SSC MATHEMATICS Guide

Illustrations & Examples

25 Chapters with explanations

2 levels of Exercise

5 Practice Sets

Fully Solved with shortcut methods

- **Head Office :** B-32, Shivalik Main Road, Malviya Nagar, New Delhi-110017
- **Sales Office :** B-48, Shivalik Main Road, Malviya Nagar, New Delhi-110017
Tel. : 011-26691021 / 26691713

Typeset by Disha DTP Team

Printed at: Repro Knowledgecast Limited, Thane

For further information about the books from DISHA,

Log on to **www.dishapublication.com** or email to **info@dishapublication.com**

CONTENTS

PRACTICE SETS

CHAPTER 1

FUNDAMENTALS

'BODMAS' RULE

A given series of calculations or operations is done in a specific order as each letter of BODMAS in order represent.

B → Brackets and order of operation of brackets is ( ), { }, [ ]
O → Of (Calculation is done the same as multiplication)
D → Division
M → Multiplication
A → Addition
S → Subtraction

So, first of all we solve the inner most brackets moving outwards. Then we perform 'of' which means multiplication, then division, addition and subtraction.

- Addition and subtraction can be done together or separately as required.
- Between any two brackets if there is not any sign of addition, subtraction and division it means we have to do multiplication

$$(20 \div 5)(7 + 3 \times 2) + 8 = 4(7 + 6) + 8$$
$$= 4 \times 13 + 8 = 52 + 8 = 60$$

BRACKETS

They are used for the grouping of things or entities. The various kind of brackets are:

(i) '–' is known as line (or bar) bracket or vinculum.
(ii) ( ) is known as parenthesis, common bracket or small bracket.
(iii) { } is known as curly bracket, brace or middle bracket.
(iv) [ ] is known as rectangular bracket or big bracket.

The order of eliminating brackets is:

(i) line bracket
(ii) small bracket (i.e., common bracket)
(iii) middle bracket (i.e., curly bracket)
(iv) big bracket (i.e., rectangular bracket)

Illustration 1: Find the value of

$$\left[5 - \left\{6 - (5 - \overline{4 - 3})\right\}\right] \text{ of } \frac{1 + \frac{1}{2}}{1 - \frac{1}{2}} \div \frac{\frac{1}{2} + \frac{1}{3}}{\frac{1}{2} - \frac{1}{3}}$$

Solution: $$\left[5 - \left\{6 - (5 - \overline{4 - 3})\right\}\right] \text{ of } \frac{1 + \frac{1}{2}}{1 - \frac{1}{2}} \div \frac{\frac{1}{2} + \frac{1}{3}}{\frac{1}{2} - \frac{1}{3}}$$

$$= [5 - \{6 - (5 - 1)\}] \text{ of } \frac{\frac{3}{2}}{\frac{1}{2}} \surd \frac{\frac{5}{6}}{\frac{1}{6}}$$

$$= \{5 - (6 - 4)\} \text{ of } \left(\frac{3}{2} \times \frac{2}{1}\right) \div \left(\frac{5}{6} \times \frac{6}{1}\right)$$

$$= (5 - 2) \text{ of } 3 \div 5$$

$$= 3 \text{ of } 3 \div 5 = 3 \times \frac{3}{5} = \frac{9}{5}$$

FACTORIAL

The product of n consecutive natural numbers (or positive integers) from 1 to n is called as the factorial 'n'. Factorial n is denoted by $n!$.

i.e.,
$$n! = 1 \times 2 \times 3 \times 4 \times 5 \times 6 \ldots (n-2)(n-1)\,n$$
$$4! = 1 \times 2 \times 3 \times 4 = 4 \times 3 \times 2 \times 1$$
$$5! = 1 \times 2 \times 3 \times 4 \times 5 = 5 \times 4 \times 3 \times 2 \times 1$$
$$6! = 1 \times 2 \times 3 \times 4 \times 5 \times 6 = 6 \times 5 \times 4 \times 3 \times 2 \times 1$$

Note: $0! = 1$ and $1! = 1$

Properties

(i) $n!$ is always an even number if $n \geq 2$.
(ii) $n!$ always ends with zero if $n \geq 5$.

ROMAN NUMBERS

In this system there are basically seven symbols used to represent the whole Roman number system. The symbols and their respective values are given below.

I = 1, V = 5, X = 10, L = 50,
C = 100, D = 500 and M = 1000

In general, the symbols in the numeral system are read from left to right, starting with the symbol representing the largest value; the same symbol cannot occur continuously more than three times; the value of the numeral is the sum of the values of the symbols.

For example LX VII = 50 + 10 + 5 + 1 + 1 = 67.

An exception to the left to the right reading occurs when a symbol of smaller value is followed immediately by a symbol of greater value, then the smaller value is subtracted from the larger. For example.

CDXL VIII = (500 – 100) + (50 – 10) + 5 + 1 + 1 + 1 = 448.

Illustration 2: The value of the numeral MCDLXIV is:

(a) 1666 (b) 664 (c) 1464 (d) 656

Solution: MCDLXIV = 1000 + (500 – 100) + 50 + 10 + (5 – 1)
= 1464

Hence (c) is the correct option.

Illustration 3: Which of the following represents the numeral for 2949

(a) MMMIXL (b) MMXMIX
(c) MMCMIL (d) MMCMXLIX

Solution: 2949 = 2000 + 900 + 40 + 9
= (1000 + 1000) + (1000 – 100) + (50 – 10) + (10 – 1)
= MMCMXLIX

Hence (d) is the correct option.

IMPORTANT CONVERSION

1 trillion = 10^{12} = 1000000000000
1 billion = 10^9 = 1000000000
1 million = 10^6 = 1000000
1 crore = 10^7 = 100 lakh
10 lakh = 10^6 = 1 million
1 lakh = 10^5 = 100000 = 100 thousand
1 thousand = 10^3 = 1000

ABSOLUTE VALUE OR MODULUS OF A NUMBER

Absolute value of a number is its numerical value irrespective of its sign.

If x be a real number N then | N | indicates the absolute value of N.

Thus | 6 | = 6, | – 6 | = 6, | 0 | = 0, | 1 | = 1, | 3.4 | = 3.4, | – 6.8 | = 6.8, etc.

| – 6 | = 6 can also be written as | – 6 | = – (– 6) = 6. Thus, if x is a negative number, then $|x| = -x$ and if x is non-negative number, then $|x| = x$

Hence $|x| = \begin{cases} x, \text{ if } x \geq 0 \\ -x, \text{ if } x < 0 \end{cases}$

PROPERTIES OF A MODULUS

(i) $|a| = |-a|$ (ii) $|ab| = |a||b|$

(iii) $\left|\frac{a}{b}\right| = \frac{|a|}{|b|}$ (iv) $|a+b| \leq |a| + |b|$

(The sign of equality holds only when the sign of a and b are same)

(v) If $|a| \leq k \Rightarrow -k \leq a \leq k$

(vi) If $|a-b| \leq k \Rightarrow -k \leq a-b \leq k \Rightarrow b-k \leq a \leq b+k$

Illustration 4: Solution of the equation $|x-2| = 5$ is

(a) 3, – 7 (b) – 3, 7
(c) 3, 6 (d) None of these

Solution: $|x-2| = 5 \Rightarrow x-2 = 5$ or $x-2 = -5$
$\Rightarrow$ $x = 7$ or $x = -3$

Hence (b) is the correct option.

Illustration 5: The minimum value of the expression $|17x-8|-9$ is

(a) 0 (b) – 9
(c) $\frac{8}{17}$ (d) none of these

Solution: The value of expression $|17x - 8| - 9$ is minimum only when $|17x - 8|$ is minimum. But the minimum value of | k | is zero.

Hence minimum value of $|17x - 8| - 9 = 0 - 9 = -9$

Hence (b) is the correct answer.

POWERS OR EXPONENTS

When a number is multiplied by itself, it gives the square of the number. i.e., $a \times a = a^2$ (Example $5 \times 5 = 5^2$)

If the same number is multiplied by itself twice we get the cube of the number i.e., $a \times a \times a = a^3$ (Example $4 \times 4 \times 4 = 4^3$)

In the same way $a \times a \times a \times a \times a = a^5$

and $a \times a \times a \times ...$ upto n times $= a^n$

There are five basic rules of powers which you should know:

If a and b are any two real numbers and m and n are positive integers, then

(i) $a^m \times a^n = a^{m+n}$ (Example: $5^3 \times 5^4 = 5^{3+4} = 5^7$)

(ii) $\frac{a^m}{a^n} = a^{m-n}$, if $m > n$ $\left(\text{Example: } \frac{6^5}{6^2} = 6^{5-2} = 6^3\right)$

$\frac{a^m}{a^n} = \frac{1}{a^{n-m}}$, if $m < n$ $\left(\text{Example: } \frac{4^3}{4^8} = \frac{1}{4^{8-3}} = \frac{1}{4^5}\right)$

and $\frac{a^m}{a^n} = a^0 = 1$, if $m = n$ $\left(\text{Example: } \frac{3^4}{3^4} = 3^{4-4} = 3^0 = 1\right)$

(iii) $(a^m)^n = a^{mn} = (a^n)^m$ (Example: $(6^2)^4 = 6^{2 \times 4} = 6^8 = (6^4)^2$

(iv) (a) $(ab)^n = a^n \cdot b^n$ (Example: $(6 \times 4)^3 = 6^3 \times 4^3$)

(b) $\left(\frac{a}{b}\right)^n = \frac{a^n}{b^n}$, $b \neq 0$ $\left(\text{Example: } \left(\frac{5}{3}\right)^4 = \frac{5^4}{3^4}\right)$

(v) $a^{-n} = \frac{1}{a^n}$ $\left(\text{Example: } 5^{-3} = \frac{1}{5^3}\right)$

(vi) For any real number a, $a^0 = 1$

Illustration 6: $\frac{5^{n+3} - 6 \times 5^{n+1}}{9 \times 5^n - 5^n \times 2^2} = ?$

Solution: $\frac{5^n \times 5^3 - 6 \times 5^n \times 5}{5^n (9 - 2^2)}$

$= \frac{5^n (5^3 - 6 \times 5)}{5^n \times 5}$

$= \frac{125 - 30}{5}$

$= \frac{95}{5} = 19$

Illustration 7: $\left\{\left(\sqrt[3]{(81)^2}\right)^{3/2}\right\}^{1/4} = ?$

Solution: $\left\{\left(\sqrt[3]{(81)^2}\right)^{3/2}\right\}^{1/4}$

$$= \left\{\left(81^2\right)^{1/3\times\frac{3}{2}}\right\}^{1/4}$$

$$= (81)^{2\times\frac{1}{3}\times\frac{3}{2}\times\frac{1}{4}} = \left(3^4\right)^{\frac{1}{4}} = 3$$

ALGEBRAIC IDENTITIES

Consider the equality $(x + 2)(x + 3) = x^2 + 5x + 6$

Let us evaluate both sides of this equality for some value of variable x say $x = 4$

LHS = $(x + 2)(x + 3) = (4 + 2)(4 + 3) = 6 \times 7 = 42$

RHS = $(4)^2 + 5 \times 4 + 6 = 16 + 20 + 6 = 42$

So for $x = 4$, LHS = RHS

Let us calculate LHS and RHS for $x = -3$

LHS = $(-3 + 2)(-3 + 3) = 0$

RHS = $(-3)2 + -(-3) + 6 = 9 - 15 + 6 = 0$

$\therefore$ for $x = -3$, LHS = RHS

If we take any value of variable x, we can find that LHS = RHS

Such an equality which is true for every value of the variable present in it is called an identity. Thus $(x + 2)(x + 3) = x^2 + 5x + 6$, is an identity.

Identities differ from equations in the following manners.

An equation is a statement of equality of two algebraic expression involving one or more variables and it is true for certain values of the variable.

For example:

$4x + 3 = x - 3$ ... (1)

$\Rightarrow$ $3x = -6 \Rightarrow x = -2$

Thus equality (1) is true only for $x = -2$, no other value of x satisfy equation (1).

Standard Identities

(i) $(a + b)^2 = a^2 + 2ab + b^2$

(ii) $(a - b)^2 = a^2 - 2ab + b^2$

(iii) $a^2 - b^2 = (a + b)(a - b)$

(iv) $(x + a)(x + b) = x^2 + (a + b)x + ab$

(v) $(a + b + c)^2 = a^2 + b^2 + c^2 + 2ab + 2bc + 2ca$

Some More Identities

We have dealt with identities involving squares. Now we will see how to handle identities involving cubes.

(i) $(a+b)^3 = a^3 + b^3 + 3a^2b + 3ab^2$

$\Rightarrow (a+b)^3 = a^3 + b^3 + 3ab(a+b)$

(ii) $(a-b)^3 = a^3 - b^3 - 3a^2b + 3ab^2$

$\Rightarrow (a-b)^3 = a^3 - b^3 - 3ab(a-b)$

(iii) $a^3 + b^3 = (a + b)(a^2 - ab + b^2)$

(iv) $a^3 - b^3 = (a - b)(a^2 + ab + b^2)$

(v) $a^3 + b^3 + c^3 - 3abc$

$= (a + b + c)(a^2 + b^2 + c^2 - ab - bc - ca)$

If $a + b + c = 0$ then $a^3 + b^3 + c^3 = 3abc$

2. Multiplication of Two Numbers Using Formulae $(a - b)(a + b) = a^2 - b^2$

If the difference between two numbers x and y is a small even number, then the smaller is express as $(a - b)$ whereas larger is expressed as $(a + b)$, then the product of x and y is found out by the formulae $x \cdot y$ i.e., $(a - b)(a + b) = a^2 - b^2$

Here a should be such that a^2 is very easily calculated.

For example:

(i) $38 \times 42 = (40 - 2) \times (40 + 2) = (40)^2 - (2)^2 = 1600 - 4 = 1596$

(ii) $66 \times 74 = (70 - 4) \times (70 + 4) = (70)^2 - (4)^2 = 4900 - 16 = 4884$

(iii) $2094 \times 2106 = (2100 - 6) \times (2100 + 6) = (2100)^2 - (6)^2 = 4410000 - 36 = 4409964$

If the difference between the two numbers is not even, still this method is used by modify as

$47 \times 54 = 47 \times 53 + 47$

$= (50 - 3) \times (50 + 3) + 47$

$= (50)^2 - (3)^2 + 47$

$= 2500 - 9 + 47 = 2538$

SQUARES

When a number is multiplied by itself, then we get the square of the number.

For example, square of $5 = 5 \times 5$ (or 5^2) = 25

Square of 2 and 3 digits numbers and cube of 2 digits numbers are very useful in CAT and CAT like competitions.

For this it is advised to learn the square of 1 to 30 as given in the table:

Number	Square	Number	Square
1	1	16	256
2	4	17	289
3	9	18	324
4	16	19	361
5	25	20	400
6	36	21	441
7	49	22	484
8	64	23	529
9	81	24	576
10	100	25	625
11	121	26	676

Number	Square	Number	Square
12	144	27	729
13	169	28	784
14	196	29	841
15	225	30	900

SQUARE ROOTS

If $b = a \times a$ or a^2, then a is called square root of b and it is represented as $\sqrt{b} = a$ or $(b)^{1/2} = a$.

Since, $16 = 4 \times 4$ or 4^2, therefore $\sqrt{16} = 4$

And $25 = 5 \times 5$ or 5^2, therefore $\sqrt{25} = 5$

There are two methods for finding the square root of a number.

(i) Prime Factorisation Method

To find the square root by this method, we first factorise the given number into prime numbers as given below for the number 3136.

2	3136
2	1568
2	784
2	392
2	196
2	98
7	49
7	7
	1

$3136 = 2 \times 2 \times 2 \times 2 \times 2 \times 2 \times 7 \times 7$

Now pair the same prime factor like

$3136 = \underline{2 \times 2} \times \underline{2 \times 2} \times \underline{2 \times 2} \times \underline{7 \times 7}$

Now product of prime numbers staken one number from each pair of prime factors is the square root of the given number

$\therefore \sqrt{3136} = 2 \times 2 \times 2 \times 7 = 56$

If we write, $3136 = (2)^6 \times (7)^2$

Then square root of 3136 is the product of prime factors 2 and 7 with the powers half of the powers raised on 2 and 7 respectively.

i.e., $\sqrt{3136} = (2)^3 \times 7 = 56$

(ii) Division Method

In this method first of all pair the digits of the given number from right side. But there may be left a single digit at the left end of the number. Further process is shown below for the number 2304.

$\sqrt{2304} = 48$

	48
4	$\overline{23}\,\overline{04}$
4	16
88	704
8	704
	×××

Illustration 8: Find the square root of 15625.

Solution: $\sqrt{15625} = 125$

	125
1	$\overline{1}\,\overline{56}\,\overline{25}$
1	1
22	56
2	44
245	1225
5	1225
	××××

CUBES

When a number multiplies itself three times, we get the cube of the number.

Cube of $4 = 4 \times 4 \times 4 = 64$

Cubes of large numbers are rarely used. It is advised to you to learn the cube of the integers from 1 to 10.

Number	1	2	3	4	5	6	7	8	9	10
Cube	1	8	27	64	125	216	343	512	729	1000

Practice Exercise

Level - I

1. $287 \times 287 + 269 \times 269 - 2 \times 287 \times 269 = ?$
   (a) 534 (b) 446
   (c) 354 (d) 324

2. If $(64)^2 - (36)^2 = 20 \times x$, then $x = ?$
   (a) 70 (b) 120
   (c) 180 (d) 140

3. If $\sqrt{3} = 1.732$ and $\sqrt{2} = 1.414$, the value of $\frac{1}{\sqrt{3}+\sqrt{2}}$ is
   (a) 0.064 (b) 0.308
   (c) 0.318 (d) 2.146

4. $\sqrt{0.01+\sqrt{0.0064}} = ?$
   (a) 0.3 (b) 0.03
   (c) $\sqrt{0.18}$ (d) None of these

5. $356 \times 936 - 356 \times 836 = ?$
   (a) 35600 (b) 34500
   (c) 9630 (d) 93600

6. The value of $\dfrac{\frac{1}{2} \div \frac{1}{2} \text{ of } \frac{1}{2}}{\frac{1}{2} + \frac{1}{2} \text{ of } \frac{1}{2}}$ is
   (a) 2/3 (b) 2
   (c) 4/3 (d) 3

7. The simplified value of

$$\frac{\left(1+\frac{1}{1+\frac{1}{100}}\right)\left(1+\frac{1}{1+\frac{1}{100}}\right)-\left(1-\frac{1}{1+\frac{1}{100}}\right)\left(1-\frac{1}{1+\frac{1}{100}}\right)}{\left(1+\frac{1}{1+\frac{1}{100}}\right)+\left(1-\frac{1}{1+\frac{1}{100}}\right)} \text{ is}$$

   (a) 100 (b) $\frac{200}{101}$
   (c) 200 (d) $\frac{202}{100}$

8. If $5^a = 3125$, then the value of $5^{(a-3)}$ is
   (a) 25 (b) 125
   (c) 625 (d) 1625

9. In a group of buffaloes and ducks, the number of legs are 24 more than twice the number of heads. What is the number of buffaloes in the group?
   (a) 6 (b) 8
   (c) 10 (d) 12

10. $\sqrt[3]{4\frac{12}{125}} = ?$
   (a) $1\frac{2}{5}$ (b) $1\frac{3}{5}$
   (c) $1\frac{4}{5}$ (d) $2\frac{2}{5}$

11. If $3^{4X-2} = 729$, then find the value of X.
   (a) 4 (b) 3
   (c) 2 (d) 5

12. What number must be added to the expression $16a^2 - 12a$ to make it a perfect square?
   (a) 9/4 (b) 11/2
   (c) 13/2 (d) 16

13. The value of $\left[\frac{1}{\sqrt{9}-\sqrt{8}}\right] - \left[\frac{1}{\sqrt{8}-\sqrt{7}}\right] + \left[\frac{1}{\sqrt{7}-\sqrt{6}}\right] - \left[\frac{1}{\sqrt{6}-\sqrt{5}}\right] + \left[\frac{1}{\sqrt{5}-\sqrt{4}}\right]$ is
   (a) 6 (b) 5
   (c) –7 (d) –6

14. Simplify: $5\sqrt[3]{250} + 7\sqrt[3]{16} - 14\sqrt[3]{54}$
   (a) $-2\sqrt[3]{2}$ (b) $-3\sqrt[3]{2}$
   (c) $3\sqrt[3]{2}$ (d) $2\sqrt[3]{3}$

15. The no. plate of a bus had peculiarity. The bus number was a perfect square. It was also a perfect square when the plate was turned upside down. The bus company had only five hundred buses numbered from 1 to 500. What was the number?
   (a) 169 (b) 36
   (c) 196 (d) Cannot say

16. If * means adding six times of second number into first number, then find the value of (1*2)*3.
   (a) 121 (b) 31
   (c) 93 (d) 91

17. If a and b are positive ingegers, such that $a^b = 125$, then $(a - b)^{a+b-4} = ?$
   (a) 16 (b) 25
   (c) 28 (d) 30

18. If $p \times q = p + q + \frac{p}{q}$, then value of $8 \times 2 = ?$
   (a) 2 (b) 10
   (c) 14 (d) 16

19. If $x*y = x^2 + y^2 - xy$, then value of 9*11 is
(a) 93 (b) 103
(c) 60.5 (d) 121

20. The least number by which we multiply to the 11760, so that we can get a perfect square number
(a) 2 (b) 3
(c) 5 (d) None of these

21. If $5\sqrt{5} \times 5^3 \div 5^{-3/2} = 5^{(a+2)}$, then value of a is
(a) 5 (b) 4
(c) 6 (d) 7

22. If difference between the $\frac{4}{5}$ of $\frac{3}{4}$ of a number and $\frac{2}{5}$ of $\frac{1}{6}$ of the same number is 648, then number is
(a) 1110 (b) 1215
(c) 1325 (d) 1440

23. If sum of two numbers is 42 and their product is 437, then find their difference.
(a) 3 (b) 4
(c) 5 (d) 7

24. 54.327 × 357.2 × 0.0057 is the same as:
(a) 5.4327 × 3.572 × 5.7
(b) 5.4327 × 3.572 × 0.57
(c) 54327 × 3572 × 0.0000057
(d) None of these

25. Write the 44000 in Roman numerals
(a) XLI (b) XLVI
(c) XLIV (d) XLVIC

26. Write LXXIX in Hindu-Arabic numerals
(a) 70000 (b) 70009
(c) 7009 (d) 700009

27. If $\frac{a+b}{b+c} = \frac{c+d}{d+a}$, then
(a) a must equal c
(b) $a+b+c+d$ must equal zero
(c) either $a = c$ or $a+b+c+d = 0$, or both
(d) $a(b+c+d) = c(a+b+d)$

28. A number lies between 300 and 400. If the number is added to the number formed by reversing the digits, the sum is 888 and if the unit's digit and the ten's digit change places, the new number exceeds the original number by 9. Find the number.
(a) 339 (b) 341
(c) 378 (d) 345

29. x and y are 2 different digits. If the sum of the two digit numbers formed by using both the digits is a perfect square, then find $x + y$.
(a) 10 (b) 11
(c) 12 (d) 13

30. Arrange the following in the decending order;
$5^{1/4}, 4^{1/3}, 6^{1/5}$.
(a) $4^{1/3}, 5^{1/4}, 6^{1/5}$ (b) $5^{1/4}, 4^{1/3}, 6^{1/5}$
(c) $6^{1/5}, 4^{1/3}, 5^{1/4}$ (d) $5^{1/4}, 4^{1/3}, 6^{1/5}$

31. If $a + b + c = 13$, $a^2 + b^2 + c^2 = 69$, then find $ab + bc + ca$.
(a) –50 (b) 50
(c) 69 (d) 75

32. If $a - 8 = b$, then determine the value of $|a-b| - |b-a|$.
(a) 16 (b) 0
(c) 4 (d) 2

33. Find the possible integral value of x, if $x^2 + |x - 1| = 1$.
(a) 1 (b) –1
(c) 0 (d) 1 and 0

DIRECTIONS (Qs. 34-49) : ***What value should come in the place of question mark (?) in the following questions ?***

34. 3.6 + 36.6 + 3.66 + 0.36 + 3.0 = ? *[SBI Clerk-June-2012]*
(a) 44.22 (b) 77.22
(c) 74.22 (d) 47.22
(e) None of these

35. 23 × 45 ÷ 15 = ? *[SBI Clerk-June-2012]*
(a) 69 (b) 65
(c) 63 (d) 71
(e) None of these

36. $4\frac{5}{6} + 7\frac{1}{2} - 5\frac{8}{11} = ?$ *[SBI Clerk-June-2012]*
(a) $2\frac{10}{33}$ (b) $6\frac{20}{33}$
(c) $2\frac{20}{33}$ (d) $6\frac{10}{33}$
(e) None of these

37. $\frac{210}{14} \times \frac{17}{15} \times ? = 4046$ *[SBI Clerk-June-2012]*
(a) 202 (b) 218
(c) 233 (d) 227
(e) None of these

38. 83% of 2350 = ? *[SBI Clerk-June-2012]*
(a) 1509.5 (b) 1950.5
(c) 1905.5 (d) 1590.5
(e) None of these

39. $\sqrt{1089} + 3 = (?)^2$ *[SBI Clerk-June-2012]*
(a) 5 (b) 6
(c) 3 (d) 8
(e) 4

40. 96 + 32 × 5 – 31 = ? *[SBI Clerk-June-2012]*
(a) 223 (b) 225
(c) 229 (d) 221
(e) None of these

41. $? \div 36 = (7)^2 - 8$ *[SBI Clerk-June-2012]*
(a) 1426 (b) 1449
(c) 1463 (d) 1476
(e) None of these

42. $\sqrt{8281} = ?$ [SBI Clerk- June-2012]
(a) 89 (b) 97
(c) 93 (d) 91
(e) 83

43. $(63)^2 - (12)^2 = ?$ [SBI Clerk-June-2012]
(a) 3528 (b) 3852
(c) 3582 (d) 3825
(e) None of these

44. $1\frac{4}{5} + 3\frac{3}{5} = ? - 4\frac{3}{10}$ [SBI Clerk-June-2012]
(a) $9\frac{7}{10}$ (b) $7\frac{7}{10}$
(c) $9\frac{3}{10}$ (d) $7\frac{9}{10}$
(e) None of these

45. $17 \times 19 \times 4 \div ? = 161.5$ [SBI Clerk-June-2012]
(a) 8 (b) 6
(c) 7 (d) 9
(e) None of these

46. $1798 \div 31 \times ? = 348$ [SBI Clerk-June-2012]
(a) 3 (b) 6
(c) 4 (d) 5
(e) None of these

47. $(9.8 \times 2.3 + 4.46) \div 3 = (3)^?$ [SBI Clerk-June-2012]
(a) 3 (b) 9
(c) 5 (d) 2
(e) None of these

48. 43% of 600 + ?% of 300 = 399 [SBI Clerk-June-2012]
(a) 45 (b) 41
(c) 42 (d) 47
(e) None of these

49. The sum of three consecutive odd numbers is 1383. What is the largest number ? [SBI Clerk-June-2012]
(a) 463 (b) 49
(c) 457 (d) 461
(e) None of these

DIRECTIONS (Qs. 50-54) : ***What approximate value should come in place of the question mark (?) in the following questions? (NOTE: You are not expected to calculate the exact value)***

50. $1504 \times 5.865 - 24.091 = ?$ [SBI Clerk-2012]
(a) 7200 (b) 9500
(c) 6950 (d) 5480
(e) 8800

51. $16.928 + 24.7582 \div 5.015 = ?$ [SBI Clerk-2012]
(a) 20 (b) 24
(c) 22 (d) 26
(e) None of these

52. $\sqrt[3]{7.938} \times (6.120)^2 - 4.9256 = ?$ [SBI Clerk-2012]
(a) 70 (b) 55
(c) 30 (d) 25
(e) 90

53. $16.046 \div 2.8 \times 0.599 = ?$ [SBI Clerk-2012]
(a) 3.5 (b) 7.9
(c) 1.9 (d) 5.6
(e) 6.2

54. $\sqrt{963} + (4.895)^2 - 9.24 = ?$ [SBI Clerk-2012]
(a) 60 (b) 35
(c) 85 (d) 45
(e) 25

DIRECTIONS (Qs. 55-69) : ***What should come in place of the question mark (?) in the following questions ?***

55. $(12 \times 19) + (13 \times 8) = (15 \times 14) + ?$ [SBI Clerk-2012]
(a) 124 (b) 122
(c) 126 (d) 128
(e) None of these

56. $\sqrt{65 \times 12 - 50 + 54} = ?$ [SBI Clerk-2012]
(a) $\sqrt{28}$ (b) 28^2
(c) 28 (d) 784
(e) None of these

57. 15% of 524 – 2% of 985 + ? = 20% of 423 [SBI Clerk-2012]
(a) 25.9 (b) 27.7
(c) 25.7 (d) 24.9
(e) None of these

58. $151 \times 8 + (228 \div 19)^2 = ?$ [SBI Clerk-2012]
(a) 1360 (b) 1354
(c) 1368 (d) 1381
(e) None of these

59. $\sqrt{1521} + \sqrt{225} = ?$ [SBI Clerk-2012]
(a) 56 (b) 58
(c) 54 (d) 62
(e) None of these

60. $38.734 + 8.638 - 5.19 = ?$ [SBI Clerk-2012]
(a) 41.971 (b) 42.179
(c) 43.072 (d) 42.182
(e) None of these

61. $7^{8.9} \div (343)^{1.7} \times (49)^{4.8} = 7^?$ [SBI Clerk-2012]
(a) 13.4 (b) 12.8
(c) 11.4 (d) 9.6
(e) None of these

62. $\sqrt[5]{512} \div \sqrt[4]{16} + \sqrt{576} = ?$ [SBI Clerk-2012]
(a) 24 (b) 31
(c) 22 (d) 18
(e) None of these

63. $(42 \times 3.2) \div (16 \times 1.5) = ?$ [SBI Clerk-2012]
(a) 5.9 (b) 5.6
(c) 6.1 (d) 4.8
(e) None of these

64. $199 + 5^3 \div 4 \times 4^2 = ?$ [SBI Clerk-2012]
(a) 969 (b) 655
(c) 966 (d) 799
(e) None of these

65. $342 \div 6 \times 28 = 1099 + ?$ [SBI Clerk-2012]

(a) 478 (b) 502
(c) 486 (d) 504
(e) None of these

66. $\frac{9.8 \times 2.5 \times 7.6}{0.5} = ?$ [SBI Clerk-2012]

(a) 384.2 (b) 379.5
(c) 364.3 (d) 372.4
(e) None of these

67. $\frac{3}{5}$ of $\frac{2}{7}$ of ? = 426 [SBI Clerk-2012]

(a) 2490 (b) 2565
(c) 2475 (d) 2485
(e) None of these

68. $3\frac{2}{5} + 1\frac{2}{9} = 4\frac{4}{5} - ?$ [SBI Clerk-2012]

(a) $\frac{8}{45}$ (b) $\frac{7}{47}$
(c) $\frac{7}{45}$ (d) $\frac{8}{51}$
(e) None of these

69. $\frac{13}{63} \div \frac{104}{14} \times \frac{52}{19} = ?$ [SBI Clerk-2012]

(a) $\frac{12}{173}$ (b) $\frac{13}{171}$
(c) $\frac{17}{171}$ (d) $\frac{18}{171}$
(e) None of these

DIRECTIONS (Qs. 70-84) : ***What should come in place of the question mark (?) in the following questions ?***

70. $\sqrt[3]{13824} \times \sqrt{?} = 864$ [SBI Clerk-2014]

(a) 1296 (b) 1156
(c) 1600 (d) 1024
(e) None of these

71. $(91)^2 + (41)^2 - \sqrt{?} = 9858$ [SBI Clerk-2014]

(a) 11236 (b) 10816
(c) 10404 (d) 9604
(e) None of these

72. $4900 \div 28 \times 444 \div 12 = ?$ [SBI Clerk-2014]

(a) 6575 (b) 6475
(c) 6455 (d) 6745
(e) None of these

73. 125% of 260 + ?% of 700 = 500 [SBI Clerk-2014]

(a) 32 (b) 56
(c) 23 (d) 46
(e) None of these

74. $3\frac{7}{11} + 7\frac{3}{11} \times 1\frac{1}{2} = ?$ [SBI Clerk-2014]

(a) $13\frac{10}{11}$ (b) $14\frac{6}{11}$
(c) $14\frac{9}{11}$ (d) $10\frac{17}{22}$
(e) None of these

75. $\frac{.23 - .023}{.0023 \div 23} = ?$ [SBI Clerk-2014]

(a) 0.207 (b) 207
(c) 2070 (d) 0.0207
(e) None of these

76. 1.05% of 2500 + 2.5% of 440 = ? [SBI Clerk-2014]

(a) 37.50 (b) 37.25
(c) 370.25 (d) 372.50
(e) None of these

77. $\frac{17 \times 4 + 4^2 \times 2}{90 \div 5 \times 12} = ?$ [SBI Clerk-2014]

(a) $\frac{25}{54}$ (b) $\frac{22}{57}$
(c) $\frac{11}{27}$ (d) $\frac{13}{27}$
(e) None of these

78. $17\frac{2}{5} \times 4\frac{5}{8} - ? = 46\frac{7}{8}$ [SBI Clerk-2014]

(a) $32\frac{3}{5}$ (b) $33\frac{3}{5}$
(c) $33\frac{2}{5}$ (d) $32\frac{2}{5}$
(e) None of these

79. 136% of 250 + ?% of 550 = 670 [SBI Clerk-2014]

(a) 64 (b) 55
(c) 56 (d) 65
(e) None of these

80. 3889 + 12.952 − ? = 3854.002 [SBI Clerk-2014]

(a) 47.95 (b) 47.752
(c) 47.095 (d) 47.932
(e) None of these

81. $(5 \times 5 \times 5 \times 5 \times 5 \times 5)^4 \times (5 \times 5)^6 \div (5)^2 = (25)^?$ [SBI Clerk-2014]

(a) 10 (b) 17
(c) 19 (d) 12
(e) None of these

82. $\frac{28 \times 5 - 15 \times 6}{7^2 + \sqrt{256} + (13)^2} = ?$ [SBI Clerk-2014]

(a) $\frac{27}{115}$ (b) $\frac{22}{117}$
(c) $\frac{25}{117}$ (d) $\frac{22}{115}$
(e) None of these

83. $1.5 \times 0.025 + (?)^2 = 0.1$ *[SBI Clerk-2014]*
(a) 0.28 (b) 0.27
(c) 0.25 (d) 0.235
(e) None of these

84. $\frac{(3.537-0.948)^2+(3.537+0.948)^2}{(3.537)^2+(.948)^2}=?$ *[SBI Clerk-2014]*
(a) 4.485 (b) 2.589
(c) 4 (d) 2
(e) None of these

DIRECTIONS (Qs. 85-89) : ***Find out the approximate value which should come in place of the question mark in the following questions. (You are not expected to find the exact value.)***

85. $\frac{(10008.99)^2}{10009.001} \times \sqrt{3589} \times 0.4987 = ?$ *[SBI Clerk-2014]*
(a) 3000 (b) 300000
(c) 3000000 (d) 5000
(e) 9000000

86. $196.1 \times 196.1 \times 196.1 \times 4.01 \times 4.001 \times 4.999 \times 4.999$
$= 196.1^3 \times 4 \times ?$ *[SBI Clerk-2014]*
(a) 100 (b) 16
(c) 10 (d) 64
(e) 32

87. $12.25 \times ? \times 21.6 = 3545.64$ *[SBI Clerk-2014]*
(a) 20 (b) 12
(c) 15 (d) 13
(e) None of these

88. ?% of 45.999 × 16% of 83.006 = 116.073
[SBI Clerk-2014]
(a) 6 (b) 24
(c) 19 (d) 30
(e) 11

89. $[(1.3)^2 \times (4.2)^2] \div 2.7 = ?$ *[SBI Clerk-2014]*
(a) 7 (b) 21
(c) 18 (d) 11
(e) 16

DIRECTIONS (Qs. 90-94) : ***What should come in place of question mark (?) in the following number series ?***

90. 3 23 43 ? 83 103 *[SBI Clerk-2014]*
(a) 33 (b) 53
(c) 63 (d) 73
(e) None of these

91. 1 9 25 49 81 ? 169 *[SBI Clerk-2014]*
(a) 100 (b) 64
(c) 81 (d) 121
(e) None of these

92. 5 6 14 45 ? *[SBI Clerk-2014]*
(a) 183 (b) 185
(c) 138 (d) 139
(e) None of these

93. 7 8 18 57 ? *[SBI Clerk-2014]*
(a) 244 (b) 174
(c) 186 (d) 226
(e) None of these

94. 1, 8, 9, ?, 25, 216, 49 *[SBI Clerk-2014]*
(a) 60 (b) 64
(c) 70 (d) 75
(e) None of these

95. Last year my age was a perfect square number. Next year it will be a cubic number. What is my present age?
[SSC-Sub. Ins.-2012]
(a) 25 years (b) 27 years
(c) 26 years (d) 24 years

96. What is the value of $(2.1)^2 \times \sqrt{0.0441}$?
[SSC-Sub. Ins.-2012]
(a) 0.9261 (b) 92.61
(c) 92.51 (d) 0.9251

97. The value of $\sqrt[3]{1372} \times \sqrt[3]{1458}$ is *[SSC-Sub. Ins.-2012]*
(a) 116 (b) 126
(c) 106 (d) 136

98. Equal amounts of water were poured into two empty jars of different capacities, which made one jar $\frac{1}{4}$ full and the other jar $\frac{1}{3}$ full. If the water in the jar with lesser capacity is then poured into the jar with greater capacity, then the part of the larger jar filled with water is *[SSC-Sub. Ins.-2012]*
(a) $\frac{1}{2}$ (b) $\frac{7}{12}$
(c) $\frac{1}{4}$ (d) $\frac{1}{3}$

99. If $\frac{5x-3}{x}+\frac{5y-3}{y}+\frac{5z-3}{z}=0$, then the value of $\frac{1}{x}+\frac{1}{y}+\frac{1}{z}$ is
[SSC-Sub. Ins.-2012]
(a) 15 (b) 3
(c) 5 (d) 10

100. Minimum value of $x^2+\frac{1}{x^2+1}-3$ is *[SSC-Sub. Ins.-2012]*
(a) −3 (b) −2
(c) 0 (d) −1

101. If $a + b = 5$, $a^2 + b^2 = 13$, the value of $a - b$ (where $a > b$) is
[SSC-Sub. Ins.-2012]
(a) 2 (b) −1
(c) 1 (d) −2

102. If $(3x - y) : (x + 5y) = 5 : 7$, then the value of $(x + y) : (x - y)$ is
[SSC-Sub. Ins.-2012]
(a) 3 : 1 (b) 1 : 3
(c) 2 : 3 (d) 3 : 2

103. The value of $1+\cfrac{1}{1+\cfrac{2}{3+\cfrac{4}{5}}}$ is: [SSC-Sub. Ins.-2013]

(a) $\frac{12}{29}$ (b) $\frac{8}{19}$
(c) $\frac{48}{29}$ (d) $\frac{2}{19}$

104. The value of

$\sqrt{19.36}+\sqrt{0.1936}+\sqrt{0.001936}+\sqrt{0.00001936}$ is: [SSC-Sub. Ins.-2013]

(a) 4.8484 (b) 4.8694
(c) 4.8884 (d) 4.8234

105. If the square of the sum of two numbers is equal to 4 times of their product. then the ratio of these numbers is : [SSC-Sub. Ins.-2013]

(a) 2 : 1 (b) 1 : 3
(c) 1 : 1 (d) 1 : 2

106. If $a^2+b^2=5ab$, then the value of $\left(\frac{a^2}{b^2}+\frac{b^2}{a^2}\right)$ is : [SSC-Sub. Ins.-2013]

(a) 32 (b) 16
(c) 23 (d) –23

107. If $x=\frac{\sqrt{3}-\sqrt{2}}{\sqrt{3}+\sqrt{2}}$ and $y=\frac{\sqrt{3}+\sqrt{2}}{\sqrt{3}-\sqrt{2}}$, then the value of x^3+y^3 is: [SSC-Sub. Ins.-2013]

(a) 950 (b) 730
(c) 650 (d) 970

108. The greatest among the following numbers $(3)^{\frac{1}{3}},(2)^{\frac{1}{2}},1,(6)^{\frac{1}{6}}$ is: [SSC-Sub. Ins.-2013]

(a) $(2)^{\frac{1}{2}}$ (b) 1
(c) $(6)^{\frac{1}{6}}$ (d) $(3)^{\frac{1}{3}}$

109. Evaluate $\frac{\sqrt{24}+\sqrt{6}}{\sqrt{24}-\sqrt{6}}$ [SSC-Sub. Ins.-2014]

(a) 2 (b) 3
(c) 4 (d) 5

110. The value of [SSC-Sub. Ins.-2014]

$3\div\left[(8-5)\div\left\{(4-2)\div\left(2+\frac{8}{13}\right)\right\}\right]$ is

(a) $\frac{15}{17}$ (b) $\frac{13}{17}$
(c) $\frac{15}{19}$ (d) $\frac{13}{19}$

111. If '+' means '÷', '×' means '–', '÷' means '×' and '–' means '+', what will be the value of the following expression?
$9+3\div 4-8\times 2=?$ [SSC-Sub. Ins.-2014]

(a) $6\frac{1}{4}$ (b) $6\frac{3}{4}$
(c) $-1\frac{3}{4}$ (d) 18

112. The next term of the sequence,

$\left(1+\frac{1}{2}\right),\left(1+\frac{1}{2}\right)\left(1+\frac{1}{3}\right),\left(1+\frac{1}{2}\right)\left(1+\frac{1}{3}\right)\left(1+\frac{1}{4}\right),\_\_$ is [SSC-Sub. Ins.-2014]

(a) 3 (b) $\left(1+\frac{1}{5}\right)$
(c) 5 (d) $\left(1+\frac{1}{2}\right)\left(1+\frac{1}{5}\right)$

113. If $a=\sqrt{6}+\sqrt{5}$, $b=\sqrt{6}-\sqrt{5}$, then $2a^2-5ab+2b^2=$ [SSC-Sub. Ins.-2014]

(a) 38 (b) 39
(c) 40 (d) 41

114. If $p=\frac{5}{18}$, then $27p^3-\frac{1}{216}-\frac{9}{2}p^2+\frac{1}{4}p$ is equal to [SSC-Sub. Ins.-2014]

(a) $\frac{4}{27}$ (b) $\frac{5}{27}$
(c) $\frac{8}{27}$ (d) $\frac{10}{27}$

115. If $x+\frac{1}{x}=2$, then $x^{2013}+\frac{1}{x^{2014}}=?$ [SSC-Sub. Ins.-2014]

(a) 0 (b) 1
(c) –1 (d) 2

116. If $a=331$, $b=336$ and $c=-667$, then the value of $a^3+b^3+c^3-3abc$ is [SSC-Sub. Ins.-2014]

(a) 1 (b) 6
(c) 3 (d) 0

117. The simplified value of

$\left(\sqrt{6}+\sqrt{10}-\sqrt{21}-\sqrt{35}\right)\left(\sqrt{6}-\sqrt{10}+\sqrt{21}-\sqrt{35}\right)$ is [SSC-Sub. Ins.-2014]

(a) 13 (b) 12
(c) 11 (d) 10

118. If $x=a-b, y=b-c, z=c-a$, then the numerical value of the algebraic expression $x^3+y^3+z^3-3xyz$ will be

(a) $a+b+c$ (b) 0
(c) $4(a+b+c)$ (d) $3abc$

119. The simplified value of $\frac{\sqrt{32}+\sqrt{48}}{\sqrt{8}+\sqrt{12}}$ is [SSC-MT-2013]

(a) 4 (b) 3
(c) 2 (d) 6

120. $\sqrt{\frac{9.5\times0.085}{0.0017\times0.19}}$ equals [SSC-MT-2013]

(a) 5 (b) 50
(c) 500 (d) 0.05

121. If $\frac{x}{b+c}=\frac{y}{c+a}=\frac{z}{a+b}$, then : [SSC 10+2-2012]

(a) $\frac{x-y}{b-a}=\frac{y-z}{c-b}=\frac{z-x}{a-c}$

(b) $\frac{x}{a}=\frac{y}{b}=\frac{z}{c}$

(c) $\frac{x-y}{c}=\frac{y-z}{b}=\frac{z-x}{a}$

(d) none of the above is true

122. If $\frac{547.527}{0.0082}=x$, then the value $\frac{547527}{82}$ is :

[SSC 10+2-2012]

(a) 10x (b) 100x
(c) $\frac{x}{100}$ (d) $\frac{x}{10}$

123. If $\sqrt[3]{3^n}=27$, then the value of n is: [SSC 10+2-2012]

(a) 9 (b) 6
(c) 1 (d) 3

124. From 9.00 AM to 2.00 PM, the temperature rose at a constant rate from 21°C to 36°C. What was the temperature at noon ? [SSC 10+2-2012]

(a) 27°C (b) 30°C
(c) 32°C (d) 28.5°C

125. If $\frac{3x+5}{5x-2}=\frac{2}{3}$, then the value of x is : [SSC 10+2-2012]

(a) 11 (b) 19
(c) 23 (d) 7

126. If the difference of two numbers is 3 and the difference of their squares is 39; then the larger number is :

[SSC 10+2-2012]

(a) 9 (b) 12
(c) 13 (d) 8

127. If $x=\sqrt{3}+\sqrt{2}$, then the value of $x^3-\frac{1}{x^3}$ is :

[SSC 10+2-2012]

(a) $14\sqrt{2}$ (b) $14\sqrt{3}$
(c) $22\sqrt{2}$ (d) $10\sqrt{2}$

128. If $a^2+b^2+c^2=2(a-b-c)-3$, then the value of $2a-3b+4c$ is [SSC 10+2-2013]

(a) 1 (b) 7
(c) 2 (d) 3

129. Let $a=\sqrt{6}-\sqrt{5}, b=\sqrt{5}-2, c=2-\sqrt{3}$.
Then point out the correct alternative among the four alternatives given below. [SSC 10+2-2013]

(a) $a<b<c$ (b) $b<a<c$
(c) $a<c<b$ (d) $b<c<a$;

130. If $a=\frac{b^2}{b-a}$ then the value of a^3+b^3 is [SSC 10+2-2013]

(a) 2 (b) 6 ab
(c) 0 (d) 1

131. If $xy+yz+zx=0$, then [SSC 10+2-2013]

$\left(\frac{1}{x^2-yz}+\frac{1}{y^2-zx}+\frac{1}{z^2-xy}\right)(x,y,z\neq0)$ is equal to

(a) 0 (b) 3
(c) 1 (d) $x+y+z$

132. The value of $\sqrt{40+\sqrt{9\sqrt{81}}}$ is [SSC 10+2-2013]

(a) 11 (b) $\sqrt{111}$
(c) 9 (d) 7

133. Which is greater $\sqrt[3]{2}$ or $\sqrt{3}$? [SSC 10+2-2013]

(A Equal (b) Cannot be compared
(c) $\sqrt[3]{2}$ (d) $\sqrt{3}$

134. If $a+b+c=9$ (where a, b, c are real numbers), then the minimum value of $a^2+b^2+c^2$ is [SSC 10+2-2013]

(a) 81 (b) 100
(c) 9 (d) 27

135. Find the value of

$3+\frac{1}{\sqrt{3}}+\frac{1}{\sqrt{3}+3}+\frac{1}{\sqrt{3}-3}$. [SSC 10+2-2013]

(a) 6 (b) 3
(c) $\frac{3}{2(\sqrt{3}+3)}$ (d) $2\sqrt{3}$

136. 'a' divides 228 leaving a remainder 18. The biggest two-digit value of 'a' is [SSC 10+2-2013]

(a) 30 (b) 70
(c) 21 (d) 35

137. A teacher wants to arrange his students in an equal number of rows and columns. If there are 1369 students, the number of students in the last row are [SSC 10+2-2014]

(a) 37 (b) 33
(c) 63 (d) 47

138. Which one of the following is true ? [SSC 10+2-2014]

(a) $\sqrt{5}+\sqrt{3}>\sqrt{6}+\sqrt{2}$
(b) $\sqrt{5}+\sqrt{3}<\sqrt{6}+\sqrt{2}$
(c) $\sqrt{5}+\sqrt{3}=\sqrt{6}+\sqrt{2}$
(d) $(\sqrt{5}+\sqrt{3})(\sqrt{6}+\sqrt{2})=1$

DIRECTIONS (139-148) : ***What will come in place of the question mark (?) in the following questions?***

139. $(3325 \div 25) \times (152 \div 16) = ?$ [*IBPS Clerk-2012*]
(a) 1269.4 (b) 1264.9
(c) 1265.3 (d) 1263.5
(e) None of these

140. $\sqrt{3136} - \sqrt{1764} = \sqrt{?}$ [*IBPS Clerk-2012*]
(a) 14 (b) $(196)^2$
(c) – 14 (d) 144
(e) None of these

141. $5\frac{1}{5} + 2\frac{2}{15} + 3\frac{2}{3} = ?$ [*IBPS Clerk-2012*]
(a) 15 (b) 13
(c) $\frac{11}{15}$ (d) 12
(e) None of these

142. $-15 - 27 - 88 - 63 + 255 = ?$ [*IBPS Clerk-2012*]
(a) 55 (b) 74
(c) 62 (d) 59
(e) None of these

143. $(2525 \times 0.25 \div 5) \times 7 = ?$ [*IBPS Clerk-2012*]
(a) 889.43 (b) 883.75
(c) 886.45 (d) 881.75
(e) None of these

144. $\frac{14}{19} \times \frac{57}{70} \times \frac{20}{21} = ?$ [*IBPS Clerk-2012*]
(a) $\frac{2}{7}$ (b) $\frac{4}{7}$
(c) $\frac{2}{9}$ (d) $\frac{3}{7}$
(e) None of these

145. 32% of 500 + 162% of 50 = ? [*IBPS Clerk-2012*]
(a) 231 (b) 245
(c) 237 (d) 247
(e) None of these

146. $45316 + 52131 - 65229 = ? + 15151$ [*IBPS Clerk-2012*]
(a) 17063 (b) 17073
(c) 17076 (d) 17067
(e) None of these

147. $\sqrt{25 - 12 + 155 + 1} = ?$ [*IBPS Clerk-2012*]
(a) 13 (b) 14
(c) 17 (d) 16
(e) None of these

148. $\frac{184 \times 4}{23\% \text{ of } 400} = ?$ [*IBPS Clerk-2012*]
(a) 7 (b) 9
(c) 8 (d) 5
(e) None of these

149. What will come in place of both the question marks (?) in the following question ? [*IBPS Clerk-2012*]

$$\frac{(?)^{4/3}}{32} = \frac{128}{?^{5/3}}$$

(a) 16 (b) 12
(c) 18 (d) 14
(e) None of these

150. If the following fractions are arranged in a descending order (from left to right), which of them will be second from the right end ? [*IBPS Clerk-2012*]

$$\frac{4}{9}, \frac{6}{13}, \frac{5}{11}, \frac{13}{16}, \frac{7}{12}$$

(a) $\frac{6}{13}$ (b) $\frac{4}{9}$
(c) $\frac{13}{16}$ (d) $\frac{7}{12}$
(e) $\frac{5}{11}$

151. A factory produces 1515 items in 3 days. How many items will they produce in a week ? [*IBPS Clerk-2012*]
(a) 3530 (b) 3553
(c) 3533 (d) 3535
(e) None of these

152. What is the **least** number that can be added to 4800 to make it a perfect square ? [*IBPS Clerk-2012*]
(a) 110 (b) 81
(c) 25 (d) 36
(e) None of these

153. If $(11)^3$ is subtracted from $(46)^2$ what will be the remainder ? [*IBPS Clerk-2012*]
(a) 787 (b) 785
(c) 781 (d) 783
(e) None of these

154. The sum of the squares of two odd numbers is 11570. The square of the smaller number is 5329. What is the other number ? [*IBPS Clerk-2012*]
(a) 73 (b) 75
(c) 78 (d) 79
(e) None of these

155. The sum of three consecutive integers is 5685. Which of the following is the correct set of these numbers ? [*IBPS Clerk-2012*]
(a) 1893, 1894, 1895 (b) 1895, 1896, 1897
(c) 1899, 1900, 1901 (d) 1897, 1898, 1899
(e) None of these

156. The product of three consecutive odd numbers is 24273. Which is the smallest number? [*IBPS Clerk-2012*]
(a) 25 (b) 29
(c) 23 (d) 37
(e) 27

DIRECTIONS (Qs. 157-171) : ***What will come in place of question mark (?) in the given question?***

157. $4\frac{1}{2}+\left(1\div 2\frac{8}{9}\right)-3\frac{1}{13}=?$ *[IBPS Clerk-2013]*

(a) $1\frac{9}{26}$ (b) $2\frac{7}{13}$

(c) $1\frac{11}{26}$ (d) $2\frac{4}{13}$

(e) $1\frac{10}{13}$

158. $\frac{6\times 136\div 8+132}{628\div 16-26.25}=?$ *[IBPS Clerk-2013]*

(a) 15 (b) 24

(c) 18 (d) 12

(e) 28

159. $\{(441)^{1/2}\times 207\times (343)^{1/3}\}\div\{(14)^2\times (529)^{1/2}\}$ *[IBPS Clerk-2013]*

(a) $6\frac{1}{2}$ (b) $5\frac{1}{2}$

(c) $5\frac{3}{4}$ (d) $6\frac{3}{4}$

(e) $6\frac{1}{4}$

160. $\left\{\sqrt{7744}\times (11)^2\right\}\div (2)^3=(?)^3$ *[IBPS Clerk-2013]*

(a) 7 (b) 9

(c) 11 (d) 13

(e) 17

161. $(4356)^{1/2}\div\frac{11}{4}=\sqrt{?}\times 6$ *[IBPS Clerk-2013]*

(a) 2 (b) 4

(c) 8 (d) 6

(e) 16

162. $\frac{3}{8}$ of $\{4624\div(564-428)\}=?$ *[IBPS Clerk-2013]*

(a) $13\frac{1}{4}$ (b) $14\frac{1}{2}$

(c) $11\frac{5}{6}$ (d) $12\frac{3}{4}$

(e) $12\frac{1}{8}$

163. $456\div 24\times 38-958+364=?$ *[IBPS Clerk-2013]*

(a) 112 (b) 154

(c) 128 (d) 136

(e) 118

164. $(43)^2+841=(?)^2+1465$ *[IBPS Clerk-2013]*

(a) 41 (b) 35

(c) 38 (d) 33

(e) 30

165. $3\frac{3}{8}\times 6\frac{5}{12}-2\frac{3}{16}\times 3\frac{1}{2}=?$ *[IBPS Clerk-2013]*

(a) 21 (b) 18

(c) 14 (d) 15

(e) 16

166. $(34.5\times 14\times 42)\div 2.8=?$ *[IBPS Clerk-2013]*

(a) 7150 (b) 7365

(c) 7245 (d) 7575

(e) 7335

167. $(216)^4\div(36)^4\times(6)^5=(6)^?$ *[IBPS Clerk-2013]*

(a) 13 (b) 11

(c) 7 (d) 9

(e) 10

168. $\frac{\sqrt{4356\times\sqrt{?}}}{\sqrt{6084}}=11$ *[IBPS Clerk-2013]*

(a) 144 (b) 196

(c) 169 (d) 81

(e) 121

169. $\left(3\frac{6}{17}\div 2\frac{7}{34}-1\frac{9}{25}\right)=(?)^2$ *[IBPS Clerk-2013]*

(a) $\frac{2}{5}$ (b) $\frac{1}{3}$

(c) $\frac{4}{5}$ (d) $\frac{1}{5}$

(e) $\frac{3}{5}$

170. $(1097.63+2197.36-2607.24)\div 3.5=?$ *[IBPS Clerk-2013]*

(a) 211.5 (b) 196.5

(c) 209.5 (d) 192.5

(e) 189.5

171. $\frac{1}{11}$ of $[(17424)^{1/2}\div(66)^2\times 3^3]=?^2$ *[IBPS Clerk-2013]*

(a) $\frac{1}{11}$ (b) $\frac{3}{11}$

(c) $\frac{2}{11}$ (d) $\frac{4}{11}$

(e) $\frac{5}{11}$

Level - II

1. Value of $999\frac{995}{999} \times 999 = ?$

(a) 990809 (b) 998996
(c) 153.6003 (d) 213.0003

2. $7892.35 \times 99.9 = ?$

(a) 753445.765 (b) 764455.765
(c) 788445.765 (d) None of these

3. How many $\frac{1}{12}$ in $18\frac{3}{4}$

(a) 522 (b) 252
(c) 225 (d) 253

4. The least possible positive number which should be added to 575 to make a perfect square number is

(a) 0 (b) 1
(c) 4 (d) None of these

5. If $a * b * c = \sqrt{\frac{(a+2)(b+3)}{(c+1)}}$, then the value of $(6 * 15 * 3)$ is

(a) 6 (b) 3
(c) 4 (d) can't be determined

6. If $x = 3+\sqrt{8}$, then $\left(x^2+\frac{1}{x^2}\right) = ?$

(a) 34 (b) 24
(c) 38 (d) 36

7. If $x^a = y^b = z^c$ and $y^2 = zx$ then the value of $\frac{1}{a}+\frac{1}{c}$ is

(a) $\frac{b}{2}$ (b) $\frac{c}{2}$
(c) $\frac{2}{b}$ (d) $2a$

8. If $\frac{2x}{1+\frac{1}{1+\frac{x}{1-x}}} = 1$, then find the value of x.

(a) $\frac{2}{3}$ (b) $\frac{3}{2}$
(c) 2 (d) $\frac{1}{2}$

9. Find the value of

$$\frac{1}{2\times 3}+\frac{1}{3\times 4}+\frac{1}{4\times 5}+\frac{1}{5\times 6}+.....+\frac{1}{9\times 10}$$

(a) $\frac{3}{2}$ (b) $\frac{2}{5}$
(c) $\frac{2}{3}$ (d) $\frac{3}{5}$

10. Find the square root of $7-2\sqrt{10}$.

(a) $\sqrt{5}+\sqrt{2}$ (b) $-\sqrt{5}-\sqrt{2}$
(c) $\pm(\sqrt{5}-\sqrt{2})$ (d) $\pm(\sqrt{5}+\sqrt{2})$

11. The product of two 2-digit numbers is 1938. If the product of their unit's digits is 28 and that of ten's digits is 15, find the larger number.

(a) 34 (b) 57
(c) 43 (d) 75

12. If $P + P! = P^3$, then the value of P is

(a) 4 (b) 6
(c) 0 (d) 5

13. For any real value of x the maximum value of $8x-3x^2$ is

(a) $\frac{8}{3}$ (b) 4
(c) 5 (d) $\frac{16}{3}$

14. If x is a number satisfying the equation

$\sqrt[3]{x+9}-\sqrt[3]{x-9} = 3$, then x^2 is between

(a) 55 and 65 (b) 65 and 75
(c) 75 and 85 (d) 85 and 95

15. The value of $\left[35.7-\left(3+\frac{1}{3+\frac{1}{3}}\right)-\left(2+\frac{1}{2+\frac{1}{2}}\right)\right]$ is

(a) 30 (b) 34.8
(c) 36.6 (d) 41.4

16. Which one of the following sets of surds is in correct sequence of ascending order of their values?

(a) $\sqrt[4]{10},\sqrt[3]{6},\sqrt{3}$ (b) $\sqrt{3},\sqrt[4]{10},\sqrt[3]{6}$
(c) $\sqrt{3},\sqrt[3]{6},\sqrt[4]{10}$ (d) $\sqrt[4]{10},\sqrt{3},\sqrt[3]{6}$

17. The last three-digits of the multiplication 12345×54321 will be

(a) 865 (b) 745
(c) 845 (d) 945

18. The sum of the two numbers is 12 and their product is 35. What is the sum of the reciprocals of these numbers ?

(a) $\frac{12}{35}$ (b) $\frac{1}{35}$
(c) $\frac{35}{8}$ (d) $\frac{7}{32}$

19. Find the value of $\left(1-\frac{1}{3}\right)\left(1-\frac{1}{4}\right)\left(1-\frac{1}{5}\right)\ldots\left(1-\frac{1}{100}\right)$.
(a) $\frac{1}{5}$ (b) $\frac{1}{10}$
(c) $\frac{1}{50}$ (d) $\frac{2}{5}$

20. An employer pays ₹20 for each day a works, and forfeits ₹ 3 for each day he is idle. At the end of 60 days, a worker gets ₹280. For how many days did the worker remain idle?
(a) 28 (b) 40
(c) 52 (d) 60

21. Simplify : $\dfrac{1}{1+\dfrac{\frac{2}{3}}{1+\frac{2}{3}+\dfrac{\frac{8}{9}}{1-\frac{2}{3}}}}$
(a) $\frac{11}{13}$ (b) $\frac{13}{15}$
(c) $\frac{13}{11}$ (d) $\frac{15}{13}$

22. The value of $\sqrt{\dfrac{(0.03)^2+(0.21)^2+(0.065)^2}{(0.003)^2+(0.021)^2+(0.0065)^2}}$ is
(a) 0.1 (b) 10
(c) 10^2 (d) 10^3

23. If $\dfrac{x^2+y^2+z^2-64}{xy-yz-zx}=-2$ and $x + y = 3z$, then the value of z is
(a) 2 (b) 3
(c) 4 (d) None of these

24. If $\sqrt{24}$ = 4.899, the value of $\sqrt{\frac{8}{3}}$ is
(a) 0.544 (b) 1.333
(c) 1.633 (d) 2.666

25. If $(X+(1/X))=4$, then the value of X^4+1/X^4 is
(a) 124 (b) 64
(c) 194 (d) Can't be determined

26. If $\sqrt{15625}=125$, then the value of $\sqrt{15625}+\sqrt{156.25}+\sqrt{1.5625}$ is
(a) 1.3875 (b) 13.875
(c) 138.75 (d) 156.25

27. A hostel has provisions for 250 students for 35 days. After 5 days, a fresh batch of 25 students was admitted to the hostel. Again after 10 days, a batch of 25 students left the hostel. How long will the remaining provisions survive?
(a) 18 days (b) 19 days
(c) 20 days (d) 17 days

28. If $\frac{97}{19}=a+\cfrac{1}{b+\frac{1}{c}}$ where a, b and c are positive integers, then what is the sum of a, b and c?
(a) 16 (b) 20
(c) 9 (d) Cannot be determined

29. If $a > 1$, then arrange the following in ascending order.
I. $\sqrt[3]{\sqrt[4]{a^3}}$ II. $\sqrt[3]{\sqrt[5]{a^4}}$
III. $\sqrt{\sqrt[3]{a}}$ IV. $\sqrt{\sqrt[5]{a^3}}$
(a) I, II, III, IV (b) I, II, IV, III
(c) IV, I, III, II (d) III, I, II, IV

30. Which of the following is correct if $A=3^{3^3}$, $B=3^{33^3}$, $C=3^{3^{33}}$ and $D=3^{333}$?
(a) $A > B = C > D$ (b) $C > A > B > D$
(c) $A > C > D > B$ (d) $C > B > D > A$

31. Find the value of x in $\sqrt{x+2\sqrt{x+2\sqrt{x+2\sqrt{3x}}}}=x$.
(a) 1 (b) 3
(c) 6 (d) 12

32. Find two numbers such that their sum, their product and the differences of their squares are equal.
(a) $\left(\frac{3+\sqrt{3}}{2}\right)$ and $\left(\frac{1+\sqrt{2}}{2}\right)$ or $\left(\frac{3+\sqrt{2}}{2}\right)$ and $\left(\frac{1+\sqrt{2}}{2}\right)$
(b) $\left(\frac{3+\sqrt{7}}{2}\right)$ and $\left(\frac{1+\sqrt{7}}{2}\right)$ or $\left(\frac{3+\sqrt{6}}{2}\right)$ and $\left(\frac{1-\sqrt{6}}{2}\right)$
(c) $\left(\frac{3-\sqrt{5}}{2}\right)$ and $\left(\frac{1-\sqrt{5}}{2}\right)$ or $\left(\frac{3+\sqrt{5}}{2}\right)$ and $\left(\frac{1+\sqrt{5}}{2}\right)$
(d) None of these

DIRECTIONS (Qs. 33-37) : ***What will come in place of the question mark (?) in the following questions ?***

33. $\sqrt{11449}\times\sqrt{6241}-(54)^2=\sqrt{?}+(74)^2$ *[IBPS-PO-2011]*
(a) 384 (b) 3721
(c) 381 (d) 3638
(e) None of these

34. $\left[\left(3\sqrt{8}+\sqrt{8}\right)\times\left(8\sqrt{8}+7\sqrt{8}\right)\right]-98=?$ *[IBPS-PO-2011]*
(a) $2\sqrt{8}$ (b) $8\sqrt{8}$
(c) 382 (d) 386
(e) None of these

35. $3463\times295-18611=?+5883$ *[IBPS-PO-2011]*
(a) 997091 (b) 997071
(c) 997090 (d) 999070
(e) None of these

36. $\frac{28}{65} \times \frac{195}{308} \div \frac{39}{44} + \frac{5}{26} = ?$ [IBPS-PO-2011]

(a) $\frac{1}{3}$ (b) 0.75

(c) $1\frac{1}{2}$ (d) $\frac{1}{2}$

(e) None of these

37. $(23.1)^2 + (48.6)^2 - (39.8)^2 = ? + 1147.69$ [IBPS-PO-2011]

(a) $(13.6)^2$ (b) $\sqrt{12.8}$

(c) 163.84 (d) 12.8

(e) None of these

DIRECTIONS (Qs. 38-42) : *What approximate value should come in place of the question mark (?) in the following questions?*

(Note : You are not expected to calculate the exact value.)

38. $\sqrt[3]{4663} + 349 = ? \div 21.003$ [IBPS-PO-2011]

(a) 7600 (b) 7650

(c) 7860 (d) 7560

(e) 7680

39. 39.897% of 4331 + 58.779% of 5003 = ? [IBPS-PO-2011]

(a) 4300 (b) 4500

(c) 4700 (d) 4900

(e) 5100

40. $59.88 \div 12.21 \times 6.35 = ?$ [IBPS-PO-2011]

(a) 10 (b) 50

(c) 30 (d) 70

(e) 90

41. $43931.03 \div 2111.02 \times 401.04 = ?$ [IBPS-PO-2011]

(a) 8800 (b) 7600

(c) 7400 (d) 9000

(e) 8300

42. $\sqrt{6354} \times 34.993 = ?$ [IBPS-PO-2011]

(a) 3000 (b) 2800

(c) 2500 (d) 3300

(e) 2600

DIRECTIONS (Qs. 43–47) : *In the following number series only one number is wrong. Find out the wrong number.*

43. 9050 5675 3478 2147 1418 1077 950 [IBPS-PO-2011]

(a) 3478 (b) 1418

(c) 5675 (d) 2147

(e) 1077

44. 7 12 40 222 1742 17390 208608 [IBPS-PO-2011]

(a) 7 (b) 12

(c) 40 (d) 1742

(e) 208608

45. 6 91 584 2935 11756 35277 70558 [IBPS-PO-2011]

(a) 91 (b) 70558

(c) 584 (d) 2935

(e) 35277

46. 1 4 25 256 3125 46656 823543 [IBPS-PO-2011]

(a) 3125 (b) 823543

(c) 46656 (d) 25

(e) 256

47. 8424 4212 2106 1051 526.5 263.25 131.625 [IBPS-PO-2011]

(a) 131.625 (b) 1051

(c) 4212 (d) 8424

(e) 263.25

48. Rubina could get equal number of ₹ 55, ₹ 85 and ₹ 105 tickets for a movie. She spents ₹ 2940 for all the tickets. How many of each did she buy? [IBPS-PO-2011]

(a) 12 (b) 14

(c) 16 (d) Cannot be determined

(e) None of these

49. Seema bought 20 pens, 8 packets of wax colours, 6 calculators and 7 pencil boxes. The price of one pen is ₹ 7, one packet of wax colour is ₹ 22, one calculator is ₹ 175 and one pencil box is ₹ 14 more than the combined price of one pen and one packet of wax colours. How much amount did Seema pay to the shopkeeper?[IBPS-PO-2011]

(a) ₹ 1,491 (b) ₹ 1,725

(c) ₹ 1,667 (d) ₹ 1,527

(e) None of these

DIRECTIONS (Qs. 50-54) : *What will come in place of the question mark (?) in the following questions ?*

50. $4003 \times 77 - 21015 = ? \times 116$ [IBPS-PO-2012]

(a) 2477 (b) 2478

(c) 2467 (d) 2476

(e) None of these

51. $\left[\left(5\sqrt{7}+\sqrt{7}\right)+\left(4\sqrt{7}+8\sqrt{7}\right)\right]-(19)^2 = ?$ [IBPS-PO-2012]

(a) 143 (b) $72\sqrt{7}$

(c) 134 (d) $70\sqrt{7}$

(e) None of these

52. $(4444 \div 40) + (645 \div 25) + (3991 \div 26) = ?$ [IBPS-PO-2012]

(a) 280.4 (b) 290.4

(c) 295.4 (d) 285.4

(e) None of these

53. $\sqrt{33124} \times \sqrt{2601} - (83)^2 = (?)^2 + (37)^2$ [IBPS-PO-2012]

(a) 37 (b) 33

(c) 34 (d) 28

(e) None of these

54. $5\frac{17}{37} \times 4\frac{51}{52} \times 11\frac{1}{7} + 2\frac{3}{4} = ?$ [IBPS-PO-2012]

(a) 303.75 (b) 305.75

(c) $303\frac{3}{4}$ (d) $305\frac{1}{4}$

(e) None of these

DIRECTIONS (Qs. 55-61) : *What approximate value should come in place of the question mark (?) in the following questions? (Note : You are not expected to calculate the exact value.)*

55. $8787 \div 343 \times \sqrt{50} = ?$ [IBPS-PO-2012]

(a) 250 (b) 140

(c) 180 (d) 100

(e) 280

56. $\sqrt[3]{54821} \times (303 \div 8) = (?)^2$ [IBPS-PO-2012]

(a) 48 (b) 38

(c) 28 (d) 18

(e) 58

57. $\frac{5}{8}$ of 4011.33 + $\frac{7}{10}$ of 3411.22 = ? [IBPS-PO-2012]

(a) 4810 (b) 4980

(c) 4890 (d) 4930

(e) 4850

58. 23% of 6783 + 57% of 8431 = ? [IBPS-PO-2012]

(a) 6460 (b) 6420

(c) 6320 (d) 6630

(e) 6360

59. $335.01 \times 244.99 \div 55 = ?$ [IBPS-PO-2012]

(a) 1490 (b) 1550

(c) 1420 (d) 1590

(e) 1400

60. Rachita enters a shop to buy ice-creams, cookies and pastries. She has to buy atleast 9 units of each. She buys more cookies than ice-creams and more pastries than cookies. She picks up a total of 32 items. How many cookies does she buy ? [IBPS-PO-2012]

(a) Either 12 or 13 (b) Either 11 or 12

(c) Either 10 or 11 (d) Either 9 or 11

(e) Either 9 or10

61. With a two digit prime number, if 18 is added, we get another prime number with digits reversed. How many such numbers are possible? [SSC CGL-2012]

(a) 2 (b) 3

(c) 0 (d) 1

62. If $x = \frac{4ab}{a+b}$, then the value of [SSC CGL-2012]

$\frac{x+2a}{x-2a} + \frac{x+2b}{x-2b}$ is

(a) a (b) b

(c) 0 (d) 2

63. If $x = 997$, $y = 998$, $z = 999$, then the value of $x^2 + y^2 + z^2 - xy - yz - zx$ will be [SSC CGL-2012]

(a) 3 (b) 9

(c) 16 (d) 4

64. If $a + b + c = 8$, then the value of [SSC CGL-2012]
$(a-4)^3 + (b-3)^3 + (c-1)^3 - 3(a-4)(b-3)(c-1)$ is

(a) 2 (b) 4

(c) 1 (d) 0

65. If $x = \sqrt{a} + \frac{1}{\sqrt{a}}$, $y = \sqrt{a} - \frac{1}{\sqrt{a}}$, then the value of $x^4 + y^4 - 2x^2y^2$ is [SSC CGL-2012]

(a) 16 (b) 20

(c) 10 (d) 5

66. If $5a + \frac{1}{3a} = 5$, then the value of $9a^2 + \frac{1}{25a^2}$ is [SSC CGL-2012]

(a) $\frac{51}{5}$ (b) $\frac{29}{5}$

(c) $\frac{52}{5}$ (d) $\frac{39}{5}$

67. If $x = 3 + 2\sqrt{2}$, then the value of $\sqrt{x} - \frac{1}{\sqrt{x}}$ is [SSC CGL-2012]

(a) $\pm 2\sqrt{2}$ (b) ± 2

(c) $\pm\sqrt{2}$ (d) $\pm\frac{1}{2}$

68. If $a + b + c = 0$, the value of [SSC CGL-2012]

$\left(\frac{a^2}{bc} + \frac{b^2}{ca} + \frac{c^2}{ab}\right)$ is

(a) 2 (b) 3

(c) 4 (d) 5

69. If a, b, c are real and $a^3 + b^3 + c^3 = 3abc$ and $a + b + c \neq 0$, then the relation between a, b, c will be [SSC CGL-2012]

(a) $a + b = c$ (b) $a + c = b$

(c) $a = b = c$ (d) $b + c = a$

70. If $a = 2$, $b = 3$, then $(a^b + b^a)^{-1}$ is [SSC CGL-2013]

(a) $\frac{1}{31}$ (b) $\frac{1}{17}$

(c) $\frac{1}{21}$ (d) $\frac{1}{13}$

71. The smallest positive integer which when multiplied by 392, gives a perfect square is [SSC CGL-2013]

(a) 2 (b) 3

(c) 5 (d) 7

72. Divide 81 into three parts so that $\frac{1}{2}$ of 1st, $\frac{1}{3}$ of 2nd and $\frac{1}{4}$ of 3rd are equal. [SSC CGL-2013]

(a) 36, 27, 18 (b) 27, 18, 36
(c) 18, 27, 36 (d) 30, 27, 24

73. The expression $x^4 - 2x^2 + k$ will be a perfect square when the value of k is [SSC CGL-2013]

(a) 1 (b) 2
(c) $\frac{1}{2}$ (d) $\frac{1}{4}$

74. If $3x - \frac{1}{4y} = 6$, then the value of $4x - \frac{1}{3y}$ is [SSC CGL-2013]

(a) 2 (b) 4
(c) 6 (d) 8

75. If $a + b + c = 0$, find the value of $\frac{a+b}{c} - \frac{2b}{c+a} + \frac{b+c}{a}$. [SSC CGL-2013]

(a) 0 (b) 1
(c) –1 (d) 2

76. If $x + \frac{4}{x} = 4$, find the value of $x^3 + \frac{4}{x^3}$. [SSC CGL-2013]

(a) 8 (b) $8\frac{1}{2}$
(c) 16 (d) $16\frac{1}{2}$

77. If $x = 3 + 2\sqrt{2}$, then the value of $\left(\sqrt{x} - \frac{1}{\sqrt{x}}\right)$ is [SSC CGL-2013]

(a) 1 (b) 2
(c) $2\sqrt{2}$ (d) $3\sqrt{3}$

78. If 'a' be a positive number, then the least value of $a + \frac{1}{a}$ is [SSC CGL-2013]

(a) 1 (b) 0
(c) 2 (d) $\frac{1}{2}$

79. Arrange the following in ascending order $3^{34}, 2^{51}, 7^{17}$, we get [SSC CGL-2014]

(a) $3^{34} > 2^{51} > 7^{17}$ (b) $7^{17} > 2^{51} > 3^{34}$
(c) $3^{34} > 7^{17} > 2^{51}$ (d) $2^{51} > 3^{34} > 7^{17}$

80. If $x = 2 + \sqrt{3}$, then $x^2 + \frac{1}{x^2}$ is equal to [SSC CGL-2014]

(a) 10 (b) 12
(c) –12 (d) 14

81. If a = 4.965, b = 2.343 and c = 2.622, then the value of $a^3 - b^3 - c^3 - 3abc$ is [SSC CGL-2014]

(a) –2 (b) –1
(c) 0 (d) 9.93^2

82. If $x + y + z = 0$, then the value of $\frac{x^2 + y^2 + z^2}{x^2 - yz}$ is [SSC CGL-2014]

(a) –1 (b) 0
(c) 1 (d) 2

83. In an examination, a boy was asked to multiply a given number by $\frac{7}{19}$. By mistake, he divided the given number by $\frac{7}{19}$ and got a result 624 more than the correct answer. The sum of digits of the given number is [SSC CGL-2014]

(a) 10 (b) 11
(c) 13 (d) 14

84. If $a^2 + b^2 + c^2 = 2a - 2b - 2$, then the value of $3a - 2b + c$ is [SSC CGL-2014]

(a) 0 (b) 3
(c) 5 (d) 2

85. If $a + b + c = 3$, $a^2 + b^2 + c^2 = 6$ and $\frac{1}{a} + \frac{1}{b} + \frac{1}{c} = 1$, where a, b, c are all non-zero, then 'abc' is equal to [SSC CGL-2014]

(a) $\frac{2}{3}$ (b) $\frac{3}{2}$
(c) $\frac{1}{2}$ (d) $\frac{1}{3}$

86. If $a^2 - 4a - 1 = 0$, $a \neq 0$, then the value of $a^2 + 3a + \frac{1}{a^2} - \frac{3}{a}$ is [SSC CGL-2014]

(a) 24 (b) 26
(c) 28 (d) 30

Hints & Solutions

Level-I

1. (d) Given Exp. = $a^2 + b^2 - 2ab$, where $a = 287$ and $b = 269 = (a-b)^2 = (287-269)^2 = (18)^2 = 324$.

2. (d) $20 \times x = (64+36)(64-36) = 100 \times 28$

$$\Rightarrow x = \frac{100 \times 28}{20} = 140.$$

3. (c) $\frac{1}{\sqrt{3}+\sqrt{2}} = \frac{1}{(\sqrt{3}+\sqrt{2})} \times \frac{(\sqrt{3}-\sqrt{2})}{(\sqrt{3}+\sqrt{2})} = \left(\frac{\sqrt{3}-\sqrt{2}}{3-2}\right)$

$$= (\sqrt{3}-\sqrt{2}) = (1.732 - 1.414) = 0.318$$

4. (a) Given expression $= \sqrt{0.01+0.08} = \sqrt{0.09} = 0.3$

5. (a) $356 \times 936 - 356 \times 836 = 356 \times (936 - 836)$
$= 356 \times 100 = 35600$

6. (a) $\dfrac{\frac{1}{2} \div \frac{1}{2} \times \frac{1}{2}}{\frac{1}{2} + \frac{1}{2} \times \frac{1}{2}} = \dfrac{\frac{1}{2} \times 2 \times \frac{1}{2}}{\frac{3}{4}} = \frac{1}{2} \times \frac{4}{3} = \frac{2}{3}$

7. (b) Given exp.

$$= \frac{a^2-b^2}{a+b} = a - b = \left(1 + \frac{1}{1+\frac{1}{100}}\right) - \left(1 - \frac{1}{1+\frac{1}{100}}\right)$$

$$= 2 \times \frac{1}{(101/100)} = 2 \times \frac{100}{101} = \frac{200}{101}$$

8. (a) $5^a = 3125 \Rightarrow 5^a = 5^5 \Rightarrow a = 5$
$\Rightarrow 5^{(a-3)} = 5^{(5-3)} = 5^2 = 25$

9. (d) Let the number of buffaloes be x and the number of ducks be y.
Then, $4x + 2y = 2(x+y) + 24 \Leftrightarrow 2x = 24 \Leftrightarrow x = 12$.

10. (b) $\sqrt[3]{4\frac{12}{125}} = \sqrt[3]{\frac{512}{125}} = \left(\frac{8 \times 8 \times 8}{5 \times 5 \times 5}\right)^{1/3} = \frac{8}{5} = 1\frac{3}{5}$

11. (c) $729 = 9^3 = 3^6$, Now $4X - 2 = 6$ or $X = 2$.

12. (a) $16a^2 - 12a = (4a)^2 - 2(4a)(3/2)$
$\therefore$ The number is $(3/2)^2 = (9/4)$.

13. (b) By rationalization we have

$$\left[\frac{1}{\sqrt{9}-\sqrt{8}}\right] = \left[\frac{1}{\sqrt{9}-\sqrt{8}}\right] \times \frac{\sqrt{9}+\sqrt{8}}{\sqrt{9}+\sqrt{8}} = \frac{\sqrt{9}+\sqrt{8}}{9-8}$$

$$= \sqrt{9} + \sqrt{8}$$

Similarly, $\left[\frac{1}{\sqrt{8}-\sqrt{7}}\right] = \sqrt{8} + \sqrt{7}$ and $\frac{1}{\sqrt{7}-\sqrt{6}} = \sqrt{7} + \sqrt{6}$

and so on. The given expression
$= (\sqrt{9}+\sqrt{8}) - (\sqrt{8}+\sqrt{7}) + (\sqrt{7}+\sqrt{6}) - (\sqrt{6}+\sqrt{5}) + (\sqrt{5}+\sqrt{4})$
$= \sqrt{9} + \sqrt{4} = 3 + 2 = 5$.

14. (b) $5\sqrt[3]{250} + 7\sqrt[3]{16} - 14\sqrt[3]{54}$

$$= 5\sqrt[3]{125 \times 2} + 7\sqrt[3]{8 \times 2} - 14\sqrt[3]{27 \times 2}$$

$$= 5 \times 5\sqrt[3]{2} + 7 \times 2\sqrt[3]{2} - 14 \times 3 \times \sqrt[3]{2}$$

$$= (25 + 14 - 42)\sqrt[3]{2} = -3\sqrt[3]{2}$$

15. (a) Work from the choices: only 169 when reversed becomes 961 and both numbers are squares.

16. (b) $1*2 = 1 + 2 \times 6 = 13$
$13*3 = 13 + 3 \times 6 = 31$

17. (a) $a^b = 125 \Rightarrow a^b = 5^3$
$\therefore a = 5, b = 3$
$(a-b)^{a+b-4} = (5-3)^{5+3-4} = 2^4 = 16$

18. (c) $p \times q = p + q + \frac{p}{q} \Rightarrow 8 \times 2 = 8 + 2 + \frac{8}{2} = 14$

19. (b) $x*y = x^2 + y^2 = xy$
$9*11 = 9^2 + 11^2 - 9 \times 11$
$= 81 + 121 - 99 = 103$

20. (d) Since the factors of 11760 are $\underline{2 \times 2} \times \underline{2 \times 2} \times 3 \times 5 \times \underline{7 \times 7}$ so we need to multiply it with 3×5 because all the factos are paired but 3 and 5 are unpaired, hence (d) is the correct choice.

21. (b) $5\sqrt{5} \times 5^3 \div 5^{-3/2} = 5^{(a+2)}$

$$5^1 \times 5^{\frac{1}{2}} \times 5^3 \times 5^{3/2} = 5^{a+2}$$

$$5^{1+\frac{1}{2}+3+\frac{3}{2}} = 5^{a+2},\ 5^{\frac{12}{2}} = 5^{a+2},\ a + 2 = 6$$

$\therefore a = 4$

22. (b) Let number be x

$$x \times \frac{4}{5} \times \frac{3}{4} - x \times \frac{2}{5} \times \frac{1}{6} = 648$$

$$\frac{3x}{5} - \frac{x}{15} = 648$$

$$\frac{9x - x}{15} = 648$$

$$8x = 648 \times 15 \Rightarrow x = \frac{648 \times 15}{8} = 81 \times 15 = 1215$$

23. (b) If sum of two is even, their difference is always even, So (b) is right answer.

24. (a) Number of decimal places in the given expression = 8
Number of decimal places in (a) = 8
Number of decimal places in (b) = 9
Number of decimal places in (c) = 7.
Clearly, the expression in (a) is the same as the given expression.

25. (c) 26. (b)

27. (c) $\frac{a+b}{b+c}=\frac{c+d}{d+a}$

$\Rightarrow ad+a^2+bd+ab=bc+c^2+bd+cd$

$\Rightarrow (a^2-c^2)+(ad-cd)+(ab-bc)=0$

$\Rightarrow (a-c)(a+c+d+b)=0$

$\Rightarrow a=c$ or $a+b+c+d=0$ or both

28. (d) Sum is 888 $\Rightarrow$ unit's digit should add up to 8. This is possible only for 4th option as "3" + "5" = "8".

29. (b) The numbers that can be formed are xy and yx. Hence $(10x+y)+(10y+x)=11(x+y)$. If this is a perfect square then $x+y=11$.

30. (a) Comparing $4^{1/3}$ and $5^{1/4}$
$(4^{1/3})^{12}$ and $(5^{1/4})^{12}$ *i.e.*, 4^4 and 5^3
$=256>125$
$\therefore\ 4^{1/3}>5^{1/4}$
Similarly, comparing $5^{1/4}$ and $6^{1/5}$
$(5^{1/4})^{20}$ and $(6^{1/5})^{20}$ *i.e.*, 5^5 and $6^4=3125>1296$
$\therefore\ 5^{1/4}>6^{1/5}$

31. (b) $(a+b+c)^2=a^2+b^2+c^2+2(ab+bc+ca)$
$\Rightarrow 2(ab+bc+ca)=(a+b+c)^2-(a^2+b^2+c^2)$
$=169-69=100$
$ab+bc+ca=50$

32. (b) $|a-b|=|8|=8 \Rightarrow |b-a|=|-8|=8$

$\Rightarrow |a-b|-|b-a|=8-8=0$

33. (d) At a value of $x=0$ we can see that the expression $x^2+|x-1|=1\rightarrow 0+1=1$. Hence, $x=0$ satisfies the given expression. Also at $x=1$, we get $1+0=1$.

34. (d) $3.6+36.6+3.66+0.36+3.0=47.22$

35. (a) $23\times 45\div 15=69$

36. (b) $4\frac{5}{6}+7\frac{1}{2}-5\frac{8}{11}=\frac{29}{6}+\frac{15}{2}-\frac{63}{11}$

$=\frac{319+495-378}{66}=\frac{436}{66}=\frac{218}{33}=6\frac{20}{33}$

37. (e) $\frac{210}{14}\times\frac{17}{15}\times ?=4046$

$?=\frac{4046\times 15\times 14}{210\times 17}=238$

38. (b) 83% of 2350 = ?

$?=\frac{83\times 2350}{100}=1950.50$

39. (b) $\sqrt{1089}+3=(?)^2$

$33+3=(6)^2$

40. (b) $96+32\times 5-31=225$

41. (d) $?\div 36=(7)^2-8$

$\frac{?}{36}=49-8$

$\therefore\ ?=36\times 41$
$=1476$

42. (d) $\sqrt{8281}=?=91$

43. (d) $(63)^2-(12)^2=3825$

44. (e) $\frac{9}{4}+\frac{18}{5}=?-\frac{43}{10}$

$?=\frac{9}{4}+\frac{18}{5}+\frac{43}{10}$

$=\frac{45+72+86}{20}=\frac{203}{20}=10\frac{3}{20}$

45. (a) $17\times 19\times 4\div ?=161.5$

$\frac{1}{?}=\frac{1615}{10\times 17\times 19\times 4}$.

$?=8$

46. (b) $1798\div 31\times ?=348$

$?=\frac{348\times 31}{1798}=6$

47. (d) $(9.8\times 2.3+4.46)\div 3=3^?$
$27\div 3=3^?\quad 3^2=3^?$
$\therefore\ ?=2$

48. (d) 43% of 600 + ?% of 300 = 399
$43\times 6+3x=399$
$3x=141$
$x=47$

49. (e) $x+(x+1)+(x+2)=1383$
$\Rightarrow 3x+3=1383$
$\Rightarrow 3x=1380$
$\Rightarrow x=\frac{1380}{3}=460$
Largest number $=x+2=462$

50. (e) $1504\times 5.865-24.091=?$
$\therefore\ ?=8796.869\approx 8800$

51. (c) $16.928+(24.7582\div 5.015)=?$
$16.928+(4.93)=?$
$\therefore\ ?=21.86\approx 22$

52. (a) $?=\sqrt[3]{7.938}\times(6.120)^2-4.9256$
$=(2\times 37.4)-4.9256$
$=74.8-4.9256$
$\approx 70\quad \therefore\ ?\approx 70$

53. (a) $16.046\div 2.8\times 0.599=?$
$(5.73)\times 0.599=?$
$\therefore\ ?=3.43\approx 3.5$

54. (d) $\sqrt{963}+(4.895)^2-9.24=?$
$31+23.9-9.24=?$
$54.91-9.24=?$
$\therefore\ ?=45.6\approx 45$

55. (b) $(12\times 19)+(13\times 8)=(15\times 14)+?$
$228+104=210+?$
$\therefore\ ?=122$

56. (c) $\sqrt{65\times 12-50+54}=?$
$\Rightarrow \sqrt{780-50+54}=?$
$\therefore\ ?=28$

57. (c) 15% of 524 – 2% of 985 + ? = 20% of 423

$\frac{15\times524}{100}-\frac{2\times985}{100}+?=\frac{20\times423}{100}$

78.6 – 19.7 – 84.6 = – ?

$\therefore$? = 25.7.

58. (e) $151\times8+(228\div19)^2=?$

$1208+(12)^2=?$

1208 + 144 = ?

$\therefore$? = 1352

59. (c) $\sqrt{1521}+\sqrt{225}=?$

39 + 15 = ?

$\therefore$? = 54

60. (d) 38.734 + 8.638 – 5.19 = ?

47.372 – 5.19 = ?

$\therefore$? = 42.182

61. (a) $7^{8.9}\div(343)^{1.7}\times(49)^{4.8}=7^?$

$\frac{7^{8.9}}{7^{3\times1.7}}\times7^{2\times4.8}=7^?,$

$\frac{7^{8.9}}{7^{5.1}}\times7^{9.6}=7^?$

$7^{8.9-5.1+9.6}=7^?,$

$\therefore$? = 13.4

62. (e) $\sqrt[3]{512}\div\sqrt[4]{16}+\sqrt{576}=?$

(8 ÷ 2) + 24 = ?

4 + 24 = ?

$\therefore$? = 28

63. (b) (42 × 3.2) ÷ (16 × 1.5) = ?

134.4 ÷ 24 = ?

$\therefore$? = 5.6

64. (e) $(199+((5^3\div4)\times4^2))=?$

$(199+(31.25\times4^2)=?$

(199 + 500) = ?

$\therefore$? = 699

65. (e) (342 ÷ 6) × 28 = 1099 + ?

57 × 28 = 1099 + ?

1596 – 1099 = ?

$\therefore$? = 497

66. (d) $\frac{9.8\times2.5\times7.6}{0.5}=?$

$=\frac{186.2}{0.5}=?$

$\therefore$? = 372.4

67. (d) $\frac{3}{5}$ of $\frac{2}{7}$ of ? = 426

? = $\frac{5}{3}$ of $\frac{7}{2}$ of 426

? = $\frac{5}{3}$ of 1491

$\therefore$? = 2485

68. (a) $3\frac{2}{5}+1\frac{2}{9}=4\frac{4}{5}-?$

$\frac{17}{5}+\frac{11}{9}=\frac{24}{5}-?$

$\frac{17}{5}+\frac{11}{9}-\frac{24}{5}=-?$

$\frac{153+55-216}{45}=-?$

$\therefore\ ?=\frac{8}{45}$

69. (b) $\left(\frac{13}{63}\div\frac{104}{14}\right)\times\frac{52}{19}=?$

$\left(\frac{13}{63}\times\frac{14}{104}\right)\times\frac{52}{19}=?$

$\therefore$? = 13/171

70. (a) $\sqrt[3]{13824}\times\sqrt{?}=864$

$\sqrt[3]{24\times24\times24}\times\sqrt{?}=864$

$\Rightarrow 24\times\sqrt{?}=864$

$\Rightarrow \sqrt{?}=\frac{864}{24}$

$\therefore$? = 36 × 36 = 1296

71. (b) $(91)^2+(41)^2-\sqrt{?}=9858$

$\Rightarrow 8281+1681-\sqrt{?}=9858$

$\Rightarrow \sqrt{?}=9962-9858=104$

$\therefore$? = 104 × 104 = 10816

72. (b) ? = 4900 ÷ 28 × 444 ÷ 12

$\Rightarrow$? = 175 × 37

$\Rightarrow$? = 6475

73. (e) 125% of 260 + ?% of 700 = 500

$\Rightarrow$?% of 700 = 500 – 125% of 260

$\Rightarrow$?% of 700 = 175

$\therefore\ ?=\frac{175\times100}{700}=25$

74. (b) $3\frac{7}{11}+7\frac{3}{11}\times1\frac{1}{2}=\frac{40}{11}+\frac{80}{11}\times\frac{3}{2}=\frac{160}{11}=14\frac{6}{11}$

75. (c) Given Expression = $\frac{0.207}{\frac{0.0023}{23}}=\frac{0.207}{0.0001}=\frac{0.2070}{0.0001}=2070.$

76. (b) ? = 1.05% of 2500 + 2.5% of 440

$\Rightarrow ?=\frac{1.05}{100}\times2500+\frac{2.5}{100}\times440$

$\Rightarrow ?=\frac{2625}{100}+\frac{1100}{100}$

$\Rightarrow ?=\frac{3725}{100}=37.25$

77. **(a)** $? = \dfrac{17 \times 4 + 4^2 \times 2}{90 \div 5 \times 12}$

$\Rightarrow ? = \dfrac{68 + 16 \times 2}{18 \times 12}$

$\Rightarrow ? = \dfrac{68 + 32}{216}$

$\Rightarrow ? = \dfrac{100}{216} = \dfrac{25}{54}$

78. **(b)** $\dfrac{87}{5} \times \dfrac{37}{8} - ? = \dfrac{375}{8}$

$\Rightarrow ? = \dfrac{3219}{40} - \dfrac{375}{8}$

$= \dfrac{3219 - 1875}{40} = \dfrac{1344}{40}$

$= \dfrac{168}{5} = 33\dfrac{3}{5}$

79. **(e)** $\dfrac{250 \times 136}{100} + \dfrac{550 \times ?}{100} = 670$

$\Rightarrow 340 + 5.5 \times ? = 670$

$\Rightarrow 5.5 \times ? = 670 - 340 = 330$

$\Rightarrow ? = \dfrac{330}{5.5} = 60$

80. **(a)** $3889 + 12.952 - ? = 3854.002$

or $? = 3889 + 12.952 - 3854.002 = 47.95$

81. **(b)** $(25)^? = (5 \times 5 \times 5 \times 5 \times 5 \times 5)^4 \times (5 \times 5)^6 \div (5)^2$

$= (25 \times 25 \times 25)^4 \times (25)^6 \div (25)^1$

$= (25^3)^4 \times (25)^6 \div 25^1 = (25)^{12} \times (25)^6 \div (25)^1$

$= (25)^{12+6-1} = (25)^{17}$

$\therefore \quad ? = 17$

82. **(c)** $? = \dfrac{28 \times 5 - 15 \times 6}{7^2 + \sqrt{256} + (13)^2}$

$\Rightarrow ? = \dfrac{140 - 90}{49 + 16 + 169}$

$\Rightarrow ? = \dfrac{50}{234} = \dfrac{25}{117}$

83. **(c)** $1.5 \times 0.025 + (?)^2 = 0.1 \Rightarrow (?)^2 = 0.1 - 1.5 \times 0.025$

$\Rightarrow (?)^2 = 0.1 - 0.0375 \Rightarrow ? = \sqrt{.0625} = 0.25$

84. **(d)** Given Expression $= \dfrac{(a-b)^2 + (a+b)^2}{(a^2+b^2)} = \dfrac{2(a^2+b^2)}{(a^2+b^2)} = 2$

85. **(b)** $? = \dfrac{(10008.99)^2}{10009.001} \times \sqrt{3589} \times 0.4987$

$= (10009) \times \sqrt{3600} \times 0.50$

$= 10009 \times 60 \times 0.50 \approx 300000$

86. **(a)** $196.1 \times 196.1 \times 196.1 \times 4.01 \times 4.01 \times 4.001 \times 4.999 \times 4.999$

$= (196.1)^3 \times 4 \times ?$

or $4 \times ? = 4.01 \times 4.001 \times 4.999 \times 4.999$ or $? = 4 \times 5 \times 5 = 100$

87. **(d)** $\because 12.25 \times ? \times 21.6 = 3545.64$

$\because ? = \dfrac{3545.64}{264.6} = 13.4 \approx 13$

88. **(c)** Let x be there in place of question mark so, $x\%$ of $45.999 \times 16\%$ of $83.006 = 116.073$.

We get, $\dfrac{x}{100} \times 46 \times \dfrac{16}{100} \times 83 = 116$

$x \times 0.46 \times 13.28 = 116$

or $\quad x \times 6.11 = 116$

$\Rightarrow \quad x = 18.98 \approx 19.$

89. **(d)** $? = \dfrac{1.69 \times 17.64}{2.7} = 11.04 \approx 11$

90. **(c)**

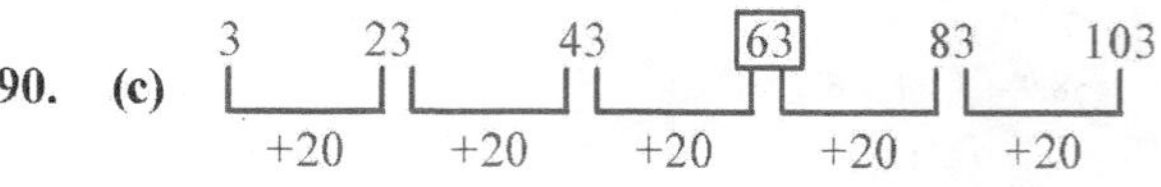

91. **(d)**

1	9	25	49	81	[121]	169
↑	↑	↑	↑	↑	↑	↑
1^2	3^2	5^2	7^2	9^2	11^2	13^2

92. **(e)** Pattern of the series would be as follows

$5 \times 1 + 1 = 6$

$6 \times 2 + 2 = 14$

$14 \times 3 + 3 = 45$

$\therefore \quad 45 \times 4 + 4 = 184$

93. **(e)** The pattern of the number series is:

$7 \times 1 + 1 = 8$

$8 \times 2 + 2 = 18$

$18 \times 3 + 3 = 57$

$57 \times 4 + 4 = \boxed{232}$

94. **(b)** Can you see that the pattern is

$1^2, 2^3, 3^2, 4^3, 5^2, 6^3, 7^2$

95. **(c)** By going options, 26 years is the present age. Present age be 26, then last year age was 25 which represents a perfect square and next year age would be 27 which represents a cubic number.

96. **(a)** Expression is $(2.1)^2 \times \sqrt{0.0441} = 4.41 \times 0.21 = 0.9261$

97. **(b)** $\sqrt[3]{1372} \times \sqrt[3]{1458}$

$= 7\sqrt[3]{4} \times 9\sqrt[3]{2} = 63 \times \sqrt[3]{4 \times 2} = 63 \times 2 = 126$

98. **(a)** Amounts of water in two jars are equal; the jar with the greater capacity is $\dfrac{1}{4}$ full, and the Jar with lesser capacity is $\dfrac{1}{3}$ full.

$\therefore$ When the water in smaller jar is poured into the larger Jar, the addition of an equal amount of water will double the amount in the larger jar, which will then be

$2\times\frac{1}{4}=\frac{1}{2}$ full.

99. (c) $\frac{5x-3}{x}+\frac{5y-3}{y}+\frac{5z-3}{z}=0$

$\frac{5x}{x}-\frac{3}{x}+\frac{5y}{y}-\frac{3}{y}+\frac{5z}{z}-\frac{3}{z}=0$

$5-\frac{3}{x}+5-\frac{3}{y}+5-\frac{3}{z}=0$

$-3\left[\frac{1}{x}+\frac{1}{y}+\frac{1}{z}\right]+15=0$

$-3\left[\frac{1}{x}+\frac{1}{y}+\frac{1}{z}\right]=-15$

$\frac{1}{x}+\frac{1}{y}+\frac{1}{z}=\frac{-15}{-3}=5$

100. (b) $x^2+\frac{1}{x^2+1}-3$

is minimum when $x=0$

$0+\frac{1}{0+1}-3=-2$

101. (c) $a+b=5$

Squaring on both sides

$(a+b)^2=(5)^2$

$a^2+b^2+2ab=25$

$13+2ab=25$

$2ab=25-13=12$ ...(1)

Again, $a^2+b^2=13$

Subtracting $(-2ab)$ from both sides

$a^2+b^2-2ab=13-2ab$

$(a-b)^2=13-12$ from equation (1)

$(a-b)^2=1$

TRICK $\Rightarrow a=3$

$b=2\ (a>b)$

$a-b=1$

102. (a) $\frac{3x-y}{x+5y}=\frac{5}{7}\Rightarrow 21x-7y=5x+25y$

$\Rightarrow 16x=32y$

$\Rightarrow x=2y$ or $\frac{x}{y}=\frac{2}{1}$, ...(1)

Now, to calculate value of $\frac{x+y}{x-y}$, divide numerator & denominator by y.

$\Rightarrow \frac{\frac{x}{y}+1}{\frac{x}{y}-1}$

Putting value of $\frac{x}{y}$ from equation (1)

$\frac{\frac{2}{1}+1}{\frac{2}{1}-1}=\frac{3}{1}$ or 3:1

103. (c) $1+\cfrac{1}{1+\cfrac{2}{\cfrac{15+4}{5}}}=1+\cfrac{1}{1+\cfrac{2\times5}{19}}$

$=1+\cfrac{1}{\cfrac{19+10}{19}}=1+\frac{19}{29}=\frac{29+19}{29}=\frac{48}{29}$

104. (c) $\sqrt{19.36}+\sqrt{0.1936}+\sqrt{0.001936}+\sqrt{0.00001936}$

$=4.4+0.44+0.044+0.0044=4.8884$

105. (c) Let the number be x and y.

According to question,

$(x+y)^2=4xy$

$\Rightarrow x^2+y^2+2xy-4xy=0$

$\Rightarrow (x-y)^2=0$

$\Rightarrow x=y$

106. (c) $a^2+b^2=5ab$

$\Rightarrow \frac{a^2+b^2}{ab}=5$

$\Rightarrow \frac{a}{b}+\frac{b}{a}=5$

On squaring both sides.

$\therefore \left(\frac{a}{b}+\frac{b}{a}\right)^2=25$

$\Rightarrow \frac{a^2}{b^2}+\frac{b^2}{a^2}+2=25$

$\Rightarrow \frac{a^2}{b^2}+\frac{b^2}{a^2}=25-2=23$

107. (d) $x=\frac{\sqrt{3}-\sqrt{2}}{\sqrt{3}+\sqrt{2}}=\frac{(\sqrt{3}-\sqrt{2})(\sqrt{3}-\sqrt{2})}{(\sqrt{3}+\sqrt{2})(\sqrt{3}-\sqrt{2})}$

$=\frac{(\sqrt{3}-\sqrt{2})^2}{3-2}=3+2-2\sqrt{3}.\sqrt{2}=5-2\sqrt{6}$

$\therefore y=\frac{\sqrt{3}+\sqrt{2}}{\sqrt{3}-\sqrt{2}}=5+2\sqrt{6}$

$\therefore x+y=5-2\sqrt{6}+5+2\sqrt{6}=10$

$xy=(5-2\sqrt{6})\cdot(5+2\sqrt{6})$

$=25-24=1$

$\therefore x^3+y^3=(x+y)^3-3xy(x+y)$

$=(10)^3-3(10)=1000-30=970$

108. (d) LCM of 3, 2 and 6 = 6

$\therefore (3)^{\frac{1}{3}} = (3^2)^{\frac{1}{6}} = (9)^{\frac{1}{6}}$

$2^{\frac{1}{2}} = (2^3)^{\frac{1}{6}} = (8)^{\frac{1}{6}}$

$(1)^{\frac{1}{6}} = 1(6)^{\frac{1}{6}} = (6)^{\frac{1}{6}}$

109. (b) $\frac{\sqrt{24}+\sqrt{6}}{\sqrt{24}-\sqrt{6}} = \frac{2\sqrt{6}+\sqrt{6}}{2\sqrt{6}-\sqrt{6}} = \frac{3\sqrt{6}}{\sqrt{6}} = 3$

110. (b) $3 \div \left[3 \div \left\{2 \div \frac{34}{13}\right\}\right]$

$3 \div \left[3 \div 2 \times \frac{13}{34}\right]$

$3 \div \left[3 \times \frac{34}{2 \times 13}\right]$

$\frac{3 \times 2 \times 13}{3 \times 34} = \frac{13}{17}$

111. (d) $9 + 3 \div 4 - 8 \times 2 = ?$

Applying rules

$9 \div 3 \times 4 + 8 - 2 = ?$

$3 \times 4 + 8 - 2 = ?$

$20 - 2 = ?$

$? = 18$

112. (a) Next term will be

$\left(1+\frac{1}{2}\right)\left(1+\frac{1}{3}\right)\left(1+\frac{1}{4}\right)\left(1+\frac{1}{5}\right)$

$= \frac{3}{2} \times \frac{4}{3} \times \frac{5}{4} \times \frac{6}{5} = 3$

113. (b) $2a^2 - 5ab + 2b^2$

$2(a^2 - 2ab + b^2) - ab$

$2(a-b)^2 - ab$

$2[\sqrt{6}+\sqrt{5}-\sqrt{6}+\sqrt{5}]^2 - (\sqrt{6}+\sqrt{5})(\sqrt{6}-\sqrt{5})$

$2 \times 4 \times 5 - 1 = 39$

114. (c) $27P^3 - \frac{1}{216} - \frac{9}{2}P^2 + \frac{1}{4}P$

$= (3P)^3 - \left(\frac{1}{6}\right)^3 - 3.(3P)^2.\frac{1}{6} + 3.3P.\left(\frac{1}{6}\right)^2$

$= \left(3P - \frac{1}{6}\right)^3$

$= \left(3 \times \frac{5}{18} - \frac{1}{6}\right)^3 = \frac{8}{27}$

115. (d) $x + \frac{1}{x} = 2$

$x^2 - 2x + 1 = 0; (x-1)^2 = 0; x = 1$

$x^{2013} + \frac{1}{x^{2014}} = 1 + 1 = 2$

116. (d) Here, $a + b + c = 0$

$\therefore a^3 + b^3 + c^3 - 3abc = 0$

117. (d)

$\left[(\sqrt{6}-\sqrt{35})+(\sqrt{10}-\sqrt{21})\right]\left[(\sqrt{6}-\sqrt{35})-(\sqrt{10}-\sqrt{21})\right]$

$= (\sqrt{6}-\sqrt{35})^2 - (\sqrt{10}-\sqrt{21})^2$

$= 6 + 35 - 2\sqrt{6}.\sqrt{35} - 10 - 21 + 2\sqrt{10}.\sqrt{21}$

$= 10 - 2\sqrt{210} + 2\sqrt{210}$

$= 10$

118. (b) $x + y + z = a - b + b - c + c - a = 0$

$\therefore x^3 + y^3 + z^3 - 3xyz = 0$

119. (c) $\frac{\sqrt{32}+\sqrt{48}}{\sqrt{8}+\sqrt{12}} = \frac{\sqrt{2\times2\times2\times2\times2}+\sqrt{2\times2\times2\times2\times3}}{\sqrt{2\times2\times2}+\sqrt{2\times2\times3}}$

$\Rightarrow \frac{4\sqrt{2}+4\sqrt{3}}{2\sqrt{2}+2\sqrt{3}} = \frac{2(2\sqrt{2}+2\sqrt{3})}{(2\sqrt{2}+2\sqrt{3})} = 2$

120. (b) $\sqrt{\frac{9.5 \times 0.085}{0.0017 \times 0.19}} = \sqrt{\frac{95}{10} \times \frac{85}{1000} \times \frac{10000}{17} \times \frac{100}{19}}$

$\Rightarrow \sqrt{5 \times 5 \times 100} = 50$

121. (a) $\frac{x}{b+c} = \frac{y}{c+a} = \frac{z}{a+b} = k(\text{say})$

So, $x = k(b+c)$

$\Rightarrow x - y = k(b+c) - k(c+a)$

$= k(b-a)$

$y = k(c+a)$

$\Rightarrow y - z = k(c+a) - k(a+b)$

$= k(c-b)$

$z = k(a+b) \Rightarrow z - x = k(a+b) - k(b+c) = k(a-c)$

So,check option (a)

$\frac{x-y}{b-a} = \frac{y-z}{c-b} = \frac{z-x}{a-c}$

$\frac{k(b-a)}{b-a} = \frac{k(c-b)}{c-b} = \frac{k(a-c)}{a-c}$

$k = k = k$

option (a) is true.

122. (d) $\frac{547.527}{0.0082} = x \Rightarrow \frac{547527}{1000} \times \frac{10000}{82} = x$

$\Rightarrow \frac{547527}{82} = \frac{x \times 1000}{10000} \Rightarrow \frac{x}{10}$

123. (a) $\left[3^n\right]^{\frac{1}{3}} = 27$

$\Rightarrow \quad 3^{\frac{n}{3}} = 3^3$

Comparing, $\frac{n}{3} = 3$

x=9

124. (b) Time difference between 9.00 A.M & 2.00 P.M = 5 hours
Temperature difference between 21°C & 36°C
= 36 – 21 = 15°C
Now, Time difference between 9.00 A.M & 12.00 Noon = 3 hrs.

In 5 hours $\xrightarrow[\text{difference}]{\text{temperature}}$ 15°C

So, In 3 hours $\xrightarrow[\text{difference}]{\text{temperature}} \left(\frac{15}{5}\times 3\right) = 9°C$

So, temperature at noon = 21 + 9 = 30°C

125. (b) $\frac{3x+5}{5x-2} = \frac{2}{3}$

$\Rightarrow \quad 9x + 15 = 10x - 4$

$\Rightarrow \quad 15 + 4 = 10x - 9x$

$\Rightarrow \quad x = 19$

126. (d) Let the numbers are x, y.

$x - y = 3$ ...(1)

$x^2 - y^2 = 39$

$\Rightarrow \quad (x-y)(x+y) = 39$

$\Rightarrow \quad x + y = 13$ ...(2)

Adding eqn (1) and (2)

$x + y + x - y = 16$

$\Rightarrow \quad x = 8$

$\therefore \quad y = 3$

Hence, 8 is the larger number.

127. (c) $x = \sqrt{3} + \sqrt{2}$

$\frac{1}{x} = \frac{1}{\sqrt{3}+\sqrt{2}} \times \frac{\sqrt{3}-\sqrt{2}}{\sqrt{3}-\sqrt{2}} = \frac{\sqrt{3}-\sqrt{2}}{3-2} = \sqrt{3}-\sqrt{2}$

Now, $x^3 - \frac{1}{x^3} = \left(\sqrt{3}+\sqrt{2}\right)^3 - \left(\sqrt{3}-\sqrt{2}\right)^3$

$= (a+b)^3 - (a-b)^3$ [Let $\sqrt{3}$ = a and $\sqrt{2}$ = b]

$= a^3 + b^3 + 3a^2b + 3b^2a - (a^3 - b^3 - 3a^2b + 3b^2a)$

$= a^3 + b^3 + 3a^2b + 3b^2a - a^3 + b^3 + 3a^2b - 3b^2a$

$= 2b^3 + 6a^2b = 2\left(\sqrt{2}\right)^3 + 6\left(\sqrt{3}\right)^2\left(\sqrt{2}\right)$

$= 4\sqrt{2} + 18\sqrt{2} = 22\sqrt{2}$

128. (a) $a^2 + b^2 + c^2 = 2(a - b - c) - 3$

$\Rightarrow \quad a^2 + b^2 + c^2 - 2(a - b - c) + 3 = 0$

$\Rightarrow \quad a^2 + b^2 + c^2 - 2a + 2b + 2c + 3 = 0$

$\Rightarrow \quad (a^2 + 1 - 2a) + (b^2 + 1 + 2b) + (c^2 + 1 + 2c) = 0$

$\Rightarrow \quad (a-1)^2 + (b+1)^2 + (c+1)^2 = 0$

This is possible when $(a-1)^2 = 0, (b+1)^2 = 0$ and $(c+1)^2 = 0.$

$\Rightarrow \quad a = 1, b = -1, c = -1$

Thus, $2a - 3b + 4c = 2(1) - 3(-1) + 4(-1)$
$= 2 + 3 - 4 = 1.$

129. (a) $\sqrt{6} = 2.44, \sqrt{5} = 2.23, \sqrt{3} = 1.73$

$a = \sqrt{6} - \sqrt{5} = 0.21$

$b = \sqrt{5} - 2 = 0.23$

$c = 2 - \sqrt{3} = 0.27$

130. (c) Given $a = \frac{b^2}{b-a}$ or $ab - a^2 = b^2$ or $ab = b^2 + a^2$

We know, $a^3 + b^3 = (a+b)(a^2 + b^2 - ab)$

$\therefore$ (a + b) (ab – ab) $\Rightarrow$ 0 (using given)

131. (a) $\frac{1}{x^2 + xy + xz} + \frac{1}{y^2 + xy + zy} + \frac{1}{z^2 + yz + zx}$

$\frac{1}{x(x+y+z)} + \frac{1}{y(x+y+z)} + \frac{1}{z(x+y+z)}$

$\frac{xy + yz + zx}{xyz(x+y+z)} = 0 \quad (\because xy + yz + zx = 0)$

132. (d) $\sqrt{40 + \sqrt{9 \times 9}} = \sqrt{49} = 7$

133. (d) $\sqrt[3]{2} = 2^{1/3}$ or $2^{\frac{1}{3}\times\frac{2}{2}} = 2^{\frac{2}{6}} = \sqrt[6]{4}$

$\sqrt{3} = 3^{\frac{1}{2}}$ or $3^{\frac{1}{2}\times\frac{3}{3}} = 3^{\frac{3}{6}} = \sqrt[6]{27}$

$\sqrt{3} > \sqrt{2}$

134. (d) $a^2 + b^2 + c^2 = (a+b+c)^2 - 2(ab + bc + ca)$

$= 9^2 - 2(ab + bc + ca)$

$a^2 + b^2 + c^2$ will be minimum if ab + bc + ca is maximum.

$ab + bc + ca$ is maximum when $a = 3, b = 3$, and $c = 3$.
[$\because$ a + b + c = 9]

$\therefore$ minimum value of $a^2 + b^2 + c^2$

$= 81 - 2(3 \times 3 + 3 \times 3 + 3 \times 3)$

$= 81 - 54 = 27$

135. (b) $3+\frac{1}{\sqrt{3}}+\frac{1}{3+\sqrt{3}}+\frac{1}{\sqrt{3}-3}$

$\Rightarrow 3+\frac{1}{\sqrt{3}}\times\frac{\sqrt{3}}{\sqrt{3}}+\frac{1}{3+\sqrt{3}}\times\frac{3-\sqrt{3}}{3-\sqrt{3}}+\frac{1}{\sqrt{3}-3}\times\frac{\sqrt{3}+3}{\sqrt{3}+3}$

$\Rightarrow \frac{3}{1}+\frac{\sqrt{3}}{3}+\frac{3-\sqrt{3}}{6}+\frac{\sqrt{3}+3}{-6}$

$\Rightarrow \frac{18+2\sqrt{3}+3-\sqrt{3}-\sqrt{3}-3}{6}$

$\Rightarrow \frac{18+2\sqrt{3}-2\sqrt{3}}{6} \quad \Rightarrow 3$

136. (b) 228 – 18 = 210 is exactly divisible biggest two digit no. 70

137. (a) If they are equal number of rows and columns then,
$\sqrt{1369}=37$

138. (a) $\sqrt{5}+\sqrt{3}>\sqrt{6}+\sqrt{2}$
Squaring both sides
$5+3+2\sqrt{15}>6+2+2\sqrt{12}$
$\sqrt{15}>\sqrt{12}$ which is true

139. (d) Given expression implies $?=\frac{3325}{25}\times\frac{152}{16}$
$=133\times9.5=1263.5$

140. (e) $\sqrt{3136}-\sqrt{1764}=\sqrt{?}$
$\Rightarrow 56-42=\sqrt{?}$
$\Rightarrow \sqrt{?}=14$
On squaring both the side
$\therefore$ $?=14\times14=196$

141. (e) $?=5+\frac{1}{5}+2+\frac{2}{15}+3+\frac{2}{3}$
$=10+\frac{1}{5}+\frac{2}{15}+\frac{2}{3}$
$=10+\frac{3+2+10}{15}=10+\frac{15}{15}$
$=10+1=11$

142. (c) $?=-15-27-88-63+255$
$=-193+255=62$

143. (b) Given expression can be written as
$?=\frac{2525\times0.25\times7}{5}=883.75$

144. (b) $?=\frac{14}{19}\times\frac{57}{70}\times\frac{20}{21}=\frac{2}{1}\times\frac{3}{10}\times\frac{20}{21}=\frac{2}{1}\times\frac{1}{1}\times\frac{2}{7}=\frac{4}{7}$

145. (e) $?=\frac{500\times32}{100}+\frac{50\times162}{100}$
$=160+81=241$

146. (d) $45316+52131-65229$
$=?+15151$
$\Rightarrow$ $32218=?+15151$
$\therefore$ $?=32218-15151=17067$

147. (a) $?=\sqrt{25-12+155+1}$
$=\sqrt{169}=13$

148. (c) $?=\frac{184\times4}{\frac{400\times23}{100}}=\frac{184\times4}{4\times23}=8$

149. (a) $?^{4/3}\times?^{5/3}=32\times128$
$\Rightarrow$ $?^3=2^5\times2^7=2^{12}$
$\therefore$ $?=(2^{12})^{1/3}=2^4=16$

150. (e) Given fractions can be written in decimal forms as
$\frac{4}{9}=0.44;\quad\frac{6}{13}=0.46;\quad\frac{5}{11}=0.45;\quad\frac{13}{16}=0.8125$
$\frac{7}{12}=0.583$
$\therefore$ Clearly,
$\frac{13}{16}>\frac{7}{12}>\frac{6}{13}>\frac{5}{11}>\frac{4}{9}$

151. (d) Number of items produced in 3 days = 1515
Number of items produced in 1 day $=\frac{1515}{3}$
Required number of items
$=\frac{1515\times7}{3}=3535$

152. (e) $4800<4900$
$\sqrt{4900}=70$
$\therefore$ Required least number
$=4900-4800=100$

153. (b) Required remainder
$=(46)^2-(11)^3$
$=2116-1331=785$

154. (d) $(\text{Larger number})^2=11570-5329=6241$
$\therefore$ Larger number $=\sqrt{6241}=79$

155. (e) Smallest number $=\frac{5685-3}{3}=1894$

156. (e) $27\times29\times31=24273$

157. (e) $4\frac{1}{2}+\left(1\div2\frac{8}{9}\right)-3\frac{1}{13}=?$

$4+\frac{1}{2}+1\times\frac{9}{26}-\left(3+\frac{1}{13}\right)$

$4+\frac{1}{2}+\frac{9}{26}-3-\frac{1}{13}$

$1+\frac{1}{2}-\frac{1}{13}+\frac{9}{26}=\frac{26+13-2+9}{26}=1\frac{10}{13}$

158. (c) $\dfrac{6\times136\div8+132}{628\div16-26.25}$

$=\dfrac{6\times136\times\frac{1}{8}+132}{628\times\frac{1}{16}-26.25}$

$=\dfrac{102+132}{39.25-26.25}=\dfrac{234}{13}=18$

159. (d) $\{(441)^{1/2}\times207\times(343)^{1/3}\}\div\{(14)^2\times(529)^{1/2}\}$
$\{(21^2)^{1/2}\times207\times(7^3)^{1/3}\}\div\{(14)^2\times(23^2)^{1/2}\}$
$(21\times207\times7)\div((14)^2\times23)$

$\dfrac{21\times207\times7}{14\times14\times23}=6\dfrac{3}{4}$

160. (c) $\left\{\sqrt{7744}\times(11)^2\right\}\div(2)^3=(?)^3$

$\left\{88\times(11)^2\right\}\div(2)^3$

$88\times(11)^2\times\dfrac{1}{8}=(11)^3$

$?=11$

161. (e) $(4356)^{\frac{1}{2}}\div\dfrac{11}{4}=\sqrt{?}\times6$

$(66^2)^{\frac{1}{2}}\times\dfrac{4}{11}$

$66\times\dfrac{4}{11}=4\times6=\sqrt{16}\times6$

$?=16$

162. (d) $\dfrac{3}{8}\text{of}\left\{4624\div(564-428)\right\}=?$

$\dfrac{3}{8}\times\left\{4624\times\dfrac{1}{136}\right\}$

$\dfrac{3}{8}\times34=12\dfrac{3}{4}$

163. (c) $456\div24\times38-958+364=?$

$=456\times\dfrac{1}{24}\times38-958+364$

$=722-958+364$

$=128$

164. (b) $(43)^2+841=(?)^2+1465$
$1849+841=(?)^2+1465$
$1225=(?)^2$
$?=35$

165. (c) $3\dfrac{3}{8}\times6\dfrac{5}{12}-2\dfrac{3}{16}\times3\dfrac{1}{2}$

$\dfrac{22}{8}\times\dfrac{77}{12}-\dfrac{35}{16}\times\dfrac{7}{2}$

$\dfrac{2079}{96}-\dfrac{245}{32}=\dfrac{2079-7365}{96}=14$

166. (c) $(34.5\times14\times42)\div2.8$

$34.5\times14\times42\times\dfrac{1}{2.8}$

$=7245$

167. (d) $(216)^4\div(36)^4\times(6)^5=(6)^?$
$(6^3)^4\div(6^2)^4\times(6)^5$

$(6^3)^4\times\dfrac{1}{6^8}\times(6)^5$

$6^{12+5-8}=6^9$

$?=9$

168. (c) $\dfrac{\sqrt{4356}\times\sqrt{?}}{\sqrt{6084}}=11$

$\dfrac{\sqrt{66\times66}\times\sqrt{?}}{\sqrt{78\times78}}=11$

$\dfrac{66\times\sqrt{?}}{78}=11$

$\sqrt{?}=\dfrac{11\times78}{66}$

$\sqrt{?}=13$

$?=169$

169. (a) $\left(3\dfrac{6}{17}\div2\dfrac{7}{34}-1\dfrac{9}{25}\right)=(?)^2$

$\dfrac{57}{17}\times\dfrac{34}{75}-\dfrac{34}{25}$

$\dfrac{19\times2}{25}-\dfrac{34}{25}=\dfrac{4}{25}=\left(\dfrac{2}{5}\right)^2$

$?=\dfrac{2}{5}$

170. (b) $(1097.63+2197.36-2607.24)\div3.5$

$(3294.99-2607.24)\times\dfrac{1}{3.5}$

$687.75\times\dfrac{1}{3.5}=196.5$

171. (b) $\dfrac{1}{11}\times\left[(17424)^{\frac{1}{2}}\times\dfrac{1}{(66)^2}\times3^3\right]$

$\dfrac{1}{11}\times\left[(132^2)^{\frac{1}{2}}\times\dfrac{1}{(66)^2}\times3^3\right]$

$\dfrac{1}{11}\times\dfrac{132}{(66)^2}\times3^3=\dfrac{2\times27^{9}}{11\times66_{33_{11}}}=\left(\dfrac{3}{11}\right)^2$

$?=11$

Level-II

1. (b) $999\frac{995}{999}\times 999=\left(999+\frac{995}{999}\right)\times 999$

$= 999\times 999+\frac{995}{999}\times 999 = 999^2+995$

$=998001+995=998996$

2. (c) 7892.35×99.9

$=\frac{789235\times 999}{1000}=\frac{789235\times(1000-1)}{1000}$

$=\frac{789235000-789235}{1000}=788445.765$

3. (c) Total number of $\frac{1}{12}=\frac{18\frac{3}{4}}{\frac{1}{12}}$

$=\frac{75}{4}\times\frac{12}{1}=225$

4. (b) This problem can't be solved by factorisation because we need not factor. So we have to solve it by division method as follows

	23.9
2	575
2	4
43	175
3	129
469	4600
9	4221

(If the number is not a perfect square then by putting decimal we can increase the zeros in pairs for further calculation.)

The result obtained is ≈ 23.9.

So by adding some number we can make it the perfect square of 24. Now since we know that $(24)^2 = 576$. So we need to add 1 ($\because 576-575=1$)

Thus (b) is the correct option.

Alternatively : Using options we can solve this problem as if we consider optin (a) then 575 itself be a perfect square but its not a perfect square. Again if we add 1 (i.e., using option (b)) we get the number 576 and then check it, we find that 576 is a perfect square. Hence (b) is correct.

Alternatively : Since we know that $(20)^2=400$ and $(25)^2=625$. It means the value of perfect square must lies in the range of 400 and 625. So we can try it manually and get that $(23)^2=529$ and $(24)^2=576$. So simply we need to add 1 to make a perfect square number.

5. (a) $6*15*3=\sqrt{\frac{(6+2)(15+3)}{(3+1)}}=\sqrt{\frac{8\times 18}{4}}=6$

6. (a) $x+\frac{1}{x}=3+\sqrt{8}+\frac{1}{3+\sqrt{8}}$

$=\frac{(3+\sqrt{8})^2+1}{(3+\sqrt{8})}=\frac{9+8+6\sqrt{8}+1}{(3+\sqrt{8})}$

$=\frac{18+6\sqrt{8}}{(3+\sqrt{8})}=\frac{6(3+\sqrt{8})}{(3+\sqrt{8})}=6,\ \left(x+\frac{1}{x}\right)^2=6^2$

$x^2+\frac{1}{x^2}+2.x.\frac{1}{x}=36,\ x^2+\frac{1}{x^2}=36-2=34$

7. (c) If $x^a=y^b=z^c$ and $y^2=zx$

Let $x^a=y^b=z^c=k$

$\Rightarrow\ x=k^{1/a},\ y=k^{1/b},\ z=k^{1/c}$

Now, $\because\ y^2=zx$

$\therefore\ (k^{1/b})^2=(k^{1/c}).(k^{1/a})$

$\Rightarrow\ k^{2/b}=k^{\frac{1}{c}+\frac{1}{a}}$

$\Rightarrow\ \frac{1}{a}+\frac{1}{c}=\frac{2}{b}$

Hence (c) is the correct option.

8. (a) We have : $\frac{2x}{1+\frac{1}{\frac{(1-x)+x}{1-x}}}=1\Leftrightarrow\frac{2x}{1+\frac{1}{[1/(1-x)]}}=1$

$\Leftrightarrow\ \frac{2x}{1+(1-x)}=1$

$\Leftrightarrow\ 2x=2-x\Leftrightarrow 3x=2\Leftrightarrow x=\frac{2}{3}$

9. (b) Given expression

$=\left(\frac{1}{2}-\frac{1}{3}\right)+\left(\frac{1}{3}-\frac{1}{4}\right)+\left(\frac{1}{4}-\frac{1}{5}\right)+\left(\frac{1}{5}-\frac{1}{6}\right)+$

$.....+\left(\frac{1}{9}-\frac{1}{10}\right)$

$=\left(\frac{1}{2}-\frac{1}{10}\right)=\frac{4}{10}=\frac{2}{5}$

10. (c) $7-2\sqrt{10}=5+2-2\sqrt{5\times 2}$

$\Rightarrow\ 7-2\sqrt{10}=(\sqrt{5})^2+(\sqrt{2})^2-2\sqrt{5}.\sqrt{2}$

$\Rightarrow\ 7-2\sqrt{10}=(\sqrt{5}-\sqrt{2})^2$

Thus the $\sqrt{7-2\sqrt{10}}=\pm(\sqrt{5}-\sqrt{2})$

11. (b) We have,
Product of unit's digits = 28
$\Rightarrow$ Product of units digits = 4 × 7
[∵ Unit's digits are one digit numbers]
$\Rightarrow$ Unit's digits are 4 and 7.
$\Rightarrow$ Product of ten's digits = 15
$\Rightarrow$ Product of ten's digits = 3 × 5
[Ten's digit are one digit numbers]
$\Rightarrow$ Ten's digits are 3 and 5.
Thus, the two numbers either 34 and 57 or 37 and 54.
Now, 34 × 57 = 34 × (50 + 7) [∵ 57 = 50 + 7]
= 34 × 50 + 34 × 7 [∵ $a \times (b + c) = a \times b + a \times c$]
= 1700 + 238 = 1938
and, 37 × 54 = 37 × (50 + 4) [∵ 54 = 50 + 4]
[∵ $a \times (b + c) = a \times b + a \times$ c]
= 1850 + 148 = 1998

12. (d) Consider $P = 5$, then

$$5 + 5! = 5^3$$
$$5 + 120 = 125$$
$$125 = 125$$

Thus (d) is correct option.

13. (d) Let $Z = 8x - 3x^2$

$$\Rightarrow Z = -3\left[x^2 - \frac{8}{3}x\right]$$

$$\Rightarrow Z = -3\left[x^2 - 2 \times x \times \frac{4}{3} + \left(\frac{4}{3}\right)^2 - \left(\frac{4}{3}\right)^2\right]$$

$$\Rightarrow Z = -3\left(x - \frac{4}{3}\right)^2 + 3 \times \left(\frac{4}{3}\right)^2$$

So the maximum value occurs when $x = \frac{4}{3}$

Maximum value $= -3 \times 0 + 3 \times \frac{16}{9} = \frac{16}{3}$

14. (c) $\sqrt[3]{x+9} - \sqrt[3]{x-9} = 3,$

$$\Rightarrow (x+9) - (x-9) - 3\,\sqrt[3]{x+9}.\sqrt[3]{x-9} = 3,$$

$$\left(\sqrt[3]{x+9} - \sqrt[3]{x-9}\right) = 3^3$$

$$\left(\because (a-b)^3 = a^3 - b^3 - 3ab(a-b)\right)$$

$$\Rightarrow 18 - 3\,\sqrt[3]{x^2 - 81} \times 3$$

$$\Rightarrow \frac{18 - 27}{9} = \sqrt[3]{x^2 - 81}$$

$$\Rightarrow -1 = x^2 - 81 \Rightarrow x^2 = 80$$

15. (a) Given expression

$$= 35.7 - \left(3 + \frac{1}{\frac{10}{3}}\right) - \left(2 + \frac{1}{\frac{5}{2}}\right)$$

$$= 35.7 - \left(3 + \frac{3}{10}\right) - \left(2 + \frac{2}{5}\right)$$

$$= 35.7 - \frac{33}{10} - \frac{12}{5} = 35.7 - \left(\frac{33}{10} + \frac{12}{5}\right)$$

$$= 35.7 - \frac{57}{10} = 35.7 - 5.7 = 30.$$

16. (b) $\sqrt[4]{10} = (10)^{1/4} = (10)^{3/12} = (1000)^{1/12}$

$\sqrt[3]{6} = (6)^{1/3} = (6)^{4/12} = (1296)^{1/12}$

$\sqrt{3} = (3)^{1/2} = (3)^{6/12} = (729)^{1/12}$

$\therefore \sqrt{3} < \sqrt[4]{10} < \sqrt[3]{6}$ is the correct order and hence (b) is correct.

17. (b) The unit's digit will be 1 × 5 = 5 (no carry over). The tens digit will be (4*1 + 5*2) = 4 (carry over 1). The hundreds digit will be (3*1 + 4*2 + 5*1) = 6 + 1 (carried over) = 7. Hence, answer is 745.

18. (a) Let the numbers be a and b. Then, $a + b = 12$ and $ab = 35$.

$$\therefore \quad \frac{a+b}{ab} = \frac{12}{35} \quad \Rightarrow \quad \left(\frac{1}{b} + \frac{1}{a}\right) = \frac{12}{35}$$

$\therefore$ Sum of reciprocals of given numbers $= \frac{12}{35}$

19. (c) Given expression

$$= \frac{2}{3} \times \frac{3}{4} \times \frac{4}{5} \times \ldots\ldots \times \frac{99}{100} = \frac{2}{100} = \frac{1}{50}$$

20. (b) Suppose the worker remained idle for x days. Then, he worked for $(60 - x)$ days.
$\therefore$ $20(60 - x) - 3x = 280 \Leftrightarrow 1200 - 23x = 280 \Leftrightarrow 23x = 920 \Leftrightarrow x = 40.$
So, the worker remained idle for 40 days.

21. (b) Given exp. $= \dfrac{1}{1 + \dfrac{\frac{2}{3}}{\dfrac{8}{\frac{5}{3} + \frac{9}{(1/3)}}}} = \dfrac{1}{1 + \dfrac{2/3}{\frac{5}{3} + \frac{8}{9} \times 3}}$

$$= \frac{1}{1 + \dfrac{2/3}{(13/3)}} = \frac{1}{1 + \dfrac{2}{13}} = \frac{13}{15}$$

22. (b) Given exp. $= \sqrt{\dfrac{(0.03)^2 + (0.21)^2 + (0.065)^2}{\left(\dfrac{0.03}{10}\right)^2 + \left(\dfrac{0.21}{10}\right)^2 + \left(\dfrac{0.065}{10}\right)^2}}$

$$= \sqrt{\frac{100\left[(0.03)^2 + (0.21)^2 + (0.065)^2\right]}{(0.03)^2 + (0.21)^2 + (0.065)^2}}$$

$= \sqrt{100} = 10.$

23. (c) Given : $x^2 + y^2 + z^2 - 64 = -2(xy - yz - zx)$ ...(i)
Now, $[x + y + (-z)]^2 = x^2 + y^2 + z^2 + 2(xy - yz - zx)$
$\Rightarrow (3z - z)^2 = x^2 + y^2 + z^2 + 2(xy - yz - zx)$
$\Rightarrow -2(xy - yz - zx) = (x^2 + y^2 + z^2) - (2z)^2$ ...(ii)
From (i) and (ii), we get:
$(2z)^2 = 64 \Leftrightarrow 4z^2 = 64 \Leftrightarrow z^2 = 16 \Leftrightarrow z = 4.$

24. (c) $\sqrt{\dfrac{8}{3}} = \sqrt{\dfrac{8\times 3}{3\times 3}} = \dfrac{\sqrt{24}}{3} = \dfrac{4.899}{3} = 1.633.$

25. (c) $\left[X + \dfrac{1}{X}\right]^2 = X^2 + \dfrac{1}{X^2} + 2 = 16$ or $X^2 + \dfrac{1}{X^2} = 14$

Now, $X^4 + \dfrac{1}{X^4} + 2 = 196$ or $X^4 + \dfrac{1}{X^4} = 194.$

26. (c) Given expression $= \sqrt{15625} + \sqrt{\dfrac{15625}{100}} + \sqrt{\dfrac{15625}{10000}}$

$= \left(125 + \dfrac{125}{10} + \dfrac{125}{100}\right) = (125 + 12.5 + 1.25) = 138.75$

27. (b) Provisions for one student $= 250 \times 35 = 8750$
250 students used provisions for 5 days.
Total provisions used by 250 students in 5 days
$= 250 \times 5 = 1250$
Remaining provision $= 8750 - 1250 = 7500$
After 5 days total number of student $= 250 + 25 = 275$
Total provisions used by 275 student in 10 days
$275 \times 10 = 2750$
Now remaining $= 7500 - 2750 = 4750$
After 15 days no. of student $= 275 - 25 = 250$
$4750 = 250 \times$ no. of extra dayus

No. of extra days $= \dfrac{4750}{250} = 19$ days

28. (a) $\dfrac{97}{19} = 5 + \dfrac{2}{19}$. Also, $\dfrac{19}{2}$ can be written as $9 + \dfrac{1}{2}$. So the values of a, b and c are 5, 9 and 2 respectively. Hence, the sum of a, b and c is 16.

29. (d) $\text{I} = \sqrt[3]{\sqrt[4]{a^3}} = \left((a^3)^{1/4}\right)^{1/3} = a^{1/4}$

$\text{II} = \sqrt[3]{\sqrt[5]{a^4}} = \left((a^4)^{1/5}\right)^{1/3} = a^{4/15}$

$\text{III} = \sqrt{\sqrt[3]{a}} = \left(a^{1/3}\right)^{1/2} = a^{1/6}$

$\text{IV} = \sqrt{\sqrt[5]{a^3}} = \left((a^3)^{1/5}\right)^{1/2} = a^{3/10}$

Now again, to compare these numbers, we need to bring the indices to a common denominator.
$\therefore$ I = a1/4 = a15/60. II = a4/15 = a16/60.
III = a1/6 = a10/60. IV = a3/10 = a18/60.
$\therefore$ The ascending order is III, I, II, IV.

30. (b) $A^{3^{3^{3^3}}} = 3^{3^{27}}$

and $C^{3^{33}} = 3^{3^{33}}$
Hence $C > A$.
Hence either (b) and (d) option is correct.

Now $A = 3^{3^{3^3}} = 3^{3^{27}}$
and $D = 3^{333}$
Hence $A > D$ (Since $3^{27} > 333$)
Thus the correct relation is $C > A > B > D$.
Hence, option (b) is correct.

31. (b) If we try to put x as 12, we get the square root of $3x$ as 6. Then the next point at which we need to remove the square root sign would be $12 + 2(6) = 24$ whose square root would be an irrational number. This leaves us with only 1 possible value ($x = 3$). Checking for this value of x we can see that the expression is satisfied as LHS = RHS.

32. (d) Solve this question through options. Also realize that $a \times b = a + b$ only occurs for the situation $2 \times 2 = 2 + 2$.
Hence, clearly the answer has to be none of these.

33. (b) $\sqrt{11449} \times \sqrt{6241} - (54)^2 = \sqrt{?} + (74)^2$

$\Rightarrow \sqrt{?} = 107 \times 79 - 2916 - 5476$
$= 8453 - 2916 - 5476 = 61$
$\therefore \quad ? = (61)^2 = 3721$

34. (c) $? = \left[\left(3\sqrt{8} + \sqrt{8}\right) \times \left(8\sqrt{8} + 7\sqrt{8}\right)\right] - 98$

$= \left(4\sqrt{8} \times 15\sqrt{8}\right) - 98 = (60 \times 8) - 98$

$= 480 - 98 = 382$

35. (a) $? + 5883 = 3463 \times 295 - 18611$
$\therefore \quad ? = 1021585 - 18611 - 5883 = 997091$

36. (d) $? = \dfrac{28}{65} \times \dfrac{195}{308} \div \dfrac{39}{44} + \dfrac{5}{26} = \dfrac{28}{65} \times \dfrac{195}{308} \times \dfrac{44}{39} + \dfrac{5}{26}$

$= \dfrac{4}{13} + \dfrac{5}{26} = \dfrac{8+5}{26} = \dfrac{13}{26} = \dfrac{1}{2}$

37. (c) $? + 1147.69 = (23.1)^2 + (48.6)^2 - (39.8)^2$
$\therefore \quad ? = 533.61 + 2361.96 - 1584.04 - 1147.69 = 163.84$

38. (e) $? \div 21.003 = \sqrt[3]{4663} + 349$
$\Rightarrow ? \div 21 = 17 + 349 = 366$
$\therefore \quad ? = 366 \times 21 = 7686 \approx 7680$

39. (c) $? = 4331 \times \frac{39.897}{100} + 5003 \times \frac{58.779}{100}$

$= 4330 \times \frac{40}{100} + 5000 \times \frac{59}{100}$

$= 1732 + 2950 = 4682 \approx 4700$

40. (c) $? = 59.88 \div 12.21 \times 6.35$

$\approx 60 \div 12 \times 6 = 60 \times \frac{1}{12} \times 6 = 30$

41. (e) $? = 43931.03 \div 2111.02 \times 401.04$

$\approx 43930 \div 2110 \times 400$

$\approx 43930 \times \frac{1}{2110} \times 400 \approx 8300$

42. (b) $? = \sqrt{6354} \times 34.993 \approx 80 \times 35 = 2800$

43. (e) The given number series is based on the following pattern:

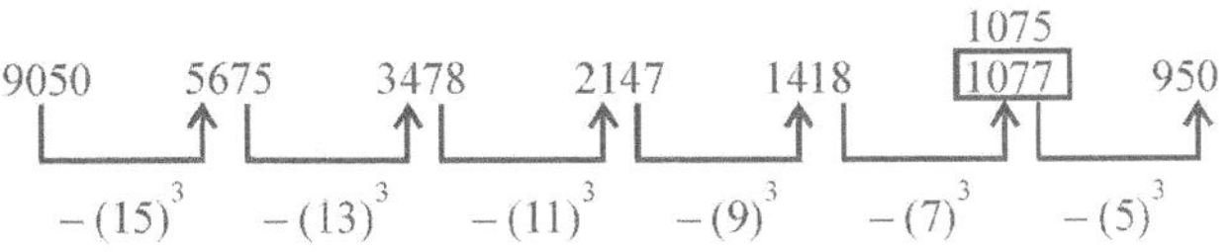

Hence, the number 1077 is wrong and it should be replaced by 1075.

44. (d) The given number series is based on the following pattern :

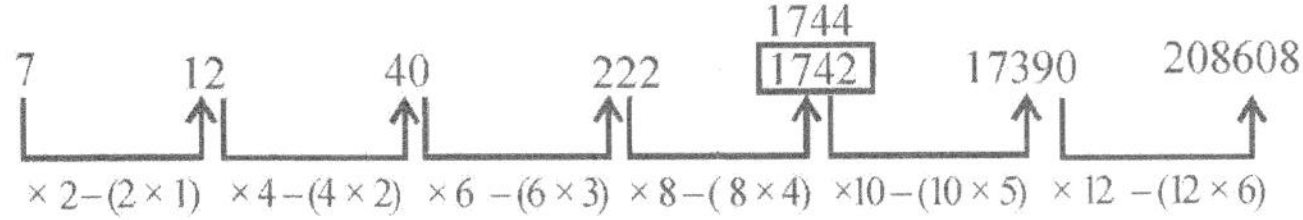

Hence, the number 1742 is wrong and it should be replaced by 1744.

45. (c) The given number series is based on the following pattern:

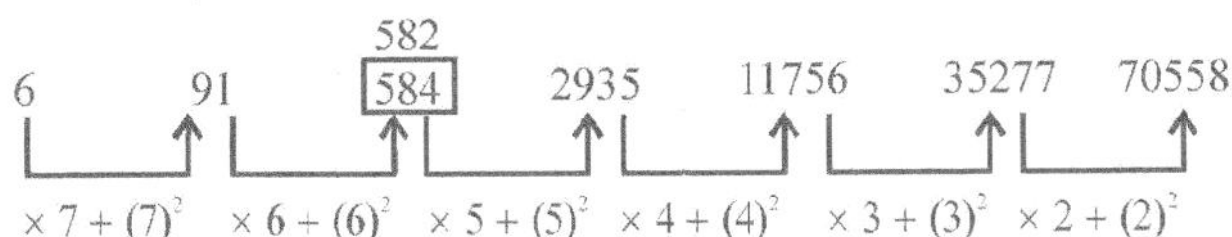

Hence, the number 584 is wrong and it should be replaced by 582.

46. (d) The given number series is based on the following pattern.

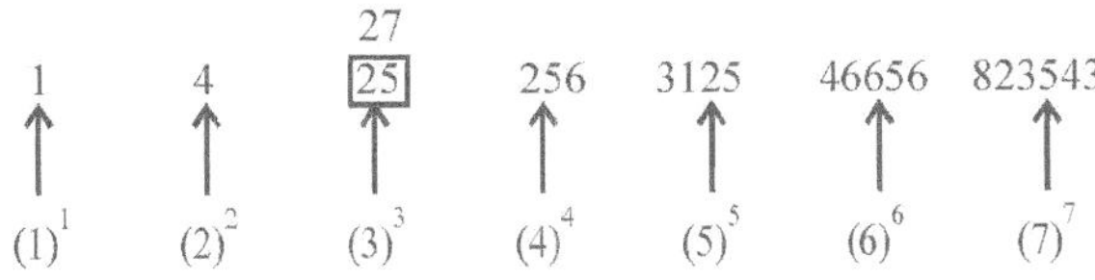

Hence, the number 25 is wrong and it should be replaced by 27.

47. (b) The given number series is based on the following pattern :

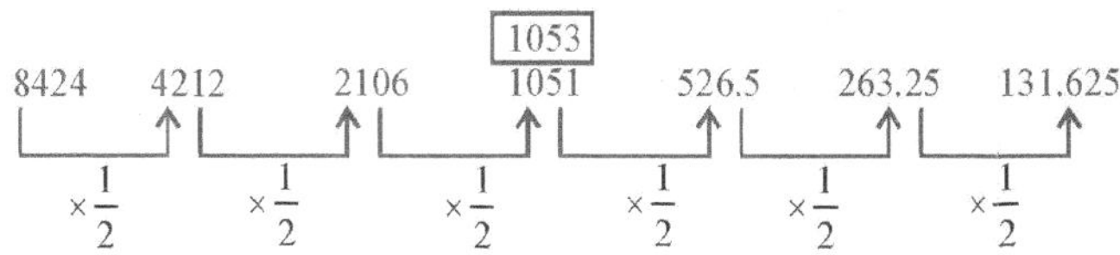

Hence, the number 1051 is wrong and it should be replaced by 1053.

48. (a) Value of one ticket of each kind = 55 + 85 + 105 = ₹245

∴ Required number of ticket of each kind

$= \frac{2940}{245} = 12$

49. (c) Cost of one pencil box = 7 + 22 + 14 = ₹43

∴ Required amount

$= (20 \times 7) + (8 \times 22) + (6 \times 175) + (7 \times 43)$

$= 140 + 176 + 1050 + 301 =$ ₹ 1667

50. (d) $4003 \times 77 - 21015 = ? \times 116$

$\Rightarrow 308231 - 21015 = ? \times 116 \Rightarrow 287216 = ? \times 116$

$\Rightarrow ? = \frac{287216}{116} = 2476$

51. (a) $\left[(5\sqrt{7} + \sqrt{7}) \times (4\sqrt{7} + 8\sqrt{7})\right] - (19)^2 = ?$

$\Rightarrow (6\sqrt{7} \times 12\sqrt{7}) - (361) = ?$

$\Rightarrow 72 \times \sqrt{7} \times \sqrt{7} - 361 = ?$

$\therefore ? = 504 - 361 = 143$

52. (b) $(4444 \div 40) + (645 \div 25) + (3991 \div 26) = ?$

$\Rightarrow ? = (111.1) + (25.8) + (153.5) \Rightarrow ? = 290.4$

53. (e) $\sqrt{33124} \times \sqrt{2601} - (83)^2 = (?)^2 + (37)^2$

$\Rightarrow (?)^2 = \sqrt{33124} \times \sqrt{2601} - (83)^2 - (37)^2$

$\Rightarrow (?)^2 = 182 \times 51 - 6889 - 1369$

$\Rightarrow (?)^2 = 9282 - 6889 - 1369$

$\Rightarrow (?)^2 = 1024$

$\therefore ? = \sqrt{1024} = 32$

54. (b) $5\frac{17}{37} \times 4\frac{51}{52} \times 11\frac{1}{7} + 2\frac{3}{4} = ?$

$\Rightarrow \left(\frac{202}{37} \times \frac{259}{52} \times \frac{78}{7}\right) + \left(\frac{11}{4}\right) = ?$

$\Rightarrow 303 + \frac{11}{4} = ?$

$\therefore ? = \frac{1223}{4} = 305.75$

55. (c) $8787 \div 343 \times \sqrt{50} = ?$

$\Rightarrow 25 \times 7 = ?$

$\therefore \quad ? = 175 \approx 180$

56. (b) $\sqrt[3]{54821} \times (303 \div 8) = (?)^2$

$\Rightarrow 38 \times 37.5 = (?)^2$

$? = \sqrt{38 \times 38}$

$? = 38$

57. (c) $\frac{5}{8}$ of $4011.33 + \frac{7}{10}$ of $3411.22 = ?$

$\Rightarrow \frac{5}{8} \times 4010 + \frac{7}{10} \times 3410 = ? \Rightarrow ? = 2506 + 2387$

$\Rightarrow ? = 4893 \approx 4890$

58. (e) 23% of 6783 + 57% of 8431 = ?

$\Rightarrow ? = 1560 + 4805$

$\therefore ? = 6365 \approx 6360$

59. (a) $335.01 \times 244.99 \div 55$

$\Rightarrow ? = \frac{335 \times 245}{55}$

$\therefore ? = 1492 \approx 1490$

60. (c) By options

(a) Either 12 or 13

then ice-creams should not be given atleast 9. This can be rejected.

(b) Either 11 or 12

Ice-cream should be atleast 9. By this combination ice cream gets less than 9.

(c) Either 10 or 11

By giving cookies 10 or 11, we get all the possible condition fulfilled.

(d) and (e), the ice-cream distribution can be more than cookies which violates our condition.

$\therefore$ option (c) is the write answer.

61. (a) Let the number be $10x + y$.

According to condition

$10x + y + 18 = 10y + x$

$y - x = 2$

So those numbers are 02, 13, 24, 35, 46, 57, 68, 79, 80

But 13 and 79 are prime numbers.

62. (c) Given, $x = \frac{4ab}{a+b}$

$\Rightarrow \frac{x}{2a} = \frac{2b}{a+b}$

Applying componendo and dividendo, we get

$\frac{x+2a}{x-2a} = \frac{2b+a+b}{2b-a-b} = \frac{a+3b}{b-a}$ ...(i)

Also, $\frac{x}{2b} = \frac{2a}{a+b}$

Applying componendo and dividendo, we get

$\frac{x+2b}{x-2b} = \frac{2a+a+b}{2a-a-b} = \frac{3a+b}{a-b}$ ...(ii)

Add (i) & (ii),

$\frac{x+2a}{x-2a} + \frac{x+2b}{x-2b} = \frac{a+3b}{b-a} + \frac{3a+b}{a-b}$

$= \frac{1}{b-a}[a + 3b - 3a - b] = \frac{2(b-a)}{(b-a)} = 2$

63. (a) $x^2 + y^2 + z^2 - xy - yz - zx$

$= \frac{2}{2}(x^2 + y^2 + z^2 - xy - yz - zx)$

$= \frac{1}{2}(2x^2 + 2y^2 + 2z^2 - 2xy - 2yz - 2zx)$

$= \frac{1}{2}(x^2 + y^2 - 2xy + y^2 + z^2 - 2yz + x^2 + z^2 - 2zx)$

$= \frac{1}{2}[(x-y)^2 + (y-z)^2 + (z-x)^2]$

$= \frac{1}{2}[(997-998)^2 + (998-999)^2 + (999-997)^2]$

$= \frac{1}{2}[1^2 + 1^2 + 2^2] = \frac{1}{2} \times 6 = 3$

64. (d) We have $x^3 + y^3 + z^3 - 3xyz = (x + y + z)(x^2 + y^2 + z^2 - xy - yz - zx)$

Here $x = a - 4, y = b - 3, z = c - 1$

So, given expression is $(x + y + z)(x^2 + y^2 + z^2 - xy - yz - zx)$

$= (a - 4 + b - 3 + c - 1)(x^2 + y^2 + z^2 - xy - yz - zx)$

$= (a + b + c - 8)(x^2 + y^2 + z^2 - xy - yz - zx)$

$= (8 - 8)(x^2 + y^2 + z^2 - xy - yz - zx)$

$= 0$

65. (a) $x^4 + y^4 - 2x^2y^2$

$\Rightarrow (x^2 - y^2)^2 \Rightarrow [(x + y)(x - y)]^2$

$\Rightarrow \left[\left(\sqrt{a} + \frac{1}{\sqrt{a}} + \sqrt{a} - \frac{1}{\sqrt{a}}\right)\left(\sqrt{a} + \frac{1}{\sqrt{a}} - \sqrt{a} + \frac{1}{\sqrt{a}}\right)\right]^2$

$\Rightarrow \left(2\sqrt{a} \times \frac{2}{\sqrt{a}}\right)^2 \Rightarrow 16$

66. (d) $5a + \frac{1}{3a} = 5$

Multiply by $\frac{3}{5}$ on both sides

$\frac{3}{5}\left(5a + \frac{1}{3a}\right) = 5 \times \frac{3}{5}$

$3a + \frac{1}{5a} = 3$

Squaring on both sides

$9a^2 + \frac{1}{25a^2} + 2 \times 3a \times \frac{1}{5a} = 9$

$\Rightarrow 9a^2 + \frac{1}{25a^2} = 9 - \frac{6}{5} = \frac{39}{5}$

67. (b) $x = 3 + 2\sqrt{2}$

$$\frac{1}{x} = \frac{1}{3+2\sqrt{2}} \times \frac{3-2\sqrt{2}}{3-2\sqrt{2}}$$

$$\frac{1}{x} = \frac{3-2\sqrt{2}}{9-8} = 3 - 2\sqrt{2}$$

$$\left(\sqrt{x} - \frac{1}{\sqrt{x}}\right)^2 = x + \frac{1}{x} - 2$$

$$\left(\sqrt{x} - \frac{1}{\sqrt{x}}\right)^2 = 3 + 2\sqrt{2} + 3 - 2\sqrt{2} - 2 = 4$$

$$\left(\sqrt{x} - \frac{1}{\sqrt{x}}\right) = \sqrt{4} = \pm 2$$

68. (b) If $a + b + c = 0$,
then $a^3 + b^3 + c^3 = 3abc$
Dividing both sides by abc

$$\frac{a^3}{abc} + \frac{b^3}{abc} + \frac{c^3}{abc} = \frac{3abc}{abc}$$

$$\frac{a^2}{bc} + \frac{b^2}{ac} + \frac{c^2}{ab} = 3$$

69. (c)

70. (b) $(a^b + b^a)^{-1} = (2^3 + 3^2)^{-1} = (8 + 9)^{-1} = (17)^{-1} = \frac{1}{17}$

71. (a) $392 \times 2 = 784 \Rightarrow (27)^2$
Hence, 2 can be multiplied by 392 which gives perfect square.

72. (c) Let lst, 2nd and 3rd part represented by x, y, z

Let $\frac{1}{2}x = \frac{1}{3}y = \frac{1}{4}z = k$

$\therefore \quad x = 2k, y = 3k, = 4k$
According to question
$x + y + z = 81$
$\Rightarrow \quad 2k + 3k + 4k = 81 \Rightarrow 9k = 81 \Rightarrow k = 9$
Hence, parts are 18, 27, 36.

73. (a) $x^4 - 2x^2 + k$
$(x^2)^2 - 2x^2 + k \Rightarrow (x^2)^2 - 2.1.x^2 + k$
For above expression to make a perfect square, the k value is equal to 1.

74. (d) $3x - \frac{1}{4y} = 6 \quad 3x = 6 + \frac{1}{4y}$

Taking 3 common on both sides

$$x = \frac{6}{3} + \frac{1}{4.3y} \Rightarrow x = 2 + \frac{1}{12y}$$

Multiplying equation by 4 on both sides

$$4x = 8 + \frac{1}{3y} \Rightarrow 4x - \frac{1}{3y} = 8$$

75. (a) $a + b + c = 0$
i.e. $a = -(b + c); b = -(c + a); c = -(a + b)$

Now, $\frac{a+b}{c} - \frac{2b}{c+a} + \frac{b+c}{a}$

$$\Rightarrow \frac{a+b}{-(a+b)} - \frac{2[-(c+a)]}{c+a} + \frac{b+c}{-(b+c)}$$

$\Rightarrow -1 + 2 - 1 = 0$

76. (b) $x + \frac{4}{x} = 4$
$x^2 + 4 = 4x \Rightarrow x^2 - 4x + 4 = 0 \Rightarrow (x - 2)^2 = 0$
$x = 2$

$$x^3 + \frac{4}{x^3} = (2)^3 + \frac{4}{(2)^3} \Rightarrow 8 + \frac{4}{8} \Rightarrow 8 + \frac{1}{2} \Rightarrow 8\frac{1}{2}$$

77. (b) $x = 3 + 2\sqrt{2}$

$x = 2 + 1 + 2\sqrt{2}$

$x = (\sqrt{2})^2 + (1)^2 + 2.1.\sqrt{2}$

$x = (\sqrt{2} + 1)^2$

$\sqrt{x} = (\sqrt{2} + 1) \qquad$ ...(1)

$$\frac{1}{\sqrt{x}} = \frac{1}{\sqrt{2}+1} \times \frac{\sqrt{2}-1}{\sqrt{2}-1} = \frac{\sqrt{2}-1}{2-1} = \sqrt{2} - 1$$

Now, $\sqrt{x} - \frac{1}{\sqrt{x}} = \sqrt{2} + 1 - (\sqrt{2} - 1) = \sqrt{2} + 1 - \sqrt{2} + 1$

$$\sqrt{x} - \frac{1}{\sqrt{x}} = 2$$

78. (c) The least value of $a + \frac{1}{a}$ is 2 where $a = 1$.

79. (a) $3^{34} = (3^2)^{17} = 9^{17}$
$2^{51} = (2^3)^{17} = 8^{17}$
Clearly, $7^{17} < 8^{17} < 9^{17}$
or $7^{17} < 2^{51} < 3^{34}$

80. (d) $x = 2 + \sqrt{3}$

$$\frac{1}{x} = \frac{1}{2+\sqrt{3}} \times \frac{2-\sqrt{3}}{2-\sqrt{3}} = 2 - \sqrt{3}$$

$$x^2 + \frac{1}{x^2} = \left(x + \frac{1}{x}\right)^2 - 2$$

$$= \left(2 + \sqrt{3} + 2 - \sqrt{3}\right)^2 - 2$$

$= 16 - 2 = 14$

81. (c) $a = 4.965 \approx 5, b = 2.343 \approx 2$
$c = 2.622$
$a - b = c$

taking cube both sides

$a^3 - b^3 - 3a^2b + 3ab^2 = c^3$

$a^3 - b^3 - c^3 - 3ab(a-b) = 0$

$a^3 - b^3 - c^3 - 3abc = 0$

82. (d) $x + y + z = 0$

$y + z = -x$

$y^2 + z^2 + 2yz = x^2$

$\Rightarrow y^2 + z^2 = x^2 - 2yz$ ...(1)

$$\frac{x^2+y^2+z^2}{x^2-yz} = \frac{x^2-2yz+x^2}{x^2-yz} = \frac{2(x^2-yz)}{x^2-yz}$$

$= 2$

83. (d) Let the number be x

$$\frac{x}{\frac{7}{19}} - \frac{7}{19} \times x = 624$$

$$x\left(\frac{19}{7} - \frac{7}{19}\right) = 624$$

$$x\left(\frac{361-49}{7 \times 19}\right) = 624$$

$$x = \frac{624 \times 133}{312}$$

$x = 266$

Sum of digits $(2 + 6 + 6) = 14$

84. (c) $a^2 + b^2 + c^2 = 2a - 2b - 2$

$(a^2 - 2a + 1) + (b^2 + 2b + 1) + c^2 = 0$

$(a-1)^2 + (b+1)^2 + c^2 = 0$

This equation is possible if

$a - 1 = 0, b + 1 = 0$ and $c = 0$

$a = 1, b = -1, c = 0$

$3a - 2b + c = 3 \times 1 - 2 \times (-1) + 0$

$= 3 + 2 = 5$

85. (b) $a + b + c = 3$

Squaring both sides

$a^2 + b^2 + c^2 + 2(ab + bc + ac) = 9$

$6 + 2(ab + bc + ca) = 9$

$ab + bc + ca = \frac{3}{2}$ ...(1)

given $\frac{1}{a} + \frac{1}{b} + \frac{1}{c} = 1$

$\Rightarrow ab + bc + ac = abc = \frac{3}{2}$ [from (1)]

86. (d) $a^2 - 4a - 1 = 0$

$a^2 - 4a = 1$

$a(a-4) = 1$

$a - 4 = \frac{1}{a}$

$a - \frac{1}{a} = 4$ ...(1)

We have $a^2 + 3a + \frac{1}{a^2} - \frac{3}{a}$

$$\left(a^2 + \frac{1}{a^2}\right) + 3\left(a - \frac{1}{a}\right)$$

$$\left(a - \frac{1}{a}\right)^2 + 3\left(a - \frac{1}{a}\right) + 2$$

$4^2 + 3 \times 4 + 2 = 30$

CHAPTER 2

NUMBER SYSTEM

Chart: Classification of Numbers

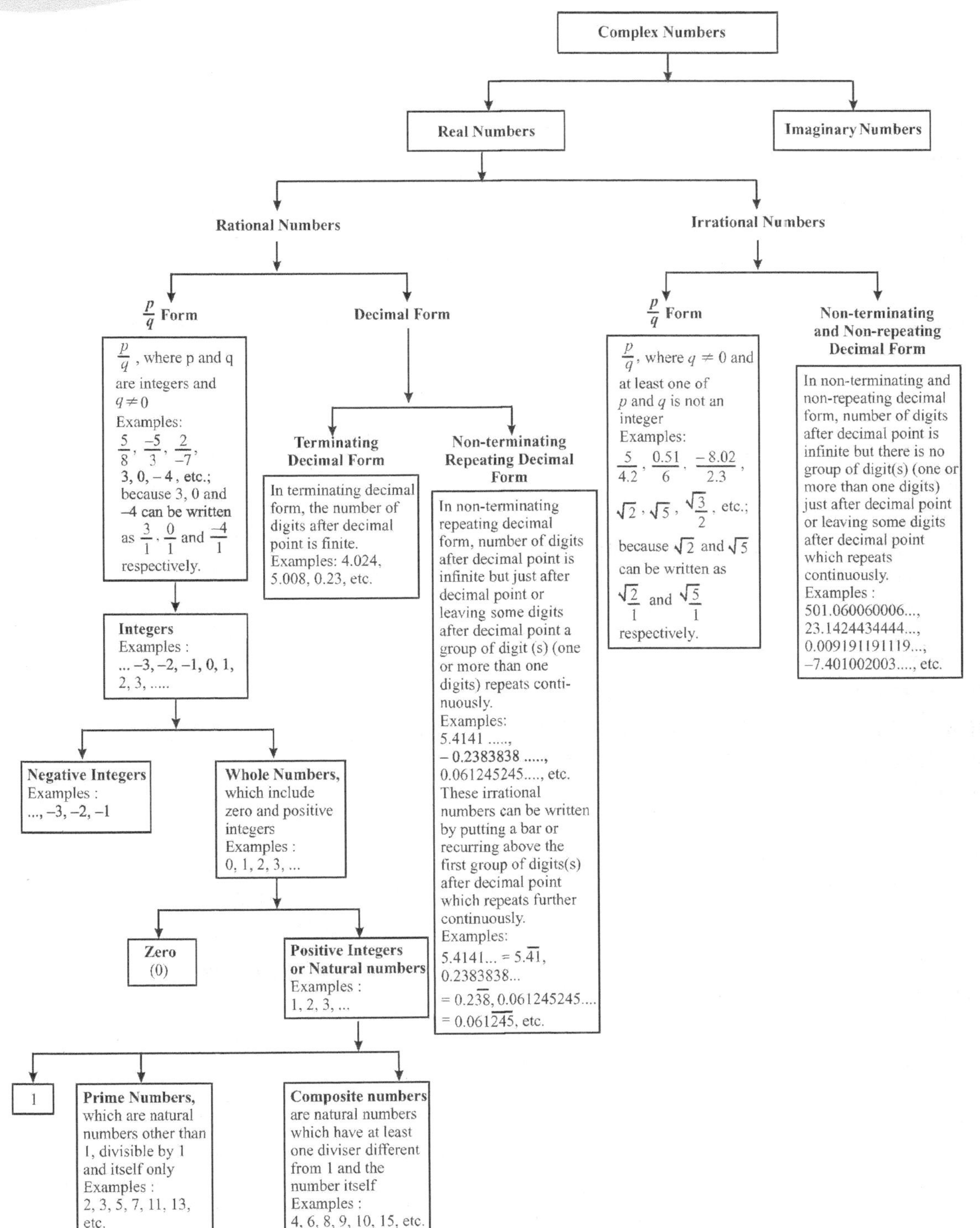

CONCEPT OF NUMBER LINE (OR NUMBER LINE)

A number line is a straight line from negative infinitive $(-, \infty)$ in left hand side to positive infinitive $(+, \infty)$ in right hand side as given:

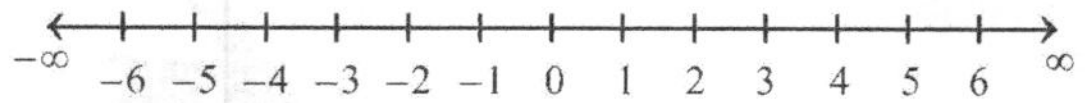

Each point on the number line represents a unique real number and each real number is denoted by a unique point on the number line.

Symbols of some special sets are:

N : the set of all natural numbers
Z : the set of all integers
Q : the set of all rational numbers
R : the set of all real numbers
Z^+ : the set of positive integers
Q^+ : the set of positive rational numbers, and
R^+ : the set of positive real numbers

The symbols for the special sets given above will be referred to throughout the text.

Even Integers

An integer divisible by 2 is called an even integer. Thus, ..., – 6, – 4, – 2, 0, 2, 4, 6, 8, 10, 12,...., etc. are all even integers. $2n$ always represents an even number, where n is an integer.

For example, by putting $n = 5$ and 8 in $2n$, we get even integer $2n$ as 10 and 16 respectively.

Odd Integers

An integer not divisible by 2 is called an odd integer.

Thus, ..., –5, –3, –1, 1, 3, 5, 7, 9, 11, 13, 15,..., etc. are all odd integers.

$(2n - 1)$ or $(2n + 1)$ always represents an odd number, where n is an integer.

For example by putting $n = 0$, 1 and 5 in $(2n - 1)$, we get odd integer $(2n - 1)$ as – 1, 1 and 9 respectively.

Properties of Positive and Negative Numbers

If n is a natural number then

$$(\text{A positive number})^{\text{natural number}} = \text{A positive number}$$
$$(\text{A negative number})^{\text{even positive number}} = \text{A positive number}$$
$$(\text{A negative number})^{\text{odd positive number}} = \text{A negative number}$$

CONVERSION OF RATIONAL NUMBER OF THE FORM NON-TERMINATING RECURRING DECIMAL INTO THE RATIONAL NUMBER OF THE FORM $\frac{p}{q}$

First write the non-terminating repeating decimal number in recurring form i.e., write

64.20132132132..... as $64.20\overline{132}$

Then using formula given below we find the required $\frac{p}{q}$ form of the given number.

Rational number in the form $\frac{p}{q}$

$$= \frac{\begin{bmatrix}\text{Complete number neglecting}\\ \text{the decimal and bar over}\\ \text{repeating digit(s)}\end{bmatrix} - \begin{bmatrix}\text{Non-recurring part of}\\ \text{the number neglecting}\\ \text{the decimal}\end{bmatrix}}{m \text{ times 9 followed by } n \text{ times 0}}$$

where m = number of recurring digits in decimal part
and n = number of non-recurring digits in decimals part

Thus, $\frac{p}{q}$ form of $64.20\overline{132} = \frac{6420132 - 6420}{99900}$

$$= \frac{6413712}{99900} = \frac{534476}{8325}$$

In short; $0.\bar{a} = \frac{a}{9}$, $0.\overline{ab} = \frac{ab}{99}$, $0.\overline{abc} = \frac{abc}{999}$, etc. and

$$0.a\bar{b} = \frac{ab - a}{90},\ 0.a\overline{bc} = \frac{abc - a}{990},\ 0.ab\bar{c} = \frac{abc - ab}{900},$$

$$0.ab\overline{cd} = \frac{abcd - ab}{9900},\ ab \cdot cd\bar{e} = \frac{abcde - abc}{990}, \text{ etc.}$$

Illustration 1: Convert $2.\overline{46102}$ in the $\frac{p}{q}$ form of rational number.

Solution: Required $\frac{p}{q}$ form $= \frac{246102 - 2}{99999} = \frac{246100}{99999}$

Illustration 2: Convert $0.167\overline{3206}$ in the $\frac{p}{q}$ form of rational number.

Solution: Required $\frac{p}{q}$ form $= \frac{1673206 - 167}{9999000} = \frac{1673039}{9999000}$

Illustration 3: Convert 31.026415555 ... into $\frac{p}{q}$ form of rational number.

Solution: First write 31.026415555... as $31.02641\bar{5}$

Now required $\frac{p}{q}$ form $= \frac{31026415 - 3102641}{900000} = \frac{27923774}{900000}$

$$= \frac{13961887}{450000}.$$

DIVISION

```
4) 275 (68
   24
   ---
    35
    32
   ---
     3
```

Here 4 is the divisor, 275 is the dividend, 68 is the quotient and 3 is the remainder. Remainder is always less than divisor.

Thus, Divisor) Dividend (Quotient
abc
Remainder

Thus,

Dividend = Divisor × Quotient + Remainder

For example, $275 = 4 \times 68 + 3$

When quotient is a whole number and remainder is zero, then dividend is divisible by divisor.

TESTS OF DIVISIBILITY

I. Divisibility by 2:

A number is divisible by 2 if its unit digit is any of 0, 2, 4, 6, 8.

Ex. 58694 is divisible by 2, while 86945 is not divisible by 2.

II. Divisible by 3:

A number is divisible by 3 only when the sum of its digits is divisible by 3.

Ex. (i) Sum of digits of the number 695421 = 27, which is divisible by 3.

$\therefore$ 695421 is divisible by 3.

(ii) Sum of digits of the number 948653 = 35, which is not divisible by 3.

$\therefore$ 948653 is not divisible by 3.

III. Divisible by 4:

A number is divisible by 4 if the number formed by its last two digits i.e. ten's and unit's digit of the given number is divisible by 4.

Ex. (i) 6879376 is divisible by 4, since 76 is divisible by 4.

(ii) 496138 is not divisible by 4, since 38 is not divisible by 4.

IV. Divisible by 5:

A number is divisible by 5 only when its unit digit is 0 or 5.

Ex. Each of the numbers 76895 and 68790 is divisible by 5.

V. Divisible by 6:

A number is divisible by 6 if it is simultaneously divisible by both 2 and 3.

Ex. 90 is divisible by 6 because it is divisible by both 2 and 3 simultaneously.

VI. Divisible by 7:

A number is divisible by 7 if and only if the difference of the number of its thousands and the remaining part of the given number is divisible by 7 respectively.

Ex. 473312 is divisible by 7, because the difference between 473 and 312 is 161, which is divisible by 7.

VII. Divisible by 8:

A number is divisible by 8 if the number formed by its last three digits i.e. hundred's, ten's and unit's digit of the given number is divisible by 8.

Ex. (i) In the number 16789352, the number formed by last 3 digits, namely 352 is divisible by 8.

$\therefore$ 16789352 is divisible by 8.

(ii) In the number 576484, the number formed by last 3 digits, namely 484 is not divisible by 8.

$\therefore$ 576484 is not divisible by 8.

VIII. Divisible by 9:

A number is divisible by 9 only when the sum of its digits is divisible by 9.

Ex. (i) Sum of digits of the number 246591 = 27, which is divisible by 9.

$\therefore$ 246591 is divisible by 9.

(ii) Sum of digits of the number 734519 = 29, which is not divisible by 9.

$\therefore$ 734519 is not divisible by 9.

IX. Divisible by 10:

A number is divisible by 10 only when its unit digit is 0.

Ex. (i) 7849320 is divisible by 10, since its unit digit is 0.

(ii) 678405 is not divisible by 10, since its unit digit is not 0.

X. Divisible by 11:

A number is divisible by 11 if the difference between the sum of its digits at odd places from right and the sum of its digits at even places also from right is either 0 or a number divisible by 11.

Ex. (i) Consider the number 29435417.

(Sum of its digits at odd places from right) – (Sum of its digits at even places from right)

$(7 + 4 + 3 + 9) - (1 + 5 + 4 + 2) = (23 - 12) = 11$, which is divisible by 11.

$\therefore$ 29435417 is divisible by 11.

(ii) Consider the number 57463822.

(Sum of its digits at odd places) – (Sum of its digits at even places)

$= (2 + 8 + 6 + 7) - (2 + 3 + 4 + 5) = (23 - 14)$

$= 9$, which is neither 0 nor divisible by 11.

$\therefore$ 57463822 is not divisible by 11.

XI. Divisible by 12:

A number is divisible by 12, if it is simultaneously divisible by both 3 and 4.

Illustration 4: Find the least value of * for which 7* 5462 is divisible by 9.

Solution: Let the required value be x. Then,

$(7 + x + 5 + 4 + 6 + 2) = (24 + x)$ should be divisible by 9.

$\Rightarrow \quad x = 3$

Illustration 5: Find the least value of * for which 4832*18 is divisible by 11.

Solution: Let the digit in place of * be x.

(Sum of digits at odd places from right) – (Sum of digits at even places from right)

$= (8 + x + 3 + 4) - (1 + 2 + 8 = (4 + x)$,

which should be divisible by 11.

$\therefore \quad x = 7$.

PRIME NUMBERS

A number other than 1 is called a prime number if it is divisible by only 1 and itself.

All prime numbers less than 100 are:

2, 3, 5, 7, 11, 13, 17, 19, 23, 29, 31, 37, 41, 43, 47, 53, 59, 61, 67, 71, 73, 79, 83, 89, 97.

Note that 2 is the smallest prime number. 2 is the only even prime number.

Smallest odd prime number is 3.

Twin Primes: A pair of prime numbers are said to be twin prime when they differ by 2. For example 3 and 5 are twin primes.

Co-primes or Relative primes: A pair of numbers are said to be co-primes or relative primes to each other if they do not have any common factor other than 1. For example 13 and 21.

Some Properties which Help in Finding Two Co-prime Numbers

(i) Two consecutive natural numbers are always co-prime.

Ex. 8 and 9 are co-prime.

Also 12 and 13 are co-prime.

(ii) Two consecutive odd integers are always co-prime.

Ex. 7, 9; 15, 17; 21, 23; etc.

(iii) Two prime numbers are always co-prime.
Ex. 19 and 23 are co-prime.
Also 29 and 41 are co-prime.

(iv) A prime number and a composite number such that the composite number is not a multiple of the prime number are always co-prime.
Ex. 7 and 15 are co-prime.

(v) Square of two co-prime numbers are always co-prime numbers.

Some Properties which Help in Finding Three Co-prime Numbers

3 numbers are co-prime to each other means all the possible pair of numbers out of these three numbers are co-prime. For example from three numbers 7, 8, 13 three pairs (7, 8), (7, 13) and (8, 13) are formed and each of these pair is a pair of co-prime. Hence, 7, 8, 13 are three co-prime numbers.

Following are some properties helping in finding three co-prime numbers:

(i) Three consecutive odd integers are always co-prime.
Ex. 9, 11, 13 are co-prime.

(ii) Three consecutive natural numbers with first one being odd are always co-primes.
Ex. 7, 8, 9 are co-prime.

(iii) Two consecutive natural numbers along with the next odd numbers are always co-primes.
Ex. 12, 13, 15 are co-prime. Also 17, 18, 19 are co-prime.

(iv) Three prime numbers are always co-prime.
Ex. 3, 11, 13 are co-prime.

To Test Whether a Given Number is Prime Number or Not

In CAT and CAT like competitions you are required to check whether a given number maximum upto 400 is prime number or not.

If you want to test whether any number is a prime number or not, take an integer equal to the square root of the given number but if square root is not an integer then take an integer just larger than the approximate square root of that number. Let it be 'x'. Test the divisibility of the given number by every prime number less than 'x'. If the given number is not divisible by any prime number less than, then the given number is prime number; otherwise it is a composite number.

Square root of 361 is 19. Prime numbers less than 19 are clearly 2, 3, 5, 7, 11, 13 and 17. Since, 361 is not divisible by any of the numbers 2, 3, 5, 7, 11, 13 and 17. Hence, 361 is a prime number.

It is advisable to learn the squared numbers of all integers from 1 to 20, which are very useful to find whether a given number is a prime or not.

Squares
$1^2 = 1$
$2^2 = 4$
$3^2 = 9$
$4^2 = 16$
$5^2 = 25$
$6^2 = 36$
$7^2 = 49$
$8^2 = 64$
$9^2 = 81$
$10^2 = 100$
$11^2 = 121$
$12^2 = 144$
$13^2 = 169$
$14^2 = 196$
$15^2 = 225$
$16^2 = 256$
$17^2 = 289$
$18^2 = 324$
$19^2 = 361$
$20^2 = 400$

From the table it is clear that if any number, say 271 lies between 256 and 289, then its square root lies between 16 and 17, because $16^2 = 256$ and $17^2 = 289$. Thus square root of the given number is not an integer. So, we take 17 as an integer just greater than the square root of the given number. Now all the prime numbers less than 17 are 2, 3, 5, 7, 11 and 13. Since 271 is not divisible by any of the numbers 2, 3, 5, 7, 11 and 13. Hence 361 is a prime number.

Illustration 6: Is 171 is a prime number ?

Solution: Square root of 171 lies between 13 and 14, because $13^2 = 169$ and $14^2 = 196$. Therefore, the integer just greater than the square root of 171 is 14.

Now prime numbers less than 14 are 2, 3, 5, 7, 11 and 13.

Since 171 is divisible by 3, therefore 171 is not a prime number.

Illustration 7: Is 167 is a prime number ?

Solution: Square root of 167 lies between 12 and 13, because $12^2 = 144$ and $13^2 = 169$. Therefore the integer just greater than the square root of 167 is 13.

Now prime numbers less than 13 are 2, 3, 5, 7 and 11.

Since 167 is not divisible by any of the prime numbers 2, 3, 5, 7 and 11; therefore 167 is a prime number.

GENERAL OR EXPANDED FORM OF 2 AND 3 DIGITS NUMBERS

(i) In a two digits number AB, A is the digit of tenth place and B is the digit of unit place, therefore AB is written using place value in expanded form as
$AB = 10A + B$
Ex. $35 = 10 \times 3 + 5$

(ii) In a three digits number ABC, A is the digit of hundred place, B is the digit of tenth place and C is the digit of unit place, therefore ABC is written using place value in expanded form as
$ABC = 100A + 10B + C$
Ex. $247 = 100 \times 2 + 10 \times 4 + 7$
These expanded forms are used in forming equations related to 2 and 3 digits numbers.

Illustration 8: In a two digit prime number, if 18 is added, we get another prime number with reversed digits. How many such numbers are possible ?

Solution: Let a two-digit number be pq.

$\therefore \quad 10p + q + 18 = 10q + p$

$\Rightarrow \quad -9p + 9q = 18 \Rightarrow q - p = 2$

Satisfying this condition and also the condition of being a prime number (pq and qp both), there are 2 numbers 13 and 79.

FACTORISATION

It is a process of representing a given number as a product of two or more prime numbers.

Here each prime number which is present in the product is called a factor of the given number.

For example, 12 is expressed in the factorised form in terms of its prime factors as $12 = 2^2 \times 3$.

Illustration 9: If $N = 2^3 \times 3^7$, then

(a) What is the smallest number that you need to multiply with N in order to make it a perfect square ?

(b) What is the smallest number that you need to divide by N in order to make it a perfect square ?

Solution:

(a) Any perfect square number in its factorised form has prime factors with even powers. So in order to make $2^3 \times 3^7$ a perfect square, the smallest number that we need to multiply it with would be 2×3 i.e. 6. The resulting perfect square will be $2^4 \times 3^8$.

(b) Similarly, in order to arrive at a perfect square by dividing the smallest number, we need to divide the number by 2×3 i.e., 6. The resulting perfect square will be $2^2 \times 3^6$.

NUMBER OF WAYS OF EXPRESSING A COMPOSITE NUMBER AS A PRODUCT OF TWO FACTORS

(i) Number of ways of expressing a composite number N which is not a perfect square as a product of two factors

$$= \frac{1}{2} \times \text{(Number of prime factors of the } N)$$

(ii) Number of ways of expressing a perfect square number M as a product of two factors $= \frac{1}{2}$ [(Number of prime factors of $M + 1$]

Illustration 10: Find the number of ways of expressing 180 as a product of two factors.

Solution: $180 = 2^2 \times 3^2 \times 5^1$

Number of factors $= (2 + 1)(2 + 1)(1 + 1) = 18$

Since 180 is not a perfect square, hence there are total $\frac{18}{2} = 9$ ways in which 180 can be expressed as a product of two factors.

Illustration 11: Find the number of ways expressing 36 as a product of two factors.

Solution: $36 = 2^2 \times 3^2$

Number of factors $= (2 + 1)(2 + 1) = 9$

Since 36 is a perfect square, hence the number of ways of expressing 36 as a product of two factors

$= \frac{9+1}{2} = 5$, as $36 = 1 \times 36, 2 \times 18, 3 \times 12, 4 \times 9$ and 6×6.

SUM OF FACTORS (OR DIVISORS) OF A COMPOSITE NUMBER

Let N be a composite number in such a way that $N = (x)^a (y)^b (z)^c$ ... where x, y, z... are prime numbers. Then, the sum of factors

(or divisors) of $N = \frac{x^{a+1}-1}{x-1} \times \frac{y^{b+1}-1}{y-1} \times \frac{z^{c+1}-1}{z-1} \ldots$

Illustration 12: What is the sum of the divisors of 60 ?

Solution: $60 = 2^2 \times 3 \times 5$

$\Rightarrow$ Sum of the divisors $= \frac{2^3-1}{2-1} \times \frac{3^2-1}{3-1} \times \frac{5^2-1}{5-1} = 168$.

SUM OF UNIT DIGITS

For given n different digits $a_1, a_2, a_3, \ldots, a_n$; the sum of the digits at unit place of all different numbers formed is

$(a_1 + a_2 + a_3 + \ldots + a_n)(n-1)!$ *i.e.*, (Sum of the digits) $(n - 1)!$

Illustration 13: Find the sum of unit digits of all different numbers formed from digits 4, 6, 7 and 9.

Solution: Required sum $= (4 + 6 + 7 + 9) - (4 - 1)!$

$= 26 - 3! = 26 - 6 = 20.$

THE LAST DIGIT FROM LEFT (i.e., UNIT DIGIT) OF ANY POWER OF A NUMBER

The last digits (from left) of the powers of any number follow a cyclic pattern i.e., they repeat after certain number of steps. If we find out after how many steps the last digit of the powers of a number repeat, then we can find out the last digit of any power of any number.

Let us look at the powers of 2:

Last digit of 2^1 is 2 .
Last digit of 2^2 is 4 .
Last digit of 2^3 is 8 .
Last digit of 2^4 is 6 .
Last digit of 2^5 is 2 .
Last digit of 2^6 is 4 .
Last digit of 2^7 is 8 .
Last digit of 2^8 is 6 .
Last digit of 2^9 is 2 .

Since last digit of 2^5 is the same as the last digit of 2^1, then onwards the last digit will start repeating, i.e., digits of $2^5, 2^6, 2^7, 2^8$ will be the same as those of $2^1, 2^2, 2^3, 2^4$. Then the last digit of 2^9 is again the same as the last digit of 2^1 and so on. Thus, we see that when power of 2 increases, the last digits repeat after every 4 steps.

In above pattern, we can see that whenever the power of 2 is a multiple of 4, the last digit of that number will be the same as the last digit of 2^4.

Suppose we want to find out the last digit of 2^{66}, we should look at a multiple of 4 which is just less than or equal to the power 66 of 2. Since 64 is a multiple of 4, the last digit of 2^{64} will be the same as the last digit of 2^4.

Then the last digits of $2^{65}, 2^{66}$ will be the same as the last digits of $2^1, 2^2$ respectively. Hence the last digit of 2^{66} is the same as the last digit of 2^2 i.e., 4.

Similarly, we can find out the last digit of 3^{75} by writing down the pattern of the powers of 3.

Last digit of 3^1 is 3.
Last digit of 3^2 is 9.
Last digit of 3^3 is 7.
Last digit of 3^4 is 1.
Last digit of 3^5 is 3.
Last digit of $3^6 = 9$
Last digit of $3^7 = 7$
Last digit of $3^8 = 1$
Last digit of $3^9 = 3$

The last digit repeats after 4 steps (like in the case of powers of 2).

Whenever the powers of 3 is a multiple of 4, the last digit of that number will be the same as the last digit of 3^4.

To find the last digit of 3^{75}, we look for a multiple of 4 which is just less than or equal to the power 75 of 3. Since, 72 is multiple of 4, the last digit of 3^{72} will be the same as that of 3^4. Hence the last digit of 3^{75} will be the same as the last digit of 3^3 i.e., 7.

Last Digit (i.e., Unit Digit) of a Product

Last digit of the product $a \times b \times c$ ... is the last digit of the product of last digits of a, b, c, ...

Illustration 14: Find the last digit of $2^{416} \times 4^{430}$.

Solution: Writing down the powers of 2 and 4 to check the pattern of the last digits, we have

We have seen that whenever the power of 2 is a multiple of 4, the last digit of that number will be the same as the last digit of 2^4.

Now, Last digit of $4^1 = 4$.
Last digit of $4^2 = 6$.
Last digit of $4^3 = 4$.
Last digit of $4^4 = 6$.

Thus last digit of any power of 4 is 4 for an odd power and 6 for an even power. The last digit of 2^{416} will be the same as 2^4 because 416 is a multiple of 4. So the last digit of 2^{416} is 6.

Last digit of 4^{430} is 6, since the power of 4 is even.

Hence the last digit of $2^{416} \times 4^{430}$ will be equal to the last digit of $6 \times 6 = 6$.

CONCEPT OF REMAINDERS

(I) Suppose the numbers $N_1, N_2, N_3, \ldots$ give quotients $Q_1, Q_2, Q_3, \ldots$ and remainder $R_1, R_2, R_3, \ldots$ when divided by a common divisor D.

Let S be the sum of $N_1, N_2, N_3 \ldots$

Therefore, $S = N_1 + N_2 + N_3 + \ldots$

$$= (D \times Q_1 + R_1) + (D \times Q_2 + R_2) + (D \times Q_3 + R_3) + \ldots$$

$$= D \times K + (R_1 + R_2 + R_3 \ldots), \quad \ldots (1)$$

where K is some number

Hence the remainder when S is divided by D is the remainder when $(R_1 + R_2 + R_3 \ldots\ldots)$ is divided by D.

(II) Suppose the numbers, $N_1, N_2, N_3, \ldots$ give quotients $Q_1, Q_2, Q_3, \ldots$ and remainders $R_1, R_2, R_3, \ldots$ respectively, when divided by a common divisor D.

Therefore $N_1 = D \times Q_1 + R_1, N_2 = D \times Q_2 + R_2,$
$N_3 = D \times Q_3 \times R_3 \ldots$ and so on.

Let P be the product of $N_1, N_2, N_3, \ldots$

Therefore,

$$P = N_1 N_2 N_3 \ldots$$

$$= (D \times Q_1 + R_1)(D \times Q_2 + R_2)(D \times Q_3 + R_3) \ldots$$

$$= D \times K + (R_1 R_2 R_3 \ldots), \quad \ldots (2)$$

where K is some number

In the above equation, since only the product $(R_1 R_2 R_3 \ldots)$ is free of D, therefore the remainder when P is divided by D is the remainder when the product $(R_1 R_2 R_3 \ldots)$ is divided by D.

Illustration 15: What is the remainder when the product 1991 × 1992 × 2000 is divided by 7 ?

Solution: The remainder when 1991, 1992 and 2000 are divided by 7 are 3, 4 and 5 respectively.

Hence the final remainder is the remainder when the product $3 \times 4 \times 5 = 60$ is divided by 7. Therefore, remainder = 4.

Illustration 16: What is the remainder when 2^{2010} is divided by 7 ?

Solution: 2^{2010} is a product $(2 \times 2 \times 2 \ldots (2010 \text{ times}))$. Since, 2 is a number less than 7, we try to convert the product into product of numbers higher than 7. Notice that $8 = 2 \times 2 \times 2$. Therefore, we convert the product in the following manner

$$2^{2010} = 8^{670} = 8 \times 8 \times 8 \ldots (670 \text{ times.})$$

The remainder when 8 is divided by 7 is 1. Hence the remainder when 8^{670} is divided by 7 is the remainder obtained when the product $1 \times 1 \times 1 \ldots$ (670 times) is divided by 7. Therefore, remainder = 1.

Illustration 17: What is the remainder when 25^{24} is divided by 9 ?

Solution: Again $25^{24} = (18 + 7)^{24} = (18 + 7)(18 + 7) \ldots$ 24 times $= 18K + 7^{24}$.

Hence, remainder when 25^{24} is divided by 9 is the remainder when 7^{24} is divided by 9.

Now, $7^{24} = 7^3 \times 7^3 \times 7^3 \ldots$ (8 times) $= 343 \times 343 \times 343 \ldots$ (8 times)

Now when 343 is divided by 9 the remainder is 1

So, the remainder when dividing $(343)^8$ by 9 means remainder when dividing $(1)^8$ by 9. So the required remainder is 1.

NUMBER OF ZEROES IN AN EXPRESSION LIKE *a* × *b* × *c* × ..., WHERE *a*, *b*, *c*,... ARE NATURAL NUMBERS

Consider an expression $8 \times 15 \times 20 \times 30 \times 40$.

The expression can be written in the standard form as :

$8 \times 15 \times 20 \times 30 \times 40$
$= (2^3) \times (3 \times 5) \times (2^2 \times 5) \times (2 \times 3 \times 5) \times (2^3 \times 5)$
$= 2^9 \times 3^2 \times 5^4$, in which base of each factor is a prime number.

A zero is formed by the product of 2 and 5 i.e. 2×5. Hence number of zeroes is equal to the number of pair(s) of 2's and 5's formed.

In the above standard form of the product there are 9 twos and 4 fives. Hence number of pairs of 2 and 5 i.e. (2×5) is 4. Hence, there will be 4 zeroes at the end of the final product.

In the same above way, we can find the number of zeroes at the end of any product given in the form of an expression like $a \times b \times c \times \ldots$, where $a, b, c, \ldots$ are natural numbers.

If there is no pair of 2 and 5 i.e. 2×5, then there is no zero at the end of the product. For example, consider the expression $9 \times 21 \times 39 \times 49$.

The given expression in standard form,

$$9 \times 21 \times 39 \times 49 = (3^2) \times (3 \times 7) \times (3 \times 13) \times (7^2)$$
$$= 3^4 \times 7^3 \times 13$$

There is no pair of 2 and 5 in the standard form of expression given as product, therefore there will be no zero at the end of the final product.

Illustration 18: Find the number of zeroes in the product $1^1 \times 2^2 \times 3^3 \times 4^4 \times 5^5 \times 6^6 \times \ldots\ldots\ldots \times 49^{49}$

Solution: Clearly the fives will be less than the twos. Hence, we need to count only the fives.

Now, $5^5 \times 10^{10} \times 15^{15} \times 20^{20} \times 25^{25} \times 30^{30} \times 35^{35} \times 40^{40} \times 45^{45}$
$= (5)^5 \times (5 \times 2)^{10} \times (5 \times 3)^{15} \times (5 \times 4)^{20} \times (5 \times 5)^{25} \times (5 \times 6)^{30} \times (5 \times 7)^{35} \times (5 \times 8)^{40} \times (5 \times 9)^{45}$

It gives us $5 + 10 + 15 + 20 + 25 \times 25 + 30 + 35 + 40 + 45$ fives *i.e.*, 825 fives

Thus the product has 825 zeroes.

BASE SYSTEM

The number system in which we carry out all calculation is decimal (base 10) system. It is called decimal system because there are 10 digits 0 to 9.

There are other number systems also depending on the number of digits contained in the base system. Some of the most common systems are Binary system, Octal system, and Hexadecimal system. A number system containing two digits 0 and 1 is called binary (base 2) system. Number system containing eight digits 0, 1, 2, 3, ..., 7 is called Octal (base 8) system.

Hexadecimal (base 16) system has 16 digits 0, 1, 2, 3, .., 9, *A*, *B*, *C*, *D*, *E*, *F*; where *A* has a value 10, *B* has a value 11 and so on.

Let a number *abcde* be written in base *p*, where *a, b, c, d* and *e* are single digits less than *p*. The value of the number abcde in base 10 = $e \times p^0 + d \times p^1 + c \times p^2 + b \times p^3 + a \times p^4$

For example, The number 7368 can be written as $8 + 6 \times 10 + 3 \times (10)^2 + 7 \times (10)^3 = 7368$ in decimal (base 10) number system.

The number 7368 in base 9 is written in decimal number system as $8 \times 9^0 + 6 \times 9 + 3 \times 9^2 + 7 \times 9^3 = 5408$

There are mainly two types of operations associated with conversion of bases: First conversion from any base to base ten and second conversion from base 10 to any base.

(i) Conversion From Any Base to Base Ten

The number $(pqrstu)_a$ (i.e., the number *pqrstu* on base *a*) is converted to base 10 by finding the value of the number.

$(pqrstu)_a = u + ta + sa^2 + ra^3 + qa^4 + pa^5$.

Here subscript '*a*' in $(pqrst)_a$ denotes the base of the number *pqrstu*.

Illustration 19: Convert $(21344)_5$ to base 10.

Solution:

$(21344)_5 = 4 \times 5^0 + 4 \times 5^1 + 3 \times 5^2 + 1 \times 5^3 + 2 \times 5^4$

$= 4 + 4 \times 5 + 3 \times 25 + 1 \times 125 + 2 \times 625 = 1474.$

(ii) Conversion From Base 10 to Any Base

A number written in base 10 can be converted to any base '*a*' by first dividing the number by '*a*' and then successively dividing the quotients by '*a*'. The remainders, written in reverse order, give the equivalent number in base '*a*'.

For example the number 238 in base 3 is found as

3	238	
	79	1
	26	1
	8	2
	2	2

Remainders

The remainders in reverse order is 22211.

Hence, 22211 is the required number in base 3.

Note: Value of a single digit number to all bases are the same. For example,

$5_4 = 5_7 = 5_8 = 5_{10}$

Addition, Subtraction and Multiplication in the Same Bases

Illustration 20: Add the numbers $(4235)_7$ and $(2354)_7$.

Solution: The numbers are written as

$$\begin{array}{r} 4\,2\,3\,5 \\ \underline{2\,3\,5\,4} \\ \hline \end{array}$$

The addition of 5 and 4 (first digit from right of both numbers) is 9 which being more than 7 would be written as $9 = 7 \times 1 + 2$. Here 1 is the quotient and 2 is the remainder when 9 is divided by 7. The remainder 2 is placed at the first place from right of the answer and the quotient 1 gets carried over to the second place from the right.

At the second place from the right $3 + 5 + 1$ (carry) $= 9 = 7 \times 1 + 2$

$$\begin{array}{cccc} & & +1 & +1 \\ 4 & 2 & 3 & 5 \\ 2 & 3 & 5 & 4 \\ \hline 6 & 6 & 2 & 2 \end{array}$$

The remainder 2 is placed at the second place from right of the answer and the quotient 1 carry over to the third place from right. In the same way, we can find the other digits of the answer.

Illustration 21: $(52)_7 + 46_8 = (?)_{10}$

(a) $(75)_{10}$ (b) $(50)_{10}$

(c) $(39)_{39}$ (d) $(28)_{10}$

Solution: (a) $(52)_7 = (5 \times 7^1 + 2 \times 7^0)_{10} = (37)_{10}$

Also, $(46)_8 = (4 \times 8^1 + 6 \times 8^0)_{10} = (38)_{10}$

Sum $= (37)_{10} + (38)_{10} = (75)_{10}$

Illustration 22: $(11)_2 + (22)_3 + (33)_4 + (44)_5 + (55)_6 + (66)_7 + (77)_8 + (88)_9 = (?)_{10}$

(a) 396 (b) 276

(c) 250 (d) 342

Solution: (b) $(11)_2 = (1 \times 2^1 + 1 \times 2^0)_{10} = (3)_{10}$

$(22)_3 = (2 \times 3^1 + 2 \times 3^0)_{10} = (8)_{10}$

$(33)_4 = (3 \times 4^1 + 3 \times 4^0)_{10} = (15)_{10}$

$(44)_5 = (4 \times 5^1 + 4 \times 5^0)_{10} = (24)_{10}$

$(55)_6 = (5 \times 6^1 + 5 \times 6^0)_{10} = (35)_{10}$

$(66)_7 = (6 \times 7^1 + 6 \times 7^0)_{10} = (48)_{10}$

$(77)_8 = (7 \times 8^1 + 7 \times 8^0)_{10} = (63)_{10}$

$(88)_9 = (8 \times 9^1 + 8 \times 9^0)_{10} = (80)_{10}$

Sum $= (3)_{10} + (8)_{10} + (15)_{10} + (24)_{10} + (35)_{10} + (48)_{10} + (63)_{10} + (80)_{10}$

$= (276)_{10}$

Illustration 23: Subtract $(247)_8$ from $(345)_8$.

Solution:

(i) 5 is less than 7. So borrow 1 from the previous digit 4. Since, we are working in octal system, so 5 become $5 + 8 = 13$. Subtract 7 from 13, you will get 6.

$$\begin{array}{r} 3\;4\;5 \\ -2\;4\;7 \\ \hline 6 \\ \hline \end{array}$$

(ii) Since, we have borrowed 1, the 4 in the first row has now become 3, which is less than the digit (4), just below it in the second row, So borrow 1 from 3 of first row. So, the 4 in first row is now becomes 3 + 8 = 11. Subtracting 4 of second row from 11, we get 7. Hence,

$$\begin{array}{r} 3\;4\;5 \\ -2\;4\;7 \\ \hline 0\;7\;6 \\ \hline \end{array}$$

FACTORS AND MULTIPLES

If one number '*a*' completely divides a second number '*b*' then 1st number '*a*' is said to be a factor of the 2nd number '*b*'. For example 3 completely divides 15, so 3 is a factor of 15; while 4 does not divide 15 completely, so 4 is not a factor of 15.

Factors of 30 are 1, 2, 3, 5, 6, 10, 15 and 30

Factors of 40 are 1, 2, 4, 5, 8, 10, 20 and 40.

If a number '*a*' is exactly divisible by *a* number '*b*' then the 1st number '*a*' is said to be a multiple of 2nd number '*b*'. For example, 35 is exactly divisible by 7, so 35 is a multiple of 7. Multiple of a number '*b*' can be written down as '*nb*' where *n* is a natural number. So multiples of 5 are 5, 10, 15, 20, 25, ...

HIGHEST COMMON FACTOR (HCF) OR GREATEST COMMON DIVISOR (GCD)

The highest (i.e. largest) number that divides two or more given numbers is called the highest common factor (HCF) of those numbers.

Consider two numbers 12 and 15.

Factors of 12 are 1, 2, 3, 4, 6, 12.

Factors of 30 are 1, 2, 3, 5, 6, 10, 15, 30.

We have some common factors out of these factors of 12 and 30, which are 1, 2, 3, 6. Out of these common factors, 6 is the highest common factor. So, 6 is called the Highest Common Factor (HCF) of 12 and 30.

Methods to Find The HCF or GCD

There are two methods to find HCF of the given numbers

(i) Prime Factorization Method

When a number is written as the product of prime numbers, then it is called the prime factorization of that number. For example, $72 = 2 \times 2 \times 2 \times 3 \times 3 = 2^3 \times 3^2$. Here, $2 \times 2 \times 2 \times 3 \times 3$ or $2^3 \times 3^2$ is called prime factorization of 72.

To find the HCF of given numbers by this methods, we perform the prime factorization of all the numbers and then check for the common prime factors. For every prime factor common to all the numbers, we choose the least index of that prime factor among the given numbers. The HCF is the product of all such prime factors with their respective least indices.

Illustration 24: Find the HCF of $36x^3y^2$ and $24x^4y$.

Solution $36x^3y^2 = 2^2.3^2.x^3.y^2$, $24x^4y = 2^3.3.x^4.y$. The least index of 2, 3, *x* and *y* in the numbers are 2, 1, 3 and 1 respectively. Hence the HCF $= 2^2.3.x^3.y = 12x^3y$.

Illustration 25: The numbers 400, 536 and 646; when divided by a number N, give the remainders of 22, 23 and 25 respectively. Find the greatest such number *N*.

Solution: *N* will be the HCF of (400 – 22), (536 – 23) and (646 – 25). Hence, *N* will be the HCF of 378, 513 and 621. Hence, *N* = 27.

Illustration 26: The HCF of two numbers is 12 and their product is 31104. How many such numbers are possible.

Solution: Let the numbers be 12*x* and 12*y*, where *x* and *y* are co-prime to each other.

Therefore, $12x \times 12y = 31104 \rightarrow xy = 216$.

Now we need to find co-prime pairs whose product is 216.

$216 = 2^3 \times 3^3$. Therefore, the co-prime pairs will be (1, 216) and (8, 27). Therefore, (12, 12 × 216) and (8 × 12, 27 × 12) are two possible numbers.

(ii) Division Method

To find the HCF of two numbers by division method, we divide the larger number by the smaller number. Then we divide the smaller number by the first remainder, then first remainder by the second remainder.. and so on, till the remainder becomes 0. The last divisor is the required HCF.

Illustration 27: Find the HCF of 288 and 1080 by the division method.

Solution:

```
288 | 1080 | 3
      864
      216 | 288 | 1
            216
             72 | 216 | 3
                  216
                    0
```

The last divisor 72 is the HCF of 288 and 1080.

Shortcut for Finding HCF or GCD

To find the HCF of any number of given numbers, first find the difference between two nearest given numbers. Then find all factors (or divisors) of this difference. Highest factor which divides all the given numbers is the HCF.

Illustration 28: Find the HCF of 12, 20 and 32.

Solution: Difference of nearest two numbers 12 and 20 = 20 – 12 = 8

All factors (or divisor) of 8 are 1, 2, 4 and 8.

1, 2 and 4 divides each of the three given numbers 12, 20 and 32. Out of 1, 2 and 4; 4 is the highest number. Hence, HCF = 4.

LEAST COMMON MULTIPLE (LCM)

The least common multiple (LCM) of two or more numbers is the lowest number which is divisible by all the given numbers.

Consider two numbers 12 and 15.

Multiples of 12 are 12, 24, 36, 48, 60, 72, 84, 96, 108, 120, 132,...

While the multiples of 15 are 15, 30, 45, 60, 75, 90, 105, 120, 135, 150,.....

Out of these series of multiples, we have some common multiples like 60, 120, 180, ..., etc. Out of these common multiples, 60 is the lowest, so 60 is called the Lowest Common Multiple (LCM) of 12 and 15.

Methods to Find The LCM

There are two methods to find the LCM.

(i) Prime Factorization Method

After performing the prime factorization of all the given numbers, we find the highest index of all the prime numbers among the given numbers. The LCM is the product of all these prime numbers with their respective highest indices because LCM must be divisible by all of the given numbers.

Illustration 29: Find the LCM of 72, 288 and 1080.

Solution:

$$72 = 2^3 \times 3^2$$
$$288 = 2^5 \times 3^2$$
$$1080 = 2^3 \times 3^3 \times 5$$

Hence, $\text{LCM} = 2^5 \times 3^3 \times 5^1 = 4320$

(ii) Division Method

To find the LCM of 5, 72, 196 and 240, we use the division method in the following way:

Check whether any prime number that divides at least two of all the given numbers. If there is no such prime number, then the product of all these numbers is the required LCM, otherwise find the smallest prime number that divides at least two of the given numbers. Here, we see that smallest prime number that divides at least two given numbers is 2.

Divide those numbers out of the given numbers by 2 which are divisible by 2 and write the quotient below it. The given number(s) that are not divisible by 2 write as it is below it and repeat this step till you do not find at least two numbers that are not divisible by any prime number.

2	5, 72, 196, 240
2	5, 36, 98, 120
2	5, 18, 49, 60
3	5, 9, 49, 30
5	5, 3, 49, 10
	1, 3, 49, 2

After that find the product of all divisors and the quotient left at the end of the division. This product is the required LCM.

Hence, LCM of the given numbers = product of all divisors and the quotient left at the end.

$$= 2 \times 2 \times 2 \times 3 \times 5 \times 3 \times 49 \times 2 = 35280$$

Illustration 30: On a traffic signal, traffic light changes its colour after every 24, 30 and 36 seconds in green, red and orange light. How many times in an hour only green and red light will change simultaneously.

Solution: LCM. of 24 and 30 = 120

So in 1 hr both green and red light will change simultaneously 3600/120 times = 30 times

LCM of 24, 30 and 36 is 360

Hence in 1 hr all three lights will change simultaneously 3600/360 times = 10 times

So in 1 hr only red and green lights will change 30 – 10 = 20 times simultaneously.

Shortcut For Finding LCM

Using idea of co-prime, you can find the LCM by the following shortcut method:

LCM of 9, 10, 15 and 36 can be written directly as $9 \times 10 \times 2$.

The logical thinking that behind it is as follows:

Step 1: If you can see a set of 2 or more co-prime numbers in the set of numbers of which you are finding the LCM, write them down by multiply them.

In the above situation, since we see that 9 and 10 are co-prime to each other, we start off writing the LCM by writing 9×10 as the first step.

Step 2: For each of the other numbers, consider what prime factor(s) of it is/are not present in the LCM (if factorised into primes) taken in step 1. In case you see some prime factors of each of the other given numbers separately are not present in the LCM (if factorised into primes) taken in step 1, such prime factors will be multiplied in the LCM taken in step 1.

Prime factorisation of $9 \times 10 = 3 \times 3 \times 2 \times 5$

Prime factorisation of $15 = 3 \times 5$

Prime factorisation of $36 = 2 \times 2 \times 3 \times 3$

Here we see that both prime factors of 15 are present in the prime factorisation of 9×10 but one prime factor 2 of 36 is not present in the LCM taken in step 1. So to find the LCM of 9, 10, 15 and 36; we multiply the LCM taken in step 1 by 2.

Thus required LCM = $9 \times 10 \times 2 = 180$

Rule For Finding HCF and LCM of Fractions

(I) HCF of two or more fractions

$$= \frac{\text{HCF of numerator of all fractions}}{\text{LCM of denominator of all fractions}}$$

(II) LCM of two or more fractions

$$= \frac{\text{LCM of numerator of all fractions}}{\text{HCF of denominator of all fractions}}$$

Illustration 31: Find the HCF and LCM of $\frac{4}{5}, \frac{6}{11}, \frac{3}{5}$.

Solution:

$$\text{HCF} = \frac{\text{HCF of } 4, 6, 3}{\text{LCM of } 5, 11, 5} = \frac{1}{55}$$

$$\text{LCM} = \frac{\text{LCM of } 4, 6, 3}{\text{HCF of } 5, 11, 5} = \frac{12}{1} = 12$$

For any two numbers, HCF × LCM = product of the two numbers

This formula is applicable only for two numbers.

For example, HCF of 288 and 1080 is 72 and LCM of these two numbers is 4320.

We can see that $72 \times 4320 = 311040 = 288 \times 1080$.

GREATEST INTEGRAL VALUE

If x be a real number, then $[x]$ indicates greatest integer equal or less than x.

If the given number is an integer, then the greatest integer gives the number itself, otherwise it gives the first integer towards the left of the number x on the number line.

For example [4] = 4, [3.4] = 3, [6.8] = 6, [–2.3] = –3, [–5.6] = –6 and so on.

Note that –3 is less than –2.3 and –6 is less than –5.6, etc.

Illustration 32: What is the value of

$$[\sqrt{1}]+[\sqrt{2}]+[\sqrt{3}]+\ldots+[\sqrt{49}]+[\sqrt{50}]$$

where $[x]$ denotes greatest integer function?

Solution: $1^2 = 1$, $2^2 = 4$, $3^2 = 9$, $4^2 = 16$, $5^2 = 25$, $6^2 = 36$, $7^2 = 49$, $8^2 = 64$

Therefore, from $[\sqrt{1}]$ to $[\sqrt{3}]$, the value will be 1, from $[\sqrt{4}]$ to $[\sqrt{8}]$ the value will be 2, from $[\sqrt{9}]$ to $[\sqrt{15}]$ the value will be 3 and so on.

Therefore, the total value

$$= 1 \times 3 + 2 \times 5 + 3 \times 7 + 4 \times 9 + 5 \times 11 + 6 \times 13 + 7 \times 2$$

$$= 3 + 10 + 21 + 36 + 55 + 78 + 14 = 217.$$

Illustration 33: What is the value of x for which $x[x] = 32$?

Solution: If the value of x is 5, $x[x] = 25$, and if the value of x is 6, then $x[x] = 36$

Therefore, the value of x lies between 5 and 6.

If x lies between 5 and 6, then $[x] = 5$.

$$\Rightarrow \quad x = \frac{28}{[x]} = \frac{32}{5} = 6.4\,.$$

Practice Exercise

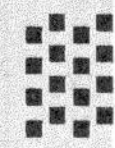

Level - I

1. The greatest number which will divide 116, 221, 356 leaving the same remainder in each case is
(a) 15 (b) 5
(c) 10 (d) 20

2. What number has to be added to 345670 in order to make it divisible by 6?
(a) 2 (b) 4
(c) 5 (d) 6

3. The least number which when divided by 35 leaves a remainder 25, when divided by 45 leaves the remainder 35 and when divided by 55 leaves 45 is
(a) 3465 (b) 3645
(c) 3655 (d) 3455

4. If n is any even number, then $n(n^2 + 20)$ is always divisible by
(a) 15 (b) 20
(c) 24 (d) 32

5. When 2^{256} is divided by 17 the remainder would be
(a) 1 (b) 16
(c) 14 (d) None of these

6. The last digit of 2137^{753} is
(a) 9 (b) 7
(c) 3 (d) 1

7. Find the least square number which is divisible by 3, 5, 6, and 9.
(a) 900 (b) 90
(c) 8100 (d) 81

8. In order that the number 1 y 3 y 6 be divisible by 11, the digit y should be
(a) 1 (b) 2
(c) 5 (d) 6

9. If n is an even natural number, then the largest natural number by which n $(n+1)(n+2)$ is divisible is
(a) 6 (b) 8
(c) 12 (d) 24

10. Which number should be added to 459045 to make it exactly divisible by 27 ?
(a) 3 (b) 9
(c) 0 (d) None of these

11. Find the last digit of the sum $19^{81} + 4^{9k}$, $K \in N$.
(a) 4 (b) 9
(c) 3 (d) Cannot be determined

12. The sum of prime numbers that are greater than 60, but less than 70 is
(a) 128 (b) 191
(c) 197 (d) 260

13. The number 311311311311311311311 is
(a) divisible by 3 but not by 11
(b) divisible by 11 but not by 3
(c) divisible by both 3 and 11
(d) neither divisible by 3 nor by 11

14. A difference between two numbers is 1365, when larger number is divided by the smaller one, the quotient is 6 and the remainder is 15. What is the smaller number?
(a) 240 (b) 360
(c) 270 (d) 295

15. If the number 517 * 324 is completely divisible by 3, then the smallest whole number in place of * will be:
(a) 0 (b) 1
(c) 2 (d) None of these

16. If the product 4864 × 9 P 2 is divisible by 12, the value of P is
(a) 2 (b) 5
(c) 6 (d) None of these

17. The largest 4-digit number exactly divisible by 88 is
(a) 9944 (b) 9768
(c) 9988 (d) 8888

18. $(x^n - a^n)$ is completely divisible by $(x + a)$, when
(a) n is any natural number
(b) n is an even natural number
(c) n is an odd natural number
(d) n is prime

19. When $0.\overline{47}$ is converted into a fraction the result is
(a) $\frac{46}{90}$ (b) $\frac{46}{99}$
(c) $\frac{47}{90}$ (d) $\frac{47}{99}$

20. Which of the following statements are true:
(i) The rational number $\frac{29}{23}$ lies to the left of zero on the number line.
(ii) The rational number $\frac{-12}{-17}$ lies to the right of zero on the number line.
(iii) The ratinal numbers $\frac{-12}{5}$ and $\frac{-7}{17}$ are on the opposite side of zero on the number line.
(v) The rational numbers $\frac{-21}{5}$ and $\frac{7}{-31}$ are on the opposite side of zero on the number line.
(a) Only (i) (b) (i) & (ii)
(c) Only (iii) (d) (i), (ii) & (iv)

21. I have a certain number of beads which lie between 600 and 900. If 2 beads are taken away the remainder can be equally divided among 3, 4, 5, 6, 7 or 12 boys. The number of beads I have
(a) 729 (b) 842
(c) 576 (d) 961

22. Find the digit at the unit's place of $(377)^{59} \times (793)^{87} \times (578)^{129} \times (99)^{99}$
(a) 1 (b) 2
(c) 7 (d) 9

23. Four different electronic devices make a beep after every 30 minutes, 1 hour, $1\frac{1}{2}$ hour and 1 hour 45 minutes respectively. All the devices beeped together at 12 noon. They will again beep together at:
(a) 12 midnight (b) 3 a.m.
(c) 6 a.m. (d) 9 a.m.

24. If N is the sum of first 13,986 prime numbers, then N is always divisible by
(a) 6 (b) 4
(c) 8 (d) None of these

25. If two numbers when divided by a certain divisor give remainder 35 and 30 respectively and when their sum is divided by the same divisor, the remainder is 20, then the divisor is
(a) 40 (b) 45
(c) 50 (d) 55

26. Find the least number which when divided by 12, leaves a remainder 7, when divided by 15, leaves a remainder 10 and when divided by 16, leaves a remainder 11
(a) 115 (b) 235
(c) 247 (d) 475

27. How many even integers n, where $100 \le n \le 200$, are divisible neither by seven nor by nine ?
(a) 40 (b) 37
(c) 39 (d) 38

28. A number is *interesting* if on adding the sum of the digits of the number and the product of the digits of the number, the result is equal to the number. What fraction of numbers between 10 and 100 (both 10 and 100 included) is *interesting?*
(a) 0.1 (b) 0.11
(c) 0.16 (d) 0.22

29. In a cricket match, Team A scored 232 runs without losing a wicket. The score consisted to byes, wides and runs scored by two opening batsmen : Ram and Shyam. The runs scored by the two batsman are 26 times wides. There are 8 more byes than wides. If the ratio of the runs scored by Ram and Shyam is 6 : 7, then the runs scored by Ram is
(a) 88 (b) 96
(c) 102 (d) 112

30. If $x+y+z=1$ and x, y, z are positive real numbers, then the least value of $\left(\frac{1}{x}-1\right)\left(\frac{1}{y}-1\right)\left(\frac{1}{z}-1\right)$ is
(a) 4 (b) 8
(c) 16 (d) None of these

31. The last digit of $3^{3^{4n}}+1$, is
(a) 0 (b) 4
(c) 8 (d) 2

32. The last digit in $(25\_)^{32}$ and $(25\_)^{33}$ both is 6. The missing digit is :
(a) 4 (b) 8
(c) 6 (d) 5

33. Which digits should come in place of * and $ if the number 62684*$ is divisible by both 8 and 5?
(a) 4, 0 (b) 0, 4
(c) 0, 0 (d) 4, 4

34. At a college football game, 4/5 of the seats in the lower deck of the stadium were sold. If 1/4 of all the seating in the stadium is located in the lower deck, and if 2/3 of all the seats in the stadium were sold, then what fraction of the unsold seats in the stadium was in the lower deck ?
(a) 3/20 (b) 1/6
(c) 1/5 (d) 1/3

35. The integers 1, 2, ..., 40 are written on a blackboard. The following operation is then repeated 39 times; In each repetition, any two numbers, say a and b, currently on the blackboard are erased and a new number $a+b-1$ is written. What will be the number left on the board at the end?
(a) 820 (b) 821
(c) 781 (d) 819

36. If $653xy$ is divisible by 80 then the value of $x+y$ is
(a) 2 (b) 3
(c) 4 (d) 6

37. How many numbers are there between 200 and 800 which are divisible by both 5 and 7?
(a) 35 (b) 16
(c) 17 (d) can't be determined

38. How many numbers are there in the set $S = \{200, 201, 202, ..., 800\}$ which are divisible by neither of 5 or 7?
(a) 411 (b) 412
(c) 410 (d) None of these

39. When a number divided by 9235, we get the quotient 888 and the remainder 222, such a least possible number is
(a) 820090 (b) 8200920
(c) 8200680 (d) None of these

40. A number which when divided by 32 leaves a remainder of 29. If this number is divided by 8 the remainder will be
(a) 0 (b) 1
(c) 5 (d) 3

41. $(0.\bar{1})^2\left[1-9(0.1\bar{6})^2\right]=?$
(a) $-\frac{1}{162}$ (b) $\frac{1}{108}$
(c) $\frac{7696}{106}$ (d) $\frac{833}{88209}$

42. A six digit number which is consisting of only one digits either 1, 2, 3, 4, 5, 6, 7, 8 or 9, *e.g.*, 111111, 222222... etc. This number is always divisible by :
(a) 7 (b) 11
(c) 13 (d) All of these

43. Product of divisors of 7056 is
(a) $(84)^{48}$ (b) $(84)^{44}$
(c) $(84)^{45}$ (d) None of these

44. The first 23 natural numbers are written in increasing order beside each other to form a single number. What is the remainder when this number is divided by 18?
(a) 1 (b) 6
(c) 12 (d) 15

45. How many positive integer values of 'a' are possible such that $\frac{a+220}{a+4}$ is an integer?
(a) 8 : 9 (b) 9 : 8
(c) 3 : 4 (d) 4 : 3

46. The sum and number of even factors of 2450.
(a) 9,3534 (b) 18,3500
(c) 12,3524 (d) 4,2453

47. Find the sum of divisors of 544 which are perfect squares.
(a) 32 (b) 64
(c) 42 (d) 21

48. Find the number of zeroes in
$100^1 \times 99^2 \times 98^3 \times 97^4 \times \ldots\ldots\ldots \times 1^{100}$
(a) 1024 (b) 250
(c) 1124 (d) 124

49. $(23)_5 + (47)_9 = (?)_8$
(a) 70 (b) 35
(c) 64 (d) 18

50. LCM of first 100 natural numbers is N. What is the LCM of first 105 natural numbers?
(a) $5! \times N$ (b) $10403\ N$
(c) $105N/103$ (d) $4\ N$

51. $N!$ is completely divisible by 13^{52}. What is sum of the digits of the smallest such number N?
(a) 11 (b) 15
(c) 16 (d) 19

52. $12^{55}/3^{11} + 8^{48}/16^{18}$ will give the digit at units place as
(a) 4 (b) 6
(c) 8 (d) 0

53. The unit digit in the expression
$36^{234}*33^{512}*39^{180} - 54^{29}*25^{123}*31^{512}$ will be
(a) 8 (b) 0
(c) 6 (d) 5

54. The last digit of the LCM of $(3^{2003} - 1)$ and $(3^{2003} + 1)$ is
(a) 8 (b) 2
(c) 4 (d) 6

55. Three persons start walking together and their steps measure 40 cm, 42 cm and 45 cm respectively. What is the minimum distance each should walk so that each can cover the same distance in complete steps?
(a) 25 m 20 cm (b) 50 m 40 cm
(c) 75 m 60 cm (d) 100 m 80 cm

56. The sum of first n odd numbers (*i.e.*, $1 + 3 + 5 + 7 + \ldots + 2n - 1$) is divisible by 11111 then the value of n is
(a) 12345 (b) 11111
(c) can't be determined (d) None of these

57. Which of the following is/are true?
(i) $43^3 - 1$ is divisible by 11
(ii) $56^2 + 1$ is divisible by 19
(iii) $50^2 - 1$ is divisible by 17
(iv) $(729)^5 - 729$ is divisible by 5
(a) (i) and (ii) (b) (iii) and (iv)
(c) (ii), (iii) and (iv) (d) (ii) and (iii)

58. The remainder when $6^{6^{6^{6^{6\ldots\infty \text{ times}}}}}$ is divided by 10 is
(a) 3 (b) 6
(c) 0 (d) can't be determined

59. The last two-digits in the multiplication
$122 \times 123 \times 125 \times 127 \times 129$ will be
(a) 20 (b) 50
(c) 30 (d) 40

60. Find $GCD\ (2^{100} - 1, 2^{120} - 1)$.
(a) $2^{20} - 1$ (b) $2^{40} - 1$
(c) $2^{60} - 1$ (d) $2^{10} - 1$

61. How many natural numbers are there which give a remainder of 41 after dividing 1997?
(a) 2 (b) 4
(c) 6 (d) None of these

62. Find the remainder when $6^{6^{6^{6^{6^{\cdot^{\cdot^{66}}}}}}}$ (100 times) when divided by 10?
(a) 6 (b) 2
(c) 4 (d) 8

63. Find the unit digit of the expression $199^{2n} + 144^{3n}$, where n is a natural number.
(a) 5 (b) 7
(c) either 5 or 7 (d) 3

64. The greatest number that can divide 140, 176, 264 leaving remainders of 4, 6, and 9 respectively is [*SSC-Sub. Ins.-2012*]
(a) 85 (b) 34
(c) 17 (d) 2

65. The ratio of two numbers is 3 : 4 and their HCF is 5. Their LCM is: [*SSC-Sub. Ins.-2013*]
(a) 10 (b) 60
(c) 15 (d) 12

66. Three tankers contain 403 litres, 434 litres, 465 litres of diesel respectively. Then the maximum capacity of a container that can measure the diesel of the three container exact number of times is [*SSC-Sub. Ins.-2014*]
(a) 31 litres (b) 62 litres
(c) 41 litres (d) 84 litres

67. L.C.M. of $\frac{2}{3}, \frac{4}{9}, \frac{5}{6}$ is [*SSC 10+2-2013*]
(a) $\frac{20}{27}$ (b) $\frac{8}{27}$
(c) $\frac{20}{3}$ (d) $\frac{10}{3}$

68. If the sum of the digits of any integer lying between 100 and 1000 is subtracted from the number, the result always is [SSC 10+2-2013]
(a) divisible by 5 (b) divisible by 6
(c) divisible by 2 (d) divisible by 9

69. If a number is as much greater than 31 as it is less than 75, then the number is. [SSC 10+2-2013]
(a) 53 (b) 106
(c) 44 (d) 74

70. The H.C.F. and L.C.M. of two numbers are 44 and 264 respectively. If the first number is divided by 2, the quotient is 44. The other number is [SSC 10+2-2014]
(a) 147 (b) 528
(c) 132 (d) 264

71. The sum of five consecutive odd numbers is 265. What is the sum of the largest number and twice the smallest number? [IBPS Clerk-2012]
(a) 156 (b) 153
(c) 155 (d) 151
(e) None of these

72. 'A', 'B' and 'C' are three consecutive even integers such that four times 'A' is equal to three times 'C'. What is the value of B ? [IBPS Clerk-2012]
(a) 12 (b) 10
(c) 16 (d) 14
(e) None of these

Level - II

1. What is the remainder obtained on dividing $34^{43} + 43^{34}$ by 7 ?
(a) 4 (b) 3 (c) 1 (d) 0

2. Two different prime numbers X and Y, both are greater than 2, then which of the following must be true ?
(a) $X - Y = 23$ (b) $X + Y \neq 87$
(c) Both (a) and (b) (d) None of these

3. What is the remainder when 1! + 2! + 3! .... + 100! is divided by 7 ?
(a) 0 (b) 5 (c) 6 (d) 3

4. On dividing 2272 as well as 875 by 3-digit number N, we get the same remainder. The sum of the digits of N is:
(a) 10 (b) 11 (c) 12 (d) 13

5. Which one of the following numbers will completely divide $(3^{25} + 3^{26} + 3^{27} + 3^{28})$?
(a) 11 (b) 16 (c) 25 (d) 30

6. There are two integers 34041 and 32506, when divided by a three-digit integer n, leave the same remainder. What is the value of n?
(a) 298 (b) 307
(c) 461 (d) can't be determined

7. After distributing the sweets equally among 25 children, 8 sweets remain. Had the number of children been 28, 22 sweets would have been left after equally distributing. What was the total number of sweets ?
(a) 328 (b) 348
(c) 358 (d) Data inadequate

8. Find the remainder when 7^{99} is divided by 2400.
(a) 1 (b) 343 (c) 49 (d) 7

9. A number N when factorized can be written as $N = p_1^4 \times p_2^3 \times p_3^7$. Find the number of perfect squares which are factors of N. (The 3 prime numbers $p_1, p_2, p_3 > 2$)
(a) 12 (b) 24 (c) 36 (d) 6

10. The number $\log_2 7$ is
(a) An integer (b) A rational number
(c) An irrational number (d) A prime number

11. Which of the following in true ?
(a) The cube of an odd integer is of the form $8q + 1$, where q is an integer .
(b) The square of an odd integer is of the form $8q + 1$, where q is an integer .
(c) The fourth power of any integer is of the form $10q + 1$, where q is an integer
(d) None of these

12. $94^3 - 23^3 - 71^3$ is atleast divisible by
(a) 71 and 23 (b) 23 and 74
(c) 71 and 94 (d) 23, 71 and 94

13. How many whole numbers between 100 and 800 contain the digit 2?
(a) 200 (b) 214 (c) 220 (d) 240

14. p, q and r are three non-negative integers such that $p + q + r = 10$. The maximum value of $pq + qr + pr + pqr$ is
(a) ≥ 40 and < 50 (b) ≥ 50 and < 60
(c) ≥ 60 and < 70 (d) ≥ 70 and < 80

15. Let a, b, c, d and e be integers such that $a = 6b = 12c$, and $2b = 9d = 12e$. Then which of the following pairs contains a number that is not an integer?
(a) $\left(\frac{a}{27}, \frac{b}{e}\right)$ (b) $\left(\frac{a}{36}, \frac{c}{e}\right)$
(c) $\left(\frac{a}{12}, \frac{bd}{18}\right)$ (d) $\left(\frac{a}{6}, \frac{c}{d}\right)$

16. If $x = (16^3 + 17^3 + 18^3 + 19^3)$, then x divided by 70 leaves a remainder of
(a) 0 (b) 1 (c) 69 (d) 35

17. Find the total number of prime factors in $2^{17} \times 6^{31} \times 7^5 \times 10^{11} \times 11^{10} \times (323)^{23}$
(a) 162 (b) 161 (c) 346 (d) 97

18. The digits of a three-digit number A are written in the reverse order to form another three-digit number B. If $B > A$ and $B-A$ is perfectly divisible by 7, then which of the following is necessarily true?
(a) $100 < A < 299$ (b) $106 < A < 305$
(c) $112 < A < 311$ (d) $118 < A < 317$

19. If $N = 1! - 2! + 3! - 4! + \ldots + 47! - 48! + 49!$, then what is the unit digit of N^N?
(a) 0 (b) 9 (c) 7 (d) 1

20. The digits of a 3-digit number in Base 4 get reversed when it is converted into Base 3. How many such numbers exist?
(a) 0 (b) 1 (c) 2 (d) 3

21. Find the remainder when $73 \times 75 \times 78 \times 57 \times 197$ is divided by 34.
(a) 22 (b) 30 (c) 15 (d) 28

22. Find the HCF of $(3^{125} - 1)$ and $(3^{35} - 1)$.
(a) 5 (b) 3
(c) $(3^5 - 1)$ (d) $(3^{35} - 1)$

23. A computer program was tested 300 times before its release. The testing was done in three stages of 100 tests each. The software failed 15 times in Stage I, 12 times in Stage II, 8 times in Stage III, 6 times in both Stage I and Stage II, 7 times in both Stage II and Stage III, 4 times in both Stage I and Stage III, and 4 times in all the three stages. How many times the software failed in a single stage only?
(a) 10 (b) 13 (c) 15 (d) 17

24. Let x denote the greatest 4-digit number which when divided by 6, 7, 8, 9 and 10 leaves a remainder of 4, 5, 6 7 and 8 respectively. Then, the sum of the four-digits of x is and 8 respectively. Then, the sum of the four-digits of x is
(a) 25 (b) 18 (c) 20 (d) 22

25. A is the set of the first 100 natural numbers. What is the minimum number of elements that should be picked from A to ensure that atleast one pair of numbers whose difference is 10 is picked?
(a) 51 (b) 55 (c) 20 (d) 11

26. The power of 45 that will exactly divide 123! is
(a) 28 (b) 30 (c) 31 (d) 59

27. What is the remainder when $_{32}32^{32}$ is divided by 7?
(a) 2 (b) 3 (c) 4 (d) 6

28. Two different two-digit natural numbers are written beside each other such that the larger number is written on the left. When the absolute difference of the two numbers is subtracted from the four-digit number so formed, the number obtained is 5481. What is the sum of the two two-digit numbers?
(a) 70 (b) 71 (c) 72 (d) 73

29. In a three-digit number, the unit digit is twice the tens digit and the tens digit is twice the hundreds digit. The same number is written as $1XY$ and $1YX$ in base 8 and base 9 respectively. Find the sum of X and Y in the decimal system.
(a) 15 (b) 7
(c) 11 (d) Cannot be determined

30. $$a+\cfrac{1}{b+\cfrac{1}{c+\cfrac{1}{d+\ldots}}}$$

If a, b, c, d etc. are positive integers, then what is the value of 'b'?
(a) 2 (b) 4 (c) 3 (d) 5

31. If m and n are positive integers such that $(m-n)^2 = \dfrac{4mn}{(m+n-1)}$, then how many pairs (m, n) are possible?
(a) 4 (b) 10 (c) 16 (d) Infinite

32. $x^2 - 3y^2 = 1376$
How many integer solutions exist for the given equation?
(a) One (b) Two (c) Four (d) Zero

33. The number of zeros at the end of the product of $222^{111} \times 35^{53} + (7!)^{6!} \times (10!)^{5!} + 42^{42} \times 25^{25}$ is
(a) 42 (b) 53
(c) 1055 (d) None of these

34. The highest power of 17 which can divide exactly the following expression :
$(18^2 - 1)(18^4 - 1)(18^6 - 1)(18^8 - 1)(18^{10} - 1) \times \ldots (18^{16} - 1)(18^{18} - 1)$ is :
(a) 1 (b) 17
(c) 9 (d) can't be determined

35. The remainder when $2^2 + 22^2 + 222^2 + 2222^2 + \ldots\ldots (222\ldots\ldots 49 \text{ twos})^2$ is divided by 9 is:
(a) 2 (b) 5 (c) 6 (d) 7

36. Find the last non-zero digit of 96!.
(a) 2 (b) 4 (c) 6 (d) 8

37. When 96 is added to a N^2, it gives another perfect square. If N is a natural no., how many distinct values of N are possible?
(a) 3 (b) 4
(c) 5 (d) None of these

38. The numbers 1 to 29 are written side by side as follows 1234567891011..........................28 29
If the number is divided by 9, then what is the remainder?
(a) 3 (b) 1
(c) 0 (d) None of these

39. The remainder when the number 123456789101112...... 484950 is divided by 16 is
(a) 3 (b) 4 (c) 5 (d) 6

40. The product of three consecutive even numbers is 4032. The product of the first and the third number is 252. What is five times the second number ? [*IBPS-PO-2012*]
(a) 80 (b) 100 (c) 60 (d) 70
(e) 90

41. What would be the sum of [*SSC CGL-2012*]
$1 + 3 + 5 + 7 + 9 + 11 + 13 + 15 + \ldots\ldots$ up to 15th term?
(a) 250 (b) 240
(c) 225 (d) 265

42. The least number which when divided by 48, 64, 90, 120 will leave the remainders 38, 54, 80, 110 respectively, is [*SSC CGL-2012*]
(a) 2870 (b) 2860 (c) 2890 (d) 2880

43. If $1^3 + 2^3 + \ldots\ldots + 9^3 = 2025$, then the approx. value of $(0.11)^3 + (0.22)^3 + \ldots\ldots + (0.99)^3$ is [*SSC CGL-2012*]
(a) 0.2695 (b) 0.3695 (c) 2.695 (d) 3.695

44. If the product of first fifty positive consecutive integers be divisible by 7^n, where n is an integer, then the largest possible value of n is [*SSC CGL-2014*]
(a) 7 (b) 8 (c) 10 (d) 5

Hints & Solutions

Level-I

1. (a) Let 'r' be the remainder ⇒ $221 - r, 116 - r, 356 - r$ are exactly divisible by that number. Now, if two numbers are divisible by a number, then so is their difference
⇒ $[(221 - r) - (116 - r)]$, $[(356 - r) - (116 - r)]$. and $[(356 - r) - (221 - r)]$ are divisible by that number
⇒ 105, 135, 240 are divisible by that number
= HCF of 105, 135, 140 = 15.

2. (a) On dividing the given number 345670 by 6, we get 4 as the remainder.
So 2 must be added to the given number.

3. (d) Since $(35 - 25) = 10$, $(45 - 35) = 10$, $(55 - 45) = 10$.
Now take the LCM of 35, 45, 55 and subtract 10 from it
⇒ $3465 - 10 = 3455$.

4. (c) $n(n^2 + 20)$ is always divisible by 24, if n is even number.

5. (a) When 2^{256} is divided by 17 then
$$\Rightarrow \frac{2^{256}}{2^4+1} \Rightarrow \frac{(2^2)^{64}}{(2^4+1)}$$
By remainder theorem when $f(x)$ is divided by $x + a$ the remainder $= f(-a)$
Here $f(1) = (2^2)^{64}$ and $x = 2^4$ and $a = 1$
∴ Remainder $= f(-1) = (-1)^{64} = 1$

6. (b) The last digit of 2137^1 is 7.
Last digit of 2137^2 is 9 .Last digit of 2137^3 is 3, the last digit of 2137^4 is 1, last digit 2137^5 is 7 and the last digit of 2137^6 is 9 and so on . Hence it form a pattern and the last digit repeats for every 5th . $753 = 4 \times 188 + 1$. Thus the last digit of 2137^{753} is the same as that of 2137^1 *i. e.*, 7.

7. (a) We have to find the least number which is divisible by 3, 5, 6 and 9 and is also a perfect square. The LCM of 3, 5, 6 and 9 is $3 \times 3 \times 2 \times 5 = 90$. Hence, the required number is $90 \times 2 \times 5 = 900$.

8 (c) Use test of 11 after putting $y = 5$.

9. (d) Out of n and $n + 2$, one is divisible by 2 and the other by 4, hence $n(n + 2)$ is divisible by 8. Also n, $n + 1$, $n + 2$ are three consecutive numbers, hence one of them is divisible by 3. Hence $n(n + 1)(n + 2)$ must be divisible by 24. This will be true for any even number n.

10. (b) Check the number for divisibility by 3.
So, $4 + 5 + 9 + 0 + 4 + 5 = 27$. Hence it is divisible by 3 and the quotient is 153015.
Now, check the quotient for divisibility by 9.
$1 + 5 + 3 + 0 + 1 + 5 = 15$
So, the number is not divisible by 9.
However, if we add 3 to the number *i.e.*, $153015 + 3 = 153018$ it would be, divisible by 9.
So, the number divisible by 27 will be $-153015 + 3 \times 3 = 459054$ *i.e.*, 9 should be added.

11. (c) Last digit in 19 – 9
$18^2 - 1$
$19^3 - 1$
for odd powers of 19
Last digit is 9 and for even it is 1
Last digit in 19^{81} is 9
Last digit in 41 is 4
42 is 6
43 is 4
for odd powers of 4
3^9k is odd irrespective of the value of k
∴ last digit in 4^9k is 4. Last digit in $19^{81} + 4^9k$ is last digit in $9 + 4$ *i.e*, in $13 = 3$

12. (a) Sum of prime numbers that are greater than 60, but less than 70 is
$61 + 67 = 128$

13. (d) 311 is repeated seven times in the number, 311 is not divisible by 3 but 311 repeated twice is not divisible by 3, but divisible by 11.
Similarly 311 repeated thrice is divisible by 3, but not by 11.
As 311 is repeated seven times, which is neither multiple of 2 nor 3.
So, number is not divisible by 3 or 11.

14. (c) $\frac{1365 - 15}{5} = 270$

15. (c) Sum of digits $= (5 + 1 + 7 + x + 3 + 2 + 4)$
$= (22 + x)$, which must be divisible by 3.
∴ $x = 2$.

16. (d) Clearly, 4864 is divisible by 4.
So, 9P2 must be divisible by 3, so, $(9 + P + 2)$ must be divisible by 3.
∴ $P = 1$.

17. (a) Largest 4-digit number = 9999

```
88 ) 9999 ( 113
     88
     119
      88
      319
      264
       55
```

Required number $= (9999 - 55) = 9944$

18. **(a)** $(x^n - a^n)$ is always divisible by $(x + a)$, when n is even natural number.

19. **(d)** $0.\overline{47} = \frac{47}{99}$.

20. **(c)**

21. **(b)** LCM of the numbers = 420. Hence there must be $(420 \times 2) + 2 = 842$ beads.

22. **(b)** Since $59 = 4 \times 14 + 3 \Rightarrow$ last digit of $(377)^{59} = 3$
$87 = 4 \times 21 + 3 \Rightarrow$ last digit of $(793)^{87} = 7$
$129 = 4 \times 32 + 1 \Rightarrow$ last digit of $(578)^{129} = 8$
$99 = 2 \times 49 + 1 \Rightarrow$ last digit of $(99)^{59} = 9$
Hence the last digit of the result is equal to the last digit of
$3 \times 7 \times 8 \times 9$, *i.e.*, 2.
$\therefore$ digit at unit's place = 2

23. **(d)** Interval after which the devices will beep together = (L.C.M. of 30, 60, 90, 105) min = 1260 min. = 21 hrs. So, the devices will again beep together 21 hrs. after 12 noon *i.e.*, at 9 a.m.

24. **(d)** N will be an odd number because N is sum of one even number (b) and 13985 odd numbers.
Hence, N will not be divisible by an even number.

25. **(b)** Divisor $= r_1 + r_2 - r_3 = 35 + 30 - 20 = 45$

26. **(b)** $12 - 7 = 5$, $15 - 10 = 5$ and $16 - 11 = 5$
Hence the desired number is 5 short for divisibility by 12, 15 and 16.
L.C.M. of 12, 15, 16 is 240
Hence the least number = 240 – 5 = 235

27. **(c)** We have to find numbers between 100 and 200 which are even and are neither divisible by 7 nor by 9.
$\therefore$ No. that are even and are divisible by 7 are 7 and no. which are even and divisible by 9 are 6.
Nos. even and divisible by 7 and 9 both are (*e.g.*, 63) is only 126 :
$\therefore$ Required answer = 7 + 6 – 1 = 12
$\therefore$ 51 – 12 = 39.

28. **(a)** Let the numbers be the form $10x + y$
According to question
$$10x + y = x + y + xy$$
$$9x = xy$$
$$\therefore \quad y = 9$$
The numbers are 19, 29, 39, 49, 59, 69, 79, 89 and 99 total of 9 numbers
Hence the required fraction $= \frac{9}{91}$
$= 0.099 \approx 0.1$

29. **(b)** Let there be w wide runs.
Byes = $w + 8$
Runs scored by batsmen = $26\,w$
Total run = 232
or $w + w + 8 + 26W = 323$
$$\Rightarrow \quad w = \frac{224}{28} = 8$$
$\therefore$ Run scored by Ram $= \frac{6}{13} \times 208 = 96$

30. **(b)** The value of the expression will be least when $x = y = z = 1/3$.
Hence, the least value $= \left(\frac{1}{1/3} - 1\right)^3$
$= 2 \times 2 \times 2 = 8.$

31. **(b)** Consider $3^{4n} = (81)^n = (1 + 80)^n = 1 + 80q, q \in N$
$\therefore \; 3^{3^{4n}} = 3^{80q+1} = (81)^{20q} . 3$
Since the last digit of $(81)^{20q}$ is 1, so the last digit of $3^{3^{4n}} + 1$ is $1 \times 3 + 1 = 4$

32. **(c)** The last digit in the number must be 6: for only numbers ending in 6, when raised to any power, result in another no. ending in 6.

33. **(a)** Since the given number is divisible by 5, so 0 or 5 must come in place of \$. But, a number ending with 5 is never divisible by 8. So, 0 will replace \$.
Now, the number formed by the last three digits is 4*0, which becomes divisible by 8, if * is replaced by 4.
Hence, digits in place of * and \$ are 4 and 0 respectively.

34. **(a)** Let total number of seats in the stadium be p; number of seats in the lower deck be x and number of seats in upper deck be y.
$\therefore \quad p = x + y, x = p/4, y = 3p/4$
Now in the lower deck, $4x/5$ seats were sold and $x/5$ seats were unsold.
No. of total seats sold in the stadium = $2p/3$.
No. of unsold seats in the lower deck = $x/5 = p/20$
No. of unsold seats in the stadium = $p/3$
$\therefore$ Required fraction $= \frac{p/20}{p/3} = \frac{3}{20}$

35. **(c)** $1 + 2 + 3 + \ldots\ldots + 40 = \frac{40 \times 41}{2} = 820$
Since at each time any two numbers a and b are erased and a single new number $(a + b - 1)$ is writen. Hence, each one is subtracted and this process is repeated 39 times. Therefore, number left on the board at the end = 820 – 39 = 781.

36. **(d)** Since $80 = 8 \times 10$ or $80 = 16 \times 5$
Thus y (*i.e.*, unit digit) must be zero.
$\therefore$ $653xy = 653x0$, where $653x0$ must be divisible by 16 or $653x$ is divisible by 8.
Thus the last 3-digit number 53x will be divisible by 8.
Hence, at $x = 6$, we get the required result.
$$\therefore \quad x + y = 6 + 0 = 6$$

37. **(c)** In the given range, the last number which is divisible by both 5 and 7. i.e., 35 is 210 and the highest number is 770. So the total number of numbers between 200 and 800 which are divisible by both 5 and 7 is
$$\left(\frac{770 - 210}{35}\right) + 1 = 17$$
Hence option (c) is correct.

38. (a) Total numbers in the set = (800 – 200) + 1 = 601

Number of numbers which are divisible by 5

$$= \frac{(800-200)}{5} + 1 = 121$$

Number of numbers which are divisible by 7

$$= \frac{(798-203)}{7} + 1 = 86$$

Number of numbers which are divisible by both 5 & 7

$$= \frac{(770-210)}{35} + 1 = 17$$

∴ Number of numbers which are either divisible by 5 or 7 or both

= (121 + 86) – 17 = 190

39. (d) Since Dividend = Divisor × Quotient + Remainder

∴ Dividend = 9235 × 888 + 222

Thus the number = 8200902

Hence (d) is the correct choice.

40. (c) Let this number be N then

$N = 32 \times Q_1 + 29$ ...(1)

Again $N = 8 \times Q_2 + R$ ...(2)

From (1) and (2)

$32Q_1 + 29 = 8Q_2 + R$ (where R is the remainder)

$8Q_2 - 32Q_1 = 29 - R$

$8(Q_2 - 4Q_1) = 29 - R$

$$\text{or}(Q_2 - 4Q_1) = \frac{29-R}{8}$$

Since Q_1, Q_2, R are integers also $Q_2 - 4Q_1$ is an integer.

Therefore 29 – R must be divisible by 8.

41. (d) $(0.\bar{1})^2\left[1-9(0.\overline{16})^2\right]$

$$= \left(\frac{1}{9}\right)^2\left[1-9\times\left(\frac{16}{99}\right)^2\right]$$

$$= \frac{1}{81}\left[1-9\times\frac{256}{9801}\right]$$

$$= \frac{1}{81}\left[1-\frac{256}{1089}\right] = \frac{1}{81}\times\frac{833}{1089} = \frac{833}{88209}$$

42. (d) Since the 7, 11 and 13 all are the factors of such a number so (d) is the correct answer.

43. (c) ∵ $7056 = 2^4 \times 3^2 \times 7^2$

∴ Number of factors/divisors of 7056

Product of factors = $(7056)^{45/2} = (84)^{45}$

Hence (c) is the correct option.

44. (d) The sum of digits of the number will be 114, which leaves a remainder of 6 when divided by 9. So when divided by 18 it would leave either 6 or 6 + 9 = 15 as the remainder.

Since the number is odd, it will leave an odd remainder only when divided by 18. So the remainder will be 15.

45. (a) $$\frac{a+220}{a+4} = \frac{a+4+216}{a+4} = 1+\frac{216}{a+4}$$

Therefore, $(a + 4)$ must be a factor of 216.

The number of factors of 216 = 16

But $(a + 4)$ cannot be equal to 1, 2, 3 and 4 as 'a' has to be a positive integer.

Total possible values = 16 – 4 = 12

46. (a) Sum of all even factors:

$(2^1)(5^0 + 5^1 + 5^2)(7^0 + 7^1 + 7^2) = 3534$

Number of even factors = 1 × 3 × 3 = 9

47. (d) Sum of divisors of 544 which are perfect square is:

$(2^0 + 2^2 + 2^4)(17^0) = 21.$

48. (c) Count the number of fives. This can get done by:

$100^1 \times 95^6 \times 90^{11} \times 85^{16} \times 80^{21} \times 75^{26} \times ... 5^{96}$

(1 + 6 + 11+ 16+ 21 +26 + 31 + 36 + 41 + 46 + ... + 96) + (1 + 26 + 51 + 76)

= 20 × 48.5 + 4 × 38.5 (Using sum of A.P. explained in the next chapter.)

= 970 + 154 = 1124.

49. (a) $(23)_5 = (2 \times 5^1 \times 3 \times 5^0)_{10} = (13)_{10} = (1 \times 8^1 + 5 \times 8^0)_8 = (15)_8$

also, $(47)_9 = (4 \times 9^1 + 7 \times 9^0)_{10} = (43)_{10}$

$= (5 \times 8^1 + 3 \times 8^0) = (53)_8$

sum $= (13)_{10} + (43)_{10} = (56)_{10} \rightarrow (70)_8$

50. (b) If we look at the numbers $100 < N \leq 105$, we see only 101 and 103 do not have their factors in N (because these are primes). So, obviously the new LCM will be $101 \times 103 \times N$.

51. (c) The number needs to be less than 13 × 52 = 676. The highest power of 13 in 676! is 56.

The power of 13 in the smallest such number needs to be exactly 52. If we subtract 13 × 3 = 39 from 676, we get 637. The number 637! will be the smallest number of type N! that is completely divisible by 1352.

The sum of the digits of 637 is 16.

52. (d) $12^{55}/3^{11} = 3^{44}.4^{55} \rightarrow 4$ as units place.

Similarly, $8^{48}/16^{18} = 2^{72} \rightarrow 6$ as the units place.

Hence, 0 is the answer.

53. (c) It can be seen that the first expression is larger than the second one. Hence, the required answer would be given by the (units digit of the first expression – units digit of the second expression) = 6 – 0 = 6.

54. (c) The given numbers are two consecutive even numbers, so their HCF = 2

Now, using LCM × HCF = Product of two numbers

LCM × 2 = (…6) × (…8)

It can be seen now that the unit digit of LCM = 4

55. **(a)** Answer is LCM of 40, 42, 45 = $2^3 \times 3^2 \times 5^1 \times 7^1$ = 2520 cm = 25.2 m.

56. **(b)** Go through option

$S_n = 1 + 3 + 5 + 7 + ... + 22221$

$S_{11111} = (11111)^2$

Hence it is divisible by 11111. Thus option (b) is correct.

57. **(b)** $(50^2 - 1) = (50 + 1)(50 - 1) = (17 \times 3) \times (7 \times 7)$ hence divisible by 17.

and $(729)^5 - 729 = 729(729^4 - 1)$

$= 729(729^2 - 1)(729^2 + 1)$

$= (729)(729 - 1)(729 + 1)(729^2 + 1)$

$= 729 \times 728 \times 730 \times (729^2 + 1)$

Hence it is divisible by 5.

58. **(b)** Since $\frac{6}{10} \to$ Remainder is 6

$\frac{6^6}{10} \to$ Remainder is 6

$\frac{6^{6^6}}{10} \to$ Remainder is 6

59. **(b)** The answer will be 50 since, 125*122 will give 50 as the last two digits.

60. **(a)** $(2^{100} - 1)$ and $(2^{120} - 1)$ will yield the *GCD* as $2^{20} - 1$.

61. **(c)** Let us assume that the quotient is Q and divisor is D. Using the condition given in question, 1997 = QD + 41

$\Rightarrow$ QD = 1956. Now we will factorize 1956 in two parts such that D (divisor) is more than 41.

62. **(a)** 6^n (where n is a natural number) will always leaves the remainder 6 when divide by 10.

63. **(c)** For any *n*, 199^{2n} has last digit as 1,

But the last digit of 144^{3n} is 4 for odd values of *n* and 6 for even values of **n**.

Therefore, last digit of the given expression is either 5 or 7.

64. **(c)** Required number = H.C.F of (140 – 4), (176 – 6) and (264 – 9) = H.C.F. of 136, 170 and 255.

```
136)255(1          17)170(10
    136                17
    ---               ----
    119)136(1           0
        119           ----
        ---
        17)119(7
           119
           ---
            ×
```

$\therefore$ Required number = 17

65. **(b)** If the numbers be 3*x* and 4*x*, then

HCF = *x* = 5

$\therefore$ Number = 15 and 20

$\therefore$ LCM = 12x = 12 × 5 = 60

66. **(a)** H.C.F. of 403, 434 and 465 is 31.

67. **(c)** LCM of $\frac{2}{3}, \frac{4}{9}, \frac{5}{6}$

$\frac{\text{LCM of}(2,4,5)}{\text{HCF of }(3,9,6)} = \frac{20}{3}$

68. **(d)** $(100x + 10y + z) - (x + y + z) = 99x + 9y$
$= 9(11x + y)$

69. **(a)** Let the number be x.

Then, $x - 31 = 75 - x$

$2x = 106$

$x = 53$

70. **(c)** First number = 2 × 44 = 88

Other number = $\frac{44 \times 264}{88} = 132$

71. **(c)** Third number = $\frac{265}{5} = 53$

$\therefore$ Smallest number = 49

Largest number = 57

$\therefore$ Required value

= 57 + 2 × 49

= 57 + 98 = 155

72. **(d)** Let A = *x*,

B = *x* + 2,

C = *x* + 4

$\therefore$ According to the question

$4x = 3(x + 4)$

$\Rightarrow$ $4x - 3x = 12 \Rightarrow x = 12$

$\therefore$ B = *x* + 2 = 12 + 2 = 14

Level-II

1. **(d)** $(34^{43} + 43^{34})/7 = [(35 - 1)^{43} + (42 + 1)^{34}/7]$.

Applying binomial theorem to $(35 - 1)^{43}$, all terms will be divisible by 35 (i.e. 7) except the last term which will be –1. Similarly, last term of $(42 + 1)^{34}$ will be +1. Therefore, $34^{43} + 43^{34}$ will leave remainder $[(-1) + (+1)] = 0$, when divided by 7.

2. **(b)** Two prime numbers greater than 2 must be odd. Sum of two odd numbers must always be even, thus, $X + Y = 87$ is not possible.

3. **(b)** 7! + 8! + 9! + 10! + ... + 100 = 7.6! + 8.7.6! + 9.8.7.6! + ... + 100! is completely divisible by 7 as each of the terms contain at least one 7 in it.

Now, 1! + 2! + 3! + 4! + 5! + 6!

= 1 + 2 + 6 + 24 + 120 + 720 = 873

which leaves a remainder of 5 when divided by 7.

4. **(a)** Clearly, (2272 – 875) = 1397, is exactly divisible by *N*.

Now, 1397 = 11 × 127

$\therefore$ The required 3-digit number is 127, the sum of whose digits is 10.

5. **(d)** $(3^{25} + 3^{26} + 3^{27} + 3^{28}) = 3^{25} \times (1 + 3 + 3^2 + 3^3)$
$= 3^{25} \times 40 = 3^{24} \times 3 \times 4 \times 10 = (3^{24} \times 4 \times 30)$, which is divisible by 30.

6. **(b)** Let the common remainder be x. Then numbers $(34041 - x)$ and $(32506 - x)$ would be completely divisible by n.
Hence the difference of the numbers $(34041 - x)$ and $(32506 - x)$ will also be divisible by n
or $(34041 - x - 32506 + x) = 1535$ will also be divisible by n.
Now, using options we find that 1535 is divisible by 307.

7. **(c)** Let the total number of sweets be $(25x + 8)$.
Then, $(25x + 8) - 22$ is divisible by 28
$\Leftrightarrow$ $(25x - 14)$ is divisible by 28 $\Leftrightarrow 28x - (3x + 14)$ is divisible by 28
$\Leftrightarrow$ $(3x + 14)$ is divisible by 28 $\Leftrightarrow x = 14$.
$\therefore$ Total number of sweets $= (25 \times 14 + 8) = 358$.

8. **(b)** $7^4/2400$ gives us a remainder of 1. Thus, the remainder of $7^{99}/2400$ would depend on the remainder of $7^3/2400 \rightarrow$ remainder $= 343$.

9. **(b)** The powers of p_1 can be 0, 2 or 4 *i.e.*, 3, powers of p_2 can be 0,2 *i.e.*, 2, Powers of p_3 can be 0, 2, 4 or 6 *i.e.*,4. Hence, a combination of these powers gives $3 \times 2 \times 4$ *i.e.*, 24 numbers. So, there are 24 perfect squares that divide N.

10. **(c)** Suppose, possible, $\log_2 7$ is rational, say p/q where p and q are integers, prime to each other.

Then, $\frac{p}{q} = \log_2 7 \Rightarrow 7 = 2^{p/q} \Rightarrow 2^p = 7^q$

which is false since L.H.S. is even and R.H.S. is odd.
Obviously $\log_2 7$ is not an integer and hence not a prime number.

11. **(b)** Square of the odd numbers can be written as
$1^n = 8 \times 0 + 1$, $3^2 = 8 \times 1 + 1$, $8 \times 3 + 1$ & so on *i. e.*, square of the odd number is of the form $8q + 1$.

12. **(d)** $94^3 - 23^3$ is divisible by $94 - 23 = 71$
$94^3 - 71^3$ is divisible by $94 - 71 = 23$
$23^3 - 71^3$ is divisible by $23 + 71 = 94$
$\therefore$ $94^3 - 23^3 - 71^3$ is divisible by 23, 71 and 94

13. **(b)** Between 100 and 199, there will be 19 numbers which contain '2'. They are as follows.
102, 112, 120 – 129 (10 numbers), 132, 142, 152, 162, 172, 182, 192.
Similar would be the case for 300 – 399, 400 – 499, 500 – 599, 600 – 699.
For 200 – 299, all 100 numbers will have 2.
$\therefore$ Total number of numbers containing '2' $= 19 \times 6 + 100 = 114 + 100 = 214$.

14. **(c)** As p, q, r are non-negative integers, the maximum will be achieved when the value of each variable is closed to each other.
i.e., p, q, r are 3, 3, 4 (not necessarily in the same order).
Hence the value of
$pq + qr + pr + pqr = 3 \times 3 + 3 \times 4 + 3 \times 4 + 3 \times 3 \times 4$
$= 9 + 12 + 12 + 36 = 69$

15. **(d)** Given $a = 6b = 12c = 27d = 36e$
Multiplied and Divide by 108 in whole expression

$$\frac{108a}{108} = \frac{108b}{18} = \frac{108c}{9} = \frac{108d}{4} = \frac{108e}{3}$$

$$\frac{1}{108}a = \frac{1}{18}b = \frac{1}{9}c = \frac{1}{4}d = \frac{1}{3}e = 1 \text{ (say)}$$

$\Rightarrow$ $a = 108, b = 18, c = 9, d = 4, e = 3$

So it is clear that $\left(\frac{a}{6}, \frac{c}{d}\right)$ contains a number $\frac{c}{d} = \left(\frac{9}{4}\right)$ which is not an integer

16. **(a)** Remember that, $a^3 + b^3 = (a + b)(a^2 + b^2 - ab)$
$x = (16^3 + 17^3 + 18^3 + 19^3)$
$x = (16^3 + 19^3) + (17^3 + 18^3)$
$x = (16 + 19)(16^2 + 19^2 - 16 \times 19) + (17 + 18)(17^2 + 18^2 - 17 \times 18)$
$x = 35[16^2 + 19^2 - 16 \times 19 + 17^2 + 18^2 - 17 \times 18]$
$x = 35 \times$ (Even number)
Hence, x is divisible by 70 and leaves remainder as zero.

17. **(a)** $2^{17} \times 6^{31} \times 7^5 \times 10^{11} \times 11^{10} \times (323)^{23}$
$= 2^{17} \times 2^{31} \times 3^{31} \times 7^5 \times 2^{11} \times 5^{11} \times 11^{10} \times 17^{23} \times 19^{23}$
$= 2^{59} \times 3^{31} \times 5^{11} \times 7^5 \times 11^{10} \times 17^{23} \times 19^{23}$
$\therefore$ Total number of prime factors
$= 59 + 31 + 11 + 5 + 10 + 23 + 23 = 162$

18. **(b)** Let the 3 digits of number A be x, y and z
Hence $A = 100x + 10y + z$
On reversing the digits of number A, we get the number B *i.e.*, zyx.
$\therefore$ $B = 100z + 10y + x$
As $B > A \Rightarrow z > x$ ...(1)
$B - A = 99z - 99x = 99(z - x)$
As 99 is not divisible by 7
so $(z - x)$ has to be divisible by 7. ...(2)
Using (1) & (2), the only possible values of z and x are (8, 1) and (9, 2)
So the minimum and maximum range of A are 108 and 299, which $\in 106 < A < 305$

19. **(d)** The unit digit of every term from 5! to 49! is 0.
Also, $1! - 2! + 3! - 4! = 1 - 2 + 6 - 24 = -19$.
Hence, the unit digit of N will be $10 - 9 = 1$.
The unit digit of N^N will also be 1.

20. **(b)** Let the 3-digit number be abc. Now according to the given condition, $(abc)_4 = (cba)_3$.
$16a + 4b + c = 9c + 3b + a$
$\Rightarrow 15a + b = 8c$
The only set of numbers which satisfies the relation given above is a = 1, $b = 1$ and $c = 2$.

21. **(a)** The remainder would be given by: $(5 \times 7 \times 10 \times 23 \times 27)/34 \to 35 \times 230 \times 27/34 \to 1 \times 26 \times 27/34 = 702/34 \to$ remainder = 22.

22. **(a)** The solution of this question is based on the rule that: The HCF of $(a^m - 1)$ and $(a^n - 1)$ is given by $(a^{\text{HCF of } m, n} - 1)$
Thus, in this question the answer is : $(3^5 - 1)$. Since 5 is the HCF of 35 and 125.]

23. **(b)** Assume that the software fails a, b, and c times in a single stage, in two stage, and in all stages respectively.
$\therefore$ $b + 3c = 6 + 7 + 4 = 17$
but $c = 4$, hence $b = 5$
Similarly, we have
$a + 2b + 3c = 15 + 12 + 8 = 35$
$a = 35 - 12 - 10 = 35 - 22 = 13$
Hence option (b)

24. **(a)** The number will be a multiple of 6, 7, 8, 9, 10
LCM of 6, 7, 8, 9, 10 = 2520
$\therefore$ Largest 4-digit number divided by this = 7560
$\therefore$ Required number = 7558
Sum of the digits of this number = 25

25. **(a)** Let's divide the first 100 natural numbers in five sets of 20 numbers each:
{1, 2, 3....20}, {21, 22, 23....40}, ...{81, 82, 83......100}. If we pick the first ten numbers from each set we will not get any pair of two numbers whose difference is 10.
However, if we pick just one more number from any of the sets, it would have a difference of 10 with one of the numbers which has already been picked.
So the answer is $10 \times 5 + 1 = 51$.

26. **(a)** $45 = 3^2 \times 5$. Hence, we need to count the number of 3^2's and 5's that can be made out of 123!.
Number of 3's = 41 + 13 + 4 + 1 = 59 $\to$ Number of 3^2's = 29
Number of 5's = 24 + 4 = 28.
The required answer is the lower of the two (viz. 28 and 29). Hence, option (a) 28 is correct.

27. **(c)** Remainder of $({}_{32}32^{32}$ divided by 7) = Remaindr of $({}_{4}32^{32}$ divided by 7)
Now find cyclicity of remainder of $({}_{4}32^{n}$ divided by 7).
Remainder when ${}_{4}32^{1}$ divided by 7 = 2
Remainder when ${}_{4}32^{2}$ divided by 7 = 4
Remainder when ${}_{4}32^{3}$ divided by 7 = 2
So, the cyclicity is 2, 4, 2, 4 and so on.
For every even value of n, remainder = 4

28. **(d)** As the larger number is written on the left, the larger number is either 54 or 55.
Let the smaller number be x.
Case I: The larger number is 54.
$5400 + x = 5481 + 54 - x$
$2x = 5535 - 5400 = 135$
(In this case x will not be a natural number.)
Case II: The larger number is 55.
$5500 + x = 5481 + 55 - x$
$2x = 5536 - 5500 = 36$
$\Rightarrow x = 18$
Hence, the required sum = 73.

29. **(c)** Let the hundreds digit be n.
The tens digit will be $2n$.
The unit digit will be $4n$.
The possible values of 'n' are 1 and 2 and hence the possible numbers are 124 and 248 respectively.
On converting 248 in base 8 and base 9, the given condition gets violated.
On converting 124 in base 8 and base 9, we get $(174)_8 = (147)_9$.
Required sum = 4 + 7 = 11.

30. **(c)** The expression can be written as $a + \frac{1}{b + x}$, where x lies in the interval (0, 1).
Since $(1.25)^3 = 1.953125$ and $(1.3)^3 = 2.197$, it can be concluded that $2^{\frac{1}{3}}$ belongs to the interval (1.25, 1.3).
Hence, $a = 1$. This implies that $\frac{1}{b+x}$ lies in the interval (0.25, 0.3). The only possible value of $b = 3$.

31. **(d)** $(m - n)^2 = \frac{4mn}{(m + n - 1)}$
$\Rightarrow (m - n)^2 (m + n - 1) = 4mn$
$\Rightarrow (m - n)^2 (m + n - 1) = (m + n)^2 - (m - n)^2$
$\Rightarrow (m - n)^2 (m + n) = (m + n)^2$
$\Rightarrow (m - n)^2 = (m + n)$
(Since, $m + n \neq$, 0)
The above equation has infinitely many solutions where m and n are positive integers.
We can put $m + n = v$ and $m - n = u$, and re-write the equation as $u^2 = v$ and then plug in different values of u and v to get different pairs of (m, n).

32. **(d)** $3y^2 = x^2 - 1376$
As we can see L.H.S. is definitely a multiple of 3 and in R.H.S. 1376 leaves a remainder of 2 when divided by 3.
There are three possibilities for x in R.H.S:
(i) If x is multiple of 3, so is x^2, and R.H.S. will leave a remainder of 1 when divided by 3.
(ii) If x is of the form $3m + 1$, x^2 will be of the form $3n + 1$ and R.H.S will leave a remainder of 2. $m, n \in, N$
(iii) If x is of the form $3m + 2$, x^2 will be of the form $3n + 1$ and R.H.S. will leave a remainder of 2. $m, n \in, N$ So R.H.S. can never be a multiple of 3, while L.H.S. is always a multiple of 3. Hence no real solution exists.

33. (a) The number of zeros at the end of $222^{111} \times 35^{53}$ is 53.
The number of zeros at the end of $(7!)^{6!} \times (10!)^{5!}$ is 960.
The number of zeros at the end of $42^{42} \times 25^{25}$ is 42.
Thus the number of zeros at the end of the whole expression is 42.

34. (c) $(18^2 - 1) = (17)(18 + 1)$
$(18^4 - 1) = (18^2 + 1)(18^2 - 1)$
$= (18^2 + 1)(18 + 1)(18 - 1)$
$(18^6 - 1) = (18^3)^2 - 1$
$= (18^3 + 1)(18^3 - 1) = 17 \times k$ etc.
Hence there will only 9 times 17 in the whole expression.

35. (c) $2^2 + 22^2 + 222^2 + 2222^2 + \ldots\ldots + (2222 \ldots\ldots 49 \text{ twos})^2$
$= 2^2 + (2)^2 + 2^2 + \ldots\ldots + 2^2$ (49 twos)
$= 4 + 4 + 4 + \ldots\ldots + 4$ (49 twos)
$= 4 \times 49 =$ last digit is 6.

36. (d) Factorize 96! into prime factors. Find the unit digit of all the factors individually and multiply to get the unit digit of 96!.
$96! = 2^{92} \times 3^{46} \times 5^{22} \times \ldots$
Now 5^{22} and 2^{22} can be eliminated, since these will result in zeroes. Find the unit digit of all the remaining.

37. (b) $N^2 + 96 = P^2$, or, $96 = P^2 - N^2$. Now factorize 96 and equate it with $(P + N)(P - N)$.

38. (a) Sum of the digits of the 'super' number
$= 1 + 2 + 3 + \ldots\ldots\ldots\ldots\ldots\ldots + 29$

$$= \frac{29}{2}.\{2 \times 1 + (29 - 1).1\}$$

$$= \frac{29}{2}.(2 + 28) = \frac{29 \times 30}{2} = 29 \times 15 = 435$$

Now, sum of digits in the number $435 = 4 + 3 + 5 = 12$ which gives a remainder of 3 when divided by 9.

39. (d) The remainder when a number is divided by 16 is given by the remainder of the last 4 digits divided by 16 (because 10000 is a multiple of 16. This principle is very similar in logic to why we look at last 2 digits for divisibility by 4 and the last 3 digits for divisibility by 8.) Thus, the required answer would be the remainder of 4950/16 which is 6.

40. (a) Let the even consecutive numbers are $2n-2, 2n, 2n+2$
$(2n - 2) \times (2n) \times (2n + 2) = 4032$ ...(1)
Product of 1st even number third even number = 252
Putting this in equation ...(1)
$252 \times 2n = 4032 \Rightarrow n = 8$
Numbers are 14, 16, 18
Five times of 2nd number is $= 5 \times 16 = 80$

41. (c) The sum forms A.P.
First term (a) = 1
Common difference (d) = 2

$$\text{Sum of 15 term} = \frac{n}{2}(2a + (n-1)d)$$

$$\text{Sum} = \frac{15}{2}(2 \times 1 + (15 - 1)2) = \frac{15}{2} \times 30 = 225$$

42. (a) Here, $(48 - 38) = 10$, $(64 - 54) = 10$, $(90 - 80) = 10$ and $(120 - 110) = 10$.
$\therefore$ Required number = (L.C.M of 48, 64, 90 and 120) – 10 = 2870

43. (c) $(0.11)^3 + (0.22)^3 + \ldots\ldots + (0.99)^3$
$= (0.11)^3\,[1^3 + 2^3 + \ldots\ldots\ 9^3]$
$= (0.11)^3 \times 2025 = 2.695$

44. (b) Product of first fifty positive consecutive integers $= 1 \times 2 \times \ldots. \times 50 = 50\,!$

$$\text{Largest possible value of n} = \left[\frac{50}{7}\right] + \left[\frac{50}{7^2}\right]$$

$= 7 + 1 = 8$

CHAPTER

3

Averages

AVERAGE

An average is a simple concept of mathematics but its uses are very common in day-to-day life. In CAT and CAT like apptitute test exams at least one question is always asked, the nature of the question asked in CAT is applied and blended with logical reasoning.

An average of a group of numbers is a number that is the best representative of the group of numbers because it tells a lot about the entire numbers of the group.

In other words an average is a measure of central tendency called arithmetic mean of a group of numbers,

The formula for finding the average is

$$\text{Average} = \frac{\text{Sum of all numbers}}{\text{Number of numbers}}$$

Thus if A_v be the average of n numbers $x_1, x_2, x_3, ..., x_n$ then

$$A_v = \frac{x_1 + x_2 + x_3 + ... + x_n}{n}$$

$\Rightarrow x_1 + x_2 + x_3 + ... + x_n = n \cdot A_v$

$\Rightarrow$ Sum of n numbers = (Number of numbers) × (Average)

Illustration 1: The average of the first nine prime numbers is:

(a) 9 **(b)** 11 **(c)** $11\frac{1}{9}$ **(d)** $11\frac{2}{9}$

Solution: (c) Average

$$= \frac{2+3+5+7+11+13+17+19+23}{9}$$

$$= \frac{100}{9} = 11\frac{1}{9}.$$

Illustration 2: The average of five consecutive odd numbers is 61. What is the difference between the highest and lowest numbers?

(a) 2 **(b)** 5

(c) 8 **(d)** Cannot be determined

Solution: (c) Let the numbers be $x, x + 2, x + 4, x + 6$ and $x + 8$.

Then, $\frac{x + (x+2) + (x+4) + (x+6) + (x+8)}{5} = 61$

or $5x + 20 = 305$ or $x = 57$.

So, required difference = (57 + 8) – 57 = 8.

WEIGHTED AVERAGE

If we have two or more groups of numbers whose individual averages are known, then combined average of all the numbers of all the groups is known as Weighted Average. Thus if there are k groups having number of numbers $n_1, n_2, n_3, ..., n_k$ with averages $A_1, A_2, A_3, ..., A_k$ respectively; then weighted average,

$$A_w = \frac{n_1 A_1 + n_2 A_2 + n_3 A_3 + ... + n_k A_k}{n_1 + n_2 + n_3 + ... + n_k}$$

Illustration 3: The average score of a cricketer in two matches is 27 and in three other matches is 32. Then find the average score in all the five matches.

Solution:

$$\text{Average in 5 matches} = \frac{2 \times 27 + 3 \times 32}{2+3} = \frac{54 + 96}{5} = 30.$$

Illustration 4: The average age of students of a class is 15.8 years. The average age of boys in the class is 16.4 years and that of the girls is 15.4 years. The ratio of the number of boys to the number of girls in the class is

(a) 1 : 2 **(b)** 2 : 3 **(c)** 3 : 4 **(d)** 3 : 5

Solution: (b) Let the number of boys in a class be x.

Let the number of girls in a class be y.

$\therefore$ Sum of the ages of the boys = $16.4\,x$

Sum of the ages of the girls = $15.4\,y$

$\therefore$ $15.8(x + y) = 16.4\,x + 15.4\,y$

$\Rightarrow 0.6\,x = 0.4\,y \Rightarrow \frac{x}{y} = \frac{2}{3}$

$\therefore$ Required ratio = 2 : 3

Illustration 5: The average age of 30 boys of a class is equal to 14 yrs. When the age of the class teacher is included the average becomes 15 yrs. Find the age of the class teacher.

Solution: Total ages of 30 boys = 14 × 30 = 420 yrs.

Total ages when class teacher is included = 15 × 31 = 465 yrs.

∴ Age of class teacher = 465 – 420 = 45 yrs.

By direct formula,

Age of new entrant

= (No. of old members) × (Increase in average) + (New average)

= 30 (15 – 14) + 15 = 45 yrs.

Illustration 6: The average weight of 45 students in a class is 52 kg. 5 of them whose average weight is 48 kg leave the class and other 5 students whose average weight is 54 kg join the class. What is the new average weight (in kg) of the class?

(a) 52.6 (b) $52\frac{2}{3}$

(c) $52\frac{1}{3}$ (d) None of these

Solution: (b) Total weight of 45 students = 45 × 52 = 2340 kg

Total weight of 5 students who leave

= 5 × 48 = 240 kg

Total weight of 5 students who join

= 5 × 54 = 270 kg

Therefore, new total weight of 45 students

= 2340 – 240 + 270 = 2370

$\Rightarrow$ New average weight $= \frac{2370}{45} = 52\frac{2}{3}$ kg.

Remember

(i) Ages and Averages

If the average age of a group of persons is x years today then after n years their average age will be $(x + n)$ years because for a group of people, 1 year is added to each person's age every year.

Similarly, n years ago their average age would have been $(x - n)$ years, because 1 year is subtracted from each person's age before every year.

(ii) Average Speed if Equal Distances are Travelled by Two Different Speeds

If a car travels at a speed S_1 from A to B and at a speed S_2 from B to A. Then

$$\text{Average speed} = \frac{2S_1 . S_2}{S_1 + S_2}$$

The above formula can be found out as follows:

If distance between A and B is d, then

$$\text{Average speed} = \frac{\text{Total distance}}{\text{Total time}} = \frac{2d}{\frac{d}{S_1} + \frac{d}{S_2}}$$

$$= \frac{2}{\frac{1}{S_1} + \frac{1}{S_2}} = \frac{2S_1 . S_2}{S_2 + S_1}$$

Similarly for three equal distances travelled by three different speeds S_1, S_2 and S_3;

$$\text{Average speed} = \frac{3S_1 . S_2 . S_3}{S_1 . S_2 + S_2 . S_3 + S_3 . S_1}$$

Let's find the average speed of a car which goes from Delhi to Panipat at a speed of 60 kmph and returns at a speed of 90 kmph.

$$\text{Average speed} = \frac{2S_1 . S_2}{S_1 + S_2} = \frac{2 \times 60 \times 90}{60 + 90} = \frac{2 \times 60 \times 90}{150}$$

= 72 kmph.

Illustration 7: The average age of a family of 6 members is 22 yrs. If the age of the youngest member be 7 yrs, then what was the average age of the family at the birth of the youngest member?

Solution: Total ages of all members = 6 × 22 = 132 yrs.

7 yrs. ago, total sum of ages = 132 – (6 × 7) = 90 yrs.

But at that time there were 5 members in the family.

∴ Average at that time = 90 ÷ 5 = 18 yrs.

Illustration 8: The average of marks obtained by 120 candidates in a certain examination is 35. If the average marks of passed candidates is 39 and that of the failed candidates is 15, what is the number of candidates who passed the examination?

Sol. Let the number of passed candidates be x.

Then total marks = 120 × 35 = $39x + (120 - x) \times 15$

or, $4200 = 39x + 1800 - 15x$ or $24x = 2400$

∴ $x = 100$

∴ number of passed candidates = 100.

Illustration 9: A cricketer has completed 10 innings and his average is 21.5 runs. How many runs must he make in his next innings so as to raise his average to 24?

Solution: Total of 10 innings = 21.5 × 10 = 215

Suppose he needs a score of x in 11th innings; then average in 11 innings $= \frac{215 + x}{11} = 24$

or, $x = 264 - 215 = 49$.

Practice Exercise

Level - I

1. The average age of 24 students and the class teacher is 16 years. If the class teacher's age is excluded, the average reduces by one year. What is the age of the class teacher?
   (a) 50 years (b) 45 years
   (c) 40 years (d) Data inadequate
2. The average age of 36 students in a group is 14 years. When teacher's age is included in it, the average increases by one. What is the teacher's age in years?
   (a) 31 (b) 36
   (c) 51 (d) cannot be determined
3. A school has 4 section of Chemistry in Class X having 40, 35, 45 and 42 students. The mean marks obtained in Chemistry test are 50, 60, 55 and 45 respectively for the 4 sections. Determine the overall average of marks per student
   (a) 50.25 (b) 52.25
   (c) 51.25 (d) 53.25
4. The average of six numbers is 3.95. The average of two of them is 3.4, while the average of the other two is 3.85. What is the average of the remaining two numbers?
   (a) 4.5 (b) 4.6
   (c) 4.7 (d) 4.8
5. The average of 5 consecutive numbers is n. If the next two numbers are also included the average will
   (a) remain the same (b) increase by 1
   (c) increase by 1.4 (d) increase by 2
6. The average of 11 numbers is 10.9. If the average of the first six numbers is 10.5 and that of the last six numbers is 11.4, then the middle number is :
   (a) 11.5 (b) 11.4
   (c) 11.3 (d) 11.0
7. The average temperature for the first four days of a week is 40.2°C and that of the last four days is 41.3°C. If the average temperature for the whole week is 40.6°C, then the temperature on the fourth day is
   (a) 40.8°C (b) 38.5°C
   (c) 41.3°C (d) 41.8°C
8. A person covers half his journey by train at 60 kmph, the remainder half by bus at 30 kmph and the rest by cycle at 10 kmph. Find his average speed during the entire journey.
   (a) 36 kmph (b) 24 kmph
   (c) 48 kmph (d) None of these
9. The marks obtained by Hare Rama in Mathematics, English and Biology are respectively 93 out of 100, 78 out of 150 and 177 out of 200. Find his average score in percent.
   (a) 87.83 (b) 86.83
   (c) 76.33 (d) 77.33
10. P is going to Delhi from Gurgaon by his car at a speed of 40 km/h. While coming back, he returns with a speed of x km/h. What should be the value of x so that his average speed during the intire journey is 80 km/h?
    (a) 160 km/h (b) 40 km/h
    (c) 120 km/h (d) It is not possible
11. The average age of Mr. and Mrs Sinha at the time of their marriage in 1972 was 23 years. On the occasion of their anniversary in 1976, they observed that the average age of their family had come down by 4 years compared to their average age at the time of their marriage. This was due to the fact that their son Vicky was born during that period. What was the age of Vicky in 1980?
    (a) 6 (b) 7
    (c) 8 (d) 5
12. A library has an average of 510 visitors of Sunday and 240 on other days. The average number of visitors per day in a month of 30 days beginning with a Sunday is
    (a) 250 (b) 276
    (c) 280 (d) 285
13. The mean of 30 values was 150. It was detected on rechecking that one value 165 was wrongly copied as 135 for the computation of the mean. Find the correct mean.
    (a) 151 (b) 149
    (c) 152 (d) None of these
14. The average of 10 numbers is 40.2. Later it is found that two numbers have been wrongly copied. The first is 18 greater than the actual number and the second number added is 13 instead of 31. Find the correct average.
    (a) 40.2 (b) 40.4
    (c) 40.6 (d) 40.8
15. In 1919, W. Rhodes, the Yorkshire cricketer, scored 891 runs for his country at an average of 34.27; in 1920, he scored 949 runs at an average of 28.75; in 1921, 1329 runs at an average of 36.70. What was his county batting average for the four years?
    (a) 36.23 (b) 37.81
    (c) 35.88 (d) 28.72
16. The average salary of all the workers in a workshop is ₹8,000. The average salary of 7 technicians is ₹12,000 and the average salary of the rest is ₹ 6,000. The total number of workers in the workshop is :
    (a) 21 (b) 20
    (c) 23 (d) 22
17. The average monthly salary of employees, consisting of officers and workers, of an organisation is ₹3000. The average salary of an officer is ₹10,000 while that of a

worker is ₹2000 per month. If there are total 400 employees in the organisation, find the number of officers.
(a) 60 (b) 50
(c) 80 (d) 40

18. Of the three numbers, the first is twice the second and the second is twice the third. The average of the reciprocal of the numbers is $\frac{7}{72}$. The numbers are
(a) 16, 8, 4 (b) 20, 10, 5
(c) 24, 12, 6 (d) 36, 18, 9

19. In a bag, there are 150 coins of ₹ 1,50 p and 25 p denominations. If the total value of coins is ₹ 150, then find how many rupees can be constituted by 50 coins.
(a) ₹ 16 (b) ₹ 20
(c) ₹ 28 (d) None of these

20. The average age of a group of persons going for picnic is 16 years. Twenty new persons with an average age of 15 years join the group on the spot due to which their average age becomes 15.5 years. The number of persons initially going for picnic is
(a) 5 (b) 10
(c) 20 (d) 30

21. The average weight of 47 balls is 4 gm. If the weight of the bag (in which the balls are kept) be included, the calculated average weight per ball increases by 0.3 gm. What is the weight of the bag?
(a) 14.8 gm (b) 15.0 gm
(c) 18.6 gm (d) None of these

22. On an average 300 people watch the movie in Sahu cinema hall on Monday, Tuesday and Wednesday and the average number of visitors on Thursday and Friday is 250. If the average number of visitors per day in the week be 400, then the average number of people who watch the movie in weekends (*i.e.,* on Saturday and Sunday) is
(a) 500 (b) 600
(c) 700 (d) None of these

23. A train travels with a speed of 20 m/s in the first 10 minutes, goes 8.5 km in the next 10 minutes, 11 km in the next 10,8.5 km in the next 10 and 6 km in the next 10 minutes. What is the average speed of the train in kilometer per hour for the journey described?
(a) 42 kmph (b) 35.8 kmph
(c) 55.2 kmph (d) 46 kmph

24. Find the average increase rate if increase in the population in the first year is 30% and that in the second year is 40%.
(a) 41 (b) 56
(c) 40 (d) 38

25. Find the average weight of four containers, if it is known that the weight of the first container is 100 kg and the total of the second, third and fourth containers' weight is defined by $f(x) = x^2 - 3/4\ (x^2)$ where $x = 100$
(a) 650 kg (b) 900 kg
(c) 750 kg (d) 450 kg

26. The average age of a group of 14 persons is 27 years and 9 months. Two persons, each 42 years old, left the group. What will be the average age of the remaining persons in the group?
(a) 26.875 years (b) 26.25 years
(c) 25.375 years (d) 25 years

27. A school has only four classes that contain 10,20,30 and 40 students respectively. The pass percentage of these classes are 20%, 30%, 60% and 100% respectively. Find the pass % of the entire school.
(a) 56% (b) 76%
(c) 34% (d) 66%

28. Find the average of $f(x)$, $g(x)$, $h(x)$, $d(x)$ at $x = 10$. $f(x) = x^2 + 2$, $g(x) = 5x^2 - 3$, $h(x) = \log x^2$ and $d(x) = (4/5)x^2$.
(a) 170 (b) 170.25
(c) 70.25 (d) 70

29. The average of 'n' numbers is z. If the number x is replaced by the number x^1, then the average becomes z^1. Find the relation between n, z, z^1, x and x^1.
(a) $\frac{z^1 - 2}{x^1 - x} = \frac{1}{n}$ (b) $\frac{x^1 - x}{z^1} = \frac{1}{n}$
(c) $\frac{z - z^1}{x - x^1} = \frac{1}{n}$ (d) $\frac{x - x^1}{z - z^1} = \frac{1}{n}$

30. A man's average expenditure for the first 4 months of the year was ₹ 251.25. For the next 5 months the average monthly expenditure was ₹ 26.27 more than what it was during the first 4 months. If the person spent ₹ 760 in all during the remaining 3 months of the year, find what percentage of his annual income of ₹ 3000 he saved in the year.
(a) 14% (b) –5.0866%
(c) 12.5% (d) None of these

31. *A* curious student of Statistics calculated the average height of all the students of his class as *A*. He also calculated the average of the average heights of all the possible pairs of students (two students taken at a time) as *B*. Further, he calculated the average of the average heights of all the possible triplets of students (three students taken at a time) as *C*. Which of the following is true of the relationship among *A*, *B* and *C*?
(a) $A + 2B = C$ (b) $A + B = 2C$
(c) $A = B = 3C$ (d) None of these

32. We write down all the digits from 1-9 side by side. Now we put '+' between as many digits as we wish to, so that the sum of numbers become 666. It is explained below
1 2 3 4 5 6 7 8 9 = 666
Now suppose we put plus signs at following places.
12 + 345 + 67 + 89 = 513
Since there are four numbers, so the average can be calculated by dividing the sum by 4. What is the average if the sum is 666?
(a) 166.5 (b) 111
(b) 133.2 (d) Cannot be determined

33. What will be the average of the followings set of scores ?
59, 84, 44, 98, 30, 40, 58 [*SBI Clerk-June-2012*]
(a) 62 (b) 66
(c) 75 (d) 52
(e) 59

34. Average of five numbers is 61. If the average of first and third number is 69 and the average of second and fourth number is 69, what is the fifth number ? [*SBI Clerk-2012*]
(a) 31 (b) 29
(c) 25 (d) 35
(e) None of these

35. Average weight of 19 men is 74 kgs, and the average weight of 38 women is 63 kgs. What is the average weight (rounded off to the nearest integer) of all the men and the women together ? [*SBI Clerk-2012*]
(a) 59 kgs. (b) 65 kgs.
(c) 69 kgs. (d) 67 kgs.
(e) 71 kgs.

36. The average age of 60 boys in a class was calculated as 12 years. It was later realised that the actual age of one of the boys in the class was 12.5 years but it was calculated as 14 years. What is the actual average age of the boys in the class ? [*SBI Clerk-2014*]
(a) 11 years (b) 11.275 years
(c) 11.50 years (d) 11.975 years
(e) None of these

37. The average of three numbers 70, *7 and 5* is 57. If * represents the same digit, then it must be [*SSC-Sub. Ins.-2012*]
(a) 3 (b) 6
(c) 4 (d) 7

38. Three years ago, the average age of a family of 8 members was 30 years. If one child is also included in the family, the present average age of the family remained the same. Then the present age of the child is [*SSC-Sub. Ins.-2012*]
(a) 3 years (b) 4 years
(c) 6 years (d) 1 year

39. The batting average for 30 inings of a cricket player is 40 runs. His highest score exceeds his lowest score by 100 runs. If these two innings are not included, the average of the remaining 28 innings is 38 runs. The lowest score of the player is: [*SSC-Sub. Ins.-2013*]
(a) 15 (b) 18
(c) 20 (d) 12

40. A boy found that the average of 20 numbers is 35 when he writes a number '61' instead of '16'. The correct average of 20 numbers is [*SSC-Sub. Ins.-2014*]
(a) 32.75 (b) 37.25
(c) 34.75 (d) 34.25

41. Out of 20 boys, 6 are each of 1 m 15 cm height, 8 are of 1 m 10 cm and rest of 1 m 12 cm. The average height of all of them is [*SSC-MT-2013*]
(a) 1 m 12 cm (b) 1 m 12.1 cm
(c) 1 m 21.1 cm (d) 1 m 21 cm

42. Average of first five prime numbers is [*SSC-MT-2013*]
(a) 3.6 (b) 5.3
(c) 5.6 (d) 5

43. A batsman in his 12th innings makes a score of 63 runs and there by increases his average scores by 2. What is his average after the 12th innings? [*SSC 10+2-2012*]
(a) 13 (b) 41
(c) 49 (d) 87

44. The average of four consecutive even numbers is 9. Find the largest number. [*SSC 10+2-2012*]
(a) 12 (b) 6
(c) 8 (d) 10

45. The average weight of 12 crewmen in a boat is increased by $\frac{1}{3}$ kg, when one of the crewmen whose weight is 55 kg is replaced by a new man. What is the weight of that new men (in kg) ? [*SSC 10+2-2012*]
(a) 58 (b) 60
(c) 57 (d) 59

46. The average of 30 numbers is 40 and that of other 40 numbers is 30. The average of all the numbers is [*SSC 10+2-2013*]
(a) 34.5 (b) $34\frac{2}{7}$
(c) 35 (d) 34

47. The average salary of all the workers in a workshop is ₹ 8,000. The average salary of 7 technicians is ₹ 12,000 and the average salary of the rest is ₹ 6,000. The total number of workers in the workshop is [*SSC 10+2-2014*]
(a) 20 (b) 21
(c) 22 (d) 23

48. 3 years ago the average age of a family of 5 members was 17 years. A baby having been born, the average age of the family is the same today. The present age of the baby is [*SSC 10+2-2014*]
(a) 1 year (b) 1½ years
(c) 2 years (d) 3 years

49. Find the average of the following set of scores : 432, 623, 209, 378, 908, 168 [*IBPS Clerk-2012*]
(a) 456 (b) 455
(c) 453 (d) 458
(e) None of these

50. The average of five numbers is 34.4. The average of the first and the second number is 46.5. The average of the fourth and the fifth number is 18. What is the third number ? [*IBPS Clerk-2012*]
(a) 45 (b) 46
(c) 42 (d) 49
(e) None of these

51. What will be the average of the following set of scores ? 78, 69, 54, 21, 94, 48, 77 [*IBPS Clerk-2012*]
(a) 63 (b) 66
(c) 67 (d) 64
(e) None of these

52. The average score of a cricketer for 13 matches is 42 runs. If his average score for the first 5 matches is 54, then what is his average score (in runs) for last 8 matches? [*IBPS Clerk-2013*]
(a) 37 (b) 39
(c) 34.5 (d) 33.5
(e) 37.5

Level - II

1. The average weight of 3 men A, B and C is 84 kg. Another man D joins the group and the average now becomes 80 kg. If another man E, whose weight is 3 kg more than that D, replaces A then the average weight of B, C, D and E becomes 75 kg. The weight of A is
(a) 70 kg (b) 72 kg
(c) 79 kg (d) 78 kg

2. In Arun's opinion, his weight is greater than 65 kg but less than 72 kg. His brother does not agree with Arun and he thinks that Arun's weight is greater than 60 kg but less than 70 kg. His mother's view is that his weight cannot be greater than 68 kg. If all of them are correct in their estimation, what is the average of different probable weights of Arun?
(a) 67 kg (b) 68 kg
(c) 69 kg (d) None of these

3. There are five boxes in a cargo hold. The weight of the first box is 200 kg and the weight of the second box is 20% higher than the weight of the third box, whose weight is 25% higher than the first box's weight. The fourth box at 350 kg is 30% lighter than the fifth box. Find the difference in the average weight of the four heaviest boxes and the four lightest boxes.
(a) 51.5 kg (b) 75 kg
(c) 37.5 kg (d) 112.5 kg

4. Of the three numbers, the average of the first and the second is greater than the average of the second and the third by 15. What is the differnce between the first and the third of the three numbers?
(a) 15 (b) 45
(c) 60 (d) None of these

5. The average monthly expenditure of Ravi was ₹1100 during the first 3 months, ₹2200 during the next 4 months and ₹4620 during the subsequent five months of the year. If the total savinig during the year was ₹ 2100, find Ravi's average monthly income.
(a) ₹ 1858 (b) ₹ 3108.33
(c) ₹ 3100 (d) None of these

6. Rajeev earns 3/2 times in January, April, July and October than his average earning of ₹ 600 per month in the rest of the month. So his savings in the January, April, July and October goes to 5/4 times that of the rest months saving of ₹ 400 per month in the year. The average expenditure of per month is:
(a) ₹ 266.66 (b) ₹ 250
(c) ₹. 233.33 (d) ₹ 433.33

7. There were five sections in MAT paper. The average score of Pooja in first 3 sections was 83 and the average in the last 3 sections was 97 and the average of all the sections (*i.e.*, whole paper) was 92, then her score in the third section was
(a) 85 (b) 92
(c) 88 (d) None of these

8. Mr. Anant Roy, the renowned author, recently got his new novel released. To his utter dismay he found that for the 1,007 pages on an average there were 2 mistakes every page. While, in the fist 612 pages there were only 434 mistakes, they seemed to increase for the latter pages. Find the average number of mistakes per page for the remaining pages.
(a) 6 (b) 4
(c) 2 (d) None of these

9. In hotel Trident, the rooms are numbered from 101 to 130 on the first floor, 221 to 260 on the second floor and 306 to 345 on the third floor. In the month of June 2012, the room occupancy was 60% on the first floor, 40% on the second floor and 75% on the third floor. If it is also known that the room charges are ₹ 200, ₹ 100 and ₹ 150 on each of the floors, then find the average income per room for the month of June 2012.
(a) ₹ 151.5 (b) ₹ 88.18
(c) ₹ 78.3 (d) ₹ 65.7

10. The average age of a couple is 25 years. The average age of the family just after the birth of the first child was 18 years. The average age of the family just after the second child was born was 15 years. The average age of the family after the third and the fourth children (who are twins) were born was 12 years. If the present average age of the family of six persons is 16 years, how old is the eldest child ?
(a) 6 years (b) 7 years
(c) 8 years (d) 9 years

11. The average monthly rainfall for a year in Guntur district is 2.7 inches, the average for the first 7 months is 1.1 inches less than the annual average. If the total rainfall for the next 4 months is 20.8 inches, then the rainfall in the last month will be
(a) 0.1 inch (b) 0.2 inch
(c) 0.4 inch (d) 0.6 inch

12. Eleven years earlier the average age of a family of 4 members was 28 years. Now the age of the same family with six members is yet the same, even when 2 children were born in this period. If they belong to the same parents and the age of the first child at the time of the birth of the younger child was same as there were total family members just after the birth of the youngest members of this family, then the present age of the youngest member of the family is
(a) 3 years (b) 5 years
(c) 6 years (d) None of these

13. The average earning of a group of persons is ₹ 50 per day. The difference between the highest earning and lowest earning of any two persons of the group is ₹ 45. If these two people are excluded the average earning of the group decreased by ₹ 1. If the minimum earning of the person in

the group lies between 42 and 47 and the number of persons initially in the group was equal to a prime number, with both its digits prime. The number of persons in the group initially was:

(a) 29 (b) 53
(c) 31 (d) None of these

14. The class X of a Vidhyalaya has four sections: A, B, C and D. The average weight of the students of A, B, C together and A, C, D together are 45 kg and 55 kg respectively, while the average weight of the students of A, B, D together and B, C, D together are 50kg and 60kg respectively. Which of the following could be the average weight of the students of all the four sections together?

(a) 47.6 kg (b) 52.5 kg
(c) 53.7 kg (d) 56.5 kg

15. The average market price of three shares x, y and z is ₹m. Shares x and y lose ₹ n each and z gains ₹$\frac{n}{2}$. As a result, the average market price of the three shares decrease by ₹ 1. The value of n is

(a) 2 (b) 3
(c) 4 (d) dependent of x

16. The average marks in English subject of a class of 24 students is 56. If the marks of three students were misread as 44, 45 and 61 of the actual marks 48, 59 and 67 respectively, then what would be the correct average? *[IBPS-PO-2011]*

(a) 56.5 (b) 59
(c) 57.5 (d) 58
(e) None of these

17. In an Entrance Examination Ritu scored 56 percent marks, Smita scored 92 percent marks and Rina scored 634 marks. The maximum marks of the examination are 875. What are the average marks scored by all the three girls together? *[IBPS-PO-2011]*

(a) 1929 (b) 815
(c) 690 (d) 643
(e) None of these

18. A batsman makes a score of 58 runs in the 15th innings and thus increases his average by 3 runs. What is the average after 15th inning? *[SSC CGL-2012]*

(a) 12 (b) 14
(c) 16 (d) 18

19. The average of 5 consecutive numbers is n. If the next two numbers are also included, the average of the 7 numbers will *[SSC CGL-2012]*

(a) increase by 2 (b) increase by 1
(c) remain the same (d) increase by 1.4

20. Out of 40 boys in a class, average weight of 30 is 60 kg and the average weight of the remaining is 56 kg. The average weight (in kilogram) of the whole class is *[SSC CGL-2013]*

(a) 58.5 (b) 58
(c) 57 (d) 59

21. The average of the first five multiples of 7 will be *[SSC CGL-2013]*

(a) 14 (b) 21
(c) 17.5 (d) 24.5

22. The average age of boys in the class is twice the number of girls in the class. The ratio of boys and girls in the class of 50 is 4 : 1. The total of the ages (in years) of the boys in the class is *[SSC CGL-2014]*

(a) 2000 (b) 2500
(c) 800 (d) 400

23. There are 100 students in 3 sections A, B and C of a class. The average marks of all the 3 sections was 84. The average of B and C was 87.5 and the average marks of A is 70. The number of students in A was *[SSC CGL-2014]*

(a) 30 (b) 35
(c) 20 (d) 25

Hints & Solutions

Level-I

1. (c) Age of the class teacher = 25 × 16 – 24 × 15
= 400 – 360 = 40 yrs.

2. (c) Age of the teacher = (37 × 15 – 36 × 14) years
= 51 years.

3. (b) Required average marks

$$= \frac{40\times50+35\times60+45\times55+42\times45}{40+35+45+42}$$

$$= \frac{2000+2100+2475+1890}{162} = \frac{8464.9997}{162}$$

= 52.25

4. (b) Sum of the remaining two numbers
= (3.95 × 6) – [(3.4 × 2) + (3.85 × 2)]
= 23.70 – (6.8 + 7.7) = 23.70 – 14.5 = 9.20

$\therefore$ Required average = $\left(\frac{9.2}{2}\right) = 4.6.$

5. (b) Check as follows,

$$\frac{1+2+3+4+5}{5} = 3$$

$$\frac{1+2+3+4+5+6+7}{7} = 4$$

6. (a) The middle number = Sum of the first six + Sum of the the last six – Sum of all the 11
= 6 × 10.5 + 6 × 10.5 – 11 × 10.9
= 63 + 68.4 – 119.9
= 11.5

7. (d) Temperature on the fourth day
= 40.2 × 4 + 41.3 × 4 – 40.6 × 7
= 160.8 + 165.2 – 284.2 = 41.8°C

8. (b) Recognise that the journey by bus and that by cycle are of equal distance. Hence, we can use the short cut illustrated earlier to solve this part of the problem.
Using the process explained above, we get average speed of the second half of the journey as
10 + 1 × 5 = 15 kmph
Then we employ the same technique for the first part and get 15 + 1 × 9 = 24 kmph

9. (d) His total score is 93 + 78 + 177 = 348 out of 450% score = 77.33.

10. (d) $80 = \frac{2.x.40}{40+x}$

$40 + x = x$
Hence, not possible

11. (b) Sum of ages of Mr. and Mrs. Sinha in 1972 = 46 years
Sum of age of their family in 1976 = 19 × 3 = 57 years
Sum of ages of Mr. and Mrs. Sinha in 1976 = (46 + 8) years
= 54 years
$\therefore$ Age of Vicky in 1980 = 57 – 54 + 4 = 7 years.

12. (d) Since the month begins with Sunday, so there will be five Sundays in the month

$\therefore$ Required average $= \left(\frac{510\times5+240\times25}{30}\right)$

$$= \frac{8550}{30} = 285$$

13. (a) Corrected mean $= \frac{150\times30-135+165}{30}$

$$= \frac{4500-135+165}{30} = \frac{4530}{30} = 151$$

14. (a) Sum of 10 numbers = 402
Corrected sum of 10 numbers
= 402 – 13 + 31 – 18 = 402

Hence, correct average $= \frac{402}{10} = 40.2$

15. (c) Find out the number of innings in each year. Then the answer will be given by:

$\frac{\text{Total runs in 4 years}}{\text{Total innings in 4 years}}$ (4270/119 = 35.88)

16. (a) Let the total no. of workers be x.
Now, 8000 x = 7 × 12000 + (x – 7) × 6000

$\Rightarrow x = \frac{42000}{2000} = 21$

17. (b) Let the number of officers be x.
Number of workers = 400 – x
$\Rightarrow$ 10000 × x + 2000(400 – x) = 3000(400)
$\Rightarrow$ 10000x + 800000 – 2000x = 12,00,000
$\Rightarrow$ 4x = 600 – 400 = 200 $\Rightarrow$ x = 50
$\therefore$ Number of officers = 50

18. (c) Let the third number be x. Then, second number = 2x. First number = 4x.

$\therefore \frac{1}{x}+\frac{1}{2x}+\frac{1}{4x} = \left(\frac{7}{72}\times3\right)$ or $\frac{7}{4x} = \frac{7}{24}$ or $4x = 24$

or x = 6
So, the numbers are 24, 12 and 6.

19. (d) For 150 coins to be of a value of ₹150, using only 25 paise, 50 paise and ₹ 1 coins, we cannot have any coins lower than the value of ₹ 1. Thus, the number of 50 paise coins would be 0. Option (d) is correct.

20. (c) Let the initial number of persons be x. Then, $16x + 20 \times 15 = 15.5(x + 20) \Leftrightarrow 0.5x = 10$ $\Leftrightarrow x = 20$.

21. (d) The average weight per ball is asked. Hence the bag does not have to be counted as the 48th item.

22. (c) $400 \times 7 = (300 \times 3) + (250 \times 2) + (n \times 2) \Rightarrow 700$.

23. (c) Find the total distance covered in each segment of 10 minutes. You will get total distance = 46 kilometers in 50 mins.

$\therefore$ Average speed $= \frac{46 \times 60}{50} = 55.2$ kmph

24. (a) $100 \to 130 \to 182$. Hence, $82/2 = 41$.

25. (a) Put $x = 100$ to get the weight of the containers. Use these weights of find average weight as $2600/4 = 650$.

26. (c) $(14 \times 27.75 - 2\ 0 + 42\ )/12 = 25.375$

27. (d) The number of pass candidates are $2 + 6 + 18 + 40 = 66$ out of a total of 100. Hence, 66%

28. (b) Put $x = 10$ in the given equations and find the average of the resultant values.

29. (c) $nz - x + x^1 = nz^1$ Simplify to get option (c) correct.

30. (b) $251.25*4 + 277.52 * 5 + 760 = 3152.6$

31. (d) Let the height of four students be 150, 160, 170, 180 cm then

$$A = \frac{150+160+170+180}{4} = \frac{660}{4} = 165\text{cm}$$

$$B = \frac{\frac{150+160}{2} + \frac{160+170}{2} + \frac{170+180}{2} + \frac{150+180}{2} + \frac{150+170}{2} + \frac{160+180}{2}}{6}$$

$$= \frac{155+165+175+165+160+170}{6} = \frac{990}{6} = 175\text{cm}$$

$$= \frac{670}{4} = 167.5$$

Similarly, $C = 167.5$

Now going through the opitons.

32. (d) We can get the sum 666 in two ways:

$1 + 2 + 3 + 4 + 567 + 89 = 666$

or $123 + 456 + 78 + 9 = 666$

So, average cannot be uniquely determined.

33. (e) Required average $= \frac{59+84+44+98+30+40+58}{7} = 59$

34. (b) Let the five no. be x_1, x_2, x_3, x_4, x_5.

Average of 5 numbers = 61

$$\frac{x_1 + x_2 + x_3 + x_4 + x_5}{5} = 61$$

$$x_1 + x_2 + x_3 + x_4 + x_5 = 305$$

Now, $\frac{x_1 + x_3}{2} = 69$

$x_1 + x_3 = 138$

$\frac{x_2 + x_4}{2} = 69$

$x_2 + x_4 = 138$

Now, $x_1 + x_3 + x_2 + x_4 + x_5 = 305$

$138 + 138 + x_5 = 305$

$x_5 = 305 - 276$

$x_5 = 29$

35. (d) Average weight of 19 men = 74 kgs

Total weight of 19 men = 74 × 19 = 1406 kgs

Average weight of 38 women = 63

Total weight of 38 women = 38 × 63 = 2394

Average weight of men and women together

$$= \frac{2394 + 1406}{38 + 19}$$

$$= \frac{3800}{57} = 66.66 \approx 67 \text{ kgs.}$$

36. (d) Let S be the sum of ages of 60 boys

Then, $12 = \frac{S}{60}$ ... (1)

New average $A = \frac{S - 14 + 12.5}{60} = \frac{S}{60} - \frac{105}{60}$

$= 12 - 0.025 = 11.975$

37. (c) Average $= \frac{70+*7+5*}{3}$

$\Rightarrow 57 = \frac{70+*7+5*}{3}$

$\Rightarrow 171 - 70 = *7+5*$

$\Rightarrow *7+5* = 101$

Then, the sum of these two numbers should be equal to 101.

Therefore by looking cptions * must be 4.

38. (c) Present age of 8 members

= (30 × 8 + 3 × 8) = 264 years

Present age of 8 members and one child

= (30 × 9) = 270 years

∴ Child's age = (270 – 264) = 6 years

39. (b) Lowest score = x
Highest score = $x + 100$
$\therefore 28 \times 38 + x + x + 100 = 30 \times 40$
$\Rightarrow 1064 + 2x + 100 = 1200$
$\Rightarrow 2x = 1200 - 1164 = 36$
$\Rightarrow x = 18$

40. (a) Sum of 20 numbers $= 20 \times 35 = 700$
Sum of 20 numbers when 61 is replace by 16
$\Rightarrow 700 - 61 + 16 = 655$

Correct average of 20 numbers $= \frac{655}{20} = 32.75$

41. (b) Average height $= \frac{6 \times (1.15) + 8 \times (1.10) + 6(1.12)}{20}$

$\Rightarrow \frac{22.42}{20} = 1.121$ or 1m 12.1cm

42. (c) First five prime numbers are 2, 3, 5, 7, 11

Average $= \frac{2+3+5+7+11}{5} = \frac{28}{5} = 5.6$

43. (b) Let the average of batsman after 11th innings = A

$$\frac{\text{Total score made by batsman at the end of 11th innings}}{11} = A$$

$\therefore$ Total score after 11th innings = 11 A

Now, $\frac{\text{Total score after 11th innings + score made in 12th innings}}{12} = A + 2$

$\Rightarrow 11A + 63 = (A + 2) \times 12$
$\Rightarrow 11A - 12A = 24 - 63$
$\Rightarrow A = 39$
12th innings average = 39 + 2 = 41

44. (a) Let the consecutive even numbers are 2n, 2n + 2, 2n + 4 and 2n + 6

Average $= \frac{2n + 2n + 2 - 2n + 4 + 2n + 6}{4}$

$8n + 12 = 4 \times 9 \Rightarrow n = 3$
Hence, the numbers are 6, 8, 10 and 12. Largest among them is 12.

45. (d) Short-cut method:
Weight of new crewmen
= Replace man weight + [No. of crew men × increased average]

$= 55 + 12 \times \frac{1}{3} = 59$ kg

46. (b) Sum of 30 numbers = 30×40 = 1200
Sum of 40 numbers = 40 × 30 = 1200

Average of 70 numbers $= \frac{1200 + 1200}{70} = \frac{2400}{70} = 34\frac{2}{7}$

47. (b) Let total number of workers be n
total salary of all workers = 8000 n
total salary of 7 technicians = 7 × 12000 = 84,000
total salary of remaining workers = (n – 7) × 6000
84000 + (n – 7) × 6000 = 8000 n
84 + 6n – 42 = 8n
42 = 2n
n = 21

48. (c) Let total age of family be S years
3 years ago, total age = S – 3 × 5 = S – 15

$\frac{S - 15}{5} = 17$

$S = 17 \times 5 + 15 = 100$
Let present age of baby be x years

$\frac{S + x}{6} = 17$

$100 + x = 17 \times 6$
$x = 102 - 100 = 2$ years

49. (c) Average $= \frac{\text{Sum of observation}}{\text{Total no. of observation}}$

Required average

$= \frac{432 + 623 + 209 + 378 + 908 + 168}{6}$

$= \frac{2718}{6} = 453$

50. (e) Third number
$= 5 \times 34.4 - 2 \times 46.5 - 2 \times 18$
$= 172 - 93 - 36 = 43$

51. (a) Required average

$= \frac{78 + 69 + 54 + 21 + 94 + 48 + 77}{7}$

$= \frac{441}{7} = 63$

52. (c) Let M_1, M_2, M_3, ... M_{13} are 13 matches played by cricket players.

$$\frac{M_1 + M_2 + M_3 + M_4 + M_5 + M_6 + M_7 + M_8 + M_9 + M_{10} + M_{11} + M_{12} + M_{13}}{13} = 42 \quad ....(1)$$

$$\frac{M_1 + M_2 + M_3 + M_4 + M_5}{5} = 54 \quad ...(2)$$

From eqns. (1) and (2)
$270 + M_6 + M_7 + M_8 + M_9 + M_{10} + M_{11} + M_{12} + M_{13} = 42 \times 13 = 546$

or, $\frac{M_6 + M_7 + M_8 + M_9 + M_{10} + M_{11} + M_{12} + M_{13}}{8}$

$= \frac{276}{8} = 34.5$

Level-II

1. **(c)** D's weight = 4 × 80 – 3 × 84 = 320 – 252 = 68.
E's weight = 68 + 3 = 71.
Now, we know that $A + B + C + D = 4 \times 80 = 320$
and $B + C + D + E = 78 \times 4 = 312$.
Hence, A's weight is 8 kg more than E's weight.
$A = 71 + 8 = 79$.

2. **(d)** Let Arun's weight be x kg.
According to Arun, $65 < x < 72$.
According to Arun's brother, $60 < x < 70$.
According to Arun's mother, $x < 68$
The values satisfying all the above conditions are 66 and 67
∴ Required average
$$= \left(\frac{66+67}{2}\right) = \left(\frac{133}{2}\right) = 66.5\text{kg}$$

3. **(b)** The weight of the boxes are 1st box → 200, 3rd box → 250 kg, 2nd box → 300 kg, 4th box → 350 and 5th box → 500 kg. Hence difference between the heavier 4 and the lighter 4 is 300. Hence, difference in the averages is 75.

4. **(d)** Let the numbers are x, y and z.
Then, $\left(\frac{x+y}{2}\right) - \left(\frac{y+z}{2}\right) = 15$ or $(x + y) - (y + z)$
$= 30$ or $x - z = 30$

5. **(b)** Required average income = (Total expenditure + total savings]/12
= [(1100 × 3 + 2200 × 4 + 4620 × 5) + 2100]/12
= 37300/12 = 3108.333

6. **(a)** Earning in the 8 months = 600 × 8 = 4800
Earning in the 4 months = $\left(600 \times \frac{3}{2}\right) \times 4 = 3600$
Total earning = ₹ 8400
Saving in 8 months = 400 × 8 = 3200
Saving in 4 months = $\left(400 \times \frac{5}{4}\right) \times 4 = 2000$
Total savings = 5200
Total expenditure for 12 months = 8400 – 5200
= 3200
Therefore average saving per month = $\frac{3200}{12}$
= 266.66

7. **(d)** $a + b + c + d + e = 5 \times 92 = 460$
$a + b + c = 3 \times 83 = 249$
$c + d + e = 3 \times 97 = 291$
∴ $c = (a + b + c) + (c + d + e) - (a + b + c + d + e)$
or $c = 540 - 460$ or $c = 80$

8. **(b)** Total mistakes = 1007 × 2 = 2014
Let x be average mistake per page for the remaining pages
$434 + 395x = 2014$
$395x = 1580$
$x = 4$

9. **(a)** The number of rooms is 18 + 16 + 30 on the three floors respectively.
Total revenues are: 18*200 + 16 *100 + 30*150 = 9700 required average = 9700/110 = 88.18.
Note here that if you could visualize here that since the number of rooms is 110 the decimal values cannot be. (c) or (d) which effectively means that options 3 and 4 are rejected.

10. **(d)** The total age of the family at the birth of first child
= 18 × 3 = 54
While the total age of the couple at marriage
= 25 × 2 = 50.
⇒ The years from marriage till the first child's birth
$= \frac{54-50}{2} = 2$ years.
The total age of family at the birth of the second child.
= 15 × 4 = 60 years.
⇒ Second child was born $= \frac{60-54}{3} = 2$ years after the first.
Similarly the twins were born $= \frac{(12\times 6)-60}{4} = 3$ years.
After the second child and today the twins are 4 years old.
(∵ average age of the family became 16 years from 12 years)
∴ Age of eldest son = 4 + 3 + 2 = 9 years.

11. **(c)** Total annual rainfall = 2.7 × 12 = 32.4 inches
Rainfall for first seven months = (2.7 – 1.1) × 7 = 11.2
Total for first 11 months = 11.2 + 20.8 = 32 inches
Rainfall for last month = 32.4 – 32 = 0.4 inches

14-17. You have to take between 25th and 30th to mean that both these dates are also included.

12. **(a)**

	No. of family members	Average	Total
Eleven years earlier	4	28	112
Presently	if 4	39	156
	6	28	168

Since it is obvious that just after the birth of the youngest member (*i.e.*, child) was 6 family members in the family. Therefore at the time of the birth of the youngest child the elder child's age was 6 years.
Now the sum of their ages
$= x + (x + 6) = 12 = (168 - 156)$
⇒ $x = 3$

13. (d) Let there be n people (initially) in the group, then the total earning of the group = $n \times 50$
Again $n \times 50 = (n-2) \times 49 + (2x + 45)$
$\Rightarrow \quad n = 2x - 53$; where x is the lowest earning of any person.
Now, since $42 < x < 47$ and $n \in$ prime numbers
Then the only possible value of $n = 37$ for $x = 45$.

14. (b) Let a, b, c, d, the number of students in section A, B, C, D respectively
then,

$$= \frac{45(a+b+c)+55(a+c+d)+50(a+b+d)+60(b+c+d)}{3(a+b+c+d)}$$

$$= 50 + \frac{5b+10c+15d}{3(a+b+c+d)}$$

Clearly, a, b, c, d are natural no. put $a = b = c = d = 1$

Then, required average $= 50 + \frac{30}{12} = 50 + 2.5 = 52.5$

15. (a) The net decrease in the average can be expressed as

$$\frac{y+y-\left(\frac{y}{2}\right)}{3} = 1 \Rightarrow y = 2$$

16. (e) Correct average

$$= \frac{(24\times56)+(48+59+67)-(44+45+61)}{24}$$

$$= \frac{1344+174-150}{24} = \frac{1368}{24} = 57$$

17. (d) Marks scored by Ritu $= 875 \times \frac{56}{100} = 490$

Marks scored by Smita $= 875 \times \frac{92}{100} = 805$

$\therefore$ Average marks scored by all the three together

$$= \frac{490+805+634}{3} = \frac{1929}{3} = 643$$

18. (c) Let average for 14 innings be x. Then,

$$\frac{14x+58}{15} = x+3 \Rightarrow 15x+45 = 14x+58 \Rightarrow x = 13$$

$\therefore$ New average $= (x + 3) = 13 + 3 = 16$ runs

19. (b) Let the numbers be $n-2$, $n-1$, n, $n+1$ and $n+2$.
Their average = n.
Next two consecutive numbers are $n + 3$ and $n + 4$.
Therefore the average of 7 consecutive numbers

$$= \frac{(n-2)+(n-1)+n+(n+1)+(n+2)+(n+3)+(n+4)}{7}$$

$$= \frac{5n+2n+7}{7} = n+1$$

20. (d) Average weight of 30 boys = 60 kg
$\Rightarrow$ Sum of weight of 30 boys = 1800
Average weight of 10 =56 kg
$\Rightarrow$ Sum of weight of 10 boys = 560
Average weight of the whole class

$$= \frac{\text{Sum of weight of all boys}}{40}$$

$$= \frac{\text{sum of weight of 30 boys + sum of weight of 10 boys}}{40}$$

$$= \frac{60\times30+56\times10}{40} = 59\text{kg}$$

21. (b) Average $= \frac{7+14+21+28+35}{5} = 21$

22. (c) Number of boys $= \frac{4}{5} \times 50 = 40$

Number of girls $= \frac{1}{5} \times 50 = 10$

Average age of boys = $2 \times 10 = 20$
Total ages of the boys = $40 \times 20 = 800$

23. (c) Total marks of all three sections = $84 \times 100 = 8400$
total marks of (B + C) = $87.5\ (n_2 + n_3)$
total marks of A = $70 \times n_1$
$n_1 + n_2 + n_3 = 100$ ...(1)
$70n_1 + 87.5\ n_2 + 87.5\ n_3 = 8400$ ...(2)
Multiplying equation (1) by 87.5 and subtract from equation (2)
We get $17.5\ n_1 = 350$
$n_1 = 20$

CHAPTER

4

ALLIGATIONS

ALLIGATION

Alligation is the simplified, faster technique to solve the problems based on weighted average. This method plays a vital role in saving the time in solving the problems related to weighted average situation.

We know that

$$\text{Weighted Average} = \frac{\text{Sum total of all numbers of all groups}}{\text{Total number of numbers in all groups together}}$$

Therefore weighted average A_w of two groups having n_1 and n_2 numbers with averages A_1 and A_2 respectively is

$$A_w = \frac{n_1 A_1 + n_2 A_2}{n_1 + n_2}$$

$$\Rightarrow (n_1 + n_2) A_w = n_1 A_1 + n_2 A_2$$

$$\Rightarrow n_1 (A_w - A_1) = n_2 (A_2 - A_w) \Rightarrow \frac{n_1}{n_2} = \frac{A_2 - A_w}{A_w - A_1}$$

Equation $\frac{n_1}{n_2} = \frac{A_2 - A_w}{A_w - A_1}$ is called Alligation Formula.

For convenient, we take $A_1 < A_2$. Hence $A_1 < A_w < A_2$.

SOLVING THE PROBLEMS OF ALLIGATIONS USING ALLIGATION FORMULA

Illustration 1: 10 kg of wheat costing ₹ 12 per kg and 15 kg of wheat costing ₹ 20 per kg are mixed. Find the average cost of the mixture per kg.

Solution: $\frac{n_1}{n_2} = \frac{A_2 - A_w}{A_w - A_1} \Rightarrow \frac{10}{15} = \frac{20 - A_w}{A_w - 12}$

$$\Rightarrow \frac{2}{3} = \frac{20 - A_w}{A_w - 12} \Rightarrow 5A_w = 84$$

$$\Rightarrow A_w = \frac{84}{5} = 16.8$$

Hence average cost of the mixture = ₹ 16.8 per kg.

Illustration 2: A mixture worth ₹ 3.25 per kg is formed by mixing two types of salts, one costing ₹ 3.10 per kg while the other ₹ 3.60 per kg. In what ratio must they have been mixed?

Solution: $\frac{n_1}{n_2} = \frac{A_2 - A_w}{A_w - A_1} \Rightarrow \frac{n_1}{n_2} = \frac{3.60 - 3.25}{3.25 - 3.10} = \frac{35}{15}$

$\Rightarrow$ $n_1 : n_2 = 7 : 3$

Hence required ratio = 7 : 3.

GRAPHICAL REPRESENTATION OF ALLIGATION– CROSS METHOD

The alligation formula $\frac{n_1}{n_2} = \frac{A_2 - A_w}{A_w - A_1}$ is graphically represented by the following cross diagram:

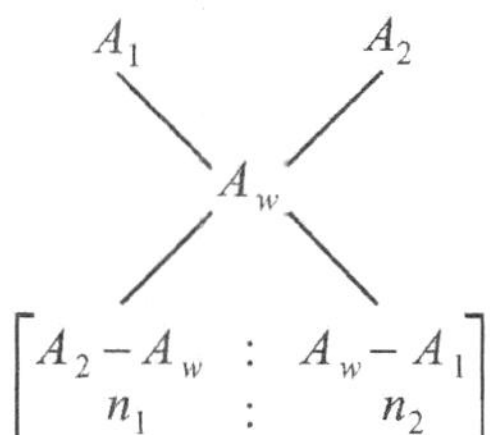

$$\begin{bmatrix} A_2 - A_w & : & A_w - A_1 \\ n_1 & : & n_2 \end{bmatrix}$$

The ratios in the bracket [ ] are equal i.e.

$n_1 : n_2 = A_2 - A_w : A_w - A_1$.

In the above graphical representation five variables A_1, A_2, A_w, n_1 and n_2 are involved.

Based on the problem situation, one of the following three cases may occur with respect to the known and the unknown out of the five variables A_1, A_2, A_w, n_1 and n_2 involved in the problem.

Case	Known	Unknown
I	(a) A_1, A_2, A_w	(a) $n_1 : n_2$
	(b) A_1, A_2, A_w, n_1	(b) n_2 and $n_1 : n_2$
II	A_1, A_2, n_1, n_2	A_w
III	A_1, A_w, n_1, n_2	A_2

Solving the problem using graphical representation of alligation is called cross method.

Let us solve some problems in each of the three cases using cross method.

Case I: When A_1, A_2, A_w are known and one of n_1 and n_2 may be also known then to find $n_1 : n_2$ and n_2 if n_1 is known OR n_1 if n_2 is known.

Illustration 3: If the average weight of the students of a class is 15kg, the average weight of the students of another class is 30kg and average weight of the students of both the classes is 25kg, then find the ratio of the number of students of the first class to the another class.

Solution:

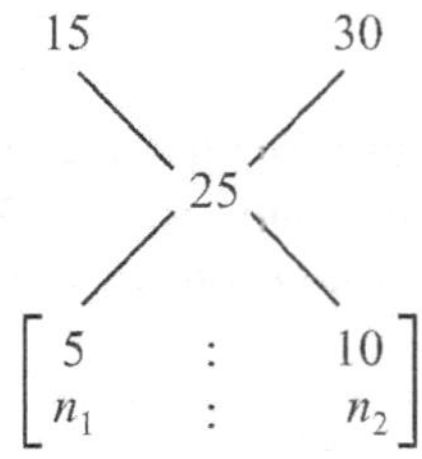

$\therefore n_1 : n_2 = 5 : 10 = 1 : 2$

Hence required ratio = 1 : 2

Case II: When A_1, A_2, n_1, n_2 are known and A_w is unknown then to find A_w.

Illustration 4: 5 kg of superior quality of sugar is mixed with 25 kg of inferior quality sugar. The price of superior quality and inferior quality sugar is ₹ 35 and ₹ 23 respectively. Find the average price per kg of the mixture.

Solution:

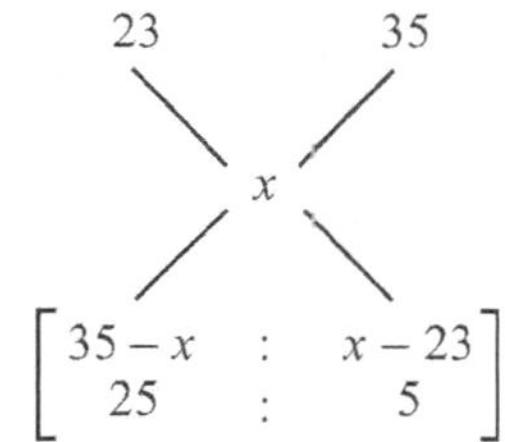

$$\therefore \quad \frac{35-x}{x-23} = \frac{25}{5}$$

$$\Rightarrow \quad 30x = 175 + 575 \Rightarrow x = \frac{750}{30} = 25$$

Hence average price per kg of the mixture = ₹ 25

Case-III: When A_1, A_w, n_1, n_2 are known and A_2 is unknown, then to find the value of A_2.

Illustration 5: The ratio of number of girls to number of boys is 1 : 2. If the average weight of the boys is 30 kg and the average weight of both the boys and girls is 25 kg, then find the average weight of the girls.

Solution:

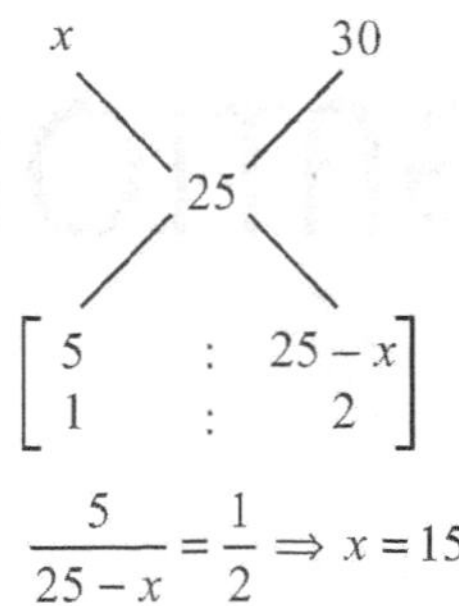

$$\frac{5}{25-x} = \frac{1}{2} \Rightarrow x = 15$$

Hence average weight of the girls = 15 kg.

THE STRAIGHT LINE APPROACH TO SOLVE THE PROBLEMS RELATED TO ALLIGATIONS

The straight line approach is actually the cross method.

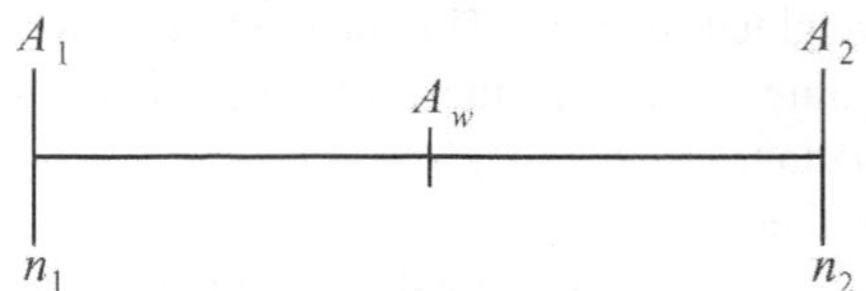

The above diagram is the straight line diagram in which the symbols A_1, A_2, A_w, n_1 and n_2 denote the same quantity as shown in cross method. Here $A_1 < A_w < A_2$.

In the above diagram,

(a) n_1 corresponds to $(A_2 - A_w)$

(b) n_2 corresponds to $(A_w - A_1)$

(c) $(n_1 + n_2)$ corresponds to $(A_2 - A_1)$

Now, we again solve the examples 3, 4 and 5 given in case-I, II and III respectively of cross method using straight line approach.

Sol. of illustration 3 by straight line approach.

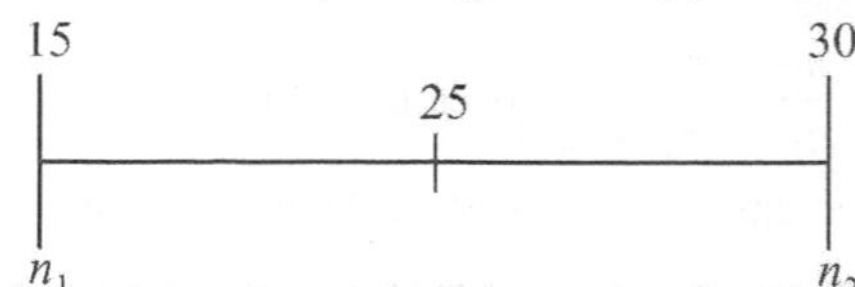

n_1 corresponds to 5 (= 30 – 25)

and n_2 corresponds to 10 (= 25 – 15)

$\therefore n_1 : n_2 = 5 : 10 = 1 : 2$

Hence required ratio = 1 : 2

Sol. of illustration 4 by straight line approach.

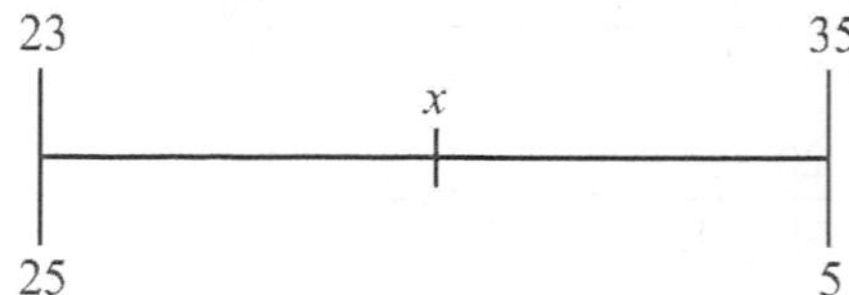

25 corresponds to $(35 - x)$

Also 5 corresponds to $(x - 23)$

$$\frac{25}{5} = \frac{35-x}{x-23}$$

$$\Rightarrow \quad x = 25$$

Hence average price per kg of mixture = ₹ 25

Sol. of illustration 5 by straight line approach.

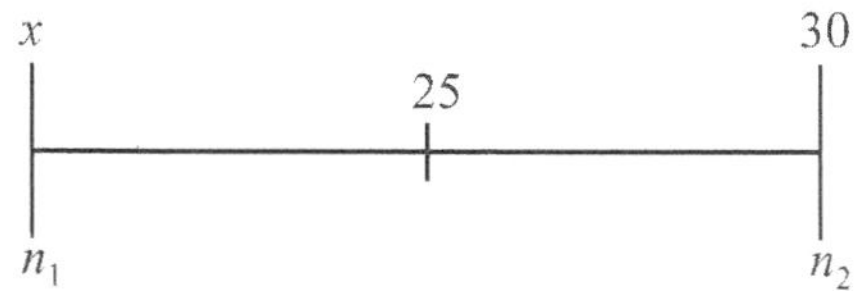

Here $n_1 : n_2 = 1 : 2$

Now, n_1 corresponds to 5 (= 30 – 25)

and n_2 corresponds to $(25 - x)$

$$\therefore \frac{n_1}{n_2} = \frac{5}{25-x} \Rightarrow \frac{1}{2} = \frac{5}{25-x} \Rightarrow x = 15$$

Hence average weight of the girls = 15 kg

RECOGNITION OF DIFFERENT SITUATIONS WHERE ALLIGATION CAN BE USED

There are many types of situations where alligation can be used, which must be recognised by the students. Here you are given some situations (or problems) which help you to recognise different alligation situations and identify A_1, A_2, n_1, n_2 and A_w in each alligation situation.

In each of the following problems

$$A_1 = 20, A_2 = 35, n_1 = 20, n_2 = 40$$

and answer as $A_w = 30$.

1. An average weight of students of a class of 40 students is 35 kg and an average weight of students of a class of 20 students is 20 kg. Find the average weight of the students of both the combined classes. (30 kg)
2. 20 litres of one variety of soda water is mixed with 40 litres of other variety of soda water. The price of first variety of soda water is ₹ 20 per litre and price of other variety of soda water is ₹ 35 per litre. Find the cost of the mixture per litre. (₹ 30)
3. A car travels at 20 km/h for 20 minutes and at 35 km/h for 40 minutes. Find the average speed of the car for the whole journey. (30 km/hr)
4. A car agency sold 20 cars at 20% profit and 40 cars at 35% profit. Find the gain percent on the sale of all these cars. (30%)
5. A trader earns a profit of 20% on 20% of his goods sold while he earns a profit of 35% on 40% of his goods sold. Find the percentage profit on whole. (30%)
6. A 40 litres mixture of water and milk contains 35% of milk and in another 20 litres of mixture of water and milk contains 20% of milk. If a new mixture is formed by mixing the both mixtures, then find the percentage of milk in new mixture. (30%)
7. A shopkeeper sold the 40% hardware at the profit of 35% and 20% software at a profit of 20%. Find the average profit% on the whole goods sold, if he sells only these two kind of things. (30%)

Some Keys to Identify A_1, A_2 & A_w and Differentiate These from n_1 and n_2

1. Normally, there are 3 averages mentioned in the problem, while there are only 2 quantities. This is not foolproof. Sometimes the question might confuse the students by giving 3 values for quantities representing n_1, n_2 and $n_1 + n_2$ respectively.
2. A_1, A_2 and A_w are always rate units, while n_1 and n_2 are quantity units.
   Rate units are like ₹ x/kg, y km/hour, etc. and corresponding quantity units are kg, hour etc.
3. The denominator of the average unit corresponds to the quantity unit (i.e., unit for n_1 and n_2).
   For example, denominator kg and hour of rate units ₹ x/kg and y km/hour are the units of quantity corresponding to rates.

A TYPICAL PROBLEM

Let's discuss the solution of a typical problem given below:

Illustration 6: A person used to draw out 20% of the honey from a jar containing 10 kg honey and replaced it with sugar solution. He has repeated the same process three times.

Find the final amount of honey left in the jar and the final ratio of honey to sugar solution finally left in the jar.

Solution: In first step: Honey drawn out 20% of 10 kg from the jar and then 2 kg sugar solution is put in the jar.

Hence after first step,

Honey remains in the jar = 10 – 20% of 10 = 10 – 2 = 8 kg and sugar solution remains in the jar = 2 kg

In second step: 20% of (8 kg honey and 2 kg sugar solution) is drawn out from the jar and then 2 kg of sugar solution is put in the jar.

⇒ 20% of 8 kg honey and 20% of 2 kg sugar solution is drawn out from the jar and then 2 kg of sugar solution is put in the container.

Thus in each step of drawing, 20% of remaining honey is drawn out.

Hence honey left in the container after second draw

= 8 – 20% of 8 = 8 – 1.6 = 6.4 kg

Honey left in the container after third draw

= 6.4 – 20% of 6.4

= 6.40 – 1.28 = 5.12 kg

Hence the final amount of the honey left in the jar = 5.12 kg

The above whole process can be shown in a single line as

10 – 20% of 10 → 8 – 20% of 8 → 6.4 – 20% of 6.4 → 5.12 kg

Now the final amount of sugar solution left in the jar

= 10 – 5.12 kg = 4.88 kg

Hence final ratio of honey to the sugar solution left in the jar

$$= \frac{5.12}{4.88} = 64 : 61.$$

Practice Exercise

Level - I

1. A mixture of certain quantity of milk with 16 litres of water is worth 90 P per litre. If pure milk be worth ₹ 1.08 per litre, how much milk is there in the mixture?
(a) 60 (b) 70
(c) 80 (d) 90

2. In my pocket there are ₹25 consisting of only the denominations of 20 paise and 50 paise. Thus there are total 80 coins in my pocket. The no. of coins of the denomination of 50 paise is
(a) 30 (b) 70
(c) 50 (d) 25

3. There are some shepherds and their sheep in a grazing field. The no. of total heads are 60 and total legs are 168 including both men and sheep. The no. of sheep is
(a) 18 (b) 26
(c) 24 (d) 36

4. If 5 kg of salt costing ₹ 5/kg and 3 kg of salt costing ₹ 4/ kg are mixed, find the average cost of the mixture per kilogram.
(a) ₹ 4.5 (b) ₹ 4.625
(c) ₹ 4.75 (d) ₹ 4.125

5. In what ratio should two qualities of coffee powder having rates of ₹ 47 per kg and ₹ 32 per kg be mixed in order to get a mixture that would have a rate of ₹ 37 per kg?
(a) 1 : 2 (b) 2 : 1
(c) 1 : 3 (d) 3 : 1

6. In what ratio should milk and water be mixed so that after selling the mixture at the cost price a profit of $16\frac{2}{3}$% is made?
(a) 1 : 2 (b) 1 : 6
(c) 2 : 3 (d) 2 : 5

7. Gold is 19 times as heavy as water and copper 9 times. In what ratio should these metals be mixed so that the mixture may be 15 times as beavy as water?
(a) 1 : 2 (b) 3 : 2
(c) 2 : 3 (d) 4 : 5

8. In a mixture of 60 litres, the ratio of milk to water is 2 : 1. If the ratio of milk to water is to be 1 : 2, then amount of water to be further added is ____________.
(a) 20 (b) 40
(c) 60 (d) 80

9. In a mixture of 45 litres, the ratio of milk and water is 4 : 1. How much water must be added to make the mixture ratio 3 : 2 ?
(a) 72 litres (b) 24 litres
(c) 15 litres (d) 1.5 litres

10. In a class of 30 students, the average weight of boys is 20 kg and the average weight of the girls is 25 kg. The fraction of boys out of the total students of the class is
(a) $\frac{4}{5}$ (b) $\frac{5}{6}$
(c) $\frac{3}{4}$ (d) Data insufficient

11. Milk and water are mixed in a vessel A in the proportion 5 : 2, and in vessel B in the proportion 8 : 5. In what proportion should quantities be taken from the two vessels so as to form a mixture in which milk and water will be in the proportion of 9 : 4?
(a) 4 : 5 (b) 5 : 7
(c) 7 : 2 (d) 7 : 9

12. A container has a capacity of 20 gallons and is full of spirit. 4 gallons of spirit is drawn out and the container is again filled with water. This process is repeated 5 times. Find out how much spirit is left in the resulting mixture finally?
(a) $6\frac{257}{525}$ gallons (b) $6\frac{346}{625}$ gallons
(c) 6.5 gallons (d) 6.25 gallons

13. A jar full of whisky contains 40% alcohol. A part of this whisky is replaced by another containing 19% alcohol and now the percentage of alcohol was found to be 26%. The quantity of whisky replaced is:
(a) $\frac{1}{3}$ (b) $\frac{2}{3}$
(c) $\frac{2}{5}$ (d) $\frac{3}{5}$

14. A dishonest grocer professes to sell pure butter at cost price, but he mixes it with adulterated fat and thereby gains 25%. Find the percentage of adulterated fat in the mixture assuming that adulterated fat is freely available.
(a) 20% (b) 25%
(c) 33.33% (d) 40%

15. A merchant purchased two qualities of pulses at the rate of ₹ 200 per quintal and ₹ 260 per quintal. In 52 quintals of the second quality, how much pulse of the first quality should be mixed so that by selling the resulting mixture at ₹ 300 per quintal, he gains a profit of 25%?
(a) 100 quintals (b) 104 quintals
(c) 26 quintals (d) None of these

16. There are two mixtures of honey and water, the quantity of honey in them being 25% and 75% of the mixture. If 2 gallons of the first are mixed with three gallons of the second, what will be the ratio of honey to water in the new mixutre?
(a) 11 : 2 (b) 11 : 9
(c) 9 : 11 (d) 2 : 11

17. Two solutions of 90% and 97% purity are mixed resulting in 21 litres of mixture of 94% purity. How much is the quantity of the first solution in the resulting mixture?
(a) 15 litres (b) 12 litres
(c) 9 litres (d) 6 litres

18. A 20 percent gain is made by selling the mixture of two types of ghee at ₹ 480 per kg. If the type costing 610 per kg was mixed with 126 kg of the other, how many kilograms of the former was mixed ?
(a) 138 kg (b) 34.5 kg
(c) 69 kg (d) Cannot be determined

19. A man makes 60 articles in the 1st hour. His efficiency decreases by 25% in the 2nd hour, increases by 40% in the 3rd hour, decreases by 33% in the 4th hour and increases by 50% in the 5th hour. If he has to work for more than 1 hour, then in which hour the average number of articles produced per hour then would be minimum ?
(a) 2nd hour (b) After 5th hour
(c) 3rd hour (d) None of these

20. There are two solutions of Sulphuric acid (acid + water) with concentration of 50% and 80% respectively. They are mixed in a certain ratio to get a 62% sulphuric acid solution. This solution is mixed with 6 liters of water to get back 50% solution. How much of the 80% solution has been used in the entire process?
(a) 15 liters (b) 12 liters
(c) 10 litres (d) None of these

21. The ratio in which a man must mix rice at ₹ 10.20 per kg and ₹ 14.40 per kg so as to make a mixture worth ₹ 12.60 per kg, is

[SSC-MT-2013]

(a) 3 : 4 (b) 4 : 3
(c) 2 : 5 (d) 18 : 24

22. Pure milk costs ₹ 16 per litre. After adding water the milkman sells the mixture ₹ 15 per litre and thereby makes a profit of 25%. In what respective ratio does he mix milk with water?

[IBPS Clerk-2013]

(a) 3 : 1 (b) 4 : 3
(c) 3 : 2 (d) 5 : 3
(e) 4 : 1

Level - II

1. 300 gm of sugar solution has 40% sugar in it. How much sugar should be added to make it 50% in the solution?
(a) 40 gm (b) 50 gm
(c) 60 gm (d) 70 gm

2. There are 65 students in a class. 39 rupees are distributed among them so that each boy gets 80 P and each girl gets 30 P. Find the number of boys and girls in that class.
(a) 45, 20 (b) 40, 25
(c) 39, 26 (d) 29, 36

3. How much water must be added to a cask which contains 40 litres of milk at cost price ₹ 3.5/litres so that the cost of milk reduces to ₹ 2/litre?
(a) 20 (b) 35
(c) 45 (d) None of these

4. A dishonest milkman professes to sell his milk at cost price but he mixes it with water and thereby gains 25%. The percentage of water in the mixture is ___________.
(a) 10% (b) 15%
(c) 20% (d) 25%

5. Jayashree purchased 150 kg of wheat of the rate of ₹ 7 per kg. She sold 50 kg at a profit of 10%. At what rate per kg should she sell the remaining to get a profit of 20% on the total deal?
(a) 6.50 (b) 8.75
(c) 7.50 (d) 9.75

6. The ratio of milk and water in 55 litres of adulterated milk is 7 : 4. How much water must be added to make the mixture's ratio 7 : 6?
(a) 5 *l* (b) 10 *l*
(c) 15 *l* (d) 25 *l*

7. From a cask full of milk, 10 litres are taken out of 50 litres and is filled with water. This was done twice. What is the quantity of milk now left in the cask?
(a) 20 litres (b) 32 litres
(c) 25 litres (d) 30 litres

8. The average weight of boys in a class is 30 kg and the average weight of girls in the same class is 20 kg. If the average weight of the whole class is 23.25 kg, what could

be the possible strength of boys and girls respectively in the same class?

(a) 14 and 26 (b) 13 and 27
(c) 17 and 27 (d) None of these

9. In what ratio should water be mixed with soda costing ₹12 per litre so as to make a profit of 25% by selling the diluted liquid at ₹13.75 per litre ?

(a) 10 : 1 (b) 11 : 1
(c) 1 : 11 (d) 12 : 1

10. Two vessels *A* and *B* of equal capacities contain mixtures of milk and water in the ratio 4 : 1 and 3 : 1, respectively. 25% of the mixture from A is taken out and added to *B*. After mixing it throughly, an equal amount is taken out from *B* and added back to *A*. The ratio of milk to water in vessel *A* after the second operation is

(a) 79 : 21 (b) 83 : 17
(c) 77 : 23 (d) 81 : 19

11. Two alloys composed of gold and silver together weight 20 kg. One lump contains 75% gold and 31.25 gm per kg silver. Another alloy contains 85% gold and 30 gm per kg silver. The total quantity of silver in two lumps is 617.5 gm. If the two lumps are melted and formed into one, what percentage of gold will it contain ?

(a) 50% (b) 89%
(c) 78% (d) 67%

12. Two vessels *A* and *B* contain spirit and water mixed in the ratio 5 : 2 and 7 : 6 respectively. Find the ratio in which these mixture be mixed to obtain a new mixture in vessel *C* containing spirit and water in the ratio 8 : 5 ?

(a) 4 : 3 (b) 3 : 4
(c) 5 : 6 (d) 7 : 9

13. Two vessels *A* and *B* contain milk and water mixed in the ratio 8 : 5 and 5 : 2 respectively. The ratio in which these two mixtures be mixed to get a new mixture containing $69\frac{3}{13}\%$ milk, is :

(a) 2 : 7 (b) 3 : 5
(c) 5 : 2 (d) 5 : 7

14. A can contains a mixture of two liquids *A* and *B* in the ratio 7 : 5. When 9 litres of mixture are drawn off and the can in filled with *B*, the ratio of *A* and *B* becomes 7 : 9. How many litres of liquid A was contained by the can initially?

(a) 10 (b) 20
(c) 21 (d) 25

15. Ram prepares solutions of alcohol in water according to customers' needs. This morning Ram has prepared 27 litres of a 12% alcohol solution and kept it ready in a 27 litre delivery container to be shipped to the customer. Just before delivery, he finds out that the customer had asked for 27 litres of 21% alcohol solution. To prepare what the customer wants, Ram replaces a portion of 12% solution by 39% solution. How many litres of 12% solution are replaced?

(a) 5 (b) 9
(c) 10 (d) 12

16. *A*, *B*, *C* subscribe together ₹ 50,000 for a business. *A* subscribes ₹4,000 more than *B* and *B* ₹ 5,000 more than *C*. Out of a total profit of ₹ 35,000, *A* receives

[*SSC CGL-2012*]

(a) ₹ 8,500 (b) ₹ 11,998
(c) ₹ 12,600 (d) ₹ 14,700

17. A vessel full of pure acid contains 10 litres of it, of which 2 litres are withdrawn. The vessel is then filled with water. Next 2 litres of the mixture are withdrawn, and again the vessel is filled up with water. The ratio of the acid left in the vessel with that of the original quantity is

[*SSC CGL-2014*]

(a) 1 : 5 (b) 4 : 5
(c) 4 : 25 (d) 16 : 25

18. Gold is 19 times as heavy as water and copper is 9 times as heavy as water. In what ratio should these be mixed to get an alloy 15 times as heavy as water? [*SSC CGL-2014*]

(a) 1 : 1 (b) 1 : 2
(c) 2 : 3 (d) 3 : 2

Hints & Solutions

Level-I

1. (c) The mean value is 90 P and the price of water is 0 P.

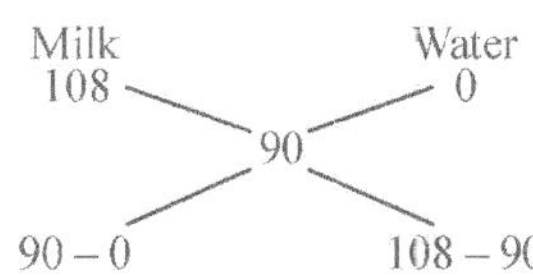

By the Alligation Rule, Milk and water are in the ratio of 5 : 1.
∴ quantity of milk in the mixture = 5 × 16 = 80 litres.

2. (a) Go through options :

$$30 \times 50 + 50 \times 20 = 2500 \text{ paise}$$

Alternatively : Since the average price of a coin

$= \frac{2500}{80} = 31.25$ paise

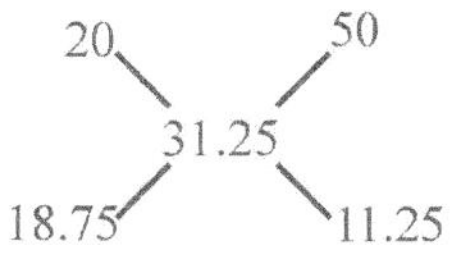

So the ratio of no. of 20 paise coins to the no. of 50 paise coins

= 18.75 : 11.25
= 75 : 45 = 5 : 3

Therefore, the no. of coins of the denominations of 50 paise is 30.

3. (c) Go through options :

$$24 \times 4 + 36 \times 2 = 168$$

Alternatively :

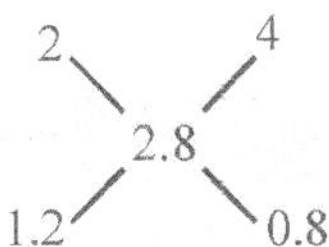

⇒ 3 : 2
Therefore, the ratio of men and sheep is 3 : 2
Alternatively : Suppose there are only men, then the no. of legs = 60 × 2 = 120.
Now since there are 48 = (168 – 120) legs extra, it means there are $24 = \left(\frac{48}{2}\right)$ sheep, since a sheep has 2 extra legs than a man has.

4. (b) Solving the following alligation figure:

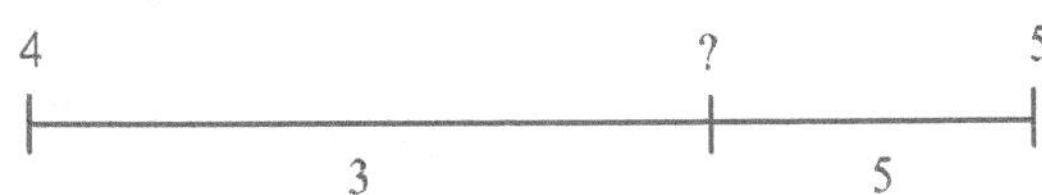

The answer would be 4.625/kg

5. (a) The ratio would be 1 : 2 as seen from the figure:

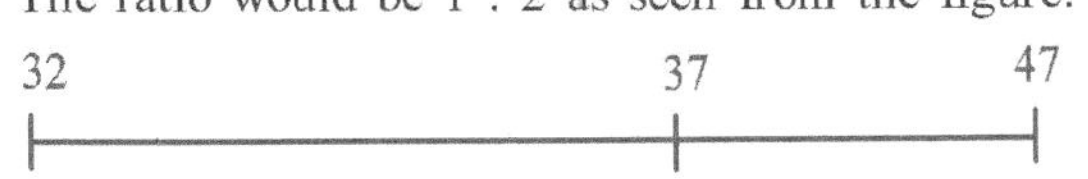

6. (b) **Short-Cut-Method :** In such questions the ratio is water : milk = $16\frac{2}{3} : 100 = 1 : 6$

7. (b)

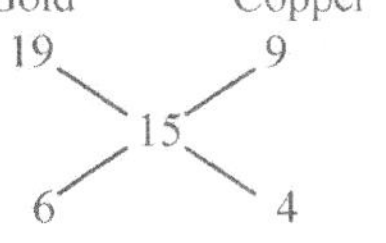

∴ Gold : Copper = 6 : 4 = 3 : 2

8. (c) Apply the alligation on fracfion of milk in each mixture.

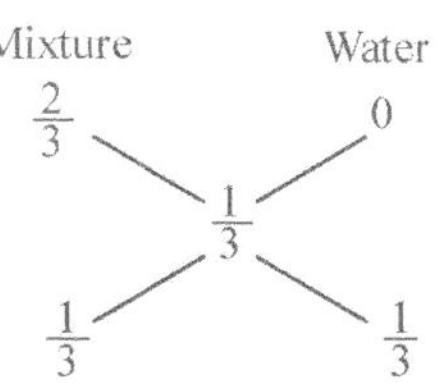

Ratio of mixture to water = 1 : 1
Therefore, if there is 60 litre of solution, 60 litres of water should be added.

9. (c) Quantity of milk = $45 \times \frac{4}{5} = 36$ litres

Quantity of water = $45 \times \frac{1}{5} = 9$ litres

Let x litres of water be added to make the ratio 3 : 2

Then, $\frac{36}{9+x} = \frac{3}{2} \Rightarrow 72 = 27 + 3x \Rightarrow x = 15$ litres

10. (d) Since we do not know either the average weight of the whole class or the ratio of no. of boys to girls.

11. (c) In vessel A, milk = $\frac{5}{7}$ of the weight of mixture

In vessel B, milk = $\frac{8}{13}$ of the weight of mixture. Now, we want to form a mixture in which milk will be $\frac{9}{13}$ of the weight of this mixture.

By alligation rule:

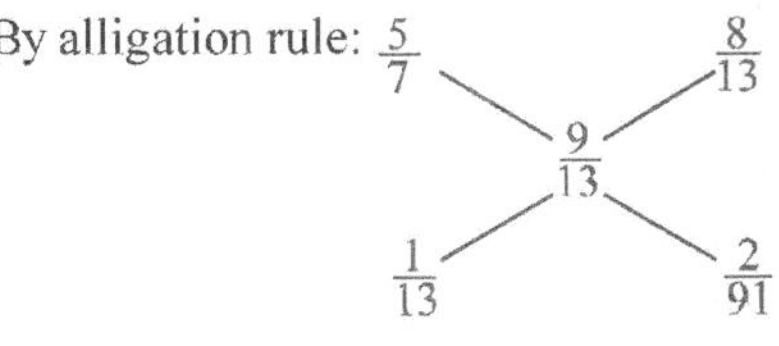

∴ required proportion is $\frac{1}{13} : \frac{2}{91} = 7 : 2$

12. (b) The amount of spirit left
= 20 × 4/5 × 4/5 × 4/5 × 4/5 × 4/5
= 4096/625 = 6 (346/625).

13. (b) By the rule of alligation, we have:

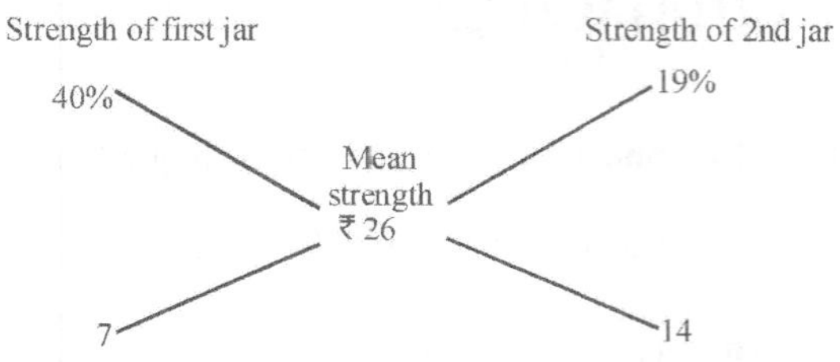

∴ So, Ratio of 1st and 2nd quantities = 7 : 14 = 1 : 2

∴ Required quantity replaced = $\frac{2}{3}$

14. (a)

Cost of adulterated fat	Average cost of the mixture to get a 25% profit by selling at CP	Cost of pure butter

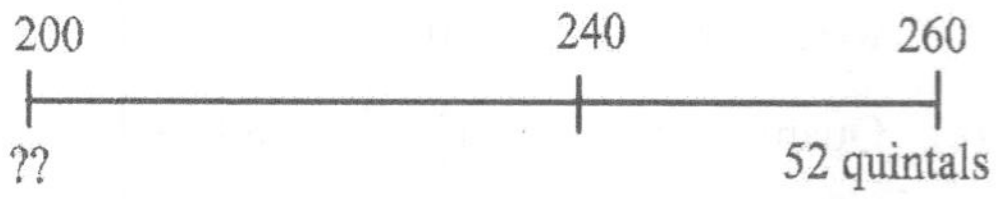

The ratio of mixing required would be 1 : 4 which means that the percentage of adulterated fat would be 20%.

15. (c) By selling at 300 if we need to get a profit of 25% it means that the cost price would be 300/1.25 = 240.

200 240 260

?? 52 quintals

Ratio of mixing required to get an average of ₹ 240 per quintal = 1 : 2

Thus, in 52 quintals of the second we need to mix 26 quintals of the first.

16. (b) The percentage of honey in the new mixture would be:
$(2 \times 25 + 3 \times 75)/5 = 275/5 = 55\%$. The ratio of honey to water in the new mixture would be
55 : 45 = 11 : 9

17. (c) 90% and 97% mixed to form 94% means that the mixing ratio is 3 : 4. The first solution would be $3 \times 21/7 = 9$ litres.

18. (d) We cannot determine the answer to this question as we do not know the price per kg of the other type of ghee. Hence, we cannot find the ratio of mixing which would be required in order to move further in this question.

19. (d) Number of articles made in 1st hour = 60
Number of articles made in 2nd hour = 45
Number of articles made in 3rd hour = 63
Number of articles made in 4th hour = 42
Number of articles made in 5th hour = 63
So, obviously articles made in 4th hour is minimum.

20. (c) Let x liters of 50% solution and y litres of 80% solutions are used

$$\frac{x}{y} = \frac{80-62}{62-50} = \frac{18}{12} = \frac{x}{y} = \frac{3}{2}$$

Solution get mixed in the ratio 3 : 2.

Now, suppose the value of acid is Z litres

$$\Rightarrow \frac{0.62z}{z+6} = \frac{1}{2}$$

$\Rightarrow$ $1.24\,Z = Z + 6$ $\Rightarrow 0.24\,Z = 6$
$Z = 25$

Hence, required rate = $\frac{2}{5} \times 25 = 10$ litres

21. (a) By the rule of alligation:

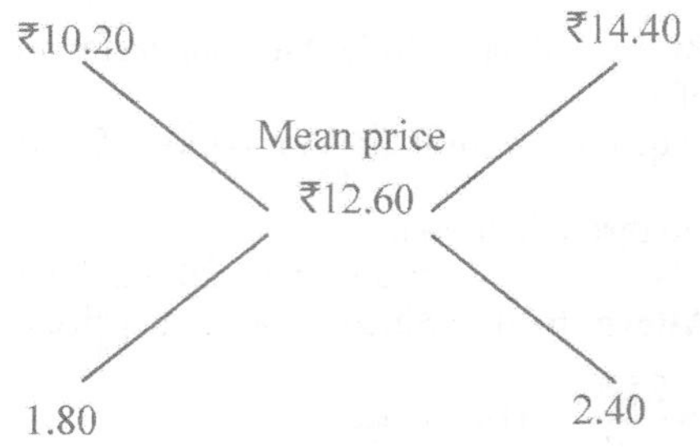

∴ Required ratio = 1.80 : 2.40 = 3 : 4.

22. (a) ∵ SP of the mixture = ₹ 15

∴ CP of the mixture = $15 \times \frac{100}{125}$ = ₹ 12

Now, by the rule of mixture,

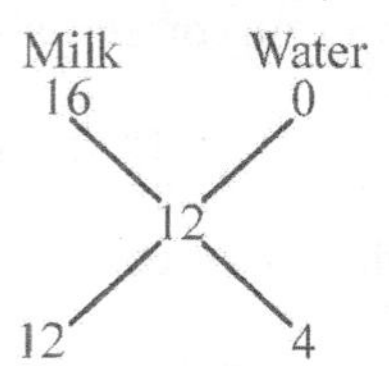

∴ Ratio of milk and water in the mixture
= 12 : 4 = 3 : 1

Level-II

1. (c) The existing solution has 40% sugar. And sugar is to be mixed; so the other solution has 100% sugar. So, by alligation method:

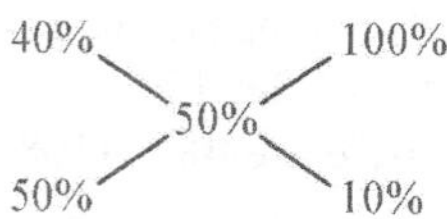

∴ The two mixtures should be added in the ratio 5 : 1.

Therefore, required sugar = $\frac{300}{5} \times 1 = 60$ gm

2. (c) Here, alligation is applicable for 'money per boy or girl".

Mean value of money per student = $\frac{3900}{65} = 60$P

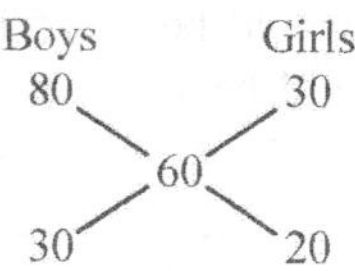

$\therefore$ Boys : Girls = 3 : 2

$\therefore$ Number of boys = $\frac{65}{3+2} \times 3 = 39$

and number of girls = 65 – 39 = 26

3. (d) This question can be solved in so many different ways. But the method of alligation method is the simplest of all the methods. We will apply the alligation on price of milk, water and mixture.

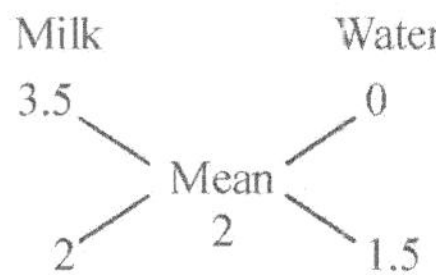

$\therefore$ ratio of milk and water should be 2 : 15 = 4 : 3

$\therefore$ added water = $\frac{40}{4} \times 3 = 30$ litres

4. (d) We will apply alligaton on % profit. If he sells the milk at CP, he gains 0%. But if he sells water at CP, he gains 100%.

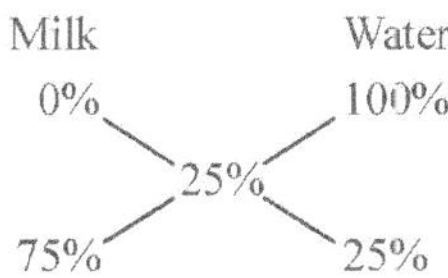

Ratio of milk to water in the mixture should be 3 : 1

$\therefore$ % of water in mixture = $\frac{1}{3+1} \times 100 = 25\%$

5. (b) Selling price of 150 kg wheat at 20% profit

$= 150 \times 7\left(\frac{120}{100}\right) =$ ₹ 1260

Selling price of 50 kg wheat at 10% profit

$= 50 \times 7\left(\frac{110}{100}\right) =$ ₹ 385

$\therefore$ Selling price per kg of remaining 100 kg wheat

$= \frac{1260 - 385}{100} =$ ₹ 8.75

6. (b) By the rule of alligation, water concentration,

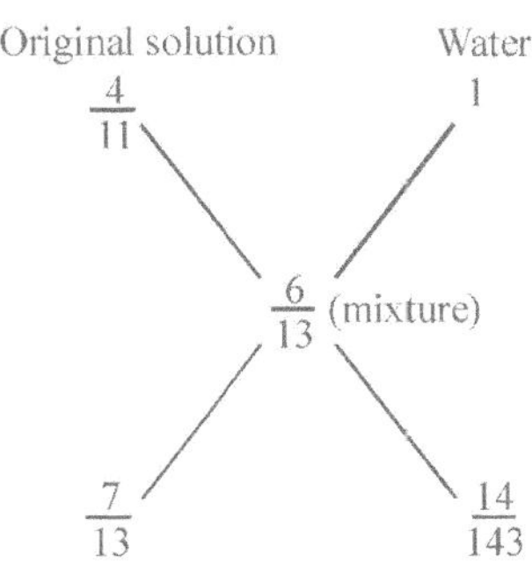

$\therefore$ water must be added to the mixture in the ratio

$\frac{14}{143} : \frac{7}{13}$ *i.e.*, 2 : 11

Quantity of water to be added $= \frac{2}{11} \times 55 = 10$ litres

7. (b) $\because$ 10 litres are withdrawn $= \frac{10}{50} = \frac{1}{5}$ of the whole

Quantity of milk after 2nd operation

$= 50\left(1 - \frac{1}{5}\right)^2 = \frac{16}{25} \times 50 = 32$ litres

8. (b) Therefore no. of boys : Number of girls = 13 : 27

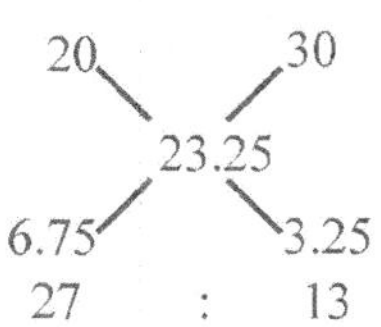

9. (c) In order to sell at a 25% profit by selling at 13.75 the cost price should be 13.75/1.25 = 11. Also since water is freely available, we can say that the ratio of water and soda must be 1:11.

10. (a) Assume there is 20 liters of the mixture in both the vessels.

In vessel *A*, milk = 16 liters and water = 4 litres

25% from *A* to *B* = milk in *B* = 15 + 4 = 19 litres

= water in *B* = 5 + 1 = 6 liters ratio = 19 : 6

Equal amount from vessel *B* to vessel *A*

= milk in $A = 12 + \frac{19}{5} = \frac{79}{5}$

= water in $A = 3 + \frac{6}{5} = \frac{21}{5}$

Hence, the ratio is 79 : 21

11. (c) Eliminating the option, we get (c) as answer because average always lies between greatest and lowest.

12. (d) Let the C.P. of spirit be ₹ 1 per litre.

Spirit in 1 litre mix. of $A = \frac{5}{7}$ litre; C.p. of a litre mix. in A = ₹ $\frac{5}{7}$.

Spirit in 1 litre mix. of $B = \frac{7}{13}$ litre; C.P. of 1 litre mix. in $B =$ ₹ $\frac{5}{13}$.

Spirit in 1 litre mix. of $C = \frac{8}{13}$ litre; Mean price = ₹ $\frac{8}{13}$.

By the rule of alligation, we have :

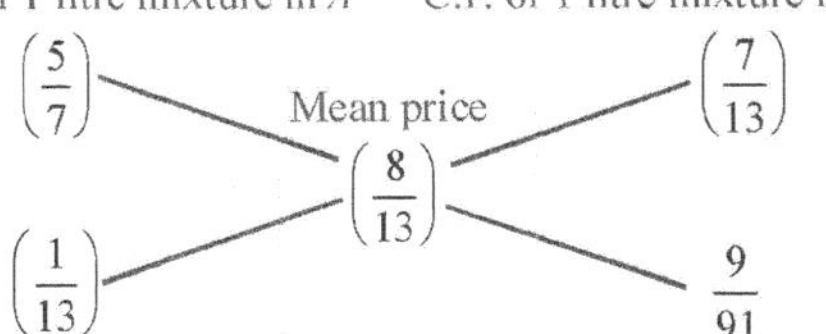

$\therefore$ Required ratio $= \frac{1}{13} : \frac{9}{13} = 7:9.$

13. **(a)** Let cost of 1 litre milk be ₹ 1.

Mlk in 1 litre mix. in $B = \frac{8}{13}$ litre, C.P. of a litre mix. in $B = ₹\ \frac{5}{7}$.

Milk in 1 litre mix. in $B = \frac{5}{7}$ litre, C.P. of 1 litre mix. in $B = ₹\ \frac{5}{7}$

Milk in 1 litre of final mix. $= \left(\frac{900}{13} \times \frac{1}{100} \times 1\right) = \frac{9}{13}$ litre;

mean price $= ₹\ \frac{9}{13}$.

By the rule of alligation, we have:

C.P. of 1 litre mixture in A C.P. of 1 litre mixture in B

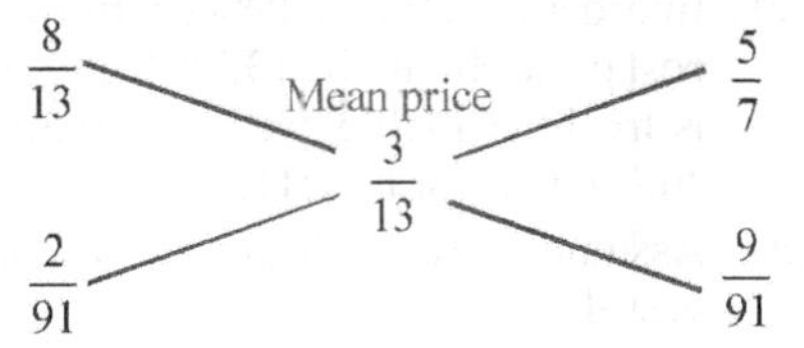

$\therefore$ Required ratio $= \frac{2}{91} : \frac{1}{13} = 2 : 7$.

14. **(c)** Suppose the can initially contains $7x$ and $5x$ litres of mixtures A and B respectively.

Quantity of A mixture left $= \left(7x - \frac{7}{12} \times 9\right)$ litres

$= \left(7x - \frac{21}{4}\right)$ litres.

Quantity of B in mixture left $= \left(5x - \frac{5}{12} \times 9\right)$ litres

$= \left(5x - \frac{15}{4}\right)$ litres.

$$\therefore \frac{\left(7x - \frac{21}{4}\right)}{\left(5x - \frac{15}{4}\right) + 9} = \frac{7}{9} \Rightarrow \frac{28x - 21}{20x + 21} = \frac{7}{9}$$

$\Rightarrow 252x - 189 = 140x + 147$

$\Rightarrow 112x = 336 \Rightarrow x = 3$.

So, the can contained 21 litre.

15. **(b)** Let Ram replaces x litres of 12% sol. with 39% solution.

Now, quality of 12% sol. in 27 litre $= \frac{27 \times 12}{100}$

$\therefore$ After replacing we have volume of 12% sol.

$$= \left(\frac{27 \times 12}{100} - \frac{12x}{100} + \frac{39x}{100}\right) = \frac{324 + 27x}{100}$$

This will be equal to 27 litre of 21% sol.

$$\therefore \frac{324 + 27x}{100} = \frac{21 \times 27}{100}$$

$$\therefore x = \frac{567 - 324}{27} = \frac{243}{27} = 9$$

Hence option (b)

16. **(d)** A = B + 4000

B = C + 5000

A + B + C = 50000

A + A – 4000 + A – 9000 = 50000

So, A = 21000

B = 17000

C = 12000

$\therefore$ A : B : C = 21000 : 17000 : 12000 = 21 : 17 : 12

A's Profit $= \frac{21}{50} \times 35000 = ₹14700$

17. **(d)** Quantity of acid left $= 10\left(1 - \frac{2}{10}\right)^2 = \frac{32}{5}$

Required ratio $= \frac{32}{5 \times 10} = \frac{16}{25} = 16 : 25$

18. **(d)** By rule of alligation, we have

Gold 19 times Copper 9 times

15 times

15 – 9 = 6 19 – 15 = 4

$\therefore$ Required ratio $= \frac{6}{4} = 3 : 2$

CHAPTER 5

PERCENTAGES

BASIC DEFINITION OF PERCENTAGE

The word per cent means per hundred or for every hundred. The symbol '%' is used for the term percent. Thus, 20 per cent is written as 20% and it means 20 out of 100.

This is written in ratio form as $\frac{20}{100}$.

The percentage value of a ratio is obtained when we multiplying the ratio by 100.

Thus percentage value of the ratio $\frac{3}{5}$ will be $\frac{3}{5}\times 100\% = 60\%$.

Illustration 1: A person saves ₹ 5,000 per month from his monthly salary ₹ 30,000. Find the percentage monthly saving of the person.

Solution: Out of monthly salary ₹ 30,000, saving is ₹ 5,000

$\Rightarrow$ Out of monthly salary ₹ 1, saving is ₹ $\frac{5,000}{30,000}$

$\Rightarrow$ Out of monthly salary ₹ 100, saving is ₹ $\frac{5,000}{30,000}\times 100$

$= ₹\ \frac{50}{3} = ₹\ 16.67$ (approx.)

Hence percentage monthly saving = 16.67% (approx.)

Illustration 2: 250 students of *ABC* school and 350 students of *XYZ* school appeared in secondary board examination conducted by CBSE in 2013. 20 students of *ABC* school and 25 students of *XYZ* school did not pass in this board examination. Students of which of the two schools *ABC* and *XYZ* have shown poor performance?

Solution: We cannot compare the performance of the students of the two schools in secondary board examination by just looking the number of students 20 of *ABC* school and 25 of *XYZ* school who did not pass in secondary board examination.

To compare the performance, you have to find the percentage of the students who did not pass the secondary board examination of each school out of those students of each school who appeared in the secondary board examination.

Percentage of the students of *ABC* school who did not pass

$$= \frac{20}{250}\times 100\% = 8\%$$

Percentage of the student of *XYZ* school who did not pass

$$= \frac{25}{350}\times 100\% = 7.1\% \text{ (approximately)}$$

Hence students of the *XYZ* school have shown poor performance.

Illustration 3: In a survey, voters of a national party *A* are increase by 2.5 lakhs and voters of national party *B* are increase by 4 lakhs in 2012. Which party *A* or *B* has grown more in 2012 ?

Solution: In first shot the answer to the question seems to be national party *B*. But actually the question can not be answered, because we don't know the just previous year's voters of each of the national party *A* and *B*.

If we had further information that in 2011, voters of national party *A* were 5 lakhs and voters of national party *B* were 10 lakhs, we can compare growth rates of two national parties.

Percentage growth rate of national party *A* in 2012

$$= \frac{250000}{500000}\times 100\% = 50\%$$

Percentage growth rate of national party *B* in 2012

$$= \frac{400000}{1000000}\times 100\% = 40\%$$

Hence, national party *A* has higher growth rate in 2012. Thus national party *A* has grown more than *B* in 2012.

In the illustrations 2 and 3, you have seen that percentage is the most powerful tool for comparing the data. 500000 and 1000000 in illustration 3 are called base values of percentage growth rate of party *A* and party *B* respectively.

Without knowing these base values, percentage growth rate of party *A* and party *B* could not be determined.

Thus percentage of anything (let X) $= \frac{\text{Value of } X}{\text{Base value of } X}\times 100$

In illustration 1, ₹ 30000 is the base value of percentage monthly saving. In illustration 2, 250 is the base value of the percentage of students of *ABC* school who did not pass and 350 is the base value of the percentage of student of *XYZ* school who did not pass.

Illustration 4: Express the following as fraction

(a) 25% **(b) $33\frac{1}{3}\%$**

Solution :

(a) $25\% = \frac{25}{100}\left(\text{Since \% means } \frac{1}{100}\right) = \frac{1}{4}$

(b) $33\frac{1}{3}\% = \frac{100}{3}\% = \frac{100}{3 \times 100} = \frac{1}{3}$

Illustration 5: 25% of a number is 80. What is the number ?

Solution:

Let the number be X. According to the given condition

$$\frac{25}{100} \times X = 80 \Rightarrow X = \frac{80 \times 100}{25} = 320.$$

Illustration 6: Express $\frac{1}{8}$ as a percentage.

Solution: $\frac{1}{8} = \frac{1}{8} \times 100\%$

$$= \frac{100}{8}\% = \frac{25}{2}\% = 12\frac{1}{2}\%$$

Illustration 7: Two third of three fifth of one eighth of a certain number is 268.50. What is 30% of the number?

(a) 1611 **(b) 1616**

(c) 1343 **(d) 594.60**

Solution: (a) Let the number be x.

According to the question $\frac{2}{3}$ of $\frac{3}{5}$ of $\frac{1}{8} \times x = 268.50$

$$\Rightarrow \frac{2}{3} \times \frac{3}{5} \times \frac{1}{8} \times x = 268.50$$

$$x = \frac{268.50 \times 3 \times 5 \times 8}{2 \times 3} = 5370$$

$$30\% \text{ of } x = \frac{30}{100} \times 5370 = 1611.00$$

Illustration 8: 4598 is 95% of ?

(a) 4800 **(b) 4840**

(c) 4850 **(d) 4880**

Solution: (b) Let 95% of $x = 4598$.

Then, $\frac{95}{100} \times x = 4598 \Rightarrow x = \left(4598 \times \frac{100}{95}\right) = 4840.$

PERCENTAGE INCREASE, PERCENTAGE DECREASE AND PERCENTAGE CHANGE

$$\text{Percentage increase} = \frac{\text{Increase}}{\text{Initial value (i.e., Base value)}} \times 100$$

$$\text{Percentage decrease} = \frac{\text{Decrease}}{\text{Initial value (i.e., Base value)}} \times 100$$

$$\text{Percentage change} = \frac{\text{Change}}{\text{Initial value (i.e., Base value)}} \times 100$$

Let income of a family in the years 2010, 2011 and 2012 are ₹ 50000, ₹ 80000 and ₹ 60000 respectively.

Here income of the family increases in 2011 but decreases in 2012.

Increase in family income in 2011 from 2010

= (Higher Income) – (Lower Income)

= (Income in 2011) – (Income in 2010)

= ₹ 80000 – ₹ 50000 = ₹ 30000

Decrease in family income in 2012 from 2011

= (Higher Income) – (Lower Income)

= (Income in 2011) – (Income in 2012)

= ₹ 80000 – ₹ 60000 = ₹ 20000

Percentage increase in family income in 2011 from 2010

$$= \frac{\left(\begin{array}{c}\text{Increase in income}\\ \text{in 2011 from 2010}\end{array}\right)}{(\text{Income in 2010})} \times 100$$

$$= \frac{30000}{50000} \times 100 = 60\%$$

Percentage decrease in family income in 2012 from 2011

$$= \frac{\left(\begin{array}{c}\text{Decrease in income}\\ \text{in 2012 from 2011}\end{array}\right)}{(\text{Income in 2011})} \times 100$$

$$= \frac{20000}{80000} \times 100 = 25\%$$

Illustration 9: Rent of the house is increased from ₹ 7000 to ₹ 7700. Express the increase in price as a percentage of the original rent.

Solution:

Increase value = ₹ 7700 – ₹ 7000 = ₹ 700

$$\text{Increase \%} = \frac{\text{Increase value}}{\text{Base value}} \times 100 = \frac{700}{7000} \times 100$$

$$= 10$$

∴ Percentage rise = 10 %.

Illustration 10: The cost of a bike last year was ₹ 19000. Its cost this year is ₹ 17000. Find the percent decrease in its cost.

Solution:

$$\% \text{ decrease} = \frac{19000 - 17000}{19000} \times 100$$

$$= \frac{2000}{19000} \times 100 = 10.5.$$

∴ Percent decrease = 10.5 %.

If the value of any thing increases, then percentage change is the percentage increase and if the value of any thing decreases, then percentage change is the percentage decrease. Thus,

Percentage change = Percentage increase, if value of any thing increases

and Percentage change = Percentage decrease, if value of anything decreases.

Practice Exercise

Level - I

1. In a public library there are 110,000 books, 40% of which are science books. It was decided to add 20,000 new books to the library. How many of the new books should be science books in order to bring the percentage of science books in the library up to 45%?
(a) 15000 (b) 1500 (c) 1450 (d) 14500
2. A batsman scored 110 runs which included 3 boundaries and 8 sixes. What percent of his total score did he make by running between the wickets ?
(a) 45% (b) $45\frac{5}{11}\%$
(c) $54\frac{6}{11}\%$ (d) 55%
3. A student secures 90%, 60% and 54% marks in test papers with 100, 150 and 200 respectively as maximum marks. The percentage of his aggregate is:
(a) 64 (b) 68
(c) 70 (d) None of these
4. If two numbers are respectively 20% and 50% of a third number, what is the percentage of the first number to the second ?
(a) 10 (b) 20 (c) 30 (d) 40
5. In an examination, 5% of the applicants were found ineligible and 85% of the eligible candidates belonged to the general category. If 4275 eligible candidates belonged to other categories, then how many candidates applied for the examination ?
(a) 30, 000 (b) 35, 000
(c) 37, 000 (d) None of these
6. Deepa decided to donate 8% of her salary to an orphanage, On the day of donation she changed her mind and donated ₹ 2240 which was 80% of what she had decided earlier. How much is Deepa's salary?
(a) ₹36000 (b) ₹42000
(c) ₹35000 (d) ₹45000
7. When the price of a radio was reduced by 20%, its sale increased by 80%. What was the net effect on the sale?
(a) 44% increase (b) 44% decrease
(c) 66% increase (d) 75% increase
8. If the price of sugar is increased by 7%, then by how much per cent should a housewife reduce her consumption of sugar, to have no extra expenditure?
(a) 7 over 107% (b) 107 over 100%
(c) 100 over 107% (d) 7%
9. A salesman's terms were changed from a flat commission of 5% on all his sales to a fixed salary of ₹ 1,000 plus 2.5% commission on all sales exceeding ₹ 4,000. If his remuneration as per the new scheme was ₹ 600 more than by the first scheme, what were his sales worth?
(a) 10,000/- (b) 11,000/- (c) 12,000/- (d) 14,000/
10. An inspector rejects 0.08% of the metres as defective. How many metres will he examine to reject 2 metres?
(a) 200m (b) 250m
(c) 2500m (d) 3000m
11. *A* invested 10% more than *B*. *B* invested 10% less than *C*. If the total sum of their investment is ₹ 14450. how much did *C* get?
(a) ₹5000 (b) ₹4800
(c) ₹5100 (d) None of these
12. A sum of ₹ 4558 is divided among *A*, *B* and *C* such that *A* receives 20% more than *C*, and *C* receives 25% less than *B*. What is A's share in the amount ?
(a) ₹1548 (b) ₹1720
(c) ₹1290 (d) ₹1345
13. In an election between two candidates, 75% of the voters cast their votes, out of which 2% of the votes were declared invalid. A candidate got 9261 votes which were 75% of total valid votes. Find the total number of votes enrolled in that election.
(a) 16080 (b) 16800
(c) 18600 (d) 16008
14. A spider climbed $62\frac{1}{2}\%$ of the height of the pole in one hour and in the next hour it covered $12\frac{1}{2}\%$ of the remaining height. If the height of the pole is 192 m, then distance climbed in second hour is:
(a) 3 m (b) 5 m
(c) 7 m (d) 9 m
15. A number is increased by 10% and then reduced by 10%. After these operations, the number:
(a) does not change (b) decreases by 1%
(c) increases by 1% (d) increases by 0.1%
16. The difference between the value of a number increased by 25% and the value of the original number decreased by 30% is 22. What is the original number ?
(a) 70 (b) 65
(c) 40 (d) 90
17. A salesman is allowed $5\frac{1}{2}\%$ discount on the total sales made by him plus a bonus of $\frac{1}{2}\%$ on the sales over ₹ 10,000. If his total earnings were ₹ 1990, then his total sales (in ₹) were:
(a) 30,000 (b) 32,000
(c) 34,000 (d) 35,000
18. If 12% of 75% is greater than 5% of a number by 75, the number is
(a) 1875 (b) 1890
(c) 1845 (d) 1860

19. Mr Yadav spends 60% of his monthly salary on consumable items and 50% of the remaining on clothes and transport. He saves the remaining amount. If his savings at the end of the year were ₹ 48456, how much amount per month would he have spent on clothes and transport?
(a) ₹4038 (b) ₹8076
(c) ₹9691.20 (d) ₹4845.60

20. In a certain year, the population of a certain town was 9000. If in the next year the population of males increases by 5% and that of the females by 8% and the total population increases to 9600, then what was the ratio of population of males and females in that given year?
(a) 4 : 5 (b) 5 : 4
(c) 2 : 3 (d) None of these

21. The salary of Raju and Ram is 20% and 30% less than the salary of Saroj respectively. By what % is the salary of Raju is more than the salary of Ram?
(a) 33.33 % (b) 50 %
(c) 15.18% (d) 14.28 %

22. Wheat is now being sold at ₹ 27 per kg. During last month its cost was ₹ 24 per kg. Find by how much per cent a family reduces its consumption so as to keep the expenditure fixed.
(a) 10.2% (b) 12.1%
(c) 12.3% (d) 11.1%

23. $33\frac{1}{3}\%$ of a man's daily output is equal to 50% of a second man's daily output. If the second man turns out 1500 screws daily, then the first man's output in terms of making screws is:
(a) 500 (b) 1000
(c) 2000 (d) 2250

24. A fraction is such that if the double of the numerator and the triple of the denominator is changed by + 10% and –30% respectively then we get 11% of 16/21. Find the fraction.
(a) $\frac{4}{25}$ (b) $\frac{2}{25}$
(c) $\frac{3}{25}$ (d) None of these

25. The entry fee in an exhinition was ₹ 10. Later this was reduced by 25%, which increased the sale of tickets by 20%. Find the percentage increase in the number of visitors.
(a) 54 (b) 57
(c) 60 (d) 66

26. An inspector rejects 0.08% of the metres as defective. How many metres will he examine to reject 2 metres?
(a) 200m (b) 250m
(c) 2500m (d) 3000m

27. If the numerator of a fraction be increased by 15% and its denominator be diminished by 8%, the value of the fraction is $\frac{15}{16}$. Find the original fraction.
(a) $\frac{1}{2}$ (b) $\frac{1}{3}$
(c) $\frac{1}{4}$ (d) $\frac{3}{4}$

28. In a class, 65% of the students are boys. On a particular day 80% of girl students were present. What was the fraction of boys who were present that day if the total number of students present that day was 70%?
(a) $\frac{2}{3}$ (b) $\frac{28}{65}$
(c) $\frac{5}{6}$ (d) $\frac{42}{65}$

29. A's income is 60% of B's income, and A's expenditure is 70% of B's expenditure. If A's income is 75% of B's expenditure, find the ratio of A's saving to B's saving.
(a) 5 : 1 (b) 1 : 5
(c) 3.5 : 1 (d) 2 : 7

30. Due to fall in manpower, the production in a factory decreases by 25%. By what per cent should the working hour be increased to restore the original production?
(a) $33\frac{1}{3}\%$ (b) 20%
(c) 25% (d) 4%

31. A number is increased by 20% and then again by 20%. By what percent should the increased number be reduced so as to get back the original number ?
(a) $19\frac{11}{31}\%$ (b) $30\frac{5}{9}\%$
(c) 40% (d) 44%

32. In the month of January, the Railway Police caught 4000 ticketless travellers. In February, the number rose by 5%. However, due to constant vigil by the Police and the total number of ticketless travellers caught in the month of April was :
(a) 3125 (b) 3255
(c) 3575 (d) 3591

33. In a mixture of milk and water the proportion of water by weight was 75%. If in the 60 gm mixture, 15 gm water was added, what would be the percentage of water ?
(a) 75% (b) 88%
(c) 90% (d) None of these

34. A shopkeeper employed a servant at a monthly salary of ₹ 1500. In addition to it, he agreed to pay him a commission of 15% on the monthly sale. How much sale in Rupees should the servant do if he wants his monthly income as ₹ 6000?
(a) ₹30000 (b) ₹415000
(c) ₹31500 (d) ₹50000

35. 720 sweets were distributed equally among children in such a way that number of sweets received by each child is 20% of the total number of children. How many sweets did each child receive ?
(a) 8 (b) 10
(c) 9 (d) 12

36. A company bought a total of 60 computers and 20 printers to modernise billing operations. If the price of each computer was three times the price of each printer then what per cent of the total cost of the purchase was the total cost of the printers ?
(a) 10% (b) 11%
(c) 15% (d) 20%

37. A city had a population of 30,00,000 in the beginning of 1999. Its average growth rate is 4% per year, but due to a massive earthquake in 2001, its population is reduced by 8% in that year. But it again maintained the same growth rate of 4% in following years. What will be the approx. population of the city at the end of 2003?
(a) 32,06,460 (b) 34,68,420
(c) 31,52,360 (d) 32,28,810

38. In a factory there are three types of machines M_1, M_2 and M_3 which produces 25%, 35% and 40% of the total products respectively. M_1, M_2 and M_3 produces 2%, 4% and 5% defective products, respectively. What is the percentage of non-defective products?
(a) 89% (b) 97.1% (c) 96.1% (d) 86.1%

39. A person saves 6% of his income. Two years later, his inocme shoots up by 15% but his savings remain the same. Find the hike in his expenditure.
(a) 15.95% (b) 15% (c) 14.8% (d) 15.5%

40. In the half yearly exam only 70% of the students were passed. Out of these (passed in half yearly) only 60% student are passed in annual exam. Out of those who did not pass the half yearly exam, 80% passed in annual exam. What percent of the students passed the annual exam?
(a) 42% (b) 56%
(c) 66% (d) None of these

41. Two vessels contain equal quantities of 40% alcohol. Anil changed the concentration of the first vessels to 50% by adding extra quantity of pure alcolol. Balu changed the concentration of the second vessels to 50% replacing a certain quantity of the solution with pure alcohol. By what percentage is the quantity of alcohol added by Anil more than that replaced by Balu?
(a) 20% (b) 25%
(c) 40% (d) Cannot be determined

42. Lagaan is levied on the 60% of the cultivated land. The revenue department collected total ₹ 3,84,4000 through the lagaan from the village of Sukhiya. Sukhiya, a very rich farmer, paid only ₹ 480 as lagaan. The percentage of total land of Sukhiya over the total taxable land of the village is:
(a) 0.15% (b) 15%
(c) 0.125% (d) None of these

43. In an election between 2 candidates, Bhiku gets 65% of the total valid votes. If the total votes were 6000, what is the number of valid votes that the other candidate Mhatre gets if 25% of the total votes were declared invalid?
(a) 1625 (b) 1575
(c) 1675 (d) 1525

44. A machine depreciates in value each year at the rate of 10% of its previous value. However, every second year there is some maintenance work so that in that particular year, depreciation is only 5% of its previous value. If at the end of the fourth year, the value of the machine stands at ₹ 1, 46, 205, then find the value of machine at the start of the first year.
(a) ₹ 1, 90, 000 (b) ₹ 2, 00, 000
(c) ₹ 1, 95, 000 (d) ₹ 2, 10, 000

45. After three successive equal percentage rise in the salary the sum of 100 rupees turned into 140 rupees and 49 paise. Find the percentage rise in the salary.
(a) 12% (b) 22%
(c) 66% (d) 82%

46. The salary of Anil and Vinit is 20% and 30% less than the salary of Dheeraj respectively. By what percentage is the salary of Anil more than the salary of Vinit?
(a) 33.33% (b) 50%
(c) 10% (d) 14.28%

47. In a certain town, at least 50% of the people read a newspaper. Among those who read a newspaper, at most 25% read more than one paper. Only one of the following statements follows from the statements given below. Which one is it?
(a) At the most 25% read exactly one newspaper
(b) At least 25% read all the newspaper
(c) At the most 37½% read exactly one newspaper
(d) At least 37½% read exactly one newspaper

48. Kajal spends 55% of her monthly income on grocery, clothes and education in the ratio of 4 : 2 : 5 respectively. If the amount spent on clothes is ₹ 5540/-, what is Kajal's monthly income? *[SBI Clerk-2012]*
(a) ₹ 55,400/- (b) ₹ 54,500/-
(c) ₹ 55,450/- (d) ₹ 55,650/-
(e) None of these

49. 35 percent of a number is two times 75 percent of another number. What is the ratio between the first and the second number respectively ? *[SBI Clerk-2012]*
(a) 35:6 (b) 31:7
(c) 23:7 (d) 32:9
(e) None of these

50. Last year there were 610 boys in a school. The number decreased by 20 percent this year. How many girls are there in the school if the number of girls is 175 percent of the total number of boys in the school this year ? *[SBI Clerk-2012]*
(a) 854 (b) 848
(c) 798 (d) 782
(e) None of these

51. Aryan got 350 marks and Vidya scored 76 percent marks in the same test. If Vidya scored 296 marks more than Aryan, what were the maximum marks of the test ?*[SBI Clerk-2012]*
(a) 650 (b) 900
(c) 850 (d) 950
(e) None of these

52. A student was awarded certain marks in an examination. However, after re-evaluation, his marks were reduced by 40% of the marks that were originally awarded to him so that the new score now became 96. How many marks did the student lose after re-evaluation ? *[SBI Clerk-2012]*
(a) 58 (b) 68
(c) 63 (d) 56
(e) 64

53. 855 candidates applied for a job, out of which 80% of the candidates were rejected. How many candidates were selected for the job ? *[SBI Clerk-2012]*
(a) 684 (b) 151
(c) 676 (d) 179
(e) None of these

54. A shopkeeper purchased 200 bulbs for ₹ 10 each. However, 5 bulbs were fused and had to be thrown away. The remaining were sold at ₹ 12 each. What will be the percentage profit ? *[SBI Clerk-2014]*
(a) 25 (b) 15
(c) 13 (d) 17
(e) None of these

55. Ajay spends 25 per cent of his salary on house rent, 5 per cent on food, 15 per cent on travel, 10 per cent on clothes and the remaining amount of ₹ 27,000 is saved. What is Ajay's income ? *[SBI Clerk-2014]*
(a) ₹60,000 (b) ₹80,500
(c) ₹60,700 (d) ₹70,500
(e) None of these

56. The salary of an employee increases every year in the month of July by 10%. If his salary in May 2000 was ₹ 15,000, his salary in October 2001 was *[SSC-Sub. Ins.-2012]*
(a) ₹ 16,500 (b) ₹18,000
(c) ₹ 18,150 (d) ₹ 19,965

57. 72% of the students of a certain class took Biology and 44% took Mathematics. If each student took Biology or Mathematics and 40 took both, the total number of students in the class was *[SSC-Sub. Ins.-2012]*
(a) 200 (b) 230
(c) 250 (d) 320

58. In a big garden 60% of the trees are coconut trees, 25% of the number of coconut trees are mango trees and 20% of the number of mango trees are apple trees. If the number of apple trees are 1500, then the number of trees in the garden is : *[SSC-Sub. Ins.-2013]*
(a) 48000 (b) 50000
(c) 51000 (c) 45000

59. If 50% of (P – Q) = 30% of (P + Q) and Q = x% of P, then the value of x is: *[SSC-Sub. Ins.-2013]*
(a) 30 (b) 25
(c) 20 (d) 50

61. In an examination 75% candidates passed in English and 60% passed in Mathematics. 25% failed in both and 240 passed the examination. Find the total number of candidates. *[SSC-Sub. Ins.-2014]*
(a) 492 (b) 300
(c) 500 (d) 400

62. If 40% of $\frac{4}{5}$ of $\frac{3}{4}$ of a number is 48, then what is 1% of the same number ? *[SSC-Sub. Ins.-2014]*
(a) 20 (b) 2
(c) 10 (d) 1

63. Two persons contested an election of Parliament. The winning candidate secured 57% of the total votes polled and won by a majority of 42,000 votes. The number of total votes polled is *[SSC-MT-2013]*
(a) 4,00,000 (b) 5,00,000
(c) 6,00,000 (d) 3,00,000

64. A number when reduced by 10% gives 30. The number is *[SSC-MT-2013]*
(a) 35 (b) $33\frac{1}{2}$
(c) $33\frac{1}{3}$ (d) 40

65. A team played 40 games in a season and won in 24 of them. What percent of games played did the team win ? *[SSC 10+2-2012]*
(a) 70% (b) 40%
(c) 60% (d) 35%

66. If 125% of x is 100, then x is : *[SSC 10+2-2012]*
(a) 80 (b) 150
(c) 400 (d) 125

67. In the annual examination Mahuya got 10% less marks than Supriyo in Mathematics. Mahuya got 81 marks. The marks of Supriyo is *[SSC 10+2-2013]*
(a) 89 (b) 90
(c) 87 (d) 88

68. Ram's income is greater than Shyam's income by 20%. Then the percent by which Shyam's income is less than Ram's income is *[SSC 10+2-2013]*
(a) $16\frac{2}{3}$ (b) $18\frac{2}{5}$
(c) $10\frac{1}{5}$ (d) $12\frac{1}{3}$

69. 1% of 1% of 25% of 1000 is *[SSC 10+2-2014]*
(a) .025 (b) .0025
(c) .25 (d) .000025

Level - II

1. A man buys a house for ₹ 100000 and rents it. He puts 12.5% of each month's rent aside for upkeep & repairs, pays ₹ 325 per year as taxes and realizes 5.5% annually on his investment. Find the monthly rent.
(a) 550 (b) 554.76
(c) 654.76 (d) 1620.45

2. When the cost of petroleum increases by 40%, a man reduces his annual consumption by 20%. Find the percentage change in his annual expenditure on petroleum.
(a) 20% (b) 16%
(c) 12% (d) 40%

3. A obtains $33\frac{1}{3}$% of the marks in a paper for which the maximum was 300. B is ahead of A by 40% of A's marks, while C is ahead of B by two-ninths of his own marks. How many marks does C get?
(a) 180 (b) 140
(c) 150 (d) 210

4. In a city, 35% of the population is composed of migrants, 20% of whom are from rural areas. Of the local population, 48% is female while this figure for rural and urban migrants is 30% and 40% respectively. If the total population of the city is 728400, what is its female population ?
(a) 324138 (b) 349680
(c) 509940 (d) None of these
5. Madan pays income tax at the rate of 10%. If his income increased by 10% and his tax rate increases to 15%. his net income would increase by ₹ 350. What is Madan's income ?
(a) ₹ 8000 (b) ₹ 10,000
(c) ₹ 12,000 (d) ₹ 14,000
6. The digit at unit place of a two-digit number is increased by 100% and the digit at ten places of the same number is increased by 50%. The new number thus formed is 19 more than the original number. What is the original number?
(a) 22 (b) 63
(c) 24 (d) None of these
7. Chunilal invests 65% in machinery, 20% in raw material and still has ₹ 1,305 cash with him. Find his total investment.
(a) ₹ 6,500 (b) ₹ 7,225
(c) ₹ 8,500 (d) None of these
8. The price of oil is increased by 25%. If the expenditure is not allowed to increase, the ratio between the reduction in consumption and the original consumption is
(a) 1 : 3 (b) 1 : 4
(c) 1 : 5 (d) 1 : 6
9. A scooter costs ₹ 25, 000 when it is brand new. At the end of each year, its value is only 80% of what it was at the beginning of the year. What is the value of the scooter at the end of 3 years?
(a) ₹ 10,000 (b) ₹ 12,500
(c) ₹ 12,800 (d) ₹ 12,000
10. If the price of sugar rises from ₹ 6 per kg to ₹ 7.50 per kg, a person, to have no increase in his expenditure on sugar, will have to reduce his consumption of sugar by
(a) 15% (b) 20%
(c) 25% (d) 30%
11. The sum of two numbers is $\frac{28}{25}$ of the first number. The second number is what percent of the first?
(a) 12% (b) 14%
(c) 16% (d) 18%
12. A positive number is by mistake divided by 6 instead of being multiplied by 6. What is the % error on the basis of correct answer?
(a) 3 (b) 97
(c) 17 (d) 83
13. p is six times as large as q. The percent that q is less than p, is :
(a) $16\frac{2}{3}$ (b) 60
(c) $83\frac{1}{3}$ (d) 90
14. 5% of income of A is equal to 15% of income of B and 10% of income of B is equal to 20% of income of C. If C's income is ₹ 2000, then the total income of A, B and C is :
(a) ₹ 6000 (b) ₹ 14,000
(c) ₹ 18,000 (d) ₹ 20,000
15. By reduction of 20% in the price of oranges, one can purchase 5 oranges more for ₹ 2.50. Find the reduced price of the oranges per dozen. Find also the original price.
(a) 120 paise, 140 paise (b) ₹ 0.8, ₹ 1.5
(c) ₹ 1.0, ₹ 1.5 (d) ₹ 1.2., ₹ 1.5
16. In a factory, producing parts of an automobile, the parts manufactured on the shop floor are required to go through quality checks, each conducted after a specific part of the processing on the raw material is completed. Only parts that are not rejected at one stage are put through subsequent stages of production and testing. If average rejection rates at these three testing stages during a month are 10%, 5% and 2% respectively, then what is the effective rejection rate for the whole plant ?
(a) 17% (b) 15.20%
(c) 84.80% (d) 16.21%
17. In some quantity of ghee, 60% is pure ghee and 40% is vanaspati. If 10 kg of pure ghee is added, then the strength of vanaspati ghee becomes 20%. The original quantity was :
(a) 10 kg (b) 15 kg
(c) 20 kg (d) 25 kg
18. The strength of a school increases and decreases every alternate year. It starts with increase by 10% and there-after the percentage of increase/decrease is the same. Which of the following is definitely true about the strength of the school in 2001 as compared to that in 1996?
(a) Increase approximately by 8%
(b) Decrease approximately by 8%
(c) Increase approximately by 20%
(d) Decrease approximately by 20%
19. $A = 10\%$ of x, $B = 10\%$ of y, $C = 10\%$ of $x + 10\%$ of y. On the basis of the above equalities, what is true in the following?
(a) A is equal to B
(b) A is greater than B
(c) B is greater than A
(d) Relation cannot be established between A and B
20. Sumitra has an average of 56% on her first 7 examinations. How much should she make on her eighth examination to obtain an average of 60% on 8 examinations?
(a) 88% (b) 78%
(c) 98% (d) Cannot be determined
21. Due to an increase of 30% in the price of eggs, 3 eggs less are availabe for ₹ 7.80. The present rate of eggs per dozen is :
(a) ₹ 8.64 (b) ₹ 8.88
(c) ₹ 9.36 (d) ₹ 10.40
22. A speaks truth in 75% and B in 80% cases. In what percentage of cases are they likely to contradict each other when narrating the same incident?
(a) 35 (b) 30
(c) 25 (d) 20

23. A's income is 60% of B's income, and A's expenditure is 70% of B's expenditure. If A's income is 75% of B's expenditure, find the ratio of A's savings to B's savings.
(a) 5 : 1 (b) 1 : 5
(c) 3.5 : 1 (d) 2 : 7

24. In a market survey, 20% opted for product *A* whereas 60% opted for product *B*. The remaining individuals were not certain. If the difference between those who opted for product *B* and those who were uncertain was 720, how many individuals were covered in the survey ?
(a) 1440 (b) 1800
(c) 3600 (d) Data inadequate

25. The income of *A* is 150% of the income of *B* and the income of *C* is 120% of the income of *A*. If the total income of *A*, *B* and *C* together is ₹ 86000, what is *C*'s income ?
(a) ₹30000 (b) ₹32000
(c) ₹20000 (d) ₹36000

26. Of the 1000 inhabitants of a town, 60% are males of whom 20% are literate. If, of all the inhabitants, 25% are literate, then what percent of the females of the town are literate ?
(a) 22.5 (b) 27.5
(c) 32.5 (d) 37.5

27. When 60% of a number A is added to another number B, B becomes 175% of its previous value. Then which of the following is true regarding the values of A and B?
(a) A > B
(b) B > A
(c) B ≥ A
(d) Either (a) or (b) can be true depending upon the values of A and B

28. In the month of January, the Railway Police caught 4000 ticketless travellers. In February, the number rise by 5%. However, due to constant vigil by the Police and the Railway staff, the number reduced by 5% and in April it further reduced by 10%. The total number of ticketless travellers caught in the month of April was:
(a) 3125 (b) 3255
(c) 3575 (d) 3591

29. The ratio of Jim's salary for October to his salary for November was 1.5 : 1.333 and the ratio of the salary for November to that for December was 2 : 2.6666. The worker got 40 rupees more for December than for October and received a bonus constituting 40 per cent of the salary for three months. Find the bonus. (Assume that the number of workdays is the same in every month.)
(a) 368.888 rupees (b) 152.5555 rupees
(c) 222.22 rupees (d) 265.6 rupees

30. King Dashratha, at his eleventh hour, called his three queens and distributed his gold in the following way: He gave 50% of his wealth to his first wife, 50% of the rest to his second wife and again 50% of the rest to his third wife. If their combined share is worth 1,30,900 kilograms of gold, find the quantity of gold King Dashratha was having initially?
(a) 1,50,000 kg (b) 1,49,600 kg
(c) 1,51,600 kg (d) 1,52,600 kg

31. After three successive raises, Aftab's salary became equal to $\frac{378}{125}$ of his initial salary. By what percentage was the salary raised the first time if the third rise was twice as high (in percentage) as the second rise was twice as high (in percentage) as the first rise?
(a) 10% (b) 15%
(c) 20% (d) 25%

32. In an assembly election at Surat, the total turnout was 80% out of which 16% of the total voters on the voting list were declared invalid. Find which of the following can be the percentage votes got by the winner of the election if the candidate who came second got 20% of the total voters on the voting list. (There were only three contestants, only one winner and the total number of voters on the voters' list was 20000.)
(a) 44.8% (b) 46.6%
(c) 48% (d) None of these

33. In the university examination last year, Rajesh scored 65% in English and 82% in History. What is the minimum percent he should score in Sociology. Which is out of 50 marks (if English and History were for 100 marks each), if he aims at getting 78% overall?
(a) 94% (b) 92%
(c) 98% (d) 96%

34. In an election of 3 candidates *A*, *B* and *C*, *A* gets 50% more votes than *B*. *A* also beats *C* by 1,80,00 votes. If it is known that B gets 5 percentage more votes than *C*, find the number of voters on the voting list (given 90% of the voters on the voting list voted and no votes were illegal)
(a) 72,000 (b) 81,000
(c) 90,000 (d) 1,00,000

35. A number is mistakenly divided by 5 instead of being multiplied by 5. Find the percentage change in the result due to this mistake.
(a) 96% (b) 95%
(c) 2400% (d) 200%

36. Ambani, a businessman, started off a business with very little capital. In the first year, he earned a profit of 50% and donated 50% of the total capital (initial capital + profit) to a charitable organisation. The same course was followed in the 2nd and 3rd years also. If at the end of three years he is left with ₹ 16, 875, then find the amount donated by him at the end of the 2nd year.
(a) ₹45, 000 (b) ₹12, 500
(c) ₹22, 500 (d) ₹20, 000

37. Arushi's project report consists of 25 pages each of 60 lines with 75 characters on each line. In case the number of lines is reduced to 55 but the number of characters is increased to 90 per lines, what is the percentage change in the number of pages. (Assume the number of pages to be a whole number.)
(a) +10% (b) +5%
(c) –8% (d) –10%

38. Recently I had gone to a locality called Shadigarh for conducting a survey about the number of married persons in the locality. The population of the locality is 7,200 and $11/18$ th of those are males and the rest females. If 40% of the males are married, find percentage of married females in the locality

(a) $48\frac{1}{7}\%$ (b) $52\frac{4}{7}\%$

(c) $62\frac{6}{7}\%$ (d) $71\frac{1}{7}\%$

39. Chintu is given a quadratic equation $ax^2 + bx + c = 0$ and is asked to make another quadratic equation from this with $a = 1$. Also one root of the second quadratic equation is same as one of the roots of the first equation but opposite in sign and the other root of the second equation is two times the second root of the first equation. Find the percentage change in the constant term of the second equation as compared to the first equation?

(a) 200% increase (b) 300% decrease

(c) 400% increase (d) 100% decrease

40. A salesgirl's terms were changed from a flat commission of 5% on all her sales to a fixed salary of ₹ 1000 plus 2.5% commission on all sales exceeding ₹4000. If her remuneration as per the new scheme was ₹600 more than that by the previous scheme, her total sales was *[SBI PO-2011]*

(a) ₹10000 (b) ₹5000

(c) ₹2000 (d) ₹12000

(e) None of these

41. Six-eleventh of a number is equal to twenty-two percent of second number. Second number is equal to the one-fourth of third number. The value of the third number is 2400. What is the 45% of first number? *[IBPS-PO-2011]*

(a) 109.8 (b) 111.7

(c) 117.6 (d) 123.4

(e) None of these

42. An HR Company employes 4800 people, out of which 45 percent are males and 60 percent of the males are either 25 years or older. How many males are employed in that HR Company who are younger than 25 years ? *[IBPS-PO-2011]*

(a) 2640 (b) 2160

(c) 1296 (d) 864

(e) None of these

43. In a test, a candidate secured 468 marks out of maximum marks 'A'. If the maximum marks 'A' were converted to 700 marks, he would have secured 336 marks. What were the maximum marks of the test? *[IBPS-PO-2011]*

(a) 775 (b) 875

(c) 975 (d) 1075

(e) None of these

44. Sum of three consecutive numbers is 2262. What is 41 % of the highest number ? *[IBPS-PO-2012]*

(a) 301.51 (b) 303.14

(c) 308.73 (d) 306.35

(e) 309.55

45. Akash scored 73 marks in subject A. He scored 56% marks in subject B and X marks in subject C. Maximum marks in each subject were 150. The overall percentage marks obtained by Akash in all the three subjects together were 54%. How many marks did he score in subject C ? *[IBPS-PO-2012]*

(a) 84 (b) 86

(c) 79 (d) 73

(e) None of these

46. In an examination, Raman scored 25 marks less than Rohit. Rohit scored 45 more marks than Sonia. Rohan scored 75 marks which is 10 more than Sonia. Ravi's score is 50 less than, maximum marks of the test. What approximate percentage of marks did Ravi score in the examination, if he gets 34 marks more than Raman? *[IBPS-PO-2013]*

(a) 90 (b) 70

(c) 80 (d) 60

(e) 85

47. Mr Giridhar spends 50% of his monthly income on household items and out of the remaining he spends 50% on transport, 25% on entertainment, 10% on sports and the remaining amount of ₹ 900 is saved. What is Mr Giridhar's monthly income? *[IBPS-PO-2013]*

(a) ₹6000 (b) ₹ 12000

(c) ₹9000 (d) Cannot be determined

(e) None of these

48. Rakesh got 273 marks in an examination and scored 5% more than the pass %. If Lokesh got 312 marks, then by what % above the pass mark did he pass the examination? *[SSC CGL-2013]*

(a) 20% (b) 27%

(c) 25% (d) 15%

49. The monthly salaries of A and B together amount to ₹ 40,000. A spends 85% of his salary and B, 95% of his salary. If now their savings are the same, then the salary (in ₹) of A is *[SSC CGL-2014]*

(a) 10,000 (b) 12,000

(c) 16,000 (d) 18,000

50. One litre of water is evaporated from 6 litres of a solution containing 5% salt. The percetange of salt in the remaining solution is *[SSC CGL-2014]*

(a) $4\frac{4}{9}\%$ (b) $5\frac{5}{7}\%$

(c) 5% (d) 6%

Hints & Solutions

Level-I

1. (d) Let X be the number of new science books. Then, Total Science books / Total books = 45%.

$$\Rightarrow \frac{\left(X+110000\times\frac{40}{100}\right)}{(20000+10000)}=\frac{45}{100} \Rightarrow X=14500.$$

2. (b) Number of runs made by running $= 110 - (3 \times 4 + 8 \times 6) = 50.$

$\therefore$ Required percentage $= \left(\frac{50}{110}\times 100\right)\%$

$$= 45\frac{5}{11}\%$$

3. (a) Total marks secured = (90% of 100 + 60% of 150 + 54% of 200)

$$= \left(\frac{90}{100}\times 100+\frac{60}{100}\times 150+\frac{54}{100}\times 200\right)$$

$= (90 + 90 + 108) = 288.$
Total maximum marks $= (100 + 150 + 200) = 450.$

$\therefore$ Aggregate Percentage $= \left(\frac{288}{450}\times 100\right)\% = 64\%$

4. (d) Let the third number be 100. Then, the first and second numbers will be 20 and 50, respectively.

Required % $= \frac{20}{50}\times 100 = 40\%$

5. (a) Let the total number of applicants be x. Number of eligible candidates = 95% of x. Eligible candidates of other categories = 15% of (95% of x)

$$= \left(\frac{15}{100}\times\frac{95}{100}\right)\times x = \frac{57}{400}x$$

$$\therefore \quad \frac{57}{400}x = 4275 \Leftrightarrow x = \left(\frac{4275\times 400}{57}\right) = 30000$$

6. (c) Let the salary of Deepa be ₹ x.
Then, 80% of 8% of x = 2240

$$\Rightarrow \quad \frac{80}{100}\times\frac{8}{100}\times x = 2240$$

$$\Rightarrow \quad x = \frac{2240\times 100\times 100}{80\times 8} = 35000$$

Hence, the salary of Deepa = ₹ 35000

7. (a) Let the original price be x and sale be of y units.
Then, the revenue collected initially $= x\times y$

Now, new price $= 0.8x$, new sale $= 1.8y$
Then, new revenue collected $= 1.44xy$

% increase in revenue $= \frac{0.44xy}{xy}\times 100$

= 44% increase

8. (a) % reduction in consumption

$$= \frac{\%\text{ change in price}}{100 + \%\text{ change in price}}\times 100$$

$$= \frac{7}{100+7}\% = \frac{7}{107}\%$$

9. (c) Let his sales be worth ₹ x. Then,
$1000 + 2.5\%\text{ of }(x-4000) = 5\%\text{ of }x + 600$

$$\Rightarrow \frac{5x}{100}-\frac{2.5(x-4000)}{100} = 1000-600$$

$$\Rightarrow 2.5x + 10000 = 40{,}000$$

$$\Rightarrow x = \frac{30{,}000}{2.5} = 12{,}000$$

10. (c) Let the inspector examined x metres, then 0.08% of $x = 2$

$$\Rightarrow \quad \frac{x\times 0.08}{100} = 2$$

or $x = \frac{200}{0.08} = 2500$ metres

11. (a) Let the investment of C = ₹ 100
Then B's investment = ₹ 90 and A's investment = ₹ 99
Sum of investment = ₹ (100 + 90 + 99) = ₹ 289

Hence, C's actual investment = ₹ $\left(\frac{14450\times 100}{289}\right)$

= ₹ 5000

12. (a) Let B get ₹ x. Then C gets = 75% of $x = \frac{3x}{4}$

and A gets = 120% of $\frac{3x}{4} = \frac{120}{100}\times\frac{3x}{4} = \frac{9x}{10}$

Now, $\frac{9x}{10}+\frac{3x}{4}+x = 4558$

$$\Rightarrow \frac{53x}{20} = 4558 \Rightarrow x = \frac{4558\times 20}{53} = 1720$$

Hence, A's share $= \frac{9x}{10} = ₹\ \frac{9\times 1720}{10} = ₹\ 1548$

13. **(b)** Let the total number of votes enrolled be x. Then, Number of votes cast = 75% of x. Valid votes = 98% of (75% of x).

$\therefore$ 75% of [98% of (75% of x)] = 9261

$$\Rightarrow \left(\frac{75}{100}\times\frac{98}{100}\times\frac{75}{100}\times x\right) = 9261$$

$$\Rightarrow x = \left(\frac{9261\times100\times100\times100}{75\times98\times75}\right) = 16800.$$

14. **(d)** Height climbed in second hour

$$= 12\frac{1}{2}\% \text{ of } \left(100-62\frac{1}{2}\right)\% \text{ of } 192 \text{ m}$$

$$= \left(\frac{25}{2}\times\frac{1}{100}\times\frac{75}{2}\times\frac{1}{100}\times192\right) m = 9 \text{ m}.$$

15. **(b)** Let the original number be 100.
Then, the new number = 100 × 1.1 × 0.9 = 99
i.e. the number decreases by 1%.

16. **(c)** Work with option, $\left(\frac{5}{4}\right)x - \left(\frac{7}{10}\right)x = 22$

Only $x = 40$ fulfil the above equation.

17. **(c)** Let the total sales be ₹ x. Then, $5\frac{1}{2}\%$ of $x + \frac{1}{2}\%$ of $(x - 10000) = 1990$

$$\Leftrightarrow \quad \frac{11}{2}\times\frac{1}{100}\times x + \frac{1}{2}\times\frac{1}{100}\times(x-10000) = 1990$$

$\Leftrightarrow \quad 12x - 10000 = 398000 \Leftrightarrow 12x = 408000$
$\Leftrightarrow x = 34000$

18. **(a)** Let the number be x,

$$\text{Then, } \frac{12}{100}\times\frac{75}{100}\times x - \frac{5}{100}\times x = 75$$

$$\Rightarrow \quad \frac{9x}{100} - \frac{5x}{100} = 75 \Rightarrow \frac{4x}{100} = 75$$

$$\Rightarrow \quad x = \frac{75\times100}{4} = 1875$$

19. **(a)** $\because$ Amount, he have spent in 1 month on clothes transport = Amount spent on saving per month
$\therefore$ Amount, spent on clothes and transport

$$= \frac{48456}{12} = ₹4038$$

20. **(a)** Let the population of males = x; then the population of females = $9000 - x$
Now, 5% of x + 8% of $(9000 - x)$
= (9600 – 9000) = 600
or $0.05x + 720 - 0.08x = 600$
or $720 - 600 = 0.08x - 0.05x$
or, $120 = 0.03x$
$\therefore$ $x = 4000$

$\therefore$ Reqd. ratio of population of males and females

$$= \frac{4000}{9000-4000} = \frac{4000}{5000} = 4:5$$

21. **(d)** Let salary of Saroj be ₹ x.

$\therefore$ Salary of Raju = $\frac{80}{100}x$

Salary of Ram = $\frac{70}{100}x$

$$\text{Required percentage} = \left(\frac{\frac{80x}{100}-\frac{70x}{100}}{\frac{70x}{100}}\right)\times100$$

$$= \frac{10x}{76x}\times100 = \frac{100}{7} = 14.28\%$$

22. **(d)** Let the family consumes 1 kg wheat
To keep expenditure at Rs. 24, its new consumption should be $\frac{24}{27} = \frac{8}{9}$ kg

$\therefore$ Percentage decrease in consumption

$$= \frac{\left(1-\frac{8}{9}\right)}{1}\times100 = 11.1\%$$

Alternative method :

$$\text{Required \%} = \frac{27-24}{27}\times100 = 11.1\%$$

23. **(d)** Let the first man's output be x.

Then, $33\frac{1}{3}\%$ of x = 50% of 1500 $\Leftrightarrow \left(\frac{100}{3}\times\frac{1}{100}\times x\right)$
$= 750 \Leftrightarrow x = 750 \times 3 = 2250.$

24. **(b)** Solve using options. 2/25 fits the requirement.

25. **(c)** 10 × 100 = 1000, 100 = no. of visitors
Now, 7.5 × No. of visitors = 1200
No. of visitors = 160

$$\text{Increase \%} = \frac{160-100}{100}\times100 = 60\%$$

26. **(c)** Let the inspector examined x metres, then 0.08% of x = 2

$$\Rightarrow \quad \frac{x\times0.08}{100} = 2$$

$$\text{or} \quad x = \frac{200}{0.08} = 2500 \text{ metres}$$

27. **(d)** Let the original fraction be $\frac{x}{y}$

Then, $\frac{115\% \text{ of } x}{92\% \text{ of } y} = \frac{15}{16} \Rightarrow \frac{115x}{92y} = \frac{15}{16}$

$\Rightarrow \quad \frac{x}{y} = \left(\frac{15}{16} \times \frac{92}{115}\right) = \frac{3}{4}$

28. **(d)** Let the class has 100 students.

$\Rightarrow$ Number of girls = 35 and number of boys = 65.

Since total number of present students = 70 and number of girls present = 80% of 35 = 28, so number of boys present = 70 – 28 = 42.

$\Rightarrow$ Required fraction = 42/65.

29. **(b)** Let 100 units be B's income and X units be B's expenditure

$\Rightarrow$ A's income = 60 units.

A's expenditure = 70X/100 units.

But 60 = 75/100 x X $\Rightarrow$ X = 80.

i.e., B's saving = (100 – 80) units = 20 units.

Hence A's saving = $60 - \frac{70}{100} \times 80$ = 4 units.

i.e., A's saving : B's saving = 4 : 20 = 1 : 5.

30. **(a)** Decrease in production is only due to decrease in manpower. Hence, manpower is decreased by 25%

Now, suppose that to restore the same production, working hours are increased by x%

Production = Manpower × Working hours

= M × W (say)

Now, M × W = (M –25% of M) × (W + x% of W)

or, $M \times W = \frac{75}{100} M \times \frac{100 + x}{100} W$

or, $100 \times 100 = 75\,(100 + x)$

or, $\frac{400}{3} = 100 + x \qquad \therefore \quad x = \frac{100}{3} = 33\frac{1}{3}\%$

31. **(b)** Let original number = 100

New number = 120% of 120% of 100

$= \left(\frac{120}{100} \times \frac{120}{100} \times 100\right) = 144.$

Decrease on 144 = 44. Decrease on 100

$= \left(\frac{44}{144} \times 100\right)\% = 30\frac{5}{9}\%$

32. **(d)** Number of ticketless travellers in April

$= 4000 \times \left(1 + \frac{5}{100}\right)\left(1 - \frac{5}{100}\right)\left(1 - \frac{10}{100}\right)$

$= \left(4000 \times \frac{21}{20} \times \frac{19}{20} \times \frac{9}{10}\right) = 3591.$

33. **(d)** Weight of water in the mixture of 60 g water

$= 60 \times \frac{75}{100} = 45\text{g}$

weight of water in the mixture of 45 g water

= 45 + 15 = 60 g

$\therefore$ Percentage of water = $\frac{60 \times 100}{75} = 80\%$

34. **(a)** Servant's commission amount

= 6000 – 1500 = ₹4500

i.e., 15% = 4500

or, $100\% = \frac{4500}{15} \times 100 =$ ₹30000

35. **(d)** Let the total number of children = x

Then, $\frac{720}{x} = 20\% \text{ of } x = \frac{20}{100} \times x = \frac{x}{5}$

$\Rightarrow \quad x^2 = 720 \times 5 = 3600$

$\Rightarrow \quad x = 60$

$\therefore$ Each child receive = $\frac{720}{60}$ = 12 sweets

36. **(a)** Suppose price of the printer = P

$\therefore$ Price of a computer = 3 P

Total cost of 60 computers = 180 P

Total cost of 20 printers = 20 P

$\therefore$ Total cost of the purchase = 200 P

Thus total cost of the printers is 10% of the total cost.

37. **(d)** Population after 2000 = 3244800

Population after 2001 = 2985216

Population at the end of 2003 = 3228810

38. **(c)** Non-defective products

$\frac{25 \times 0.98 + 35 \times 0.96 + 40 \times 0.95}{100} \times 100 = 96.1\%$

39. **(a)** On ₹ 100 he saves ₹ 6. On 115 he still saves ₹ 6. percentage increase of 15 on 94 = 15.95%

40. **(c)** **Half yearly exam**

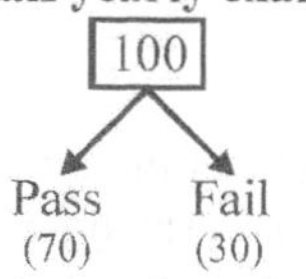

Annual exam

$\underbrace{70 \times 0.6}_{42} + \underbrace{30 \times 0.8}_{24}$

$\therefore$ Total pass in annual exam = 42 + 24 = 66

41. **(a)** Solution = 100 ml and Alcohol = 40 ml

For first vessel

$\frac{40 + x}{100 + x} = \frac{1}{2},$

so, x = 20ml

For second vessel

$$\frac{40+\frac{3}{5}y}{100+\frac{2}{5}y}=\frac{1}{2},$$

so, $y = 25$ml

Required percentage $= \frac{5}{25}\times 100 = 20\%$

42. **(d)** Total land of Sukhiya $= \frac{480x}{0.6} = 800\ x$

$\therefore$ Cultivated land of village $= 384000\,x$

$\therefore$ Required percentage $= \frac{800x}{384000}\times 100 = 0.20833.$

43. **(b)** Total votes = 6000. Valid votes = 75% of 6000 = 4500. Bhiku gets 65% of 4500 votes and Mhatre gets 35% of 4500. Hence, Mhatre gets: 0.35 × 4500 = 1575 votes.

44. **(b)** Solve using options. Checking for option (b), gives us:

$200000 \to 180000 \to 171000 \to 153900 \to 146205$

(by consecutively decreasing 200000 by 10% and 5% alternately)

45. **(a)** Solve through trial and error using the options. 12% (option (a)) is the only value that fits the situation.

46. **(d)** Salary of Dheeraj = ₹ 100

Salary of Anil = ₹ 80

Salary of Vinit = ₹ 70

Required percent $= \frac{10}{70}\times 100 = 14.28\%$

47. **(d)** Let population = 100

At least 50 people read a newspaper

At most 12.5 people read more than a newspaper

Hence, at least 37.5 people read only one newspaper.

48. **(a)** Let monthly income be y

Let money spent on grocery, clothes and education be $4x, 2x, 5x$

Money spent of clothes = ₹ 5540 = $2x$

$x = 2770$

Now $4x + 2x + 5x = 11x = 11 \times 2770$

$= 30470 = 55\%$ of y

$$y = \frac{30470\times 100}{55}$$

$y =$ ₹55,400

49. **(e)** Let the two number be x and y.

$35\%\,x = 2 \times 75\%\,y$

$35\%\,x = 150\%\,y$

$$\frac{35}{100}\times x = \frac{150}{100}\times y$$

$35x = 150y$

$$\frac{x}{y} = \frac{\cancel{150}^{30}}{\cancel{35}^{7}},\ x:y = 30:7$$

50. **(a)** No. of boys, last year = 610

20% of 610 = 122

No. of boys, current year = 610 – 122 = 488

No. of girls = 175% of 488

$= \frac{175\times 488}{100} = 854$ girls

51. **(c)** Let maximum marks of test $= x$

Vidya marks = 350 + 296 = 646 = 76% of x

$$x = \frac{646\times 100}{76} = 850$$

52. **(e)** Let initial marks of student $= x$

After Re-evaluation marks reduced by 40% of x

New score = 60% of x = 96

$$= \frac{60}{100}\times x = 96$$

$$x = \frac{96\times 100}{60}$$

$x = 160$

Marks lose = 160 – 96 = 64.

53. **(e)** No. of candidates selected for job = 20% of 855

$$= \frac{20\times 855}{100} = 171$$

54. **(d)** Total cost price = 200 × 10 = ₹ 2000

Total selling price = 12 × 195 = ₹2340

$\therefore$ Profit per cent $= \frac{2340-2000}{2000}\times 100 = 17\%$

55. **(a)** Saving percentage = (100 – 55)% = 45%

If the income of Ajay be ₹ x, then,

$$\frac{45\times x}{100} = 27000$$

$\Rightarrow x = \frac{27000\times 100}{45} =$ ₹60000

56. **(c)** Salary in May 2000 = ₹ 15000

Salary in July 2000 ⇒ 15000 + 10% of 15000 = ₹16500

Salary in October 2001 = 16500 + 10% of 16500 = ₹18150

57. **(c)** Let the total number of students in the class be x.

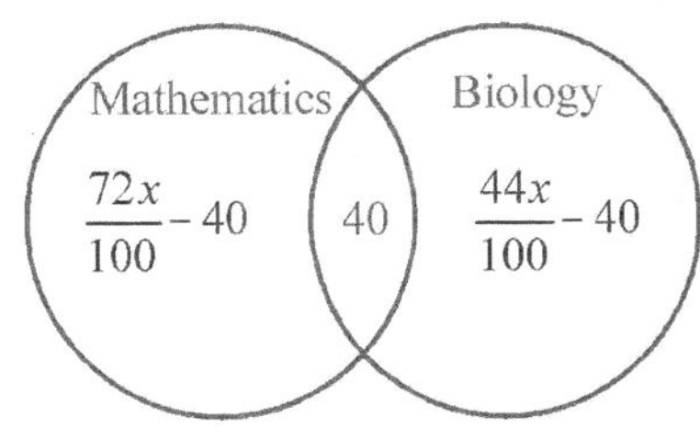

$$\because \frac{72x}{100} - 40 + 40 + \frac{44x}{100} - 40 = x$$

$$\Rightarrow \frac{72x}{100} x + \frac{44x}{100} - x = 40$$

$$\Rightarrow \frac{16x}{100} = 40 \Rightarrow x = \frac{40 \times 100}{16}$$

$$\Rightarrow x = 250$$

58. **(c)** If the number of trees in the garden be x, then

$$x \times \frac{60}{100} \times \frac{25}{100} \times \frac{20}{100} = 1500$$

$$\Rightarrow x \times \frac{3}{5} \times \frac{1}{4} \times \frac{1}{5} = 1500$$

$$\Rightarrow x = \frac{1500 \times 5 \times 4 \times 5}{3} = 50000$$

59. **(b)** $\frac{P - Q}{2} = (P + Q) \times \frac{30}{100}$

$\Rightarrow 5(P - Q) = (P + Q) \times 3$

$\Rightarrow 5P - 3P = 5Q + 3Q$

$\Rightarrow 2P = 8Q \Rightarrow P = 4Q$

$$Q = \frac{x}{100} \times P$$

$$Q = \frac{x}{100} \times 4Q$$

$$\Rightarrow \frac{4x}{100} \Rightarrow x = 25$$

60. **(d)** Let the total number of students be x.

Let A and B represent the sets of students who passed in English and Mathematics respectively.

Then, number of students passed in one or both the subjects

$= n(A \cup B) = n(A) + n(B) - n(A \cap B)$

$= 75\% \text{ of } x + 60\% \text{ of } x - (x - 25\% \text{ of } x)$

$$= \frac{3}{4}x + \frac{3}{5}x - \frac{3}{4}x = \frac{3}{5}x$$

So, $\frac{3}{5}x = 240$

$$x = \frac{240 \times 5}{3} = 400$$

61. **(b)** $\frac{40}{100} \times \frac{4}{5} \times \frac{3}{4} \times x = 48$

$$\frac{6}{25}x = 48$$

$$x = \frac{48 \times 25}{6} = 200$$

1% of 200 is 2.

62. **(d)** % of votes secured by second candidate

$= (100 - 57) = 43\%$

Let total votes polled be x.

According to question,

$(57 - 43)\% \text{ of } x = 42000$

$14\% \text{ of } x = 42000 \Rightarrow x = 3,00,000$

63. **(c)** Let the number is x.

According to question

$x - 10\% \text{ of } x = 30$

$$x - \frac{10}{100}x = 30$$

$$\left(\frac{100 - 10}{100}\right)x = 30$$

$$x = \frac{30 \times 100}{90} = 33\frac{1}{3}$$

Hence, the number is $33\frac{1}{3}$

64. **(c)** Required percentage $= \frac{24}{40} \times 100 = 60\%$

65. **(a)** $\frac{125}{100} \times x = 100$

$$\Rightarrow x = \frac{100 \times 100}{125} = 80$$

66. **(b)** Marks of Supriyo = x marks

Accoding to question

Mahuya marks = Supriyo marks – 10% of Supriyo marks

$$81 = x - 10\% \text{ of } x \Rightarrow x\left(1 - \frac{10}{100}\right)$$

$$81 = \frac{9}{10}x \Rightarrow \frac{810}{9} = x$$

$\therefore x = 90$ marks

67. **(a)** $R = S + 0.2S = 1.2S$

$$\left(\frac{R - S}{R}\right) \times 100 = \left(1 - \frac{S}{R}\right) \times 100$$

$$= \left(1 - \frac{1}{1.2}\right) \times 100$$

$$= \frac{100}{6} = \frac{50}{3} = 16\frac{2}{3}$$

68. **(a)** $\frac{1}{100} \times \frac{1}{100} \times \frac{25}{100} \times 1000 = 0.025$

Level-II

1. (b) We have 5.5% of 100000
= Rent – 12.5% of Rent – 325.

$\Rightarrow \frac{5500}{12} = \text{Rent} - \frac{\text{Rent}}{8} - \frac{325}{12}$

$\Rightarrow \frac{5500}{12} + \frac{325}{12} = \frac{7}{8} \times \text{Rent}.$

$\Rightarrow \text{Rent} = \frac{5500}{12} \times \frac{8}{7} = ₹554.76$ per month.

2. (c) **First expenditure:** Suppose 100 litres of petroleum at 100 units of money per litre, then total expenditure = 100 × 100 units of money = 10000 units of money.
Second expenditure: Now 80 litres of petroleum at 140 units of money per litre, total expenditure = 80 × 140 units of money = 11200 units.
$\Rightarrow$ Expenditure increases by

$\frac{11200-10000}{10000} \times 100 = 12\%$

Short-cut: $Exp_1 = PX$, $Exp_2 = 1.4P(0.8X) = 1.12PX$.
$\Rightarrow$ Directly we see, answer = 12%.

3. (a) A's marks $= 300 \times \frac{33\frac{1}{3}}{100} = 100.$

B's marks = 100 × (1 + 40/100) = 140.
$\Rightarrow$ C is ahead of B by 2/9 of his own marks i.e.
7/9 of C's marks = 140
$\Rightarrow$ C's marks = 140 × 9/7 = 180.

4. (d) Migrants = 35% of 728400 $= \left(\frac{35x}{100} \times 728400\right)$

= 254940.
Local population = (728400 – 254940) = 473460
Rural population = 20% of 473460 = 94692.
Urban population = (254940 – 94692) = 160248.
$\therefore$ Female population = 48% of 473460 + 30% of 94692 + 40% of 160248

$= \left(\frac{48}{100} \times 473460 + \frac{30}{100} \times 94692 + \frac{40}{100} \times 160248\right)$

= 227260.8 + 28407.6 + 64099.2 = 896660.

5. (b) Let Madan's income be ₹ x.
Then, Net income = (100 – 10)% of ₹ x

$= 90\%$ of ₹ x = ₹ $\frac{9x}{10}$.

New net income = 85% of 110% of ₹ x

$= ₹\left(\frac{85}{100} \times \frac{110}{100} \times x\right) = ₹\frac{187}{200}x$

$\therefore \quad \frac{187x}{200} - \frac{9x}{10} = 350 \Rightarrow \frac{7x}{200} = 350$

$\Rightarrow x = \left(\frac{350 \times 200}{7}\right) = 10000.$

6. (d) Working with options, we have

	Original number	New number	Difference
(a)	22	34	12
(b)	63	96	33
(c)	24	38	14

Obviously, (d) is the correct option.

7. (d) Let he had originally ₹ x. Then
65% of x + 20 % of x + 1305 = x
$0.65x + 0.2x + 1305 = x$
$\Rightarrow 0.15x = 1305 \Rightarrow x = ₹8700$
$\therefore$ His total investment = 65% of 8700 + 20% of 8700
= 85% of 8700 = ₹ 7395

8. (c) Let original consumption be 1 unit costing ₹ 100
New cost = ₹125. New consumption

$= \left(\frac{1}{125} \times 100\right) = \frac{4}{5}$ unit.

$\therefore \quad \frac{\text{Reduction in consumption}}{\text{Original consumption}} = \frac{\left(1-\frac{4}{5}\right)}{1} = \frac{1}{5}$

i.e., 1 : 5.

9. (c) After first year, the value of the scooter
= ₹ 20,000
After second year, the value of scooter = ₹ 16,000
After third year, the value of scooter = ₹ 12,800

10. (b) Let original consumption = 100 kg
and new consumption = x kg
So, 100 × 6 = x × 7.50 $\Leftrightarrow x$ = 80 kg
$\therefore$ Reduction is consumption = 20%

11. (a) Let the numbers be x and y. Then,

$x + y = \frac{28}{25}x \Rightarrow y = \frac{28}{25}x - x \Rightarrow y = \frac{3}{25}x$

$\Rightarrow \frac{y}{x} = \left(\frac{3}{25} \times 100\right)\% = 12\%.$

12. (b) Let the number be x. Then,

$\%\text{ error} = \frac{6x - x/6}{6x} \times 100 = \frac{35}{36} \times 100 = 97.2\%$

13. (c) $p = 6q$. So, q is less than p by $5q$.

$\therefore \quad$ Required percentage $= \left(\frac{5q}{p} \times 100\right)\%$

$= \left(\frac{5q}{6q} \times 100\right)\% = 83\frac{1}{3}\%$

14. (c) $\frac{5}{100}A=\frac{15}{100}B$ and $\frac{10}{100}B=\frac{20}{100}C \Rightarrow A = 3B$ and $B = 2C = 2 \times 2000 = 4000$.

$\therefore \quad A = 3 \times 4000 = 12000.$

Hence, $A + B + C = (12000 + 4000 + 2000) = 18000$.

15. (d) Let original price be ₹ x per orange. Then,

Reduced rate $= (1 - 0.2)x =$ ₹ $0.8x$

$$\therefore \frac{2.50}{0.8x}-\frac{2.50}{x}=5$$

$$\Rightarrow \frac{25}{8x}-\frac{2.5}{x}=5 \Rightarrow x=\frac{1}{8}$$

$\therefore$ Original price of oranges per dozen $\frac{1}{8}\times 12 =$ ₹ 1.5

and Reduced price = ₹ $(0.8)(1.5)$ = ₹ 1.2

16. (d) Let the total no. of parts produced at initial stage be 100. Then after three successive percentage rejections of 10%, 5% and 2%, we have

$100 \times 0.9 \times 0.95 \times 0.98 = 83.79$

Therefore, a single effective rejection

$= 100 - 83.79 = 16.21$

17. (a) Let the original quantity be x kg. Vanaspati ghee in

$$x \text{ kg} = \left(\frac{40}{100}x\right)\text{kg} = \left(\frac{2x}{5}\right)\text{kg.}$$

$$\text{Now, } \frac{\frac{2x}{5}}{x+10}=\frac{20}{100} \Leftrightarrow \frac{2x}{5x+50}=\frac{1}{5} \Leftrightarrow 5x = 50$$

$\Leftrightarrow x = 10$

18. (a) Let the strength of school was x in 1998

$\therefore$ strength in 2001 will be

$$= x\frac{110}{100}\times\frac{90}{100}\times\frac{110}{100}\times\frac{90}{100}\times\frac{110}{100}=1.07811x$$

$\therefore$ increment $= 1.07811x - x = 0.07811x$

$\therefore$ % increase $= 7.811 \approx 8\%$

19. (d) The given information gives no indication regarding the comparison of x and y.

20. (d) Since the weightage of eighth examination is not known, hence can not be determined.

21. (c) Let the original price per egg be ₹ x. Then, increased

price = ₹ $\left(\frac{130}{100}x\right)$

$$\therefore \quad \frac{7.80}{x}-\frac{7.80}{\frac{130}{100}x}=3 \Leftrightarrow \frac{7.80}{x}-\frac{780}{130x}=3$$

$\Leftrightarrow \quad 1014 - 780 = 3 \times 130x \Leftrightarrow 390x = 234$

$\Leftrightarrow \quad x = 0.6$

So, present price per dozen = ₹ $\left(12\times\frac{130}{100}\times 0.6\right)$

= ₹ 9.36.

22. (a) Let the truth spoken by A and B be p_1 and p_2 respectively, *i.e.*, $p_1=\frac{3}{4}$ and $p_2=\frac{4}{5}$

They will contradict each other only when one speaks truth and the other is lying.

i.e., $\frac{3}{4}\times\frac{1}{5}+\frac{4}{5}\times\frac{1}{4}=\frac{3}{20}+\frac{4}{20}=\frac{7}{20}=\frac{35}{100}$ *i.e.*, 35%

23. (b) Let B's Income = ₹ x

A's Income = ₹ $\frac{3}{5}x$

And B's expenditure = ₹ y

A's expenditure = ₹ $\frac{7}{10}y$

Also, $\frac{3}{5}x=\frac{3}{4}\cdot\frac{7}{10}y$

$$\frac{\text{A'savings}}{\text{B'savings}}=\frac{x-y}{\frac{3}{5}x-\frac{7}{10}y}=\frac{\frac{7}{8}y-y}{\frac{3}{5}\cdot\frac{7}{8}y-\frac{7}{10}y}=\frac{-y/8}{\frac{21y}{40}-\frac{7}{10}y}$$

$=\frac{5}{25}\approx 1:5$

24. (b) Percentage of uncertain individuals

$= [100 - (20 + 60)]\ \% = 20\%$

$\therefore \quad 60\%$ of $x - 20\%$ of $x = 720 \Leftrightarrow 40\%$ of $x = 720$

$$\Leftrightarrow \frac{40}{100}x = 720 \Leftrightarrow x=\left(\frac{720\times 100}{40}\right) = 1800.$$

25. (d) Suppose Income of B = ₹ x

Income of $A=\frac{150}{100}\times x=$ ₹ $\frac{3x}{2}$

Income of $C=\frac{120}{100}\times\frac{3x}{2}$

$\frac{6}{5}\times\frac{3x}{2}=\frac{9x}{5}$

$$\therefore \quad x+\frac{3x}{2}+\frac{9x}{5}=86000$$

$$\frac{10x+15x+18x}{10}=86000$$

$43x = 860000$

$x = 20000$

So, income of C $=\frac{9}{5}\times 20000 =$ ₹ 36000

26. (c) Number of males = 60% of 1000 = 600.

Number of females = (1000 – 600) = 400.

Number of literates = 25% of 1000 = 250.

Number of literate males = 20% of 600 = 120

Number of literate females = (250 – 120) = 130

∴ Required percentage = $\left(\frac{130}{400}\times 100\right)\% = 32.5\%$

27. (d) B + 60% of A = 175% of B → 60% of A = 75% of B.

i.e. 0.6A = 0.75B

A/B = 5/4

Apparently it seems that A is bigger, but if you consider A and B to be negative the opposite would be true.

Hence, option (d) is correct

28. (d) Number of ticketless travellers in April

$= 4000\times\left(1+\frac{5}{100}\right)\left(1-\frac{5}{100}\right)\left(1-\frac{10}{100}\right)$

$= \left(4000\times\frac{21}{20}\times\frac{19}{20}\times\frac{9}{10}\right) = 3591.$

29. (d) October : November : December = 9 : 8 : 10.666 since, he got ₹40 more in December than October, we can conclude that 1.666 = 40 → 1= 24.

Thus, total Bonus for the three months is:

0.4 × 27.666 × 24 = 265.6

30. (b) The total wealth given would be 50% + 25% (which is got by 50% of the remaining 50%) + 12.5% (which is got by 50% of the remaining 25%). Thus, the total wealth given by him would be equivalent to 87.5% of the total. Since, this is equal to 130900 kilograms of gold, the total gold would be:

130900 × 8/7 = 149600.

31. (c) $\frac{378}{125} = 3+\frac{3}{125} = 302.4\%$

Let original salary be ₹ 100

And now going through option, we get (c) as answer.

32. (d) Out of a total of 100% votes; 80% voted. 16% were invalid and 20% went to the second placed candidate. This means that the maximum the winner can get is 44%. Options a, b and c are greater than 44% and hence cannot be correct. Hence, none of these.

33. (d) Rajesh's scores in each area is 65 and 82 respectively out of 100 each. Since, the exam is of a total of 250 marks (100 + 100 + 50) he needs a total of 195 marks in order to get his target of 78% overall. Thus, he should score 195 – 65 – 82 = 195 –147 = 48 marks in Sociology which would mean 96%

34. (d) The only values that fit this situation are *C* 25%, *B* 30%, and *A* 45%. These are the percentage of votes polled. (Note: these values can be got either through trial and error or through solving $c + c + 5 + 1.5(c + 5) = 100\%$

Then, 20% is 18000 (the difference between A & C.) Hence, 90000 people must have voted and 100000 people must have been on the voter's list.

35. (a) Let the number be N. Then, 5N should be the correct outcome. But instead the value got is 0.2N. Change in value = 5N – 0.2N = 4.8N. The percentage change in the value = 4.8N × 100/5N = 96%

36. (c) 100 → 150 → 75 (yr. 1) → 112.5 → 56.25 (yr. 2) → 84.375 → 42.1875

Now, 42.1875 = ₹ 16,875

Hence, 1 → 400

Also year 2 donation is 56.25 × 400 = 22500

37. (c) Total characters in her report = 25 × 60 × 75

Let the new no. of pages be *n*

Then: $n \times 55 \times 90 = 25 \times 60 \times 75$

$n = 22.72$

This means that her report would require 23 pages. A drop of 8% in terms of the pages.

38. (c) No. of males = $\frac{11}{18}\times 7200 = 4400$

No. of males married = $\frac{40}{100}\times 4400 = 1760$

No. of females married = 1760

Required percentage = $\frac{1760}{2800}\times 100 = 62\frac{6}{7}\%$

39. (b) Let the equation be

$x^2 - 2x + 1 = 0$ ...(1)

and $x^2 - x - 2 = 0$ ...(2)

Required percentage = $\frac{1-(-2)}{1}\times 100 = 300\%$

40. (d) $\frac{5x}{100}+600 = 1000+\frac{5}{200}(x-4000)$

$\frac{5x}{100}-\frac{5x}{200} = 300$

$5x = 200\times 300$

$\Rightarrow x = 12000$

41. (e) According to question

Third number = 2400

∴ Second number = $2400\times\frac{1}{4} = 600$

Again,

First number $\times\frac{6}{11}$ = Second number $\times\frac{22}{100}$

∴ First number = $600\times\frac{22}{100}\times\frac{11}{6} = 242$

∴ 45% of the first number = $242\times\frac{45}{100} = 108.9$

42. (d) Required number = $4800\times\frac{45}{100}\times\frac{40}{100} = 864$

43. **(c)** Suppose the maximum mark of the test be x.

Then, $\frac{468\times100}{x}=\frac{336\times100}{700}$

$\therefore\ x=\frac{468\times100\times700}{336\times100}=975$

44. **(e)** Let the numbers are $x, x+1, x+2$

sum of three consecutive numbers = 2262

$x+x+1+x+2=2262$

$3x+3=2262$

$3x=2259$

$x=753$

Number are 753, 754, 755

$\therefore$ 41% of 755 = 309.55

45. **(b)** Marks is subject B = 56% of 150 = 84

Total marks obtained = 54 % of Total marks

$=\frac{54}{100}\times450$ [$\because$ Maximum marks in each subject is 150]

$=243$

Total marks obtained = A + B + C

$243=73+84+X$

$X=86$

46. **(b)** Rohan's marks = 75

Sonia's marks = 65

Rohit's marks = 65 + 45 = 110

Raman's marks = 110 – 25 = 85

Ravi got marks = 85 + 34 = 119

Total maximum marks = 119 + 50 = 169

Percentage of Ravi's mark $=\frac{119}{169}\times100\%=70.4\%=70\%$

47. **(b)** Let total monthly income of Mr. Giridhar be ₹ x.

According to question,

$\therefore\ x\times\frac{50}{100}\times\frac{15}{100}=900$

x = ₹ 12000

Hence, monthly income of Mr. Giridhar = ₹ 12000.

48. **(a)** Let passing marks be represented by p.

$p\times1.05=273$

$p=260$

Lokesh passing % $=\frac{312-260}{260}\times100=20\%$

49. **(a)** Let the monthly salary of A be x, monthly salary of B is (40000 – x).

Savings of A = (100 – 85)% of x = 0.15x

Savings of B = (100 – 95)% of (40000 – x)

= 0.05 (40000 – x)

$0.15x=0.05(40000-x)$

$\Rightarrow 0.15x+0.05x=40000\times0.05$

$\Rightarrow 0.2x=2000$

$\Rightarrow x=10000$

50. **(d)** Quantity of salt = 5% of 6l = 300 ml

Quantity of water = 6000 ml – 300 ml = 5700 ml

Quantity of water left after evaporation

= (5700 – 100) ml = 4700 ml

% of salt $=\frac{300\text{ ml}}{(4700+300)\text{ ml}}\times100=6\%$

CHAPTER 6

PROFIT, LOSS AND DISCOUNT

TOTAL COST PRICE (CP)

The total amount paid or expended in either purchasing an object (or a service) or producing an object (or a service) is known as its Total Cost Price of that object (or the service) for purchaser or producer respectively.

Total cost price is subdivided into three parts as given below:

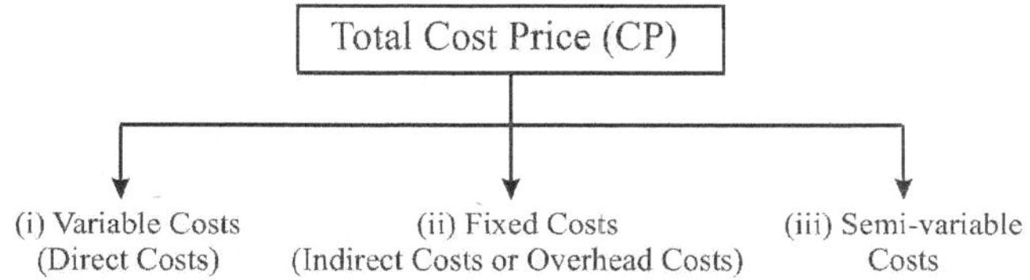

(I) Variable Costs (Direct Costs)

It is that part of the total cost that varies directly with the number of units of objects (or services) purchased or produced. For example price of raw material used in producing one unit of product. Wages to labour in producing one unit of the product when the wages are given on a piece rate basis, price per unit of an object at which a trader bought it, etc.

If price of raw material used for producing one unit of product is = ₹ 20. Then price of raw material used for producing five units of products = ₹ 20 × 5 = ₹ 100

If price of a note-book at which a trader bought it = ₹ 50

Then price of 10 note-books, which the trader paid to buy them = ₹ 50 × 10 = ₹ 500

Here, we clearly see that price of raw material varies directly with the number of product produced. Also price of note books that the trader paid to buy them varies directly with the number of note-books bought. Hence, they are the direct costs.

(II) Fixed Costs (Indirect Costs or Overhead Costs)

This is the part of the total cost which incurred irrespective of the number of items produced or purchased for sale. For example, irrespective of the number of units of a product produced or purchased for sale, the rent of the premises, salary of the employees, etc. are fixed and hence are fixed costs.

(III) Semi-variable Costs

Some costs like rent behave as fixed costs under normal circumstances but have to be increased after a certain level of production or purchase to sell, because the company needs additional space to accommodate the increased products and increased work load.

Here the rent is not fixed after a certain level therefore in the case mentioned here rent is neither completely fixed nor completely variable and hence rent is semi-variable costs.

Semi-variable costs are also a part of the total cost.

Thus

Total Cost Price = (Variable Costs) + (Fixed Costs) + (Semi-variable costs)

In most of the problems; Fixed costs and Semi-variable costs are neither given nor are to be found out for these problems,

Total Cost Price = Variable Costs Price

Total Cost Price is simply called Cost Price.

SELLING PRICE (SP)

When a person (or an agency) P sells an object or a service to another person (or agency) Q at a price S, then P is called seller, Q is called purchaser and S is called Selling Price of the seller. S is also called variable cost of the purchaser. If there is no fixed cost and semi-variable cost of the purchaser, then S is called Total Cost Price or simply called cost price of the purchaser.

PROFIT (OR GAIN) AND LOSS

(1) When selling price is more than cost price (i.e. SP > CP) then profit has been incurred.

(i)
$$\text{Profit} = \text{SP} - \text{CP}$$
$$\text{SP} = \text{CP} + \text{Profit}$$
$$\text{CP} = \text{SP} - \text{Profit}$$

(ii)
$$\text{Percentage Profit (or profit percent)} = \frac{\text{Profit}}{\text{Cost Price}} \times 100\%$$

Percentage profit means profit when cost price is ₹ 100. Percentage profit is always calculated on CP unless otherwise stated.

To understand the percentage profit clearly, suppose cost price (CP) and selling price (SP) of a book are ₹ 500 and ₹ 700

Profit = SP – CP = ₹ 700 – ₹ 500 = ₹ 200

Here, we see that, when CP is ₹ 500, then profit = ₹ 200

$\Rightarrow$ When CP will be ₹ 1, then profit = ₹ $\frac{200}{500}$

$\Rightarrow$ When CP will be ₹ 100, then profit = ₹ $\frac{200}{500} \times 100$

or percentage profit = $\frac{200}{500} \times 100\% = 40\%$

Hence, percentage profit (or profit per cent)

$$= \frac{\text{Profit}}{\text{Cost Price}} \times 100\%$$

Here profit and cost price means total profit and total cost price respectively.

$\Rightarrow$ $\boxed{\text{Profit} = \frac{\text{CP} \times \text{Profit Per cent}}{100}}$, in terms of profit percent

(iii) SP = CP + Profit

If we substitute the value of profit in term of profit percent then

$$\text{SP} = \text{CP} + \frac{\text{CP} \times \text{Profit Percent}}{100}$$

$$\Rightarrow \boxed{\text{SP} = \frac{\text{CP}\ (100 + \text{Profit Percent})}{100}}$$

$$\Rightarrow \text{CP} = \frac{\text{SP} \times 100}{(100 + \text{Profit Percent})}$$

(2) When Selling Price is less than Cost Price (i.e. SP < CP), then loss has been incurred.

(i)

Loss = CP – SP
SP = CP – Loss
CP = SP + Loss

(ii) Percentage Loss (or Loss Percent) = $\frac{\text{Loss}}{\text{CP}} \times 100$, here loss and CP means total loss and total CP respectively.

Percentage loss means loss when cost price is ₹ 100.

Percentage loss is always calculated on CP unless otherwise stated.

Loss in terms of loss percent,

$$\boxed{\text{Loss} = \frac{\text{CP} \times \text{Loss Percent}}{100}}$$

(iii) Since, SP = CP – Loss

In terms of loss percent,

$$\text{SP} = \text{CP} - \frac{\text{CP} \times \text{Loss Percent}}{100}$$

$$\Rightarrow \boxed{\text{SP} = \frac{\text{CP}\ (100 - \text{Loss Percent})}{100}}$$

$$\Rightarrow \boxed{\text{CP} = \frac{\text{SP} \times 100}{100 - \text{Loss Percent}}}$$

Illustration 1: If the cost price is 96% of the selling price, then what is the profit percent?

(a) 4.5% **(b) 4.2%**

(c) 4% **(d) 3.8%**

Solution: (b) Let S.P. = ₹ 100. Then, C.P. = ₹ 96; Profit = ₹ 4.

$\therefore$ Profit % $= \left(\frac{4}{96} \times 100\right)\% = \frac{25}{6}\% = 4.17\%. \approx 4.2\%$.

Illustration 2: Arun got ₹ 0.70 as gain over ₹ 70. Find his gain percent.

(a) 1% **(b) 0.01%**

(b) 0.1% **(d) 7%**

Solution: (a) Gain % = $\frac{0.70}{70} \times 100 = 1\%$.

Illustration 3: Vishal buys an old bike for ₹ 4700 and spends ₹ 800 on its repairs, then he sells it for ₹ 5800. Find his gain percent.

(a) 5.2% **(b) $4\frac{4}{7}\%$**

(c) 5% **(d) $5\frac{5}{11}\%$**

Solution: (d) Total C.P. for Vishal = 4700 + 800 = ₹ 5500

S.P. = ₹ 5800

$\therefore$ Gain = 5800 – 5500 = ₹ 300

Gain % = $\frac{300}{5500} \times 100 = 5\frac{5}{11}\%$.

Illustration 4: *P* buys some toffees at 6 for a rupee and sells them at 4 for a rupee. Find his gain percent.

Solution: LCM of 6 and 4 is 12

CP of 12 toffees = $\frac{1}{6} \times 12$ = ₹ 2

SP of 12 toffees = $\frac{1}{4} \times 12$ = ₹ 3

Gain = 3 – 2 = ₹ 1

$\therefore$ Gain % = $\frac{1}{2} \times 100 = 50\%$.

MARKED PRICE, LIST PRICE, DISCOUNT AND SUCCESSIVE DISCOUNTS

(i) Marked Price (MP)

In big shops and departmental stores, every article is tagged with a card and its price is written on the card. This is called the marked price of the article. Mark price of an article is the retail price, which is decided by the retail shopkeeper. So the marked price of the same article can be different on different shops.

(ii) List Price

When a manufacturer decides the retail prices of its different products, then these retail prices are either printed on the products or a list of retail prices of different products is sent to all its retail shopkeepers. Since the list price is decided by the manufacturer and not by its retail shopkeeper, therefore it is the same at all retail shops.

(iii) Discount

In order to increase the sale or clear the old stock, sometimes the shopkeepers offer a certain percentage of rebate on the marked price or list price. This rebate is known as discount. Discount is always given on marked price or list price. Hence

Selling price = (Marked price or List price) – (Discount)

Illustration 5: After allowing a discount of $\frac{15}{2}$% on marked price, an article is sold for ₹ 555. Find its M.P.

Solution: Let M.P. = ₹ 100

$$\text{Discount} = \frac{15}{2}\%$$

$$\text{S.P.} = 100 - 7.50 = 92.50$$

If SP is ₹ 92.50, then M.P. = ₹ 100

If SP is ₹ 555, then M.P. = $\frac{100}{92.50} \times 555$ = ₹ 600.

Illustration 6: A garment dealer allows his customers 10% discount on marked price of the goods and still makes a profit of 25%. Find the cost price of a shirt if it is marked at ₹ 1250.

Solution:

Marked Price (MP) = ₹ 1250,

Discount = 10%, Profit = 25%

Let Cost Price (C.P.) = ₹ 'x'

Selling Price (S.P.) = M.P. – Discount

SP = 1250 – 10 % of 1250

SP = 1250 – 125 = ₹ 1125

Now, $$\% \text{ Profit} = \frac{SP - CP}{CP} \times 100$$

$$25 = \left(\frac{1125 - x}{x}\right) \times 100$$

$$\Rightarrow \quad x = 1125 \times 4 - 4x$$

$$\Rightarrow \quad 5x = 1125 \times 4$$

$$\Rightarrow \quad x = \frac{1125 \times 4}{5} = 225 \times 4$$

$$\therefore \quad x = ₹\,900$$

C.P. = ₹ 900.

Illustration 7: What price should Neha mark on a sari which cost her ₹ 3000, so as to gain 20% after allowing a discount of 10%?

Solution:

Let Marked Price (MP) of sari = ₹ x,

Discount = 10%

C.P. = ₹ 3000, % gain = 20%

SP = MP – Discount

$$SP = x - 10\% \text{ of } x = x - \frac{10}{100}x = \frac{9x}{10}$$

$$\% \text{ gain} = \frac{SP - CP}{CP} \times 100$$

$$20 = \frac{\frac{9x}{10} - 3000}{3000} \times 100 \Rightarrow 20 \times 30 = \frac{9x}{10} - 3000$$

$$\Rightarrow \quad \frac{9x}{10} = 3600$$

$$\Rightarrow \quad x = \frac{3600 \times 10}{9} = 4000$$

∴ MP of Sari = ₹ 4000.

Successive Discounts

If two or more discounts are allowed one after the other then such discounts are known as successive discounts or discounts in series.

Suppose a discount of 15% is given, then on the reduced price a discount of 10% is also given. In such a case, we say that the successive discounts of 15% and 10% are given.

Illustration 8: Find the single discount equivalent to two successive discounts of 20% and 10%.

Solution:

$$\underset{(CP)}{100} \xrightarrow[-20]{20\% \downarrow} 80 \xrightarrow[-8]{10\% \downarrow} \underset{(SP)}{72}$$

(First discount) (Second discount)

∴ Single discount equivalent to two given successive discounts

= (100 – 72)% = 28%.

Illustration 9: M.P. of a bed is ₹ 7500. The shopkeeper allows successive discounts of 8%, 5% and 2% on it. What is the net selling price ?

Solution:

M.P. of bed = ₹ 7500

$$7500 \xrightarrow[-600]{8\% \downarrow} 6900 \xrightarrow[-345]{5\% \downarrow} 6555 \xrightarrow[-131.10]{2\% \downarrow} 6423.90$$

∴ Net selling price = ₹ 6423.90

Practice Exercise 

Level - I

1. By selling 12 marbles for a rupee, a shopkeeper loses 20%. In order to gain 20% in the transaction, he should sell the marbles at the rate of how many marbles for a rupee?
   (a) 8 (b) 6
   (c) 4 (d) 3

2. Three successive discounts of 10%, 12% and 15% amount to a single discount of:
   (a) 36.28% (b) 34.68%
   (c) 37% (d) None of these

3. A reduction of 20% in the price of sugar enables a purchaser to obtain $2\frac{1}{2}$ kg more for ₹ 160. Find the original price per kg of sugar.
   (a) ₹ 12 (b) ₹ 20
   (c) ₹ 16 (d) ₹ 18

4. Two motor cars were sold for ₹ 9,900 each, gaining 10% on one and losing 10% on the other. The gain or loss per cent in the whole transaction is :
   (a) Neither loss no gain (b) $\frac{1}{99}$% gain
   (c) $\frac{100}{99}$% profit (d) 1% loss

5. A cycle agent buys 30 bicycles, of which 8 are first grade and the rest are second grade for ₹ 3150. Find at what price he must sell the first grade bicycles so that if he sells the second grade bicycles at third quarter of the price, he may make a profit of 40% on both the types of transactions ?
   (a) ₹ 200 (b) ₹ 240
   (c) ₹ 180 (d) ₹ 210

6. A dairyman pays ₹ 6.4 per litre of milk. He adds water and sells the mixture at ₹ 8 per litre, thereby making 37.5% profit. The proportion of water to milk received by the customers is :
   (a) 1 : 5 (b) 1 : 10
   (c) 1 : 20 (d) 1 : 12

7. The cost price of 20 articles is the same as the selling price of x articles. If the profit is 25%, then the value of x is
   (a) 25 (b) 18
   (c) 16 (d) 15

8. A departmental store receives a shipment of 1,000 shirts, for which it pays ₹ 9,000. The store sells the shirts at a price 80 per cent above the cost for one month, after which it reduces the price of the shirts to 20 per cent above the cost. The store sells 750 shirts for one month and 50 per cent of the remaining shirts afterwards. How much gross income did the sales of the shirts generate ?
   (a) ₹ 10,000 (b) ₹ 10,800
   (c) ₹ 12,150 (d) ₹ 13,500

9. A company blends two varieties of tea from two different tea gardens, one variety costing ₹ 20 per kg and other ₹ 25 per kg, in the ratio 5 : 4. He sells the blended tea at ₹ 23 per kg. Find his profit per cent :
   (a) 5% profit (b) 3.5% loss
   (c) 3.5% profit (d) No profit, no loss

10. An article is listed at ₹ 65. A customer bought this article for ₹ 56.16 and got two successive discounts of which the first one is 10%. The other rate of discount of this scheme that was allowed by the shopkeeper was :
   (a) 3% (b) 4%
   (c) 6% (d) 2%

11. Three partners altogether invested ₹ 1,14,000 in a business. At the end of the year, one got ₹ 337.50, the second ₹ 1,125.00 and the third, ₹ 675 as profit. What is the percentage of profit ?
   (a) 5.8% (b) 4.8%
   (c) 1.8% (d) 3.8%

12. A shopkeepers sells an article at $12\frac{1}{2}$% loss. If he sells it for ₹ 92.50 more, then he gains 6%. What is the cost price of the article?
   (a) ₹ 510 (b) ₹ 500
   (c) ₹ 575 (d) ₹ 600

13. Ramesh purchased a bicycle for ₹ 5,200 and spent ₹ 800 on its repairs. He had to sell it for ₹ 5,500. Find his profit or loss per cent.
   (a) ₹ 844.37 (b) ₹ 488.47
   (c) ₹ 588.47 (d) None of these

14. Dhiraj purchased 150 kg of rice. He sold $\frac{1}{3}$rd of it at 10% loss. At what per cent of profit must he sell the remaining rice so that he can make 10% profit on the whole?
   (a) 20% (b) 15%
   (c) 10% (d) None of these

15. A grocer purchased 20 kg of rice at the rate of ₹ 15 per kg and 30 kg of rice at the rate of ₹ 13 per kg. At what price per kg should he sell the mixture to earn $33\frac{1}{3}$% profit on the cost price?
   (a) ₹ 28.00 (b) ₹ 20.00
   (c) ₹ 18.40 (d) ₹ 17.40

16. A builder purchased a plot of land for ₹ 80 lakh and constructed a five-storey building inclusive of ground floor on it. How much should he charge for each flat to make 25% profit on his investment on land, if there are five flats on each storey?
   (a) ₹ 50000 (b) ₹ 100000
   (c) ₹ 500000 (d) None of these

17. The difference between a discount of 35% and two successive discounts of 20% and 20% on a certain bill was ₹ 22. Find the amount of the bill.
(a) ₹ 1,100 (b) ₹ 200
(c) ₹ 2,200 (d) None of these

18. A grocer purchased 80 kg of sugar at ₹ 13.50 per kg and mixed it with 120 kg sugar at ₹ 16 per kg. At what rate should he sell the mixture to gain 16%?
(a) ₹ 17 per kg (b) ₹ 17.40 per kg
(c) ₹ 16.5 per kg (d) ₹ 16 per kg

19. A sells a tube to B at a profit of 20% and B sells it to C at profit of 25 %. If C pays ₹ 225 for it, what did A pay for it?
(a) ₹ 100 (b) ₹ 125
(c) ₹ 150 (d) ₹ 175

20. Prabhu purchased 30 kg of rice at the rate of ₹ 17.50 per kg and another 30 kg rice at a certain rate. He mixed the two and sold the entire quantity at the rate of ₹ 18.60 per kg and made 20 per cent overall profit. At what price per kg did he purchase the lot of another 30 kg rice?
(a) ₹ 14.50 (b) ₹ 12.50
(c) ₹ 15.50 (d) ₹ 13.50

21. A trader marks his goods at such a price that he can deduct 15% for cash and yet make 20% profit. Find the marked price of an item which costs him ₹ 90
(a) ₹ $135\frac{11}{13}$ (b) ₹ $105\frac{3}{21}$
(c) ₹ $127\frac{1}{17}$ (d) ₹ $95\frac{1}{21}$

22. A trader wants 10% profit on the selling price of a product whereas his expenses amount to 15% on sales. What should be his rate of mark up on an article costing ₹ 9?
(a) 20% (b) $66\frac{2}{3}\%$
(c) 30% (d) $\frac{100}{3}\%$

23. An article is listed at ₹ 65. A customer bought this article for ₹ 56.16 and got two successive discounts of which the first one is 10%. The other rate of discount of this scheme that was allowed by the shopkeeper was
(a) 3% (b) 4%
(c) 6% (d) 2%

24. The sale price of an article including the sales tax is ₹ 616. The rate of sales tax is 10%. If the shopkeeper has made a profit of 12%, then the cost price of the article is :
(a) ₹ 500 (b) ₹ 515
(c) ₹ 550 (d) ₹ 600

25. A man sold two watches for ₹ 1000 each. On one he gains 25% and on the other 20% loss. Find how much % does he gain or lose in the whole transaction?
(a) $\frac{100}{41}\%$ loss (b) $\frac{100}{41}\%$ gain
(c) No gain, no loss (d) Cannot be determined

26. The cost price of 20 articles is equal to the selling price of 25 articles. The loss percent in the transaction is
(a) 5 (b) 20
(c) 25 (d) 30

27. Rajni purchased a mobile phone and a refrigerator for ₹ 12000 and ₹ 10000 respectively. She sold the first at a loss of 12% and the second at a profit of 8%. What is her overall loss/profit?
(a) loss of ₹ 280 (b) profit of ₹ 2160
(c) loss of ₹ 240 (d) None of these

28. A property dealer sells a house for ₹ 6,30,000 and in the bargain makes a profit of 5%. Had he sold it for ₹ 5,00,000, then what percentage of loss or gain he would have made?
(a) $2\frac{1}{4}\%$ gain (b) 10% loss
(c) $12\frac{1}{2}\%$ loss (d) $16\frac{2}{3}\%$ loss

29. A manufacturer sells a car to a dealer at a profit of 50%, the dealer sells it to a customer at a profit fo 20% and the customer sells it to a friend for ₹ 288000 at a loss of 20%. Find the cost of manufacturer.
(a) 200000 (b) 300000
(c) 400000 (d) 50000

30. A dishonest dealer professes to sell his goods at cost price, but he uses a weight of 960 gm for the kg weight. Find his gain percent.
(a) 2.8% (b) $4\frac{1}{6}\%$
(c) 4.16% (d) $3\frac{1}{3}\%$

31. A shopkeeper sold an article offering a disount of 5% and earned a profit of 23.5%. What would have been the percentage of profit earned if no discount was offered?
(a) 24.5 (b) 28.5
(c) 30 (d) None of these

32. A man sells an article at 5% profit. If he had bought it at 5% less and sold if for ₹ 1 less, he would have gained 10%. Find the cost price.
(a) 100 (b) 150
(c) 200 (d) 250

33. It is known that the shopkeeper takes a discount of 10% from his supplier and he disregards this discount while marking up (*i.e.*, he marks up at the undiscounted price), find the percentage profit for the shopkeeper if there is no other change from the previous problem.
(a) 32% (b) 36.66%
(c) 40.33% (d) 46.66%

34. A shopkeeper marks up his goods by 40% and gives a discount of 10%. Apart from this, he uses a faulty balance also, which reads 1000 gm for 800 gm. What is his net profit percentage?
(a) 57.5% (b) 63.5%
(c) 42.5% (d) 36.5%

35. A supplier sells 20 pencils at the marked price of 16 pens to a retailer. The retailer, in turn, sells them at the marked price. What is the percentage profit or percentage loss of the retailer?
(a) Loss 25% (b) Profit 25%
(c) Loss 20% (d) Profit 20%

36. A milkman defrauds by means of a false measure to the tune of 20% in buying and also defrauds to the tune of 25% in selling. Find his overall % gain.
(a) 15% (b) 30%
(c) 50% (d) 45%

37. A businessman, while selling 20 articles. loses the cost price of 5 articles. Had he purchased the 20 articles for 25% less and sold them for $33\frac{1}{3}$% more than the original selling price, what is his gain?
(a) 5% (b) 75%
(c) $33\frac{1}{3}$% (d) 45%

38. $\frac{2}{3}$ of a consignment was sold at 6 % profit and the rest at a loss of 3 %. If there was an overall profit of ₹ 540, find the value of the consignment.
(a) ₹ 15,000 (b) ₹ 18000
(c) ₹ 35000 (d) ₹ 45000

39. The ratio between the sale price and the cost price of an article is 7 : 5. What is the ratio between the profit and the cost price of that article?
(a) 2 : 7 (b) 5 : 2
(c) 7 : 9 (d) None of these

40. The percentage profit earned by selling an article for ₹ 1920 is equal to the percentage loss incurred by selling the same article for ₹ 1280. At what price should the article be sold to make 25% profit?
(a) ₹ 2000 (b) ₹ 2200
(c) ₹ 2400 (d) None of these

41. The profit by selling an item was 25%. If the item was marked 40% above the selling price then what is the ratio of the marked price to the cost price of the item?
(a) $\frac{5}{4}$ (b) $\frac{7}{4}$
(c) $\frac{3}{4}$ (d) $\frac{1}{4}$

42. Two dealers X and Y selling the same model of refrigerator mark them under the same selling prices. X gives successive discounts of 25% and 5% and Y gives successive discounts of 16% and 12%. From whom is it more profitable to purchase the refrigerator?
(a) From Y
(b) From X
(c) Indifferent between the two
(d) Cannot be determined

43. A shopkeeper marks up his goods by 20% and then gives a discount of 20%. Besides he cheats both his supplier and customer by 100 grams *i.e.*, he takes 1100 gram from his supplier and sells only 900 grams to his customer. What is his net profit percentage?
(a) 24.5% (b) 17.33%
(c) 25% (d) 32.5%

44. Amit brought two cars. He then sold the first car at 10% profit and the second one at 25% profit. The selling price of the second car is 25% more than the selling price of the first car. What is the approximate profit per cent in both the cars together?
(a) 17.85% (b) 18.36%
(c) 16.19% (d) Cannot be determined

45. A trader mixes three varieties of groundnuts costing ₹ 50, ₹ 20 and ₹ 30 per kg in the ratio 2 : 4 : 3 in terms of weight, and sells the mixture of ₹ 33 per kg. What percentage of profit does he make?
(a) 8% (b) 9%
(c) 10% (d) None of these

46. A manufacturer sells a pair of glasses to a wholesale dealer at a profit of 18%. The wholesaler sells the same to a retailer at a profit of 20%. The retailer in turn sells them to a customer for ₹ 30.09, thereby earning a profit of 25%. The cost price for the manufacturer is
(a) ₹ 16 (b) ₹ 20
(c) ₹ 17 (d) ₹ 24

47. Samant bought a microwave oven and paid 10% less than the original price. He sold it with 30% profit on the price he had paid. What percentage of profit did Samant earn on the original price?
(a) 17% (b) 20%
(c) 27% (d) 32%

48. If 5% more is gained by selling an article for ₹ 350 than by selling it for ₹ 340 the cost of the article is:
(a) ₹ 50 (b) ₹ 160
(c) ₹ 200 (d) ₹ 225

49. A discount of 15% on one article is the same as a discount of 20% on another article. The costs of the two articles can be:
(a) ₹ 40, ₹ 20 (b) ₹ 60, ₹ 40
(c) ₹ 80, ₹ 60 (d) ₹ 60, ₹ 40

50. A shopkeeper earns a profit of 12% on selling a book at 10% discount on the printed price. The ratio of the cost price to the printed price of the book is:
(a) 45 : 56 (b) 50 : 61
(c) 55 : 69 (d) 99 : 125

51. By selling a watch at a profit of 10 per cent, a man got ₹ 15 more than half its price. What is the price of the watch?
(a) 10 (b) 15
(c) 25 (d) 5

52. A bookseller marks his books at an advance of 69% on the actual cost of production. He allows a discount of 15% and also given a copy free for every dozen sold at a time. What rate per cent profit does the bookseller make, if books are sold in lots of 12 ?
(a) 32.6 (b) 47.5
(c) 24.9 (d) None of these

53. 21 articles were bought for ₹ 6531 and sold for ₹ 9954. How much was the approximate profit percentage per article ? [SBI Clerk-June-2012]
(a) 56% (b) 43%
(c) 52% (d) 49%
(e) 61%

54. Meera purchased 23 bracelets at the rate of ₹160 per bracelet. At what rate per bracelet should she sell the bracelets so that profit earned is 15% ? [SBI Clerk-2012]
(a) ₹ 184/- (b) ₹ 186/-
(c) ₹ 192/- (d) ₹ 198/-
(e) None of these

55. The profit earned after selling a pair of shoes for ₹ 2,033 is the same as loss incurred after selling the same pair of shoes for ₹ 1,063. What is the cost of the shoes ? [SBI Clerk-2014]
(a) ₹ 1,650 (b) ₹ 1,548
(c) ₹ 1,532 (d) Cannot be determined
(e) None of these

56. Rahul bought two cycles for a total sum of ₹ 1,500. He sold one cylce at 20% loss and the other cycle at 20% gain. If the selling price of both the cycles is the same, find the cost price of the two cycles. [SSC-Sub. Ins.-2012]
(a) ₹ 500, ₹ 1,000 (b) ₹ 600, ₹ 900
(c) ₹ 750 each (d) ₹ 550, ₹ 950

57. A bookseller makes 8% profit after selling the book at 10% discount. The ratio of the cost price to the marked price is [SSC-Sub. Ins.-2012]
(a) 4 : 5 (b) 5 : 4
(c) 5 : 6 (d) 6 : 5

58. By selling an article for ₹ 21,000, a man gains 5%. To get a profit of 15%, he has to sell it for [SSC-Sub. Ins.-2012]
(a) ₹ 19,800 (b) ₹ 20,700
(c) ₹23,000 (d) ₹25,000

59. A single discount equivalent to the following three successive discounts of 30%; 20% and 10% is given by : [SSC-Sub. Ins.-2013]
(a) 49.6% (b) 50.60%
(c) 49.40% (d) 50.40%

60. A man sold 250 chairs and had a gain equal to selling price of 50 chairs. His profit per cent is: [SSC-Sub. Ins.-2013]
(a) 20% (b) 25%
(c) 50% (d) 15%

61. An article was sold at 16% gain. Had it been sold for ₹ 200 more, the gain would have been 20%. Then the cost price of the article is : [SSC-Sub. Ins.-2013]
(a) ₹ 5000 (b) ₹4800
(c) ₹4500 (d) ₹5200

62. A shopkeeper marked the selling price of his goods in such a way that after giving a discount of 10% he gains 17%. How much per cent above the cost price is the marked price? [SSC-Sub. Ins.-2013]
(a) 36% (b) 27%
(c) 30% (d) 40%

63. A trader who marks his goods up to 50% offered a discount of 20%. What % profit the trader makes after offering the payment? [SSC-Sub. Ins.-2014]
(a) 30% (b) 70%
(c) 20% (d) 50%

64. A retailer buys a sewing machine at a discount of 15% and sells it for ₹1955. Thus he makes a profit of 15%. The discount is [SSC-Sub. Ins.-2014]
(a) ₹270 (b) ₹290
(c) ₹300 (d) ₹310

65. A tea-merchant professes to sell tea at cost price but uses a false weight of 900 gram for a kilogram. The profit percent in his transaction is [SSC-Sub. Ins.-2014]
(a) $11\frac{1}{9}\%$ (b) 10%
(c) $9\frac{1}{11}\%$ (d) 15%

66. Mahesh earned a profit of 20% by selling 60 apples at the rate of ₹42.50 for 5 apples. Then the total cost, at which the apples were bought is [SSC-Sub. Ins.-2014]
(a) ₹452 (b) ₹425
(c) ₹450 (d) ₹485

67. A merchant purchases a wrist watch for ₹ 450 and fixes its list price in such a way that after allowing a discount of 10%, he earns a profit of 20%. Then the list price of the watch is [SSC-MT-2013]
(a) ₹600 (b) ₹650
(c) ₹700 (d) ₹550

68. Two successive discounts of 70% and 30% are equivalent to a single discount of [SSC-MT-2013]
(a) 89% (b) 75%
(c) 79% (d) 100%

69. A merchant allows a discount of 10% on marked price for the cash payment. To make a profit of 17%, he must mark his goods higher than their cost price by [SSC-MT-2013]
(a) 30% (b) 33%
(c) 40% (d) 27%

70. A dishonest grocer sells rice at a profit of 10% and also uses weights which are 20% less than the marked weight. The total gain earned by him will be [SSC-MT-2013]
(a) 35% (b) 37.5%
(c) 40% (d) 30.5%

71. The cost price of a radio is ₹ 600. 5% of the cost price is charged towards transportation. After adding that, if the net profit to be made is 15%, then the selling price of the radio must be [SSC-MT-2013]
(a) ₹684.50 (b) ₹704.50
(c) ₹724.50 (d) ₹664.50

72. By selling a fan for ₹ 600, a man loses 10%. To make a gain of 20%, the selling price of the fan should be [SSC-MT-2013]
(a) ₹800 (b) ₹900
(c) ₹1000 (d) ₹700

73. A man sold two articles at ₹375 each. On one, he gains 25% and on the other, he loses 25%. The gain or loss% on the whole transaction is : [SSC 10+2-2012]

(a) 6% (b) $4\frac{1}{6}\%$

(c) ₹50 (d) $6\frac{1}{4}\%$

74. A bought an article, paying 5% less than the original price. A sold it with 20% profit on the price he had paid. What percent of profit did A earn on the original price ? [SSC 10+2-2012]

(a) 10 (b) 13

(c) 14 (d) $\frac{17}{2}$

75. The profit percent of a bookseller if he sells book at marked price after enjoying a commission of 25% on marked price will be: [SSC 10+2-2012]

(a) 30% (b) 25%

(c) 20% (d) $33\frac{1}{3}\%$

76. The Banker's discount on a bill due 6 months hence at 16% per annum is ₹216. The true discount is: [SSC 10+2-2012]

(a) ₹212 (b) ₹180

(c) ₹210 (d) ₹200

77. The printed price of a book is ₹320. A retailer pays ₹244.80 for it. He gets successive discounts of 10% and an another rate. His second rate is : [SSC 10+2-2012]

(a) 15% (b) 16%

(c) 14% (d) 12%

78. A sells an article to B at a gain of 10%, B sells it to C at a gain of 5%. If C pays ₹462 for it, what did it cost to A ? [SSC 10+2-2012]

(a) ₹500 (b) ₹450

(c) ₹600 (d) ₹400

79. A shopkeeper blends two varieties of tea costing ₹18 and ₹13 per 100 gm in the ratio 7:3. He sells the blended variety at the rate of ₹18.15 per 100 gm. His percentage gain in the transaction is [SSC 10+2-2013]

(a) 8% (b) 10%

(c) 12% (d) 14%

80. A got 30% concession on the label price of an article sold for ₹8,750 with 25% profit on the price he bought. The label price was [SSC 10+2-2013]

(a) 10,000 (b) 13,000

(c) 16,000 (d) 12,000

81. The cost price of a book is ₹150. At what price should it be sold to gain 20% ? [SSC 10+2-2013]

(a) ₹80 (b) ₹120

(c) ₹180 (d) ₹100

82. If books bought at prices ranging from ₹150 to ₹300 are sold at prices ranging from ₹250 to ₹350, what is the greatest possible profit that might be made in selling 15 books ? [SSC 10+2-2013]

(a) ₹3,000 (b) Cannot be determined

(c) ₹750 (d) ₹4,250

83. The marked price of a saree is ₹200. After allowing a discount of 20% on the marked price, the shopkeeper makes a profit of ₹16. Find the gain percent. [SSC 10+2-2014]

(a) $11\frac{1}{9}\%$ (b) $9\frac{1}{11}\%$

(c) 11% (d) 8%

84. The marked price of an item is twice the cost price. For a gain of 15%, the discount should be [SSC 10+2-2014]

(a) 7.5% (b) 20.5%

(c) 32.5% (d) 42.5%

85. A man sold his watch at a loss of 5%. Had he sold it for ₹56.25 more, he would have gained 10%. What is the cost price of the watch (in ₹) ? [SSC 10+2-2014]

(a) 370 (b) 365

(c) 375 (d) 390

86. A merchant bought some goods worth ₹6000 and sold half of them at 12% profit. At what profit per cent should he sell the remaining goods to make and overall profit of 18%? [IBPS Clerk-2013]

(a) 24 (b) 28

(c) 18 (d) 20

(e) 26

Level - II

1. A cash payment that will settle a bill for 250 chairs at ₹50 per chair less 20% and 15% with a further discount of 5% on cash payment is

(a) ₹8075 (b) ₹7025

(c) ₹8500 (d) None of these

2. An oil refinery takes 1000 L of crude oil as input and after refining for 1 h gives certain amount of output oil X L. This can be sold in the market at a profit of ₹30 per L. If this oil is further refined for $\frac{1}{2}$ h, it gives oil Y L. This can be sold at a profit of ₹50 per L. Output and input ratio at both the stages is 90%. The maximum amount that can be earned from 1000 L of crude input is

(a) ₹40000 (b) ₹30000

(c) ₹27000 (d) ₹40500

3. A manufacturer sells a pair of glasses to a wholesale dealer at a profit of 18%. The wholesaler sells the same to a retailer at a profit of 20%. The retailer in turn sells them to a customer for ₹30.09, thereby earning a profit of 25%. The cost price for the manufacturer is

(a) ₹15 (b) ₹16

(c) ₹17 (d) ₹18

4. A dealer offers a cash discount of 20% and still makes a profit of 20%, when he further allows 16 articles to a dozen to a particularly sticky bargainer. How much per cent above the cost price were his wares listed?
(a) 100% (b) 80%
(c) 75% (d) 66 2/3%

5. Instead of a metre scale cloth merchant uses a 120 cm scale while buying but uses an 80 cm scale while selling the same cloth. If he offers a discount of 20 per cent of cash payment, what is his overall per cent profit?
(a) 20% (b) 25%
(c) 40% (d) 15%

6. A book is sold at profit of ₹ 20, which is 10% of its cost price. If its C.P. is increased by 50% and it is still sold at a profit of 10%, then find the new profit.
(a) ₹ 30 (b) ₹ 50
(c) ₹ 60 (d) ₹ 300

7. A fruitseller sells mangoes at the rate of ₹ 9 per kg and thereby loses 20%. At what price per kg, he should have sold them to make a profit of 5%?
(a) ₹ 11.81 (b) ₹ 12
(c) ₹ 12.25 (d) ₹ 12.31

8. A man would gain 20% by selling a chair for ₹ 47.5 and would gain 15% by selling a table for ₹ 57.5. He sells the chair for ₹ 36, what is the least price for which he must sell the table to avoid any loss on the two together
(a) ₹ 50.2 (b) ₹ 55.8
(c) ₹ 60 (d) ₹ 53.6

9. By selling 5 dozen mangoes for ₹ 156 it was found that $\frac{3}{10}$th of the outlay was gained. What should the retail price per mango be in order to gain 60% ?
(a) ₹ 4 (b) ₹ 2
(c) ₹ 3.2 (d) ₹ 4.2

10. An article is sold at 20 % profit. If its CP and SP are less by ₹ 10 and ₹ 5 respectively the percentage of profit increases by 10 %. Find the cost price.
(a) ₹ 40 (b) ₹ 80
(c) ₹ 60 (d) ₹ 50

11. A man purchases two clocks A and B at a total cost of ₹ 650. He sells A with 20% profit and B at a loss of 25% and gets the same selling price for both the clocks. What are the purchasing prices of A and B respectively?
(a) ₹ 225; ₹ 425 (b) ₹ 250; ₹ 400
(c) ₹ 275; ₹ 375 (d) ₹ 300; ₹ 350

12. A person purchases 100 pens at a discount of 10%. The net amount of money spent by the person to purchase the pens is ₹ 600. The selling expenses incurred by the person are 15% on the net cost price. What should be the selling price for 100 pens in order to earn a profit of 25%?
(a) ₹ 802.50 (b) ₹ 811.25
(c) ₹ 862.50 (d) ₹ 875

13. A milkman buys milk contained in 10 vessels of equal size. If he sells his milk at ₹ 5 a litre, he loses ₹ 200; if he sells it at ₹ 6 a litre, he would gain ₹ 150 on the whole. Find the number of litres contained in each vessel.
(a) 20 litres (b) 30 litres
(c) 25 litres (d) 35 litres

14. A shopkeeper purchased a table marked at ₹ 200 at successive discount of 10% and 15% respectively. He spent ₹ 7 on transportation and sold the table for ₹ 200. Find his gain %?
(a) No loss or gain (b) 25%
(c) 30% (d) 40%

15. A man buys 2 dozen bananas at ₹ 16 per dozen. After selling 18 bananas at the rate of ₹ 12 per dozen, the shopkeeper reduced the rate of ₹ 4 pre dozen. The percent loss is:
(a) 25.2% (b) 32.4%
(c) 36.5% (d) 37.5%

16. A space research company wants to sell its two products A and B. If the product A is sold at 20% loss and the product B at 30% gain, the company will not lose anything. If the product A is sold at 15% loss and the product B at 15% gain, the company will lose ₹ 6 million in the deal. What is the cost of product B ?
(a) ₹ 140 million (b) ₹ 120 million
(c) ₹ 100 million (d) ₹ 80 million

17. Two-third of a consignment was sold at a profit of 5% and the remainder at a loss of 2 %. If the total profit was ₹ 400, the value of the consignment (in rupees)
(a) 20,000 (b) 15,000
(c) 12, 000 (d) 10, 000

18. A dealer sold a radio at a loss of 2.5%. Had he sold it for ₹ 100 more, he would have gained $7\frac{1}{2}$ %. In order to gain $12\frac{1}{2}$ %, he should sell it for:
(a) ₹ 850 (b) ₹ 925
(c) ₹ 1, 080 (d) ₹ 1, 125

19. The raw material and manufacturing cost formed individually 70% and 30% of the total cost and the profit percentage is 14.28% of the raw material. If the cost of raw material increase by 20% and the cost of manufacturing is increased by 40% and the selling price is increased by 80%, then the new profit percentage is :
(a) 57% (b) 65.8%
(c) 60% (d) can't determined

20. A person purchased a cupboard and a cot for ₹ 18,000. He sold the cupboard at a profit of 20% and the cot at a profit of 30%. If his total profit was 25.833%, find the cost price of the cupboard.
(a) ₹ 10,500 (b) ₹ 12,000
(c) ₹ 7500 (d) ₹ 10,000

21. *A* sells a car priced at ₹ 36,000. He gives a discount of 8% on the first ₹ 20,000 and 5% on the remaining ₹ 16,000. His competitor *B* sells a car of the same make, priced at ₹ 36,000. If he wants to be competitive what percent discount should *B* offer on the marked price.
(a) 5% (b) 5.5%
(c) 6.66% (d) 8.33%

22. In a certain store, the profit is 320% of the cost. If the cost increases by 25% but the selling price remains constant, approximately what percentage of the selling price is the profit?
(a) 30% (b) 70%
(c) 100% (d) 250%

23. A shopkeeper bought 150 calculators at the rate of ₹ 250 per calculator. He spent ₹ 2500 on transportation and packing. If the marked price of calculator is ₹ 320 per calculator and the shopkeeper gives a discount of 5% on the marked price then what will be the percentage profit gained by the shopkeeper?
(a) 20% (b) 14%
(c) 15% (d) 16%

24. A firm of readymade garments makes both men's and women's shirts. Its average profit is 6% of the sales. Its profit in men's shirts average 8% of the sales and women's shirts comprise 60% of the output. The average profit per sale rupee in women shirts is
(a) 0.0466 (b) 0.0666
(c) 0.0166 (d) 0.0366

25. A tradesman marks his goods at 25% above cost price and allows discount of 12.5 per cent for cash payment. What profit per cent does he make ?
(a) $9\frac{3}{8}$ (b) $9\frac{1}{8}$
(c) $9\frac{5}{8}$ (d) $9\frac{7}{8}$

26. A bookseller sells a book at a profit of 10%. If the had bought it at 4% less and sold it for ₹ 6 more, he would have gained $18\frac{3}{4}$ per cent. What did it cost him?
(a) 120 (b) 130
(c) 140 (d) 150

27. A watch passes through three hands and each gains 25%. If the third sells it for ₹ 250, what did the first pay for it?
(a) 128 (b) 130
(c) 145 (d) 150

28. I loss 9 per cent by selling pencils at the rate of 15 a rupee. How many for a rupee must I sell them to gain 5 per cent?
(a) 10 (b) 13
(c) 15 (d) 18

29. A tradesman marks an article at ₹ 205 more than the cost price. He allows a discount of 10% on the marked price. Find the profit per cent if the cost price is ₹ x.
(a) $\dfrac{\left[\dfrac{x}{(18450)}-10\right]}{x}$ (b) $\dfrac{[(18450)]-10x}{x}$
(c) $\dfrac{\left[\dfrac{x}{(18450)}-100\right]}{x}$ (d) 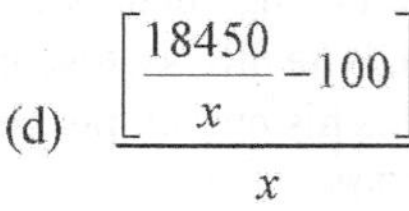

30. A manufacturer makes a profit of 15% by selling a colour TV for ₹ 5750. If the cost of manufacturing increases by 30% and the price paid by the retailer is increased by 20%, find the profit percent made by the manufacturer.
(a) 6(2/13)% (b) 4(8/13)%
(c) 6(1/13)% (d) 7(4/13)%

31. The profit earned when an article is sold for ₹ 800 is 20 times the loss incurred when it is sold for ₹ 275. At what price should the article be sold if it is desired to make a profit of 25%
(a) ₹ 300 (b) ₹ 350
(c) ₹ 375 (d) ₹ 400

32. Each of A and B sold their article at ₹ 1818 but A incurred a loss of 10% while B gained by 1%. What is the ratio of cost price of the articles of A to that of B?
(a) 101 : 90 (b) 85 : 89
(c) 81 : 75 (d) None of these

33. A manufacturer of a certain item can sell all he can produce at the selling price of ₹ 60 each. It costs him ₹ 40 in materials and labour to produce each item and he has overhead expenses of ₹ 3,000 per week in order to operate that plant. The number of units he should produce and sell in order to make a profit of at least ₹ 1,000 per week is
(a) 300 (b) 250
(c) 400 (d) 200

34. Dolly goes to a shop to purchase a doll priced at ₹ 400. She is offered 4 discount options by the shopkeeper. Which of these options should she opt for to gain maximum advantage of the discount offered?
(a) Single discount of 30%
(b) 2 successive discounts of 15% each
(c) 2 successive discounts of 20% and 10%
(d) 2 successive discounts of 20% and 12%

35. A trader sells goods to a customer at a profit of k% over the cost price, besides it he cheats his customer by giving 880 g only instead of 1 kg. Thus his overall profit percentage is 25%. Find the value of k?
(a) 8.33% (b) 8.25%
(c) 10% (d) 12.5%

36. A, B and C invest in the ratio of 3 : 4 : 5. The percentage of return on their investments are in the ratio of 6 : 5 : 4. Find the total earnings, if B earns ₹ 250 more than A :
(a) ₹ 6000 (b) ₹ 7250
(c) ₹ 5000 (d) None of these

37. A car mechanic purchased four old cars for ₹ 1 lakh. He spent total 2 lakh in the maintenance and repairing of these four cars. What is the average sale price of the rest three cars to get 50% total profit if he has already sold one of the four cars at ₹ 1.2 lakh?
(a) 1.5 lakh (b) 1.1 lakh
(c) 1.2 lakh (d) 1.65 lakh

38. A person purchases 90 clocks and sells 40 clocks at a gain of 10% and 50 clocks at gain of 20%. If he sold all of them at a uniform profit of 15%, then he would have got ₹ 40 less. The cost price of each clock is:
(a) ₹ 50 (b) ₹ 60
(c) ₹ 80 (d) ₹ 90

39. A tradesman fixed his selling price of goods at 30% above the cost price. He sells half the stock at this price, one quarter of his stock at a discount of 15% on the original selling price and rest at a discount of 30% on the original selling price. Find the gain percent altogether.
(a) 14.875% (b) 15.375%
(c) 15.575% (d) 16.375%

40. Cheap and Best, a kirana shop bought some apples at 4 per rupee and an equal number at 5 per rupee. He then sold the entire quantity at 9 for 2 rupees. What is his percentage profit or loss?
(a) 1.23% loss (b) 6.66%
(c) 8.888% (d) No profit no loss

41. Amar sold his moped to Bharat at 20% profit and Bharat sold it to Sridhar at 10% profit. Sridhar sold the same to a mechanic and received ₹ 2, 316. If Amar had sold the same moped to the mechanic and receive the same amount the mechanic paid to Sridhar, what profit percentage would Amar have made?
(a) 52% (b) 48%
(c) 33.3% (d) Cannot be determined

42. A dishonest dealer prefers to sell his goods at cost price but uses less weight for a kg weight and gains $4\frac{1}{6}\%$. What does he use for a kg weight? [*SBI PO-2011*]
(a) 950 gm (b) 980 gm
(c) 960 gm (d) 840 gm
(e) None of these

43. An article was purchased for ₹ 78,350/-. Its price was marked up by 30%. It was sold at a discount of 20% on the marked up price. What was the profit percent on the cost price ?
(a) 4 (b) 7 [*IBPS-PO-2012*]
(c) 5 (d) 3
(e) 6

44. A businessman allows a discount of 10% on the written price. How much above the cost price must he mark his goods to make a profit of 17%? [*SSC CGL-2012*]
(a) 30% (b) 20%
(c) 27% (d) 18%

45. A man sold an article at a loss of 20%. If he sells the article for ₹ 12 more, he would have gained 10%. The cost price of the article is [*SSC CGL-2012*]
(a) ₹ 60 (b) ₹ 40
(c) ₹ 30 (d) ₹ 22

46. A trader has a weighing balance that shows 1,200 gm for a kilogram. He further marks up his cost price by 10%. Then the net profit percentage is [*SSC CGL-2012*]
(a) 32% (b) 23%
(c) 31.75% (d) 23.5%

47. Kabir buys an article with 25% discount on its marked price. He makes a profit of 10% by selling it at ₹ 660. The marked price is [*SSC CGL-2013*]
(a) ₹ 600 (b) ₹ 685
(c) ₹ 700 (d) ₹ 800

48. On the eve of Gandhi Jayanti, Gandhi Ashram declared a 25% discount on silk. If selling price of a silk saree is ₹ 525, what is its marked price ? [*SSC CGL-2013*]
(a) ₹ 700 (b) ₹ 725
(c) ₹ 750 (d) ₹ 775

49. A shopkeeper marks an article at a price which gives a profit of 25%. After allowing certain discount, the profit reduces to $12\frac{1}{2}\%$. The discount percent is [*SSC CGL-2013*]
(a) 12% (b) 12.5%
(c) 10% (d) 20%

50. A CD was sold at a profit of $12\frac{1}{2}\%$. If it had been sold at a profit of 15%, it would have gained him ₹ 10 more. The cost prices of CD is (in ₹) [*SSC CGL-2013*]
(a) 450 (b) 500
(c) 400 (d) 550

51. A shopkeeper allows 10% discount on goods when he sells without credit. Cost price of his goods is 80% of his selling price. If he sells his goods by cash, then his profit is
(a) 50% (b) 70% [*SSC CGL-2014*]
(c) 25% (d) 40%

52. A dealer of scientific instruments allows 20% discount on the marked price of the instruments and still makes a profit of 25%. If his gain over the sale of an instrument is ₹ 150, find the marked price of the instrument. [*SSC CGL-2014*]
(a) ₹ 938.50 (b) ₹ 940
(c) ₹ 938 (d) ₹ 937.50

53. Ram bought a T.V. with 20% discount on the labelled price. Had he bought it with 30% discount he would have saved ₹ 800. The value of the T.V. set that he bought is [*SSC CGL-2014*]
(a) ₹ 5,000 (b) ₹ 8,000
(c) ₹ 9,000 (d) ₹ 10,000

54. A sold an article to B at 20% profit and B sold it to C at 15% loss. If A sold it to C at the selling price of B, then A would make [*SSC CGL-2014*]
(a) 5% profit (b) 2% profit
(c) 2% loss (d) 5% loss

55. The cost price of 8 books is equal to the selling price of 6 books. The percentage of gain is [*SSC CGL-2014*]
(a) $4\frac{1}{6}$ (b) $33\frac{1}{3}$
(c) 2 (d) 25

Hints & Solutions

Level-I

1. (a) Given SP of 12 marbles = ₹ 1, loss = 20%

$\therefore$ CP of 12 marbles = ₹ $\frac{1}{0.8}$ = ₹ 1.25

Now, SP of 12 marbles at a gain of 20%

$= \text{CP} \times 1.2 = 1.25 \times 1.2 =$ ₹ 1.5

This implies that in order to gain 20%, he should sell 12 marbles for ₹ 1.5.

$\Rightarrow$ For ₹ 1, he should sell $\frac{12}{1.5} = 8$ marbles.

2. (d) Applying successive discounts of 10%, 12% and 15% on 100, we get $100 \times 0.9 \times 0.88 \times 0.85 = 67.32$

$\Rightarrow$ Single discount $= 100 - 67.32 = 32.68$

Hence, none of the given options is correct.

3. (c) Total amount used for purchasing = ₹ 160. A reduction of 20% in the price means, now a person gets 5/2 kg for ₹ 32 and this is the present price of the sugar.

$\therefore$ Present price per kg $= \frac{32}{5} \times 2 =$ ₹ 12.8

Let the original price be ₹ x. Then new price is arrived after reduction of 20% on it.

$\Rightarrow x \times 0.8 = 12.8$ or $x =$ ₹ 16.

4. (d) If any two transactions of SP is the same and also gain % and loss % are same then there is always a loss

$\therefore$ loss % $= \left(\frac{\text{Common gain or loss\%}}{10}\right)^2 = \left(\frac{10}{10}\right)^2 = 1\%$

5. (c) Let he sells first grade cycle at a rate of ₹ z per bicycle.

Then, $8z + \frac{22 \times 3}{4} z = 3150 \times 1.4$

$\Rightarrow 32z + 66z = 17640$

$\Rightarrow 98z = 17640 \Rightarrow z = 180$

He should sell the first grade bicycles at a rate of ₹ 180.

6. (b) Let the quantity of milk purchased be x litres and quantity of water added to it be y litres. Then ratio of water to milk will be $y : x$.

Now, CP = $6.4x$ and SP = $8(x + y)$

and profit % = 37.5%

$\Rightarrow 8(x + y) = 6.4x \times 1.375$

$\Rightarrow 8x + 8y = 8.8x$

$\Rightarrow 8y = 0.8x$ or $\frac{x}{y} = \frac{80}{8} = \frac{10}{1}$

$\Rightarrow y : x = 1 : 10$

7. (c) CP of 20 articles = SP of x articles = 1 (say)

Therefore, CP of 1 article $= \frac{1}{20}$,

And SP of 1 article $= \frac{1}{x}$

Now, gain% $= \frac{\frac{1}{x} - \frac{1}{20}}{\frac{1}{20}} = \frac{25}{100} \Rightarrow \left(\frac{20 - x}{20x}\right) \times 20 = \frac{25}{100}$

$\Rightarrow 80 - 4x = x \Rightarrow 5x = 80 \Rightarrow x = 16$

8. (d) 750 × (180% of ₹ 9) + 125 × (120% of ₹ 9)

= 750 × 16.20 + 125 × 10.80 = 12150 + 1350 = ₹ 13500

9. (c) Let the quantity of two varieties of tea be $5x$ kg and $4x$ kg, respectively.

Now, SP = $23 \times 9x = 207x$

and CP = $20 \times 5x + 25 \times 4x = 200x$

Profit % $= \frac{7x}{200x} \times 100 = 3.5\%$

10. (b) Price of the article after first discount, $65 - \left(65 \times \frac{10}{100}\right)$

= ₹ 58.5

Therefore, the second discount

$= \frac{58.5 - 56.16}{58.5} \times 100 = 4\%$

11. (c) Total profit = 337.50 + 1125.00 + 675 = ₹ 2137.50

Percentage profit $= \frac{2137.50}{114000} \times 100 = 1.8\%$

12. (b) S.P. = 100 – 12.5 = ₹ 87.5

S.P. after 6% gain = ₹ 106

Difference = ₹ 18.5

$\therefore$ C.P. $= \frac{92.5}{18.5} \times 100 =$ ₹ 500

13. (d) Cost price of bicycle = ₹ 5,200.

He spent ₹ 800 on it's repairs

$\therefore$ C.P. = 5200 + 800 = 6000.

Selling price = ₹ 5, 500

$\therefore$ Loss = 6000 – 5500 = 500

Hence, loss % $= \frac{500}{6000} \times 100 = 8.33\%$

14. (a) Let the C.P. of 150 kg of rice be ₹ 150.

$\therefore$ S.P. of 50 kg of rice at 10%

loss $= \frac{90}{100} \times 50 =$ ₹ 45

For 10% of gain on the whole.

$$S.P. = 150 \times \frac{110}{100} = ₹165$$

∴ 100 kg rice should be sold for ₹ 120.

∴ Per cent gain = 20

15. **(c)** $CP = 20 \times 15 + 30 \times 13 = ₹690$

$$\therefore \quad SP = \frac{4}{3} \text{ of } 690 \times \frac{1}{50} = ₹18.40$$

16. **(d)** We do not know the total investment of builder, because in the question construction cost is not given. Hence, 'none of these' is the answer.

17. **(c)** Successive discount $= 20\% + \frac{20 \times 80}{100}$

$= 20 + 16 = 36\%$

Difference in discount = 36 – 35 = 1%

∴ Bill amount = 22 × 100 = ₹2200

18. **(b)** C.P. of 200 kg of mixture = ₹ (80 × 13.50 + 120 × 16)

= ₹ 3000.

$$S.P. = 116\% \text{ of } ₹3000 = ₹\left(\frac{116}{100} \times 3000\right) = ₹\ 3480.$$

$$\therefore \text{ Rate of S.P. of the mixture} = ₹\left(\frac{3480}{200}\right) \text{ per kg}$$

= ₹ 17.40 per kg.

19. **(c)** Let A paid = ₹ x

125 % of 120% of x = 225

$$\Rightarrow \frac{125}{100} \times \frac{120}{100} \times x = 225$$

$$\Rightarrow x = \frac{225 \times 100 \times 100}{125 \times 120} = ₹\ 150$$

20. **(d)** Let he purchase of ₹ x/kg.

$$\therefore \ (525 + 30x) \times \frac{120}{100} = 60 \times 18.60$$

$\Rightarrow x = ₹\ 13.5 / \text{kg}.$

21. **(c)** SP = 90 × 1.2 = ₹ 108

$$\text{Marked price} = \frac{108}{0.85} = ₹127.05$$

22. **(d)** Let the SP of the article be ₹ x

Expenses = 15% of x = ₹ $0.15x$

Profit = 10% of x = ₹ $0.10x$

CP = ₹ 9 (given)

Therefore, $9 + 0.15x + 0.1x = x \Rightarrow x = 12$

$$\therefore \ \% \text{ increase for marked price} = \frac{12 - 9}{9} \times 100$$

$$= \frac{100}{3}\%$$

23. **(b)** Price of the article after first discount

65 – 6.5 = ₹ 58.5

Therefore, the second discount

$$= \frac{58.5 - 56.16}{58.5} \times 100 = 4\%$$

24. **(a)** Let the CP of the article be ₹ x

Then, SP = $x \times 1.12 \times 1.1$

Now, $x \times 1.12 \times 1.1 = 616$

$$\Rightarrow x = \frac{616}{1.232} = ₹500$$

25. **(b)** When $S_1 = S_2$, then

overall % gain or % loss

$$= \left[100 - \frac{2(100 + x_1)(100 - x_2)}{(100 + x_1) + (100 - x_2)}\right]\%$$

$$= \left(100 - \frac{2(125)(80)}{(125) + (80)}\right)\% = \left(100 - \frac{2 \times 125 \times 80}{205}\right)\%$$

$$= \frac{100}{41}\% \text{ gain } (\because \text{ it is +ve})$$

26. **(c)** Let C.P. of 1 article = ₹ 1

then C.P. of 25 articles = ₹ 25

and S.P. of 25 articles = ₹ 20

$$\therefore \text{ loss } \% = \frac{25 - 20}{20} \times 100 = 25\%$$

27. **(d)** Total cost price of mobile phone and refrigerator

= ₹ (12000 + 10000) = ₹ 22000

SP of mobile phone = (88% of 12000)

$$= ₹\left(\frac{88}{100} \times 12000\right) = ₹\ 10560$$

SP of refrigerator = 108% of 10000

$$= ₹\left(\frac{108}{100} \times 10000\right) = ₹\ 10800$$

Total SP of both the articles = ₹ (10560 + 10800)

= ₹ 21360

Loss = ₹ (22000 – 21360) = ₹ 640.

28. **(d)** $C.P. = ₹\left(\frac{100}{105} \times 630000\right) = ₹\ 600000.$

$$\therefore \text{ Required loss } \% = \left(\frac{100000}{600000} \times 100\right)\% = 16\frac{2}{3}\%.$$

29. **(a)** Let ₹ X be the C.P. of the manufacturer of the car

$$\Rightarrow X \times \frac{150}{100} \times \frac{120}{100} \times \frac{80}{100} = 288000$$

$\Rightarrow X = 200000$

30. **(c)** Supposing the goods cost the dealer ₹ 1 for the kg., he sells for ₹ 1 goods which cost him ₹ 0.96.

Gain on ₹ 0.96 = 0.04;

$$\% \text{ Gain} = \frac{0.04 \times 100}{0.96} = 4\frac{1}{6}\% = 4.16\%$$

31. **(c)** Let C.P. be ₹ 100. Then, S.P. = ₹ 123.50.

Let marked price be ₹ x. Then, $\frac{95}{100}x = 123.50 \Rightarrow x$

$= ₹\left(\frac{12350}{95}\right) = ₹130.$

Now, S.P. = ₹ 130, C.P. = ₹100

∴ Profit % = 30%

32. **(c)** Let cost price = x

Then we have, $x\left(\frac{95}{100}\right)\left(\frac{110}{100}\right) = x\left(\frac{105}{100}\right) - 1$

or, $x = \frac{100 \times 100}{105 \times 100 - 95 \times 100} \times 200$

∴ Cost price = ₹ 200

33. **(d)** The new situation is
Buying:
1100 grams for ₹ 900
Hence, 1320 grams for ₹ 1080
Selling: 900 grams for ₹ 1080

Profit % = $\frac{420}{900} \times 100 = 46.66\%$

34. **(a)** Let us assume his CP/1000 gm = ₹ 100
So, his SP/kg (800 gm) = ₹ 126
So, his CP/800 gm = ₹ 80
So, profit = ₹ 46
So profit percentage = 46/80 × 100 = 57.5%

35. **(b)** MP of 1 Pencil = ₹ 1
For supplier, SP of 20 pencils = ₹ 16
For retailer, SP of 20 pencils = ₹ 20

Profit percentage = $\frac{4}{16} \times 100 = 25\%$

36. **(c)** The milkman defrauds 20% in buying and also defrauds 25% in selling, so his overall % gain will be

$$\left[\frac{(100+20\%)(100+25\%)}{100} - 100\right]\% = 50\%$$

37. **(c)** Let the price of 1 article = ₹ 1
⇒ Loss = 20 C.P – 20 S.P.
⇒ 5C.P. = 20 C.P. – 20 S.P. ⇒ 20 S.P = 15 C.P
⇒ CP_1 of 20 articles = ₹ 20
⇒ SP_1 of 20 articles = ₹ 15
Also given that , had he purchased the 20 articles for 25% less and sold them for $33\frac{1}{3}\%$ more, then
⇒ CP_2 of 20 articles = ₹ 15
⇒ CP_2 of 20 articles = ₹ 20

$$\therefore \text{ Gain percentage} = \frac{20-15}{15} \times 100 = 33\frac{1}{3}\%$$

38. **(b)** Value of consignment = $\frac{540 \times 100}{\frac{2}{3} \times 6 + \frac{1}{3}(-3)} = \frac{540 \times 100}{4-1}$ = ₹18,000

39. **(d)** Let C.P. = ₹ $5x$ and S.P. = ₹ $7x$. Then, Gain = ₹ $2x$
∴ Required ratio $2x : 5x = 2 : 5$

40. **(a)** Let C.P. be ₹ x.

Then, $\frac{1920 - x}{x} \times 100 = \frac{x - 1280}{x} \times 100$

$\Rightarrow 1920 - x = x - 1280 \Rightarrow 2x = 3200 \Rightarrow x = 1600.$

∴ Required S.P. = 125% of ₹1600

$= ₹\left(\frac{125}{100} \times 1600\right) = ₹\ 2000$

41. **(b)** Let the cost price of an item = ₹ 100,
then, selling price = ₹ 125
(∵ Profit by selling is 25%)
Now, marked price is 40% above the selling price

$\therefore \text{M.P} = 125 + 125 \times \frac{40}{100}$

$\Rightarrow \text{Marked price} = 125\left(1 + \frac{40}{100}\right) = 175$

Hence, $\frac{\text{Marked price}}{\text{Cost price}} = \frac{175}{100} = \frac{7}{4}$

42. **(b)** Assume marked price for both to be 100.
X's selling price = 100 ×0.75×0.95 = 71.25
Y's selling price = 100 × 0.84 × 0.88 = 73.92.
Buying from '*X*' is more profitable.

43. **(b)** Assume his CP = ₹ 1000/1100 gm
MP = ₹ 1200 and SP = ₹ 960/900 gm
So, SP/1100 gm = ₹ 1173.33
So, profit = ₹ 173.33
Profit percentage = 17.33%

44. **(a)** $100 \xrightarrow{10\%} \text{First profit} \xrightarrow{25\%} \text{Net profit}$

45. **(c)** Suppose he bought 2 kg, 4 kg and 3 kg of the three varieties. C.P. of 9 kg = ₹ (2 × 50 + 4 × 20 + 3 × 30) = ₹ 270.
S.P. of 9 kg = ₹ (9 × 33) = ₹ 297.

$\therefore$ Profit % = $\left(\frac{27}{270} \times 100\right)\% = 10\%.$

46. **(c)** Let the cost price for the manufacturer be ₹x.
Then, 125% of 120% of 118% of x = 30.09.

$\Rightarrow \frac{125}{100} \times \frac{120}{100} \times \frac{118}{100} x = \frac{3009}{100} \Rightarrow \frac{177}{100} x = \frac{3009}{100}$

$\Rightarrow x = \left(\frac{3009}{177}\right) = 17.$

47. (a) Let original price = ₹ 100.

Then C.P. = ₹ 90, S.P. = 130% of ₹ 90 = ₹$\left(\frac{130}{100}\times 90\right)$

= ₹ 117.

∴ Required percentage = (117 – 100)% = 17%.

48. (c) Let C.P. be ₹ x. Then, 5% of x = (350 – 340) = 10

$\Rightarrow \frac{x}{20} = 10 \Rightarrow x = 200.$

49. (c) Let the costs of the two articles be x and y. Then,

15% of x = 20% of y $\Rightarrow \frac{x}{y} = \frac{20}{15} = \frac{4}{3}$.

So, x and y must be in the ratio of 4 : 3.

₹ 80, ₹ 60.

50. (a) Let cost price be ₹ 100. The, S.P. = ₹ 112.

Let printed price be ₹ x.

90% of x = 112 $\Rightarrow x = \left(\frac{112\times 100}{90}\right)$ = ₹ $\frac{1120}{9}$

∴ Required ratio = 100 : $\frac{1120}{9}$ = 900 : 1120 = 45 : 56

51. (c) Price = ₹ X

SP = ₹ 1.1 x = 0.5 x + 15, So, 0.6 x = 15

So, x = 25

52. (a) CP = ₹ X/dozen = 0.833/copy

So, MP = 1.69 x/dozen

SP = ₹ 1.4365 x/13 copies = 0.1105 x/copy

So, profit = 32.6%

53. (c) Cost price per article = 311

Selling price per article $= \frac{9954}{21} = 474$

Profit per article $= \frac{474-311}{311}\times 100 = 52\%$

54. (a) Cost of 23 bracelet purchased at rate of ₹ 160/bracelet = ₹ 23 × 160 = ₹ 3680

If profit earned is 15%, then

Profit amount = $\frac{3680\times 15}{100}$ = ₹ 552

Total amount Meera have after selling 23 bracelets = 3680 + 552 = 4232

S.P. of one bracelet = $\frac{4232}{23}$ = ₹ 184

55. (b) Le the CP of the shoes be ₹ x.

∴ 2033 – x = x – 1063

⇒ 2x = 2033 + 1063 = 3096

⇒ x = $\frac{3096}{2}$ = ₹ 1548

56. (b) 2 cycles – 1500

By options,

600	900
↓	↓
+20% → 720	–20% → 720

57. (c)

C.P.	M.P.		S.P.
X	100	10% discount	90

Let cost price (C.P.) = ₹ x

& Marked price (M.P.) = ₹ 100

$x \times \frac{108}{100} = 90$

$x = \frac{90\times 100}{108}$

New cost price : cost price

$\frac{90\times 100}{108} : 100 = 5 : 6$

58. (c) C.P. = $\frac{100}{(100+5)}\times 21000$

C.P = ₹ 20000

New profit = 15%

New S.P = $\frac{(100+15)}{100}\times 20000$ = ₹ 230000

∴ To get 15% profit he has to sell an article at ₹ 23000.

59. (a) Single equivalent discount for 30% and 20%

$= \left(30+20-\frac{30\times 20}{100}\right) = 44\%$

Single equivalent discount for 44% and 10%

$= \left(44+10-\frac{44\times 10}{100}\right) = 49.6\%$

60. (b) Percentage profit $= \frac{250-200}{200}\times 100 = 25\%$

61. (a) If the C.P. of article be ₹ x, then

$x\times\frac{116}{100} + 200 = \frac{x\times 120}{100}$

$\Rightarrow x\times\frac{4}{100} = 200$

$\Rightarrow x\times\frac{200\times 100}{4}$ = ₹ 5000

62. (c) C.P. of the article = ₹ 100 and marked price = ₹ x

∴ $x\times\frac{90}{100} = 117$

$\Rightarrow x = \frac{117\times 100}{90}$ = 130 = 30% above C.P.

63. (c) Let cost price of goods be ₹ 100
Trader mark up at 50% more i.e. 150

Selling price of goods $= 150 - \frac{20}{100} \times 150 = 120$

Profit % $= \frac{120-100}{100} \times 100 = 20$

64. (c) Let original price of sewing machine be ₹ x

Retailer bought it at $x - \frac{15}{100}x = 0.85x$

$0.85x + \frac{15}{100} \times 0.85x = 1955$

$1.15 \times 0.85x = 1955$

$x = \frac{1955 \times 10000}{115 \times 85} = 2000$

Discount is $\frac{15}{100} \times 2000 =$ ₹ 300

65. (a) Profit % $= \frac{1000-900}{900} \times 100 = 11\frac{1}{9}\%$

66. (b) Selling price of 5 apples = ₹ 42.50

Selling price of 60 apples $= \frac{42.5}{5} \times 60 = 510$

C.P + Profit = S.P

$C.P + \frac{20}{100} \times C.P = 510$

$C.P. = \frac{510}{120} \times 100 =$ ₹ 425

67. (a) C.P. = ₹ 450; profit = 20%

$\therefore$ S.P. $= \frac{(100+20)}{100} \times 450 =$ ₹540

Let the list price of the wrist watch be ₹x.

Then discount @ 10% = ₹ $x \times \frac{10}{100} =$ ₹$\frac{x}{10}$

$\therefore$ S.P. $= x - \frac{x}{10} =$ ₹$\frac{9}{10}x$

According to quesiton, $\frac{9x}{10} = 540$

$x = \frac{540 \times 10}{9} =$ ₹600

68. (c) Single discount $= x + y + \frac{xy}{100}$

$= -70 - 30 + \frac{(-70 \times -30)}{100}$

$= -100 + 21 = -79\%$

'–' denotes discount. Hence, single discount equivalent to 79%

69. (a) Solving this type of question by short cut.

Net profit% $= x + y + \frac{xy}{100}$

$17\% = -10 + y + \frac{(-10) \times y}{100}$ [∴ '–' for discount]

$27 = y - \frac{y}{10} \Rightarrow 27 = \frac{10y - y}{10}$

$27 \times 10 = 9y$

$y = 30\%$

Hence, he must mark his goods 30% higher than their cost price.

70. (b) Let us consider a packet or rice marked 1kg. It's actual weight is 80% of 1000 gm = 800 gm
Let C.P. of each gm be ₹1.
Then, C.P. of this packet = ₹ 800
S.P. of this packet = 110% of C.P. of 1kg

$= \frac{110}{100} \times 1000 =$ ₹1110

$\therefore$ Gain % $= \frac{(1100-800)}{1100} \times 100 = 37.5\%$

71. (c) C.P. of a radio = ₹ 600
New C.P. after adding transportation charges

= ₹(600 + 5% of 600) = ₹$\left(600 + \frac{5}{100} \times 600\right)$ = ₹630

S.P. $= \left(\frac{100 + \text{Profit}\%}{100}\right) \times C.P$

$= \frac{100+15}{100} \times 630 = \frac{115}{10} \times 63 =$ ₹724.50

72. (a) $S.P = \frac{(100 - loss\%)}{100} \times C.P_1$

$600 = \frac{(100-10)}{100} \times CP_1$

$\therefore CP_1 = \frac{100 \times 600}{90}$

To make a gain of 20%, the S.P. of fan should be

$\frac{(100 + \text{gain}\%)}{100} \times CP_1$

$\therefore S.P = \frac{(100+20)}{100} \times \frac{100 \times 600}{90} = \frac{120}{90} \times 600 = 800$

Hence, S.P. should be ₹ 800.

73. (d) In such type of question,

Required % loss $= \frac{(25)^2}{100}\%$

$= \frac{625}{100}\% = 6.25\% = 6\frac{1}{4}\%$

74. (c) Required % earned by A

$$=\left\{100\times\frac{(100-5)}{100}\times\frac{(100+20)}{100}-100\right\}\%$$

$$=\left\{100\times\frac{95}{100}\times\frac{120}{100}-100\right\}\%$$

$=(114-100)\%=14\%$

75. (d) Let MP = 100

So, SP = 100 – 25% of 100

= 100 – 25 = 75

$$\text{So, Profit percent} = \frac{100-75}{75}\times 100$$

$$=\frac{25}{75}\times 100\% = \frac{1}{3}\times 100 = 33\frac{1}{3}\%$$

76. (d) Let x = true discount

So, 216 = x + 8% of x

= x + 0.08x = 1.08x

$$x=\frac{216}{1.08}=₹\,200$$

77. (a) Printed price = ₹ 320, $d_1 = 10\%$

Let $d_2 = x\%$, Amount actually paid = ₹ 244.80

$$\text{So, amount actually paid} = \left(1-\frac{10}{100}\right)\left(1-\frac{x}{100}\right)\times 320$$

$$244.80=\frac{90}{100}\times\frac{(100-x)}{100}\times 320$$

$$100-x=\frac{244.80\times100\times100}{90\times320}=85\Rightarrow x=15\%$$

78. (d) Cost to A $\xrightarrow{10\%\uparrow}$ cost to B $\xrightarrow{5\%\uparrow}$ cost to C

$$\Rightarrow \text{Cost to A}\times\frac{110}{100}\times\frac{105}{100}=₹\,462$$

$$\Rightarrow \text{Cost to A}=\frac{462\times100\times100}{110\times105}=₹\,400$$

79. (b) CP of 1000 gm tea = 18 × 7 + 13 × 3

= ₹ (126 + 39) = ₹ 165

CP of 100 g = ₹ 16.5

S.P. of 100 g = ₹ 18.15

Profit = ₹(18.15 – 16.5) = ₹ 1.65

$$\%\text{ gain}=\frac{1.65}{16.5}\times 100 = 10\%$$

80. (a) Let the labelled price be ₹ x

$$\text{Now, C.P}=\frac{100}{(100+\text{profit }\%)}\times S.P$$

$$\text{C.P}=\frac{100}{(100+25)}\times 8750 = ₹\,7000$$

Now, (1 – 30% concession) label price = C.P

$$\left(1-\frac{30}{100}\right)x = 7000$$

$$\frac{70}{100}x = 7000$$

$$x=\frac{7000\times100}{70}$$

$$x=₹\,10,000$$

81. (c) $₹150+\frac{20}{100}\times150=₹180$

S.P = ₹180

82. (a) Least cost price = ₹(150 × 15) = ₹2250

Greatest selling price = ₹(350 × 15) = ₹5250

Required profit = ₹(5250 – 2250) = ₹3000

83. (a) Selling price = Marked price – Discount

= 200 – 20% of 200 = 160

Cost Price = 160 – 16 = 144

$$\text{Gain}\%=\frac{16}{144}\times100=\frac{100}{9}\%=11\frac{1}{9}\%$$

84. (d) Marked Price, M = 2C, where C is cost price for 15% gain, S.P. = $C+\frac{15}{100}C=1.15C$

Let discount be x%

$$2C-\frac{x}{100}\times 2C=1.15C\Rightarrow x=42.5\%$$

85. (c) Let Cost Price of watch be ₹ x

$$\text{S.P}=x-\frac{5}{100}x=.95x$$

If S.P = 0.95x + 56.25 then profit = 10%

$$\frac{0.95x+56.25-x}{x}\times100=10$$

$$\frac{56.25}{x}-0.05=\frac{1}{10}$$

$$\frac{56.25}{x}=\frac{1}{10}+\frac{5}{100}=\frac{3}{20}$$

$$x=56.25\times\frac{20}{3}=375$$

86. (a) Profit on all the goods = 18% of 6000 = ₹ 1080

Profit on half of the goods = 12% of 3000 = ₹ 360

∴ Profit on remaining half of the objects

= 1080 – 360 = ₹ 720

$$\text{Hence, required profit percentage}=\frac{720}{3000}\times100\%$$

= 24 %

Level-II

1. (a) Cost of one chair = ₹ 50

$\therefore$ Amount paid for 1 chair = $50 \times \frac{100-20}{100} \times \frac{100-15}{100}$

($\because$ Given, 20% and 15 % discount on chair)

$= 50 \times \frac{80}{100} \times \frac{85}{100} = 34$

cost of 250 chair at the rate of 34 per chair is

$250 \times 34 =$ ₹ 8500

Further , 5% discount on ₹ 8500 is

$8500 \times \frac{5}{100} =$ ₹ 425

$\therefore$ Total payment after discount

$= 8500 - 425 = 8075$

2. (d) First case,

(Refining for one hr)

Input = 1000 L

Output = $1000 \times \frac{90}{100} = 900\text{L} \Rightarrow x = 900\text{L}$

Profit = $900 \times 30 = 27000$

Second case

(Refining for $\frac{1}{2}$ hr).

Input = 9000 L

Output $y = 900 \times \frac{90}{100} = 810$ L

Profit = $810 \times 50 =$ ₹ 40500

3. (c) Let the cost price of manufactures is = P

Selling price of manufacturer = $P + P \times \frac{18}{100} = \frac{59P}{50}$

Wholesaler selling price = $\frac{59P}{50} + \frac{59P}{50} \times \frac{20}{100}$

$= \frac{59P}{50} + \frac{59P}{250} = \frac{354P}{250}$

Retailer selling price = $\frac{354P}{250} + \frac{354P}{250} \times \frac{25}{100}$

$= \frac{354P}{250} + \frac{177P}{500} = \frac{805P}{500}$

Now, $\frac{805P}{500} = 30.09$

$\Rightarrow P = 17$

Short

$P = \left(\frac{100}{118} \times \frac{100}{120} \times \frac{100}{125} \times 30.09\right) = 17$

4. (a) If listed price of article be ₹ 100 then discounted price be ₹ 80 (since discount = 20%)

After offering 16 articles to a dozen

Price of 16 articles = 80×12

Price of one article $= \frac{80 \times 12}{16} = 60$

Profit = 20 %

Cost price x (say) = $60 - x \times \frac{20}{100}$

$\Rightarrow \quad x + \frac{x}{5} = 60 \Rightarrow x = 60 \times \frac{5}{6} = 50$

Per cent above the cost price $= \frac{100-50}{50} \times 100 = 100\%$

5. (a) Let the cost of cloth per cm be ₹ x

As he uses 120 cm scale, so, he has 120 cm cloth cost incurred = $100x$. While selling he uses 80 cm scale, so actually he charges for $\frac{100}{80} \times 120 = 150$ cm of cloth

Amount obtained after 20% discount

$= 0.8x \times 150 = 120x$

$\therefore$ Profit = $\frac{20x}{100x} \times 100 = 20\%$

6. (a) 10% of cost price = ₹ 20

$\therefore$ Original cost of book = ₹ 200

Now, revised cost = 200×1.5 = ₹ 300

$\therefore$ Profit = 10% of C.P. = ₹ 30

7. (a) $80 : 9 = 105 : x$ or $x = \left(\frac{9 \times 105}{80}\right) = 11.81$

Hence, S.P. per k.g = ₹ 11.81.

8. (d) C.P. of the chair = ₹ $47.5 \times \frac{100}{120}$ = ₹ 39.6

C.P. of the table = ₹ $57.5 \times \frac{100}{115}$ = ₹ 50

$\Rightarrow$ The required S.P. for the table

= ₹ 39.6 + ₹ 50 − ₹ 36 = ₹ 53.6

9. (c) Let CP of 5 dozen mangoes by ₹ x. SP = ₹ 156 and Gain = $0.3\,x$

$\therefore\ 156 - x = 0.3\,x \qquad \therefore\ x = 120$

$\therefore$ SP of 60 mangoes = 120×1.6 = ₹ 192

$\therefore$ SP per mango = ₹ 3.2.

10. (b) $\frac{10(20+10)-100(5-10)}{10} = \frac{800}{10} =$ ₹ 80

11. (b) Let the cost price of clocks A and B be 'a' and $(650-a)$ respectively.

Selling price for A = Selling price for B

$$a\left(1+\frac{20}{100}\right)=(650-a)\left(1-\frac{25}{100}\right)$$

$$\left(\frac{120a}{100}\right)=(650-a)\left(\frac{75}{100}\right)$$

$$a=250$$

Cost price for $B = 650-250 = 400$

12. (c) Cost price = Money spent by the person to purchase + selling expenses

$$600+\left(600\times\frac{15}{100}\right)=690$$

Hence, selling price = $690\left(1+\frac{25}{100}\right)=862.50$

13. (d) Difference in rupees by increasing the price by rupees 1 is ₹350. That means that the quantity of milk is 350 litre. Now, 10 equal containers will become 35 litre per container.

14. (b) $\text{C.P.}=200\left(1-\frac{10}{100}\right)\left(1-\frac{15}{100}\right)+7$

$$=\frac{200\times90\times85}{100\times100}+7=₹153+7=₹160$$

$\therefore$ $\%\text{ gain}=\frac{200-160}{160}\times100=\frac{40}{160}\times100=25\%$

15. (d) C.P. = ₹ (16 × 2) = 32. S.P. = ₹ (12 × 1.5 + 4 × 0.5) = ₹ (18 + 2) = ₹ 20.

$\therefore$ $\text{Loss\%}=\left(\frac{12}{32}\times100\right)\%=37.5\%$

16. (d) Since, selling price of both the products is same

$\therefore$ % loss = % gain

$\Rightarrow$ 20% of A = 30% of $B \Rightarrow A/B = 3/2$

Let cost of product $A = 3x$ and cost of product $B = 2x$.

According to the question,

$$3x\times\frac{15}{100}-2x\times\frac{15}{100}=6$$

$\Rightarrow$ $45x-30x=600 \Rightarrow x=\frac{600}{15}=40$

Hence, cost of product $B = 2\times40 = 80$ million

17. (b) Let the value of consignment be x.

when $\frac{2}{3}$rd of consignment was sold at a profit of 5%,

then Profit = $\frac{\frac{2}{3}x\times5}{100}$

When the remaining which is $\frac{x}{3}$ consignment was sold at a loss of 2%, then according to Ques, we have

$$\frac{\frac{2x}{3}\times5}{100}-\frac{\frac{x}{3}\times2}{100}=400 \quad (\because \text{Total profit}=400)$$

$\Rightarrow 10x-2x=120000 \Rightarrow x=15{,}000$

18. (d) Let the cost price of radio be ₹ 1000.

Dealer sold it at a loss of 2.5%

i.e. selling price = 997.5

When he sold it ₹ 100 more, then selling price = 1007.5

$$\left(\because \text{gain}=7\frac{1}{2}\%\right)$$

Now, In order to gain $12\frac{1}{2}\%$,

Selling price will be $1000\times\frac{25}{200}+1000=1125$.

19. (a) RM + MC = Total cost

Total cost + Profit = Sale price

70 + 30 = 100   100 + 10 = 110 ⟶ + 80%

84 + 42 = 126   126 + 72 = 198

Therefore profit % = $\frac{72}{126}\times100=57.14\%$

20. (c) The following alligation visualization would help us solve the problem:

CUPBOARD		COT
20% profit	25.833% profit	30% profit

Ratio of cost of cupboard to cost of COT
= 4.1666 : 5.8333
= 25 : 35
= 5 : 7

Cost of cupboard = 5 × 18000/12 = 7500.

21. (c) The total discount offered by A = 8% on 20000 + 5% on 16000 = 1600 + 800 = 2400.

If B wants to be as competitive, he should also offer a discount of ₹ 2400 on 3600. Discount percentage = 2400 × 100/36000 = 6.66% discount.

22. (b) Let the original cost price be ₹ 100.

Then, profit = ₹ 320 and SP = ₹ 420

New CP = ₹ 125

New profit = SP – New CP = ₹ 295

$\Rightarrow$ Required profit = $\frac{295}{420}\times100=70.23\%\approx70\%$

23. (b) CP of 150 calculators = 150 × 250 = ₹ 37,500.

$\therefore$ total CP = 37,500 + 2500 = ₹ 40,000

Marked price of 150 calculators = 150 × 320 = ₹48,000

Selling price after discount = $48000 \times \frac{95}{100}$

= ₹45,600

∴ Percentage profit = $\frac{45,600-40,000}{40,000} \times 100 = 14\%$

24. **(a)** Women's shirt comprise 60% of the output.

∴ Men's shirts comprise (100 – 60) = 40% of the out put.

∴ Average profit from men's shirt = 8% of 40 = 3.2 out of 40

Overall average profit = 6 out of 100

∴ Average profit from women's shirts = 2.8 out of 60 i.e. 0.0466 out of each shirt.

25. **(a)** If the CP is 100, marked price = 125.

But discount to the cash purchase

= 12.5% of ₹125 = ₹$15\frac{5}{8}$

Now the reduced price he gets from the cash purchaser

$= 125 - 15\frac{5}{8} = 109\frac{3}{8}$

i.e., Gain percent = $109\frac{3}{8} - 100 = 9\frac{3}{8}$

26. **(d)** Let the actual cost price be ₹100

Actual selling price at 10% profit = ₹110

Supposed cost price at 4% less = ₹96

Supposed selling price at $18\frac{3}{4}$% profit

$= ₹96 \times \frac{118\frac{3}{4}}{100} = ₹114.$

∴ The difference in the selling price

= ₹114 – ₹110 = ₹4

Now, use the unitary method.

If the difference is ₹4. the cost price = ₹100

If the difference is ₹6, the cost price

$= ₹\frac{100}{4} \times 6 = ₹150.$

27. **(a)** By Rule fraction:

First purchased for $250\left(\frac{100}{125}\right)\left(\frac{100}{125}\right)\left(\frac{100}{125}\right)$

$= 250\left(\frac{4}{5}\right)\left(\frac{4}{5}\right)\left(\frac{4}{5}\right) = ₹128$

28. **(b)** By the rule of fraction:

He purchased $15\left(\frac{100-9}{100}\right)$ for a rupee.

Now to gain 5%, he must sell $15\left(\frac{91}{100}\right)\left(\frac{100}{105}\right) = 13$ for a rupee

29. **(b)** Cost price = x

Marked price = $x + 205$

Selling price = $0.9x + 184.5$

Percentage profit = $[(-0.1x + 184.5)x \times 100.$

$= \frac{18450 - 10x}{x}$

30. **(a)** Original Cost Price = ₹5000

New Cost Price = 1.3 × 5000 = ₹6500

Price paid by retailer = 1.2 × 5750 = ₹6900

Profit percentage = (400/6500)×100 = 6(2/13)%

31. **(c)** The interpretation of the first statement is that if the loss at 275 is 1L, the profit at 800 is 20L.

Thus, 21L = 800 – 275 = 525 → L = 25.

Thus, the cost price of the item is ₹300.

To get a profit of 25%, the selling price should be 1.25 × 300 = 375.

32. **(a)** CP of A = $\frac{1818}{0.9} = 2020$

CP of B = $\frac{1818}{1.01} = 1800$

$\frac{\text{CP of A}}{\text{CP of B}} = \frac{2020}{1800} = \frac{101}{90} = 101 : 90$

33. **(d)** Let x be no. of units.

∴ Profit per unit $x = (60 - 40)x = 20x$.

Now, additional cost = 3000

∴ To make a profist of at least ₹1000 we have

$20x - 3000 = 1000 \Rightarrow 20x = 4000$

$\Rightarrow x = 200.$

34. **(a)** She should opt for a straight discount of 30% as that gives her the maximum benefit.

35. **(c)** Profit % = $\frac{25}{100} = \frac{120 + k\text{(Profit)}}{880\ \text{(Sale)}} \Rightarrow k = 100$

Therefore, net profit % = $\frac{100}{1000} \times 100 = 10\%$

36. **(b)**

	A	B	C
Investment	$3x$	$4x$	$5x$
Rate of return	$6y\%$	$5y\%$	$4y\%$
Return	$\frac{18xy}{100}$	$\frac{20xy}{100}$	$\frac{20xy}{100}$

Total = (18 + 20 + 20) = $\frac{58xy}{100}$

B's earnings – A's earnings = $\frac{2xy}{100} = 250$

Total earning = $\frac{58xy}{100} = 7250$

37. (b) Total cost of 4 cars = 1 + 2 = 3 lakh

Total SP of 4 cars = 3 × 1.5 = 4.5 lakh

SP of car = 1.2 lakh

SP of rest 3 cars = 4.5 – 1.2 = 3.3 lakh

Average SP of all the 3 cars = 1.1 lakh

38. (c) Let C.P. of each clock be ₹ x. Then, C.P. of 90 clocks = ₹ $90x$.

$\therefore$ $[(110\% \text{ of } 40x) + (120\% \text{ of } 50x)] - (115\% \text{ of } 90x) = 40$

$\Rightarrow 44x + 60x - 103.5x = 40 \Rightarrow 0.5x = 40 \Rightarrow x = 80$

39. (b) Total cost (assume) = 100.

Recovered amount = 65 + 0.85 × 32.5 + 0.7 × 32.5

= 65 + 27.625 + 22.75 = 115.375

Hence, profit percent = 15.375%

40. (a) Assume he bought 20 apples each. Net investment ⇒ ₹ 5 + ₹ 4 = ₹ 9 for 40 apples. He would sell 40 apples @ (40 × 2)/9 = ₹ 8.888 → Loss of ₹ 0.111 on ₹ 9 investment

Loss percentage = 1.23%

41. (d) Amar–100, Bharat–120, Sridhar–132

No profit or loss is mentioned about the deal between Sridhar and the mechanic. So the answer cannot be determined.

42. (c) $100 \times \frac{1000}{x} - 100 = \frac{25}{6} \Rightarrow x = 960$ gm

43. (a) Article purchased = ₹ 78350

Marked price = ₹$\left(78350 \times \frac{130}{100}\right)$ = ₹101855

After Discount Price of Article

= ₹$\left(101855 \times \frac{80}{100}\right)$ = ₹81484

Profit Percentage = $\frac{\text{Profit}}{\text{Cost Price}} \times 100$

$\Rightarrow \frac{81484 - 78350}{81484} \times 100 = 3.8 \approx 4\%$

44. (a) Let CP = ₹100

Then, S.P = ₹117

Let marked price be Rs x.

Then, 90% of $x = 117 \Rightarrow x = \left(\frac{117 \times 100}{90}\right) = 130$

$\therefore$ Marked price = 30% above C.P.

45. (b) $S.P = C.P\left(\frac{80}{100}\right) \Rightarrow S.P = \frac{4}{5}C.P$ ...(1)

$S.P + 12 = C.P\left(\frac{110}{100}\right) \Rightarrow S.P = \frac{11}{10}C.P - 12$ ...(2)

From eqs. (1) and (2)

$\frac{4}{5}C.P = \frac{11}{10}C.P - 12$

$\Rightarrow \frac{11}{10}C.P - \frac{4}{5}C.P = 12 \Rightarrow C.P = ₹40$

46. (a) The trader professes to sell 1200 kg but sells only 1000 kg.

So profit = 20%

Markup = 10%

Total profit $= 10 + 20 + \frac{10 \times 20}{100} = 32\%$

47. (d) Let the marked price be ₹ x.

$\because$ C.P. $= (x - 25\% \text{ of } x) = \frac{3}{4}x$

$\Rightarrow$ S.P. $= \left(\frac{3x}{4} + 10\% \text{ of } \frac{3x}{4}\right) = \frac{33}{40}x$

But, $\frac{33}{40}x = 660 \Rightarrow x = 800$.

48. (a) Let the marked price be ₹ x.

$\because$ S.P. $= (x - 25\% \text{ of } x) = \frac{3}{4}x$

But, S.P = ₹ 525

$\therefore \frac{3}{4}x = 525 \Rightarrow x = 700$

49. (c) **Short cut method :**

Net profit = Profit + Discount + $\frac{\text{Profit} \times \text{Discount}}{100}$

$\frac{25}{2} = 25 - \text{Discount} - \frac{25 \times \text{Discount}}{100}$

('–' to represent discount)

$\frac{25}{2} - 25 = \frac{-5}{4}$ Discount

$\therefore$ Discount % = 10%

50. (c) **Ist case :**

$S.P = \frac{100 + \text{Profit\%}}{100} \times C.P. \Rightarrow S.P. = \frac{100 + \frac{25}{2} \times C.P}{100}$

$\Rightarrow S.P = \frac{112.5}{100}CP$ ...(1)

IInd case :

$$S.P = \frac{100 + \text{Profit } \%}{100} \times C.P. \Rightarrow (S.P + 10) = \frac{100 + 15}{100} \times C.P.$$

$$\Rightarrow \quad (S.P+10) = \frac{115}{100} C.P \qquad ...(2)$$

Dividing equation (1) by (2)

$$\frac{S.P}{S.P+10} = \frac{112.5}{100}(C.P) \times \frac{100}{115(C.P)}$$

$$S.P = \left(\frac{112.5}{115}\right)(S.P + 10)$$

115 S.P = 112.5 SP + 1125

S.P = 450

$$\therefore \quad C.P = \frac{S.P \times 100}{112.5} = \frac{450 \times 100}{112.5} = 400$$

51. **(c)** Let marked price of goods be ₹ 100.

$$\text{Selling price of goods} = 100 - \frac{10}{100} \times 100 = ₹\,90$$

Cost price of goods is 80% of its selling price

$$C.P. = \frac{80}{100} \times 90 = 72$$

Profit on goods = (90 – 72) = ₹ 18

$$\text{Profit } \% = \frac{18}{72} \times 100 = 25\%$$

52. **(a)** Let marked price of the instrument be ₹ x

$$\text{Selling price, S.P.} = x - \frac{20}{100} x = 0.8x$$

$$\text{Cost price, C.P.} = C.P. + \frac{25}{100} C.P. = 0.8x$$

$$C.P. = \frac{0.8 \times 100}{125} = \frac{16}{25} x$$

$$x = \frac{25}{16} C.P.$$

$$\text{Given that } \frac{25}{100} C.P = 150$$

$$\Rightarrow C.P. = \frac{150 \times 100}{25} = 600$$

$$\text{Marked price } x = \frac{25}{16} \times 6{,}000 = ₹\,938.50$$

53. **(b)** Let labelled price of T.V. be ₹ x

$$\text{Price after 20\% discount, } x - \frac{20}{100} x = 0.8x$$

$$\text{Price after 30\% discount, } x - \frac{30}{100} x = 0.7x$$

According to question

0.8x – 0.7x = 800

x = 800 × 10 = 8000

54. **(b)** Let ₹ 100 be the cost price for A.

S.P. for A = 100 + 20% of 100 = 120

S.P. for B = 120 – 15% of 120 = 102

$$\text{Profit } \% = \frac{102 - 100}{100} \times 100 = 2\%$$

55. **(b)** 8 C.P = 6 S.P

$$S.P = \frac{8}{6} C.P = \frac{4}{3} C.P$$

$$\text{gain } \% = \frac{\frac{4}{3} C.P. - C.P.}{C.P.} \times 100 = 33\frac{1}{3}\%$$

CHAPTER 7

INTEREST

INTEREST

If an agency (i.e. an individual, a firm or a bank etc.) borrow some money from any other agency, then the first agency is called the *borrower* and the second agency is called the *lender*. The borrowed money is called the *principal*.

If the borrower has to pay some additional money together with the borrowed money for the benefit of using borrowed money for a certain time period is called *loan period*, then this additional money is called the *interest* and the principal together with the interest is called the *amount* (i.e. Amount = Principal + Interest). When we deposite money in a bank, we earn interest, interest is calculated according to an agreement which specifies the rate of interest. Generally the rate of interest is taken as "percent per annum" which means "per ₹ 100 per year". For example, a rate of 10% per annum means ₹ 10 on ₹ 100 for 1 year.

SIMPLE INTEREST (S.I.)

If the principal remains the same for whole loan period, then the interest is called the simple interest.

$$\text{S.I.} = \frac{\text{PRT}}{100}$$

where P = Principal, R = Rate of interest in percent per annum, T = Loan period (or whole time period in years)

In the formula of simple interest, by putting the value of any three unknowns out of the four unknowns *S.I., P, R, T*; you can find the remaining fourth unknown.

Simple rate of interest is generally written as rate of interest only i.e. if it is not mentioned whether the interest is simple or compound, then we should assume it as simple interest.

Illustration 1: At what rate percent by simple interest, will a sum of money double itself in 5 years 4 months ?

Solution: Let P = ₹ x

Then A = ₹ $2x$

$\therefore$ S.I. = $A - P$ = ₹ $2x$ – ₹ x = ₹ x

$$T = 5 \text{ years } 4 \text{ months} = 5\frac{4}{12} \text{ years}$$

$$= 5\frac{1}{3} \text{ year} = \frac{16}{3} \text{ years}$$

Let R be the rate percent per annum.

Using $R = \frac{\text{S.I.} \times 100}{\text{P} \times \text{T}}$, We get $R = \frac{x \times 100}{x \times \frac{16}{3}} = \frac{300}{16} = 18.75$.

Hence required rate = 18.75 % p.a.

Illustration 2: Find the SI on ₹ 1800 from 21st Feb. 2003 to 12th April 2003 at 7.3% rate per annum.

Solution: P = ₹ 1800; R = 7.3%; I = ?

No. of days = 7 + 31 + 12 = 50 days.

$$T = \frac{50}{365} \text{ years.}$$

$$I = \frac{PTR}{100} = \frac{1800 \times \frac{50}{365} \times 7.3}{100} = ₹\ 18.$$

COMPOUND INTEREST (C.I.)

If the borrower and the lender agree to fix up a certain interval of time (a year, a half year or a quarter of a year etc.) called conversion period, so that the amount (= principal + interest) at the end of an conversion period becomes the principal for the next conversion period, then the total interest over the whole loan period calculated in this way is called the compound interest.

Note: The main difference between the simple interest and the compound interest is that the principal in the case of simple interest remains constant throughout the loan period whereas in the case of compound interest, the principal changes periodically (i.e. after each conversion period) throughout the loan period.

Rate of interest is always given annually but it can be compounded annually, half yearly, quaterly or monthly.

Interest compounded annually means conversion period is one year and hence amount at the end of every one year becomes the principal for the next conversion period.

Interest compounded half yearly means conversion period is half year and hence amount at the end of every half year becomes the principal for the next conversion period.

Interest compounded quarterly means conversion period is a quarter of a year and hence amount at the end of every quarter of a year becomes the principal for the next conversion period.

Similarly, interest compounded monthly means conversion period is one month and hence amount at the end of every one month becomes the principal for the next conversion period.

1. Computation of Compound Interest When Interest is Compounded Annually

$$A = P\left(1+\frac{r}{100}\right)^n$$

$$\therefore \quad \text{C.I.} = A - P = P\left[\left(1+\frac{r}{100}\right)^n - 1\right]$$

Here A is the amount,
P is the principal, r is the rate of interest in percent per conversion period and n is the number of conversion periods in the whole loan period.

In the formula of compound interest by putting the value of any three unknowns out of the four unknowns A, P, r and n; you can find the remaining fourth unknown.

Illustration 3: Roohi deposited ₹ 7,000 in a finance company for 3 years at an interest of 15% per annum compounded annually. What is the compound interest and the amount that Roohi gets after 3 years ?

Solution: Principal, P = ₹ 7000, $n = 3$ years, $r = 15\%$ per annum

Amount of C.I.

$$A = P\left(1+\frac{r}{100}\right)^n = 7000\left(1+\frac{15}{100}\right)^3 = 7000\left(\frac{115}{100}\right)^3$$

$$= 7000 \times \frac{115}{100} \times \frac{115}{100} \times \frac{115}{100} = 10646.125 = ₹\ 10646.13$$

$$= ₹\ 10646 \text{ (approx.)}$$

Compound interest = A – P = 10646 – 7000 = ₹ 3646

2. Computation of Compound Interest When Interest is Compounded *k* Times Every Year

If r be the rate of interest in percent per year, then the rate of interest in percent per conversion period is $\frac{r}{k}$.

If n be the number of years in the whole loan period (or whole time period), then the number of conversion period is nk.

$$\therefore \quad A = P\left(1+\frac{r}{100k}\right)^{nk}$$

$$\text{and} \quad \text{C.I.} = P\left[\left(1+\frac{r}{100k}\right)^{nk} - 1\right]$$

(a) In case of interest compounded half-yearly, $k = 2$

$$\therefore \quad A = P\left(1+\frac{r}{100\times 2}\right)^{2n}$$

$$\text{and} \quad \text{C.I.} = P\left[\left(1+\frac{r}{100\times 2}\right)^{2n} - 1\right]$$

(b) In case of interest compounded quarterly, $k = 4$

$$\therefore \quad A = P\left(1+\frac{r}{100\times 4}\right)^{4n}$$

$$\text{and} \quad \text{C.I.} = P\left[\left(1+\frac{r}{100\times 4}\right)^{4n} - 1\right]$$

(c) In case of interest compounded monthly, $k = 12$

$$\therefore \quad A = P\left(1+\frac{r}{100\times 12}\right)^{12n}$$

$$\text{and} \quad \text{C.I.} = P\left[\left(1+\frac{r}{100\times 12}\right)^{12n} - 1\right]$$

Illustration 4: A sum of money is lent out at compound interest rate of 20 % per annum for 2 years. It would fetch ₹ 482 more if interest is compounded half-yearly. Find the sum.

Solution: Suppose the sum is ₹ P.

C.I. when interest is compounded yearly

$$= P\left[1+\frac{20}{100}\right]^2 - P$$

C.I. when interest is compounded half-yearly

$$= P\left[1+\frac{10}{100}\right]^4$$

Now, we have,

$$P\left[1+\frac{10}{100}\right]^4 - P\left[1+\frac{20}{100}\right]^2 = 482$$

$$\Rightarrow \quad P\left[\{1.1\}^4 - \{1.2\}^2\right] = 482$$

$$\Rightarrow \quad P\left[\left\{(1.1)^2 - (1.2)\right\}\left\{(1.1)^2 + (1.2)\right\}\right] = 482$$

$$\Rightarrow \quad P\,[\{1.21 - 1.2\}\{1.21 + 1.2\}] = 482$$

$$\Rightarrow \quad P\,[(0.01)(2.41)] = 482$$

$$\therefore \quad P = \frac{482}{2.41\times 0.01} = ₹\ 20{,}000$$

Illustration 5: Lussy deposited ₹ 7500 in a bank which pays him 12% interest per annum compounded quarterly. What is the amount which she will receive after 9 months?

Solution: Here, P = ₹ 7500, $r = 12\%$ per annum and $n = 9$ months $= \frac{9}{12}$ years $= \frac{3}{4}$ years.

$$\therefore \quad \text{Amount after 9 months} = P\left(1+\frac{r}{400}\right)^{4n}$$

$$= ₹\,7500 \times \left(1+\frac{12}{400}\right)^{4\times\frac{3}{4}} = ₹\,7500 \times \left(1+\frac{3}{100}\right)^3$$

$$= ₹\,7500 \times \frac{103}{100} \times \frac{103}{100} \times \frac{103}{100} = ₹\,8195.45$$

Illustration 6: A sum of money placed at compound interest doubles itself in 4 years. In how many years will it amount to eight times itself ?

Solution: We have

$$P\left(1+\frac{r}{100}\right)^4 = 2P$$

$$\therefore \quad \left(1+\frac{r}{100}\right)^4 = 2$$

Cubing both sides, we get

$$\left(1+\frac{r}{100}\right)^{12} = 2^3 = 8$$

$$\therefore \quad P\left(1+\frac{r}{100}\right)^{12} = 8P$$

Hence required time = 12 years.

Shortcut Approach:

x becomes $2x$ in 4 yrs.

$2x$ becomes $4x$ in 4 yrs.

$4x$ becomes $8x$ in 4 yrs.

Thus x becomes $8x$ in 4 + 4 + 4 = 12 yrs.

Remember the follow result

If a sum becomes x times in y years at compound interest then it will be $(x)^n$ times in ny years

Thus if a sum becomes 3 times in y years at compound interest, it will be $(3)^2$ times in $2 \times 3 = 6$ years.

Illustration 7: If a sum deposited at compound interest becomes double in 4 years, when will it be 4 times at the same rate of interest ?

Solution: Using the above conclusion, we say that the sum will be $(2)^2$ times in $2 \times 4 = 8$ years.

---

3. Computation of Compound Interest When Interest is Compounded Annually but Rate of Interest in Percent being Different for Different Years

$$A = P\left(1+\frac{R_1}{100}\right)\left(1+\frac{R_2}{100}\right)...\left(1+\frac{R_n}{100}\right),$$

where R_1, R_2, ..., R_n are rate of interest in percent per year for different years.

Illustration 8: Ram Singh bought a refrigerator for ₹ 4000 on credit. The rate of interest for the first year is 5% and of the second years is 15%. How much will it cost him if he pays the amount after two years.

Solution: Here, P = ₹ 4000, R_1 = 5% per annum and R_2 = 15% per annum.

$$\therefore \quad \text{Amount after 2 years} = P\left(1+\frac{R_1}{100}\right)\left(1+\frac{R_2}{100}\right)$$

$$= ₹\, 4000 \times \left(1+\frac{5}{100}\right)\left(1+\frac{15}{100}\right)$$

$$= ₹\, 4000 \times \left(1+\frac{1}{20}\right)\left(1+\frac{3}{20}\right)$$

$$= ₹\, 4000 \times \frac{21}{20} \times \frac{23}{20} = ₹\, 4830$$

Thus, the refrigerator will cost ₹ 4830 to Ram Singh.

---

4. If *P* be the value of an article (or population of a town or a country etc.) at a certain time and *R*% per annum is the rate of depreciation, then the value *A* at the end of *n* years is given by

$$A = P\left(1-\frac{R}{100}\right)^n$$

Illustration 9: The population of a town 2 years ago was 62500. Due to migration to cities, it decreases every year at the rate of 4% per annum. Find its present population.

Solution: We have,

Population two years ago = 62500

Rate of decrease of population = 4% per annum.

$$\therefore \quad \text{Present population} = 62500 \times \left(1-\frac{4}{100}\right)^2$$

$$= 62500 \times \left(1-\frac{1}{25}\right)^2$$

$$= 62500 \times \left(\frac{24}{25}\right)^2$$

$$= 62500 \times \frac{24}{25} \times \frac{24}{25} = 57600$$

Hence, present population = 57600

---

5. If P be the population of a country (or value of an article etc.) at a certain time, which increases at the Rate R_1 % per year for first n_1 years and decreases at the rate of R_2 % per year for next n_2 years, then the population at the end of $(n_1 + n_2)$ years is given by

$$A = P\left(1+\frac{R_1}{100}\right)^{n_1} \cdot \left(1-\frac{R_2}{100}\right)^{n_2}$$

This formula can be extended for more than 2 different periods and rates.

Illustration 10: 10000 workers were employed to construct a river bridge in four years. At the end of first year, 10% workers were retrenched. At the end of the second year, 5% of the workers at the begining of the second year were retrenched. However to complete the project in time, the number of workers was increased by 10% at the end of the third year. How many workers were working during the fourth year ?

Solution: We have,

Initial number of workers = 10000

Reduction of workers at the end of first year = 10%

Reduction of workers at the end of second year = 5%

Increase of workers at the end of third year = 10%

$\therefore$ Number of workers working during the fourth year

$$= 10000\left(1-\frac{10}{100}\right)\left(1-\frac{5}{100}\right)\left(1+\frac{10}{100}\right)$$

$$= 10000 \times \frac{9}{10} \times \frac{19}{20} \times \frac{11}{10} = 9405$$

Hence, the number of workers working during the fourth year was 9405.

Practice Exercise

Level - I

1. To buy furniture for a new apartment, Sylvia Chang borrowed ₹ 5000 at 11% per annum simple interest for 11 months. How much interest will she pay?
(a) 500 (b) 504.17
(c) 6050 (d) 605

2. Find the compound interest on ₹ 18,750 in 2 years the rate of interest being 4% for the first year and 8% for the second year.
(a) 2310 (b) 1130
(c) 3120 (d) None of these

3. At a simple interest ₹ 800 becomes ₹ 956 in three years. If the interest rate, is increased by 3%, how much would ₹ 800 become in three years?
(a) ₹ 1020.80 (b) ₹ 1004
(c) ₹ 1028 (d) Data inadequate

4. On retirement, a person gets 1.53 lakhs of his provident fund which he invests in a scheme at 20% p.a. His monthly income from this scheme will be
(a) ₹ 2,450 (b) ₹ 2,500
(c) ₹ 2,550 (d) ₹ 2,600

5. In how many minimum number of complete years, the interest on ₹ 212.50 P at 3% per annum will be in exact number of rupees?
(a) 6 (b) 8
(c) 9 (d) 7

6. A scooter costs ₹ 25, 000 when it is brand new. At the end of each year, its value is only 80% of what it was at the beginning of the year. What is the value of the scooter at the end of 3 years?
(a) ₹ 10,000 (b) ₹ 12,500
(c) ₹ 12,800 (d) ₹ 12,000

7. Village A has a population of 6800, which is decreasing at the rate of 120 per year. Village B has a population of 4200, which is increasing at the rate of 80 per year. In how many years will the population of the two villages will become equal ?
(a) 9 (b) 11
(c) 13 (d) 16

8. A person invested some amount at the rate of 12% simple interest and a certain amount at the rate of 10% simple interest. He received yearly interest of ₹ 130. But if he had interchanged the amounts invested, he would have received ₹ 4 more as interest. How much did he invest at 12% simple interest ?
(a) ₹ 700 (b) ₹ 500
(c) ₹ 800 (d) ₹ 400

9. A certain amount is lent at x% p.a. simple interest for two years. Instead, if the amount was lent at $2x$% p.a. simple interest for 'y' more years, then the interest would have been five times the earlier interest. What is the value of y?
(a) 2 years (b) 3 years
(c) 4 years (d) 5 years

10. A certain sum of money triple itself in 8 years. In how many years it will be five times?
(a) 22 years (b) 16 years
(c) 20 years (d) 24 years

11. The difference between CI and SI on a certain sum of money at 10% per annum for 3 years is ₹ 620. Find the principal if it is known that the interest is compounded annually.
(a) ₹ 200, 000 (b) ₹ 20,000
(c) ₹ 10,000 (d) ₹ 100, 000

12. An amount of ₹ 12820 due 3 years hence, is fully repaid in three annual instalments starting after 1 year The first instalment is 1/2 the second instalment and the second instalment is 2/3 of the third instalment. If the rate of interest is 10% per annum, find the first instalment.
(a) ₹ 2400 (b) ₹ 1800
(c) ₹ 2000 (d) ₹ 2500

13. What will be the ratio of simple interest earned by certain amount at the same rate of interest for 6 years and that for 9 years?
(a) 1 : 3 (b) 1 : 4
(c) 2 : 3 (d) None of these

14. A man borrows ₹ 6000 at 5% interest, on reducing balance, at the start of the year. If he repays ₹ 1200 at the end of each year, find the amount of loan outstanding, (in ₹), at the beginning of the third year.
(a) 3162.75 (b) 4155.00
(c) 4155.00 (d) 5100.00

15. Two equal sums were lent, one at the rate of 11% p.a. for five years and the other at the rate of 8% p.a. for six years, both under simple interest. If the difference in interest accrued in the two cases is ₹ 1008. find the sum.
(a) ₹ 11,200 (b) ₹ 5,600
(c) ₹ 12,600 (d) ₹ 14,400

16. A sum is invested at compound interest payable annually. The interest in two successive years was ₹ 225 and ₹ 236.25. Find the rate of interest
(a) 2% (b) 3%
(c) 5% (d) 9%

17. A person borrowed ₹ 500 at 3% per annum S.I. and ₹ 600 at $4\frac{1}{2}$% per annum on the agreement that the whole sum, will be returned only when the total interest becomes ₹ 126. The number of years, after which the borrowed sum is to be returned, is :
(a) 2 (b) 3
(c) 4 (d) 5

18. A bank offers 5% compound interest calculated on half-yearly basis. A customer deposits ₹ 1600 each on 1st January and 1st July of a year. At the end of the year, the amount he would have gained by way of interest is
(a) ₹ 120 (b) ₹ 121
(c) ₹ 122 (d) ₹ 123

19. A sum of money invested at simple interest triples itself in 8 years. How many times will it become in 20 years time?
(a) 8 times (b) 7 times
(c) 6 times (d) 9 times

20. The population of a city is 200,000. If the annual birth rate and the annual death rate are 6% and 3% respectively, then calculate the population of the city after 2 years.
(a) 212,090 (b) 206,090
(c) 212,000 (d) 212,180

21. The population of Bangalore was 1283575 on 1 January 2001 and the growth rate of population was 10% in the last year and 5% in the years prior to it, the only exception being 1999 when because of a huge exodus there was a decline of 20% in population. What was the population of January 1, 1995 ?
(a) 1,000,000 (b) 1,200,000
(c) 1,250,000 (d) 1,500,000

22. A person bought a motorbike under the following scheme: Down payment of ₹ 15,000 and the rest amount at 8% per annum for 2 years. In this way, he paid ₹ 28,920 in total. Find the actual price of the motorbike. (Assume simple interest).
(a) ₹ 26,000 (b) ₹ 27,000
(c) ₹ 27,200 (d) ₹ 26,500

23. The ratio of the amount for two years under C.I. annually and for one year under S.I. is 6 : 5. When the rate of interest is same, then the value of rate of interest is
(a) 12.5% (b) 18%
(c) 20% (d) 16.66%

24. Mr. Bajaj invested $\frac{1}{7}$ of his total investment at 4% and $\frac{1}{2}$ at 5% and rest at 6% for the one year and received total interest of ₹ 730. What is the total sum invested?
(a) ₹ 70000 (b) ₹ 14000
(c) ₹ 24000 (d) ₹ 38000

25. Akram Ali left an amount of ₹ 340000 to be divided between his two sons aged 10 years and 12 years such that both of them would get an equal amount when each attain 18 years age. What is the share of elder brother if the whole amount was invested at 10% simple interest ?
(a) 120000 (b) 140000
(c) 160000 (d) 180000

26. A Sonata watch is sold for ₹ 440 cash or for ₹ 200 cash down payment together with ₹ 244 to be paid after one month. Find the rate of interest charged in the instalment scheme
(a) 10% (b) 15%
(c) 20% (d) 25%

27. What will be the compound interest on a sum of ₹ 7500/- at 4 p.c.p.a. in 2 years ? *[SBI Clerk-June-2012]*
(a) ₹ 618/- (b) ₹ 612/-
(c) ₹ 624/- (d) ₹ 606/-
(e) ₹ 621/-

28. In how many years will ₹ 4600 amount to ₹ 5428 at 3 p.c.p.a. simple interest ? *[SBI Clerk-June-2012]*
(a) 3 (b) 5
(c) 6 (d) 4
(e) None of these

29. A sum of money becomes eight times in 3 years if the rate is compounded annually. In how much time, the same amount at the same compound interest rate will become sixteen times ? *[SBI Clerk-2014]*
(a) 6 years (b) 4 years
(c) 8 years (d) 5 years
(e) None of these

30. If the compound interest on a certain sum of money for 3 years at 10% p.a. be ₹ 993, what would be the simple interest ? *[SBI Clerk-2014]*
(a) ₹ 800 (b) ₹ 950
(c) ₹ 900 (d) ₹ 1000
(e) None of these

31. Mahesh starts work as a sales representative on an annual salary of ₹ 1,60,000. If he receives a 15% pay-rise each year, the number of years he has worked for the company, when his annual salary became ₹ 2,79,841 is *[SSC-Sub. Ins.-2012]*
(a) 2 (b) 3
(c) 4 (d) 5

32. A sum of money placed at compound interest doubles itself in 5 years. It will amount to eight times of itself in : *[SSC-Sub. Ins.-2013]*
(a) 15 years (b) 12 years
(c) 10 years (d) 20 years

33. A sum amounts double in 8 years by simple interest. Then the rate of simple interest p.a. is *[SSC-Sub. Ins.-2014]*
(a) 10% (b) 12.5%
(c) 15% (d) 20%

34. Rekha invested a sum of ₹ 12000 at 5% per annum compound interest. She received an amount of ₹ 13230 after n years. Find n. *[SSC-Sub. Ins.-2014]*
(a) 2.8 years (b) 3.0 years
(c) 2.5 years (d) 2.0 years

35. A sum of money becomes 1.331 times in 3 years as compound interest. The rate of interest is [SSC-MT-2013]
(a) 50% (b) 8%
(c) 7.5% (d) 10%

36. A person deposited ₹ 500 for 4 years and ₹ 600 for 3 years at the same rate of simple interest in a bank. Altogether he received ₹ 190 as interest. The rate of simple interest per annum was [SSC-MT-2013]
(a) 3% (b) 4%
(c) 5% (d) 2%

37. The difference between the interests received from two different banks on ₹ 500 for 2 years is ₹ 2.50. The difference between their rates is: [SSC 10+2-2012]
(a) 0.5% (b) 2.5%
(c) 0.25% (d) 1%

38. A principal of ₹ 10,000, after 2 years compounded annually, the rate of interest being 10% per annum during the first year and 12% per annum during the second year (in rupees) will amount to: [SSC 10+2-2012]
(a) 12,000 (b) 12,320
(c) 12,500 (d) 11,320

39. A sum becomes ₹ 2,916 in 2 years at 8% per annum compound interest. The simple interest at 9% per annum for 3 years on the same amount will be [SSC 10+2-2013]
(a) ₹ 625 (b) ₹ 600
(c) ₹ 675 (d) ₹ 650

40. The population of a village increases by 5% annually. If its present population is 4410, then its population 2 years ago was [SSC 10+2-2014]
(a) 4500 (b) 4000
(c) 3800 (d) 3500

41. A sum of ₹ 210 was taken as a loan. This is to be paid back in two equal instalments. If the rate of interest be 10% compounded annually, then the value of each instalment is [SSC 10+2-2014]
(a) ₹ 127 (b) ₹ 121
(c) ₹ 210 (d) ₹ 225

42. ₹ 64,000 will amount to ₹ 68,921 at 5% per annum and interest payable half yearly in [SSC 10+2-2014]
(a) $3\frac{1}{2}$ years (b) 2 years
(c) $2\frac{1}{2}$ years (d) $1\frac{1}{2}$ years

43. If the simple interest and compound interest at the same rate of certain amount for 2 years are ₹ 400 & ₹ 420 respectively, then the rate of interest is [SSC 10+2-2014]
(a) 12% (b) 8%
(c) 10% (d) 11%

44. What would be the compound interest obtained on an amount of ₹ 1,210 at the rate of 6. p.c.p.a. after a year ? [IBPS Clerk-2012]
(a) ₹ 70.50 (b) ₹ 74.60
(c) ₹ 73.80 (d) ₹ 72.60
(e) None of these

45. What is the difference between the simple and compound interest earned from a sum of ₹ 13,033 at a rate of 13 percent per annum for a period of 3 years (rounded off to 2 digits after decimal) ? [IBPS Clerk-2012]
(a) ₹ 5,082.87 (b) ₹ 689.41
(c) ₹ 5,772.28 (d) ₹ 680.94
(e) None of these

46. ₹ 58,750 amounts to ₹ 79,900 in four years at simple interest. What is the rate of interest paid? [IBPS Clerk-2012]
(a) 14 (b) 13
(c) 12 (d) 16
(e) 9

47. How much will a sum of ₹ 12,000 deposited at a rate of 9% per annum (simple interest) for 13 years amount to? [IBPS Clerk-2012]
(a) ₹ 14,040 (b) ₹ 20,650
(c) ₹ 13,404 (d) ₹ 27,800
(e) ₹ 26,040

48. Simple interest on a sum of money for 4 yrs at 7 pcpa is ₹ 3584. What would be the compound interest (compounded annually) on the same amount of money for 2 yr at 4 pcpa? [IBPS Clerk-2013]
(a) ₹ 1162.32 (b) ₹ 1098.72
(c) ₹ 992.38 (d) ₹ 1231.76
(e) ₹ 1044.48

Level - II

1. The comound interest on a certain sum for 2 years at 10% per annum is ₹ 1260. The simple interest on the same sum for double the time at half the rate per cent per annum is
(a) ₹ 1200 (b) ₹ 1160
(c) ₹ 1208 (c) ₹ 1175

2. The simple interest on a sum of money will be ₹ 300 after 5 years. In the next 5 years principal is trepled, what will be the total interest at the end of the 10th year?
(a) 1200 (b) 900
(c) 600 (d) 1500

3. A person lent a certain sum of money at 4% simple interest; and in 8 years the interest amounted to ₹ 340 less than the sum lent. Find the sum lent.
(a) 500 (b) 600
(c) 1000 (d) 1500

4. A sum was put at simple interest at a certain rate for 2 years. Had it been put at 1% higher rate, it would have fetched ₹ 24 more? The sum is
(a) 1200 (b) 1500
(c) 1800 (d) 2000

5. A sum of money at compound interest amounts in two years to ₹ 2809, and in three years to ₹ 2977.54. Find the rate of interest and the original sum
(a) 2000 (b) 2100
(c) 2200 (d) 2500

6. Consider the following statements
If a sum of money is lent at simple interest, then the
I. money gets doubled in 5 years if the rate of interest is $16\frac{2}{3}\%$.
II. money gets doubled in 5 years if the rate of interest is 20%.
III. money becomes four times in 10 years if it gets doubled in 5 years.
Of these statements,
(a) I and III are correct (b) II alone is correct
(c) III alone is correct (d) II and III are correct

7. Simple interest on a certain amount is $\frac{9}{16}$ of the principal. If the numbers representing the rate of interest in percent and time in years be equal, then time, for which the principal is lent out, is
(a) $5\frac{1}{2}$ years (b) $6\frac{1}{2}$ years
(c) 7 years (d) $7\frac{1}{2}$ years

8. If the rate increases by 2%, the simple interest received on a sum of money increases by ₹ 108. If the time period is increased by 2 years, the simple interest on the same sum increases by ₹ 180. The sum is :
(a) ₹ 1800 (b) ₹ 3600
(c) ₹ 5400 (d) Data inadequate

9. A man lends ₹ 10,000 in four parts. If he gets 8% on ₹ 2000; $7\frac{1}{2}\%$ on ₹ 4000 and $8\frac{1}{2}\%$ on ₹ 1400; what percent must he get for the remainder, if his average annual interest is 8.13% ?
(a) 7% (b) 9%
(c) $9\frac{1}{4}\%$ (d) $10\frac{1}{2}\%$

10. A man borrows ₹ 12,500 at 20% compound interest. At the end of every year he pays ₹ 2000 as part repayment. How much does he still owe after three such instalments?
(a) ₹ 12,000 (b) ₹ 12,864
(c) ₹ 15,600 (d) None of these

11. A part of ₹ 38,800 is lent out at 6% per six months. The rest of the amount is lent out at 5% per annum after one year. The ratio of interest after 3 years from the time when first amount was lent out is 5 : 4. Find the second part that was lent out at 5%
(a) ₹ 26, 600 (b) ₹ 28,800
(c) ₹ 27,500 (d) ₹ 28,000

12. The difference between C.I. and S.I. on a certain sum of money at 10% per annum for 3 years is ₹ 620. Find the principal if it is known that the interest is compounded annually.
(a) ₹ 200,000 (b) ₹ 20,000
(c) ₹ 10,000 (d) ₹ 100,000

13. The population of towns A and B is the ratio of 1 : 4. For the next 2 years, the population of A would increase and that of B would decrease by the same percentage every year. After 2 years, their population became equal. What is the percentage change in the population?
(a) 33.33% (b) 66.66%
(c) 25% (d) Not possible

14. If the population of a town at the beginning of a year was 1530000, and the birth rate was 53.2, while the death rate was 31.2 per 1000 of the population, then the net increase in the population at the end of the year was
(a) 336600 (b) 363600
(c) 366300 (d) 330000

15. Arun borrowed a sum of money from Jayant at the rate of 8% per annum simple interest for the first four years, 10% per annum for the next six years and 12% per annum for the period beyond ten years. If he pays a total of ₹ 12,160 as interest only at the end of 15 years, how much money did he borrow?
(a) ₹ 8000 (b) ₹ 10,000
(c) ₹ 12,000 (d) ₹ 9,000

16. What will be the difference in simple and compound interest on ₹ 2000 after three years at the rate of 10 percent per annum?
(a) ₹ 160 (b) ₹ 42
(c) ₹ 62 (d) ₹ 20

17. Aniket deposited two parts of a sum of ₹ 25000 in different banks at the rates of 15% per annum and 18% per annum respectively. In one year he got ₹ 4050 as the total interest. What was the amount deposited at the rate of 18% per annum?
(a) ₹ 9000 (b) ₹ 18000
(c) ₹ 15000 (d) None of these

18. Mr. X invested an amount for 2 years at 15 percent per annum at simple interest. Had the interest been compounded, he would have earned ₹ 450/- more as interest. What was the amount invested?
(a) ₹ 22000 (b) ₹ 24000
(c) ₹ 25000 (d) None of these

19. Mr Sridharan invested money in two schemes A and B, offering compound interest at 8 percent per annum and 9 percent per annum respectively. If the total amount of interest accrued through the two schemes together in two years was ₹ 4818.30 and the total amount invested was ₹ 27,000, what was the amount invested in Scheme A ?
(a) ₹ 15,000 (b) ₹ 13,500
(c) ₹ 12,000 (d) Cannot be determined

20. Parameshwaran invested an amount of ₹ 12,000 at the simple interest rate of 10 percent per annum and another amount at the simple interest rate of 20 percent per annum. The total interest earned at the end of one year on the total amount invested became 14 percent per annum. Find the total amount invested.
(a) ₹ 22,000 (b) ₹ 25,000
(c) ₹ 20,000 (d) ₹ 24,000

21. A father left a will of ₹ 68,000 to be divided between his two sons aged 10 years and 12 years such that they may get equal amount when each attains the age of 18 years If the money is reckoned at 10% p.a., find how much each gets at the time of the will.
(a) ₹ 30,000, ₹ 38,000 (b) ₹ 28,000, ₹ 40,000
(c) ₹ 32,000, ₹ 36,000 (d) Cannot be determined

22. Two equal sums of money were invested, one at 4% and the other at 4.5%. At the end of 7 years, the simple interest received from the latter exceeded to that received from the former by ₹ 31.50. Each sum was
(a) ₹ 1,200 (b) ₹ 600
(c) ₹ 750 (d) ₹ 900

23. A sum of ₹ 725 is lent in the beginning of a year at a certain rate of interest. After 8 months, a sum of ₹ 362.50 more is lent but at the rate twice the former. At the end of the year, ₹ 33.50 is earned as interest from both the loans. What was the original rate of interest?
(a) 3.6% (b) 4.5%
(c) 5% (d) 3.46%

24. David invested certain amount in three different schemes *A*, *B* and *C* with the rate of interest 10% p.a., 12% p.a. and 15% p.a. respectively. If the the total interest accrued in one year was ₹ 3200 and the amount invested in Scheme *C* was 150 % of the amount invested in Scheme *A* and 240% of the amount invested in Scheme *B*, what was the amount invested in Scheme *B*?
(a) ₹5000 (b) ₹6500
(c) ₹8000 (d) cannot be determined

25. Subash purchased a refrigerator on the terms that he is required to pay ₹ 1,500 cash down payment followed by ₹ 1,020 at the end of first year, ₹ 1,003 at the end of second year and ₹ 990 at the end of third year. Interest is charged at the rate of 10% per annum. Calculate the cash price
(a) ₹ 3,000 (b) ₹ 2,000
(c) ₹ 4,000 (d) ₹ 5,000

26. A owes *B* ₹ 1,573, payable $1\frac{1}{2}$ years hence. Also *B* owes *A* ₹ 1,444.50, payable 6 months hence. If they want to settle the account forthwith, keeping 14% as the rate of interest, then who should pay whom and how much ?
(a) *A* to *B*, ₹ 28.50 (b) *B* to *A*, ₹ 37.50
(c) *A* to *B*, ₹ 50 (d) *B* to *A*, ₹ 50

27. Seema invested an amount of ₹ 16,000 for two years on compound interest and received an amount of ₹ 17,640 on maturity. What is the rate of interest ?
(a) 5% pa (b) 8% pa
(c) 4% pa (d) Data inadequate

28. A finance company declares that, at a certain compound interest rate, a sum of money deposited by anyone will become 8 times in three years. If the same amount is deposited at the same compound rate of interest, then in how many year will it become 16 times ?
(a) 5 years (b) 4 years
(c) 6 years (d) 7 years

29. Two friends *A* and *B* jointly lent out ₹ 81,600 at 4% per annum compound interest. After 2 years *A* gets the same amount as *B* gets after 3 years. The investment made by *B* was
(a) ₹ 40,000 (b) ₹ 30,000
(c) ₹ 45,000 (d) ₹ 38,000

30. A money-lender, lends a part of his money at 10% per annum and the rest at 15% per annum. His annual income is ₹ 1900. However, if he had interchanged the rate of interest on the two sums, he would have earned ₹ 200 more. The amount lent will fetch what 15%?
(a) ₹ 6000 (b) ₹ 4000
(c) ₹ 10000 (d) ₹ 4400

31. The simple interest on a sum of money is $\frac{1}{9}$th of the principal, and the number of years is equal to the rate per cent per annum. Find the rate per cent.
(a) $3\frac{1}{3}$% (b) 3%
(c) 10% (d) None of these

32. Amin borrowed some money from Vishwas. The rate of interest for first two years is 8% p.a., for the next three years is 11 % p.a. and for the period beyond 5 years 14% p.a. Vishwas got an amount of ₹ 10920 as an interest at the end of eight years. Then what amount was borrowed by Amin'?
(a) ₹ 12000 (b) ₹ 15000
(c) ₹ 1400 (d) None of these

33. The simple interest accrued on an amount of ₹ 22,500 at the end of four years is ₹ 10,800. What would be the compound interest accrued on the same amount at the same rate at the end of two years ? *[IBPS-PO-2011]*
(a) ₹ 16,908 (b) ₹ 5,724
(c) ₹ 28,224 (d) ₹ 8,586
(e) None of these

34. What is the difference between the simple and compound interest on ₹ 7,300/- at the rate of 6 p.c.p.a. in 2 years ? *[IBPS-PO-2012]*
(a) ₹ 29.37/- (b) ₹ 26.28/-
(c) ₹ 31.41/- (d) ₹ 23.22/-
(e) ₹ 21.34/-

35. The simple interest accrued on a sum of certain principal in 8 yr at the rate of 13% per year is ₹ 6500. What would be the compound interest accrued on that principal at the rate of 8% per year in 2 yrs? *[IBPS-PO-2013]*
(a) ₹ 1040 (b) ₹ 1020
(c) ₹ 1060 (d) ₹ 1200
(e) None of these

36. The difference between C. I. (Compound Interest) and S.I. (Simple Interest) on a sum of ₹ 4,000 for 2 years at 5% p.a. payable yearly is *[SSC CGL-2012]*
(a) ₹ 20 (b) ₹ 10
(c) ₹ 50 (d) ₹ 60

37. Two equal sums were borrowed at 8% simple interest per annum for 2 years and 3 years respectively. The difference in the interest was ₹ 56. The sums borrowed were *[SSC CGL-2013]*
(a) ₹ 800 (b) ₹ 700
(c) ₹ 560 (d) ₹ 350

Hints & Solutions

Level-I

1. (b) From the formula, $I = Prt$, with $P = 5000$, $r = .11$, and $t = 11/12$ (in years). The total interest she will pay is $I = 5000(.11)(11/12) = 504.17$
or ₹504.17

2. (a) After first year the amount
$$= 18750\left(1+\frac{4}{100}\right) = 18750\left(\frac{104}{100}\right)$$
After 2nd year the amount $= 18750\left(\frac{104}{100}\right)\left(\frac{108}{100}\right)$
$$= 18750\left(\frac{26}{25}\right)\left(\frac{27}{25}\right) = 21060$$
$\therefore$ C.I. = 21060 – 18,750 = ₹2310.

3. (c) Rate of interest $= \frac{956-800}{3\times 800}\times 100 = 6.50\%$
$\therefore$ Amount $= 800 + \frac{800\times 9.5\times 3}{100}$
= 800 + 228 = ₹ 1028

4. (c) Let S.I. = ₹x
$$= \frac{1.53\times 10^5\times 20}{100} = 30600$$
Monthly income $= \frac{30600}{12}$ = ₹ 2550

5. (b) Interest for one year = ₹ $212.50\times\frac{3}{100}\times 1$ = ₹ $\frac{51}{8}$
Thus in 8 years, the interest is ₹ 51.

6. (c) After first year, the value of the scooter
$= \frac{25000\times 80}{100}$ = ₹ 20,000
After second year, the value of scooter = ₹ 16,000
After third year, the value of scooter = ₹ 12,800

7. (c) Checking with options, we find that after 13 years, population of the village $A = 6800 - 120\times 13 = 5240$
And that of village $B = 4200 + 80\times 13 = 5240$

8. (b) Let amount invested at 12% be x and amount invested at 10% be y.
According to question
$$130 = \frac{x\times 12\times 1}{100} + \frac{y\times 10\times 1}{100}$$
$\Rightarrow 13000 = 12x + 10y$ ...(1)
And $134 = \frac{x\times 10\times 1}{100} + \frac{y\times 12\times 1}{100}$
$\Rightarrow 13400 = 10x + 12y$ (2)
From equations (1) and (2)
$x = 500$

9. (b) Let the sum be ₹ x.
$$ATQ = \frac{100\times 2x\times(y+2)}{100} = \frac{5\times 100\times x\times 2}{100}$$
$\Rightarrow y = 3$ years

10. (b) It triple itself in 8 years, which makes interest equal to 200% of principal.
So, 200% is added in 8 years
Hence, 400% which makes the whole amount equal to five times of the principal, which will be added in 16 years

11. (b) Go through trial and error of the options. You will get:
20000×(1.3) = 26000 (@ simple interest)
20000 × 1.1×1.1×1.1 = 26620 @ compound interest.
Thus 20000 is the correct answer.

12. (c) Solve using options. Option (c) fits the situation as:
12820 = 2000 + 2 years interest on 2000 + 4000 + 1 years interest on 4000 + 6000 (use 10% compound interest for calculation of interest) →
12820 = 2000 + 420 + 4000 + 400 + 6000.
Thus, option (c) fits the situation perfectly.

13. (c) Let the principal be P and rate of interest be R%.
$$\therefore \text{Required ratio} = \left[\frac{\left(\frac{P\times R\times 6}{100}\right)}{\left(\frac{P\times R\times 9}{100}\right)}\right] = \frac{6PR}{9PR} = \frac{6}{9} = 2:3.$$

14. (c) The amount man gets after one year
$$= 6000 + \frac{6000\times 5\times 1}{100} - 1200$$
= 6000 + 300 – 1200 = 5100
$\therefore$ Amount after two years *i.e.*, at the beginning of the third year
$$= 5100 + \frac{5100\times 5\times 1}{100} - 1200 = 5100 + 255 - 1200 = 4155$$
Hence option (c)

15. (d) Let the sum be ₹ x.
$$\therefore \frac{x\times 11\times 5}{100} - \frac{x\times 8\times 6}{100} = 1008$$
$$\Rightarrow \frac{7x}{100} = 1008$$
$\Rightarrow x = 14400$

16. (c) Difference in interest = 236.25 – 225 = ₹ 11.25
This difference is the simple interest over ₹ 225 for one year. Hence, rate of interest
$$= \frac{11.25\times 100}{225\times 1} = 5\%$$

17. **(b)** Let the time be x years. Then,

$$\left(\frac{500\times3\times x}{100}\right)+\left(\frac{600\times9\times x}{100\times2}\right)=126$$

$\Leftrightarrow 15x + 27x = 126 \Leftrightarrow 42x = 126 \Leftrightarrow x = 3.$

$\therefore$ Required time = 3 years

18. **(b)** Amount = ₹ $\left[1600\times\left(1+\frac{5}{2\times100}\right)^2 +1600\times\left(1+\frac{5}{2\times100}\right)\right]$

= ₹ $\left[1600\times\frac{41}{40}\times\frac{41}{40}+1600\times\frac{41}{40}\right]$

= ₹ $\left[1600\times\frac{41}{40}\left(\frac{41}{40}+1\right)\right]$ = ₹ $\left(\frac{1600\times41\times81}{40\times40}\right)$

= ₹ 3321.

C.I. = ₹ (3321 – 3200) = ₹ 121

19. **(c)** Tripling in 8 years means that the interest earned in 8 years is equal to 200% of the capital value. Thus, interest per year (simple interest) is 25% of the capital. In 20 years, total interest earned = 500% of the capital and hence the capital would become 6 times it's original value.

20. **(d)** The yearly increase in the population is 3%. Thus, the population would increase by 3% each year. 200000 would become 206000 while 206000 would become 212180.

21. **(b)** Solve through options to see that the value of 1200000 fits the given situation.

22. **(b)** Solve using options. If the price is 27000, the interest on 12000 (after subtracting the down payment) would be 16% of 12000 = 1920. Hence, the total amount paid would be 28920.

23. **(c)** On the second year (in terms of C.I.) is

$$\frac{P\left(1+\frac{r}{100}\right)^2}{\left(P+\frac{Pr}{100}\right)}=\frac{6}{5}\quad\Rightarrow\left(1+\frac{r}{100}\right)=\frac{6}{5}$$

$\Rightarrow r = 20\%$

24. **(b)** Let the principal be x, then

$$\frac{\left(\frac{x}{7}\times4+\frac{x}{2}\times5+\frac{5x}{14}\times6\right)}{100}=730$$

$\Rightarrow x = 14000$

Alternatively : Go through suitable options. Choose any middlemost option so that if the choosen option is not correct, then you can determine that whether you have to increase or decrease the value of the choices given.

25. **(d)** Go through options

$$1.8+\frac{1.8\times6\times10}{100}=1.6+\frac{1.6\times8\times10}{100}$$

Hence (d) is correct.

Alternatively : $P_1+\frac{P_1\times6\times10}{100}=P_2+\frac{P_2\times8\times10}{100}$

$\Rightarrow \quad \frac{P_1}{P_2}=\frac{9}{8}$

Share of elder brother = $\frac{340000\times9}{17}$ = ₹180000

26. **(c)** Principal for next month = 440 – 200 = 240

Amount paid after next month = 244

Therefore interest charged at ₹ 240 = 4

$\therefore \quad 4=\frac{240\times r\times1}{12\times100}$

$r = 20\%$ per annum

27. **(b)** Total amount = $7500\left(1+\frac{4}{100}\right)^2$

$= 7500\left(\frac{26}{25}\times\frac{26}{25}\right)=8112$

C I = Total amount – sum

= 8112 – 7500 = ₹612

28. **(c)** $P+\frac{p\times r\times t}{100}=5428$

$\frac{4600\times3\times t}{100}=5428-4600=828$

$\Rightarrow t=\frac{828\times100}{4600\times3}=6$ years

29. **(b)** Let the sum of money be ₹ x.

Now, $8x = x\left(1+\frac{r}{100}\right)^3$

or, $\left(1+\frac{r}{100}\right)^3=(2)^3$ or $1+\frac{r}{100}=2$

Again, let the sum becomes 16 times in n years. Then,

$16x = x\left(1+\frac{r}{100}\right)^n$

$\Rightarrow \; 16=2^n$ or $2^4=2^n$ or $n=4$

30. **(c)** Let Principal = ₹ P

$P\left(1+\frac{10}{100}\right)^3-P=993 \Rightarrow \left(\frac{11}{10}\times\frac{11}{10}\times\frac{11}{10}-1\right)P=993$

$\Rightarrow \left(\frac{1331-1000}{1000}\right)P=993$ or ,

$$P=\frac{993\times1000}{331}=3000$$

$\therefore$ Simple interest $=₹\left(\frac{3000\times3\times10}{100}\right)=₹900$

31. (c) Initial salary = 160000
15% pay-rise each year
After 1 year salary = 160000 + 15% of 160000
= 160000 + 24000 = ₹ 184000
After 2 years salary = 184000 + 15% of 184000
= 184000 + 27600 = ₹211600
After 3 years salary = 211600 + 15% of 211600
= 211600 + 31740 = ₹243340
After 4 years salary = ₹ 279841

32. (a) $A=P\left(1+\frac{R}{100}\right)^T$

$$\Rightarrow 2=1\left(1+\frac{R}{100}\right)^5$$

Cubing both sides.

$$2^3=1\left(1+\frac{R}{100}\right)^{15}$$

Therefore, T = 15 years.

33. (b) Let P be the principle amount and R be rate of interest.

$$2P=P+\frac{P\times R\times8}{100}$$

$$R=\frac{100}{8}=12.5\%$$

34. (d) P = ₹ 12000, Rate = 5 %, Time (n) = ?, Amount = 13230

$$A=P\left(1+\frac{R}{100}\right)^T\Rightarrow13230=12000\left(1+\frac{5}{100}\right)^n$$

$$\Rightarrow\frac{13230}{12000}=\left(\frac{21}{20}\right)^n\Rightarrow\frac{1323}{1200}=\left(\frac{21}{20}\right)^n$$

$$\Rightarrow\frac{441}{400}=\left(\frac{21}{20}\right)^n$$

$$\left(\frac{21}{20}\right)^2=\left(\frac{21}{20}\right)^n$$

$\therefore$ n = 2 years

35. (d) $P\left(1+\frac{R}{100}\right)^3=1.331P\Rightarrow\left(1+\frac{R}{100}\right)^3=1.331$

$$\left(1+\frac{R}{100}\right)^3=\left(\frac{11}{10}\right)^3$$

$$1+\frac{R}{100}=\frac{11}{10}\Rightarrow\frac{R}{100}=\frac{11}{10}-1\Rightarrow\frac{R}{100}=\frac{1}{10}$$

$R=10\%$

36. (c) S.I (Simple Interest)

$$=\frac{\text{Principle}_1\times\text{Rate}\times\text{Time}_1}{100}+\frac{\text{Principle}_1+\text{Rate}\times\text{Time}_1}{100}$$

$$190=\frac{500\times R\times4}{100}+\frac{600\times R\times3}{100}$$

$$190=20R+18R\Rightarrow38R=190\Rightarrow R=5\%$$

37. (c) Let r_1 and r_2 are the rates of interests.
So, the difference in S.I

$$=\frac{\text{principal}\times\text{time}\times\text{difference between the rates of interests}}{100}$$

$$\Rightarrow 2.50=\frac{500\times2\times(r_1-r_2)}{100}$$

So, $(r_1-r_2)=\frac{2.50\times100}{500\times2}=0.25$

38. (b) $A=P\left(1+\frac{r_1}{100}\right)\left(1+\frac{r_2}{100}\right)$

$$A=10000\left(1+\frac{10}{100}\right)\left(1+\frac{12}{100}\right)$$

$$A=10000\left(\frac{110}{100}\right)\left(\frac{112}{100}\right)$$

A = 12320

39. (c) $2916=P\left(1+\frac{8}{100}\right)^2$

$$P=\frac{2916}{(1.08)^2}=2500$$

$$S.I=\frac{2500\times9\times3}{100}=675$$

40. (b) Population 2 years ago $=\frac{4410}{\left(1+\frac{5}{100}\right)^2}=\frac{4410}{441}\times400$

= 4000

41. (a) Principal (P) = 210
Ratio (R) = 10%
Loan has to be paid in the instalments i.e., it take two years to pay.

$$CI=P\left(1+\frac{R}{100}\right)^2$$

$$=210\left(1+\frac{10}{100}\right)^2\Rightarrow210\times\frac{11}{10}\times\frac{11}{10}=254$$

So, equal instalment $=\frac{254}{2}=127$

42. (d) For Halfyearly, $A = P\left(1+\frac{\frac{R}{2}}{100}\right)^{2n}$

$$68921 = 64000\left(1+\frac{5}{200}\right)^{2n}$$

$$\frac{68921}{64000} = \left(\frac{41}{40}\right)^{2n}$$

$$\left(\frac{41}{40}\right)^{3} = \left(\frac{41}{40}\right)^{2n}$$

$$n = \frac{3}{2} = 1\frac{1}{2} \text{ years}$$

43. (c) Let P be the Principal amount and R be the rate of interest

$$400 = \frac{P \times R \times 2}{100}$$

$$PR = 20000 \text{ or } P = \frac{20000}{R}$$

For 2 years, $C.I - S.I. = P\frac{R^2}{(100)^2}$

$$420 - 400 = \frac{20000}{R} \times \frac{R^2}{10000}$$

$20 = 2R$

$R = 10$

44. (d) $C.I. = P\left[\left(1+\frac{R}{100}\right)^{T} - 1\right] = 1210\left[\left(1+\frac{6}{100}\right)^{1} - 1\right]$

$$1210\left[1+\frac{3}{50}-1\right] = \frac{1210 \times 3}{50} = \frac{363}{5} = ₹\ 72.60$$

45. (b) $S.I. = \frac{13033 \times 13 \times 3}{100} = ₹5082.87$

$$C.I. = 13033\left[\left(1+\frac{13}{100}\right)^{3} - 1\right]$$

$= 13033 \times 0.44$

$= ₹5772.28$

Difference $= 5772.28 - 5082.87$

$= ₹689.41$

46. (e) $S.I. = 79900 - 58750 = ₹21150$

$$\text{Rate} = \frac{S.I. \times 100}{\text{Principal} \times \text{Time}}$$

$$= \frac{21150 \times 100}{58750 \times 4} = 9\% \text{ per annum}$$

47. (e) $S.I. = \frac{12000 \times 9 \times 13}{100}$

$= ₹14040$

∴ Amount $= 12000 + 14040$

$= ₹26040$

48. (e) Using $SI = \frac{Prt}{100}$

$$3584 = \frac{P \times 7 \times 4}{100}$$

$\Rightarrow P = ₹\ 12800$

Now, amount got by CI

$$A = P\left(1+\frac{r}{100}\right)^{n}$$

$$= 12800\left(1+\frac{4}{100}\right)^{2} = 12800 \times 1.04 \times 1.04$$

$= ₹13844.48$

Hence, $CI = A - P = 13844.48 - 12800 = ₹1044.48$

Level-II

1. (a) Let the sum be ₹ P. Then,

$$\Rightarrow \left[P\left(1+\frac{10}{100}\right)^{2} - P\right] = 1260$$

$$\Rightarrow \left[P\left(\frac{11}{10}\right)^{2} - 1\right] = 1260$$

∴ Sum = ₹ 6000

So, $S.I. = ₹\left(\frac{6000 \times 4 \times 5}{100}\right) = ₹\ 1200$

2. (a) Simple interest for 5 years = ₹ 300

Now, when principal is trebled, the simple interest for 5 years will also treble the simple interest on original principal for the same period. Thus, S.I. for last 5 years when principal is trebled.

= 3 × 300 = ₹ 900

∴ Total SI for 10 years = 300 + 900 = ₹1200

3. (a) Let the sum be ₹ x.

∴ Interest $= \frac{x \times 8 \times 4}{100} = \frac{32x}{100}$

$$x - \frac{32x}{100} = \frac{68x}{100}$$

When interest is $\frac{68x}{100}$ less, the sum is ₹ x.

∴ When interest is ₹ 340 less, the sum is

$$\frac{x}{68x} \times 100 \times 340 = ₹\ 500$$

Direct Formula:

$$\text{Sum} = \frac{100}{100-8\times4}\times340 = \frac{100\times340}{68} = ₹\ 500$$

4. (a) $\text{Sum} = \dfrac{\text{Difference in interest}\times100}{\text{Times}\times\text{Difference in rate}}$

$$= \frac{24\times100}{2\times1} = ₹\,1200$$

5. (d) Difference in amounts = 2977.54 – 2809 = ₹ 168.54
Now, we see that ₹ 168.54 is the interest on ₹ 2809 in one year (it is either simple or compound interest because both are the same for a year).

Hence, rate of interest $= \dfrac{168.54\times100}{2809} = 6\%$

Now, for the original sum,

$$2809 = x\left(1+\frac{6}{100}\right)^2$$

$$\text{Or, } 2809 = x\left(\frac{53}{50}\right)^2$$

$$\therefore\quad x = \frac{2809\times50\times50}{53\times53} = ₹\ 2500$$

6. (b) Let sum be x. Then, S.I. = x.

I. $\text{Time} = \dfrac{100\times x}{x\times\dfrac{50}{3}} = 6$ years (False)

II. $\text{Time} = \dfrac{100\times x}{x\times20} = 5$ years (True)

III. Suppose sum = x. Then, S.I. = x and Time = 5 years.

$$\text{Rate} = \left(\frac{100\times x}{x\times5}\right)\% = 20\%$$

Now, sum = x, S.I. = $3x$ and Rate = 20%.

$$\therefore\quad \text{Time} = \left(\frac{100\times3x}{x\times20}\right) \text{ years} = 15 \text{ years (False)}$$

So, II alone is correct.

7. (d) Let sum = x. Then, S.I. = $\dfrac{9}{16}x$.

Let rate = R% and time = R years.

$$\therefore\quad \left(\frac{x\times R\times R}{100}\right) = \frac{9x}{16} \Leftrightarrow R^2 = \frac{900}{16}$$

$$\Leftrightarrow R = \frac{30}{4} = 7\frac{1}{2}.$$

Hence, time = $7\frac{1}{2}$ years.

8. (d) Let the sum be ₹ x, rate be R% p.a. and time be T years.

$$\text{Then, } \left[\frac{x\times(R+2)\times T}{100}\right]-\left(\frac{x\times R\times T}{100}\right) = 108$$

$\Leftrightarrow 2xT = 10800$ ...(i)

$$\left[\frac{x\times R\times(T+2)}{100}\right]-\left(\frac{x\times R\times T}{100}\right) = 180$$

$\Leftrightarrow 2xR = 18000$ ...(ii)

Clearly, from (i) and (ii), we cannot find the value of x.

So, the data is inadequate.

9. (b) Let the required rate be R. Then,

$$\left(\frac{20000\times8\times1}{100}\right)+\left(4000\times\frac{15}{2}\times\frac{1}{100}\right) + \left(1400\times\frac{17}{2}\times\frac{1}{100}\right) + \left(2600\times R\times\frac{1}{100}\right)$$

$$= \left(\frac{813}{10000}\times10000\right)$$

$\Leftrightarrow 160 + 300 + 119 + 26R = 813 \Leftrightarrow R = 9.$

10. (d) $\text{Balance} = ₹\left[\left\{12500\times\left(1+\dfrac{20}{100}\right)^3\right\} - \left\{2000\times\left(1+\dfrac{20}{100}\right)^2 + 2000\times\left(1+\dfrac{20}{100}\right)+2000\right\}\right]$

$$= ₹\left[\left(12500\times\frac{5}{6}\times\frac{6}{5}\times\frac{6}{5}\right) - \left(2000\times\frac{6}{5}\times\frac{6}{5}+2000\times\frac{6}{5}+2000\right)\right]$$

$= ₹\,[21600 - (2880 + 2400 + 2000)] = ₹14320.$

11. (b) $\dfrac{F\times(0.06)\times6}{(38800-F)\times0.05\times2} = 5/4$

where F is the first part.
$1.44F = 19400 - 0.5F$
$F = 19400/1.94 = 10000$
Thus, the second part = 38800 – 10000 = 28800

12. (b) Go through trial and error of the options. You will get:
20000 × (1.3) = 26000 @ simple interest)
20000 × 1.1 × 1.1 × 1.1 = 26620 @ compound interest.
Thus, 20000 is the correct answer.

13. (a) $x\left(1+\frac{r}{100}\right)^2 = 4x\left(1-\frac{r}{100}\right)^2$

$\left(1+\frac{r}{100}\right) = 2\left(1-\frac{r}{100}\right)$

$\frac{3r}{100} = 1$

$r = \frac{100}{3} = 33\frac{1}{3}\% = 33.33\%$

14. (a) The original population was 1530 thousand
$\Rightarrow$ No. of births was 53.2% of 1530 thousand $= 813960$
$\Rightarrow$ No. of deaths was 31.2% of 1530 thousand $= 47360$
$\Rightarrow$ Net increase in population $= 813960 - 47360 = 336600$

Examination method : Net increase $= (53.2 - 31.2)\%$ of total

15. (a) Let the Principal $= P$

Then $\frac{P\times8\times4}{100} + \frac{P\times10\times6}{100} + \frac{P\times12\times5}{100}$

$= 12160$

$\Rightarrow 152P = 12160 \times 100$

or $\frac{12160\times100}{152} = ₹8000$

16. (c) **For 3 years:**

$$\text{Diff.} = \frac{\text{Sum}\times(\text{rate})^2(300+\text{rate})}{(100)^3}$$

$$= \frac{2000\times10\times10\times310}{100\times100\times100} = ₹62$$

17. (d) Let the amount deposited at the rate of 15% per annum be ₹ x.
15% of x + 18% of $(25000 - x) = 4050$
or, 15% of x + 18% of 25000 – 18% of $x = 4050$
or, 3% of $x = 4500 - 4050 = 450 \Rightarrow x = ₹15000$
$\therefore$ Amount deposited at 18%
$= (25000 - 15000 =)$ ₹ 10000

18. (d) $\frac{30p}{100} + 450 = \left[p\left(1+\frac{15}{100}\right) - p\right]$

$\Rightarrow p = ₹20{,}000.$

19. (c) Let, in scheme A, Sridharan invest ₹ x.
Then, his investment in scheme $B = ₹(27000 - x)$.
Now,

$x\left(1+\frac{8}{100}\right)^2 + (27000-x)\left(1+\frac{9}{100}\right)^2$

$-27000 = 4818.30$

or, $x(1.08)^2 + (27000-x)(1.09)^2 = 31818.30$
or, $1.1664x + 32078.7 - 1.1881x = 31818.30$
or, $0.0217x = 260.4$

or, $x = \frac{260.4}{0.0217} = ₹12000$

20. (c) Let the amount invested at 20% rate be ₹ x. According to the question,

$$12000\times\frac{10}{100} + x\times\frac{20}{100} = (12000+x)\times\frac{14}{100}$$

or, $1200 + \frac{x}{5} = 1680 + \frac{7}{50}x$

or, $\frac{x}{5} - \frac{7}{50}x = 480$

or, $\frac{3}{50}x = 480$

$\therefore$ $x = ₹\,8000$

$\therefore$ Total amount invested ₹ $= (12000 + 8000)$
$= ₹\,20000$

21. (c) Let one gets $= ₹\,x$
then, second gets $= ₹\,(68{,}000 - x)$
Given : $A_1 = A_2$

$$x + \frac{x\times10\times8}{100} = (68{,}000-x) + \frac{(68000-x)\times10\times6}{100}$$

$\Rightarrow x[100+80] = (68{,}000-x)[100+60]$

$\Rightarrow \frac{180x}{160} = 68{,}000 - x$

$\Rightarrow 34x = 68000\times16 \Rightarrow x = ₹32{,}000$

$\therefore$ second gets $= ₹\,36{,}000$

22. (d) Difference of $S.I. =$ ₹ $\sqrt{31.50}$
Let each sum be ₹ x. Then

$$\frac{x\times4\frac{1}{2}\times7}{100} - \frac{x\times4\times7}{100} = 31.50$$

or $\frac{7x}{100}\times\frac{1}{2} = \frac{63}{2}$

or $x = ₹\,900$

23. (d) Let the original rate be $R\%$. Then, new rate $= (2R)\%$.

$\therefore \left(\frac{725\times R\times1}{100}\right) + \left(\frac{362.50\times2R\times1}{100\times3}\right) = 33.50$

$\Rightarrow (2175 + 725)R = 33.50\times100\times3 = 10050$

$\Rightarrow R = \frac{10050}{2900} = 3.46\%$

24. **(a)** Let x, y and z be the amounts invested in schemes A, B and C respectively. Then,

$$\left(\frac{x\times 10\times 1}{100}\right)+\left(\frac{y\times 12\times 1}{100}\right)+\left(\frac{z\times 15\times 1}{100}\right)=3200$$

$\Rightarrow 10x+12y+15z=320000$ ..... (1)

Now, $z = 240\%$ of $y = \frac{12}{5}y$ ..... (2)

And, $z = 150\%$ of $x = \frac{3}{2}x$

$\Rightarrow x=\frac{2}{3}z=\left(\frac{2}{3}\times\frac{12}{5}\right)y=\frac{8}{5}y$ ..... (3)

From (1), (2) and (3), we have :

$16y+12y+36y=320000 \Rightarrow 64y=320000 \Rightarrow y=5000.$

$\therefore$ Sum invested in scheme B = ₹ 5000.

25. **(c)** Cash down payment = ₹ 1500

Let ₹ x becomes ₹ 1020 at the end of first year.

Then, $1020 = x\left(1+\frac{10}{100}\right)$

or $x=\frac{1020\times 100}{110}$ =₹.927.27

Similarly, $1003 = y\left(1+\frac{10}{100}\right)^2$

or $y=\frac{1003\times 20\times 20}{22\times 22}$ =₹ 828.92

and $z=\frac{990\times 20\times 20\times 20}{22\times 22\times 22}$ =₹ 743.80

Hence, CP = 1500 + 927.27 + 828.92 + 743.80

= 3999.99 or ₹ 4000.

26. **(d)** 14% in 1.5 yrs will be 21%

in 6 months will be 7%

$\therefore$ A's debt $=\frac{1573\times 100}{121}$ =₹ 1300

B's debt $=\frac{1444.5\times 100}{107}$ =₹ 1350

Hence, B must pay ₹ 50 to A.

27. **(a)** Amount = ₹ 17640, Principal = ₹ 16000

Time = 2 yrs, Rate = R

$$17640=16000\left(1+\frac{R}{100}\right)^2$$

$$\Rightarrow \frac{17640}{16000}=\left(1+\frac{R}{100}\right)^2 \Rightarrow 1.1025=\left(1+\frac{R}{100}\right)^2$$

$$\Rightarrow 1+\frac{R}{100}=1.05 \Rightarrow \frac{R}{100}=1.05-1=0.05$$

$\Rightarrow R = 5\%$

28. **(b)** Given $8P = P\left(1+\frac{r}{100}\right)^3$

Where P = Principal amount,

r = Compound interest rate

$\Rightarrow r = 100\%$

$\therefore$ let the time in which the principal amount becomes 16 times be n

Then $16P = P\left(1+\frac{100}{100}\right)^n$

$\Rightarrow 16 = 2^n \Rightarrow n = 4$yrs.

29. **(a)** Let A lent ₹ x and B lent ₹ y

Since, A and B together lent out ₹ 81600

$\therefore x+y=81,600$

Now, given (r) Rate = 4%

$$\therefore\ 1+r=1+\frac{4}{100}=\frac{26}{25}$$

According to the question, we have

$$\frac{x}{y}=\left(\frac{26}{25}\right)^{3-2}=\frac{26}{25}$$

$\therefore$ Investment made by $B = 81600\times\frac{25}{51} = 40,000$

30. **(a)** Let the amount lent = P_1 at 15% and P_2 at 10%

According first condition.

$$\frac{P_1\times 15\times 1}{100}+\frac{P_2\times 10\times 1}{100}=1900$$

$15P_1+10P_2=1900\times 100$ ....(1)

According to second condition.

$$\frac{P_1\times 10\times 1}{100}+\frac{P_2\times 15\times 1}{100}=1900+200=2100$$

$10P_1+15P_2=2100\times 100$ ....(2) × 10

$15P_1+10P_2=1900\times 100$ ....(1) × 15

$100P_1+150P_2=2100000$

$225P_1+150P_2=2850000$

$-\quad\quad -\quad\quad -$

$125P_1=750000$

$P_1=6000$

31. **(a)** Let principal = P, time = t years, rate = t

Then, $\frac{Ptt}{100}=\frac{P}{9}$

$\therefore\ t^2=\frac{100}{9}$ $\quad\therefore\ t=\frac{10}{3}=3\frac{1}{3}$

$\therefore$ rate = $3\frac{1}{3}\%$

Direct formula:

Rate = time = $\sqrt{100\times\frac{1}{9}}=\frac{10}{3}=3\frac{1}{3}\%$

32. (a) Let 'x' be the amount borrowed by Amin.

$\therefore \frac{x\times2\times8}{100}+\frac{x\times3\times11}{100}+\frac{x\times3\times14}{100}=10920$

or, $\frac{91}{100}x=10920$ or $x=\frac{10920\times100}{91}=12000$

33. (b) Rate of simple interest = $\frac{\text{Interest}\times100}{\text{Principal}\times\text{Time}}$

$=\frac{10800\times100}{22500\times4}=12\%$

Compound interest = Principal $\left[\left(1+\frac{\text{rate}}{100}\right)^{\text{time}}-1\right]$

$=22500\left[\left(1+\frac{12}{100}\right)^2-1\right]$

$=22500\left[\left(1+\frac{3}{25}\right)^2-1\right]$

$=22500\left[\left(\frac{28}{25}\right)^2-1\right]=22500\left(\frac{784}{625}-1\right)$

$=22500\times\frac{159}{625}=$ ₹ 5724

34. (b) Required difference = $P\left(\frac{R}{100}\right)^2$

$=7300\times\left(\frac{6}{100}\right)^2=$ ₹26.28

35. (a) $P=\frac{SI\times100}{R\times T}=\frac{6500\times100}{8\times13}=6250$

$CI=6250\left(1+\frac{8}{100}\right)^2-6250=$ ₹ 1040

36. (b) Required difference = $\frac{PR^2}{(100)^2}$

$\Rightarrow\frac{4000\times5\times5}{100\times100}=$ ₹10

37. (b) Let principal be represented by P.

Ist Case :

$S.I.=\frac{P\times R\times T}{100}=\frac{P\times8\times2}{100}$

IInd Case :

$S.I=\frac{P\times R\times T}{100}=\frac{P\times8\times3}{100}$

According to question

$\frac{P\times8\times3}{100}-\frac{P\times8\times2}{100}=56$

$\frac{P\times8}{100}=56\Rightarrow P=\frac{56\times100}{8}=700$

CHAPTER

8

RATIO, PROPORTION AND VARIATION

RATIO

Ratio is the comparison between two quantities in terms of their magnitudes. The ratio of two quantities is equivalent to a fraction that one quantity is of the other.

For example, let Swati has 5 note books and Priya has 7 note books. Then the ratio of the number of books that have with Swati to the number of books that have with Priya is 5 is to 7. This ratio is expressed as 5 : 7 or $\frac{5}{7}$, which is a quotient of 5 and 7. Ratio of any two numbers a and b is expressed as $a : b$ or $\frac{a}{b}$. The numbers that form the ratio is called the terms of the ratio. The numerator of the ratio is called the antecedent and the denominator is called the consequent of the ratio.

DECIMAL AND PERCENTAGE VALUE OF A RATIO

A ratio can be expressed in decimal and percentage.

Decimal value of $\frac{3}{5} = 0.6$

To express the value of a ratio as a percentage, we multiply the ratio by 100.

Hence $\frac{3}{5} = \frac{3}{5} \times 100\% = 60\%$

To find the decimal value of any ratio, you may calculate the percentage value using the percentage rule (discussed in the chapter Percentage) and then shift the decimal point 2 places towards left. Hence the decimal value of a ratio whose percentage value is 54.82% will be 0.5482.

PROPERTIES OF RATIOS

(i) $\frac{a_1}{b_1} = \frac{a_2}{b_2} = \frac{a_3}{b_3} = \ldots = \frac{a_1 + a_2 + a_3 + \ldots}{b_1 + b_2 + b_3 + \ldots}$

This means that if two or more ratios are equal, then the ratio whose numerator is the sum of the numerators of all the ratios and denominator is the sum of the denominators of all the ratios is equal to the original ratio.

Since $\frac{35}{50} = \frac{7}{10}$

$\therefore \quad \frac{35}{50} = \frac{7}{10} = \frac{35 + 7}{50 + 10} = \frac{42}{60}$

(ii) If $\frac{a_1}{b_1}, \frac{a_2}{b_2}, \frac{a_3}{b_3}, \ldots, \frac{a_n}{b_n}$ are unequal ratios (or fractions), then $\frac{a_1 + a_2 + a_3 + \ldots + a_n}{b_1 + b_2 + b_3 + \ldots + b_n}$ lies between the lowest and the highest of these ratios.

(iii) If the ratio $\frac{a}{b} > 1$ and k is a positive number, then

$$\frac{a + k}{b + k} < \frac{a}{b} \text{ and } \frac{a - k}{b - k} > \frac{a}{b}$$

Similarly, if $\frac{a}{b} < 1$, then

$$\frac{a + k}{b + k} > \frac{a}{b} \text{ and } \frac{a - k}{b - k} < \frac{a}{b}$$

(iv) If $\frac{c}{d} > \frac{a}{b}$, then $\frac{a + c}{b + d} > \frac{a}{b}$

and if $\frac{c}{d} < \frac{a}{b}$, then $\frac{a + c}{b + d} < \frac{a}{b}$

Illustration 1: Salaries of Rajesh and Sunil are in the ratio of 2 : 3. If the salary of each one is increased by ₹ 4000 the new ratio becomes 40 : 57. What is Sunil's present salary ?

(a) ₹ 17000 (b) ₹ 20000

(c) ₹ 25500 (d) None of these

Solution: (d) Let the salaries of Rajesh and Sunil be ₹ $2x$ and ₹ $3x$ respectively.

Then, $\frac{2x + 4000}{3x + 4000} = \frac{40}{57}$

or $114x + 228000 = 120x + 160000$

or $6x = 68000$

or $3x =$ ₹ 34000

Illustration 2: The ratio between the present ages of P and Q is 5 : 8. After four years, the ratio between their ages will be 2 : 3. What is Q's age at present ?

(a) 36 years (b) 20 years

(c) 24 years (d) None of these

Solution: (d) $\frac{P}{Q}=\frac{5}{8}$ or $P=\frac{5Q}{8}$ ... (1)

$$\frac{P+4}{Q+4}=\frac{2}{3}$$

or $3P+12=2Q+8$

or $2Q-3P=4$ ... (2)

Putting value of P from eq. (1),

$$2Q-3\times\frac{5}{8}Q=4 \Rightarrow Q=32.$$

USES OF RATIOS

(i) As a Bridge between three or more Quantities

If $a:b=N_1:D_1$

$b:c=N_2:D_2$

$c:d=N_3:D_3$

and $d:e=N_4:D_4$

Then $a:b:c:d:e=N_1N_2N_3N_4:D_1N_2N_3N_4:D_1D_2N_3N_4:D_1D_2D_3N_4:D_1D_2D_3D_4$

Here

a is correspond to the product of all four numerators ($N_1N_2N_3N_4$)

b is correspond to the first denominator and the last three numerators ($D_1N_2N_3N_4$)

c is correspond to the first two denominators and the last two numerators ($D_1D_2N_3N_4$)

d is correspond to the first three denominators and the last numerators ($D_1D_2D_3N_4$)

e is correspond to the product of all four denominators ($D_1D_2D_3D_4$)

This method is applied for any three or more ratios.

This can be understood by following illustrations:

Illustration 3: Ratio of the age of A and B is 3 : 5 and ratio of the age of B and C is 4 : 7. Find the ratio of the age of A and C.

Solution: $A:B=3:5$; $B:C=4:7$

$\Rightarrow$ $A:B:C=3\times4:5\times4:5\times7=12:20:35$

Here

A is correspond to the product of both numerators (3×4)

B is correspond to the product of first denominator and second numerator (5×4)

and C is correspond to the product of both denominators (5×7)

Hence ratio of the age of A and C = 12 : 35

Conventional Method

LCM of 5 and 4 (the two values corresponding B's amount) is 20.

Now convert B's value in both ratio to 20.

Hence $A:B=3\times4:5\times4=12:20$

$B:C=4\times5:7\times5=20:35$

$\Rightarrow$ $A:B:C=12:20:35$

$\Rightarrow$ $A:C=12:35$

This conventional method will be long for more than three ratios.

Illustration 4: If $A:B=4:5$; $B:C=3:7$; $C:D=6:7$; $D:E=12:17$ then find the value of ratio $A:E$.

Solution: $A:B:C:D:E=(4\times3\times6\times12):(5\times3\times6\times12):(5\times7\times6\times12):(5\times7\times7\times12):(5\times7\times7\times17)$

$\therefore$ $A:E=(4\times3\times6\times12):(5\times7\times7\times17)=864\times4165$

Note that here we have found the ratio of $A:E$ directly without finding the consolidate ratio ($A:B:C:D:E$) of A, B, C, D and E.

COMPARISON OF RATIOS

The value of a ratio is directly related to the value of numerator but inversely related to the value of denominator i.e. if (only numerator decrease)/(only denominator increases)/(numerator decreases and denominator increases) then the value of the ratio decreases and vice-versa.

There are eight cases in which we have to compare two ratios. In six out of these eight cases, we can easily compare the two ratios by keeping the above mentioned facts related to ratios in mind as shown in the following table.

S.No.	Cases	Comparison of Ratios	Comparison of Ratios (Example)
(i)	Numerator : Decreases Denominator : Fixed	(First Ratio) > (Second Ratio)	$\frac{5}{8}>\frac{3}{8}$
(ii)	Numerator : Increases Denominator : Fixed	(First Ratio) < (Second Ratio)	$\frac{4}{9}<\frac{7}{9}$
(iii)	Numerator : Fixed Denominator : Decreases	(First Ratio) < (Second Ratio)	$\frac{6}{7}<\frac{6}{5}$
(iv)	Numerator : Fixed Denominator : Increases	(First Ratio) > (Second Ratio)	$\frac{5}{8}>\frac{5}{9}$
(v)	Numerator : Decreases Denominator : Increases	(First Ratio) > (Second Ratio)	$\frac{6}{7}>\frac{5}{8}$
(vi)	Numerator : Increases Denominator : Decreases	(First Ratio) < (Second Ratio)	$\frac{3}{7}<\frac{5}{4}$

In the remaining two cases, we cannot compare the two ratios just by looking them.

The remaining two cases are

(vii) Numerator : Decreasing
Denominator : Decreasing

(viii) Numerator : Increasing
Denominator : Increasing

In both the remaining two cases (vii) and (viii), we can compare the two ratios by any one of the following two methods.

Method-I: Cross Multiplication Method

$$\frac{a}{b} > \frac{c}{d}, \text{ if } ad > bc$$

and $$\frac{a}{b} < \frac{c}{d}, \text{ if } ad < bc$$

For example $\frac{6}{7} > \frac{3}{5}$ because $6 \times 5 > 7 \times 3$

and $\frac{4}{5} < \frac{7}{8}$ because $4 \times 8 < 5 \times 7$

Method-II: Denominator Equating Method

By making the denominator of each ratio equal to the LCM of the denominators of both ratios, we can compare the two ratios by checking their numerators.

Illustration 5: Which of the two ratios $\frac{5}{6}$ and $\frac{8}{9}$ is greater.

Solution: LCM of 6 and 9 = 18

$$\frac{5}{6} = \frac{5 \times 3}{6 \times 3} = \frac{15}{18}$$

$$\frac{8}{9} = \frac{8 \times 2}{9 \times 2} = \frac{16}{18}$$

Since numerator of second ratio is greater than the numerator of first ratio,

$$\therefore \quad \frac{16}{18} > \frac{15}{18} \Rightarrow \frac{8}{9} > \frac{5}{6}$$

PROPORTION

When two ratios are equal, the four quantities composing them are said to be proportionals. Hence, if $\frac{a}{b} = \frac{c}{d}$, then a, b, c, d are in proportional and is written as

$$a : b :: c : d$$

The terms a and d are called extremes while the terms b and c are called the means.

$$a : b :: c : d \Rightarrow \frac{a}{b} = \frac{c}{d} \Rightarrow ad = bc$$

Hence product of extremes = Product of means

Illustration 6: What must be added to each of the four numbers 10, 18, 22, 38 so that they become in proportion ?

Solution: Let the number to be added to each of the four numbers be x.

By the given condition, we get

$$(10 + x) : (18 + x) :: (22 + x) : (38 + x)$$

$$\Rightarrow (10 + x)(38 + x) = (18 + x)(22 + x)$$

$$\Rightarrow \quad 380 + 48x + x^2 = 396 + 40x + x^2$$

Cancelling x^2 from both sides, we get

$$380 + 48x = 396 + 40x$$

$$\Rightarrow 48x - 40x = 396 - 380$$

$$\Rightarrow \quad 8x = 16 \Rightarrow x = \frac{16}{8} = 2$$

Therefore, 2 should be added to each of the four given numbers.

Continue Proportion

(i) If $\frac{a}{b} = \frac{b}{c}$, then $a, b, c,$ are said to be in continue proportion and vice-versa.

Now $\frac{a}{b} = \frac{b}{c} \Rightarrow ac = b^2$

Here b is called mean proportional and c is called third proportional of a and b.

(ii) If a, b, c and d are in continue proportion, then

$$\frac{a}{b} = \frac{b}{c} = \frac{c}{d}$$

Also if $\frac{a}{b} = \frac{b}{c} = \frac{c}{d} = k$ (let), a constant

Then $c = dk$

$b = ck = dk \,.\, k = dk^2$

$a = bk = dk^2 \,.\, k = dk^3$

PROPERTIES OF PROPORTION

(i) Invertendo: If $\frac{a}{b} = \frac{c}{d}$, then $\frac{b}{a} = \frac{d}{c}$

(ii) Alternando: If $\frac{a}{b} = \frac{c}{d}$, then $\frac{a}{c} = \frac{b}{d}$

(iii) Componendo: If $\frac{a}{b} = \frac{c}{d}$, then $\frac{a+b}{b} = \frac{c+d}{d}$

(iv) Dividendo: If $\frac{a}{b} = \frac{c}{d}$, then $\frac{a-b}{b} = \frac{c-d}{d}$

(v) Componendo and Dividendo: If $\frac{a}{b} = \frac{c}{d}$, then

$$\frac{a+b}{a-b} = \frac{c+d}{c-d}$$

Illustration 7: Find the value of $\frac{x+a}{x-a} + \frac{x+b}{x-b}$, if $x = \frac{2ab}{a+b}$

Solution: $x = \frac{2ab}{a+b} \Rightarrow \frac{x}{a} = \frac{2b}{a+b}$

By componendo – dividendo,

$$\frac{x+a}{x-a}=\frac{3b+a}{b-a}$$

Similarly, $$\frac{x}{b}=\frac{2a}{a+b}$$

$$\Rightarrow \quad \frac{x+b}{x-b}=\frac{3a+b}{a-b}$$

$$\therefore \quad \frac{x+a}{x-a}+\frac{x+b}{x-b}=\frac{3b+a}{b-a}+\frac{3a+b}{a-b}$$

$$=\frac{-(3b+a)}{a-b}+\frac{3a+b}{a-b}=\frac{2a-2b}{a-b}=2.$$

VARIATIONS

We come across many situations in our day to day life where we see change in one quantity bringing change in the other quantity. For example:

(a) If the number of items purchased increases, its cost also increases.

(b) If the number of workers working to complete a job increases then days required to complete the job will decrease.

Here we observe that change in one quantity leads to change in other quantity. This is called variation.

TYPES OF VARIATIONS

There are three types of variations: Direct variation, Indirect variation and Compound variation.

(i) Direct Variations

There is a direct variation in two quantities if they are related in such a way that an increase in one causes an increase in the other in the same ratio or a decrease in one causes a decrease in the other in the same ratio. This means that if one quantity becomes double then the other quantity also becomes double and if one quantity becomes half then the other quantity also becomes half etc. In other words if x and y are two variables then y varies directly with x if the ratio $\frac{y}{x}$ is a constant.

'y varies directly with x' is represented as $y \propto x$

y varies directly as x is simply say that y varies as x.

Here symbol '$\propto$' means 'varies as'.

The representation $y \propto x$ can be converted to an equation $y = kx$, where k is a positive constant and called constant of proportionality.

Hence $\frac{y}{x}$ = constant

The equation $\frac{y}{x} = k$, means all ratios of a value of y with their corresponding value of x are equal.

If y_1, y_2 are two values of y corresponding to two values x_1 and x_2 of x, then $\frac{y_1}{x_1}=\frac{y_2}{x_2}$.

Graph

If y varies directly as x, then graph between x and y will be as shown below:

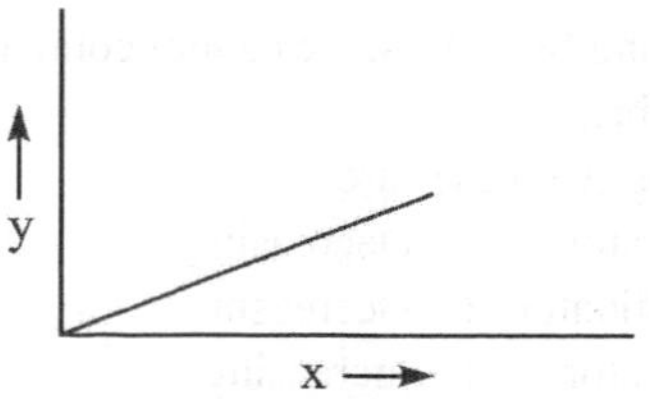

Some Examples of Direct Variations

- Number of persons $\propto$ Amount of work done
  More number of persons, more work.
- Number of days $\propto$ Amount of work
  More days, More work
- Working rate $\propto$ Amount of work
  More working rate, more work
- Efficiency of worker $\propto$ Amount of work
  More efficient worker, More work.

Illustration 8: A machine takes 5 hours to cut 120 tools. How many tools will it cut in 20 hours?

Solution: Here more time, more number of tools i.e. time and number of tools cut vary directly.

Let number of tools cut in 20 hours be 'x', then

$$\frac{5}{120}=\frac{20}{x} \qquad \left(\because \frac{y_1}{x_1}=\frac{y_2}{x_2}\right)$$

$$\Rightarrow \quad x=\frac{20\times 120}{5}$$

$$x = 480$$

Hence required number of tools = 480.

(ii) Inverse Variations

There is an inverse variation in two quantities if they are so related that an increase in one causes a decrease in the other in the same ratio or vice-versa. This means that if one quantity becomes double then other quantity becomes half and if one quantity becomes one third then other quantity becomes thrice etc.

In other words if x and y are variables then y varies inversely with x, if xy is a constant.

'y varies inversely with x' is represented as $y \propto \frac{1}{x}$.

Here symbol '$\propto$' means 'varies as'. The representation $y \propto \frac{1}{x}$ can be converted to an equation $y = \frac{k}{x}$ or $xy = k$, where k is a positive constant, called constant of proportionality.

$$\Rightarrow \quad xy = \text{constant}$$

The equation xy = constant, means all products of a value of y and their corresponding value of x are equal. That is if y_1, y_2 are two values of y corresponding to the values x_1, x_2 of x respectively, then

$$x_1 y_1 = x_2 y_2$$

Graph

If y varies inversely as x, then graph between x and y will be as shown below:

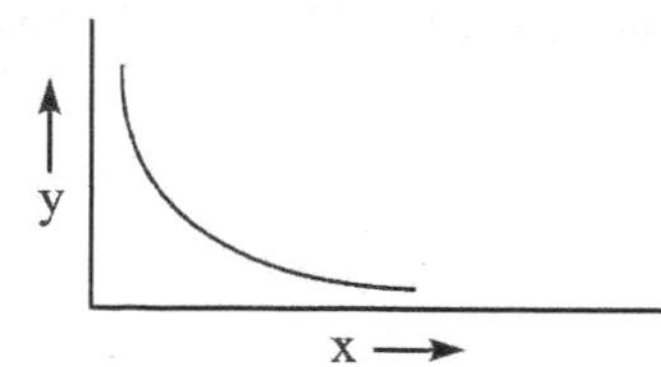

Illustration 9: If 900 persons can finish the construction of a building in 40 days, how many persons are needed to complete the construction of building in 25 days.

Solution: Let the required number of persons be 'x'. As the number of days required to complete the job is less, so more number of persons will be required. It is a case of inverse variation.

So $900 \times 40 = x \times 25$

$$\Rightarrow \quad x = \frac{900 \times 40}{25} = 1440$$

Hence required number of persons = 1440.

COMPOUND VARIATIONS

In real life, there are many situations which involve more than one variation, *i.e.* change in one quantity depends on changes in two or more quantities either directly or inversely or by both.

Let x, y and z are variables, *i.e.* $y \propto x$

(a) y varies directly as x when z is constant, i.e., $y \propto x$ and y varies directly as z when x is constant, i.e. $y \propto z$, then we say that y varies directly as the product of x and z.

Thus $y \propto xz$

or $y = k(xz)$, k is a positive constant

(b) y varies directly as x when z is constant, i.e. $y \propto x$ and y varies inversely as z when x is constant i.e. $y \propto \frac{1}{z}$, then

$y \propto \frac{x}{z}$ or $y = k\left(\frac{x}{z}\right)$, where k is a positive constant.

(c) y varies inversely as x when z is constant i.e. $y \propto \frac{1}{x}$ and y varies inversely as z when x is constant then $y \propto \frac{1}{xz}$ or

$y = \frac{k}{xz}$, where k is a positive constant.

Illustration 10: 25 horses eat 5 bags of corn in 12 days, how many bags of corn will 10 horses eat in 18 days ?

Solution: Here three quantities : number of horses (h), number of bags (b) and number of days (d) are involved.

Number of bags increases as number of hourses increases. Also, number of bags increases as number of days increases.

Hence $b \propto hd \Rightarrow \frac{b}{hd}$ = constant

$$\Rightarrow \quad \frac{b_1}{h_1 d_1} = \frac{b_2}{h_2 d_2} \Rightarrow b_2 = \frac{b_1 h_2 d_2}{h_1 d_1}$$

$$\therefore \quad b_2 = \frac{5 \times 10 \times 18}{25 \times 12} = 3$$

Hence number of bags required by 10 horses in 18 days = 3 bags.

PARTNERSHIP

A partnership is an association of two or more persons who invest their money in order to carry on a certain business.

A partner who manages the business is called the **working partner** and the one who simply invests the money is called the **sleeping partner.**

Partnership is of two kinds :

(i) Simple (ii) Compound.

Simple partnership : If the capitals is of the partners are invested for the same period, the partnership is called simple.

Compound partnership : If the capitals of the partners are invested for different lengths of time, the partnership is called compound.

☒ If the period of investment is same for each partner, then the profit or loss is divided in the ratio of their investments.

If A and B are partners in a business investing for same period, then

$$\frac{\text{Investment of A}}{\text{Investment of B}} = \frac{\text{Profit of A}}{\text{Profit of B}} \text{ or } = \frac{\text{Loss of A}}{\text{Loss of B}}$$

If A, B and C are partners in a business, then

Investment of A : Investment of B : Investment of C

= Profit of A : Profit of B : Profit of C, or

= Loss of A : Loss of B : Loss of C

Illustration 11: Three partner Rahul, Puneet and Chandan invest ₹ 1600, ₹ 1800 and ₹ 2300 respectively in a business. How should they divide a profit of ₹ 399 ?

Solution: Profit is to be divided in the ratio 16 : 18 : 23

Rahul's share of profit $= \frac{16}{16+18+23} \times 399$

$= \frac{16}{57} \times 399 = ₹\,112$

Puneet's share of profit $= \frac{18}{57} \times 399 = ₹\,126$

Chandan's share of profit $= \frac{23}{57} \times 399 = ₹\,161$

Illustration 12: A, B and C enter into a partnership by investing 1500, 2500 and 3000 rupees, respectively. A as manager gets one-tenth of the total profit and the remaining profit is divided among the three in the ratio of their investment. If A's total share is ₹ 369, find the shares of B and C.

Solution: If total profit is x, then

A's share $= \frac{1}{10}x + \frac{15}{15+25+30}$ of the balance $\frac{9}{10}x$

$\Rightarrow \frac{1}{10}x + \frac{27x}{140} = 369$

$\Rightarrow 14x + 27x = 369 \times 140$

$\Rightarrow x = \frac{369 \times 140}{41} = 9 \times 140 = 1260$

B's share $= \frac{5}{14} \times \frac{9}{10} \times 1260 = ₹\,405$

C's share $= \frac{6}{14} \times \frac{9}{10} \times 1260 = ₹\,486$

Illustration 13: A and B invested in the ratio 3 : 2 in a business. If 5% of the total profit goes to charity and A's share is ₹ 855, find the total profit.

Solution: Let the total profit be ₹ 100.

Then, ₹ 5 goes to charity.

Now, ₹ 95 is divided in the ratio 3 : 2.

∴ A's share $= \frac{95}{3+2} \times 3 =$ ₹ 57

But A's actual share is ₹ 855.

∴ Actual total profit $= 855\left(\frac{100}{57}\right) =$ ₹ 1500

In a group of n persons invested different different amount for different period then their profit or loss ratio is

$$At_1 : Bt_2 : Ct_3 : Dt_4 .... : Xt_n$$

[Here first person invested amount A for t_1 period, second person invested amount B for t_2 period and so on.]

Illustration 14: A and B start a business. A invests ₹ 600 more than B for 4 months and B for 5 months. A's share is ₹ 48 more than that of B, out of a total profit of ₹ 528. Find the capital contributed by each.

Solution: B's profit $= \frac{528-48}{2} =$ ₹240

A's profit = 528 – 240 = ₹ 288

$$\frac{\text{A's capital} \times 4}{\text{B's capital} \times 5} = \frac{288}{240} = \frac{6}{5}$$

$$\therefore \frac{\text{A's capital}}{\text{B's capital}} = \frac{6}{5} \times \frac{5}{4} = \frac{3}{2}$$

$$\Rightarrow \frac{\text{B's capital} + 600}{\text{B's capital}} = \frac{3}{2}$$

⇒ B's capital = ₹ 1200 and A's capital = ₹ 1800

Illustration 15: Three persons A, B, C rent the grazing of a park for ₹ 570. A puts in 126 oxen in the park for 3 months, B puts in 162 oxen for 5 months and C puts in 216 oxen for 4 months. What part of the rent should each person pay ?

Solution: Monthly equivalent rent of A = 126 × 3 = 378

Monthly equivalent rent of B = 162 × 5 = 810

Monthly equivalent rent of C = 216 × 4 = 864

∴ Rent is to be divided in the ratio 378 : 810 : 864, i.e. 7 : 15 : 16

∴ A would have to pay $\frac{7}{7+15+16}$ of the rent

$= \frac{7}{38}$ of the rent $= \frac{7}{38} \times 570 =$ ₹ 105

∴ B would have to pay $\frac{15}{38}$ of the rent $= \frac{15}{38} \times 570$

= ₹ 225

and C would have to pay $\frac{16}{38}$, i.e. $\frac{8}{19}$ of the rent

$= \frac{8}{19} \times 570 =$ ₹ 240

Illustration 16: Shekhar started a business investing ₹ 25,000 in 1999. In 2000, he invested an additional amount of ₹ 10,000 and Rajeev joined him with an amount of ₹ 35,000. In 2002, Shekhar invested another additional amount of ₹ 10,000 and Jatin joined them with an amount of ₹ 35,000. What will be Rajeev's share in the profit of ₹ 1,50,000 earned at the end of 3 years from the start of the business in 1999 ?

(a) ₹ 45,000 (b) ₹ 50,000

(c) ₹ 70,000 (d) ₹ 75,000

Solution: (b) Ratio of Shekhar, Rajeev and Jatin's investments

= 25000 × 36 + 10000 × 24 + 10000 × 12 : 35000 × 24 : 35000 × 12,

= 25 × 36 + 10 × 24 + 10 × 12 : 35 × 24 : 35 × 12

= 25 × 3 + 10 × 2 + 10 × 1 : 35 × 2 : 35 × 1

= 75 + 20 + 10 : 70 : 35

= 105 : 70 : 35, i.e. 3 : 2 : 1

∴ Rajeev's share in the profit $= \frac{2}{6} \times 150000 =$ ₹ 50000

Practice Exercise

Level - I

1. Find the value of $\frac{x+a}{x-a}+\frac{x+b}{x-b}$, if $x=\frac{2ab}{a+b}$.
   (a) -2 (b) 2
   (c) 1 (d) -1
2. A certain sum of money was divided among A, B and C in a certain way. C got half as much as A and B together got. A got one third of what B and C together got. What is the ratio of A's share to that of C's share?
   (a) 1:4 (b) 3:4
   (c) 4:1 (d) 3:5
3. Two numbers are in the ratio of 3 : 4. If 5 is subtracted from each, the resulting numbers are in the ratio 2 : 3. Find the numbers
   (a) 12, 16 (b) 24, 32
   (c) 60, 80 (d) 15, 20
4. The wages of labourers in a factory increased in the ratio 22 : 25 and there was a reduction in their number in the ratio 15 : 11. Find the original wage bill if the present bill is ₹ 5000.
   (a) ₹2500 (b) ₹3000
   (c) ₹5000 (d) ₹6000
5. Which of the following numbers should be added to 11, 15, 17 and 23 so that they are in proportion?
   (a) 2 (b) 3
   (c) 5 (d) 1
6. Find the forth proportional to $12X^3, 9aX^2, 8a^3X$.
   (a) $4a^3$ (b) $6a^4$
   (c) $5a$ (d) $7a^5$
7. Vijay decides to leave 100 acres of his land to his three daughters Vijaya, Sunanda and Ansuya in the proportion of one-third, one-fourth and one-fifth respectively. But Vijaya suddenly expires. Now how should Vijay divide the land between Sunanda and Anusuya?
   (a) $\frac{500}{9}, \frac{400}{9}$ (b) $\frac{450}{8}, \frac{350}{8}$
   (c) $\frac{420}{7}, \frac{280}{7}$ (d) $\frac{320}{7}, \frac{380}{7}$
8. Find $a:b:c$, if $6a=9b=10c$.
   (a) 12:10:8 (b) 15:4:3
   (c) 15:18:9 (d) 15:10:9
9. What is the least integer which when added to both terms of the ratio 5 : 9 will make a ratio greater than 7 : 10?
   (a) 6 (b) 8
   (c) 5 (d) 7
10. If $a:b=2:3, b:c=3:4, c:d=4:5$, find $a:b:c:d$.
   (a) 5:4:3:2 (b) 30:20:15:12
   (c) 2:3:4:6 (d) 2:3:4:5
11. ₹ 1220 is divided, among A, B, C and D, such that B's share is $\frac{5}{9}$ th of A's; C's share is $\frac{7}{10}$ th of B's and D has $\frac{1}{3}$ as much as B and C together. Find A's share.
   (a) ₹ 540 (b) ₹ 802
   (c) ₹ 100 (d) ₹ 650
12. In an examination, there are five subjects and each has the same maximum. A boy's marks are in the ratio 3 : 4 : 5 : 6 : 7 and his aggregate is $\frac{3}{5}$ th of the full marks. In how many subjects did he get more than 50% marks?
   (a) 1 (b) 2
   (c) 3 (d) 4
13. Three friends started a business of renting out air conditioners by investing ₹ 20000, ₹ 24000 and ₹ 16000, respectively. C gets 20% of total profit for repair and maintenance of the air conditioner. If in a particular year, C gets ₹ 487.50 less than the total earnings of the other two, then the total profit for the year is :
   (a) ₹2812.50 (b) ₹3625.50
   (c) ₹4515.00 (d) None of these
14. The ratio of the prices of two houses A and B was 4 : 5 last year. This year, the price of A is increased by 25% and that of B by ₹ 50000. If their prices are now in the ratio 9 : 10, the price of A last year was :
   (a) ₹3,60,000 (b) ₹4,50,000
   (c) ₹4,80,000 (d) ₹5,00,000
15. The dimensions of a rectangular room when increased by 4 metres are in the ratio of 4 : 3 and when decreased by 4 metres, are in the ratio of 2 : 1. The dimensions of the room are
   (a) 6 m and 4 m (b) 12 m and 8 m
   (c) 16 m and 12 m (d) 24 m and 16 m
16. The sum of three numbers is 98. If the ratio of the first to the second is 2 : 3 and that of the second to the third is 5 : 8, then the second number is:
   (a) 20 (b) 30
   (c) 38 (d) 48
17. Two numbers are such as that square of one is 224 less than 8 times the square of the other. If the numbers are in the ratio of 3 : 4, they are
   (a) 12, 16 (b) 6, 8
   (c) 9, 12 (d) None of these

18. Tea worth ₹ 126 per kg and ₹ 135 per kg are mixed with a third variety in the ratio 1 : 1 :2. If the mixture is worth ₹ 153 per kg, then the price of the third variety per kg is

(a) ₹169.50 (b) ₹170
(c) ₹175.50 (d) ₹180

19. In a mixture of 45 litres, the ratio of milk and water is 3 : 2. How much water must be added to make the ratio 9 : 11?

(a) 10 litres (b) 15 litres
(c) 17 litres (d) 20 litres

20. The ratio of the rate of flow of water in pipes varies inversely as the square of the radii of the pipes. What is the ratio of the rates of flow in two pipes of diameters 2 cm and 4 cm, respectively?

(a) 1:2 (b) 2:1
(c) 1:8 (d) 4:1

21. Given that 24 carat gold is pure gold. 18 carat gold is $\frac{3}{4}$ pure gold and 20 carat gold is $\frac{5}{6}$ pure gold. The ratio of the pure gold in 18 carat gold to the pure gold in 20 carat gold is :

(a) 3:8 (b) 9:10
(c) 15:24 (d) 8:5

22. If $\frac{y}{x-z} = \frac{y+x}{z} = \frac{x}{y}$, then find $x : y : z$.

(a) 1:2:3 (b) 3:2:1
(c) 4:2:3 (d) 2:4:7

23. Salaries of A, B and C were in the ratio 3 : 5 : 7, respectively. If their salaries were increased by 50%, 60% and 50% respectively, what will be the new ratio of the their respective new salaries?

(a) 4:5:7 (b) 3:6:7
(c) 4:15:18 (d) 9:16:21

24. The average score of boys in an examination of a school is 71 and that of the girls is 73. The average score of the whole school in that examination is 71.8. Find the ratio of the number of boys to the number of girls that appeared in the examination.

(a) 4:5 (b) 3:2
(c) 3:5 (d) 5:2

25. Two casks of 48 L and 42 L are filled with mixtures of wine and water, the proportions in the two casks being respectively 13 : 7 and 18 : 17. If the contents of the two casks be mixed and 20 L of water is added to the whole, what will be the proportion of wine to water in the resultant solution?

(a) 21:31 (b) 12:13
(c) 13:12 (d) None of these

26. What amounts (in litres) of 90% and 97% pure acid solutions are mixed to obtain 21 L of 95% pure acid solution?

(a) 6 and 15 L (b) 14 and 15 L
(c) 12 and 15 L (d) 13 and 12 L

27. Arvind began a business with ₹ 550 and was joined afterwards by Brij with ₹ 330. When did Brij join, if the profits at the end of the year were divided in the ratio 10 : 3?

(a) After 4 months (b) After 6 months
(c) After 4.5 months (d) None of these

28. A, B and C are partners. A receives 9/10 of the profit and B and C share the remaining profit equally. A's income is increased by ₹ 270 when the profit rises from 12 to 15%. Find the capital invested by B and C each

(a) ₹5000 (b) ₹1000
(c) ₹500 (d) ₹1500

29. A fort had provision of food for 150 men for 45 days. After 10 days, 25 men left the fort. The number of days for which the remaining food will last, is

(a) $29\frac{1}{5}$ (b) $37\frac{1}{4}$
(c) 42 (d) 54

30. In a mixture of 45 L, the ratio of milk and water is 2 : 1. If this ratio is to be 3 : 2, the quantity of water to be further added is

(a) 3 L (b) 5 L
(c) 8 L (d) None of these

31. If 40% of a number is equal to two-third of another number, what is the ratio of first number to the second number?

(a) 2 : 5 (b) 3 : 7
(c) 5 : 3 (d) 7 : 3

32. If the cost of printing a book of 320 leaves with 21 lines on each page and on an average 11 words in each line is ₹ 19, find the cost of printing a book with 297 leaves, 28 lines on each page and 10 words in each line.

(a) ₹ $22\frac{3}{8}$ (b) ₹ $20\frac{3}{8}$
(c) ₹ $21\frac{3}{8}$ (d) ₹ $21\frac{3}{4}$

33. A and B entered into a partnership with investments of ₹ 15000 and ₹ 40000 respectively. Aftere 3 months A left from the business, at the same time C joins with ₹ 30000. At the end of 9 months, they got ₹ 7800 as profit. Find the share of B.

(a) ₹4800 (b) ₹ 600
(c) ₹2400 (d) ₹ 1200

34. The third proportional to $(x^2 - y^2)$ and $(x - y)$ is :

(a) $(x + y)$ (b) $(x - y)$
(c) $\frac{x+y}{x-y}$ (d) $\frac{x-y}{x+y}$

35. The sides of a triangle are in the ratio $\frac{1}{2}:\frac{1}{3}:\frac{1}{4}$ and its perimeter is 104 cm. The length of the longest side is

(a) 52 cm (b) 48 cm
(c) 32 cm (d) 26 cm

36. Three friends *A*, *B* and *C* started a business by investing a sum of money in the ratio of 5 : 7 : 6. After 6 months *C* withdraws half of his capital. If the sum invested by '*A*' is ₹ 40,000, out of a total annual profit of ₹ 33,000, *C*'s share will be
(a) ₹9,000 (b) ₹12,000
(c) ₹11,000 (d) ₹10,000

37. The numbers of students speaking English and Hindi are in the ratio of 4:5. If the number of students speaking English increased by 35% and that speaking Hindi increased by 20%, what would be the new respective ratio?
(a) 19:20 (b) 7:8
(c) 8:9 (d) 9:10

38. The ratio of males and females in a city is 7 : 8 and the percentage of children among males and females is 25% and 20% respectively. If the number of adult females in the city is 156800 what is the total population?
(a) 245000 (b) 367500
(c) 196000 (d) 171500

39. *A*, *B* and *C* started a business with a total investment of ₹ 72000. *A* invests ₹ 6000 more than *B* and *B* invests ₹ 3000 less than *C*. If the total profit at the end of a years is ₹ 8640, find A's share.
(a) ₹3240 (b) ₹2520
(c) ₹2880 (d) ₹3360

40. *A*, *B* and *C* enter into a partnership. They invest ₹ 40,000, ₹ 80,000 and ₹ 1,20,000 respectively. At the end of the first year, *B* withdraws ₹ 40,000, while at the end of the second year, *C* withdraws ₹ 80,000. In what ratio will the profit be shared at the end of 3 years?
(a) 2 : 3 : 5 (b) 3 : 4 : 7
(c) 4 : 5 : 9 (d) None of these

41. Incomes of two companies *A* and *B* are in the ratio of 5 : 8. Had the income of company *A* been more by ₹ 25 lakh, the ratio of their incomes would have been 5 : 4. What is the income of company *B*?
(a) ₹80 lakh (b) ₹50 lakh
(c) ₹40 lakh (d) ₹60 lakh

42. Abhishek started a business investing ₹ 50,000. After one year he invested another ₹ 30,000 and Sudin also joined him with a capital of ₹ 70,000. If the profit earned in three years from the starting of business was ₹ 87,500, then find the share of Sudin in the profit.
(a) ₹37,500 (b) ₹35,000
(c) ₹38,281 (d) ₹52,500

43. In 1 kg mixture of sand and iron, 20% is iron. How much sand should be added so that the proportion of iron becomes 10%?
(a) 1 kg (b) 200 gms
(c) 800 gms (d) 1.8 kg

44. *A* started a business with ₹ 21,000 and is joined afterwards by *B* with ₹ 36,000. After how many months did *B* join if the profits at the end of the year are divided equally?
(a) 3 (b) 4
(c) 5 (d) 6

45. When 30 percent of a number is added to another number the second number increases to its 140 per cent. What is the ratio between the first and the second number?
(a) 3:4 (b) 4:3
(c) 3:2 (d) None of these

46. The ratio of number of ladies to gents at a party was 1 : 2, but when 2 ladies and 2 gents left, the ratio became 1 : 3. How many people were originally present at the party?
(a) 6 (b) 9
(c) 12 (d) 10

47. A bag contains an equal number of one rupee, 50 paise and 25 paise coins respectively. If the total value is ₹ 35, how many coins of each type are there?
(a) 20 coins (b) 30 coins
(c) 28 coins (d) 25 coins

48. *A* and *B* invest ₹ 3,000 and ₹ 4,000 in a business. *A* receives ₹ 10 per month out of the profit as a remuneration for running the business and the rest of profit is divided in proportion to the investments. If in a year '*A*' totally receives ₹ 390, what does *B* receive?
(a) ₹375 (b) ₹360
(c) ₹350 (d) ₹260

49. If $f(x)=\frac{(x+1)}{(x-1)}$, then the ratio of x to $f(y)$ where $y=f(x)$ is
(a) $x:y$ (b) $x^2:y^2$
(c) 1:1 (d) $y:x$

50. If $\frac{a}{b+c}=\frac{b}{c+a}=\frac{c}{a+b}$, then each fraction is equal to
(a) $(a+b+c)^2$ (b) 1/2
(c) 1/4 (d) 0

51. If $a:b=c:d$ then the value of $\frac{a^2+b^2}{c^2+d^2}$ is
(a) 1/2 (b) $\frac{a+b}{c+d}$
(c) $\frac{a-b}{c-d}$ (d) $\frac{ab}{cd}$

52. In Ramnagar Colony, the ratio of school going children to non-school going children is 5 : 4. If in the next year, the number of non-school going children is increased by 20%, making it 35,400, what is the new ratio of school going children to non-school going children?
(a) 4:5 (b) 3:2
(c) 25:24 (d) None of these

53. If ₹ 1066 is divided among *A*, *B*, *C* and *D* such that $A:B=3:4$, $B:C=5:6$ and $C:D=7:5$, who will get the maximum?
(a) B (b) A
(c) C (d) D

54. The ratio of the ages of A and B seven years ago was 3 : 4 respectively. The ratio of their ages nine years from now will be 7 : 8 respectively. What is B's age at present ? [SBI Clerk-June-2012]
(a) 16 years (b) 19 years
(c) 28 years (d) 23 years
(e) None of these

55. The respective ratio between the present ages of father, mother and daughter is 7 : 6 : 2. The difference between mother's and the daughter's age is 24 years. What is the father's age at present ? [SBI Clerk-2012]
(a) 43 years (b) 42 years
(c) 39 years (d) 38 years
(e) None of these

56. Number of students studying in colleges *A* and *B* are in the ratio of 3 : 4 respectively. If 50 more students join college *A* and there is no change in the number of students in college *B*, the respective ratio becomes 5 : 6. What is the number of students in college *B* ? [SBI Clerk-2014]
(a) 450 (b) 500
(c) 400 (d) 600
(e) None of these

57. A certain sum of money is distributed to A and B in the ratio 2 : 5. If A received ₹100, then the money received by B is [SSC-Sub. Ins.-2012]
(a) ₹200 (b) ₹150
(c) ₹250 (d) ₹300

58. A man leaves ₹ 12,600 to be divided among 7 sons, 3 daughters and 5 nephews. If each daughter receives three times as much as each nephew and each son seven times as much as each nephew, then each daughter's share is [SSC-Sub. Ins.-2012]
(a) ₹ 700 (b) ₹ 650
(c) ₹ 600 (d) ₹ 750

59. The proportion of acid and water in three samples is 2 : 1, 3 : 2 and 5 : 3. A mixture containing equal quantities of all three samples is made. The ratio of water and acid in the mixture is : [SSC-Sub. Ins.-2013]
(a) 120 : 133 (b) 227 : 133
(c) 227 : 120 (d) 133 : 227

60. If x : y ::2 : 3 and 2 : x :: 4 : 8 the value of y is [SSC-Sub. Ins.-2014]
(a) 6 (b) 8
(c) 4 (d) 12

61. ₹ 730 were divided among A, B, C in such a way that if A gets ₹ 3, then B gets ₹ 4 and if B gets ₹ 3.50 then C gets ₹ 3. The share of B exceeds that of C by [SSC-Sub. Ins.-2014]
(a) ₹ 30 (b) ₹ 40
(c) ₹ 70 (d) ₹ 210

62. A certain amount of money is divided among x, y and z. If x receives 25% more than y and y receives 25% less than z, then x : y : z is equal to [SSC-MT-2013]
(a) 12 : 10 : 11 (b) 14 : 12 : 13
(c) 15 : 12 : 16 (d) 10 : 9 : 12

63. If 10% of *x* is the same as 20% of *y*, then *x* : *y* is [SSC-MT-2013]
(a) 5 : 1 (b) 1 : 2
(c) 3 : 1 (d) 2 : 1

64. In a school, the ratio of boys to girls is 4 : 3 and the ratio of girls to teachers is 8 : 1. The ratio of student to teachers is : [SSC 10+2-2012]
(s) 56 : 3 (b) 55 : 1
(c) 49 : 3 (d) 56 : 1

65. A, B and C are batsmen. The ratio of the runs scored by them in a certain match are given below:
A : B = 5 : 3 and B : C = 4 : 5. In all they scored 564 runs. The number of runs scored by B is: [SSC 10+2-2012]
(a) 124 (b) 104
(c) 114 (d) 144

66. The ratio of age of two boys is 5 : 6. After two years the ratio will be 7 : 8. The ratio of their ages after 12 years will be [SSC 10+2-2013]
(a) 11/12 (b) 22/24
(c) 15/16 (d) 17/18

67. A invests ₹ 64,000 in a business. After few months B joined him with ₹ 48,000. At the end of year, the total profit was divided between them in the ratio 2 : 1. After how many months did B join ? [SSC 10+2-2013]
(a) 7 (b) 8
(c) 4 (d) 6

68. If $\frac{x}{y}=\frac{4}{5}$, then the value of $\left(\frac{4}{7}+\frac{2y-x}{2y+x}\right)$ is [SSC 10+2-2014]
(a) $\frac{3}{7}$ (b) $1\frac{1}{7}$
(c) 1 (d) 2

69. Ram left $\frac{1}{3}$ of his property to his widow and $\frac{3}{5}$ of the remainder to his daughter. He gave the rest to his son who received ₹ 6,400. How much was his original property worth? [SSC 10+2-2014]
(a) ₹ 16,000 (b) ₹ 32,000
(c) ₹ 24,000 (d) ₹ 1,600

70. A total profit of ₹ 3,600 is to be distributed amongst A, B and C such that A : B = 5 : 4 and B : C = 8 : 9. The share of C in the profit is [SSC 10+2-2014]
(a) ₹ 1,200 (b) ₹ 1,500
(c) ₹ 1,650 (d) ₹ 1,700

71. Three friends divide ₹ 624 among themselves in the ratio $\frac{1}{2}:\frac{1}{3}:\frac{1}{4}$; the share of the third friend is [SSC 10+2-2014]
(a) ₹ 288 (b) ₹ 192
(c) ₹ 148 (d) ₹ 144

72. Two numbers are in the ratio 3 : 5. If 9 is subtracted from each, the new numbers are in the ratio 12 : 23. The small number is [SSC 10+2-2014]
(a) 27 (b) 33
(c) 49 (d) 55

73. If x : y = 5 : 2, then (8x + 9y) : (8x + 2y) is [SSC 10+2-2014]
(a) 22 : 29 (b) 26 : 61
(c) 29 : 22 (d) 61 : 26

74. A is twice as fast as B and B is thrice as fast as C is. The journey covered by C in 1½ hours will be covered by A in [SSC 10+2-2014]
(a) 15 minutes (b) 20 minutes
(c) 30 minutes (d) 1 hour

75. If $\frac{x}{xa+yb+zc}=\frac{y}{ya+zb+xc}=\frac{z}{xa+xb+yc}$ and $x+y+z \neq 0$, then each ratio is [SSC 10+2-2014]
(a) $\frac{1}{a-b-c}$ (b) $\frac{1}{a+b-c}$
(c) $\frac{1}{a-b+c}$ (d) $\frac{1}{a+b+c}$

76. In the expression xy^2, the values of x and y are each decreased by 25%. The value of the expression is decreased by [SSC 10+2-2014]
(a) $\frac{37}{64}$ of its value (b) $\frac{1}{2}$ of its value
(c) $\frac{27}{64}$ of its value (d) $\frac{3}{4}$ of its value

77. If the numerator of a fraction is increased by 300% and the denominator is increased by 200%, the resultant fraction is $\frac{4}{15}$. What is the original fraction ? [IBPS Clerk-2012]
(a) $\frac{3}{5}$ (b) $\frac{4}{5}$
(c) $\frac{2}{5}$ (d) $\frac{1}{5}$
(e) None of these

78. The ratio between Gloria's and Sara's present ages is 4 : 7 respectively. Two years ago the ratio between their ages was 1 : 2 respectively. What will be Sara's age three years hence ? [IBPS Clerk-2012]
(a) 17 years (b) 14 years
(c) 11 years (d) 8 years
(e) None of these

79. The respective ratio of salaries of A and B is 8 : 7. If the salary of B increases by 20% and the salary of A increases by 21%, the new ratio becomes 121 : 105 respectively. What is A's salary? [IBPS Clerk-2013]
(a) ₹22560 (b) ₹21600
(c) ₹20640 (d) ₹23040
(e) Cannot be determined

80. A, B and C started a business by investing ₹ 12800, ₹ 16800 and ₹ 9600 respectively. If after 8 months B received ₹13125 as his share of profit, what amount did C get as his share of profit? [IBPS Clerk-2013]
(a) ₹7800 (b) ₹7150
(c) ₹7750 (d) ₹8250
(e) ₹7500

81. 12 yrs hence the ratio between the ages of A and B will be 3 :4 respectively. The present age of A is $3\frac{3}{4}$ times of C's present age. If C's present age is 10 yrs, then what is B's present age? (in years) [IBPS Clerk-2013]
(a) 48 (b) 46
(c) 60 (d) 54
(e) 36

82. M, N, O and P divided ₹ 44352 among themselves. M took $\frac{3}{8}$th of the money, N took $\frac{1}{6}$th of the remaining amount and rest was divided among O and P in the ratio of 3 : 4 respectively. How much did O get as his share? [IBPS Clerk-2013]
(a) ₹9600 (b) ₹10600
(c) ₹10300 (d) ₹8700
(e) ₹9900

83. A and B are two numbers. 6 times of square of B is 540 more than the square of A. If the respective ratio between A and B is 3 : 2, what is the value of B? [IBPS Clerk-2013]
(a) 10 (b) 12
(c) 16 (d) 8
(e) 14

Level - II

1. A man completes 5/8 of a job in 10 days. At this rate, how many more days will it take him to finish the job?
(a) 5 (b) 6
(c) 7 (d) $7\frac{1}{2}$

2. ₹ 1104 is divided between 3 men, 4 women and 6 boys, so that the share of a man, a woman and a boy are in the proportion of 3 : 2 : 1. How much does each boy get?
(a) ₹ 48 (b) ₹ 64
(c) ₹ 96 (d) Cannot be determined

3. Seats of Physics, Chemistry and Mathematics in a school are in the ratio 4 : 5 : 6. There is a proposal to increase these seats by 75 in each department. What were the total number of seats in the school finally?
(a) 600 (b) 750
(c) 900 (d) None of these

4. 60 kg of an alloy A is mixed with 100 kg of alloy B. If alloy A has lead and tin in the ratio 3 : 2 and alloy B has tin and copper in the ratio 1 : 4, then the amount of tin in the new alloy is

(a) 36 kg (b) 44 kg
(c) 53 kg (d) 80 kg

5. A, B and C started a business. A invests $\frac{1}{2}$ capital for $\frac{1}{4}$ time, B invests $\frac{1}{8}$ capital for $\frac{1}{2}$ time and C invests the remaining capital for whole time. Find the share of B in the total profit of ₹ 9900.

(a) ₹ 2200 (b) ₹ 1100
(c) ₹ 6600 (d) ₹ 4400

6. Two jars having a capacity of 3 and 5 litres respectively are filled with mixtures of milk and water. In the smaller jar 25% of the mixture is milk and in the larger 25% of the mixture is water. The jars are emptied into a 10 litre cask whose remaining capacity is filled up with water. Find the percentage of milk in the cask.

(a) 55% (b) 50%
(c) 45% (d) None of these

7. A, B, C subscribe ₹ 50,000 for a business. A subscribes ₹ 4000 more than B and ₹ 5000 more than C. Out of a total profit of ₹ 35,000, A receives :

(a) ₹ 8,400 (b) ₹ 11,900
(c) ₹ 13,600 (d) ₹ 14,700

8. There is a ratio of 5 :4 between two numbers. If 40 percent of the first number is 12 then what would be the 50 percent of the second number?

(a) 12 (b) 24
(c) 18 (d) None of these

9. In a partnership, A invests $\frac{1}{6}$ of the capital for $\frac{1}{6}$ of the time, B invests $\frac{1}{3}$ of the capital for $\frac{1}{3}$ of the time and C, the rest of the capital for whole time. Find A's share of the total profit of ₹ 2,300.

(a) ₹ 100 (b) ₹ 200
(c) ₹ 300 (d) ₹ 400

10. A and B rent a pasture for 10 months; A puts in 80 cows for 7 months. How many can B put in for the remaining 3 months, if he pays half as much again as A?

(a) 120 (b) 180
(c) 200 (d) 280

11. The resistance of a wire is proportional to its length and inversely proportional to the square of its radius. Two wires of the same material have the same resistance and their radii are in the ratio 9 : 8. If the length of the first wire is 162 cms., find the length of the other.

(a) 64 cm. (b) 120 cm.
(c) 128 cm. (d) 132 cm.

12. A diamond falls and breaks into three pieces whose weights are in the ratio 1 : 3 : 6. The value of the diamond is proportional to the square of its weight. If the original value is ₹ 30,000, What is the loss in the in the value due to the breakage?

(a) ₹ 13, 800 (b) ₹ 16,200
(c) ₹ 18, 600 (d) ₹ 19, 400

13. When a bus started from the first stop, the number of male passengers to the number of female passengers was 3 : 1. At the stop 16 passengers get down and 6 more female passengers get into. Now the ratio of the male to female passengers becomes 2 : 1. What was the total number of passengers in the bus when it started from the first stop?

(a) 64 (b) 48
(c) 54 (d) 72

14. In three vessels, the ratio of water and milk is 6 : 7, 5 : 9 and 8 : 7, respectively. If the mixtures of the three vessels are mixed together, then what will be the ratio of water and milk?

(a) 2431 : 3781 (b) 3691 : 4499
(c) 4381 : 5469 (d) None of these

15. In two alloys, the ratio of iron and copper is 4 : 3 and 6 : 1, respectively. If 14 kg of the first alloy and 42 kg of the second alloy is mixed together to form a new alloy, then what will be the ratio of iron to copper in the new alloy?

(a) 11 : 3 (b) 11 : 8
(c) 8 : 1 (d) None of these

16. Mixture of milk and water has been kept in two separate containers. Ratio of milk to water in one of the containers is 5 : 1 and that in the other container is 7 : 2. In what ratio should the mixtures of these two containers be added together so that the quantity of milk in the new mixture may become 80%?

(a) 3 : 2 (b) 2 : 3
(c) 4 : 5 (d) None of these

17. Three containers of capacity 20 L, 5 L and 9 L contain mixture of milk and water with milk concentrations 90%, 80% and 70% respectively. The contents of three containers are emptied into a large vessel. What is the approximate ratio of milk to water in the resultant mixture?

(a) 3 : 1 (b) 4 : 1
(c) 5 : 1 (d) 2 : 1

18. Ratio of the earnings (in ₹) of A and B is 4 : 7. If the earnings of A increase by 50% and those of B decrease by 25%, the new ratio of their earnings becomes 8 : 7. How much is A earning?

(a) ₹ 28000 (b) ₹ 21000
(c) ₹ 26000 (d) Data inadequate

19. In the famous Bhojpur island, there are four men for every three women and five children for every three men. How many children are there in the island if it has 531 women?

(a) 454 (b) 1180
(c) 1070 (d) 389

20. If $a/b = 1/3$, $b/c = 2$, $c/d = 1/2$, $d/e = 3$ and $e/f = 1/4$, then what is the value of abc/def?
(a) 3/8 (b) 27/8
(c) 3/4 (d) 27/4

21. When the numerator and the denominator of a fraction are increased by 1 and 2 respectively, the fraction becomes $\frac{2}{3}$, and when the numerator and the denominator of the same fraction are increased by 2 and 3 respectively, the fraction becomes $\frac{5}{7}$. What is the original fraction? [*SBI PO-2011*]
(a) $\frac{5}{6}$ (b) $\frac{3}{4}$
(c) $\frac{3}{5}$ (d) $\frac{6}{7}$
(e) None of these

22. When X is subtracted from the numbers 9,15 and 27, the remainders are in continued proportion. What is the value of X ? [*IBPS-PO-2012*]
(a) 8 (b) 6
(c) 4 (d) 5
(e) None of these

23. A certain amount was to be distributed among A, B and C in the ratio 2 :3 :4 respectively, but was erroneously distributed in the ratio 7:2:5 respectively. As a result of this, B got ₹40 less. What is the amount ? [*IBPS-PO-2012*]
(a) ₹210/- (b) ₹270/-
(c) ₹230/- (d) ₹280/-
(e) None of these

24. ₹ 73,689/- are divided between A and B in the ratio 4 : 7. What is the difference between thrice the share of A and twice the share of B ? [*IBPS-PO-2012*]
(a) ₹36,699/- (b) ₹46,893/-
(c) ₹20,097/- (d) ₹26,796/-
(e) ₹ 13,398/-

25. If the numerator of a fraction is increased by 20% and the denominator is increased by 25%, the fraction obtained is $\frac{3}{5}$. What was the original fraction? [*IBPS-PO-2013*]
(a) $\frac{5}{7}$ (b) $\frac{4}{7}$
(c) $\frac{3}{8}$ (d) Cannot be determined
(e) None of these

26. The respective ratio between the present ages of son, mother, father and grandfather is 2 : 7 : 8 : 12. The average age of son and mother is 27 yrs. What will be mother's age after 7 yrs? [*IBPS-PO-2013*]
(a) 40 yrs (b) 41 yrs
(c) 48 yrs (d) 49 yrs
(e) None of these

27. The prize money of ₹ 1,800 is divided among 3 students A, B and C in such a way that 4 times the share of A is equal to 6 times the share of B, which is equal to 3 times the share of C. Then A's share is [*SSC CGL-2013*]
(a) ₹400 (b) ₹600
(c) ₹700 (d) ₹800

28. A man borrowed some money from a private organisation at 5% simple interest per annum. He lended 50% of this money to another person at 10% compound interest per annum and thereby the man made a profit of ₹ 13,205 in 4 years. The man borrowed [*SSC CGL-2014*]
(a) ₹80,000 (b) ₹1,00,000
(c) ₹1,20,000 (d) ₹1,50,000

29. A, B and C enter into a partnership with their capitals in the $\frac{7}{2}:\frac{4}{3}:\frac{6}{5}$. After 4 months, A increases his share 50%. If the total profit at the end of the year was ₹ 2,16,000, then B's share in the profit was [*SSC CGL-2014*]
(a) ₹22,000 (b) ₹24,000
(c) ₹30,000 (d) ₹40,000

Hints & Solutions

Level-I

1. (b) $x = \frac{2ab}{a+b} \Rightarrow \frac{x}{a} = \frac{2b}{a+b}$

$\Rightarrow \frac{x+a}{x-a} = \frac{3b+a}{b-a}$ (componendo di videndo)

Similarly, $\frac{x}{b} = \frac{2a}{a+b} \Rightarrow \frac{x+b}{x-b} = \frac{3a+b}{a-b}$

$\therefore \frac{x+a}{x-a} + \frac{x+b}{x-a} = \frac{3b+a}{b-a} + \frac{3a+b}{a-b}$

$= -\frac{(3b+a)}{a-b} + \frac{3a+b}{a-b} = \frac{2a-2b}{a-b} = 2$

2. (b) Let us represent their shares by the corresponding letter of their names.
$A+B=2C$ and $B+C=3A$.
$\Rightarrow A+3A-C=2C$ (since $B=3A-C$)
$4A=3C \Rightarrow A:C=3:4$

3. (d) Let $3X$ and $4X$ be the numbers $\Rightarrow \frac{3X-5}{4X-5} = \frac{2}{3}$

$\Rightarrow 9X-15=8X-10 \Rightarrow X=5$
$\Rightarrow$ The required numbers are 15 and 20.

4. (d)

	Original	Present
Wages	$22x$	$25x$
Number	$15y$	$11y$

Ratio of total wages $= \frac{22x \times 15y}{25x \times 11y} = \frac{6}{5}$

If the present bill is ₹ 5000, the original was ₹ 6000.

5. (d) Let x is to be added
$(11+x):(15+x)=(17+x):(23+x)$

$\Rightarrow \frac{11+x}{15+x} = \frac{17+x}{23+x} \Rightarrow x=1$

6. (b) Let r be the 4th proportional.

Then $\frac{12X^3}{9aX^2} = \frac{8a^3X}{r} \Rightarrow r = 6a^4$

7. (a) The 100 acres should no be divided between Sunanda and Ansuya in the ratio

$\frac{1}{4}:\frac{1}{5}$ i.e. $5:4$

So, Sunanda gets $\frac{500}{9}$ acres and Ansuya gets $\frac{400}{9}$

8. (d) $\frac{a}{b} = \frac{9}{6} = 3:2 = 15:10$

$\frac{b}{c} = \frac{10}{9} = 10:9$

Hence, $a:b:c = 15:10:9$

9. (c) If x is the integer, $\frac{5+x}{9+x} > \frac{7}{10}$

$\therefore 50+10x > 63+7x$
$\therefore 3x > 13$

$\therefore x > \frac{13}{3}$

The least integer greater than $\frac{13}{3}$ is 5.

10. (d) Obviously the ratio is 2 : 3 : 4 : 5

11. (a) If A's share is 1, B's share $= \frac{5}{9} \times 1 = \frac{5}{9}$

C's share $= \frac{7}{10} \times \frac{5}{9} = \frac{7}{18}$;

D's share $= \frac{1}{3}\left(\frac{5}{9}+\frac{7}{18}\right) = \frac{17}{54}$

$\therefore A:B:C:D = 1:\frac{5}{9}:\frac{7}{18}:\frac{17}{54} = 54:30:21:17.$

$\therefore A$'s share $= \frac{54}{122} \times 1220 = $ Rs.540.

12. (c) If the maximum for each paper is 100, total marks = 500

and his aggregate $= \frac{3}{5} \times 500 = 300$. which when divided

in the given ratio gives marks 36, 48, 60, 72 and 84 and so there are 3 subjects in which he gets more than 50.

13. (a) 80% of the total profit is divided in the ratio
$20000:24000:16000 = 5:6:4$
80% of total profit $= 5x+6x+4x = 15x$

Total profit $= \frac{15x}{80\%} = 18.75x$

$\therefore$ Share of C in profit $= 4x + 20\%$ of $18.75x$
$= 4x + 3.75x = 7.75x$

Share of A in profit $= 5x$
Share of B in profit $= 6x$
$\therefore (6x+5x) - 7.75x = 487.50$
$3.25x = 487.50 \Rightarrow x = 150$
$\therefore$ Total profit $= 18.75 \times 150 =$ ₹2812.50.

14. (a) Let the prices of two houses A and B be ₹ $4x$ and ₹ $5x$, respectively for the last year.
Then, the prices of A this year = ₹ $(1.25 \times 4x)$ and that of B = ₹ $(5x+50{,}000)$
This year, Ratio of their prices = 9 : 10

$\therefore \frac{1.25 \times 4x}{5x+50{,}000} = \frac{9}{10}$

$\Rightarrow 50x - 45x = 450000 \Rightarrow 5x = 4{,}50{,}000$
$\Rightarrow x = 90{,}000$
Hence, the price of A last year was
$4x =$ ₹ 3,60,000

15. (b) Let the length and breadth of the rectangular room be ℓ and b.

We have, $\frac{\ell+4}{b+4}=\frac{4}{3}$

$\Rightarrow 3\ell+12=4b+16$

$\Rightarrow 3\ell-4b=4$ ...(1)

Again, we have $\frac{\ell-4}{b-4}=\frac{2}{1} \Rightarrow \ell-4=2b-8$

$\Rightarrow \ell-2b=-4$ ...(2)

Solving (1) and (2), we get $\ell=12$ and b = 8.

16. (b) Let A, B and C be the first, second and third nos. respectively.

Then, $A:B=2:3$ and $B:C=5:8$

Consider, $A:B=2:3=2\times5:3\times5=10:15$

and $B:C=5:8=5\times3:8\times3=15:24$

$\therefore$ $A:B:C=10:15:24$

Let the required number be $10x$, $15x$ and $24x$.

Given, sum of three numbers = 98

Then,

$\therefore$ $10x+15x+24x=98$

$\Rightarrow 49x=98 \Rightarrow x=2$

$\Rightarrow$ Second number $=15x=15\times2=30$

17. (b) Given, ratio of numbers is 3 : 4

$\therefore$ The numbers are $3x$ and $4x$.

Now, according to the question

$16x^2=8(3x)^2-224$

$\Rightarrow 16x^2=72x^2-224 \Rightarrow 56x^2=224$

$\Rightarrow x=2$

$\therefore$ Required numbers = 6, 8

18. (c) Let the third type of tea is priced at ₹ x per kg. Also suppose that the three types of tea mixed together are 1, 1 and 2 kg, respectively.

Now, $\frac{126\times1+135\times1+2x}{1+1+2}=153$

$\Rightarrow \frac{261+2x}{4}=153 \Rightarrow 261+2x=612$

$\Rightarrow x=\frac{351}{2}=$ ₹ 175.5 per kg.

19. (b) Quantity of milk $=\frac{3}{5}\times45=27$ litres

Quantity of water $=\frac{2}{5}\times45=18$ litres

Let x litres of water be added to make the ratio 9 : 11.

$\therefore$ $\frac{18+x}{27}=\frac{11}{9} \Rightarrow 18+x=33 \Rightarrow x=15l$

20. (d) Radius of the two pipes are 1 cm and 2 cm.

Square of the radii of the pipes are 1 and 4.

$\therefore$ Required ratio of rates of flow in the two pipes $=1:\frac{1}{4}$

$=4:1$

21. (b) 18 carat gold $=\frac{3}{4}$ pure gold $=\frac{3}{4}\times24$

= 18 carat pure gold

20 carat gold $=\frac{5}{6}$ pure gold $=\frac{5}{6}\times24$

= 20 carat pure gold

$\therefore$ Required ratio = 18 : 20 = 9 : 10

22. (c) We have, $\frac{y}{x-z}=\frac{y+x}{z}$

$\Rightarrow yz=xy+x^2-yz-xz$ ....(1)

Also, $\frac{x}{y}=\frac{y}{x-z} \Rightarrow x^2-xz=y^2$ ...(2)

From (1) and (2), we have

$yz=xy-yz+y^2$

$\Rightarrow 2yz=xy+y^2$

$\therefore$ $2z=x+y$ ...(3)

Checking with the options, we find that the values given in option c satisfies the equation (3)

23. (d) Increased ratio of their respective salaries

$=3\times\frac{150}{100}:5\times\frac{160}{100}:7\times\frac{150}{100}$

$=\frac{9}{2}:8:\frac{21}{2}=9:16:21$

24. (b) Let the no. of the boys and girls that appeared in the examination be x and y, respectively.

Now, $71.8=\frac{71x+73y}{x+y}$

$\Rightarrow 71.8x+71.8y=71x+73y$

$\Rightarrow (71.8-71)x=(73-71.8)y \Rightarrow 0.8x=1.2y$

$\Rightarrow \frac{x}{y}=\frac{1.2}{0.8}=\frac{3}{2}$

25. (b) In first cask,

Quantity of water $=\frac{7}{20}\times48=16.8L$

Quantity of wine $=\frac{13}{20}\times48=31.2L$

In second cask,

Quantity of water $=\frac{17}{35}\times42=20.6L$

Quantity of wine $=\frac{18}{35}\times42=21.6L$

Now after mixing:

Total quantity of wine = 52.8 L

Quantity of water = 57.2 L

Ratio after mixing $=\frac{52.8}{57.2}=\frac{528}{572}=\frac{12}{13}$

= 12 : 13.

26. **(a)** 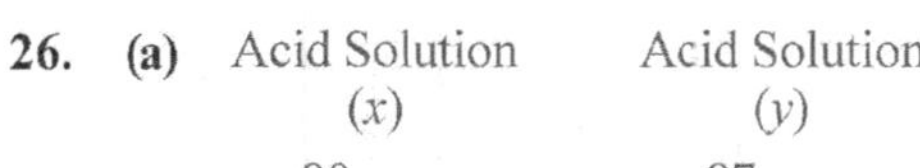
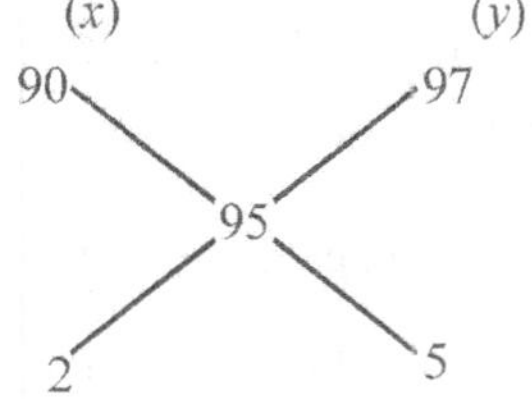

$\therefore$ Amount of the solution x and y in ratio 2 : 5

Amount of acid in solution $x = \frac{2}{7} \times 21 = 6\text{L}$

Amount of acid in solution $y = \frac{5}{7} \times 21 = 15\text{L}$

6L acid in x and 15L in Y

27. **(b)** Profit ratio = 10 : 3
Time ratio = $t_1 : t_2$
Cost ratio = 550 : 330

$550\,t_1 : 330\,t_2 = 10 : 3$

$$\frac{550t_1}{330t_2} = \frac{10}{3}$$

$$\frac{t_1}{t_2} = 2$$

given t_1 = 12 months
$\Rightarrow$ t_2 = 6 months

28. **(c)** Let the profit = x

Profit of $A = \frac{9x}{10}$, Remaining profit $= \frac{x}{10}$

Profit of $B = \frac{x}{20}$

Profit of $C = \frac{x}{20}$

Ratio of profit $= \frac{9}{10} : \frac{1}{20} : \frac{1}{20}$

= 18 : 1 : 1

A's income is increased by ₹270. When profit rises 3%

Investment of $A = \frac{270}{3} \times 100 =$ ₹9000.

If investment of A, B and C = $18x$, x and x
$18x = 9000$
$x = 500$
B investment = ₹500.
C investment = ₹500.

29. **(c)** After 10 days : 150 men had food for 35 days.
Suppose 125 men had food for x days. Now,
Less men, More days (Indirect Proportion)
Then,
men days

men	days
↑ 150	35 ↓
125	x

$\therefore 125 : 150 :: 35 : x \Rightarrow 125 \times x = 150 \times 35$

$$\Rightarrow x = \frac{150 \times 35}{125} \Rightarrow x = 42.$$

Hence, the remaining food will last for 42 days.

30. **(b)** Quantity of milk $= \frac{2}{3} \times 45 = 30$ L

Quantity of water $= \frac{1}{3} \times 45 = 15$ L

Let the required quantity of water to be added be x litre.

Then, $\frac{30}{15 + x} = \frac{3}{2}$

$\Rightarrow$ $3\,(15 + x) = 30 \times 2 = 60$
$\Rightarrow$ $x = 5\text{L}$

31. **(c)** Let 40% of A $= \frac{2}{3}B$. Then, $\frac{40A}{100} = \frac{2B}{3} \Leftrightarrow \frac{2A}{5} = \frac{2B}{3}$

$$\Leftrightarrow \frac{A}{B} = \left(\frac{2}{3} \times \frac{5}{2}\right) = \frac{5}{3}$$

$\therefore$ $A : B = 5 : 3$

32. **(c)** Less leaves, less cost
(Direct Proportion)
More lines, more cost (Direct Proportion)
Less words, less cost (Direct Proportion)

$$\left.\begin{array}{ll} \text{leaves} & 320:297 \\ \text{lines} & 21:28 \\ \text{words} & 11:10 \end{array}\right\} :: 19 : x$$

$\therefore 320 \times 21 \times 11 \times x = 297 \times 28 \times 10 \times 19$

$$\Rightarrow x = \frac{171}{8} = 21\frac{3}{8}$$

33. **(a)** Ratio of capitals of A, B and C
= (15000 × 3) : (40000 × 9) : (30000 × 6)
= 1 : 8 : 4

B's share = ₹$\left(\frac{8}{13} \times 7800\right)$ = ₹ 4800

34. **(d)** Let the third proportional to $(x^2 - y^2)$ and $(x - y)$ be z. Then,
$(x^2 - y^2) : (x - y) :: (x - y) : z \Leftrightarrow (x^2 - y^2) \times z$
$= (x - y)^2 \Leftrightarrow z = \frac{(x-y)^2}{(x^2 - y^2)} = \frac{(x-y)}{(x+y)}$

35. **(b)** Ratio of sides $= \frac{1}{2} : \frac{1}{3} : \frac{1}{4} = 6 : 4 : 3$

Largest side $= \left(104 \times \frac{6}{13}\right)$ cm = 48 cm

36. **(a)** Sum invested by A, B and C is

$5 \times 12 : 7 \times 12 : 6 \times 6 + 3 \times 6$
or, 60 : 84 : 54 or, 10 : 14 : 9

$\therefore$ Share of $C = \frac{9}{33} \times 33,000 =$ ₹9,000

37. (d) Required ratio $= \frac{4 \times 135}{5 \times 120} = 9:10$

38. (b) Number of females $= 156800 \times \frac{100}{80} = 196000$

$\therefore$ Number of males $= \frac{7}{8} \times 196000 = 171500$

$\therefore$ Total population $= 196000 + 171500 = 367500$

39. (a) Let C's investment = ₹ x.

B's investment = ₹ $(x - 3000)$

A's investment = ₹ $(x - 3000 + 6000)$ = ₹ $(x + 3000)$

Now, $(A + B + C)$'s investment = ₹ 72000

$\Rightarrow x + (x - 3000) + (x + 3000) = 72000$

$\Rightarrow 3x = 72000 \Rightarrow x = 24000$

Hence, A's investment = ₹ 27000

B's investment = ₹ 21000

C's investment = ₹ 24000

Ratio of the capitals of A, B and C

$= 27000 : 21000 : 24000$

$= 9 : 7 : 8$

A's share = ₹ $\left(\frac{9}{24} \times 8640\right)$ = ₹ 3240

40. (b) $A : B : C = (40000 \times 36) : (80000 \times 12 + 40000 \times 24) : (120000 \times 24 + 40000 \times 12)$

$= 144 : 192 : 336 = 3 : 4 : 7$

41. (c) Let the incomes of two companies A and B be $5x$ and $8x$ respectively.

From the question,

$\frac{5x + 25}{8x} = \frac{5}{4} \Rightarrow 20x + 100 = 40x \therefore x = 5$

$\therefore$ Income of company $B = 8x$ = ₹ 40 lakh

42. (b) Ratio of Abhishek and Sudin for one month

$= (50,000 \times 36) + (30,000 \times 24) : (70,000 \times 24)$

$= (18,00,000 + 7,20,000) : 16,80,000 = 3 : 2$

Hence share of Sudin in the profit earned from the business.

$= \frac{87,500}{(3+2)} \times 2 =$ ₹35,000.

43. (a) In 1 kg mixture quantity of iron = 200 gm

Let x gm sand should be added, then

10% of $(1000 + x) = 200$

$\therefore x = 1000$ gm = 1 kg

44. (c) Suppose B joined after x months.

Then, $21000 \times 12 = 36000 \times (12 - x) \Leftrightarrow 36x = 180$

$\Leftrightarrow x = 5$.

Hence, B joined after 5 months.

45. (b) Let the first and the second numbers be x and y respect then

$y + 30\%$ of $x = 140\%$ of y

or, $y + 0.3x = 1.4y$

or, $0.3x = 0.4y$

$\therefore x : y = 0.4 : 0.3 = 4 : 3$

46. (c) Let number of ladies $= x$

then, number of gents $= 2x$

Now, $\frac{x-2}{2x-2} = \frac{1}{3} \Rightarrow 3x - 6 = 2x - 2$

$\Rightarrow x = 4$

$\therefore$ Total number of people originally present $= 4 + 8 = 12$

47. (a) Let number of each type of coin $= x$. Then,

$1 \times x + .50 \times x + .25x = 35$

$\Rightarrow 1.75x = 35 \Rightarrow x = 20$ coins

48. (b) In a year, for A, total amount as a remuneration

$= 10 \times 12 =$ ₹ 120

$\therefore$ Amount of A's profit $= 390 - 120 =$ ₹ 270

Ratio of investment $= 3 : 4$

Let total profit = ₹ x

Then, B's profit = ₹ $(x - 270)$

$\therefore \frac{3}{3+4} \times x = 270 \Rightarrow x = 630$

$\therefore B$'s profit $= 630 - 270 =$ ₹ 360

49. (c) Let $x = 5$

Then $f(x) = 6/4 = 1.5 = y$

And $f(y) = 2.5/0.5 = 5$.

Thus, the ratio of $x : f(y) = 1 : 1$

Note: Even if you take some other value of y, you would still get the same answer.

50. (b) The given condition has a, b and c symmetrically placed. Thus, if we use $a = b = c = 2$ (say) we get each fraction as 1/2.

51. (d) $1 : 2 = 3 : 6$, so $(a^2 + b^2)/(c^2 + d^2) = 5/45 = 1/9$

From the given options, only ab/cd gives us this value.

52. (c) $5 : 4 \rightarrow 5 : 4.8 \rightarrow 25 : 24$

Option (c) is correct.

53. (c) Since $A : B = 3 : 4$ ...(1)

$B : C = 5 : 6$ ...(2)

and $C : D = 7 : 5$ ...(3)

Therefore, by, proportionating, (1) and (2)

$A : B = 3 \times 5 : 4 \times 5 = 15 : 20$

$B : C = 20 : 24$ and $C : D = 7 : 5$

Hence, $A : B : C = 15 : 20 : 24$ ...(4)

Now, $A : B : C = 15 \times 7 : 20 \times 7 : 24 \times 7$

$= 105 : 140 : 168$

$\therefore C : D = 24 \times 7 : 24 \times 5 = 168 : 120$

[By proportionating (3) and (4)]

Hence, $A : B : C : D = 105 : 140 : 168 : 120$

Hence, C gets the maximum share.

54. (d) Let the present age of A $= x$ and B $= y$ years

According to first condition

$\frac{x-7}{y-7} = \frac{3}{4} \Rightarrow 4x - 28 = 3y - 21 \Rightarrow 4x - 3y = 7$ .... (i)

According to second condition

$\frac{x+9}{y+9} = \frac{7}{8} \Rightarrow 8x + 72 = 7y + 63$

$\Rightarrow 7y - 8x = 9$ .... (ii)

$8x - 6y = 14$

$7y - 8x = 9$

$y = 23$ years.

55. (b) Let present age of father, mother and daughter be $7x, 6x, 2x$

$6x - 2x = 24$

$4x = 24$

$x = 6$

Father age = $7x = 42$ years.

56. (d) Let total number of students in college $A = 3x$ and total number of students in college $B = 4x$

After 50 more students join college A

New Ratio $= \dfrac{3x+50}{4x} = \dfrac{5}{6}$

$\Rightarrow \quad 18x + 300 = 20x$

$\Rightarrow \quad 2x = 300$

$\Rightarrow \quad x = \dfrac{300}{2} = 150$

Total number of students in college $B = 4x = 4 \times 150 = 600$

57. (c) Let 'x' be the total sum of money.

Money received by A $= \dfrac{2}{7} \times x$

$\Rightarrow 100 = \dfrac{2}{7} \times x \Rightarrow x = ₹350$

$\therefore$ Money received by B $= \dfrac{5}{7} \times 350 = ₹250$

58. (c) $7S + 3D + 5N = ₹12600$ ...(1)

According to Question

$D = 3N$

$S = 7N$

Putting these values in equation (1)

$49N + 9N + 5N = 12600$

$\Rightarrow 63N = 12600$

$\Rightarrow N = 200$

Each daughter's share

= 3 × Nephew's share

$= 3 \times 200 = ₹600$

59. (b) Required ratio $= \left(\dfrac{2}{3}+\dfrac{3}{5}+\dfrac{5}{8}\right) : \left(\dfrac{1}{3}+\dfrac{2}{5}+\dfrac{3}{8}\right)$

$= \left(\dfrac{80+72+75}{120}\right) : \left(\dfrac{40+48+45}{120}\right) = 227 : 133$

60. (a) $\dfrac{x}{y} = \dfrac{2}{3}; \ \dfrac{2}{x} = \dfrac{4}{8}$

$x = 4$

$y = \dfrac{3}{2}x = \dfrac{3}{2} \times 4 = 6$

61. (b) $\dfrac{A}{B} = \dfrac{3}{4}; \ \dfrac{B}{C} = \dfrac{3.5}{3} = \dfrac{7}{6}$

and $A + B + C = 730$

$\dfrac{3}{4}B + B + \dfrac{6}{7}B = 730$

$B\left(\dfrac{3}{4} + 1 + \dfrac{6}{7}\right) = 730$

$B = \dfrac{730 \times 28}{73} = 280$

$C = \dfrac{6}{7} \times B = \dfrac{6}{7} \times 280 = 240$

B exceeds that of C by $(280 - 240) = ₹40$

62. (c) $x = x + \dfrac{125}{100}y$ or $\dfrac{x}{y} = \dfrac{5}{4}$ or $x : y = 5 : 4$

$y = \dfrac{75}{100}z$ or $\dfrac{y}{z} = \dfrac{3}{4}$ or $y : z = 3 : 4$

x : y : z

5 : 4

3 : 4

15 : 12 : 16

Then, $x : y : z$ is equal to 15: 12: 16

63. (d) 10% of x = 20% of y

$\dfrac{10}{100}x = \dfrac{20}{100}y$

$x = 2y$

or $\dfrac{x}{y} = \dfrac{2}{1}$

or $x : y = 2 : 1$

64. (a) boys : girls   girls : teacher

4 : 3   8 : 1

So, boys : girls : teacher

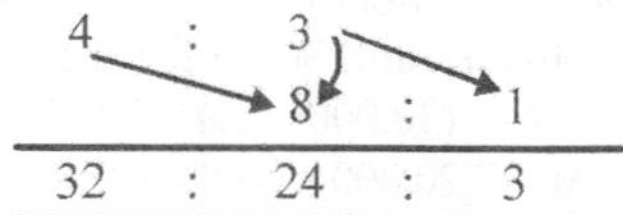

So, Student : teacher

$\Rightarrow$ (boys + girls) : teacher

(32 + 24) : 3

56 : 3

65. (d) $A : B = 5 : 3$

$B : C = 4 : 5$

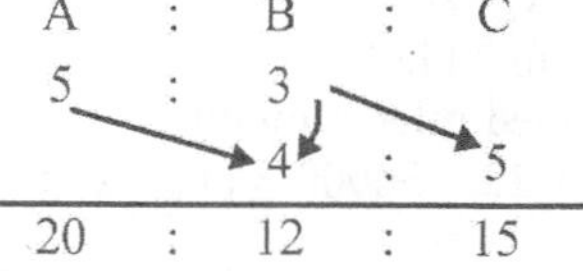

No. of runs scored by B $= \dfrac{12}{47} \times 564 = 144$

66. (d) $\dfrac{A}{B} = \dfrac{5}{6} \Rightarrow B = \dfrac{6}{5}A$ ...(1)

$\dfrac{A+2}{B+2} = \dfrac{7}{8} \Rightarrow 8A + 16 = 7B + 14 \Rightarrow 7B - 8A = 2$ ...(2)

From (1) and (2), A = 5, B = 6

$$\frac{5+12}{6+12} = \frac{17}{18}$$

67. (c) Suppose, B Joined after x month
Then B's money was invested for (12 – x) months
∴ According to question

$$\frac{64000 \times 12}{48000 \times (12-x)} = \frac{2}{1}$$

$$\frac{16}{12-x} = \frac{2}{1} \Rightarrow 16 = 24 - 2x$$

$2x = 24 - 16 = 8 \Rightarrow x = 4$
Hence, B joined after 4 months

68. (c) $$\frac{4}{7} + \frac{\frac{2y}{y} - \frac{x}{y}}{\frac{2y}{y} + \frac{x}{y}} = \frac{4}{7} + \frac{2 - \frac{4}{5}}{2 + \frac{4}{5}} = \frac{4}{7} + \frac{6}{14} = 1$$

69. (c) Let original property worth ₹ x

Property left for Ram's widow = $\frac{x}{3}$

Property left for his daughter = $\frac{3}{5} \times \frac{2x}{3} = \frac{2x}{5}$

Remaining property = $x - \left(\frac{x}{3} + \frac{2x}{5}\right) = \frac{4x}{15}$

$$\frac{4x}{15} = 6,400$$

$$\Rightarrow x = \frac{6,400 \times 15}{4} = 24,000$$

70. (a) A : B = 5 : 4, B : C = 8 : 9
A : B : C = 5 × 8 : 4 × 8 : 4 × 9 = 40 : 32 : 36
A : B : C = 10 : 8 : 9

Share of C in the profit = $\frac{9}{10+8+9} \times 3600$ = ₹ 1,200

71. (d) $\frac{1}{2} : \frac{1}{3} : \frac{1}{4} = 6 : 4 : 3$

Share of third friend = $\frac{3}{6+4+3} \times 624$

$= \frac{3}{13} \times 624$ = ₹ 144

72. (b) Let two numbers be 3x and 5x

$$\frac{3x-9}{5x-9} = \frac{12}{23}$$

$23(3x - 9) = 12(5x - 9)$
$69x - 207 = 60x - 108$
$9x = 99$
$x = 11$
Hence, the small number will be 3 × 11 = 33

73. (c) $\frac{x}{y} = \frac{5}{2}$

$$\frac{8x+9y}{8x+2y} = \frac{8\frac{x}{y} + \frac{9y}{y}}{8\frac{x}{y} + \frac{2y}{y}} = \frac{8 \times \frac{5}{2} + 9}{8 \times \frac{5}{2} + 2} = \frac{29}{22}$$

74. (a) Let C's speed = x km/h
Then, B's speed = 3x km/h
and A's speed = 6x km/h
Ratio of speeds of A, B, C = 6x : 3x : x = 6 : 3 : 1

Ratio of time taken $= \frac{1}{6} : \frac{1}{3} : 1$
$= 1 : 2 : 6$

C takes 90 minutes
Hence, 6x = 90
x = 15 minutes
Hence, A should take 15 minutes.

75. (d)

76. (a) $K = xy^2$
$K' = (x - 25\% \text{ of } x)(y - 25\% \text{ of } y)^2$

$$= \left(\frac{3x}{4}\right)\left(\frac{3y}{4}\right)^2 = \frac{27}{64}xy^2$$

Value decreased by $K - K' = \left(1 - \frac{27}{64}\right)xy^2$

$$= \frac{37}{64}xy^2$$

77. (d) Let the original fraction be $\frac{x}{y}$.

According to the question

$$\therefore \quad \frac{x \times 400}{y \times 300} = \frac{4}{15}$$

$$\Rightarrow \quad \frac{x}{y} = \frac{4}{15} \times \frac{3}{4} = \frac{1}{5}$$

78. (a) Let Gloria's and Sara's present ages be $4x$ and $7x$ years respectively.
Two years ago,

$$\frac{4x-2}{7x-2} = \frac{1}{2}$$

⇒ $8x - 4 = 7x - 2$
⇒ $x = 2$
∴ Sara's age three years hence = $7x + 3$
= 17 years

79. (e) $$\frac{\text{A's salary}}{\text{B's salary}} = \frac{8}{7}$$

A's salary = 8x
B's salary = 7x

Now, A's salary = $8x + 8x \times \frac{21}{100} = 8x + 1.68x = 9.68x$

Now B's salary = $7x + 7x \times \frac{20}{100} = 7x + 1.4x = 8.4x$

$\frac{9.68x}{8.4x} = \frac{121}{105}$

$\frac{9.68}{8.4} = \frac{121}{105}$

Here x is cancelled. So, salary of A can't be calculated.

80. (e) A's investment = 12800
B's investment = 16800
C's investment = 9600
B's profit = 13125
Investment ratio
A : B : C = 128 : 168 : 96 = 16 : 21 : 12
A, B and C profit 16x, 21x, 12x
21x = 13125
x = 625
C's profit = 12x = 12 × 625 = 7500

81. (d) $\frac{A+12}{B+12} = \frac{3}{4}$

$A = \frac{15}{4}C$

$A = \frac{15}{4} \times 10 = 37.5$

$\frac{37.5+12}{B+12} = \frac{3}{4}$

B = 54

82. (e) M's share = $44352 \times \frac{3}{8} = 16632$

Remaining after M's share = 27720

N's share = $27720 \times \frac{1}{6} = 4620$

Remaining after M & N's share = 23100

$\frac{O}{P} = \frac{3}{4} \Rightarrow$ O's share = $23100 \times \frac{3}{7} = 9900$

83. (b) $6B^2 = A^2 + 540$

$\frac{A}{B} = \frac{3}{2}$

$A = \frac{3B}{2}$

$6B^2 = \frac{9B^2}{4} + 540$

$3.75\,B^2 = 540$

$B = \sqrt{144} = 12$

Level-II

1. (b) Work done = $\frac{5}{8}$. Balance work = $\left(1-\frac{5}{8}\right) = \frac{3}{8}$.

Less work, Less days (Direct Proportion)
Let the required number of days be x. Then,

Work	days
↑ 5/8	10 ↑
3/8	x

Then, $\frac{5}{8} : \frac{3}{8} :: 10 : x \Rightarrow \frac{5}{8} \times x = \frac{3}{8} \times 10$

$\Rightarrow x = \left(\frac{3}{8} \times 10 \times \frac{8}{5}\right) = 6$.

2. (a) Let each boy gets x, so the women gets $2x$ and a man gets $3x$.
Now, $(3 \times 3x) + (4 \times 2x) + (6 \times x) = 1104$
$\Rightarrow 23x = 1104 \Rightarrow x = 48$
∴ Each boy gets ₹ 48.

3. (d) Let the number of seats in Physics, Chemistry and Mathematics be $4x$, $5x$ and $6x$.
New ratio of seats = $(4x + 75) : (5x + 75) : (6x + 75)$
∴ The given data is insufficient.

4. (b) Quantity of tin in 60 kg of A = $\left(60 \times \frac{2}{5}\right)$ kg = 24 kg

Quantity of tin in 100 kg of B = $\left(100 \times \frac{1}{5}\right)$ kg = 20 kg

Quantity of tin in the new alloy = (24 + 20) kg = 44 kg.

5. (b) C's capital = $1 - \left(\frac{1}{2} + \frac{1}{8}\right) = 1 - \frac{5}{8} = \frac{3}{8}$

Ratio of capitals of A, B and C

$= \left(\frac{1}{2} \times \frac{1}{4}\right) : \left(\frac{1}{8} \times \frac{1}{2}\right) : \left(\frac{3}{8} \times 1\right)$

$= \frac{1}{8} : \frac{1}{16} : \frac{3}{8} = 2 : 1 : 6$

B's share = ₹ $\left(\frac{1}{9} \times 9900\right)$ = ₹ 1100

6. (c) There will be a total of 4.5 litres of milk (25% of 3 + 75% of 5) giving a total of 4.5. Hence , 45%.

7. (d) Let $C = x$. Then, $B = x + 5000$ and
$A = x + 5000 + 4000 = x + 9000$.
So, $x + x + 5000 + x + 9000 = 50000 \Leftrightarrow 3x = 36000$
$\Leftrightarrow x = 12000$.
$A : B : C = 21000 : 17000 : 12000 = 21 : 17 : 12$

∴ A's share = ₹ $\left(35000 \times \frac{21}{50}\right)$ = ₹ 14,700.

8. (a) $\frac{a}{b}=\frac{5}{4}, b=\frac{4}{5}a$ Given (40% of $a=$) $\frac{2}{5}a=12$

$\therefore \quad a=5\times 6$ and $b=\frac{4}{5}\times 5\times 6=24$

$\therefore \quad 50\%$ of $b=\frac{24}{2}=12$

9. (a) Remaining capital $=1-\left(\frac{1}{6}+\frac{1}{3}\right)=\frac{1}{2}$

Ratio of their profit

$$=\frac{1}{6}\times\left[\frac{1}{6}\times 12\right]:\frac{1}{3}\times\left[\frac{1}{3}\times 12\right]:\frac{1}{2}\times 12$$

$$=\frac{1}{3}:\frac{4}{3}:6=1:4:18$$

$\therefore$ A's share $=\frac{1}{1+4+18}\times 2300=$ ₹ 100

10. (d) Let B puts $= x$ cows

then amount paid by $B=\frac{3}{2}\times$ amount paid by A.

$$\therefore \frac{80\times 7}{x\times 3}=\frac{\text{amount paid by } A}{3/2\times \text{amount paid by } A}$$

$$\Rightarrow x=\frac{80\times 7\times 3}{3\times 2}=280 \text{ cows}$$

11. (c) If R is the resistance, l is the length and r is the radius.

$$R\,\alpha\,\frac{\ell}{r^2}$$

$$\therefore R=\frac{k\ell}{r^2} \quad \text{(where } k \text{ is a constant)}$$

$$\therefore \frac{R_1}{R_2}=\frac{\frac{k\times 162}{81}}{\frac{k\times \ell}{64}}\text{; But } R_1=R_2.$$

$$\therefore \frac{k\times 162}{81}=k\times\frac{\ell}{64} \quad \therefore \frac{162}{81}=\frac{\ell}{64}$$

$\therefore \ell=128$ cms.

12. (b) Let the weight of the three pieces be x, $3x$ and $6x$ and value (V) α square of W

$$V\,\alpha\,W^2\Rightarrow\frac{V}{W^2}=\text{constant}=\frac{30000}{(10x)^2}$$

$\therefore$ loss due to breakage

$$=\frac{30000(10x)^2-(x)^2+(6x)^2}{(10x)^2}=\frac{3000x54x^2}{100x^2}=16200$$

13. (a) Let initially, the number of males and females in the bus be $3x$ and x, respectively.

$\therefore \quad$ Total no. of passengers $=4x$

At the first stop, the number of male passengers

$$=(4x-16)\times\frac{3}{4}=3x-12$$

and female passengers $=(4x-16)\times\frac{1}{4}+6=x+2$

Now, $\frac{3x-12}{x+2}=\frac{2}{1}\Rightarrow 3x-12=2x+4\Rightarrow x=16$

Hence, the total numbers of passengers, initially $=16\times 4=64$

14. (b)

	Water	Milk	Total.
1st vessel	6	7	13
2nd vessel	5	9	14
3rd vessel	8	7	15

LCM of 13, 14 & 15 = 2730

Increase value of total to 2730 as follows.

1st vessel	1260	1470	2730
2nd vessel	975	1755	2730
3rd vessel	1456	1274	2730
Total	3691	4499	8190

$\therefore$ Required ratio $=\frac{3691}{4499}$

15. (a)

	Iron		Copper	
Alloy I	8	:	6	14 kg.
Alloy II	36	:	6	42 kg.
	44	:	12	56 kg

16. (b) Let x and y be two containers.

Ratio of milk to water in container $x=5:1$

and ratio of milk to water in container $y=7:2$

It is given that quantity of milk should be 80% in new mixture. This means that quantity of water will be 20% that quantity of water will be 20%

$\therefore$ Ratio of milk to water in new mixture $=80:20=4:1$.

$$\therefore \frac{5}{6}x+\frac{7}{9}y:\frac{1}{6}x+\frac{2y}{9}=4:1$$

$$\Rightarrow \frac{\frac{5x}{6}+\frac{7y}{9}}{\frac{x}{6}+\frac{2y}{9}}=\frac{4}{1}$$

$$\Rightarrow \frac{15x+14y}{3x+4y}=\frac{4}{1}\Rightarrow 15x+14y=12x+16y$$

$$\Rightarrow 3x=2y\Rightarrow x:y=2:3$$

17. (c) In first container,

Quantity of milk $=20\times\frac{90}{100}=18$L

Quantity of water = 2L

In second container,

Quantity of milk $=5\times\frac{80}{100}=4$L

Quantity of water = 1 L

In third container,

Quantity of milk $= 9 \times \frac{70}{100} = 6.3L$

Quantity of water = 2.7 L
After mixing :
total milk = 18 + 4 + 6.3 = 28.3 L
total water = 2 + 1 + 2.7 = 5.7 L
Ratio milk and water after mixing = 28. 3 L : 5.7 L = 5 : 1 (approx)

18. (d) Ratio of earning of A and B is 4 : 7
A earning = ₹ $4x$.
B earning = ₹ $7x$
A earning increase by 50%

Now, A earning $= 4x + 4x \times \frac{50}{100} = 6x$

B earning decreases by 25%

Now B earning $= 7x - 7x \times \frac{25}{100} = \frac{21x}{4}$

Now, Ratio $= \frac{6x}{\frac{21x}{4}} = 8:7$

Data inadequate

19. (b) Women : Men = 3 : 4
Men : Children = 3 : 5
→ Women : Men : Children = 9 : 12 : 20
In the ratio, 9 → 531 Women
Thus, 20 → 1180 children.

20. (a) $a : b : c = 2 : 6 : 3$
$a : b : c : d : e : f = 6 : 18 : 9 : 18 : 6 : 24$
$abc/def = 3/8$

21. (b) $\frac{x+1}{y+2} = \frac{2}{3} \Rightarrow 3x - 2y = 1$

$\frac{x+2}{y+3} = \frac{5}{7} \Rightarrow 7x - 5y = 1$

or, $3x - 2y = 7x - 5y \Rightarrow 3y = 4x \Rightarrow \frac{x}{y} = \frac{3}{4}$

22. (e) 9, 15, 27
$9 - x, 15 - x, 27 - x$

$\frac{15-x}{9-x} = \frac{27-x}{15-x}$

$\Rightarrow (15 - x)^2 = (27 - x)(9 - x)$
$\Rightarrow 225 + x^2 - 30x = 243 - 9x - 27x + x^2$
$\Rightarrow -30x + 9x + 27x = 243 - 225$
$\Rightarrow 6x = 18 \Rightarrow x = 3$

23. (a) Let amount of B = ₹ x

B's Share without error $= \frac{\text{B's ratio}}{\text{Total ratio}} \times \text{Total Amount}$

$x = \frac{3}{9} \times \text{Total Amount}$ ...(1)

B's share after error $= \frac{\text{B's new ratio}}{\text{Total new ratio}} \times \text{Total Amount}$

$x - 40 = \frac{2}{14} \times \text{Total Amount}$ ...(2)

From equations, (1) and (2)
$3x = 7(x-40)$
$3x - 7x = -280$
$\therefore x = 70$
Total Amount = 7 (70 – 40) = ₹ 210

24. (e) A and B ratio is 4 : 7
$\Rightarrow 4x + 7x = 73689$
$\Rightarrow 11x = 73689$
$\Rightarrow x = 6699$
Share of A = 4x = ₹ 26796
Share of B = 7x = ₹ 46893
Difference = twice of share B – thrice of share A
$= 2 \times 46893 - 3 \times 26796 =$ ₹ 13398

25. (e) Let fraction be $\frac{x}{y}$.

$\therefore$ According to the question, $\frac{x \times 120\%}{y \times 125\%} = \frac{3}{5}$

$\Rightarrow \frac{x}{y} = \frac{3}{5} \times \frac{125}{120} = \frac{5}{8}$

26. (d) Total age of son and mother
$2x + 7x = 2 \times 27$
$9x = 54$
$x = 6$
$\therefore$ Mother's age after 7 yr $= 7x + 7 = 7 \times 6 + 7 = 49$ yrs]

27. (b) $4A = 6B \Rightarrow 2A = 3B \Rightarrow A : B = 3 : 2$
$B = 3C \Rightarrow 2B = C \Rightarrow B : C = 1 : 2$

A	B	C
3	2	
	1	2
3	2	4

A's share $= \frac{3}{(3+2+4)} \times 1800 = \frac{3}{9} \times 1800 = 600$

28. (b) **29. (d)**

CHAPTER

9

TIME AND WORK

CONCEPT OF EFFICIENCY

Efficiency means rate of doing work. This means that more the efficiency, less will be the number of days required to complete a certain work and less the efficiency, more will be the number of days required to complete a certain work.

Aliza is twice as efficient as Binny.

$\Rightarrow$ Aliza does twice as much work as Binny in the same time interval

$\Rightarrow$ Aliza will require half the time as required by Binny to do the same work.

CONCEPT OF NEGATIVE WORK

Suppose two persons A and B are working to build a wall while C is working to demolish the wall. If we consider the work as the building of the wall, then breaking the wall (by C) is negative work.

The concept of negative work generally appears in the problems based on pipes and cisterns, where there are inlet pipes and outlet pipes/leaks, which are working against each other.

If we consider the work of filling a tank, the inlet pipe does positive work while the outlet pipe/leaks does negative work.

Illustration 1: *A* can build a wall in 15 days and *B* can build it in 10 days, while *C* can completely demolish the wall in 12 days. If they start working at the same time, in how many days will the work be completed.

Solution: Work per day by $A = \frac{1}{15}$

Work per day by $B = \frac{1}{10}$

Work per day by $C = -\frac{1}{12}$

(negative sign is taken for negative work)

The net combined work per day by A, B and C

$$= \frac{1}{15} + \frac{1}{10} - \frac{1}{12} = \frac{1}{12}$$

Since, Total work done = (Work done per day) × (No. of days required to complete the work)

$\therefore$ No. of days required to complete the work

$$= \frac{\text{Total work done}}{\text{Work done per day}} = \frac{1}{\frac{1}{12}} = 12$$

CONCEPT OF MAN-DAYS

If 'M' men working together can complete a work in 'D' days, then the product of number of men (M) and number of days (D) i.e. $M \times D$ is known as the number of MAN-DAYS.

Number of man days to complete a specific task always remains constant.

Suppose 30 persons working together for 20 days to complete a job, then the total work done is equal to (30 × 20 = 600) man-days. If we change the number of days in which the work is to be completed, then the other factor i.e. the number of persons will change accordingly, so that the product of the factors becomes equal to 600 man-days.

WORK DONE

Consider a whole work as the unit work.

1. Work Done by Two Persons

Let A can do a whole work in x days and B can do the same one unit work in y days.

Hence work done by A in one day $= \frac{1}{x}$

and work done by B in one day $= \frac{1}{y}$

Then work done in one day when A and B work together

$$= \frac{1}{x} + \frac{1}{y}$$

$$= \frac{y+x}{xy} \text{ or } \frac{x+y}{xy}$$

Whole work = (Work done in one day) × (Number of days required to complete the whole work)

Hence, number of days required to complete the whole work $= \dfrac{\text{Whole work}}{\text{Work done in one day}}$

$\Rightarrow$ Number of days required to complete the whole work when A and B are working together

$= \dfrac{1}{\frac{x+y}{xy}} = \dfrac{xy}{x+y}$, because a whole work is considered as one unit of work.

Illustration 2: If A can do a work in 10 days and B can do the same work in 15 days, then how many days will they take to complete the work both while working together?

Solution: Work done by A in one day $= \dfrac{1}{10}$

Work done by B in one day $= \dfrac{1}{15}$

Work done in one day when A and B work together

$$= \frac{1}{10} + \frac{1}{15} = \frac{3+2}{30} = \frac{5}{30} = \frac{1}{6}$$

Hence required number of days $= \dfrac{1}{\frac{1}{6}} = 6$

We can find the required number of days directly by using the formula,

Number of days $= \dfrac{xy}{x+y}$, derived above

$$= \frac{10 \times 15}{10+15} = \frac{150}{25} = 6.$$

Illustration 3: 'A' completes a work in 12 days. 'B' completes the same work in 15 days. 'A' started working alone and after 3 days B joined him. How many days will they now take together to complete the remaining work?

(a) 5 **(b) 8**

(c) 6 **(d) 4**

Solution: (a) Work done by 'A' in 3 days $= \dfrac{1}{12} \times 3 = \dfrac{1}{4}$

$\therefore$ Remaining work $= 1 - \dfrac{1}{4} = \dfrac{3}{4}$

Work done by A and B together $= \dfrac{12 \times 15}{27} = \dfrac{20}{3}$

$\therefore$ Remaining work done by A and B together in

$$= \frac{3}{4} \times \frac{20}{3} = 5 \text{ days}$$

2. Work Done by Three Persons

As we derived the formula for two persons, you can also derived the formula for three persons in the same way.

If A, B, C can do a work in x, y and z days respectively, then all of them working together can finish the work in $\dfrac{xyz}{xy + yz + zx}$ days.

Illustration 4: If A, B, C can do a work in 12, 15 and 20 days respectively, then how many days will they take to complete the work when all the three work together.

Solution:

Required number of days $= \dfrac{xyz}{xy + yz + zx}$

$$= \frac{12 \times 15 \times 20}{12 \times 15 + 15 \times 20 + 20 \times 12}$$

$$= \frac{3600}{180 + 300 + 240} = \frac{3600}{720} = 5$$

3. If A and B Together Can do a Work in x Days and A Alone can do it in y Days, then B alone can do the Work in $\frac{xy}{y-x}$ Days

Illustration 5: A and B can do a work in 8 days and A alone can do it in 12 days. In how many days can B alone do it?

Solution:

Work done by A and B working together in one day $= \dfrac{1}{8}$

Work done by A in one day $= \dfrac{1}{12}$

$\therefore$ Work done by B in one day $= \dfrac{1}{8} - \dfrac{1}{12} = \dfrac{3-2}{24} = \dfrac{1}{24}$

Hence number of days in which B alone can do the whole work

$$= \frac{1}{\frac{1}{24}} = 1 \times \frac{24}{1} = 24$$

You can find required number of days directly by using the above formula as

Required number of days $= \dfrac{xy}{y-x} = \dfrac{8 \times 12}{12 - 8} = \dfrac{8 \times 12}{4} = 24.$

EXTENSION OF THE CONCEPT OF TIME AND WORK

1. Pipes and Cisterns

Problems related to Pipes and Cisterns are almost the same as those of Time and Work. Statement 'pipes A and B can fill a tank in 2 hours and 3 hours working individually' is similar to the statement 'A and B can do a work in 2 hours and 3 hours respectively working individually'.

If a pipe fills a tank in 3 hours, then the pipe fills $\dfrac{1}{3}$rd of the same tank in 1 hour.

The only difference with the pipes and cisterns problems is that there are inlets as well as outlets. Inlet is a pipe connected with a tank (or a cistern or a reservoir) that fills it. Outlet is a pipe connected with a tank (or a cistern or a reservoir) that empties it.

Hence, if we consider filling the tank by inlet as positive work, then empting the tank by outlet will be considered as negative work.

(a) Let a pipe fill a tank in x hours and another pipe can empty the full tank in y hours. Then the net part of the tank filled in 1 hour, when both the pipes are opened, if x is less than y.

$$= \frac{1}{x} - \frac{1}{y} = \frac{y-x}{xy}$$

$\therefore$ time taken to fill the tank, when both the pipes are opened

$$= \frac{1}{\frac{y-x}{xy}} = \frac{xy}{y-x}.$$

(b) Let a pipe fill a tank in x hours while another fills the same tank in y hours but a third one empties the full tank in z hours. If all the three pipes are opened together, then the

net part of the tank filled in 1 hour $= \frac{1}{x} + \frac{1}{y} - \frac{1}{z}$

$$= \frac{yz + zx - xy}{xyz}$$

$\therefore$ time taken to fill the tank $= \frac{xyz}{yz + zx - xy}$

(c) Let a pipe fill a tank in x hours but due to the leak in the bottom, the tank is filled in y hours and when the tank is filled, the time taken by the leak to empty the tank is z hours.

Net part of the tank filled in 1 hour by the pipe when there is the leak in the bottom $= \frac{1}{x} - \frac{1}{z} = \frac{z-x}{xz}$

Since the tank will be filled completely in y hours by the pipe when there is the leak in the bottom, therefore

$$\left(\frac{z-x}{xz}\right) \times y = 1 \Rightarrow y = \frac{xz}{z-x} \Rightarrow yz - xy = xz$$

$$\Rightarrow \quad z(y-x) = xy \Rightarrow z = \frac{xy}{y-x}$$

Hence, if a pipe can fill a tank in x hours but due to the leak in the bottom, the tank is filled in y hours, then the fully filled tank will be emptied in $\frac{xy}{y-x}$ hours.

(d) Let a pipe A fill a tank in x hrs while pipe B can fill the tank in y hrs alone. When both the pipes are opened together, then time required to fill the tank $= \frac{xy}{x+y}$ hrs.

(e) Let pipes A, B and C fill a tank alone in x, y and z hrs respectively. When all the three pipes open together, then time required to fill the tank $= \frac{xyz}{xy + yz + zx}$ hrs.

Illustration 6: If a pipe fills a tank in 4 hrs and another pipe can empty the full tank in 6 hrs. When both the pipes are opened together, then find the time required to completely fill the tank.

Solution: Required time $= \frac{xy}{y-x}$

Here $x = 4, \ y = 6$

$\therefore$ Required time $= \frac{4 \times 6}{6-4} = 12$ hrs.

Illustration 7: Pipe A can fill a tank in 6 hrs while pipe B alone can fill it in 5 hrs and pipe C can empty the full tank in 8 hrs. If all the pipes are opened together, how much time will be needed to completely fill the tank?

Solution: Required time

$$= \frac{xyz}{yz + zx - xy} = \frac{6 \times 5 \times 8}{5 \times 8 + 8 \times 6 - 6 \times 5}$$

$$= \frac{6 \times 5 \times 8}{58} = \frac{120}{29} = 4\frac{4}{29} \text{ days.}$$

Illustration 8: A pipe can fill a tank in 10 hrs. Due to a leak in the bottom, it is filled in 15 hrs. If the tank is full, how much time will the leak take to empty it.

Solution: Required time $= \frac{xy}{y-x} = \frac{10 \times 15}{15 - 10} = 30$ hrs.

Illustration 9: If three pipes A, B and C can fill the tank alone in 5, 6 and 8 hrs, then when all the three pipes are opened together, find the time to fill the tank completely.

Solution: Required time

$$= \frac{xyz}{xy + yz + zx} = \frac{5 \times 6 \times 8}{5 \times 6 + 6 \times 8 + 8 \times 5}$$

$$= \frac{240}{30 + 48 + 40} = \frac{240}{118} = 2\frac{2}{59} \text{ hrs}$$

3. Alternate Work

In some problems two or more people of different efficiencies work alternatively or in some particular pattern. You can understand the method to solve these types of problems through the following illustration.

Illustration 10: Sanjeev can build a wall in 20 days and Parveen can demolish the same wall in 30 days. If they work on alternate days with Sanjeev starting the job on the 1st day, then in how many days will the wall be built for the first time ?

Solution: Let us assume the total units of work

= 60 units (i.e. LCM of 20 and 30)

So, the wall built by Sanjeev in one day = 3 units

And wall demolished by Parveen in one day = 2 units

So, effectively in two days, total wall built = 1 unit

Now, they work on alternate days, so days taken to built 57 units = 57 days

On 58th day Sanjeev will add another 3 units and so completing the construction of wall in 58 days.

(This problem can be understood well with another very traditional problem—A frog climbs up a pole 3 inches in 1 minute and slips 2 inches in next minute. If height of the pole is 120 inches, then how much time is taken by the frog to reach the top of the pole ?)

Practice Exercise

Level - I

1. A and B together can do a job in 12 days. B alone can finish it in 28 days. In how many days can A alone finish the work?
   (a) 21 days (b) 19 days
   (c) 20 days (d) None of these

2. A can do $\frac{3}{4}$ of a work in 12 days. In how many days can he finish $\frac{1}{8}$ of the work?
   (a) 6 days (b) 5 days
   (c) 3 days (d) 2 days

3. A can finish a work in 18 days and B can do the same work in half the time taken by A. Then, working together, what part of the same work they can finish in a day?
   (a) $\frac{1}{6}$ (b) $\frac{1}{9}$ (c) $\frac{2}{5}$ (d) $\frac{2}{7}$

4. A man is twice as fast as a woman. Together the man and the woman do the piece of work in 8 days. In how many days each will do the work if engaged alone?
   (a) man-14 days, woman-28 days
   (b) man-12 days, woman-24 days
   (c) man-10 days, woman-20 days
   (d) None of these

5. A is 30% more efficient than B. How much time will they, working together, take to complete a job which A along could have done in 23 days?
   (a) 11 days (b) 13 days
   (c) $20\frac{3}{17}$ days (d) None of these

6. A contractor undertakes to built a walls in 50 days. He employs 50 peoples for the same. However after 25 days he finds that only 40% of the work is complete. How many more man need to be employed to complete the work in time?
   (a) 25 (b) 30 (c) 35 (d) 20

7. 12 men complete a work in 18 days. Six days after they had started working, 4 men joined them. How many days will all of them take to complete the remaining work?
   (a) 10 days (b) 12 days
   (c) 15 days (d) 9 days

8. A man, a woman or a boy can do a job in 20 days, 30 days or 60 days respectively. How many boys must assist 2 men and 8 women to do the work in 2 days?
   (a) 15 boys (b) 8 boys
   (c) 10 boys (d) None of these

9. 10 men can complete a piece of work in 15 days and 15 women can complete the same work in 12 days. If all the 10 men and 15 women work together, in how many days will the work get completed?
   (a) 6 (b) $6\frac{1}{3}$ (c) $6\frac{2}{3}$ (d) $7\frac{2}{3}$

10. After working for 8 days, Anil finds that only $\frac{1}{3}$ of the work has been done. He employs Rakesh who is 60% efficient as Anil. How many more days will Anil take to complete the job?
   (a) 15 days (b) 12 days
   (c) 10 days (d) 8 days

11. A can knit a pair of socks in 3 days. B can knit the same thing in 6 days. If they are knitting together, in how many days will they knit two pairs of socks?
   (a) 4 days (b) 2 days
   (c) $4\frac{1}{2}$ days (d) 3 days

12. A can build up a wall in 8 days while B can break it in 3 days. A has worked for 4 days and then B joined to work with A for another 2 days only. In how many days will A alone build up the remaining part of wall?
   (a) $13\frac{1}{3}$ days (b) $7\frac{1}{3}$ days
   (c) $6\frac{1}{3}$ days (d) 7 days

13. Sakshi can do a piece of work in 20 days. Tanya is 25% more efficient than Sakshi. The number of days taken by Tanya to do the same piece of work is
   (a) 15 (b) 16 (c) 18 (d) 25

14. Three men, four women and six children can complete a work in seven days. A woman does double the work a man does and a child does half the work a man does. How many women alone can complete this work in 7 days?
   (a) 7 (b) 8
   (c) 12 (d) Cannot be determined

15. Sunil and Pradeep can complete a work in 5 days and 15 days respectively. They both work for one day and then Sunil leaves. In how many days in the remaining work completed by Pradeep ?
   (a) 11 days (b) 12 days
   (c) 15 days (d) 8 days

16. Suresh can finish a piece of work by himself in 42 days. Mahesh, who is $\frac{1}{5}$ times more efficient as Suresh, requires X days to finish the work by working all by himself. Then what is the value of X ?
   (a) 25 days (b) 30 days
   (c) 35 days (d) 20 days

17. If 6 BSF or 10 CRPF companies can demolish a terrorist outfit in Kashmir in 2 days, find how long will 4 BSF and 9 CRPF companies take to do the same ?
   (a) 1.27 days (b) 2.27 days
   (c) 3.27 days (d) 4.27 days

18. 2 men and 3 boys can do a piece of work in 10 days while 3 men and 2 boys can do the same work in 8 days. In how many days can 2 men and 1 body to the work ?

(a) $12\frac{1}{2}$ days (b) $11\frac{1}{2}$ days

(c) $15\frac{1}{2}$ days (d) $13\frac{1}{2}$ days

19. Three pumps working 8 hours a day can empty a tank in 2 days. How many hours a day must 4 pumps work to empty the tank in 1 day.

(a) 10 hours (b) 12 hours

(c) 8 hours (d) None of these

20. If 18 binders bind 900 books in 10 days, how many binders will be required to bind 660 books in 12 days ?

(a) 14 (b) 13 (c) 22 (d) 11

21. If 27 men take 15 days to mow 225 hectares of grass, how long will 33 men take to mow 165 hectare ?

(a) 9 days (b) 18 days

(c) 6 days (d) 12 days

22. *X* and *Y* can do a piece of work in 72 days. *Y* and *Z* can do it in 120 days. *X* and *Z* can do it in 90 days. In how many days all the three together can do the work ?

(a) 100 days (b) 150 days

(c) 60 days (d) 80 days

23. If 6 men and 8 boys can do a piece of work in 10 days and 26 men and 48 boys can do the same work in 2 days, the time taken by 15 men and 20 boys to do the same type of work will be

(a) 6 days (b) 4 days

(c) 8 days (d) 7 days

24. The work done by man, a woman and a boy are in the ratio 3 : 2 : 1. There are 24 men, 20 women and 16 boys in a factory whose weekly wages amount to ₹ 224. What will be the yearly wages of 27 men, 40 women and 15 boys?

(a) ₹ 16366 (b) ₹ 16466

(c) ₹ 16066 (d) ₹ 16016

25. Two pipes can fill a cistern in 6 minutes and 7 minutes respectively. Both the pipes are opened alternatively for 1 minute each. In what time will they fill the cistern.

(a) 6 minutes (b) $6\frac{2}{3}$ minutes

(c) $6\frac{3}{7}$ minutes (d) $3\frac{1}{2}$ minutes

26. Three pipes *A*, *B* and *C* can fill a tank from empty to full in 30 minutes, 20 minutes and 10 minutes respectively. When the tank is empty, all the three pipes are opened. *A*, *B* and *C* discharge chemical solutions *P*, *Q* and *R* respectively. What is the proportion of solution *R* in the liquid in the tank after 3 minutes?

(a) $\frac{3}{11}$ (b) $\frac{6}{11}$ (c) $\frac{4}{11}$ (d) $\frac{7}{11}$

27. *A* and *B* can finish a work in 10 days while *B* and *C* can do it in 18 days. *A* started the work, worked for 5 days, then *B* worked for 10 days and the remaining work was finished by *C* in 15 days. In how many days could *C* alone have finished the whole work ?

(a) 30 days (b) 15 days

(c) 45 days (d) 24 days

28. A certain number of men can do a work in 60 days. If there were 8 men more it could be finished in 10 days less. How many men are there ?

(a) 75 men (b) 40 men

(c) 48 men (d) 45 men

29. *A* and *B* can do a job in 16 days and 12 days respectively. *B* has started the work alone 4 days before finishing the job, *A* joins *B*. How many days has B worked alone?

(a) 6 days (b) 4 days

(c) 5 days (d) 7 days

30. Two pipes *A* and *B* when working alone can fill a tank in 36 min. and 45 min. respectively. A waste pipe *C* can empty the tank in 30 min. First *A* and *B* are opened. After 7 min., *C* is also opened. In how much time will the tank be full ?

(a) 1/60 (b) 1/30

(c) 7/20 (d) 13/20

31. *A* can do a piece of work in 25 days and *B* in 20 days. They work together for 5 days and then *A* goes away. In how many days will *B* finish the remaining work ?

(a) 17 days (b) 11 days

(c) 10 days (d) 15 days

32. 12 men complete a work in 18 days. Six days after they had started working, 4 men joined them. How many days will all of them take to complete the remaining work ?

(a) 10 days (b) 12 days

(c) 15 days (d) 9 days

33. *A* can do a piece of work in 10 days, while *B* alone can do it in 15 days. They work together for 5 days and the rest of the work is done by *C* in 2 days. If they get ₹ 450 for the whole work, how should they divide the money ?

(a) ₹ 225, ₹ 150, ₹ 75 (b) ₹ 250, ₹ 100, ₹ 100

(c) ₹ 200, ₹ 150, ₹ 100 (d) ₹ 175, ₹ 175, ₹ 100

34. *A* can do some work in 24 days, *B* can do it in 32 days and *C* can do it in 60 days. They start working together. *A* left after 6 days and *B* left after working for 8 days. How many more days are required to complete the whole work?

(a) 30 (b) 25 (c) 22 (d) 20

35. Mayank can do 50% more work than Shishu in the same time. Shishu alone can do a piece of work in 30 hours. Shishu starts working and he had already worked for 12 hours when Mayank joins him. How many hours should Shishu and Mayank work together to complete the remaining work?

(a) 6 (b) 12 (c) 4.8 (d) 9.6

36. In a fort there was sufficient food for 200 soldiers for 31 days. After 27 days 120 soldiers left the fort. For how many extra days will the rest of the food last for the remaining soldiers?

(a) 12 days (b) 10 days

(c) 8 days (d) 6 days

37. Sambhu can do 1/2 of the work in 8 days while kalu can do 1/3 of the work in 6 days. How long will it take for both of them to finish the work?

(a) $\frac{88}{17}$ days (b) $\frac{144}{17}$ days

(c) $\frac{72}{17}$ days (d) 8 days

38. A and B can do a piece of work in 45 and 40 days respectively. They began the work together, but A leaves after some days and B finished the remaining work in 23 days. After how many days did A leave

(a) 7 days (b) 8 days
(c) 9 days (d) 11 days

39. There is sufficient food for 400 men for 31 days. After 28 days, 280 men leave the place. For how many days will the rest of the food last for the rest of the men?

(a) 10 days (b) 12 days
(c) 16 days (d) 20 days

40. A tyre has two punctures. The first puncture along would have made the tyre flat in 9 minutes and the second alone would have done it in 6 minutes. If air leaks out at a constant rate, how long does it take both the punctures together to make it flat?

(a) $1\frac{1}{2}$ minutes (b) $3\frac{1}{2}$ minutes

(c) $3\frac{3}{5}$ minutes (d) $4\frac{1}{4}$ minutes

41. 12 men and 16 boys can do a piece of work in 5 days, 13 men and 24 boys can do it in 4 days. Then the ratio of daily work done by a man to that of a boy is

(a) 2 : 1 (b) 3 : 1 (c) 3 : 2 (d) 5 : 4

42. Two taps can fill a tank in 12 and 18 minutes respectively. Both are kept open for 2 minutes and the first is turned off. In how many minutes more will the tank be filled ?

(a) 15 min. (b) 20 min.
(c) 11 min. (d) 13 min.

43. A cistern normally takes 6 hours to be filled by a tap but because of a leak, 2 hours more. In how many hours will the leak empty a full cistern ?

(a) 20 hrs (b) 24 hrs
(c) 26 hrs (d) None of these

44. If 3 men or 4 women can reap a field in 43 days, how long will 7 men and 5 women take to reap it?

(a) 7 days (b) 11 days
(c) 12 days (d) 16 days

45. If m man can do a work in r days, then the number of days taken by $(m + n)$ men to do it is:

(a) $\frac{m+n}{mn}$ (b) $\frac{m+n}{mr}$ (c) $\frac{mr}{(m+n)}$ (d) $\frac{(m+n)r}{mn}$

46. Pipes A and B can fill a tank in 5 and 6 hours, respectively. Pipe C can empty it in 12 hours. The tank is half full. All the three pipes are in operation simultaneously. After how much time, the tank will be full ? [*SBI Clerk-2014*]

(a) $3\frac{9}{17}$ h (b) 11 h (c) $2\frac{8}{11}$ h (d) $1\frac{13}{17}$ h

(e) None of these

47. If 10 men or 18 boys can do a work in 15 days, then the number of days required by 15 men and 33 boys to do twice the work is [*SSC-Sub. Ins.-2012*]

(a) $4\frac{1}{2}$ (b) 8 (c) 9 (d) 36

48. In a fort, there was sufficient food for 200 soldiers for 31 days. After 27 days, 120 soldiers left the fort. For how many extra days will be rest of the food last for the remaining soldiers? [*SSC-Sub. Ins.-2012*]

(a) 10 days (b) 6 days
(c) 4 days (d) 12 days

49. A can do as much work as B and C together can do. A and B can together do a piece of work in 9 hours 36 minutes and C can do it in 48 hours. The time (in hours) that B needs to do the work alone, is : [*SSC-Sub. Ins.-2013*]

(a) 18 (b) 24 (c) 30 (d) 12

50. 3 men and 7 women can do a job in 5 days. While 4 men and 6 women can do it in 4 days. The number of days required for a group of 10 women working together, at the same rate as before, to finish the same job is: [*SSC-Sub. Ins.-2013*]

(a) 30 (b) 36 (c) 40 (d) 20

51. A can do $\frac{7}{8}$ of work in 28 days, B can do $\frac{5}{6}$ of the same work in 20 days. The number of days they will take to complete if they do it together is [*SSC-Sub. Ins.-2014*]

(a) $15\frac{3}{7}$ days (b) $17\frac{3}{5}$ days

(c) $14\frac{5}{7}$ days (d) $13\frac{5}{7}$ days

52. Seventy-five men are employed to lay down a railway line in 3 months. Due to certain emergency conditions, the work was to be finished in 18 days. How many more men should be employed to complete the work in the desired time ? [*SSC-Sub. Ins.-2014*]

(a) 300 (b) 325 (c) 350 (d) 375

53. Two pipes A and B can fill a tank in 6 hours and 4 hours respectively. If they are opened on alternate hours and if pipe A is opened first, then the tank shall be full in [*SSC-MT-2013*]

(a) $4\frac{1}{2}$ hrs (b) 5 hrs (c) $5\frac{1}{2}$ hrs (d) 6 hrs

54. A, B and C can do a piece of work in 10, 12 and 15 days respectively. A leaves 5 days before the completion of the work and B leaves 2 days after A. The whole work lasts for [*SSC-MT-2013*]

(a) 7 days (b) 6 days
(c) 12 days (d) 13 days

55. A can do a piece of work in 20 days which B can do in 12 days. B worked at it for 9 days. A can finish the remaining work in : [*SSC 10+2-2012*]
(a) 5 days (b) 7 days
(c) 11 days (d) 3 days

56. A man walks 'a' km in 'b' hours. The time taken to walk 200 metres is: [*SSC 10+2-2012*]
(a) $\frac{200b}{a}$ hours (b) $\frac{b}{5a}$ hours
(c) $\frac{b}{a}$ hours (d) $\frac{ab}{200}$ hours

57. A is thrice as good a workman as B and takes 60 days less than B for doing a job. The time in which they can do it together is: [*SSC 10+2-2012*]
(a) 15 days (b) 30 days
(c) $22\frac{1}{2}$ days (d) 60 days

58. A can do a work in 20 days and B can do the same work in 30 days. In how many days can A and B together do the work? [*SSC 10+2-2013*]
(a) 15 (b) 16 (c) 10 (d) 12

59. A and B working separately can do a piece of work in 9 and 15 days respectively. If they work for a day alternatively, with A beginning, then the work will completed in [*SSC 10+2-2014*]
(a) 10 days (b) 11 days
(c) 9 days (d) 12 days

60. Two pipes A and B can fill a tank in 36 min. and 45 min. respectively. Another pipe C can empty the tank in 30 min. First A and B are opened. After 7 minutes, C is also opened. The tank is filled up in [*SSC 10+2-2014*]
(a) 39 min. (b) 46 min.
(c) 40 min. (d) 45 min.

61. 9 women can complete a piece of work in 19 days. How many days will 18 women take to complete the same piece of work? [*IBPS Clerk-2012*]
(a) 12 days (b) 6.5 days
(c) 9 days (d) 8.5 days
(e) None of these

62. Two pipes can full a tank in 10 h and 16 h respectively. A third pipe can empty the tank in 32 h. If all the three pipes function simultaneously, then in how much time the tank will be full? (in hours) [*IBPS Clerk-2013*]
(a) $7\frac{11}{21}$ (b) $7\frac{13}{21}$ (c) $8\frac{4}{21}$ (d) $6\frac{5}{14}$
(e) $8\frac{9}{14}$

63. 56 workers can finish a piece of work in 14 days. If the work is to be completed in 8 days, then how many extra workers are required? [*IBPS Clerk-2013*]
(a) 36 (b) 48
(c) 44 (d) 42
(e) 32

Level - II

1. A pipe can fill a tank in 15 minutes and another one in 10 minutes. A third pipe can empty the tank in 5 minutes. The first two pipes are kept open for 4 minutes in the beginning and then the third pipe is also opened. In what time will the tank be emptied ?
(a) 35 min (b) 15 min
(c) 20 min (d) Cannot be emptied

2. Filling pipe, if opened alone, takes 5 minutes to fill a cistern. Suddenly, during the course of fillling, the waste pipe (which is of similar size and flow as of fill pipe) is opened for 2 minutes, then the cistern will be filled in
(a) $3\frac{1}{7}$ min (b) $3\frac{1}{3}$ min
(c) 5 min (d) 7 min

3. Three taps *A*, *B* and *C* can fill a tank in 12, 15 and 20 hours respectively. If *A* is open all the time and *B* and *C* are open for one hour each alternately, then the tank will be full in :
(a) 6 hrs. (b) $6\frac{2}{3}$ hrs.
(c) 7 hrs. (d) $7\frac{1}{2}$ hrs.

4. 1 man or 2 women or 3 boys can do a work in 44 days. Then, in how many days will 1 man, 1 woman and 1 boy do the work?
(a) 12 days (b) 24 days
(c) 18 days (d) 36 days

5. *A*, *B* and *C* can do a work in 8, 16 and 24 days respectively. They all begin together. *A* continues to work till it is finished, *C* leaving off 2 days and *B* one day before its completion. In what time is the work finished?
(a) 3 days (b) 4 days
(c) 5 days (d) 8 days

6. Two pipes *A* and *B* can fill a tank in 24 minutes and 32 minutes respectively. If both the pipes are opened simultaneously, after how much time should *B* be closed so that the tank is full in 18 minutes?
(a) 6 min. (b) 8 min.
(c) 12 min. (d) 14 min.

7. A contractor undertook to do a piece of work in 9 days. He employed certain number of laboures but 6 of them were absent from the very first day and the rest could finish the work in only 15 days. Find the number of men originally employed .
(a) 15 (b) 6 (c) 13 (d) 9

8. After working for 8 days, Anil finds that only $\frac{1}{3}$ of the work has been done. He employs Rakesh who is 60 % efficient as Anil. How many more days will Anil take to complete the job?
(a) 15 days (b) 12 days
(c) 10 days (d) 8 days

9. A can build up a wall in 8 days while B can break it in 3 days. A has worked for 4 days and then B joined to work with A for another 2 days only. In how many days will A alone build up the remaining part of wall?
(a) $13\frac{1}{3}$ days (b) $7\frac{1}{3}$ days
(c) $6\frac{1}{3}$ days (d) 7 days

10. A cistern has two taps which fill it in 12 minutes and 15 minutes respectively. There is also a waste pipe in the cistern. When all the three are opened, the empty cistern is full in 20 minutes. How long will the waste pipe take to empty the full cistern ?
(a) 10 min (b) 12 min
(c) 15 min (d) None of these

11. A pump can be operated both for filling a tank and for emptying it. The capacity of tank is 2400 m^3. The emptying capacity of the pump is 10 m^3 per minute higher than its filling capacity. Consequently, the pump needs 8 minutes less to empty the tank to fill it. Find the filling capacity of pump.
(a) 50 m^3/min (b) 60 m^3/min
(c) 58 m^3/min (c) None of these

12. A tank is filled in 5 hours by three pipes *A*, *B* and *C*. The pipe *C* is twice as fast as *B* and *B* is twice as fast as *A*. How much time will pipe *A* alone take to fill the tank ?
(a) 20 hrs (b) 25 hrs
(c) 35 hrs (d) Cannot be determind

13. Two pipes *A* and *B* can fill a tank in 15 hours and 20 hours respectively while a third pipe *C* can empty the full tank in 25 hours. All the three pipes are opened in the begining. After 10 hours, *C* is closed. In how much time, will the tank be full?
(a) 12 hrs (b) 13 hrs
(c) 16 hrs (d) 18 hrs

14. 4 men and 6 women finish a job in 8 days, while 3 men and 7 women finish in 10 days. In how many days will 10 women finish it?
(a) 20 days (b) 30 days
(c) 40 days (d) 50 days

15. *A* can do a work in 25 days and *B* can do the same work in 20 days. They work together for 5 days and then *A* goes away. In how many days will *B* finish the work?
(a) 9 days (b) 11 days
(c) 15 days (d) 20 days

16. There is leak in the bottom of a tank. This leak can empty a full tank in 8 hours. When the tank is full, a tap is opened into the tank which admits 6 litres per hour and the tank is now emptied in 12 hours. What is the capacity of the tank?
(a) 28.8 litres (b) 36 litres
(c) 144 litres (d) Can't be determined

17. A company has a job to prepare certain no. of cans and there are three machines *A*, *B* & *C* for this job. *A* can complete the job in 3 days, *B* can complete the job in 4 days and *C* can complete the job in 6 days. How many days the company will take to complete job if all the machines are used simultaneously?
(a) 4 days (b) 4/3 days
(c) 3 days (d) 12 days

18. 3 small pumps and a large pump are filling a tank. Each of the three small pumps works at 2/3rd the rate of the large pump. If all 4 pumps work at the same time, they should fill the tank in what fraction of the time that it would have taken the large pump alone?
(a) 4/7 (b) 1/3 (c) 2/3 (d) 3/4

19. *A* and *B* can do a job in 15 days and 10 days, respectively. They began the work together but *A* leaves after some days and *B* finished the remaining job in 5 days. After how many days did *A* leave?
(a) 2 days (b) 3 days
(c) 1 day (d) None of these

20. If 12 men or 15 women or 18 boys can do a piece of work in 15 days of 8 hours each, find how many men assisted by 5 women and 6 boys will finish the same work in 16 days of 9 hours each.
(a) 6 men (b) 2 men
(c) 8 men (d) 4 men

21. The work done by a man, a woman and a child is in the ratio of 3 : 2 : 1. There are 20 men, 30 women and 36 children in a factory. Their weekly wages amount to ₹ 780, which is divided in the ratio of work done by the men, women and children. What will be the wages of 15 men, 21 women and 30 children for 2 weeks?
(a) ₹ 585 (b) ₹ 292.5
(c) ₹ 1170 (d) ₹ 900

22. *x* is 3 times as faster as *y* and is able to complete the work in 40 days less than *y*. Then the time in which they can complete the work together?
(a) 15 days (b) 10 days
(c) $7\frac{1}{2}$ days (d) 5 days

23. The Bubna dam has four inlets. Through the first three inlets, the dam can be filled in 12 minutes; through the second, the third and the fourth inlet, it can be filled in 15 minutes; and through the first and the fourth inlet, in 20 minutes. How much time will it take all the four inlets to fill up the dam?
(a) 8 min (b) 10 min
(c) 12 min (d) None of these

24. Two pipes can fill a cistern in 14 and 16 hours respectively. The pipes are opened simultaneously and it is found that due to leakage in the bottom of the cistern, it takes 32 minutes extra for the cistern to be filled up. When the cistern is full, in what time will the leak empty it?

(a) 114 h (b) 112 h (c) 100 h (d) 80 h

25. A student studying the weather for d days observed that (i) it rained on 7 days. morning or afternoon; (ii) when it rained in the afternoon, it was clear in the morning; (iii) there were five clear afternoons and (iv) there were six clear morning. Then d equals

(a) 3 (b) 7 (c) 11 (d) 9

26. If 6 BSF or 10 CRPF companies can demolish a terrorist outfit in Kashmir in 2 days, find how long will 4 BSF and 9 CRPF companies take to do the same ?

(a) 1.27 days (b) 2.27 days
(c) 3.27 days (d) 4.27 days

27. Three pumps working 8 hours a day can empty a tank in 2 days. How many hours a day must 4 pumps work to empty the tank in 1 day ?

(a) 10 hours (b) 12 hours
(c) 8 hours (d) None of these

28. A group of men decided to do a job in 4 days. But since 20 men dropped out every day, the job completed at the end of the 7th day. How many men were there at the beginning?

(a) 240 (b) 140 (c) 280 (d) 150

29. The total number of men, women and children working in a factory is 18. They earn ₹ 4000 in a day. If the sum of the wages of all men, all women and all children is in the ratio of 18 : 10 : 12 and if the wages of an individual man, woman and child is in the ratio 6 : 5 : 3, then how much a woman earn in a day?

(a) ₹ 400 (b) ₹ 250 (c) ₹ 150 (d) ₹ 120

30. Raju can do a piece of work in 10 days, Vicky in 12 days and Tinku in 15 days. They all start the work together, but Raju leaves after 2 days and Vicky leaves 3 days before the work is completed. In how many days is the work completed?

(a) 5 days (b) 6 days (c) 7 days (d) 8 days

31. A can do a piece of work in 10 days and B is 25% more efficient than A. In what time will the work be finished if A and B work together? [*SBI PO-2011*]

(a) $4\frac{4}{9}$ days (b) $5\frac{5}{7}$ days

(c) 5 days (d) $6\frac{2}{3}$ days

(e) None of these

32. A and B together can complete a task in 20 days. B and C together can complete the same task in 30 days. A and C together can complete the same task in 40 days. What is the respective ratio of the number of days taken by A when completing the same task alone to the number of days taken by C when completing the same task alone ?

[*IBPS-PO-2012*]

(a) 2 : 5 (b) 2 : 7 (c) 3 : 7 (d) 1 : 5
(e) 3 : 5

33. 8 men and 4 women together can complete a piece of work in 6 days. The work done by a man in one day is double the work done by a woman in one day. If 8 men and 4 women started working and after 2 days 4 men left and 4 new women joined, in how many more days will the work be completed?

[*IBPS-PO-2013*]

(a) 5 days (b) 8 days
(c) 6 days (d) 4 days
(e) 9 days

34. X and Y can do a piece of work in 30 days. They work together for 6 days and then X quits and Y finishes the work in 32 more days. In how many days can Y do the piece of work alone? [*SSC CGL-2012*]

(a) 30 days (b) 32 days
(c) 34 days (d) 40 days

35. A can do a certain work in the same time in which B and C together can do it. If A and B together could do it in 10 days and C alone in 50 days, then B alone could do it in

[*SSC CGL-2013*]

(a) 15 days (b) 20 days
(c) 25 days (d) 30 days

36. A can do a piece of work in 6 days. B can do the same work in 15 days. How long would both of them take to do the same work ? [*SSC CGL-2013*]

(a) 2 days (b) 4 days
(c) 6 days (d) 8 days

37. 12 men construct 1.5 km of road in 7 days. 28 men will construct 12 km of roads in [*SSC CGL-2013*]

(a) 20 days (b) 24 days
(c) 28 days (d) 38 days

38. A, B and C together can do a piece of work in 40 days. After working with B and C for 16 days, A leaves and then B and C complete the remaining work in 40 days more. A alone could do the work in [*SSC CGL-2014*]

(a) 80 days (b) 90 days
(c) 100 days (d) 120 days

39. Three pipes A, B and C can fill a tank in 6 hours. After working it together for 2 hours, C is closed and A and B can fill the remaining part in 7 hours. The number of hours taken by C alone to fill the tank is [*SSC CGL-2014*]

(a) 10 (b) 12
(c) 14 (d) 16

40. Pratibha is thrice as efficient as Sonia and is therefore able to finish a piece of work in 60 days less than Sonia. Pratibha and Sonia can individually complete the work respectively in [*SSC CGL-2014*]

(a) 30, 60 days (b) 60, 90 days
(c) 30, 90 days (d) 40, 120 days

41. 40 men can finish a piece of work in 60 days. After some days, 10 men leave the work so that the work is finished in 70 days. The number of days after which 10 men left the work is [*SSC CGL-2014*]

(a) 20 days (b) 25 days
(c) 30 days (d) 40 days

Hints & Solutions

Level-I

1. (a) $(A + B)$'s 1 day's work $= \frac{1}{12}$th part of whole work.

B's 1 day's work $= \frac{1}{28}$th part of whole work.

$\therefore$ A's 1 day's work $= \frac{1}{12} - \frac{1}{28} = \frac{1}{21}$th part of whole work.

$\therefore$ A alone can finish the work in 21 days

2. (d) $\because$ A can do $\frac{3}{4}$ of the work in 12 days

$\therefore$ A can do $\frac{1}{8}$ of the work in $12 \times \frac{4}{13} \times \frac{1}{8}$ days $= 2$ days

3. (a) A's 1 day's work $= \frac{1}{18}$ and B's 1 day's work $= \frac{1}{9}$.

$\therefore$ $(A + B)$'s 1 day's work $= \left(\frac{1}{18} + \frac{1}{9}\right) = \frac{1}{6}$.

4. (b) Let the man alone do the work in x days.
Then, the woman alone do the work in $2x$ days.

Their one day's work $= \frac{1}{8}$th part of whole work

i.e., $\frac{1}{x} + \frac{1}{2x} = \frac{1}{8}$

$\Rightarrow x = 12$ days

$\therefore$ man takes 12 days and woman $2x = 24$ days.

5. (b) Ratio of times taken by A and $B = 100 : 130 = 10 : 13$.
Suppose B takes x days to do the work.

Then, $10 : 13 :: 23 : x \Rightarrow x = \left(\frac{23 \times 13}{10}\right) \Rightarrow x = \frac{299}{10}$.

A's 1 day's work $= \frac{1}{23}$; B's 1 days work $= \frac{10}{299}$.

$(A + B)$'s 1 day's work $= \left(\frac{1}{23} + \frac{10}{299}\right) = \frac{23}{299} = \frac{1}{13}$.

$\therefore$ A and B together can complete the job in 13 days.

6. (a) 50 men complete 0.4 work in 25 days.

Applying the work rule, $m_1 \times d_1 \times w_2 = m_2 \times d_2 \times w_1$ we have,

$50 \times 25 \times 0.6 = m_2 \times 25 \times 0.4$

or $m_2 = \frac{50 \times 25 \times 0.6}{25 \times 0.4} = 75$ men

Number of additional men required $= (75 - 50) = 25$

7. (d) In 1 day, work done by 12 men $= \frac{1}{18}$

In 6 days, work done by 12 men $= \frac{6}{18} = \frac{1}{3}$

Remaining work $= \frac{2}{3}$

Now, $m_1 \times d_1 \times w_2 = m_2 \times d_2 \times w_1$

or $12 \times 18 \times \frac{2}{3} = 16 \times d_2 \times 1$

or $d_2 = \frac{4 \times 18 \times 2}{16} = 9$ days

8. (b) Man's two day's work $= 2 \times \frac{1}{20}$th work $= \frac{1}{10}$th work

Woman's two days's work

$= 2 \times \frac{1}{30}$th work $= \frac{1}{15}$th work

Boy's two day's work $= 2 \times \frac{1}{60}$th work $= \frac{1}{30}$th work

Now, let 2 men, 8 women and x boys can complete work in 2 days. Then ,

2 men's work + 8 women's work + x boy's work $=1$

$2\left(\frac{1}{10}\right) + 8\left(\frac{1}{15}\right) + x\left(\frac{1}{30}\right) = 1$

$\Rightarrow x = \left(1 - \frac{1}{5} - \frac{8}{15}\right) \times 30 \Rightarrow x = 8$ boys

9. (c) 10 men's 1 day's work $= \frac{1}{15}$;

15 women's 1 day's work $= \frac{1}{12}$.

(10 men + 15 women)'s 1 day's work

$= \left(\frac{1}{15} + \frac{1}{12}\right) = \frac{9}{60} = \frac{3}{20}$.

$\therefore$ 10 men and 15 women will complete the work in $\frac{20}{3} = 6\frac{2}{3}$ days.

10. (c) In 8 days, Anil does $= \frac{1}{3}$rd work.

$\therefore$ in 1 day, he does $= \frac{1}{24}$th work.

$\therefore$ Rakesh's one day's work $= 60\%$ of $\frac{1}{24} = \frac{1}{40}$th work.

Remaining work $= 1 - \frac{1}{3} = \frac{2}{3}$

(Anil and Rakesh)'s one day's work

$= \frac{1}{24} + \frac{1}{40} = \frac{1}{15}$th work

Now, $\frac{1}{15}$th work is done by them in one day

$\therefore \frac{2}{3}$rd work is done by them in $15 \times \frac{2}{3} = 10$ days

11. (a) A's one day's work $= \frac{1}{3}$rd work.

B's one day's work $= \frac{1}{6}$th work.

$(A + B)$'s one day's work $= \frac{1}{3} + \frac{1}{6} = \frac{1}{2}$nd work

$\therefore$ A and B together can complete the work (knit a pair of socks) in 2 days.

$\therefore$ They together knit two pair of socks in 4 days.

12. (b) A's one day's work $= \frac{1}{8}$th work

B's one day's work $= \frac{1}{3}$rd work

$\therefore$ A's 4 day's work $= 4 \times \frac{1}{8} = \frac{1}{2}$nd work

$\therefore$ In next two days, total wall $= \frac{1}{2} + 2\left(\frac{1}{8}\right) - 2\left(\frac{1}{3}\right)$

$= \frac{1}{12}$th wall

Remaining wall $= 1 - \frac{1}{12} = \frac{11}{12}$th

Now, $\frac{1}{8}$th wall is built up by A in one day.

$\therefore \frac{11}{12}$th wall is built up by A in $8 \times \frac{11}{12} = 7\frac{1}{3}$ days.

13. (b) Sakshi's one day's work $= \frac{1}{20}$th work

Tanya's one day's work

$= \frac{1}{20} + 25\%$ of $\frac{1}{20} = \frac{1}{16}$th work

Hence, Tanya takes 16 days to complete the work.

14. (a) Let 1 woman's 1 day's work $= x$.

Then, 1 man's 1 day's work $= \frac{x}{2}$

and 1 child's 1 day's work $= \frac{x}{4}$.

So, $\left(\frac{3x}{2} + 4x + \frac{6x}{4}\right) = \frac{1}{7} \Rightarrow x = \left(\frac{1}{7} \times \frac{4}{28}\right) = \frac{1}{49}$.

$\therefore$ 1 woman alone can complete the work in 49 days.

So, to complete the work in 7 days, women required

$= \left(\frac{49}{7}\right) = 7.$

15. (a) Sunil takes 5 days and Pradeep takes 15 days to do the work.

In a day they would complete $\frac{1}{5} + \frac{1}{15}$ i.e., $\frac{4}{15}^{th}$ work.

The remaining 11/15th work would be completed by Pradeep in $\frac{11}{15} \times 15$ i.e. 11 days.

16. (c) Suresh, working alone 42 days = 1 unit of work. Mahesh is 1/5 times more efficient that Suresh. So Mahesh is 6/5 times as efficient as Suresh. Hence Mahesh should require 5/6th of the time, the time taken by Suresh.

Therefore time taken by Mahesh = 5/6 × 42 = 35 days.

17. (a) Given 6 BSF = 10 CRPF $\Rightarrow$ 4 BSF + 9 CRPF

$= 4 + (9 \times 6/10)$ BSF $= \frac{94}{10}$ BSF

Now work = 6 × 2 BSF days $= \frac{94}{10} \times X$ BSF days

We have $6 \times 2 = \frac{94}{10} \times X \Rightarrow X = 1.27$ days

18. (a) Let 1 man's 1 days' work $= x$ and 1 boy's 1 day's work $= y$

Then, $2x + 3y = \frac{1}{10}$ and $3x + 2y = \frac{1}{8}$

Solving, we get : $x = \frac{7}{200}$ and $y = \frac{1}{100}$

$\therefore$ (2 men + 1 boy)'s 1 day's work

$= \left(2 \times \frac{7}{200} + 1 \times \frac{1}{100}\right) = \frac{16}{200} = \frac{2}{25}$

So, 2 men and 1 boy together can finish the work in $12\frac{1}{2}$ days.

19. (b) Let the required number of working hours/day $= x$

More pumps, less working hrs per day (Indirect)

Less days, more working hrs per day (Indirect)

$\left.\begin{array}{l}\text{Pumps } 4:3 \\ \text{Days } 1:2\end{array}\right\} :: 8 : x$

$\therefore 4 \times 1 \times x = 3 \times 2 \times 8$

$\Rightarrow x = \frac{3 \times 2 \times 8}{4} = 12$

20. (d) Let required number of binders be 'x'

Less books, less binders (direct)

More days, less binders (indirect)

$$\left.\begin{matrix}\text{Books} & 900 & :660 \\ \text{Days} & 12 & :10\end{matrix}\right\}::18:x$$

$900 \times 12 \times x = 660 \times 10 \times 18$

$$x = \frac{660 \times 10 \times 18}{900 \times 12} = 11$$

21. (a) 27 men mow 225 hectares in 15 days

$\therefore$ 1 man mow 225 hectares in (15×27) days (indirect)

$\therefore$ 1 man mow 1 hectares in $\frac{15 \times 27}{225}$ days (direct)

1 man mow 165 hectares in $\frac{15 \times 27}{225} \times 165$ days (direct)

$\therefore$ 33 men mow 165 hectares in $\frac{15 \times 27 \times 165}{225 \times 33} = 9$ days

22. (c) $(X + Y)$'s one day work $= \frac{1}{72}$

$(Y + Z)$'s one day work $= \frac{1}{120}$

$(Z + X)$'s one day work $= \frac{1}{90}$

$\therefore$ $2(X + Y + Z)$'s one day work $= \frac{1}{72} + \frac{1}{120} + \frac{1}{90}$

$= \frac{5+3+4}{360} = \frac{12}{360} = \frac{1}{30}$

$\therefore$ $(X + Y + Z)$'s one day work $= \frac{1}{2} \times \frac{1}{30} = \frac{1}{60}$

$\therefore$ They will complete the work in 60 days.

23. (b) Given $(6M + 8B) \times 10 = (26M + 48B) \times 2$

$\Rightarrow 60M + 80B = 52M + 96B$

$\Rightarrow 8M = 16B$

$\Rightarrow 1M = 2B$

$\therefore 15M + 20B = 30B + 20B = 50B$

$6M + 8B = 12B + 8B = 20B$

Now	Boys	Days
	20	10
	50↓	x (Let)

$$\therefore \quad x = \frac{20 \times 10}{50} = 4 \text{ days}$$

24. (d) 1 Man = 3 Boys and 1 Woman = 2 Boys

$\therefore$ 24 Men + 20 Women + 16 Boys

$= (24 \times 3) + (20 \times 2) + 16$

$= 72 + 40 + 16$

$= 128$ Boys

27 Men + 40 Women + 15 Boys $= (27 \times 3) + (40 \times 2) + 15$

$= 81 + 80 + 15 = 176$ Boys.

Now,

No. of Boys	Duration	Wages
128 ↑	1 ↑	224
176	52	x (Let)

$$\therefore \quad x = \frac{176}{128} \times \frac{52}{1} \times 224$$

$x = ₹\ 16,016$

25. (c) Part of the cistern filled by first pipe in 1 minute $= \frac{1}{6}$

Part of the cistern filled by second pipe in 2 minutes $= \frac{1}{7}$

Part of the cistern filled in first 2 minutes $= \frac{1}{6} + \frac{1}{7} = \frac{13}{42}$

Part of the cistern filled in 6 minutes $= \frac{3 \times 13}{42} = \frac{39}{42}$

Remaining part $= 1 - \frac{39}{42} = \frac{3}{42} = \frac{1}{14}$

$\therefore$ Time taken to fill $\frac{1}{14}$ parts $= \frac{6}{14} = \frac{3}{7}$

$\therefore$ Total time $= 6 + \frac{3}{7} = 6\frac{3}{7}$ minutes

26. (b) Part filled by $(A + B + C)$ in 3 minutes

$$= 3\left(\frac{1}{30} + \frac{1}{20} + \frac{1}{10}\right) = 3 \times \frac{11}{60} = \frac{11}{20}$$

Part filled by C in 3 minutes $= \frac{3}{10}$

$$\therefore \text{Required ratio} = \frac{\frac{3}{10}}{\frac{11}{20}} = \frac{3}{10} \times \frac{20}{11} = \frac{6}{11}$$

27. (c) Let C completes the work in x days.

Work done by $(A + B)$ in 1 day $= \frac{1}{10}$

Work done by $(B + C)$ in 1 day $= \frac{1}{18}$

A's 5 days' work + B's 10 days' work + C's 15 days' work = 1

or $(A + B)$'s 5 days' work + $(B + C)$'s 5 days' work + C's 10 days' work = 1

or $\frac{5}{10} + \frac{5}{18} + \frac{10}{x} = 1$

$\therefore \quad x = 45$ days

28. (b) We have :

x men to the work in 60 days and $(x + 8)$ men do the work in

$(60 - 10 =)$ 50 days.

Then by "basic formula", $60x = 50(x + 8)$

$$\therefore x = \frac{50 \times 8}{10} = 40 \text{ men.}$$

29. **(c)** A's one day's work = $\frac{1}{16}$th work

B's one day's work $= \frac{1}{12}$th work

Let the number of days B has worked alone = x days. Then,

A's amount of work + B's amount of work = 1

$$\Rightarrow 4\left(\frac{1}{16}\right) + (x+4)\left(\frac{1}{12}\right) = 1$$

$$\Rightarrow \frac{1}{4} + \frac{x+4}{12} = 1 \Rightarrow x = \frac{3}{4} \times 12 - 4 \Rightarrow x = 5 \text{ days}$$

30. **(a)** Part filled in 7 min. $= 7 \times \left(\frac{1}{36} + \frac{1}{45}\right) = \frac{7}{20}$

Remaining part $= \left(1 - \frac{7}{20}\right) = \frac{13}{20}$

Part filled by $(A + B + C)$ in 1 min.

$$= \left(\frac{1}{36} + \frac{1}{45} - \frac{1}{30}\right) = \frac{1}{60}.$$

31. **(b)** $(A + B)$'s 5 days' work

$$= 5\left(\frac{1}{25} + \frac{1}{20}\right) = \frac{45}{100} = \frac{9}{20}$$

Remaining work $= \left(1 - \frac{9}{20}\right) = \frac{11}{20}$

$\frac{11}{20}$ of the work would be finished by B in

$$\frac{\frac{11}{20}}{\frac{1}{20}} = 11 \text{days.}$$

32. **(d)** In 1 day, work done by 12 men $= \frac{1}{18}$

In 6 days, work done by 12 men $= \frac{6}{18} = \frac{1}{3}$

Remaining work $= \frac{2}{3}$

Now, $m_1 \times d_1 \times w_2 = m_2 \times d_2 \times w_1$

or $\quad 12 \times 18 \times \frac{2}{3} = 16 \times d_2 \times 1$

or $\quad d_2 = \frac{4 \times 18 \times 2}{16} = 9 \text{ days}$

33. **(a)** Work done by A and B in 5 days $= \left(\frac{1}{10} + \frac{1}{15}\right) \times 5 = \frac{5}{6}$

Work remaining $= 1 - \frac{5}{6} = \frac{1}{6}$

∴ C alone can do the work in 6 × 2 = 12 days

Ratio of their share work $= \frac{5}{10} : \frac{5}{15} : \frac{2}{12} = 3:2:1$

Share of wages = ₹ 225, ₹ 150, ₹ 75.

34. **(c)** In 6 days A would do 25% of the work and in 8 days B would do 25% of the work himself. So C has to complete 50% of the work by himself.
In all C would require 30 days to do 50% of the work.
So, he would require 22 more days.

35. **(d)** Ratio of efficiency of Mayank and Shishu = 3/2
So ratio of time taken by Mayank and Shishu = 2/3
So if Shishu takens 30 hours, then Mayank will take 20 hours
Shishu in 6 hours = 1/5 the work.
Remaining work = 1 – 1/5 = 4/5the work,

Shishu and Mayank together $= \frac{1}{20} + \frac{1}{30} = \frac{1}{12}$

So required time $= \frac{4/5}{1/12} = 9.6 \text{hours}$

36. **(d)** After 27 days, food left = 4 × 200 = 800 soldier days worth of food. Since, now there are only 80 soldiers, this food would last for 800/80 = 10 days. Number of extra days for which the food lasts = 10 – 4 = 6 days.

37. **(b)** Sambhu requires 16 days to do the work while Kalu requires 18 days to do the work.
$(1/16 + 1/18) \times n = 1$
$\rightarrow n = 288/34 = 144/17$

38. **(c)** $n(1/45 + 1/40) + 23/40 = 1 \rightarrow n = 9$

39. **(a)** The rest of the food will last for (31 – 28 =) 3 days if nobody leaves the place.

Thus, the rest of the food will last for $3\left(\frac{400}{120}\right)$ days for the 120 men left.

$$\therefore = 3\left(\frac{400}{120}\right) = 10 \text{ days}$$

40. **(c)** 1 minute's work of both the punctures $= \left(\frac{1}{9} + \frac{1}{6}\right) = \frac{5}{18}.$

So, both the punctures will make the tyre flat in

$\frac{18}{5} = 3\frac{3}{5}$ min.

41. **(a)** Let 1 man's 1 days work = x
1 boy's 1 day's work = y

$$12x + 16y = \frac{1}{5}$$

$$13x + 24y = \frac{1}{4}$$

Solving these two equation we get,

$x = \frac{1}{100}, y = \frac{1}{200}$

Required ratio = 2 : 1

42. (d) Part filled by first tap in one min $= \frac{1}{12}$th

Part filled by second tap in one min $= \frac{1}{18}$th

Now, $2\left[\frac{1}{12} + \frac{1}{18}\right]$ + unfilled part = 1

$\Rightarrow$ unfilled part $= \frac{13}{18}$th

$\because \frac{1}{18}$th part of tank is filled by second tap in 1min.

$\therefore \frac{13}{18}$th part of tank is filled by second tap in 1 min.

$= 18 \times \frac{13}{18}$ min $= 13$ min.

43. (b) $\because$ Cistern fill in 6 hours.

$\therefore$ in 1 hour, filled part $= \frac{1}{6}$th

Now, due to leakage, filled part in 1 hour $= \frac{1}{8}$th

Part of the cistern emptied, due to leakage in 1 hour

$= \frac{1}{6} - \frac{1}{8} = \frac{1}{24}$th

$\therefore$ The leakage will empty the full cistern in 24 hrs.

44. (c) 3 men reap $\frac{1}{43}$ rd of the field in 1 day.

$\therefore$ 1 man reaps $\frac{1}{43 \times 3}$ rd of the field in 1 day.

4 women reap $\frac{1}{43}$ rd of the field in 1 day.

$\therefore$ 1 woman reaps $\frac{1}{43 \times 3}$ th of the field in 1 day.

$\therefore$ 7 men and 5 women reap $\left(\frac{7}{43 \times 3} + \frac{5}{43 \times 4}\right)$

$= \frac{1}{12}$ th of the field in 1 day

$\therefore$ 7 men and 5 women will reap the whole field in 12 days.

45. (c) $M_1 \times D_1 = M_2 \times D_2$

$m \times r = (m + n) \times D_2$

$D_2 = \frac{mr}{(m+n)}$

46. (d) Part of the tank filled by the three pipes working simultaneously in one hour is $= \frac{1}{5} + \frac{1}{6} - \frac{1}{12} = \frac{17}{60}$

i.e. it takes $\frac{60}{17}$ hours to fill up the tank completely.

Now, $\frac{1}{2}$ of the tank is filled with all the pipes open, simultaneously together in $\frac{60}{17} \times \frac{1}{2} = 1\frac{13}{17}$ hours

47. (c) 10 men in 15 days

$\Rightarrow$ 1 man can do the work in 150 days

$\Rightarrow$ 1 man can do twice the work in 300 days

Similarly, 18 boys in 15 days

$\Rightarrow$ 1 boy can do the work in 270 days

$\Rightarrow$ 1 boy can do twice the work in 540 days

Now, if there are 15 men and 33 boys trying to do twice the work then

$\left(15 \times \frac{1}{300}\right) + \left(33 \times \frac{1}{540}\right)$

$= \frac{1}{20} + \frac{11}{180} = \frac{9+11}{180} = \frac{20}{180} = \frac{1}{9}$

$\Rightarrow$ It will take 9 days for 15 men and 33 Boys to do twice the work.

48. (b) Ratio of new number of persons in fort : original number of persons in fort = 80 : 200 = 2 : 5

Hence the food will last for 5/2 days of the original (4 days = 31 days – 27 days)

$= \frac{5}{2} \times 4 = 10$ days

So, extra days = 6 days

49. (b) 9 hours 36 minutes

$= 9 + \frac{36}{60} = 9\frac{3}{5}$ hours $= \frac{48}{5}$ hours

(A + B)'s 1 hour's work

$= \frac{5}{48}$ hours

C's 1 hour's work $= \frac{1}{48}$

(A + B + C)'s 1 hour's work $= \frac{5}{48} + \frac{1}{48} = \frac{1}{8}$ ...(1)

A's 1 hours work = (B + C)'s 1 hour's work ...(2)

2 ×A's 1 hour's work $= \frac{1}{8}$

A's 1 hour's work $= \frac{1}{16}$

$\therefore$ B's 1 hour's work $= \frac{5}{48} - \frac{1}{16} = \frac{5-3}{48} = \frac{1}{24}$

$\therefore$ B alone will finish the work in 24 hours

50. (d) 3×5 men + 7×5 women
$= 4 \times 4$ men + 6×4 women
$\Rightarrow$ 16 men – 15 men = 35 women – 24 women
$\Rightarrow$ 1 man = 11 women
$\therefore$ 3 men + 7 women = 40 women
Now, $M_1 D_1 = M_2 D_2$
$\Rightarrow 40 \times 5 = 10 \times D_2$
$\Rightarrow D_2 = 20$ days

51. (d) A can complete whole work in $\frac{28}{7} \times 8 = 32$ days

B can complete whole work in $\frac{20 \times 6}{5} = 24$ days

A and B together can complete whole work in

$$\frac{32 \times 24}{32 + 24} = \frac{32 \times 24}{56} = \frac{96}{7} = 13\frac{5}{7} \text{days}$$

52. (a) More the no. of men less time they take to complete work.
Let x men are added

$$\frac{75}{75+x} = \frac{18}{90} \quad \text{(Inverse Proportion)}$$

$$\frac{75}{75+x} = \frac{1}{5}$$

$375 - 75 = x$
$x = 300$

53. (b) A's work in 1 hour $= \frac{1}{6}$

B's work in 1 hour $= \frac{1}{4}$

(A + B)'s 2 hour's work when opened alternately

$$= \left(\frac{1}{6} + \frac{1}{4}\right) = \frac{5}{12}$$

(A + B)'s 4 hour's work when opened alternately

$$= \frac{10}{12} = \frac{5}{6}$$

Remaining part $= \left(1 - \frac{5}{6}\right) = \frac{1}{6}$

Now, it is A's turn and $\frac{1}{6}$ part is filled by A in 1 hour.

$\therefore$ Total time taken to fill the tank = (4 + 1) hrs. = 5 hrs.

54. (a) Suppose, the work was finished in x days. Then,
A's (x – 5) day's work + B's (x – 3) day's work + C's x day's work = 1.

$$\Rightarrow \frac{x-5}{10} + \frac{x-3}{12} + \frac{x}{15} = 1$$

$\Rightarrow 6(x - 5) + 5(x - 3) + 4x = 60.$
$\Rightarrow 6x - 30 + 5x - 15 + 4x = 60$
$\Rightarrow 15x = 60 + 30 + 15$
$\Rightarrow 15x = 105 \Rightarrow x = 7$ days.

55. (a) B's 1 day work $= \frac{1}{12}$

B's 9 day's work $= \frac{9}{12} = \frac{3}{4}$

Remaining work $= 1 - \frac{3}{4} = \frac{1}{4}$

$\therefore$ A can finish this work in $\frac{20}{4}$ days = 5 days

56. (b) 1 km = 1000 m
$D = S \times T$
a km = S × b hr

$$S = \frac{a \times 1000}{b}$$

Now, D = 200 m

Time taken $= \frac{D}{S}$

$$= \frac{200}{a \times 1000} \times b = \frac{b}{5a} \text{hrs}$$

57. (c) If A can finish a work = x days
B will do this work = 3x days
From question, $A - B = 3x - x = 60 \Rightarrow x = 30$
A = 30 days, B = 90 days

A's 1 day work + B's 1 day work $= \frac{1}{30} + \frac{1}{90} = \frac{4}{90}$

So, A and B working together can complete work

$$= \frac{90}{4} = 22.5 \text{ days}$$

58. (d) A's 1 day's work $= \frac{1}{20}$

B's 1 day's work $= \frac{1}{30}$

(A + B)'s 1 day's work $= \left(\frac{1}{20} + \frac{1}{30}\right) = \frac{5}{60}$

$\therefore$ Both A and B will finish the work in $\frac{60}{5}$ = 12 days.

59. (b) Two days work $= \frac{1}{9} + \frac{1}{15} = \frac{5+3}{45} = \frac{8}{45}$

Ten days work $= 5 \times \frac{8}{45} = \frac{40}{45} = \frac{8}{9}$

Remaining work $= 1 - \frac{8}{9} = \frac{1}{9}$ which is done by A on 11th day.

Hence, the work will be completed in 11 days.

60. (a) In one minute (A + B) can together fill $\frac{1}{36}+\frac{1}{45}=\frac{1}{20}$ part.

In 7 minutes part of tank filled = $\frac{7}{20}$

remaining part $=1-\frac{7}{20}=\frac{13}{20}$

In 8th minutes, part filled by A, B and C altogether

$=\frac{1}{36}+\frac{1}{45}-\frac{1}{30}=\frac{1}{20}-\frac{1}{30}=\frac{1}{60}$

$\frac{13}{20}$ part of tank filled by (A + B + C)

$=60\times\frac{13}{20}=39$ minutes

61. (e) $M_1D_1=M_2D_2$

$\Rightarrow 9\times 19=18\times D_2$

$\Rightarrow D_2=\frac{9\times 19}{18}=9.5$ days

62. (b) 10 hr A pipe $\rightarrow 1$

16 hr B pipe $\rightarrow 1$

32 hr C pipe $\rightarrow 1$

$\frac{1}{10}+\frac{1}{16}-\frac{1}{32}=\frac{21}{160}$

$\frac{160}{21}=7\frac{13}{21}$ hr

63. (d) Here, $M_1=56$, $D_1=14$, $M_2=?$, $D_2=8$

Using

$M_1D_1=M_2D_2$,

$56\times 14=M_2\times 8$

$\Rightarrow M_2=98$

Hence, extra workers to be required

$=98-56=42$

Level-II

1. (c) Proportion of the volume of the tank filled by both the pipes in 4 min $=4\left(\frac{1}{15}+\frac{1}{10}\right)=\frac{2}{3}$rd of the tank.

Volume of the tank filled by all the pipes working together $=\frac{1}{15}+\frac{1}{10}-\frac{1}{5}=\frac{-1}{30}$

i.e., $\frac{1}{30}$ tank is emptied in 1 min.

$\therefore \frac{2}{3}$rd of the tank can be emptied in $\frac{2\times 30}{3}=20$ min

2. (d) Since, flow of waste pipe = flow of filling pipe.

$\Rightarrow$ Filled part in one min = emptied part in one min.

$\therefore$ After opening the waste pipe for 2 min, cistern will be full in (5 + 2) = 7 min.

3. (c) $(A+B)$'s 1 hour's work $=\left(\frac{1}{12}+\frac{1}{15}\right)=\frac{9}{60}=\frac{3}{20}$

$(A+C)$'s 1 hour's work $=\left(\frac{1}{12}+\frac{1}{20}\right)=\frac{8}{60}=\frac{2}{15}$

Part filled in 2 hrs $=\left(\frac{3}{20}+\frac{2}{15}\right)=\frac{17}{60}$

Part filled in 6 hrs $=\left(3\times\frac{17}{60}\right)=\frac{17}{20}$

Remaining part $=\left(1-\frac{17}{20}\right)=\frac{3}{20}$

Now, it is the turn of A and B and $\frac{3}{20}$ part is filled by A and B in 1 hour.

$\therefore$ Total time taken to fill the tank = (6 + 1) hrs = 7 hrs.

4. (b) Thus, by our extended formula, number of required days

$=\frac{1}{\frac{1}{44\times 1}+\frac{1}{44\times 2}+\frac{1}{44\times 3}}=\frac{44\times 1\times 2\times 3}{6+3+2}=24$ days

5. (c) Let the work be finished in x days.

Then, A's x day's work + B's $(x-1)$ day's work + C's $(x-2)$ day's work = 1

or, $\frac{x}{8}+\frac{x-1}{16}+\frac{x-2}{24}=1$

or, $\frac{6x+3x-3+2x-4}{48}=1$

or, $11x=55$

$\therefore x=5$ days

6. (b) Let B be closed after x minutes. Then, part filled by $(A+B)$ in x min. + part filled by A in $(18-x)$ min = 1.

$\therefore\quad x\left(\frac{1}{24}+\frac{1}{32}\right)+(18-x)\times\frac{1}{24}=1$

or, $\frac{7x}{96}+\frac{18-x}{24}=1$ or, $7x+4(18-x)=96$

or, $3x=24\quad\therefore x=8$.

So, B should be closed after 8 min.

Direct Formula:

Pipe B should be closed after $\left(1-\frac{18}{24}\right)\times 32=8$ min.

7. (a) Let the number of men originally employed be x.

$9x=15(x-6)$

or $\quad x=15$

8. (c) In 8 days, Anil does $=\frac{1}{3}$rd work.

$\therefore$ in 1 day, he does $=\frac{1}{24}$th work.

$\therefore$ Rakesh's one day's work = 60% of $\frac{1}{24}=\frac{1}{40}$th work.

Remaining work $=1-\frac{1}{3}=\frac{2}{3}$

(Anil and Rakesh)'s one day's work

$=\frac{1}{24}+\frac{1}{40}=\frac{1}{15}$th work

Now, $\frac{1}{15}$th work is done by them in one day.

$\therefore \frac{2}{3}$rd work is done by them in $15\times\frac{2}{3}=10$ days

9. (b) A's one day's work $=\frac{1}{8}$th work

B's one day's work $=\frac{1}{3}$rd work

$\therefore$ A's 4 day's work $=4\times\frac{1}{8}=\frac{1}{2}$nd work

$\therefore$ In next two days, total wall $=\frac{1}{2}+2\left(\frac{1}{8}\right)-2\left(\frac{1}{3}\right)$

$=\frac{1}{12}$th wall

Remaining wall $=1-\frac{1}{12}=\frac{11}{12}$th

Now, $\frac{1}{8}$th wall is built up by A in one day.

$\therefore \frac{11}{12}$th wall is built up by A in $8\times\frac{11}{12}=7\frac{1}{3}$ days.

10. (a) Work done by the waste pipe in 1 minutes

$=\frac{1}{20}-\left(\frac{1}{12}+\frac{1}{15}\right)=-\frac{1}{10}$ [–ve sign means emptying]

$\therefore$ Waste pipe will empty the full cistern in 10 minutes.

11. (a) Let the filling capacity of pump be x m^3/min.
Then, emptying capacity of pump = $(x+10)$ m^3/min.

$\therefore \frac{2400}{x}-\frac{2400}{x+10}=8$

$\Rightarrow x^2+10x-3000=0$

$\Rightarrow (x-50)(x+60)=0 \Rightarrow x=50$ m^3/min.

12. (c) Suppose pipe A alone takes x hours to fill the tank.

Then, pipes B and C will take $\frac{x}{2}$ and $\frac{x}{4}$ hours respectively to fill the tank.

$\therefore \frac{1}{x}+\frac{2}{x}+\frac{4}{x}=\frac{1}{5}\Rightarrow\frac{7}{x}=\frac{1}{5}\Rightarrow x=35$ hrs.

13. (a) Part filled in 10 hours $=10\left(\frac{1}{15}+\frac{1}{20}-\frac{1}{25}\right)=\frac{23}{30}$.

Remaining part $=\left(1-\frac{23}{30}\right)=\frac{7}{30}$.

$(A+B)$'s 1 hour's work $=\left(\frac{1}{15}+\frac{1}{20}\right)=\frac{7}{60}$.

$\frac{7}{60}:\frac{7}{30}::1:x$ or $x=\left(\frac{7}{30}\times1\times\frac{60}{7}\right)=2$ hours.

$\therefore$ The tank will be full in (10 + 2) hrs = 12 hrs.

14. (c) **Method I.** Considering one day's work:

$4m+6w=\frac{1}{8}$ .....(1)

$3m+7w=\frac{1}{10}$ ....(2)

(1) × 3 – (2) × 4 gives

$18w-28w=\frac{3}{8}-\frac{4}{10}$ or, $10w=\frac{1}{40}$

$\therefore$ 10 women can do the work in 40 days.

Method II. We find that
$8(4m+6w)=10(3m+7w)$
or, $2m=22w$
$\therefore 4m=44w$
$\therefore$ 4 men + 6 women = 50 women do in 8 days

$\therefore$ 10 women do in $\frac{8\times50}{10}=40$ days

15. (b) $A+B$ can do the work in 5 days $=5\left[\frac{1}{25}+\frac{1}{20}\right]$

$=\frac{5\times45}{25\times20}=\frac{9}{20}$

Rest of the work $=1-\frac{9}{20}=\frac{11}{20}$

B will do the rest of the work in $20\times\frac{11}{20}=11$ days.

16. (c) Let the capacity of tank be x litres

In one hour tank empties $=\frac{1}{8}$ of $x=\frac{x}{8}$ litre

In one hour, tap admits 6 litres
after opening tap tank is emptied in 12 hours.

So in one hour tank empties by $\frac{1}{12}$ of $x=\frac{x}{12}$ litres.

Therefore equation becomes $6-\frac{x}{8}=-\frac{x}{12}$

$$6=\frac{x}{8}-\frac{x}{12}=\frac{4x}{96}=\frac{x}{24}$$

$\therefore$ $x = 144$ litres

17. (b) Let work done by A in one day be a, similarly, for B, b and for C, c

So, $3a = 1$, $4b = 1$, $6c = 1$ [Total work be 1 unit]

So, Total work done by the 3 Machines in one day

$$=\frac{1}{3}+\frac{1}{4}+\frac{1}{6}=\frac{3}{4}$$

Therefore, time taken to complete the work is

$$\frac{1}{3/4}=\frac{4}{3}\text{ days.}$$

18. (b) Suppose large pump takes t hours to fill a tank

$\therefore$ 1 hour work of large pump fills $=\frac{1}{t}$ part

1 hour work of each small pump fills $=\frac{1}{t}\times\frac{2}{3}$

1 hour work of all 4 pumps fill $=\frac{1}{t}+3\times\frac{2}{3t}=\frac{3}{t}$

Therefore, $\frac{3}{t}$ part is filled by all 4 pumps in 1 hour

$\therefore$ Whole tank would be filled in $1\times\frac{t}{3}=\frac{t}{3}h$ this is 1/3 of the time taken by large pump i.e., t hour

19. (b) A's one day's work $=\frac{1}{15}$th work.

B's one day's work $=\frac{1}{10}$th work.

$(A + B)$'s one day's work $=\frac{1}{15}+\frac{1}{10}=\frac{1}{6}$th work.

Let A left after x days.

$\therefore$ $(A + B)$'s x days' work $=\frac{x}{6}$th work.

Remaining work $=1-\frac{x}{6}=\frac{6-x}{6}$th work.

Now, in 5 days, work done by $B=\frac{6-x}{6}$th work.

$\therefore$ In 1 day work done by $B=\frac{6-x}{30}$th work

and $\frac{6-x}{30}=\frac{1}{10}$

$\therefore$ $x = 3$ days

20. (b) Given 12 men ≡ 15 women ≡ 18 boys

$\therefore$ 1 Man ≡ 1.5 boys, 1 woman = 6/5 boys.

Now, $5W + 6B = 12B$.

Required answer is calculated as follows :

Total no. of boys reqd. = 18 × [(15/16) × (8/9)]

= 15 boys

The number of boys already present = 12.

Hence, 3 boys more required.

But 3 boys = 2 men.

So, 2 men are required.

21. (c)

	Men	Women	Children
Work	3	2	1
Numbers	20	30	36

Ratio of wages = (3 × 20) : (2 × 20) : (1 × 36) = 5 : 5 : 3

Total wages of men $=\frac{5}{13}\times780$ = ₹ 300

$\therefore$ Wages of a man = ₹ 15

Similarly, wages of woman = ₹ 10

and wages of child = ₹ 5

Total waves of 15 men, 21 women and 30 children

= 15 × 15 + 21 × 10 + 30 × 5 = 585

Total wages for 2 weeks = ₹ 1170

22. (a) If x complete a work in x days. y will do the same task in $3x$ days.

$3x - x = 40$

$\Rightarrow$ $x = 20$

y will finish the task in 60 days

$(x + y)$'s 1 days work

$$=\frac{1}{20}+\frac{1}{60}=\frac{1}{15}$$

Both of them will complete the work in 15 days.

23. (b) Let the inlets be A, B, C and D.

$A + B + C = 8.33$ %

$B + C + D = 6.66\%$

$A + D = 5\%$

Thus $2A + 2B + 2C + 2D = 20\%$

and $A + B + C + D = 10\%$

→ 10 minutes would be required to fill the tank completely.

24. (b) The 32 minutes extra represents the extra time taken by the pipes due to the leak.

Normal time for the pipes → n × (1/14 + 1/16) = 1 → n = 112/15 = 7 hrs 28 minutes.

Thus, with 32 minutes extra, the pipes would take 8 hours to fill the tank.

Thus, 8(1/14 + 1/16) – 8 × (1/L) = 1 → 8/L

= 8(15/112) – 1

1/L = 15/112 – 1/8

= 1/112.

Thus, L = 112 hours.

25. (d) Let x = Number of days it rained in the morning and had clear afternoons.

y = Number of days it rained in the afternoon and had clear mornings.

z = Number of days it rained in the morning or afternoon

So according to question, $x + y = 7$

$x + z = 5$

$y + z = 6$

Adding all three equations, $x + y + z = 9$

So, d = 9 days

26. **(a)** Given 6 BSF ≡ 10 CRPF ⇒ 4 BSF + 9 CRPF

$= 4 + (9 \times 6/10)$ BSF $= \frac{94}{10}$ BSF

Now work $= 6 \times 2$ BSF days $= \frac{94}{10} \times X$ BSF days

We have $6 \times 2 \equiv \frac{94}{10} \times X \Rightarrow X = 1.27$ days

27. **(b)** Let the required number of working hours/day $= x$

More pumps, less working hrs per day (Indirect)

Less days, more working hrs per day (Indirect)

$$\left.\begin{array}{ll}\text{Pumps} & 4:3\\ \text{Days} & 1:2\end{array}\right\} :: 8 : x$$

$\therefore 4 \times 1 \times x = 3 \times 2 \times 8$

$\Rightarrow x = \frac{3 \times 2 \times 8}{4} = 12$

28. **(b)** Go through option

$140 \times 4 = (140 + 120 + 100 + ... + 20)$

$560 = 560$

Alternatively: Let n be the initial number of worker then

$n \times 4 = n + (n - 20) + (n - 40) + ... + (n - 120)$

$4n = 7n - 420$

$\Rightarrow 3n = 420$

$\Rightarrow n = 140$ workers

29. **(b)** Ratio of number of men, women and children

$= \frac{18}{6} : \frac{10}{5} : \frac{12}{3} = 3x : 2x : 4x$

$\therefore (3x + 2x + 4x) = 18$

$\therefore x = 2$

Therefore, number of women = 4

Share of all women $= \frac{10}{40} \times 4000 =$ ₹ 1000

($\because$ 18 + 10 + 12 = 40)

$\therefore$ Share of each woman $= \frac{1000}{4}$ ₹ 250

30. **(c)** Raju = 10%, Vicky = 8.33% and Tinku = 6.66%. Hence, total work for a day if all three work = 25%. In 2 days they will complete, 50% work. On the third day onwards Raju doesn't work. The rate of work will become 15%. Also, since Vicky leaves 3 days before the actual completion of the work, Tinku works alone for the last 3 days (and must have done the last 6.66 × 3 = 20% work alone). This would mean that Vicky leaves after 80% work is done. Thus, Vicky and Tinku must be doing 30% work together over two days. Hence, total time required = 2 days (all three) + 2 days (Vicky and Tinku) + 3 days (Tinku alone).

31. **(a)** Time taken by B $= 10 \times \frac{100}{125} = 8$ days

Required answer $= \frac{8 \times 10}{18} = 4\frac{4}{9}$ days

Alternatively :

Reqd. days $= 10 \div (1 + 1.25) = \frac{10}{2.25} = \frac{40}{9} = 4\frac{4}{9}$

32. (d) $(A + B)$ 1 day's work $= \frac{1}{20}$ ...(1)

$(B + C)$ 1 day's work $= \frac{1}{30}$ ...(2)

$(C + A)$ 1 day's work $= \frac{1}{40}$ ...(3)

Adding eqs. (1), (2) and (3)

$2(A + B + C) = \frac{1}{20} + \frac{1}{30} + \frac{1}{40}$

$2(A + B + C) = \frac{6 + 4 + 3}{120}$

$\Rightarrow (A + B + C)$ 1 day work together $= \frac{13}{240}$

A' Alone 1 day's work = $(A + B + C)$ 1 day's work – $(B + C)$ 1 day's work

$A = \frac{13}{240} - \frac{1}{30} \Rightarrow \frac{13 - 8}{240} = \frac{5}{240}$

Number of days taken by $A = \frac{240}{5}$ days

C' Alone 1 day's work = $(A + B + C)$ 1 day's work – $(A + B)$ 1' day's work

$\Rightarrow \frac{13}{240} - \frac{1}{20} \Rightarrow \frac{13 - 12}{240} = \frac{1}{240}$

Number of days taken by $C = \frac{240}{1}$ days

Required Ratio $\frac{240}{5} : \frac{240}{1} \Rightarrow 1 : 5$

33. (a) 1M = 2W

(8M + 4W) × (6 days – 2 days) = (4M + 8W) × x days

$[M_1D_1 = M_2D_2]$

⇒ (8 × 2W + 4W) × (6 – 2) days = (4 × 2W + 8W) × x days

⇒ (16 + 4)W × 4 days = 16W × x days

$\Rightarrow x = \frac{20 \times 4}{16} = 5$ days

34. **(d)** $(x + y)$'s 6 days' work $= \left(\frac{1}{30} \times 6\right) = \frac{1}{5}$.

Remaining work $= \left(1 - \frac{1}{5}\right) = \frac{4}{5}$

Now, $\frac{4}{5}$ work is done by y in 32 days.

Whole work will be done by y in $\left(32 \times \frac{5}{4}\right) = 40$ days.

35. **(c)** (A + B)'s 1 day's work $= \frac{1}{10}$; C's 1day's work $= \frac{1}{50}$

(A + B + C)'s 1 day's work $= \left(\frac{1}{10} + \frac{1}{50}\right) = \frac{6}{10} = \frac{3}{25}$ ...(1)

Also, A's 1 day's work = (B + C)'s 1 day's work ...(2)

From (1) and (2), we get : 2 × (A's 1 day's work) $= \frac{3}{25}$

$\Rightarrow$ A's 1 day's work $= \frac{3}{50}$

$\therefore$ B's 1 day's work $= \left(\frac{1}{10} - \frac{3}{50}\right) = \frac{2}{50} = \frac{1}{25}$

So, B alone could do the work in 25 days.

36. **(c)** A's 1 day's work $= \frac{1}{10}$ and B's 1 day's work $= \frac{1}{15}$

$\therefore$ (A + B)'s 1 day's work $= \left(\frac{1}{10} + \frac{1}{15}\right) = \frac{1}{6}$

So both together will finish the work in 6 days.

37. **(b)** Let the required number of days be x.
Then, more men, more km (Direct proportion)
more days, more km (Direct proportion)
men

$$\left.\begin{array}{l}\text{Men } 12 : 28 \\ \text{Days } 7 : x\end{array}\right\} :: 1.5 : 12$$

$\therefore$ $12 \times 7 \times 12 = 28 \times x \times 1.5$

$x = \frac{12 \times 7 \times 12}{28 \times 1.5} = 24$

38. **(c)** (A + B + C)'s 1 day's work $= \left(\frac{1}{40}\right)^{th}$ part of whole work

(A + B + C)'s 16 days work $= \frac{16}{40} = \frac{2}{5}$ of whole work

(B + C) completes remaining work in 40 days (B + C) completes $\left(\frac{3}{5}\right)^{th}$ part of work in 40 days.

(B + C) completes whole work in $\frac{40 \times 5}{3} = \frac{200}{3}$ days.

$\frac{1}{A} + \frac{1}{B} + \frac{1}{C} = \frac{1}{40}$

$\frac{1}{A} + \frac{3}{200} = \frac{1}{40}$

$\frac{1}{A} = \frac{1}{40} - \frac{1}{200} = \frac{4}{200}$

$\frac{1}{A} = \frac{1}{100}$

A alone can complete whole work in 100 days.

39. **(c)** $\frac{1}{A} + \frac{1}{B} + \frac{1}{C} = \frac{1}{6}$

(A + B + C) can do $\frac{2}{6} = \frac{1}{3}$ part of work in 2 days.

Remaining work $= 1 - \frac{1}{3} = \frac{2}{3}$

In one hour (A + B) can do $\frac{2}{3 \times 7}$ part of work

$\frac{1}{C} = \frac{1}{6} - \left(\frac{1}{B} + \frac{1}{C}\right)$

$\frac{1}{C} = \frac{1}{6} - \frac{2}{21} = \frac{3}{42}$

C = 14 hours

40. **(c)** Let Pratibha can finish the work in x days then, Sonia can finish the same work in 3x days
According to question
$3x - x = 60$
$2x = 60 \Rightarrow x = 30$
Pratibha and Sonia can individually complete the work in 30 days and 90 days respectively.

41. **(c)** Let 10 men left the work after 10 days.
Then, $M \times D = M \times x + (M - 10)(70 - x)$
$40 \times 60 = 40x + 30(70 - x)$
$\Rightarrow 2400 = 2100 + 10x \Rightarrow 10x = 300 \Rightarrow x = 10$ days

CHAPTER

10

TIME, SPEED AND DISTANCE

MOTION OR MOVEMENT

The relation between speed (S), distance (D) and time (T) is given below :

Distance = Speed × Time

or, Speed × Time = Distance *i.e.* S × T = D

In the above relation, the unit used for measuring the distance (D) covered during the motion and the unit of time (T) i.e. duration to cover the distance (D) will be the same as in numerator and denominator respectively of the unit used for the speed.

CONVERSION OF KMPH (KILOMETER PER HOUR) TO M/S (METRE PER SECOND) AND VICE-VERSA

$$1 \text{ kmph or } 1 \text{ km/h} = \frac{1 \text{ km}}{1 \text{ hr}} = \frac{1000 \text{ m}}{60 \times 60 \text{ sec}} = \frac{5 \text{ m}}{18 \text{ sec}} = \frac{5}{18} \text{ m/s}$$

$$\Rightarrow \quad x \text{ kmph} = \frac{5x}{18} \text{ m/s} \quad \text{and vice-versa } x$$

$$\text{m/s} = \frac{18x}{5} \text{ kmph} \quad \text{or} \quad \frac{18x}{5} \text{ km/h}$$

i.e. to convert km/hr to m/sec, multiply by $\frac{5}{18}$ and to convert m/sec to km/hr multiply by $\frac{18}{5}$.

Illustration 1: Convert 90 km/h into m/s.

Solution: 90 km/h = $90 \times \frac{5}{18} = 25$ m/s.

Illustration 2: The driver of a Maruti car driving at the speed of 68 km/h locates a bus 40 metres ahead of him. After 10 seconds, the bus is 60 metres behind. The speed of the bus is

(a) 30 km/h **(b) 32 km/h**

(c) 25 km/h **(d) 38 km/h**

Solution: (b) Let speed of Bus = S_B km/h.

Now, in 10 sec., car covers the relative distance

= (60 + 40) m = 100 m

$\therefore$ Relative speed of car = $\frac{100}{10} = 10$ m/s

$= 10 \times \frac{18}{5} = 36$ km/h

$\therefore \quad 68 - S_B = 36$

$\Rightarrow \quad S = 32$ km/h

AVERAGE SPEED

Average speed is defined as the ratio of total distance covered to the total time taken by an object *i.e.*

$$\text{Average speed} = \frac{\text{Total distance travelled}}{\text{Total time taken}}$$

If an object travels $d_1, d_2, d_3, ..., d_n$ distances with different speeds $s_1, s_2, s_3, ..., s_n$ in time $t_1, t_2, t_3, ..., t_n$ respectively; then average speed (S_a) is given by

$$\boxed{S_a = \frac{d_1 + d_2 + d_3 + ... + d_n}{t_1 + t_2 + t_3 + ... + t_n}} \quad ... (1)$$

Since, Distance = Speed × Time

$$\therefore \ d_1 = s_1 t_1, \ d_2 = s_2 t_2, \ d_3 = s_3 t_3, ..., d_n = s_n t_n$$

Hence from (1),

$$\boxed{S_a = \frac{s_1 t_1 + s_2 t_2 + s_3 t_3 + ... + s_n t_n}{t_1 + t_2 + t_3 + ... + t_n}}$$

Since $\quad \text{Time} = \frac{\text{Distance}}{\text{Speed}}$

$$\therefore \quad t_1 = \frac{d_1}{s_1}, \ t_2 = \frac{d_2}{s_2}, \ t_3 = \frac{d_3}{s_3}, ..., t_n = \frac{d_n}{s_n}$$

Hence from (1),

$$\boxed{S_a = \frac{d_1 + d_2 + d_3 + ... + d_n}{\frac{d_1}{s_1} + \frac{d_2}{s_2} + \frac{d_3}{s_3} + ... + \frac{d_n}{s_n}}}$$

Special Cases

In chapter of Averages, we studied that

(i) If with two different speeds s_1 and s_2 the same distance d is covered, then

$$\text{Average Speed} = \frac{2 s_1 \cdot s_2}{s_1 + s_2}$$

(ii) If with three different speeds s_1, s_2 and s_3 the same distance d is covered, then

$$\text{Average Speed} = \frac{3s_1 \cdot s_2 \cdot s_3}{s_1 \cdot s_2 + s_2 \cdot s_3 + s_3 \cdot s_1}.$$

Illustration 3: A car moves 300 km at a speed of 45 km/h and then it increases its speed to 60 km/h to travel another 500 km. Find average speed of car.

Solution:

$$\text{Average speed} = \frac{d_1 + d_2}{\frac{d_1}{s_1} + \frac{d_2}{s_2}} = \frac{300 + 500}{\frac{300}{45} + \frac{500}{60}} = \frac{800}{\frac{45}{3}} = \frac{160}{3}$$

$$= 53\frac{1}{3} \text{ km/h}$$

Illustration 4: ***A*** **covers 1/3rd of the journey at the speed of 10 km/h and half of the rest at the speed of 20 km/h and rest at the speed of 30 km/h. What is the average speed of** ***A*** **?**

Solution:

Distance covered at 10 km/h = 1/3rd of the whole journey

Distance covered at 20 km/h = $\left(\left(1 - \frac{1}{3}\right) \times \frac{1}{2} = \frac{1}{3}\right)$ rd of the whole journey

Distance covered at 30 km/h = $\left(1 - \frac{1}{3} - \frac{1}{3} = \frac{1}{3}\right)$ rd of the whole journey

Since the distances covered with each of the three given speeds are the same, therefore

$$\text{Average speed} = \frac{3s_1 \cdot s_2 \cdot s_3}{s_1 \cdot s_2 + s_2 \cdot s_3 + s_3 \cdot s_1}$$

$$= \frac{3 \times 10 \times 20 \times 30}{10 \times 20 + 20 \times 30 + 30 \times 10}$$

$$= 16\frac{4}{11} \text{ km/h.}$$

Illustration 5: A man makes his upward journey at 16 km/h and downward journey at 28 km/h. What is his average speed ?

(a) 32 km/h **(b) 56 km/h**

(c) 20.36 km/h **(d) 22 km/h**

Solution: (c) Let the distance travelled during both upward and down-ward journey be x km.

$$\text{Average speed} = \frac{\text{Total distance covered}}{\text{Total time taken}}$$

$$= \frac{x + x}{\frac{x}{16} + \frac{x}{28}} = \frac{2}{\frac{28+16}{28 \times 16}}$$

$$= \frac{2 \times 28 \times 16}{44} = 20.36 \text{ km/h}$$

RELATIVE SPEED

Generally, when we talk about the speed of a body, we mean the speed of the body with respect to a stationary point (or object), which we have already discussed. In many cases, we need to determine the speed of a body with respect to an independent moving point (or body). In such cases, we have to take into account the speed of the independent body with respect to which we want to find the speed of another body.

The speed of a body 'A' with respect to an independent moving body 'B' is called relative speed of the body A with respect to the body 'B'.

Formulae of Relative Speed

(i) If two bodies are moving in opposite directions at speeds s_1 and s_2 respectively, then relative speed of any one body with respect to other body is $(s_1 + s_2)$.

(ii) If two bodies are moving in the same direction at speeds s_1 and s_2 respectively, then relative speed of any one body with respect to other body is given by

$s_1 - s_2$, when s_1 is greater than s_2

and $s_2 - s_1$ when s_2 is greater than s_1.

Illustration 6: A car ***X*** **starts from Delhi and another car** ***Y*** **starts from Moradabad at the same time to meet each other. Speed of car** ***X*** **is 40 km/h while speed of car** ***Y*** **is 50 km/h. If the distance between Delhi and Moradabad is 210 kms, when will they meet ?**

Solution: Effective speed = Relative speed = 40 + 50

= 90 km/h

$$\text{Time taken} = \frac{210}{90} = 2\frac{1}{3} \text{ hrs.}$$

TO AND FRO MOTION IN A STRAIGHT LINE BETWEEN TWO POINTS *A* AND *B*

Two and fro motion in a straight line between two points A and B means motion of one or more bodies between two fixed points A and B such that when any body reached at any end point A or B, they start moving towards the opposite end point.

1. When two bodies start moving towards each other from two points *A* and *B*

(a) If distance between A and B is D, then the two bodies together have to cover D unit of distance for the first meeting.

D

A B

(b) For the next number of meeting (*i.e.* second, third, fourth meeting and so on) both A and B together have to cover $2D$ distance more from the previous meeting.

D

A B

Hence to meet the fifth time they have to cover together $D + (4 \times 2D) = 9D$ unit of distances. Similarly for the ninth meeting they have to cover together $D + (8 \times 2D) = 17D$ units of distance. Thus, for the nth meeting they have to cover together $D + (n - 1) \times 2D$ *i.e.* $(2n - 1)D$ units of distance.

(c) At any point of time ratio of the distances covered by the two bodies will be equal to the ratio of their speeds.

2. When two bodies start moving towards the same direction from the point A

(a) Since the faster body reaches the next end (or opposite end) first than the slower body and the faster body starts returning before the slower body reaches the same opposite end and hence the two bodies meet somewhere between the two ends. For the first meeting after they start to move they have to cover $2D$ distance, where D is the distance between two particular end points (*i.e.* A and B)

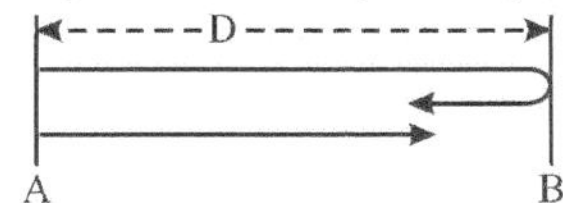

(b) For every subsequent meeting they have to cover together $2D$ unit distance more from the previous meeting.

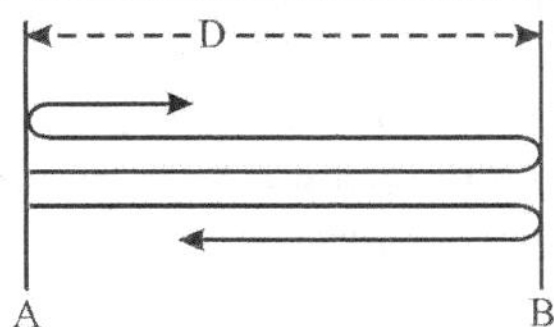

Thus, for the nth meeting they have to cover together $(n \times 2D)$ units of distance.

(c) At any point of time ratio of the distances covered by the two bodies will be equal to the ratio of their speeds.

Illustration 7: Two runners Shiva and Abhishek start running to and fro between opposite ends A and B of a straight road towards each other from A and B respectively. They meet first time at a point $0.75D$ from A, where D is the distance between A and B. Find the point of their 6th meeting.

Solution: At the time when Shiva and Abhishek meet first time,

Ratio of their speeds = Ratio of distance covered by them

$= 0.75 : 0.25$

$= 3 : 1$

Total distance covered by Shiva and Abhishek together till they meet at 6th time $= D + 5 \times 2D = 11D$

Total distance covered by Shiva till he meets Abhishek 6th time $= \frac{3}{3+1} \times 11D = 8.25D$

After covering a distance of $8.25D$, Shiva will be at a point at a distance of $0.25D$ from A or $0.75D$ from B.

CONCEPT RELATED TO MOTION OF TRAINS

The following things need to be kept in mind before solving questions on trains.

(i) For the train is crossing a moving object, the speed of the train has to be taken as the relative speed with respect to the object.

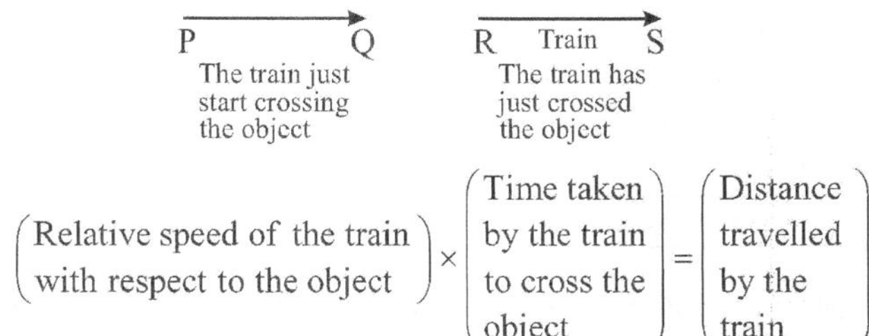

$$\begin{pmatrix}\text{Relative speed of the train} \\ \text{with respect to the object}\end{pmatrix} \times \begin{pmatrix}\text{Time taken} \\ \text{by the train} \\ \text{to cross the} \\ \text{object}\end{pmatrix} = \begin{pmatrix}\text{Distance} \\ \text{travelled} \\ \text{by the} \\ \text{train}\end{pmatrix}$$

(ii) For object moving in opposite direction of the train,

$$\begin{pmatrix}\text{Relative speed of the train} \\ \text{with respect to the object}\end{pmatrix} = \begin{pmatrix}\text{Speed of} \\ \text{the train}\end{pmatrix} + \begin{pmatrix}\text{Speed of} \\ \text{the object}\end{pmatrix}$$

(iii) For object moving in the same direction of the train,

$$\text{(a)} \begin{pmatrix}\text{Relative speed of the train} \\ \text{with respect to the object}\end{pmatrix} = \begin{pmatrix}\text{Speed of} \\ \text{the train}\end{pmatrix} - \begin{pmatrix}\text{Speed of} \\ \text{the object}\end{pmatrix}$$

(b) (Distance travelled by the train when crossing the object)

= Distance travelled by the engine from Q to S

$= QR + RS$

$= AB + RS$

= Length of the object + Length of the train

In the case of a train crossing a man, tree or a pole, the length of the man, tree or pole is actually its diameter (or width) which is generally considered as negligible *i.e.* a man, a tree, a pole or a point etc. has no length.

S. No.	Situations	Basic Formulae	Expended Form of Basic Formulae	Expended Formulae in Symbolic Form
1.	When a train crossing a moving object with length in opposite direction	Relative Speed × Time = Distance	$\left[\begin{pmatrix}\text{Speed} \\ \text{of the} \\ \text{train}\end{pmatrix} + \begin{pmatrix}\text{Speed} \\ \text{of the} \\ \text{object}\end{pmatrix}\right] \times \begin{pmatrix}\text{Time taken by} \\ \text{the train to cross} \\ \text{the moving object}\end{pmatrix} = \begin{pmatrix}\text{Length} \\ \text{of the} \\ \text{train}\end{pmatrix} + \begin{pmatrix}\text{Length} \\ \text{of the} \\ \text{object}\end{pmatrix}$	$(S_T + S_0) \times t = (L_T + L_0)$
2.	When a train crossing a moving object with length in the same direction	Relative Speed × Time = Distance	$\left[\begin{pmatrix}\text{Speed} \\ \text{of the} \\ \text{train}\end{pmatrix} - \begin{pmatrix}\text{Speed} \\ \text{of the} \\ \text{object}\end{pmatrix}\right] \times \begin{pmatrix}\text{Time taken by} \\ \text{the train to cross} \\ \text{the moving object}\end{pmatrix} = \begin{pmatrix}\text{Length} \\ \text{of the} \\ \text{train}\end{pmatrix} + \begin{pmatrix}\text{Length} \\ \text{of the} \\ \text{object}\end{pmatrix}$	$(S_T - S_0) \times t = (L_T + L_0)$
3.	When a train crossing a moving object without length like a man, a tree, a pole, a point etc. in opposite direction	Relative Speed × Time = Distance	$\left[\begin{pmatrix}\text{Speed} \\ \text{of the} \\ \text{train}\end{pmatrix} + \begin{pmatrix}\text{Speed} \\ \text{of the} \\ \text{object}\end{pmatrix}\right] \times \begin{pmatrix}\text{Time taken by} \\ \text{the train to cross} \\ \text{the moving object}\end{pmatrix} = \begin{pmatrix}\text{Length} \\ \text{of the} \\ \text{train}\end{pmatrix}$	$(S_T + S_0) \times t = L_T$

S. No.	Situations	Basic Formulae	Expended Form of Basic Formulae	Expended Formulae in Symbolic Form
4.	When a train crossing a moving object without length in the same direction	Relative Speed × Time = Distance	[(Speed of the train) − (Speed of the object)] × (Time taken by the train to cross the moving object) = (Length of the train)	$(S_t - S_0) \times t = L_T$
5.	When a train crossing a stationary object with length	Speed × Time = Distance	(Speed of the train) × (Time taken to cross the stationary object) = [(Length of the train) + (Length of the object)]	$S_T \times t = L_T + L_0$
6.	When a train crossing a stationary object without length	Speed × Time = Distance	(Speed of the train) × (Time taken to cross the stationary object) = (Length of the train)	$S_T \times t = L_T$
S_T = Speed of the train, S_0 = Speed of the object, L_T = Length of the train, L_0 = Length of the object, t = time taken by the train to cross the object				

Illustration 8: A train passes an electric pole in 10 seconds and a platform 120 m long in 18 seconds. Find the length of the train.

Solution: Let the length of the train be = x m

$$\text{Speed} = \frac{x}{10} = \frac{120 + x}{18} \Rightarrow x = 150$$

Hence length of the train = 150 m.

Illustration 9: A train of length 100 m takes 1/6 hour to pass over another train 150 m long coming from the opposite direction. If the speed of first train is 60 km/h, then find speed of the second train.

Solution: Let speed of the second train be x km/h.

$$\text{Relative Speed} = \text{Sum of speed of two trains}$$

$$= (60 + x) \text{ km/h} = (60 + x)\frac{5}{18} \text{ m/s}$$

$$\text{Time} = \frac{\text{Sum of length of two trains}}{\text{Relative Speed}}$$

$$\Rightarrow \quad 10 = \frac{250 \times 18}{(60 + x) \times 5} \Rightarrow x = 30 \text{ km/h.}$$

Illustration 10: Two trains 137 metres and 163 metres in length are running towards each other on parallel lines, one at the rate of 42 kmph and another at 48 kmph. In what time will they be clear of each other from the moment they meet?

(a) 10 sec **(b)** 12 sec
(c) 14 sec **(d)** cannot be determined

Solution: (b) Relative speed of the trains

$$= (42 + 48) \text{ kmph} = 90 \text{ kmph}$$

$$= \left(90 \times \frac{5}{18}\right) \text{ m/sec} = 25 \text{ m/sec.}$$

Time taken by the trains to pass each other

= Time taken to cover (137 + 163) m at 25 m/sec

$$= \left(\frac{300}{25}\right) \text{ sec} = 12 \text{ seconds.}$$

Illustration 11: A train 110 m in length travels at 60 km/h. How much time does the train take in passing a man walking at 6 km/h against the train?

(a) 6 s **(b)** 12 s
(c) 10 s **(d)** 18 s

Solution: (a) Relative speeds of the train and the man

$$= (60 + 6) = 66 \text{ km/h} = \frac{66 \times 5}{18} \text{ m/s}$$

Distance = 110 m

Therefore, time taken in passing the men

$$= \frac{110 \times 18}{66 \times 5} = 6\text{s}$$

BOATS AND STREAMS

In still water, a boat moves with its own speed which is called speed of the boat in still water (S_B).

When the boat is moving against the flow of the water or with the flow of the water, the speed of movement of the boat depends on the speed of flow of water [*i.e.* speed of stream (S_S)].

Speed of the boat moving against the flow of water (*i.e.* moving in upstream)

= Speed of boat in still water – Speed of stream

$= S_B - S_S$

Speed of the boat moving with flow of water (*i.e.* moving in downstream)

= Speed of boat in still water + Speed of stream

$= S_B + S_S$

The basic formula used for solving the problems of boats and streams is

Speed × Time = Distance

Illustration 12: A man can row a boat in downstream at 12 km/h and in upstream at 8 km/h. Find the speed of the boat that the man can row in still water.

Solution:

$$S_B + S_S = 12 \quad \text{... (1)}$$

$$S_B - S_S = 8 \quad \text{... (2)}$$

On adding (1) and (2), we get

$$2S_B = 20 \Rightarrow S_B = 10$$

Hence speed of boat in still water = 10 km/h.

Illustration 13: A boat covers 48 km in upstream and 72 km in downstream in 12 hours, while it covers 72 km in upstream and 48 km in downstream in 13 hours. Find the speed of the stream.

Solution: $\frac{48}{S_B - S_S} + \frac{72}{S_B + S_S} = 12$

$$\frac{72}{S_B - S_S} + \frac{48}{S_B + S_S} = 13$$

Let $\frac{1}{S_B - S_S} = x$ and $\frac{1}{S_B + S_S} = y$

Then $48x + 72y = 12$ ... (1)

and $72x + 48y = 13$ ... (2)

On adding (1) and (2),

$$120x + 120y = 25 \Rightarrow x + y = \frac{5}{24} \quad \text{... (3)}$$

On subtracting (2) from (1),

$$24y - 24x = -1 \Rightarrow x - y = \frac{1}{24} \quad \text{... (4)}$$

On adding (3) and (4),

$$2x = \frac{1}{4} \Rightarrow x = \frac{1}{8} \Rightarrow S_B - S_S = 8 \quad \text{... (5)}$$

On subtracting (4) from (3),

$$2y = \frac{1}{6} \Rightarrow y = \frac{1}{12} \Rightarrow S_B + S_S = 12 \quad \text{... (6)}$$

Subtracting (5) from (6),

$$2S_S = 4 \Rightarrow S_S = 2$$

Hence speed of stream = 2 km /h.

Illustration 14: A motor boat takes 12 hours to go downstream and it takes 24 hours to return the same distance. Find the ratio of the speed of boat in still water to the speed of stream.

Solution: Distance = Speed × Time

Distance travelled in downstream

= Distance travelled in upstream

$$(S_B + S_S) \times 12 = (S_B - S_S) \times 24$$

$$\Rightarrow S_B + S_S = 2S_B - 2S_S$$

$$\Rightarrow 3S_S = S_B \Rightarrow \frac{S_B}{S_S} = \frac{3}{1} \Rightarrow S_B : S_S = 3 : 1$$

Hence required ratio = 3 : 1.

CIRCULAR MOTION

When two bodies start moving from a place on a circular track simultaneously in the same direction, the faster body keeps increasing the distance by which the slower body is behind the faster body. When the distance by which the faster body is in front of the slower body becomes equal to the circumference of the track, the faster body meets the slower body first time *i.e.* faster body comes in line with the slower body.

(i) When two bodies are moving in the opposite directions, their relative speed is equal to the sum of their individual speeds.

(ii) When two bodies are moving in the same direction, their relative speed is equal to the difference of the speeds of the two bodies.

First Meeting

(i) Let A and B are two runners.

Time taken by A and B to meet for the first time

$$= \frac{\text{Circumference of the circular Track}}{\text{Relative speed}}$$

(ii) When there are more than two runners, suppose A is the fastest runner and A meets B first time in time t_{AB}, A meets C first time in time t_{AC}, A meets D first time in time t_{AD} and so on. Then time taken by all of them to meet for the first time is the LCM of t_{AB}, t_{AC}, t_{AD}, etc.

First Meeting at the Starting Point

Let A take, t_A time, B takes t_B time, C takes t_C times and so on, to complete one round, then the time taken to meet all the runners for the first time at the starting point

= LCM of t_A, t_B, t_C etc.

Illustration 15: The jogging track in a sports complex is 726 metres in circumference. Pradeep and his wife start from the same point and walk in opposite directions at 4.5 km/h and 3.75 km/h, respectively. They will meet for the first time in

(a) 5.5 min **(b) 6.0 min**

(c) 5.28 min **(d) 4.9 min**

Solution: (c) Let the husband and the wife meet after x minutes 4500 metres are covered by Pradeep in 60 minutes.

In x minutes, he will cover $\frac{4500}{60}x$ metres.

Similarly,

In x minutes, his wife will cover $\frac{3750}{60}x$ m.

Now, $\frac{4500}{60}x + \frac{3750}{60}x = 726$

$$\Rightarrow x = \frac{726 \times 60}{8250} = 5.28 \text{min}$$

Illustration 16: *A* and *B* run on a circular track of circumference 800 m in the opposite direction. Speeds of *A* and *B* are 50 m/s and 30 m/s respectively. Initially *A* and *B* are diametrically opposite to each other.

(i) When will they meet for the first time ?

(ii) What is the ratio of distances covered by each one to meet for the first time ?

Solution:

(i) Relative speed of A with respect to $B = 50 + 30 = 80$ m/s

Initially A and B are diametrically opposite to each other means B is 400 m ahead of A in the race.

Time taken by A to meet B first time $= \frac{400}{80} = 5$ s

(ii) To meet second time A and B have to cover 800 m

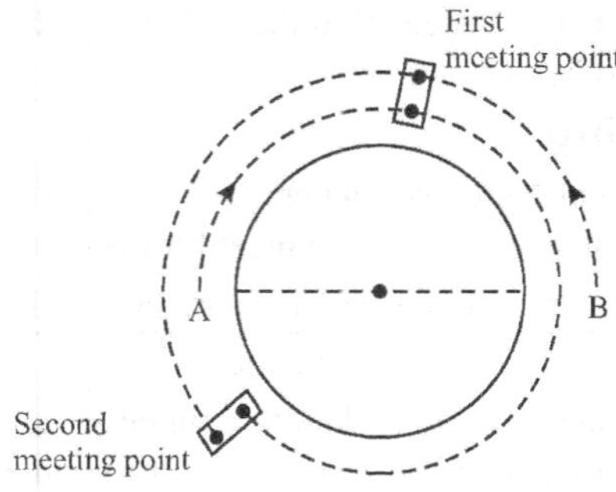

Hence time taken to meet second time = $\frac{800}{80}$ = 10 seconds

CLOCKS

Problems on clocks are based on the movement of the minute hand and hour hand. We consider the dial of a clock as a circular track having a circumference of 60 km. minute hand and hour hand are two runners running with the speed of 60 km/h and 5 km/hr respectively in the same direction. Hence relative speed of minute hand with respect to hour hand is 55 km/h. This means that for every hour elapsed, the minute hand goes 55 km more than the hour hand.

Degree Concept of a Clock

Total angle subtended at the centre of a clock = 360°

Angle made by hour hand at the centre = 30° per hour

= 0.5° per minute

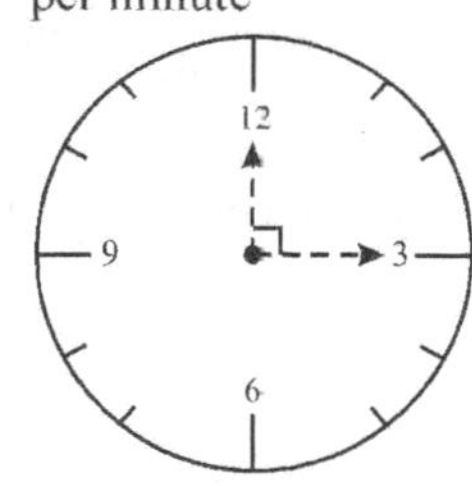

Angle made by minute hand at the centre = 360° per hour

= 6° per minute

Number of Right Angles and Straight Angles Formed by Minute Hand and Hour Hand

A right angle is formed by hour hand and minute hand when distance between tip of hour hand and tip of minute hand is 15 km. A straight line is formed by hour hand and minute hand when distance between their tips is 30 km.

A clock makes two right angles in every hour. Thus there are 2 right angles between marked 1 to 2, 2 to 3, 3 to 4 and so on the dial.

Two straight lines are formed by hour hand and minute hand in every hour.

Thus two straight lines are formed by hour hand and minute hand between marked 1 to 2, 2 to 3, 3 to 4 and so on.

(iii) Hour hand and minute hand of a clock are together after every $65\frac{5}{11}$ minutes. So, if hour hand and minute hand of a clock are meeting in less than $65\frac{5}{11}$ minutes, then the clock is running fast and if hour hand and minute hand are meeting in more than $65\frac{5}{11}$ minutes,then clock is running slow.

Illustration 17: Between 5 O' clock and 6 O' clock, when hour hand and minute hand of a clock overlap each other ?

Solution: At 5 O' clock, distance between tips of two hands = 25 km

Relative speed = 55 km/h

Required time to overlap the two hands

$= \frac{25 \text{ km}}{55 \text{ km/h}} = \frac{5}{11}\text{h}$

$= \frac{5 \times 60}{11}\text{min}$

$= 27 \text{ min} + \frac{3 \times 60}{11} \text{ sec}$

= 27 min + 16 sec.

= 27 minutes 16 seconds.

Illustration 18: Mrs. Veena Gupta goes for marketing between 5 P.M. and 6 P.M. When she comes back, she finds that the hour hand and the minute hand have interchanged their positions. For how much time was she out of her house ?

Solution: Since two hands are interchange their positions, so sum of the angles subtended at the centre by hour hand and minute hand = 360°

Let us suppose that she was out of house for 't' minutes.

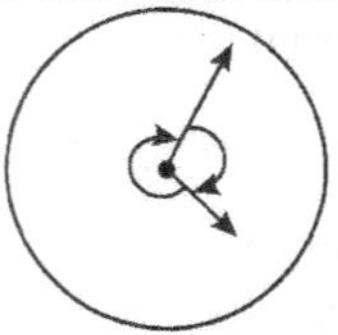

So, the sum of the angles subtended at the centre by the hour hand and minute hand = $(0.5 \times t)° + (6t)°$

$\because \quad 0.5t + 6t = 360$

$\Rightarrow \quad 6.5t = 360 \Rightarrow t = 55.4$ (app.)

Hence required time = 55.4 minutes.

CALENDAR

INTRODUCTION

The solar year consists of 365 days, 5 hrs 48 minutes, 48 seconds. In 47 BC, Julius Ceasar arranged a calendar known as the Julian calendar in which a year was taken as $365\frac{1}{4}$ days and in order to get rid of the odd quarter of a day, an extra day was added once in every fourth year and this was called as leap year or Bissextile. Nowadays, the calendar, which is mostly used, is arranged by Pope Gregory XII and known as Gregorian calendar.

In India, number of calendars were being used till recently. In 1952, the Government adopted the National Calendar based on Saka era with Chaitra as its first month. In an ordinary year, Chaitra 1 falling on March 22 of Gregorian Calendar and in a leap year it falls on March 21.

Remember

★ In an ordinary year,
1 year = 365 days = 52 weeks + 1 day

★ In a leap year,
1 year = 366 days = 52 weeks + 2 days

NOTE : *First January 1 A.D. was Monday. So we must count days from Sunday.*

★ 100 years or one century contains 76 ordinary years and 24 leap years.

$\Rightarrow$ [76 × 52 weeks + 76 odd days] + [24 × 52 weeks + 24 × 2 odd days]

= (76 + 24) × 52 weeks + (76 + 48) odd days

= 100 × 52 weeks + 124 odd days

= 100 × 52 weeks + (17 × 7 + 5) odd days

= (100 × 52 + 17) weeks + 5 odd days

∴ 100 years contain 5 odd days.

Similarly, 200 years contain 3 odd days,
300 years contain 1 odd days,
400 years contain 0 odd days.

Year whose non-zero numbers are multiple of 4 contains no odd days; like 800, 1200, 1600 etc.

The number of odd days in months

The month with 31 days contains (4 × 7 + 3) ie. 3 odd days and the month with 30 days contains (4 × 7 + 2) ie. 2 odd days.

NOTE : *February in an ordinary year gives no odd days, but in a leap year gives one odd day.*

Illustration 19: What day of the week was 15th August 1949?

Sol. 15th August 1949 means
1948 complete years + first 7 months of the year 1949 + 15 days of August.

1600 years give no odd days.
300 years give 1 odd day.
48 years give {48 + 12} = 60 = 4 odd days.
[∵ For ordinary years → 48 odd days and for leap year 1 more day (48 ÷ 4) = 12 odd days; 60 = 7 × 8 + 4]
From 1st January to 15th August 1949
Odd days :
January – 3
Feburary – 0
March – 3
April – 2
May – 3
June – 2
July – 3
August – 1
17 ⇒ 3 odd days.
∴ 15th August 1949 → 1 + 4 + 3 = 8 = 1 odd day.

This means that 15th Aug. fell on 1st day. Therefore, the required day was Monday.

Illustration 20: How many times does the 29th day of the month occur in 400 consecutive years?

Sol. In 400 consecutive years, there are 97 leap years. Hence, in 400 consecutive years, February has the 29th day 97 times and the remaining eleven months have the 29th day 400 × 1100 = 4400 times
∴ The 29th day of the month occurs (4400 + 97) or 4497 times.

Illustration 21: Today is 5th February. The day of the week is Tuesday. This is a leap year. What will be the day of the week on this date after 5 years?

Sol. This is a leap year. So, next 3 years will give one odd day each. Then leap year gives 2 odd days and then again next year give 1 odd day.
Therefore (3 + 2 + 1) = 6 odd days will be there.
Hence the day of the week will be 6 odd days beyond Tuesday, i.e., it will be Monday.

Illustration 22: What day of the week was 20th June 1837 ?

Sol. 20th June 1837 means 1836 complete years + first 5 months of the year 1837 + 20 days of June.
1600 years give no odd days.
200 years give 3 odd days.
36 years give (36 + 9) or 3 odd days.
1836 years give 6 odd days.
From 1st January to 20th June there are 3 odd days.
Odd days :

January	: 3
February	: 0
March	: 3
April	: 2
May	: 3
June	: 6

	17

Therefore, the total number of odd days = (6 + 3) or 2 odd days.
This means that the 20th of June fell on the 2nd day commencing from Monday. Therefore, the required day was Tuesday.

Practice Exercise

Level - I

1. A certain distance is covered by a train with a certain speed. If half the distance is covered in double time, then the ratio of this speed to that of the original one is
   (a) 1 : 4 (b) 4 : 1
   (c) 1 : 2 (d) 2 : 1
2. A man makes his upward journey at 16 km/h and downward journey at 28 km/h. What is his average speed ?
   (a) 32 km/h (b) 56 km/h
   (c) 20.36 km/h (d) 22 km/h
3. Sound is said to travel in air at about 1100 feet per second. A man hears the axe striking the tree, $\frac{11}{5}$ seconds after he sees it strike the tree. How far is the man from the wood chopper?
   (a) 2197 ft (b) 2420 ft
   (c) 2500 ft (d) 2629 ft
4. A salesman travels a distance of 50 km in 2 hours and 30 minutes. How much faster, in kilometres per hour, on an average, must he travel to make such a trip in $\frac{5}{6}$ hour less time?
   (a) 10 (b) 20
   (c) 30 (d) None of these
5. Two persons *A* and *B* started from two different places towards each other. If the ratio of their speed be 3 : 5, then what is the ratio of distance covered by *A* and *B* respectively till the point of meeting?
   (a) 1 : 2 (b) 3 : 4
   (c) 3 : 5 (d) 5 : 3
6. If a man travels at 30 km/h, he reaches his destination late by 10 minutes but if he travels at 42 km/h then he reaches 10 minutes earlier. The distance travelled by him is
   (a) 30 km (b) 35 km
   (c) 45 km (d) 36 km
7. Two trains each of 120 m in length, run in opposite directions with a velocity of 40 m/s and 20 m/s respectively. How long will it take for the tail ends of the two trains to meet each other during the course of their journey?
   (a) 20 s (b) 3 s
   (c) 4 s (d) 5 s
8. Two trains starting at the same time from two stations, 200 km apart and going in opposite directions, cross each other at a distance of 110 km from one of them. What is the ratio of their speeds?
   (a) 11 : 20 (b) 9 : 20
   (c) 11 : 9 (d) 19 : 20
9. Two runner start running together for a certain distance, one at 8 km/h and another at 5 km/h. The former arrives one and half an hour, before the latter. The distance (in km) is:
   (a) 12 (b) 20
   (c) 25 (d) 36
10. A can complete a journey in 10 hours. He travels first half of the journey at the rate of 21 km/hr and second half at the rate of 24 km/hr. Find the total journey in km.
    (a) 220 km (b) 224 km
    (c) 230 km (d) 234 km
11. A train is moving at a speed of 132 km/h. If the length of the train is 110 metres, how long will it take to cross a railway platform, 165 metres long ?
    (a) 5 s (b) 7.5 s
    (c) 10 s (d) 15 s
12. A person travels equal distances with speeds of 3km/hr, 4 km/hr and 5km/hr and takes a total time of 47 minutes. The total distance (in km) is:
    (a) 2 (b) 3
    (c) 4 (d) 5
13. *A* and *B* travel the same distance at 9 km/h and 10 km/h respectively. If *A* takes 20 minutes longer than *B*, the distance travelled by each is:
    (a) 16 (b) 20
    (c) 30 (d) None of these
14. A passenger train takes two hours less for a journey of 300 km if its speed is increased by 5 km/h from its normal speed. The normal speed of the train is
    (a) 35 km/h (b) 50 km/h
    (c) 25 km/h (d) 30 km/h
15. A gun is fired at a distance of 3.32 km from Chauhan. He hears its sound 10 seconds later. Find the speed of the sound.
    (a) 301 m/s (b) 302 m/s
    (c) 332 m/s (d) 340 m/s
16. A walks around a circular field at the rate of one round per hour while B runs around it at the rate of six rounds per hour. They start in the same direction from the same point at 7.30 a.m. They shall first cross each other at:
    (a) 7.42 a.m. (b) 7.48 a.m.
    (c) 8.10 a.m. (d) 8.30 a.m.
17. A car driver travels from the plains to a hill station, which are 200 km apart at an average speed of 40 km/h. In the return trip he covers the same distance at an average speed of 20 km/h. The average speed of the car over the entire distance of 400 km is
    (a) 16.56 km/h (b) 17.89 km/h
    (c) 26.67 km/h (d) 35 km/h

18. Two trains of equal lengths are running on parallel tracks in the same direction at 46 km/h and 36 km/h, respectively. The faster train passes the slower train in 36 sec. The length of each train is

(a) 50 m (b) 80 m
(c) 72 m (d) 82 m

19. In a 800 m race around a stadium having the circumference of 200 m, the top runner meets the last runner on the 5th minute of the race. If the top runner runs at twice the speed of the last runner, what is the time taken by the top runner to finish the race ?

(a) 20 min (b) 15 min
(c) 10 min (d) 5 min

20. Excluding stoppages, the speed of a train is 45 km/h and including stoppages, it is 36 km/h. For how many minutes does the train stop per hour ?

(a) 10 min. (b) 12 min.
(c) 15 min. (d) 18 min.

21. The driving wheel of a locomotive engine, 2.1 m in radius, makes 75 revolutions in one minute. Find the speed of the train in km/h.

(a) 60 km/h (b) 59.4 km/h
(c) 61.5 km/h (d) None of these

22. A train covers 180 km distance in 4 hours. Another train covers the same distance in 1 hour less. What is the difference in the distances covered by these trains in one hour ?

(a) 45 km (b) 9 km
(c) 40 km (d) None of these

23. Speed of a speed-boat when moving in the direction parallel to the direction of the current is 16 km/hr. Speed of the current is 3 km/hr. So the speed of the boat against the current will be (in km/hr)

(a) 22 (b) 9.5
(c) 10 (d) None of these

24. A plane left 30 minutes later than the scheduled time and in order to reach the destination 1500 km away in time, it had to increase the speed by 250 km/h from the usual speed. Find its usual speed.

(a) 720 km/h (b) 740 km/h
(c) 730 km/h (d) 750 km/h

25. Two trains are 2 km apart and their lengths are 200 m and 300 m. They are approaching towards each other with a speed of 20 m/s and 30 m/s, respectively. After how much time will they cross each other?

(a) 50 s (b) 100 s
(c) 25/3 s (d) 150 s

26. A train 300 m long is running at a speed of 90 km/hr. How many seconds will it take to cross a 200 m long train running in the opposite direction at a speed of 60 km/hr?

(a) $7\frac{1}{5}$ (b) 60
(c) 12 (d) 20

27. A boat travels upstream from B to A and downsteam from A to B in 3 hours. If the speed of the boat in still water is 9 km/hr and the speed of the current is 3 km/hr, the distance between A and B is

(a) 4 km (b) 8 km
(c) 6 km (d) 12 km

28. A motor boat can travel at 10 km/h in still water. It traveled 91 km downstream in a river and then returned, taking altogether 20 hours. Find the rate of flow of the river.

(a) 6 km/hr (b) 5 km/hr
(c) 8 km/hr (d) 3 km/hr

29. Two men starting from the same place walk at the rate of 5 km/h and 5.5 km/h respectively. What time will they take to be 8.5 km apart, if they walk in the same direction?

(a) 16 h (b) 8 h 30 min
(c) 4h / 5min (d) 17 h

30. Speed of a boat in standing water is 9 km/h and the speed of the stream is 1.5 kmlh. A man rows to a place at a distance of 105 km and comes back to the starting point. The total time taken by him is

(a) 20 h (b) 18 h
(c) 16 h (d) 24 h

31. An aeroplane travels distances 2500 km, 1200km and 500km at the rate of 500 km/hr, 400 km/hr, and 250 km/hr, respectively. The average speed is

(a) 420 km/hr (b) 405 km/hr
(c) 410 km/hr (d) 575 km/hr

32. There are 20 poles with a constant distance between each pole. A car takes 24 seconds to reach the 12th pole . How much time will it take to reach the last pole?

(a) 25.25 s (b) 17.45 s
(c) 35.75 s (d) 41.45 s

33. A man walks half of the journey at 4 km/h by cycle does one third of journey at 12 km/h and rides the remainder journey in a horse cart at 9 km/h, thus completing the whole journey in 6 hours and 12 minutes. The length of the journey is

(a) 36 km (b) $\frac{1332}{67}$ km
(c) 40 km (d) 28 km

34. A train covers 180 km distance in 4 hours. Another train covers the same distance in 1 hour less. What is the difference in the distances covered by these trains in one hour ?

(a) 45 km (b) 9 km
(c) 40 km (d) None of these

35. The jogging track in a sports complex is 726 metres in circumference. Pradeep and his wife start from the same point and walk in opposite directions at 4.5 km/h and 3.75 km/h, respectively. They will meet for the first time in

(a) 5.5 min (b) 6.0 min
(c) 5.28 min (d) 4.9 min

36. A boat goes 24 km upstream and 28 km downstream in 6 hours. It goes 30km upstream and 21 km downstream in 6 hours and 30 minutes. The speed of the boat in still water is :

(a) 10 km/h (b) 4 km/h
(c) 14 km/h (d) 6km/h

37. Two trains for Mumbai leave Delhi at 6 a.m. and 6 : 45 am and travel at 100 kmph and 136 kmph respectively. How many kilometres from Delhi will the two trains be together
(a) 262.4 km (b) 260 km
(c) 283.33 km (d) 275 km

38. A 200 m-long train passes a 350 m long platform in 5 s. If a man is walking at a speed of 4 m/s along the track and the train is 100 m away from him, how much time will it take to reach the man?
(a) Less than 1 s (b) 1.04 s
(c) More than 2s (d) Data insufficient

39. A clock gains 15 minutes per day. It is set right at 12 noon. What time will it show at 4.00 am, the next day?
(a) 4 : 10 am (b) 4 : 45 am
(c) 4 : 20 am (d) 5 : 00 am

40. During a journey of 80 km a train covers first 60km with a speed of 40 km/h and completes the remaining distance with a speed of 20 km/h. What is the average speed of the train during the whole journey?
(a) 30 km/h (b) 32 km/h
(c) 36 km/h (d) 40 km/h

41. *A* travels from *B* to *C*, a distance of 250 miles, in 5.5 hours. He returns to *B* in 4 hours 40 minutes. His average speed is
(a) 44 (b) 46
(c) 48 (d) 50

42. A circular running path is 726 metres in circumference. Two men start from the same point and walk in opposite directions at 3.75 km/h and 4.5 km/h, respectively. When will they meet for the first time ?
(a) After 5.5 min (b) After 6.0 min
(c) After 5.28 min (d) After 4.9 min

43. *R* and *S* start walking each other at 10 AM at the speeds of 3 km/hr and 4 km/hr respectively. They were initially 17.5 km apart. At what time do they meet?
(a) 2 : 30 PM (b) 11 : 30 AM
(c) 1 : 30 PM (d) 12 : 30 PM

44. A person travels from *P* to *Q* at a speed of 40 kmph and returns by increasing his speed by 50%. What is his average speed for both the trips?
(a) 36 kmph (b) 45 kmph
(c) 48 kmph (d) 50 kmph

45. A car travels first half distance between two places with a speed of 40 km/h and the rest of the half distance with a speed of 60 km/h. The average speed of the car is
(a) 48 km/h (b) 37 km/h
(c) 44 km/h (d) None of these

46. Two cyclists start on a circular track from a given point but in opposite directions with speeds of 7 m/sec and 8 m/sec respectively. If the circumference of the circle is 300 metres, after what time will they meet at the starting point ?
(a) 100 sec (b) 20 sec
(c) 300 sec (d) 200 sec

47. If a train runs at 40 kmph, it reaches its destination late by 11 minutes but if it runs at 50 kmph, it is late by 5 minutes only. The correct time for the train to complete its journey is:
(a) 13 min. (b) 15 min.
(c) 19 min. (d) 21 min.

48. A man while returning from his factory, travels 2/3 of the distance by bus and $\frac{3}{4}$ of the rest by car, and the remaining by foot. If he travels 2 km on foot, find the distance covered by him.
(a) 24 km (b) 22 km
(c) 28 km (d) 26 km

49. A car driver, driving in a fog, passes a pedestrian who was walking at the rate of 2 km/hr in the same direction. The pedestrian could see the car for 6 minutes and it was visible to him up to a distance of 0.6 km. What was the speed of the car?
(a) 15 km/hr (b) 30 km/hr
(c) 20 km/hr (d) 8 km/hr

50. A plane left 30 min later than its scheduled time to reach its destination 1500 km away. In order to reach in time it increases its speed by 250 km/h. What is its original speed?
(a) 1000 km/h (b) 750 km/h
(c) 600 km/h (d) 800 km/h

51. Bombay Express left Delhi for Bombay at 14.30 hrs, travelling at a speed of 60 kmph and Rajdhani Express left Delhi for Bombay on the same day at 16.30 hrs, travelling at a speed of 80 kmph. How far away from Delhi will the two trains meet?
(a) 120 km (b) 360 km
(c) 480 km (d) 500 km

52. A person can swim at a speed of 9 km per hour in still water. If the speed of the stream is 6 km per hour, then how long does he take to swim up to a distance of 9 km and return at the starting point?
(a) 2 hours (b) $2\frac{1}{2}$ hours
(c) $3\frac{3}{5}$ hours (d) $3\frac{3}{4}$ hours

53. A thief goes away with a Maruti car at a speed of 40 km/h. The theft has been discovered after half an hour and the owner sets off in another car at 50 km/h. When will the owner overtake the thief from the start.
(a) $2\frac{1}{2}$ hours (b) 2 hr 20 min
(c) 1 hr 45 min (d) cannot be determined

54. A man is walking at a speed of 10 km per hour. After every kilometre, he takes rest for 5 minutes. How much time will he take to cover a distance of 5 kilometres?
(a) 48 min. (b) 50 min.
(c) 45 min. (d) 55 min.

55. One-fourth of a certain journey is covered at the rate of 25 km/h, one-third at the rate of 30 km/h and the rest at 50 km/h. Find the average speed for the whole journey.
(a) 600/53 km/h (b) 1200/53 km/h
(c) 1800/53 km/h (d) 1600/53 km/h

56. A railway passenger counts the telegraph poles on the rail road as he passes them. The telegraph poles are at a distance of 50 meters. What will be his count in 4 hours if the speed of the train is 45 km per hour?
(a) 2500 (b) 600
(c) 3600 (d) 5000

57. A long distance runner runs 9 laps of a 400 metres track everyday. His timings (in minutes) for four consecutive days are 88, 96, 89 and 87 resplectively. On an average, how many metres/minute does the runner cover ?
(a) 40 m/min (b) 45 m/min
(c) 38 m/min (d) 49 m/min

58. A dog starts chasing to a cat 2 hours later. It takes 2 hours to dog to catch the cat. If the speed of the dog is 30 km/h, what is the speed of cat?
(a) 10 km/h (b) 15 km/h
(c) 20 km/h (d) Can't be determined

59. A truck covers a distance of 368 km at a certain speed in 8 hours. How much time would a car take at an average speed which is 18 km/hr more than that of the speed of the truck to cover a distance which is 16 km more than that travelled by the truck ? *[SBI Clerk-2012]*
(a) 7 hrs (b) 5 hrs
(c) 6 hrs (d) 8 hrs
(e) None of these

60. An aeroplane takes off 30 minutes later than the scheduled time and in order to reach its destination 1500 km away in time, it has to increase its speed by 250 km/h from its usual speed. Find its usual speed. *[SBI Clerk-2014]*
(a) 1000 km/h (b) 750 km/h
(c) 850 km/h (d) 650 km/h
(e) None of these

61. A man walks a certain distance and rides back taking a total time of 37 minutes. He could walk both ways in 55 minutes. How long would he take to ride both ways? *[SSC-Sub. Ins.-2012]*
(a) 9.5 minutes (b) 18 minutes
(c) 19 minutes (d) 20 minutes

62. On a journey across Kolkata, a taxi averages 50 km per hour for 50% of the distance. 40 km per hour for 40% of it and 20 km per hour for the remaining. The average speed in km/hour, for the whole journey is :*[SSC-Sub. Ins.-2013]*
(a) 42 (b) 40
(c) 35 (d) 45

63. A train 270 metre long is running at a speed of 36 km per hour, then it will cross a bridge of length 180 metres in : *[SSC-Sub. Ins.-2013]*
(a) 40 sec (b) 45 sec
(c) 50 sec (d) 35 sec

64. A student goes to school at the rate of $\frac{5}{2}$ km/hr and reaches 6 minutes late. If he travels at the speed of 3 km/hr, he reaches 10 minutes earlier. The distance of the school is *[SSC-Sub. Ins.-2014]*
(a) 45 km (b) 20 km
(c) 10 km (d) 4 km

65. A train 50 metre long passes a platform 100 metre long in 10 sec. The speed of the train in km/hr is *[SSC-Sub. Ins.-2014]*
(a) 10 (b) 54
(c) 15 (d) 100

66. Walking at a speed of 5 km/hr, a man reaches his office 6 minutes late. Walking at 6 km/hr, he reaches there 2 minutes early. The distance of his office is *[SSC-MT-2013]*
(a) 2 km (b) 3 km
(c) 4 km (d) 3.5 km

67. Two trains 108 m and 112 m in length are running towards each other on the parallel lines at a speed of 45 km/hr and 54 km/hr respectively. To cross each other after they meet, it will take *[SSC-MT-2013]*
(a) 10 sec (b) 12 sec
(c) 9 sec (d) 8 sec

68. A train 100 metres long meets a man going in opposite Directions at 5 km/hr and passes him in $7\frac{1}{5}$ seconds. What is the speed of the train in km/hr ? *[SSC 10+2-2012]*
(a) 45 km/hr (b) 60 km/hr
(c) 55 km/hr (d) 50 km/hr

69. A train is moving at a speed of 80 km/h and covers a certain distance in 4.5 hours. The speed of the train to cover the same distance in 4 hours is *[SSC 10+2-2013]*
(a) 90 km/h (b) 100 km/h
(c) 70 km/h (d) 85 km/h

70. A horse take 2½ seconds to complete a round around a circular field. If the speed of the horse was 66 m/sec, then the radius of the field is, *[SSC 10+2-2014]*
[Given $\pi = \frac{22}{7}$]
(a) 25.62 m (b) 26.52 m
(c) 25.26 m (d) 26.25 m

71. The area of a square park is 25 sq. km. The time taken to complete a round of the field once, at a speed of 3 km/hour is *[SSC 10+2-2014]*
(a) 4 hours 60 minutes (b) 4 hours 50 minutes
(c) 6 hours 40 minutes (d) 5 hours 40 minutes

72. The speed of a man is $\frac{3}{4}$th the speed of a bicycle. The bicycle covers 192 m in 8 seconds. How much time will the man take to cover 54 m ? *[IBPS Clerk-2012]*
(a) 3 seconds (b) 4 seconds
(c) 7 seconds (d) 5 seconds
(e) None of these

73. A bus covers 572 kms in 13 hours. What is the speed of the bus? *[IBPS Clerk-2012]*
(a) 40 km/hr (b) 44 km/hr
(c) 43 km/hr (d) 47 km/hr
(e) None of these

74. A 210 m long train takes 6 s to cross a man running at 9 km/h in a direction opposite to that of the train. What is the speed of the train? (in km/h) *[IBPS Clerk-2013]*
(a) 127 (b) 121
(c) 117 (d) 108
(e) 111

Level - II

1. A and B can run 200 m in 22 seconds and 25 seconds, respectively. How far is B from the finishing line when A reaches in ?
   (a) 8 m (b) 12 m
   (c) 16 m (d) 24 m

2. If a man walks at the rate of 5 kmph, he misses a train by 7 minutes. However, if he walks at the rate of 6 kmph, he reaches the station 5 minutes before the arrival of the train. Find the distance covered by him to reach the station.
   (a) 4 km (b) 6 km
   (c) 5 km (d) 7 km

3. The speed of a car increases by 2 kms after every one hour. If the distance travelled in the first one hour was 35 kms, what was the total distance travelled in 12 hours?
   (a) 456 kms (b) 482 kms
   (c) 552 kms (d) None of these

4. It takes eight hours for a 600 km journey, if 120 km is done by train and the rest by car. It takes 20 minutes more, if 200 km is done by train and the rest by car. The ratio of the speed of the train to that of the speed of the car is
   (a) 4 : 3 (b) 3 : 4
   (c) 3 : 2 (d) 2 : 3

5. A person has to cover a distance of 6 km in 45 minutes, If he covers one-half of the cistance in two-thirds of the total time; to cover the remaining distance in the remaining time, his speed (in km/hr) must be:
   (a) 6 (b) 8
   (c) 12 (d) 15

6. A car travels the first one-third of a certain distance with a speed of 10 km/hr, the next one-third distance with a speed of 20 km/hr, and the last one-third distance with a speed of 60 km/hr. The average speed of the car for the whole journey is
   (a) 18 km/hr (b) 24 km/hr
   (c) 30 km/hr (d) 36 km/hr

7. A train starts from Delhi at 6 : 00 AM and reaches Ambala Cantt at 10 AM. The other train starts from Ambala Cantt at 8 AM and reaches Delhi at 11:30 PM. If the distance between Delhi and Ambala Cantt. is 200 km, then at what time did the two trains meet each other ?
   (a) 8 : 56 AM (b) 8 : 46 AM
   (c) 7 : 56 AM (d) 8 : 30 AM

8. Rahul can row a certain distance downstream in 6 hours and return the same distance in 9 hours. If the speed of Rahul in still water is 12 km/hr, find the speed of the stream.
   (a) 2 km/hr (b) 2.4 km/hr
   (c) 3 km/hr (d) Data inadequate

9. A man can row 4.5 km/hr in still water and he finds that it takes him twice as long to row up as to row down the river. Find the rate of the stream.
   (a) 1.5 km/hr (b) 2 km/hr
   (c) 2.5 km/hr (d) 1.75 km/hr

10. A man sitting in a train travelling at the rate of 50 km/hr observes that it takes 9 sec for a goods train travelling in the opposite direction to pass him. If the goods train is 187.5 m long, find its speed.
   (a) 40 km/hr (b) 25 km/hr
   (c) 35 km/hr (d) 36 km/hr

11. Two trains are moving in opposite directions at speeds of 60 km/hour and 90 km/hour. Their lengths are 1.10 km and 0.9 km respectively. The time taken by the slower train to cross the faster train in seconds is
   (a) 36 (b) 49
   (c) 45 (d) 48

12. It takes eight hours for a 600 km journey, if 120 km is done by tain and the rest by car. It takes 20 minutes more, if 200 km is done by train and the rest by car. The ratio of the speed of the train to that of the car is
   (a) 2 : 3 (b) 3 : 2
   (c) 3 : 4 (d) 4 : 3

13. The distance between two cities A and B is 330 km. A train starts from A at 8 a.m. and travels towards B at 60 km/hr. Another train starts from B at 9 a.m. and travels towards A at 75 km/hr. At what time do they meet?
   (a) 10 a.m. (b) 10.30 a.m.
   (c) 11 a.m. (d) 11.30 a.m.

14. A motorcyclist covered two thirds of a total journey at his usual speed. He covered the remaining distance at three fourth of his usual speed. As a result, he arrived 30 minutes later than the time he would have taken at usual speed. If the total journey was 180 km, the what is his usual speed?
   (a) 40 kmph (b) 36 kmph
   (c) 30 kmph (d) 32 kmph

15. A man can row a certain distance against the stream in 6 hours. However, he would take two hours less to cover the same distance with the current. If the speed of the current is 2 kmph, then what is the rowing speed in still water?
   (a) 10 kmph (b) 12 kmph
   (c) 14 kmph (d) 8 kmph

16. If I walk at 4 km/h, I miss the bus by 10 minutes. If I walk at 5 km/h, I reach 5 minutes before the arrival of the bus. How far I walk to reach the bus stand ?
   (a) 5 km (b) 4.5 km
   (c) $5\frac{1}{4}$ km / h (d) Cannot be determined

17. A man covers a certain distance on a toy train. If the train moved 4 km/h faster, it would take 30 minutes less. If it moved 2 km/h slower, it would have taken 20 minutes more. Find the distance.
(a) 60 km (b) 58 km
(c) 55 km (d) 50 km

18. An aeroplane flies along the four sides of a square at the speeds of 200, 400, 600 and 800 km/h. Find the average speed of the plane around the field.
(a) 384 km/h (b) 370 km/h
(c) 368 km/h (d) None of these

19. A thief steals a car at 2 : 30 p.m. and drives it at 60 kmph. The theft is discovered at 3 p.m. and the owner sets off in another car at 75 kmph. When will he overtake the thief?
(a) 4 : 30 p.m. (b) 4 : 45 p.m.
(c) 5 p.m. (d) 5 : 15 p.m.

20. Points A and B are 70 km apart on a highway. One car starts form A and the another one from B at the same time. If they travel in the same direction, they meet in 7 hours. But if they travel towards each other, they meet in one hour. The speeds of the two cars are, respectively.
(a) 45 and 25 km/h (b) 70 and 10 km/h
(c) 40 and 30 km/h (d) 60 and 40 km/h

21. A river 3 m deep and 40 m wide is flowing at the rate of 2 km per hour. How much water (in litres) will fall into the sea in a minute?
(a) 4,00,000 (b) 40,00,000
(c) 40,000 (d) 4,000

22. A dog sees a cat. It estimates that the cat is 25 leaps away. The cat sees the dog and starts running with the dog in hot pursuit. If in every minute, the dog makes 5 leaps and the cat makes 6 leaps and one leap of the dog is equal to 2 leaps of the cat. Find the time in which the cat is caught by the dog (assume an open field with no trees)
(a) 12 minutes (b) 15 minutes
(c) 12.5 minutes (d) None of these

23. A group of soldiers are marching with a speed of 5 m/s. The distance between the first and the last row of soldiers is 100 m. A dog starts running from the last row and moves towards the first row, turns and comes back to the last row. If the dog has travelled 400 m, the speed of the dog is
(a) $5\sqrt{2}$ m/s (b) $3\sqrt{5}$ m/s
(c) $6\sqrt{5}$ m/s (d) $6\sqrt{2}$ m/s

24. Ram runs 7/4 times as fast as Sham, If Ram gives Sham a start of 300 m, how far must the winning post be if both Ram and Sham have to end the race at the same time?
(a) 1400 m (b) 700 m
(c) 350 m (d) 210 m

25. A watch, which gains time uniformly, was 5 minutes behind the correct time when it showed 11:55 AM on Monday. It was 10 minutes ahead of the correct time when it showed 06:10 PM on the next day. When did the watch show the correct time?
(a) 6 AM, Tuesday (b) 6 PM, Monday
(c) 2 PM, Tuesday (d) 10 PM, Monday

26. With an average speed of 40 km/h, a train reaches its destination in time. If it goes with an average speed of 35 km/h, it is late by 15 minutes. The length of the total journey is:
(a) 40 km (b) 70 km
(c) 30 km (d) 80 km

27. A student rides on a bicycle at 8 km/h and reaches his school 2.5 minutes late. The next day he increases his speed to 10 km/h and reaches the school 5 minutes early. How far is the school from his house?
(a) 1.25 km (b) 8 km
(c) 5 km (d) 10 km

28. Two rockets approach each other, one at 42000 mph and the other at 18000 mph. They start 3256 miles apart. How far are they apart (in miles) 1 minute before impact ?
(a) 1628 (b) 1000
(c) 826 (d) 1200

29. Two guns were fired form the same place at an interval of 10 minutes and 30 seconds, but a person in the train approaching the place hears the second shot 10 minutes after the first. The speed of the train (in km/hr), supposing that speed travels at 330 metres per second, is
(a) 19.8 (b) 58.6
(c) 59.4 (d) 111.80

30. Train A running at 60 km/h leaves Mumbai for Delhi at 6 p.m. Train B running at 90 km/h also leaves for Delhi at 9 p.m. Train C leaves Delhi for Mumbai at 9 p.m. If all the three trains meet at the same time between Mumbai and Delhi, then what is the speed of train C, if distance between Delhi and Mumbai is 1260 km ?
(a) 60 km/h (b) 90 km/h
(c) 120 km/h (d) 135 km/h

31. A boat, while going downstream in a river covered a distance of 50 mile at an average speed of 60 miles per hour. While returning, because of the water resistance, it took one hour fifteen minutes to cover the same distance . What was the average speed of the boat during the whole journey?
(a) 40 mph (b) 48 mph
(c) 50 mph (d) 55 mph

32. A man takes 5 hour 45 min. in walking to a certian place and riding back. He would have gained 2 hours by riding both ways. The time he would take to walk both ways, is
(a) 3 hrs 45 min (b) 7 hrs 30 min
(c) 7 hrs 45 min (d) 11 hrs 45 min

33. A boatman rows to a place 45 km distant and back in 20 hours. He finds that he can row 12 km with the stream in same time as 4 km against the stream . Find the speed of the stream.
(a) 3 km/hr (b) 2.5 km/hr
(c) 4 km/hr (d) Cannot be determined

34. A man goes 15 metres due west and then 8 metres due north. How far is he from the starting point?
(a) 19 metres (b) 16 metres
(c) 17 metres (d) 15 metres

35. Two trains, 130 m and 110 m long, are going in the same direction. The faster train takes one minute to pass the other completely. If they are moving in opposite directions, they pass each other completely in 3 seconds. Find the speed of each train.
(a) 38 m/sec, 36 m/sec (b) 42 m/sec, 38 m/sec
(c) 36 m/sec, 42 m/sec (d) None of these

36. A man who can swim 48 m/min in still water swims 200 m against the current and 200 m with the current. If the difference between those two times is 10 minutes, find the speed of the current.
(a) 30 m/min (b) 29 m/min
(c) 31 m/min (d) 32 m/min

37. A man who can swim 48 m/min in still water swims 200 m against the current and 200 m with the current. If the difference between those two times is 10 min, what is the speed of the current?
(a) 30 m/min (b) 31 m/min
(c) 29 m/min (d) 32 m/min

38. A man walks a certain distance and rides back in $6\frac{1}{4}$h. He can walk both ways in $7\frac{3}{4}$h. How long it would take to ride both ways ?
(a) 5 hours (b) $4\frac{1}{2}$hours
(c) $4\frac{3}{4}$hours (d) 6 hours

39. An accurate clock shows 8 o'clock in the morning. Through how many degrees will the hour hand rotate when the clock shows 2 o'clock in the afternoon?
(a) 144° (b) 150°
(c) 168° (d) 180°

40. A dog after travelling 50 km meets a swami who counsels him to go slower. He then proceeds at 3/4 of his former speed and arrives at his destination 35 minutes late. Had the meeting occurred 24 km further the dog would have reached its destination 25 minutes late. The speed of the dog is
(a) 48 km/h (b) 36 km/h
(c) 54 km/h (d) 58 km/h

41. *A*, *B*, and *C* are three participants in a kilometer race. If *A* can give *B* a start of 40 metres and *B* can give *C* a start of 25 metres, how many metres of a start can *A* give to *C*?
(a) 60 m (b) 64 m
(c) 62 m (d) 66 m

42. A monkey ascends a greased pole 12 metres high. He ascends 2 metres in first minute and slips down 1 metre in the alternate minute. In which minute, he reaches the top ?
(a) 21st (b) 22nd
(c) 23rd (d) 24th

43. Mallah can row 40 km upstream and 55 km downstream in 13 h and 30 km upstream and 44 km downstrean in 10 hours. What is the speed of Mallah in still water?
(a) 6 km/h (b) 12 km/h
(c) 3 km/h (d) 8 km/h

44. A passenger sitting in a train of length 100 m, which is running with speed of 60 km/h passing through two bridges, notices that he crosses the first bridge and the second bridge in time intervals which are in the ratio of 7 : 4 respectively. If the length of first bridge be 280 m, then the length of second bridge is:
(a) 490 m (b) 220 m
(c) 160 m (d) Can't be determined

45. A tiger is 50 of its own leaps behind a deer. The tiger takes 5 leaps per minute to the deer's 4. If the tiger and the deer cover 8 m and 5 m per leap respectively, what distance will the tiger have to run before it catches the deer?
(a) 600 m (b) 700 m
(c) 800 m (d) 1000 m

46. A candle of 6 cm long burns at the rate of 5 cm in 5 h and another candle of 8 cm long burns at the rate of 6 cm in 4h. What is the time required by each candle to remain of equal lengths after burning for some hours, when they start to burn simultaneously with uniform rate of burning?
(a) 1 h (b) 1.5 h
(c) 2 h (d) None of these

47. Two persons start from the opposite ends of a 90 km straight track and run to and fro between the two ends. The speed of first person is 30 m/s and the speed of other is 125/6 m/s. They continue their motion for 10 hours. How many times they pass each other?
(a) 10 (b) 9
(c) 12 (d) None of these

48. At what time after 3:10 am, the acute angle made by the minute and hour-hand is double to that of a 3:10 am, for the first time?
(a) 4 h 43 min (b) 3 h 48 min
(c) $3\text{h}\frac{320}{11}\text{min}$ (d) None of these

49. A swiss watch is being shown in a museum which has a very peculiar property. It gains as much in the day as it loses during night between 8 pm to 8 am. In a week how many times will the clock show the correct time?
(a) 6 times (b) 14 times
(c) 7 times (d) 8 times

50. The metro service has a train going from Mumbai to Pune and Pune to Mumbai every hour, the first one at 6 a.m. The trip from one city to other takes 4½ hours, and all trains travel at the same speed. How many trains will you pass while going from Mumbai to Pune if you start at 12 noon?

(a) 8 (b) 10
(c) 9 (d) 13

51. A wall clock gains 2 minutes in 12 hours, while a table clock loses 2 minutes in 36 hours; both are set right at noon on Tuesday. The correct time when they both show the same time next would be

(a) 12 : 30 night (b) 12 noon
(c) 1 : 30 night (d) 12 night

52. A watch loses 2/3% time during the 1st week and gains 1/3% time during the next week. If on a Sunday noon, it showed the right time, what time will it show at noon on the Saturday after the next.

(a) 11 : 26 : 24 a.m. (b) 10 : 52 : 18 a.m.
(c) 10 : 52 : 48 a.m. (d) 11 : 36 : 24 a.m.

53. The time taken by a man to walk five times around the boundary of a square field having 16 hectares as area, at the rate of 5 km per hour is [*SBI PO-2011*]

(a) 16 minutes (b) 24 minutes
(c) 48 minutes (d) 96 minutes
(e) None of these

54. A wheel of a motorbike has radius 35 cm. How many revolutions per minute must the wheel make so that the speed of the bike is 33 km/hr? [*SBI PO-2011*]

(a) 300 (b) 250
(c) 200 (d) 220
(e) None of these

55. An aeroplane flies with an average speed of 756 km/h. A helicopter takes 48 h to cover twice the distance covered by aeroplane in 9 h. How much distance will the helicopter cover in 18 h? (Assuming that flights are non-stop and moving with uniform speed.) [*IBPS-PO-2013*]

(a) 5010 km (b) 4875 km
(c) 5760 km (d) 5103 km
(e) None of these

56. A car covers four successive 6 km stretches at speeds of 25 kmph, 50 kmph, 75 kmph and 150 kmph respectively. Its average speed over this distance is [*SSC CGL-2012*]

(a) 25 kmph (b) 50 kmph
(c) 75 kmph (d) 150 kmph

57. A wheel rotates 3.5 times in one second. What time (in seconds) does the wheel take to rotate 55 radian of angle? [*SSC CGL-2012*]

(a) 1.5 (b) 2.5
(c) 3.5 (d) 4.5

58. Anil calculated that it will take 45 minutes to cover a distance of 60 km by his car. How long will it take to cover the same distance if the speed of his car is reduced by 15 km/hr? [*SSC CGL-2013*]

(a) 36 min (b) 55.38 min
(c) 48 min (d) 40 min

59. A train 100 metres long moving at a speed of 50 km/hr. crosses a train 120 metres long coming from opposite direction in 6 sec. The speed of the second train is [*SSC CGL-2013*]

(a) 60 km/hr. (b) 82 km/hr.
(c) 70 km/hr. (d) 74 km/hr.

60. It takes 8 hours for a 600 km journey, if 120 km is done by train and the rest by car. It takes 20 minutes more if 200 km is down by train and the rest by car. The ratio of the speed of the train to that of the car is [*SSC CGL-2014*]

(a) 2 : 3 (b) 3 : 2
(c) 3 : 4 (d) 4 : 3

61. If a train runs at 70 km/hour, it reaches its destination late by 12 minutes. But if it runs at 80 km/hour, it is late by 3 minutes. The correct time to cover the journey is [*SSC CGL-2014*]

(a) 58 minutes (b) 2 hours
(c) 1 hour (d) 59 minutes

62. On a journey across Kolkata, a taxi averages 40 kmph for 60% of distance, 30 kmph for 20% of the distance, and 10 kmph for the remainder. The average speed of the whole journey is [*SSC CGL-2014*]

(a) 25 kmph (b) 26 kmph
(c) 24 kmph (d) 30 kmph

Hints & Solutions

Level-I

1. (a) Let a distance x be covered in time t.

$$\text{Required ratio} = \frac{\frac{x/2}{2t}}{\frac{x}{t}} = \frac{1}{4} = 1:4$$

2. (c) Let the distance travelled during both upward and downward journey be x km.

$$\text{Average speed} = \frac{\text{Total distance covered}}{\text{Total time taken}}$$

$$= \frac{x+x}{\frac{x}{16}+\frac{x}{28}} = \frac{2}{\frac{28+16}{28\times16}}$$

$$= \frac{2\times28\times16}{44} = 20.36 \text{ km/h}$$

3. (b) Distance $= \left(1100 \times \frac{11}{5}\right)$ feet = 2420 feet.

4. (a) Time required = (2 hrs 30 min – 50 min) = 1hr 40 min

$$= 1\frac{2}{3} \text{ hrs}$$

$\therefore$ Required speed $= \left(50 \times \frac{3}{5}\right)$ km/hr = 30 km/hr.

Original speed $= \left(50 \times \frac{2}{5}\right)$ km/hr = 20 km/hr.

$\therefore$ Difference in speed = (30 – 20) km/hr = 10 km/hr.

5. (c) When time is constant the distance covered by A and B will be in the ratio of their speeds, respectively.

6. (b) Let the distance travelled be x km.
Then, the correct time at a speed of 30 km/h

$$= \frac{x}{30} - \frac{10}{60} \text{ and}$$

the correct time at a speed of 42 km/h $= \frac{x}{42} + \frac{10}{60}$

Now, $\frac{x}{30} - \frac{10}{60} = \frac{x}{42} + \frac{10}{60}$

or $\frac{x}{30} - \frac{x}{42} = \frac{2}{6}$ or $\frac{12x}{1260} = \frac{2}{6}$ or $x = 35$ km

7. (c) Relative speed of the trains = (40 + 20) = 60 m/s
Distance = (120 + 120) = 240 m

Time taken by trains to cross each other completely

$$= \frac{240}{60} = 4 \text{ s}$$

$\therefore$ Larger the no. of cogs (tooth of wheel) of wheel, lesser will be that no. of revolution made by it.

8. (c) Let the speed of trains be x km/h and y km/h, respectively.
When the trains cross each other, time taken by both the trains will be equal.

i.e. $\frac{110}{x} = \frac{90}{y} \Rightarrow \frac{x}{y} = \frac{110}{90} \Rightarrow x:y = 11:9$

9. (b) Required distance $= \frac{S_1 S_2}{(S_1 \sim S_2)} \times$ Time difference

$$= \frac{8\times5}{3}\times\frac{3}{2} = 20\text{km}$$

10. (b) Let the total distance be x km. Then,

$$\frac{\frac{1}{2}x}{21} + \frac{\frac{1}{2}x}{24} = 10 \Rightarrow \frac{x}{21} + \frac{x}{24} = 20$$

$$\Rightarrow 15x = 168 \times 20 \Rightarrow x = \left(\frac{168\times20}{15}\right) = 224 \text{ km.}$$

11. (b) Speed of the train = 132 km/h $= \frac{132\times5}{18}$ m/s
Distance = (110 + 165) = 275 m
Time required to cross the railway platform

$$= \frac{275\times18}{132\times5} = 7.5\text{s}$$

12. (b) Let the total distance be $3x$ km.

Then, $\frac{x}{3} + \frac{x}{4} + \frac{x}{5} = \frac{47}{60} \Leftrightarrow \frac{47x}{60} = \frac{47}{60} \Leftrightarrow x = 1.$

$\therefore$ Total distance = (3 × 1) km = 3 km.

13. (c) $\frac{x}{9} - \frac{x}{10} = \frac{20}{60}$

or, $\frac{10x-9x}{90} = \frac{20}{60}$

$\therefore x = 30$ km

14. (c) Let the normal speed = x km/h
Then, the new speed = $(x + 5)$ km/h.

Now, $\frac{300}{x} - 2 = \frac{300}{(x+5)}$ or $\frac{300}{x} - \frac{300}{(x+5)} = 2$

Checking with options, we see that $x = 25$ km/h.

15. **(c)** Distance between Chauhan and the gun

$= 3.32 \times 1000 = 3320$ m

Time taken = 10 s

$\Rightarrow$ Speed $= \frac{3320}{10} = 332\,\text{m/s}$

16. **(a)** Since A and B move in the same direction along the circle, so they will first meet each other when there is a difference of one round between the two.

Relative speed of A and $B = (6 - 1) = 5$ rounds per hour.

Time taken to complete one round at this speed

$= \frac{1}{5}$ hr = 12 min.

$\therefore$ They meet at 7:42 a.m.

17. **(c)** Average speed $= \frac{\text{Total distance covered}}{\text{Total time taken}}$

$$= \frac{2\times200}{\frac{200}{40}+\frac{200}{20}} = \frac{2\times40\times20}{40+20}$$

$$= \frac{2\times40\times20}{60} = \frac{80}{3} = 26.67 \text{ km/h.}$$

18. **(a)** Let the length of each train be x metres.

Then, the total distance covered $= (x + x) = 2x$ m

Relative speed $= (46 - 36) = 10$ km/h $= \frac{10\times5}{18}$ m/s

Now, $36 = \frac{2x\times18}{50}$

or $x = 50$ m

19. **(c)** After 5 minutes (before meeting), the top runner covers 2 rounds i.e., 400 m and the last runner covers 1 round *i.e.*, 200 m.

$\therefore$ Top runner covers 800 m race in 10 minutes.

20. **(b)** Due to stoppges the train travels $(45 - 36) = 9$ km less in an hour than it could have travelled without stoppages.

Thus train stops per hour for $\frac{9}{45}\times60 = 12$ min.

21. **(b)** Distance travelled by the train in 1 hour

$= 2\times\frac{22}{7}\times2.1\times75\times60\,\text{m.} = \frac{132\times450}{1000} = 59.4$ km

i.e. speed of the train = 59.4 km/h.

22. **(d)** First train's speed is 45km/hr.

$\left(\text{Using speed} = \frac{\text{Distance}}{\text{Time}}\right)$

Second train's speed is 60km/hr.

Difference in the distance covered by these trains in 1 hr. is 15 km.

23. **(c)** Speed of speed-boat $= 16 - 3 = 13$ km/hr.

$\therefore$ Speed of boat against the current $= 13 - 3$

$= 10$ km/hr.

24. **(d)** Let the usual speed be x km/hr, then

$$\frac{1500}{x} - \frac{1500}{x+250} = \frac{1}{2}$$

$\Rightarrow x = 750$ km/hr

25. **(a)** Relative velocity = 20 + 30 = 50 m/s.

Distance = 2.5 kms. = 2500 m.

$t = 2500/50 = 50$ s.

26. **(c)** Relative speed = 90 + 60 = 150 km/hr.

Total distance to be covered = 300 + 200 = 500 m

Time required $= \frac{500}{150\times1000}\times3600 = 12$ sec.

27. **(d)** Required distance between A and B

$$= \frac{3\left((9)^2-(3)^2\right)}{2(9)} = \frac{3(81-9)}{18} = \frac{72}{6} = 12\,\text{km.}$$

28. **(d)** Total distance covered $= 2 \times 91$ km = 182 km

Time taken = 20 hours

$\therefore$ Average speed $= \frac{182}{20} = 9.1$ km/h

Let the speed of flow of the river = x km/hr

then, $\frac{10^2 - x^2}{10} = 9.1 \Rightarrow 100 - 91 = x^2 \Rightarrow x = \pm3$

Hence, rate of flow of the river = 3 km/h

29. **(d)** Relative speed = 5.5 – 5 = 0.5 km/h.

Required time $= \frac{8.5}{0.5} = 17$ h

30. **(d)** x (speed of boat in standing water) = 9 km/hr

speed of stream = 1.5 km.h

Total time taken by him $= \frac{105}{10.5} + \frac{105}{7.5}$

= 10 + 14 = 24 h

31. **(a)** Given, distances are 2500 km, 1200 km and 500 km.

Given, speeds are 500 km/h, 400 km/h and 250 km/h

$\therefore$ Total time $= \frac{2500}{500} + \frac{1200}{400} + \frac{500}{250}$

$= 5 + 3 + 2 = 10$ hr.

$\therefore$ Average speed $= \frac{\text{Total distance}}{\text{Total time}}$

$$= \frac{2500+1200+500}{10} = \frac{4200}{10}$$

= 420 km/hr

32. **(d)** Let the distance between each pole be x m.

Then, the distance up to 12th pole = 11 x m

Speed $= \frac{11x}{24}$ m/s

Time taken to covers the total distance of $19x$

$$= \frac{19x \times 24}{11x} = 41.45\text{s}$$

33. (a) Let the length of the journey = x km.

$\therefore$ Journey rides by horse cart $= x\left(1 - \frac{1}{2} - \frac{1}{3}\right)$

$$= \frac{1}{6}x \text{ km.}$$

Then, total time taken to complete journey $= \frac{31}{5}$ hr

$$\Rightarrow t_1 + t_2 + t_3 = \frac{31}{5}$$

$$\Rightarrow \frac{x}{2} \times \frac{1}{4} + \frac{x}{3} \times \frac{1}{12} + \frac{x}{6 \times 9} = \frac{31}{5}$$

$$\Rightarrow x = \frac{31}{5} \times \frac{216}{37} = 36.2\text{km} \approx 36\text{km}$$

34. (d) Required difference $= \frac{180}{3} - \frac{180}{4} = 15$ km

35. (c) Let the husband and the wife meet after x minutes. 4500 metres are covered by Pradeep in 60 minutes.

In x minutes, he will cover $\frac{4500}{60}x$ metres.

Similarily,

In x minutes, his wife will cover $\frac{3750}{60}x$ m.

Now, $\frac{4500}{60}x + \frac{3750}{60}x = 726$

$$\Rightarrow x = \frac{726 \times 60}{8250} = 5.28 \text{ min}$$

36. (a) Let speed of the boat in still water be x km/h and speed of the current be y km/h.
Then, upstream speed = $(x - y)$ km/h
and downstream speed = $(x + y)$ km/h

Now, $\frac{24}{(x-y)} + \frac{28}{(x+y)} = 6$ ...(1)

and $\frac{30}{(x-y)} + \frac{21}{(x+y)} = \frac{13}{2}$ ...(2)

Solving (1) and (2), we have
$x = 10$ km/h and $y = 4$ km/h

37. (c) The train that leaves at 6 am would be 75 km ahead of the other train when it starts. Also, the relative speed being 36 kmph, the distance from Mumbai would be (75/36) × 136 = 283.33 km

38. (a) The train can cover (200 + 350) m distane in five seconds which means the speed of the train is 110 m/s. Relative speed of man and trian is 114 m/s. To cover the distance of 100 metre, it will take less than one second.

39. (a) The clock gains 15 min in 24 hours.
Therefore, in 16 hours, it will gain 10 minutes.
Hence, the time shown by the clock will be 4.10 am.

40. (b) Average speed $= \frac{\text{Total distance}}{\text{Total time}}$

$$= \frac{80}{\frac{60}{40} + \frac{20}{20}} = \frac{80}{2.5} = 32\text{ km/h}$$

41. (d) Total distance = 250 × 2 = 500 km

Total time $= 5\frac{1}{2}\text{hrs} + 4\frac{2}{3}\text{hrs} = 10\frac{1}{6}\text{hrs}$

Average speed $= \frac{\text{Total distance}}{\text{Total time}} = \frac{500}{10\frac{1}{6}} = \frac{3000}{61}\text{hrs}$

= 49.18 hours ≈ 50 hours (approx.)

42. (c) Their relative speeds $= (4.5 + 3.75) = 8.25$ km/h

Distance = 726 metres $= \frac{726}{1000} = 0.726\text{km}$

Required time $= \frac{0.726}{8.25} \times 60 = 5.28$ min

43. (d) Since they are moving in opposite direction, therefore their relative speed will be 4 + 3 = 7 km/hr.

Time $= \frac{d}{s} = \frac{17.5}{7} = 2.5\text{hrs.}$

(where d is distance and s is speed).
$\therefore$ They should meet at 12.30 PM.

44. (c) Speed on return trip = 150% of 40 = 60 kmph.

$\therefore$ Average speed $= \left(\frac{2 \times 40 \times 60}{40 + 60}\right)$ km/hr

$$= \left(\frac{4800}{100}\right) \text{ km/hr} = 48 \text{ km/hr.}$$

45. (a) Average speed $= \frac{2 \times V_1 \times V_2}{V_1 + V_2} = \frac{2 \times 40 \times 60}{40 + 60}$

= 48 km/h

46. (c) The speeds of the two cyclists are different Hence, when one of the cyclist has covered one round more than the other cyclist, only then they will meet at the starting point.
$\therefore$ Time when the two cyclists will meet
= 300 m × (difference in speeds)
= 300 × (8 – 7) sec = 300 seconds.

47. (c) Let the correct time to complete the journey be x min.

Distance covered in $(x + 11)$ min. at 40 kmph
= Distance covered in $(x + 5)$ min. at 50 kmph

$\therefore \frac{(x+11)}{60} \times 40 = \frac{(x+5)}{60} \times 50 \Leftrightarrow x = 19$ min.

48. (a) Let x be the total distance.

$\therefore$ According to the question,

Distance covered by him on foot = $\frac{1}{3} x \times \frac{1}{4} = \frac{x}{12}$

But given he travels on foot = 2 km

$\therefore \frac{x}{12} = 2 \Rightarrow x = 24$ km.

49. (d) Let speed of car = x km / hr

Let speed of pedestrian = y = 2km / hr

$\therefore$ Relative speed = $(x - 2)$ km / hr

$\therefore$ According to the question,

$(x-2) \times \frac{6}{60} = 0.6 \Rightarrow x - 2 = 6 \Rightarrow x = 8$ km / h

50. (b) Let the original time be T hours and original speed be x km/h

$$\frac{1500}{x} = T \quad \ldots(1)$$

$$\frac{1500}{x+250} = T - \frac{30}{60} \quad \ldots(2)$$

Solving equations (1) and (2), we get

Speed of plane = x = 750 or – 1000 (not possible)

$\therefore$ x = 750 km/h

51. (c) Suppose they meet x hours after 14.30 hrs.

Then, $60x = 80 (x - 2)$ or $x = 8$.

$\therefore$ Required distance = $(60 \times 8) = 480$ km.

52. (c) Total time taken

$= \left(\frac{9}{9+6} + \frac{9}{9-6}\right)$ hour

$= \left(\frac{3}{5} + 3\right)$ hours $= 3\frac{3}{5}$ hours

53. (a) Distance to be covered by the thief and by the owner is same.

Let after time 't', owner catches the thief.

$\therefore 40 \times t = 50\left(t - \frac{1}{2}\right)$

$\Rightarrow 10t = 25 \Rightarrow t = \frac{5}{2}$ hr $= 2\frac{1}{2}$ hr

54. (b) Rest time = Number of rest × Time for each rest

= 4 × 5 = 20 minutes

Total time to cover 5 km

$= \left(\frac{5}{10} \times 60\right)$ minutes + 20 minutes = 50 minutes.

55. (c) Assume that the distance is 120 km. Hence, 30 km is covered @ 25 kmph, 40@30 kmph and so on. Then average speed is 120/total time

56. (c) Time taken to cross a pole = $\frac{50}{1000} \times \frac{1}{45}$ hr

$\therefore$ No. of counts = $\frac{4 \times 1000 \times 45}{50} = 80 \times 45 = 3600$.

57. (a) Average speed $= \frac{\text{Total distance}}{\text{Total time}}$

$= \frac{400 \times 4 \times 9}{88 + 96 + 89 + 87} = \frac{400 \times 4 \times 9}{360}$

= 40 metres /minutes

58. (b) Time = $\frac{\text{Distance advanced}}{\text{Relative speed}}$

$2 = \frac{2 \times x}{(30 - x)}$

$\Rightarrow$ x = 15 km/h

59. (c) Average speed of truck = 368/8 = 46 km/hr

Average speed of car = 46 + 18 = 64 km/hr

Distance travelled by car = 368 + 16 = 384 km

Time taken by car = 384/64 = 6 hrs.

60. (b) Let the usual speed of the aeroplane be x km/h.

Then, $\frac{1500}{x} - \frac{1}{2} = \frac{1500}{(x+250)}$

$\frac{1500}{x} - \frac{1500}{x+250} = \frac{1}{2} \Rightarrow \frac{1500x + 3750000 - 1500x}{x(x+250)} = \frac{1}{2}$

$\Rightarrow 750000 = x^2 + 250x$ or $x^2 + 250x - 750000 = 0$

$\Rightarrow x^2 + 1000\,x - 750\,x - 75000 = 0$

$\Rightarrow (x + 1000)(x - 750) = 0 \Rightarrow x = 750, -1000$

Speed cannot be negative

We get x = 750 km/h

61. (c) To walk both ways, duration = 55 minutes

$\therefore$ To walk one way, duration = $\frac{55}{2}$ minutes

To walk one way + To ride one way = 37 minutes

$\therefore$ To ride both ways $= 2 \times \frac{19}{2} = 19$ minutes

62. (b) Let total distance be 100 km.

Total time = $\frac{50}{50} + \frac{40}{40} + \frac{10}{20} = 1 + 1 + \frac{1}{2} = \frac{5}{2}$ hours

$\therefore$ Average speed = $\frac{100 \times 2}{5} = 40$ kmph

63. (b) Speed of train = 36 kmph

$= \left(36 \times \frac{5}{18}\right)$ m/sec. = 10 m/sec.

Required time = $\frac{\text{length of train + Bridge}}{\text{Speed of train}}$

$= \frac{270 + 180}{10} = 45$ seconds

64. **(d)** Let original time taken by student be x hours.

$$\frac{5}{2}\times\left(x+\frac{6}{60}\right)=3\times\left(x-\frac{10}{60}\right)$$

$$5x+\frac{1}{2}=6x-1$$

$$x=\frac{3}{2}\text{hours}$$

$$\therefore \text{distance of school}=\frac{5}{2}\times\left(\frac{3}{2}+\frac{1}{10}\right)=4\text{km}$$

65. **(b)** Total distance = (100 + 50) m = 150 m

$$\text{Speed of the train}=\frac{150\times3600}{1000\times10}=54\text{ km/hr}$$

66. **(c)** Let the required distnace be x km.
Difference in the times taken at two speeds

$$=8\text{ min}=\frac{8}{60}\text{ hr}=\frac{2}{15}\text{ hr}$$

$$\therefore \frac{x}{5}-\frac{x}{6}=\frac{2}{15}$$

$$\frac{6x-5x}{30}=\frac{2}{15}$$

$$x=\frac{2}{15}\times30=4$$

Hence, the required distance is 4 km.

67. **(d)** Relative speed

$$=(45+54)=99\text{km / hr}=\frac{99\times5}{18}\text{m / sec}$$

Distance covered in crossing each other
= (108+112) = 220m

$$\text{Required time}=\frac{220}{99}\times\frac{18}{5}=8\text{sec}$$

68. **(a)** Let speed of train = x km/hr
Distance travelled by train
= Relative speed of train × Time

$$100\text{ m}=(x+5)\text{ km/hr}\times\frac{36}{5}\text{ seconds}$$

$$\frac{100}{1000}\text{km}=(x+5)\times\left(\frac{36}{5}\times\frac{1}{3600}\right)\text{hrs}$$

$$\Rightarrow x+5=50$$

$$\therefore x=45\text{ km/hr}$$

69. **(a)** Distance covered in 4.5 h = 80 × 4.5 = 360 km

$$\text{Speed}=\frac{360}{4\text{h}}\text{km}=90\text{ km/h}$$

70. **(d)** Total distance covered by horse in $2\frac{1}{2}$ seconds

$$=66\times\frac{5}{2}=165\text{m}$$

$$\text{Radius of the field}=\frac{165}{2\pi}=\frac{165\times7}{2\times22}=26.25\text{m}$$

71. **(c)** Side of square park = $\sqrt{25}$km = 5 km
Perimeter of park = 4 × 5 = 20 km

$$\text{Time taken}=\frac{20\text{km}}{3\text{ km / h}}=6\text{ hours 40 minutes}$$

72. **(a)** $\text{Speed of bicycle}=\frac{\text{Distance}}{\text{Time}}$

$$=\frac{192}{8}=24\text{ metre/second}$$

$$\therefore \text{Speed of man}=24\times\frac{3}{4}$$

= 18 metre/second

$$\therefore \text{Required time}=\frac{\text{Distance}}{\text{Speed}}$$

$$=\frac{54}{18}=3\text{ seconds}$$

73. **(b)** $\text{Speed of bus}=\frac{\text{Distance}}{\text{Time}}$

$$=\frac{572}{13}=44\text{ kmph}$$

74. **(c)** V be the speed of train

$$V+9=\frac{210}{6\times1000}\times3600$$

$$V+9=126$$

$$V=117\text{ km/h}$$

Level-II

1. (d) When A covers 200 metres, B covers

$$200\times\frac{22}{25}=176\,\text{m}$$

So, B is $(200-176)=24$ m far away from the end point when A reaches in.

2. (b) Let the required distance be x km.

Difference in the times taken at two speeds

$$=12\text{ min}=\frac{1}{5}\text{hr}.$$

$$\therefore\ \frac{x}{5}-\frac{x}{6}=\frac{1}{5}\Leftrightarrow 6x-5x=6\Leftrightarrow x=6$$

Hence, the required distance is 6 km.

3. (c) Total distance travelled in 12 hours = (35 + 37 + 39 + ... upto 12 terms).

This is an A.P. with first term, $a=35$, number of terms, $n=12$, common difference. $d=2$.

$$\therefore\ \text{Required distance}=\frac{12}{2}[2\times 35+(12-1)\times 2]$$

$$=\ 6(70+22)=552\text{ km}.$$

4. (b) Let the speed of the train and the car be x km/h and y km/h, respectively.

$$\text{Now, }\frac{120}{x}+\frac{480}{y}=8 \qquad \text{...(1)}$$

$$\text{and }\frac{200}{x}+\frac{400}{y}=\frac{25}{3} \qquad \text{...(2)}$$

From (1), $120y+480x=8xy$ and ...(3)

From (2), $200y+400x=\frac{25}{3}xy$ ...(4)

From (3) and (4),

$$\frac{120y+480x}{8}=\frac{3(200y+400x)}{25}$$

or $15y+60x=24y+48x$

or $12x=9y$ or $\frac{x}{y}=\frac{3}{4}$ or $x:y=3:4$

5. (c) Remaining distance = 3 km and Remaining time

$$=\left(\frac{1}{3}\times 45\right)\text{ min}=15\text{ min}=\frac{1}{4}\text{ hour}.$$

$\therefore$ Required speed = (3×4) km/hr = 12 km/hr.

6. (a) Let the whole distance travelled be x km and the average speed of the car for the whole journey be y km/hr.

$$\text{Then, }\frac{(x/3)}{40}+\frac{(x/3)}{20}+\frac{(x/3)}{60}=\frac{x}{y}$$

$$\Leftrightarrow\ \frac{x}{30}+\frac{x}{60}+\frac{x}{180}=\frac{x}{y}$$

$$\Leftrightarrow\ \frac{1}{18}y=1$$

$\Leftrightarrow\ y=18$ km/hr.

7. (a) Speed of first train = 50 km/hr.

Speed of second train $=\frac{400}{7}$ km/hr.

At 8:00 AM distance between two trains is 100 kms.

Relative velocity

$$=50+\frac{400}{7}=\frac{350+400}{7}=\frac{750}{7}\text{km/h}$$

Time taken $=\frac{100\times 7}{750}\times 60=56$ min. Hence, the two trains meet each other at 8:56 AM.

8. (b) Let the speed of the stream be x km/hr and distance travelled be S km. Then,

$$\frac{S}{12+x}=6\text{ and }\frac{S}{12-x}=9$$

$$\Rightarrow\frac{12-x}{12+x}=\frac{6}{9}\Rightarrow 108-9x=72+6x$$

$$\Rightarrow 15x=36\Rightarrow x=\frac{36}{15}=2.4\text{ km/hr}.$$

9. (a) If the rate of the stream is x, then $2(4.5-x)=4.5+x$

$\Rightarrow 9-2x=4.5+x\Rightarrow 3x=4.5\Rightarrow x=1.5$ km/hr

10. (b) Distance covered = 187.5m, Time = 9 secs

$$\text{Relative speed}=\frac{187.5}{9}\times\frac{3600}{1000}=75\text{ km/hr}$$

As the trains are travelling in opposite directions, speed of goods train = 75 – 50 = 25 km/hr.

11. (d) Relative speed of both trains = 60 + 90 = 150 km / h

Total distance = 1.10 + 0.9 = 2 km

$$\therefore\ \text{Required time}=\frac{2\times 60\times 60}{150}=48\text{ seconds}.$$

12. (c) Let the speed of the train be x km/hr and that of the car be y km/hr.

$$\text{Then, }\frac{120}{x}+\frac{480}{y}=8\text{ or }\frac{1}{x}+\frac{4}{y}=\frac{1}{15} \qquad \text{...(1)}$$

And, $\frac{200}{x}+\frac{400}{y}=\frac{25}{3}$ or $\frac{1}{x}+\frac{2}{y}=\frac{1}{24}$ ...(2)

Solving (1) and (2), we get $x = 60$ and $y = 80$.

∴ Ratio of speeds = 60 : 80 = 3 : 4.

13. (c) Suppose they meet x hrs after 8 a.m. Then,

(Distance moved by first in x hrs) + [Distance moved by second in $(x - 1)$ hrs] = 330

$\therefore 60x + 75(x - 1) = 330$

$\Rightarrow x = 3$

So, they meet at (8 + 3), i.e. 11 a.m.

14. (a) Total journey = 180 km

$\frac{1}{3}$rd of journey $=\frac{180}{3}=60$ km.

If usual speed be x kmph, then

$$\frac{60}{\frac{3x}{4}}-\frac{60}{x}=\frac{1}{2}$$

$$\Rightarrow \frac{80}{x}-\frac{60}{x}=\frac{1}{2}$$

$$\Rightarrow \frac{80}{x}-\frac{60}{x}=\frac{1}{2}$$

$\Rightarrow x = 40$ kmph

15. (a) If the rowing speed in still water be x kmph, and the distance be y km, then

$$\frac{y}{x-2}=6$$

$\Rightarrow y = 6(x - 2)$ ...(1)

and, $\frac{y}{x+2}=4$

$\Rightarrow y = 4(x + 2)$ ...(2)

$\Rightarrow 6(x - 2) = 4(x + 2)$

$\Rightarrow x = 10$ kmph

16. (a) d = product of speed $\left[\frac{\text{difference of time}}{\text{difference of speed}}\right]$

$d=\frac{4\times5}{60}\left[\frac{10-(-5)}{5-4}\right]$ [Here, –ve sign indicates before the schedule time]

$\Rightarrow d = 5$ km

17. (a) Let the distance be x km. Let speed of train be y km/h. Then by question, we have

$\frac{x}{y+4}=\frac{x}{y}-\frac{30}{60}$ ...(1)

and $\frac{x}{y-2}=\frac{x}{y}+\frac{20}{60}$ ...(2)

On solving (1) and (2), we get $x = 3y$

Put $x = 3y$ in (1) we get

$$\frac{3y}{y+4}=3-\frac{1}{2} \Rightarrow y=20$$

Hence, distance = 20 × 3 = 60 km.

18. (a) Let each side of the square be x km and let the average speed of the plane around the field be y km/h. Then,

$$\frac{x}{200}+\frac{x}{400}+\frac{x}{600}+\frac{x}{800}=\frac{4x}{y}$$

$$\Rightarrow \frac{25x}{2400}=\frac{4x}{y} \Rightarrow y=\left(\frac{2400\times4}{25}\right)=384.$$

∴ Average speed = 384 km/h.

19. (c) Here, distance to be covered by the thief and by the owner is same.

Let after 2 : 30 p. m., owner catches the thief in t hrs.

$$\text{Then, } 60\times t=75\left(t-\frac{1}{2}\right) \Rightarrow t=\frac{5}{2}\text{ hrs}$$

So, the thief is overtaken at 5 p.m.

20. (c) Let the speed of the cars be x km/h and y km/h, respectively.

Their relative speeds when they are moving in same direction = $(x - y)$ km/h.

Their relative speeds when they are in opposite directions = $(x + y)$ km/h.

Now, $\frac{70}{x+y}=1$ or $x + y = 70$ ...(1)

and $\frac{70}{(x-y)}=7$ or $x - y = 10$ ...(2)

Solving (1) and (2), we have

$x = 40$ km/h and $y = 30$ km/h.

21. (b) Volume of water flowed in an hour

= 2000 × 40 × 3 cubic metre = 240000 cubic metre

∴ volume of water flowed in 1 minute

$=\frac{240000}{60}=4000$ cubic metre = 40,00,000 litre

22. (c) Initial distance = 25 dog leaps.

Per minute → dog makes 5 dog leaps

Per minute → Cat makes 6 cat leaps = 3 dog leaps.

Relative speed = 2 dog leaps/minutes.

An initial distance of 25 dog leaps would get covered in 12.5 minutes.

23. **(a)** Form the equations first and then use the options.

24. **(b)**

	Ram	:	Sham
Speed	7	:	4
Time	4	:	7
Distance	4	:	7

Now, $7x - 4x = 300$

Means $x = 100$

Therefore, the winning post is $7 \times 100 = 700$ m away from the starting point

25. **(d)** The watch gains (5 + 10) = 15 min in 30 hours (12 Noon to 6 PM next day). This means that it will show the correct time when it gains 5 min in 10 hours or at 10 PM on Monday.

26. **(b)** The train needs to travel 15 minutes extra @35 kmph. Hence, it is behind by 8.75 kms. The rate of losing distance is 5 kmph. Hence, the train must have travelled for 8.75/5 = 1 hour 45 minutes. @40 kmph → 70 km.

Alternatively, you can also see that 12.5% drop in speed results in 14.28% increase in time. Hence, total time required is 105 minutes @ 40 kmph → 70 kilometers.

Alterntively, solve through options.

27. **(c)** Let the distance between the school and the home be x km.

Then, $\frac{x}{8} - \frac{2.5}{60} = \frac{x}{10} + \frac{5}{60}$ or $\frac{x}{8} - \frac{x}{10} = \frac{5}{60} + \frac{2.5}{60}$

or $\frac{2x}{80} = \frac{7.5}{60}$ or $x = \frac{7.5 \times 80}{2 \times 60} = 5$ km

28. **(b)** Relative speed of rockets

$= (42000 + 18000) = 60000$ mile/h

It means both of them together cover a distance of 60000 miles between themselves in 60 minutes or 1000 miles in 1 minute.

Hence, they should be 1000 miles apart, 1 minute before impact.

29. **(c)** Let the speed of the train be x m/sec. Then,

Distance travelled by the train in 10 min. = Distance travelled by sound in 30 sec.

$\Leftrightarrow x \times 10 \times 60 = 330 \times 30$

$\Leftrightarrow x = 16.5.$

$\therefore$ Speed of the train $= 16.5$ m/sec $= \left(16.5 \times \frac{18}{5}\right)$ km/hr

$= 59.4$ km/hr

30. **(c)** Let the speed of train C be x km/h.

At 9 p.m. the train A will have covered a distance of 180 km.

For trains A and B relative speed $= (90 - 60) = 30$ km/h

Distance between them = 180 km

Time after which they meet $= \frac{180}{30} = 6$ hrs

For trains A and C relative speeds $= (60 + x)$ km/h

Distance between them = 1080 km.

Time after which they meet $= \frac{1080}{(60 + x)}$ hrs

As the time of meeting of all the three trains is the same, we have $\frac{1080}{(60 + x)} = 6$

or $x = 120$ km/h

31. **(b)** Time taken by the boat during downstream journey $= \frac{50}{60} = \frac{5}{6}$ h

Time taken by the boat in upstream journey $= \frac{5}{4}$ h

Average speed $= \frac{2 \times 50}{\frac{5}{6} + \frac{5}{4}} = \frac{100 \times 24}{50} = 48$ mph

32. **(c)** Let the distance be x km. Then,

(Time taken to walk x km) + (Time taken to ride x km)

$= \frac{23}{4}$ hrs.

$\Rightarrow$ (Time taken to walk $2x$ km) + (Time taken to ride $2x$ km) $= \frac{23}{2}$ hrs.

But, time taken to ride $2x$ km $= \frac{15}{4}$ hrs.

$\therefore$ Time taken to walk $2x$ km $= \left(\frac{23}{2} - \frac{15}{4}\right)$ hrs $= \frac{31}{4}$ hrs

= 7 hrs 45 min.

33. **(a)** Let the speed of the boatman be x km/hr and that of stream by y km/hr. Then

$\frac{12}{x + y} = \frac{4}{x - y}$

$\Rightarrow 12x - 12y = 4x + 4y$

$\Rightarrow 8x = 16y \Rightarrow x = 2y$

Now $\frac{45}{x+y} + \frac{45}{x-y} = 20$

$\Rightarrow 45 + 135 = 60y \Rightarrow 180 = 60y \Rightarrow y = 3$km/hr.

34. (c) Required distance

$= \sqrt{8^2 + 15^2}$

$= \sqrt{64 + 225}$

$= \sqrt{289} = 17$ m

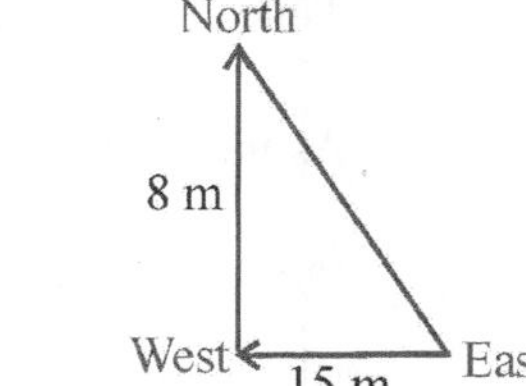

35. (b) Let the Speed of faster train be x and speed of slower train be y.

Now, when both the train move in same direction their relative speed = $x - y$

Now, total distance covered = 130 + 110 = 240

Now, distance = speed × time

$\therefore 240 = (x - y) \times 60$ ($\because$ 1 min = 60 sec)

$\Rightarrow x - y = 4$ ...(1)

When the trains move in opposite direction then their relative speed = $x + y$

$\therefore 240 = (x + y) \times 3$

$\Rightarrow 80 = x + y$ ...(2)

on solving eqs (1) and (2), we get $x = 42$ m/sec and $y = 38$ m/sec

36. (d) Let v_m = velocity of man = 48 m/min

Let v_c = velocity of current

then t_1 = time taken to travel 200 m against the current.

i.e., $t_1 = \frac{200}{v_m - v_c}$ ....(1)

and t_2 time taken to travel 200 m with the current

i.e., $t_2 = \frac{200}{v_m + v_c}$ ....(2)

Given : $t_1 - t_2 = 10$ min

$\therefore \frac{200}{v_m - v_c} - \frac{200}{v_m + v_c} = 10$

$\Rightarrow v_m^2 - v_c^2 = 40v_c \Rightarrow v_c^2 + 40v_c - (48)^2 = 0$

$\Rightarrow v_c = 32, -72$

Hence, speed of the current = 32 ($\because v_c \neq -72$).

37. (d) Let speed of current = v.m/min

$\frac{200}{48 - v} - \frac{200}{48 + v} = 10$

$20(48 + v) - 20(48 - v) = 48^2 - v^2$

$40v = 48^2 - v^2$

$v^2 + 40v - 2304 = 0$

$v = 32$ m/min.

38. (c) We know that, the relation in time taken with two different modes of transport is

$t_{\text{walk both}} + t_{\text{ride both}} = 2(t_{\text{walk}} + t_{\text{ride}})$

$\frac{31}{4} + t_{\text{ride both}} = 2 \times \frac{25}{4}$

$\Rightarrow t_{\text{ride both}} = \frac{25}{2} - \frac{31}{4} = \frac{19}{4} = 4\frac{3}{4}$ hrs

39. (d) Time difference between 8 am and 2 pm = 6 hrs.

Angle traced by the hour hand in 6 hours

$= \left(\frac{360}{12} \times 6\right)^\circ = 180^\circ$

40. (a) The dog loses 1/3rd of his normal time from the meeting point. (Thus normal time = 35 × 3 = 105 minutes)

If the meeting occurred 24 km further, the dog loses 25 minutes.

This means that the normal time for the new distance would be 75 minutes. Thus, normally the dog would cover this distance of 24 km in 30 minutes.

Thus, normal speed = 48 km/hr.

41. (b) When A covers 1000 m, B covers 960 m.

Similarly, when B covers 1000 m, C covers 975 m.

$\therefore$ When B covers 960 m, C covers $\frac{975}{1000} \times 960 = 936$ m.

Thus, A can give a start to C by a distance

= (1000 – 936) m = 64 m.

42. (a) In 2 minutes, he ascends = 1 metre

$\therefore$ 10 metres, he ascends in 20 minutes.

$\therefore$ He reaches the top in 21st minute.

43. (d) $\frac{40}{(B-S)}+\frac{55}{(B+S)}=13$

$\frac{30}{(B-S)}+\frac{44}{(B+S)}=10$

On solving these, we get $B = 8$ km/h, $S = 5$ km/h

$\therefore$ speed of Mallah in still water = 8 km/h

44. (c) Note here the length of the train in which passenger is travelling is not considered since we are concerned with the passenger instead of train. So, the length of the bridge will be directly proportional to the time taken by the passenger respectively.

$t \rightarrow$ Time

$l \rightarrow$ Length of bridge

Therefore, $\frac{t_1}{t_2}=\frac{l_1}{l_2}$

$\frac{7}{4}=\frac{280}{2}$

$\Rightarrow$ $x = 160$ m

45. (c) Speed of tiger = 40 m/min

Speed of deer = 20 m/min

Relative speed = 40 – 20 = 20 m/min

Difference in distances = 50 × 8 = 400 m

$\therefore$ Time taken in overtaking (or catching) $=\frac{400}{20}=20$ min

$\therefore$ Distance travelled in 20 min = 20 × 40 = 800 m

46. (d) $(6 - x) = (8 - 1.5x)$

$\Rightarrow$ $x = 4$ cm

So, it will take 4 hours to burn in such a way that they remain equal in length.

47. (c) The speeds of two persons is 108 km/h and 75 km/h. The first person covers 1080 km in 10 hours and thus he makes 12 rounds. Thus, he will pass over another person 12 times in any one of the direction.

48. (c) Angle between two hands at 3 : 10 am

= (90 + 5) – 60 = 35°

So, the required angle = 70°, after 3:10 am

Total time required to make 70° angle when minute-hand is ahead of hour-hand.

$=\frac{90+70}{11/2}=\frac{320}{11}$ min

So at 3h $\frac{320}{11}$ min the required angle will be formed.

49. (d) $(n + 1)$ times in n days.

50. (c) If you start at 12 noon, you would reach at 4 : 30 PM. You would be able to meet the train which left Mumbai at 8 AM, 9 AM, 10 AM, 11 AM, 12 Noon, 1 PM, 2 PM, 3 PM and 4 PM – a total of 9 trains.

51. (b) In 36 hours, there would be a gap of 8 minutes. The two watches would show the same time when the gap would be exactly 12 hours or 720 minutes.

The no. of 36 hour time frames required to create this gap = 720/8 = 90.

Total time= 90 × 36 = 3240 hours. Since this is divisible by 24, the watches would show 12 noon.

52. (c) The net time loss is 1/3% of 168 hours.

53. (d) Perimeter $= 4\times\sqrt{160000} = 1600$ m

$\text{Time}=\frac{\text{Distance}}{\text{Speed}}=\frac{1600\times5\times60}{5000}=96 \text{ min}$

54. (b) Circumference $=\frac{44}{7}\times35=220$ cm

Distance travelled in 1 minute $=\frac{33000}{60}=550$ m

Required no. of revolutions $=\frac{550\times100}{220}=250$

55. (d) Distance covered by the aeroplane in 9 h

= 9 × 756 = 6804 km

Speed of helicopter $=\frac{2\times6804}{48}=283.5$ km/h

$\therefore$ Distance covered by helicopter in 18 h

= 283.5 × 18 = 5103 km

56. (b) $\text{Average speed}=\frac{\text{Total distance covered}}{\text{Total time taken}}$

$=\frac{6+6+6+6}{\frac{6}{25}+\frac{6}{50}+\frac{6}{75}+\frac{6}{150}}\Rightarrow\frac{24}{6\left[\frac{1}{25}+\frac{1}{50}+\frac{1}{75}+\frac{1}{150}\right]}$

$=\frac{24\times300}{6\times24}\Rightarrow 50$ km/hr

57. (b) Radian covered in one second $=2\times\frac{22}{7}\times3.5$

Time required to covered 55 radian $=\frac{55}{2\times\frac{22}{7}\times3.5}=2.5$

58. (b) $D = S \times T$

$60 = S\times\left(\frac{45}{60}\right)$hr

$S=\frac{60\times60}{45}\Rightarrow 80$km/hr

Now, new speed = 80 – 15 = 65 km/hr.

$\therefore \quad \text{Time} = \frac{\text{Distance}}{\text{Speed}} = \frac{60}{65}\text{ hr.}$

or $\frac{60}{65} \times 60 \text{ min} = 55.38 \text{min}.$

Hence, time to taken by car to travel same distance is 55.38 min.

59. (b) Let speed of the second train = x km/hr.

Relative speed of trains = $(50 + x)$ km/hr.

Distance travelled by trains = (100 + 120) = 220 metres

Distance = Speed × Time

$$\left(\frac{220}{1000}\right)\text{km} = (50 + x)\text{ km/hr.} \times \left(\frac{6}{3600}\right)\text{hr}$$

$$50 + x = \frac{220 \times 3600}{1000 \times 6}$$

$50 + x = 132$

$x = 132 - 50 = 82$ km/hr

60. (c) Let T be the speed of train and C be the speed of car.

$$\frac{120}{T} + \frac{480}{C} = 8 \Rightarrow \frac{1}{T} + \frac{4}{C} = \frac{1}{15} \quad ...(1)$$

$$\frac{200}{T} + \frac{400}{C} = 8 + \frac{20}{60} \Rightarrow \frac{1}{T} + \frac{2}{C} = \frac{1}{24} \quad ...(2)$$

Subtracting (2) from (1)

$$\frac{2}{C}(2-1) = \frac{1}{15} - \frac{1}{24}$$

$$\frac{2}{C} = \frac{1}{40} \Rightarrow C = 80$$

$$\frac{1}{T} = \frac{1}{15} - \frac{4}{80}$$

$$\frac{1}{T} = \frac{1}{60} \Rightarrow T = 60$$

Required ratio = 60 : 80 = 3 : 4

61. (c) Let correct time to cover journey be t hours

$$70\left(t + \frac{12}{60}\right) = 80\left(t + \frac{3}{60}\right)$$

70t + 14 = 80t+ 4

10t = 10

t = 1 hour

62. (c) Let total distance be d.

$$\text{time taken for 60\% distance} = \frac{0.6d}{40} = \frac{3d}{200}\text{h}$$

$$\text{time taken for 20\% distance} = \frac{0.2d}{30} = \frac{d}{150}\text{h}$$

time taken for remaining 20% distance

$$= \frac{0.2d}{10} = \frac{d}{50}\text{h}$$

$$\text{average speed} = \frac{d}{\frac{3d}{200} + \frac{d}{150} + \frac{d}{50}}$$

$$= \frac{200 \times 150 \times 50}{22500 + 10000 + 30000} = \frac{200 \times 150 \times 50}{62500}$$

= 24 kmph

CHAPTER 11

PROGRESSIONS

ARITHMETIC PROGRESSIONS (A.P.)

A sequence of numbers which are either continuously increased or continuously decreased by a common difference found by subtracting any term of the sequence from the next term.

The following sequences of numbers are arithmeticprogressions:

(i) 5, 8, 11, 14, ...

(ii) – 6, – 1, 4, 9, 14, ...

(iii) 10, 7, 4, 1, – 2, – 5, ...

(iv) $p, p + q, p + 2q, p + 3q, ...$

In the arithmetic progression (i); 5, 8, 11 and 14 are first term, second term, third term and fourth term respectively. Common difference of this A.P. is found out either by subtracting 5 from 8, 8 from 11 or 11 from 14. Thus common difference = 3. Similarly, common difference of arithmetic progression (ii), (iii) and (iv) are 5, – 3 and q respectively. First term and common difference of an A.P. are denoted by a and d respectively. Hence

d of (i) A.P.= 3, d of (ii) A.P.= 5,

d of (iii) A.P. = – 3 and d of (iv) A.P. = q

n^{th} TERM OF AN A.P.

To find an A.P. if first term and common difference are given, we add the common difference to first term to get the second term and add the common difference to second term to get the third term and so on.

The standard form of an A.P. is

$$a, a + d, a + 2d, a + 3d, ...$$

Here 'a' is the first term and 'd' is the common difference. Also we see that coefficient of d is always less by one than the position of that term in the A.P. Thus n^{th} term of the A.P. is given by

$$T_n = a + (n - 1)\, d \qquad ...(1)$$

This equation (1) is used as a formula to find any term of the A.P.

If l be the last term of a sequence containing n terms, then

$$l = T_n = a + (n - 1)\, d$$

To find any particular term of any A.P., generally we put the value of a, n and d in the formula (i) and then calculate the required term.

For example to find the 25th term of the A.P. 6, 10, 14, 18, ... ; using the formula (i), we put the value of $a = 6$, $n = 25$ and $d = 4$ in formula and calculate as

Illustration 1: In an A.P. if $a = -7.2$, $d = 3.6$, $a_n = 7.2$, then find the value of n.

Solution: $a_n = a + (n - 1)\, d$

$\Rightarrow \quad 7.2 = -7.2 + (n - 1)(3.6)$

$\Rightarrow \quad 14.4 = (n - 1)(3.6)$

$\Rightarrow \quad n - 1 = 4 \Rightarrow n = 5.$

Illustration 2: Which term of the A.P. 21, 42, 63, ... is 420 ?

Solution: $420 = a_n = a + (n - 1)\, d$

[Here $a = 21$, $d = 42 - 21 = 21$]

$= 21 + (n - 1)\, 21$

$= 21n$

$\therefore \quad n = \dfrac{420}{21} = 20$

$\therefore$ required term is 20th term.

Illustration 3: Is – 150 a term of the A.P. 11, 8, 5, 2, ... ?

Solution: Here $a = 11$, $d = -3$

$-150 = a_n = a + (n - 1)\, d$

$= 11 + (n - 1)(-3)$

$= 11 - 3n + 3$

$= 14 - 3n$

$3n = 14 + 150$

$n = \dfrac{164}{3} = 54\dfrac{2}{3}$,

which is not possible because n is +ve integer.

$\therefore$ – 150 is not a term of the given A.P.

SUM OF FIRST n TERMS OF AN A.P.

Sum of first n terms means sum of from first term to n^{th} term.

Consider an A.P. whose first term and common difference are 'a' and 'd' respectively. Sum of first n terms S_n of this A.P. is given by

$$S_n = \frac{n}{2}\,[2a + (n - 1)\, d] \qquad ...(1)$$

If last term of an A.P. containing n terms be l, then n^{th} term $= l = a + (n-1)\,d$.

$$\therefore \quad S_n = \frac{n}{2}\,[2a + (n-1)\,d] = \frac{n}{2}\,[a + \{a + (n-1)\,d\}]$$

$$\Rightarrow \quad S_n = \frac{n}{2}\,(a + l) \qquad ...(2)$$

CONSIDERING THE TERMS IN AN A.P.

If sum of three consecutive terms of an A.P. is given, then if required consider the three consecutive terms as $(a - d)$, a and $(a + d)$. This reduces one unknown d thereby making the solution easier.

Similarly, we consider the four consecutive terms as $(a - 3d)$, $(a - d)$, $(a + d)$, $(a + 3d)$ and five consecutive terms as $(a - 2d)$, $(a - d)$, a, $(a + d)$ and $(a + 2d)$; if their sums are given otherwise consider three terms as a, $a + d$, $a + 2d$; four terms as a, $a + d$, $a + 2d$, $a + 3d$ and five terms as a, $a + d$, $a + 2d$, $a + 3d$, $a + 4d$.

USEFUL RESULTS

(i) **(a)** Sum of first n natural numbers

$$= 1 + 2 + 3 + ... + n = \frac{n\,(n+1)}{2}$$

(b) Sum of first n odd natural numbers

$$= 1 + 3 + 5 + ... + (2n - 1) = n^2$$

(c) Sum of first n even natural numbers

$$= 2 + 4 + 6 + ... + 2n = n\,(n + 1)$$

(d) Sum of odd numbers $\le n$

$$= \begin{cases} \left(\dfrac{n+1}{2}\right)^2, & \text{if } n \text{ is odd} \\ \left(\dfrac{n}{2}\right)^2, & \text{if } n \text{ is even} \end{cases}$$

(e) Sum of even numbers $\le n$

$$= \begin{cases} \dfrac{n}{2}\left(\dfrac{n}{2} + 1\right), & \text{if } n \text{ is odd} \\ \left(\dfrac{n-1}{2}\right)\left(\dfrac{n+1}{2}\right), & \text{if } n \text{ is even} \end{cases}$$

(ii) **(a)** Sum of squares of first n natural numbers

$$= 1^2 + 2^2 + 3^2 + ... + n^2 = \frac{n\,(n+1)\,(2n+1)}{6}$$

(b) Sum of cubes of first n natural numbers

$$= 1^3 + 2^3 + 3^3 + ... + n^3 = \left[\frac{n\,(n+1)}{2}\right]^2$$

= Square of the sum of first n natural numbers.

(iii) **(a)** $T_n = S_n - S_{n-1}$

(b) For A.P., $d = S_2 - 2S_1$

(iv) **(a)** In an A.P., the sum of terms equidistant from the beginning and end is constant and equal to the sum of first term and last term.

(b) If in an A.P. sum of p terms is equal to sum of q terms, then sum of $(p + q)$ terms is zero.

(c) If in an A.P., p^{th} term is q and q^{th} term is p then n^{th} term is $(p + q - n)$.

(d) If in an A.P., sum of p terms is q and sum of q terms is p, then sum of $(p + q)$ terms is $-(p + q)$.

Practice Exercise

Level - I

1. If the 4th term of an arithmetic progression is 14 and 12th term is 70, then the first term is
(a) – 10 (b) – 7
(c) +7 (d) + 10

2. Find the sum of all numbers in between 10–50 excluding all those numbers which are divisible by 8. (include 10 and 50 for counting.)
(a) 1070 (b) 1220
(c) 1320 (d) 1160

3. How many 3-digit numbers are completely divisible by 6?
(a) 149 (b) 150
(c) 151 (d) 166

4. $(11^2 + 12^2 + 13^2 + ... + 20^2) = ?$
(a) 385 (b) 2485
(c) 2870 (d) 3255

5. On March 1st 2016, Sherry saved ₹ 1. Everyday starting from March 2nd 2016, he save ₹ 1 more than the previous day. Find the first date after March 1st 2016 at the end of which his total savings will be a perfect square.
(a) 17th March 2016 (b) 18th April 2016
(c) 26th March 2016 (d) None of these

6. A man arranges to pay off a debt of ₹ 3,600 in 40 annual instalments which form an AP. When 30 of the instalments are paid, he dies leaving one-third of the debt unpaid. Find the value of the first instalment.
(a) 55 (b) 53
(c) 51 (d) 49

7. A number 15 is divided into three parts which are in AP and the sum of their squares is 83. Find the smallest number.
(a) 5 (b) 3
(c) 6 (d) 8

8. A boy agrees to work at the rate of one rupee on the first day, two rupees on the second day, four rupees on the third day and so on. How much will the boy get if he starts working on the 1st of February and finishes on the 20the of February ?
(a) 2^{20} (b) $2^{20} - 1$
(c) $2^{19} - 1$ (d) 2^{19}

9. What is the sum of all the two-digit numbers which when divided by 7 gives a remainder of 3?
(a) 94 (b) 676
(c) 696 (d) None of these

10. The sum of the 6th and 15th terms of an arithmetic progression is equal to the sum of 7th, 10th and 12th terms of the same progression. Which term of the series should necessarily be equal to zero ?
(a) 10th (b) 8th
(c) 1st (d) None of these

11. If the mth term of an AP is $1/n$ and nth term is $1/m$, then find the sum to mn terms.
(a) $(mn - 1)/4$ (b) $(mn + 1)/4$
(c) $(mn + 1)/2$ (d) $(mn - 1)/2$

12. Find the value of $1 - 2 - 3 + 2 - 3 - 4 + ... +$ upto 100 terms.
(a) –694 (b) –626
(c) –624 (d) –549

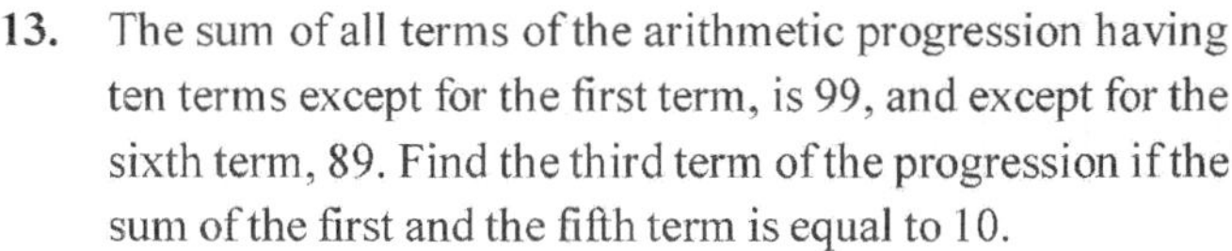

13. The sum of all terms of the arithmetic progression having ten terms except for the first term, is 99, and except for the sixth term, 89. Find the third term of the progression if the sum of the first and the fifth term is equal to 10.
(a) 15 (b) 5
(c) 8 (d) 10

14. There are 4 terms in an A.P. such that the sum of two means is 110 and product of their extremes is 2125. The 3rd term is *[SSC-Sub. Ins.-2012]*
(a) 65 (b) 75
(c) 55 (d) 45

15. The first term of an Arithmetic Progression is 22 and the last term is –11. If the sum is 66, the number of terms in the sequence are : *[SSC 10+2-2014]*
(a) 10 (b) 12
(c) 9 (d) 8

16. The sum of the squares of three consecutive natural numbers is 194. The sum of the numbers is *[SSC 10+2-2014]*
(a) 24 (b) 27
(c) 21 (d) 30

Level - II

1. Each of the series 13 + 15 + 17 + ... and 14 + 17 + 20 + ... is continued to 100 terms. Find how many terms are identical between the two series?
(a) 35 (b) 34
(c) 32 (d) 33

2. The sum of thirty-two consecutive natural numbers is a perfect square. What is the least possible sum of the smallest and the largest of the thirty-two numbers?
(a) 81 (b) 36
(c) 49 (d) 64

3. The middle term of arithmetic series 3, 7, 11...147, is
(a) 71 (b) 75
(c) 79 (d) 83

4. If a man saves ₹ 4 more each year than he did the year before and if he saves ₹ 20 in the first year, after how many years will his savings be more than ₹ 1000 altogether?
(a) 19 years (b) 20 years
(c) 21 years (d) 18 years

5. What is the maximum sum of the terms in the arithmetic progression 25, 24½, 24, ...?
(a) 637½ (b) 625
(c) 662½ (d) 650

6. $\left(1-\frac{1}{n}\right)+\left(1-\frac{2}{n}\right)+\left(1-\frac{3}{n}\right)+\ldots$ upto n terms = ?
(a) $\frac{1}{2}n$ (b) $\frac{1}{2}(n-1)$
(c) $\frac{1}{2}n(n-1)$ (d) None of these

7. If $1^3 + 2^3 + ... + 9^3 = 2025$, then the value of $(0.11)^3 + (0.22)^3 + ... + (0.99)^3$ is close to:
(a) 0.2695 (b) 0.3695
(c) 2.695 (d) 3.695

8. How many terms are identical in the two APs 1,3, 5,... up to 120 terms and 3, 6, 9, ... up to 80 terms ?
(a) 38 (b) 39
(c) 40 (d) 41

9. If the sum of the first $2n$ terms of the AP 2, 5, 8 ....is equal to the sum of first n terms of the AP 57, 59, 6 ..., then what is the value of n?
(a) 7 (b) 9
(c) 11 (d) 13

10. If the positive real numbers a, b and c are in Arithmetic Progression, such that $abc = 4$, then minimum possible value of b is :
(a) $2^{\frac{3}{2}}$ (b) $2^{\frac{2}{3}}$
(c) $2^{\frac{1}{3}}$ (d) None of these

11. After striking a floor a rubber ball rebounds $(7/8)^{th}$ of the height from which it has fallen. Find the total distance that it travels before coming to rest, if it is gently dropped from a height of 420 meters?
(a) 2940 (b) 6300
(c) 1080 (d) 3360

12. On the ground 12 stones are placed. The distance between the first and the second is 1 metre, between second and 3rd 3 m, between 3rd and 4th 5 m, and so on. How far will a boy have to run to touch the last stone if he starts from the first?
(a) 144m (b) 121m [*SBI PO-2011*]
(c) 132m (d) 110m
(e) None of these

Hints & Solutions

Level-I

1. (a) $a_4 = a + (4-1) \times d$

$14 = a + 3d \Rightarrow a = 14 - 3d$ ...(1)

Also, $70 = a + 11d$ ...(2)

After putting the value of a from equation (1) in equation (2)

$14 - 3d + 11d = 70$

$8d = 70 - 14$

$\therefore d = 8$

$\therefore a = 14 - 24 = -10$

2. (a) The answer will be given by:

[10 + 11 + 12 + ... + 50] – [16 + 24 + ... + 48]

= 41 × 30 – 32 × 5

= 1230 – 160 = 1070.

3. (b) 3-digit numbers divisible by 6 are

102, 108, 114, ..., 996

This is an A.P. in which $a = 102$, $d = 6$ and $\ell = 996$

Let the number of terms be n. Then $t_n = 996$.

$\therefore \quad a + (n-1)d = 996 \Rightarrow 102 + (n-1) \times 6 = 996$

$\Rightarrow \quad 6 \times (n-1) = 894 \Rightarrow (n-1) = 149 \Rightarrow n = 150$

$\therefore$ Number of terms = 150.

4. (b) $(11^2 + 12^2 + 13^2 + ... + 20^2) = (1^2 + 2^2 + ... + 30^2) - (1^2 + 2^2 + .. + 10^2)$

$$= \frac{20 \times 21 \times 41}{6} - \frac{10 \times 11 \times 21}{6}$$

$$\left[\because (1^2 + 2^2 + ... + n^2) = \frac{1}{6}n(n+1)(2n+1)\right]$$

= 2870 – 385 = 2485

5. (d) $n(n+1)/2$ should be a perfect square. The first value of n when this occurs would be for $n = 8$. Thus, on the 8th of March the required condition would come ture.

6. (c) Sum of 40 instalments $= S_{40} = 3600 = 20(2a + 39d)$

or $2a + 39d = 180$ ...(1)

Sum of 30 instalments $= S_{30} = 2400 = 15(2a + 29d)$

or $2a + 29d = 160$ ...(2)

From (1) and (2), we get $a = 51$ and $d = 2$

The value of first instalment = ₹ 51

7. (b) The three parts are 3, 5 and 7 since $3^2 + 5^2 + 7^2 = 83$. Since, we want the smallest number, the answer would be 3.

8. (b) Sum of a G.P. with first term 1 and common ratio 2 and number of terms 20.

$$\frac{1 \times (2^{20} - 1)}{(2-1)} = 2^{20} - 1$$

9. (b) This series is like → 10, 17, 21, ...94.

Here $n = 13$, $d = 7$ and $a = 10$

Using the formula for the sum

$$S_n = \frac{n}{2}\left[2a + (n-1)d\right], \text{sum} = 676$$

Alternatively, using the average method,

average = (1st number + last number)/2

$$\text{Average} = \frac{10 + 94}{2} = 52$$

So, the sum = average × number of numbers

= 52 × 13 = 676

10. (b) Let the first term and common difference of the AP be a and d, respectively.

Now, $(a + 5d) + (a + 14d)$

$= (a + 6d) + (a + 9d) + (a + 11d)$

or $\quad 2a + 19d = 3a + 26d$

or $a + 7d = 0$

i.e., 8th term is 0.

11. (c) The A.P. will become

1/6, 1/3, 1/2, 2/3, 5/6, 1

or in decimal terms, 0.166, 0.333, 0.5, 0.666, 0.833, 1

Sum to 6 terms = 3.5

Check the option with m = 2 and n = 3. Only option (c) gives 3.5.

12. (b) The first 100 terms of this series can be viewed as:

$(1-2-3) + (2-3-4) + ... + (33-34-35) + 34$

The first 33 terms fo the above series (indicated inside the brackets) will give an A.P: – 4, – 5, –6 ... –36

Sum of this A.P. = 33 × –20 = –660

Answer = –660 + 34 = – 626

13. (b) Sum of the first term and the fifth term = 10

or $\quad a + a + 4d = 10$

or $\quad a + 2d = 5$ (1)

and, the sum of all terms of the A.P. except for the 1st term = 99

or $\quad 9a + 45d = 99$

$a + 5d = 11$ ...(2)

Solve (1) and (2) we $a = 1$, $b = 2$ to get the answer.

14. (a) Let the 4 terms in A.P are a – 3d, a – d, a + d, a + 3d

According to question

a – d + a + d = 110 ... (1)

(a – 3d) (a + 3d) = 2125 ... (2)

From equation (1)

a – d + a + d = 110

2a = 110 ⇒ a = 55

From equation (2)

(a – 3d) (a + 3d) = 2125

$\Rightarrow a^2 - 9d^2 = 2125$

$\Rightarrow (55)^2 - 9d^2 = 2125$
$\Rightarrow 3025 - 9d^2 = 2125$
$\Rightarrow 900 = 9d^2$
$\Rightarrow d^2 = 100$
$\Rightarrow d = 10$
$\therefore\ a = 55, d = +10$
series would be :
25, 45, 65, 85
IIIrd term would be 65.

15. (b) The sum of Arithmetic Progression is given by

$$s = \frac{n}{2}(a + l)$$

$$66 = \frac{n}{2}(22 - 11) = \frac{n}{2} \times 11$$

$n = 12$

16. (a) $(a-1)^2 + a^2 + (a+1)^2 = 194$
$(a^2 - 2a + 1) + a^2 + (a^2 + 2a + 1) = 194$
$3a^2 + 2 = 194$
$3a^2 = 192$
$a^2 = 64$
$a = 8$
Number are 7, 8, 9
Sum of numbers = 7 + 8 + 9 = 24

Level-II

1. (d) The two series till their hundredth terms are 13, 15, 17 .... 211 and 14, 17, 20 ... 311. The common terms of the series would be given by the series 17, 23, 29 . ... 209. The number of terms in this series of common terms would be 192/6 + 1 = 33.

2. (c) Let the numbers be a, $a + 1$, $a + 2$, ..., $a + 31$.

Sum of these numbers $32a + \frac{31 \times 32}{2} = 16(2a + 31)$

As 16 is a perfect square, the least possible value of $2a + 31 = 49$.

Therefore, $a = 9$ and $a + 31 = 40$.

The least possible sum = 49.

3. (b) 3, 7, 11 147
It is an arithmetic series whose
first term, $a = 3$
last term, $x_n = 147$
common difference, $d = 4$
$x_n = a + (n-1)d$
$147 = 3 + (n-1) \times 4$

$$n - 1 = \frac{147 - 3}{4}$$

$n - 1 = 36, n = 37$

The given series consists of 37 terms. Therefore, its middle term will be

$$\frac{37+1}{2} = 19\text{th term}$$

$x_{19} = 3 + (19-1)4$
$= 3 + 18 \times 4 = 75$

$\therefore$ The middle term of the given arithmetic series is 75.

4. (a) We need the sum of the series 20 + 24 + 28 to cross 1000. Trying out the options, we can see that in 20 years the sum of his savings would be: 20 + 24 + 28 + ... + 96. The sum of this series would be 20 × 58 = 1160. If we remove the 20th year we will get the series for savings for 19 years. The series would be 20 + 24 + 28 + ... 92. Sum of the series would be 1160 – 96 = 1064. If we remove the 19th years's savings the savings would be 1064 – 92 which would go below 1000. Thus, after 19 years his savings would cross 1000.

5. (a) The maximum sum would occur when we take the sum of all the positive terms of the series.
The series 25, 24.5, 24, 23.5, 23, ... 1, 0.5, 0 has 51 terms. The sum of the series would be given by: $n \times$ average = 51 × 12.5 = 637.5

6. (b) Given sum = (1 + 1 + 1 + ... to n terms)

$$-\left(\frac{1}{n} + \frac{2}{n} + \frac{3}{n} + \text{....to } n \text{ terms}\right)$$

$$= n - \frac{n}{2}\left(\frac{1}{n} + 1\right) \qquad \left[\because \ell = \text{nth terms} = \frac{n}{n} = 1\right]$$

$$= n - \frac{n+1}{2} = \frac{1}{2}(n-1)$$

7. (c) $(0.11)^3 + (0.22)^3 + ... + (0.99)^3$
$= (0.11)^3 (1^3 + 2^3 + ... + 9^3)$
$= 0.001331 \times 2025 = 2.695275 = 2.695.$

8. (c) The first series is 3, 5, 7 239
While the second series is 3, 6, 9 ... 240
Hence, the last common term is 237.

Thus our answer becomes $\frac{237-3}{6} + 1 = 40$

9. (c) The equation can be written as:

$$\frac{2n}{2}\left[4 + (2n-1)3\right] = \frac{n}{2}\left[114 + (n-1)2\right]$$

$\therefore\quad n = 11$

10. (b) Product of three numbers a, b and c in A.P. is 4
i.e., $abc = 4$, a constant.
Hence, the minimum possible value of $b = (4)^{1/3}$
$= (2)^{2/3}$

11. (b) The sum of the total distance it travels would be given by the infinite sum of the series:
420 × 8/1 + 367.5 × 8/1 = 3360 + 2940 = 6300.

12. (b)

1	2	3	4
1 m	3 m	5 m	....

Sum of first n odd no. = n^2
Ans = $(12 - 1)^2 = 121$

CHAPTER 12

LINEAR EQUATIONS

LINEAR EQUATIONS

Many times in mathematics, we have to find the value of an unknown. In this case we represent the unknown by using some letters like p, q, r, x, y etc. These letters are then called as the variable representations of the unknown quantity.

Let's see a problem: A man says, "I am thinking of a number, when I divide it by 3 and then add 5, my answer is twice the number thought of ". Find the number.

Although you do not have the actual number in your mind, you can still move ahead to solve the problem by assuming a variable to represent the number.

The information given in the problem related to the number ultimately will give the value of the unknown i.e., the number in this particular problem. See the process involved in solving the above problem:

Let the number be x.

On dividing the number x by 3, we get $\frac{x}{3}$.

On adding 5 to $\frac{x}{3}$, we get $\frac{x}{3}+5$.

According to the information given in the problem, $\frac{x}{3}+5$ is twice the number i.e., $2x$.

$$\therefore \quad \frac{x}{3}+5 = 2x$$

$$\Rightarrow \quad 5 = 2x - \frac{x}{3}$$

$$\Rightarrow \quad 5 = \frac{6x-x}{3}$$

$$\Rightarrow \quad 5 = \frac{5x}{3}$$

$$\Rightarrow \quad 5 \times 3 = 5x$$

$$\Rightarrow \quad \frac{5\times 3}{5} = x \Rightarrow x = 3$$

Hence, required number = 3

Here, '$\frac{x}{3}+5=2x$' is the mathematical statement of equality involving the variable x.

Each mathematical statement of equality involving any number of variables is called an equation. Note that in the above equation there is a single variable x, but according to the given and required information, you may have to suppose more than one variable to move ahead to solve the problem and hence, an equation may have one or more than one variable. If all the variables in the equation are in numerator, no product or quotient (of the expressions including variable(s)) is available in the equation and the power of each variable is unity, then the equation is called linear equation.

Linear equations are commonly used in CAT and Cat like Apptitute tests.

See the following illustration, whose solutions will be found out by converting the statements of the problems into linear equation(s).

Illustration 1: Find the two consecutive even numbers whose sum is 76.

Solution: Let one of the two consecutive even numbers be x.

As we know that the difference between any two consecutive even number is always 2. Therefore the next consecutive even number will be $(x + 2)$.

According to the question, sum of the two consecutive even numbers is 76.

$$\therefore \quad x + (x + 2) = 76$$

$$\Rightarrow \quad 2x + 2 = 76, \Rightarrow 2x = 76 - 2 = 74$$

$$\Rightarrow \quad x = \frac{74}{2} = 37$$

Hence the two consecutive numbers are 37 and 39.

Note that 'the difference between any two consecutive even numbers is always 2' is an information related to the variable x is an extra information because it is not given in the problem, but without this information, we would not form the equation required for solving the problem. Thus you must use the extra information, which helps in formation of equation, if needed.

Illustration 2: Sanjay starts his job with a certain monthly salary and earns a fixed increment every year. If his salary was ₹ 31,000 after four years of service and ₹ 40,000 after 10 years, find his initial salary and annual increment.

Solution: Let the initial salary be ₹ x and fixed increment every year be ₹ y.

$\therefore \quad x + 4y = 31000 \quad ...(1)$

and $\quad x + 10y = 40000 \quad ...(2)$

On subtracting equation (1) from (2), we get

$$6y = 9000 \Rightarrow y = 1500$$

Now putting the value of y in equation (1), we get

$$x + 6000 = 31000 \Rightarrow x = 25000$$

Hence initial salary = ₹ 25000

and fixed annual increment = ₹ 1500.

Illustration 3: If a number is decreased by 4 and divided by 6, the result is 8. What would be the result if 2 was subtracted from the number and then it was divided by 5?

(a) $9\frac{2}{3}$ (b) 10

(c) $10\frac{1}{5}$ (d) $11\frac{1}{5}$

Solution: (b) Let the number be x. Then,

$$\frac{x-4}{6} = 8 \Rightarrow x - 4 = 48 \Rightarrow x = 52$$

$$\therefore \quad \frac{x-2}{5} = \frac{52-2}{5} = \frac{50}{5} = 10.$$

Illustration 4: If three numbers are added in pairs, the sums equal 10, 19 and 21. The numbers are

(a) 4, 6, 10 (b) 6, 4, 15

(c) 3, 5, 10 (d) 2, 5, 15

Solution: (b) Let the numbers be x, y and z. Then,

$x + y = 10$ ..... (1) $\quad y + z = 19$ ..... (2) $\quad x + z = 21$ ..... (3)

Adding (1), (2) and (3), we get : $2(x + y + z) = 50$

or $\quad x + y + z = 25.$

Thus, $\quad x = 25 - 19 = 6; y = 25 - 21 = 4;$

$z = 25 - 10 = 15.$

Hence, the required numbers are 6, 4 and 15.

Illustration 5: If the sum of two numbers is 42 and their product in 437, then find the absolute difference between the numbers.

(a) 4 (b) 7

(c) 9 (d) Cannot be determined

Solution: (a) Let the numbers be x and y. Then, $x + y = 42$ and $xy = 437$.

$$x - y = \sqrt{(x+y)^2 - 4xy} = \sqrt{(42)^2 - 4 \times 437}$$

$$= \sqrt{1764 - 1748} = \sqrt{16} = 4.$$

$\therefore$ Required difference = 4.

Note that depending upon the number of variables in a problem, a linear equation may have one, two or even more variables. But to get the value of the variables the number of equations should be always equal to the number of variables.

STEPS TO BE FOLLOWED TO SOLVE A WORD PROBLEM USING LINEAR EQUATION(S)

Step (i): Read the problem carefully and note what is/are given and what is/are required.

Step (ii): Denote the unknown quantity by some letters, say p, q, r, x, y etc.

Step (iii): Translate the statements of the problem into mathematical statements i.e., equations using the condition(s) given in the problem and extra information(s) related to the variable(s) derived from the statement(s) in the problem.

Step (IV): Solve the equation(s) for the unknown(s).

Step (V): Check whether the solution satisfies the equation(s).

Most of the time in solving the word problem you get struck. It could be due to one or more of the following four reasons:

Reason (i): You are not able to interpret one or more statements in the problem. In this case you concentrate on developing your ability to decode the mathematical meaning of the statement(s) in the problems.

Reason (ii): You have either not used all the information given in the problem or have used them in the incorrect order.

In such a case, go back to the problem and try to identify each statement and see whether you have utilized it or not. If you have already used all the information, then check whether you have used the information given in the problem in the correct order.

Reason (iii): Even though you might have used all the information given in the problem, you have not utilized some of the information completely.

In such a case, you need to review each part of each information given in the problem and look at whether any additional details can be derived out from the same informations. If derived any additional details, use them in forming or solving the equation(s). Sometimes a statement can be used for more than one perspective. In this case, if you have used that statement for one perspective, then using it in the other perspective will solve the problem.

Reason (iv): You are struck because the problem does not have a solution. In such a case, check the solution once and if it is correct go back to reason (i), (ii) and (iii).

Illustration 6: Find the two odd numbers whose sum is 12.

Solution: Let the two odd numbers are x and y.

Then $x + y = 12$

There is no other information about the two variable x and y.

Hence, there will be no other equation between the variable x and y. So, we can not find the exact solution of the problem. The equation formed above yields a set of possibilities for the value of x and y as (1, 11), (3, 9), (5, 7), (7, 5), (9, 3), (11, 1). One of these possibilities has to be the correct answer.

Illustration 7: A piece of wire is 80 metres long. It is cut into three pieces. The longest piece is 3 times as long as the middle-sized and the shortest piece is 46 metres shorter than the longest piece. Find the length of the shortest piece (in metres).

Solution: Let the length of the longest piece = a metres

Length of middle-sized piece = b metres

Since sum of the length of three pieces of wire = 80 metres

$\therefore$ length of shortest piece = $80 - (a + b)$ metres

Now $a = 3b$ ...(1)

and $80 - (a + b) = a - 46$ ...(2)

From (1) and (2),

$$80 - \left(a + \frac{a}{3}\right) = a - 46$$

$$\Rightarrow \quad 80 - \frac{3a + a}{3} = a - 46$$

$$\Rightarrow \quad 80 + 46 = a + \frac{4a}{3}$$

$$\Rightarrow \quad \frac{7a}{3} = 126 \Rightarrow a = 126 \times \frac{3}{7} = 54$$

$$\therefore \quad b = \frac{a}{3} = \frac{54}{3} = 18,$$

and $80 - (a + b) = 80 - (54 + 18) = 8$

Hence length of shortest piece = 8 metres.

Illustration 8: Mohan took five papers in an examination, where each paper was of 250 marks. His marks in these papers were in the ratio 6 : 8 : 10 : 12 : 15. In all papers together, Mohan obtained 70% of the total marks. Then find the number of papers in which he got more than 80% marks.

Solution: Ratio of marks obtained in five papers are

6 : 8 : 10 : 12 : 15.

Let marks obtained in five papers are $6x$, $8x$, $10x$, $12x$ and $15x$.

$$\therefore \; 6x + 8x + 10x + 12x + 15x = 5 \times 250 \times \frac{70}{100}$$

$$\Rightarrow 51x = 125 \times 7 \Rightarrow x = \frac{125 \times 7}{51} = 17 \text{ (approx.)}$$

Now $80\% \text{ of } 250 = 250 \times \frac{80}{100} = 200$

Now $10x = 170$, $12x = 12 \times 17 = 204$

Hence $6x, 8x, 10x < 200$

and $12x, 15x > 200$

Therefore Mohan got more than 80% in only two subjects.

Illustration 9: The sum of the digits of a two digit number is 16. If the number formed by reversing the digits is less than the original number by 18. Find the original number.

Solution: Let unit digit be x. Then tens digit = $16 - x$

$$\therefore \quad \text{Original number} = 10 \times (16 - x) + x = 160 - 9x.$$

On reversing the digits, we have x at the tens place and $(16 - x)$ at the unit place.

$$\therefore \quad \text{New number} = 10x + (16 - x) = 9x + 16$$

$$\text{Original number} - \text{New number} = 18$$

$$(160 - 9x) - (9x + 16) = 18$$

$$160 - 18x - 16 = 18$$

$$-18x + 144 = 18$$

$$-18x = 18 - 144 \Rightarrow 18x = 126$$

$$\Rightarrow \quad x = 7$$

$\therefore$ In the original number, we have unit digit = 7

Tens digit = 16 – 7 = 9

Thus, original number = 97

Illustration 10: The denominator of a rational number is greater than its numerator by 4. If 4 is subtracted from the numerator and 2 is added to its denominator, the new number becomes $\frac{1}{6}$. Find the original number.

Solution: Let the numerator be x.

Then, denominator = $x + 4$

$$\therefore \quad \frac{x - 4}{x + 4 + 2} = \frac{1}{6}$$

$$\Rightarrow \quad \frac{x - 4}{x + 6} = \frac{1}{6}$$

$$\Rightarrow \quad 6(x - 4) = x + 6$$

$$\Rightarrow \quad 6x - 24 = x + 6 \Rightarrow 5x = 30$$

$$\therefore \quad x = 6$$

Thus, Numerator = 6, Denominator = 6 + 4 = 10.

Hence the original number = $\frac{6}{10}$.

Illustration 11: A man covers a distance of 33 km in $3\frac{1}{2}$ hours; partly on foot at the rate of 4 km/hr and partly on bicycle at the rate of 10 km/hr. Find the distance covered on foot.

Solution: Let the distance covered on foot be x km.

$\therefore$ Distance covered on bicycle = $(33 - x)$ km

$$\therefore \text{ Time taken on foot} = \frac{\text{Distance}}{\text{Speed}} = \frac{x}{4} \text{ hr.}$$

$$\therefore \text{ Time taken on bicycle} = \frac{33 - x}{10} \text{ hr.}$$

$$\text{The total time taken} = \frac{7}{2} \text{ hr.}$$

$$\frac{x}{4} + \frac{33 - x}{10} = \frac{7}{2}$$

$$\frac{5x + 66 - 2x}{20} = \frac{7}{2}$$

$$6x + 132 = 140$$

$$6x = 140 - 132$$
$$6x = 8$$
$$x = \frac{8}{6} = 1.33 \text{ km.}$$

$\therefore$ The distance covered on foot is 1.33 km.

Illustration 12: The total age of A and B is 12 years more than the total age of B and C. C is how many years younger than A?

(a) 12 **(b) 24**
(c) C is elder than A **(d) Data inadequate**

Solution: (a) $(A + B) - (B + C) = 12 \Rightarrow A - C = 12.$

C is 12 year younger than A.

Illustration 13: The sum of four numbers is 64. If you add 3 to the first number, 3 is subtracted from the second number, the third is multiplied by 3 and the fourth is divided by 3, then all the results are equal. What is the difference between the largest and the smallest of the original numbers?

(a) 21 **(b) 27**
(c) 32 **(d) Cannot be determined**

Solution: (c) Let the four numbers be A, B, C and D.

Let $A + 3 = B - 3 = 3C = D/3 = x$ (let)

Then, $A = x - 3$, $B = x + 3$, $C = x/3$ and $D = 3x$.

$A + B + C + D = 64 \Rightarrow (x - 3) + (x + 3) + x/3 + 3x = 64$

$\Rightarrow 5x + x/3 = 64 \Rightarrow 16x = 192 \Rightarrow x = 12$

Thus, the numbers are 9, 15, 4 and 36.

$\therefore$ Required difference = (36 – 4) = 32.

CONDITION OF CONSISTENCY AND INCONSISTANCY OF A PAIR OF SIMULTANEOUS LINEAR EQUATIONS IN TWO VARIABLES

Let $a_1x + b_1y = c_1$ .... (i)

and $a_2x + b_2y = c_2$ .... (ii)

are a pair of simultaneous linear equations in two various x and y. Here a_1, b_1, c_1 and a_2, b_2, c_2 are coefficient of x, coefficient of y and real constants in equation (i) and (ii) respectively.

(a) Consistent with unique solution :

The two equations (i) and (ii) have an unique solution,

if $\frac{a_1}{a_2} \neq \frac{b_1}{b_2}$

(b) Consistent with infinite many solution :

The two equations (i) and (ii) have infinite many solution,

if $\frac{a_1}{a_2} = \frac{b_1}{b_2} = \frac{c_1}{c_2}$

(c) Inconsistent :

The two equations (i) and (ii) have no solution, if

$$\frac{a_1}{a_2} = \frac{b_1}{b_2} \neq \frac{c_1}{c_2}$$

Illustration 14: Find the value of k for which the system of linear equation:
$kx + 4y = k - 4$, $16x + ky = k$, has infinite many solutions.

Solution:

$kx + 4y = k - 4$ .... (1)

$16x + ky = k$ .... (2)

$a_1 = k,\ b_1 = 4,\ c_1 = -(k - 4)$

$a_2 = 16,\ b_2 = k,\ c_2 = -k$

Here condition for infinite solutions is

$$\frac{a_1}{a_2} = \frac{b_1}{b_2} = \frac{c_1}{c_2}$$

$$\Rightarrow \frac{k}{16} = \frac{4}{k} = \frac{k-4}{k}$$

$$\Rightarrow \frac{k}{16} = \frac{4}{k} \Rightarrow k^2 = 64 \quad x = \pm 8$$

Also, $\frac{4}{k} = \frac{k-4}{k} \Rightarrow 4k = k^2 - 4k$

$\Rightarrow k^2 - 8k = 0 \Rightarrow k(k - 8) = 0$

$\Rightarrow k = 0$ or $k = 8$ but $k = 0$ is not possible otherwise equation will be one variable.

$\therefore$ $k = 8$ is correct value for many solution.

Illustration 15: Check wehether the following given pair of equations has no solution, unique solution or infinite solutions.

$3x + 4y = 8$

$9x + 12y = 24$

Solution:

For these two equations

$a_1 = 3,\ a_2 = 9,\ b_1 = 4,\ b_2 = 12,\ c_1 = -8,\ c_2 = -24,$

$$\therefore \frac{a_1}{a_2} = \frac{b_1}{b_2} = \frac{c_1}{c_2}$$

Since, $\frac{3}{4} = \frac{4}{12} = \frac{-8}{-24}$

The above pair of equations will have infinite solutions.

Illustration 16: Find the value of P for which the given system of equations has only solution (i.e., unique solution)

$Px - y = 2$; $6x - 2y = 3$

Solution:

$Px - y = 2$ .... (1)

$6x - 2y = 3$ .... (2)

$a_1 = P,\ b_1 = -1,\ c_1 = -2$

$a_2 = 6,\ b_2 = -2,\ c_2 = -3$

Condition for unique solution is $\frac{a_1}{a_2} \neq \frac{b_1}{b_2}$

$$\Rightarrow \frac{P}{6} \neq \frac{+1}{+2} \Rightarrow P \neq \frac{6}{2} \Rightarrow P \neq 3$$

$\therefore$ P can have all real values except 3.

Practice Exercise

Level - I

1. If $(x-3)(2x+1)=0$, then the possible values of $2x+1$ are:
   (a) 0 only (b) 0 and 3
   (c) $-\frac{1}{2}$ and 3 (d) 0 and 7
2. Father is 5 years older than the mother and mother's age now is thrice the age of the daughter. The daughter is now 10 years old. What was father's age when the daughter was born?
   (a) 20 years (b) 15 years
   (c) 25 years (d) 30 years
3. A father told his son, "I was as old as you are at present, at the time of your birth," If the father is 38 years old now, what was the son's age five years back ?
   (a) 19 years (b) 14 years
   (c) 38 years (d) 33 years
4. When 24 is subtracted from a number, it reduces to its four-seventh. What is the sum of the digits of that number ?
   (a) 1 (b) 9
   (c) 11 (d) Data inadequate
5. If the sum of one-half and one-fifth of a number exceeds
   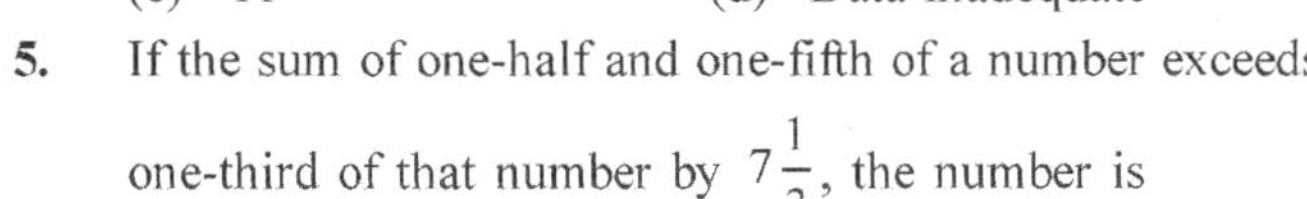
   one-third of that number by $7\frac{1}{3}$, the number is
   (a) 15 (b) 18
   (c) 20 (d) 30
6. A driver's income consists of his salary and tips. During one week his tips were 5/4 of his salary. What fraction of his income came from tips ?
   (a) $\frac{4}{9}$ (b) $\frac{5}{9}$
   (c) $\frac{5}{8}$ (d) $\frac{5}{4}$
7. In a certain party, there was a bowl of rice for every two guests, a bowl of broth for every three of them and a bowl of meat for every four of them. If in all there were 65 bowls of food, then how many guests were there in the party ?
   (a) 65 (b) 24
   (c) 60 (d) 48
8. Two numbers are such that the square of one is 224 less than 8 times the square of the other. If the numbers be in the ratio of 3 : 4, the numbers are
   (a) 36 (b) 48
   (c) 56 (d) 64
9. Ram and Mohan are friends. Each has some money. If Ram gives ₹ 30 to Mohan, then Mohan will have twice the money left with Ram. But if Mohan gives ₹ 10 to Ram, then Ram will have thrice as much as is left with Mohan. How much money does each have ?
   (a) ₹ 62, ₹ 34 (b) ₹ 6, ₹ 2
   (c) ₹ 170, ₹ 124 (d) ₹ 43, ₹ 26
10. The sum of two numbers is 25 and their difference is 13. Find their product.
    (a) 104 (b) 114
    (c) 315 (d) 325
11. A person on tour has ₹ 360 for his daily expenses. He decides to extend his tour programme by 4 days which leads to cutting down daily expenses by ₹ 3 a day. The number of days of his tour programme is
    (a) 15 (b) 20
    (c) 18 (d) 16
12. The difference between the squares of two numbers is 256000 and the sum of the numbers is 1000. The numbers are:
    (a) 600, 400 (b) 628, 372
    (c) 640, 360 (d) None of these
13. The sum of three consecutive odd numbers is 20 more than the first of these numbers. What is the middle number ?
    (a) 7 (b) 9
    (c) 11 (d) Data inadequate
14. The autorickshaw fare consists of a fixed charge together with the charge for the distance covered. For a journey of 10 km, the charge paid is ₹ 85 and for a journey of 15 km, the charge paid is ₹ 120. The fare for a journey of 25 km will be
    (a) ₹ 175 (b) ₹ 190
    (c) ₹ 180 (d) ₹ 225
15. The denominator of a rational number is greater than its numerator by 4. If 4 is subtracted from the numerator and 2 is added to its denominator, the new number becomes $\frac{1}{6}$. Find the original number.
    (a) $\frac{1}{6}$ (b) $\frac{6}{10}$
    (c) $\frac{10}{6}$ (d) 6
16. The present ages of Vikas and Vishal are in the ratio 15 : 8. After ten years, their ages will be in the ratio 5 : 3. Find their present ages.
    (a) 60 years, 32 years (b) 32 years, 60 years
    (c) 15 years, 8 years (d) 8 years, 15 years

17. The sum of three consecutive multiples of 3 is 72. What is the largest number ?
(a) 21 (b) 24
(c) 27 (d) 36

18. Two-fifths of one-fourth of three-sevenths of a number is 15. What is half of that number?
(a) 96 (b) 196
(c) 94 (d) None of these

19. The sum of the ages of a father and his son is 4 times the age of the son. If the average age of the father and the son is 28 years, what is the son's age?
(a) 14 years (b) 16 years
(c) 12 years (d) Data inadequate

20. The product of two numbers is 192 and the sum of these two numbers is 28. What is the smaller of these two numbers?
(a) 16 (b) 14
(c) 12 (d) 18

21. The sum of three consecutive even numbers is 14 less than one-fourth of 176. What is the middle number?
(a) 8 (b) 10
(c) 6 (d) Data inadequate

22. The difference between the numerator and the denominator of a fraction is 5. If 5 is added to its denominator, the fraction is decreased by $1\frac{1}{4}$. Find the value of the fraction.
(a) $\frac{1}{6}$ (b) $2\frac{1}{4}$
(c) $3\frac{1}{4}$ (d) 6

23. The sum of three numbers is 300. If the ratio between first and second be 5 : 9 and that between second and third be 9 : 11, then second number is
(a) 12 (b) 60
(c) 108 (d) 132

24. When 20 is subtracted from a number, it reduces to seven-twelve of the number. What is the sum of the digit of the number?
(a) 40 (b) 44
(c) 46 (d) 48

25. If the number obtained on the interchanging the digits of two-digit number is 18 more than the original number and the sum of the digits is 8, then what is the original number?
(a) 50 (b) 51
(c) 52 (d) 53

26. There are two numbers such that sum of twice the first number and thrice the second number is 100 and the sum of thrice the first number and twice the second number is 120. Which is the larger number?
(a) 32 (b) 12
(c) 14 (d) 35

27. There are two number such that the sum of twice the first number and thrice the second number is 300 and the sum of thrice the first number and twice the second number is 265. What is the larger number?
(a) 24 (b) 39
(c) 85 (d) 74

28. If the digits of a two-digit number are interchanged, the number formed is greater than the orginal number by 45. If the difference between the digits is 5, then what is the orginal number?
(a) 16 (b) 27
(c) 38 (d) Cannot be determined

29. Krishna has some hens and some goats. If the total number animal heads are 81 and the total number of animnal legs are 234, how many goats does Krishna have?
(a) 45 (b) 24
(c) 36 (d) Cannot be determined

30. The average age of father and his son is 22 years. The ratio of their ages is 10 : 1 respectively. What is the age of the son?
(a) 24 (b) 4
(c) 40 (d) 14

31. The sum of third, fourth and fifth part of a number exceeds half of the number by 34. Find the number.
(a) 60 (b) 120
(c) 30 (d) None of these

32. A series of books was published at seven years interval. When the seventh book was issued, the sum of the publication year was 13,524. When was the first book published?
(a) 1932 (b) 1942
(c) 1911 (d) 1917

33. In a two-digit number the digit in the unit's place is three times the digit in the tenth's place. The sum of the digits is equal to 8. Then, what is the number ?
(a) 20 (b) 26
(c) 39 (d) 13

34. The number obtained by interchanging the two digits of a two-digit number is lesser than the original number by 54. If the sum of the two-digit number is 10, then what is the original number ?
(a) 28 (b) 39
(c) 82 (d) Cannot be determined

35. The age of the father 5 years ago was 5 times the age of his son. At present the father's age is 3 times that of his son. What is the present age of the father?
(a) 33 years (b) 30 years
(c) 45 years (d) None of these

36. If the numerator of a fraction is increased by 150% and denominator of the fraction is increased by 350%. The resultant fraction is 25/31. What is the original fraction?
(a) $\frac{11}{7}$ (b) $\frac{11}{15}$
(c) $\frac{15}{17}$ (d) $\frac{13}{15}$

37. The denominator of a fraction is 2 more than thrice its numerator. If the numerator as well as denominator is increased by one, the fraction becomes 1/3. What was the original fraction?

(a) $\frac{4}{13}$ (b) $\frac{3}{11}$

(c) $\frac{5}{13}$ (d) $\frac{5}{11}$

38. Smita was asked to multiply a certain number by 36. She multiplied it by 63 instead and got an answer of 3834 more than the correct one. What was the number to be multiplied?

(a) 152 (b) 126

(c) 142 (d) 148

39. Ravi has spent a quarter $\left(\frac{1}{4}\right)$ of his life as a boy, one-fifth $\left(\frac{1}{5}\right)$ as a youth, one-third $\left(\frac{1}{3}\right)$ as man and thirteen (13) years in old age. What is his present age?

(a) 70 years (b) 80 years

(c) 60 years (d) 65 years

40. In a group of equal number of cows and herdsmen the number of legs was 28 less than four times the number of heads. The number of herdsmen was

(a) 7 (b) 28

(c) 21 (d) 14

41. The ratio of the present ages of a mother and daughter is 7 : 1. Four years ago the ratio of their ages was 19 : 1. What will be the mother's age four years from now?

(a) 42 years (b) 38 years

(c) 46 years (d) 36 years

42. A number when subtracted by $\frac{1}{7}$ of itself gives the same value as the sum of all the angles of a triangle. What is the number?

(a) 224 (b) 210

(c) 140 (d) 350

43. Farah got married 8 years ago. Today her age is $1\frac{2}{7}$ times her age at the time of her marriage. At present her daughter's age is one-sixth of her age. What was her daughter's age 3 years ago?

(a) 6 years (b) 4 years

(c) 3 years (d) None of these

44. There are some parrots and some tigers in a forest. If the total number of animal heads in the forest is 858 and the total number of animal legs is 1,846, what is the number of parrots in the forest?

(a) 845 (b) 833

(c) 800 (d) 793

45. The ratio between a two-digit number and the sum of the digits of that number is 4 : 1. If the digit in the unit's place is 3 more than the digit in the ten's place, what is the number?

(a) 36 (b) 63

(c) 39 (d) 93

46. The ratio of two numbers is 4 : 7. If each of these numbers increases by 30, their ratio will become 5 : 8 . What is the average of these two numbers?

(a) 135 (b) 145

(c) 155 (d) 165

47. A number of two digits has 3 for its unit's digit, and the sum of digits is $\frac{1}{7}$ of the number itself, The number is

(a) 43 (b) 53

(c) 63 (d) 73

48. A number is doubled and 9 is added. If the resultant is trebled, it becomes 75. What is that number?

(a) 3.5 (b) 6

(c) 8 (d) None of these

49. The difference between a two-digit number and the number obtained by interchanging the position of its digits is 36. What is the difference between the two digits of that number?

(a) 3 (b) 4

(c) 9 (d) Cannot be determined

50. 54 is to be divided into two parts such that the sum of 10 times the first and 22 times the second is 780. The bigger part is:

(a) 24 (b) 34

(c) 30 (d) 32

51. The sum of five whole numbers is 146. If m is the largest of the five numbers, then which is the smallest value that m can have

(a) 30 (b) 35

(c) 28 (d) 27

52. A man has ₹ 480 in the denominations of one-rupee notes, five-rupee notes and ten-rupee notes. The number of notes of each denomination is equal. What is the total number of notes that he has ?

(a) 45 (b) 60

(c) 75 (d) 90

53. If the numerator of a fraction is increased by 200% and the denominator is increased by 200%, then resultant fraction is $2\frac{4}{5}$. What is the original fraction?

(a) 4/7 (b) 13/12

(c) 11/12 (d) None of these

54. Cost of 36 pens and 42 pencils is ₹ 460/-. What is the cost of 18 pens and 21 pencils ? *[SBI Clerk-June-2012]*

(a) ₹ 230/- (b) ₹ 203/-

(c) ₹ 302/- (d) ₹ 320/-

(e) None of these

55. Bus fare between Raipur and Mirpur for one adult is six times the fare for one child. If an adult's bus fare is ₹ 114/, how much amount will be paid by 4 adults and 5 children together for travelling the same distance ? [*SBI Clerk-2012*]

(a) ₹ 505/- (b) ₹ 551/-
(c) ₹ 572/- (d) ₹ 560/-
(e) None of these

56. The sum of the two digits of a two-digit number is 15 and the difference between the two digits of the two digit number is 3. What is the product of the two digits of the two digit number ? [*SBI Clerk-2014*]

(a) 72 (b) 56
(c) 54 (d) Cannot be determined
(e) None of these

57. The shaded region represents [*SSC-Sub. Ins.-2012*]

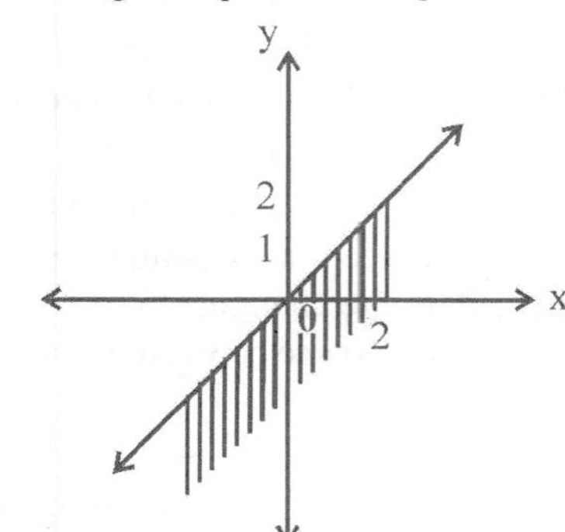

(a) $y \le x$ (b) $y \ge -x$
(c) $y \ge x$ (d) $y \le -x$

58. The graph of $y = x + |x|$ is given by [*SSC-Sub. Ins.-2012*]

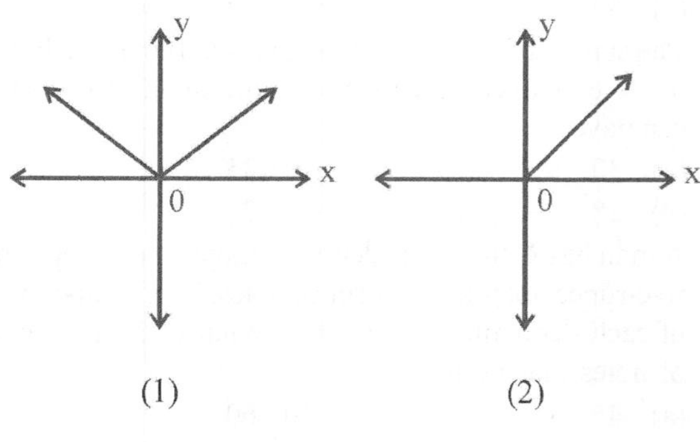

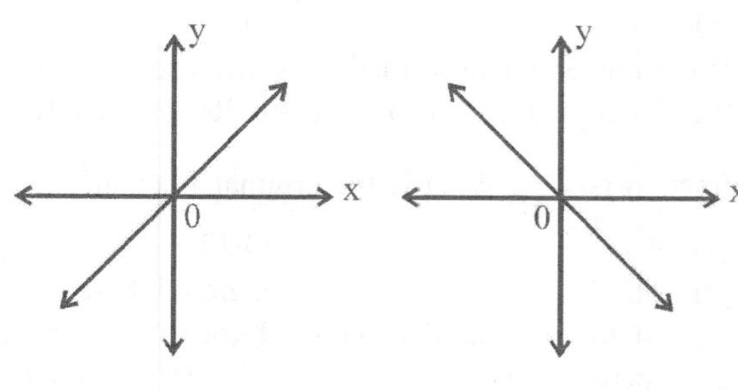

(a) 1 (b) 2
(c) 3 (d) 4

59. The equation of this graph is [*SSC-Sub. Ins.-2012*]

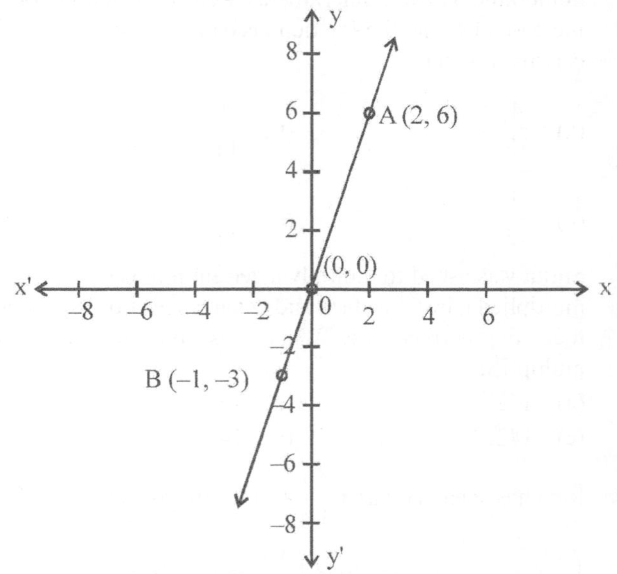

(a) $y = -x$ (b) $y = -3x$
(c) $y = x$ (d) $y = 3x$

60. The linear equation such that each point on its graph has an ordinate four times its abscissa is: [*SSC-Sub. Ins.-2013*]

(a) $y + 4x = 0$ (b) $y = 4x$
(c) $x = 4y$ (d) $x + 4y = 0$

61. A man ordered 4 pairs of black socks and some pairs of brown socks. The price of a black socks is double that of a brown pair. While preparing the bill the clerk interchanged the number of black and brown pairs by mistake which increased the bill by 50%. The ratio of the number of black and brown pairs of socks in the original order was : [*SSC-Sub. Ins.-2013*]

(a) 2 : 1 (b) 1 : 4
(c) 1 : 2 (d) 4 : 1

62. 11 friends went to a hotel and decided to pay the bill amount equally. But 10 of them could pay ₹ 60 each, as a result 11th has to pay ₹ 50 extra than his share. Find the amount paid by him. [*SSC-Sub. Ins.-2014*]

(a) ₹ 105 (b) ₹ 110
(c) ₹ 115 (d) ₹ 120

63. The present ages of two persons are 36 and 50 years respectively, if after *n* years the ratio of their ages will be 3 : 4, then the value of *n* is [*SSC-MT-2013*]

(a) 3 (b) 4
(c) 7 (d) 6

64. Number of solutions of the two equations $4x - y = 2$ and $2y - 8x + 4 = 0$ is [*SSC 10+2-2013*]

(a) infinitely many (b) zero
(c) one (d) two

65. Divide 50 into two parts so that the sum of their reciprocals is 1/12. *[SSC 10+2-2013]*
(a) 28, 22 (b) 35, 15
(c) 20, 30 (d) 24, 36

66. In a two-digit number, the digit at the unit's place is 1 less than twice the digit at the ten's place. If the digits at unit's and ten's place are interchanged, the difference between the new and the original number is less than the original number by 20. The original number is *[SSC 10+2-2013]*
(a) 47 (b) 59
(c) 23 (d) 35

67. The cost of 5 pens and 8 pencils is ₹ 31. What would be the cost of 15 pens and 24 pencils ? *[IBPS Clerk-2012]*
(a) ₹ 93 (b) ₹ 99
(c) ₹ 96 (d) Cannot be determined
(e) None of these

68. Joel purchased 40 notebooks at the rate of ₹ 18 per notebook and 55 pencils at the rate of ₹ 8 per pencil. What is the total amount that he paid to the shopkeeper ?*[IBPS Clerk-2012]*
(a) ₹ 1,165 (b) ₹ 1,160
(c) ₹ 1,166 (d) ₹ 1,161
(e) None of these

Level - II

1. The sum of the digits of a three-digit number is 16. If the tens digit of the number is 3 times the units digit and the units digit is one-fourth of the hundredth digit, then what is the number ?
(a) 446 (b) 561
(c) 682 (d) 862

2. A two digit number is such that the product of its digits is 14. When 45 is added to the number, then the digits interchange their places. Find the number.
(a) 72 (b) 27
(c) 37 (d) 14

3. When Ranjeev was born, his father was 32 years older than his brother and his mother was 25 years older than his sister. If Ranjeev's brother is 6 years older than Ranjeev and his mother is 3 years younger than his father, how old was Ranjeev's sister when he was born?
(a) 15 years (b) 14 years
(c) 7 years (d) 10 years

4. In an exercise room some discs of denominations 2 kg and 5 kg are kept for weightlifting. If the total number of discs is 21 and the weight of all the discs of 5 kg is equal to the weight of all the discs of 2 kg, find the weight of all the discs together.
(a) 80 kg (b) 90 kg
(c) 56 kg (d) None of these

5. One-third of Ramesh's marks in Arithmetic is equal to half his marks in English. If he gets 150 marks in the two subjects together, how many marks has he got in English?
(a) 60 (b) 120
(c) 30 (d) 50

6. The sum of four numbers is 64. If you add 3 to the first number, 3 is subtracted from the second number, the third is multiplied by 3 and the fourth is divided by 3, then all the results are equal. What is the difference between the largest and the smallest of the original numbers?
(a) 21 (b) 27
(c) 32 (d) Cannot be determined

7. In a family, a couple has a son and a daughter. The age of the father is three times of his daughter and the age of the son is half of his mother. The wife is nine years younger to her husband and the brother is seven years older than his sister. What is the age of the mother?
(a) 40 years (b) 50 years
(c) 45 years (d) 60 years

8. The sum of the numerator and denominator of a fraction is 11. If 1 is added to the numerator and 2 is subtracted from the denominator it becomes 3/2. The fraction is
(a) $\frac{3}{8}$ (b) $\frac{5}{6}$
(c) $\frac{7}{4}$ (d) $\frac{9}{2}$

9. In an objective examination of 90 questions, 5 marks are allotted for every correct answer and 2 marks are deducted for every wrong answer. After attempting all the 90 questions a student got a total of 387 marks. Find the number of questions that he attempted wrong.
(a) 36 (b) 18
(c) 9 (d) 27

10. Two different natural numbers are such that, their product is less than their sum. Then one of the number must be
(a) 3 (b) 1
(c) 2 (d) 0

11. Out of total number of students in a college 12% are interested in sports. $\frac{3}{4}$th the total number of students are interested in dancing. 10% of the total number of students are interested in singing and the remaining 15 students are not interested in any of the activities. What is the total number of students in the college?
(a) 450 (b) 500
(c) 600 (d) Cannot be determined

12. A number consists of two digits such that the digit in the ten's place is less by 2 than the digit in the unit's place. Three times the number added to $\frac{6}{7}$ times the number obtained by reversing the digits equals 108. The sum of digits in the number is :
(a) 8 (b) 9
(c) 6 (d) 7

13. When the numerator and the denominator of a fraction are increased by 1 and 2 respectively, the fraction becomes $\frac{2}{3}$, and when the numerator and the denominator of the same fraction are increased by 2 and 3 respectively, the fraction becomes $\frac{5}{7}$. What is the original fraction?
(a) $\frac{5}{6}$ (b) $\frac{3}{4}$
(c) $\frac{3}{5}$ (d) $\frac{6}{7}$

14. If three numbers are added in pairs, the sums equal 10, 19 and 21. Find the numbers.
(a) 6, 4, 15 (b) 1, 9, 12
(c) 9, 10, 2 (d) 5, 6, 10

15. Find the number of positive integer solutions of the equation $\frac{2}{x}+\frac{15}{y}=5$.
(a) 0 (b) 1
(c) 2 (d) 3

16. Of the three numbers, the sum of the first two is 45; the sum of the second and the third is 55 and the sum of the third and thrice the first is 90. The third number is
(a) 20 (b) 25
(c) 30 (d) 3

17. One of the angles of a triangle is two-third angle of sum of adjacent angles of parallelogram. Remaining angles of the triangle are in ratio 5 : 7 respectively. What is the value of second largest angle of the triangle?
(a) 25° (b) 40°
(c) 35° (d) Cannot be determined

18. Ramola's monthly income is three times Ravina's monthly income. Ravina's monthly income is fifteen percent more than Ruchira's monthly income. Ruchira's monthly income is ₹ 32,000. What is Ramola's annual income ? *[IBPS-PO-2011]*
(a) ₹ 1,10,400 (b) ₹ 13,24,800
(c) ₹ 36,800 (d) ₹ 52,200
(e) None of these

19. The respective ratio between the present age of Manisha and Deepali is 5 : X. Manisha is 9 years younger than Parineeta. Parineeta's age after 9 years will be 33 years. The difference between Deepali's and Manisha's age is same as the present age of Parineeta. What will come in place of X? *[IBPS-PO-2011]*
(a) 23 (b) 39
(c) 15 (d) Cannot be determined
(e) None of these

20. The fare of a bus is ₹ X for the first five kilometers and ₹ 13/- per kilometer thereafter. If a passenger pays ₹ 2402/- for a journey of 187 kilometers, what is the value of X ? *[IBPS-PO-2012]*
(a) ₹ 29/- (b) ₹ 39/-
(c) ₹ 36/- (d) ₹ 31/-
(e) None of these

21. The sum of the ages of 4 members of a family 5 years ago was 94 years. Today, when the daughter has been married off and replaced by a daughter-in-law, the sum of their ages is 92. Assuming that there has been no other change in the family structure and all the people are alive, what is the difference in the sum ages of three members and the daughter-in-law? *[IBPS-PO-2012]*
(a) 22 years (b) 11 years
(c) 25 years (d) 19 years
(e) 15 years

22. If the positions of the digits of a two-digit number are interchanged, the number obtained is smaller than the original number by 27. If the digits of the number are in the ratio of 1 : 2, what is the original number? *[IBPS-PO-2013]*
(a) 36 (b) 63
(c) 48 (d) Cannot be determined
(e) None of these

23. 465 coins consists of 1 rupee, 50 paise and 25 paise coins. Their values are in the ratio 5: 3 : 1. The number of each type of coins respectively is *[SSC CGL-2012]*
(a) 155, 186, 124 (b) 154, 187, 124
(c) 154, 185, 126 (d) 150, 140, 175

24. If $a = 0$, $b \neq 0$, $c \neq 0$, then the equation $ax + by + c = 0$ represents a line parallel to *[SSC CGL-2013]*
(a) $x + y = 0$ (b) x-axis
(c) y-axis (d) None of these

25. For what value of k, the system of equations kx + 2y = 2 and 3x + y = 1 will be coincident? *[SSC CGL-2014]*
(a) 2 (b) 3
(c) 5 (d) 6

Hints & Solutions

Level-I

1. (d) $(x-3)(2x+1)=0 \Rightarrow (x-3)=0$ or $(2x+1)=0$
when $x-3=0, x=3$
when $2x+1=0 \Rightarrow x=-\frac{1}{2}$
When $x=3$, then $(2x+1)=7$ and when $x=-\frac{1}{2}$, then $2x+1=0$,
Possible values of $(2x+1)$ are 0 and 7.

2. (c) Let father's, mother's and daughter's present age be F, M, D respectively.
We have, $F=M+5, M=3D$ and $D=10$
$\Rightarrow M=3\times 10=30$ years and $F=30+5=35$ years
The father's age at the time of birth of the daughter $=35-10=25$ years

3. (b) Let the present age of the son be x years, then
$x=38-x$ or $x=19$ years
Five years back, son's age $=x-5=19-5=14$ years

4. (c) Let the number be x, Then,
$$x-24=\frac{4}{7}x \Leftrightarrow x-\frac{4}{7}x=24 \Leftrightarrow \frac{3}{7}x=24$$
$$\Leftrightarrow x=\left(\frac{24\times 7}{3}\right)=56.$$

5. (c) Let the number be x.
$$\text{Then, }\left(\frac{1}{2}x+\frac{1}{5}x\right)-\frac{1}{3}x=\frac{22}{3} \Leftrightarrow \frac{11x}{3}$$
$$=\frac{22}{3} \Leftrightarrow x=\left(\frac{22\times 30}{3\times 11}\right)=20$$

6. (b) Let the salary of the driver be ₹ x.
Then, his income during one week $= x+\frac{5}{4}x=\frac{9x}{4}$
$$\text{Required fraction}=\frac{\frac{5}{4}x}{\frac{9}{4}x}=\frac{5}{9}$$

7. (c) Let the number of rice bowls be x,
the number of broth bowls be y
and the number of meat bowls be z.
Now, $x+y+z=65$ ...(1)
and $2x=3y=4z$ ...(2)
From (1) and (2), we have $x=30, y=20, z=15$
Thus, the total number of guests $=2x=3y=4z=60$

8. (c) Let the numbers be $4x$ and $7x$. Then, $\frac{4x+4}{7x+4}=\frac{3}{5}$
$\Leftrightarrow 5(4x+4)=3(7x+4) \Leftrightarrow x=8$
$\therefore$ Larger number $=7x=56$

9. (a) We have, $(M+30)=2(R-30)$
or $M+30=2R-60$
or $M-2R=-90$ ...(1)
Again $(R+10)=3(M-10)$
or $R+10=3M-30$
or $R-3M=-30-10$
or $R-3M=-40$ ...(2)
Solving (1) and (2), we have $M=34$ and $R=62$.

10. (b) Let the numbers be x and y. Then, $x+y=25$ and $x-y=13$.
$\Rightarrow xy=114$

11. (b) Person's daily expenses = ₹ x
Number of days tour last $=y$ days
So, $x\times y=360$ ...(1)
$(x-3)(y+4)=360$ ... (2)
Solving equations (1) and (2), we get
$y=20$ or -24 (not possible)
$\therefore$ $y=20$ days

12. (b) Let the numbers be x and y. Then, $x^2-y^2=256000$ and $x+y=1000$.
On dividing, we get : $x+y=256$.
Solving $x+y=1000$ and $x-y=256$.
We get : $x=628$ and $y=372$.

13. (b) Let the numbers be x, $x+2$ and $x+4$
Then, $x+(x+2)+(x+4)=x+20 \Leftrightarrow 2x=14$
$\Leftrightarrow x=7$.
$\therefore$ Middle number $=x+2=9$.

14. (b) Let fixed charge = ₹ x
and charge for 1 km is ₹ y
$x+10y=85$
$x+15y=120$
$-\quad-\quad-$
$-5y=-35$
$y=$ ₹ 7 per km
$x=$ ₹ 15
Charges for 25 km $=15+25\times 7=$ ₹ 190

15. (b) Let the numerator be x.
Then, denominator $=x+4$
$$\therefore \quad \frac{x-4}{x+4+2}=\frac{1}{6}$$
$$\Rightarrow \quad \frac{x-4}{x+6}=\frac{1}{6}$$
$\Rightarrow$ $6(x-4)=x+6$
$\Rightarrow$ $6x-24=x+6 \Rightarrow$ $5x=30$
$\therefore$ $x=6$
Thus, Numerator = 6, Denominator $=6+4=10$.
Hence, the original number $=\frac{6}{10}$.

16. (a) Let the present ages of Vikas and Vishal be $15x$ years and $8x$ years.
After 10 years,
Vikas's age = $15x + 10$ and
Vishal's age = $8x + 10$

$\therefore \quad \frac{15x+10}{8x+10} = \frac{5}{3}$

$\Rightarrow \quad 3(15x + 10) = 5(8x + 10)$

$\Rightarrow \quad 45x + 30 = 40x + 50$

$\Rightarrow \quad 5x = 20 \quad \Rightarrow x = \frac{20}{5} = 4$

$\therefore$ Present age of Vikas = $15x = 15 \times 4 = 60$ years
Present age of Vishal = $8x = 8 \times 4 = 32$ years.

17. (c) Let the numbers be $3x$, $3x + 3$ and $3x + 6$
Then, $3x + (3x + 3) + (3x + 6) = 72$
$\Leftrightarrow 9x = 63 \Leftrightarrow x = 7.$
$\therefore$ Largest number = $3x + 6 = 27$

18. (d) $\frac{2}{5} \times \frac{1}{4} \times \frac{3}{7} \times x = 15$

$\therefore \quad \frac{x}{2} = \frac{5 \times 7 \times 2 \times 5}{2} = 25 \times 7 = 175$

19. (a) $F + S = 4S$
or, $F = 3S \Rightarrow F : S = 3 : 1$
The ages of father and son = 56 years

$\therefore$ Son's age $= \frac{1}{4} \times 56 = 14$ years

20. (c) Let the two numbers be x and y.
$\therefore \quad xy = 192, x + y = 28$ ...(1)
$\therefore \quad (x - y)^2 = (x + y)^2 - 4xy = 784 - 768 = 16$
$\therefore \quad x - y = 4$ ...(2)
Combining (1) and (2), $x = 16$, and $y = 12$.

21. (b) Let the middle no. = x

$(x - 2) + x + (x + 2) = \frac{176}{4} - 14$

or $3x = \frac{120}{4}$ or, $x = 10$

22. (b) Let the denominator be x. Then, numerator = $x + 5$.

Now, $\frac{x+5}{x} - \frac{x+5}{x+5} = \frac{5}{4} \Leftrightarrow \frac{x+5}{x} = \frac{5}{4} + 1 = \frac{9}{4} = 2\frac{1}{4}$

So, the fraction is $2\frac{1}{4}$

23. (c) Let the three numbers be $5x$, $9x$ and $11x$ respectively.
Then, $25x = 300 \Rightarrow x = 12$
So, the second number is $9x = 9 \times 12 = 108$

24. (d) Let the number be x. Then,

$x - 20 = \frac{7x}{12} \Rightarrow x - \frac{7x}{12} = 20 \Rightarrow \frac{5x}{12} = 20$

$\Rightarrow x = 48$

25. (d) Let the unit's digit be y and the ten's digit be x.
Then, the number is $10x + y$.
Interchanging the numbers, the new number is $10y + x$.
Then, $10x + y = 10y + x + 18 \Rightarrow 9x - 9y = 18$
$\Rightarrow x - y = 2$ and given $x + y = 8$
Solving $x = 5, y = 3$
Then, the original number is 53.

26. (a) Let the two numbers be x and y. Then,
$2x + 3y = 100$ ...(1)
and $3x + 2y = 120$ ...(2)
Solving eqs (1) and (2), we get $y = 12$ and $x = 32$
So, the larger of the numbers is 32.

27. (d) Let the two numbers be x and y respectively. Then,
$2x + 3y = 300$ ...(1)
and $3x + 2y = 265$ ...(2)
Solving eqs (1) and (2), we get $x = 39$ and $y = 74$
So, the larger number is 74.

28. (a) From the options, the required two-digit number is 16.

29. (c) $h + g = 81$ ...(1)
and $2h + 4g = 234$ ...(2)
Solving eqs (1) and (2), we get $h = 45$ and $g = 36$

30. (c) Suppose the age of son is x years.
Therefore, age of father = $10x$ years
According to question

$\frac{10x + x}{2} = 22$

$\Rightarrow 11x = 44$

$\therefore x = \frac{44}{11} = 4$ years

Age of father = $10 \times 4 = 40$ years

31. (b) Let X be the given number. Then
$X/3 + X/4 + X/5 - X/2 = 34.$
Solving this, we get $X = 120$.

32. (c) Let first book published in year x
According to question
$x + x + 7 + x + 14 + x + 21 + x + 28 + x + 35 + x + 42 = 13524$
$147 + 7x = 13524$
$7x = 13524 - 147 = 13377$

$\therefore x = \frac{13377}{7} = 1911$

33. (b) Let the ten's digit be x. Then, the unit's digit is $3x$.
Then, $x + 3x = 8 \Rightarrow x = 2$. So, ten's digit is 2 and unit's digit is 6. So, number is 26.

34. (c) From the option, 82 is the right choice as
$82 - 28 = 54$

35. (b) Let the present age of father = x year and Son's present age = y years.
5 years ago, father's age = $x - 5$ and
Son's age = $y - 5$
According to the question,
$x - 5 = 5(y - 5)$ .... (1)
and $x = 3y$ ....(2)
$\therefore$ From eqs (1) and (2), we have
$y = 10$ and $x = 30$ years.
Hence, father's present age = 30 years.

36. (c) Let the original fraction be $\frac{x}{y}$.

Then, $\frac{\frac{250}{100}\times x}{\frac{450}{100}\times y}=\frac{25}{51}\Rightarrow\frac{250x}{450y}=\frac{25}{51}$

$\Rightarrow \quad \frac{x}{y}=\frac{450\times 25}{250\times 51}=\frac{15}{17}$

37. (b) By trial and error method.

38. (c) Let the number be x.
$63x - 36x = 3834 \Rightarrow 27x = 38834 \Rightarrow x = 142$

39. (c) Suppose his present age is x years.
According to question

$\frac{x}{4}+\frac{x}{5}+\frac{x}{3}=x-13$

$\Rightarrow \quad \frac{15x+12x+20x}{60}=x-13$

$\Rightarrow \quad 47x = 60x - 780$

$\Rightarrow \quad 60x - 47x = 780$

$\Rightarrow \quad 13x = 780$

$\therefore \quad x=\frac{780}{13}$ = 60 years

40. (d) Suppose the number of cows = x
Therefore, the number of herdsmen = x
According to question,
$4 \times 2x - 28 = x \times 2 + x \times 4$

$\Rightarrow \quad 8x - 28 = 2x + 4x$

$\Rightarrow \quad 8x - 6x = 27$

$\therefore \quad x=\frac{28}{2}=14$

41. (c) Let the present age of mother and daughter be $7x$ and x.
Then,

$\frac{7x-4}{x-4}=\frac{19}{1}\Rightarrow 7x-4$

$=19x - 76$

$= 12x = 72$

$= x = 6$

$\therefore$ mother's age 4 yrs hence = $7 \times 6 + 4 = 46$ yrs.

42. (b) $\frac{6x}{7}=180\Rightarrow x=210$

43. (c) Age of Farah = $x=(x-8)\times\frac{9}{7}$

$\Rightarrow x=\frac{9(x-8)}{7}$

$\Rightarrow 7x = 9x - 72$

$\Rightarrow 2x = -72$

$\Rightarrow x=\frac{72}{2}$ = 36 years

Present age of her daughter = $\frac{36}{6}$ = 6 years

$\therefore$ Age of daughter 3 years ago
$= 6 - 3 =$ 3 years

44. (d) Let the number of parrots be p and the number of tigers be t. Then

$p + t = 858$ ...(1)

$2p + 4t = 1846$ ...(2)

After rearranging equation (2), we get

$p + 2t = 923$ ...(3)

Solving (1) & (2) we get

$t = 65$ & $p = 793$

45. (a) Let the ten's digit be x. Then, units digit = $(x + 3)$.
Sum of the digits = $x + (x + 3) = 2x + 3$.
Number = $10x + (x + 3) = 11x + 3$.

$\therefore \quad \frac{11x+3}{2x+3}=\frac{4}{1} \Leftrightarrow 11x + 3 = 4(2x + 3)$

$\Leftrightarrow 3x = 9 \quad \Leftrightarrow x = 3$

Hence, required number = $11x + 3 = 36$

46. (d) Let the numbers be x and y.

$\therefore \quad \frac{x}{y}=\frac{4}{7}$

$\therefore \quad 7x = 4y$ ... (1)

$\frac{x+30}{y+30}=\frac{5}{8}$

$\therefore \quad 8x - 5y = -90$ ...(2)

From eqn (2), $32x - 20y = -360$

From eqn (1), $35x = 20y$

$\therefore \quad 32x - 35x = -360$

$\therefore \quad x=\frac{360}{3}=120$

$y = 210$

$\therefore$ Average $=\frac{330}{2}=165$

47. (c) Let the ten's digit be x. Then, number = $10x + 3$ and sum of digits = $(x + 3)$.

So, $(x + 3) = \frac{1}{7}(10x + 3) \Leftrightarrow 7x + 21 = 10x + 3$

$\Leftrightarrow 3x = 18 \Leftrightarrow x = 6$.

Hence, the number is 63.

48. (c) Let the number be x. Then, $3(2x + 9) = 75$

$\Rightarrow 2x + 9 = 25 \Rightarrow 2x = 16 \Rightarrow x = 8$.

49. (b) Let the ten's digits be x and unit's digit be y.
Then, $(10x + y) - (10y + x) = 36 \Leftrightarrow 9(x - y) = 36$

$\Leftrightarrow x - y = 4$.

50. (b) Let the two parts be $(54 - x)$ and x.
Then, $10(54 - x) + 22x = 780 \Leftrightarrow 12x = 240$

$\Leftrightarrow x = 20$.

$\therefore$ Bigger part = $(54 - x) = 34$.

51. (a) Greatest of the five numbers will be least if remaining four numbers are less then m and as large as possible

$\Rightarrow$ The remaining four numbers are same.

$4(m - 1) + m = 146$

$\Rightarrow \quad m = 30$

52. (d) Let number of notes of each denomination be x.
Then, $x + 5x + 10x = 480 \Leftrightarrow 16x = 480 \Leftrightarrow x = 30$.
Hence, total number of notes $= 3x = 90$.

53. (d) Let the original fraction be $\frac{x}{y}$. Then,

$$\frac{\frac{300x}{100}}{\frac{300y}{100}} = \frac{14}{5} \Rightarrow \frac{x}{y} = \frac{14}{5}$$

54. (a) Let one pen cost be ₹ x and one pencil cost be ₹ y
$36x + 42y = 460$ .... (1)
$18x + 21y = ?$
Divided eq. (1) by 2
$18x + 21y = 230$

55. (b) Let adult fare be x and child fare be y
$x = 6y$
$x = 114$
$y = 114/6 = 19$
Now $4x + 5y$
$= 4 \times 114 + 5 \times 19 = 456 + 95 =$ ₹ 551 /-

56. (c) Let the nubmer be $10x + y$ where $x > y$.
According to the question,
$x + y = 15$
and $x - y = 3$
Solving both the equations,
$x = 9, y = 6$
$\therefore x \times y = 9 \times 6 = 54$

57. (a) 58. (b) 59. (d)

60. (b) $y = 4x$,
When, $x = 1, y = 4$

61. (b) Number of brown socks $= x$
Price of brown socks = ₹ y per pair
Price of black socks = ₹ $2y$ per pair

$$\therefore 4y + x \times 2y = \frac{150}{100}(4 \times 2y + xy)$$

$$\Rightarrow 4 + 2x = \frac{3}{2}(8 + x)$$

$\Rightarrow 8 + 4x = 24 + 3x$
$\Rightarrow x = 24 - 8 = 16$
$\therefore$ Requried ratio $= 4 : 16 = 1 : 4$

62. (c) Let total bill would be ₹ x

Each one decided to pay = ₹$\left(\frac{x}{11}\right)$

10 friends could pay $10 \times 60 =$ ₹ 600
According to question,

$$600 + \frac{x}{11} + 50 = x$$

$$650 = x - \frac{x}{11} = \frac{10x}{11}$$

$$x = \frac{650 \times 11}{10} = 715$$

Amount paid by 11th friend $= \frac{715}{11} + 50 =$ ₹ 115

63. (d) According to question

$$\frac{36 + n}{50 + n} = \frac{3}{4}$$

$36 \times 4 + 4n = 50 \times 3 + 3n$
$4n - 3n = 150 - 144$
$n = 6$

64. (a) Given $4x - y = 2$ or $4x - y - 2 = 0$
and $2y - 8x + 4 = 0$ or $-8x + 2y + 4 = 0$
Therefore $a_1 = 4; b_1 = -1; c_1 = -2$
$a_2 = -8; b_2 = 2; c_2 = 4$

Now, $\frac{a_1}{a_2} = \frac{b_1}{b_2} = \frac{c_1}{c_2} \Rightarrow \frac{4}{-8} = \frac{-1}{2} = \frac{-2}{4}$

This is true only when system of equations has infinitely many solutions.

65. (c) $\frac{1}{x} + \frac{1}{50 - x} = \frac{1}{12}$
$x^2 - 50x + 600 = 0$
$x^2 - 30x - 20x + 600 = 0$
$x(x - 30) - 20(x - 30) = 0$
$x = 30, 20$

66. (a) Since two digit number $= 10x + y$
According to question $\rightarrow y = 2x - 1$ ..(i)
When digits are interchanged then new number
$= 10y + x$
then original number – [new number – original number] = 20
$\Rightarrow 10x + y - [10y + x - (10x + y)] = 20$
$\Rightarrow 10x + y - 10y - x + 10x + y = 20$
$19x - 8y = 20$
$19x - 8(2x - 1) = 20$ (Using eq. (i))
$19x - 16x + 8 = 20$
$3x = 12 \Rightarrow x = 4$
From (i) $y = 2 \times 4 - 1 = y = 7$
$\therefore$ original number $= 10x + y = 10 \times 4 + 7 = 47$

67. (a) Cost of 5 pens + 8 pencils = ₹ 31
On multiplying by 3
15 pens + 24 pencils
$= 3 \times 31 =$ ₹ 93

68. (b) Amount paid
= ₹ $(40 \times 18 + 55 \times 8)$
= ₹ $(720 + 440)$
= ₹ 1160

Level-II

1. (d) Let a, b and c be the digits at the hundredth, tens and units places, respectively.
Now, $a + b + c = 16$, ...(1)
$b = 3c$ ...(2)

and $c = \frac{1}{4}a$ ...(3)

From (2) and (3),

$b = \frac{3}{4}a$ ...(4)

From (1), (3) and (4), we have

$$a+\frac{3}{4}a+\frac{1}{4}a=16$$

or $a=8$, $b=6$ and $c=2$

Hence, the three digit number is 862.

2. (b) Let the digit at units place be y and that at the tens place be x.

Number $= 10x + y$

We have, $xy = 14$ and $10x + y + 45 = 10y + x$

$\Rightarrow 9x - 9y = -45$ or $x - y = -5$ ...(1)

Now, $(x + y)^2 = (x - y)^2 + 4xy = (-5)^2 + 4 \times 14$
$= 25 + 56 = 81$

$\Rightarrow x + y = 9$ ...(2)

From (1) + (2)

$2x = 4$ or $x = 2$

$\Rightarrow$ Number = 27.

3. (d) Let present age of Ranjeev $= x$ years

Present age of Ranjeev's brother $= (x + 6)$ years

Present age of Ranjeev's father $= (x + 6 + 32)$years
$= (x + 38)$ years

$\therefore$ Present age of Ranjeev's mother $= (x + 38 - 3)$ years
$= (x + 35)$ years

$\therefore$ Present age of Ranjeev's sister $= (x + 35 - 25)$ years
$= (x + 10)$ years

$\therefore$ Age of Ranjeev's sister when he was born
$= (x + 10 - x) = 10$ years.

4. (d) Let the total number of discs of 2 kg and 5 kg be 'a' and 'b' respectively.

Then, $a + b = 21$ and $5b = 2a$

Solving the above two equations, we get $a = 15$, $b = 6$

$\therefore$ Weight of all discs together
$= 15 \times 2 + 6 \times 5 = 60$ kg

5. (a) $\frac{1}{3}A = \frac{E}{2}$

$\Rightarrow \frac{A}{3} - \frac{E}{2} = 0$

$\Rightarrow 2A - 3E = 0$ ...(1)

$A + E = 150$ ...(2)

From equations (1) and (2)

$E = 60$

6. (c) Let the four numbers be A, B, C and D.

Let $A + 3 = B - 3 = 3C = D/3 = x$.

Then, $A = x - 3$, $B = x + 3$, $C = x/3$ and $D = 3x$.

$A + B + C + D = 64$

$\Rightarrow (x - 3) + (x + 3) + x/3 + 3x = 64$

$\Rightarrow 5x + x/3 = 64 \Rightarrow 16x = 192 \Rightarrow x = 12$

Thus, the numbers are 9, 15, 4 and 36.

$\therefore$ Required difference $= (36 - 4) = 32$.

7. (d) Suppose husband's age be H years.

Then wife's age $W = H - 9$

Son's age $S = \frac{H-9}{2}$

Daughter's age $D = \frac{H}{3}$

According to question,

$\frac{H}{3}+7=\frac{H-9}{2} \Rightarrow \quad 2H+42=3H-27$

$\Rightarrow H = 42 + 27 = 69$

$\therefore W = 60$.

Solving through option (c).

45+9 = 54 H.

D = 18
S = 22.5 → Difference is 4.5 years, so this is incorrect

Solving through option (d) matches all conditions.

8. (b) Let the fraction is $\frac{a}{b}$, then $a + b = 11$

and $\frac{a+1}{b-2}=\frac{3}{2} \Rightarrow 2a+2=3b-6$

$\Rightarrow 2a - 3b = -8$

Solving both $a + b = 11$ and $2a - 3b = -8$

$\Rightarrow a = 5, b = 6 \quad \therefore$ fraction $= \frac{5}{6}$

9. (c) Let the number of questions that he attempted wrongly be n, so the questions attempted correctly will be $(90 - n)$.

According to the question,

$5 \times (90 - n) + (-2) \times n = 387 \Rightarrow 450 - 7n = 387$

$\Rightarrow n = 9$

10. (b) Since, $1 \times x < 1 + x$, So, one of the number is 1.

11. (b) Let 'x' be the total number of students in college

$$x-\left[\frac{12x}{100}+\frac{3x}{4}+\frac{10x}{100}\right]=15$$

$$x-\left[\frac{48x+300x+40x}{400}\right]=15 \quad \therefore x=500$$

12. (c) Let the unit's digit be x.

$\therefore$ Ten's digit $= x - 2$

$\therefore$ Number $= 10(x-2) + x = 10x - 20 + x = 11x - 20$

New number obtained after reversing the digits
$= 10x + x - 2 = 11x - 2$

According to the question,

$3(11x - 20) + \frac{6}{7}(11x - 2) = 108$

$\Rightarrow (11x-20)+\frac{2}{7}(11x-2)=36$

$\Rightarrow 77x - 140 + 22x - 4 = 252$

$\Rightarrow 99x = 252 + 144 \Rightarrow x = \frac{396}{99} = 4$

$\therefore$ Number $= 11x - 20 = 11 \times 4 - 20 = 24$

$\therefore$ Sum of digits $= 2 + 4 = 6$

13. (b) $\frac{x+1}{y+2}=\frac{2}{3} \Rightarrow 3x-2y=1$

$\frac{x+2}{y+3}=\frac{5}{7} \Rightarrow 7x-5y=1$

or, $3x - 2y = 7x - 5y \Rightarrow 3y = 4x \Rightarrow \frac{x}{y}=\frac{3}{4}$

14. (a) Let the numbers be x, y and z. Then,

$x + y = 10$ ...(1)

$y + z = 19$ ...(2)

$x + z = 21$ ...(3)

Adding (1), (2) and (3), we get : $2(x + y + z) = 50$ or $(x + y + z) = 25$.

Thus, $x = (25 - 19) = 6$; $y = (25 - 21) = 4$; $z = (25 - 10) = 15$.

Hence, the required numbers are 6, 4 and 15.

15. (b) We have $\frac{2}{x}+\frac{15}{y}=5$

$\Rightarrow 2y + 15x = 5xy$

$\Rightarrow 5xy - 2y - 15x = 0$

$\Rightarrow (y - 3)(5x - 2) = 6$

Now, 6 can be written as 2×3, -2×-3, 1×6 or -1×-6.

The only possible case is $5x \,.\, 2 = 3$ and $y \,.\, 3 = 2$.

Therefore, $x = 1$ and $y = 5$.

16. (c) Let the numbers be x, y and z. Then, $x + y = 45$, $y + z = 55$ and $3x + z = 90$

$\Rightarrow$ $y = 45 - x$, $z = 55 - y = 55 - (45 - x) = 10 + x$.

$\therefore$ $3x + 10 + x = 90$ or $x = 20$.

$y = (45 - 20) = 25$ and $z = (10 + 20) = 30$.

$\therefore$ Third number = 30.

17. (c) An angle is a triangle

$= \frac{2}{3}\times 180° = 120°$

Remaining $180° - 120° = 60°$ is the ratio of 5 : 7.

So, $5x + 7x = 60$

$12x = 60$

$x = 5$

So, angles are $5 \times 5 = 25°$

and $7 \times 5 = 35°$

and 120°

So, value of second largest angle of triangle is 35°.

18. (b) Ravina's monthly income

$= 32000 \times \frac{100+15}{100} = 32000 \times \frac{115}{100} = ₹\,36800$

= Ramola's annual income = $36800 \times 3 \times 12$

= ₹ 1324800

19. (e) According to the question

Present age of Parineeta = 33 – 9 = 24 years

Present age of Manisha = 24 – 9 = 15 years

Present age of Deepali = 24 + 15 = 39 years

$\because$ 5 : X = 15 : 39

$\therefore$ $X = \frac{5\times 39}{15} = 13$

20. (c) ₹[(x for first 5 km) + 13 × remaining kms] = Total pay

₹x + ₹ 13 × 182 = ₹ 2402

$x + 2366 = 2402$

x = ₹ 36

21. (a) Let the 4 members are x_1, x_2, x_3, daughter

Sum of 4 members five years ago

$= x_1 + x_2 + x_3$ + daughter = 94

After 5 years,

$x_1 + x_2 + x_3$ + daughter = 114 ...(1)

daughter + daughter in law = 92

Daughter = 92 – daughter in law

Put this eqn. ...(1)

$x_1 + x_2 + x_3$ + 92 – Daughter in law = 114

$x_1 + x_2 + x_3$ = 22 + Daughter in law

So, the required difference is 22 years.

22. (b) Let one's digit = x

ten's digit = 2x

Number = 10 (2x) + x = 21 x

After interchange the digit number = 12 x

$\therefore$ 21x – 12x = 27

9x = 27

x = 3

$\therefore$ one's digit = 3

Ten's digit = 2 × 3 = 6

Number = 10 × 6 + 3 = 63

23. (a) The ratio of number of coins = 5 : 6 : 4

$\therefore$ The number of one rupee coins $= \frac{465}{5+6+4}\times 5 = 155$

The number of 50 paise coins $= \frac{465}{5+6+4}\times 6 = 186$

The number of 25 paise coins $= \frac{465}{5+6+4}\times 4 = 124$

24. (b) It $a = 0$, $b \neq 0$, $c \neq 0$, then equation $ax + by + c = 0$ represents a line parallel to x-axis.

25. (d) Kx + 2y = 2 ...(1)

3x + y = 1 ...(2)

divide eqn (1) by (2)

$\frac{K}{2}+y = 1$

for system of equation to be coincident

$\frac{K}{2} = 3$

K = 6

CHAPTER 13

FUNCTIONS

INTRODUCTION

Function in mathematics is an equation or rule that defines a relationship between the two variables; one of them is dependent variable and other is independent variable. This chapter is very important from the point of view of CAT and other equivalent aptitude tests. The number of questions being asked from this topic is almost constant. Basically on an average 3–4 problems are asked from this chapter. A deep understanding of the concepts of this chapter is required to solve the problems.

FUNCTION

A function is a rule which relates two or more than two variables. Out of these variables one is dependent variable and others are independent variables. If y is dependent variable and x is independent variable, then the function is symbolically expressed as

$$y = f(x)$$

$y = f(x)$ is read as y is the function of x. But f denotes the rule by which y varies with x.

In the function $y = f(x)$, there is a unique real value of y for each real value of x. A set D of all real values of x for which the value of y is a unique real value is called domain of the function $y = f(x)$. A set R of all unique real values of y corresponding to each value of x from set D is called Range of the function $y = f(x)$.

The concept of the function can be easily understood by the following examples:

(i) The function between diameter d of a circle and radius r is

$$d = 2r$$

Here d is a dependent variable and r is an independent variable, because d and r both are variable but value of d is dependent upon the value of r.

Here domain is a set of all positive real values, because value of r cannot be non-positive and for each positive real value of r, the value of d is a unique positive real number.

Range is also a set of all positive real values, because the diameter,which is twice the length of the radius will be all the positive real numbers for all positive real value of r.

(ii) The function between the volume V of a cuboid with its side length x is

$$V = x^3$$

Here V is dependent variable and x is independent variable

Domain = Set of all positive real numbers.

Range = Set of all positive real numbers.

(iii) The function between the area A of the circle with its radius r is

$$A = \pi r^2$$

Here A and r are dependent and independent variables respectively.

Since value of r can be any positive real number and for all positive real values of r, values of A will be all positive real numbers, hence

Domain = Set of all positive real numbers.

Range = Set of all positive real numbers.

(iv) For the function $y = x^2$, y is a dependent variable and x is an independent variable,

Domain = Set of all real numbers

But Range = Set of all non-negative real numbers, because value of y cannot be negative for any value of x for the given function.

Illustration 1: If $f(x) = -2x + 7$ and $g(x) = x^2 - 5x + 6$, find $f(3), f(-4), g(2)$, and $g(-1)$.

Solution:

$f(x) = -2x + 7$,	$g(x) = x^2 - 5x + 6$
$f(3) = -2(3) + 7 = 1$	$g(2) = 2^2 - 5(2) + 6 = 0$
$f(-4) = -2(-4) + 7 = 15$	$g(-1) = (-1)^2 - 5(-1) + 6 = 12$

RULES FOR FINDING THE DOMAIN OF A FUNCTION

1. Domain of Algebraic Functions

(i) Denominator should be non-zero

For the function $y = \dfrac{2x}{x-3}$, the value of x can be any real number but can not be 3, because for $x = 3$, denominator of the function will be zero.

Hence domain of the function is the set of all real numbers except 3 i.e. domain = $R - \{3\}$.

(ii) Expression under the even root (i.e. square root, fourth root, etc.) should be non-negative.

For the function $y = \sqrt{5-x}$,

$$5 - x \geq 0 \Rightarrow x \leq 5$$

Hence domain = Set of all real numbers which are equal or less than 5.

2. Domain of Logarithmic Functions

$\log_b a$ is defined when $a > 0$, $b > 0$ but $b \neq 1$.

For the function $y = \log_2 (x - 4)$

$$x - 4 > 0 \Rightarrow x > 4$$

Hence domain = Set of all real numbers greater than 4.

3. Domain of Exponential Functions

a^x is defined for all real values of x, where $a > 0$.

For the function $y = (3x - 2)^x$,

$$3x - 2 > 0 \Rightarrow x > \frac{2}{3}$$

Hence domain = Set of all real numbers greater than $\frac{2}{3}$.

Note: If a and b are two real numbers such that $a > b$, then

(i) Interval $[a, b]$ means all real numbers equal or greater than a but equal or less than b.

(ii) Interval $[a, b)$ means all real numbers equal or greater than a but less than b.

(iii) Interval $(a, b]$ means all real numbers greater than a but equal or less than b.

(iv) Interval (a, b) means all real numbers greater than a but less than b.

(v) $(a, b) \cup (c, d)$ means all real numbers greater than a but less than b or greater than c but less than d.

Illustration 2: The domain of the function $f(x) = \dfrac{1}{\sqrt{x^2 - 3x + 2}}$ is

(a) $(-\infty, 1)$
(b) $(-\infty, 1) \cup (2, \infty)$
(c) $(-\infty, 1] \cup [2, \infty)$
(d) $(2, \infty)$

Solution: (b) For $f(x)$ to be defined, we must have

$x^2 - 3x + 2 = (x - 1)(x - 2) > 0 \Rightarrow x < 1$ or > 2

Domain of $f = (-\infty, 1) \cup (2, \infty)$.

METHODS OF REPRESENTATION OF FUNCTIONS

A function is represented mainly in three ways as given below.

1. Analytical Representation

When a function is represented by a uniform equation for the entire domain or by several equations which are different for different parts of the domain.

For example

(a) $y = 5x^2 + 2x$

This is the uniform function for entire domain

(b) $y = \begin{cases} x^2 + 4, & \text{if } x \leq 2 \\ x - 3, & \text{if } x > 2 \end{cases}$

This is the function which is represented by two equations which are different for different parts of the domain as given above.

2. Tabular Representation

When a function is represented by a sequence of values of the independent variable with the corresponding values of the dependent variable, then this representation is called Tabular representation of the function.

For example,

(a)

x	1	2	3	4	5	6
y	1	4	9	16	25	36

(b)

θ	0°	30°	45°	60°	90°
sin θ	0	1/2	$1/\sqrt{2}$	$\sqrt{3}/2$	1

3. Graphical Representation

When a function is represented by a graph taking different values of dependent variable along x-axis and corresponding values of independent variable along y-axis in a cartesian plane, then this representation of function is called graphical representation of function.

For example

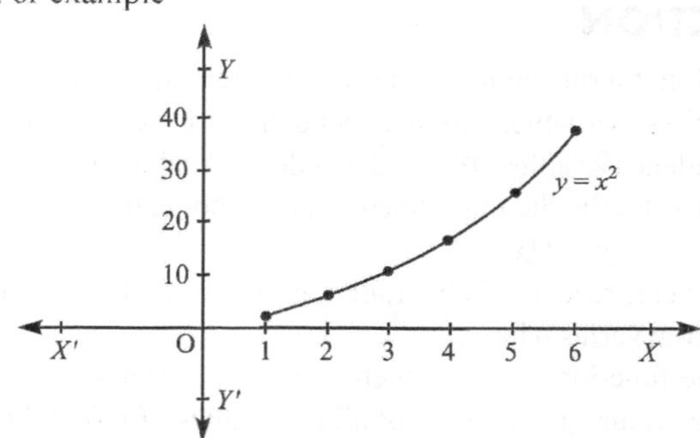

SOME SPECIAL FUNCTIONS

1. Even and Odd Functions

(i) Even functions: If a function $y = f(x)$ be such that $f(-x) = f(x)$, then the function is called an even function. Graph of the even function $y = f(x)$ is symmetrical about the y-axis. For example the graph of even function $y = x^2$ is symmetrical about y-axis.

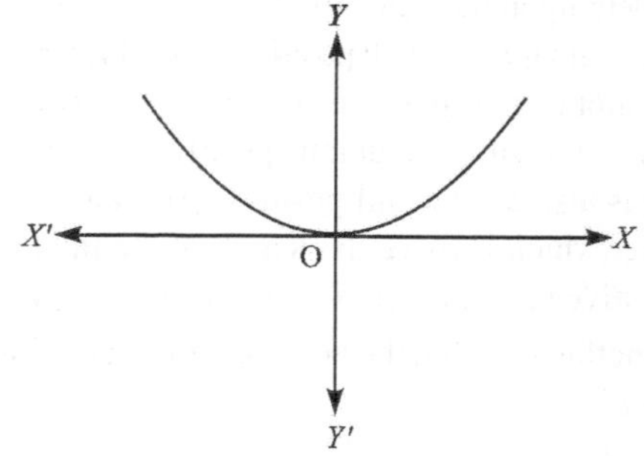

However, if y is independent variable and x is dependent variable, then the even function $x = f(y)$ is symmetrical about the x-axis.

Sum, difference, product and quotient of even functions are also even.

(ii) **Odd functions:** If a function $y = f(x)$ is such that $f(-x) = -f(x)$, then the function is called an odd function.

For example graph of the odd function $y = x^3$ is shown in the figure.

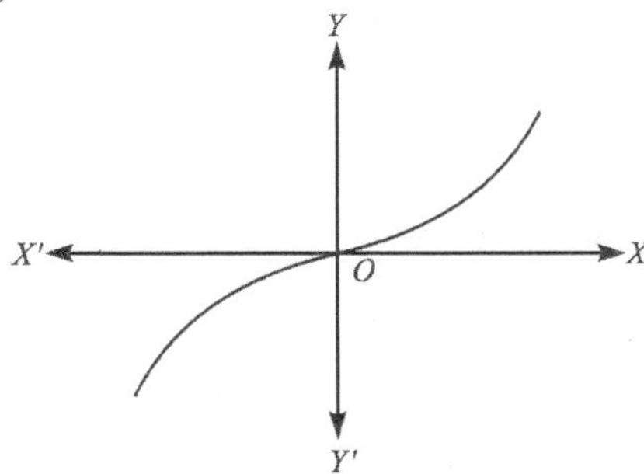

Graph of odd functions are two-fold graphs i.e., on folding the graph paper twice, once along x-axis and then along y-axis, one part of the graph overlaps the other part of the graph.

Some examples of odd functions are $y = x^3 - 2x$, $y = x^5$, $y = x^3 + \frac{1}{x}$, etc.

- Sum and difference of two odd functions is odd function.
- Product of two odd functions is an even function.
- Sum of even and odd function is neither even nor odd function.
- Product of an even and an odd function is odd function.
- Every function can be expressed as the sum of an even function and an odd function.
- A function may be even, odd or neither even nor odd.

For example $4x^3 + 3x^2 + 5$ is neither an even function nor an odd function.

Illustration 3: The function $f(x) = x\dfrac{a^x - 1}{a^x + 1}$ is odd or even ?

Solution:

$$\text{Since } f(-x) = -x.\frac{a^{-x} - 1}{a^{-x} + 1} = -x.\frac{1 - a^x}{1 + a^x} = x\frac{a^x - 1}{a^x + 1} = f(x)$$

$\therefore f(x)$ is an even function.

2. Modulus Function

$$f(x) = |x|$$

or

$$f(x) = \begin{cases} x, & \text{if } x \geq 0 \\ -x, & \text{if } x < 0 \end{cases}$$

Domain = Set of all real numbers

Range = Set of all non-negative real numbers

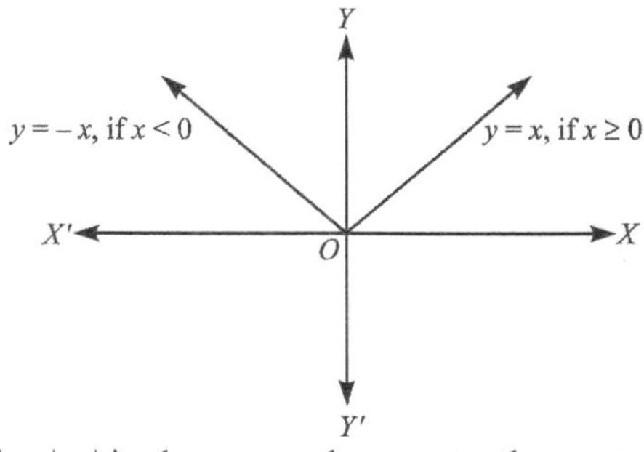

Note that $|x|$ is always equal or greater than zero

i.e. $|x| \geq 0$

For example,

$|0| = 0$

$|5| = 5$, since $5 > 0$

$|-5| = -(-5) = 5$, since $-5 < 0$

Illustration 4: If $|6x - 4| = 5$, find the value of x.

Solution: **Case–I;** $6x - 4 = 5$, if $6x - 4 \geq 0$

$\Rightarrow$ $x = \frac{3}{2}$, if $x \geq \frac{2}{3}$

Case–II: $-(6x - 4) = 5$, if $6x - 4 < 0$

$\Rightarrow$ $6x = -1$, if $x < \frac{2}{3}$

$\Rightarrow$ $x = -\frac{1}{6}$, if $x < \frac{2}{3}$

Illustration 5: Find the value of x if $2x^2 + 6|x| + 3 = 0$.

Solution: Since $2x^2$ and $6|x|$ is non-negative and 3 is positive, therefore their sum cannot be equal to zero.

Hence, there is no value of x for which $2x^2 + 6|x| + 3 = 0$

3. Composite Function

If two or more functions are composed into one function, then the resulting function is called composite function.

For example, if

$y = f(x)$ and $y = g(x)$ are two functions then

$f(g(x))$ and $g(f(x))$ are composite functions

Let $f(x) = 2x - 3$ and $g(x) = -3x^2$

Then $f(g(x)) = 2(-3x^2) - 3 = -6x^2 - 3$

and $g(f(x)) = -3(2x - 3)^2$

$f(g(x))$ and $g(f(x))$ are also written as $f'g(x)$ and $g'f(x)$ respectively

Illustration 6: Given $f(x) = 2x + 1$ and $g(x) = x^2 + 2x - 1$, find $(f - g)(x)$. Then evaluate the difference when $x = 2$.

Solution: The difference of the functions f and g is given by

$$(f - g)(x) = f(x) - g(x)$$
$$= (2x + 1) - (x^2 + 2x - 1) = -x^2 + 2.$$

When $x = 2$, the value of this difference is

$$(f - g)(2) = -(2)^2 + 2 = -2.$$

Practice Exercise

Level - I

1. A function f is defined by $f(x)=x+\frac{1}{x}$. Consider the following.
(1) $(f(x))^2=f(x^2)+2$
(2) $(f(x))^3=f(x^3)+3f(x)$
Which of the above is/are correct?
(a) 1 only (b) 2 only
(c) Both 1 and 2 (d) Neither 1 nor 2

2. What is the range of the function $f(x)=\frac{|x|}{x}, x\neq 0$?
(a) Set of all real numbers (b) Set of all integers
(c) $\{-1, 1\}$ (d) $\{-1, 0, 1\}$

3. The domain of the function $f(x)=\frac{\sqrt{(x+1)(x-3)}}{x-2}$ is
(a) $[-1, 2)\cup[3,\infty)$ (b) $(-1, 2)\cup[3,\infty)$
(c) $[-1, 2]\cup[3,\infty)$ (d) None of these

4. If $f(x)=\sqrt{x^3}$, then $f(3x)$ will be equal to
(a) $\sqrt{3x^3}$ (b) $3\sqrt{x^3}$
(c) $3\sqrt{(3x^3)}$ (d) $3\sqrt{x^5}$

5. If $f(x)=e^x$, then the value of $7f(x)$ will be equal to
(a) e^{7x} (b) $7e^x$
(c) $7e^{7x}$ (d) e^x

6. If $f(x)=\frac{x+1}{x-1}, x\neq 1$, find $f(f(f(f(f(2)))))$
(a) 2 (b) 3
(c) 4 (d) 6

7. Find fof if $f(t)=t/(1+t^2)^{1/2}$.
(a) $1/(1+2t^2)^{1/2}$ (b) $t/(1+2t^2)^{1/2}$
(c) $(1+2t^2)$ (d) None of these

8. $f(x)=3x^2$, $g(x)=h(x)=3x^3+3$. The value of $f(x)\,g(x)$ differ from the corresponding values of $h(x)$ approximately by what value
(a) 9 (b) 5
(c) 3 (d) Cannot be determined

9. If $f(x)=|x|$ and $g(x)=[x]$, then value of
$fog\left(-\frac{1}{4}\right)+gof\left(-\frac{1}{4}\right)$ is
(a) 0 (b) 1
(c) -1 (d) 1/4

10. If $f(x)$ is an even function, then the graph $y=f(x)$ will be symmetrical about
(a) x-axis (b) y-axis
(c) Both the axes (d) None of these

11. The domain of definition of $y=\left[\log_{10}\left(\frac{5x-x^2}{4}\right)\right]^{1/2}$ is
(a) $[1, 4]$ (b) $[-4, -1]$
(c) $[0, 5]$ (d) $[-1, 5]$

12. If $f(t)=\sqrt{t}$, $g(t)=t/4$ and $h(t)=4t-8$, then the formula for $g(f(h(t)))$ will be
(a) $\frac{\sqrt{t-2}}{4}$ (b) $2\sqrt{t}-8$
(c) $\frac{\sqrt{(4t-8)}}{4}$ (d) $\frac{\sqrt{(t-8)}}{4}$

13. If $f(x)=5x^3$ and $g(x)=3x^5$, then $f(x).g(x)$ will be
(a) Even function (b) Odd function
(c) Both (d) None of these

14. If $f(x)=\begin{cases}1-x, & 0\le x\le 2\\ x-1, & 2\le x\le 4\\ 1, & 4\le x\le 6\end{cases}$; then find
$f(0)+f\left(\frac{1}{2}\right)+f(1)+f\left(\frac{45}{18}\right)$
(a) 1 (b) 2
(c) 3 (d) None of these

15. Given $f(x)=\log\left(\frac{1+x}{1-x}\right)$ and $g(x)=\frac{3x+x^3}{1+3x^2}$, then $fog(x)$ is
(a) $-f(x)$ (b) $3f(x)$
(c) $[f(x)]^3$ (d) None of these

16. If $3f(x)+5f\left(\frac{1}{x}\right)=\frac{1}{x}-3,\ \forall\ x\neq 0\in R$, then $f(x)=$
(a) $\frac{1}{16}\left(\frac{3}{x}+5x-6\right)$ (b) $\frac{1}{16}\left(-\frac{3}{x}+5x-6\right)$
(c) $\frac{1}{14}\left(-\frac{3}{x}+5x+6\right)$ (d) None of these

17. Which of the following is not an even function?
(a) $f(x) = e^x + e^{-x}$ (b) $f(x) = e^x - e^{-x}$
(c) $f(x) = e^{2x} + e^{-2x}$ (d) None of these

18. Let $f(x) = |x-2| + |x-3| + |x-4|$ and $g(x) = f(x+1)$. Then
(a) $g(x)$ is an even function
(b) $g(x)$ is an odd function
(c) $g(x)$ is neither even nor odd
(d) None of these

19. Find the value of $f(f(-2))$, if $f(x) = \dfrac{x}{x+1}$
(a) 3/2 (b) 4/3
(c) 2/3 (d) None of these

20. Find the value of $f(f(f(3))) + f(f(1))$, if

$$f(x) = \begin{cases} \dfrac{x}{x+1}; & \text{if } x \text{ is an integer} \\ \dfrac{1}{x-(x)}; & \text{if is not an integer} \end{cases}$$

(a) 4 (b) 5
(c) 6 (d) 7

21. Let $f(x)$ be a function satisfying $f(x)f(y) = f(xy)$ for all real x, y. If $f(2) = 4$, then what is the value of $f\left(\dfrac{1}{2}\right)$?
(a) 0 (b) $\dfrac{1}{4}$
(c) $\dfrac{1}{2}$ (d) cannot be determined

22. Which of the following functions is an odd function?
(a) $2^{-x.x}$ (b) $2^{x-x.x.x.x}$
(c) Both (a) and (b) (d) Neither (a) nor (b)

23. If $f(t) = t^2 + 2$ and $g(t) = (1/t) + 2$, then for $t = 2, f[g(t)] - g[f(t)] = ?$
(a) 1.2 (b) 2.6
(c) 4.34 (d) None of these

24. Given $f(t) = kt + 1$ and $g(t) = 3t + 2$. If $fog = gof$, find k.
(a) 2 (b) 3
(c) 5 (d) 4

25. If $f(x) = e^x$ and $g(x) = \log_e x$, then value of fog will be
(a) x (b) 0
(c) 1 (d) e

Level - II

1. Which of the following two functions are identical?
(i) $f(x) = x^2/x$ (ii) $g(x) = \left(\sqrt{x}\right)^2$
(iii) $h(x) = x$
(a) (i) and (ii) (b) (ii) and (iii)
(c) (i) and (iii) (d) None of these

2. If $f(x) = \log x^4$ and $g(x) = 4 \log x$, then the domain for which $f(x)$ and $g(x)$ are identical?
(a) $(-\infty, \infty)$ (b) $[0, \infty)$
(c) $(0, \infty)$ (d) None of these

3. If $f(x) = x^3 - 4x + p$, and $f(0)$ and $f(1)$ are of opposite signs, then which of the following is necessarily true?
(a) $-1 < p < 2$ (b) $0 < p < 3$
(c) $-2 < p < 1$ (d) $-3 < p < 0$

4. If $f(x)$ is a function satisfying $f(x).f(1/x) = f(x) + f(1/x)$ and $f(4) = 65$, what will be the value of $f(6)$?
(a) 37 (b) 217
(c) 64 (d) None of these

Hints & Solutions

Level-I

1. (c) $f(x^2)+2 = x^2 + \frac{1}{x^2} + 2$

$$= \left(x+\frac{1}{x}\right)^2 = \{f(x)\}^2$$

and $f(x^3)+3f(x)$

$$= x^3 + \frac{1}{x^3} + 3\left(x+\frac{1}{x}\right) = \left(x+\frac{1}{x}\right)^3 = \{f(x)\}^3$$

Thus, both 1 and 2 are correct.

2. (c) As we know

$$|x| = \begin{cases} x & \text{if } x > 0 \\ -x & \text{if } x < 0 \end{cases}$$

$$\therefore f(x) = \frac{|x|}{x} = \begin{cases} \frac{x}{x} & \text{if } x > 0 \\ \frac{-x}{x} & \text{if } x < 0 \end{cases}$$

$$= \begin{cases} 1 & \text{if } x > 0 \\ -1 & \text{if } x < 0 \end{cases}$$

Hence, range = {–1, 1}.

3. (a)

4. (c) $f(x) = \sqrt{x^3} \Rightarrow f(3x) = \sqrt{(3x)^3} = 3\sqrt{3x^3}$

5. (b) $7f(x) = 7e^x$.

6. (b) $f(2) = \frac{2+1}{2-1} = 3$

$$f(f(2)) = f(3) = \frac{3+1}{3-1} = 2$$

$$f(f(f(2))) = f(f(3)) = f(2) = \frac{2+1}{2-1} = 3$$

$$f(f(f(f(2)))) = f(3) = \frac{3+1}{3-1} = 2$$

$$f(f(f(f(f(2))))) = f(2) = \frac{2+1}{2-1} = 3$$

7. (b) $f(f(t)) = f[t/(1+t^2)^{1/2}] = t/(1+2t^2)^{1/2}$

8. (c) $h(x) = 3x^3 + 3 = (3x^2)(x) + 3 = f(x)\,g(x) + 3$

Thus, for every x, the corresponding values of $f(x)$ $g(x)$ and $h(x)$ differ by 3.

9. (b) $\because fog\left(-\frac{1}{4}\right) = f\left[g\left(-\frac{1}{4}\right)\right] = f(-1) = 1$

and $gof\left(-\frac{1}{4}\right) = g\left[f\left(-\frac{1}{4}\right)\right] = g\left(\frac{1}{4}\right) = [1/4] = 0$

$\therefore$ Required value = 1 + 0 = 1

10. (b) y-axis by definition.

11. (a) $\frac{5x-x^2}{4} \geq 1 \Rightarrow 1 \leq x \leq 4$

12. (c) $g(f(h(t))) = g(f(4t-8)) = g\left(\sqrt{4t-8}\right)$

$$= \frac{\sqrt{4t-8}}{4}$$

13. (a) $f(x).g(x) = 15x^8$, which is an even function. Thus, option (a) is corect.

14. (c) $f(0) = 1 - 0 = 1,\ f\left(\frac{1}{2}\right) = 1 - \frac{1}{2} = \frac{1}{2} = 0.5$

$$f(1) = 1 - 1 = 0,\ f\left(\frac{45}{18}\right) = 2.5 - 1 = 1.5$$

$$\therefore f(0) + f\left(\frac{1}{2}\right) + f(1) + f\left(\frac{45}{18}\right) = 1 + 0.5 + 0 + 1.5 = 3$$

15. (b) We have $(fog)(x) = f(g(x))$

$$= \log\left\{\frac{1+g(x)}{1-g(x)}\right\} \quad \left[\text{since } f(x) = \log\left(\frac{1+x}{1-x}\right)\right]$$

$$= \log\left\{\frac{1+\left(\frac{3x+x^3}{1+3x^2}\right)}{1-\left(\frac{3x+x^3}{1+3x^2}\right)}\right\} \quad [\text{Substituting for } g(x)]$$

$$= \log\left\{\frac{1+3x^2+3x+x^3}{1+3x^2-3x-x^3}\right\}$$

$$= \log\left\{\frac{(1+x)^3}{(1-x)^3}\right\} = 3\log\left(\frac{1+x}{1-x}\right) = 3f(x)$$

16. (b) We have, $3f(x)+5f\left(\frac{1}{x}\right)=\frac{1}{x}-3, \forall\ x(\neq 0)\in R$ ...(1)

$\Rightarrow\ 3f\left(\frac{1}{x}\right)+5f(x)=x-3$ ...(2)

$\left[\text{Replacing } x \text{ by } \frac{1}{x}\right]$

Multiplying (1) by 3 and (2) by 5 and subtracting, we get

$\Rightarrow\ f(x)=\frac{1}{16}\left(-\frac{3}{x}+5x-6\right), \forall\ x(\neq 0)\in R$.

17. (b) Is not even since $e^x - e^{-x} \neq e^{-x} - e^x$.

18. (c) $g(x) = f(x+1) = |x-2+1| + |x-3+1| + |x-4+1|$
$= |x-1| + |x-2| + |x-3|$

19. (c) $f(-2)=\frac{-2}{-2+1}=2$

$f(f(-2)) = f(2) = \frac{2}{2+1}=\frac{2}{3}$

20. (b) $f(3)=\frac{3}{3+1}=\frac{3}{4}$

$f(f(3)) = f\left(\frac{3}{4}\right)=\frac{1}{\frac{3}{4}-\left[\frac{3}{4}\right]}=\frac{1}{\frac{3}{4}-0}=\frac{4}{3}$

$f(f(f(3))) = f\left(\frac{4}{3}\right)=\frac{1}{\frac{4}{3}-\left[\frac{4}{3}\right]}=\frac{1}{\frac{1}{3}}=3$

$f(1)=\frac{1}{1+1}=\frac{1}{2}$

$f(f(1))=\frac{1}{\frac{1}{2}-\left[\frac{1}{2}\right]}=2 \Rightarrow 3+2=5$

21. (b) $f(x).f(y)=f(x.y)$

$\Rightarrow\ p(0).p(1)=p(0)$

$\therefore\ p(1)=1$

Now, $p(2).\ p\left(\frac{1}{2}\right)=p(1)$

$\Rightarrow\ 4\times p\left(\frac{1}{2}\right)=1$

$\therefore\ p\left(\frac{1}{2}\right)=\frac{1}{4}$.

22. (d) Neither $2^{-x.x}$ nor $2^{x-x.x.x.x}$ is an odd function as for neither of them is $f(x) = -f(-x)$

23. (d) $f(g(t)) - g(f(t)) = f(2.5) - g(6) = 8.25 - 2.166$
$= 6.0833$.

24. (a) $fog = f(3t+2) = K(3t+2)+1$

$gof = g(kt+1) = 3(kt+1)+2$

$K(3t+2)+1 = 3(kt+1)+2$

$\Rightarrow\ 2k+1=5$

$\Rightarrow\ k=2$

25. (a) $fog = f(\log_e x) = e^{\log_e x} = x$.

Level-II

1. (d) For two functions to be identical, their domains should be equal.

Checking the domains of $f(x)$, $g(x)$ and $h(x)$,

$f(x) = x^2/x$, x should not be equal to zero.

So, domain will be all real numbers except at $x = 0$.

$g(x)=\left(\sqrt{x}\right)^2$, x should be non-negaive.

So, domain will be all positive real numbers.

$h(x) = x$, x is defined every where,

So, we can see that none of them have the same domain.

2. (c) Domain $f(x)$ is $R - \{0\}$ *i.e.,* $(-\infty,0)\cup(0,\infty)$

and Domain $g(x)$ is R^+ *i.e.,* $(0,\infty)$

$\therefore$ Common domain of $f(x)$ and $g(x)$ is $(0,\infty)$

Hence, if $x\in(0,\infty)$, then $f(x) = g(x)$

3. (b) $f(x) = x^3 - 4x + p$

$f(0) = p$

Let $p > 0$ ....(1)

$f(1) = p - 3$ (which will be negative)

$\Rightarrow p - 3 < 0$ or $p < 3$ ....(2)

From (1) and (2)

$0 < p < 3.$

Again let $p < 0$ (3), then $p - 3 > 0$ (iv)

From (3) and (4) :

$3 < p < 0$

which is not possible

4. **(b)** We have $f(x).f(1/x) = f(x) + f(1/x)$

$\Rightarrow f(1/x)\,[f(x) - 1] = f(x)$

For $x = 4$, we have $f(1/4)\,[f(4) - 1] = f(4)$

$\Rightarrow f(1/4)\,[64] = 54$

$\Rightarrow f(1/4) = 65/64 = 1/64 + 1$

This mean $f(x) = x^3 + 1$

For $f(6)$ we have $f(6) = 216 + 1 = 217$

CHAPTER 14

QUADRATIC AND CUBIC EQUATIONS

QUADRATIC POLYNOMIALS

An expression in the form of $ax^2 + bx + c$, where a, b, c are real numbers but $a \neq 0$, is called a quadratic polynomial. For examples $2x^2 - 5x + 3$, $-x^2 + 2x$, $3x^2 - 7$, $\sqrt{2}x^2 + 7x + 2$, etc.

QUADRATIC EQUATIONS

A quadratic expression when equated to zero is called a quadratic equation. Hence an equation in the form of $ax^2 + bx + c = 0$, where a, b, c are real numbers and $a \neq 0$, is called a quadratic equation. For examples,

$$2x^2 - 5x + 3 = 0,\ -x^2 + 2x = 0,$$

$$3x^2 - 7 = 0 \text{ and } \sqrt{2}x^2 + 7x + 2 = 0, \text{ etc.}$$

Illustration 1: Which of the following is not a quadratic equation?

(a) $x^2 - 2x + 2(3 - x) = 0$
(b) $x(x + 1) + 1 = (x - 2)(x - 5)$
(c) $(2x - 1)(x - 3) = (x + 5)(x - 1)$
(d) $x^3 - 4x^2 - x + 1 = (x - 2)^3$

Solution: (b) Hint: $x(x + 1) + 1 = (x - 2)(x - 5)$

$\Rightarrow x^2 + x + 1 = x^2 - 7x + 10$

$\Rightarrow 8x - 9 = 0$, which is not a quadratic equation.

Discriminant (D)

For the quadratic equation $ax^2 + bx + c = 0$,

$$D = b^2 - 4ac$$

Here, D is the symbol of discriminant.

Roots or Solution of a Quadratic Equation

(i) If $D > 0$, then the quadratic equation $ax^2 + bx + c = 0$ has two distinct roots given by

$$\alpha = \frac{-b + \sqrt{D}}{2a} \text{ and } \beta = \frac{-b - \sqrt{D}}{2a}$$

Here α and β are symbols of roots of the quadratic equation.

(ii) If $D = 0$, then the quadratic equation $ax^2 + bx + c = 0$ has two equal roots given by

$$\alpha = \beta = -\frac{b}{2a}$$

Illustration 2: If $ax^2 + bx + c = 0$ has equal roots, then $c =$

(a) $-\frac{b}{2a}$ (b) $\frac{b}{2a}$

(c) $-\frac{b^2}{4a}$ (d) $\frac{b^2}{4a}$

Solution: (d) $ax^2 + bx + c = 0$ has equal roots if disc. $b^2 - 4ac = 0$

$\Rightarrow \quad b^2 = 4ac$

$\Rightarrow \quad c = \frac{b^2}{4a}$

Illustration 3: If $x^2 + 4x + k = 0$ has real roots, then

(a) $k \geq 4$ **(b)** $k \leq 4$
(c) $k \leq 0$ **(d)** $k \geq 0$

Solution: (b) Since $x^2 + 4x + k = 0$ has real roots.

$\therefore$ Disc. $(4)^2 - 4k \geq 0$

$\Rightarrow 16 - 4k \geq 0$

$\Rightarrow 4k \leq 16$

$\Rightarrow k \leq 4$

Properties of Quadratic Equations and Their Roots

(i) If D is a perfect square then roots are rational otherwise irrational.

(ii) If $p + \sqrt{q}$ is one root of a quadratic equation, then their conjugate $p - \sqrt{q}$ must be the other root and vice-versa, where p is rational and $\sqrt{q}$ is a surd.

(iii) If a quadratic equation in x has more than two roots, then it is an identity in x.

SUM AND PRODUCT OF ROOTS

If α and β are the roots of a quadratic equation $ax^2 + bx + c = 0$, Then,

$$\text{Sum of roots},\quad \alpha + \beta = -\frac{b}{a} = -\frac{\text{coefficient of } x}{\text{coefficient of } x^2}$$

$$\text{Product of roots},\quad \alpha\beta = \frac{c}{a} = \frac{\text{constant term}}{\text{coefficient of } x^2}$$

Illustration 4: Find the sum and product of roots of $-2x^2 + 3x - 5 = 0$.

Solution: Sum of roots $= -\frac{b}{a} = -\frac{3}{-2} = \frac{3}{2}$

Product of roots $= \frac{c}{a} = \frac{-5}{-2} = \frac{5}{2}$

FORMATION OF AN EQUATION WITH GIVEN ROOTS

If α and β are the roots of a quadratic equation, then the quadratic equation will be

$$x^2 - (\alpha + \beta)x + \alpha.\beta = 0$$

i.e., x^2 – (Sum of the roots) x + Product of the roots = 0

Illustration 5: If α and β are the roots of the equation $3x^2 - x + 4 = 0$, then find the quadratic equation whose roots are $\frac{1}{\alpha}$ and $\frac{1}{\beta}$.

Solution: $\alpha + \beta = -\frac{-1}{3} = \frac{1}{3}, \alpha.\beta = \frac{4}{3}$

Now, $\frac{1}{\alpha} + \frac{1}{\beta} = \frac{a+b}{ab}$

$$= \frac{\frac{1}{3}}{\frac{4}{3}} = \frac{1}{4}$$

$$\frac{1}{\alpha} \cdot \frac{1}{\beta} = \frac{1}{ab} = \frac{1}{\frac{4}{3}} = \frac{3}{4}$$

Hence required quadratic equation,

$$x^2 - \frac{1}{4}x + \frac{3}{4} = 0$$

$$\Rightarrow \quad 4x^2 - x + 3 = 0$$

Practice Exercise

Level - I

1. Which of the following is a quadratic equation ?

(a) $x^{\frac{1}{2}}+2x+3=0$

(b) $(x-1)(x+4)=x^2+1$

(c) $x^4-3x+5=0$

(d) $(2x+1)(3x-4)=2x^2+3$

2. Solve $x-\frac{1}{x}=1\frac{1}{2}$

(a) $-\frac{1}{2},2$ (b) $\frac{1}{2},2$

(c) $\frac{1}{2},\frac{2}{3}$ (d) None of these

3. If $2x^2-7xy+3y^2=0$, then the value of $x:y$ is

(a) 3 : 2 (b) 2 : 3

(c) 3 : 1 or 1 : 2 (d) 5 : 6

4. Father's age is 4 less than five times the age of his son and the product of their ages is 288. Find the father's age.

(a) 40 years (b) 36 years

(c) 26 years (d) 42 years

5. The sum of a rational number and its reciprocal is $\frac{13}{6}$, find the number.

(a) $\frac{2}{3}$ or $\frac{3}{2}$ (b) $\frac{3}{4}$ or $\frac{4}{3}$

(c) $\frac{2}{5}$ or $\frac{5}{2}$ (d) None of these

6. $\sqrt{6+\sqrt{6+\sqrt{6+\ldots\ldots\ldots}}}=?$

(a) 2.3 (b) 3

(c) 6 (d) 6.3

7. If $x^2+2=2x$, then the value of $x^4-x^3+x^2+2$ is

(a) 1 (b) 0

(c) -1 (d) $\sqrt{2}$

8. Minimum value of $x^2+\frac{1}{x^2+1}-3$ is

(a) 0 (b) -1

(c) -3 (d) -2

9. One root of $x^2+kx-8=0$ is square of the other. Then the value of k is

(a) 2 (b) 8

(c) -8 (d) -2

10. If the roots x_1 and x_2 of the quadratic equation $x^2-2x+c=0$ also satisfy the equation $7x_2-4x_1=47$, then which of the following is true?

(a) $c=-15$ (b) $x_1=-5, x_2=3$

(c) $x_1=4.5, x_2=-2.5$ (d) None of these

11. For what value of k, are the roots of the quadratic equation $(k+1)x^2-2(k-1)x+1=0$ real and equal?

(a) $k=0$ only (b) $k=-3$ only

(c) $k=0$ or $k=3$ (d) $k=0$ or $k=-3$

12. If the roots of the equation

$$(a^2+b^2)x^2-2ab(a+c)x+(b^2+c^2)=0$$

are equal, then which one of the following is correct?

(a) $2b=a+c$ (b) $b^2=ac$

(c) $b+c=2a$ (d) $b=ac$

13. If α and β are the roots of the equation $x^2-2x+4=0$, then what is the value of $\alpha^3+\beta^3$?

(a) 16 (b) -16

(c) 8 (d) -8

14. If p and q are the roots of the equation $x^2-px+q=0$, then what are the values of p and q respectively?

(a) 1, 0 (b) 0, 1

(c) -2, 0 (d) -2, 1

15. What is the value of $\sqrt{5\sqrt{5\sqrt{5\sqrt{5\ldots\infty}}}}$?

(a) 5 (b) $\sqrt{5}$

(c) 1 (d) $(5)^{1/4}$

16. If r and s are roots of $x^2+px+q=0$, then what is the value of $\frac{1}{r^2}+\frac{1}{s^2}$?

(a) p^2-4q (b) $\frac{p^2-4q}{2}$

(c) $\frac{p^2-4q}{q^2}$ (d) $\frac{p^2-2q}{q^2}$

17. Find the solution of $\frac{2x}{x-3}+\frac{1}{2x+3}+\frac{3x+9}{(x-3)(2x+3)}=0$

(a) 0 (b) -1

(c) 3 (d) -3

18. If the roots of $x^2-kx+1=0$ are non-real, then

(a) $-3<k<3$ (b) $-2<k<2$

(c) $k>2$ (d) $k<-2$

19. If $ax^2+bx+c=0$ has real and different roots, then

(a) $b^2-4ac=0$ (b) $b^2-4ac>0$

(c) $b^2-4ac<0$ (d) $b^2-4ac\le 0$

20. If $\sqrt{3x^2+x+5} = x-3$, then the given equator has ...... solution/solutions.

(a) $x=-4$ (b) $x=\frac{1}{2}$

(c) $x=-4, \frac{1}{2}$ (both) (d) No solution

21. The sum of two numbers p and q is 18 and the sum of their reciprocals is $\frac{1}{4}$. Then the numbers are

(a) 10, 8 (b) 12, 6

(c) 9, 9 (d) 14, 4

22. If the roots of the equation $x^2-bx+c=0$ differ by 2, then which of the following is true?

(a) $c^2=4(c+1)$ (b) $b^2=4c+4$

(c) $c^2=b+4$ (d) $b^2=4(c+2)$

23. The sum of a number and its reciprocal is one-fifth of 26. What is the sum of that number and its square?

(a) 3 (b) 4

(c) 5 (d) 6

24. Two numbers are such that the square of greater number is 504 less than 8 times the square of the other. If the numbers are in the ratio 3 : 4. Find the number.

(a) 15 and 20 (b) 6 and 8

(c) 12 and 16 (d) 9 and 12

25. The equation $x+\sqrt{x-2}=4$ has

(a) two real roots and one imaginary root

(b) one real and one imaginary root

(c) two imaginary roots

(d) one real root

26. The equation $\sqrt{x+10}-\frac{6}{\sqrt{x+10}}=5$ has

(a) an extraneous root between –5 and –1

(b) an extraneous root between –10 and –6

(c) two extraneous roots

(d) a real root between 20 and 25

[An extraneous root means a root which does not satisfy the equation.]

27. If $\log_{10}(x^2-3x+6)=1$, then the value of x is

(a) 10 or 2 (b) 4 or –2

(c) 4 only (d) 4 or –1

28. The roots of the equation $2\sqrt{x}+2x^{-\frac{1}{2}}=5$ can be found by solving

(a) $4x^2-25x+4=0$ (b) $4x^2+25x-4=0$

(c) $4x^2-17x+4=0$ (d) None of these

29. Two numbers whose sum is 6 and the absolute value of whose difference is 8 are the roots of the equation

(a) $x^2-6x+7=0$ (b) $x^2-6x-7=0$

(c) $x^2+6x-8=0$ (d) $x^2-6x+8=0$

30. The roots of the equation $x^2+2\sqrt{3}x+3=0$ are

(a) real and equal (b) rational and equal

(c) rational and unequal (d) imaginary

31. The roots of the equation $ax^2+bx+c=0$ will be reciprocal if

(a) $a=b$ (b) $a=bc$

(c) $c=a$ (d) $b=c$

32. If $\frac{b}{x-a}=\frac{x+a}{b}$ then the value of x in terms of a and b is

(a) $\pm\sqrt{a^2+b^2}$ (b) $+\sqrt{a^2+b^2}$

(c) $-\sqrt{a^2+b^2}$ (d) None of these

33. For what value of b and c would the equation $x^2+bx+c=0$ have roots equal to b and c.

(a) (0, 0) (b) (1, –2)

(c) (1, 2) (d) Both (a) and (b)

34. One of the factors of the expression $4\sqrt{3}x^2+5x-2\sqrt{3}$ is: *[SSC-Sub. Ins.-2013]*

(a) $4x+\sqrt{3}$ (b) $4x+3$

(c) $4x-3$ (d) $4x-\sqrt{3}$

35. If $x+\frac{1}{x}=3$, then the value of $\frac{3x^2-4x+3}{x^2-x+1}$ is *[SSC 10+2-2014]*

(a) $\frac{4}{3}$ (b) $\frac{3}{2}$

(c) $\frac{5}{2}$ (d) $\frac{5}{3}$

36. If $x=p+\frac{1}{p}$ and $y=p-\frac{1}{p}$, then value of $x^4-2x^2y^2+y^4$ is *[SSC 10+2-2014]*

(a) 24 (b) 4

(c) 16 (d) 8

37. If $x=3+2\sqrt{2}$, then $\frac{x^6+x^4+x^2+1}{x^3}$ is equal to *[SSC 10+2-2014]*

(a) 216 (b) 192

(c) 198 (d) 204

38. A certain number of capsules were purchased for ₹ 216. 15 more capsules could have been purchased in the same amount if each capsule was cheaper by ₹ 10. What was the number of capsules purchased? *[IBPS Clerk-2013]*

(a) 6 (b) 14

(c) 8 (d) 12

(e) 9

Level - II

1. The discriminant of $ax^2 - 2\sqrt{2}x + c = 0$ with a, c are real constants is zero. The roots must be
   (a) equal and integral (b) rational and equal
   (c) real and equal (d) imaginary
2. If one root of the equation $ax^2 + bx + c = 0$ is three times the other, then ______
   (a) $b^2 = 16\,ac$ (b) $b^2 = ac$
   (c) $3b^2 = 16\,ac$ (d) None of these
3. If the product of roots of the equation $x^2 - 3(2a+4)x + a^2 + 18a + 81 = 0$ is unity, then a can take the values as
   (a) $3, -6$ (b) $10, -8$
   (c) $-10, -8$ (d) $-10, -6$
4. If the roots of the equation $(a^2 + b^2)x^2 - 2(ac + bd)x + (c^2 + d^2) = 0$ are equal, then which of the following is true?
   (a) $ab = cd$ (b) $ad = bc$
   (c) $ad = \sqrt{bc}$ (d) $ab = \sqrt{cd}$
5. For what values of c in the equation $2x^2 - (c^3 + 8c - 1)x + c^2 - 4c = 0$ the roots of the equation would be opposite to signs?
   (a) $c \in (0,4)$ (b) $c \in (-4,0)$
   (c) $c \in (0,3)$ (d) $c \in (-4,4)$
6. If $x^2 - 3x + 2$ is a factor of $x^4 - ax^2 + b = 0$, then the values of a and b are
   (a) $-5, -4$ (b) $5, 4$
   (c) $-5, 4$ (d) $5, -4$
7. If α and β are the roots of the quadratic equation $ax^2 + bx + c = 0$, then the value of $\frac{\alpha^2}{\beta} + \frac{\beta^2}{\alpha}$ is
   (a) $\frac{3bc - a^3}{b^2c}$ (b) $\frac{3abc - b^3}{a^2c}$
   (c) $\frac{3abc - b^2}{a^3c}$ (d) $\frac{ab - b^2c}{2b^2c}$
8. If a, b are the two roots of a quadratic equation such that $a + b = 24$ and $a - b = 8$, then the quadratic equation having a and b as its roots is
   (a) $x^2 + 2x + 8 = 0$ (b) $x^2 - 4x + 8 = 0$
   (c) $x^2 - 24x + 128 = 0$ (d) $2x^2 + 8x + 9 = 0$
9. If $m + \frac{1}{m-2} = 4$ then, what is value of
   $$(m-2)^2 + \frac{1}{(m-2)^2} = ?$$
   (a) -2 (b) 0
   (c) 2 (d) 4
10. If $x^2 + y^2 + \frac{1}{x^2} + \frac{1}{y^2} = 4$, then the value of $x^2 + y^2$ is
    (a) 2 (b) 4
    (c) 8 (d) 16
11. Let x, y be two positive numbers such that $x + y = 1$. Then, the minimum value of $\left(x + \frac{1}{x}\right)^2 + \left(y + \frac{1}{y}\right)^2$ is
    (a) 12 (b) 20
    (c) 12.5 (d) 13.3
12. Solve the simultaneous equations
    $$\sqrt{\frac{x}{y}} + \sqrt{\frac{y}{x}} = \frac{5}{2}; x + y = 10$$
    (a) 8, 6 (b) 8, 2
    (c) 4, 6 (d) 5, 5
13. If roots of an equation $ax^2 + bx + c = 0$ are positive, then which one of the following is correct?
    (a) Signs of a and c should be like
    (b) Signs of b and c should be like
    (c) Signs of a and b should be like
    (d) None of the above
14. If the sum of the squares of the roots of $x^2 - (p-2)x - (p+1) = 0$ $(p \in R)$ is 5, then what is the value of p?
    (a) 0 (b) -1
    (c) 1 (d) $\frac{3}{2}$
15. If α and β are the roots of the equation $x^2 + 6x + 1 = 0$, then what is $|\alpha - \beta|$ equal to?
    (a) 6 (b) $3\sqrt{2}$
    (c) $4\sqrt{2}$ (d) 12
16. If $\frac{1}{2 - \sqrt{-2}}$ is one of the roots of $ax^2 + bx + c = 0$, where a, b, c are real, then what are the values of a, b, c respectively?
    (a) $6, -4, 1$ (b) $4, 6, -1$
    (c) $3, -2, 1$ (d) $6, 4, 1$
17. If α and β are the roots of the equation $ax^2 + bx + c = 0$, then the equation whose roots are $\frac{1}{\alpha + \beta}, \frac{1}{\alpha} + \frac{1}{\beta}$ is equal to
    (a) $acx^2 + (a^2 + bc)x + bc = 0$
    (b) $bcx^2 + (b^2 + ac)x + ab = 0$
    (c) $abx^2 + (c^2 + ab)x + ca = 0$
    (d) None of these
18. Find the roots of the equation $a^3x^2 + abcx + c^3 = 0$
    (a) $\alpha^2\beta, \beta^2\alpha$ (b) α, β^2
    (c) $\alpha^2\beta, \beta\alpha$ (d) $\alpha^3\beta, \beta^3\alpha$

19. A natural number when increased by 12, equals 160 times its reciprocal. Find the number.
(a) 3 (b) 5
(c) 8 (d) 16

20. Solve: $\frac{1}{a+b+x} = \frac{1}{a}+\frac{1}{b}+\frac{1}{x}$; $a \neq 0, x \neq 0$
(a) a, b (b) $-a, b$
(c) $0, a$ (d) $-a, -b$

21. Which is not true?
(a) Every quadratic polynomial can have at most two zeros.
(b) Some quadratic polynomials do not have any zero. [*i.e.* real zero]
(c) Some quadratic polynomials may have only one zero. [*i.e.* one real zero]
(d) Every quadratic polynomial which has two zeros.

22. The expression $a^2 + ab + b^2$ is_____ for $a < 0, b < 0$
(a) $\neq 0$ (b) < 0
(c) > 0 (d) $= 0$

23. For what value of c the quadratic equation $x^2 - (c + 6)x + 2(2c - 1) = 0$ has sum of the roots as half of their product?
(a) 5 (b) −4
(c) 7 (d) 3

24. If α and β are the roots of the equation $ax^2 + bx + c = 0$, then the equation whose roots are $\alpha + \frac{1}{\beta}$ and $\beta + \frac{1}{\alpha}$ is
(a) $abx^2 + b(c + a)x + (c + a)^2 = 0$
(b) $(c + a)x^2 + b(c + a)x + ac = 0$
(c) $cax^2 + b(c + a)x + (c + a)^2 = 0$
(d) $cax^2 + b(c + a)x + c(c + a)^2 = 0$

25. If $x^2 + ax + b$ leaves the same remainder 5 when divided by $x - 1$ or $x + 1$, then the values of a and b are respectively
(a) 0 and 4 (b) 3 and 0
(c) 0 and 5 (d) 4 and 0

26. The condition that both the roots of quadratic equation $ax^2 + bx + c = 0$ are positive is
(a) a and c have an opposite sign that of b
(b) b and c have an opposite sign that of a
(c) a and b have an opposite sign that of c
(d) None of these

27. If the equation $x^2 - bx + 1 = 0$ does not possess real roots, then which one of the following is correct?
(a) $-3 < b < 3$ (b) $-2 < b < 2$
(c) $b > 2$ (d) $b < -2$

28. If the roots of the quadratic equation $3x^2 - 5x + p = 0$ are real and unequal, then which one of the following is correct?
(a) $p = 25/12$ (b) $p < 25/12$
(c) $p > 25/12$ (d) $p \leq 25/12$

29. If the roots of the equation $x^3 - ax^2 + bx - c = 0$ are three consecutive integers, then what is the smallest possible value of b?
(a) $-\frac{1}{\sqrt{3}}$ (b) −1
(c) 0 (d) 1

30. If α, β are the roots of the equation $2x^2 - 3x - 6 = 0$, find the equation whose roots are $\alpha^2 + 2$ and $\beta^2 + 2$.
(a) $4x^2 + 49x + 118 = 0$ (b) $4x^2 - 49x + 118 = 0$
(c) $4x^2 - 49x - 118 = 0$ (d) $4x^2 + 49x - 118 = 0$

31. Sum of the areas of two squares is 468 m^2. If the difference of their perimeters is 24 m, find the sides of the two squares.
(a) 9 m, 6 m (b) 18 m, 12 m
(c) 18 m, 6 m (d) 9 m, 12 m

32. The sum of the ages of Puneet and his father is 45 years and the product of their ages is 126. What is the age of Puneet ? [*SSC CGL-2013*]
(a) 3 years (b) 5 years
(c) 10 years (d) 45 years

Hints & Solutions

Level-I

1. (d) Equations in options (a) and (c) are not quadratic equations as in (a) max. power of x is fractional and in (c), it is not 2 in any of the terms.

For option (b), $(x-1)(x+4) = x^2+1$

or $x^2+4x-x-4 = x^2+1$

or $3x-5=0$

which is not a quadratic equations but a linear.

For option (d), $(2x+1)(3x-4) = 2x^2+3$

or $6x^2-8x+3x-4 = 2x^2+3$

or $4x^2-5x-7=0$

which is clearly a quadratic equation.

2. (a) $x-\frac{1}{x}=1\frac{1}{2} \Rightarrow \frac{x^2-1}{x}=\frac{3}{2}$

$\Rightarrow 2(x^2-1)=3x \Rightarrow 2x^2-2=3x$

$\Rightarrow 2x^2-3x-2=0$

$\Rightarrow 2x^2-4x+x-2=0$

$\Rightarrow 2x(x-2)+1(x-2)=0$

$\Rightarrow (2x+1)(x-2)=0$

Either $2x+1=0$ or $x-2=0$

$\Rightarrow 2x=-1$ or $x=2$

$\Rightarrow x=\frac{-1}{2}$ or $x=2$

$\therefore x=\frac{-1}{2}$, 2 are solutions.

3. (c) $2x^2-7xy+3y^2=0$

$$2\left(\frac{x}{y}\right)^2-7\left(\frac{x}{y}\right)+3=0$$

$$\frac{x}{y}=\frac{-b\pm\sqrt{b^2-4ac}}{2a}=\frac{7\pm\sqrt{49-24}}{2\times 2}=\frac{7\pm 5}{4}=3,\frac{1}{2}$$

$\Rightarrow \frac{x}{y}=\frac{3}{1}$ or $\frac{x}{y}=\frac{1}{2}$

4. (b) Let the son's age be x years.

So, father's age $= 5x-4$ years.

$\therefore x(5x-4)=288$

$\Rightarrow 5x^2-4x-288=0 \Rightarrow 5x^2-40x+36x-288=0$

$\Rightarrow 5x(x-8)+36(x-8)=0$

$\Rightarrow (5x+36)(x-8)=0$

Either $x-8=0$ or $5x+36=0 \Rightarrow x=8$ or $x=\frac{-36}{5}$

x cannot be negative; therefore, $x=8$ is the solution.

$\therefore$ Son's age = 8 years and Father's age = $5x-4$ = 36 years.

5. (a) Let the number be x.

Then, $x+\frac{1}{x}=\frac{13}{6} \Rightarrow \frac{x^2+1}{x}=\frac{13}{6} \Rightarrow 6x^2-13x+6=0$

$\Rightarrow 6x^2-9x-4x+6=0 \Rightarrow (3x-2)(2x-3)=0$

$\Rightarrow x=\frac{2}{3}$ or $x=\frac{3}{2}$.

Hence, the required number is $\frac{2}{3}$ or $\frac{3}{2}$.

6. (b) $\sqrt{6+\sqrt{6+\sqrt{6+\ldots\ldots\ldots\ldots}}}$

$\sqrt{6+x}=x$

$6+x=x^2$

$x^2-x-6=0$

$x^2-3x+2x-6=0$

$(x-3)(x+2)=0$

$x=3$

7. (b) $x^2+2=2x \Rightarrow x^2-2x+2=0$

$$\begin{array}{r|l} x^2-2x+2 & x^4-x^3+x^2+2 \;(x^2+x+1 \\ & x^4-2x^3+2x^2 \\ & -\quad +\quad - \\ \hline & x^3-x^2+2 \\ & x^3-2x^2+2x \\ & -\quad +\quad - \\ \hline & x^2-2x+2 \\ & x^2-2x+2 \\ \hline & 0 \end{array}$$

$\therefore x^4-x^3+x^2+2 = (x^2-2x+2)(x^2+x+1)=0$

8. (d) $x^2 \ge 0$

$\therefore$ Minimum value

$=0+\frac{1}{1}-3=-2$

9. (d) Given $x^2+kx-8=0$

Let a and b be the roots of given equation and $b=a^2$ (given)

Sum of roots $= a+b=-k=a+a^2$ .....(1)

Product of roots $= ab=-8=a^3 \Rightarrow a=-2$

Using $a=-2$ in (1), $-k=-2+4=2$ or $k=-2$

10. (a) $7x_2 - 4x_1 = 47$
$x_1 + x_2 = 2$
Solving $11x_2 = 55$
$x_2 = 5$ & $x_1 = -3$
$\therefore c = -15$

11. (c) Since, the roots of the equation $(k+1)x^2 - 2(k-1)x + 1 = 0$ are real and equal.
$\therefore \{-2(k-1)\}^2 - 4(k+1) = 0 \quad (\because b^2 - 4ac = 0)$
$\Rightarrow 4(k^2 - 2k + 1) - 4(k+1) = 0$
$\Rightarrow k^2 - 2k + 1 - 1 = 0$
$\Rightarrow k^2 - 3k = 0$
$\Rightarrow k = 0, k = 3$

12. (b) Since roots of the given equation are equal.
$\therefore D = 0$
On solving we get $b^2 = ac$

13. (b) Use $\alpha^3 + \beta^3 = (\alpha+\beta)^3 - 3\alpha\beta(\alpha+\beta)$

14. (a)

15. (a) Let $x = \sqrt{5\sqrt{5\sqrt{5...\infty}}}$
$\Rightarrow x^2 = 5x \Rightarrow x = 0, 5$

16. (d) $\frac{1}{r^2} + \frac{1}{s^2} = \frac{s^2 + r^2}{(rs)^2} = \frac{(s+r)^2 - 2sr}{(rs)^2} = \frac{p^2 - 2q}{q^2}$

17. (b) Clearly, the given equation is valid if $x - 3 \neq 0$ and $2x + 3 \neq 0$ i.e.,. when $x \neq \frac{-3}{2}, 3$

Now, $\frac{2x}{x-3} + \frac{1}{2x+3} + \frac{3x+9}{(x-3)(2x+3)} = 0$
$\Rightarrow 2x(2x+3) + (x-3) + 3x + 9 = 0$
[Multiplying throughout by $(x-3)(2x+3)$]
$\Rightarrow 4x^2 + 6x + x - 3 + 3x + 9 = 0$
$\Rightarrow 4x^2 + 10x + 6 = 0$
$\Rightarrow 2x^2 + 5x + 3 = 0$
$\Rightarrow 2x^2 + 2x + 3x + 3 = 0$
$\Rightarrow 2x(x+1) + 3(x+1) = 0$
$\Rightarrow (2x+3)(x+1) = 0$
$\Rightarrow x + 1 = 0 \Rightarrow x = -1 \; [\because 2x + 3 \neq 0]$
Hence, $x = -1$ is the only solution of the given equation.

18. (b) Since the roots of $x^2 - kx - 4 = 0$ are non-real.
$\therefore$ Disc., $(-k^2) - 4 < 0 \Rightarrow k^2 - 4 < 0$
$\Rightarrow k^2 < 4 \Rightarrow |k| < 2 \Rightarrow -2 < k < 2$

19. (b)

20. (d) Square both sides, we shall get $x = -4, \frac{1}{2}$. But both of them do not satisfy the given equation.

21. (b) $p + q = 18$ ...(1)

and $\frac{1}{p} + \frac{1}{q} = \frac{1}{4}$ ...(2) (Given)

i.e $\frac{p+q}{pq} = \frac{1}{4} \Rightarrow \frac{18}{pq} = \frac{1}{4}$
$\Rightarrow pq = 72$ ...(3)
From (1) and (3), $p(18 - p) = 72$
$\Rightarrow p^2 - 18p + 72 = 0 \Rightarrow (p-6)(p-12) = 0$
$\Rightarrow p = 6, 12$ when $p = 6, q = 12$; when $p = 12, q = 6$
Hence the numbers are 12, 6.

22. (b) Let the roots be α and $\alpha + 2$.
Then $\alpha + \alpha + 2 = b \Rightarrow \alpha = (b-2)/2$ (1)
and $\alpha(\alpha+2) = c \Rightarrow \alpha^2 + 2\alpha = c$ (2)
Putting the value of α from (1) in (2).
$((b-2)/2)^2 + 2(b-2)/2) = c$
$\Rightarrow (b^2 + 4 - 4b)/4 + b - 2 = c$
$\Rightarrow b^2 + 4 - 8 = 4c$
$\Rightarrow b^2 = 4c + 4$

23. (c) Let the number be x. Then,

$x + \frac{1}{x} = \frac{26}{5}$

$\Rightarrow \frac{x^2 + 1}{x} = \frac{26}{5}$
$\Rightarrow 5x^2 - 26x + 5 = -0$
$\Rightarrow (x-5)(5x-1) = 0$
$\Rightarrow x = 5$ or $\frac{1}{5}$

24. (d)

25. (d) $x + \sqrt{x-2} = 4$
$\sqrt{x-2} = 4 - x$
Squaring on the both sides
$x - 2 = 16 + x^2 - 8x$
$x^2 - 9x + 18 = 0$
$(x-6)(x-3) = 0$
$x = 6$ or 3
But by checking, only $x = 3$ satisfies the equation.

26. (b) $\sqrt{x+10} - \frac{6}{\sqrt{x+10}} = 5$

$x + 10 - 6 = 5\sqrt{x+10}$
$x + 4 = 5\sqrt{x+10}$
Squaring on both sides,
$x^2 + 8x + 16 = 25x + 250$
$x^2 - 17x - 234 = 0$
$x^2 - 26x + 9x - 234 = 0$
$x(x-26) + 9(x-26) = 0$
$(x-26)(x+9) = 0$
$x = 26$ (or) -9
Here $x = -9$ is not satisfying. So it is extraneous.

27. **(d)** $\log_{10}(x^2-3x+6)=1$

$x^2-3x+6=10^1$

$x^2-3x-4=0$

$(x-4)(x+1)=0$

$x=4$ or -1

28. **(c)** $2\sqrt{x}+\frac{2}{\sqrt{x}}=5$

$2x+2=5\sqrt{x}$

$\Rightarrow 4x^2+8x+4=25x$

$\Rightarrow 4x^2-17x+4=0$

29. **(b)** Let α and β are the roots

$\alpha+\beta=6$

$\alpha-\beta=8$

$2\alpha=14$

$\alpha=7$

$\beta=-1$

$\alpha+\beta=6, \alpha\beta=-7$

The quadratic equation is $x^2-6x-7=0$

30. **(a)** $b^2-4ac=(2\sqrt{3})^2-4(1)(3)=0$. So the roots are real and equal.

31. **(c)** Since roots are reciprocal,

product of the roots $=1 \Rightarrow \frac{c}{a}=1$

$\Rightarrow c=a.$

32. **(a)** $\frac{b}{x-a}=\frac{x+a}{b}$

$x^2-a^2=b^2$

$x^2=b^2+a^2$

$x=\pm\sqrt{a^2+b^2}$

33. **(d)** Solve using options. It can be seen that $b=0$ and $c=0$ the condition is satisfied. It is also satisfied at $b=1$ and $c=-2$.

34. **(d)** $4\sqrt{3}\,x^2+5x-2\sqrt{3}$

$=4\sqrt{3}\,x^2+8x-3x-2\sqrt{3}$

$=4x(\sqrt{3}x+2)-\sqrt{3}(\sqrt{3x}+2)$

$=\left(4x-\sqrt{3}\right)\left(\sqrt{3x}+2\right)$

35. **(c)** $\frac{3x^2-4x+3}{x^2-x+1}=\frac{\frac{3x^2}{x}-\frac{4x}{x}+\frac{3}{x}}{\frac{x^2}{x}-\frac{x}{x}+\frac{1}{x}}$

$$\frac{3\left(x+\frac{1}{x}\right)-4}{\left(x+\frac{1}{x}\right)-1}=\frac{3\times3-4}{3-1}=\frac{5}{2}$$

36. **(c)** $x^4-2x^2y^2+y^4=(x^2-y^2)^2=[(x+y)(x-y)]^2$

$$=\left(2p\times\frac{2}{p}\right)^2=16$$

37. **(d)** We have, $x=3+2\sqrt{2}$

$$\frac{1}{x}=\frac{1}{3+2\sqrt{2}}\times\frac{3-2\sqrt{2}}{3-2\sqrt{2}}=3-2\sqrt{2}$$

$$x+\frac{1}{x}=6$$

$$\frac{x^6+x^4+x^2+1}{x^3}=x^3+x+\frac{1}{x}+\frac{1}{x^3}$$

$$=\left(x^3+\frac{1}{x^3}\right)+\left(x+\frac{1}{x}\right)$$

$$=\left(x+\frac{1}{x}\right)\left(x^2+\frac{1}{x^2}-1\right)+\left(x+\frac{1}{x}\right)$$

$$=\left(x+\frac{1}{x}\right)\left[\left(x+\frac{1}{x}\right)^2-3\right]+\left(x+\frac{1}{x}\right)$$

$=6[6^2-3]+6=198+6=204$

38. **(d)** Let x be the price of one capsule

y be the total number of capsule.

$xy=216$ ...(1)

$(x-10)(y+15)=216$ ...(2)

From eqs (1) and (2)

$$\left(\frac{216}{y}-10\right)(y+15)=216$$

$$(216-10y)(y+15)=216y$$

$$216y+216\times15-10y^2-150y=216y$$

$$216y+3240-10y^2-150y=216y$$

$$-10y^2-150y+3240=0$$

$$y^2+15y-324=0$$

$$y=12,-27$$

Number of capsules cannot be negative.

Level-II

1. **(c)** $ax^2-2\sqrt{2}x+c=0$

$(2\sqrt{2})^2-4ac=0$

$4ac=8$

$ac=2$

$c=\frac{2}{a}$

Let α, β be the roots.

$\alpha+\beta\ \frac{2\sqrt{2}}{a},\ \alpha\beta=\frac{c}{a}=\frac{2}{a^2}$

$(\alpha-\beta)^2=(\alpha+\beta)^2-4\alpha\beta$

$=\frac{8}{a^2}-\frac{8}{a^2}=0$

$\alpha=\beta$

So, $\alpha=\beta=\frac{\sqrt{2}}{a}$

Hence the roots are real and equal.

2. **(c)** Let α, 3α are the roots.

$\alpha+3\alpha=\frac{-b}{a}\Rightarrow 4\alpha=\frac{-b}{a}$

$\Rightarrow\alpha=\frac{-b}{4a}$ ...(1)

$\alpha\times 3\alpha=\frac{c}{a}\Rightarrow 3\alpha^2=\frac{c}{a}$

$\frac{3b^2}{16a^2}=\frac{c}{a}$ [by (1)]

$3b^2=16ac.$

3. **(c)** The product of the roots is given by: $(a^2+18a+81)/1$.
Since product is unity we get: $a^2+18a+81=1$
Thus, $a^2+18a+80=0$
Solving, we get $a=-10$ and $a=-8$.

4. **(b)** Solve this by assuming each option to be true and then check whether the given expression has equal roots for the option under check.
Thus, if we check for option (b).
$ad=bc$.
We assume $a=6$, $d=4$ $b=12$ $c=2$ (any set of values that satisfies $ad=bc$)
Then $(a^2+b^2)x^2-2(ac+bc)x+(c^2+d^2)=0$
$180x^2-120x+20=0$
We can see that this has equal roots. Thus, option (b) is a possible answer. The same way if we check for *a*, *c* and *d* we see that none of them gives us equal roots and can be rejected.

5. **(a)** For the roots to be opposite in sign, the product of roots should be negative.
$(c^2-4c)/2<0\Rightarrow 0<c<4$

6. **(b)** $x^2-3x+2=0$ gives its roots as $x=1, 2$. Put these values in the equation and then use the options.

7. **(b)** Here, $\alpha+\beta=-\frac{b}{a}$ and $\alpha\beta=\frac{c}{a}$

Thus, $\frac{\alpha^2}{\beta}+\frac{\beta^2}{\alpha}=\frac{\alpha^3+\beta^3}{\alpha\beta}$

$=\frac{(\alpha+\beta)\ (\alpha^2-\alpha\beta+\beta^2)}{\alpha\beta}$ ...(1)

Now, $(\alpha^2+\beta^2-\alpha\beta)=[(\alpha+\beta)^2-2\alpha\beta-\alpha\beta]$

$=[(\alpha+\beta)^2-3\alpha\beta]$

Hence (1) becomes

$$\Rightarrow\frac{(\alpha+\beta)[(\alpha+\beta)^2-3\alpha\beta)]}{\alpha\beta}=\frac{\frac{-b}{a}\left[\frac{b^2}{a^2}-\frac{3c}{a}\right]}{\frac{c}{a}}$$

$$=\frac{-b}{c}\left[\frac{b^2-3ac}{a^2}\right]=\frac{3abc-b^3}{a^2c}$$

8. **(c)** $a+b=24$ and $a-b=8$
$\Rightarrow a=16$ and $b=8\Rightarrow ab=16\times 8=128$
A quadratic equation with roots *a* and *b* is
$x^2-(a+b)x+ab=0$ or $x^2-24x+128=0$

9. **(c)** $m+\frac{1}{m-2}=4$
$m^2-2m-3=0$
$(m-3)(m+1)=0$
$m=3$
$m-2=1$

Now $(m-2)^2+\frac{1}{(m-2)^2}$

$=1^2+\frac{1}{1^2}=2$

10. **(a)** $x^2+y^2+\frac{1}{x^2}+\frac{1}{y^2}-4=0$

$\Rightarrow\ x^2+\frac{1}{x^2}-2+y^2+\frac{1}{y^2}-2=0$

$\Rightarrow\ \left(x-\frac{1}{x}\right)^2+\left(y-\frac{1}{y}\right)^2=0$

$\Rightarrow\ x-\frac{1}{x}=0$

$\Rightarrow\ x^2-1=0\Rightarrow x=1$

Similarly,
$y=1$
$\therefore\ x^2+y^2=1+1=2$

11. **(c)** Given, $x+y=1$

Then, $\left(x+\frac{1}{x}\right)^2+\left(y+\frac{1}{y}\right)^2=x^2+y^2+\frac{1}{x^2}+\frac{1}{y^2}+4$

Minimum value of x^2+y^2 occur when $x=y$
$[\because x+y=1]$

Put $x = y = \frac{1}{2}$

Minimum value $= \left(\frac{5}{2}\right)^2 + \left(\frac{5}{2}\right)^2 = \frac{25}{2} = 12.5$

12. **(b)** We have $\sqrt{\frac{x}{y}} + \sqrt{\frac{y}{x}} = \frac{5}{2}$ ...(1)

and $x + y = 10$ ...(2)

Now, $\sqrt{\frac{x}{y}} + \sqrt{\frac{y}{x}} = \frac{5}{2} \Rightarrow \frac{x+y}{\sqrt{xy}} = \frac{5}{2}$

$\Rightarrow \frac{10}{\sqrt{xy}} = \frac{5}{2}$ [using eq. (2)]

$\Rightarrow \sqrt{xy} = 4 \Rightarrow xy = 16$

Thus, the given system of simultaneous equations reduces to

$x + y = 10$ and $xy = 16$

$\Rightarrow y = 10 - x$

and $xy = 16$

$\Rightarrow x(10 - x) = 16$

$\Rightarrow x^2 - 10x + 16 = 0$

$\Rightarrow (x-2)(x-8) = 0 \Rightarrow x = 2$ or $x = 8$

Now, $x = 2$ and $x + y = 10 \Rightarrow y = 8$

and $x = 8$ and $x + y = 10 \Rightarrow y = 2$

Hence, the required solution are $x = 2, y = 8$ and $x = 8, y = 2$

13. **(a)** If roots of an equation $ax^2 + bx + c = 0$ are positive, then signs of a and c should be like.

14. **(c)** Let α and β be the roots of $x^2 - (p-2)x - (p+1) = 0$

Then, $\alpha + \beta = p - 2$

and $\alpha\beta = -(p+1)$

$\therefore \alpha^2 + \beta^2 = 5$

$\Rightarrow (\alpha+\beta)^2 - 2\alpha\beta = 5$

$\Rightarrow (p-2)^2 + 2(p+1) = 5$

$\Rightarrow p^2 - 4p + 4 + 2p + 2 = 5$

$\Rightarrow p^2 - 2p + 1 = 0 \Rightarrow (p-1)^2 = 0$

$\Rightarrow p = 1$

15. **(c)** $\because$ α and β are the roots of the equation $x^2 + 6x + 1 = 0$

$\therefore \alpha + \beta = -6$ and $\alpha\beta = 1$

Now, $(\alpha - \beta)^2 = (\alpha+\beta)^2 - 4\alpha\beta$

$= (-6)^2 - 4$

$= 36 - 4 = 32$

$\Rightarrow |\alpha - \beta| = \sqrt{32} = 4\sqrt{2}$

16. **(a)** The given root is

$= \frac{1}{2 - \sqrt{-2}} = \frac{2 + \sqrt{2i}}{6}$

$\therefore$ Another root $= \frac{2 - \sqrt{2i}}{6}$

Now, find sum and product of the roots and put in $x^2 -$ (sum of the roots) x + (multiplication of the roots) = 0

17. **(b)** $S = \frac{1}{\alpha+\beta} + \frac{\alpha+\beta}{\alpha\beta} = -\frac{a}{b} - \frac{b}{c} = -\frac{(ac + b^2)}{bc}$

$P = \frac{1}{\alpha+\beta} \cdot \frac{\alpha+\beta}{\alpha\beta}$

$= \frac{1}{\alpha\beta} = \frac{a}{c}$

Put the values of P and S in $x^2 - Sx + P = 0$, we get the required result.

18. **(a)** Dividing the equation $a^3x^2 + abcx + c^3 = 0$ by c^2, we get

$a\left(\frac{ax}{c}\right)^2 + b\left(\frac{ax}{c}\right) + c = 0$

$\Rightarrow \frac{ax}{c} = \alpha, \beta$

$\Rightarrow x = \frac{c}{a}\alpha, \frac{c}{a}\beta$

$\Rightarrow x = \alpha^2\beta, \alpha\beta^2$

$[\because \frac{c}{a} = \alpha\beta$ = product of roots]

Hence, $\alpha^2\beta$ and $\alpha\beta^2$ are the roots of the equation $a^3x^2 + abcx + c^3 = 0$.

19. **(b)** Let the natural number be $= x$.

By the given condition: $x + 12 = \frac{160}{x} (x \neq 0)$

$\Rightarrow x^2 + 12x - 160 = 0 \Rightarrow x = -\frac{12 \pm \sqrt{144 + 640}}{2}$

$= -\frac{12 \pm \sqrt{784}}{2} = \frac{-12 \pm 28}{2} = -\frac{40}{2}$ or $\frac{16}{2}$

$= -10$ or 5. But x is a natural number $\therefore x = 5$.

20. **(d)** $\frac{1}{a+b+x} = \frac{1}{a} + \frac{1}{b} + \frac{1}{x} \Rightarrow \frac{1}{a+b+x} - \frac{1}{x} = \frac{1}{a} + \frac{1}{b}$

$\Rightarrow \frac{x - a - b - x}{(a+b+x)x} = \frac{b-a}{ab}$

$\Rightarrow \frac{-(a+b)}{(x^2 + ax + bx)} = \frac{(a+b)}{ab}$

$\Rightarrow \frac{-1}{x^2 + ax + bx} = \frac{1}{ab}$

$\Rightarrow x^2 + ax + bx = -ab \Rightarrow x(x+a) + b(x+a) = 0$

$\Rightarrow (x+a)(x+b) = 0 \Rightarrow x = -a$ or $x = -b$

21. **(d)** (a) is clearly true.

(b) $x^2 + 1$ is a quadratic polynomial which has no real value of x for which $x^2 + 1$ is zero.

$[\therefore x^2 \geq 0 \Rightarrow x^2 + 1 > 0$ for all real $x]$ $\therefore$ (b) is true.

(c) The quadratic polynomial

$x^2 - 2x + 1 = (x-1)^2$ has only one zero i.e. 1

$\therefore$ (c) is true.

$[\because (x-1)^2 > 0$ at $x \neq 1$ and for $x = 1, (x-1)^2 = 0]$

(d) is false $[\because$ of (b), (c)]

Hence (d) holds.

22. **(c)** For a, b negative the given expression will always be positive since, a^2, b^2 and ab are all positive.

23. **(c)** $(c+6) = 1/2 \times 2(2c-1) \Rightarrow c + 6 = 2c - 1 \Rightarrow c = 7$

24. **(c)** Assume any equation:

Say $x^2 - 5x + 6 = 0$

The roots are 2, 3.

We are now looking for the equation, whose roots are: $(2 + 1/3) = 2.33$ and $(3 + 1/2) = 3.5$.

Also $a = 1, b = -5$ and $c = 6$.

Put these values in each option to see which gives 2.33 and 3.5 as its roots.

25. **(c)** $f(x) = x^2 + ax + b$

$f(1) = f(-1) = 5$

$\Rightarrow a + b = -a + b = 5$

$\Rightarrow a = 0, b = 5$

26. **(a)** For both the roots: (α, β) to be positive

$\alpha + \beta > 0$ and $\alpha\beta > 0$

$$\Rightarrow \frac{-b}{a} > 0 \text{ and } \frac{c}{a} > 0$$

i.e., b and a are of opposite sign and c and a are of same sign.

27. **(b)** Given quadratic equation is $x^2 - bx + 1 = 0$

It has no real roots. It means, equation has imaginary roots.

Which is possible when $B^2 - 4AC < 0$

Here, $B = -b, A = 1, C = 1$

$$\Rightarrow b^2 - 4 < 0 \Rightarrow b^2 < 4 \Rightarrow -2 < b < 2$$

28. **(b)** The given equation is,

$3x^2 - 5x + p = 0$

We have, $a = 3, b = -5, c = p$

$D = b^2 - 4ac = 25 - 12p$

For Real and unequal, $D > 0$

$\therefore 25 - 12p > 0$

$$\Rightarrow 25 > 12p \Rightarrow p < \frac{25}{12}$$

29. **(b)** Let roots are $(n-1)$, n and $(n+1)$

Sum of the roots $= b$

$(n-1)n + n(n+1) + (n+1)(n-1) = b$

$\Rightarrow n^2 - n + n^2 + n + n^2 - 1 = b$

$\Rightarrow 3n^2 - 1 = b$

The value of b will be minimum when the value of n^2 is minimum i.e., $n^2 = 0$

Hence, minimum value of $b = -1$.

30. **(b)** Since, α, β are root of the equation $2x^2 - 3x - 6 = 0$

$$\therefore \alpha + \beta = \frac{3}{2} \text{ and } \alpha\beta = -3$$

$$\Rightarrow \alpha^2 + \beta^2 = (\alpha + \beta)^2 - 2\alpha\beta$$

$$= \frac{9}{4} + 6 = \frac{33}{4}$$

Now, $(\alpha^2 + 2) + (\beta^2 + 2) = (\alpha^2 + \beta^2) + 4$

$$= \frac{33}{4} + 4 = \frac{49}{4}$$

and $(\alpha^2 + 2)(\beta^2 + 2) = \alpha^2\beta^2 + 2(\alpha^2 + \beta^2) + 4$

$$= (-3)^2 + 2\left(\frac{33}{4}\right) + 4 = \frac{59}{2}$$

So, the equation whose roots are $\alpha^2 + 2$ and $\beta^2 + 2$ is $x^2 - x\{(\alpha^2 + 2) + (\beta^2 + 2)\} + (\alpha^2 + 2)(\beta^2 + 2) = 0$

$$\Rightarrow x^2 - \frac{49}{4}x + \frac{59}{2} = 0$$

$\Rightarrow 4x^2 - 49x + 118 = 0$

31. **(b)** Let first square has side x, $\therefore$ Area $= x^2$, Perimeter $= 4x$ and let second square has side y,

$\therefore$ Area $= y^2$, Perimeter $= 4y$

Let $x > y$ so that $4x > 4y$

Given, $x^2 + y^2 = 468$ ...(1)

and $4x - 4y = 24 \Rightarrow x - y = 6 \Rightarrow y = x - 6$ ...(2)

Using (2) in (1), we get $x^2 + (x-6)^2 = 468$

$\Rightarrow x^2 + x^2 - 12x + 36 = 468 \Rightarrow 2x^2 - 12x - 432 = 0$

$$\Rightarrow x^2 - 6x - 216 = 0 \Rightarrow x = \frac{6 \pm \sqrt{36 + 864}}{2} = \frac{6 \pm \sqrt{900}}{2}$$

$$= \frac{6 \pm 30}{2} = \frac{36}{2}, \frac{-24}{2} = 18, -12$$

But x being length cannot be negative $\therefore x = 18$

put $x = 18$ in (2), we get $y = x - 6 = 18 - 6 = 12$

$\therefore$ sides of the two squares $= x, y = 18$ m, 12 m

32. **(a)** Let Puneet's age $= x$ yrs.

Let Puneet's father age $= y$ yr.

$x + y = 45 \Rightarrow y = (45 - x)$

$xy = 126$

Putting the value of y.

$(x)(45 - x) = 126$

$45x - x^2 = 126$

$x^2 - 45x + 126 = 0$

$x^2 - 42x - 3x + 126 = 0$

$x(x - 42) - 3(x - 42) = 0$

$x = 3, x = 42$

Hence, Puneet's age in 3yrs.

CHAPTER 15

INEQUALITIES

INEQUALITY

Two real numbers, two algebraic expressions or an algebraic expression and a real number related by the symbol >, <, ≥ or ≤ form an inequality.

'>' means 'greater than'. Hence $a > b$ read as a is greater than b.

'<' means 'less than'. Hence $a < b$ read as a is less than b.

'≥' means 'greater than or equal to'. Hence $a \geq b$ is read as a is greater than or equal to b.

'≤' means 'less than or equal to'. Hence $a \leq b$ is read as a is less than or equal to b.

TYPES OF INEQUALITIES

1. Numerical Inequalities
   Inequalities which does not contain any variable are called numerical, inequalities.
2. Literal Inequalities
   Inequalities which does not contain any variable are called literal inequalities. For examples, $8 > 6$, $-7 < 0$, etc.
3. An inequality may contain more than one variable. For examples $2xy < 8$, $x + 3y \geq 20$, etc.
   An inequality in one variable may be linear, quadratic or cubic etc. For examples $2x + 5 < 10$, $x^2 + 4x + 3 \geq 0$, $-x^3 + 2x^2 - 4 \leq 8$, etc.
4. Strict Inequalities
   Inequalities involving the symbol '>' or '<' are called strict inequalities.
5. Slack Inequalities
   Inequalities involving the symbol '≥' or '≤' are called slack inequalities.

   $a \geq b$ means $a > b$ or $a = b$

   $a \leq b$ means $a < b$ or $a = b$

   Note that simultaneous relation between any three different quantities a, b and c will be either $a < b < c$, $a < b \leq c$, $a \leq b < c$ or $a \leq b \leq c$

SOME PROPERTIES OF INEQUALITY

(i) If $a > b$, then evidently $b < a$ i.e. if the sides of an equality be transposed, the sign of equality must be reversed.

(ii) Sign of inequality does not change when equal numbers added to (or subtracted from) both sides of an inequality.

i.e. $a > b$

$\Rightarrow$ $a + 5 > b + 5$

and also $a - 4 > b - 4$

(iii) Sign of inequality does not change when both sides of an inequality can be multiplied (or divided) by the same positive number. But when both sides are multiplied or divided by a negative number, then the sign of inequality is reversed.

i.e. $a \leq b$

$\Rightarrow$ $3a \leq 3b$

and also $\frac{a}{5} \leq \frac{b}{5}$.

But $-3a \geq -3b$

and also $\frac{a}{-5} \geq \frac{b}{-5}$

(iv) If $a > b$ and $b > c$, then $a > c$. Since $5 > 4$ and $4 > 2$, therefore $5 > 2$.

(v) If $a > b > 0$ then $\frac{1}{a} < \frac{1}{b}$

Since $6 > 2 > 0$, therefore $\frac{1}{6} < \frac{1}{2}$.

(vi) If $a > b > 0$ and $n > 0$ then $a^n > b^n$ and $(a)^{1/n} > (b)^{1/n}$

Since $3 > 2 > 0$ and $4 > 0$, therefore $(3)^4 > (2)^4$ and also $(3)^{1/4} > (2)^{1/4}$

(vii) If $x > y > 0$ and $a > 1$, then $a^x > a^y$

Since $5 > 3 > 0$ and $6 > 1$, therefore $(6)^5 > (6)^3$

(viii) If $x > y > 0$ and $0 < a < 1$ then $a^x < a^y$

Since $6 > 4 > 0$ and $0 < \frac{2}{3} < 1$, therefore $\left(\frac{2}{3}\right)^6 < \left(\frac{2}{3}\right)^4$.

IMPORTANT RESULTS

(i) Square of any real number is always equal or greater than 0. i.e. if a is a real number, then $a^2 \geq 0$.

(ii) For any real number a,

$$|a| \geq 0$$

(iii) If a is a positive real number and $|x| \leq a$, then

$$-a \leq x \leq a$$

(iv) If a is a positive real number and $|x| \geq a$, then

$$x \leq -a \text{ or } x \geq a$$

(v) $|a+b| \leq |a| + |b|$

In general

$|a_1 + a_2 + a_3 + ... + a_n| \leq |a_1| + |a_2| + |a_3| + ... + |a_n|$

(vi) $|a-b| \geq |a| - |b|$

(vii) $a^2 + b^2 \geq 2ab$

NOTATION AND RANGES

If a, b, c, d are four numbers such that $a < b < c < d$, then

(i) $x \in (a, b)$ means $a < x < b$

(ii) $x \in [a, b]$ means $a \leq x \leq b$

(iii) $x \in [a, b)$ means $a \leq x < b$

(iv) $x \in (a, b]$ means $a < x \leq b$

(v) $x \in (a, b) \cup (c, d)$ means $a < x < b$ or $c < x < d$

SOLUTIONS OF LINEAR INEQUALITIES IN ONE UNKNOWN

Inequalities of the form $ax + b > 0$, $ax + b \geq 0$, $ax + b < 0$ and $ax + b \leq 0$ are called linear inequalities.

(i) $ax + b > 0$

$\Rightarrow \quad x > -\dfrac{b}{a}$, if $a > 0$

and $x < -\dfrac{b}{a}$, if $a < 0$

(ii) $ax + b \geq 0$

$\Rightarrow \quad x \geq -\dfrac{b}{a}$, if $a > 0$

and $x \leq -\dfrac{b}{a}$, if $a < 0$

(iii) $ax + b < 0$

$\Rightarrow \quad x < -\dfrac{b}{a}$, if $a > 0$

and $x > -\dfrac{b}{a}$, if $a < 0$

(iv) $ax + b \leq 0$

$\Rightarrow \quad x \leq -\dfrac{b}{a}$, if $a > 0$

and $x \geq -\dfrac{b}{a}$, if $a < 0$

Illustration 1: Solve $2(x-3) + 4 \geq 4 - x$

Solution: $2(x-3) + 4 \geq 4 - x$

$\Rightarrow \quad 2x - 6 + 4 \geq 4 - x \Rightarrow 2x + x - 2 \geq 4$

$\Rightarrow \quad 3x \geq 4 + 2 \Rightarrow 3x \geq 6$

$\Rightarrow \quad x \geq \dfrac{6}{3} \Rightarrow x \geq 2$

This solution can also be written as $x \in [2, \infty)$.

Illustration 2: Solve $3(x+4) + 1 < 2(3x+1) + 15$

Solution: $3(x+4) + 1 < 2(3x+1) + 15$

$\Rightarrow \quad 3x + 12 + 1 < 6x + 2 + 15$

$\Rightarrow \quad 3x - 6x < 17 - 13$

$\Rightarrow \quad -3x < 4$

$\Rightarrow \quad x > \dfrac{4}{-3} \Rightarrow x > -\dfrac{4}{3}$

This solution can also be written as $x \in \left(-\dfrac{4}{3}, \infty\right)$.

Illustration 3: Solve the following inequations:

$$\frac{2x+4}{x-1} \geq 5$$

Solution: We have,

$\dfrac{2x+4}{x-1} \geq 5$

$\Rightarrow \quad \dfrac{2x+4}{x-1} - 5 \geq 0$

$\Rightarrow \quad \dfrac{x-3}{x-1} \leq 0$ [Dividing both sides by 3]

$\Rightarrow \quad 1 < x \leq 3$

$\Rightarrow \quad x \in (1, 3]$

Hence, the solution set of the given inequation is (1,3].

Illustration 4: Solve: $-5 \leq \dfrac{2-3x}{4} \leq 9$.

Solution: We have,

$-5 \leq \dfrac{2-3x}{4} \leq 9$

$\Rightarrow \quad \dfrac{22}{3} \geq x \geq \dfrac{-34}{3}$

$\Rightarrow \quad \dfrac{-34}{3} \leq x \leq \dfrac{22}{3}$

$\Rightarrow \quad x \in [-34/3, 22/3]$

Hence, the interval [–34/3, 22/3] is the solution set of the given system of inequations.

INEQUALITIES CONTAINING A MODULUS

(i)
- If $a > 0$, then $|x| \leq a \Rightarrow -a \leq x \leq a$
- If $a > 0$, then $|x| < a \Rightarrow -a < x < a$

(ii)
- If $a > 0$, then $|x| \geq a \Rightarrow x \leq -a$ and $x \geq a$
- If $a > 0$,then $|x| > a \Rightarrow x < -a$ and $x > a$
- If $a < 0$, then $|x| \geq a \Rightarrow x \leq a$ and $x \geq -a$
- If $a < 0$, then $|x| > a \Rightarrow x < a$ and $x > -a$

Illustration 5: Solve $|x-3| \geq 4$

Solution: $|x-3| \geq 4$

$\Rightarrow \quad (x-3) \leq -4$ and $(x-3) \geq 4$

$\Rightarrow \quad x \leq -4+3$ and $x \geq 4+3$

$\Rightarrow \quad x \leq -1$ and $x \geq 7$

i.e. $\quad x \in (-\infty, -1] \cup [7, \infty)$

Illustration 5: Solve $|5-4x| < -2$

Solution: $|5-4x| < -2$

$\Rightarrow \quad (5-4x) < -2$ and $(5-4x) > 2$

$\Rightarrow \quad -4x < -2-5$ and $-4x > 2-5$

$\Rightarrow \quad -4x < -7$ and $-4x > -3$

$\Rightarrow \quad 4x > 7$ and $4x < 3$

$\Rightarrow \quad x > \dfrac{7}{4}$ and $x < \dfrac{3}{4}$

i.e. $\quad x \in \left(-\infty, \dfrac{3}{4}\right) \cup \left(\dfrac{7}{4}, \infty\right)$

Practice Exercise

Level - I

1. If $0 < x < 5$ and $1 < y < 2$, then which of the following is true?
   (a) $x + y < 0$ (b) $-3 < 2x - 3y < 4$
   (c) $-6 < 2x - 3y < 7$ (d) $-3 < 3x - y < 2$

2. Which of the following is the solution set of $|2x - 3| < 7$?
   (a) $\{x : -5 < x < 2\}$ (b) $\{x : -5 < x < 5\}$
   (c) $\{x : -2 < x < 5\}$ (d) $\{x : x < -5 \text{ or } x > 2\}$

3. $|x^2 - 4x| < 5$
   (a) $-1 \le x \le 5$ (b) $1 \le x \le 5$
   (c) $-1 \le x \le 1$ (d) $-1 < x < 5$

4. If $1 \le x \le 3$ and $2 \le y \le 4$, what is the maximum value of $\left(\frac{x}{y}\right)$?
   (a) $\frac{2}{3}$ (b) 4
   (c) $\frac{3}{2}$ (d) 2

5. The solution set of $\frac{1}{4} + \frac{1}{x} < \frac{1}{2} + \frac{1}{3}$, when $x \in R$, is:
   (a) $\frac{1}{x} < \frac{7}{12}$ (b) $x \ge \frac{12}{7}$
   (c) $x < \frac{14}{7}$ (d) $\frac{1}{x} > \frac{3}{2}$

6. $x > 2, y > -1$, then which of the following holds good ?
   (a) $xy > -2$ (b) $xy < -2$
   (c) $x > -\frac{2}{y}$ (d) None of these

7. If $x > 5$ and $y < -1$, then which of the following statements is true?
   (a) $(x + 4y) > 1$ (b) $x > -4y$
   (c) $-4x < 5y$ (d) None of these

8. If a and b are negative, and c is positive, which of the following statements is/are true?
   I. $a - b < a - c$ II. if $a < b$, then $\frac{a}{c} < \frac{b}{c}$
   III. $\frac{1}{b} < \frac{1}{c}$
   (a) I only (b) II only
   (c) III only (d) II and III only

9. If $x + y > 5$ and $x - y > 3$, then which of the following gives all possible values of x?
   (a) $x > 3$ (b) $x > 4$
   (c) $x > 5$ (d) $x < 5$

Level - II

1. If $x \in R$, and $\alpha = \frac{x^2}{(1+x^4)}$, then
   (a) $0 \le \alpha \le 2$ (b) $0 \le \alpha \le 1$
   (c) $0 \le \alpha \le \frac{1}{4}$ (d) $0 \le \alpha \le \frac{1}{2}$

2. If $6 \ge x \ge -2$ and $4 \ge y \ge -4$, find the limits for $\frac{y}{x}$, where x and y are non-zero integers.
   (a) $\frac{y}{x} \ge 2, \frac{y}{x} \le \frac{2}{3}$ (b) $\frac{y}{x} \ge \frac{-2}{3}, \frac{y}{x} \le 2$
   (c) $\frac{y}{x} \ge \frac{-2}{3}, \frac{y}{x} \le \frac{1}{4}$ (d) $\frac{y}{x} \ge -4, \frac{y}{x} \le 4$

3. The shaded portion of figure shows the graph of which of the following ?

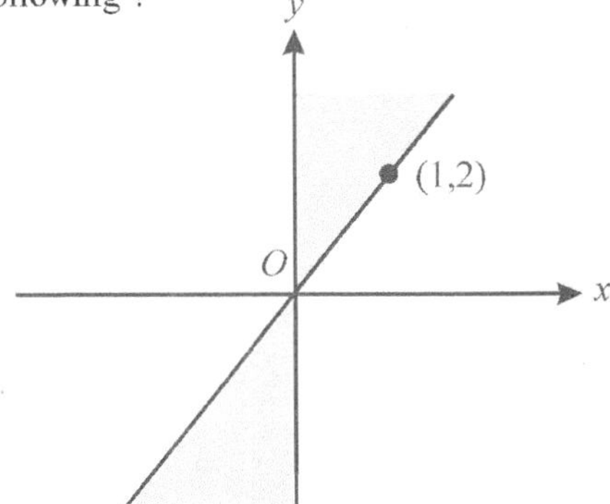

   (a) $x(y - 2x) \ge 0$ (b) $x(y - 2x) \le 0$
   (c) $x\left(y + \frac{1}{2}x\right) \ge 0$ (d) $x\left(y - \frac{1}{2}x\right) \le 0$

4. For the real numbers p, q, r, x, y, let $p<x<q$ and $p<y<r$. Which one of the following is correct?

(a) $p<x<y<r$ (b) $p<x<q<r$

(c) $p<y<x<q$ (d) None of these

5. Given that $-1 \le v \le 1, -2 \le u \le -0.5$ and $-2 \le z \le -0.5$ and $w = \frac{vz}{u}$, then which of the following is necessarily true?

(a) $-0.5 \le w \le 2$ (b) $-4 \le w \le 4$

(c) $-4 \le w \le 2$ (d) $-2 \le w \le -0.5$

6. If $|b| \ge 1$ and $x = -|a|b$, then which one of the following is necessarily true?

(a) $a - xb < 0$ (b) $a - xb \ge 0$

(c) $a - xb > 0$ (d) $a - xb \le 0$

7. The number of solutions of the equation $2x + y = 40$ where both x and y are positive integers and $x \le y$

(a) 7 (b) 13

(c) 14 (d) 18

8. If a, b and c are three real numbers, then which of the following is NOT true?

(a) $|a+b| \le |a|+|b|$

(b) $|a-b| \le |a|+|b|$

(c) $|a-b| \le |a|-|b|$

(d) $|a-c| \le |a-b|+|b-c|$

9. x and y are real numbers satisfying the conditions $2<x<3$ and $-8<y<-7$. Which of the following expressions will have the least value?

(a) x^2y (b) xy^2

(c) $5xy$ (d) None of these

Hints & Solutions

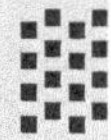

Level-I

1. **(c)** $0 < x < 5$ ...(1)
$0 < 2x < 10$ ...(2) (multiply (1) by 2)
$1 < y < 2$ ...(3)
$-6 < -3y < -3$ ...(4) (multiply (3) by –3)
Subtracting (4) from (2)
$0 - 6 < 2x - 3y < 10 - 3$ *i.e.,* $-6 < 2x - 3y < 7$.

2. **(c)** If the expression between the absolute value bars is positive. It's less than +7 or, if the expression between the bars is negative, it's greater than –7. In other words, $2x - 3$ is between –7 and +7
$-7 < 2x - 3 < 7$
$-4 < 2x < 10$
$-2 < x < 5$

3. **(d)** At $x = 0$, inequality is satisfied, option (b) is rejected.
At $x = 2$, inequality is satisfied, option (c) is rejected.
At $x = 5$, LHS = RHS.
At $x = -1$, LHS = RHS.
Thus, option (d) is correct.

4. **(c)** $\text{Max}\left(\frac{x}{y}\right) = \frac{\text{Max}(x)}{\text{Min}(y)} = \frac{3}{2}$

5. **(a)** $\frac{1}{4} + \frac{1}{x} < \frac{5}{6}$ $\quad \therefore \frac{1}{x} < \frac{5}{6} - \frac{1}{4}$

$\therefore \frac{1}{x} < \frac{7}{12}$.

6. **(d)** Given, $y > -1$ *i.e.,* if y is positive no. then product of x and y also positive.
But any option does not give xy is +ve.
By putting different values of x and y, we see that none of these three hold good.

7. **(d)** $x > 5$ and $y < -1 \Rightarrow 4y < -4$
(i) $x > 5$ and $4y < -4$ so $x + 4y < 1$
(ii) Let $x > -4y$ be true $\Rightarrow 4y < -4$ or $-4y > 4$
So, $x > 4$, which is not true as given $x > 5$.
So, $x > -4y$ is not necessarily true.
(iii) $x > 5 \Rightarrow -4x < -20$ and $5y < -5$
It is not necessary that $-4x < 5y$ as $-4x$ can be greater than $5y$, since $5y < -5$.
Hence, none of the options is true.

8. **(d)** Let $a = x$, $b = -y$ and $c = z$
Statement I : $-x + y < -x - z$
So I is not true.
Statement II : $\frac{-x}{z} < \frac{-y}{z}$
Since $x < y$ so II is true
Statement III : $\frac{1}{-y} < \frac{1}{z}$
Since y is negative and z is positive.
So III is true
Hence, statements II and III are true.

9. **(b)** $x + y > 5$ ...(1)
$x - y > 3$ ...(2)
Adding inequations (1) and (2), we get
$2x > 8$ i.e. $x > 4$

Level-II

1. **(d)** For real x, we have $x^2/(1+x^4) \geq 0$ also
$1 + x^4 - 2x^2 = (1 - x^2)^2 \geq 0$
$\Rightarrow 1 + x^4 \geq 2x^2$
$\therefore \left(\frac{x^2}{1+x^4}\right) \leq \frac{1}{2}$

So $0 \leq \frac{x^2}{1+x^4} \leq \frac{1}{2} \Rightarrow 0 \leq x \leq \frac{1}{2}$

2. **(d)**

y	4	3	2	1	–1	–3	–4
x	6	5	4	3	1	–1	–2

Hence, minimum value of $\frac{y}{x} = \frac{4}{-1} = -4$

and maximum value of $\frac{y}{x} = \frac{4}{1} = 4$

$\therefore \frac{y}{x} \geq -4$ and $\frac{y}{x} \leq 4$.

3. **(a)** Each of the answer choices in the form of the product of two factors on the left and a "≥ 0" or " ≤ 0" on the right.
The product will be negative when the two factors have opposite signs, and it will be positive when the factors have the same sign. Choice (1), for exampled, has *a* "≥ 0", so you'll be looking other factors to have the same sign.
Either : $x \geq 0$ and $y - 2x \geq 0 \Rightarrow x \geq 0$ and $y \geq 2x$.
or $x \leq 0$ and $y - 2x \leq 0 \Rightarrow x \leq 0$ and $y \leq 2x$

The graph of $x \geq 0$ and $y \geq 2x$ looks like this :

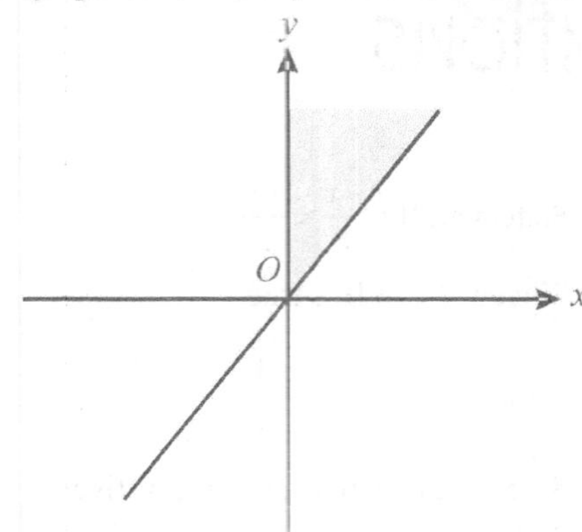

The graph of $x \leq 0$ and $y \leq 2x$ looks like this

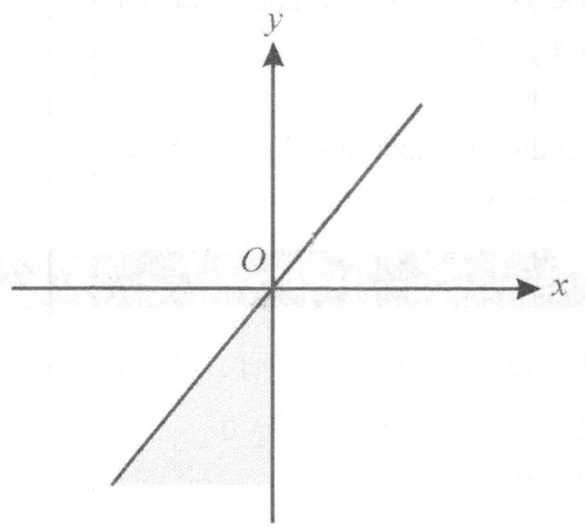

Together they make the graph in the figure.

4. **(b)** For the real number p, q, r, x and y
$p < x < q$ and $p < y < r$
$p < x < q < r$

5. **(b)** Substitute the extreme values in the given equation :
$v = 1,\ u = -0.5,\ z = -2.$

Then $w = \dfrac{vz}{u} = 4.$

Only option (b) gives this.

6. **(b)** $|b| \geq 1 \Rightarrow b \geq 1$ or $b \leq -1$

$x = -|a|b$ (given)

Consider $a - xb = a - (-|a|b)b$

$= a + |a|b^2 \geq 0$ since, $b^2 \geq 1$

7. **(b)** $2x + y = 40;\ x \leq y;\ x, y \in I^+$
This problem can be solved by putting various values for x and y. Starting from $x = 1$. The above equation can be solved till $x = 13$.
At $x = 13$, $y = 14$ which is $> x$. But above this value of x, it becomes greater than y so the condition $x \leq y$ is violated.

8. **(c)** This can be checked by taking arbitrary values of a and b in the given terms. Taking $a = 2$ and $b = 3$, we conclude that (c) is not true.

9. **(c)** $2 < x < 3$ and $-8 < y < -7$
$4 < x^2 < 9$
and $-8 < y < -7$

$-32 < x^2 y < -63$

While $-80 < 5xy < -105$

Hence, $5xy$ is the least because xy^2 is positive

CHAPTER 16

LOGARITHMS

DEFINITION

If $x = a^m$, then $\log_a x = m$, where 'a' and 'x' both are positive real numbers but 'a' not equal to 1 i.e., $a, x > 0$, but $a \neq 1$.

Here log is the short form of logarithm.

$\log_a x$ is read as log of x to the base a.

For example,

(i) Since, $10 = 10^1$, $100 = 10^2$, $1000 = 10^3$, etc.

Hence, $\log_{10}10 = 1$, $\log_{10}100 = 2$, $\log_{10}1000 = 3$, etc.

(ii) Since, $8 = 2^3$, $16 = 2^4$, $32 = 2^5$, etc.

Hence, $\log_2 8 = 3$, $\log_2 16 = 4$, $\log_2 32 = 5$, etc.

(iii) Since, $\frac{1}{8} = (2)^{-3}$, $\frac{1}{16} = (2)^{-4}$, etc.

Hence, $\log_2\left(\frac{1}{8}\right) = -3$, $\log_2\left(\frac{1}{16}\right) = -4$, etc.

(iv) Since, $0.01 = (10)^{-2}$, $0.001 = (10)^{-3}$, etc.

Hence $\log_{10}(0.01) = -2$, $\log_{10}(0.001) = -3$, etc.

LAWS OF LOGARITHM

(i) $\log_a(m \times n) = \log_a m + \log_a n$

In general,

$\log_a(m \times n \times p \times ....) = \log_a m + \log_a n + \log_a p + ...$

For example:

$\log_2(4 \times 5 \times 6) = \log_2 4 + \log_2 5 + \log_2 6$

Note that

$\log_a m + \log_a n \neq \log_a(m + n)$

(ii) $\log_a\left(\frac{m}{n}\right) = \log_a m - \log_a n$

For example:

$\log_4\left(\frac{8}{15}\right) = \log_4 8 - \log_4 15$

Note that

$\log_a m - \log_a n \neq \log_a(m - n)$

(iii) $\log_a(m)^n = n \log_a m$

For example:

$\log_3(5)^4 = 4\log_3 5$

(iv) $\log_b a = \frac{\log_c a}{\log_c b}$ [Change of base rule]

For example,

$\log_5 20 = \frac{\log_2 20}{\log_2 5} = \frac{\log_4 20}{\log_4 5} = \frac{\log_7 20}{\log_7 5} = ...$ etc.

(v) $\log_b a = \frac{1}{\log_a b}$

For example,

$\log_{10} 100 = \frac{1}{\log_{100} 10}$

(vi) $\log_b a \, . \, \log_c b = \log_c a$ [Chain Rule]

In general,

$\log_b a \, . \, \log_c b \, . \, \log_d c ....\log_n m = \log_n a$

For example,

$\log_{24} 256 \, . \, \log_{10} 24 \, . \, \log_2 10 = \log_2 256$

Illustration 1: $\log_a 4 + \log_a 16 + \log_a 64 + \log_a 256 = 10$. Then a = ?

(a) 4 **(b) 2**
(c) 8 **(d) 5**

Solution: (a) The given expression is:

$\log_a(4 \times 16 \times 64 \times 256) = 10$

i.e. $\log_a 4^{10} = 10$

Thus, $a = 4$.

Illustration 2: Find x if $\log x = \log 1.5 + \log 12$

(a) 12 **(b) 8**
(c) 18 **(d) 15**

Solution: (c) $\log x = \log 18 \Rightarrow x = 18$

Illustration 3: Find x, if $\log(2x - 2) - \log(11.66 - x) = 1 + \log 3$

(a) 452/32 **(b) 350/32**
(c) 11 **(d) 11.33**

Solution: (c) $\log(2x - 2)/(11.66 - x) = \log 30$

$\Rightarrow (2x - 2)/(11.66 - x) = 30$

$2x - 2 = 350 - 30x$

Hence, $32x = 352 \Rightarrow x = 11$.

Illustration 4: Solve for x:

$$\log \frac{75}{35} + 2\log \frac{7}{5} - \log \frac{105}{x} - \log \frac{13}{25} = 0$$

(a) 90 **(b) 65**
(c) 13 **(d) 45**

Solution: (c) (75/35) × (49/25) × (x/105) × (25/13) = 1
⇒ $x = 13$

SOME IMPORTANT PROPERTIES

(i) $x = a^m \Rightarrow \log_a x = m$
and $\log_a x = m \Rightarrow x = a^m$
Here equation $x = a^m$ is in exponential form and equation $\log_a x = m$ is in logarithmic form.

(ii) If base of log is not mentioned, then we assume the base as 10.
∴ $\log m = \log_{10} m$
log to the base 10 is called common log.

(iii) Since, $10000 = (100)^2 = (10)^4$
∴ $\log_{100} 10000 = 2$, $\log_{10} 10000 = 4$
Thus value of log of a number on different bases is different i.e., value of log of a number depends on its base.

(iv) (a) Since, $a = a^1$, hence $\log_a a = 1$
For example, $\log_5 5 = 1$, $\log_{10} 10 = 1$
Thus log of any number to the same base is always 1.
(b) Since, $1 = a^0$, hence, $\log_a 1 = 0$
For example, $\log_8 1 = 0$
Thus log of 1 to any base always equal to 0.

(v) $a^{(\log_a x)} = x$
For example,

$$20^{(\log_{20} 50)} = 50$$

(vi) (a) log of zero and negative numbers is not defined.
(b) Base of log is always positive but not equal to 1.

Illustration 5: Find x, if $0.01^x = 2$

(a) log 2/2 **(b) 2/log 2**
(c) – 2/log 2 **(d) – log 2/2**

Solution: (d) $x = \log_{0.01} 2 = -\log 2/2$.

Illustration 6: If log 3 = .4771, find log $(.81)^2$ × log $\left(\frac{27}{10}\right)^{\frac{2}{3}}$ ÷ log 9.

(a) 2.689 **(b) – 0.0552**
(c) 2.2402 **(d) 2.702**

Solution: (b) 2 log (81/100) × 2/3 log (27/10) ÷ log 9
= 2 [log 3^4 – log 100] × 2/3 [(log 3^3 – log 10)] ÷ 2 log 3
= 2 [log 3^4 – log 100] × 2/3 [(3log 3 – 1)] ÷ 2 log 3
Substitute log 3 = 0.4771 ⇒ – 0.0552.

Practice Exercise

Level - I

1. Find the value of $\log_5 10 \times \log_{10} 15 \times \log_{15} 20 \times \log_{20} 25$.
   (a) 5/2 (b) 5
   (c) 2 (d) $\log\left(\frac{5}{2}\right)$
2. If $\log_3 a = 4$, find value of a.
   (a) 27 (b) 3
   (c) 9 (d) 81
3. Find the value of $\log\frac{9}{8} - \log\frac{27}{32} + \log\frac{3}{4}$
   (a) 0 (b) 1
   (c) 3 (d) log (3/4)
4. Evaluate : $3^{2-\log_3 5}$
   (a) $\frac{9}{5}$ (b) 45
   (c) 5/9 (d) $9\log_{35}$
5. The value of $\left[\frac{1}{\log_{xy}(xyz)} + \frac{1}{\log_{yz}(xyz)} + \frac{1}{\log_{zx}(xyz)}\right]$ is equal to
   (a) 1 (b) 2
   (c) 3 (d) 4
6. If $\log_2 [\log_3 (\log_2 x)] = 1$, then x is equal to
   (a) 512 (b) 128
   (c) 12 (d) 0
7. Find the value of $\log_{27} \frac{1}{81}$
   (a) – 4/3 (b) – 3
   (c) – 1 (d) – 1/3
8. Find the value of $\frac{8\log_8 8}{2\log_{\sqrt{8}} 8}$
   (a) 1 (b) 2
   (c) 3 (d) 4
9. $\log_3 (5 + x) + \log_8 8 = 2^2$
   (a) 22 (b) 33
   (c) 11 (d) 44
10. $\log 216\sqrt{6}$ to the base 6 is equal to
    (a) 3 (b) 3/2
    (c) 7/2 (d) None of these
11. If $\log_k x \log_5 k = 3$, then find the value of x.
    (a) k^5 (b) $5k^3$
    (c) 243 (d) 125
12. $\log_a\left(\frac{m}{n}\right)$ is equal to
    (a) $\log_a(m-n)$ (b) $\log_a m - \log_a n$
    (c) $\frac{(\log_a m)}{n}$ (d) $\log_a m \div \log_a n$
13. If $\log_5\left[\log_3\left(\log_2 x\right)\right] = 1$ then x is
    (a) 2^{234} (b) 243
    (c) 2^{243} (d) None of these
14. The value of $\left[3\log\left(\frac{81}{80}\right) + 5\log\left(\frac{25}{24}\right) + 7\log\left(\frac{16}{15}\right)\right]$ is
    (a) log 3 (b) log 5
    (c) log 7 (d) log 2
15. If $\log_{10} a + \log_{10} b = c$, then the value of a is
    (a) bc (b) $\frac{c}{b}$
    (c) $\frac{(10)^c}{b}$ (d) $\frac{10b}{c}$
16. If $\log_y x = 8$ and $\log_{10y} 16x = 4$, then find the value of y.
    (a) 1 (b) 2
    (c) 3 (d) 5
17. log 0.0867 = ?
    (a) log 8.67 + 2 (b) log 8.67 – 2
    (c) $\frac{\log 867}{1000}$ (d) – 2 log 8.67
18. Find x, if $0.01^x = 2$
    (a) log 2/2 (b) 2/log 2
    (c) – 2/log2 (d) – log 2/2
19. If $2^x.3^{2x} = 100$, then the value of x is
    (log 2 = 0.3010, log 3 = 0.4771)
    (a) 2.3 (b) 1.59
    (c) 1.8 (d) 1.41
20. If $\log_{10} a = b$, then find the value of 10^{3b} in terms of a.
    (a) a^3 (b) $3a$
    (a) $a \times 1000$ (d) $a \times 100$
21. If log 3 = 0.4771, find $\log (0.81)^2 \times \log\left(\frac{27}{10}\right)^{\frac{2}{3}} \div \log 9$.
    (a) 2.689 (b) – 0.0552
    (c) 2.2402 (d) 2.702

22. $\log_{10} 10 + \log_{10} 10^2 + \ldots\ldots + \log_{10} 10^n$

(a) $n^2 + 1$ (b) $n^2 - 1$

(c) $\left(\frac{n^2+n}{3}\right)$ (d) $\frac{n^2+n}{2}$

23. If a, b and c are distinct positive number ($\neq 1$) such that $(\log_b a \log_c a - \log_a a) + (\log_a b \log_c b - \log_b b) + (\log_a c \log_b c - \log_c c) = 0$. What is the value of abc?

(a) 1

(b) 0

(c) -1

(d) None of these

24. What is the value of x in the following expression $\log_{3/4} \log_2 (x^2 + 7) \log_{1/4} (x^2 + 7)^{-1} = -2$?

(a) $+3$ (b) -3

(c) ± 3 (d) None of these

Level - II

1. If $\log_{10} 2 = 0.3010$, then the value of $\log_{10} 80$ is :

(a) 1.9030 (b) 1.6020

(c) 3.9030 (d) 2.9030

2. The value of $\log_{2\sqrt{3}} (1728)$ is

(a) 3 (b) 5

(c) 6 (d) 9

3. If $\log 2 = 0.30103$, then the number of digits in 4^{50} is

(a) 30 (b) 31

(c) 100 (d) 200

4. If $\log_7 \log_5 (\sqrt{x} + 5 + \sqrt{x}) = 0$, find the value of x.

(a) 1 (b) 0

(c) 2 (d) None of these

5. If $\log_3 [\log_3 [\log_3 x]] = \log_3 3$, then what is the value of x?

(a) 3 (b) 27

(c) 3^9 (d) 3^{27}

6. What is $\log\left(a + \sqrt{a^2+1}\right) + \log\left(\frac{1}{a + \sqrt{a^2+1}}\right)$ is equal to?

(a) 1 (b) 0

(c) 2 (d) $\frac{1}{2}$

7. $\frac{1}{(\log_a bc)+1} + \frac{1}{(\log_b ac)+1} + \frac{1}{(\log_c ab)+1}$ is equal to

(a) 1 (b) 2

(c) 0 (d) abc

8. If $p = \log_3 5$ and $q = \log_{17} 25$, which one of the following is correct?

(a) $p < q$ (b) $p = q$

(c) $p > q$ (d) can't say

9. If $\log_{10} x = a$, $\log_{10} y = b$ and $\log_{10} z = c$, then antilog $(pa + qb - rc) = ?$

(a) $\frac{pxqy}{rz}$ (b) $px + qy - rz$

(c) $\frac{x^p y^q}{z^r}$ (d) $x^p y^q z^r$

10. If a, b, c are three consecutive integers, then $\log (ac + 1)$ has the value

(a) $\log b$ (b) $(\log b)^2$

(c) $2 \log b$ (d) $\log 2b$

11. Find the value of $\left(7^3\right)^{-2\log_7 8}$

(a) 8^{-7} (b) 6^{-8}

(c) 8^{-6} (d) None of these

12. If $(\log_3 x)^2 + \log_3 x < 2$, then which one of the following is correct ?

(a) $0 < x < \frac{1}{9}$ (b) $\frac{1}{9} < x < 3$

(c) $3 < x < \infty$ (d) $\frac{1}{9} \leq x \leq 3$

13. If $\log_{10} x - \log_{10} \sqrt{x} = 2 \log_x 10$, then a possible value of x is given by

(a) 10 (b) 1/100

(c) 1/1000 (d) None of these

14. What is the value of $\frac{\log_{27} 9 \times \log_{16} 64}{\log_4 \sqrt{2}}$?

(a) $\frac{1}{6}$ (b) $\frac{1}{4}$

(c) 8 (d) 4

15. If $(\log_x x)(\log_3 2x)(\log_{2x} y) = \log_x x^2$, then what is the value of y ?

(a) 9/2 (b) 9

(c) 18 (d) 27

16. What is the value of $\log_{10}\left(\frac{9}{8}\right) - \log_{10}\left(\frac{27}{32}\right) + \log_{10}\left(\frac{3}{4}\right)$?

(a) 3 (b) 2

(c) 1 (d) 0

17. If $\log_{10} x, \log_{10} y, \log_{10} z$ are in AP then x, y, z are in

(a) AP (b) GP
(c) HP (d) None of these

18. Find the value of $\frac{\log\sqrt{27} + \log\sqrt{8} - \log\sqrt{125}}{\log 6 - \log 5}$

(a) $\frac{2}{3}$ (b) $\frac{1}{3}$
(c) $\frac{3}{2}$ (d) None of these

19. Find the value of x and y respectively for

$\log_{10}\left(x^2y^3\right) = 7$ and $\log_{10}(x/y) = 1$

(a) $x = 10, y = 100$ (b) $x = 100, y = 10$
(c) $x = 10, y = 20$ (d) None of these

20. What is the value of $\log_3 2. \log_4 3. \log_5 4....\log_{16} 15$?

(a) 1/2 (b) 1/3
(c) 2/3 (d) 1/4

21. If $\log_4 5 = a$ and $\log_5 6 = b$ then what is the value of $\log_3 2$?

(a) $\frac{1}{2a+1}$ (b) $\frac{1}{2b+1}$
(c) $2ab + 1$ (d) $\frac{1}{2ab-1}$

22. What is the value of x if

$\log_3 x + \log_9 x + \log_{27} x + \log_{81} x = \frac{25}{4}$?

(a) 9 (b) 27
(c) 81 (d) None of these

23. What is the value of $\log_{32} 27 \times \log_{243} 8$?

(a) $\frac{\log 9}{\log 4}$ (b) $\frac{\log 3}{\log 2}$
(c) $\log 27$ (d) None of these

24. $\log a^n / b^n + \log b^n / c^n + \log c^n / a^n$

(a) 1 (b) n
(c) 0 (d) 2

Hints & Solutions

Level-I

1. (c) $\log_5 10 \times \log_{10} 15 \times \log_{15} 20 \times \log_{20} 25$.
$= (\log 10/\log 5) \times (\log 15/\log 10) \times (\log 20/\log 15) \times (\log 25/\log 20)$
$= \log 25/\log 5 = 2\log 5/\log 5 = 2$.

2. (d) $\because\ \log_3 a = 4\ \therefore\ 3^4 = a \Rightarrow a = 81$

3. (a) Given $\log\frac{9}{8} - \log\frac{27}{32} + \log\frac{3}{4} = \log\left(\frac{9}{8} \div \frac{27}{32}\right) + \log\frac{3}{4}$

$= \log\left(\frac{9}{8} \times \frac{32}{27} \times \frac{3}{4}\right) = \log 1 = 0 \qquad \because\ \log_a 1 = 0$

4. (a) Given : $3^{2-\log_3 5} = 3^2.3^{-\log_3 5} \qquad (\because a^{m+n} = a^m.a^n)$

$= 9.3^{\log_3 5^{-1}} = 9 \times 5^{-1} = \frac{9}{5}$

5. (b) Given expression
$= \log_{xyz}(xy) + \log_{xyz}(yz) + \log_{xyz}(zx)$
$= \log_{xyz}(xy \times yz \times zx) = \log_{xyz}(xyz)^2$
$= 2\log_{xyz}(xyz) = 2 \times 1 = 2$

6. (a) $\log_2[\log_3(\log_2 x)] = 1 = \log_2 2$
$\Rightarrow \log_3(\log_2 x) = 2$
$\Rightarrow \log_2 x = 3^2 = 9$
$\Rightarrow x = 2^9 = 512$

7. (a) Let $\log_{27}\left(\frac{1}{81}\right) = x$

$\therefore\ (27)^x = \frac{1}{81}$

$\therefore\ 3^{3x} = 3^{-4} \Rightarrow 3x = -4 \Rightarrow x = -\frac{4}{3}$

8. (b) $\frac{8\log_8 8}{2\log_{\sqrt{8}} 8} = \frac{8 \times 1}{2\log_{\sqrt{8}}\left(\sqrt{8}\right)^2} = \frac{8}{4\log_{\sqrt{8}}\sqrt{8}} = \frac{8}{4} = 2$

9. (a) $\log_3(5+x) + \log_8 8 = 2^2$
$\log_3(5+x) + 1 = 4$
$\log_3(5+x) = 3$
$3^3 = 5 + x$
$5 + x = 27$
$x = 27 - 5 = 22$.

10. (c) $\log_6 216\sqrt{6} = \log_6(6)^3(6)^{1/2} = \log_6(6)^{7/2}$

$= \frac{7}{2}\log_6 6 = \frac{7}{2} \quad (\because \log_a a = 1)$

11. (d) Given, $\log_5 k \log_k x = 3$

$\frac{\log k}{\log 5}.\frac{\log x}{\log k} = 3 \Rightarrow \frac{\log x}{\log 5} = 3$

$\Rightarrow \log x = 3\log 5 \Rightarrow \log x = \log 5^3$
$\Rightarrow x = 5^3 \Rightarrow x = 125$

12. (b) $\log_a\left(\frac{m}{n}\right) = \log_a m - \log_a n$

13. (c) $\log_5\left[\log_3\left(\log_2 x\right)\right] = 1 = \log_5 5$

$\Rightarrow \log_3(\log_2 x) = 5 = \log_3 3^5$

$\Rightarrow \log_2 x = 3^5 = 243$

$\Rightarrow 2^{243} = x$

14. (d) $3\log\frac{81}{80} + 5\log\frac{25}{24} + 7\log\frac{16}{15}$

$= \log\left[\left(\frac{81}{80}\right)^3 \times \left(\frac{25}{24}\right)^5 \times \left(\frac{16}{15}\right)^7\right]$

$= \log\left(\frac{3^{12} \times 5^{10} \times 2^{28}}{2^{12} \times 5^3 \times 2^{15} \times 3^5 \times 3^7 \times 5^7}\right)$

$= \log 2$

15. (c) $\log_{10} a + \log_{10} b = c$

$\Rightarrow \log_{10}(ab) = c$

$\Rightarrow 10^c = ab$

$\Rightarrow a = \frac{(10)^c}{b}$

16. (d) $\log_y x = 8 \Rightarrow y^8 = x$ ...(1)

$\log_{10y} 16x = 4 \Rightarrow 10^4 y^4 = 16x$ ...(2)

Dividing (2) by (1) $10^4 y^{-4} = 16 \Rightarrow y = 5$

17. (b) $\log 0.0867 = \log(8.67/100) = \log 8.67 - \log 100$
$\log 8.67 - 2$

18. (d) $x = \log_{0.01} 2 = -\log 2/2$.

19. (b) $2^x.3^{2x} = 100$
$\Rightarrow x\log 2 + 2x\log 3 = \log 100$
$\Rightarrow x(0.3010 + 2 \times 0.4771) = 2$

$\Rightarrow x = \frac{1}{1.2552} = 1.59$

20. (a) $\log_{10} a = b \Rightarrow 10^b = a \Rightarrow$ By definition of logs.
Thus $10^{3b} = (10^b)^3 = a^3$.

21. (b) $2\log(81/100) \times 2/3\log(27/10) \div \log 9$
$= 2[\log 3^4 - \log 100] \times 2/3[(\log 3^3 - \log 10)] \div 2\log 3$
$= 2[\log 3^4 - \log 100] \times 2/3[(3\log 3 - 1)] \div 2\log 3$
Substitute $\log 3 = 0.4771 \Rightarrow -0.0552$

22. (d) $\log_{10} 10 + \log_{10} 10^2 + \ldots\ldots\ldots + \log_{10} 10^n$

$= 1 + 2 + 3 + \ldots\ldots + n = \frac{n(n+1)}{2}$

23. **(a)** $\log_b a \log_e a + \log_a b \log_a b + \log_a c \log_b c = 3$

$$\Rightarrow \frac{(\log a)^2}{\log b.\log c} + \frac{(\log b)^2}{\log a \log c} + \frac{(\log c)^2}{\log a \log b} = 3$$

$$\Rightarrow \frac{(\log a)^3 + (\log b)^3 + (\log c)^3}{\log a.\log b.\log c} = 3$$

$$\Rightarrow (\log a)^3 + (\log b)^3 + (\log c)^3 = 3\log a.\log b.\log c$$

We know, $x^3 + y^3 = z^3 = 3xyz$ when $x + y + z = 0$

$\log a + \log b + \log c = 0 \Rightarrow \log abc = 0 \Rightarrow abc = 1.$

24. **(c)** Go through the options.

Level-II

1. **(a)** $\log_{10} 80 = \log_{10}(8 \times 10) = \log_{10} 8 + \log_{10} 10$

$= \log_{10} 2^3 + 1$

$= (3 \log_{10} 2) + 1 = (3 \times 0.3010) + 1 = 1.9030$

2. **(c)** Let $\log_{2\sqrt{3}}(1728) = x$.

Then, $(2\sqrt{3})^x = 1728 = (12)^3$

$$= \left[(2\sqrt{3})^x\right]^3 = (2\sqrt{3})^6.$$

$\therefore x = 6$, *i.e.*, $\log_{2\sqrt{3}}(1728) = 6$.

3. **(b)** $\log 4^{50} = 50 \log 4 = 50 \log 2^2$

$= (50 \times 2) \log 2 = 100 \times \log 2$

$= (100 \times 0.30103) = 30.103$

So the number of digits = 31.

4. **(b)** $\log_7 \log_5(\sqrt{x} + 5 + \sqrt{x}) = 0$

use $\log_a x = b$

$\Rightarrow a^b = x$

$\therefore \quad \log_5(\sqrt{x} + 5 + \sqrt{x}) = 7^0 = 1$

$\sqrt{x} + 5 + \sqrt{x} = 5^1 = 5 \Rightarrow 2\sqrt{x} = 0 \therefore x = 0$

5. **(d)** Consider $\log_3[\log_3[\log_3 x]] = \log_3 3$

$\Rightarrow \log_3[\log_3 x] = 3$

$\Rightarrow \log_3 x = 3^3$

$\Rightarrow \log_3 x = 27 \Rightarrow x = 3^{27}$

6. **(b)** Let $\log(a + \sqrt{a^2 + 1}) + \log\left(\frac{1}{a + \sqrt{a^2 + 1}}\right)$

$= \log(a + \sqrt{a^2 + 1}) + \log 1 - \log(a + \sqrt{a^2 + 1})$

$= \log(a + \sqrt{a^2 + 1}) - \log(a + \sqrt{a^2 + 1})$

$= 0$

7. **(a)** $$\frac{1}{(\log_a bc) + 1} + \frac{1}{(\log_b ac) + 1} + \frac{1}{(\log_c ab) + 1}$$

$$= \frac{1}{\log_a bc + \log_a a} + \frac{1}{\log_b ac + \log_b b} + \frac{1}{\log_c ab + \log_c c}$$

$$= \frac{1}{\log_a abc} + \frac{1}{\log_b abc} + \frac{1}{\log_c abc}$$

$= \log_{abc} a + \log_{abc} b + \log_{abc} c$

$\log_{abc} abc = 1$

8. **(c)** $q = \log_{17}(5)^2 = 2 \log_{17} 5$

$$\Rightarrow \frac{1}{q} = \frac{1}{2}\log_5 17$$

And $\frac{1}{p} = \log_5 3 = \frac{1}{2} \times (2 \log_5 3)$

$= \frac{1}{2}(\log_5 9)$

$\therefore \quad \frac{1}{p} < \frac{1}{q} \Rightarrow p > q$

9. **(c)** $(pa + qb - rc) = p \log_{10} x + q \log_{10} y - r \log_{10} z$

$= \log_{10}(x^p) + \log_{10}(y^q) - \log_{10}(z^r)$

$$= \log_{10}\left(\frac{x^p y^q}{z^r}\right)$$

$$\Rightarrow \text{antilog}\,(pa + qb - rc) = \frac{x^p y^q}{z^r}$$

10. **(c)** a, b, c are consecutive integers

$\therefore b = a + 1$ and $c = a + 2$

$\therefore \log(ac + 1) = \log[a(a + 2) + 1]$

$= \log[(b - 1)(b - 1 + 2) + 1] \qquad [\because a = b - 1]$

$= \log b^2 = 2 \log b$

11. **(c)** $(7^3)^{-2\log_7 8} = 7^{-6\log_7 8} = 7^{(\log_7 8^{-6})}$

$$= 8^{-6} = \frac{1}{8^6}$$

12. **(b)** Given equation is $(\log_3 x)^2 + \log_3 x < 2$

$\Rightarrow (\log_3 x)^2 + (\log_3 x) - 2 < 0$

$\Rightarrow (\log_3 x + 2)(\log_3 x - 1) < 0$

$\Rightarrow -2 < \log_3 x < 1$

$\Rightarrow \log_3 3^{-2} < \log_3 x < \log_3 3$

$\Rightarrow \frac{1}{9} < x < 3$

13. **(b)** Let $\log_{10} x - \log_{10}\sqrt{x} = 2\log_x 10$

$\Rightarrow \frac{1}{2}\log_{10} x = 2\log_x 10 \Rightarrow \log_{10} x = \log_x 10^4$

$\Rightarrow \frac{\log_{10} x}{\log_x 10} = 4 \Rightarrow (\log_{10} x)^2 = 4$

$\Rightarrow \log_{10} x = \pm 2$

$\Rightarrow x = 10^2$ or 10^{-2}

14. (d) The given logarithm expression

$$\frac{\log_{27} 9 \log_{16} 64}{\log_4 \sqrt{2}}$$

is simplified as :

$$\frac{\log 9}{\log 27} \times \frac{\log 64}{\log 16} \times \frac{\log 4}{\log \sqrt{2}}$$

$$= \frac{2\log 3}{3\log 3} \times \frac{6\log 2}{4\log 2} \times \frac{2\log 2}{\frac{1}{2}\log 2}$$

$$= \frac{2}{3} \times \frac{6}{4} \times 4 = 4$$

15. (b) $(\log_x x)(\log_3 2x)(\log_{2x} y) = \log_x x^2$

$\Rightarrow 1 (\log_3 2x)(\log_{2x} y) = 2 \qquad (\because \log_x x^2) = 2\log_x x)$

$$\Rightarrow \left(\frac{\log 2x}{\log 3}\right)\left(\frac{\log y}{\log 2x}\right) = 2$$

$$\Rightarrow \frac{\log y}{\log 3} = 2 \Rightarrow \log y = 2\log 3$$

$\Rightarrow \log y = \log 3^2 \Rightarrow y = 3^2 \Rightarrow y = 9$

16. (d) Consider, $\log\frac{9}{8} - \log\frac{27}{32} + \log\frac{3}{4}$

$$= \log\left(\frac{9}{8} \times \frac{32}{27}\right) + \log\frac{3}{4}$$

$$= \log\left(\frac{4}{3}\right) + \log\frac{3}{4} = \log\left(\frac{4}{3} \times \frac{3}{4}\right) = \log 1 = 0$$

17. (b) $\log_{10}x, \log_{10}y, \log_{10}z$ are in AP

$\therefore \quad 2\log_{10} y = \log_{10} x + \log_{10} z$

$\Rightarrow \quad \log_{10} y^2 = \log_{10}(xz)$

$\Rightarrow \quad y^2 = xz$

$\Rightarrow \quad x, y, z$ are in GP

18. (c) $$\frac{\log\sqrt{27} + \log\sqrt{8} - \log\sqrt{125}}{\log 6 - \log 5}$$

$$= \frac{\log\left(3^{3/2} \times 2^{3/2}\right) - \log(5)^{3/2}}{\log 6 - \log 5}$$

$$= \frac{\log 6^{3/2} - \log 5^{3/2}}{\log 6 - \log 5}$$

$$= \frac{\frac{3}{2}\log\left(\frac{6}{5}\right)}{\log\left(\frac{6}{5}\right)} = \frac{3}{2}$$

19. (b) Best way is to go through options

Alternatively: $\log_{10} x^2y^3 = 7$

$\Rightarrow \quad x^2y^3 = 10^7 \qquad \ldots(1)$

and $\log_{10}\left(\frac{x}{y}\right) = 1$

$\Rightarrow \quad \frac{x}{y} = 10 \qquad \ldots(2)$

$$\therefore \quad \frac{x^2y^3}{(x/y)^2} = \frac{10^7}{(10)^2} \Rightarrow y^5 = 10^5$$

$\Rightarrow \quad y = 10 \therefore x = 100$

20. (d) $\log_3 2, \log_4 3. \log_5 4 \ldots\log_{16} 15$

$$= \frac{\log 2}{\log 3} \cdot \frac{\log 3}{\log 4} \cdot \frac{\log 4}{\log 5} \ldots\ldots \frac{\log 15}{\log 16} = \frac{\log 2}{\log 16} = \frac{\log 2}{\log 2^4}$$

$$= \frac{\log 2}{4\log 2} = \frac{1}{4}$$

21. (d) $\log_4 5 = a$ and $\log_5 6 = b$

$\Rightarrow \log_4 5 \times \log_5 6 = ab$

$\Rightarrow \log_4 6 = ab \Rightarrow \frac{1}{2}\log_2 6 = ab$

$\Rightarrow (1 + \log_2 3) = 2ab$

22. (b) $\log_3 x + \log_9 x + \log_{27} x + \log_{81} x = \frac{25}{4}$

$$\Rightarrow \log_3 x + \frac{1}{2}\log_3 x + \frac{1}{3}\log_3 x + \frac{1}{4}\log_3 x = \frac{25}{4}$$

$$\Rightarrow \log_3 x\ [1 + 1/2 + 1/3 + 1/4] = \frac{25}{4}$$

$$\Rightarrow \frac{25}{2} \times \log_3 x = \frac{25}{4}$$

$\Rightarrow \log_3 x = 3 \Rightarrow x = 27$

23. (d) $\log_{32} 27 \times \log_{243} 8 = \log_{2^5} 3^3 \times \log_{3^5} 2^3$

$$= \frac{3}{5}\log_2 3 \times \frac{3}{5}\log_3 2$$

$$= \left(\frac{3}{5}\right)^2 \log_2 3 \times \log_3 2 = \frac{9}{25}$$

24. (c) $\log\left(a^nb^nc^n / a^nb^nc^n\right) = \log 1 = 0$

CHAPTER 17

SET THEORY

SETS

A set is a well- defined collection of different objects.

In everyday life, we often speak about the collection of objects of particular kind such as a cricket team, the rivers of India, the vowels in the English alphabet etc. Each of these collection is well-defined collection of objects in the sense that we can definitely decide whether a given particular object belongs to a given collection or not. For example, we say that 10 does not belongs to the given collection of all odd natural numbers. On the other hand, 15 belongs to this given collection.

Note that

(i) Objects, elements and members of a set are synonymous terms.

(ii) Sets are usually denoted by capital letters A, B, C, D, E, F, etc.

(iii) The elements of a set are represented by small letters a, b, c, d, e, f, etc.

(iv) Each element in a set comes only once i.e. repetation of any element is not allowed.

If a is an element of a set A, we say that "a belongs to A". The Greek symbol $\in$(epsilon) is used to denote the phrase 'belongs to'.

Thus, we write $a \in A$. If 'b' is not an element of a set A, we write $b \notin A$ and read "b does not belong to A".

If V be the set of vowels of English alphabet, then $a \in V$ but $b \notin V$. In the set P of prime factors of 30, $3 \in P$ but $15 \notin P$.

REPRESENTATIONS OF SETS

There are two methods of representing a set:

(i) Roster or tabular form (ii) Set-builder form.

Roster or Tabular Form

(i) In roster form, all the elements of a set are listed within a bracket { } and separated by commas. For example, the set of all even positive integers less than 7 is described in roster form as {2, 4, 6}.

(ii) In roster form, the order in which the elements are listed is immaterial.

Set-builder Form

The set $\{a, e, i, o, u\}$ in roster form can be written as set in builder form as $\{x : x$ is a vowel of English alphabet$\}$. Here the set written in set builder form is read as 'x' is an element of the set such that x is a vowel of English alphabet'. Here the colon (:) read as 'such that'. In set-builder, a common property which posses all the elements of the set is written after colon (:).

Statement	Roster form	Set-builder form
(1) The set of currencies used in USA, England, Japan, Germany and Russia.	{Dollar, Pound, Yen, Euro, Rouble}	$\{x : x$ is the currencies used in USA, England, Japan, Germany and Russia $\}$
(2) The set of Capital of Kerala, Karnataka, Tamilnadu, Andhra Pradesh and Gujarat	{Tiruvananthapuram, Banglore, Chennai, Hyderabad and Gandhi Nagar}	$\{x : x$ is the capitals of Kerala, Karnataka, Tamilnadu, Andhra Pradesh and Gujarat$\}$
(3) The set of all distinct letters used in the word student.	$\{s, t, u, d, \underline{e}, n\}$	$\{x : x$ is the distinct letters used in the word student.$\}$
(4) The set of all the states of India beginning with the letter A.	{Andhra Pradesh, Arunachal Pradesh, Assam}	$\{x : x$ is the state of India beginning with the letter A$\}$
(5) The set of six presidents of India since 1980.	{Neelam Sanjeeva Reddy, Gyani Zail Singh, Radha Swami Venkat Raman, Dr. Shankar Dayal Sharma, K.R. Narayan, A.P.J. Abdul Kalam}	$\{x : x$ is the presidents of India since 1980$\}$
(6) The set of all natural numbers between 11 and 15.	{12, 13, 14}	$\{x : x \in N, 11 < x < 15\}$

Illustration 1: Write the set $X = \left\{1, \frac{1}{4}, \frac{1}{9}, \frac{1}{16}, \frac{1}{25}, \ldots\right\}$ in the set-builder form.

Solution:

We observe that the elements of set X are the reciprocals of the squares of all natural numbers. So, the set X in set builder form is

$$X = \left\{\frac{1}{n^2} : n \in N\right\}.$$

Illustration 2: Write the following intervals in set builder form

(i) (–3, 0) (ii) [6, 12] (iii) (6, 12] (iv) [–23, 5)

Solution:

The following intervals are written in set builder form as :

(i) (–3, 0) is an open interval which does not include both – 3 and 0. So, it can be shown in the set builder form as : $\{x : x \in R, -3 < x < 0\}$.

(ii) [6, 12] is a closed interval which includes both 6 and 12. So it can be shown in the set builder form as $\{x : x \in R, 6 \leq x \leq 12\}$

(iii) (6, 12] is an interval open at the first end and closed at the second end i.e. it excludes 6 but includes 12. So it is shown in the set builder form as : $\{x : x \in R, 6 < x \leq 12\}$

(iv) [–23, 5) is an interval closed at the first end point but open at the second end point. It means that the interval includes –23 but excludes 5. It is written in the set builder form as : $\{x : x \in \text{R}, -23 \leq x < 5\}$.

STANDARD SYMBOLS OF SOME SPECIAL SETS

N : Set of all natural numbers
Z : Set of all integers
Q : Set of all rational numbers
R : Set of all real numbers
Z⁺ : Set of all positive integers
Q⁺ : Set of all positive rational numbers, and
R⁺ : Set of all positive real numbers.

The symbols for the special sets given above will be referred throughout the chapter.

TYPES OF SETS

Empty Set

A set which does not contain any element is called an empty set, null set or void set.

The empty set is denoted by the symbol ϕ or {}.

Given below are few examples of empty sets.

(i) If $A = \{x : 1 < x < 2, x \text{ is a natural number}\}$, then A is the empty set, because there is no natural number between 1 and 2.

(ii) If $B = \{x : x^2 - 2 = 0 \text{ and } x \text{ is rational number}\}$, then B is the empty set, because the equation $x^2 - 2 = 0$ is not satisfied by any rational value of x.

(iii) If $C = \{x : x \text{ is an even prime number greater than } 2\}$, then C is the empty set, because 2 is the only even prime number.

(iv) If $D = \{x : x^2 = 4, x \text{ is odd}\}$, then D is the empty set, because the equation $x^2 = 4$ is not satisfied by any odd value of x.

Equal Sets

Two sets A and B are said to be equal if they have exactly the same elements and we write $A = B$. Otherwise, the sets are said to be unequal and we write $A \neq B$.

(i) Let $A = \{1, 2, 3, 4\}$ and $B = \{3, 1, 4, 2\}$, then $A = B$, because elements of both sets are the same. Only order of the elements in the two sets is different but it is not considered in a set.

(ii) Let A be the set of prime numbers less than 6 and P the set of prime factors of 30. Then A and P are equal, since 2, 3 and 5 are the only prime factors of 30 and also these are less the only prime numbers than 6.

Illustration 3: Find the pairs of equal sets, from the following sets, if any, giving reasons:

$A = \{0\}$, $B = \{x : x > 15 \text{ and } x < 5\}$, $C = \{x : x - 5 = 0\}$, $D = \{x : x^2 = 25\}$

$E = \{x : x$ is an integral positive root of the equation $x^2 - 2x - 15 = 0\}$.

Solution: We have,

$A = \{0\}$,

$B = \{x : x > 15 \text{ and } x < 5\} = \phi$,

$C = \{x : x - 5 = 0\} = \{5\}$,

$D = \{x : x^2 = 25\} = \{-5, 5\}$,

and $E = \{5\}$.

Clearly, $C = E$.

SUBSETS

Set A is said to be a subset of a set B if every element of set A is also an element of set B. Here set B is called superset of set A. A is a subset of B, is represented ACB. Thus $A \subset B$ if whenever $a \in A$, then $a \in B$. It is often convenient to use the symbol "$\Rightarrow$" which means implies. Using this symbol, we can write the definition of subset as follows: $A \subset B$ if $a \in A \Rightarrow a \in B$.

We read the above statement as "A is a subset of B if a is an element of A implies a is also an element of B". If A is not a subset of B, we write $A \not\subset B$. For example:

(i) The set Q of rational numbers is a subset of the set R of real numbers, so we write $Q \subset R$.

(ii) If A is the set of all divisors of 56 and B the set of all prime divisors of 56, then B is a subset of A so we write $B \subset A$.

(iii) Let $A = \{1, 3, 5\}$ and B $\{x : x$ is an odd natural number less than $6\}$. Then $A \subset B$ and $B \subset A$ and hence $A = B$.

(iv) Let $A = \{a, e, i, o, u\}$ and $B = \{a, b, c, d\}$. Then A is not a subset of B. Also B is not a subset of A.

Important Points about Subsets

(i) Every set is a subset of itself.

B Empty set is a subset of every set.

(iii) Total number of subsets of a finite set containing n element is 2^n.

UNIVERSAL SET

If there are some sets under consideration, and out of these sets, there is a set which is the superset of all other given sets i.e., all other sets under consideration are subsets of this set. Such a set is known as the universal set, denoted by U.

For example,

(i) In the context of human population studies, the universal set consists of all the people in the world.

(ii) If {1, 2, 3, 4}, {2, 5, 6}, {1, 3, 7, 8, 9} and {1, 2, 3, 4, 5, 6, 7, 8, 9} are the sets under consideration, then set {1, 2, 3, 4, 5, 6, 7, 8, 9} can be considered as universal set because all other three sets are the subsets of this set.

VENN DIAGRAMS

In order to illustrate universal sets, subsets and certain operations on sets in a clear and simple way, we use geometric figures. These figures are called Venn-Diagrams. In Venn Diagrams, a universal set is represented by a rectangle and any other set is represented by a circle.

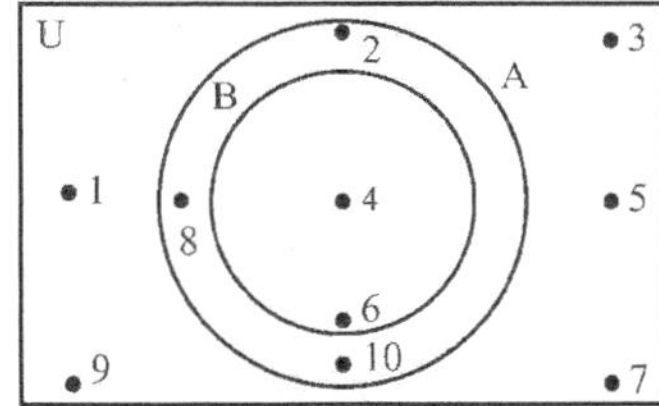

In the Venn-diagrams, the elements of the sets are written in their respective circles.

In the Venn-diagrams, $U = \{1, 2, 3, ..., 10\}$ is the universal set of which $A = \{2, 4, 6, 8, 10\}$ and $B = \{4, 6\}$ are subsets, and also $B \subset A$.

OPERATION ON SETS

Union of Sets

Union of two sets A and B is the set which consists of all those elements which are either in A or in B (including those which are in both sets A and B). In symbols, we write

$$A \cup B = \{x : x \in A \text{ or } x \in B\}$$

The union of two sets A and B can be represented by a Venn diagram as shown in figure by shaded portion

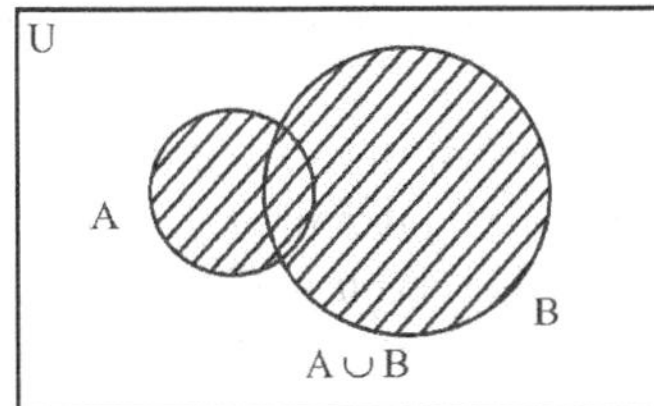

Intersection of Sets

The intersection of two sets A and B is the set of all those elements which belong to both sets A and B. Symbolically, we write

$$A \cap B = \{x : x \in A \text{ and } x \in B\}.$$

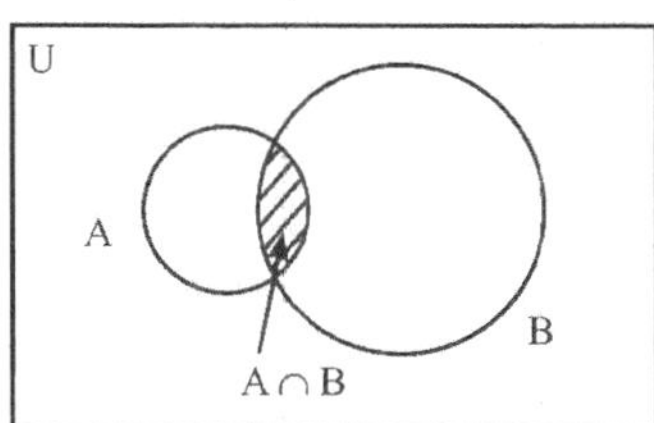

The shaded portion in figure indicates the intersection of sets A and B.

Difference of Sets

The difference of the sets A and B (in the order A minus B) is the set of elements which belong to A but not to B. Symbolically, we write $A - B$ and read as "A minus B".

In the set builder notation, we can write

$$A - B = \{x : x \in A \text{ but } x \notin B\}$$

The difference of two sets A and B is represented in Venn diagram by shaded portion.

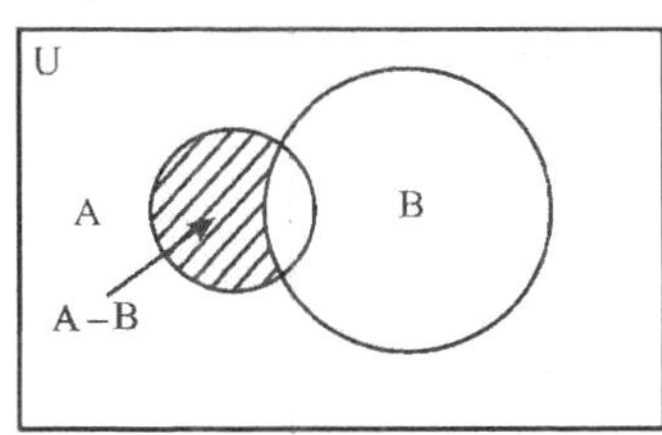

Complements of a Set

Let U be the universal set and A be a subset of U. Then the complement of A is the set of all elements of U which are not the elements of set A. Symbolically, we write A' or A^c to denote the complement of set A.

Thus, $A' = \{x : x \in U \text{ but } x \notin A\}$.

Obviously $A' = U - A$

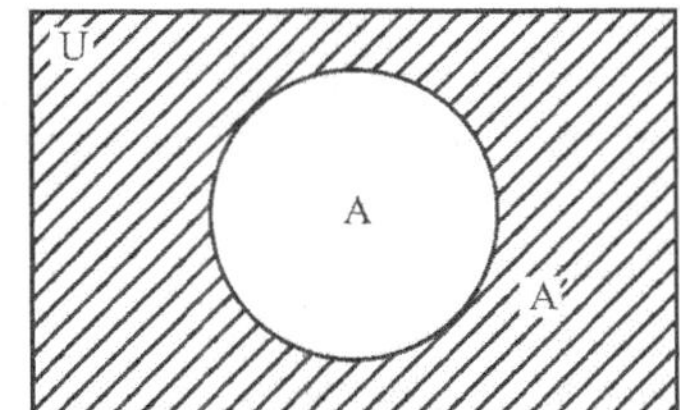

Complement of set A i.e. A' is represented in Venn diagram by shaded region.

Some Properties of Complement of a Set

1. Complement laws:
   (i) $A \cup A' = U$
   (ii) $A \cap A' = \phi$
2. De Morgan's law:
   (i) $(A \cup B)' = A' \cap B'$
   (ii) $(A \cap B)' = A' \cup B'$
3. Law of double complementation: $(A')' = A$
4. Laws of empty set and universal set: $\phi' = U$ and $U' = \phi$.

These laws can be verified by using Venn diagrams.

Illustration 4: If $A = \{x : x = 3n, n \in Z\}$ and $B = \{x : x = 4n, n \in Z\}$, then find $A \cap B$.

Solution: We have,

$x \in A \cap B \Leftrightarrow x = 3n, n \in Z$ and $x = 4n, n \in Z$

$\Leftrightarrow x$ is a multiple of 3 and x is a multiple of 4

$\Leftrightarrow x$ is a multiple of 3 and 4 both

$\Leftrightarrow x$ is a multiple of 12.

$\Leftrightarrow x = 12n, n \in Z$

Hence, $A \cap B = \{x : x = 12n, n \in Z\}$.

If A and B are two sets, then $A \cap B = A$, if $A \subset B$ and $A \cap B = B$, if $B \subset A$.

Illustration 5: Let $U = \{1, 2, 3, 4, 5, 6, 7, 8, 9\}$, $A = \{1, 2, 3, 4\}$, $B = \{2, 4, 6, 8\}$ and $C = \{3, 4, 5, 6\}$. Find

(i) A' **(ii) $(A \cup B)'$**
(iii) $(A')'$ **(iv) $(B - C)'$**

Solution:

(i) $\{5, 6, 7, 8, 9\}$ (ii) $\{5, 7, 9\}$
(iii) A (iv) $\{1, 3, 4, 5, 6, 7, 9\}$

Illustration 6: Find the union of each of the following P airs of sets:

(i) $A = \{x : x$ is a natural number and $1 < x \leq 6\}$
$B = \{x : x$ is a natural number and $6 < x \leq 10\}$

(ii) $A = \{1, 2, 3\}$, $B = \phi$.

Solution:

(i) $A = \{x : x$ is a natural number and $1 < x \leq 6\}$

$\Rightarrow \quad A = \{2, 3, 4, 5, 6\}$

$B = \{x : x$ is a natural number and $6 < x \leq 10\}$

$\Rightarrow \quad B = \{7, 8, 9, 10\}$

$\therefore \quad A \cup B = \{2, 3, 4, 5, 6\} \cup \{7, 8, 9, 10\}$

$\Rightarrow \quad A \cup B = \{2, 3, 4, 5, 6, 7, 8, 9, 10\}$

(ii) We have, $A = \{1, 2, 3\}$, $B = \phi$

$\Rightarrow \quad A \cup B = \{1, 2, 3\} \cup \phi$

$\Rightarrow \quad A \cup B = \{1, 2, 3\}$

Illustration 7: If $A = \{x : x = 3n, n \in Z\}$ and $B \{x : x = 4n, n \in Z\}$, then find $(A \cap B)$.

Solution:

Let $x \in (A \cap B) \Leftrightarrow x \in A$ and $x \in B$

$\Leftrightarrow x$ is a multiple of 3 and x is a multiple of 4.

$\Leftrightarrow x$ is a multiple of 3 and 4 both

$\Leftrightarrow x$ is a multiple of 12

$\Leftrightarrow x = 12n, n \in Z$

Hence $A \cap B = \{x : x = 12n, n \in Z\}$

DISJOINT SETS

If A and B are two sets such that $A \cap B = \phi$, then A and B are called disjoint sets.

For example, let $A = \{2, 4, 6, 8\}$ and $B = \{1, 3, 5, 7\}$. Here A and B are disjoint sets, because there is no element common to both sets A and B.

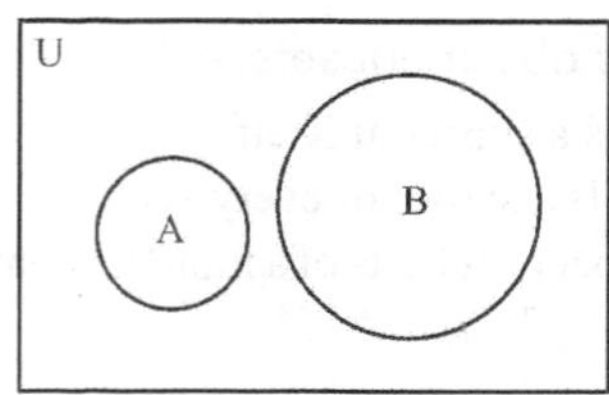

In the Venn diagram, A and B are disjoint sets.

CARDINAL NUMBER

Number of element in a set A is called cardinal number of set A. It is represented by $n(A)$. If $A = \{a, b, c, d, e, f\}$, then $n(A) = 6$

1. If A and B are finite sets then
$$n(A \cup B) = n(A) + n(B) - n(A \cap B)$$
2. If A, B and C are three finite sets, then
$$n(A \cup B \cup C) = n(A) + n(B) + n(C) - n(A \cap B) - n(B \cap C) - n(C \cap A) + n(A \cap B \cap C)$$

Illustration 8: In a political survey, 78% of the politicians favour at least one proposal, 50% of them are in favour of proposal A, 30% are in favour of proposal B and 20% are in favour of proposal C. 5% are in favour of all three proposals. what is the percentage of people favouring more than one proposal?

(a) 16 **(b) 17**
(c) 18 **(d) 19**

Solution: (b)

$n(A \cup B \cup C) = n(A) + n(B) + n(C) - n(A \cap B) - n(B \cap C) - n(6A \cap C) + n(A \cap B \cap C)$

or $78 = 50 + 30 + 20 - \Sigma n(A \cap B) + 5$

or $\Sigma n(A \cap B) = 27$

This includes $n(A \cap B \cap C)$ three times.

$\therefore$ Percentage of people favouring more than one proposal $= 27 - 5 \times 2 = 17$

Illustration 9: If X and Y are two sets such that $X \cup Y$ has 50 elements, X has 28 elements and Y has 32 elements, how many elements does $X \cap Y$ have?

Solution:

Given that $n(X \cup Y) = 50, \quad n(X) = 28,$
$n(Y) = 32, \quad n(X \cap Y) = ?$

By using the formula,

$n(X \cup Y) = n(X) + n(Y) - n(X \cap Y),$

We find that $n(X \cap Y) = n(X) + n(Y) - n(X \cup Y)$
$= 28 + 32 - 50 = 10$

Illustration 10: In a class of 35 students, 24 like to play cricket and 16 like to play football. Also, each student likes to play at least one of the two games. How many students like to play both cricket and football?

Solution: Let X be the set of students who like to play cricket and Y be the set of students who like to play football. Then $X \cup Y$ is the set of students who like to play at least one of the two games, and $X \cap Y$ is the set of students who like to play both games.

Given $n(X) = 24$, $n(Y) = 16$, $n(X \cup Y) = 35$, $n(X \cap Y) = ?$

Using the formula $n(X \cup Y) = n(X) - n(Y) - n(X \cap Y)$,

We get $35 = 24 + 16 - n(X \cap Y)$

Thus, $n(X \cap Y) = 5$ i.e., 5 students like to play both games.

SITUATION BASED VENN DIAGRAMS

1. Suppose set C represents the people who like cricket and F represents the people who like football.

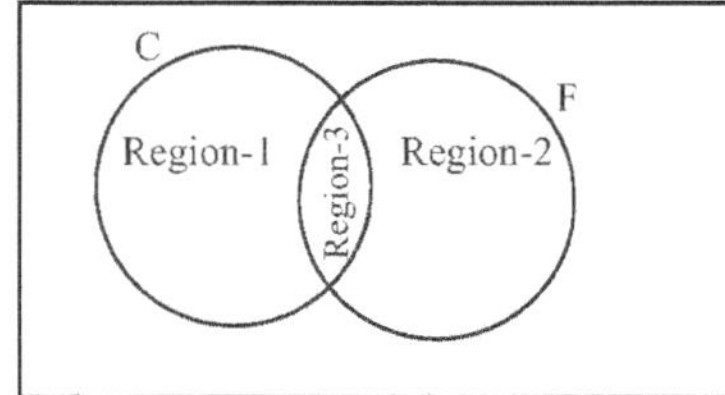

In the above Venn-diagram,

Region- 1: Represents the people who like cricket only (means people who like cricket but not football.)

Region- 2: Represents the people who like football only (means people who like football but not cricket.)

Region- 3: Represents the people who like both cricket and football.

The people who like both cricket and football is represented by the common shaded region of set A and set B in the Venn diagram.

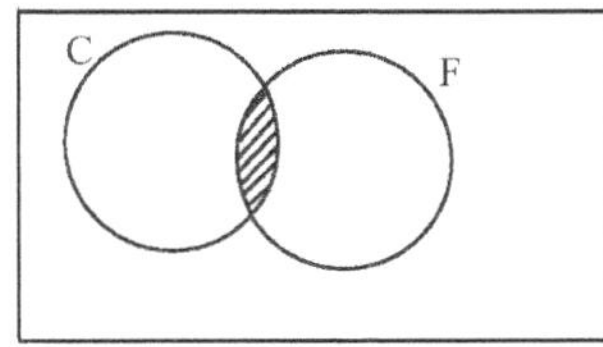

2. Let M represent the students who passed in mathematics, E represents the students who passed in English and S represents the students who passed in Science. Then students who passed in both Mathematics and English are represented by common region of the sets M and E.

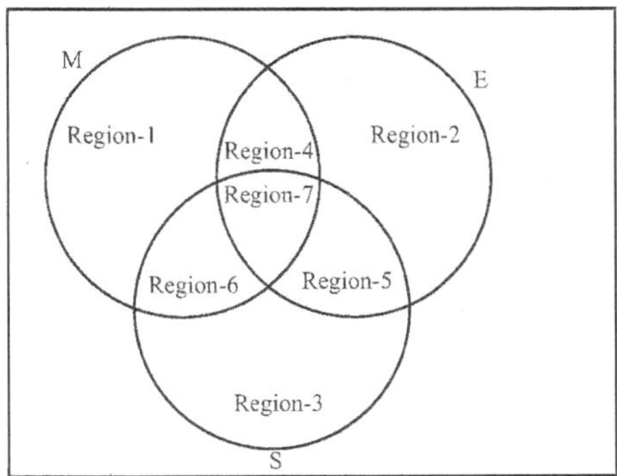

Students who passed in both English and Science are represented by the common region of set E and S. Students who passed in both Science and Mathematics represented by the common region of set S and M. Students who passed in both Mathematics and English are represented by the common region of sets M and E. Students who passed in all the three subjects, Mathematics, English and Science are represented by common region of all the three sets M, E and S.

Region- 1: Represents the students who passed in Mathematics only (means the students who passed in Mathematics but not passed in English and Science).

Region- 2: Represents the students who passed in English only (means the students who passed in English but not passed in Science and mathematics).

Region- 3: Represents the students who passed in Science only (means the students who passed in science but not passed in Mathematics and English).

Region- 4: Represents the students who passed in both Mathematics and English only (means the students who passed in both Mathematics and English but not in Science).

Region- 5: Represents the students who passed in both English and Science only (means the students who passed in both English and Science but not passed in Mathematics)

Region- 6: Represents the students who passed in both Science and Mathematics only (means the students who passed in both Science and Mathematics, but not passed in English).

Region- 7: Represents the students who passed in all the three subjects Mathematics, English and Science.

Note that

- Students who passed in Mathematics are represented by the sum of the regions 1, 4, 6 and 7.
- Students who passed in English are represented by the sum of the regions 2, 4, 5 and 7.
- Students who passed in Science are represented by the sum of the regions 3, 5, 6 and 7.
- Students who passed in both Mathematics and English are represented by the sum of the regions 4 and 7.
- Students who passed in both English and Science are represented by the sum of the regions 5 and 7.
- Students who passed in both Science and Mathematics are represented by the sum of the regions 6 and 7.

Illustration 11: Draw the appropriate Venn diagram for each of the following:

(i) $(A \cup B)'$ (ii) $A' \cap B'$ (iii) $(A \cap B)'$ (iv) $A' \cup B'$

Solution:

(i) $(A \cup B)'$ is represented by the shaded region.

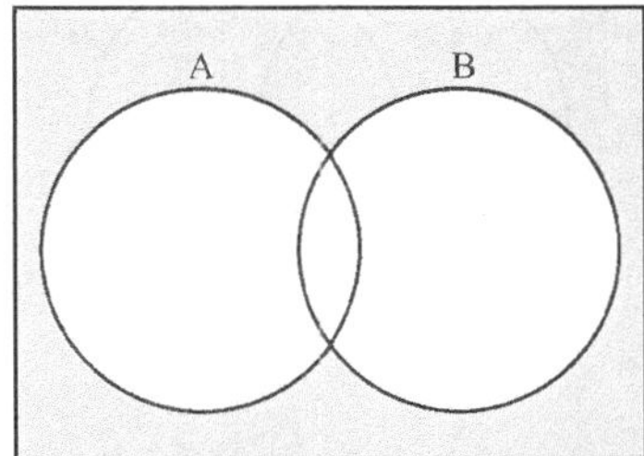

(ii) $A' \cap B'$ is represented by the shaded region.

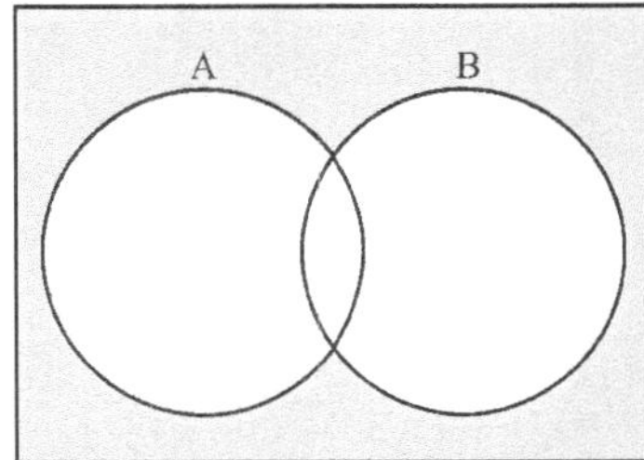

(iii) $(A \cap B)'$ is represented by the shaded region.

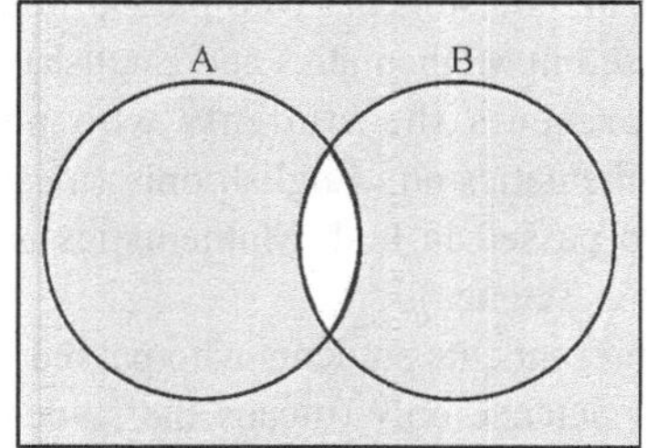

(iv) $A' \cup B'$ is represented by the shaded region.

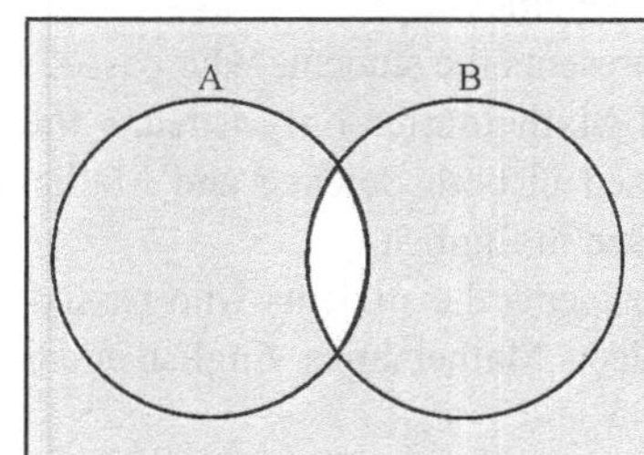

Illustration 12: Out of 10000 people surveyed, 3700 liked city *A*, 4000 liked city *B* and 5000 liked city *C*. 700 people liked *A* and *B* 1200 liked *A* and *C* and 1000, liked *B* and *C*. Each person liked at least one city. Then find

(A) The number of people liking all the three cities.

(B) The number of persons liking at least two cities as a % of number of people liking exactly one city.

(C) The number of persons liking exactly two cities as a percentage of the number of people liking at least one city.

(D) The number of persons liking *A* and *B* but not *C*.

Solution:

Refer the figure given

$n(A \cup B \cup C) = n(A) + n(B) + n(C) - n(A \cap B \cap C)$

$\Rightarrow$ $10000 = 3700 + 4000 + 5000 - 700 - 1000 - 1200 + d$

$\Rightarrow$ $d = 200$

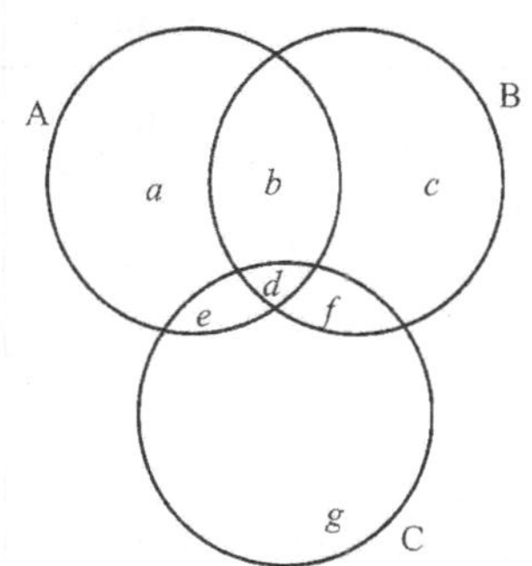

Once the value of d is known, all other values will be determined fastly.

e.g. $b + d = 700$ (given) $\Rightarrow$ $b = 500$

Similarly $e = 1000, f = 800, a = 2000, c = 2500, g = 3000$

A. $d = 200$.

B. At least two cities $b + d + e + f = 2500$

Exactly one city $a + c + g = 7500$

$\Rightarrow \% = 2500/7500 \times 100\ \% = 33.33\%$

C. Exactly two cities $= b + e + f = 2300$

At least one city = 10000

$\Rightarrow$ Required % = 23 %.

D. $b = 500$.

Illustration 13: In a survey of 100 students, the number of students studying the various languages were found to be: English only 18, English but not Hindi 23, English and Sanskrit 8, English 26, Sanskrit 48, Sanskrit and Hindi 8, no language 24. Find:

(i) How many students were studying Hindi?

(ii) How many students were studying English and Hindi?

Solution:

We have, $a = 18$, $a + b = 23$, $d + e = 8$, $a + b + d + e = 26$, $d + e + f + g = 48$, $e + f = 8$, $a + b + c + d + e + f + g = 100 - 24 = 76$

$\therefore$ $a = 18, b = 0, c = 10, d = 5, e = 3, f = 5$ and $g = 35$

(i) $n(H) = b + c + e + f = 18$

(ii) $n(H \cap E) = b + e = 3$

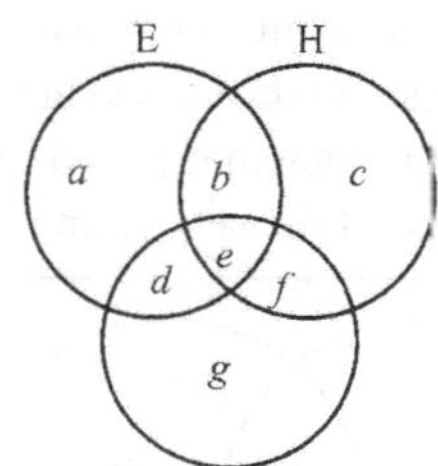

Practice Exercise

Level - I

1. In a group of students, 100 students know Hindi, 50 know English and 25 know both. Each of the students knows either Hindi or English. How many students are there in the group?
(a) 100 (b) 115
(c) 110 (d) 125
2. If X and Y are two sets such that $n(X) = 17, n(Y) = 23$ and $n(X \cup Y) = 38$, then $n(X \cap Y)$
(a) 2 (b) 1
(c) 3 (d) 4
3. If X and Y are two sets such that X has 40 elements, $X \cup Y$ has 60 elements and $X \cap Y$ has 10 elements, how many elements does Y have ?
(a) 25 (b) 35
(c) 30 (d) 40
4. Let $S = \{0, 1, 5, 4, 7\}$. Then the total number of subsets of S is
(a) 64 (b) 32
(c) 40 (d) 20
5. If two sets A and B are having 99 elements in common, then the number of elements common to each of the sets $A \times B$ and $B \times A$ are
(a) 2^{99} (b) 99^2
(c) 100 (d) 19
6. In an examination out of 100 students, 75 passed in English 60 passed in Mathematics and 45 passed in both English and Mathematics. What is the number of students passed in exactly one of the two subjects?
(a) 45 (b) 60
(c) 75 (d) 90
7. If $X = \{4^n - 3n - 1 : n \in N\}$ and $Y = \{9(n-1) : n \in N\}$, then precisely:
(a) $X \subset Y$ (b) $X \subseteq Y$
(c) $X = Y$ (d) $X \supseteq Y$
8. $A \cup B = A$, then
(a) $A \subset B$ (b) $B \subset A$
(c) $A \not\subset B$ and $B \not\subset A$ (d) None of these

Level - II

DIRECTIONS (Qs. 1–3) : ***Read the information given below and answer the questions that follow.***

The result of an exam is given below. Out of 1000 students who appeared
(i) 658 failed in Physics
(ii) 166 failed in Physics and Chemistry
(iii) 372 failed in Chemistry, 434 failed in Physics and Maths
(iv) 590 failed in Maths, 126 failed in Maths & Chemistry.

Find the number of people who failed in (assuming that none is passed in all subjects).

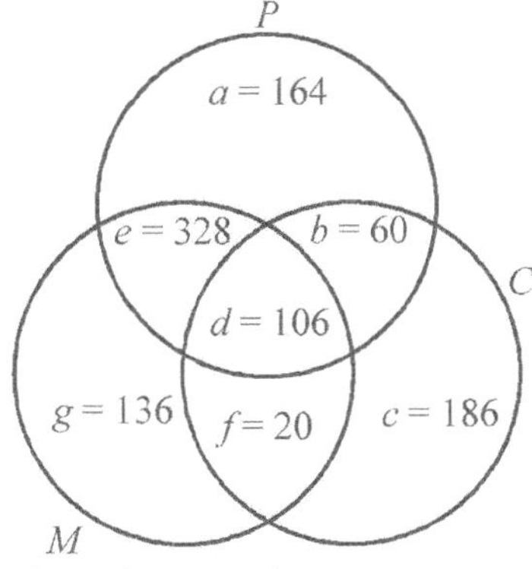

We have the following equations :
$a + b + c + d + e + f + g = 1000$
$a + b + d + e = 658, b + d = 166$
$b + d + c + f = 372.$
$d + e = 434$ as in the figure.
$d + e + f + g = 590, d + f = 126.$ Find the values.

1. Chemistry but not in Physics.
(a) 318 (b) 198
(c) 213 (d) 206
2. Physics or Maths but not in Chemistry.
(a) 558 (b) 718
(c) 628 (d) None of these
3. Physics but neither Chemistry nor Maths.
(a) 164 (b) 228
(c) 196 (d) None of these
4. 70 per cent of the employees in a multinational corporation have VCD players, 75 per cent have microwave ovens, 80 per cent have ACs and 85 per cent have washing machines. At least what percentage of employees has all four gadgets?
(a) 15 (b) 5
(c) 10 (d) Cannot be determined
5. If $A = \{1, 2, 5, 6\}$ and $B = \{1, 2, 3\}$, then $(A \times B) \cap (B \times A)$ is equal to?
(a) $\{(1,1), (2,1), (6,1), (3,2)\}$
(b) $\{(1,1), (1,2), (2,1), (2,2)\}$
(c) $\{(1,1), (2,2)\}$
(d) $\{(1,1), (1,2), (2,5), (2,6)\}$
6. Which one of the following is a null set ?
(a) $\{0\}$ (b) $\{\{\{\}\}\}$
(c) $\{\{\}\}$ (d) $\{x \mid x^2 + 1 = 0, x \in R\}$
7. In a certain office, 72% of the workers prefer tea and 44% prefer coffee. If each of them prefers tea or coffee and 40 like both, the total number of workers in the office is :
(a) 200 (b) 240
(c) 250 (d) 320

8. A survey show that 63% of the Indians like cheese whereas 76% like apples. If x% of the Indians like both cheese and apples, then find the range of x.
(a) $0 \le x \le 23\%$ (b) $0 \le x \le 39\%$
(c) $4 \le x \le 35\%$ (d) $6 \le x \le 33\%$

9. $A \equiv \{x \mid x \text{ is a prime number} \le 100\}$
$B \equiv \{x \mid x \text{ is an odd number } \le 100\}$
What is the ratio of the number of subsets of set A to set B ?
(a) 2^{25} (b) 2^{-25}
(c) 2 (d) $\frac{50^2}{25^2}$

10. If A and B are two sets such that A has 12 elements, B has 17 elements, and $A \cup B$ has 21 elements, how many elements does $A \cap B$ have ?
(a) 7 (b) 8
(c) 9 (d) 10

11. In an examination 70% of the candidates passed in English, 65% in Mathematics, 27 % failed in both the subjects. Find the total number of candidates.
(a) 200 (b) 400
(c) 300 (d) 100

12. If the set A has p elements, B has q elements, then the number of elements in $A \times B$ is
(a) $p+q+1$ (b) pq
(c) p^2 (d) $p+q$

13. Let $A = \{(n, 2n) : n \in N\}$ and $B = \{(2n, 3n) : n \in N\}$. Then $A \cap B$ equal to ?
(a) $\{(n, 6n) : n \in N\}$ (b) $\{(2n, 6n) : n \in N\}$
(c) $\{(n, 3n) : n \in N\}$ (d) ϕ

14. If $n(A) = 115, n(B) = 326, n(A - B) = 47$, then what is $n(A \cup B)$ equal to?
(a) 373 (b) 165
(c) 370 (d) 394

For the next Four (15–18) questions that follow:

In a city, three daily newspapers A, B, C are published, 42% read A; 51% read B; 68% read C; 30% read A and B; 28% read B and C; 36% read A and C; 8% do not read any of the three newspapers.

15. What is the percentage of persons who read all the three papers?
(a) 20% (b) 25%
(c) 30% (d) 40%

16. What is the percentage of persons who read only two papers?
(a) 19% (b) 31%
(c) 44% (d) None of these

17. What is the percentage of persons who read only one paper?
(a) 38% (b) 48%
(c) 51% (d) None of these

18. What is the percentage of persons who read only A but neither B nor C?
(a) 4% (b) 3%
(c) 1% (d) None of these

19. If A and B are any two sets, then what is $A \cap (A \cup B)$ equal to?
(a) Complement of A (b) Complement of B
(c) B (d) A

20. 40% of the people read newspaper X, 50% read newspaper Y and 10% read both the papers. What percentage of the people read neither newspaper?
(a) 10% (b) 15%
(c) 20% (d) 25%

DIRECTIONS (Qs. 21–24) : *Read the passage below and solve the questions based on it.*

5% of the passengers do not like coffee, tea and lassi and 10% like all the three, 20% like coffee and tea, 25% like lassi and coffee and 25% like lassi and tea. 55% like coffee, 50% like tea, and 50% like lassi.

21. The passengers who like only coffee is greater than the passengers who like only lassi by
(a) 25% (b) 100%
(c) 75% (d) 0%

22. The percentage of passengers who like both tea and lassi but not coffee, is
(a) 15 (b) 25
(c) 40 (d) 75

23. The percentage passengers who like at least 2 of the coffee, tea and lassi, is
(a) 30 (b) 45
(c) 50 (d) 60

24. If the number of passengers is 180, then the number of passengers who like lassi only, is
(a) 10 (b) 18
(c) 27 (d) 36

25. In a town three newspapers A, B and C are published. 42% of the people in that town read A, 68% read B, 51% read C, 30% read A and B, 28% read B and C, 36% A and C and 18% do not read any paper. Find the % of population of town that reads all the three. *[SBI PO-2011]*
(a) 15% (b) 25%
(c) 20% (d) 35%
(e) None of these

DIRECTIONS (Qs. 26 - 29) : *Answer these questions on the basis of the information given below :*

In a survey of 1000 boys conducted in an area, it is found that 65% play Cricket, 48% play Football and 40% play Hockey. Of the total, 30% play both Football and Cricket, 25% play Football and Hockey, while 24% play Cricket and Hockey. Only 5% do not play any of the three games. *[SBI PO-2011]*

26. Find the number of players who play Football but not Hockey.
(a) 180 (b) 230
(c) 350 (d) Can't be determined
(e) None of these

27. How many play all the three above-mentioned games?
(a) 180 (b) 240
(c) 230 (d) 210
(e) None of these

28. How many play Hockey but neither Cricket nor Football?
(a) 140 (b) 320
(c) 120 (d) Can't be determined
(e) None of these

29. Find the percentage of players who play only Football.
(a) 12 (b) 14
(c) 32 (d) 18
(e) None of these

Hints & Solutions

Level-I

1. (d) Total number of students = 100 + 50 – 25 = 125
2. (a) 3. (c) 4. (b) 5. (b)
6. (a) Total number of students = 100

Let E denote the students who have passed in English.

Let M denote the students who have passed in Maths.

$\therefore n(E) = 75, n(M) = 60$ and $n(E \cap M) = 45$

we know $n(E \cup M) = n(E) + n(M) - n(E \cap M)$

$= 75 + 60 - 45 = 90$

Required number of students = 90 – 45 = 45

7. (a) 8. (b)

Level-II

1. (d) Chemistry but not Physics = $c + f = 206$.
2. (c) Physics and Maths but not Chemistry = $a + e + g = 628$.
3. (a) Physics but neither Maths nor Chemistry = $a = 164$.
4. (c) Employees who doesn't have VCD = 100 – 70 = 30%

Employees who doesn't have MWO = 100 – 75 = 25%

Employees who doesn't have AC = 100 – 80 = 20%

Employees who doesn't have WM = 100 – 85 = 15%

$\therefore$ Total employees who doesn't have atleast one of the four equipments = 30 + 25 + 20 + 15 = 90%

$\therefore$ Percentage of employees having all four gadgets = 100 – 90 = 10%.

5. (b) Let $A = \{1, 2, 5, 6\}$ and $B = \{1, 2, 3\}$

$\therefore A \times B = \{(1, 1), (1, 2), (1, 3), (2, 1), (2, 2), (2, 3), (5, 1), (5, 2), (5, 3), (6, 1), (6, 2), (6, 3)\}$

and $B \times A = \{(1, 1), (1, 2), (1, 5), (1, 6), (2, 1), (2, 2), (2, 5), (2, 6), (3, 1), (3, 2), (3, 5), (3, 6)\}$

$\Rightarrow (A \times B) \cap (B \times A)$

$= \{(1, 1), (1, 2), (2, 1), (2, 2)\}$

6. (d) Consider the set given in option 'd'.

$\{x \mid x^2 + 1 = 0, x \in R\}$

Let $x^2 + 1 = 0 \Rightarrow x^2 = -1 \Rightarrow x = \pm i$ which is complex.

But $x \in R$. Hence for, any $x \in R$, $x^2 + 1$ can not be zero.

7. (c) Let total number be x. Then

$n(A) = \frac{72}{100}x = \frac{18x}{25}$, $n(B) = \frac{44}{100}x = \frac{11x}{25}$ and

$n(A \cap B) = 40$ $n(A \cap B) = n(A) + n(B) - n(A \cup B)$

$\Rightarrow x = \frac{18x}{25} + \frac{11x}{26} - 40 \Rightarrow \frac{29x}{25} - x = 40$

$\Rightarrow \frac{4x}{25} = 40 \Rightarrow x = 250$

8. (b) $n(C) = 63\%$

$n(A) = 76\%$

$n(C \cup A) = n(C) + n(A) - n(C \cap A)$

$100\% = 63\% + 76\% - X\%$

$X\% = 39\%$

9. (b) We know that there are 25 prime number below 100.

$n(A) = 25$

The total number of subsets of sets $A = 2^{25}$

There are 50 odd numbers below 100.

The total number of subsets of $B = 2^{50}$

Required ratio = $\frac{2^{25}}{2^{50}} = 2^{-25}$.

10. (b) $n(A) = 12, n(B) = 17, n(A \cup B) = 21$

$(A \cup B) = n(A) + n(B) - n(A \cap B)$

$21 = 12 + 17 - n(A \cap B)$ or $n(A \cap B)$

$= 12 + 17 - 21 = 8$

$\Rightarrow A \cap B$ has 8 elements.

11. (b) Let the set E and M represent students who passed in English and Mathematics respectively.

$n(E \cup M) = (100–27)\% = 73\%$

$n(E \cup M) = n(E) + n(M) - n(E \cap M)$

$73\% = 70\% + 65\% - x\%$

$x\% = 62\%$

Now. $62\% \equiv 248$

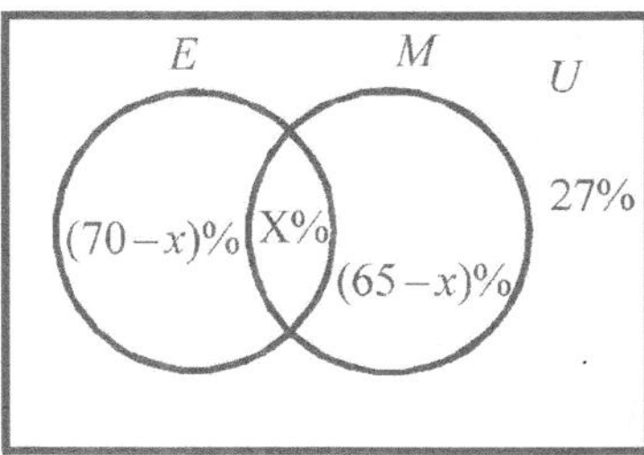

$\therefore$ Total number of candidates = $\frac{248 \times 100}{62} = 400$

12. (b) As A has p elements and B has q elements so, $A \times B$ has pq elements.

13. (d) $A = \{(n, 2n) : n \in N\}$ and $B = \{(2n, 3n)\} : n \in N$

Listing few members of each set

$A = \{(1, 2), (2, 4), (3, 6), ...\}$

$B = \{(2, 3), (4, 6), (6, 9) ......\}$

There is no member common to both these sets, hence. $A \cap B = \phi$

14. (a) We know, for two sets A and B

$A - B = A - (A \cap B)$

$\therefore n(A - B) = n(A) - n(A \cap B)$

Given, $n(A) = 115, n(B) = 326$ and $n(A - B) = 47$

$\Rightarrow 47 = 115 - n(A \cap B)$

$\Rightarrow n(A \cap B) = 68$

Consider $n(A \cup B) = n(A) + n(B) - n(A \cap B)$

$= 115 + 326 - 68 = 373$

(15-18).

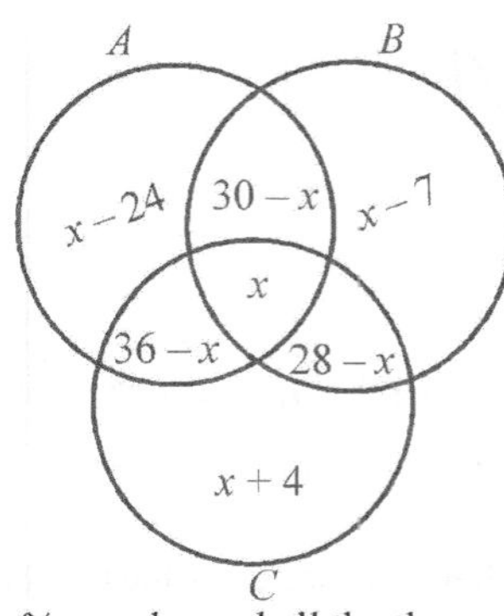

Let x % people read all the three newspapers.
Since 8% people do not read any newspapers.

$\therefore \quad (x-24)+(x-7)+(x+4)+(30-x)+(36-x)+(28-x)+x=92$

$\Rightarrow \quad x+98-31=92$

$\Rightarrow \quad x=92-67=25$

15. **(b)** Hence people who read all the three newspapers = 25%

16. **(d)** $(30-x)+(36-x)+(28-x)=94-3x$
$=98-3\times 25=23$
Hence percentage of people who read only two newspapers = 23%

17. **(b)** $(x-24)+(x-7)+(x+4)=3x-27$
$=3\times 25-27=48$
Hence percentage of people who read only one newspaper = 48%

18. **(c)** $x-24=25-24=1$
Hence percentage of people who read only Newspaper A but neither B nor C = 1%

19. **(d)**

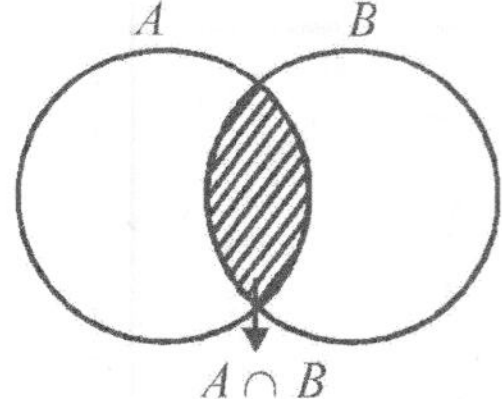

$A\cap(A\cup B)=(A\cap A)\cup(A\cap B)=A\cup(A\cap B)$
$=A$ (By diagram)

Thus, $A\cap(A\cup B)=A$

20. **(c)** $n(A)=40, n(B)=50, n(A\cap B)=10.$
$n(A\cup B)=n(A)+n(B)-n(A\cap B)=40+50-10=80.$
$\therefore$ Percentage reading either or both newspapers = 80%.
Hence, percentage reading neither newspaper
$=(100-80)\%=20\%$

(21 – 24)

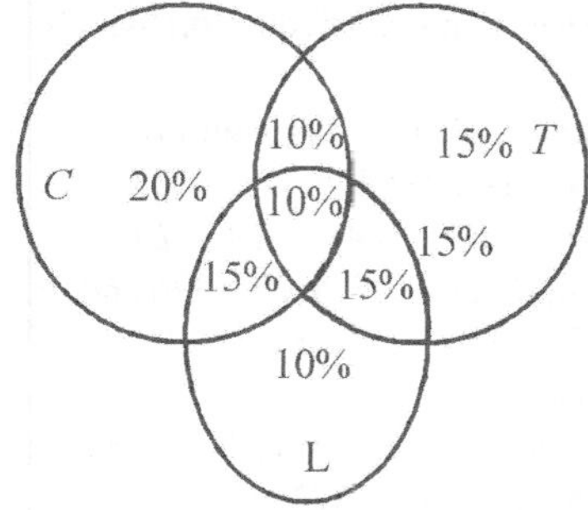

Where C = Coffee, T = Tea and L = Lassi

21. **(b)** The passengers who like only coffee = 20% and the passengers who like only lassi = 10%
Required passengers = 100%

22. **(a)** It can be seen that the percentage of passengers who like both tea and lassi but not coffee = 15%. This is the figure representing this area

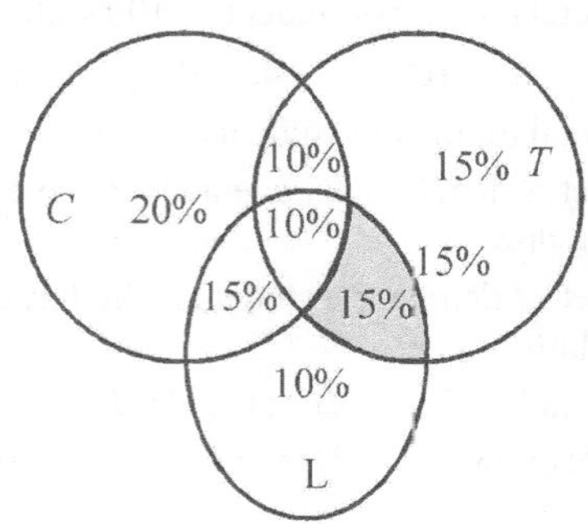

23. **(c)** The percentage of passengers who like at least 2 of the coffee, tea and lassi can be seen in the below figure:

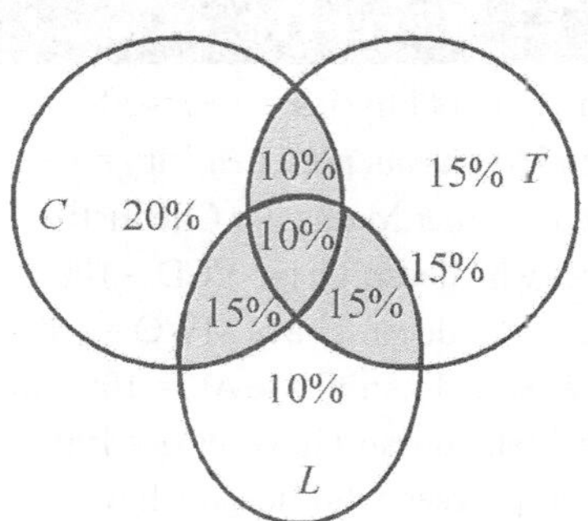

24. **(b)** 10% of the people like only lassi. So, the number of persons = 18

25. **(a)** $n(A\cup B\cup C)=n(A)+n(B)+n(C)-n(A\cap B)$
$-n(A\cap C)-n(B\cap C)+n(A\cap B\cap C)$
$\Rightarrow 100-18=42+68+51-30-28-36+x$
$\Rightarrow x=15$

(26-29) :

$n(C)=650, n(F)=480, n(H)=400$
$n(C\cap F)=300, n(F\cap H)=250, n(C\cap H)=240$
and $n(C\cup F\cup H)=(100-5)\%$ of $1000=950$
Since, $n(C\cup F\cup H)=n(C)+n(F)+n(H)-n(C\cap F)-n(F\cap H)-n(C\cap H)+n(C\cap F\cap H)$
$\Rightarrow n(C\cap F\cap H)=210$

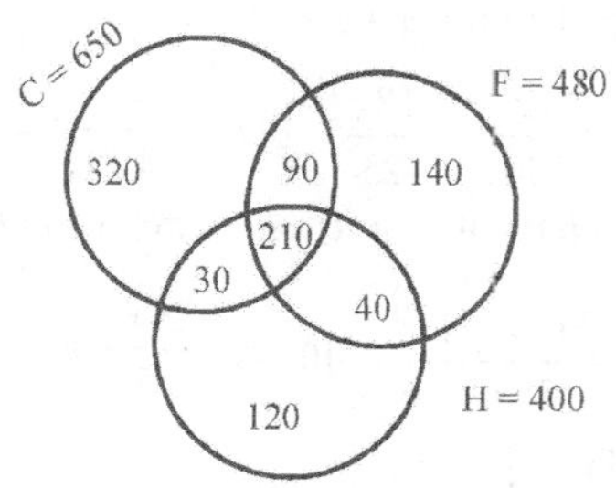

26. **(b)** **27.** **(d)** **28.** **(c)** **29.** **(b)**

CHAPTER 18

GEOMETRY

POINTS, LINES, LINE SEGMENT, RAY AND PLANE

Point: A point is like a dot marked by a very sharp pencil on a plane paper. A point is named by a capital letter like *P*. In the figure *P* is a point. Length, breadth and height of a point are negligible and hence cannot be measured.

Line: A line is defined as a group of points. Which are straight one after another. Each line is extended infinitely in two directions. Examples:

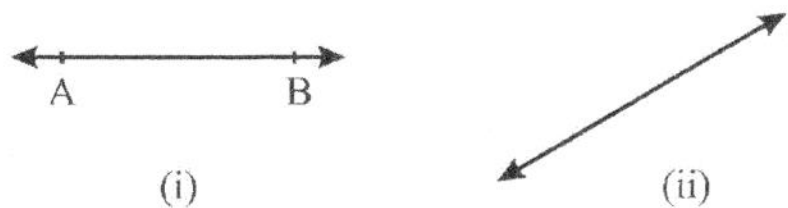

A line is named by either any two points on it or by a single small letter. In figure (i), *AB* is a line. In figure (ii), *l* is a line.

Arrows on both sides of a line indicate that the line is extended both sides infinitely. A line has only length. It does not have any width or height.

Line Segment: If a part of the line is cut out, then this cut out piece of the line is called a line segment. A line segment has no arrow at its any end.

This means that no line segment is extended infinitely in any direction.

Ray: A ray is a part of a line extended infinitely in any one direction only. Example:

A ray is named by two points, one of which is the end point on the ray called initial point and other point is any point on the ray.

In the figure, *AB* is a ray. The point *A* is called the initial point. Arrow of the ray indicates that the ray is extended infinitely towards arrow head.

Plane: It is a flat surface extended infinitely. It has only length and breadth but no thickness. Surface of a black board, surface of a wall, surface of a table are some examples of parts of planes because they are flat surfaces but not extended infinitely.

LINES AND ANGLES

Intersecting Lines: If two or more lines intersect each other, then they are called intersecting lines. In the figure *AB* and *CD* are intersecting lines.

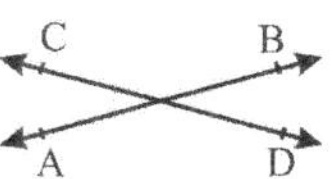

If two lines intersect at right angles, then two lines are called perpendicular lines

In the following figure *AB* and *CD* are perpendicular lines.

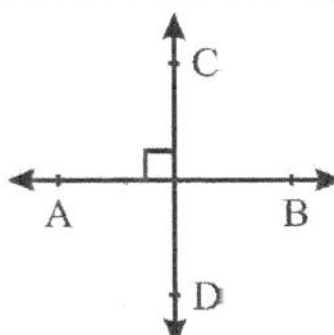

Symbolically it is represented as AB ⊥ BC or BC ⊥ AB.

Concurrent Lines: If three or more lines pass through a point, then they are called concurrent lines and the point through which these all lines pass is called point of concurrent.

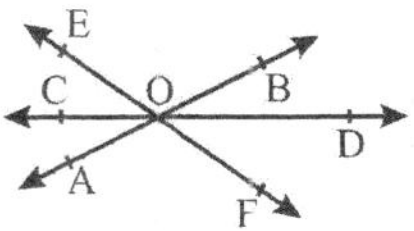

In the figure, *AB*, *CD* and *EF* are concurrent lines and point *O* is the point of concurrent.

Parallel Lines: Two straight lines are parallel if they lie in the same plane and do not intersect even if they produced.

Perpendicular distances between two parallel lines are the same at all places.

A B

C D

In the figure *AB* and *CD* are parallel lines.

Symbol for parallel lines is ||. Hence parallel lines *AB* and *CD* represented symbolically as *AB* || *CD*.

Transversal Line: A line which intersects two or more given lines at distinct points is called a transversal of the given lines.

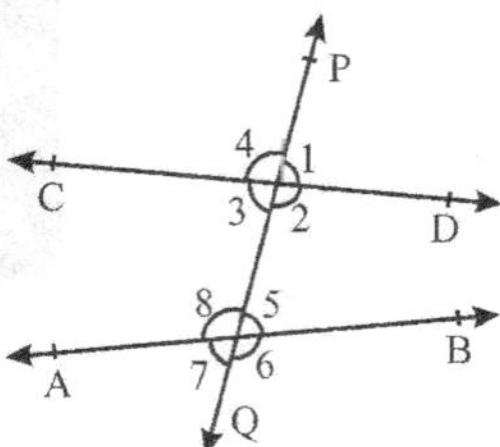

In figure straight lines AB and CD are intersected by a transversal PQ.

(i) **Corresponding angles:** In the figure $\angle 1$ and $\angle 5$, $\angle 4$ and $\angle 8$, $\angle 2$ and $\angle 6$, $\angle 3$ and $\angle 7$ are four pairs of corresponding angles.

(ii) **Alternate interior angles:** $\angle 3$ and $\angle 5$, $\angle 2$ and $\angle 8$, are two pairs of alternate interior angles.

(iii) **Alternate exterior angles:** $\angle 1$ and $\angle 7$, $\angle 4$ and $\angle 6$ are two pairs of alternate exterior angles.

(iv) **Consecutive interior angles:** In the figure, $\angle 2$ and $\angle 5$, $\angle 5$ and $\angle 8$, $\angle 8$ and $\angle 3$, $\angle 3$ and $\angle 2$ are four pairs of consecutive interior angles.

Interior angles on the same side of a transversal are called cointerior angles. In the fig. $\angle 2$ and $\angle 5$, $\angle 3$ and $\angle 8$ are two pairs of cointerior angles.

When a transversal intersects two parallel lines:

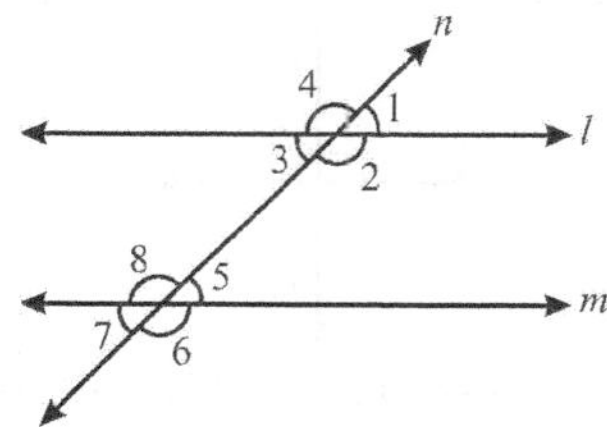

In the figure two parallel lines l and m are intersected by a transversal line n, then

(a) Two angles of each pair of corresponding angles are equal i.e. $\angle 1 = \angle 5$; $\angle 2 = \angle 6$; $\angle 4 = \angle 8$; $\angle 3 = \angle 7$

(b) Two angles of each pair of alternate interior angles are equal i.e.
$\angle 2 = \angle 8$; $\angle 3 = \angle 5$

(c) Two angles of each pair of alternate exterior angles are equal i.e.
$\angle 1 = \angle 7$; $\angle 4 = \angle 6$

(d) Any two consecutive interior angles are supplementary. i.e. their sum is 180°. Hence
$\angle 2 + \angle 5 = 180°$; $\angle 5 + \angle 8 = 180°$; $\angle 8 + \angle 3 = 180°$; $\angle 3 + \angle = 180°$

Note that

(i) If two angles of any pair of corresponding angles are equal, then the two lines are parallel.

(ii) If two angles of any pair of alternate interior angles are equal, then the two lines are parallel.

(iii) If two angles of any pair of alternate exterior angles are equal, then the two lines are parallel.

(iv) If any two consecutive interior angles are supplementary (i.e. their sum is 180°), then the two lines are parallel.

Acute angle: An angle is said to be acute angle if it is less than 90°.

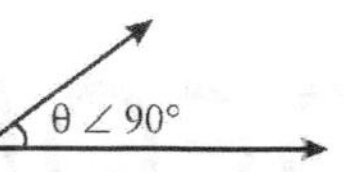

Here $0° \angle \theta \angle 90°$, hence θ is acute angle.

Right angle: An angle is said to be right angle if it is of 90°.

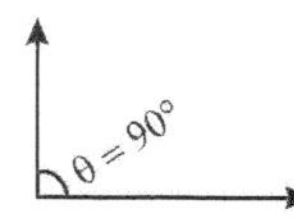

Here θ is right angle.

Obtuse angle: An angle is said to be obtuse angle if it is of more than 90°.

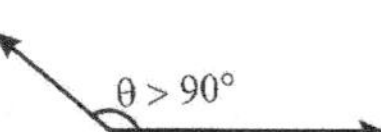

Here θ is obtuse angle.

Straight angle: An angle is said to be straight angle if it is of 180°.

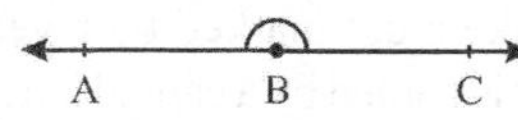

Here θ is a straight angle.

Reflex angle: An angle is said to be reflex angle if it is of greater than 180°.

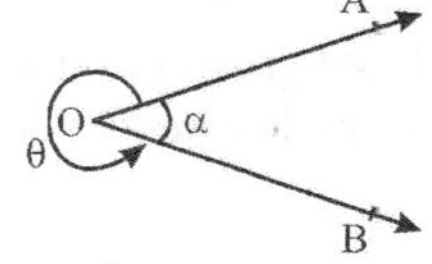

Here θ is the reflex angle.

Reflex angle θ is written as

$$\theta = 360° - \angle AOB \text{ (or } 360° - \alpha)$$

Here $\angle AOB$ or α is less than 180°

Complementary angles: Two angles, the sum of whose measures is 90°, are called the complementary angles.

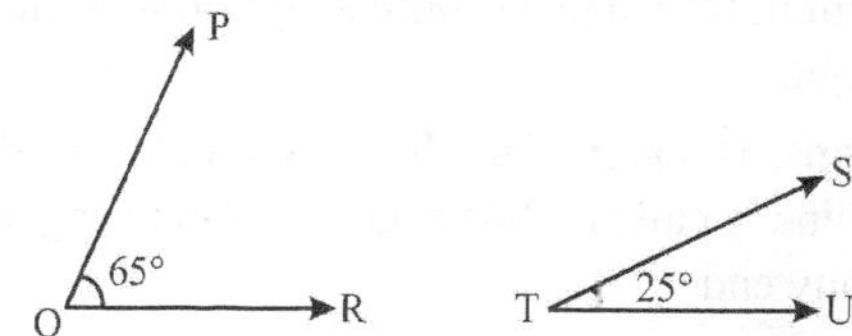

$\angle PQR$ and $\angle STU$ are complementary angles.

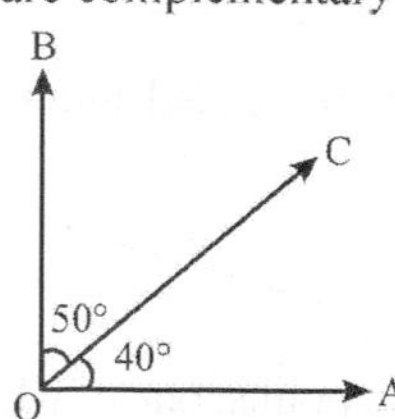

In figure $\angle AOC$ and $\angle BOC$ are also complementary angles.

Supplementary angles: Two angles, the sum of whose measures is 180°, are called the supplementary angles.

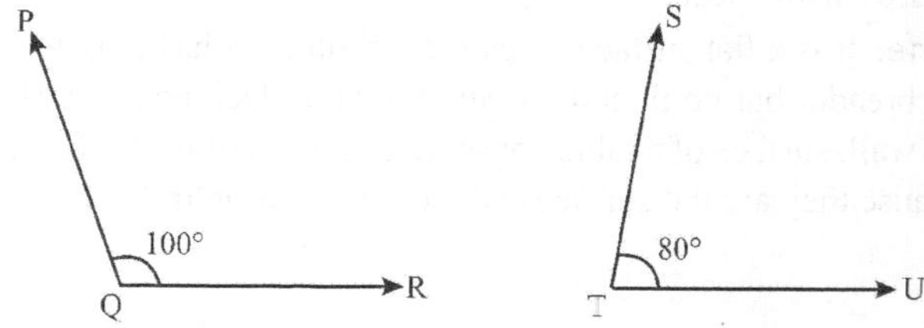

In figure, $\angle PQR$ and $\angle STU$ are supplementary angles.

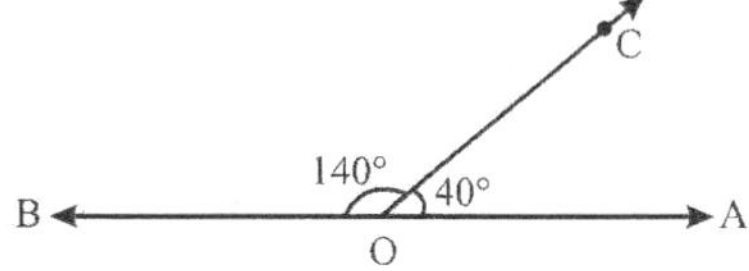

In figure, ∠AOC and ∠BOC are also supplementary angles.

Adjacent angles: Two angles are called adjacent angles, if

(i) they have the same vertex
(ii) they have a common arm and
(iii) non-common arms are on either side of the common arm.

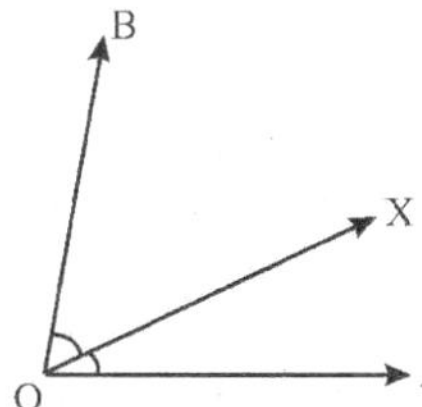

In figure, $\angle AOX$ and $\angle BOX$ are adjacent angles because O is the common vertex, OX is common arm, non-common arm OA and OB are on either side of OX.

Linear pair of angles: Two adjacent angles are said to form a linear pair of angles, if their non common arms are two opposite rays. In other words if the sum of two adjacent angles is 180°, then they are said to form a linear pair of angles.

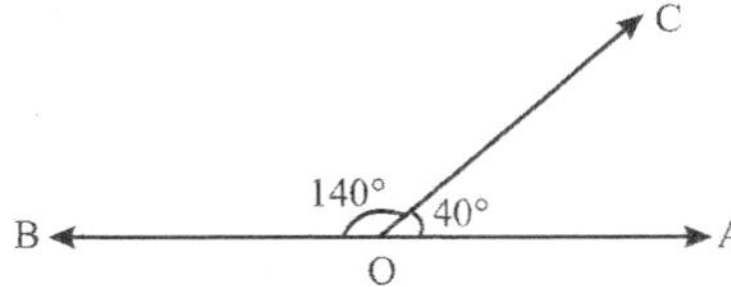

In figure, $\angle AOC$ and $\angle BOC$ are linear pair angles.

Vertically opposite angles: Two angles are called a pair of vertically opposite angles, if their arms form two intersecting lines.

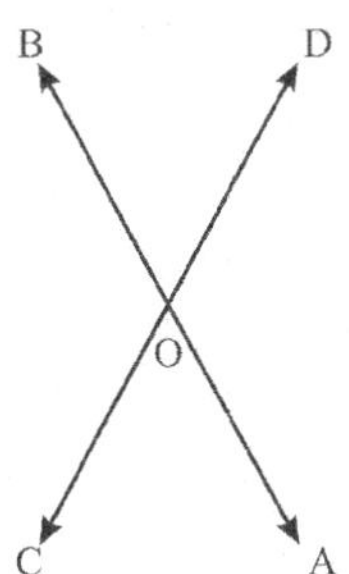

In figure, $\angle AOC$ and $\angle BOD$ form a pair of vertically opposite angles. Also $\angle AOD$ and $\angle BOC$ from a pair of vertically opposite angles.

Angles on one side of a line at a point on the line: Sum of all the angles on any one side of a line at a point on the line is always 180°.

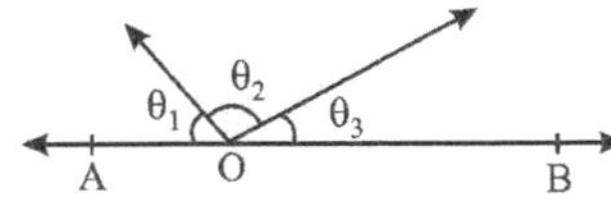

Here AOB is a straight line, hence in figure, $\theta_1 + \theta_2 + \theta_3 = 180°$.

Angle around a point: Sum of all the angles around a point is always 360°.

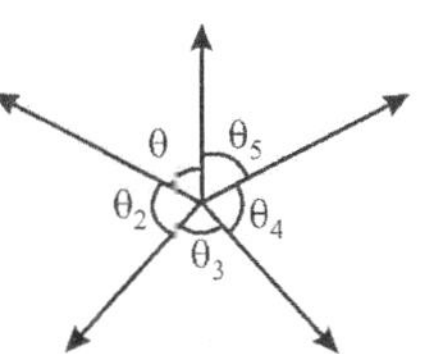

Here θ_1, θ_2, θ_3, θ_4 and θ_5 are the angles around a point. Hence $\theta_1 + \theta_2 + \theta_3 + \theta_4 + \theta_5 = 360°$

Angle bisector: An angle bisector is a ray which bisects the angle whose initial point be the vertex of the angle.

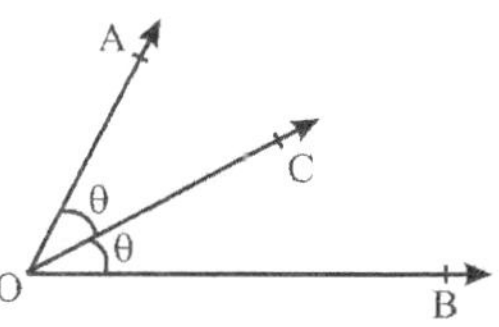

Since $\angle AOC = \angle BOC = \theta$

Hence ray OC is the bisector of $\angle AOB$.

Illustration 1: Three straight lines, *X*, *Y* and *Z* are parallel and the angles are as shown in the figure above. What is ΔAFB equal to?

(a) 20° **(b) 15°**
(c) 30° **(d) 10°**

Solution: (b) $\angle CDE = 180° - 125° = 55°$

In ΔDCE,

$\angle CED = 180° - 55° - 80° = 45°$

and $\angle ABF = 30°$ (vertically opposite)

Also, $\angle ABF = \angle BFM = 30°$ (alternate angle)

and, $\angle DEF = \angle EFM$ (alternate angle)

$\angle EFM = 45°$

$\Rightarrow \angle EFB + \angle BFM = 45° \Rightarrow \angle EFB = 45° - 30°$

$\Rightarrow \angle AFB = 15°$

Illustration 2: In figure, if $AB \parallel CD$, $CD \parallel EF$ and $y : z = 3 : 7$, $x = ?$

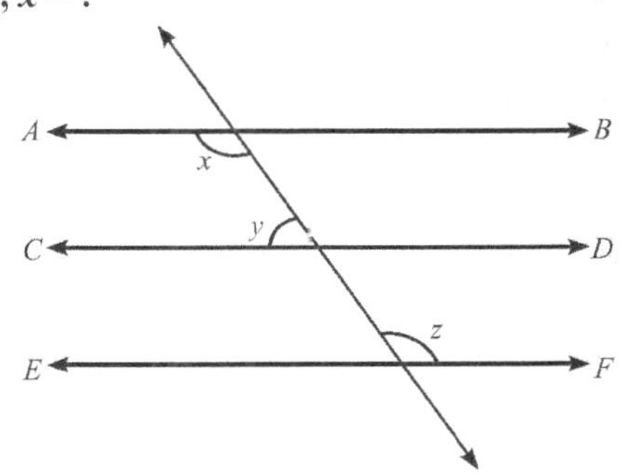

(a) 112° **(b) 116°**
(c) 96° **(d) 126°**

Solution: (d) As $y + z = 180°$, $\therefore y = 54°$

$$x + y = 180°$$
$$x = 180 - 54 = 126°$$

Illustration 3: In the ΔPQR, PS is the bisector of $\angle P$ and $PT \perp QR$, then $\angle TPS$ is equal to

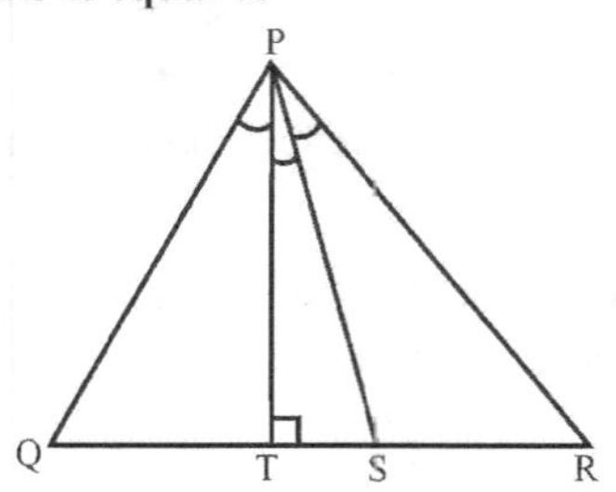

(a) $\angle Q + \angle R$ (b) $90° + \frac{1}{2}\angle Q$

(c) $90° - \frac{1}{2}\angle R$ (d) $\frac{1}{2}(\angle Q - \angle R)$

Solution: (d) PS is the bisector of $\angle QPR$

$\therefore$ $\angle 1 + \angle 2 = \angle 3$ ...(1)

$\Rightarrow$ $\angle Q = 90° - \angle 1$

$\angle R = 90° - \angle 2 - \angle 3$

So, $\angle Q - \angle R = [90° - \angle 1] - [90° - \angle 2 - \angle 3]$

$\Rightarrow$ $\angle Q - \angle R = \angle 2 + \angle 3 - \angle 1$

$= \angle 2 + (\angle 1 + \angle 2) - \angle 1$ [From Eq. (1)]

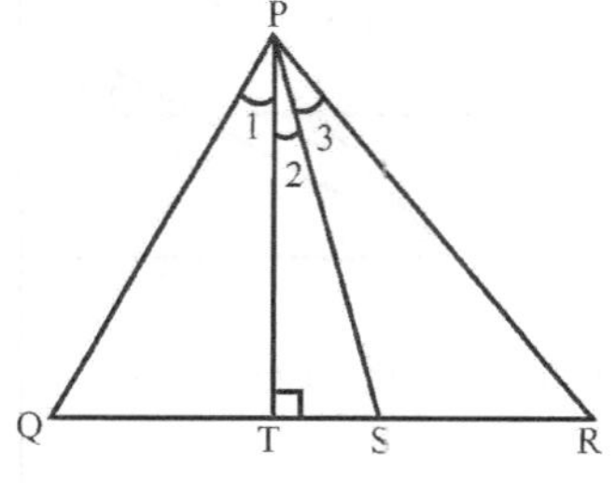

$\Rightarrow$ $\angle Q - \angle R = 2\angle 2 \Rightarrow \frac{1}{2}(\angle Q - \angle R) = \angle TPS$

POLYGONS

Polygons are closed plane figures formed by series of line segments, e.g. triangles, rectangles, etc.

Polygons can also be classified into convex and concave polygons.

A convex polygon is a polygon in which any line segment joining any two points of the polygon always lies completely inside the polygon, otherwise the polygon is concave polygon.

ABCDE is a convex polygon because any line segment joining any two points of the polygon completely lies inside the polygon.

FGHIJK is a concave polygon because line segment joining two points *R* and *S* of the polygon does not lie completely inside the polygon.

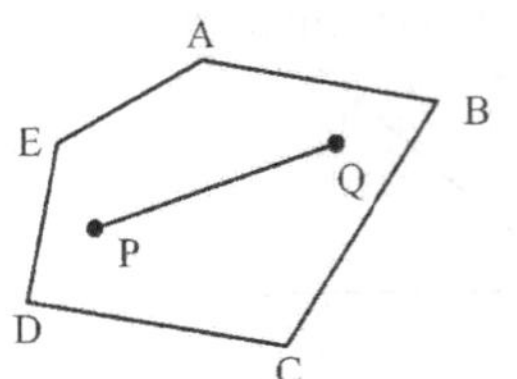

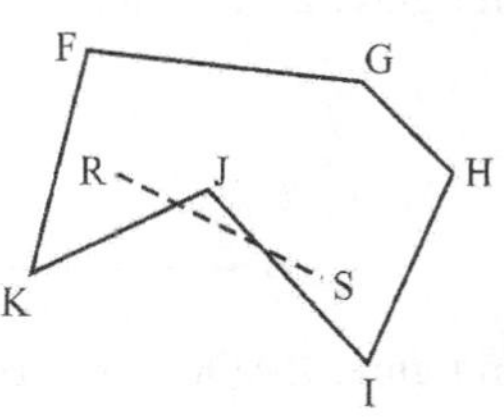

Convex polygons can be classified into regular and irregular polygons.

(a) Regular polygon: A convex polygon whose all the sides are equal and also all the angles equal is called a regular polygon.
A regular polygon is simply called polygon.

(b) Irregular Polygon: A convex polygon in which all the sides are not equal or all the angles are not of the same measure is called an irregular polygon.

Polygons can also be divided on the basis of number of sides they have

No. of sides of the polygon	Name of the polygon
3	Triangle
4	Quadrilateral
5	Pentagon
6	Hexagon
7	Heptagon
8	Octagon
9	Nonagon
10	Decagon
:	:
etc.	etc.

Interior and Exterior Angles of a Polygon

An angle inside a polygon between any two adjacent sides at a vertex of the polygon is called an interior angle of the polygon. An angle outside a polygon made by a side of the polygon with the its adjacent side produced is called an exterior angle of the polygon.

In the figure *ABCDEF* is a polygon.

$\angle FAB$, $\angle ABC$, $\angle BCD$, $\angle CDE$, $\angle DEF$ and $\angle EFA$ are interior angles of the polygon *ABCDEF*.

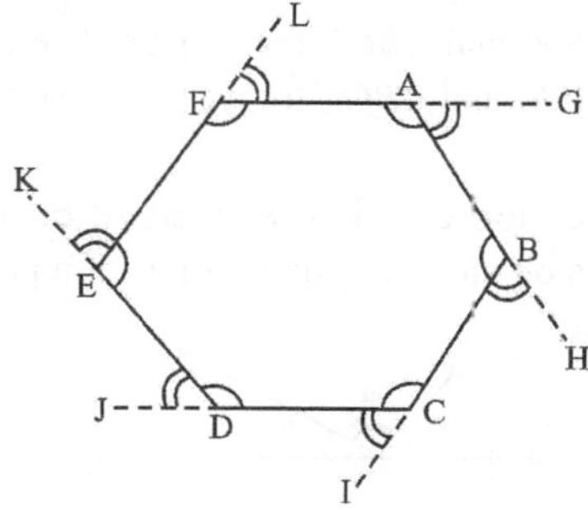

$\angle BAG$, $\angle CBH$, $\angle DCI$, $\angle EDJ$, $\angle FEK$ and $\angle AFL$ are exterior angles of the polygon *ABCDEF*.

Diagonals of a Polygon

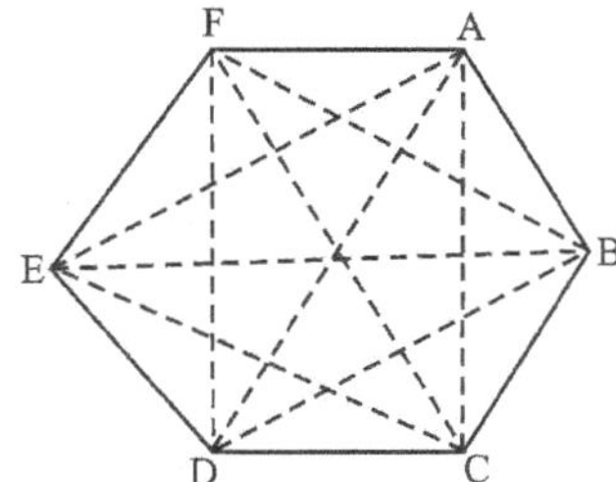

A diagonal of a polygon is a line segment connecting two non-consecutive vertices of the Polygon.

In the figure, diagonals are drawn by dotted line segments.

Properties of Polygons

(i) Sum of all the interior angles of a polygon with 'n' sides $= (n-2)\,180°$

(ii) Sum of all the exterior angles of a polygon = 360°

$\angle 1 + \angle 2 + \angle 3 + \angle 4 + \angle 5 + \angle 6 = 360°$

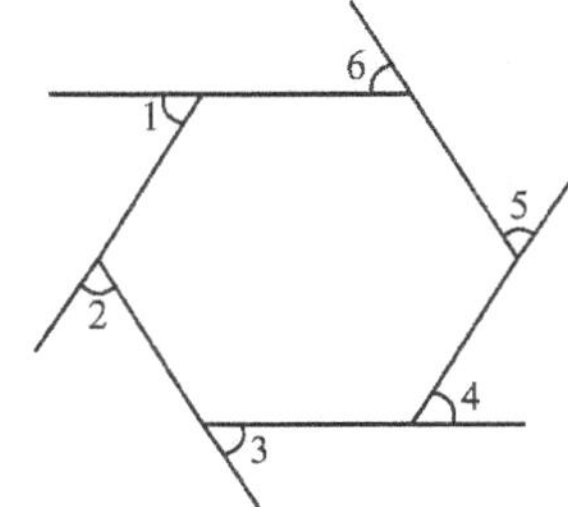

(iii) Perimeter of a regular polygon with a side length of $a = n \times a$

(iv) No. of sides of a regular polygon = an exterior

$$= \frac{360°}{\text{An exterior angle}}$$

(v) Number of diagonals of a polygon with n sides

$$= \frac{n\,(n-3)}{2}$$

Illustration 4: An interior angle of a regular polygon is 135°. Find the number of sides of the polygon.

Solution: Since interior angle of the regular polygon = 135°, hence exterior angle = 180° – 135° = 45°

$$\therefore \quad \text{No. of sides} = \frac{360°}{\text{An exterior angle}} = \frac{360°}{45°} = 8$$

$\therefore$ No. of sides = 8

Illustration 5: An interior angle of a regular polygon is 100° more than its an exterior angle. Find the number of sides the polygon.

Solution: Let measure of each exterior angle be x°.

Then measure of each interior angle $= (x + 100)$

Now $x + (x + 100) = 180$

$\Rightarrow \quad 2x = 80 \Rightarrow x = 40$

$$\text{Now} \quad \text{number of sides} = \frac{360}{\text{An exterior angle}} = \frac{360}{40} = 9.$$

TRIANGLES

A triangle is a convex polygon having three sides.

A triangle is represented by the symbol Δ.

Triangles can be classified on the basis of their sides or angles.

On the basis of sides, triangles are of the following types

(a) **Equilateral triangle:** All the three sides are equal

(b) **Isosceles triangle:** Two sides are equal

(c) **Scalene triangle:** All the three sides are unequal.

On the basis of angles, triangles are of the following types

(a) **Acute angled triangle:** Each interior angle is less than 90°.

(b) **Right angled triangle:** One of the interior angle is equal 90°.

(c) **Obtuse angled triangle:** One of the interior angle is more than 90°.

BASIC PROPERTIES AND SOME IMPORTANT THEOREMS OF TRIANGLES

1. Sum of measures of the interior angles of a triangle is 180°.

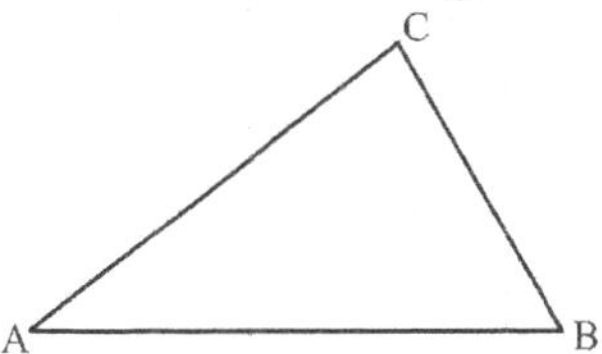

In ΔABC, $\angle CAB + \angle ABC + \angle ACB = 180°$

or $\angle A + \angle B + \angle C = 180°$

2. The exterior angle of a triangle is equal to the sum of the opposite (not adjacent) interior angles

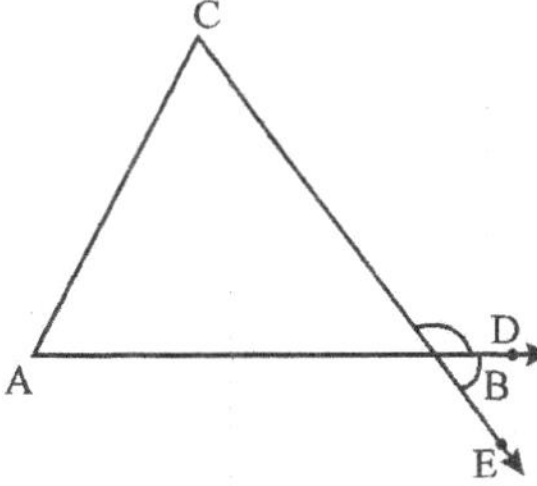

In ΔABC, $\angle CBD = \angle A + \angle C = \angle ABE$

3. Sum of the lengths of any two sides of a triangle is greater than the length of the third side.

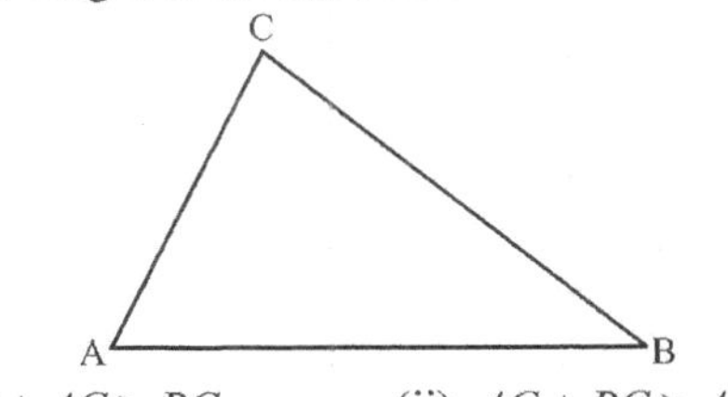

(i) $AB + AC > BC$ (ii) $AC + BC > AB$

(iii) $AB + BC > AC$

4. Difference between the lengths of any two sides of a triangle is smaller than the length of the third side.

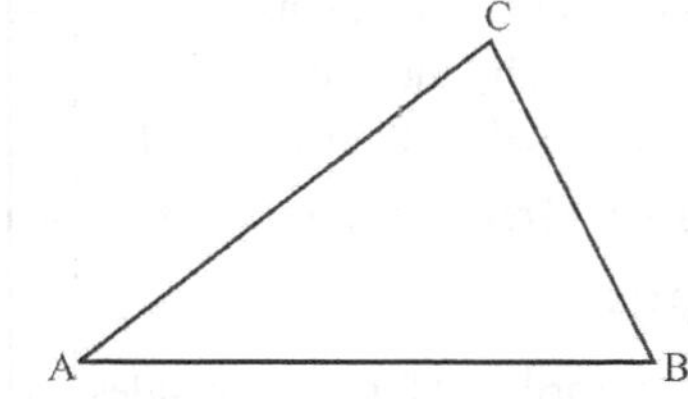

(i) $|AB - BC| < AC$ (ii) $|AC - AB| < BC$
(iii) $|AC - BC| < AB$

5. In any triangle, side opposite to greatest angle is largest and side opposite to smallest angle is smallest.

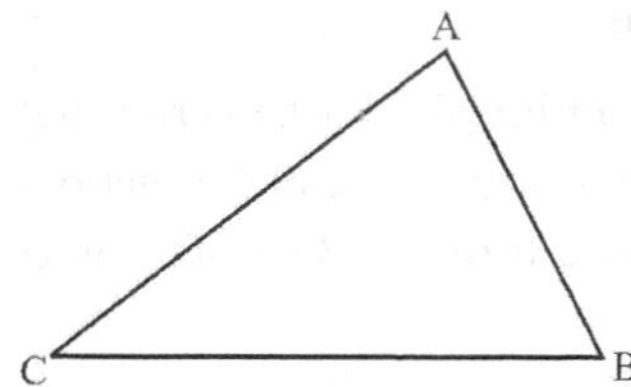

In ΔABC, if $\angle A > \angle B > \angle C$, then BC is the largest side and AB is the smallest side.

6. In any triangle line joining any vertex to the mid point of its opposite side is called a median of the opposite side of the triangle.

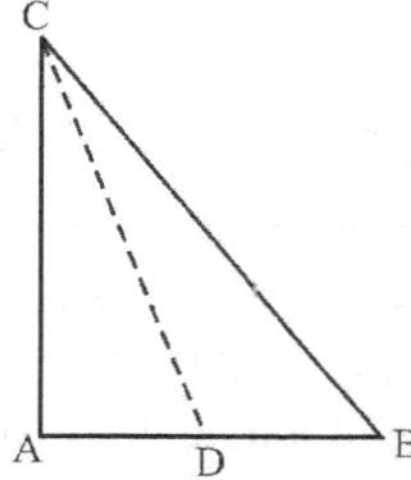

In $\Delta ABCD$ is the mid point of AB
Hence CD is a median of ΔABC.
A triangle can have 3 medians.
Any median of a triangle divides the triangle into two triangles of equal areas.

7. Sides opposite to equal angles in a triangle are equal.

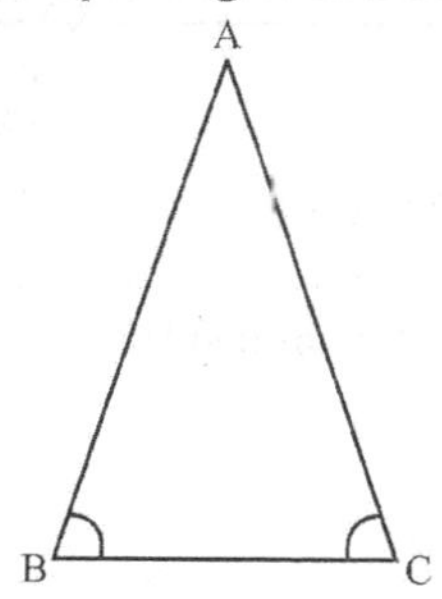

In ΔABC, $\angle B = \angle C$
$\therefore$ $AB = AC$
Converse of this property is also true.

8. In an isosceles triangle, if a perpendicular is drawn to unequal side from its opposite vertex, then
(a) The perpendicular is the median
(b) The perpendicular bisects the vertex angle.

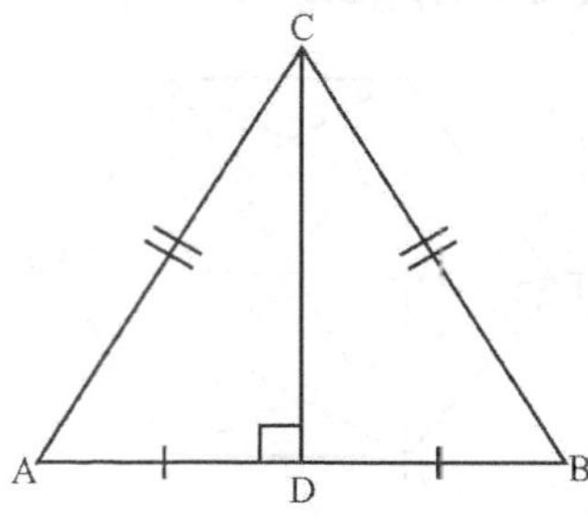

ΔABC is an isosceles triangle in which $AC = BC$.
CD is perpendicular to AB, hence CD is a median and $\angle ACD = \angle BCD$

9. In a right angled triangle, the line joining the vertex of the right angle to the mid point of the hypotenuse is half the length of the hypotenuse.

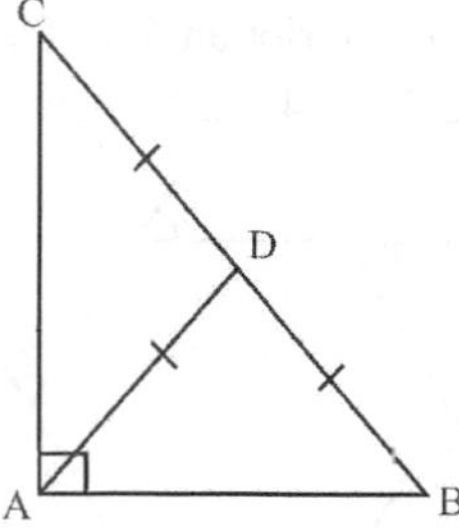

In ΔABC, $\angle BAC = 90°$ and D is the mid point of BC, then

$$AD = \frac{1}{2}BC = BD = CD$$

10. **Mid-point theorem:** In any triangle, line segment joining the mid points of any two sides is parallel to the third side and equal to half of the length of third side.

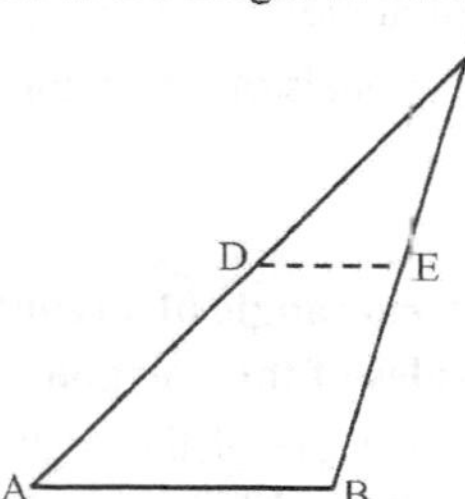

In ΔABC, D and E are mid points of sides AC and BC, then

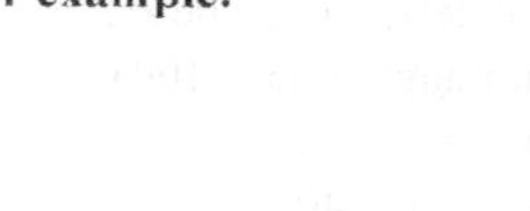

DE is parallel to AB i.e. $DE \parallel AB$ and $DE = \frac{1}{2} AB$

11. **Angle Bisector Theorem:** Bisector of an angle (internal or external) of a triangle divides the opposite side (internally or externally) in the ratio of the sides containing the angle.
For example:

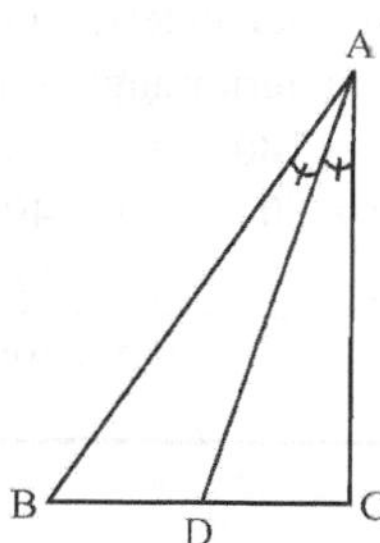

In figure AD is the bisector of exterior $\angle BAC$

$\therefore \quad \frac{AB}{AC} = \frac{BD}{DC}$

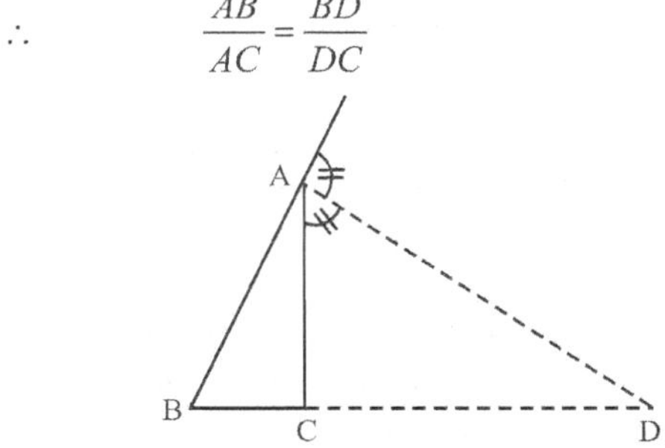

In figure AD is the bisector of exterior $\angle BAC$.

$\therefore \quad \frac{AB}{AC} = \frac{BD}{DC}$

Converse of the angle bisector theorem is also true.

12. **Pythagoras Theorem:** In a right angled triangle.
Square of longest or hypotenuse = Sum of square of other two sides.

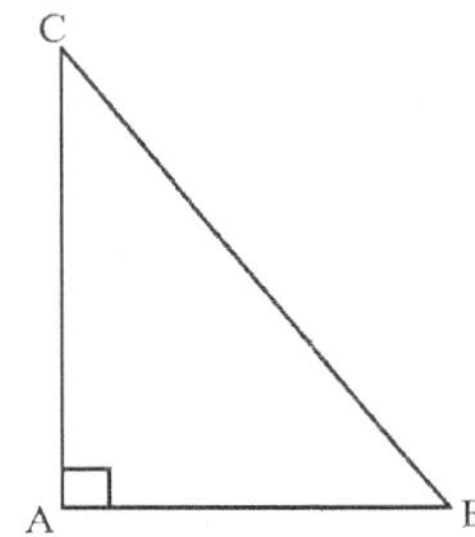

In figure ΔABC is a triangle right angled at A.

$\therefore \quad (BC)^2 = (AB)^2 + (AC)^2$

Converse of this theorem is also true.

13. **Basic Proportionality Theorem (BPT):** If a line is drawn parallel to one side of a triangle which intersects the other two sides in distinct points, the other two sides are divided in the same ratio.

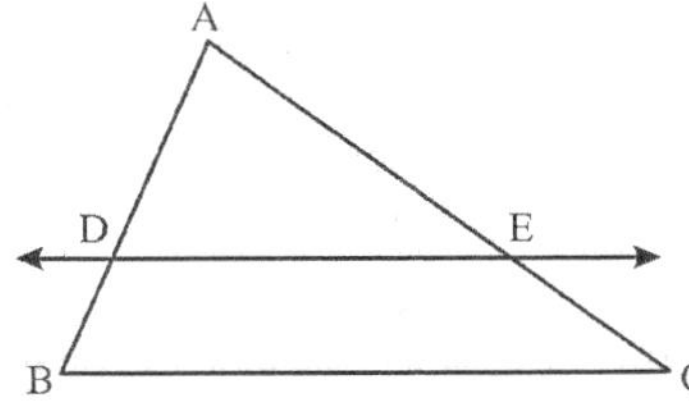

In ΔABC, $DE \parallel BC$,

Then, $\frac{AD}{DB} = \frac{AE}{EC}$

This theorem is also known as Thalse theorem.
Converse of this theorem is also true.

Illustration 6: In a triangle ABC, $\angle A = x$, $\angle B = y$, and $\angle C = y + 20$.

If $4x - y = 10$, then the triangle is :

(a) Right-angled **(b) Obtuse-angled**
(c) Equilateral **(d) None of these**

Solution: (a) We have, $x + y + (y + 20) = 180$

or $\quad x + 2y = 160 \quad$ ...(1)

and $\quad 4x - y = 10 \quad$ ...(2)

From (1) and (2), $y = 70$, $x = 20$

Angles of the triangles are 20°, 70°, 90°. Hence the triangle is a right angled.

Illustration 7: In the given figure, $CD \parallel AB$. Find y.

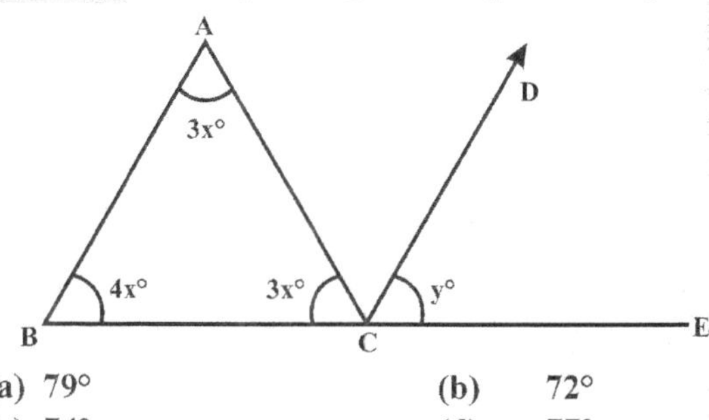

(a) 79° **(b) 72°**
(c) 74° **(d) 77°**

Solution: (b) In ΔABC,

$\angle ABC + \angle BCA + \angle CAB = 180°$

$\Rightarrow \quad 4x + 3x + 3x = 180° \Rightarrow 10°x = 180° \Rightarrow x = 18°$

Now, $\quad \angle ABC = \angle DCE$

(corresponding angles are equal)

$\Rightarrow \quad \angle DCE = 4x° \Rightarrow y = 4 \times 18° = 72°$

Illustration 8: In the adjoining figure, AE is the bisector of exterior $\angle CAD$ meeting BC produced in E. If $AB = 10$ cm, $AC = 6$ cm and $BC = 12$ cm, then CE is equal to

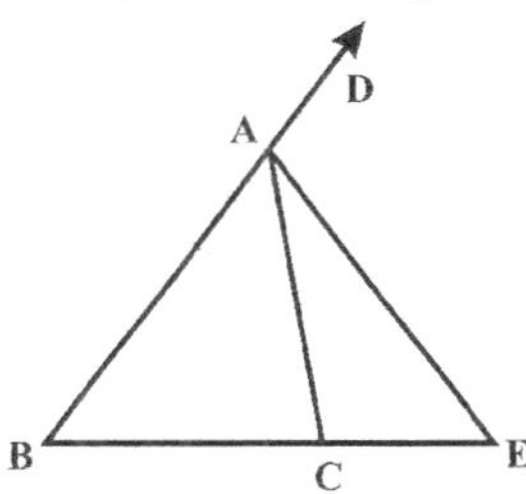

(a) 6 cm **(b) 12 cm**
(c) 18 cm **(d) 20 cm**

Solution: (c) $\frac{BE}{CE} = \frac{AB}{AC}$ as AE is an exterior angle bisector.

Let $\quad CE = x$, $BE = BC + EC = 12 + x$

$\Rightarrow \quad \frac{12 + x}{x} = \frac{10}{6} \quad \Rightarrow (12 + x)\, 6 = 10x$

$\Rightarrow \quad 72 + 6x = 10x \Rightarrow 4x = 72 \Rightarrow x = 18$ cm

Illustration 9: OB and OC are respectively the bisectors of $\angle ABC$ and $\angle ACB$. Then, $\angle BOC$ is equal to

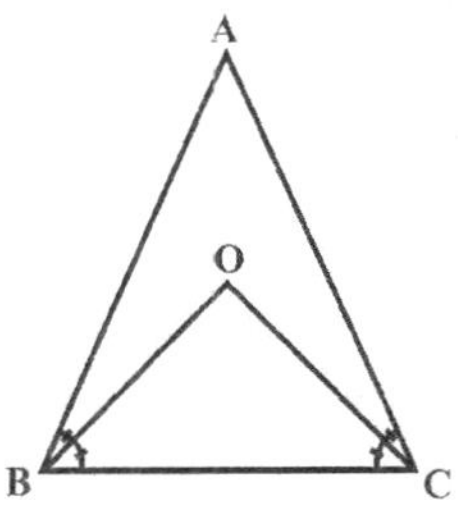

(a) $90° - \frac{1}{2}\angle A$ **(b)** $90° + \angle A$

(c) $90° + \frac{1}{2}\angle A$ **(d)** $180° - \frac{1}{2}\angle A$

Solution: (c) In ΔBOC,

$\angle 1 + \angle 2 + \angle BOC = 180°$

$\angle A + \angle B + \angle C = 180°. \quad$ .... (1)

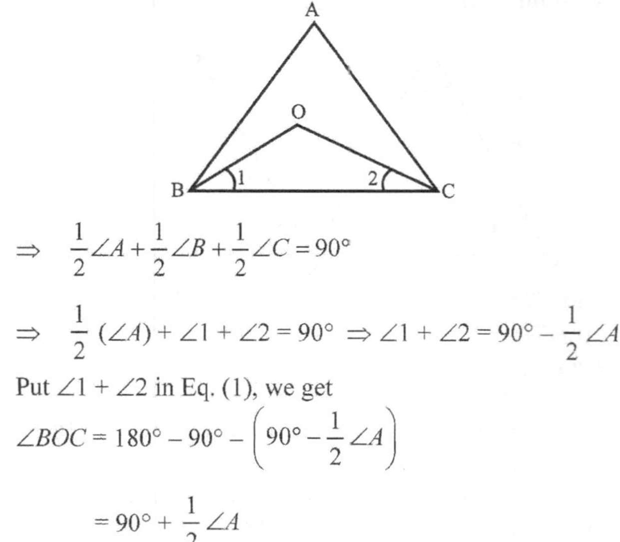

$$\Rightarrow \frac{1}{2}\angle A + \frac{1}{2}\angle B + \frac{1}{2}\angle C = 90°$$

$$\Rightarrow \frac{1}{2}(\angle A) + \angle 1 + \angle 2 = 90° \Rightarrow \angle 1 + \angle 2 = 90° - \frac{1}{2}\angle A$$

Put $\angle 1 + \angle 2$ in Eq. (1), we get

$$\angle BOC = 180° - 90° - \left(90° - \frac{1}{2}\angle A\right)$$

$$= 90° + \frac{1}{2}\angle A$$

IMPORTANT TERMS RELATED TO A TRIANGLE

1. **Medians and Centroid:** We know that a line segment joining the mid point of a side of a triangle to its opposite vertex is called a median.

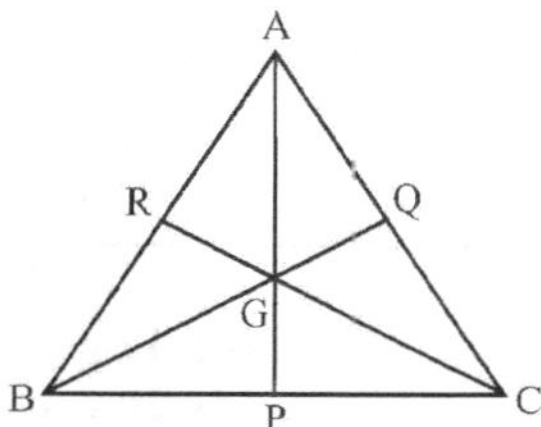

AP, BQ and CR are medians of ΔABC where P, Q and R are mid points of sides BC, CA and AB respectively.

(i) Three medians of a triangle on concurrent. The point of concurrent of three medians is called Centroid of the triangle denoted by G.

(ii) Centroid of the triangle divides each median in the ratio 2 : 1

i.e. $AG : GP = BG : GQ = CG : GR = 2 : 1$, where G is the centroid of ΔABC.

2. **Altitudes and Orthocentre:** A perpendicular drawn from any vertex of a triangle to its opposite side is called altitude of the triangle. There are three altitudes of a triangle.

In the figure, AP, BQ and CR are altitudes of ΔABC.

The altitudes of a triangle are concurrent (meet at a point) and the point of concurrency of altitudes is called Orthocentre of the triangle, denoted by O.

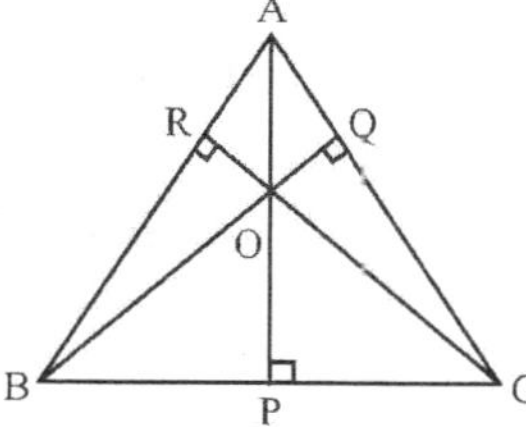

In figure, AP, BQ and CR meet at O, hence O is the orthocentre of the triangle ABC.

Note: The angle made by any side at the orthocentre and at the vertex opposite to the side are supplementary angle.

Hence, $\angle BAC + \angle BOC = \angle ABC + \angle AOC = \angle ACB + \angle AOB = 180°$.

3. **Perpendicular Bisectors and Circumcentre:** A line which is perpendicular to a side of a triangle and also bisects the side is called a perpendicular bisector of the side.

(i) Perpendicular bisectors of sides of a triangle are concurrent and the point of concurrency is called circumcentre of the triangle, denoted by 'C'.

(ii) The circumcentre of a triangle is centre of the circle that circumscribes the triangle.

(iii) Angle formed by any side of the triangle at the circumcentre is twice the vertical angle opposite to the side.

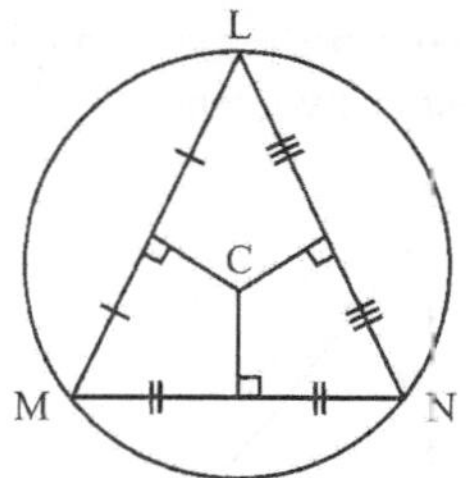

In figure, perpendicular bisectors of sides LM, MN and NL of ΔLMN meets at C. Hence C is the circumcentre of the triangle LMN.

$$\angle MCN = 2\ \angle MLN.$$

4. **Angle Bisectors and Incentre:** Lines bisecting the interior angles of a triangle are called angle bisectors of triangle.

(i) Angle bisectors of a triangle are concurrent and the point of concurrency is called Incentre of the triangle, denoted by I.

(ii) With I as centre and radius equal to length of the perpendicular drawn from I to any side, a circle can be drawn touching the three sides of the triangle. So this is called incircle of the triangle. Incentre is equidistant from all the sides of the triangle.

(iii) Angle formed by any side at the incentre is always 90° more than half the vertex angle opposite to the side.

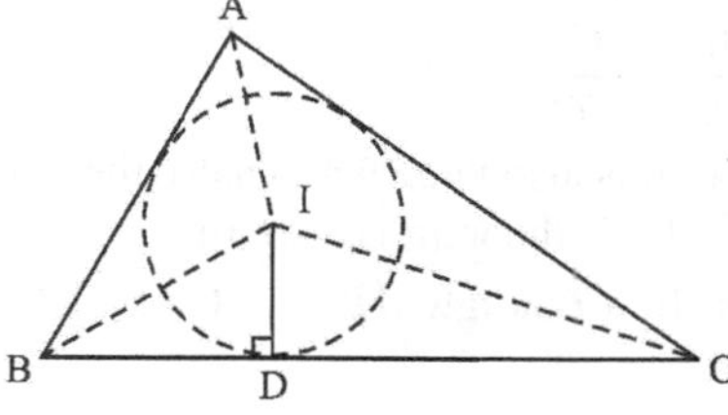

In figure AI, BI, CI are angle bisectors of ΔABC.

Hence I is the incentre of the ΔABC and

$$\angle BIC = 90° + \frac{1}{2}\angle A,\ \angle AIC = 90° + \frac{1}{2}\angle B$$

and $$\angle AIB = 90° + \frac{1}{2}\angle C$$

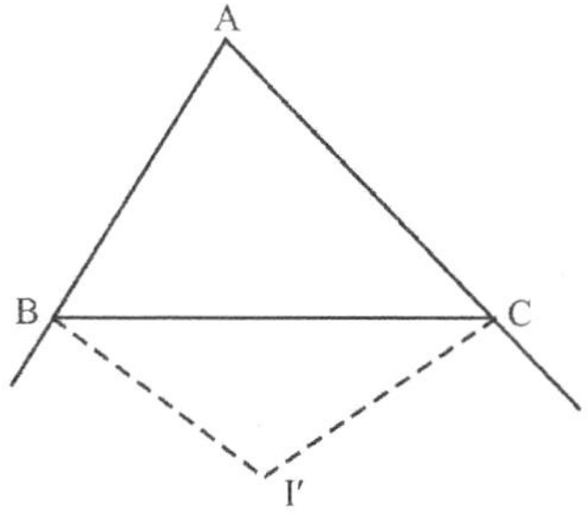

If BI' and CI' be the angle bisectors of exterior angles at B and C, then

$$\angle BI'C = 90° - \frac{1}{2}\angle A.$$

Illustration 10: If in the given figure $\angle PQR = 90°$, O is the centroid of ΔPQR, $PQ = 5$ cm and $QR = 12$ cm, then OQ is equal to

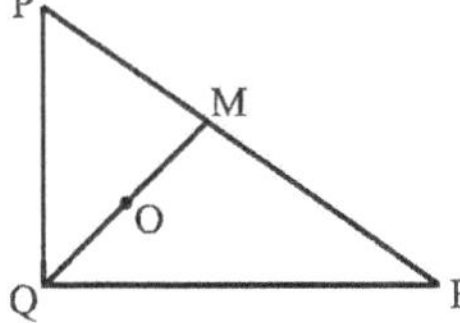

(a) $3\frac{1}{2}$ **cm** **(b)** $4\frac{1}{3}$ **cm**

(c) $4\frac{1}{2}$ **cm** **(d)** $5\frac{1}{3}$ **cm**

Solution: (b) By Pythagoras theorem,

$$PR = \sqrt{PQ^2 + QR^2} = \sqrt{5^2 + 12^2} = 13 \text{ cm}$$

$\therefore$ O is centroid $\Rightarrow$ QM is median and M is mid-point of PR.

$$QM = PM = \frac{13}{2}$$

$\therefore$ Centroid divides median in ratio 2 : 1.

$$\therefore OQ = \frac{2}{3}QM = \frac{2}{3} \times \frac{13}{2} = \frac{13}{3} \quad \therefore OQ = 4\frac{1}{3} \text{ cm}$$

CONGRUENCY OF TWO TRIANGLES

Two triangles are congruent if they are of the same shape and size i.e .if any one of them can be made to superpose on the other it will cover exactly.

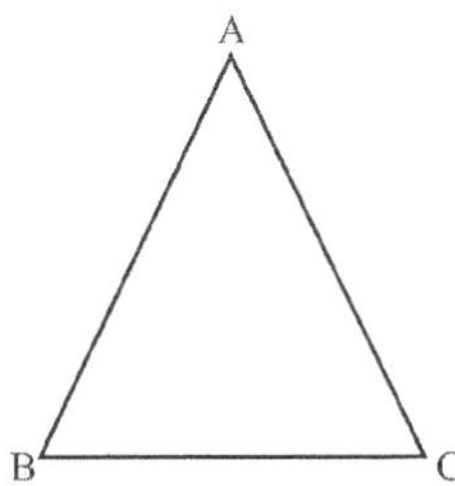

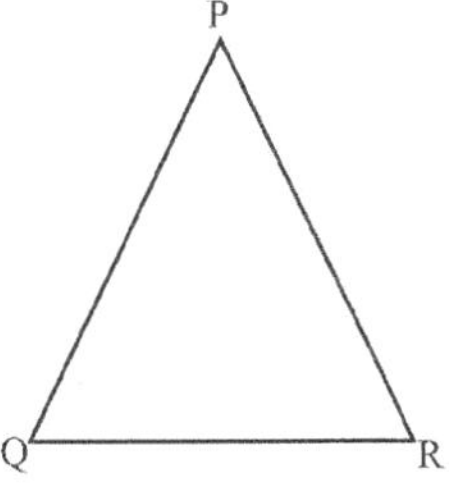

If two triangles ABC and PQR are congruent then 6 elements (i.e. three sides and three angles) of one triangle are equal to corresponding 6 elements of other triangle.

(i) $\angle A = \angle P$, $\angle B = \angle Q$, $\angle C = \angle R$

(ii) $AB = PQ$, $BC = QR$, $AC = PR$

This is symbolically written as $\Delta ABC \cong \Delta PQR$

Note: In two congruent triangles, sides opposite to equal angles are corresponding sides and angles opposite to equal sides are corresponding angles.

Conditions of Congruency

There are 4 conditions of congruency of two triangles.

1. **SAS (Side-Angle-Side) Congruency:** If two sides and the included angle between these two sides of one triangle is equal to corresponding two sides and included angle between these two sides of another triangle, then the two triangles are congruent.

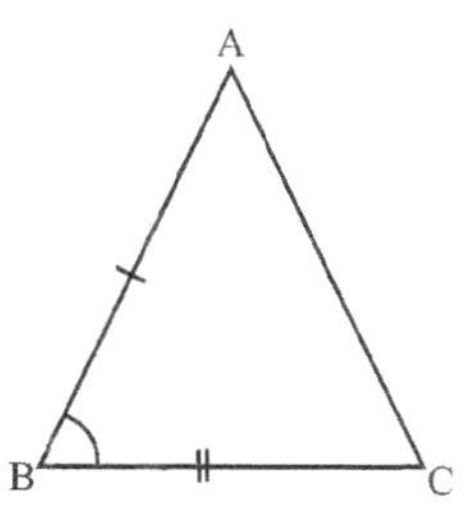

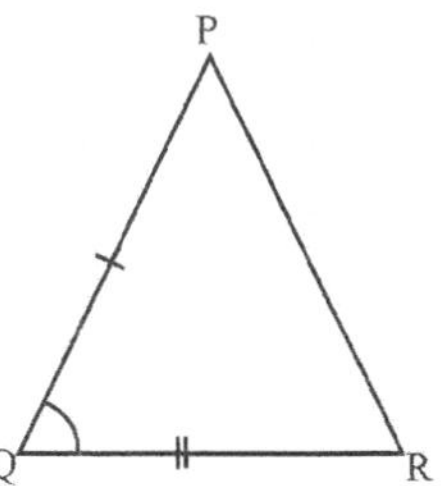

In ΔABC and ΔPQR

$$AB = PQ,$$
$$BC = QR$$
and $$\angle ABC = \angle PQR$$
$$\therefore \quad \Delta ABC \cong \Delta PQR \quad \text{[by SAS congruency]}$$

Here $\cong$ is the sign of congruency.

2. **ASA (Angle-Side-Angle) Congruency:** If two angles and included side between these two angles of one triangle are equal to corresponding angles and included side between these two angles of another triangle, then two triangles are congruent.

In ΔABC and ΔPQR

$$\angle A = \angle P$$
$$\angle B = \angle Q$$
$$AB = PQ$$
$$\therefore \quad \Delta ABC \cong \Delta PQR \quad \text{[by ASA congruency]}$$

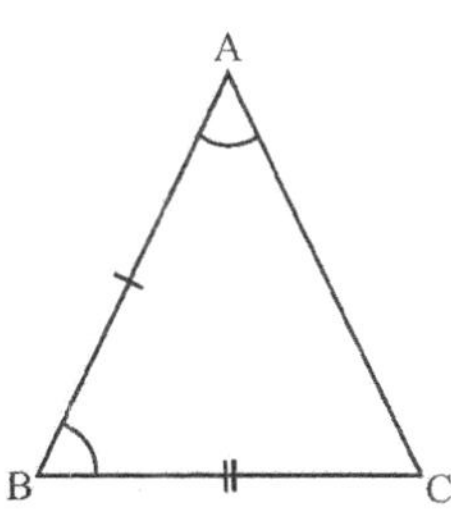

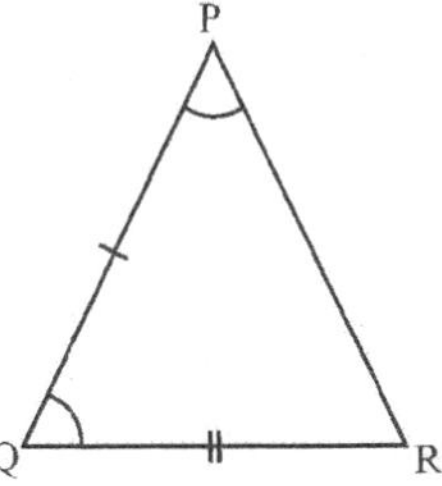

3. **SSS (Side-Side-Side) Congruency:** If three sides of one triangle are equal to corresponding three sides of another triangle, the two triangles are congruent.

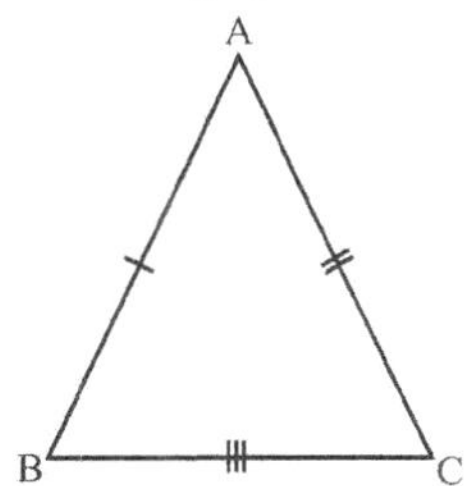

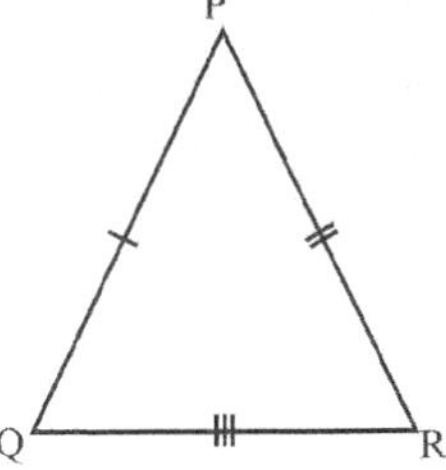

In ΔABC and ΔPQR

$$AB = PQ$$
$$BC = QR$$
$$CA = RP$$

$\therefore \quad \Delta ABC \cong \Delta PQR$ [by SSS congruency]

4. **RHS (Rightangle-Hypotenuse-Side) Congruency:** Two right angled triangles are congruent to each other if hypotenuse and one side of one triangle are equal to hypotenuse and corresponding side of another triangle.

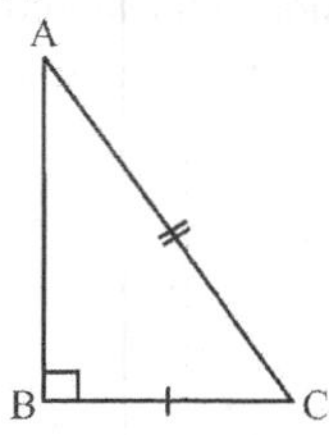

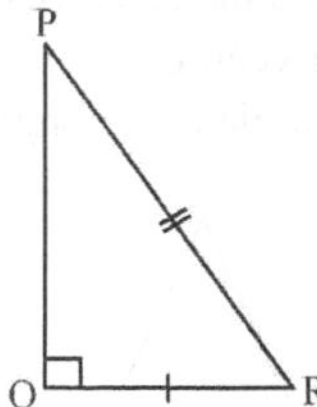

In ΔABC and ΔPQR

$$\angle ABC = \angle PQR = 90°$$
$$AC = PR$$
$$BC = QR$$

$\therefore \quad \Delta ABC \cong \Delta PQR$ [by RHS congruency]

SIMILARITY OF TWO TRIANGLES

Two triangles are said to be similar, if their shapes are the same but their size may or may not be equal.

When two triangles are similar, then

(i) all the corresponding angles are equal and

(ii) all the corresponding sides are in the same ratio (or proportion)

Note: In two similar triangles, sides opposite to equal angles are called corresponding sides. And angles opposite to side proportional to each other are called corresponding angles.

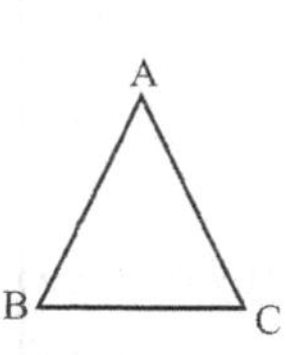

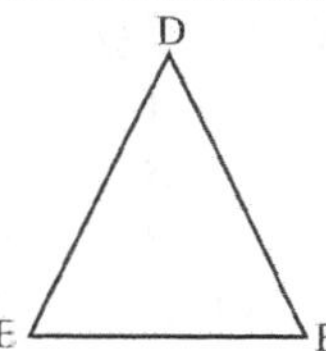

If ΔABC and ΔDEF are similar, then

$$\angle A = \angle D$$
$$\angle B = \angle E$$
$$\angle C = \angle F$$

and $\dfrac{AB}{DE} = \dfrac{BC}{EF} = \dfrac{CA}{FD}$

$\Delta ABC \sim \Delta DEF$, read as triangle ABC is similar to triangle DEF. Here ~ is the sign of similarity.

Conditions of Similarity

There are 4 conditions of similarity.

1. **AAA (Angle–Angle–Angle) Similarity:** Two triangles are said to be similar, if their all corresponding angles are equal. For example:

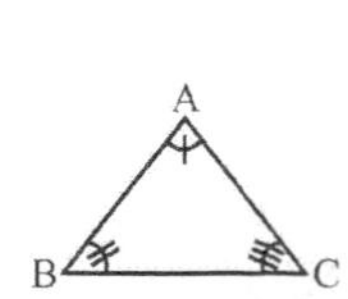

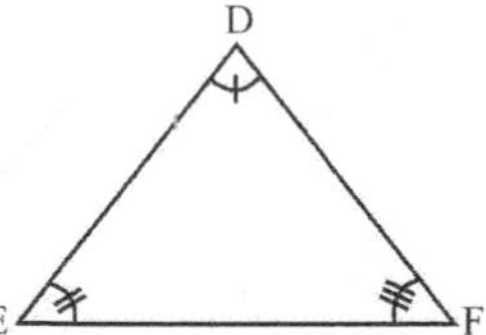

In ΔABC and ΔDEF, if

$$\angle A = \angle D$$
$$\angle B = \angle E$$
$$\angle C = \angle F$$

Then $\Delta ABC \sim \Delta DEF$ [By AAA Similarity]

Corollary AA (Angle-Angle) Similarity: If two angles of one triangle are respectively equal to two angles of another triangles, then two triangles are similar.

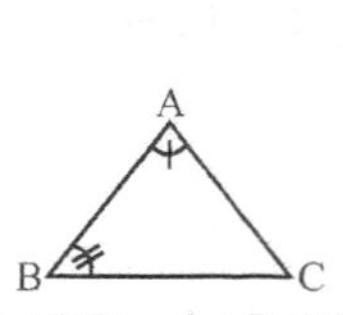

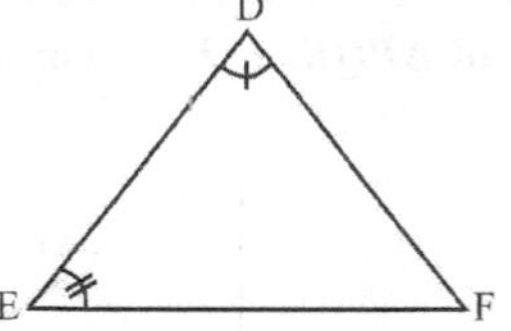

In ΔABC and ΔDEF, if

$$\angle A = \angle D$$
$$\angle B = \angle E$$

then $\Delta ABC \sim \Delta DEF$ [By AA Similarity]

2. **SSS (Side–Side–Side) Similarity:** Two triangles are said to be similar, if sides of one triangle are proportional (or in the same ratio of) to the sides of the other triangle:

For example:

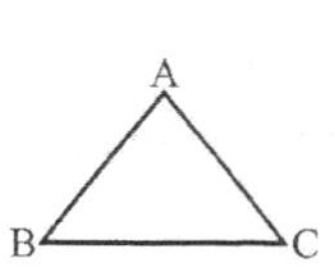

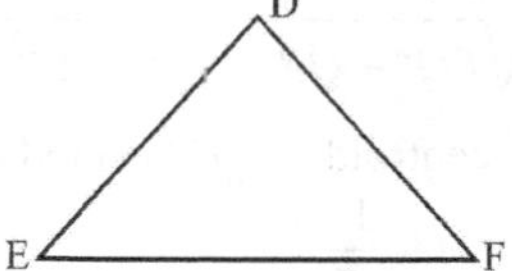

In ΔABC and ΔDEF, if

$$\frac{AB}{DE} = \frac{BC}{EF} = \frac{CA}{FD}$$

Then $\Delta ABC \sim \Delta DEF$ [By SSS Similarity]

3. **SAS (Side–Angle–Side) Similarity:** Two triangles are said to be similar if two sides of a triangle are proportional to the two sides of the other triangle and the angles included between these sides of two triangles are equal.

For example:

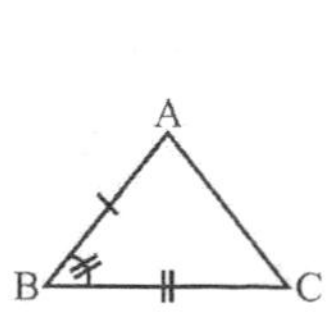

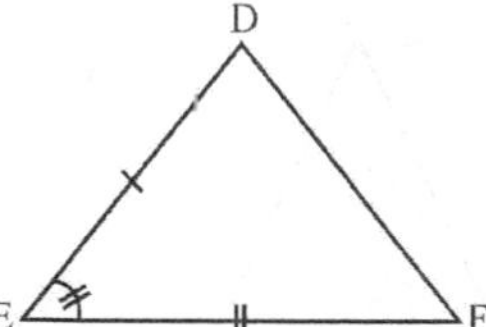

In ΔABC and ΔDEF, if

$$\frac{AB}{DE} = \frac{BC}{EF}$$

and $\angle B = \angle E$

Then, $\Delta ABC \sim \Delta DEF$ [By SAS Similarity]

4. **RHS (Rightangle-Hypotenuse-Side) Similarity:** Two triangles are said to be similar if one angle of both triangle is right angle and hypotenuse of both triangles are proportional to any one other side of both triangles respectively.

For example:

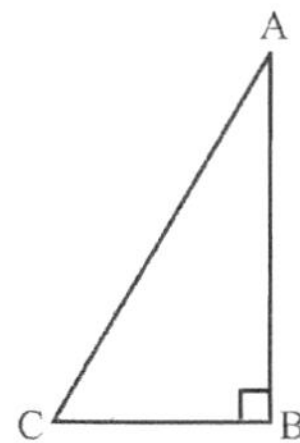

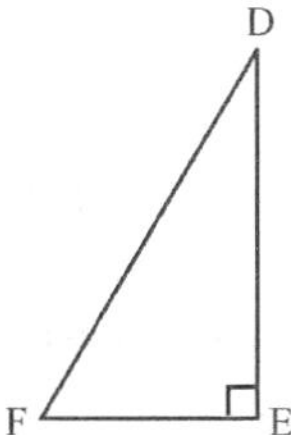

In ΔABC and ΔDEF, if

$$\angle B = \angle E\ [= 90°]$$

$$\frac{AC}{DF} = \frac{AB}{DE}$$

Then $\Delta ABC \sim \Delta DEF$ [By RHS similarity]

Note: In similar triangles,

Ratio of medians = Ratio of corresponding heights
= Ratio of circumeradii
= Ratio of inradii

Theorem

If two triangles are similar, then ratio of areas of two similar triangle is equal to the ratio of square of corresponding sides.

Illustration 11: D and E are the points on the sides AB and AC respectively of a ΔABC and AD = 8 cm, DB = 12 cm, AE = 6 cm and EC = 9 cm, then BC is equal to

(a) $\frac{2}{5}DE$ **(b)** $\frac{5}{2}DE$

(c) $\frac{3}{2}DE$ **(d)** $\frac{2}{3}DE$

Solution: (b) As in ΔADE and ΔABC

$$\frac{AD}{AB} = \frac{8}{20} = \frac{2}{5}, \frac{AE}{AC} = \frac{6}{15} = \frac{2}{5}$$

So, $\frac{AD}{AB} = \frac{AE}{EC}$

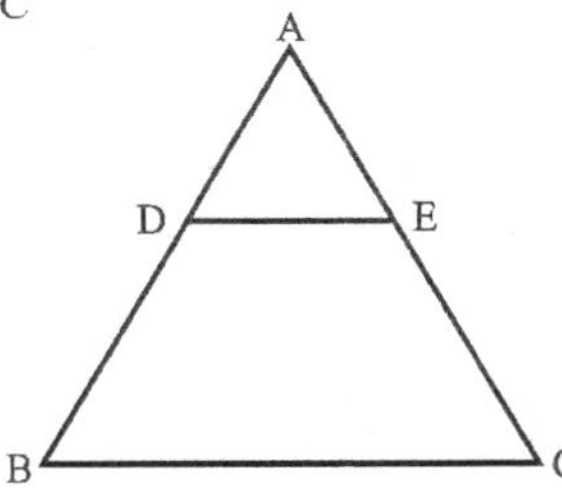

and $\angle A = \angle A$ (common)

$\Delta ADE \sim \Delta ABC$

$$\therefore \quad \frac{DE}{BC} = \frac{AD}{AB} \Rightarrow \frac{DE}{BC} = \frac{2}{5}$$

$$\Rightarrow \quad BC = \frac{5}{2}DE$$

Illustration 12: In a right angled ΔABC in which $\angle A$ = 90°. If $AD \perp BC$, then the correct statement is

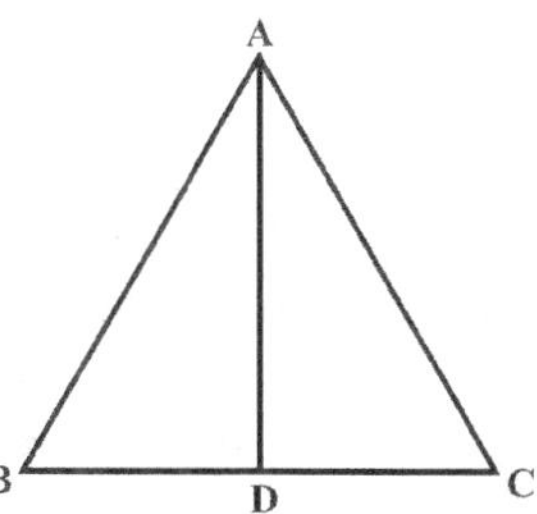

(a) $AB^2 = BD \times DC$ **(b)** $AB^2 = BD \times AD$

(c) $AB^2 = BC \times DC$ **(d)** $AB^2 = BC \times BD$

Solution: (d) Clearly, $\Delta ABD \sim \Delta CBA$

$$\Rightarrow \quad \frac{AB}{BD} = \frac{CB}{BA}$$

$$\Rightarrow \quad AB^2 = BC \times BD$$

Illustration 13: From the adjoining diagram, calculate

(i) AB **(ii) AP**

Solution: In ΔAPC and ΔABC,

$$\angle ACP = \angle ABC$$

$$\angle A = \angle A$$

$$\Rightarrow \quad \Delta ACP \sim \Delta ABC \Rightarrow \frac{AP}{AC} = \frac{PC}{BC} = \frac{AC}{AB}$$

$$\therefore \quad \frac{AP}{6} = \frac{8}{10} = \frac{6}{AB}$$

$$\Rightarrow \quad AP = 6 \times \frac{8}{10} = 4.8 \text{ and } AB = \frac{60}{8} = 7.5$$

$$\Rightarrow \quad AP = 4.8 \text{ cm and } AB = 7.5 \text{ cm}$$

Illustration 14: In the adjoining figure, $DE \parallel BC$ and $AD : DB = 4 : 3$

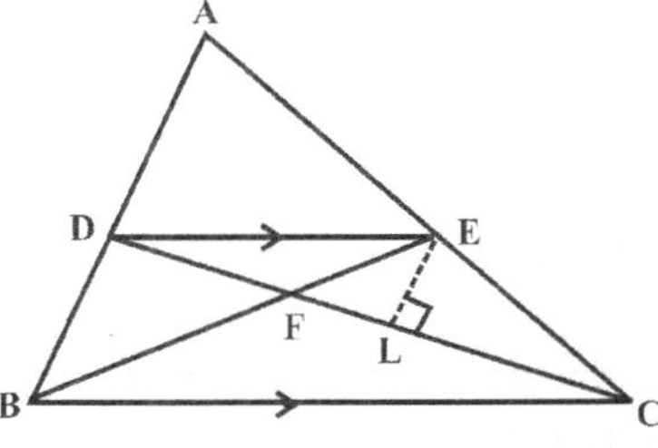

Find $\frac{AD}{AB}$ and then $\frac{DE}{BC}$

Solution: Since the sides of similar triangles are proportional, we have

$$\frac{AD}{AB} = \frac{DE}{BC}$$

But, $\frac{AD}{DB} = \frac{4}{3} \Rightarrow \frac{AD}{AD+DB} = \frac{4}{4+3} \Rightarrow \frac{AD}{AB} = \frac{4}{7}$

$\therefore \quad \frac{DE}{BC} = \frac{AD}{AB} = \frac{4}{7}$

Illustration 15: In the given figure, *DE* parallel to *BC*. If *AD* = 2 cm, *DB* = 3 cm and *AC* = 6 cm, then *AE* is

(a) 2.4 cm (b) 1.2 cm

(c) 3.4 cm (d) 4.8 cm

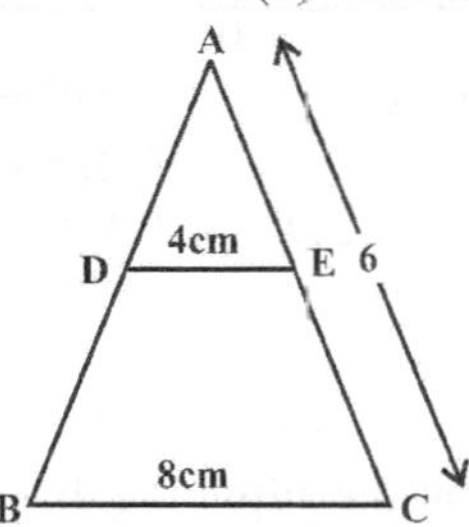

Solution: (a) The triangles *ADE* and *ABC* are similar.

$$\Rightarrow \quad \frac{AD}{AB} = \frac{AE}{AC}$$

$$\text{or} \quad \frac{2}{5} = \frac{AE}{6}$$

$$\therefore \quad AE = \frac{12}{5}$$

$$= 2.4\text{cm}$$

Illustration 16: The perimeters of two similar triangles *ABC* and *PQR* are 36 cm, and 24 cm, respectively. If *PQ* = 10 cm, then the length of *AB* is :

(a) 16 cm (b) 12 cm

(c) 14 cm (d) 15 cm

Solution: (d)

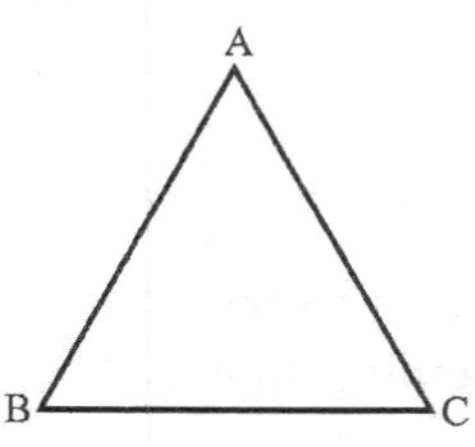

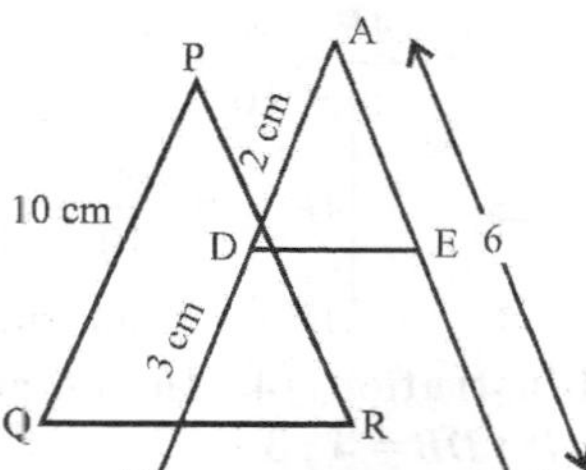

ΔABC and ΔPQR are similar.

$$\frac{AB}{PQ} = \frac{\text{Perimeter of } \Delta ABC}{\text{Perimeter of } \Delta PQR} \Rightarrow \frac{AB}{PQ} = \frac{36}{24}$$

$$\text{or} \quad AB = \frac{36}{24} \times 10 = 15$$

QUADRILATERALS

Quadrilateral is a plane figure bounded by four straight lines. The line segment which joins the opposite vertices of a quadrilateral is called diagonal of the quadrilateral. In figure, *PQRS* is a quadrilateral and *PR*, *QS* are its two diagonals.

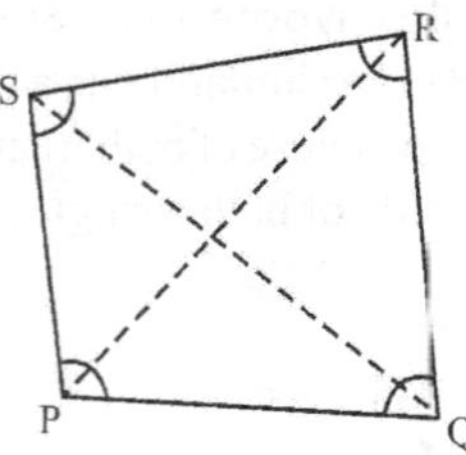

Sum of angles of a quadrilateral = 360°

i.e. $\angle P + \angle Q + \angle R + \angle S = 360°$

Types of Quadrilaterals

1. **Parallelogram:** A parallelogram is a quadrilateral with opposite sides parallel and equal.

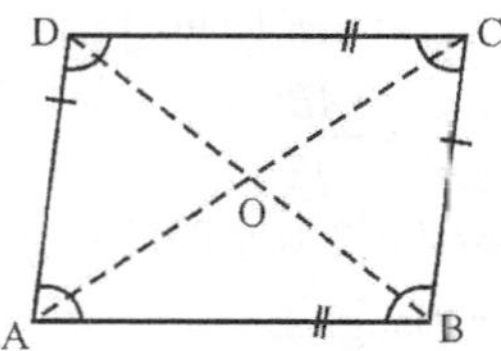

In figure, *ABCD* is a parallelogram in which *AC* and *BD* as diagonals which intersect each other at *O*.

Properties:

(i) Opposite sides are equal i.e.

$$AB = DC, \quad AD = BC$$

(ii) Opposite sides are parallel i.e.

$$AB \parallel DC \text{ and } AD \parallel BC$$

(iii) Opposite angles are equal i.e.

$$\angle BAD = \angle BCD \text{ and } \angle ABC = \angle ADC$$

(iv) Diagonals bisect each other, i.e.

$$OA = OC, \quad OB = OD$$

(v) Sum of pair of consecutive angles is 180° i.e.,

$$\angle A + \angle B = 180°, \quad \angle B + \angle C = 180°,$$
$$\angle C + \angle D = 180°, \angle D + \angle A = 180°.$$

2. **Rectangle:** A rectangle is a parallelogram with all angles equal to 90°.

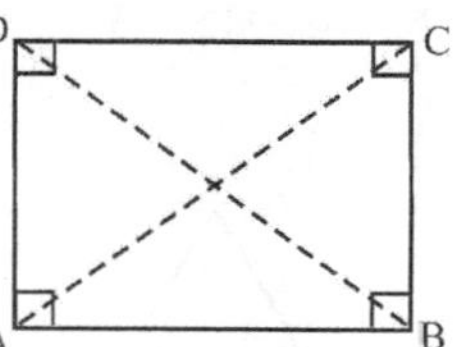

In figure, $\angle A = \angle B = \angle C = \angle D = 90°$

Properties:

(i) In a rectangle

Length of diagonal, are equal i.e.

$$AC = \sqrt{AB^2 + BC^2} = BD$$

(ii) In a rectangle diagonals bisect each other.

(iii) All rectangles are parallelogram but all parallelograms are not rectangles.

3. **Rhombus:** A parallelogram is a rhombus if its all sides are equal.

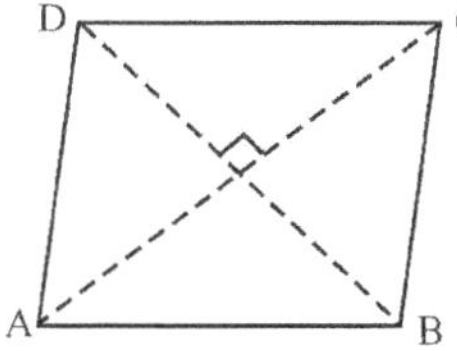

In rhombus $ABCD$, $AB = BC = CD = DA$

Properties:

(i) In a rhombus diagonals bisect each other at right angles i.e. angle between AC and DB is 90°.

(ii) All rhombus are parallelogram but all parallelograms are not rhombus.

4. **Square:** A parallelogram is a square if all the four sides are equal and also all the four angles are equal (i.e. 90°).

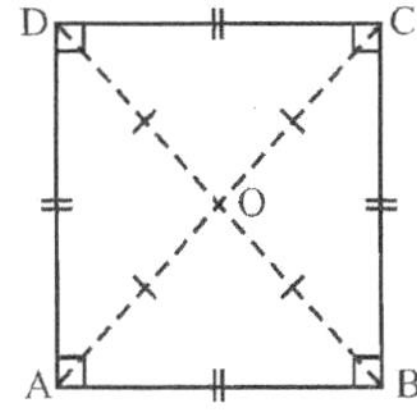

In figure, $ABCD$ is a square in which $AB = BC + CD = DA$ and $\angle A = \angle B = \angle C = \angle D = 90°$

Properties:

(i) In a square diagonals are equal i.e. $AC = BD$

(ii) In a square diagonals bisect each other at right angle, i.e.

$$OA = OC, \quad OB = OD$$

and $\angle AOB = \angle BOC = \angle COD = \angle DOA = 90°$.

(iii) All square are rhombus but rhombus may or may not be a square.

5. **Trpaezium:** A quadrilateral is a trapezium if one pair of opposite sides are parallel.

In trapezium $ABCD$, $AB \parallel DC$.

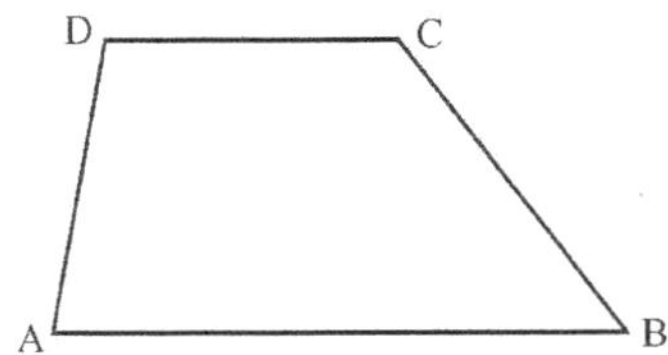

If lateral sides (i.e. non-parallel sides) of a trapezium are equal, then it is called isosceles trapezium.

Properties of isosceles trapezium

In the figure ABCD is an isosceles trapezium, then

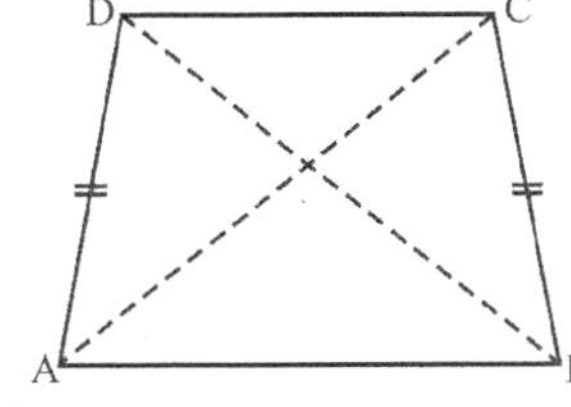

(i) $AB \parallel DC$

(ii) $AD = BC$

(iii) Diagonals are equal i.e. $AC = BD$

Diagonal Properties of all Parallelograms

Sr. No.	Diagonal Properties	Type of Parallelogram			
		Parallelogram	Rectangle	Rhombus	Square
1	Diagonals bisect each other	✓	✓	✓	✓
2	Diagonals are equal	×	✓	×	✓
3	Diagonals are at 90° to each other	×	×	✓	✓

CIRCLES

A circle is a locus i.e. path of a point in a plane which moves in such a way that its distance from a fixed point always remains constant.

In figure, 'O' is the fixed point and P is a moving point in the same plane. The path traced by P is called a circle. Fixed point O is the centre of the circle and the constant distance OP is called radius of the circle.

A diameter is a line segment passing through the centre and joins the two points on the circle in the figure.

AB is the diameter as it passes through the centre and joins the two points on the circle. Diameter = 2 × radius.

A circle divides the plane in which it lies into three parts.

(i) Inside the circle, called interior of the circle

(ii) The circle

(iii) Outside the circle, called the exterior of the circle.

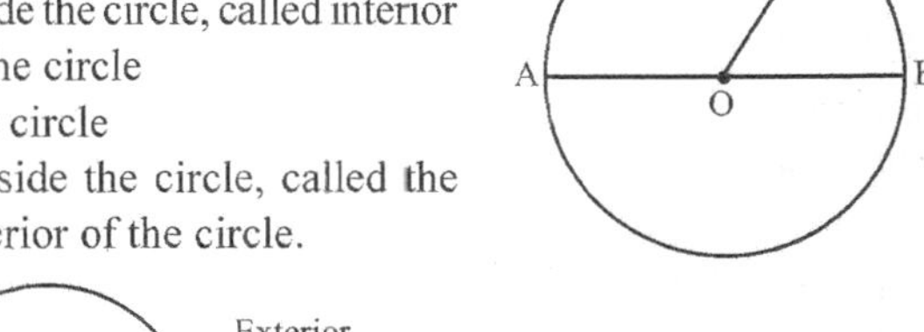

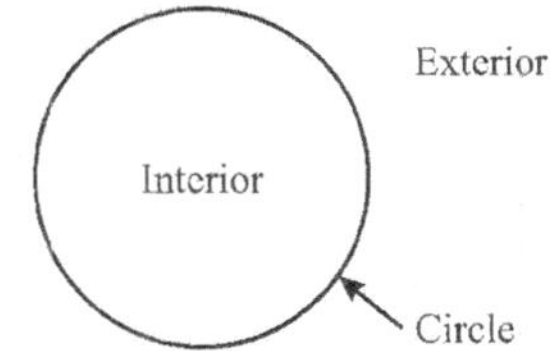

The circle and its interior make up the circular region.

Circumference

Length of a complete circle is called its circumference.

In figure, AB is tangent to circle of radius 'r', which touches the circle at point P.

P is called the point of contact of tangent to the circle. Radius through the point of contact is always perpendicular to the tangent at the point of contact i.e. $OP \perp AB$.

Chord: A line segment joining any two points on the circle is called chord of the circle. A chord which passes through the centre is the diameter of the circle.

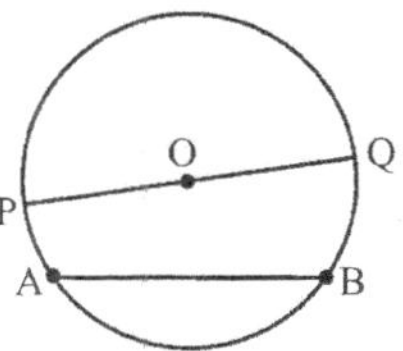

In the figure, O is the centre of the circle.

AB and PQ both are chords.

But PQ is the diameter (longest chord) also.

Arcs: A piece of a circle between two points is called an arc.

Consider two points M and N on the circle. We find that there are two pieces of circle between M and N. One is longer and other is smaller.

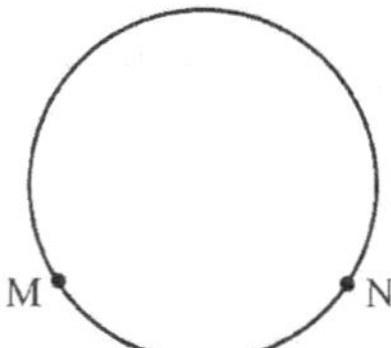

The longer piece is called major arc and smaller piece is called minor arc

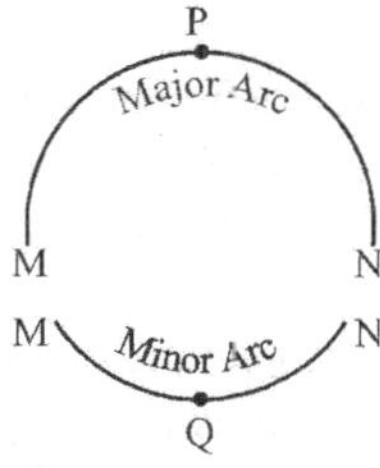

Major arc is denoted by $\widehat{MPN}$ and minor arc is denoted by $\widehat{MQN}$.

When M and N are ends of a diameter then both the arcs are equal and both are called semicircle.

Segment: The region between a chord and an arc of a circle is called a segment.

There are two segments corresponding to two arcs, major segment and minor segment. Major segment is the segment enclosed by major arc. Centre of the circle lies in the major segment. Minor segment is the segment enclosed by minor arc. Centre of the circle does not lie in the minor segment.

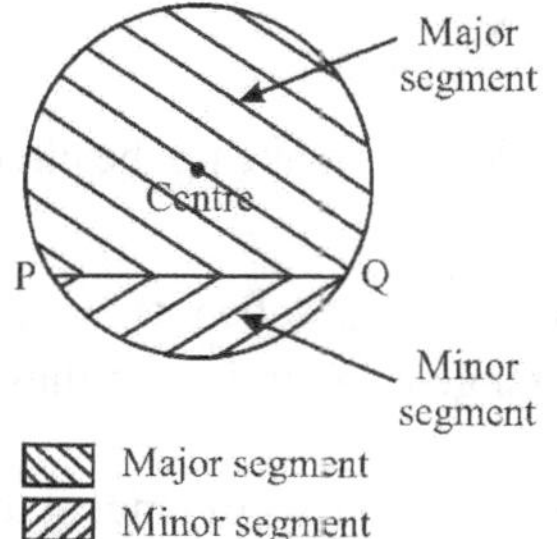

If two arcs are equal, then both segments are semi-circles.

Sector: The region between an arc and the two radii joining the centre to the end point of the arc is called a sector. There are two sectors Minor and Major Sectors.

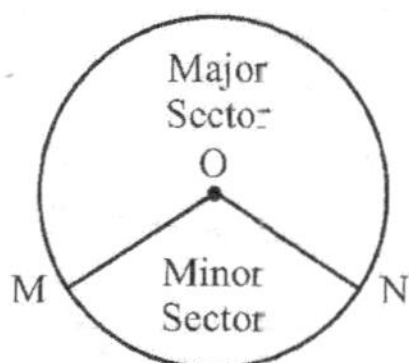

The sector which is larger than semicircular region is called major sector and the region less than the semicircular region is called minor sector.

If both sectors are equal, then each sector is a semi-circle.

Tangent: A tangent is a straight line which touches the circumference of a circle at only one point. A tangent does not intersect the circumference, if produced infinitely on either sides.

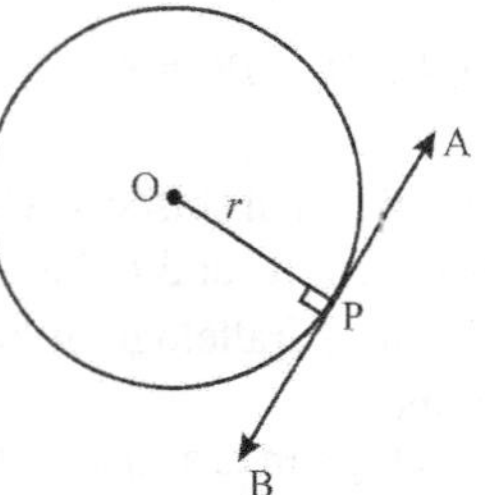

Secant: A secant is a straight line of infinite length which intersects the circumference of a circle at two different points. In figure, AB is a secant.

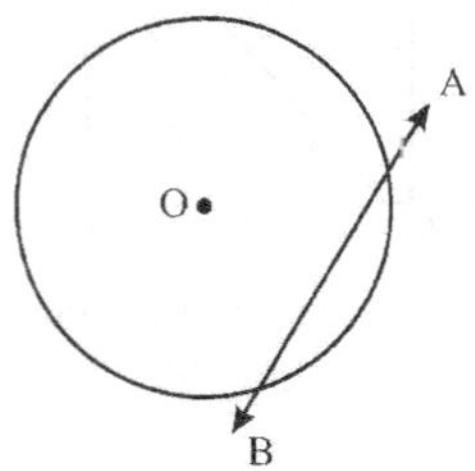

Basic Properties of a Circle

1. Equal chord of a circle subtend equal angles at the centre.

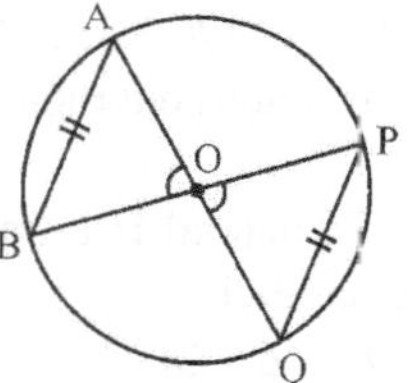

If $AB = PQ$, then $\angle AOB = \angle POQ$

The converse is also true.

2. The perpendicular from the centre of a circle to a chord of the circle bisects the chord.

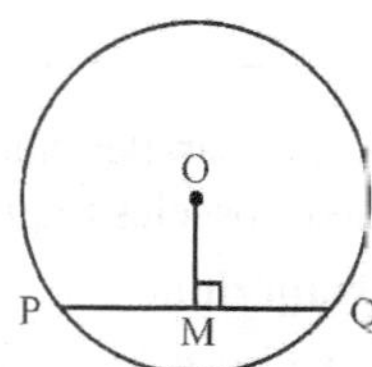

In figure, PQ is chord of a circle with centre 'O', OM is perpendicular to PQ therefore $PM = MQ$. The converse is also true.

3. One and only one circle can pass through given three non-collinear points.

If three or more points lie on a line, then they are called collinear points otherwise called non-collinear points.

4. Equal chords of a circle are equidistant from the centre of the circle.

In the figure, if $AB = CD$, then $OP = OQ$

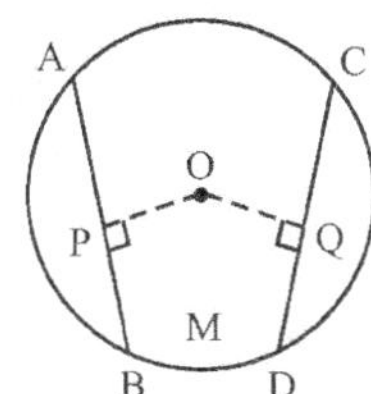

The converse is also true.

5. Two equal chords have equal corresponding arcs.

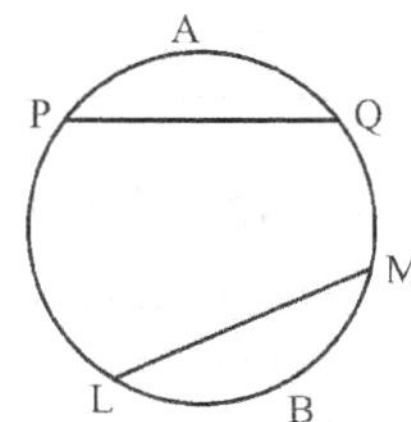

If $PQ = LM$ then

(a) $\overset{\frown}{PAQ} = \overset{\frown}{LBM}$ (Minor Arc)

(b) $\overset{\frown}{PBQ} = \overset{\frown}{LAM}$ (Major Arc)

6. The greater of the two chords is nearer to the centre.

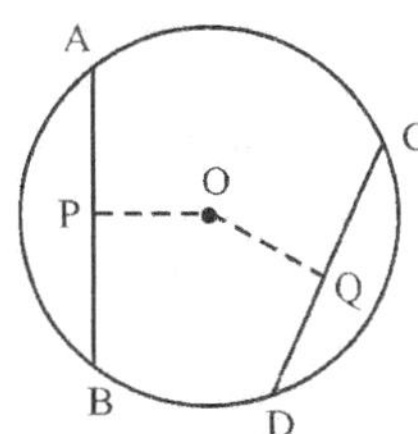

If $AB > CD$, then $OP < OQ$

7. The angle subtended by an arc at the centre is double the angle subtended by it at any point on the remaining part of the circle.

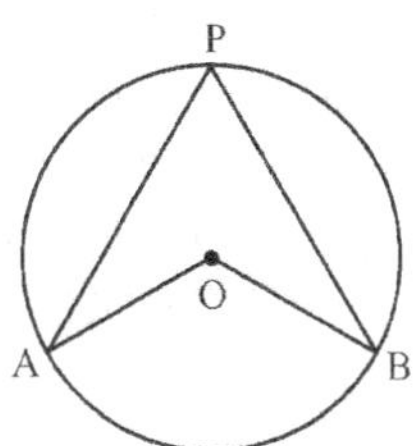

minor arc $\overset{\frown}{AB}$ subtend $\angle AOB$ at the centre O and also subtend $\angle APB$ at point P (situated on remaining part of circle). So $\angle AOB = 2\ \angle APB$

8. Angle in a semicircle is a right angle.

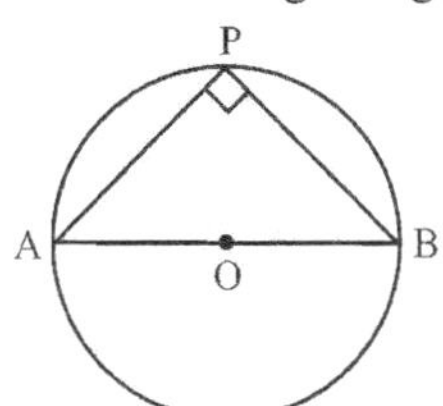

In figure, AOB is a diameter, hence $AOBPA$ is a semicircle, therefore $\angle APB = 90°$.

9. Angles in the same segment of a circle are equal. $\angle ACB$, $\angle ADB$, $\angle AEB$ are in the same segment $ACDEBA$ of the circle.

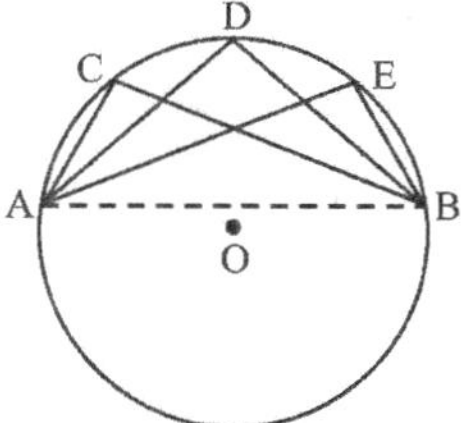

$\therefore\ \angle ACB = \angle ADB = \angle AEB.$

10. If in a plane a line segment joining two points subtends equal angles at two other points lying on the same side of a line containing the line segment, the four points lie on a circle i.e. they are concyclic.

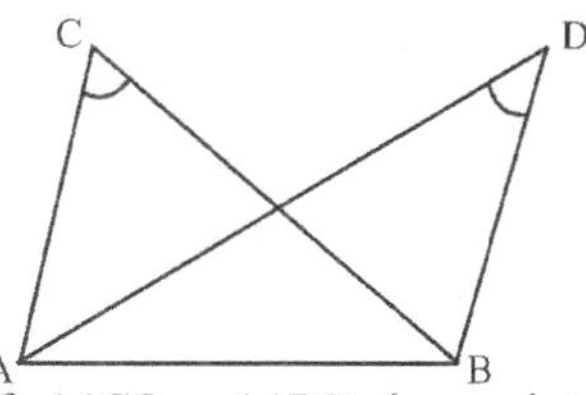

In figure, if $\angle ACB = \angle ADB$, then points A, B, D, C lie on a circle.

11. The sum of either pair of opposite angles of a cyclic quadrilateral is 180°.
A cyclic quadrilateral is the quadrilateral whose four vertices are concyclic i.e. the four vertices lie on a circle. In figure, $ABCD$ is a cyclic quadrilateral,

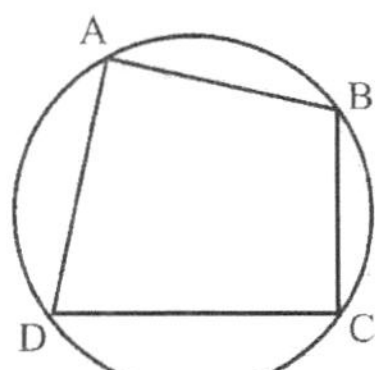

$\therefore\ \angle A + \angle C = 180°$ and $\angle B + \angle D = 180°$

The converse is also true.

12. If a side of a cyclic quadrilateral is produced the exterior angle so formed is equal to the interior opposite angle.

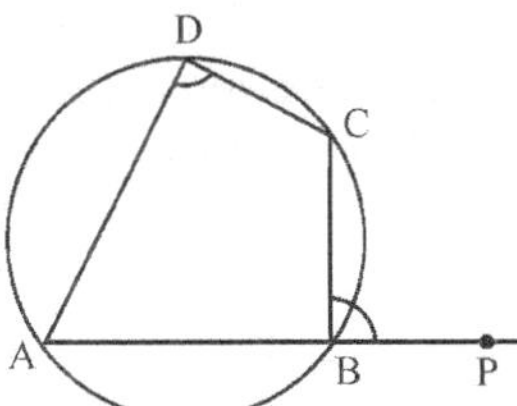

In figure, $ABCD$ is a cyclic quadrilateral,

$\therefore\ \angle CBP = \angle CDA$

13. Two circles C_1 with centre O_1, radius r_1 and C_2 with centre O_2, radius r_2 will touch

(a) Externally, if and only if $O_1O_2 = r_1 + r_2$

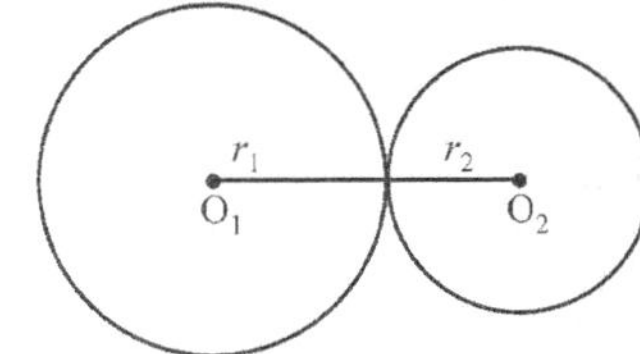

(b) Internally, if and only if $O_1O_2 = |r_1 - r_2|$

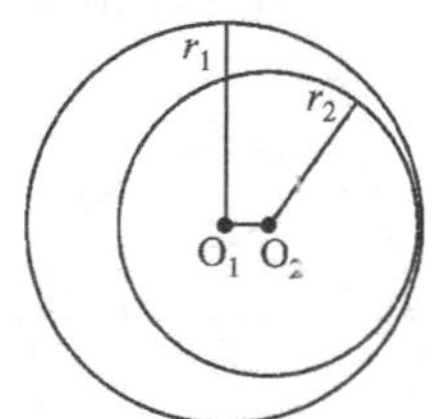

14. Two circles are congruent if their radii are equal.

Illustration 17: In a circle of radius 17 cm, two parallel chords are drawn on opposite sides of a diameter. The distance between the chords is 23 cm. If length of one chord is 16 cm, then the length of the other one is :

(a) 15 cm **(b) 23 cm**
(c) 30 cm **(d) 34 cm**

Solution: (c) Let PQ and RS be two parallel chords of the circle on the opposite sides of the diameter AB = 16 cm

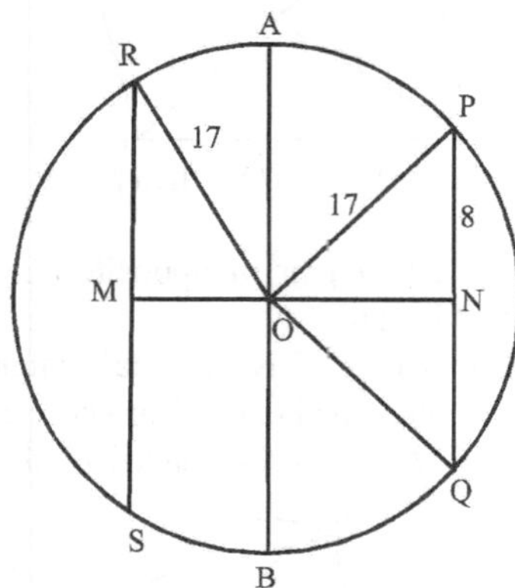

Now, PN = 8 (Since ON is the perpendicular bisector)

In ΔPON

$$ON^2 = OP^2 - PN^2$$
$$= (17)^2 - (8)^2 = 289 - 64 = 225$$

or $ON = 15 \Rightarrow \therefore OM = 23 - 15 = 8$

In ΔORM,

$$RM^2 = OR^2 - OM^2$$
$$17^2 - 8^2 = 289 - 64 = 225$$

or $RM = 15 \Rightarrow RS = 15 \times 2 = 30$ cm

Illustration 18: In the cyclic quadrilateral $ABCD$, $\angle BCD$ = 120°, m (arc DZC) = 7°, find DAB and m (arc CXB).

(a) 60°, 70° **(b) 60°, 40°**
(c) 60°, 50° **(d) 60°,60°**

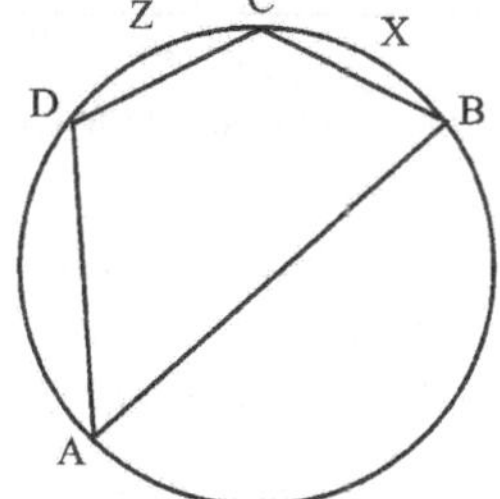

Solution: (c) m $\angle DAB$ + 180° – 120° = 60° ......(Opposite angles of a cyclic quadrilateral)

m (arc BCD) = 2m $\angle DAB$ = 120°.

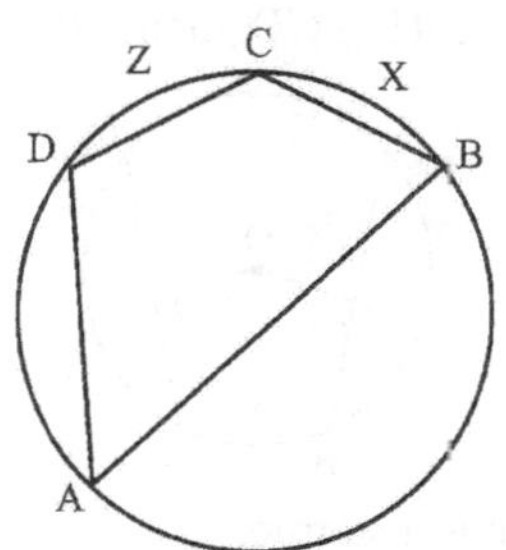

$\therefore$ m (arc CXB) = m (BCD) – m (arc DZC)
= 120° – 70° = 50° .

BASIC PYTHAGOREAN TRIPLETS

A Pythagorean triplet is a set of three natural numbers a, b and c, which are length of the sides of a right angled triangle.

Hence, if $a^2 + b^2 = c^2$, $b^2 + c^2 = a^2$ or $c^2 + a^2 = b^2$, then the set of natural numbers a, b and c is a Pythagorean triplet.

Since $3^2 + 4^2 = 5^2$, hence 3, 4, 5 form a Pythagorean triplet.

General Rule To Find Pythagorean Triplet: If r and s are two natural numbers such that $r > s$, $r - s$ is odd and GCD of r and s is 1, then the Pythagorean triplet a, b, c are defined by

$$a = r^2 - s^2, \quad b = 2rs \quad \text{and} \quad c = r^2 + s^2.$$

Note: If each term of any Pythagorean triplet is multiplied or divided by such a positive number that the products or quotients obtained respectively are natural numbers then the new products or quotients are also form Pythagorean triplets.
Since 3, 4, 5 form a Pythagorean triplet, therefore 9, 12 and 15 also form a Pythagorean triplet.

DETERMINATION OF NATURE OF TRIANGLE

Let length of three sides of a triangle are a, b and c.

- If c be the length of longest side and $c^2 = a^2 + b^2$, then the triangle is right-angled triangle.
- If c be the length of longest side and $c^2 > a^2 + b^2$, then the triangle is an obtuse-angled triangle.
- If c be the length of longest side and $c^2 < a^2 + b^2$, then the triangle is an acute-angled triangle.

IMPORTANT POINTS

1. In ΔABC right angled at A, if AD is perpendicular to BC.

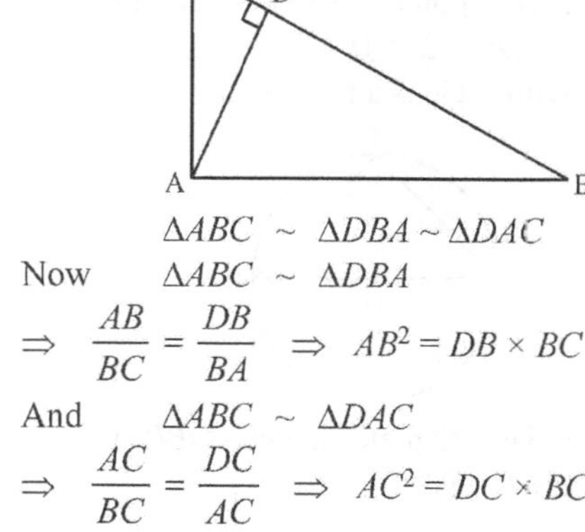

$$\Delta ABC \sim \Delta DBA \sim \Delta DAC$$

Now $\Delta ABC \sim \Delta DBA$

$$\Rightarrow \frac{AB}{BC} = \frac{DB}{BA} \Rightarrow AB^2 = DB \times BC$$

And $\Delta ABC \sim \Delta DAC$

$$\Rightarrow \frac{AC}{BC} = \frac{DC}{AC} \Rightarrow AC^2 = DC \times BC$$

And $\Delta DBA \sim \Delta DAC$

$$\Rightarrow \frac{DA}{DB} = \frac{DC}{DA} \Rightarrow DA^2 = DB \times DC$$

2. In a cyclic quadrilateral, product of the diagonals is equal to the sum of the products to the opposite sides,

$$AC \times BD = (AD \times BC) + (AB \times CD)$$

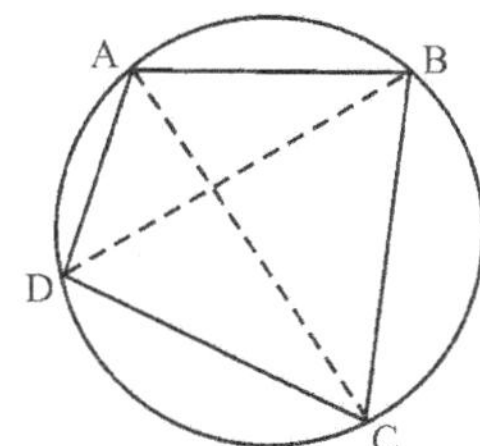

3. Bisectors of the angles of a parallelogram or a rectangle form a rectangle.

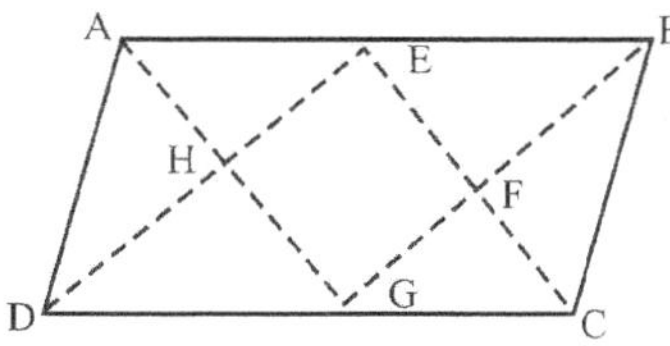

In parallelogram *ABCD*, *AG*, *BG*, *CE* and *DE* are the bisectors of $\angle A$, $\angle B$, $\angle C$ and $\angle D$ respectively. Hence in the figures *EFGH* is a rectangle.

4. A parallelogram inscribed in a circle is a rectangle. In figure, *ABCD* is a rectangle.

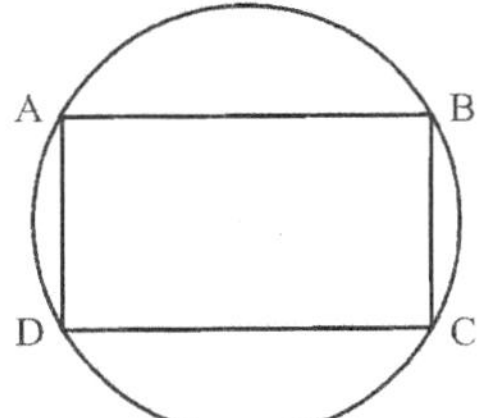

5. A parallelogram circumscribed a circle is a rhombus. In figure, *ABCD* is a rhombus.

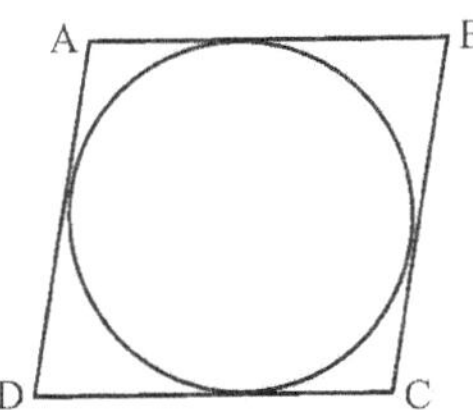

6. Median of a trapezium is the line segment joining mid-points of non-parallel sides of the trapezium. In the figure *E* and *F* are the mid points of non-parallel sides *AB* and *CD* respectively. Hence *EF* is the median of trapezium *ABCD*.

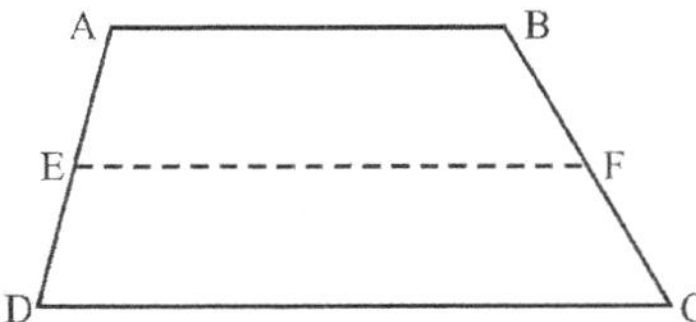

$$EF = \frac{1}{2}(AB + CD)$$

Also $$EF = \frac{a \times (\text{AB}) + b \times (\text{DC})}{\text{AD}},$$

where $AE = a$ and $ED = b$

7. Perpendicular bisectors of two chords of a cricle intersect at its centre of the circle.

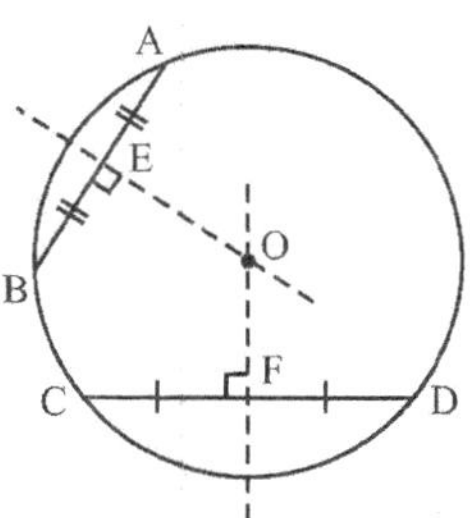

In figure, *OE* and *OF* are perpendicular bisectors of chords *AB* and *CD*, *OE* and *OF* meet at point *O*. Hence *O* is the centre of the circle.

8. If two circles intersect each other at two points then the line through the centres is the perpendicular bisectors of the common chord.

In figure, two circles with centre *P* and *Q* intersect each other at two points *A* and *B*.

Hence *AB* is the common chord of the two circles.

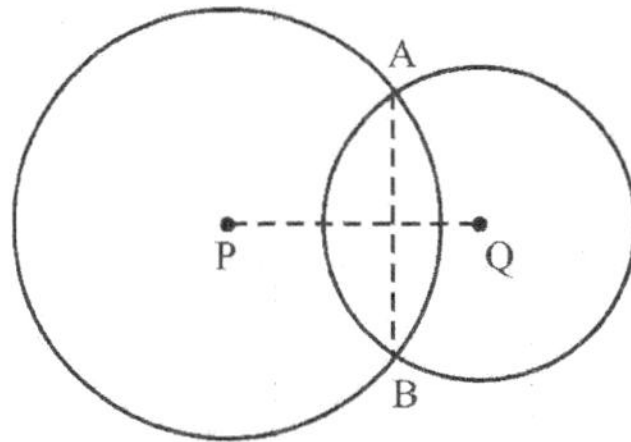

Therefore, *PQ* is the perpendicular bisector of common chord *AB*.

9. Equal chords of a circle or congruent circles are equidistant from the centre.

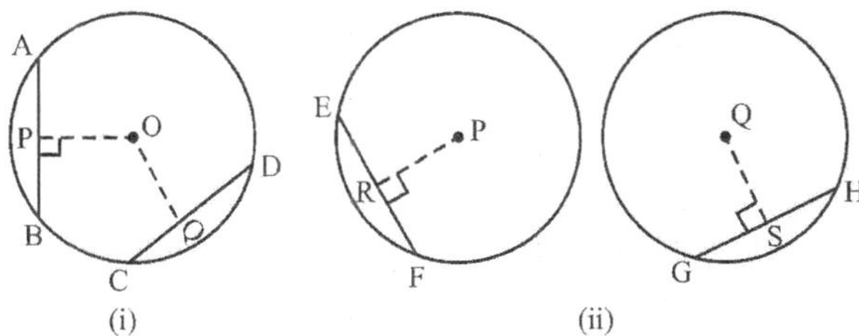

(i) (ii)

In figure (i), *AB* and *CD* are two equal chords of a circle, therefore their perpendicular distances *OP* and *OQ* respectively from the centre *O* are equal.

In figure (ii), two circles are congruent i.e. their radii are equal. *EF* and *GH* are two equal chords. Hence their perpendicular distances from centre *P* and *Q* respectively are equal.

10. If a circle touches all the four sides of a quadrilateral then the sum of the two opposite sides is equal to the sum of other two.

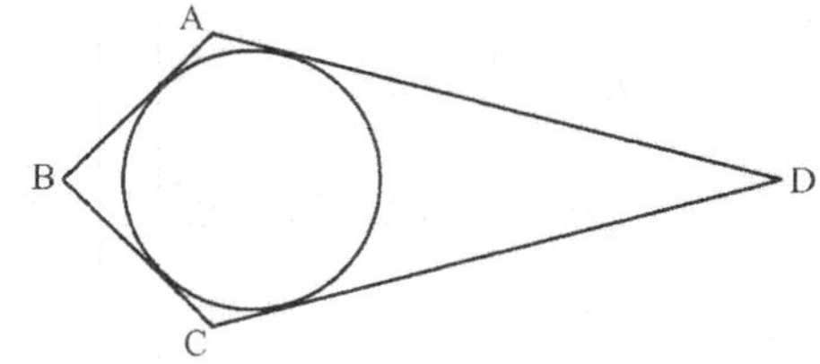

$AB + DC = AD + BC$

11. In two concentric circles, if a chord of the larger circle is also tangent to the smaller circle, then the chord is bisected at the point of contact.

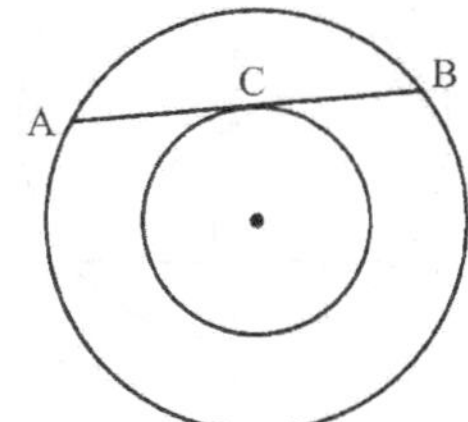

Hence in the figure, $AC = CB$

12. Length of two tangents from an exterior point to a circle are equal.

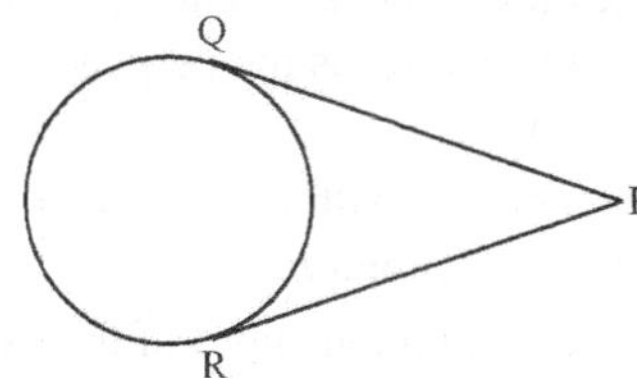

In figure PQ and PR are two tangents drawn from an exterior point to a circle.

$\therefore \quad PQ = PR$

13. **Direct common tangent:** A tangent to two circles are such that the two circles lies on the same side of the tangent, then the tangent is called direct tangent to the two circles.

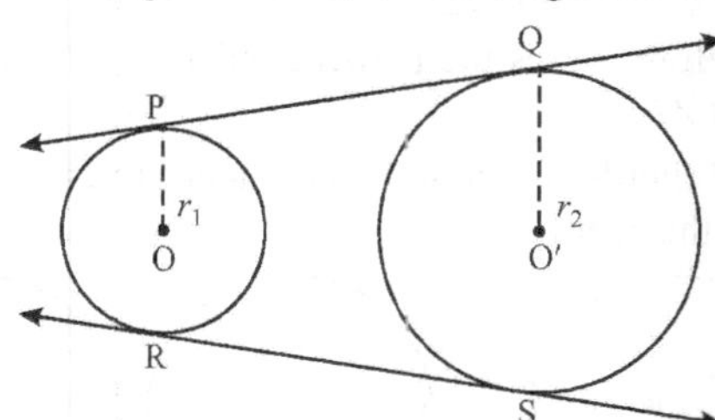

In the figure, PQ and RS are two direct common tangent to the same two circles. Length of these two common tangents to the same two circles are equal.

i.e. $\quad PQ = RS$

Also $\quad PQ = RS = \sqrt{(OO')^2 + (r_2 - r_1)^2}$

Here O, O' are the centres and r_1, r_2 are the radii of the two circles respectively. Also $r_2 > r_1$.

14. **Indirect or Transverse Common Tangent:** If a tangent to two circles is such that the two circles lie on opposite sides of the tangent, then the tangent is called indirect tangent.

Length of two indirect tangents to two circles is equal.

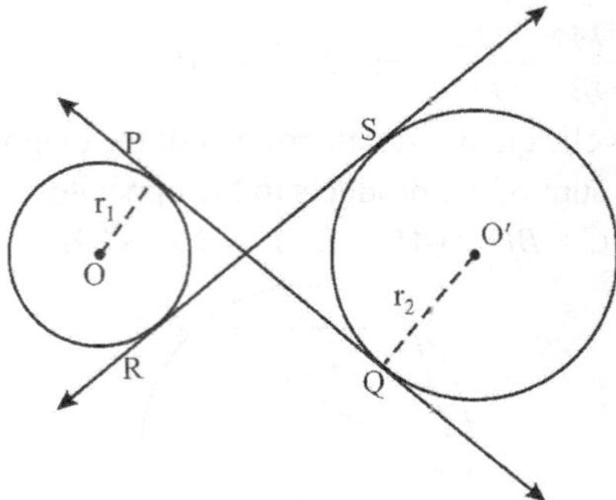

In the figure, PQ and RS are two indirect common tangents to the same two circles.

$\therefore \quad PQ = RS$

Also $\quad PQ = RS = \sqrt{(OO')^2 - (r_1 + r_2)^2}$

Here O, O' are centres r_1, r_2 are radii of the two circles respectively.

15. **Star:** A star has a shape like given in the figure.

If a star has n sides, then

Sum of its all angles = $(n - 4) \times 180°$.

16. In a triangle, the sum of the square of any two sides of a triangle is equal to twice the sum of the square of the median to the third side and square of half the third side.

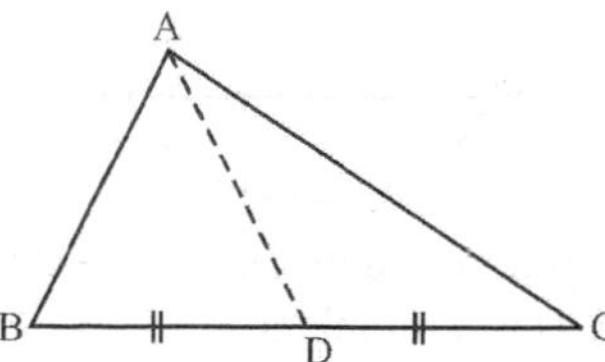

In the figure, AD is the median.

$$\therefore \quad AB^2 + AC^2 = 2\left[AD^2 + \left(\frac{BC}{2}\right)^2\right]$$

17. In a triangle,

$$3 \times \begin{pmatrix}\text{Sum of square of}\\ \text{three sides of}\\ \text{a triangle}\end{pmatrix} = 4 \times \begin{pmatrix}\text{Sum of the square of}\\ \text{three medians of}\\ \text{the triangle}\end{pmatrix}$$

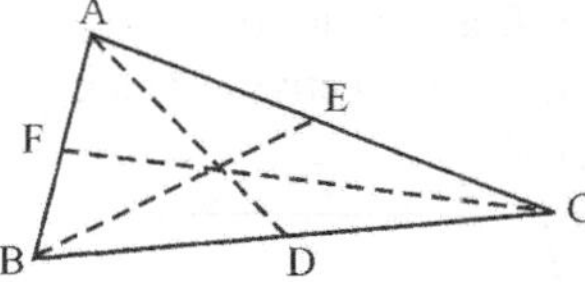

In figure AD, BE and CF are medians of ΔABC.

$\therefore \quad 3 \times (AB^2 + BC^2 + CA^2) = 4 \times (AD^2 + BE^2 + CF^2)$

18. In the figure given below, if P is any point inside the rectangle $ABCD$, then

$$PA^2 + PC^2 = PB^2 + PD^2$$

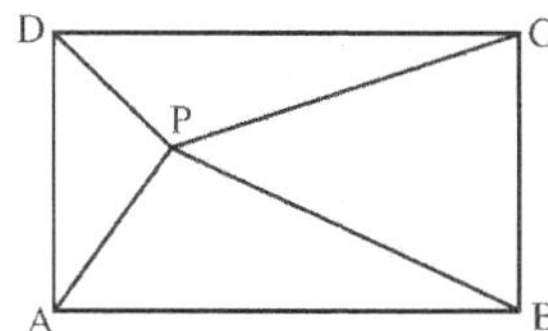

19. Diagonals of a trapezium divide each other in the ratio of the parallel sides of the trapezium. In trapezium $ABCD$, $AB \parallel DC$

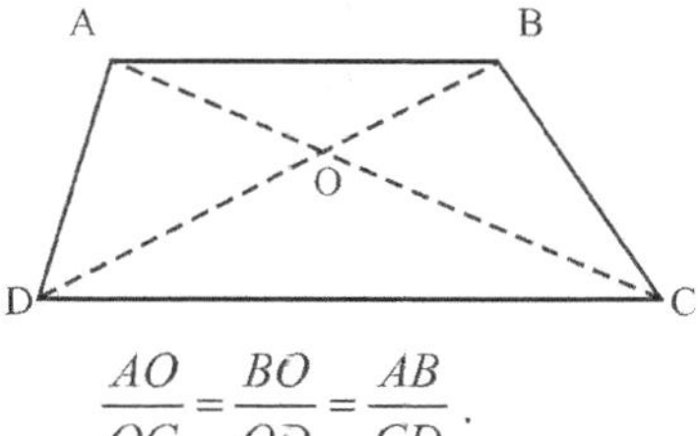

$$\therefore \quad \frac{AO}{OC} = \frac{BO}{OD} = \frac{AB}{CD}.$$

20. If a trapezium is inscribed inside a circle, then the trapezium is an isosceles trapezium i.e. its non-parallel sides are equal.

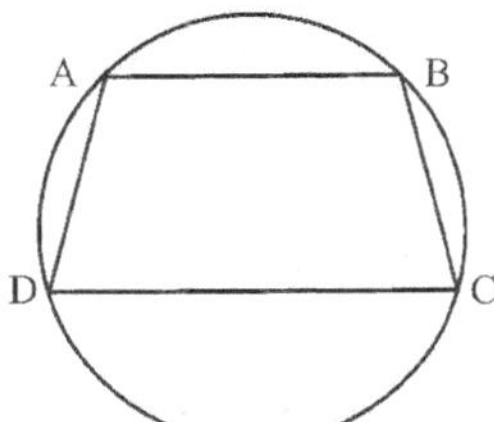

In the figure, $ABCD$ is a trapezium in which $AB \parallel CD$

$\therefore \quad AD = BC$

21. Area of triangles on the same base and lie between the same pair of parallel lines are equal.

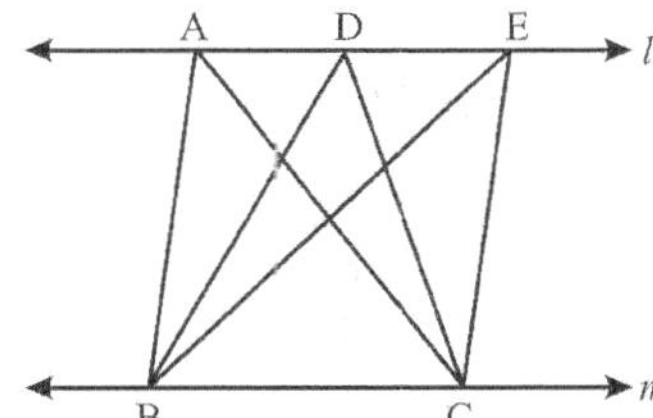

In the figure, ΔABC, ΔDBC and ΔEBC are on the same base BC and lie between the same pair of parallel lines l and m.

$\therefore$ area of ΔABC = area of ΔDBC = area of ΔEBC.

22. If a parallelogram and a triangle are on the same base and lie between the same pair of parallel lines, then area of the parallelogram is twice the area of the triangle.

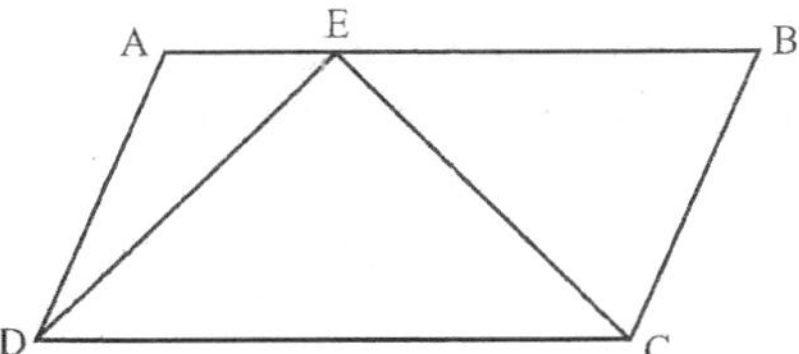

In the figure, $ABCD$ a parallelogram and EDC a triangle are on the same base and lie between the same pair of parallel lines AB and CD.

$\therefore$ area of parallelogram $ABCD = 2 \times$ (area of ΔEDC).

23. Concentric circles: Two or more circles in a plane are said to be concentric, if they have the same centre.

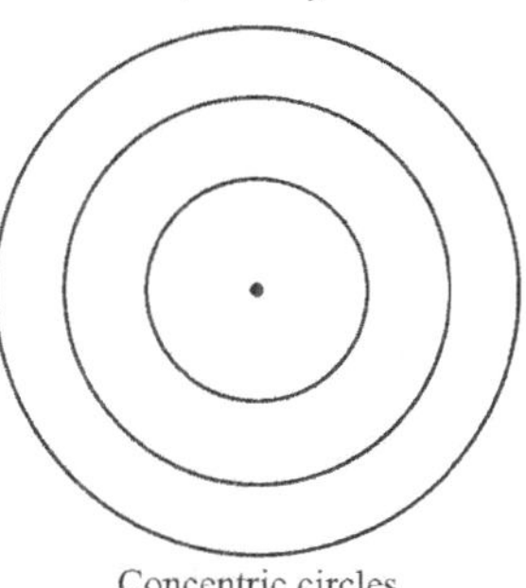

Concentric circles

24. Intercepts made by three or more parallel lines on two or more lines are in the same ratios.

In the figure three parallel lines AD, BE and CF made intercepts AB, BC and DE, EF on two lines AC and DF respectively.

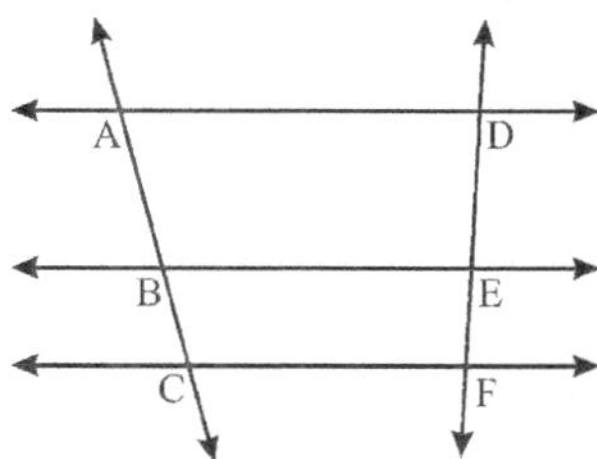

$$\therefore \quad \frac{AB}{BC} = \frac{DE}{EF}$$

25. (a) In an equilateral triangle centroid, incentre, circumcentre, orthocentre coincide at the same point.

(b) Circumradius = 2 × in radius

26. A parallelogram is a rectangle if its diagonals are equal.

27. If two chords AB and CD of a circle intersect inside a circle (or outside a circle when produced) at point E, then $AE \times EB = CE \times ED$.

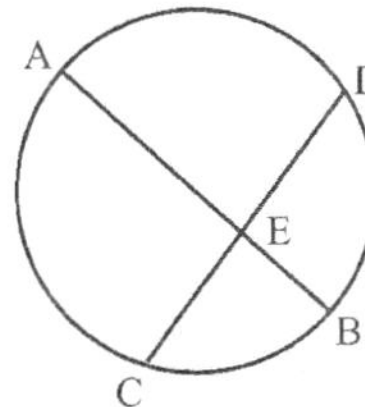

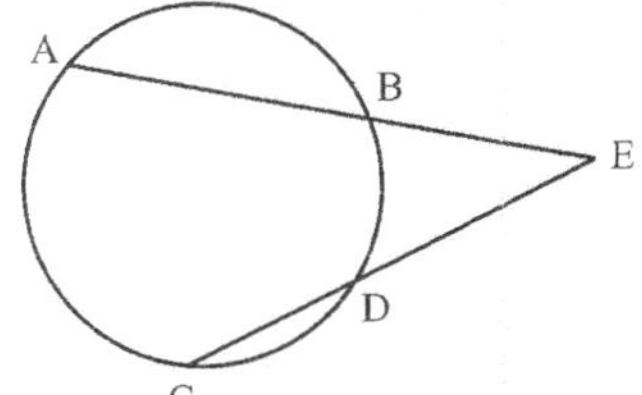

28. If PB is a secant which intersects the circle at A and B and PT is a tangent at T to the circle, then

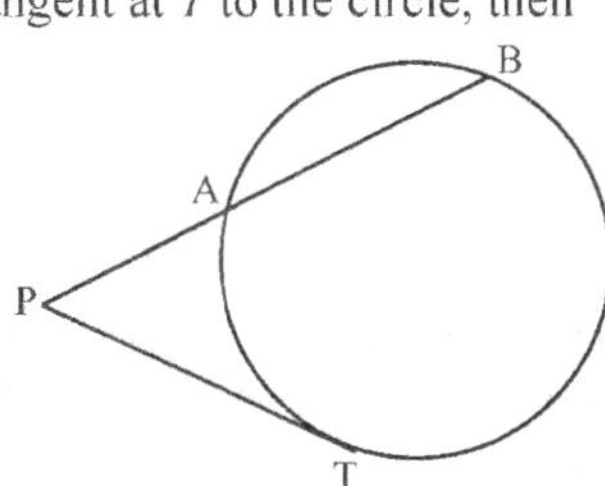

$$PA \times PB = PT^2$$

29. Angles in the alternate segment:

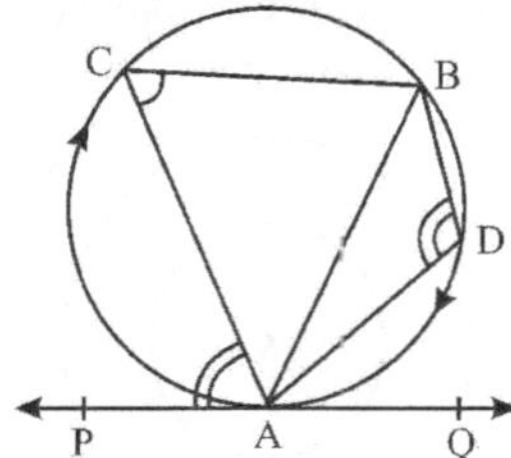

In the figure, AB is a chord of a circle. PQ is a tangent at an end point A of the chord to the circle. C is any point on arc AB and D is any point on arc BA.

$\angle BAQ$ and $\angle ACB$ are angles in the alternate segments

$\angle BAP$ and $\angle ADB$ are angles in the alternate segments.

Angles in the alternate segments of a circle are equal i.e.

$$\angle BAQ = \angle ACB \text{ and } \angle BAP = \angle ADB$$

Illustration 19: In the given figure, chords AB and CD of a circle intersect externally at P. If $AB = 6$ cm, $CD = 3$ cm and $PD = 5$ cm, then $PB = ?$

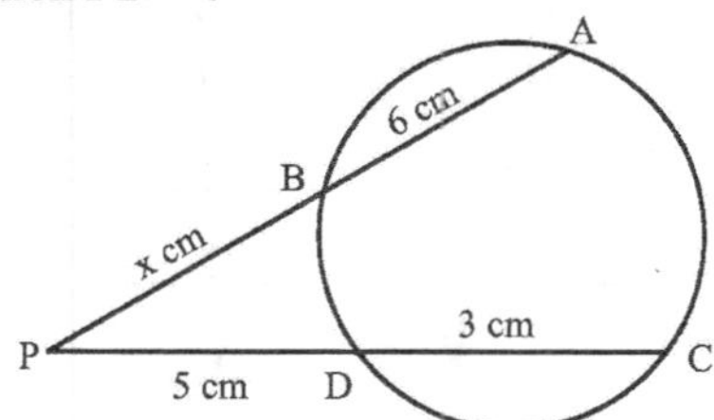

(a) 5 cm **(b) 6.25 cm**
(c) 6 cm **(d) 4 cm**

Solution: (d) $PA \times PB = PC \times PD$ (According to property of circle)

$\Rightarrow \quad (x + 6) \times x = 8 \times 5$

$\Rightarrow \quad x^2 + 6x - 40 = 0$

$\Rightarrow \quad (x + 10)(x - 4) = 0 \Rightarrow x = 4$

$\therefore \quad PB = 4$ cm

Illustration 20: In the given figure, PAB is a secant and PT is a tangent to the circle from P. If $PT = 5$ cm, $PA = 4$ cm and $AB = x$ cm, then x is equal to

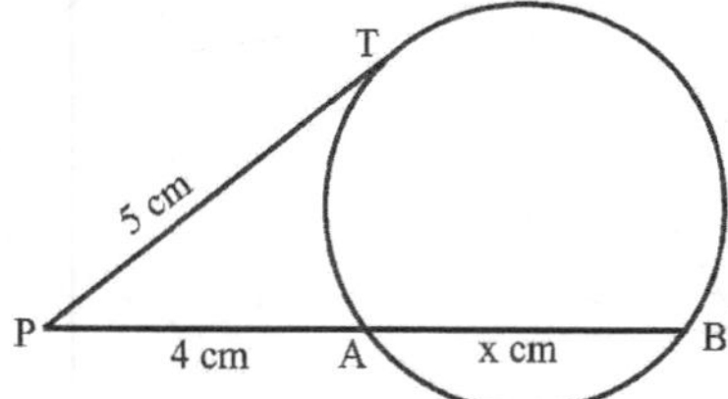

(a) 2.5 cm **(b) 2.6 cm**
(c) 2.25 cm **(d) 2.75 cm**

Solution: (c) $PA \times PB = PT^2 \Rightarrow 4 \times (4 + x) = 25$

$$\Rightarrow \quad 4 + x = \frac{25}{4} = 6.25 \Rightarrow x = 2.25 \text{ cm}$$

Illustration 21: Two equal circles pass through each other's centre. If the radius of each circle is 5 cm, what is the length of the common chord?

(a) $5\sqrt{3}$ **(b) $10\sqrt{3}$**
(c) $\frac{5\sqrt{3}}{2}$ **(d) 5**

Solution: (a)

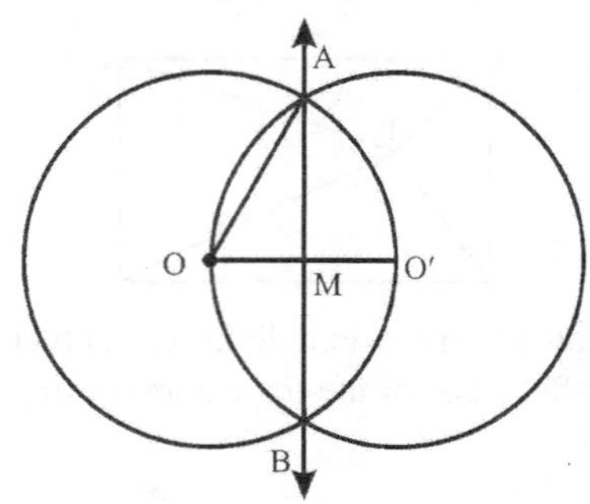

Given, distance between the centres of two circle = 5 cm

$$OO' = 5 \text{ cm}$$

$$\therefore \quad OM = \frac{5}{2} \text{ cm}$$

In ΔOAM,

$$OA^2 = OM^2 + AM^2$$

$$(5)^2 = \left(\frac{5}{2}\right)^2 + AM^2$$

$$AM = \sqrt{25 - \frac{25}{4}} = \frac{5\sqrt{3}}{2} \text{ cm}$$

$\therefore$ The length of common chord, $AB = 2 \times AM$

$$= 2 \times \frac{5\sqrt{3}}{2} = 5\sqrt{3} \text{ cm}$$

Illustration 22: The radius of a circle is 13 cm and xy is a chord which is at a distance of 12 cm from the centre. The length of the chord is

(a) 12 cm **(b) 10 cm**
(c) 20 cm **(d) 15 cm**

Solution: (b) From figure,

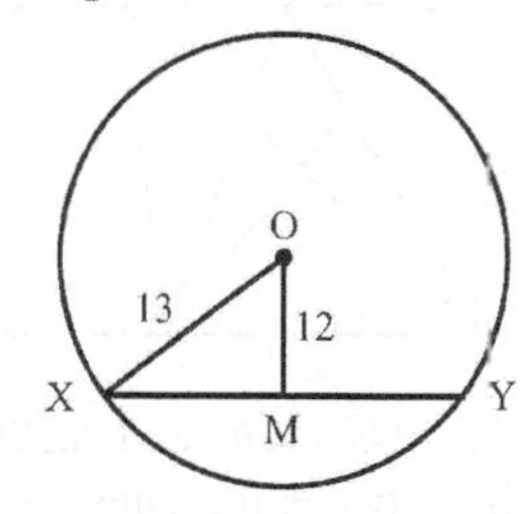

$$XM = \sqrt{13^2 - 12^2}$$

$$= \sqrt{169 - 144} = 5$$

$\therefore$ Length of the chord $= 2 \times XM$

$$= 2 \times 5 = 10 \text{ cm}$$

Illustration 23: Two circles of radii 10 cm and 8 cm. intersect and length of the common chord is 12 cm. Find the distance between their centres.

(a) 13.8 cm **(b) 13.29 cm**
(c) 13.2 cm **(d) 12.19 cm**

Solution: **(b)** Here, $OP = 10$ cm; $O'P = 8$ cm

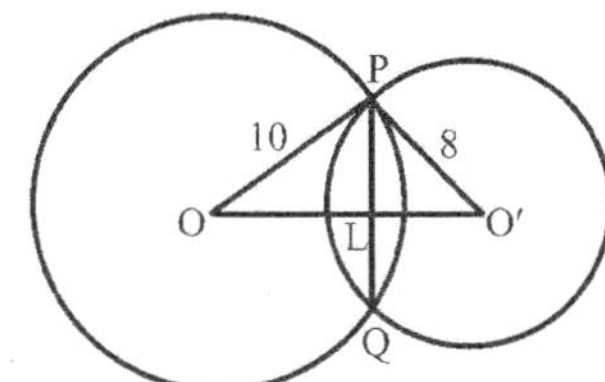

$PQ = 12$ cm

$\therefore \quad PL = 1/2\ PQ \Rightarrow PL = \frac{1}{2} \times 12 \Rightarrow PL = 6$ cm

In rt. ΔOLP, $OP^2 = OL^2 + LP^2$ (using Pythagoras theorem)

$\Rightarrow \quad (10)^2 = OL^2 + (6)^2 \Rightarrow OL^2 = 64; OL = 8$

In $\Delta O'LP$, $(O'L)^2 = O'P^2 - LP^2 = 64 - 36 = 28$

$O'L^2 = 28 \Rightarrow O'L = \sqrt{28}$

$O'L = 5.29$ cm

$\therefore \quad OO' = OL + O'L = 8 + 5.29$

$OO' = 13.29$ cm

LOCUS

The locus of a point is the path traced out by a moving point under given geometrical conditions. Alternatively, the locus is the set of all those points which satisfy the given geometrical conditions.

The plural of locus is loci and is read as 'Losai'.

The Locus of a Point in Different Conditions

(i) The locus of a point which is equidistant from two fixed points is the perpendicular bisector of the line segment joining the two fixed points.

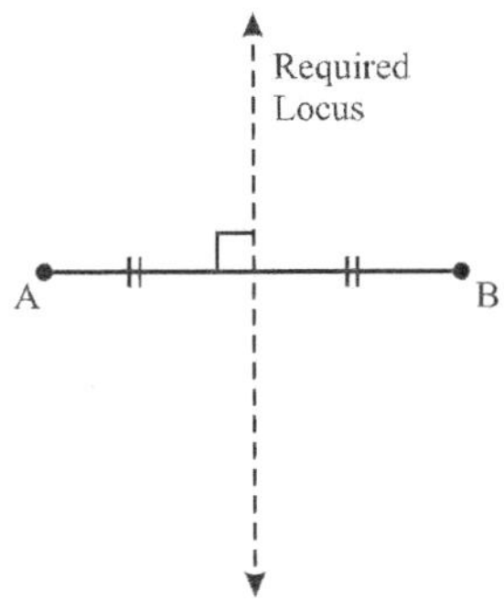

(ii) The locus of a point which is equidistant from two intersecting straight lines is a pair of straight lines which bisect the angles between the two given lines.

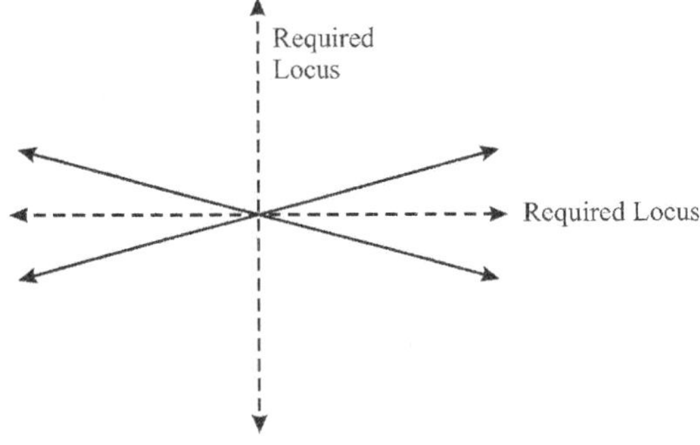

(iii) The locus of a point equidistant from two given parallel straight lines is a straight line parallel to the given straight lines and midway between them.

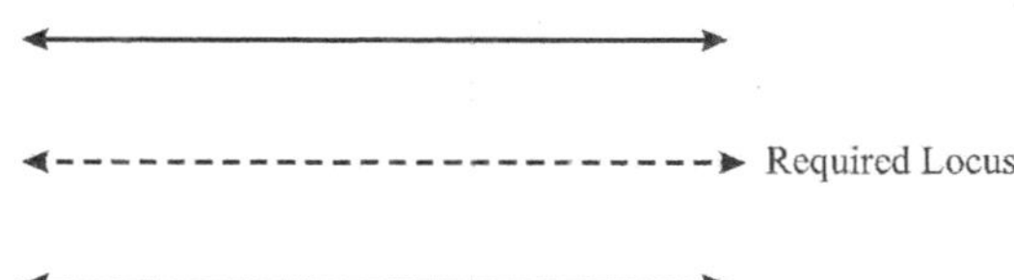

(iv) The locus of a point which is equidistant from a fixed point in a plane is a circle.

(v) The locus of a point, which is at a given distance from a given straight line, is a pair of parallel straight lines either side to the given line at a given distance from it.

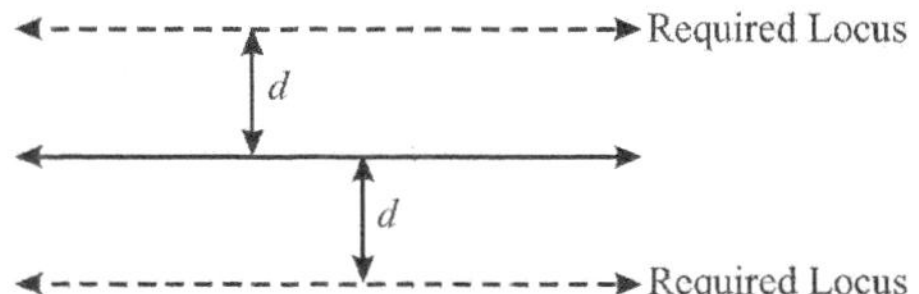

Here d is the given distance.

(vi) The locus of the centre of a wheel moving on a straight horizontal road, is a straight line parallel to the road and at a height equal to the radius of the wheel.

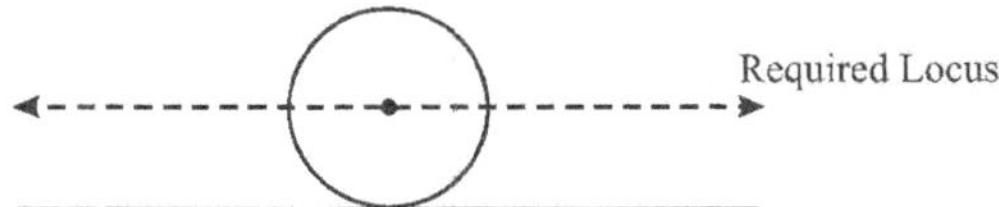

(vii) The locus of mid-points of all parallel chords of a circle, is the diameter of the circle which is perpendicular to the given parallel chords.

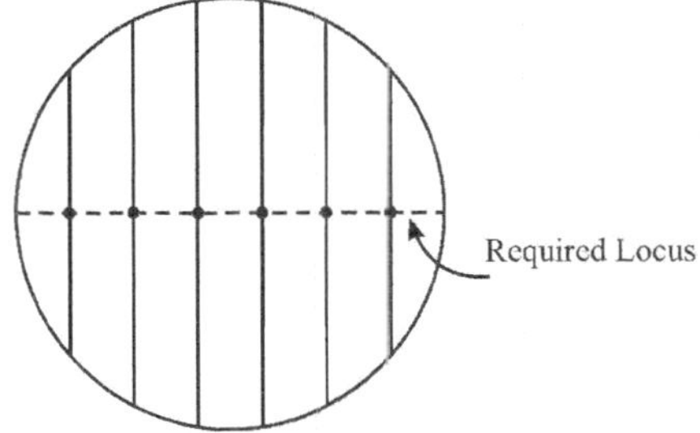

(viii) The locus of a point which is equidistant from two concentric circles is the circumference of the circle concentric with the given circles and midway between them.

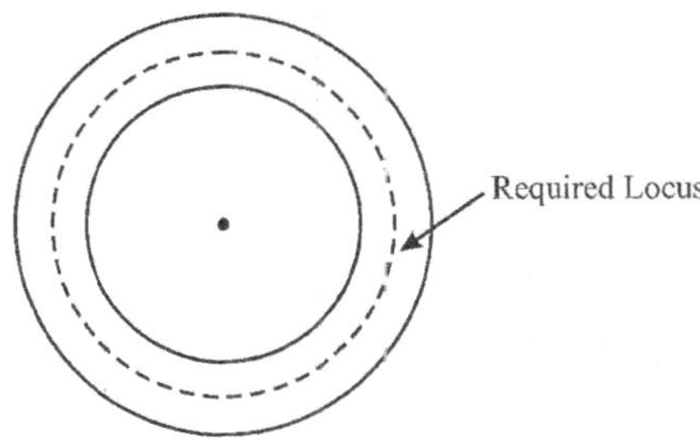

(ix) If A and B are two fixed points, then the locus of a point P such that ∠APB = 90°, is the circle with AB as diameter.

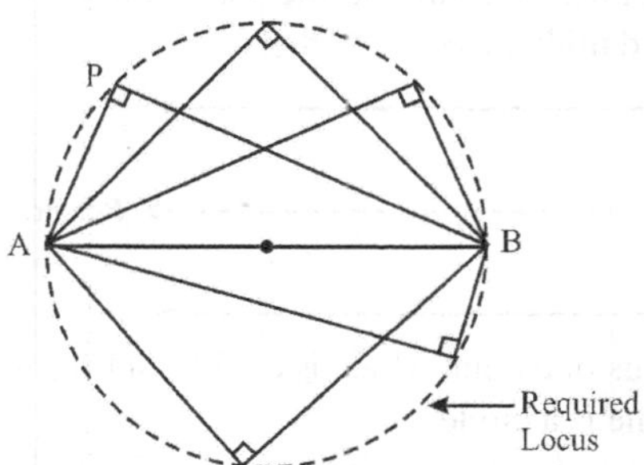

(x) The locus of midpoints of all equal chords of a circle is the circumference of the circle concentric with the given circle and radius equal to the distance of equal chords from the centre of the given circle.

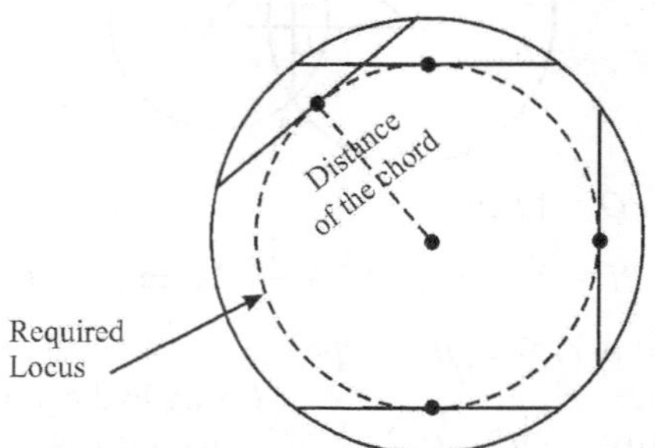

Practice Exercise

Level - I

1. In triangle ABC, angle B is a right angle. If (AC) is 6 cm, and D is the mid-point of side AC. The length of BD is

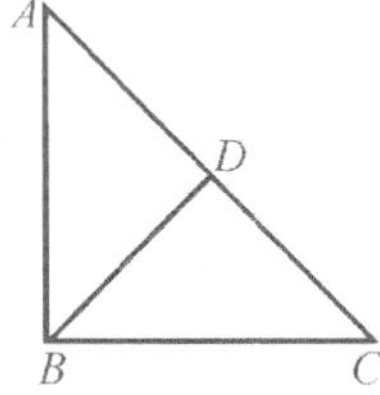

(a) 4 cm (b) $\sqrt{6}$cm
(c) 3 cm (d) 3.5 cm

2. $AB \perp BC$ and $BD \perp AC$. And CE bisects the angle C. $\angle A = 30°$. The, what is $\angle CED$.

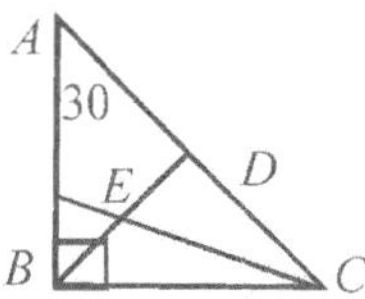

(a) 30° (b) 60°
(c) 45° (d) 65°

3. If two parallel lines are cut by two distinct transversals, then the quadrilateral formed by these four lines will always be a :
(a) parallelogram (b) rhombus
(c) square (d) trapezium

4. In the adjoining the figure, points A, B, C and D lie on the circle. $AD = 24$ and $BC = 12$. What is the ratio of the area of the triangle CBE to that of the triangle ADE

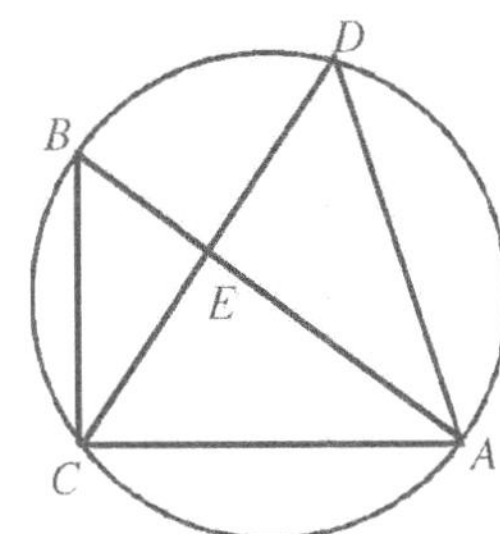

(a) 1 : 4 (b) 1: 2
(c) 1 : 3 (d) Insufficient data

5. In ΔABC, AD is the bisector of $\angle A$ if $AC = 4.2$ cm., $DC = 6$ cm., $BC = 10$ cm., find AB.
(a) 2.8 cm (b) 2.7 cm
(c) 3.4 cm (d) 2.6 cm

6. How many sides a regular polygon has with its sum of interior angles eight times its sum of exterior angles?
(a) 16 (b) 24
(c) 18 (d) 30

7. A point P is 26 cm away form the centre O of a circle and the length PT of the tangent draw from P to the circle is 10cm. Find radius of the circle
(a) 2.4 cm (b) 3.2 cm
(c) 2.2 cm (d) 4.2 cm

8. In the given figure, $AB \parallel CD$, $\angle BAE = 45°$, $\angle DCE = 50°$ and $\angle CED = x$, then find the value of x.

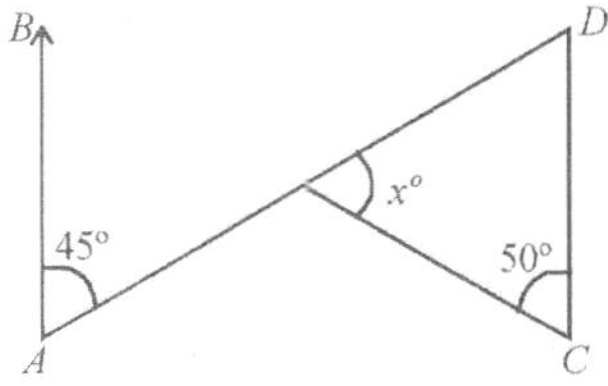

(a) 85° (b) 95°
(c) 60° (d) 20°

9. Given the adjoining figure. Find a, b, c

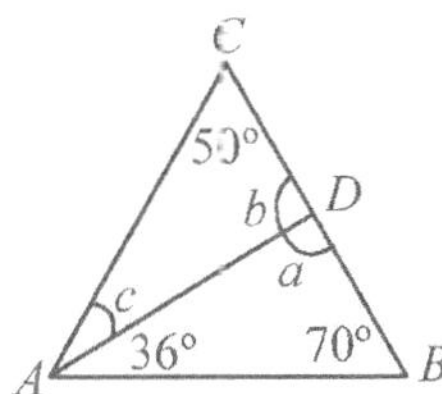

(a) 74°, 106°, 20° (b) 90°, 20°, 24°
(c) 60°, 30°, 24° (d) 106°, 24°, 74°

10. In the figure given below, AB is a diametre of the semicircle $APQB$, centre O, $\angle POQ = 48°$ cuts BP at X, calculate $\angle AXP$.

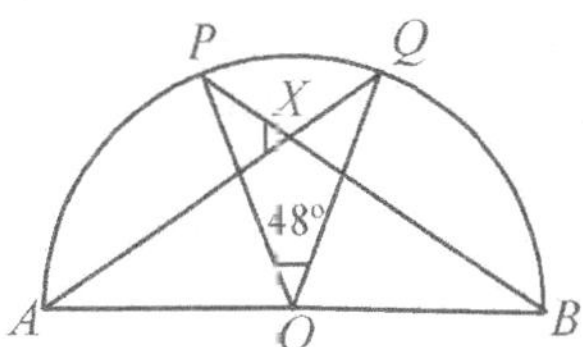

(a) 50° (b) 55°
(c) 66° (d) 40°

11. In the figure, if $\frac{NT}{AB} = \frac{9}{5}$ and if $MB = 10$, find MN.

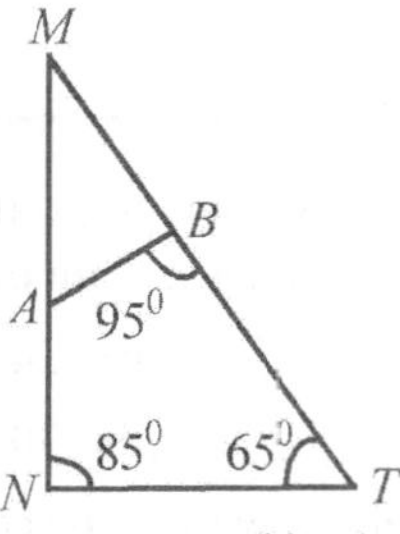

(a) 5 (b) 4
(c) 28 (d) 18

12. In the given figure, $AB \parallel CD$, $\angle ABO = 40°$ and $\angle CDO = 30°$. If $\angle DOB = x$, then find the value of x.

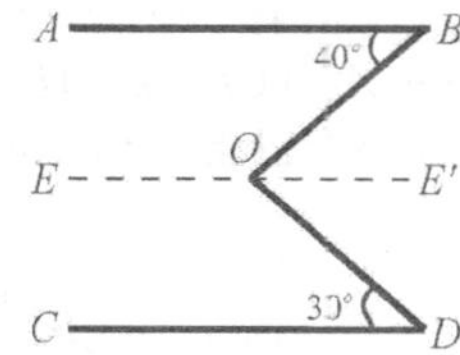

(a) 10° (b) 70°
(c) 110° (d) 20°

13. M and N are points on the sides PQ and PR respectively of a ΔPQR. For each of the following cases state whether MN is parallel to QR

A. $PM = 4$, $QM = 4.5$, $PN = 4$, $NR = 4.5$

B. $PQ = 1.28$, $PR = 2.56$, $PM = 0.16$, $PN = 0.32$

(a) only in case A
(b) only in case B
(c) both in the case A & B
(d) None of these

14. Given the adjoining figure. Find a, b, c.

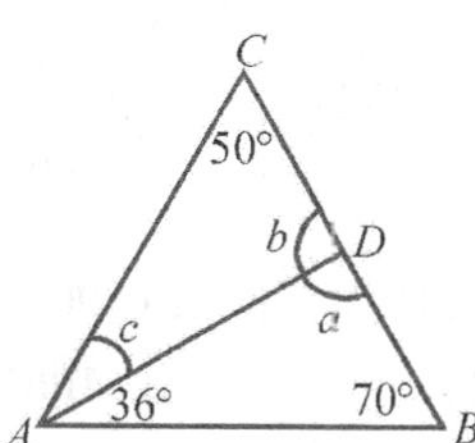

(a) 74°, 106°, 24° (b) 90°, 20°, 24°
(c) 60°, 30°, 24° (d) 106°, 24°, 74°

15. The perimeters of two similar Δs ABC and PQR are respectively 36 cm and 24 cm. If $PQ = 10$ cm, then AB is equal to

(a) 5 cm (b) 10 cm
(c) 15 cm (d) 9 cm

16. In the triangle ABC, AD bisects $\angle BAC$, $BC = 6.4$, $AB = 5$ and $AC = 3$, then the length of BD is equal to

(a) 3.5 (b) 5.5
(c) 3.2 (d) 4

17. In the given figure, $m\angle EDC = 54°$, $m\angle DCA = 40°$. Find x, y and z respectively.

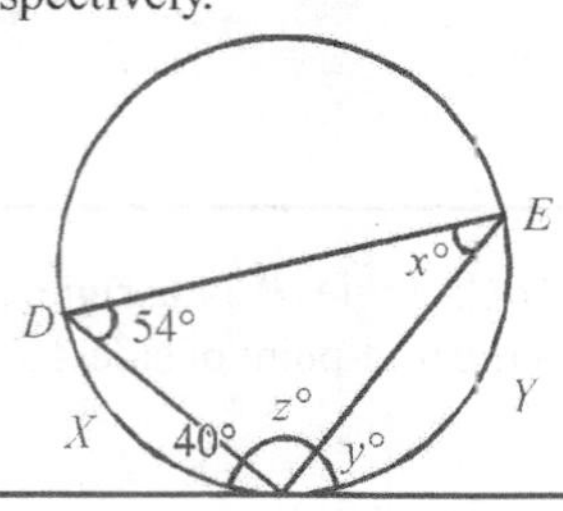

(a) 20°, 27°, 86° (b) 40°, 54°, 86°
(c) 20°, 27°, 43° (d) 40°, 54°, 43°

18. In the adjoining figure, $ABCD$ is a cyclic quadrilateral. If AB is a diameter, $BC = CD$ and $\angle ABD = 40°$, find the measure of $\angle DBC$.

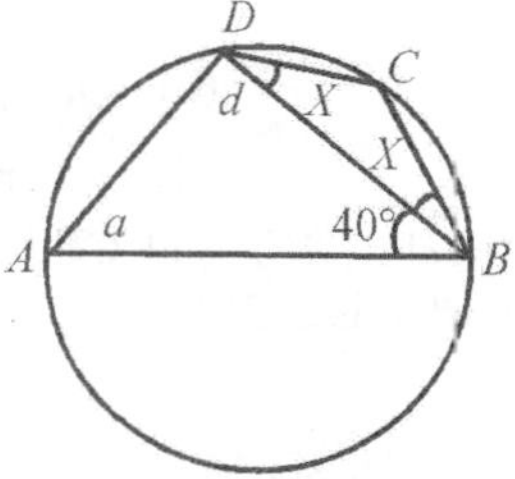

(a) 65 (b) 25
(c) 45 (d) 60

19. In the cyclic quadrilateral $ABCD$, $\angle BCD = 120°$, m (arc DZC) = 70°, find $\angle DAB$ and m(arc CXB).

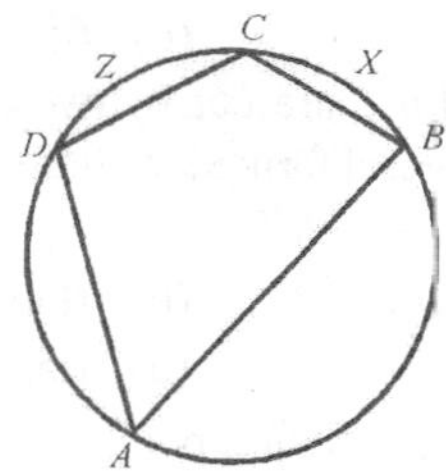

(a) 60°, 70° (b) 60°, 40°
(c) 60°, 50° (d) 60°, 60°

20. In the above figure (ii), angle c is –

(a) 270° (b) 70°
(c) 105° (d) 45°

21. If two tangents inclined at an angle 60° are drawn to a circle of radius 3 cm, then length of each tangent is equal to

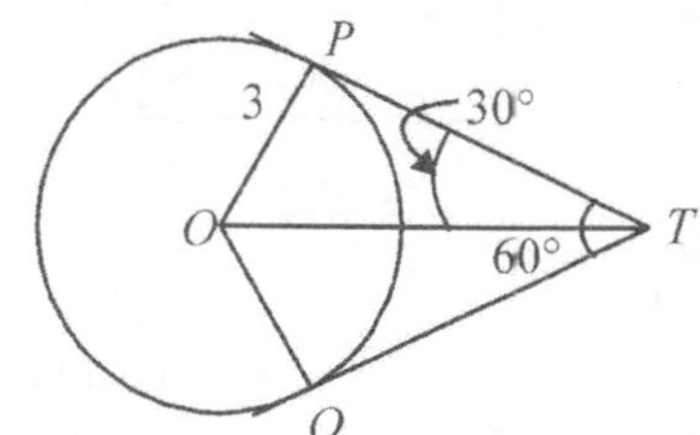

(a) $\frac{3}{2}\sqrt{2}$ cm (b) 6 cm
(c) 3 cm (d) $3\sqrt{3}$ cm

22. In the given fig. PQ is a chord of a circle and PT is the tangent at P such that $\angle QPT = 60°$. Then $\angle PRQ$ is equal to

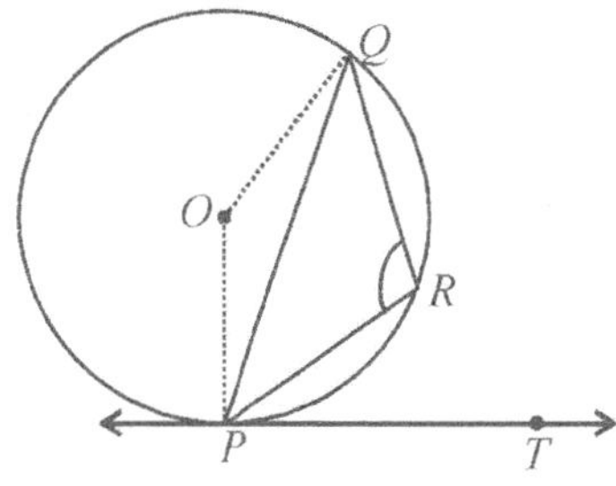

(a) 135° (b) 150°
(c) 120° (d) 110°

23. If four sides of a quadrilateral $ABCD$ are tangential to a circle, then.
(a) AC + AD = BD + CD (b) AD + BC = AB + CD
(c) AB + CD = AC + BC (d) AC + AD = BC + DB

24. In the given figure, $AB \parallel CD$, $\angle ALC = 60°$, EC is the bisector of $\angle LCD$ and $EF \parallel AB$. Then, find the measure of $\angle CEF$.

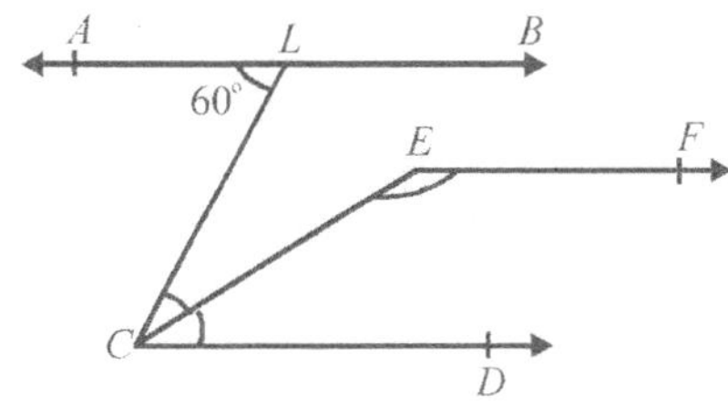

(a) 80° (b) 130°
(c) 120° (d) 150°

25. D, E, F are midpoints of BC, CA and AB respectively. G, H, I are midpoints of FE, FD, DE respectively. Areas of ΔDHI and ΔAFE are in the ratio

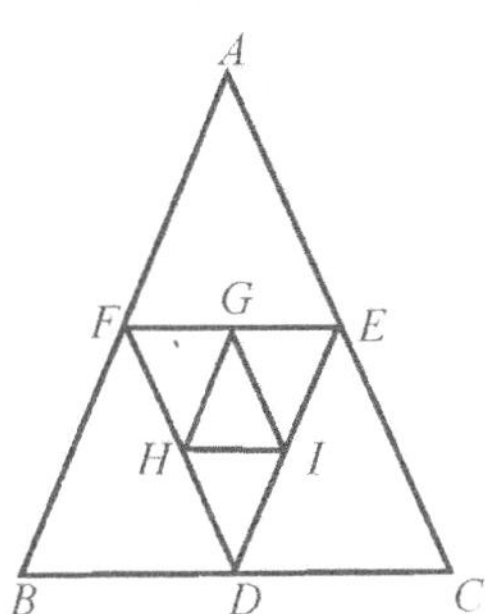

(a) 1 : 3 (b) 1 : 4
(c) 1 : 9 (d) 1 : 16

26. John wishes to determine the distance between two objects A and B, but there is an obstacle between the two objects which prevents him from making a direct measurement. He designed an ingenious way to overcome this difficulty. First, he fixes a pole at convenient point O so that from O, both ends are visible. Then he fixes another pole at a point D on the line AO (produced) such that $AO = DO$. In a similar way, he fixes a third pole at a point C on the line BO (produced) such that $BO = CO$. Then he measures CD and finds that $CD = 170$ cm. Find the distance between the objects A and B.
(a) 90 cm (b) 170 cm
(c) 140 cm (d) 150 cm

27. In the adjoining figure, $ABCD$ is a cyclic quadrilateral. Then $r + s$ is equal to

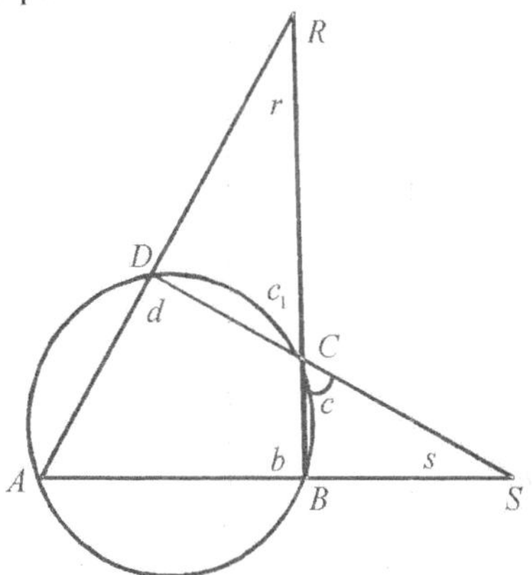

(a) 180° (b) $2c$
(c) $180° + 2c$ (d) $180° - 2c$

28. P is the centre of the circle
$m \angle ACB = 65°$.
Find m (are AXB)
(a) 105°
(b) 115°
(c) 65°
(d) 245°

29. The centroid, circumcenter, orthocenter in a triangle–
(a) are always coincident.
(b) are always collinear.
(c) are always the inside the triangular area.
(d) always coincide in a equilateral triangle and otherwise collinear.

30. In the given figure $AB \parallel CD$ and $AC \parallel BD$. If $\angle EAC = 40°$, $\angle FDG = 55°$, $\angle HAB = x°$, then find the value of x.

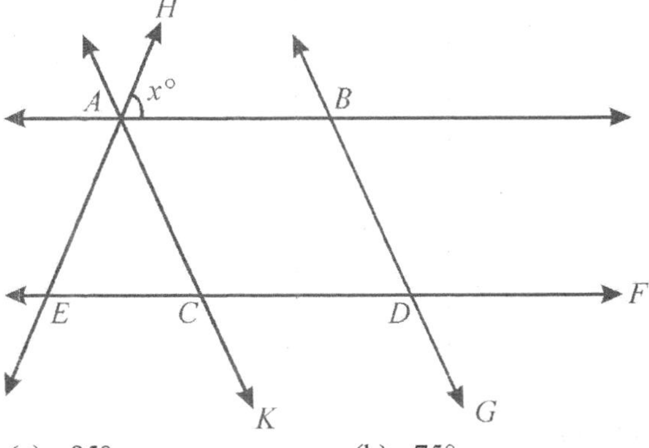

(a) 85° (b) 75°
(c) 65° (d) 55°

31. Which one of the following cannot be the ratio of angles in a right angled triangle?
(a) 1 : 2 : 3 (b) 1 : 1 : 2
(c) 1 : 3 : 6 (d) None of these

32. In ΔABC, $AB \perp BC$ and $BD \perp AC$. And CE bisects the angle C. $\angle A = 30°$. What is $\angle CED$?

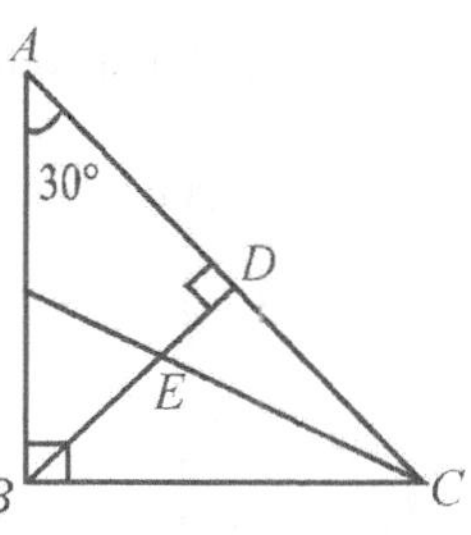

(a) 30° (b) 60°
(c) 45° (d) 65°

33. In the adjoining figure $ABCD$ is a rectangle and $DF = CF$ also, $AE = 3BE$. What is the value of $\angle EOF$, if $\angle DFO = 28°$ and $\angle AEO = 42°$?

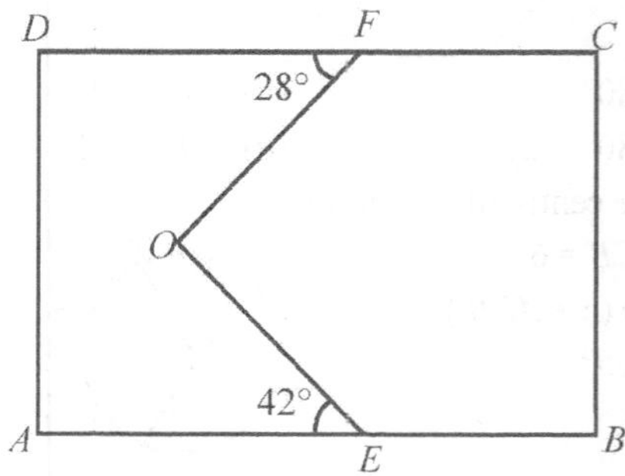

(a) 14° (b) 42°
(c) 70° (d) 90°

34. Each interior angle of a regular polygon exceeds its exterior angle by 132°. How many sides does the polygon have?
(a) 9 (b) 15
(c) 12 (d) None of these

35. In a triangle ABC, O is the centre of incircle PQR, $\angle BAC = 65°$, $\angle BCA = 75°$, find $\angle ROQ$:

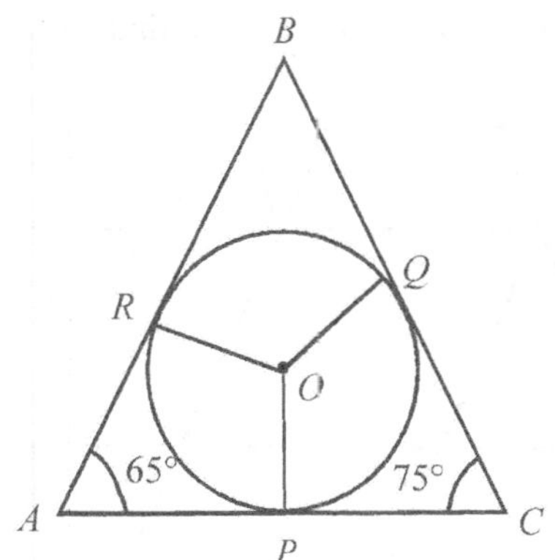

(a) 80° (b) 120°
(c) 140° (d) can't be determined

36. ABC and CDE are right angled triangle. $\angle ABC = \angle CDE = 90°$. D lies on AC and E lies on BC. $AB = 24$ cm, $BC = 60$ cm. If $DE = 10$ cm, then CD is:

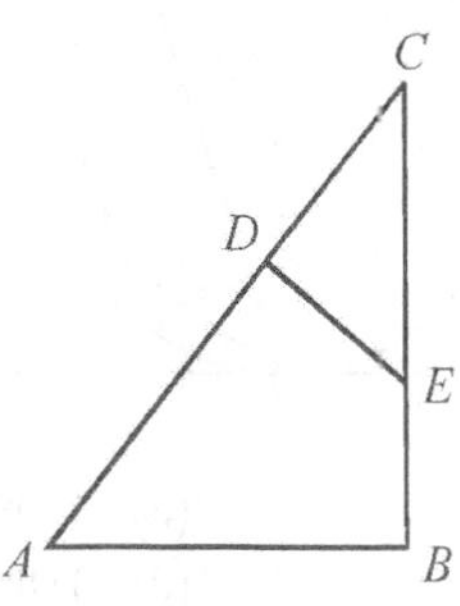

(a) 28 cm (b) 35 cm
(c) 25 cm (d) can't be determined

37. The largest angle of a triangle of sides 7 cm, 5 cm and 3 cm is
(a) 45° (b) 60°
(c) 90° (d) 1200

38. $ABCD$ is a paralellogram in which $\angle B = 70°$. Find the number of points X in the plane of the parallelogram such that it is equidistant from its vertices.
(a) zero (b) one
(c) two (d) three

39. $PQRS$ is trapezium, in which PQ is parallel to RS, and $PQ = 3\,(RS)$. The diagonal of the trapezium intersect each other at X, then the ratio of ΔPXQ and ΔRXS is
(a) 6 : 1 (b) 3 : 1
(c) 9 : 1 (d) 7 : 1

40. C is the midpoint of DE. Area of parallelogram $ABCD = 16$ sq. cm. Find the area of Δ $BCDE$.
(a) 8 sq.cm (b) 16 sq. cm
(c) 32 sq. cm (d) 24 sq. cm

41. What are the respective value of x, y and z in the given rectangle $ABCD$?

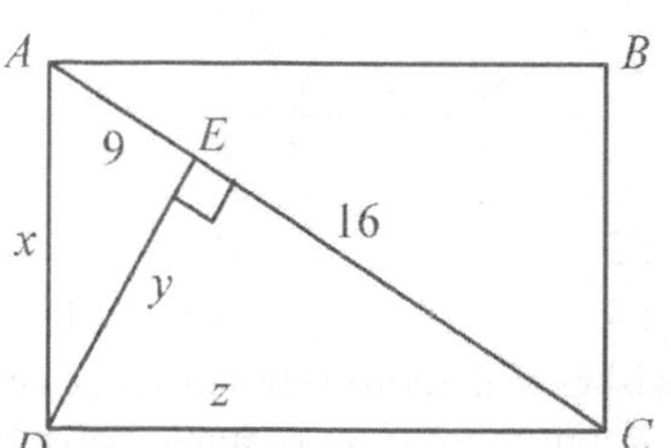

(a) 15, 12, 20 (b) 12, 15, 20
(c) 8, 10, 12 (d) None of these

42. In the figure (not drawn to scale) given below, if $AD = CD = BC$, and $\angle BCE = 96°$, how much is $\angle DBC$?

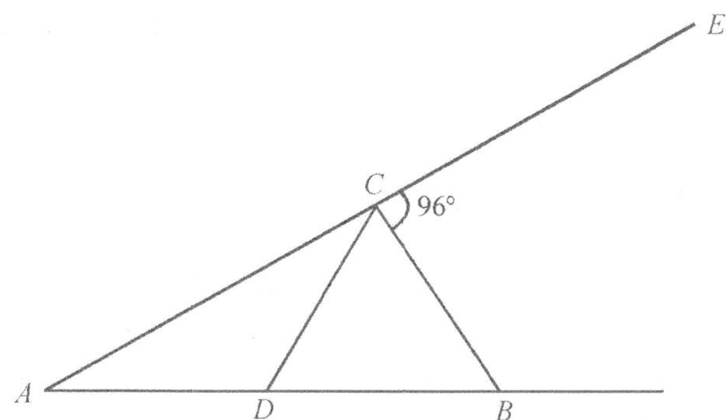

(a) 32° (b) 84°
(c) 64° (d) Cannot be determined

43. In a trapezium $ABCD, AB \parallel CD$ and $AD = BC$. If P is point of intersection of diagonals AC and BD, then all of the following is wrong except.

(a) $PA.PB = PC.PD$ (b) $PA.PC = PB.PD$
(c) $PA.AB = PD.DC$ (d) $PA.PD = AB.DC$

44. Find $\angle BOA$.

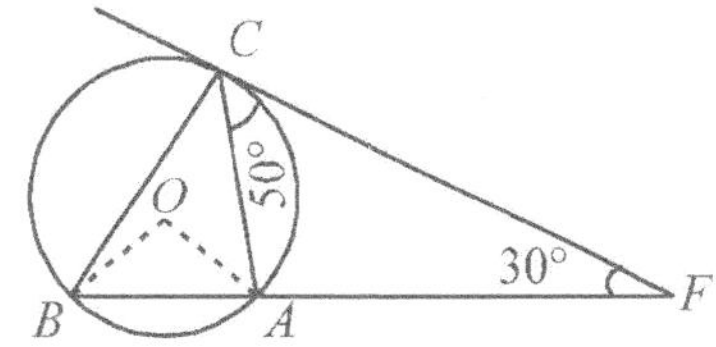

(a) 100° (b) 150°
(c) 80° (d) Indeterminate

45. ABCD is a quadrilateral in which diagonal BD = 64 cm, AL ⊥ BD and CM ⊥ BD, such that AL = 13.2 cm and CM = 16.8 cm. The area of the quadrilateral ABCD in square centimetres is [SSC-Sub. Ins.-2012]

(a) 537.6 (b) 960.0
(c) 422.4 (d) 690.0

46. ABCDEF is a regular hexagon of side 2 feet. The area, in square feet of the rectangle BCEF is [SSC-Sub. Ins.-2012]

(a) 4 (b) $4\sqrt{3}$
(c) 8 (d) $4+4\sqrt{3}$

47. The area of the shaded region in the following graph is [SSC-Sub. Ins.-2012]

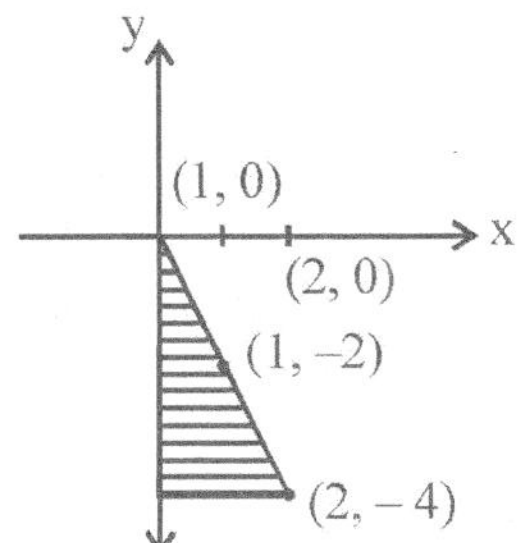

(a) 2 units (b) 4 units
(c) 6 units (d) 8 units

48. In ΔABC, ∠B = 60°, ∠C = 40°. If AD bisects ∠BAC and AE ⊥ BC, then ∠EAD is [SSC-Sub. Ins.-2012]

(a) 40° (b) 80°
(c) 10° (d) 20°

49. In the figure below, if AB ∥ CD and CE ⊥ ED, then the value of x is [SSC-Sub. Ins.-2012]

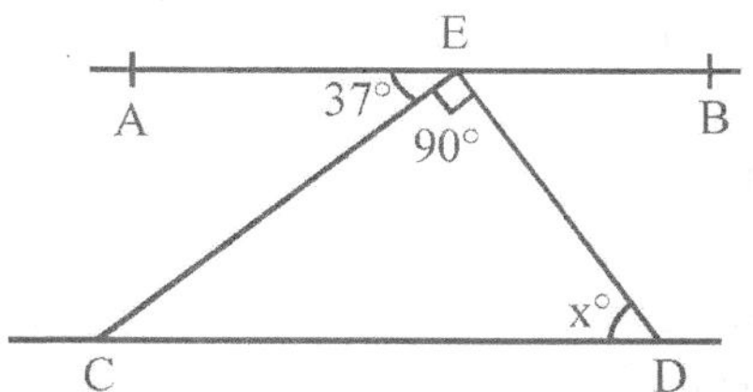

(a) 37 (b) 45
(c) 53 (d) 63

50. PA and PB are two tangents drawn from an external point P to a circle with centre O where the points A and B are the points of contact. The quadrilateral OAPB must be [SSC-Sub. Ins.-2012]

(a) a square (b) concylic
(c) a rectangle (d) a rhombus

51. G is the centroid of ΔABC. If AG = BC, then ∠BGC is [SSC-Sub. Ins.-2012]

(a) 60° (b) 120°
(c) 90° (d) 30°

52. In the following figure, if OA = 10 and AC = 16, then OB must be [SSC-Sub. Ins.-2012]

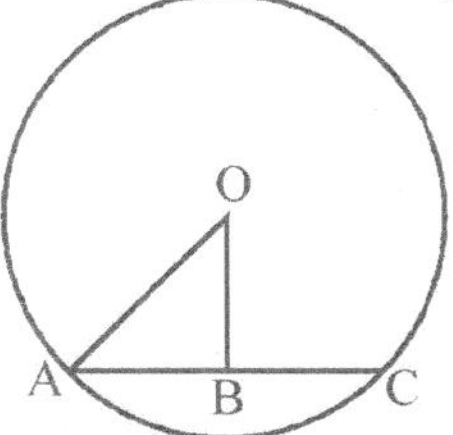

(a) 3 (b) 4
(c) 5 (d) 6

53. Triangle PQR circumscribes a circle with centre O and radius r cm such that ∠PQR = 90°. If PQ = 3 cm, QR= 4 cm, then the value of r is : [SSC-Sub. Ins.-2013]

(a) 2 (b) 1.5
(c) 2.5 (d) 1

54. In the following figure. AB be diameter of a circle whose centre is O. If ∠AOE = 150°. ∠DAO = 51° then the measure of ∠CBE is: [SSC-Sub. Ins.-2013]

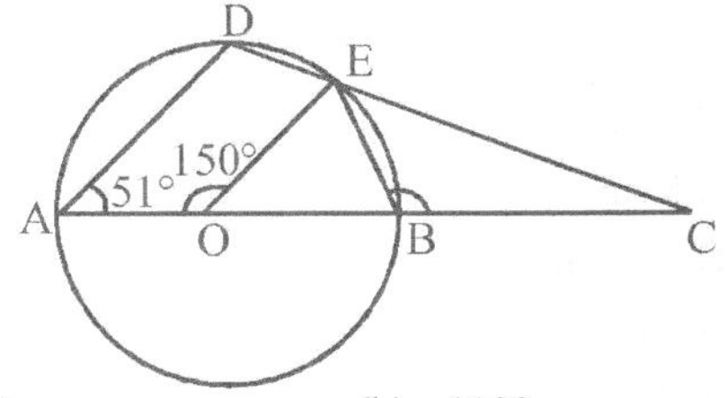

(a) 115° (b) 110°
(c) 105° (d) 120°

55. The areas of two similar triangles ABC and DEF are 20 cm^2 and 45 cm^2 respectively. If AB = 5 cm. then DE is equal to: *[SSC-Sub. Ins.-2013]*

(a) 6.5 cm (b) 7.5 cm
(c) 8.5 cm (d) 5.5 cm

56. In a triangle ABC, BC is produced to D so that CD = AC. If $\angle BAD = 111°$ and $\angle ACB = 80°$, then the measure of $\angle ABC$ is: *[SSC-Sub. Ins.-2013]*

(a) 31° (b) 33°
(c) 35° (d) 29°

57. In ΔABC, $\angle A + \angle B = 145°$ and $\angle C + 2\angle B = 180°$. State which one of the following relations is true? *[SSC-Sub. Ins.-2013]*

(a) CA = AB (b) CA < AB
(c) BC > AB (d) CA > AB

58. In a Δ ABC, $\frac{AB}{AC} = \frac{BD}{DC}$, $\angle B = 70°$ and $\angle C = 50°$, then $\angle BAD =$ *[SSC-Sub. Ins.-2014]*

(a) 60° (b) 20°
(c) 30° (d) 50°

59. In a Δ ABC, AD, BE and CF are three medians. The perimeter of ΔABC is always *[SSC-Sub. Ins.-2014]*

(a) equal to $\left(\overline{AD} + \overline{BE} + \overline{CF}\right)$

(b) greater than $\left(\overline{AD} + \overline{BE} + \overline{CF}\right)$

(c) less than $\left(\overline{AD} + \overline{BE} + \overline{CF}\right)$

(d) None of these

60. In a Δ ABC, $\overline{AD}$, $\overline{BE}$ and $\overline{CF}$ are three medians. Then the ratio $\left(\overline{AD} + \overline{BE} + \overline{CF}\right) : \left(\overline{AB} + \overline{AC} + \overline{BC}\right)$ is *[SSC-Sub. Ins.-2014]*

(a) equal to $\frac{3}{4}$ (b) less than $\frac{3}{4}$

(c) greater than $\frac{3}{4}$ (d) equal to $\frac{1}{2}$

61. Two circles with radii 25 cm and 9 cm touch each other externally. The length of the direct common tangent is *[SSC-Sub. Ins.-2014]*

(a) 34 cm (b) 30 cm
(c) 36 cm (d) 32 cm

62. If AB = 5 cm, AC = 12 and $AB \perp AC$, then the radius of the circumcircle of ΔABC is *[SSC-Sub. Ins.-2014]*

(a) 6.5 cm (b) 6 cm
(c) 5 cm (d) 7 cm

63. ABC is a right angled triangle, right angled at C and p is the length of the perpendicular from C on AB. If a, b and c are the lengths of the sides BC, CA and AB respectively, then *[SSC 10+2-2012]*

(a) $\frac{1}{p^2} = \frac{1}{b^2} - \frac{1}{a^2}$ (b) $\frac{1}{p^2} = \frac{1}{a^2} + \frac{1}{b^2}$

(c) $\frac{1}{p^2} + \frac{1}{a^2} = \frac{1}{b^2}$ (d) $\frac{1}{p^2} = \frac{1}{a^2} - \frac{1}{b^2}$

64. From a point P, two tangents PA and PB are drawn to a circle with centre O. If OP is equal to diameter of the circle, then $\angle APB$ is *[SSC 10+2-2013]*

(a) 60° (b) 45°
(c) 90° (d) 30°

65. A chord 12 cm long is drawn in a circle of diameter 20 cm. The distance of the chord from the centre is *[SSC 10+2-2013]*

(a) 16 cm (b) 8 cm
(c) 6 cm (d) 10 cm

66. If in Δ ABC, $\angle ABC = 5\angle ACB$ and $\angle BAC = 3\angle ACB$, then $\angle ABC =$ *[SSC 10+2-2013]*

(a) 120° (b) 130°
(c) 80° (d) 100°

67. The perpendiculars, drawn from the vertices to the opposite sides of a triangle, meet at the point whose name is *[SSC 10+2-2013]*

(a) orthocentre (b) incentre
(c) circumcentre (d) centroid

68. In ΔABC, D and E are two points on the sides AB and AC respectively so that DE||BC and $\frac{AD}{BD} = \frac{2}{3}$. Then $\frac{\text{the area of trapezium DECB}}{\text{the area of } \Delta ABC}$ is equal to *[SSC 10+2-2014]*

(a) $\frac{5}{9}$ (b) $\frac{21}{25}$

(c) $1\frac{4}{5}$ (d) $5\frac{1}{4}$

69. One of the angles of a parallelogram is 45°. What will be the sum of the larger angle and twice the smaller angle of the parallelogram ? *[IBPS Clerk-2012]*

(a) 228° (b) 224°
(c) 225° (d) 222°
(e) None of these

Level - II

1. Here XY has been divided into 5 congruent segments and semicircles have been drawn. But suppose XY were divided into millions of congruent segments and semicircles were drawn, what would the sum of the lengths of the arcs be?

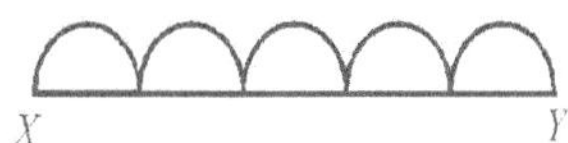

(a) $2YX$ (b) $5XY$
(c) XY (d) None of these

2. In the adjoining figure, chord AD and BC of a circle are produced to meet at P, $PA = 10$ cm, $PB = 8$ cm, $PC = 5$ cm, $AC = 6$ cm. Find BD, PD.

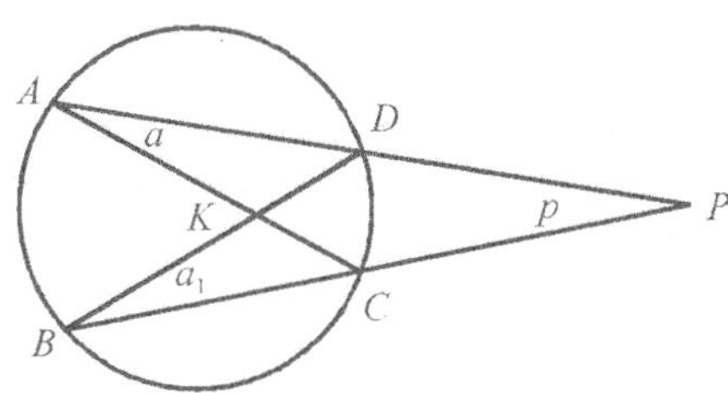

(a) 5.8, 3 (b) 3.8, 5
(c) 2.8, 6 (d) 4.8, 4

3. In the adjoining figure the circles touches the side of the quadrilateral $ABCD$. If $AB = p$, express $(AD + BC)$ in terms of p and

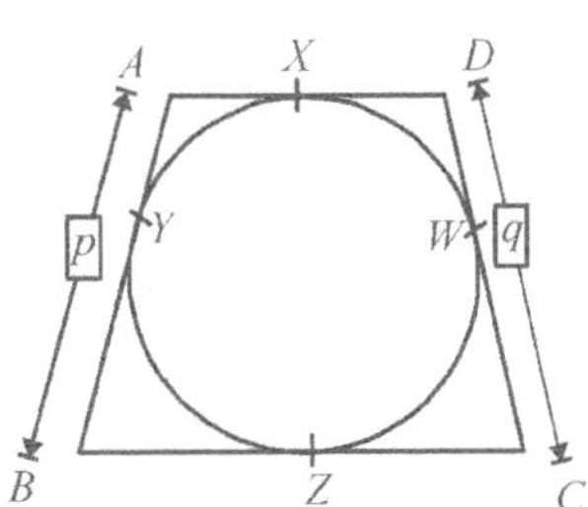

(a) $p + q$ (b) $\frac{1}{2}p + q$
(c) $2(p - q)$ (d) $3(p - q)$

4. In the figure given below, AB is a diametre of the semi-circle $APQB$, centre O, $\angle POQ = 48°$ cuts BP at X, calculate $\angle AXP$.

(a) 50°
(b) 55°
(c) 66°
(d) 40°

5. In the figure, if $PS = 360$, find PQ, QR and RS.

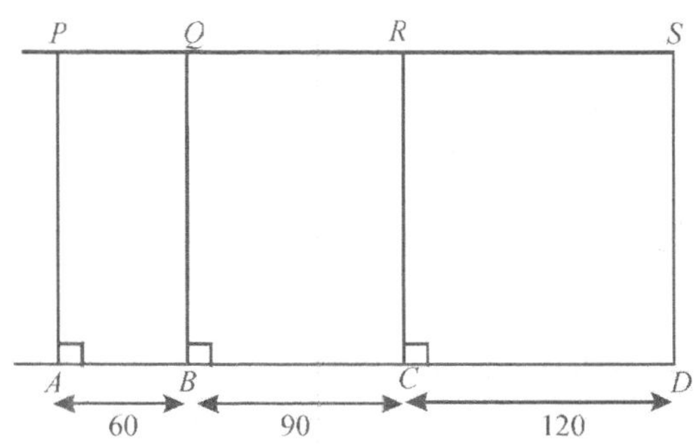

(a) 150° (b) 160°
(c) 180° (d) 190°

6. If $ABCD$ is a square and BCE is an equilateral triangle, what is the measure of the angle DEC?

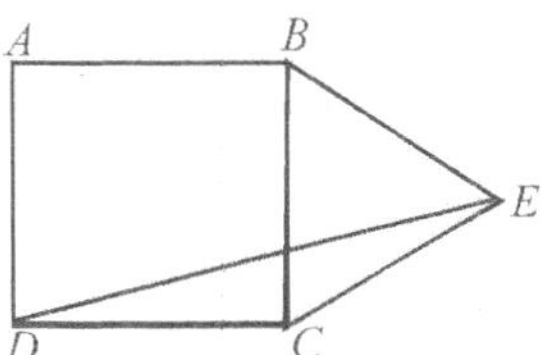

(a) 15° (b) 30°
(c) 20° (d) 45°

7. Based on the figure below, what is the value of x, if $y = 10$

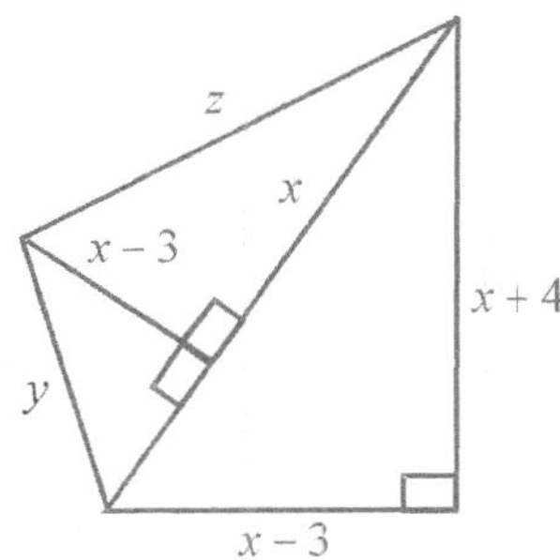

(a) 10 (b) 11
(c) 12 (d) None of these

DIRECTIONS (Qs. 8–10) : ***Answer the questions on the basis of the information given below.***

In the adjoining figure, I and II are circles with centers P and Q respectively. The two circle touch each other and have a common tangent that touches them at points R and S respectively. This common tangent meets the line joining P and Q at O. The diameters of I and II are in the rartio 4 : 3. It is also known that the length of PO is 28 cm.

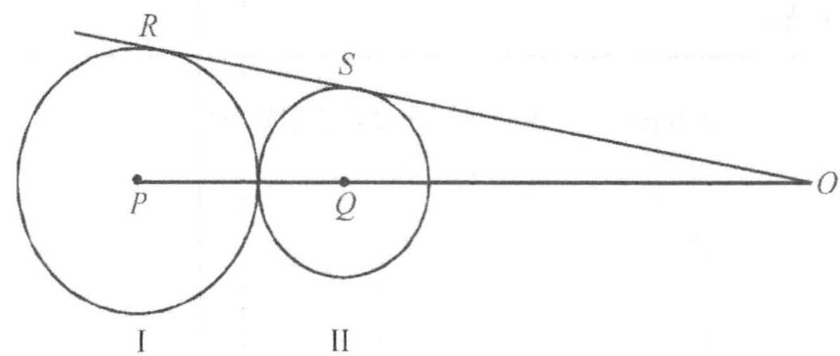

8. What is the ratio of the length of PQ to that of QO?

(a) 1 : 4 (b) 1 : 3
(c) 3 : 8 (d) 3 : 4

9. What is the radius of the circle II?

(a) 2 cm (b) 3 cm
(c) 4 cm (d) 5 cm

10. The length of SO is

(a) $8\sqrt{3}$ cm (b) $10\sqrt{3}$ cm
(c) $12\sqrt{3}$ cm (d) $14\sqrt{3}$ cm

11. What is the inradius of the incircle shown in the figure?

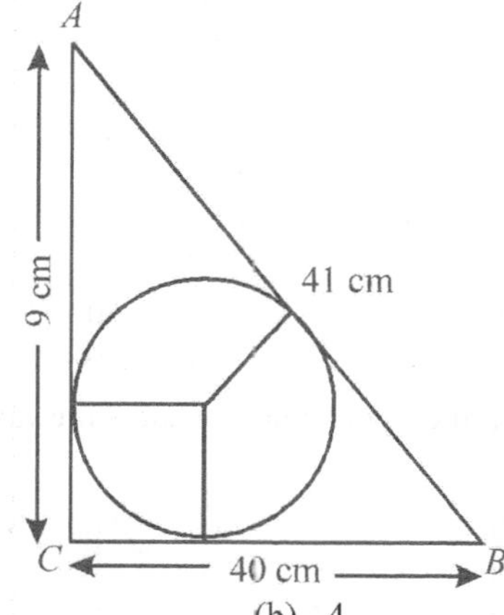

(a) 9 cm (b) 4
(c) can't be determined (d) None of these

12. In a circle O is the centre and $\angle COD$ is right angle. $AC = BD$ and CD is the tangent at P. What is the value of $AC + CP$, if the radius of the circle is 1 metre?

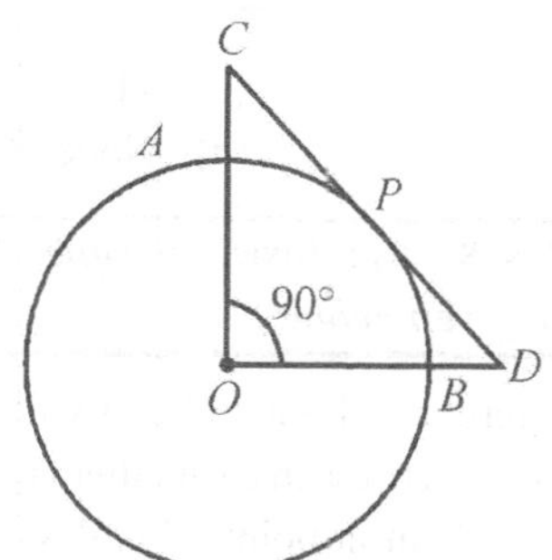

(a) 105 cm (b) 141.4 cm
(c) 138.6 cm (d) can't be determined

13. In the triangle ABC, MN is parallel to AB. Area of trapezium $ABNM$ is twice the area of triangle CMN. What is ratio of $CM : AM$?

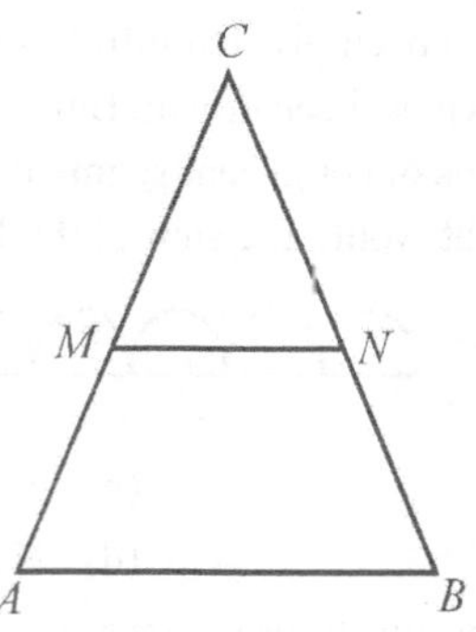

(a) $\frac{1}{\sqrt{3}+1}$ (b) $\frac{\sqrt{3}-1}{2}$
(c) $\frac{\sqrt{3}+1}{2}$ (d) None of these

14. ABC is a triangle in which $\angle CAB = 80°$ and $\angle ABC = 50°$, AE, BF and CD are the altitudes and O is the orthocentre. What is the value of $\angle AOB$?

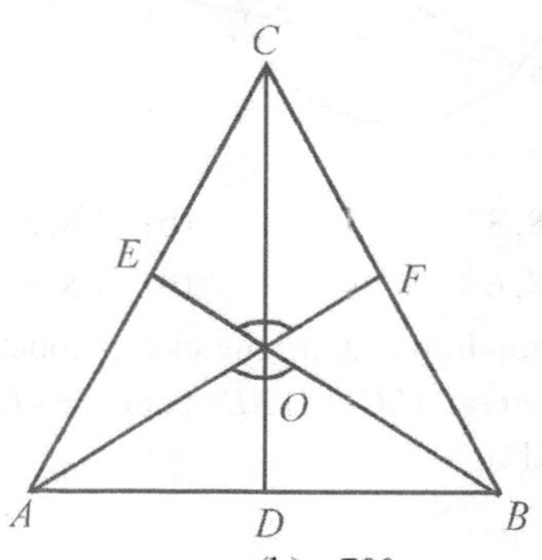

(a) 65° (b) 70°
(c) 50° (d) 130°

15. In the given diagram O is the centre of the circle and CD is a tangent. $\angle CAB$ and $\angle ACD$ are supplementary to each other $\angle OAC$ 30°. Find the value of $\angle OCB$:

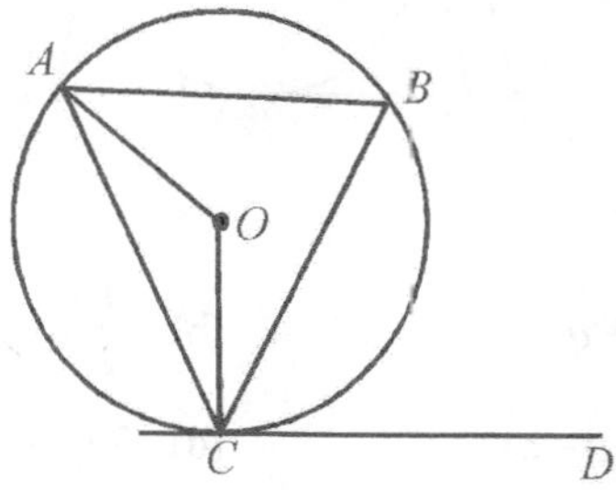

(a) 30° (b) 20°
(c) 60° (d) None of these

16. The sides of a triangle are in the ratio of $\frac{1}{2}:\frac{1}{3}:\frac{1}{4}$. If the perimeter is 52 cm, then the length of the smallest side is

(a) 9 cm (b) 10 cm
(c) 11 cm (d) 12 cm

17. The ratio of the area of a square to that of the square drawn on its diagonal is

(a) 1 : 4 (b) 2 : 1
(c) 1 : 2 (d) 1 : 3

18. PQ is a tangential to circles with centers A and B at P and Q respectively. If AB = 10 cm. and PQ = 8 cm, find the radius of the bigger circle. Given that area of triangle APO is four times the area of triangle OQB –

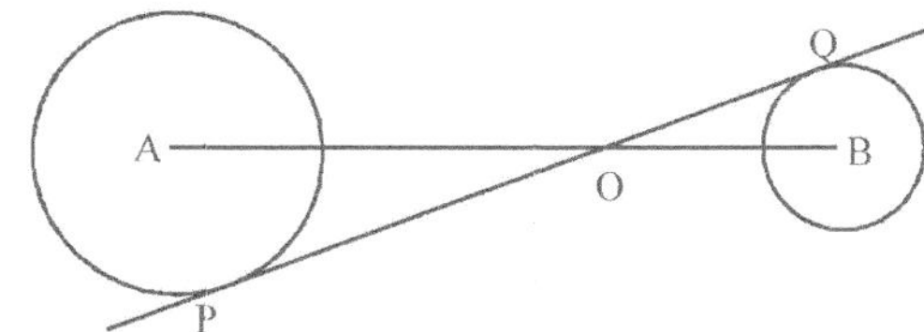

(a) 2 cm (b) 4 cm
(c) 6 cm (d) 8cm

19. Two circles touch each other internally. Their radii are 2 cm and 3 cm. The biggest chord of the outer circle which is outside the inner circle is of length

(a) $2\sqrt{2}$ cm (b) $3\sqrt{2}$ cm
(c) $2\sqrt{3}$ cm (d) $4\sqrt{2}$ cm

20. The sum of the interior angles of a polygon is 1620°. The number of sides of the polygon are :

(a) 9 (b) 11
(c) 15 (d) 12

21. In ΔABC, $DE \,||\, BC$ and $\frac{AD}{DB} = \frac{3}{5}$. If AC = 5.6 cm, find AE.

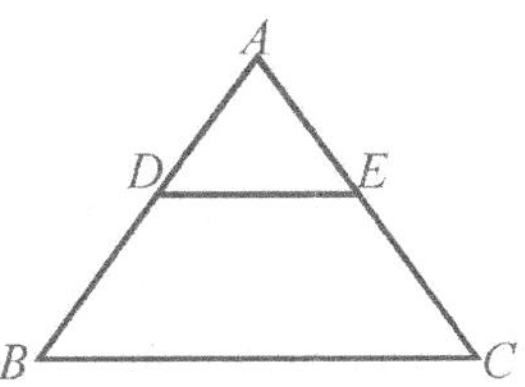

(a) 2.1 cm (b) 3.1 cm
(c) 1.2 cm (d) 2.3 cm

22. If one of the diagonals of a rhombus is equal to its side, then the diagonals of the rhomhus are in the ratio:

(a) $\sqrt{3} : 1$ (b) $\sqrt{2} : 1$
(c) 3 : 1 (d) 2 : 1

23. If $ABCD$ is a square and BCE is an equilateral triangle, what is the measure of the angle DEC?

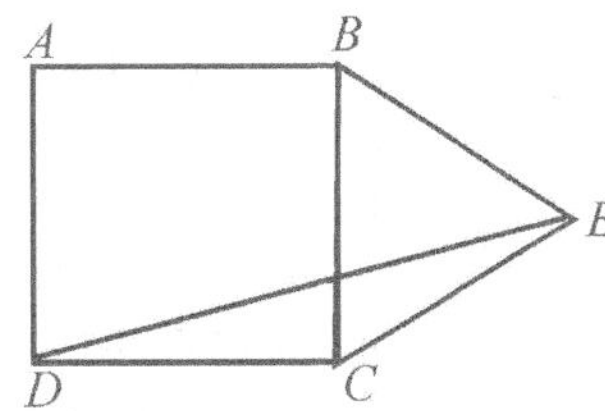

(a) 15° (b) 30°
(c) 20° (d) 45°

24. $ABCD$ is a square, F is the mid-point of AB and E is a point on BC such that BE is one-third of BC. If area of ΔFBE = 108 m^2, then the length of AC is :

(a) 63 m (b) $36\sqrt{2}$ m
(c) $63\sqrt{2}$ m (d) $72\sqrt{2}$ m

25. Arc ADC is a semicircle and $DB \perp AC$. If AB = 9 and BC = 4, find DB.

(a) 6 (b) 8
(c) 10 (d) 12

26. In the figure below, which of the following is the relationship between 'x' and 'y' if the equal circles shown are tangents to each other and to the sides of the rectangle

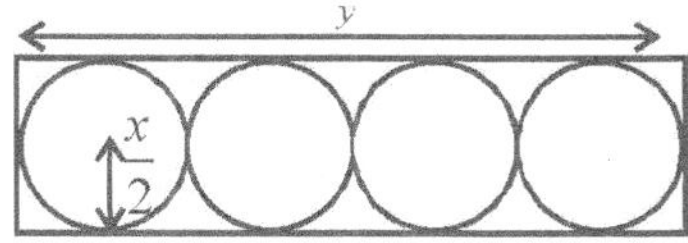

(a) $x = \frac{1}{4}y$ (b) $x = \frac{y}{\pi}$
(c) $x = \pi y^2$ (d) $x = 2\pi y$

27. In the given figure given below, E is the mid-point of AB and F is the midpoint of AD. if the area of $FAEC$ is 13, what is the area of $ABCD$?

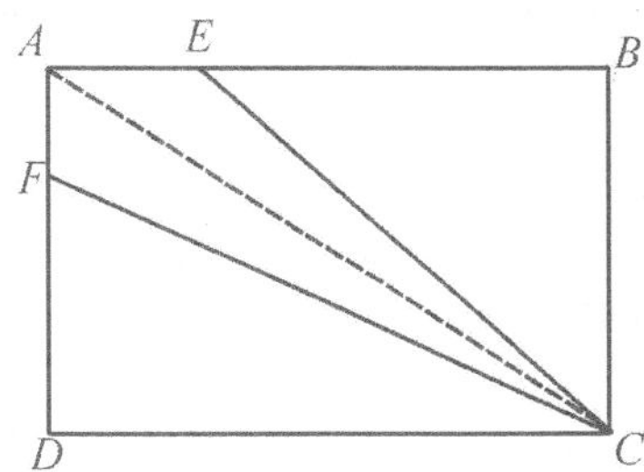

(a) 19.5 (b) 26
(c) 39 (d) None of these

28. In the given figure, $\angle ABC$ and $\angle DEF$ are two angles such that $BA \perp ED$ and $EF \perp BC$, then find value of $\angle ABC + \angle DEF$.

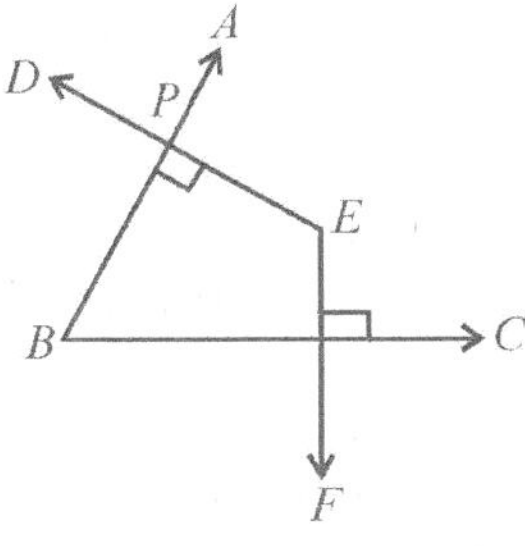

(a) 120° (b) 180°
(c) 150° (d) 210°

29. In the cyclic quadrilateral $ABCD$ $BCD = 120°$, m (arc DZC) $= 7°$, find DAB and m (arc CXB).

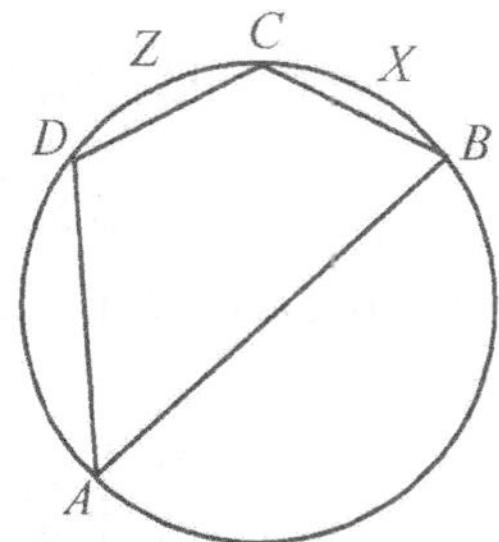

(a) 60°, 70° (b) 60°, 40°
(c) 60°, 50° (d) 60°,60°

30. In the given figure, AB is chord of the circle with centre O, BT is tangent to the circle. The values of x and y are

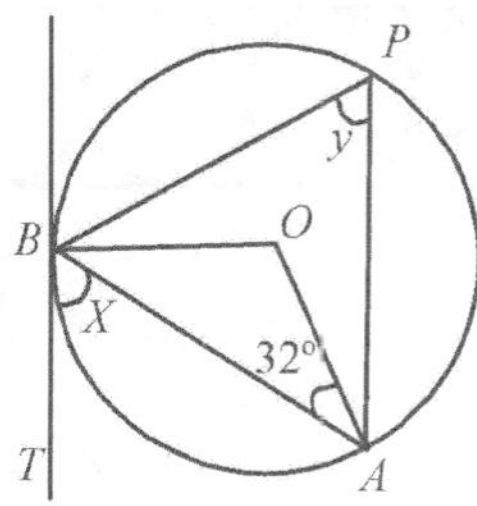

(a) 52°, 52° (b) 58°, 52°
(c) 58°, 58° (d) 60°, 64°

31. The distance between two parallel chords of length 8 cm each in a circle of diameter 10 cm is
(a) 6 cm (b) 7 cm
(c) 8 cm (d) 5.5 cm

32. The internal bisectors of the angles B and C of a triangle ABC meet at O. Then find the measure of $\angle BOC$.
(a) $90° - \frac{\angle A}{2}$ (b) $180° - \frac{\angle A}{2}$
(c) $90° + \frac{\angle A}{2}$ (d) $180° + \frac{\angle A}{2}$

33. In a ΔABC, angle C is 68°, the perpendicular bisector of AB at R meets BC at P. If $\angle PAC = 42°$ then $\angle ABC$ is equal to
(a) 45° (b) 42°
(c) 35° (d) 34°

34. A chord of length 14 cm is at a distance of 6 cm from the centre of a circle. Find the length of another chord at a distance of 2 cm from the centre of the circle.
(a) 18 cm (b) 16 cm
(c) 10 cm (d) 12 cm

35. In the adjoining figure x is a point on diameter AB of the circle with centre o, such that $AX = 9$ cm, $XB = 5$ cm. Find the radius of the circle (centre Y) which touches the diameter at X and touches the circle, centre O, internally at Z.

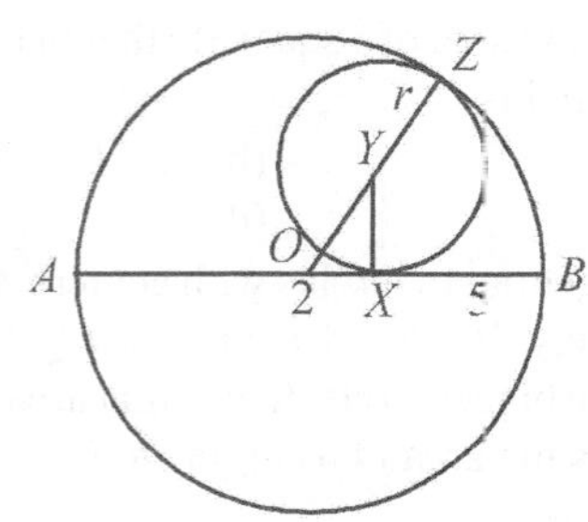

(a) $3\frac{3}{14}$ cm (b) $3\frac{1}{14}$ cm.
(c) $1\frac{1}{14}$ cm. (d) $2\frac{3}{14}$ cm.

36. In ΔABC, $AB = AC = 8$, PR and PQ are parallel to lines AC and AB respectively. P is the midpoint of BC. Find the perimeter of □$PRAQ$.
(a) 16 (b) 18
(c) 20 (d) 12

37. The height of the hexagon whose side is a

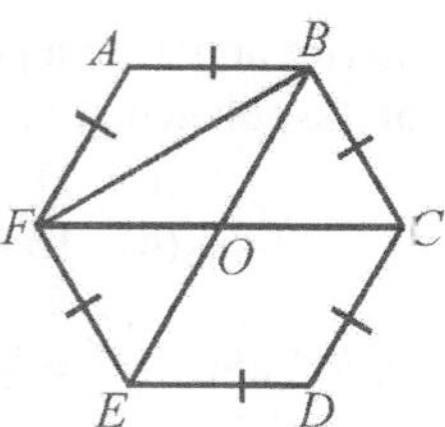

(a) $\frac{3\sqrt{3}}{2}a$ (b) $\frac{3\sqrt{3}}{4}$
(c) $\sqrt{3}\,a$ (d) None of these

38. In ΔABC, $AB = 8$, $AC = 6$, Altitude $AD = 4.8$ AE is the diameter of the circumcircle. Find the circumradius.

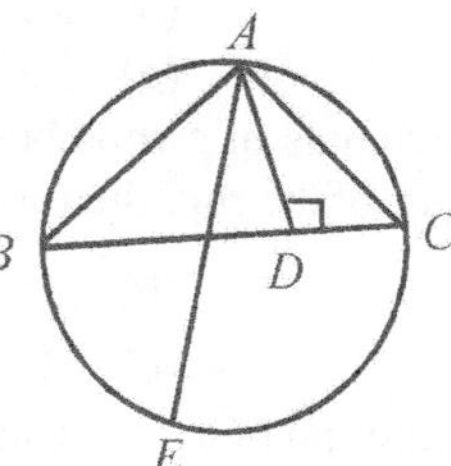

(a) 5 (b) 10
(c) 15 (d) Cannot be determined

39. The length of a ladder is exactly equal to the height of the wall it is resting against. If lower end of the ladder is kept on a stool of height 3 m and the stool is kept 9 m away from the wall the upper end of the ladder coincides with the tip of the wall. Then, the height of the wall is
(a) 12 m. (b) 15 m.
(c) 18 m. (d) 11 m.

40. Three circles, each of radius 20 and centres at P, Q, R. further, $AB = 5$, $CD = 10$ and $EF = 12$. What is the perimeter of the triangle PQR?

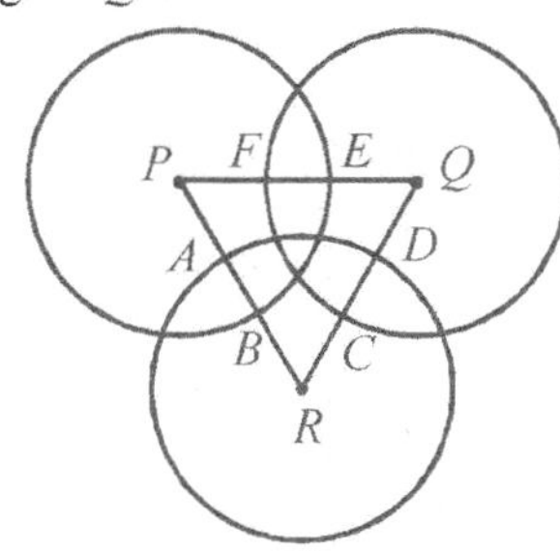

(a) 120 (b) 66
(c) 93 (d) 87

41. In the diagram given below, $\angle ABD = \angle CDB = \angle PQD = 90°$. If $AB : CD = 3 : 1$, the ratio of $CD : PQ$ is

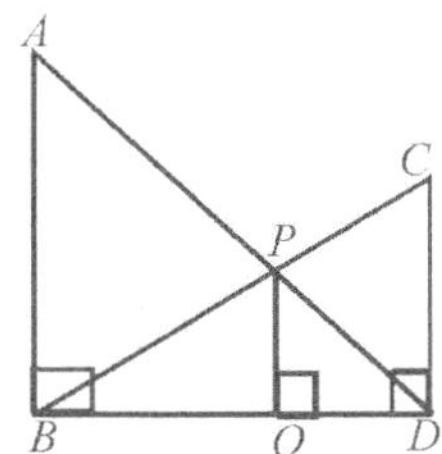

(a) 1 : 0.69 (b) 1 : 0.75
(c) 1 : 072 (d) None of these

42. What is the distance in cm between two parallel chords of lengths 32 cm and 24 cm in a circle of radius 20 cm?
(a) 1 or 7 (b) 2 or 14
(c) 3 or 21 (d) 4 or 28

43. In the adjoining figure O is the centre of the circle. The radius OP bisects a rectangle $ABCD$, at right angle. $DM = NC = 2$ cm and $AR = SB = 1$ cm and $KS = 4$ cm and $OP = 5$ cm. What is the area of the rectangle?

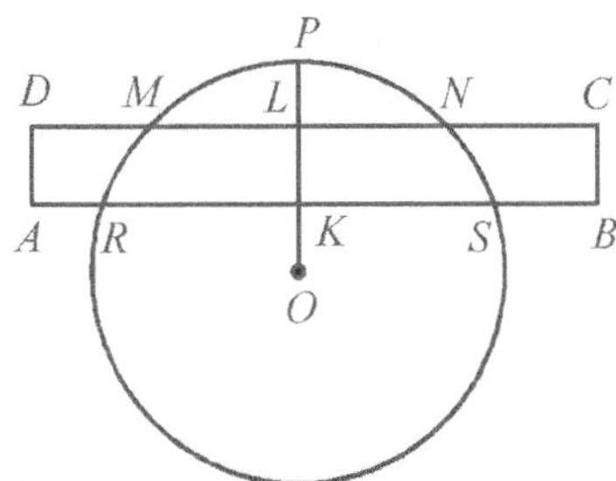

(a) 8 cm^2 (b) 10 cm^2
(c) 12 cm^2 (d) None of these

44. There are two circles each with radius 5 cm. Tangent AB is 26 cm. The length of tangent CD is:

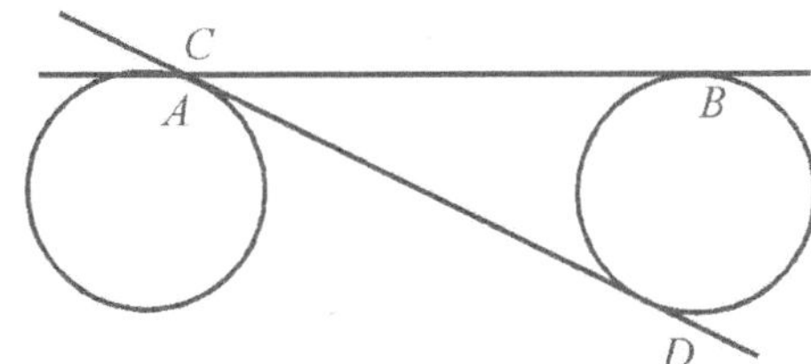

(a) 15 cm (b) 21 cm
(c) 24 cm (d) can't be determined

45. In the given triangle ABC, the length of sides AB and AC is same (i.e., $b = c$) and $60° < A < 90°$, then the possible length of BC is

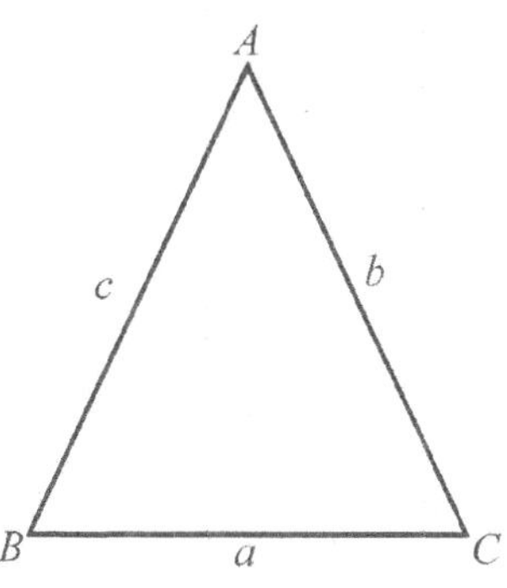

(a) $b < a < 2b$ (b) $\frac{c}{3} < a < 3a$

(c) $b < a < b\sqrt{3}$ (d) $c < a < c\sqrt{2}$

46. The angles of a triangle are in the ratio of 4 : 1 : 1. Then the ratio of sine of the largest angle to the smallest angle is the largest side to the perimeter is
[sin 120° = sin 60°]

(a) $\frac{2}{3}$ (b) $\frac{1}{2+\sqrt{3}}$

(c) $\frac{\sqrt{3}}{1}$ (d) $\frac{2}{1+\sqrt{3}}$

47. What is the sum of all the angles of a 9 pointed star (i.e., $\angle 1 + \angle 2 + \angle 3 + ... \angle 8 + \angle 9$):
(a) 909° (b) 900°
(c) 720° (d) 540°

48. A smaller circle touches internally to a larger circle at A and passes through the centre of the larger circle. O is the centre of the larger circle and BA, OA are of the diameters of the larger and smaller circles respectively. Chord AC intersects the smaller circle at a point D. If $AC = 12$ cm, then AD is:

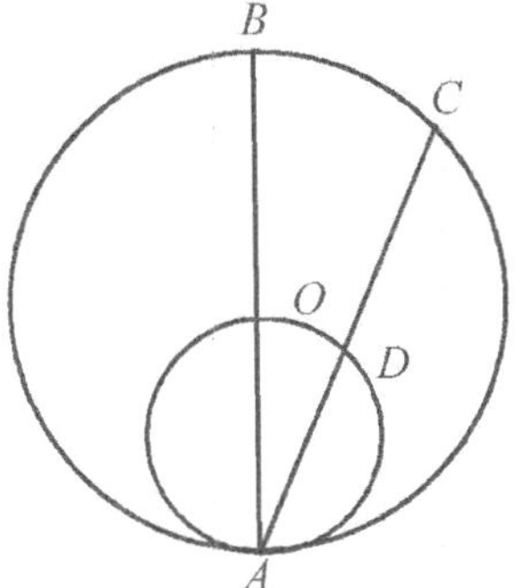

(a) 4 cm (b) 6 cm
(c) 5.6 cm (d) Data insufficient

49. Two circles $C(O, r)$ and $C(O', r')$ intersect at two points A and B and O lies on $C(O', r')$. A tangent CD is drawn to the circle $C(O', r')$ at A. Then

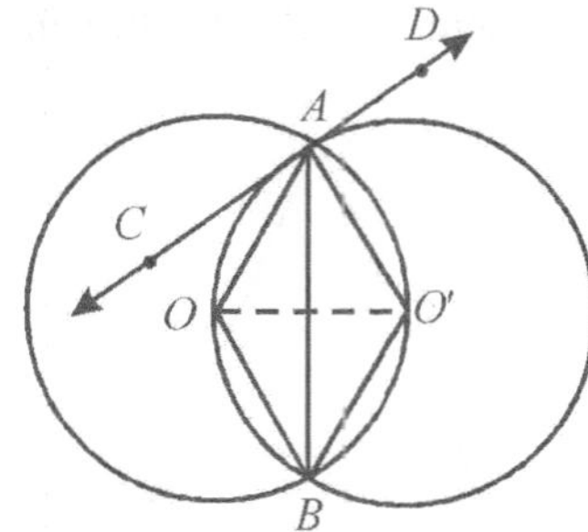

(a) $\angle OAC = \angle OAB$ (b) $\angle OAB = \angle AO'O$
(c) $\angle AO'B = \angle AOB$ (d) $\angle OAC = \angle AOB$

50. Find the perimeter of the given figure.

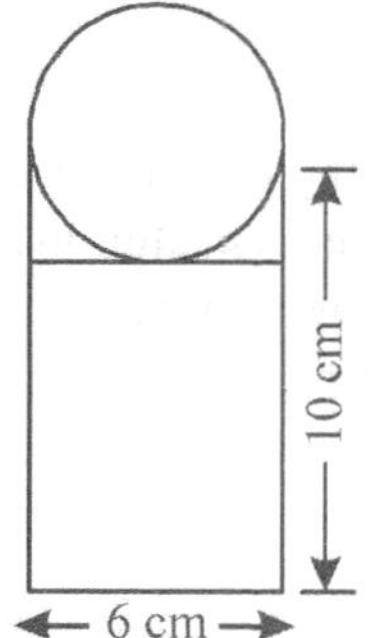

(a) $(32 + 3\pi)$ cm (b) $(36 + 6\pi)$ cm
(c) $(46 + 3\pi)$ cm (d) $(26 + 3\pi)$ cm

51. ΔABC has sides AB, AC measuring 2001 and 1002 units respectively. How many such triangles are possible with all integral sides?
(a) 2001 (b) 1002
(c) 2003 (d) 1004

52. One of the angles of a quadrilateral is thrice the smaller angle of a parallelogram. The respective ratio between the adjacent angles of the parallelogram is 4:5. Remaining three angles of the quadrilateral are in ratio 4 : 11: 9 respectively. What is the sum of the largest and the smallest angles of the quadrilateral? [*IBPS-PO-2013*]
(a) 255° (b) 260°
(c) 265° (d) 270°
(e) None of these

53. Two circles intersect each other at P and Q. PA and PB are two diameters. Then $\angle AQB$ is [*SSC CGL-2012*]
(a) 120° (b) 135°
(c) 160° (d) 180°

54. O is the centre of the circle passing through the points A, B and C such that $\angle BAO = 30°$, $\angle BCO = 40°$ and $\angle AOC = x°$. What is the value of x ? [*SSC CGL-2012*]
(a) 70° (b) 140°
(c) 210° (d) 280°

55. A and B are centres of the two circles whose radii are 5 cm and 2 cm respectively. The direct common tangents to the circles meet AB extended at P. Then P divides AB. [*SSC CGL-2012*]
(a) externally in the ratio 5 : 2
(b) internally in the ratio 2 : 5
(c) internally in the ratio 5 : 2
(d) externally in the ratio 7 : 2

56. A, B, P are three points on a circle having centre O. If $\angle OAP = 25°$ and $\angle OBP = 35°$, then the measure of $\angle AOB$ is [*SSC CGL-2013*]
(a) 120° (b) 60°
(c) 75° (d) 150°

57. Side $\overline{BC}$ of ΔABC is produced to D. If $\angle ACD = 140°$ and $\angle ABC = 3\angle BAC$, then find $\angle A$. [*SSC CGL-2013*]
(a) 55° (b) 45°
(c) 40° (d) 35°

58. The length of tangent (upto the point of contact) drawn from an external point P to a circle of radius 5 cm is 12 cm. The distance of P from the centre of the circle is [*SSC CGL-2013*]
(a) 11 cm (b) 12 cm
(c) 13 cm (d) 14 cm

59. ABCD is a cyclic quadrilateral, AB is a diameter of the circle. If $\angle ACD = 50°$, the value of $\angle BAD$ is [*SSC CGL-2013*]
(a) 30° (b) 40°
(c) 50° (d) 60°

60. Two circles of equal radii touch externally at a point P. From a point T on the tangent at P, tangents TQ and TR are drawn to the circles with points of contact Q and R respectively. The relation of TQ and TR is [*SSC CGL-2013*]
(a) TQ < TR (b) TQ > TR
(c) TQ = 2TR (d) TQ = TR

61. When two circles touch externally, the number of common tangents are [*SSC CGL-2013*]
(a) 4 (b) 3
(c) 2 (d) 1

62. D and E are the mid-points of AB and AC of ΔABC. If $\angle A = 80°$, $\angle C = 35°$, then $\angle EDB$ is equal to [*SSC CGL-2013*]
(a) 100° (b) 115°
(c) 120° (d) 125°

Hints & Solutions

Level-I

1. **(c)** In a right angled Δ, the length of the median is $\frac{1}{2}$ the length of the hypotenuse. Hence $BD = \frac{1}{2} AC = 3$ cm.

2. **(b)** In $\Delta ABC, \angle C = 180 - 90 - 30 = 60°$

$\therefore \Delta DCE = \frac{60}{2} = 30°$

Again in ΔDEC, $\Delta CED = 180 - 90 - 30 = 60°$

3. **(d)** The quadrilateral obtained will always be a trapeziam as it has two lines which are always parallel to each other.

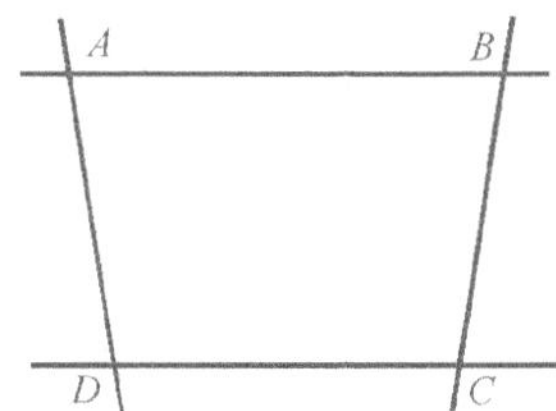

4. **(a)** $AD = 24$, $BC = 12$

In ΔBCE & ΔADE

since $\angle CBA = \angle CDA$ (Angles by same arc)

$\angle BCE = \angle DAE$ (Angles by same arc)

$\angle BEC = \angle DEA$ (Opp. angles)

$\therefore \angle BCE$ & $\angle DAE$ are similar Δs

with sides in the ratio 1 : 2

Ratio of area = 1:4 (i.e square of sides)

5. **(a)**

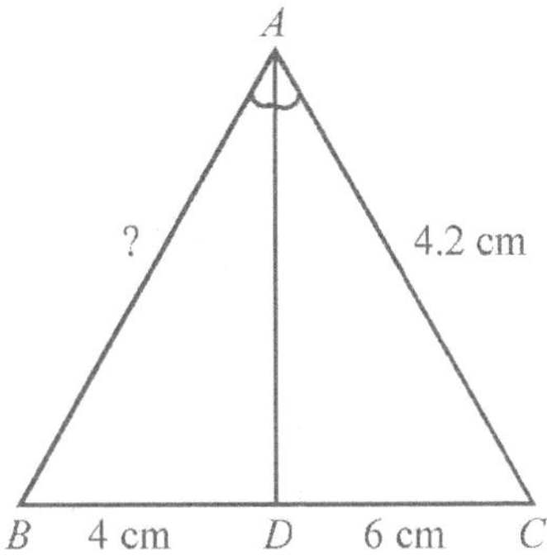

$\Delta ABD \sim \Delta ACD$

$\frac{AC}{DC} = \frac{AB}{BD} \Rightarrow \frac{4.2}{6} = \frac{AB}{4}$

$\therefore AB = 2.8$ cm

6. **(c)** Let n be the number of sides of the polygon

Now, sum of interior angles = 8 × sum of exterior angles

i.e. $(2n - 4) \times \frac{\pi}{2} = 8 \times 2\pi$

or $(2n - 4) = 32$

or $n = 18$

7. **(a)** 2.4 cm

8. **(a)** $\angle EDC = \angle BAD = 45°$ (alternate angles)

$\therefore x = DEC = 180° - (50° + 45°) = 85°$.

9. **(a)** $a + 36° + 70° = 180°$ (sum of angles of triangle)

$\Rightarrow a = 180° - 36° - 70° = 74°$

$b = 36° + 70°$(Ext. angle of triangle) = 106°

$c = a - 50°$ (Ext. angle of triangle) $= 74° - 50°$ $= 24°$.

10. **(c)** $b = \frac{1}{2}(48°)$

($\angle$at centre = 2 at circumference on same PQ) 24°

$\angle AQB = 90°$ ($\angle$In semi-circle)

$\angle QXB = 180° - 90° - 24°$ ($\angle$sum of Δ) = 66°

11. **(d)** $\angle MBA = 180° - 95° = 85°$

$\angle AMB = \angle TMN$...(Same angles with different names)

$\therefore \Delta MBA - \Delta MNT$......($AA$ test for similarity)

$\frac{MB}{MN} = \frac{AB}{NT}$ ....(proportional sides)

$\frac{10}{MN} = \frac{5}{9} \quad \therefore MN = \frac{90}{5} = 18.$

12. **(b)** Through O draw EOE' parallel to AB & so to CD.

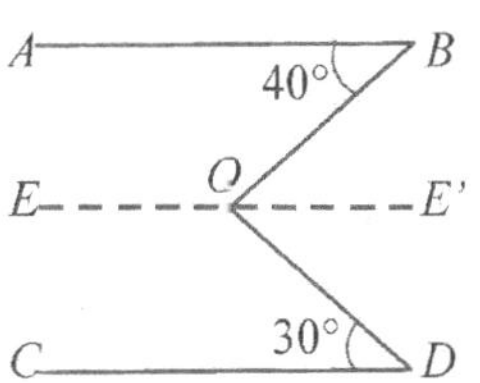

$\therefore \angle BOE' = \angle ABO = 40°$ (alternate angles)

$\angle E'OD = \angle CDO = 30°$ (alternate angles)

$\therefore \angle BOD = (40° + 30°) = 70°$. So, $x = 70$.

13. **(c)** The triangle PQR is isosceles

$\Rightarrow MN \parallel QR$ by converse of Proportionality Theorem.

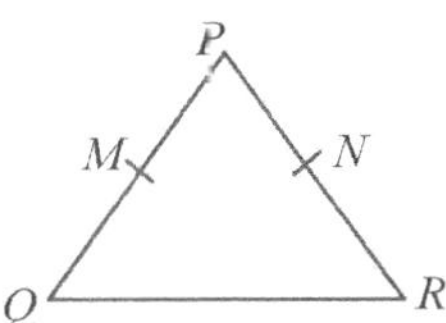

(b) Again by Converse of Proportionality theorem, $MN \parallel QR$.

14. (a) a + 36° + 70° = 180° (sum of angles of triangle)
⇒ a = 180° – 36° – 70° = 74°
b = 36° + 70° (Ext. angle of triangle) = 106°
c = a – 50° (Ext. angle of triangle) = 74° – 50°
= 24°.

15. (c) Perimeter of ΔABC = 36 cm.
Perimeter of ΔPQR = 24 cm and PQ = 10 cm.
We have to find AB. Perimeter of $\Delta ABC = AB + BC + AC$.
Perimeter of $\Delta PQR = PQ + QR + PR$. Since $\Delta ABC \sim \Delta PQR$.

$$\therefore \frac{AB}{PQ} = \frac{BC}{QR} = \frac{AC}{PQ} = \frac{AB+BC+AC}{PQ+QR+PR} = \frac{36}{24}$$

$$\Rightarrow \frac{AB}{PQ} = \frac{36}{24} \Rightarrow AB = \frac{36}{24} \times PQ = \frac{36}{24} \times 10 = 15 \text{ cm.}$$

16. (d) AD is the bisector of $\angle A$.

$$\therefore \frac{AB}{AC} = \frac{BD}{DC} = \frac{5}{3}$$

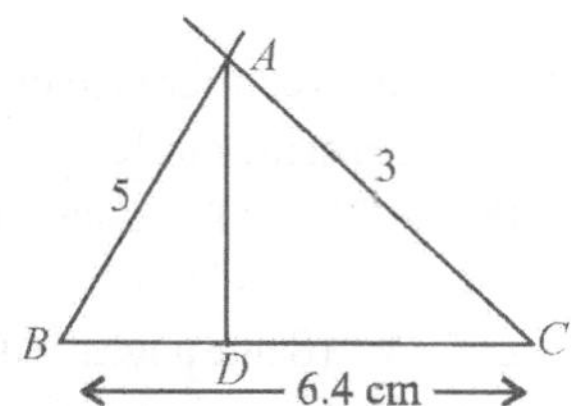

$$\Rightarrow \frac{DC}{BD} = \frac{3}{5} \Rightarrow \frac{DC+BD}{BD} = \frac{3+5}{5}$$

$$\Rightarrow \frac{BC}{BD} = \frac{8}{5} \Rightarrow BD = BC \times \frac{5}{8} = 6.4 \times \frac{5}{8} = 4$$

17. (b) $m\angle ACD = m\angle DEC$
$\therefore m\angle DEC = x = 40°$
$\therefore m\angle ECB = m\angle EDC$
$\therefore m\angle ECB = y = 54°$
$54° + x + z = 180°$ .... (sum of all the angles of a triangle)
$54° + 40° + z = 180°$
$\therefore z = 86°$

18. (b) In ΔBCD, $BC = CD$, $\angle BDC = \angle CBD = x$
In cyclic quadrilateral $ABCD$, $\angle ABC + \angle ADC = 180°$
$40° + x + 90° + x = 180° \Rightarrow x = 25°$.

19. (c) $m\angle DAB = 180° – 120° = 60°$ ...(opposite angles of a cyclic quadrilateral) $m(\text{arc } BCD) = 2\, m\angle DAB = 120°$.

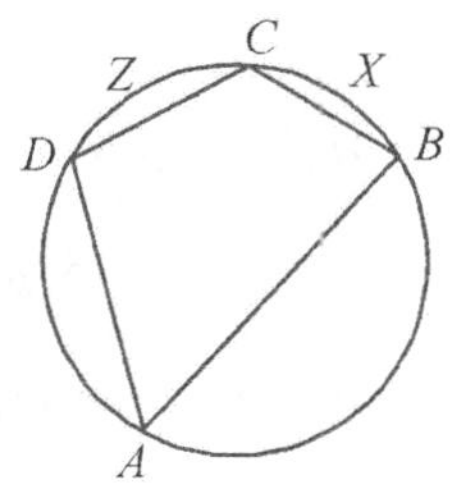

$\therefore m(\text{arc } CXB) = m(\text{arc } BCD) – m(\text{arc } DZC)$
$= 120° – 70° = 50°$.

20. (d)

21. (d) $\frac{OP}{PT} = \tan 30° = \frac{1}{\sqrt{3}} \Rightarrow PT = \sqrt{3}\; OP = 3\sqrt{3}$ cm.

22. (c) $\angle OPQ = \angle OQP = 30°$, i.e., $\angle POQ = 120°$.
Also,

$$\angle PRQ = \frac{1}{2} \text{ reflex } \angle POQ$$

23. (b) Since $ABCD$ is a quadrilateral
Again AP, AQ are tangents to the circle from the point A.

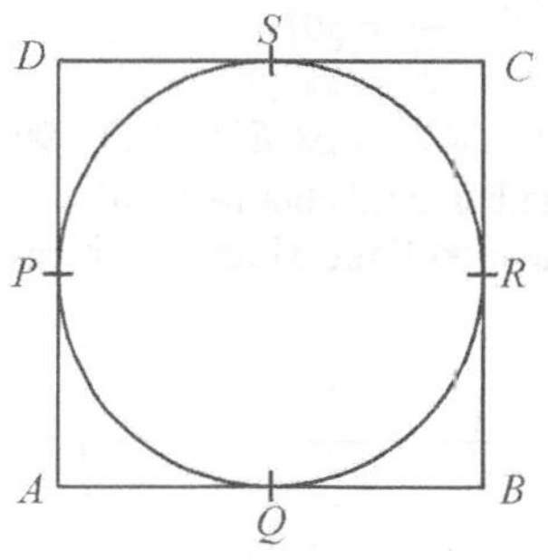

$\therefore \quad AP = AQ$
Similarly $BR = BQ$
$CR = CS$
$DP = DS$
$\therefore \quad (AP + DP) + (BR + CR) = AQ + DS + BQ + CS$
$= (AQ + BQ) + (CS + DS)$
$\Rightarrow \quad AD + BC = AB + CD$

24. (d) $\angle LCD = \angle ALC = 60°$ (alternate angles)
$\angle DCE = \frac{1}{2}\angle LCD = 30°$. ($EC$ is the angle bisector)
$\therefore \quad \angle FEC = (180° – 30°) = 150°$.

25. (b) We have area of triangle $AFE = A/4$. (If A = Area of triangle ABC) and area of triangle $DHI = (A/4)/4 = A/16$. Hence, ratio = 1 : 4.

26. (b) In ΔAOB and ΔCOD

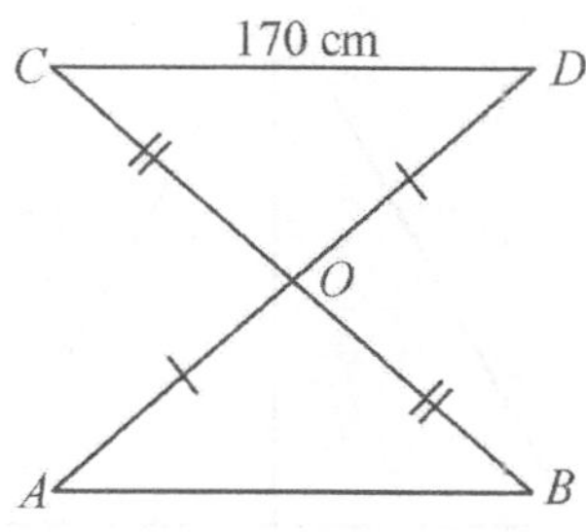

$AO = OD$, $BO = OC$
$\angle AOB = \angle COD$ (vertically opposite angles)
$\therefore \quad \Delta AOB \cong \Delta COD$
$\therefore \quad AB = CD = 170$ cm.

27. (d) $c = c_1$ (Vert. opp. $\angle$ s). $b = c + s$ (Ext. $\angle$).
$d = c_1 + r$ (Ext. $\angle$)

But $b + d = 180°$ (Opp. $\angle s$, cyclic quad.)

$\Rightarrow \quad c + s + c_1 + r = 180°$

$\Rightarrow \quad r + s + 2c = 180° \Rightarrow r + s = 180° - 2c.$

28. **(b)** $m\angle PAC = m\angle PBC = 90°$

....(Tangent perpendicularity theorem)

$m\angle PAC + m\angle PBC + m\angle ACB = 360°$

$\therefore \quad m\angle APB = 360 - (90 + 90 + 65) = 115°$

$\therefore \quad m(AXB) = 115°.$

29. **(d)** Basic concept

30. **(a)** $\angle DCK = \angle FDG = 55°$ (corr. $\angle s$)

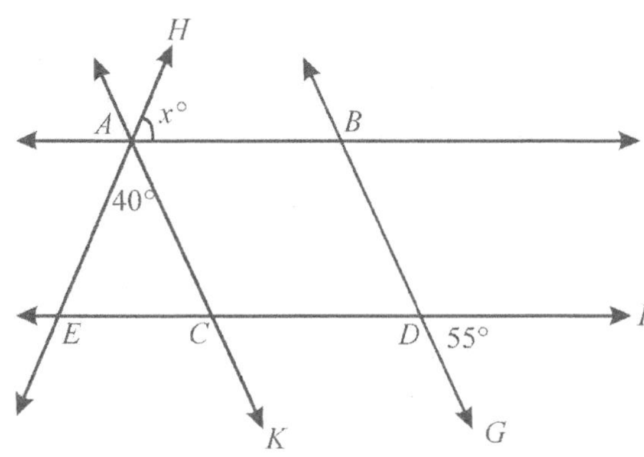

$\therefore \angle ACE = 180° - (\angle EAC + \angle ACE)$

$\therefore \angle HAB = \angle AEC = 85°$ (corr. $\angle s$)

Hence, $x = 85°$

31. **(c)** Clearly option (a) shows the angles would be 30, 60 and 90. It can be the ratio of angle in a right angled triangle.

Option (b) shows the angles would be 45, 45 and 90, then it can be the ratio of angle in a right angled triangle.

But option (c) cannot form the ratio of angles of right angled triangle.

32. **(b)** In $\Delta ABC, \angle C = 180 - 90 - 30 = 60°$

$\therefore \angle DCE = \frac{60}{2} = 30°$

Again in $\Delta DEC, \angle CED = 180 - 90 - 30 = 60°$

33. **(c)** $\angle DFO = \angle FOM$

and $\angle AEO = \angle EOM$ (since $CD \parallel AB$)

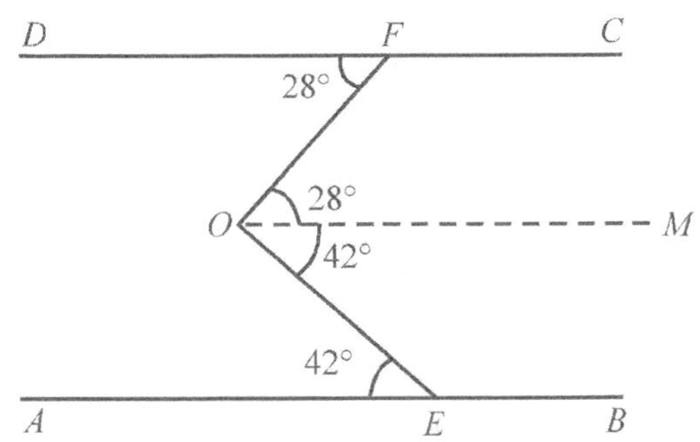

$\therefore \quad \angle FOE = (28° + 42°) = 70°$

34. **(b)** Go through option for quicker answer

Exterior angle $= \frac{360}{15} = 24°$ (for $n = 15$)

$\therefore$ Interior angle $= 180° - 24° = 156°$

$\therefore$ Interior – Exterior $= 156 - 24 = 132°$

Hence, option (b) is correct.

35. **(c)** $\angle ABC = 180 - (65 + 75) = 40°$

$\angle ORB = \angle OQB = 90°$

$\therefore \quad \angle ROQ = 360 - (90 + 90 - 40)$

$\therefore \quad \angle ROQ = 140°$

36. **(c)** ΔABC is similar to ΔEDC

$$\therefore \quad \frac{AB}{ED} = \frac{BC}{DC} = \frac{AC}{EC}$$

$$\therefore \quad \frac{AB}{DE} = \frac{BC}{DC} \Rightarrow \frac{24}{10} = \frac{60}{DC}$$

$\Rightarrow \quad DC = 25$ cm

37. **(d)** Clearly, triangle is obtuse, So (d) is the correct option.

38. **(a)** No such point is possible

39. **(c)**

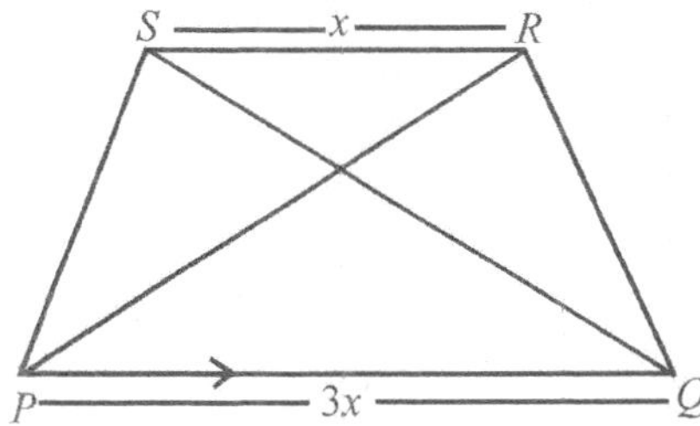

$$\frac{ar(\Delta P \times Q)}{(\Delta R \times S)} = \frac{PQ^2}{RS^2} = \frac{(3x)^2}{x^2} = 9:1$$

40. **(a)** The parallelogram $ABCD$ and ΔBCE lies between the same parallel lines AB and DE and has base of equal length. $\therefore A(\Delta BCE) = \frac{1}{2} A(\square ABCD) = \frac{1}{2} \times 16 = 8$ sq. cm.

41. **(a)** Form the figure given in the question ,we get $x^2 - y^2 = 81, x^2 + y^2 = 625$ and $y^2 + 256 = z^2$

Form the option the only triplet satisfying the three equations is 15, 12, 20

42. **(c)**

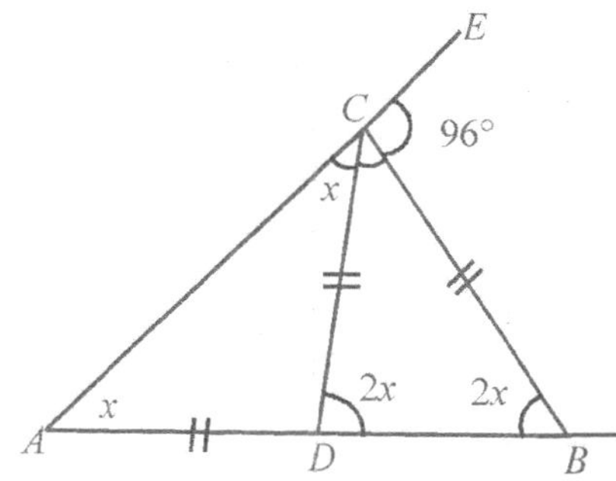

Let $\angle CAD = \angle ACD = x$

At point C,
$x + (180° - 4x) + 96° = 180°$
$\Rightarrow 180° - 3x + 96° = 180°$
$\therefore\ x = 32°$
Hence, $\angle DBC = 2 \times 32 = 64°$

43. (b)

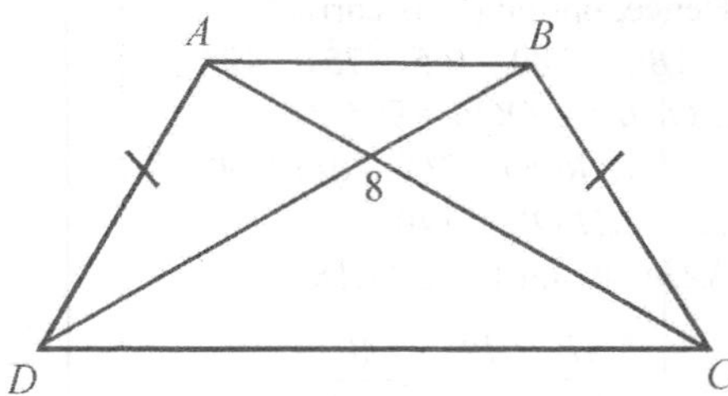

$\Delta APD \sim \Delta BPC$

$\therefore \quad \frac{PA}{PB} = \frac{PD}{PC}$

i.e., $PA.\ PC = PB.\ PD.$
$\therefore$ option (b)

44. (a) $\angle CAF = 100°$. Hence $\angle BAC = 80°$
Also, $\angle OCA = (90 – ACF) = 90 - 50 = 40° = \angle OAC$
(Since the triangle OCA is isosceles)
Hence $\angle OAB = 40°$
In isosceles ΔOAB, $\angle OBA$ will also be 40°
Hence, $\angle BOA = 180 - 40 - 40 = 100°$

45. (b)

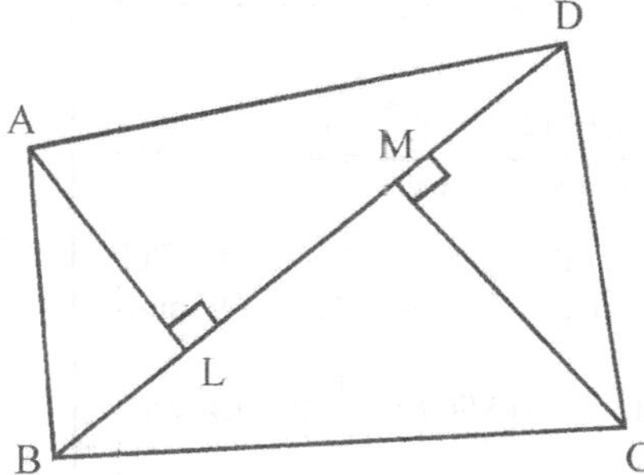

Given :
BD = 64 cm
AL = 13.2 cm
CM = 16.8 cm
So, Area (ABCD) = Area (ΔABD) + Area (ΔBCD)

$= \frac{1}{2} \times AL \times BD + \frac{1}{2} \times CM \times BD$

$= \frac{1}{2} \times BD \times (AL + CM)$

$= \frac{64}{2}(13.2 + 16.8)$

$= 32 \times 30 = 960\ cm^2$

46. (b)

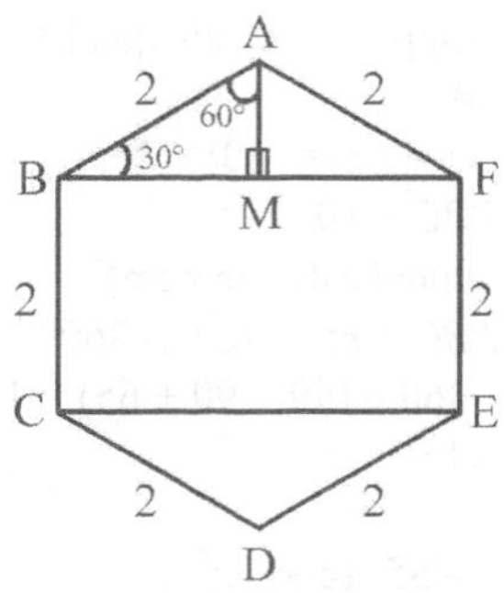

Given BC & EF are each 2 feet. Since area of rectangle is length × width.
To find out BF or CE, Take ΔABF. It has two equal sides (AB = AF), so the perpendicular from A to line BF divides ABF into two congruent Δs.
So, each of the two triangles is 30°-60°-90° right angle Δ with hypotenuse 2.

In ΔABM $\cos 30° = \frac{BM}{AB} \Rightarrow \frac{\sqrt{3}}{2} = \frac{BM}{2} \Rightarrow BM = \sqrt{3}$

So, $BF = 2 \times BM = 2\sqrt{3}$

Area of rectangle $= 2\sqrt{3} \times 2 = 4\sqrt{3}$

47. (b)

48. (c)

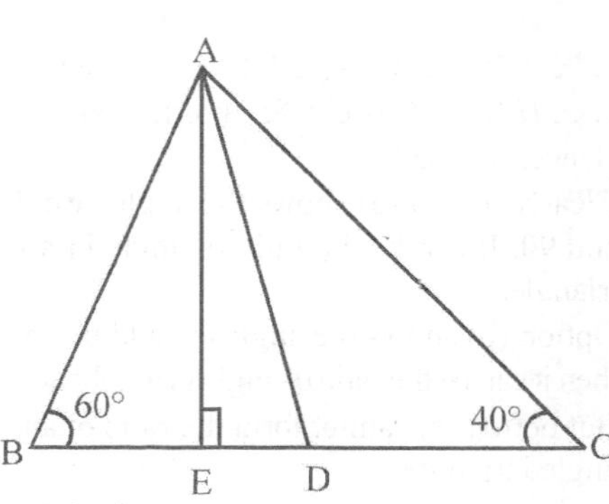

In ΔABC,
$\angle A + \angle B + \angle C = 180°$
$\angle A + 60° + 40° = 180°$
$\angle A = 180° - 60° - 40° = 80°$
AD bisects $\angle$BAC
$\therefore\ \angle A = \angle BAD + \angle DAC$
$\angle BAD = \angle DAC = 40°$
Now, In ΔABE
$\angle B + \angle E + \angle BAE = 180°$
$60° + 90° + \angle BAE = 180°$
$\angle BAE = 30°$
$\therefore\ \angle EAD = \angle BAD - \angle BAE$
$= 40° - 30° = 10°$

49. (c) $\angle AEC = \angle ECD$ (Alternate interior angles as AB || CD)
In ΔCED,
$\angle ECD + \angle CED + x° = 180°$
(Sum of angles of Δ are 180°)
$37° + 90° + x° = 180°$
$x° = 180° - 37° - 90°$
$x° = 53°$

50. (b)

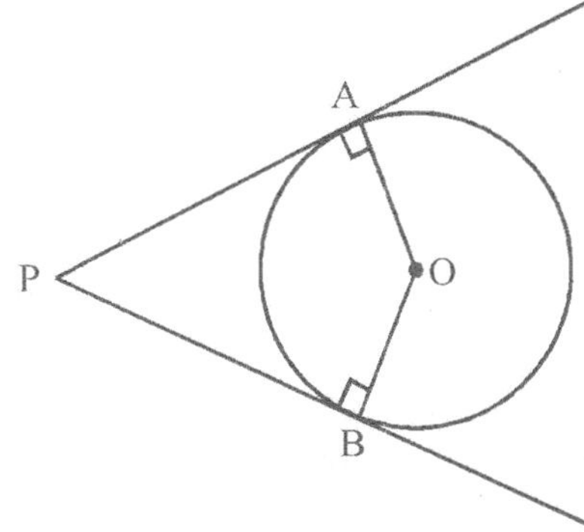

OAPB is concyclic because ∠A + ∠B = 180°
& ∠O + ∠P = 180°

51. (c)

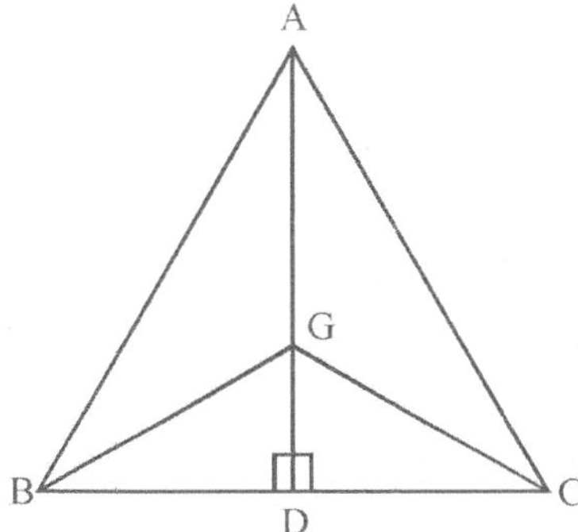

AG = BC (Given)
BD = DC (given) AD is median
So, GD = BD = DC
ΔBCD & ΔGCD are both isosceles Δ.
Then ∠BGC = 90°

52. (d)

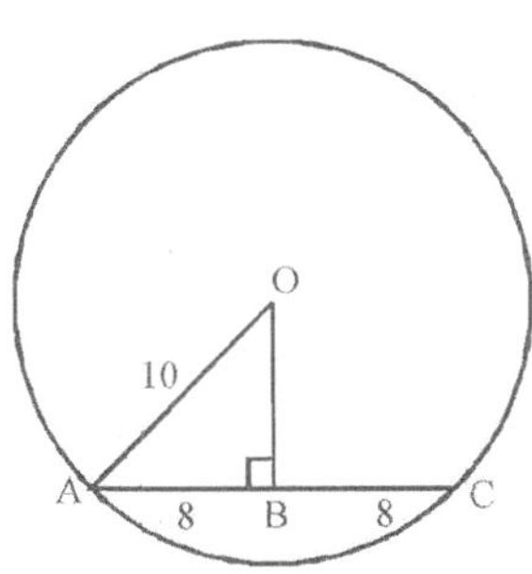

In OAB,
$OA^2 = OB^2 + AB^2$

[∴ $AB = \frac{1}{2}$ AC, because line drawn from centre to a chord bisect & perpendicular to it]
$(10)^2 = (OB)^2 + (8)^2$
$100 - 64 = OB^2$
$OB^2 = 36$
OB = 6

53. (d) $PR^2 = PQ^2 + PR^2 = 3^2 + 4^2 = 25$

$\therefore PR = \sqrt{25} = 5$ cm

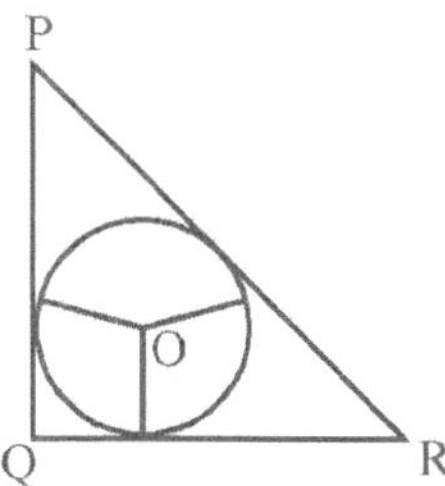

$$r = \frac{\text{Area of triangle}}{\text{Semi-perimeter of triangle}}$$

$$= \frac{\frac{1}{2} \times 3 \times 4}{\frac{3+4+5}{2}} = \frac{6}{6} = 1\text{cm}$$

54. (c)

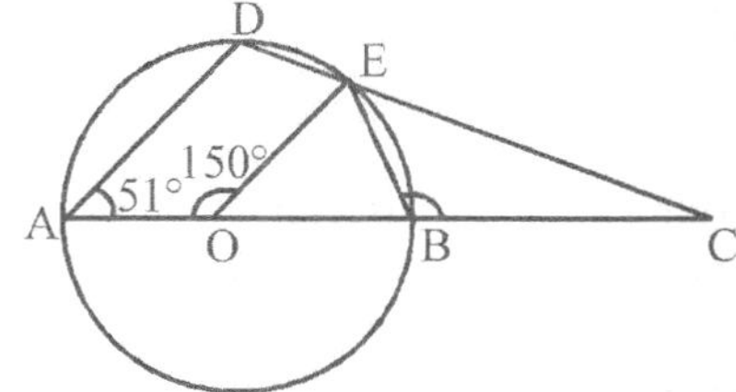

∠AOE = 150°
∠DAO = 51°
∠EOB = 180° – 150° = 30°
OE = OB

$\therefore \angle OEB = \angle OBE = \frac{150}{2} = 75°$

$\therefore \angle CBE = 180° - 75° = 105°$

55. (b) $\frac{\Delta ABC}{\Delta DEF} = \frac{AB^2}{DE^2}$

$\Rightarrow \frac{20}{45} = \frac{25}{DE^2}$

$\Rightarrow DE^2 = \frac{45 \times 25}{20} = \frac{225}{4}$

$\therefore DE = \frac{15}{2} = 7.5$ cm

56. (d) ∠ACB = 80°
∠ACD = 180° – 80° = 100°
∵ AC = CD
∴ ∠CAD = ∠CDA

$= \frac{80}{2} = 40°$

∠BAC = 111° – 40°
= 71°
∠ABC = 180° – 71° – 80° = 29°

57. (d)

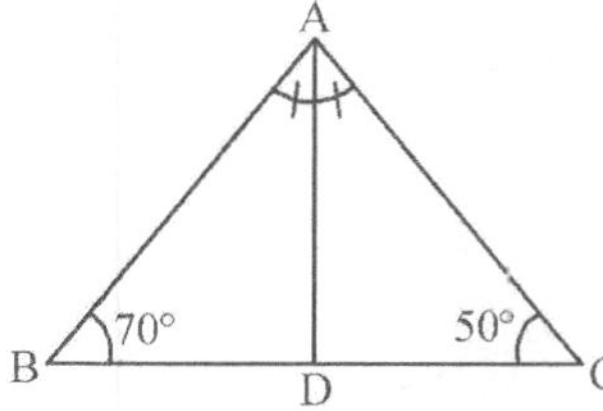

$\angle A + \angle B = 145°$

$\angle C + 180° - 145° = 35°$

$\angle C + 2\angle B = 180°$

$\Rightarrow 2\angle B = 180° - 35° = 145°$

$\Rightarrow \angle B = \frac{145}{2} = 72.5° = \angle A$

$\angle B > \angle C$

$\therefore AC > AB$

58. (c)

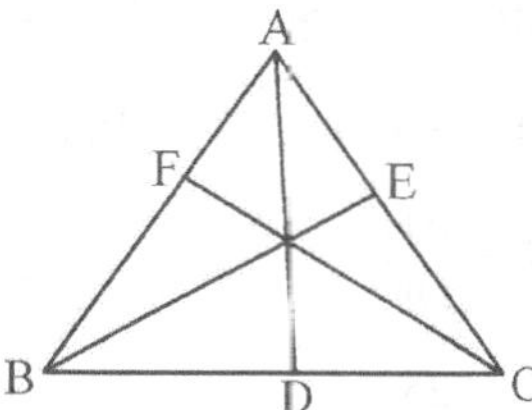

According to angle bisector theorem : The angle bisctor, like segment AD, divides the sides of the triangle proportionally.

In Δ ABC

$\angle A + \angle B + \angle C = 180°$

$\angle A = 180° - 70° - 50° = 60°$

$\angle BAD = \frac{60}{2} = 30°$

59. (b)

Let ABC be the triangle and D, E and F are midpoints of BC, CA and AB respectively.

Hence, in Δ ABD, AD is median

$AB + AC > 2\,AD$

Similarly, we get

$BC + AC > 2\,CF$

$BC + AB > 2\,BE$

On adding the above inequations, we get

$(AB + AC + BC + AC + BC + AB) > 2(AD + BE + CF)$

$2\,(AB + AC + BC) > 2\,(AD + BE + CF)$

$\therefore AB + BC + BC > AD + BE + CF$

Thus, the perimeter of triangle is greater than the sum of the medians.

60. (c)

61. (b)

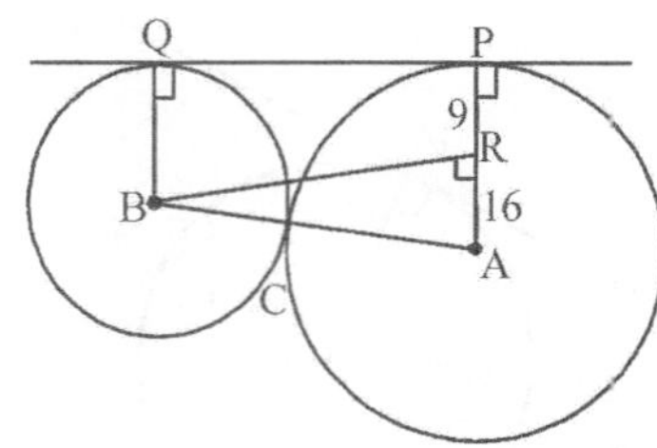

Let the two circles with centre A, B and radii 25 cm and 9 cm touch each other externally at point C. Then

$AB = AC + CB$

$= 25 + 9 = 34$ cm

Let PQ be the direct common tangent i.e. $BQ \perp PQ$ and $AP \perp PQ$. Draw $BR \perp AP$. Then BRQP is a rectangle. (Tangent ⊥ radius at pt. of contact)

In Δ ABR

$AB^2 = AR^2 + BR^2$

$(34)^2 = (16)^2 + (BR)^2$

$BR^2 = 1156 - 256 = 900$

$BR = \sqrt{900} = 30$ cm

62. (a) In Δ ABC,

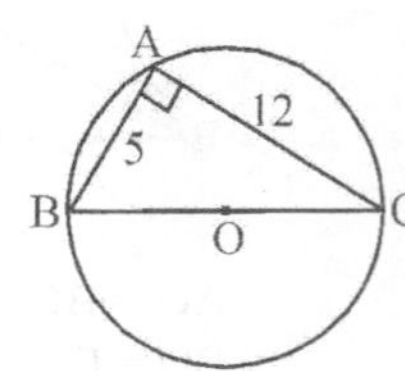

$BC^2 = AB^2 + AC^2$

$BC^2 = (5)^2 + (12)^2$

$BC^2 = 25 + 144$

$BC^2 = 169$

$BC = \sqrt{169} = 13$ cm

Radius of triangle $= \frac{BC}{2} = \frac{13}{2} = 6.5$ cm

63. (b) Here,

$\angle ACB = 90°$

$\angle ADC = 90°$

$\angle BDC = 90°$

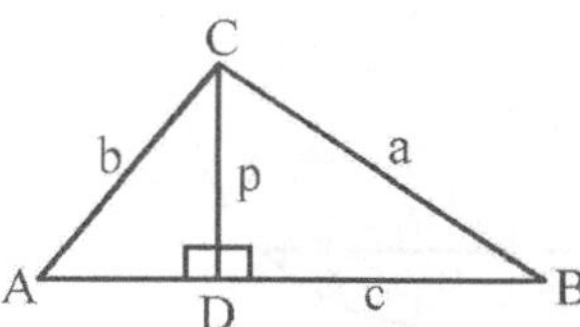

Triangles ACB, ADC and BDC are right angle triangles.

Here, Area of ΔABC = Area of ΔADC + Area of ΔBDC

$\Rightarrow \frac{1}{2} a \times b = \frac{1}{2} \times p \times AD + \frac{1}{2} \times p \times DB$

$\Rightarrow ab = p\,(AD + DB)$

$\Rightarrow ab = pc \Rightarrow c = \frac{ab}{p}$ ... (1)

Now, In ΔABC,

$$c^2 = a^2 + b^2 \left(\frac{ab}{p}\right)^2 = a^2 + b^2$$

$$\Rightarrow \frac{a^2b^2}{p^2} = a^2 + b^2$$

$$\frac{1}{p^2} = \frac{1}{a^2} + \frac{1}{b^2}$$

64. (a) Given OP = 2r = Diameter of circle

$(\because OA \perp PA \,\&\, OB \perp PB)$

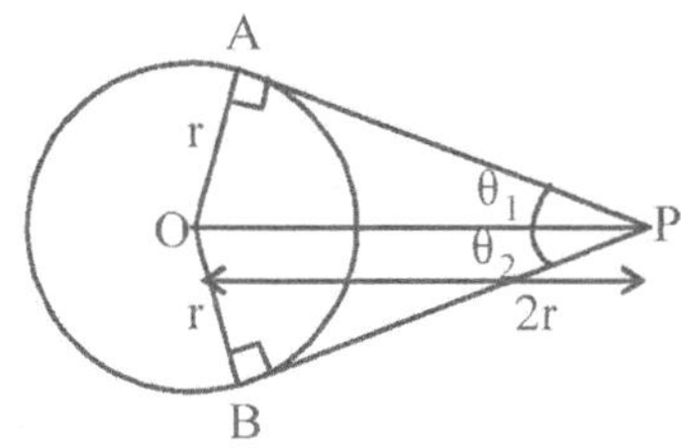

$\therefore$ In Δ OAP, $\sin\theta_1 = \frac{r}{2r} = \frac{1}{2}$

$\sin\theta_1 = \sin 30° \Rightarrow \theta_1 = 30°$

Similary, in ΔOBP, $\sin\theta_2 = \frac{r}{2r} = \frac{1}{2}$

$\sin\theta_2 = \sin 30° \Rightarrow \theta_2 = 30°$

$\therefore \angle APB = \theta_1 + \theta_2 = 30° + 30° = 60°$

65. (b) Given, AB = 12 cm; CD = 20 cm

OE = ?

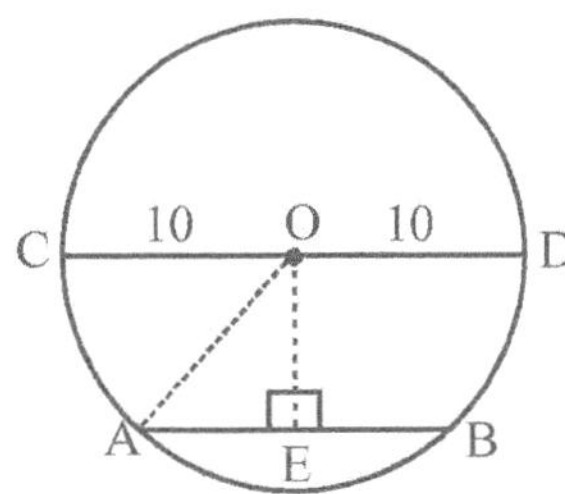

Now, $AE = EB$ = 6cm (The line drawn from centre of circle to the chord bisect the chord)

In ΔOAE, By phythagoras theorem

$(OA)^2 = (OE)^2 + (AE)^2 \Rightarrow (10)^2 = (OE)^2 + (6)^2$

$100 - 36 = (OE)^2 \Rightarrow 64 = OE^2 \Rightarrow OE = 8$ cm

66. (d) $\angle A + \angle B + \angle C = 180°$

$3\angle C + 5\angle C + \angle C = 180°$

$9\angle C = 180°$

$\angle C = 20°$

$\angle B = 100°$

67. (a)

68. (b)

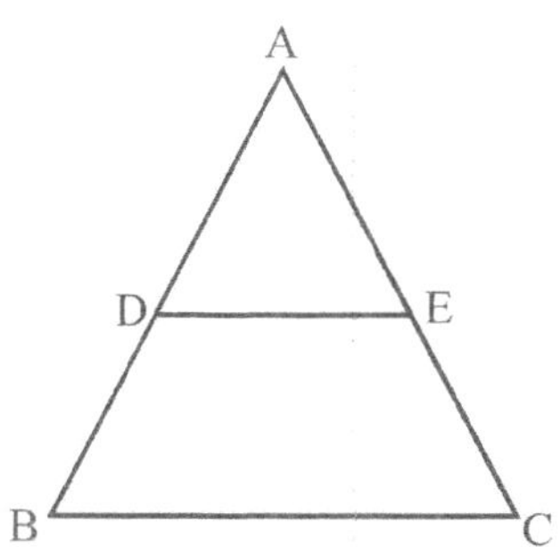

Since DE is parallel to BC

ΔADE ≅ ΔABC

$$\frac{ar(\Delta ABC)}{ar(\Delta ADE)} = \frac{(AB)^2}{(AD)^2} = \frac{25}{4}$$

$$\frac{ar(DECB)}{ar(ADE)} + \frac{ar(ADE)}{ar(ADE)} = \frac{25}{4}$$

$$\frac{ar(DECB)}{ar(ADE)} = \frac{25}{4} - 1 = \frac{21}{4} = 5\frac{1}{4}$$

69. (c) Second angle of parallelogram

= 180° – 45° = 135°

$\therefore$ Required value

= 135 + 2 × 45

= 135 + 90 = 225°

Level-II

1. (c) Should be XY since you divide XY into millions of congruent portions, each portion which is the diameter of the semicircle is very small. So the sum of all the arcs should be XY.

2. (d) In Triangles ACP and BDP; $a = a_1$ (∠ in same seg.); $p = p$ (common)

$\angle ACP = \angle BDP$ (3rd ∠ of triangle)

$\Rightarrow$ Triangle ACP ~ Triangle BDP (A.A.A.)

$BD/BP = AC/AP$ (corr. sides of ~ triangles)

$\Rightarrow$ $BD/8 = 6/10 \Rightarrow BD = 4.8$ cm

$PD/BP = PC/AP$ (Corr. sides) $PD/8 = 5/10$

$\Rightarrow$ $PD = 40/10 = 4$ cm.

3. (a) Let $AY = AY = a$

$BY = BZ = b$

$CZ = CW = c$ (tangents from ext. pt.)

$DW = DX = d$

$AD + BC = a + d + b + c = a + b + d + c = p + q.$

4. (c) $b = \frac{1}{2}(48°)$

(∠at centre = 2 at circumference on same PQ) 24°

$\angle AQB = 90°$ (∠ in semi- circle)

$\angle QXB = 180° - 90° - 24°$ (∠sum of Δ) = 66°

5. (b) PA, AB, RC and SD are perpendicular to AD. Hence they area parallel. So, the intercepts are proportional.

$\therefore \frac{AB}{BD} = \frac{PQ}{QS} \quad \therefore \frac{60}{210} = \frac{x}{360-x}$

$\therefore \frac{2}{7} = \frac{x}{360-x} \quad \therefore x = \frac{720}{9} = 80$

$\therefore PQ = 80 \quad \therefore QS = 360 - 80 = 280$

Again, $\frac{BC}{CD} = \frac{QR}{RS} \quad \therefore \frac{90}{120} = \frac{y}{280-y}$

$\therefore \frac{3}{4} = \frac{y}{280-y} \quad \therefore 7y = 280 \times 3 \;\therefore y = 120$

$\therefore QR = 120$

$\therefore SR = 280 - 120 = 160$

Another method: $60 : 90 : 120 = 2 : 3 : 4$

$\therefore$ Divide 360 in the ratio $2 : 3 : 4$

$\Rightarrow PQ = 80$, $QR = 120$ and $RS = 160$

6. (a)

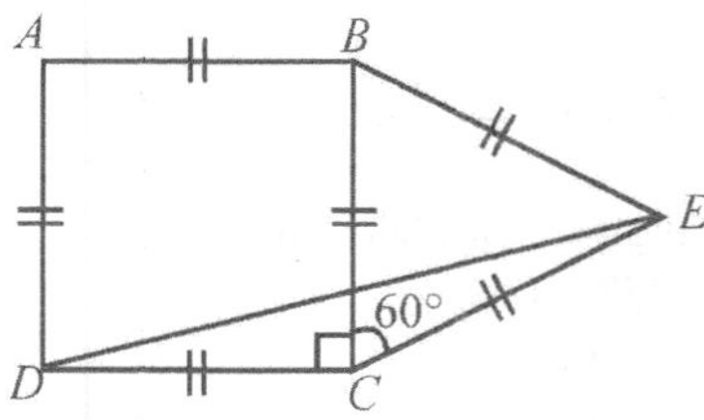

In $\Delta DEC, \angle DCE = 90 + 60 = 150°$

$\angle CDE = \angle DEC = \frac{180-150}{2} = 15°$

7. (b)

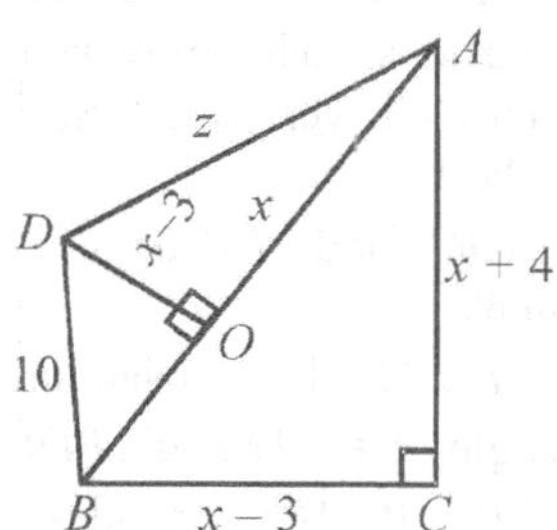

$AB^2 = (x+4)^2 + (x-3)^2 = 2x^2 + 25 + 2x$

Since solving this equation is very difficult. So, it is a better approach (Time saving) to put the values given in the options and try tc find out a solution.

Hence, trying out we get 11 as the value of x.

8-10.

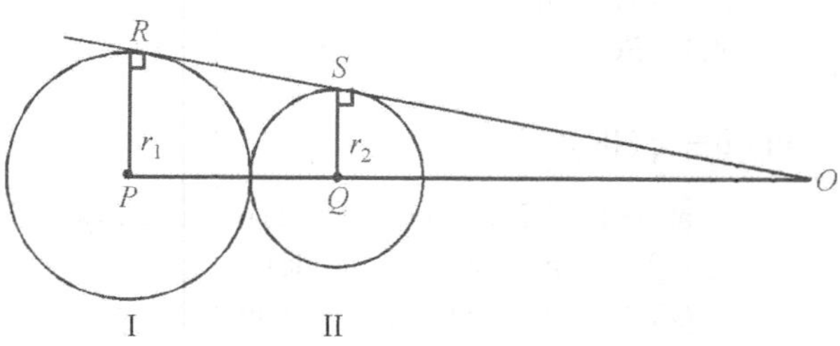

8. (b) In ΔSOQ and ΔROP

$\angle O$ is common

$\angle S = \angle R = 90°$ (tangent at circle)

$\therefore \Delta SOQ \sim \Delta ROP$

$\Rightarrow \frac{RP}{SQ} = \frac{OP}{OQ} = \frac{PQ+OQ}{OQ} = \frac{PQ}{OQ} + 1$

$\Rightarrow \frac{4}{3} = \frac{PQ}{OQ} + 1$ or $\frac{PQ}{OQ} = \frac{4}{3} - 1 = \frac{1}{3}$

$\Rightarrow PQ = 7$ and $OQ = 21$

$\therefore$ Required ratio $= \frac{7}{21} = \frac{1}{3}$

9. (b) $PQ = r_1 + r_2 = 7$

As the ratio of radii is 4 : 3.

So, the only value which satisfies the radii of circle II = 3

10. (c) In ΔSOQ,

$\Rightarrow SO^2 + SQ^2 = OQ^2$

$\Rightarrow SO^2 = 21^2 - 3^2 = (21-3)(21+3) = 18 \times 24 = 432$

$\Rightarrow SO = 12\sqrt{3}$

11. (b) Inradius of right angled triangle

$= \frac{AB + BC - AB}{2}$

$= \frac{9+40-41}{2} = 4$ cm

12. (b)

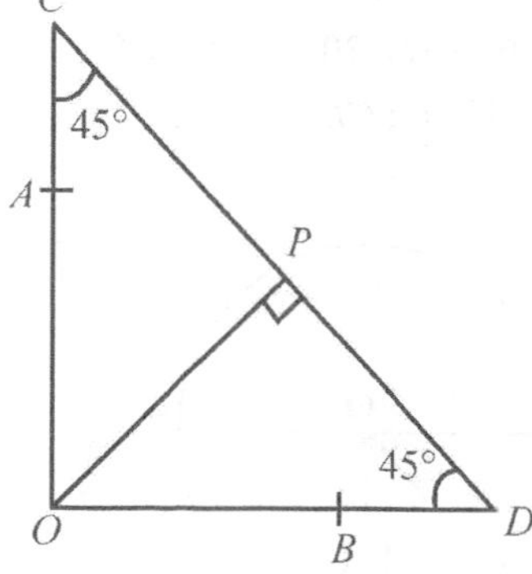

$OC = OD$ and $OA = OP = OB$

$OP = 1$ m

$\therefore \quad PC = 1$ m

$OC = \sqrt{2}$m

$\therefore \quad AC = OC - OA$

$= \left(\sqrt{2} - 1\right)$ m

and $AC + CP = \left(\sqrt{2} - 1\right) + 1$

$= \sqrt{2}$ m

$= 1.414$ m $= 141.4$ cm

13. (c) $\dfrac{ar(\Delta CMN)}{ar(ABNM)} = \dfrac{1}{2}$

$\therefore \quad \dfrac{ar(\Delta CMN)}{ar(\Delta CAB)} = \dfrac{1}{3}$

$\Rightarrow \quad \dfrac{MN}{AB} = \dfrac{CM}{CA} = \dfrac{1}{\sqrt{3}}$

$\Rightarrow \quad \dfrac{CM}{MA} = \dfrac{1}{\sqrt{3}-1} = \dfrac{\sqrt{3}+1}{2} \qquad MA = (CA - CM)$

14. (d) $\angle ACB = 50°$
$\angle CFO = \angle CEO = 90°$
$\therefore \quad \angle FOE = 360° - (90° + 90° + 50°) = 130°$
but $\angle AOB = \angle FOE = 130°$

15. (a) $\angle OCD = 90°$
$\angle OAC = \angle OCA = 30°$
$\angle ACD = \angle ACO + \angle OCD = 30° + 90° = 120°$
$\therefore \quad \angle BAC = 180° - 120° = 60°$
$\Rightarrow \quad \angle BCD = 60° \qquad (\angle BCD = \angle BAC)$
$\Rightarrow \quad \angle OCB = \angle OCB - \angle BCD$
$= 90° - 60° = 30°$

16. (d) $\dfrac{1}{2} : \dfrac{1}{3} : \dfrac{1}{4} = 6 : 4 : 3$
$6x + 4x + 3x = 52$, or $13x = 52x$, or $x = 4$
Required length = 12 cm.

17. (c) Required ratio $= \dfrac{a^2}{\left(\sqrt{2}a\right)^2} = 1 : 2$

18. (b) Δ's APO and BQO are similar
($\angle APO = \angle BQO = 90°$, tangent is $\perp r$ to radius
$\angle AOP = \angle QOB$, vertically opposite angles).
$\therefore AO : OB :: 2 : 1$ and $OP : OQ :: 2 : 1$, $AB = 10$

$\Rightarrow AO = \dfrac{2}{3} \times AB = \dfrac{20}{3}$ and $OP = \dfrac{2}{3} \times PQ = \dfrac{16}{3}$

$AP^2 = OA^2 - OP^2$ (In $\Delta\ OAP$, $\angle APO = 90°$)

$= \dfrac{1}{3^2}(20^2 - 16^2) = \dfrac{1}{3^2} \times 144$

$\therefore \ AP = \dfrac{12}{3} = 4$

19. (d)

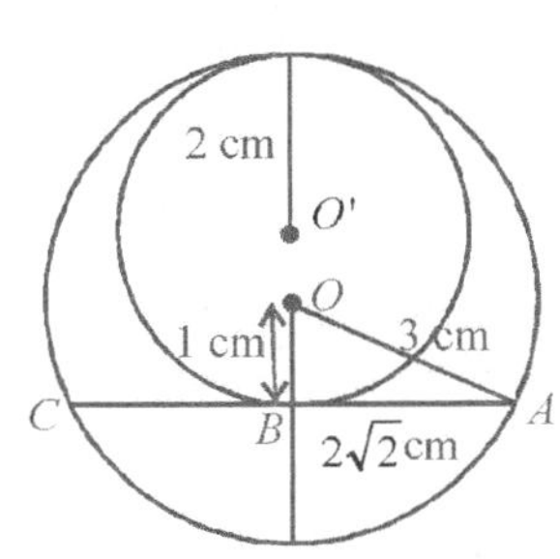

$AB = \sqrt{3^2 - 1^2} = 2\sqrt{2}$ cm

$\therefore \quad AC = 4\sqrt{2}$ cm

20. (b) The sum of the interior angles of a polygon of n sides is given by the expression $(2n - 4)\ \dfrac{\pi}{2}$

$\Rightarrow \ (2n-4) \times \dfrac{\pi}{2} = 1620 \times \dfrac{\pi}{180}$

$(2n-4) = \dfrac{1620 \times 2}{180} = 18$

or $2n = 22$
or $n = 11$

21. (a) In ΔABC, $DE \parallel BC$
By applying basic Proportionality theorem,

$\dfrac{AD}{DB} = \dfrac{AE}{EC}$

But $\dfrac{AD}{DB} = \dfrac{3}{5}$ (Given)

$\therefore \ \dfrac{AE}{EC} = \dfrac{3}{5}$ or $\dfrac{AE}{EC + AE} = \dfrac{3}{5+3}$ or $\dfrac{AE}{AC} = \dfrac{3}{8}$

or $\dfrac{AE}{5.6} = \dfrac{3}{8} \Rightarrow 8AE = 3 \times 5.6$

$\Rightarrow \quad AE = 3 \times 5.6 / 8$

$\therefore \ AE = 2.1$ cm.

22. (a) Let the diagonals of the rhombus be x and y and the its sides be x.

Now, $x^2 = \left(\dfrac{x}{2}\right)^2 + \left(\dfrac{y}{2}\right)^2$

or $\quad x^2 - \dfrac{x^2}{4} = \dfrac{y^2}{4}$

$3x^2 = y^2$

or $\quad \dfrac{x}{y} = \dfrac{1}{\sqrt{3}}$ or $y : x = \sqrt{3} : 1$

23. (a)

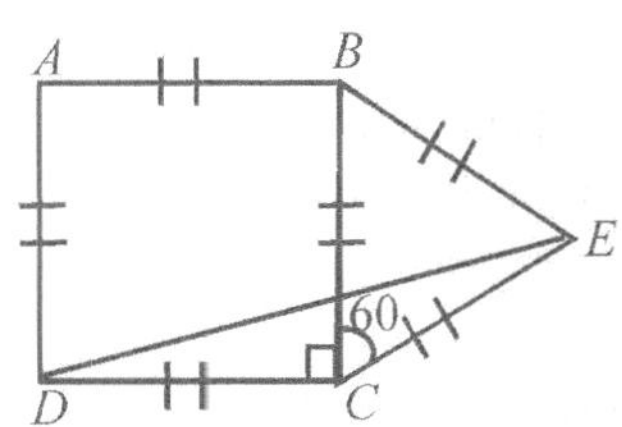

In $\Delta DEC, \angle DCE = 90° + 60° = 150°$

$\angle CDE = \angle DEC = \dfrac{180 - 150}{2} = 15°$

24. **(b)** Let the side of the square be x, then

$BE = \frac{x}{3}$ and $BF = \frac{x}{2}$

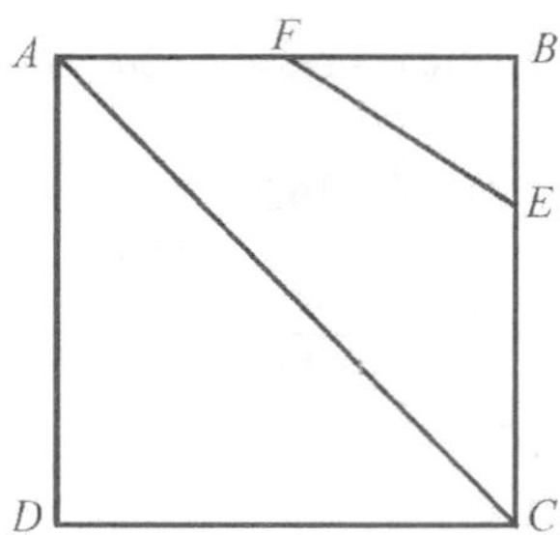

Area of $\Delta FEB = \frac{1}{2} \times \frac{x}{3} \times \frac{x}{2} = \frac{x^2}{12}$

Now, $\frac{x^2}{12} = 108$

$\Rightarrow \quad x^2 = 108 \times 12 = 1296$

In ΔADC, we have

$AC^2 = AD^2 + DC^2$

$= x^2 + x^2 = 2x^2$

$= 2 \times 1296 = 2592$

or $\quad AC = \sqrt{2592} = 36\sqrt{2}$

25. **(a)** $m \angle ADC = 90°$

(Angle subtended by the diameter on a circle is 90°)

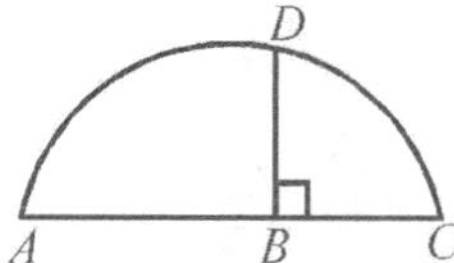

$\therefore$ ΔADC is a right angled triangle.

$\therefore (DB)^2 = l \times BC.$

(DB is the perpendicular to the hypotenuse)

$= 9 \times 4 = 36$

$\therefore DB = 6$

26. **(a)** Diameter of circle $= x$

$\therefore y = 4x \quad \therefore x = \frac{1}{4}y.$

27. **(b)** As F is the mid-point of AD, CF is the median of the triangle ACD to the side AD.

Hence area of the triangle FCD

= area of the triangle ACF.

Similarly area of triangle BCE = area of triangle ACE.

$\therefore$ Area of $ABCD$ = Area of ($CDF + CFA + ACE + BCE$)

$= 2$ Area ($CFA + ACE$) $= 2 \times 13 = 26$ sq. units.

28. **(b)** Since the sum of all the angle of a quadrilateral is 360°

We have $\angle ABC + \angle BQE + \angle DEF + \angle EPB = 360°$

$\therefore \angle ABC + \angle DEF = 180° \quad [\because BPE = EQB = 90°]$

29. **(c)** $m \angle DAB + 180° - 120° = 60°$

(Opposite angles of a cyclic quadrilateral)

m (arc BCD) = 2m $\angle DAB = 120°$.

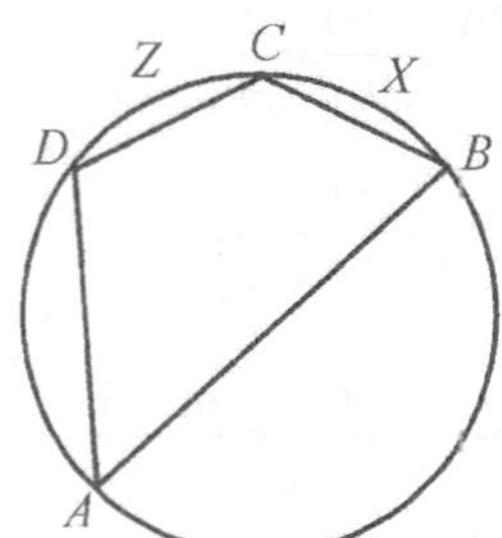

$\therefore$ m (arc CXB) = m (BCD) – m (arc DZC)

$= 120° - 70° = 50°$.

30. **(c)** Given AB is a circle and BT is a tangent, $\angle BAO = 32°$

Here, $\angle OBT = 90°$

[$\because$ Tangent is $\perp$ to the radius at the point of contact]

$OA = OB$ [Radii of the same circle]

$\therefore \angle OBA = \angle OAB = 32°$

[Angles opposite to equal side are equal]

$\therefore \angle OBT = \angle OBA + \angle ABT = 90°$ or $32° + x = 90°$.

$\angle x = 90° - 32° = 58°$.

Also, $\angle AOB = 180° - \angle OAB - \angle OBA$

$= 180° - 32° - 32° = 116°$

Now $Y = \frac{1}{2} AOB$

[Angle formed at the center of a circle is double the angle formed in the remaining part of the circle]

$= \frac{1}{2} \times 116° = 58°$.

31. **(a)**

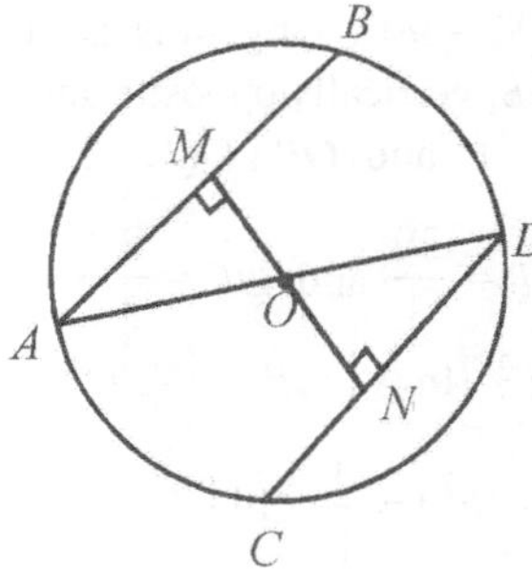

Two parallel chords AB & CD & $AB = CD = 8$ cm

Diameter of circle $= AD = 10$ cm.

$\therefore \quad$ radius $= AO = OD = \frac{10}{2} = 5$ cm

$AM = MB = \frac{AB}{2} = 4$ cm.

ΔAOM is right angle Δ,

$AO^2 = AM^2 + OM^2$

$5^2 = 4^2 + OM^2$

$OM^2 = 25 - 16 = 9$

$\Rightarrow \quad OM = 3$ cm.

Similarly,

$OM = ON = 3$ cm

$\therefore$ Distance between parallel chords $= MN$

$= OM + ON$

$= 3 + 3 = 6$ cm

32. (c) $\angle A + \angle B + \angle C = \angle 180°$

$$\Rightarrow \frac{1}{2}\angle B + \frac{1}{2}\angle C = 90 - \frac{1}{2}\angle A$$

$$\Rightarrow \frac{1}{2}\angle B + \frac{1}{2}\angle C + \angle BOC = 180°$$

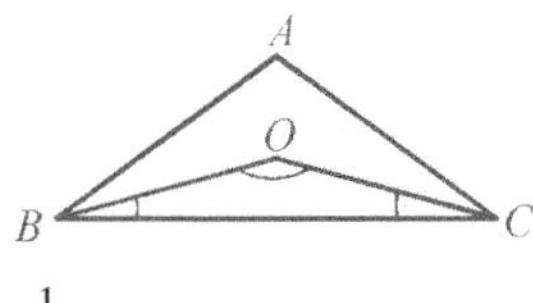

$$\Rightarrow 90° - \frac{1}{2}\angle A + \angle BOC = 180°$$

$$\Rightarrow \angle BOC = 180° - \left(90° - \frac{\angle A}{2}\right) = \left(90° + \frac{\angle A}{2}\right)$$

33. (c) $\angle APB = 42° + 68° = 110°$

(Exterior angle of a triangle is equal to sum of opposite interior $\angle s$).

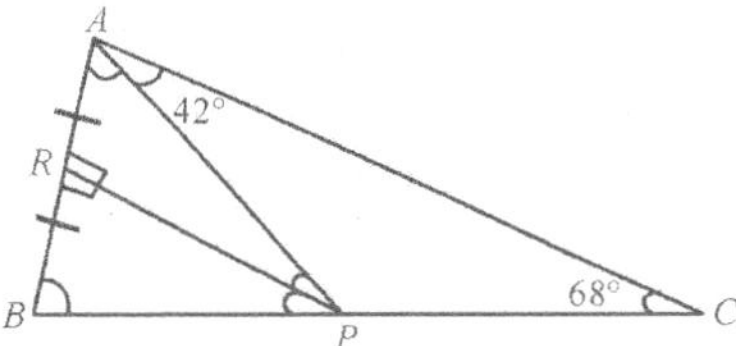

$\Delta APR \cong \Delta BPR$ [SAS condition]

$$\therefore \angle RPB = \angle RPA = \frac{110°}{2} = 55°$$

$\therefore$ In ΔBRP, $\angle ABC = 90° - 55° = 35°$.

34. (a) Let AB be the chord of length 14 cm.

at a distance of 6 cm from the centre O.

Draw $OE \perp AB$. Then, $BE = 7$ cm and $OE = 6$ cm.

$\therefore OB^2 = OE^2 + BE^2 = (6^2 + 7^2) = 85$.

Let CD be the chord at a distance of 2 cm from O.

Now, $OF = 2$ cm, $OD^2 = OB^2 = 85$.

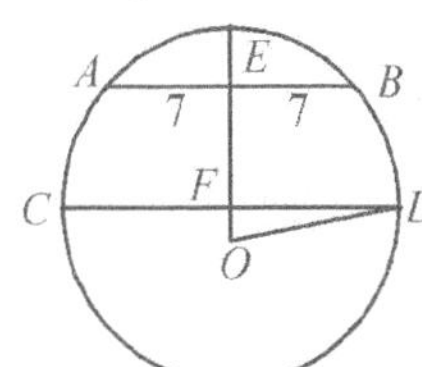

$\therefore FD = 9$ cm.

$\therefore CD = 2 \times FD = 18$ cm.

35. (a) Let $YX = YZ = r$ (same radii); OYZ is a straight line (contact of circles)

$YX \perp AB$ (Tangent $\perp$ to radius);

$AX = 9$, $XB = 5$ (given)

$\Rightarrow AB = 14$, $OB = OZ = 7$(Same radii)

$OX + 7 - 5 = 2$

In triangle OXY, $OY = 7 - r$; $YX + r$, $OX = 2$

$\Rightarrow OY^2 = YX^2 + OX^2$ (Pythagoras' Theorem)

$(7 - r)^2 = r^2 + 2^2 \quad 49 - 14r + r^2 = r^2 + 4$

$14r = 45 \quad r = 3\frac{3}{14}$ cm.

36. (a) R is the midpoint of side. AB and Q is the midpoint of side AC.. (midpoint theorem)

$$\therefore RP = \frac{1}{2}AC = \frac{1}{2}AB$$

$$PQ = \frac{1}{2}AB$$

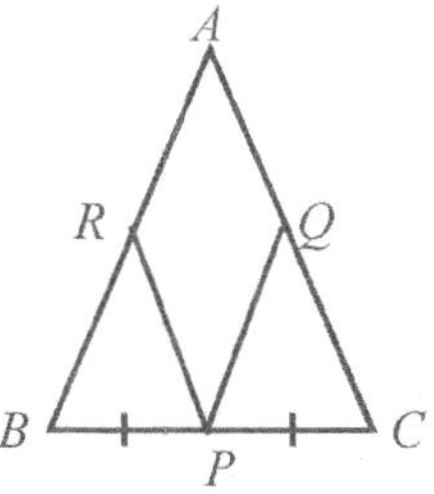

Perimeter of $\square PRQ = 2(RP + PQ)$

$$= 2\left(\frac{1}{2}AB + \frac{1}{2}AB\right)$$

$= 2AB = 2 \times 8 = 16$

37. (c) The given hexagon is regular hexagon.

$\therefore AF = AO = FO = OE = FE$

Height of the hexagon = distance AE.

$$= \frac{\sqrt{3}}{2}a + \frac{\sqrt{3}}{2}a = \sqrt{3}\,a.$$

38. (a) In ΔABD and ΔAEC

$m\angle ADB = m\angle ACE$ (both 90°)

$m\angle ABC = m\angle ABC$ (intercept the same area AC)

$\therefore \quad \Delta ABD - \Delta AEC$ (AA similarity)

$$\frac{AB}{AE} = \frac{AD}{AC} \quad \text{(Proportional sides)}$$

$$\therefore \quad AE = \frac{AB}{AD} \times AC = \frac{8}{4.8} \times 6 = 10$$

$\therefore$ Circumradius = 5,

39. (b)

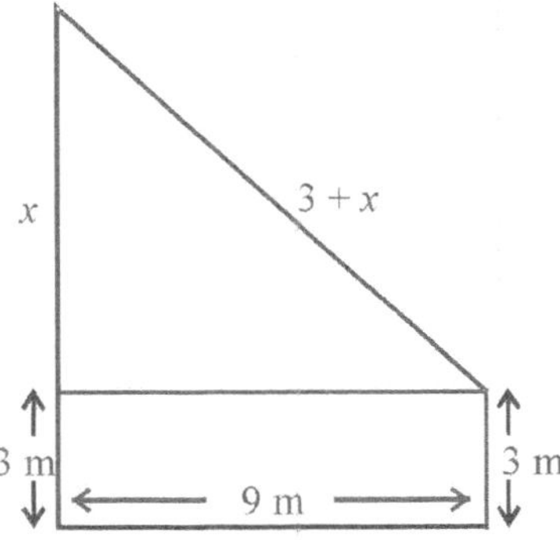

Using Pythagoras theorem, $x^2 + 81 = (3 + x)^2$

$\Rightarrow x^2 + 81 = 9 + x^2 + 6x \Rightarrow 6x = 72 \Rightarrow x = 12\text{m}$

Height of wall = 12 + 3 = 15 m

40. (c) $PR = PB + AR - 5 = 20 + 20 - 5$ [$\because AB = 5$ cm]

So, perimeter = $PR + PQ + QR$

$= 20 + (20 - 5) + 20 + (20 - 10) + 20 + (20 - 12)$

$= 35 + 30 + 28 = 93$

41. (b) Using the quality of similar triangles, $\frac{CP}{PB} = \frac{CD}{AB} = \frac{1}{3}$;

In $\Delta\, BPQ$ and BCD, $\frac{CD}{PQ} = \frac{BC}{BP} = \frac{4}{3} = 1:0.75$

42. (d) Remember that a perpendicular from the centre to a chord divides it into two equal parts.

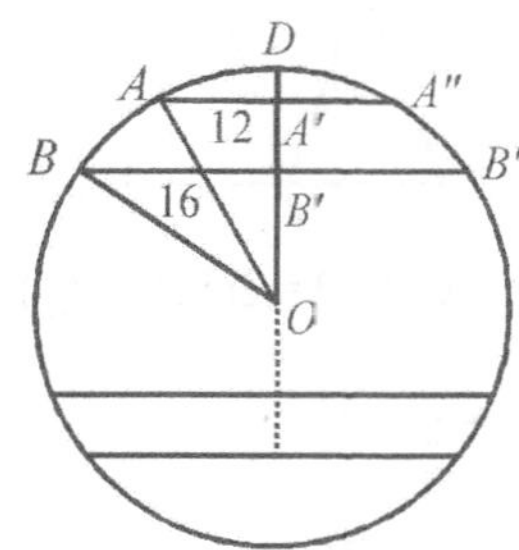

In $\Delta\, OBB'$,

$OB^2 = BB'^2 + OB'^2$

$\Rightarrow 20^2 = 16^2 + OB'^2$

$\Rightarrow OB' = \sqrt{20^2 - 16^2} = 12$

Similarly in $\Delta\, OAA'$, $OA' = \sqrt{20^2 - 12^2} = 16$

$\therefore$ Distance between the two parallel chords

= 16 – 12 = 4 cm or 16 + 12 = 28 cm

43. (b) $(OS)^2 = (OK) + (KS)^2$

$25 = OK^2 + 16 \Rightarrow OK = 3$

and $(OS)^2 = (OL)^2 + (LN)^2$

$25 = (OL)^2 + 9$

$\Rightarrow OL = 4$ cm

$\therefore KL = OL - OK = 1$ cm

$\therefore$ Area of rectangle = 1 × 10 = 10 cm^2

44. (c) $AB = PQ = 26$ cm

and $PO = OQ = 13$ CM

$CO = \sqrt{(PO)^2 - (PC)^2}$

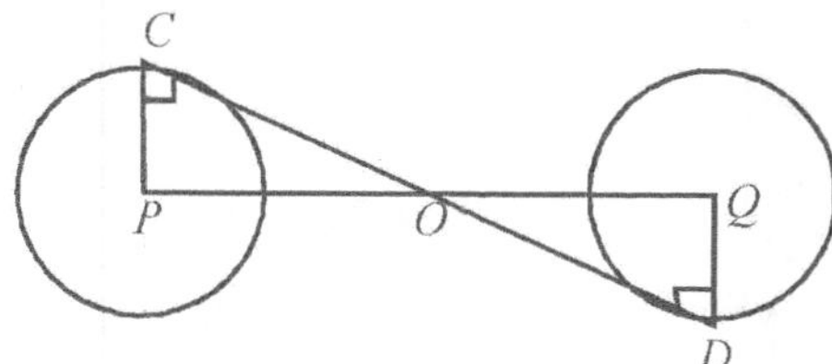

$CO = \sqrt{(13)^2 - (5)^2}$

$CO = 12$ cm

$CD = 2CO = 24$ cm

Alternatively: Solve by using the formula of tangents.

45. (d) At $\angle A = 60°$, $BC = b = c$

and at $\angle A = 90°$, $BC = \sqrt{2}b = \sqrt{2}c$

$\therefore 60° < A < 90°$, $BC = c < a < c\sqrt{2}$

46. (c) $\angle A : \angle B : \angle C = 4 : 1 : 1$

Hence we can suppose

$\angle A = 4x$, $\angle B = x$, $\angle C = x$

$\therefore 4x + x + x = 180$

$x = 30$

$\therefore A = 120, B = 30, C = 30$

Now, $\frac{\sin A}{\sin B} = \frac{\sin 120°}{\sin 30°} = \frac{\sin 60°}{\sin 30°} = \frac{\frac{\sqrt{3}}{2}}{\frac{1}{2}} = \frac{\sqrt{3}}{1}$

47. (b) $9 \times 180 - 2 \times 360$

$= 180 \times 5 = 900°$

$\because \left\{\begin{matrix} n \times 180 - 2 \times 360 \\ = 180(n-4) \end{matrix}\right\}$

48. (b) $\angle ADO$ is a right angle (angle of semicircle)

Again when OD is perpendicular on the chord AC and OD passes through the centre of circle ABC, then it must bisect the chord AC at D.

$\therefore AD = CD = 6$ cm

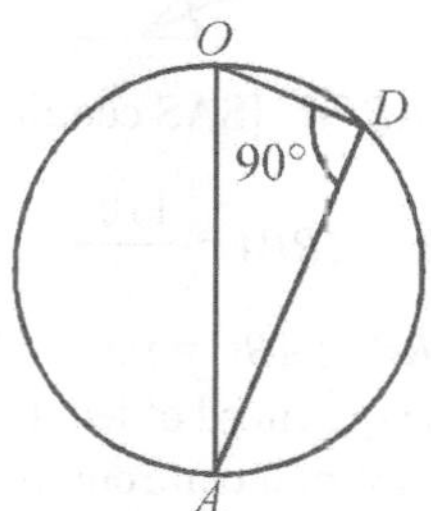

49. (a) $OB = OA$ – radius of circle

$\Rightarrow \angle CAO = \angle OBA$

(angles in alternate segments are equal)

Now, if $\angle CAO = \angle OBA$

$\therefore \angle OAC = \angle OAB$

$\therefore$ option (a) is correct

50. (d) Perimeter of the figure = $10 + 10 + 6 + 3\pi$

$= 26 + 3\pi$ cm

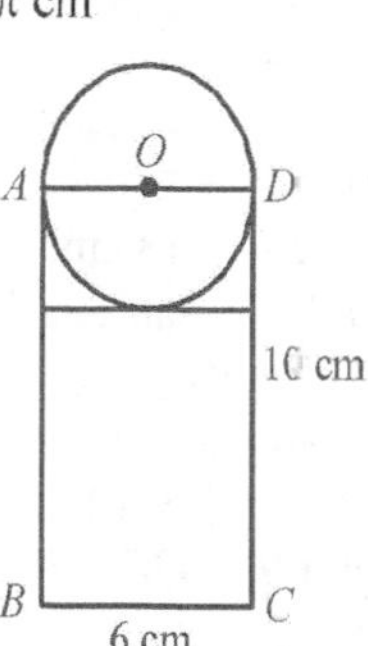

51. (c) Value of BC will lie in between 999 and 3003. Hence $999 < BC < 3003$.
So, the total values possible for $BC = 2003$.

52. (b) Let the adjacent angles of the parallelogram be 4x and 5x.
Then, $4x + 5x = 180 \Rightarrow 9x = 180 \Rightarrow x = 20$
One angle of quadrilateral $= 3 \times 80° = 240°$
Again, sum of angles of quadrilateral
$4y + 11y + 9y + 240° = 360°$
$24y = 120° \Rightarrow y = 5$
Hence, the sum of the largest and the smallest angles of the quadrilateral $= 4 \times 5 + 240 = 260°$

53. (d)
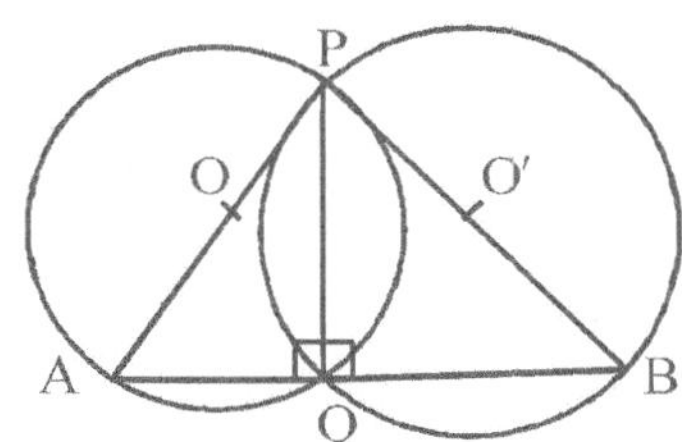

$\angle AQP = \frac{\pi}{2}$ (Angle in the semicircle is 90°)

$\angle BQP = \frac{\pi}{2}$ (Angle in the semicircle is 90°)

$\angle AQB = \angle AQP + \angle BQP = \frac{\pi}{2} + \frac{\pi}{2} \Rightarrow \pi$ or 180°
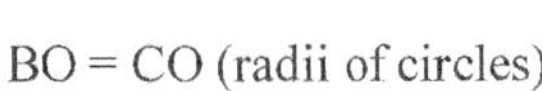

54. (b) In ΔAOB
AO = BO (radii of circles)
$\therefore \angle ABO = \angle BAO = 30°$
In ΔBOC

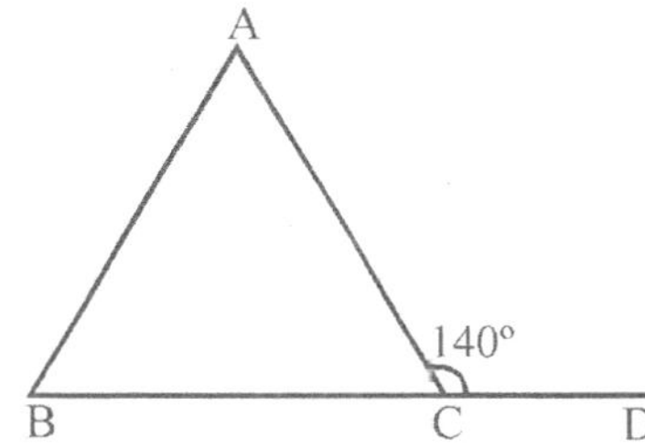

BO = CO (radii of circles)
$\therefore \angle BCO = \angle OBC = 40°$
$\angle ABC = \angle ABO + \angle OBC$
$\angle ABC = 30° + 40° = 70°$
$2 \times \angle ABC = \angle AOC \Rightarrow x° = 140$

55. (a)
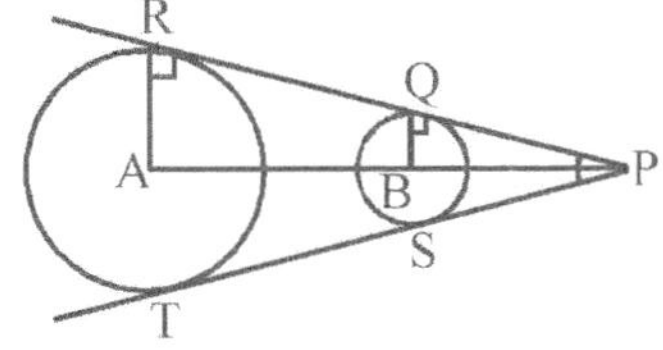

Δ PQB and Δ PRA are similar triangle by AAA criteria.

$\therefore \frac{AP}{BP} = \frac{AR}{BQ} = \frac{5}{2}$

P divides AB externally in the ratio of 5 : 2

56. (a) In ΔOBP.
OB = OP ($\because$ radius)

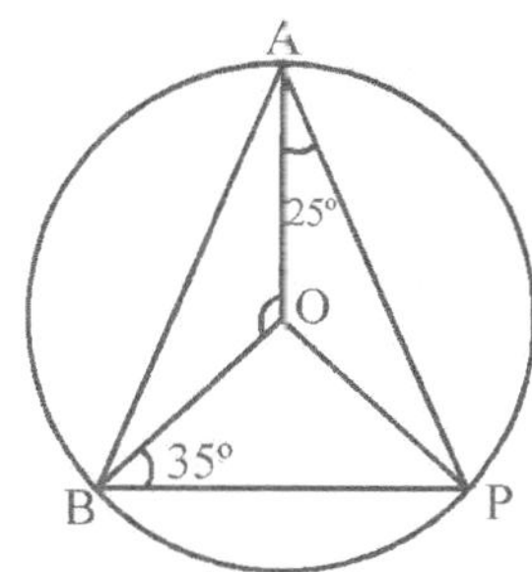

$\therefore \angle OBP = \angle OPB = 35°$
In Δ AOP
OA = OP ($\because$ radius)
$\therefore \angle OAP = \angle OPA = 25°$
Now, $\angle APB = \angle OPA + \angle OPB$
$= 25° + 35° = 60°$
Hence, $\angle AOB = 2\angle APB$
(Angle be substended by are at centre is twice)
$= 2 \times 60° = 120°$

57. (d)
A
140°
B C D

$\angle ACB + \angle ACD = 180°$ (linear pair)
$\therefore \angle ACB = 180° - 140° = 40°$
In Δ ABC
$\angle BAC + \angle ABC + \angle ACB = 180°$
$\angle BAC + 3\angle BAC + 40° = 180°$
$4\angle BAC = 180° - 40°$

$\angle BAC = \frac{140}{4} = 35°$

58. (c)
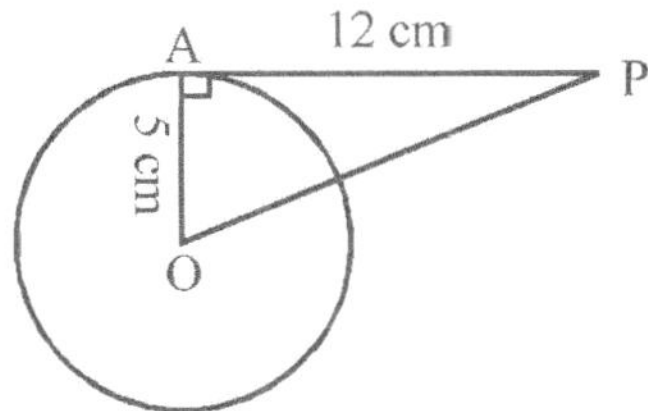

AP is a tangent and OA is a radius.
Therefore, OA is ⊥ at AP.
So, In Δ OAP
$OP^2 = 5^2 + 12^2$
$OP^2 = 25 + 144 = 169$
OP = 13 cm

59. (b) In Δ ABC, $\angle$ ACB = 90°

$\therefore$ $\angle$ACB + $\angle$ACD

$\Rightarrow$ 90° + 50° = 140°

As angle mode by triangle in semicircle is equal to 90°.

$\therefore$ In quad. ABCD,

$\angle$BAD + $\angle$BCD = 180°

(Sum of opposite angles of a cyclic quad. is equal to 180°)

$\angle$ BAD = 180° – 140° = 40°

60. (d)

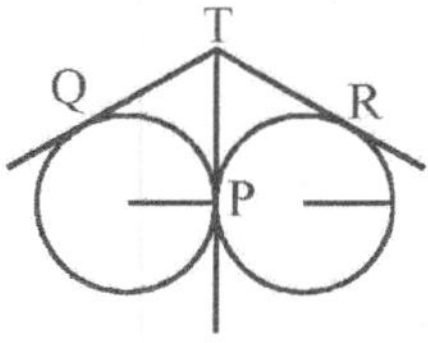

TP = TQ [The length of tangents drawn from an external point to a circle are equal]

Similarly, TP = TR

Using both equation, we get

TQ = TR

The relation of TQ and TR is TQ = TR.

61. (b)

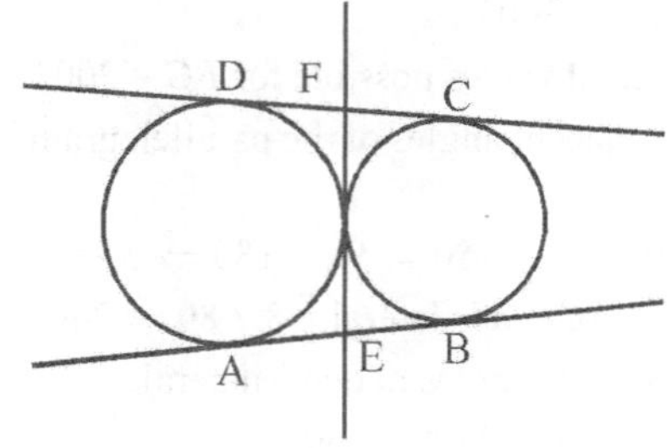

There are three common tangents AB, CD and EF.

62. (b) DE is parallel to BC

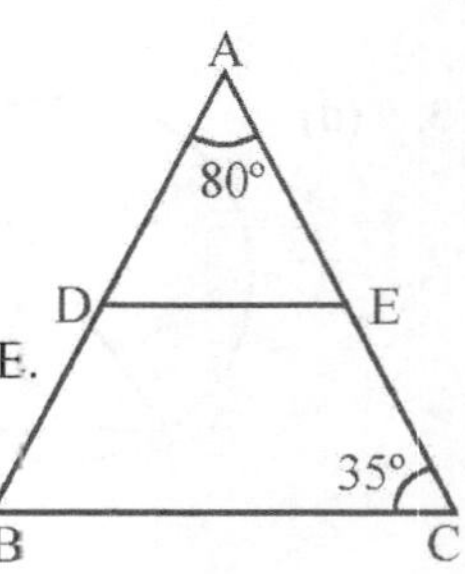

So $\angle$ AED = $\angle$C = 35°

Since $\angle$A = 80°

Then $\angle$ ADE = 65°

$\angle$ EDB is supplement to $\angle$ADE.

So, $\angle$ EDB = 180° – $\angle$ ADE

= 180° – 65° = 115°

CHAPTER 19

MENSURATION

BASIC CONVERSION OF UNITS

(i) Length:

1 m = 10 dm = 100 cm = 1000 mm

1 dm = 10 cm = 100 mm

1 cm = 10 mm

1 feet (ft) = 12 inches

1 inch = 2.54 cm

1 yard (y) = 3 feet (ft)

1 m = 1.094 yard (y) = 39.37 inches

1 yard (y) = 0.914 metre (m)

1 km = 1000 m = $\frac{5}{8}$ miles

1 mile = 1760 yards (y) = 5280 feet (ft)

1 nautical mile (knot) = 6080 feet (ft)

(ii) Surface Area:

Surface areas are measured in square units.

1 square metre = 1m × 1m = 100 cm × 100 cm = 10000 cm^2

1 square yard = 1y × 1y = 3 ft × 3 ft = 9 ft^2

1 acre = 4047 m^2 (approx.)

1 hectare = 10000 m^2

(iii) Mass:

1 kg = 1000 grams (g) = 2.2 pounds (approx.)

1 gram = 10 miligram (mg)

1 quintal = 100 kg

1 tonne = 10 quintal = 1000 kg

(iv) Volume:

Volumes are measured in cubic units.

1 litre = 1000 cm^3 or cc

1 m^3 = 10000 litres (= $10^4\, l$) = 10^7 cm^3

Note that

$\sqrt{2} = 1.414$, $\sqrt{3} = 1.732$, $\sqrt{5} = 2.236$,

$\sqrt{6} = 2.45$, $\pi = \frac{22}{7}$ or 3.14

PLANE FIGURES

We have already dealt with plane figures (Triangles, Quadrilaterals and Circles) in geometry chapter. In this chapter, we will deal with perimeter and area of plane figures.

Perimeter: The perimeter of a plane geometrical figure is the total length of sides (or boundary) enclosing the figure. Units of measuring perimeter can be cm, m, km, etc.

Area: The area of any figure is the amount of surface enclosed within its bounding lines. Area is always expressed in square units.

AREA OF A TRIANGLE

1. If in a triangle, we draw a perpendicular AP from vertex A on opposite side BC then AP is called altitude (or height) of the triangle ABC corresponding to base BC.

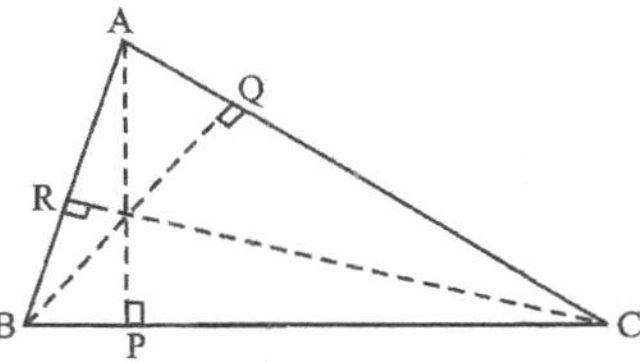

Similarly, BQ and CR are altitude of ΔABC corresponding to, bases AC and AB respectively.

$$\text{Area} = \frac{1}{2} \times \text{base} \times \text{corresponding altitude}$$

$$\text{Area of } \Delta ABC = \frac{1}{2} \times BC \times AP = \frac{1}{2} \times AC \times BQ = \frac{1}{2} \times AB \times CR$$

Note that in ΔKLM, LN is the perpendicular on KM produced.

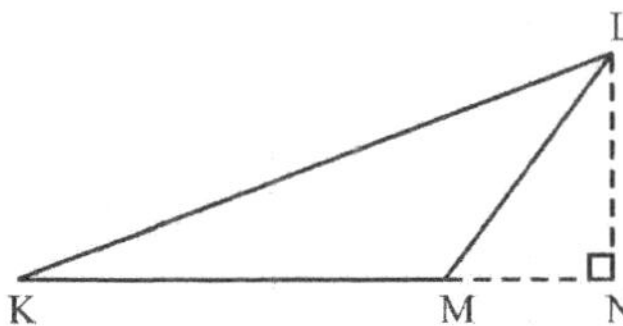

Here, LN is the altitude corresponding to the base KM of ΔKLM.

$\therefore$ Area of $\Delta KLM = \frac{1}{2} \times KM \times LN$

2. Let in ΔABC, $BC = a$, $AC = b$ and $AB = c$; then perimeter of $\Delta ABC = a + b + c$

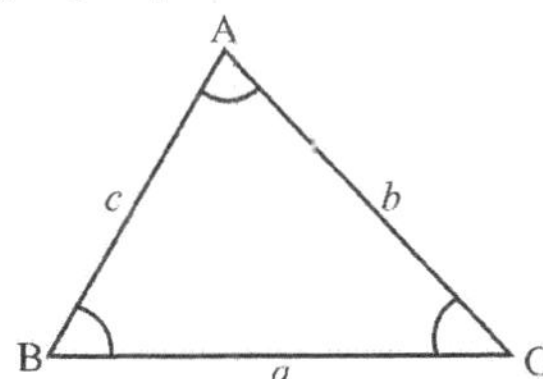

Semi-perimeter of $\Delta\, ABC$'s $s = \dfrac{a+b+c}{2}$

Area of $\Delta\, ABC = \sqrt{s(s-a)(s-b)(s-c)}$ (Heron's formula)

3. Area of $\Delta\, ABC = \dfrac{1}{2} \times$ (Product of two sides) $\times$ (Sine of the included angle)

$$= \frac{1}{2}\, ac \sin B \text{ or } \frac{1}{2}\, ab \sin C \text{ or } \frac{1}{2}\, bc \sin A$$

Note that $\sin 30° = \dfrac{1}{2}$, $\sin 45° = \dfrac{1}{\sqrt{2}}$,

$\sin 60° = \dfrac{\sqrt{3}}{2}$, $\sin 90° = 1$

Area of an Equilateral Triangle

Since, $\Delta\, ABC$ is an equilateral triangle.

$\therefore \quad AB = BC = CA = a$ (say)

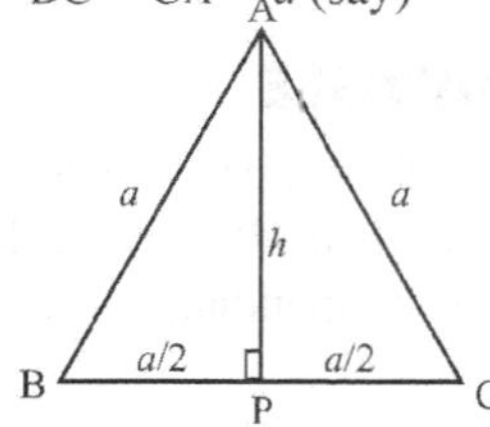

From $\Delta\, APC$,

$$AP^2 = AC^2 - PC^2 = a^2 - \frac{a^2}{4} = \frac{3a^2}{4}$$

$$AP = \frac{\sqrt{3}}{2} a \Rightarrow h = \frac{\sqrt{3}}{2} a$$

Area of an equilateral $\Delta = \dfrac{1}{2} \times a \times \dfrac{\sqrt{3}}{2} a = \dfrac{\sqrt{3}}{4} a^2$,

where a is the length of its one side

Note that

(i) among all the triangles that can be formed with a given perimeter, the equilateral triangle will have the maximum area.

(ii) For a given area of triangle, the perimeter of equilateral triangle is minimum.

Area of Incircle and Circumcircle of a Triangle

(i) If a circle touches all the three sides of a triangle, then it is called incircle of the triangle.

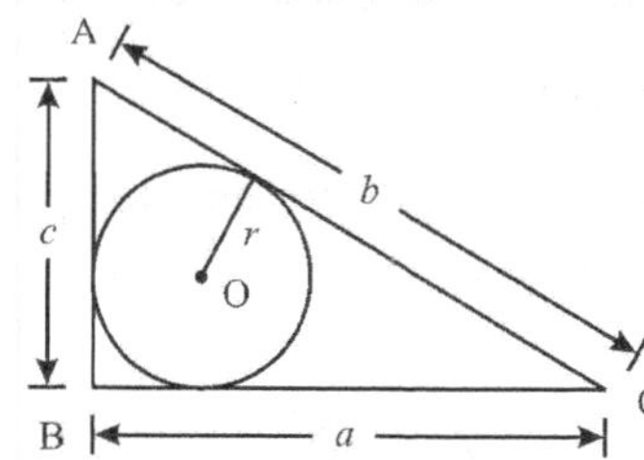

Area of incircle of a triangle $= r.\, s$, where r is the radius of the incircle and s is the half of the perimeter of the triangle. If a, b, c are the length of the sides of $\Delta\, ABC$, then

$$s = \frac{a+b+c}{2}$$

For an equilateral triangle,

$$r = \frac{\text{Length of a side of the triangle}}{2\sqrt{3}} = \frac{h}{3},$$

where h is the height of the triangle.

(ii) If a circle passess through the vertices of a triangle, then the circle is called circumcircle of the triangle.

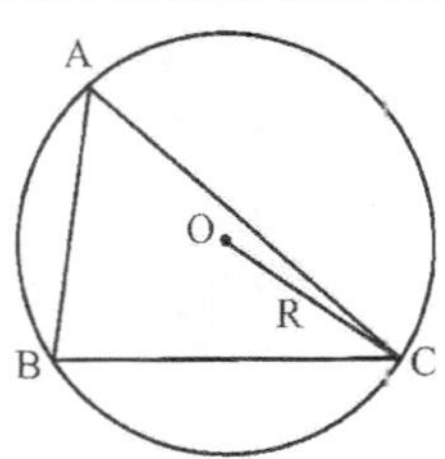

Area of the circumcircle $= \dfrac{abc}{4R}$, where R is the radius of the circumcircle and a, b, c are the length of sides of the triangle.

For an equilateral triangle,

$$R = \frac{\text{Length of a side of the triangle}}{\sqrt{3}} = \frac{2h}{3},$$

where h is the height or altitude of the equilateral triangle.

Hence for an equilateral triangle, $R = 2r$.

Note that an equilateral triangle inscribed in a circle will have the maximum area compared to other triangles inscribed in the same circle.

AREA OF A QUADRILATERAL

1. Area of quadrilateral $ABCD$

$= \dfrac{1}{2} \times$ (Length of the longest diagonal) $\times$ (Sum of length of perpendicular to the longest diagonal from its opposite vertices)

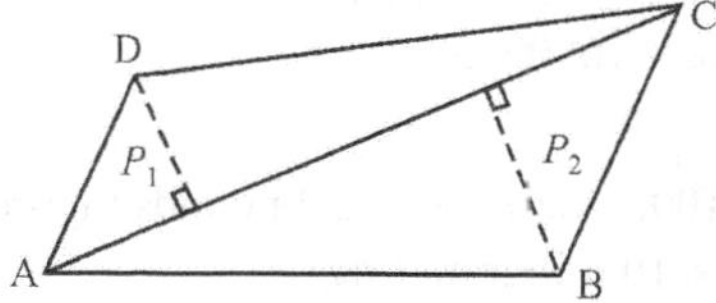

$= \dfrac{1}{2} \times d \times (p_1 + p_2)$, where $d = AC$ (i.e. longest diagonal)

2. If length of four sides and one of its diagonals of quadrilateral $ABCD$ are given, then

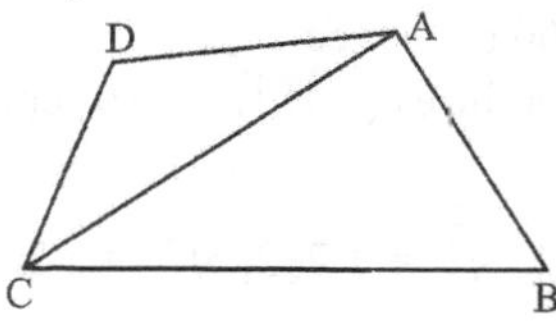

Area of the quadrilateral $ABCD$

$= $ Area of $\Delta\, ABC +$ Area of $\Delta\, ADC$

3. Area of circumscribed quadrilateral

$$= \sqrt{(s-a)(s-b)(s-c)(s-d)}$$

where $\quad s = \dfrac{a+b+c+d}{2}$ and a, b, c, d are length of sides of quadrilateral $ABCD$.

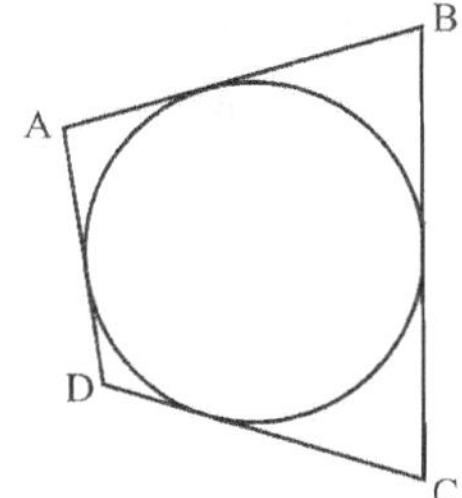

4. If θ_1 and θ_2 are the angles between the diagonals of a quadrilateral, then

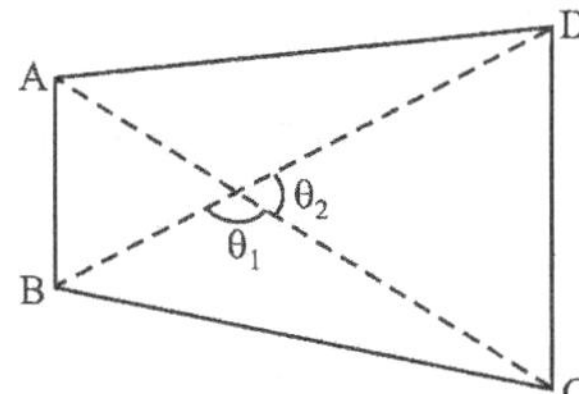

Area of the quadrilateral $= \frac{1}{2}\ d_1 d_2 \sin\theta_1$ or $\frac{1}{2}\ d_1 d_2 \sin\theta_2$

Here d_1 and d_2 are the length of the diagonals of the quadrilateral.

Area of a Parallelogram

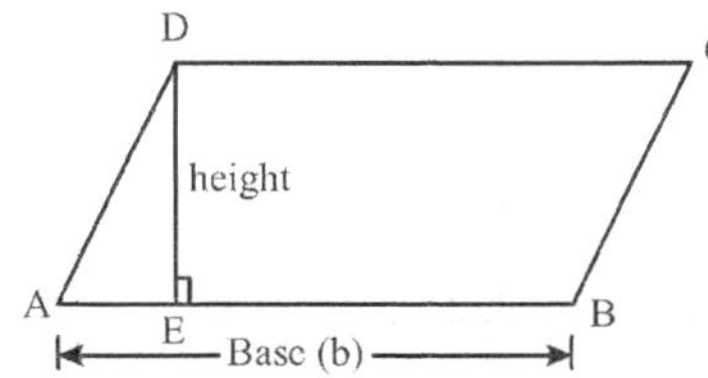

Area of parallelogram = Base × Corresponding height

$$A = b \times h$$

Perimeter of a parallelogram $= 2(a + b)$, where a and b are length of adjacent sides.

If θ be the angle between any two adjacent sides of a parallelogram whose length are a and b, then

Area of parallelogram $= ab \sin\theta$

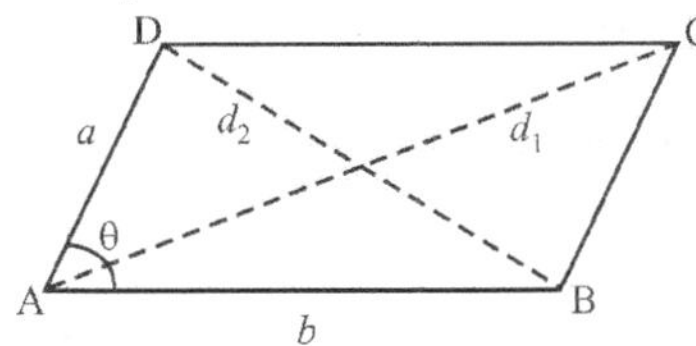

Note that in a parallelogram sum of squares of two diagonals

= 2 (sum of squares of two adjacent sides)

i.e., $d_1^2 + d_2^2 = 2\,(a^2 + b^2)$

Area of a Rectangle

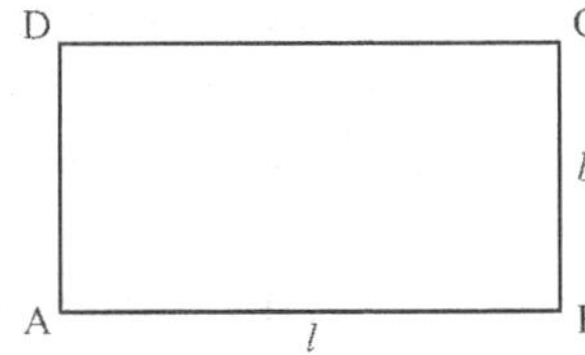

Area of a rectangle = Length × Breadth $= l \times b$

[If any one side and diagonal is given]

Perimeter of a rectangle $= 2(l + b)$

Area of a Square

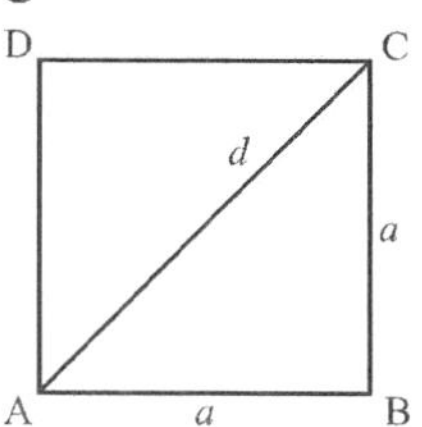

Area of square = side × side $= a \times a = a^2$

Length of diagonal $(d) = a\sqrt{2}$ (by Pythagoras theoram)

Hence area of the square $= \left(\frac{d}{\sqrt{2}}\right)^2 = \frac{d^2}{2}$

Perimeter of square = 4 × side = $4 \times a$

For a given perimeter of a rectangle, a square has maximum area.

Note that the side of a square is the diameter of the inscribed circle and diagonal of the square is the diameter of the circumscribing circle.

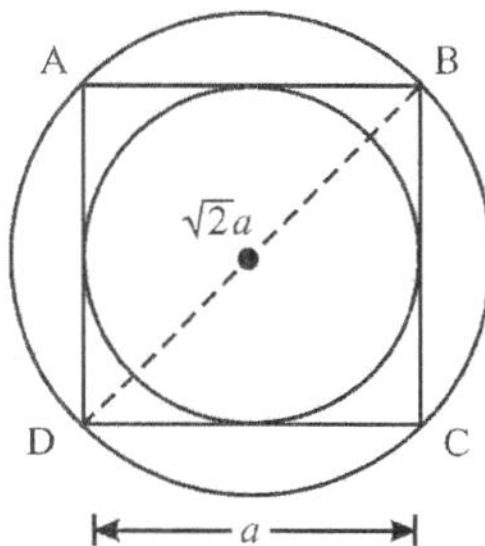

Hence inradius $= \frac{a}{2}$ and circumradius $= \frac{\sqrt{2}a}{2} = \frac{a}{\sqrt{2}}$

Area of a Rhombus

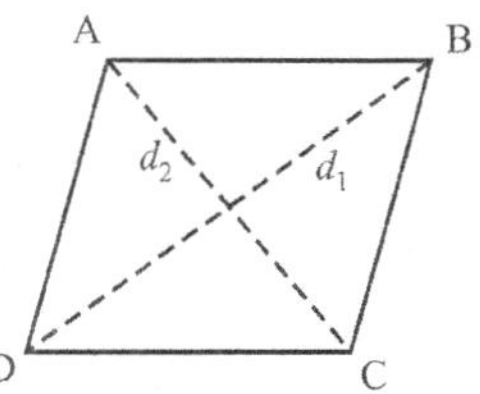

Area of a rhombus $= \frac{1}{2}$ × product of diagonals

$= \frac{1}{2} \times d_1 \times d_2$

Area of a Trapezium

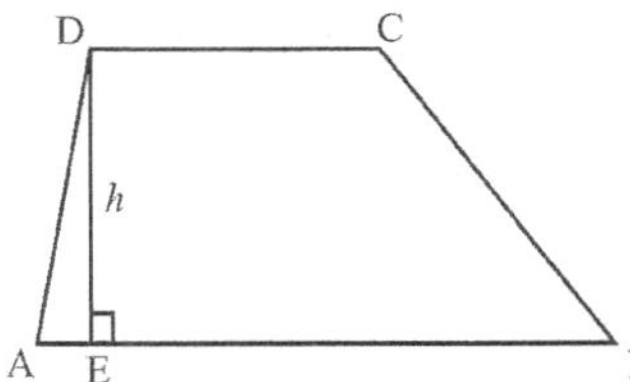

Distance between parallel sides of a trapezium is called height of trapezium.

In fig. $ABCD$ is a trapezium, whose sides AB and CD are parallel,

$DE = h$ = Height of the trapezium
= Distance between || sides.

Area of trapezium = $\frac{1}{2}$ (sum of || sides) × height

$= \frac{1}{2} \times (AB + CD) \times DE$

Area of a Cyclic Quadrilateral

For a given quadrilateral *ABCD* inscribed in a circle with sides measuring *a*, *b*, *c*, and *d*;

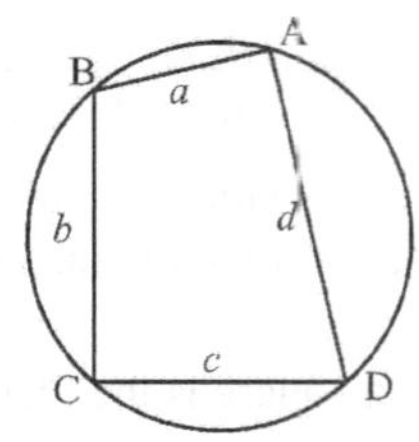

$$\text{Area} = \sqrt{(s-a)(s-b)(s-c)(s-d)}$$

where $s = \frac{a+b+c+d}{2}$

Illustration 1: A rectangular parking space is marked out by painting three of its sides. If the length of the unpainted side is 9 feet, and the sum of the lengths of the painted sides is 37 feet, then what is the area of the parking space in square feet?

(a) 46 **(b)** 81
(c) 126 **(d)** 252

Solution: (c) Clearly, we have : $l = 9$ and $l + 2b = 37$ or $b = 14$.
∴ Area = $(l \times b) = (9 \times 14)$ sq. ft. = 126 sq. ft.

Illustration 2: A square carpet with an area 169 m^2 must have 2 metres cut-off one of its edges in order to be a perfect fit for a rectangular room. What is the area of rectangular room?

(a) 180 m^2 **(b)** 164 m^2
(c) 152 m^2 **(d)** 143 m^2

Solution: (d) Side of square carpet $\sqrt{\text{Area}} = \sqrt{169} = 13$ m
After cutting of one side,
Measure of one side = 13 – 2 = 11 m
and other side = 13 m (remain same)
∴ Area of rectangular room = 13 × 11 = 143 m^2

Illustration 3: The ratio between the length and the breadth of a rectangular park is 3 : 2. If a man cycling along the boundary of the park at the speed of 12 km/hr completes one round in 8 minutes, then the area of the park (in sq. m) is:

(a) 15360 **(b)** 153600
(c) 30720 **(d)** 307200

Solution: (b) Perimeter = Distance covered in 8 min.

$$= \left(\frac{12000}{6} \times 8\right) \text{m} = 1600 \text{ m}.$$

Let length = $3x$ metres and breadth = $2x$ metres.
Then, $2(3x + 2x) = 1600$ or $x = 160$.
∴ Length = 480 m and Breadth = 320 m.
∴ Area = (480 × 320) m^2 = 153600 m^2.

Illustration 4: The length and breadth of a playground are 36m and 21 m respectively. Poles are required to be fixed all along the boundary at a distance 3m apart. The number of poles required will be

(a) 39 **(b)** 38
(c) 37 **(d)** 40

Solution: (b) Given, playground is rectangular.
Length = 36 m, Breadth = 21 m
Now, perimeter of playground = 2(21 + 36) = 114
Now, poles are fixed along the boundary at a distance 3 m.
∴ Required no. of poles = $\frac{114}{3} = 38$.

Illustration 5: Find the area of the trapezium *ABCD*.

(a) $5/2\,(13 + 2\sqrt{3})$ **(b)** $\frac{5\sqrt{3}\,(13 + 5\sqrt{3})}{2}$
(c) $13\,(13 + 2\sqrt{3})$ **(d)** None of these

Solution: (d)

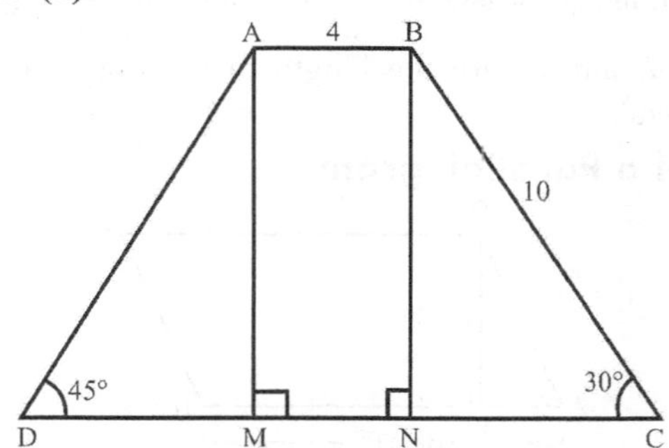

AB and *DC* are the parallel sides
Height = *AM* = *BN*
AB = *MN* = 4
Δ*BNC* and Δ*AMD* are right angled triangles

In Δ*BNC* ⇒ $\sin 30 = \frac{BN}{10} \Rightarrow BN = 5$

Using Pythagoras theorem $NC = \sqrt{10^2 - 5^2} = 5\sqrt{3}$

In Δ*ADM*; *AM* = 5; $\tan 45 = \frac{AM}{DM} = 1 \frac{5}{DM}$

$\Rightarrow DM = 5$

Area of trapezium ⇒ $\frac{1}{2}$ (Sum of parallel sides) × height

$$\Rightarrow \frac{1}{2}(4 + 4 + 5\sqrt{3} + 5) \times 5 = \frac{5(13 + 5\sqrt{3})}{2}$$

Illustration 6: Two goats tethered to diagonally opposite vertices of a field formed by joining the mid-points of the adjacent sides of another square field of side $20\sqrt{2}$. What is the total grazing area of the two goats?

(a) $10\,\pi\text{m}^2$ **(b)** $50(\sqrt{2} - 1)\,\pi\text{m}^2$
(c) $100\pi\,(3 - 2\sqrt{2})\text{m}^2$ **(d)** $200\pi\,(2 - \sqrt{2})\text{m}^2$

Solution: (a)

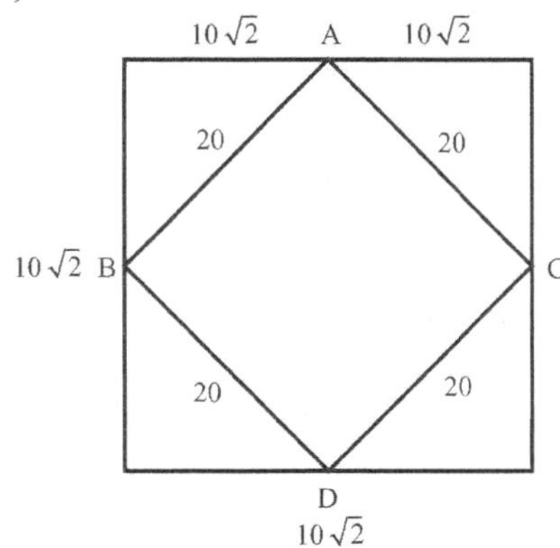

The length of rope of goat = $10\sqrt{2}$ m

Then the two goats will graze an area = Area of a semicircle with radius $10\sqrt{2}$ m.

So total area grazed = $\frac{\pi r^2}{2} \Rightarrow 100\ \pi m^2$

AREA OF A REGULAR HEXAGON

Area = $\frac{3\sqrt{3}}{2}a^2$, where 'a' is the length of each side of the regular hexagon.

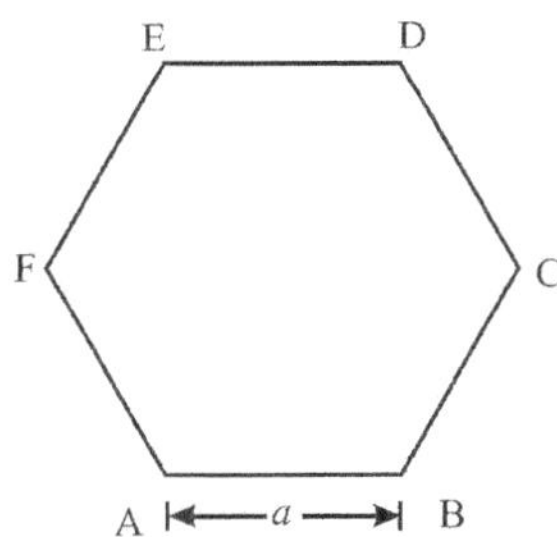

Diagonals of a hexagon divide it into six equelateral triangle. Hence, radius of the circumcircle of the hexagon

= Length of a side of the hexagon = a

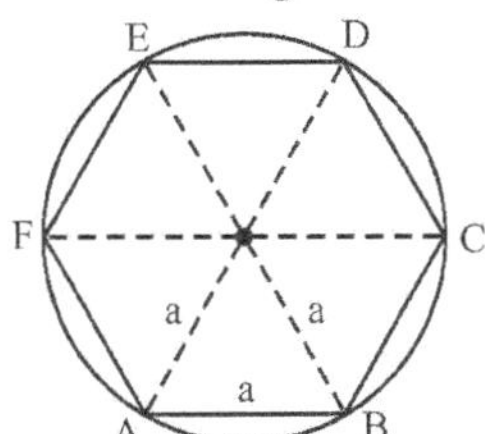

PATHS

1. Pathways Running Across the Middle of a Rectangle

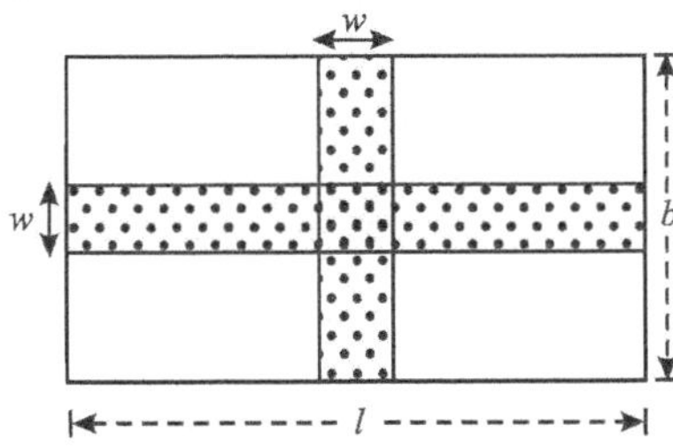

Area of the path = $l.w + b.w - w.w$

$= (l + b - w).w$

Perimeter of the path

$= 2l + 2b - 4w$

$= 2(l + b - 2w)$

Here w is the width of the path.

2. Pathways Outside a Rectangle

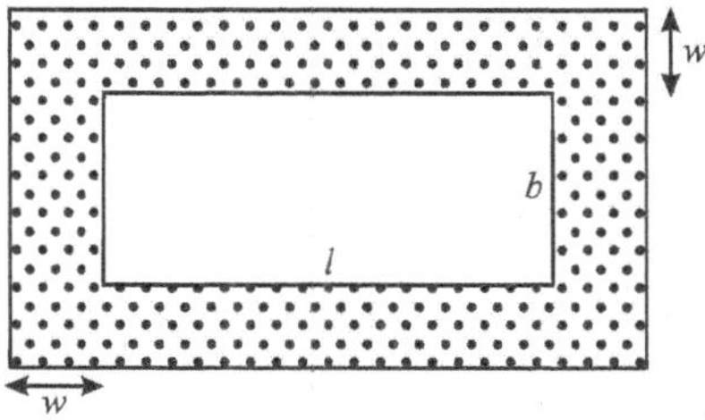

Area of path $= 2(lw) + 2(b.w) + 4(w.w)$

$= (l + b + 2w)2w$

Perimeter of path

= (Internal perimeter) + (External perimeter)

$= 2(l + b) + 2(l + b + 4w)$

$= 4(l + b + 2w)$

Here w is the width of the path.

3. Pathway Inside a Rectangle

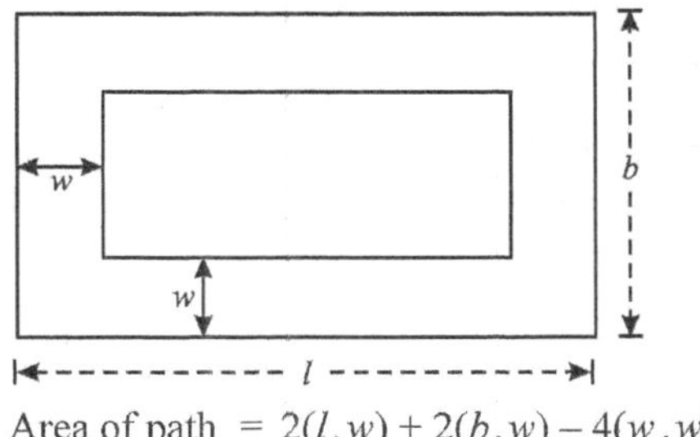

Area of path $= 2(l.w) + 2(b.w) - 4(w.w)$

$= (l + b - 2w).2w$

Perimeter of path

= Length of outer path + Length of inner path

$= 2(l + b) + 2(l + b - 4w)$

$= 4(l + b - 2w)$

AREA RELATED TO A CIRCLE

Circle

Set of all points in a plane which are at a fixed distance from a fixed point in the same plane is called a circle.

The fixed point is called centre of the circle and the fixed distance is called radius of the circle.

Circumference or perimeter of a circle of radius r is

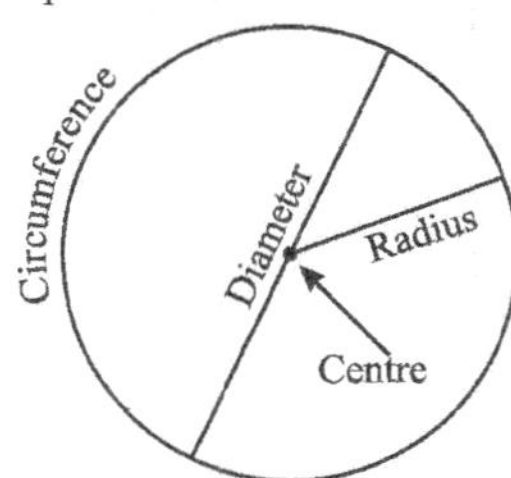

$c = 2\pi r = \pi d \quad (2r = d = \text{diameter})$

Area of the circle $= \pi r^2 = \frac{\pi d^2}{4} = \frac{c^2}{4\pi} = \frac{1}{2} \times c \times r$

Circular Ring

Region enclosed between two concentric circles of different radii in a plane is called a ring.

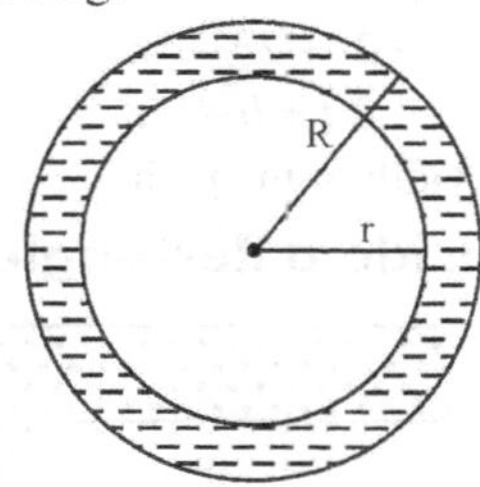

Area of the ring = $\pi R^2 - \pi r^2 = \pi (R^2 - r^2)$

Circumference of the ring

= (External circumference) + (Internal circumference)

$= 2\pi R + 2\pi r = 2\pi(R + r)$

Semi-circle

A semi-circle is a figure enclosed by a diameter and one half of the circumference of the circle.

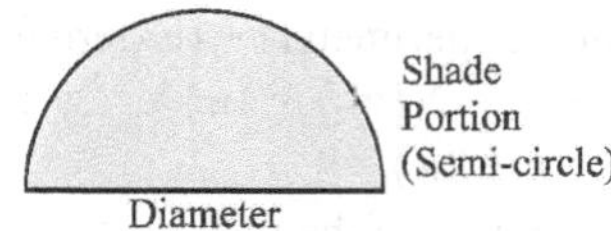

Area of the semi-circle = $\dfrac{\pi r^2}{2}$

Circumference of the semi-circle = $\pi r + 2r = r(\pi + 2)$

Sector of a Circle

Sector of a circle is the portion of a circle enclosed by two radii and an arc of the circle. *OACB* is a sector of the circle.

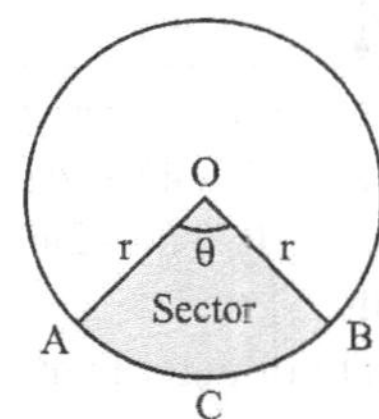

Length of arc *ACB* (which make angle θ at the centre)

$$= (2\pi r) \times \frac{\theta}{360} = \frac{\pi r\theta}{180}$$

Perimeter of the sector *OACB* = $2r + \dfrac{\pi r\theta}{180}$

Area of sector *OACB* = $(\pi r^2) \times \dfrac{\theta}{360}$

Segment of a Circle

A segment of a circle is a region enclosed by a chord and an arc of the circle.

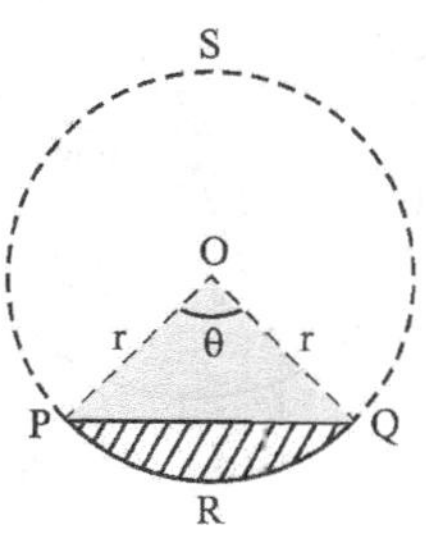

Any chord of a circle which is not a diameter divides the circle into two segments, one of which is the major segment and other is minor segment.

Perimeter of the segment *PRQP*

= Length of the arc *PRQ* + Length of *PQ*

$$= \frac{\pi r\theta}{180} + 2r\sin\frac{\theta}{2}$$

Area of (minor) segment *PQR*

= Area of sector *OPRQO* – Area of ΔOPQ

Area of (major) segment *PSQ*

= Area of circle – Area of segment *PQR*

Illustration 7: A circular grass lawn of 35 metres in radius has a path 7 metres wide running around it on the outside. Find the area of path.

(a) 1694 m² **(b) 1700 m²**
(c) 1598 m² **(d) None of these**

Solution: (a) Radius of a circular grass lawn (without path) = 35 m

$\therefore$ Area $= \pi r^2 = \pi (35)^2$

Radius of a circular grass lawn (with path)

$= 35 + 7 = 42$ m

$\therefore$ Area $= \pi r^2 = \pi (42)^2$

$\therefore$ Area of path $= \pi(42)^2 - \pi(35)^2$

$= \pi(42^2 - 35^2)$

$= \pi(42 + 35)(42 - 35)$

$= \pi \times 77 \times 7 = \dfrac{22}{7} \times 77 \times 7 = 1694 \text{ m}^2$

Illustration 8: A wire can be bent in the form of a circle of radius 56 cm. If it is bent in the form of a square, then its area will be:

(a) 3520 cm² **(b) 6400 cm²**
(c) 7744 cm² **(d) 8800 cm²**

Solution: (c) Length of wire = $2\pi \times R = \left(2 \times \dfrac{22}{7} \times 56\right)$ cm

= 352 cm.

Side of the square = $\dfrac{352}{4}$ cm = 88 cm.

Area of the square = $(88 \times 88) \text{ cm}^2 = 7744 \text{ cm}^2$.

Illustration 9: There are two concentric circular tracks of radii 100 m and 102 m, respectively. A runs on the inner track and goes once round on the inner track in 1 min 30 sec, while B runs on the outer track in 1 min 32 sec. Who runs faster?

(a) Both A and B are equal
(b) A
(c) B
(d) None of these

Solution: (b) Radius of the inner track = 100 m

and time = 1 min 30 sec ≡ 90 sec.

Also, Radius of the outer track = 102 m

and time = 1 min 32 sec ≡ 92 sec.

Now, speed of *A* who runs on the inner track

$$= \frac{2\pi(100)}{90} = \frac{20\pi}{9} = 6.98$$

And speed of B who runs on the outer track

$$= \frac{2\pi(102)}{90} = \frac{51\pi}{23} = 6.96$$

Since, speed of $A >$ speed of B

$\therefore$ A runs faster than B.

Illustration 10: A rectangular plate is of 6 m breadth and 12 m length. Two apertures of 2 m diameter each and one apertures of 1 m diameter have been made with the help of a gas cutter. What is the area of the remaining portion of the plate?

(a) 68.5 sq. m. **(b) 62.5 sq m**

(c) 64.5 sq. m **(d) None of these**

Solution: (c) Given, Length = 12 m and Breadth = 6 m

$\therefore$ Area of rectangular plate = $12 \times 6 = 72$ m^2

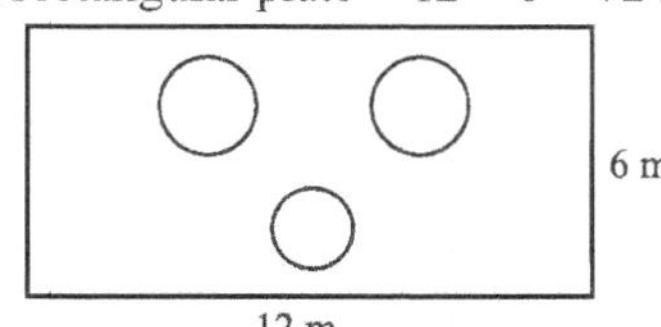

Since, two apertures of 3 m diameter each have been made from this plate.

$\therefore$ Area of these two apertures $= \pi(1)^2 + \pi(1)^2$

$$= \pi + \pi = 2\pi$$

Area of 1 aperture of 1m diameter $= \pi\left(\frac{1}{2}\right)^2 = \frac{\pi}{4}$

$\therefore$ Total area of aperture $= 2\pi + \frac{\pi}{4} = \frac{9\pi}{4} = \frac{9}{4} \times \frac{22}{7}$

$$= \frac{99}{14}$$

$\therefore$ Area of the remaining portion of the plate

$$= 72 - \frac{99}{14} \text{ sq. m} = \frac{909}{14} \text{ sq. m} \approx 64.5 \text{ sq.m}$$

Illustration 11: In the adjoining figure, *AOBCA* represents a quadrant of a circle of radius 3.5 cm with centre *O*. Calculate the area of the shaded portion.

(a) 35 cm^2 **(b) 7.875 cm^2**

(c) 9.625 cm^2 **(d) 6.125 cm^2**

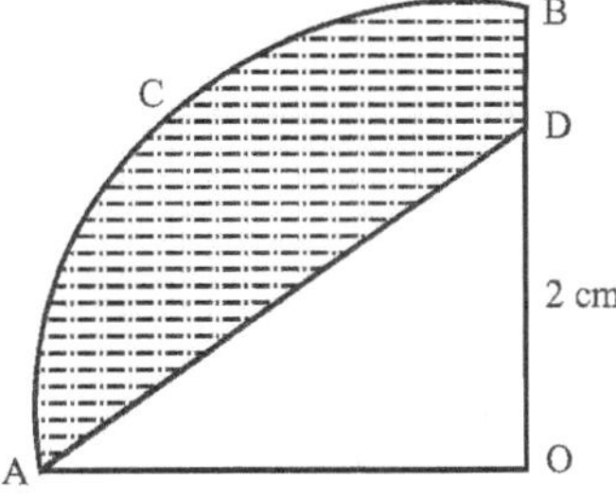

Solution: (d)

Area of shaded portion = Area of quadrant – Area of triangle

$$\Rightarrow \frac{\pi r^2}{4} - \frac{1}{2} \times 3.5 \times 2 = \frac{3.14 \times (3.5)^2}{4} - 3.5$$

$\Rightarrow 6.125$ cm^2

Illustration 12: Find the perimeter and area of the shaded portion of the adjoining diagram:

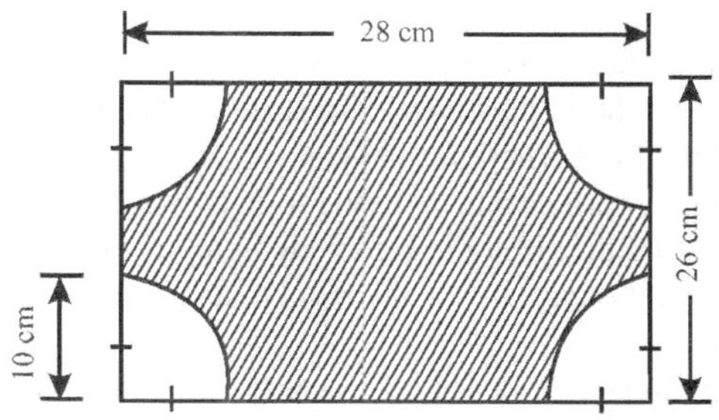

(a) 90.8 cm, 414 cm^2 **(b) 181.6 cm, 423.7 cm^2**

(c) 90.8 cm, 827.4 cm^2 **(d) 181.6 cm, 827.4 cm^2**

Solution: (a)

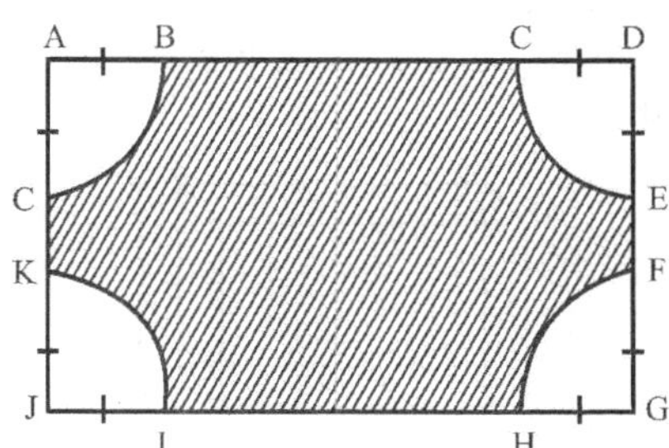

KJ = radius of semicircles = 10 cm

4 quadrants of equal radius = 1 circle of that radius

Area of shaded portion $\Rightarrow$ Area of rectangle – Area of circle

$(28 \times 26) - (3.14 \times 102) \Rightarrow 414$ cm^2

$BC = 28 - (10 + 10) = 8$ and $EF = 26 - (10 + 10) = 6$

Perimeter of shaded portion = 28 cm + $2\pi r$

Answer $\Rightarrow$ 414 cm^2 = Area and

Perimeter = 90.8

Illustration 13: *ABDC* is a circle and circles are drawn with *AO*, *CO*, *DO* and *OB* as diameters. Areas *E* and *F* are shaded *E*/*F* is equal to

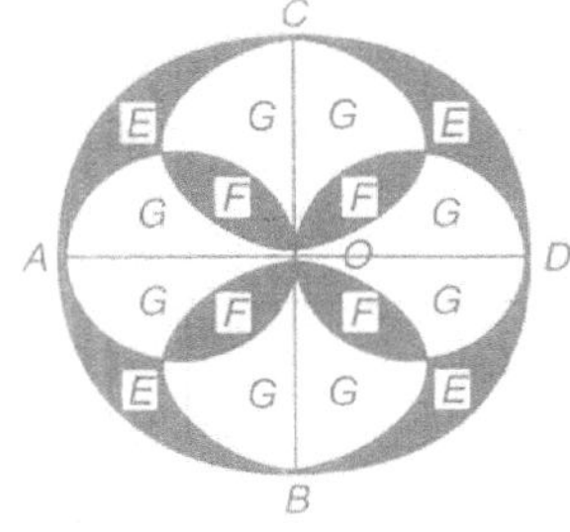

(a) 1/1 **(b) 1/2**

(c) 1/2 **(d) π/4**

Solution: (a)

$AO = CO = DO = OB$ = radius of bigger circle = r (let)

Then area of $(G + F) = \frac{\pi r^2}{2}$

Area of $2(G + F) = \pi r^2$. Also area of $2G + F + E = \pi r^2$

i.e. $2G + F + F = 2G + F + E \Rightarrow F = E$

So the ratio of areas E and F = 1 : 1

SURFACE AREA AND VOLUME OF SOLIDS

Solid

A solid body has three dimensions namely length, breadth (or width) and height (or thickness). The surfaces that bind it are called faces and the lines where faces meet are called edges.

The area of the surface that binds the solid is called its surface area.

We measure the size of a solid body in terms of its volume.

The amount of space that any solid body occupies is called its volume.

Surface areas are measured in square units and volumes are measured in cubic units.

Cuboid

A cuboid is like a three dimensional box. It is defined by its length (l), breadth (b) and height (h). A cuboid can also be visualised as a room. It has six rectangular faces. It is also called rectangular parallelopiped.

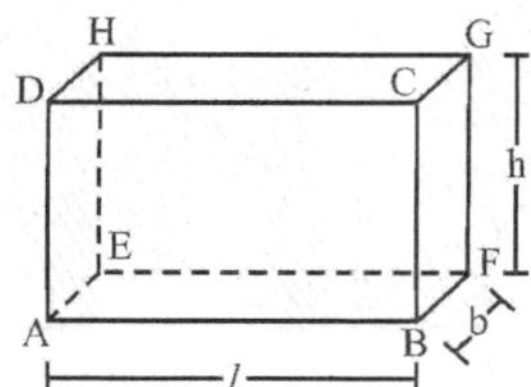

A cuboid is shown in the figure with length 'l', breadth 'b' and height 'h'. 'd' denotes the length of a diagonal (AG, CE, BH or DF) of the cuboid.

Total surface area of a cuboid = $2(lb + bh + hl)$

Lateral surface area (i.e., total area excluding area of the base and top) = $2h(l + b)$

Length of a diagonal of a cuboid = $\sqrt{l^2 + b^2 + h^2}$

Volume of a cuboid = Space occupied by cuboid
= Area of base × height
= $(l \times b) \times h = lbh$

Cube

A cube is a cuboid whose all edges are equal i.e.,

length = breadth = height = a (say)

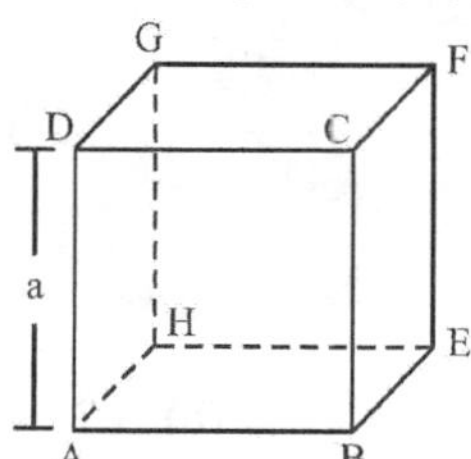

Area of each face of the cube is a^2 square units.

Total surface area of the cuboid = Area of 6 square faces of the cube
$= 6 \times a^2 = 6a^2$

Lateral surface area of cube i.e., total surface area excluding top and bottom faces = $4a^2$

Length of diagonal (d) of the cube

$$= \sqrt{a^2 + a^2 + a^2}$$

$$= \sqrt{3a^2} = a\sqrt{3}$$

Volume of the cube (V) = Base area × Height
$= a^2 \times a = a^3$

Note that if a cube of the maximum volume is inscribed in a sphere of radius 'r', then the edge of the cube = $\frac{2r}{\sqrt{3}}$

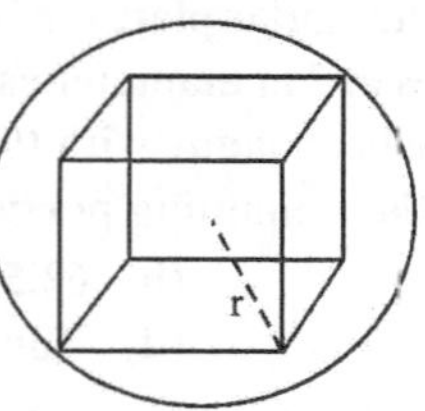

Cylinder

A cylinder is a solid object with circular ends of equal radius and the line joining their centres perpendicular to them. This line is called axis of the cylinder. The length of axis between centres of two circular ends is called height of the cylinder.

In the figure, a cylinder with circular ends each of radius r and height h is shown.

Curved surface area of a cylinder
= Circumference of base × height
$= 2\pi r \times h = 2\pi rh$

If cylinder is closed at both the ends then total surface area of the cylinder
= Curved surface area + Area of circular ends
$= 2\pi rh + 2 \times \pi r^2 = 2\pi r(h + r)$

Volume of the cylinder (V) = Base area × Height
$= \pi r^2 \times h = \pi r^2 h$

- Note that a cylinder can be generated by rotating a rectangle by fixing one of its sides.
- The curved surface of a cylinder is also called lateral surface.

Hollow Cylinder

A hollow cylinder is like a pipe.

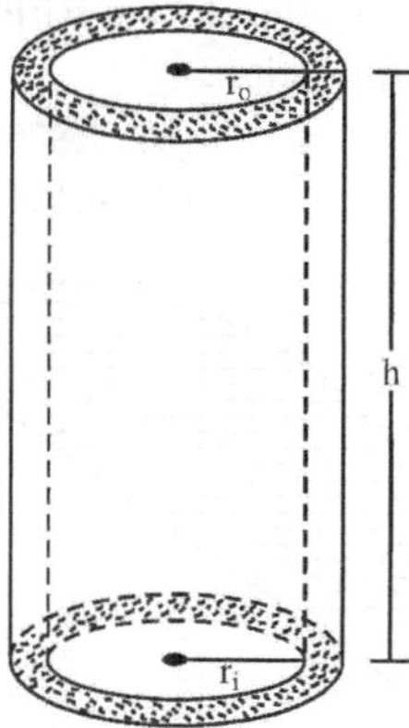

Inner radius = r_i and outer radius = r_o

Hence $r_o - r_i$ = thickness of material of the cylinder.

Let length or height of the cylinder = h,

Curved surface area (C.S.A) of the hollow cylinder
= Outer curved surface area of the cylinder
+ Inner curved surface area of the cylinder
$= 2\pi r_o h + 2\pi r_i h = 2\pi h(r_o + r_i)$

Total surface area of hollow cylinder

= C.S.A. of hollow cylinder

+ Area of 2 circular end rings.

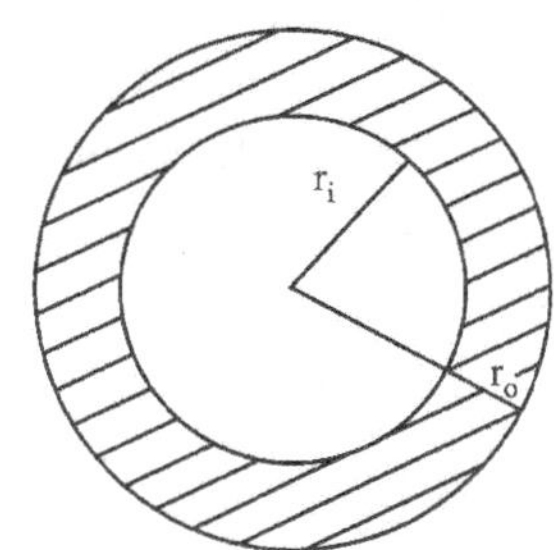

(one end of the pipe)

$= 2\pi h (r_o + r_i) + 2\pi (r_o^2 - r_i^2)$

$= 2\pi (r_o + r_i)(h + r_o + r_i)$

Volume of hollow cylinder = Volume of the material used in making the cylinder

$= \pi (r_o^2 - r_i^2)h$

Cone

A cone is a solid obtained by rotating a strip in the shape of a right angled triangle about its height. It has a circular base and a slanting lateral curved surface that converges at a point. Its dimensions are defined by the radius of the base (r), the height (h) and slant height (l).

A structure similar to cone is the ice-cream cone.

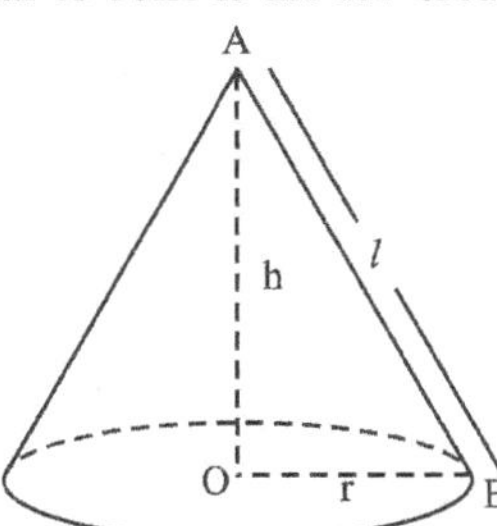

Height (AO) of cone is always perpendicular to base radius (OB) of the cone.

Slant height (l) = $\sqrt{h^2 + r^2}$

Volume of cone = $\frac{1}{3} \times$ base area $\times$ height = $\frac{1}{3} \times \pi r^2 \times h$

Curved surface area (C.S.A.) = πrl

Total surface area (T.S.A.) = C.S.A. + Base area

$= \pi rl + \pi r^2 = \pi r(l + r)$

When a conical cup of paper (hollow cylinder) is unrolled, it forms a sector of a circle

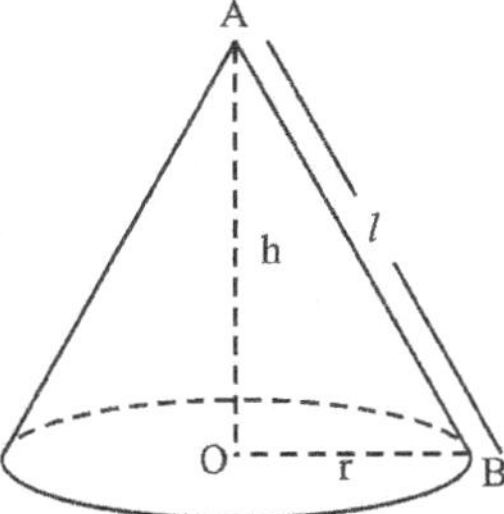

Conical cup of paper

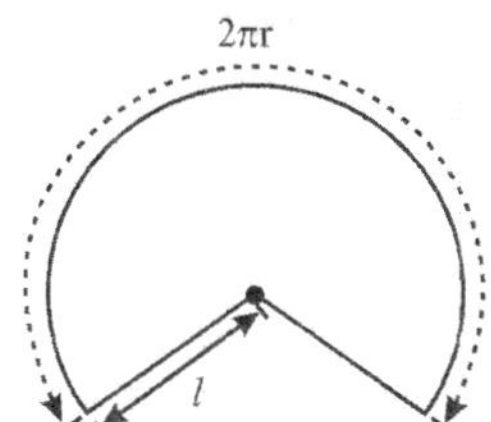

Unrolled conical cup, which is a sector of a circle.

Radius of this sector is equal to slant height of the cone.

Length of curved edge of this sector is equal to the circumference of the base of the cone.

Sphere

A sphere is formed by revolving a semi-circle about its diameter. It has one curved surface which is such that all points on it are equidistant from a fixed point within it, called the centre.

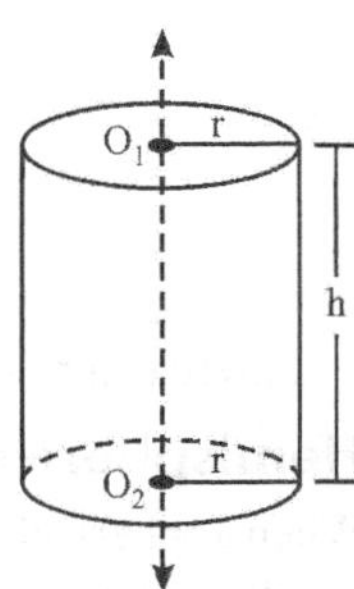

Length of a line segment joining the centre to any point of the curved surface is called the radius (r) of the sphere.

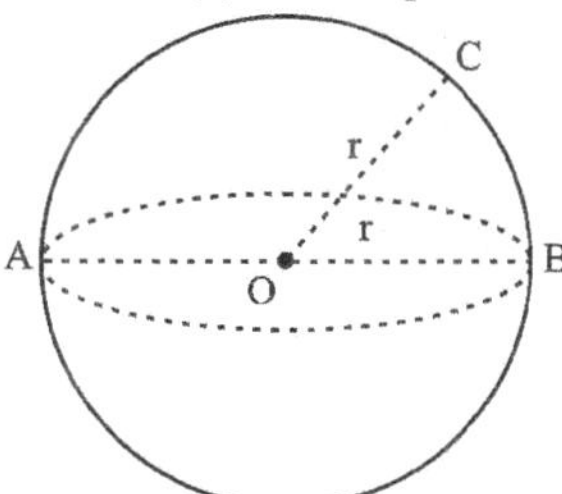

Any line segment passing through the centre and joining two points on the curved surface is called the diameter (d) of the sphere.

Centre = O

Radius = $OC = OA = OB = r$,

Diameter = AB

$= d = 2r$

Surface area of a sphere = $4\pi r^2$

Volume of a sphere (V) = $\frac{4}{3}\pi r^3$

Hemisphere

A plane through the centre of the sphere cuts the sphere into two equal parts. Each part is called a hemisphere.

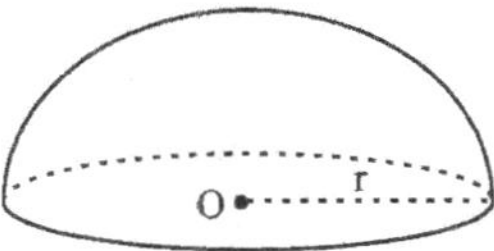

Volume of a hemisphere = $\frac{2}{3}\pi r^3$

Curved surface area (C.S.A.) of a hemisphere = $2\pi r^2$

Total surface area (T.S.A.) of a hemisphere

= C.S.A. + Base area

$= 2\pi r^2 + \pi r^2 = 3\pi r^2$

Note that if a sphere is inscribed in a cylinder then the volume of the sphere is $\frac{2}{3}$rd of the volume of the cylinder.

Hollow Sphere or Spherical Shell

A rubber ball is an example of hollow sphere. In the rubber ball air is filled inside it. Thickness of the rubber in the ball is uniform. If outer and inner radii are R and r, then thickness of rubber or material used in hollow sphere = $R - r$.

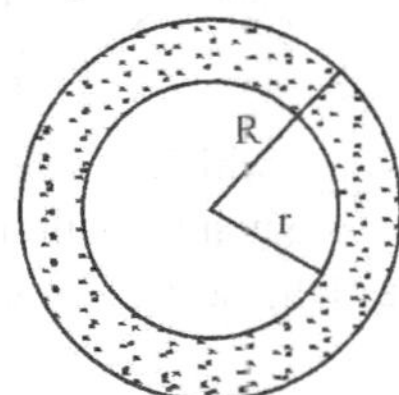

Volume of the rubber or material used in hollow sphere

= External volume – Internal volume

$$= \frac{4}{3}\pi R^3 - \frac{4}{3}\pi r^3$$

$$= \frac{4}{3}\pi(R^3 - r^3)$$

External surface area = $4\pi R^2$.

Hemispherical Bowl

When a spherical shell is cut off in two equal parts, then each part is called a hemispherical bowl as shown in the figure.

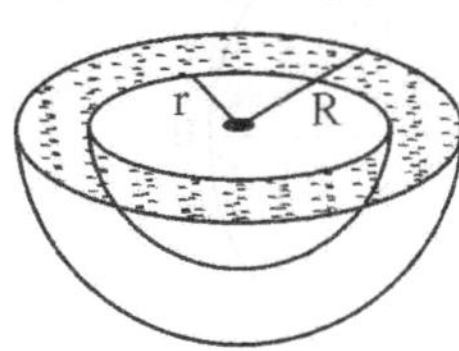

If R and r are external and internal radii of the hemisphere respectively, then

Volume of the material used in the hemispherical bowl

= External volume – Internal volume

$$= \frac{2}{3}\pi R^3 - \frac{2}{3}\pi r^3 = \frac{2}{3}\pi(R^3 - r^3)$$

External curved surface area = $2\pi R^2$

Internal surface area = $2\pi r^2$

Area of the cross-sectional ring = $\pi R^2 - \pi r^2 = \pi(R^2 - r^2)$

Total surface area

= (External curved surface area) + (Internal curved surface area) + (Area of cross-sectional ring)

$$= 2\pi R^2 + 2\pi r^2 + \pi(R^2 - r^2)$$

$$= \pi(3R^2 + r^2)$$

Illustration 14: If the radius of a sphere is increased by 2 cm, then its surface area increases by 352 cm2. The radius of the sphere before the increase was:

(a) 3 cm **(b) 4 cm**

(c) 5 cm **(d) 6 cm**

Solution: (d) $4\pi (r + 2)^2 - 4\pi r^2 = 352$

$$\Rightarrow (r + 2)^2 - r^2 = \left(352 \times \frac{7}{22} \times \frac{1}{4}\right) = 28.$$

$$\Rightarrow (r + 2 + r)(r + 2 - r) = 28$$

$$\Rightarrow 2r + 2 = \frac{28}{2} \Rightarrow 2r + 2 = 14 \Rightarrow r = 6 \text{ cm}$$

Illustration 15: A cylindrical bucket of height 36 cm and radius 21 cm is filled with sand. The bucket is emptied on the ground and a conical heap of sand is formed, the height of the heap being 12 cm. The radius of the heap at the base is :

(a) 63 cm **(b) 53 cm**

(c) 56 cm **(d) 66 cm**

Solution: (a) Volume of the bucket = volume of the sand emptied

Volume of sand = $\pi (21)^2 \times 36$

Let r be the radius of the conical heap.

Then, $\frac{1}{3}\pi r^2 \times 12 = \pi (21)^2 \times 36$

or $r^2 = (21)^2 \times 9$ or $r = 21 \times 3 = 63$ cm

Illustration 16: The length of the longest rod that can be placed in a room which is 12 m long, 9 m broad and 8 m high is

(a) 27 m **(b) 19 m**

(c) 17 m **(d) 13 m**

Solution: (c) Required length = length of the diagonal

$$= \sqrt{12^2 + 9^2 + 8^2} = \sqrt{144 + 81 + 64} = \sqrt{289} = 17 \text{ m}$$

Illustration 17: The internal measurements of a box with lid are 115 × 75 ×735 cm3 and the wood of which it is made is 2.5 cm thick. Find the volume of wood.

(a) 82,125 cm³ **(b) 70,054 cm³**

(c) 78,514 cm³ **(d) None of these**

Solution: (a) Internal volume = $115 \times 75 \times 35 = 3{,}01{,}875$ cm³

External volume = $(115 + 2 \times 2.5) \times (75 + 2 \times 2.5) \times (35 + 2 \times 2.5)$

$= 120 \times 80 \times 40 = 3{,}84{,}000$ cm³

∴ Volume of wood = External volume – Internal volume

$= 3{,}84{,}000 - 3{,}01{,}875 = 82{,}125$ cm³

Illustration 18: A rectangular tank is 225 m by 162 m at the base. With what speed must water flow into it through an aperture 60 cm by 45 cm that the level may be raised 20 cm in 5 hours ?

(a) 5000 m/hr **(b) 5400 m/hr**

(c) 5200 m/hr **(d) 5600 m/hr**

Solution: (b) Required speed of flow of water

$$= \frac{225 \times 162 \times 20}{5 \times 100} = \frac{60}{100} \times \frac{45}{100} \times h$$

∴ $h = 5400$

Illustration 19: A metallic sheets is of rectangular shape with dimensions 48 cm × 36 cm. From each one of its corners, a square of 8 cm is cut off. An open box is made of the remaining sheet. Find the volume of the box

(a) 5110 cm³ **(b) 5130 cm³**

(c) 5120 cm³ **(d) 5140 cm³**

Solution: (c) Volume of the box made of the remaining sheet

$= 32 \times 20 \times 8 = 5120$ cm³

48 cm
8 cm 8 cm
8 cm 8 cm
36 cm 20 cm
8 cm 8 cm
8 cm 8 cm
8
32 cm

Illustration 20: The capacity of a cylindrical tank is 246.4 litres. If the height is 4 metres, what is the diameter of the base?

(a) 1.4 m **(b) 2.8 m**

(c) 14 m **(d) None of these**

Solution: (d) Volume of the tank = 246.4 litres = 246400 cm^3.

Let the radius of the base be r cm. Then,

$$\left(\frac{22}{7} \times r^2 \times 400\right) = 246400$$

$$\Rightarrow r^2 = \left(\frac{246400 \times 7}{22 \times 400}\right) = 196 \Rightarrow r = 14.$$

$\therefore$ Diameter of the base = $2r$ = 28 cm = .28 m

Prism

A 'prism' is a solid having identical and parallel top and bottom (or base) faces. These identical faces are regular polygon of any number of sides. The side faces of a prism are rectangular and are known as lateral faces. Number of lateral faces is equal to the number of sides in the base.

Here are some example of prisms

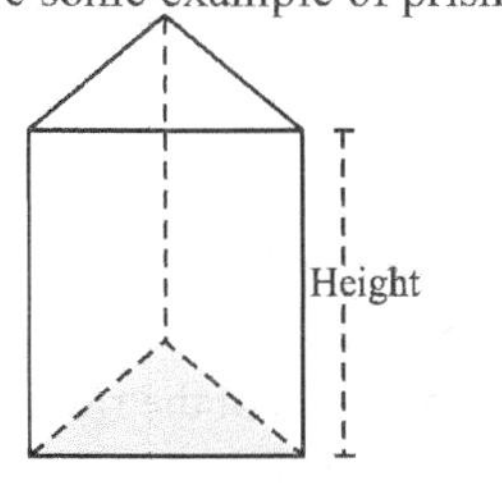

Triangular base prism

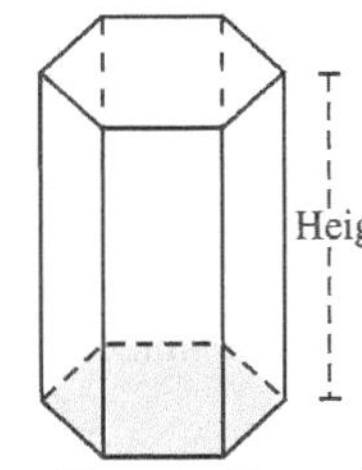

Hexagonal base prism

Lateral surface area of the prism

= (Perimeter of the base) × (Height)

Total surface area of the prism

= (Surface area of the top and bottom) + (Lateral surface area)

= 2 × Area of the base + Perimeter of base × Height

Volume of the prism = (Area of base) × (Height)

The actual formula used to find the surface area and volume will depend upon the number of sides in the base of the prism.

Pyramid

It is a three-dimensional body made up of a regular polygon shaped base and triangular lateral faces that meet at a point called vertex, which is also called the apex of the pyramid.

The number of triangular faces is equal to the number of sides in the base.

For example: A pyramid with a square base has four triangular faces, whereas a pyramid with a hexagonal face is made up of six triangular faces, and so on.

Lower face is called the base and the perpendicular distance of the vertex (or top) from the base is called the height or altitude of the pyramid.

The altitude of a lateral face of a pyramid is the slant height, which is the perpendicular distance of the vertex (or top) from the mid-point of any side of the base.

The lateral surface area of a regular pyramid is the sum of the areas of its lateral faces.

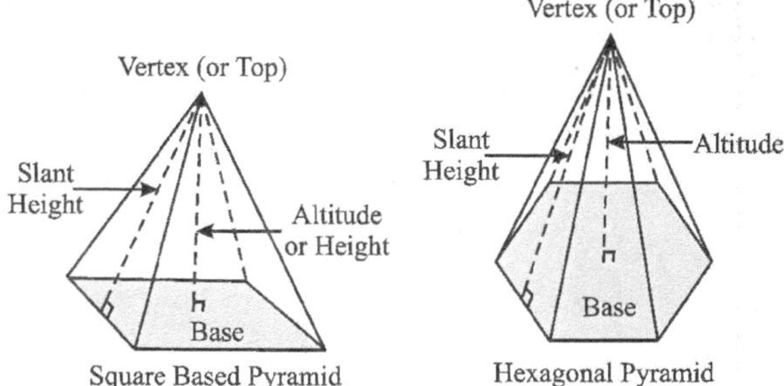

Square Based Pyramid Hexagonal Pyramid

Lateral surface area of a pyramid

$$= \frac{1}{2} \times (\text{Area of the base}) \times (\text{Slant height})$$

Total surface area of a pyramid

$$= \frac{1}{2} \times (\text{Perimeter of the base}) \times (\text{Slant height}) + (\text{Area of the base})$$

Volume of a pyramid = $\frac{1}{3}$ × Area of base × Height

Illustration 21: Find the lateral surface area of a regular pyramid with triangular base, if each edge of the base measures 8 cm and slant height is 5 cm.

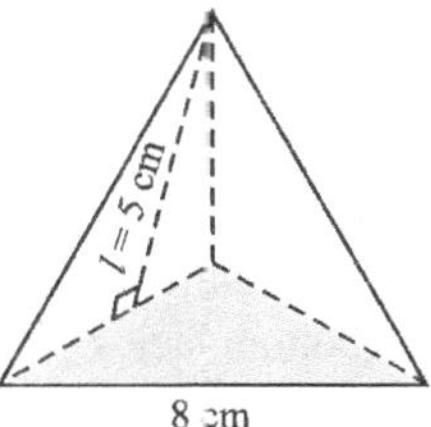

Solution: The perimeter of the base is the sum of the sides,

$$p = 3.(8) = 24 \text{ cm}$$

$$\text{L.S.A.} = \frac{1}{2} \times (24) \times (5) = 60 \text{ cm}^2$$

Illustration 22: Find the total surface area of a pyramid with a square base if each side of the base measures 16 cm, the slant height of a side is 17 cm and the altitude is 15 cm.

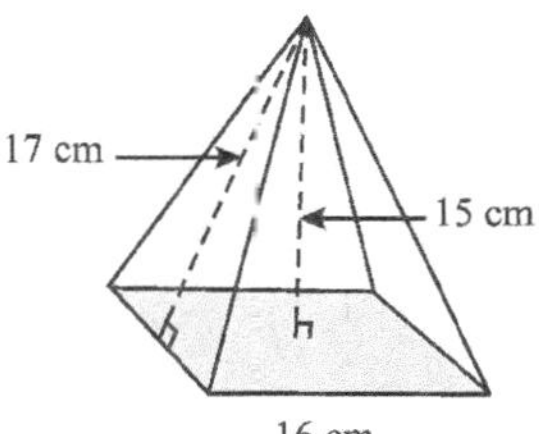

Solution: The perimeter of the base,

$$p = 4 \times 16 = 64 \text{ cm}$$

The area of the base

$$= 16^2 = 256 \text{ cm}^2$$

$$\text{T.S.A.} = \frac{1}{2}(64)(17) + 256$$

$$= 544 + 256 = 800 \text{ cm}^2$$

Frustum of a Cone

When top portion of a cone cut off by a plane parallel to the base of it, the left-over part is called the frustum of the cone.

In the figure, r and R are the radius of two ends, h is the height and l is the slant height of the frustum of cone.

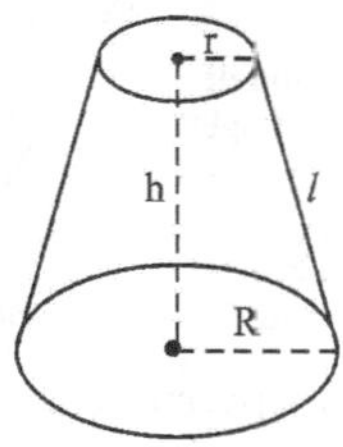

Slant height, $l = \sqrt{(R-r)^2 + h^2}$

Curved surface area $= \pi(R + r)l$

Total surface area

$= (\text{Curved surface area}) + (\text{Area of two circular ends})$

$= \pi(R + r)l + \pi R^2 + \pi r^2$

$= \pi(Rl + rl + R^2 + r^2)$

Height of the original cone $= \dfrac{Rh}{R - r}$

Volume of the frustum of cone

$$= \frac{\pi h}{3}\left(R^2 + r^2 + Rr\right)$$

Frustum of a Pyramid

When top portion of a pyramid is cut off by a plane parallel to the base of it, the left-over part is called the frustum of the pyramid.

If A_1, A_2 are of top and bottom face, P_1 and P_2 are the perimeters of top and bottom face, h is the height and l is the slant height of the frustum of the pyramid, then

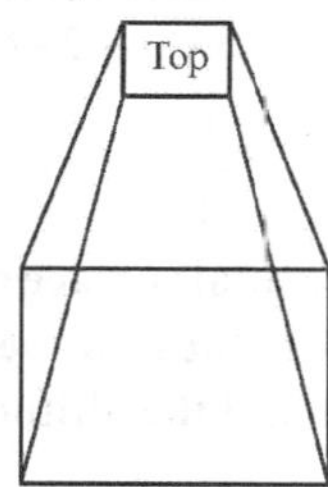

Lateral surface area $= \dfrac{1}{2}(P_1 + P_2)l$

Total surface area $=$ Lateral surface area $+ A_1 + A_2$

$$= \frac{1}{2}(P_1 + P_2)l + A_1 + A_2$$

$$\text{Volume} = \frac{1}{3}h(A_1 - A_2 + \sqrt{A_1 . A_2})$$

Tetrahedron (Only Shape)

A tetrahedron is a solid object which has 4 faces. All the faces of a tetrahedron are equilateral triangles. A tetrahedron has 4 vertices and 6 edges.

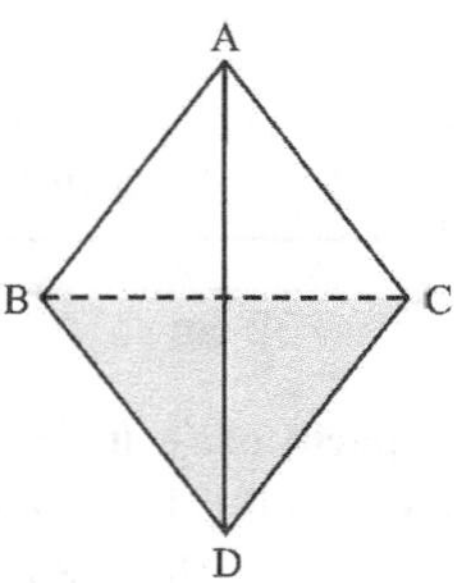

EULER'S RULE

For any regular shape solid (like cuboid, cube, cylinder, etc)

Number of faces (F) + Number of vertices (V)

= Number of edges (E) + 2

i.e., $F + V = E + 2$

SOME OTHER IMPORTANT CONCEPTS

1. In the figure ABC is a triangle right angled at B. Three semi-circles are drawn taking the three sides AB, BC and CA as diameter. The region enclosed by the three semi-circles is shaded.

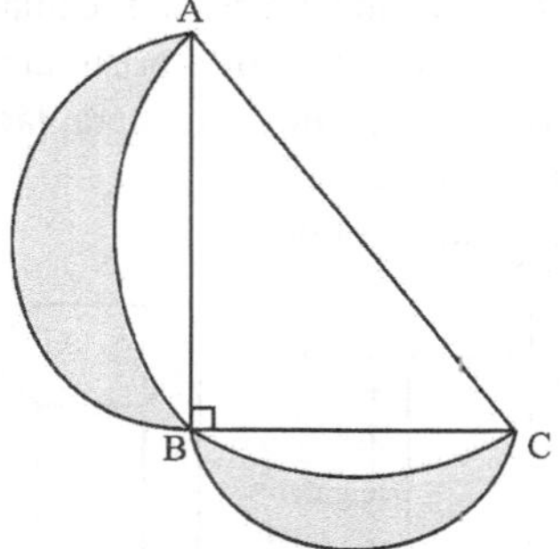

Area of the shaded region = Area of the right angled triangle.

2. In the figure given below all triangles are equilateral triangles and circles are inscribed in these triangles. If the side of triangle $ABC = a$, then the side of triangle $DEF = \dfrac{a}{2}$ and the side of triangle $GHI = \dfrac{a}{4}$

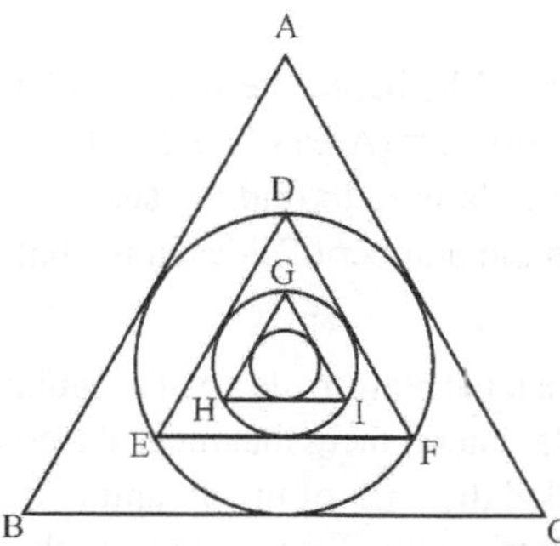

Thus length of a side of an inner triangle is half the length of immediate outer triangle. Similarly the radius of an inner circle is half the radius of immediate outer circle.

Practice Exercise

Level - I

1. The side and the height of a rhombus are 13 and 20 cms respectively. Find the area.
(a) 260 cm^2 (b) 275 cm^2
(c) 290 cm^2 (d) None of these
2. The circumference of a circle is 44 metres. Find the area of the circle.
(a) 154 m^2 (b) 160 m^2
(c) 175 m^2 (d) 168 m^2
3. The length and breadth of a rectangle are in the ratio 9 : 5. If its area is 720 m^2, find its perimeter.
(a) 112 metre (b) 115 metre
(c) 110 metre (d) 118 metre
4. How many squares are there in a 5 inch by 5 inch square grid, if the grid is made up one inch by one inch squares?
(a) 50 (b) 150 (c) 55 (d) 25
5. If the ratio of areas of two squares is 9 : 1, the ratio of their perimeter is :
(a) 9 : 1 (b) 3 : 4 (c) 3 : 1 (d) 1 : 3
6. A circle and a rectangle have the same perimeter. The sides of the rectangle are 18 cm and 26 cm. What is the area of the circle ?
(a) 88 cm^2 (b) 154 cm^2
(c) 1250 cm^2 (d) 616 cm^2
7. If the perimeter and diagonal of a rectangle are 14 and 5 cms respectively, find its area.
(a) 12 cm^2 (b) 16 cm^2
(c) 20 cm^2 (d) 24 cm^2
8. When the circumference and area of a circle are numerically equal, then the diameter is numerically equal to
(a) area (b) circumference
(c) 4 (d) 2 π
9. In a parallelogram, the length of one diagonal and the perpendicular dropped on that diagonal are 30 and 20 metres respectively. Find its area.
(a) 600 m^2 (b) 540 m^2
(c) 680 m^2 (d) 574 m^2
10. The area of a triangle is 615 m^2. If one of its sides is 123 metre, find the length of the perpendicular dropped on that side from opposite vertex.
(a) 15 metres (b) 12 metres
(c) 10 metres (d) None of these
11. How many plants will be there in a circular bed whose outer edge measure 30 cms, allowing 4 cm^2 for each plant ?
(a) 18 (b) 750
(c) 24 (d) 120
12. A square carpet with an area 169 m^2 must have 2 metres cut-off one of its edges in order to be a perfect fit for a rectanguilar room. What is the area of rectangular room?
(a) 180 m^2 (b) 164 m^2
(c) 152 m^2 (d) 143 m^2
13. If the area of a circle decreases by 36%, then the radius of a circle decreases by
(a) 20% (b) 18%
(c) 36% (d) 64%
14. The altitude drawn to the base of an isosceles triangle is 8 cm and the perimeter is 32 cm. The area of the triangle is
(a) 72 cm^2 (b) 60 cm^2
(c) 66 cm^2 (d) None of these
15. The area of a square field is 576 km^2. How long will it take for a horse to run around at the speed of 12 km/h ?
(a) 12 h (b) 10 h
(c) 8 h (d) 6 h
16. Four equal circles are described about the four corners of a square so that each touches two of the others. If a side of the square is 14 cm, then the area enclosed between the circumferences of the circles is :
(a) 24 cm^2 (b) 42 cm^2
(c) 154 cm^2 (d) 196 cm^2
17. The ratio between the length and the breadth of a rectangular park is 3 : 2. If a man cycling along the boundary of the park at the speed of 12km / hr completes one round in 8 minutes, then the area of the park (in sq. m) is:
(a) 15360 (b) 153600
(c) 30720 (d) 307200
18. A wire can be bent in the form of a circle of radius 56 cm. If it is bent in the form of a square, then its area will be:
(a) 3520 cm^2 (b) 6400 cm^2
(c) 7744 cm^2 (d) 8800 cm^2
19. The length of a room is double its breadth. The cost of colouring the ceiling at ₹ 25 per sq. m is ₹ 5,000 and the cost of painting the four walls at ₹ 240 per sq. m is ₹ 64,800. Find the height of the room.
(a) 4.5 m (b) 4 m (c) 3.5 m (d) 5 m
20. A metal cube of edge 12 cm is melted and formed into three smaller cubes. If the edges of two smaller cubes are 6 cm and 8 cm, then find the edge of the third smaller cube.
(a) 10 cm (b) 14 cm
(c) 12 cm (d) 16 cm
21. A well 22.5 deep and of diameter 7 m has to be dug out. Find the cost of plastering its inner curved surface at ₹ 3 per sq. metre.
(a) ₹ 1465 (b) ₹ 1485
(c) ₹ 1475 (d) ₹ 1495
22. The surface area of a cube is 150 m^2. The length of its diagonal is
(a) $5\sqrt{3}$ m (b) 5 m
(c) $\frac{10}{\sqrt{3}}$ m (d) 15 m

23. The length of the longest rod that can be placed in a room which is 12 m long, 9 m broad and 8 m high is
(a) 27 m (b) 19 m
(c) 17 m (d) 13 m

24. If the volume of a sphere is divided by its surface area, the result is 27 cms. The radius of the sphere is
(a) 9 cms (b) 27 cms
(c) 81 cms (d) 243 cms

25. The volume of water measured on a rectangular field 500 m × 300 m is 3000 m^3. Find the depth (amount) of rain that has fallen.
(a) 2 cms (b) 3 cms
(c) 4 cms (d) 3.5 cms

26. How many spherical bullets can be made out of a lead cylinder 28 cm high and with base radius 6 cm, each bullet being 1.5 cm in diameter?
(a) 1845 (b) 1824
(c) 1792 (d) 1752

27. Water flows out through a circular pipe whose internal diameter is 2 cm, at the rate of 6 metres per second into a cylindrical tank, the radius of whose base is 60 cm. By how much will the level of water rise in 30 minutes?
(a) 2 m (b) 4 m
(c) 3 m (d) 5 m

28. A spherical ball of lead, 3 cm in diameter, is melted and recast into three spherical balls. The diameter of two of these balls are 1.5 cm and 2 cm respectively. The diameter of the third ball is
(a) 2.5 cm (b) 2.66 cm
(c) 3 cm (d) 3.5 cm

29. A cube of 384 cm^2 surface area is melt to make x number of small cubes each of 96 mm^2 surface area. The value of x is
(a) 80,000 (b) 8
(c) 8,000 (d) 800

30. The capacity of a cylindrical tank is 246.4 litres. If the height is 4 metres, what is the diameter of the base?
(a) 1.4 m (b) 2.8 m
(c) 14 m (d) None of these

31. A conical cavity is drilled in a circular cylinder of 15 cm height and 16 cm base diameter. The height and the base diameter of the cone are same as those of the cylinder. Determine the total surface area of the remaining solid.
(a) 440 π cm^2 (b) 215π cm^2
(c) 542 π cm^2 (d) 376 π cm^2

32. If the radius of a sphere is increased by 2 cm, then its surface area increases by 352 cm^2. The radius of the sphere before the increase was:
(a) 3 cm (b) 4 cm
(c) 5 cm (d) 6 cm

33. A hollow sphere of internal and external diameters 4 cm and 8 cm respectively is melted into a cone of base diamater 8 cm. The height of the cone is:
(a) 12 cm (b) 14 cm
(c) 15 cm (d) 18 cm

34. The length and breadth of a playground are 36m and 21 m respectively. Poles are required to be fixed all along the boundary at a distance 3m apart. The number of poles required will be
(a) 39 (b) 38 (c) 37 (d) 40

35. The length of a rectangular field is double its width. Inside the field there is a square-shaped pond 8 m long. If the area of the pond is 1/8 of the area of the field, what is the length of the field?
(a) 32 m (b) 16 m (c) 64 m (d) 20 m

36. A horse is tethered to one corner of a rectangular grassy field 40 m by 24 m with a rope 14 m long. Over how much area of the field can it graze?
(a) 154 cm^2 (b) 308 m^2
(c) 150 m^2 (d) None of these

37. The length of a cold storage is double its breadth. Its height is 3 metres. The area of its four walls (including the doors) is 108 m^2. Find its volume.
(a) 215 m^3 (b) 216 m^3
(c) 217 m^3 (d) 218 m^3

38. The cost of the paint is ₹ 36.50 per kg. If 1 kg of paint covers 16 square feet, how much will it cost to paint outside of a cube having 8 feet each side?
(a) ₹ 692 (b) ₹ 768
(c) ₹ 876 (d) ₹ 972

39. A cuboidal block of 6 cm × 9 cm × 12 cm is cut up into an exact number of equal cubes. The least possible number of cubes will be:
(a) 6 (b) 9
(c) 24 (d) 30

40. A semicircular sheet of paper of diameter 28 cm is bent to cover the exterior surface of an open conical ice-cream cup. The depth of the ice-cream cup is
(a) 10.12 cm (b) 8.12 cm
(c) 12.12 cm (d) 13.27 cm

41. How many squares are there in a 5 inch by 5 inch square grid, if the grid is made up one inch by one inch squares ?
(a) 50 (b) 150
(c) 55 (d) 25

42. A solid wooden toy in the shape of a right circular cone is mounted on a hemisphere. If the radius of the hemisphere is 4.2 cm and the total height of the toy is 10.2 cm, find the volume of the wooden toy.

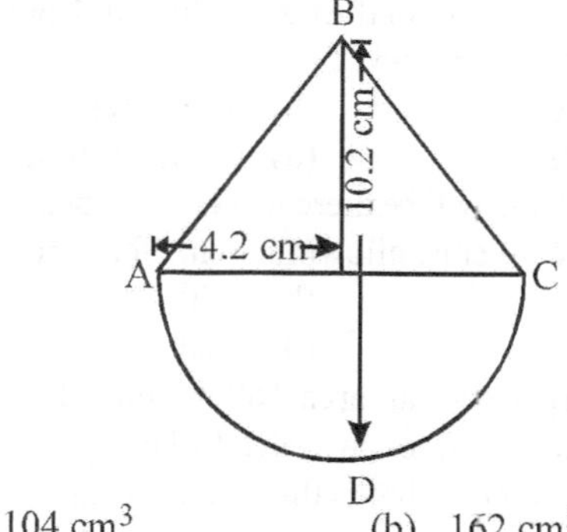

(a) 104 cm^3 (b) 162 cm^3
(c) 427 cm^3 (d) 266 cm^3

43. The dimensions of a field are 20 m by 9 m. A pit 10 m long, 4.5 m wide and 3 m deep is dug in one corner of the field and the earth removed has been evenly spread over the remaining area of the field. What will be the rise in the height of field as a result of this operation?
(a) 1 m (b) 2 m (c) 3 m (d) 4 m

44. In a triangle ABC, points P, Q and R are the mid-points of the sides AB, BC and CA respectively. If the area of the triangle ABC is 20 sq. units, find the area of the triangle PQR
(a) 10 sq. units (b) 5.3 sq. units
(c) 5 sq. units (d) None of these

45. From a circular sheet of paper with a radius of 20 cm, four circles of radius 5 cm each are cut out. What is the ratio of the areas of uncut to the cut portion?
(a) 1 : 3 (b) 4 : 1 (c) 3 : 1 (d) 4 : 3

46. The figure shows a circle of diameter AB and radius 6.5 cm. If chord CA is 5 cm long, find the area of triangle ABC

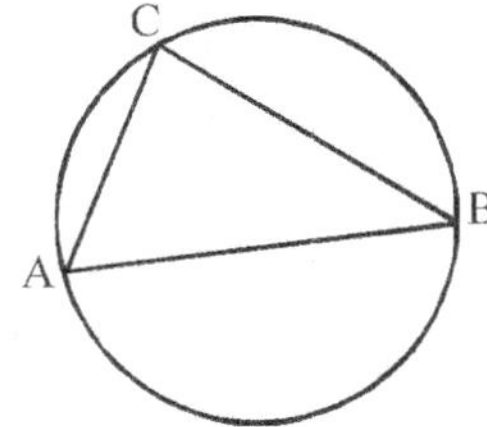

(a) 60 sq. cm. (b) 30 sq. cm.
(c) 40 sq. cm. (d) 52 sq. cm.

47. The sides of a triangle are 5, 12 and 13 units respectively. A rectangle is constructed which is equal in area to the triangle and has a width of 10 units. Then the perimeter of the rectangle is
(a) 30 (b) 26
(c) 13 (d) None of these

48. One diagonal of a rhombus is 24 cm whose side is 13 cm. Find the area of the rhombus.
(a) 25 sq. cm (b) 312 sq. cm.
(c) 125 sq. cm. (d) 120 sq. cm.

49. The radius of the incircle in the given diagram will be

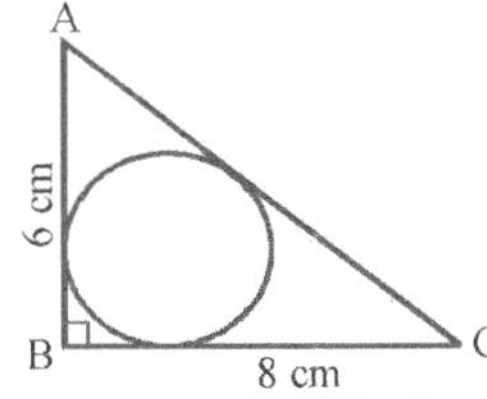

(a) 1.8 cm (b) 2 cm
(c) 2.5 cm (d) 3.6 cm

50. If a rectangular paper of length 6 cm. and width 3 cm. is rolled to form a cylinder with height equal to the width of the paper, then its base radius is –
(a) $\frac{6}{\pi}$cm (b) $\frac{3}{\pi}$cm
(c) $\frac{3}{2\pi}$cm (d) $\frac{9}{2\pi}$cm

51. A hollow spherical shell is made of metal of density 4.8 g/cm^3, If its internal and external radii are 10 cm and 12 cm respectively, find the weight of the shell
(a) 15.24 kg (b) 12.84 kg
(c) 14.64 kg (d) None of these

52. The area of the circle that can be inscribed in a square of side 6 cm is
(a) 36π cm^2 (b) 18π cm^2
(c) 12π cm^2 (d) 9π cm^2

53. Circumference of a sector of angle $p°$ of a circle with radius R is
(a) $\frac{p}{180}\times 2\pi R$ (b) $\frac{p}{180}\times \pi R^2$
(c) $\frac{p}{360}\times 2\pi R$ (d) $\frac{p}{720}\times 2\pi R^2$

54. Three circles with centres A, B and C and with unit radii touch each other at O, P and Q. Find the area of the shaded region .

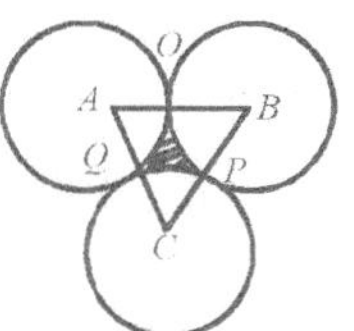

(a) 0.16 sq. units (b) 1.21 sq. units
(c) 0.03 sq. units (d) 0.32 units

55. The inside perimeter of a practice running track with semi-circular ends and straight parallel sides is 312 m. The length of the straight portion of the track is 90 m. If the track has a uniform width of 2 m throughout, find its area.

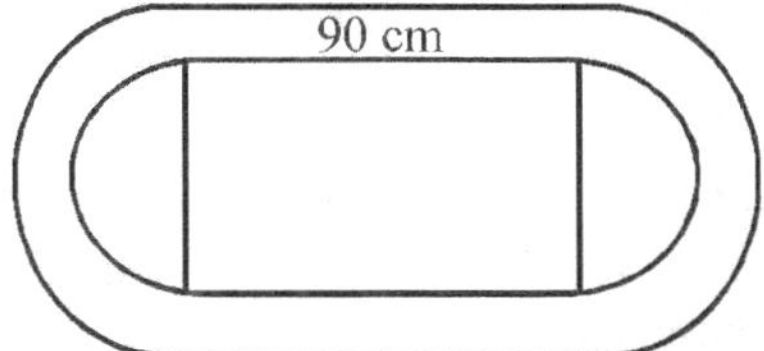

(a) 5166 m^2 (b) 5802.57 m^2
(c) 636.57 m^2 (d) 1273.14 m^2

56. The circumference of a circle is 792 meters. What will be its radius ? *[SBI Clerk-June-2012]*
(a) 120 metres (b) 133 metres
(c) 145 metres (d) 136 metres
(e) None of these

57. The area of a rectangle is 1209 square metres. Its length measures 39 metres. How much is its perimeter ? *[SBI Clerk-June-2012]*
(a) 122 metres (b) 134 metres
(c) 148 metres (d) 144 metres
(e) None of these

58. Area of rectangular field is 3584 m^2 and the length and the breadth are in the ratio 7 : 2 respectively. What is the perimeter of the rectangle ? *[SBI Clerk-2012]*
(a) 246 m (b) 292 m
(c) 286 m (d) 288 m
(e) None of these

59. The base of a triangle is 2 cm more than twice its altitude. If the area is 12 sq. cm, its altitude will be [SSC-Sub. Ins.-2012]
(a) 6 cm (b) 5 cm
(c) 4 cm (d) 3 cm

60. If the volume and the surface area of a sphere are numerically equal, then the numerical value of the radius of the sphere is [SSC-Sub. Ins.-2012]
(a) 1 (b) 2
(c) 3 (d) 4

61. The area of a semi-circular field is 308 sq. m; then taking $\pi = \frac{22}{7}$, the length of the railing to surround it has to be [SSC-Sub. Ins.-2012]
(a) 44 m (b) 72 m
(c) 88 m (d) 80 m

62. Volume of a right circular cone is numerically equal to its slant surface area. Then value of $\left(\frac{1}{h^2}+\frac{1}{r^2}\right)$, where h and r are height and radius of the cone respectively, is [SSC-Sub. Ins.-2012]
(a) 9 units (b) $\frac{1}{9}$ unit
(c) 4 units (d) $\frac{1}{4}$ unit

63. If the numerical value of the volume of a right circular cylinder and its curved surface area are equal, then its radius is [SSC-Sub. Ins.-2012]
(a) 2 units (b) 4 units
(c) 3 units (d) 6 units

64. A solid right circular cylinder and a solid hemisphere stand on equal bases and have the same height. The ratio of their whole surface areas is: [SSC-Sub. Ins.-2013]
(a) 3: 2 (b) 3: 4
(c) 4: 3 (d) 2: 3

65. If area of an equilateral triangle is a and height b, then value of $\frac{b^2}{a}$ is: [SSC-Sub. Ins.-2013]
(a) 3 (b) $\frac{1}{3}$
(c) $\sqrt{3}$ (d) $\frac{1}{\sqrt{3}}$

66. A copper sphere of diameter 18 cm is drawn into a wire of diameter 4 mm. The length of the wire, in metre, is : [SSC-Sub. Ins.-2013]
(a) 2.43 (b) 243
(c) 2430 (d) 24.3

67. Water flows at the rate of 10 metres per minute from a cylindrical pipe 5 mm in diameter. How long it lake to fill up a conical vessel whose diameter at the base is 30 cm and depth 24 cm? [SSC-Sub. Ins.-2013]
(a) 28 minutes 48 seconds
(b) 51 minutes 12 seconds
(c) 51 minutes 24 seconds
(d) 28 minutes 36 seconds

68. If the volumes of two right circular cones are in the ratio 4 : 1 and their diameters are in the ratio 5 : 4 then the ratio of their heights is : [SSC-Sub. Ins.-2013]
(a) 25: 16 (b) 25: 64
(c) 64 : 25 (d) 16: 25

69. Three circles of equal radius 'a' cm touch each other. The area of the shaded region is : [SSC-Sub. Ins.-2013]

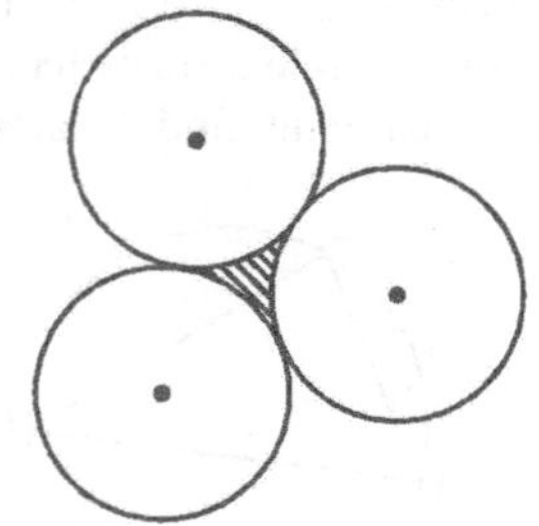

(a) $\left(\frac{\sqrt{3}+\pi}{2}\right)a^2$sq.cm (b) $\left(\frac{6\sqrt{3}-\pi}{2}\right)a^2$sq.cm
(c) $(\sqrt{3}-\pi)a^2$sq.cm (d) $\left(\frac{2\sqrt{3}-\pi}{2}\right)a^2$sq.cm

70. The radius of a right circular cone is 3 cm and its height is 4 cm. The total surface area of the cone is [SSC-Sub. Ins.-2014]
(a) 48.4 sq.cm (b) 64.4 sq.cm
(c) 96.4 sq.cm (d) 75.4 sq.cm

71. A wooden box of dimension 8 metre × 7 metre × 6 metre is to carry rectangular boxes of dimensions 8 cm × 7 cm × 6 cm. The maximum number of boxes that can be carried in 1 wooden box is [SSC-Sub. Ins.-2014]
(a) 7500000 (b) 9800000
(c) 1200000 (d) 1000000

72. Two circular cylinders of equal volume have their heights in the ratio 1 : 2; Ratio of their radii is (Take $\pi = \frac{22}{7}$) [SSC-Sub. Ins.-2014]
(a) 1 : 4 (b) $1:\sqrt{2}$
(c) $\sqrt{2}:1$ (d) 1 : 2

73. A rectangular piece of paper of dimensions 22 cm by 12 cm is rolled along its length to form a cylinder. The volume (in cm^3) of the cylinder so formed is (use $\pi = \frac{22}{7}$) [SSC-Sub. Ins.-2014]
(a) 562 (b) 412
(c) 462 (d) 362

74. A sphere is placed inside a right circular cylinder so as to touch the top, base and the lateral surface of the cylinder. If the radius of the sphere is R, the volume of the cylinder is [SSC-Sub. Ins.-2014]

(a) $2\pi R^3$ (b) $4\pi r^3$

(c) $8\pi R^3$ (d) $\frac{8}{3}\pi R^3$

75. A godown is 15 m long and 12 m broad. The sum of the areas of the floor and the ceiling is equal to the sum of areas of the four walls. The volume (in m^3) of the godown is: [SSC-Sub. Ins.-2014]

(a) 900 (b) 1200

(c) 1800 (d) 720

76. A circle is inscribed in an equilateral triangle and a square is inscribed in that circle. The ratio of the areas of the triangle and the square is [SSC-MT-2013]

(a) $3\sqrt{3}:1$ (b) $\sqrt{3}:4$

(c) $\sqrt{3}:8$ (d) $3\sqrt{3}:2$

77. If the sum of the length, breadth and height of a rectangular parallelopiped is 24 cm and the length of its diagonal is 15 cm, then its total surface area is [SSC-MT-2013]

(a) 351 cm^2 (b) 256 cm^2

(c) 265 cm^2 (d) 315 cm^2

78. Diagonal of a cube is $6\sqrt{3}$ cm. Ratio of its total surface area and volume (numerically) is: [SSC 10+2-2012]

(a) 2 : 1 (b) 1 : 6

(c) 1 : 1 (d) 1 : 2

79. The minute hand of a big wall-clock is 35 cm long. Taking $\pi = \frac{22}{7}$, length of the arc, its extremity moves in 18 seconds is : [SSC 10+2-2012]

(a) 11 cm (b) 1.1 cm

(c) 6.6 cm (d) 6 cm

80. The length of the two sides forming the right angle of a right-angled triangle are 6 cm and 8 cm. The length of its circum-radius is : [SSC 10+2-2012]

(a) 5 cm (b) 7 cm

(c) 6 cm (d) 10 cm

81. The length of radius of a circumcircle of a triangle having sides 3 cm, 4 cm and 5 cm is: [SSC 10+2-2012]

(a) 2 cm (b) 2.5 cm

(c) 3 cm (d) 1.5 cm

82. The length and breadth of a square are increased by 30% and 20% respectively. The area of the rectangle so formed exceeds the area of the square by: [SSC 10+2-2012]

(a) 46% (b) 66%

(c) 42% (d) 56%

83. The volume of a cubical box is 3.375 cubic meters. The length of edge of the box is: [SSC 10+2-2012]

(a) 75 cm (b) 1.5 m

(c) 1.125 m (d) 2.5 m

84. The length of a minute hand of a clock is 7 cm. The area swept by the minute hand in 30 minutes is: [SSC 10+2-2012]

(a) 210 sq. cm (b) 154 sq. cm

(c) 77 sq. cm (d) 147 sq. cm

85. The perimeter of a semi-circular area is 18 cm, then the radius is : (using $\pi = \frac{22}{7}$) [SSC 10+2-2012]

(a) $5\frac{1}{3}$ cm (b) $3\frac{1}{2}$ cm

(c) 6 cm (d) 4 cm

86. The circumference of the base of a 16 cm height solid cone is 33 cm. What is the volume of the cone in cm^3 ? [SSC 10+2-2012]

(a) 1028 (b) 616

(c) 462 (d) 828

87. The ratio of the edges of rectangular parallelopiped is 1 : 2 : 3 and its volume is 1296 cubic cm. The area of the whole surface in sq. cm is : [SSC 10+2-2012]

(a) 696 (b) 792

(c) 824 (d) 548

88. The base of a right pyramid is an equilateral triangle of side $10\sqrt{3}$ cm. If the total surface area of the pyramid is $270\sqrt{3}$ sq. cm, its height is [SSC 10+2-2013]

(a) 12 cm (b) $12\sqrt{3}$ cm

(c) 10 cm (d) $10\sqrt{3}$ cm

89. The volumes of a cylinder and a cone are in the ratio 3 : 1. Find their diameters and then compare them when their heights are equal. [SSC 10+2-2013]

(a) Diameter of cylinder < Diameter of cone

(b) Diameter of cylinder = 2 times of diameter of cone

(c) Diameter of cylinder = Diameter of cone

(d) Diameter of cylinder > Diameter of cone

90. A square of side 3 cm is cut off from each corner of a rectangular sheet of length 24 cm and breadth 18 cm and the remaining sheet is folded to form an open rectangular box. The surface area of the box is [SSC 10+2-2013]

(a) 423 cm^2 (b) 468 cm^2

(c) 396 cm^2 (d) 612 cm^2

91. The sides of a triangle are 16 cm, 12 cm and 20 cm. Find the area [SSC 10+2-2013]

(a) 81 cm^2 (b) 64 cm^2

(c) 112 cm^2 (d) 96 cm^2

92. What is the height of a cylinder that has the same volume and radius as a sphere of diameter 12 cm? [SSC 10+2-2013]

(a) 8 cm (b) 7 cm

(c) 10 cm (d) 9 cm

93. The volume of air in a room is 204 m^3 . The height of the room is 6 m. What is the floor area of the room? [SSC 10+2-2013]

(a) 34 m^2 (b) 32 m^2

(c) 46 m^2 (d) 44 m^2

94. If the total surface area of a cube is 96 cm^2, its volume is [SSC 10+2-2013]

(a) 36 cm^3 (b) 56 cm^3
(c) 16 cm^3 (d) 64 cm^3

95. 360 sq. cm and 250 sq. cm are the areas of two similar triangles. If the length of one of the sides of the first triangle be 8 cm, then the length of the corresponding side of the second triangle is [SSC 10+2-2013]

(a) 6 cm (b) $6\frac{1}{5}$ cm
(c) $6\frac{1}{3}$ cm (d) $6\frac{2}{3}$ cm

96. The length and breadth of a rectangle are doubled. Percentage increase in area is [SSC 10+2-2013]

(a) 400% (b) 150%
(c) 200% (d) 300%

97. The base of a right prism is a triangle whose perimeter is 28 cm and the inradius of the triangle is 4 cm. If the volume of the prism is 366 cc, then its height is [SSC 10+2-2013]

(a) 4 cm (b) 8 cm
(c) 6 cm (d) None of these

98. The difference between the circumference and the radius of a circle is 37 cm. The area of circle is (Take $\pi = \frac{22}{7}$) [SSC 10+2-2013]

(a) 154 sq. cm (b) 259 sq. cm
(c) 148 sq. cm (d) 111 sq. cm

99. The sum of the interior angles of a polygon is 1444°. The number of sides of the polygon is [SSC 10+2-2014]

(a) 6 (b) 9
(c) 10 (d) 12

100. If the sides of a right angled triangle are three consecutive integers, then the length of the smallest side is [SSC 10+2-2014]

(a) 3 units (b) 2 units
(c) 4 units (d) 5 units

101. If the three medians of a triangle are same, then the triangle is [SSC 10+2-2014]

(a) equilateral (b) isosceles
(c) right-angled (d) obtuse-angled

102. The perimeters of two similar triangles ΔABC and ΔPQR are 36 cm and 24 cm respectively. If PQ = 10 cm, the AB is [SSC 10+2-2014]

(a) 15 cm (b) 12 cm
(c) 14 cm (d) 26 cm

103. Two circles intersect each other at the points A and B. A straight line parallel to AB intersects the circles at C, D, E and F. If CD = 4.5 cm, then the measure of EF is [SSC 10+2-2014]

(a) 1.50 cm (b) 2.25 cm
(c) 4.50 cm (d) 9.00 cm

104. Area of a regular hexagon with side 'a' is [SSC 10+2-2014]

(a) $\frac{3\sqrt{3}}{4}a^2$ sq. unit (b) $\frac{12}{2\sqrt{3}}a^2$ sq. unit
(c) $\frac{9}{2\sqrt{3}}a^2$ sq. unit (d) $\frac{6}{\sqrt{2}}a^2$ sq. unit

105. If the sum of the dimensions of a rectangular parallelepiped is 24 cm and the length of the diagonal is 15 cm, then the total surface area of it is [SSC 10+2-2014]

(a) 420 cm^2 (b) 275 cm^2
(c) 351 cm^2 (d) 378 cm^2

106. A flask in the shape of a right circular cone of height 24 cm is filled with water. The water is poured in a right circular cylindrical flask whose radius is $\frac{1}{3}^{rd}$ of the radius of the base of the circular cone. Then the height of the water in the cylindrical flask is [SSC 10+2-2014]

(a) 32 cm (b) 24 cm
(c) 48 cm (d) 72 cm

107. The external fencing of a circular path around a circular plot of land is 33 m more than its interior fencing. The width of the path around the plot is [SSC 10+2-2014]

(a) 5.52 m (b) 5.25 m
(c) 2.55 m (d) 2.25 m

108. A vessel is in the form of an inverted cone. Its height is 11 cm and radius of its top, which is open, is 2.5 cm. It is filled with water upto the rim. When lead shots, each of which is a sphere of radius 0.25 cm are dropped into the vessel, 2/5 of the water flows out. The number of lead shots dropped into the vessel is [SSC 10+2-2014]

(a) 880 (b) 440
(c) 220 (d) 110

109. A plot of 1800 sq. ft. is available at the rate of ₹ 630 per sq. ft. If 45% of the total cost of the plot is to be paid at the time of booking it, how much is the booking amount? [IBPS Clerk-2012]

(a) ₹ 11,34,000 (b) ₹ 5,10,300
(c) ₹ 6,03,000 (d) ₹ 6,00,300
(e) None of these

110. 1/3rd the diagonal of a square is $3\sqrt{2}$ m. What is the measure of the side of the concerned square? [IBPS Clerk-2013]

(a) 12 m (b) 9 m
(c) 18 m (d) 6 m
(e) 7 m

111. The perimeter of a rectangle whose length is 6 m more than its breadth is 84 m. What would be the area of a triangle whose base is equal to the diagonal of the rectangle and whose height is equal to the length of the rectangle? (in m^2) [IBPS Clerk-2013]

(a) 324 (b) 372
(c) 360 (d) 364
(e) 348

Level - II

1. 2 cm of rain has fallen on a sq. km of land. Assuming that 50% of the raindrops could have been collected and contained in a pool having a 100 m × 10 m base, by what level would the water level in the pool have increased?
(a) 15 m (b) 20 m (c) 10 m (d) 25 m
2. A right circular solid cylinder of base radius 4 cm and vertical height 22.5 cm is melted to form 8 equal solid spheres. If there is a process loss of 20% during such formation, then what is the radius of each of the solid sphere so formed?
(a) 2 cm (b) 3 cm (c) 2.5 cm (d) 3.5 cm
3. If the radius of a circle is diminished by 10%, the area is diminished by
(a) 36% (b) 20% (c) 19% (d) 10%
4. A landowner increased the length and breadth of a rectangular plot by 10% and 20% respectively. Find the percentage change in the cost of the plot.
(a) 35% (b) 33% (c) 22.22% (d) 32%
5. In the given diagram a rope is wound round the outside of a circular drum whose diameter is 70 cm and a bucket is tied to the other end of the rope. Find the number of revolutions made by the drum if the bucket is raised by 11 m.

(a) 10 (b) 2.5 (c) 5 (d) 5.5
6. *ABCD* is a square of area 4, which is divided into four non overlapping triangles as shown in the fig. Then the sum of the perimeters of the triangles is

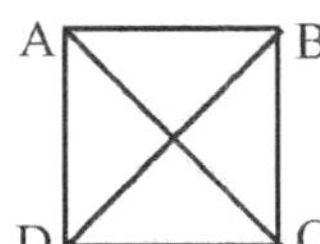

(a) $8(2+\sqrt{2})$ (b) $8(1+\sqrt{2})$
(c) $4(1+\sqrt{2})$ (d) $4(2+\sqrt{2})$
7. A cone, a hemisphere and a cylinder stand on equal bases and have the same height, the height being equal to the radius of the circular base. Their total surface areas are in the ratio:
(a) $(\sqrt{2}+1):3:4$ (b) $(\sqrt{3}+1):3:4$
(c) $\sqrt{2}:3:4$ (d) $\sqrt{3}:7:8$
8. Four identical coins are placed in a square. For each coin, the ratio of area to circumference is same as the ratio of circumference to area.

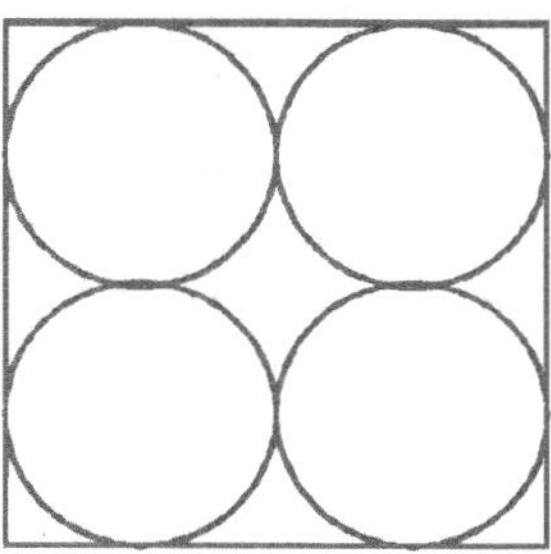

Then, find the area of the square that is not covered by the coins
(a) $16(\pi - 1)$ (b) $16(8 - \pi)$
(c) $16(4 - \pi)$ (d) $16\left(4-\frac{\pi}{2}\right)$
9. The figure shows a circle of diameter *AB* and radius 6.5 cm. If chord *CA* is 5 cm. long, find the area of ΔABC

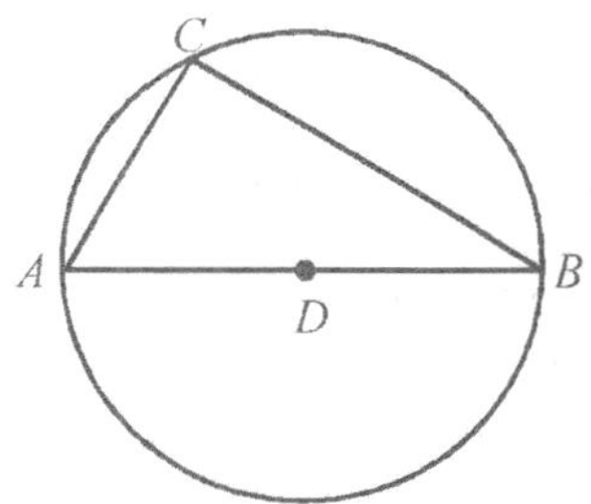

(a) 60 sq. cm (b) 30 sq. cm
(c) 40 sq. cm (d) 52 sq. cm.
10. A slab of ice 8 inches in length, 11 inches in breadth, and 2 inches thick was melted and resolidified in the form of a rod of 8 inches diameter. The length of such a rod, in inches, in nearest to
(a) 3 (b) 3.5
(c) 4 (d) 4.5
11. A passage 12 m long, 3m high and 4 m wide has two doors of 2.5 m by 1.5 m and a window of 2 m by 0.60 m. The cost of colouring the walls and ceiling at ₹ 15 per sq. m is
(a) ₹ 1023 (b) ₹ 432
(c) ₹ 2029.5 (d) ₹ 1635
12. A pipes each of 3 inch diameter are to be replaced by a single pipe discharging the same quantity of water. What should be the diameter of the single pipe, if the speed of water is the same ?
(a) 6 inch (b) 3 inch
(c) 9 inch (d) 12 inch

13. A sphere is melted and half of the molten liquid is used to form 11 identical cubes, whereas the remaining half is used to form 7 identical smaller spheres. The ratio of the side of the cube to the radius of the new small sphere is
(a) $(4/3)^{1/3}$ (b) $(8/3)^{1/3}$ (c) $(3)^{1/3}$ (d) 2

14. Find the area of an isosceles triangle whose equal sides are 8 cm each and the third side is 10 cm ?
(a) 10 cm^2 (b) 48 cm^2
(c) $5\sqrt{39}$ cm^2 (d) $10\sqrt{10}$ cm^2

15. In the figure given below, *ABCD* is a square of side 4 cm. Two quadrants of a circle with *B* and *D* as centres are draw. The radius of each of the quadrants is 4cm. What is the area of the shaded portion?

(a) 4.56 sq. cm (b) 9.12 sq. cm
(c) 13.68 sq. cm (d) 7.76 sq.cm

16. Find the volume and the total surface area of a solid right pyramid of its height is 4 cm, and its square base is of side 6 cm.
(a) 86 sq. cm. (b) 90 sq. cm.
(c) 80 sq. cm. (d) 96 sq. cm.

17. The radius of the incircle of triangle when sides are 18, 24 and 3 cms is
(a) 2 cm. (b) 4 cm. (c) 6 cm. (d) 9 cm.

18. The sides of a triangle are 21, 20 and 13 cm. Find the area of the larger triangle into which the given triangle is divided by the perpendicular upon the longest side from the opposite vertex.
(a) 72 cm^2 (b) 96 cm^2 (c) 168 cm^2 (d) 144 cm^2

19. A solid sphere of radius 6 cm is melted into a hollow cylinder of uniform thickness. If the external radius of the base of the cylinder is 5 cm and its height is 32 cm, find the uniform thickness of the cylinder.
(a) 2 cm (b) 3 cm (c) 1 cm (d) 3.5 cm

20. Find the sum of the areas of the shaded sectors given that *ABCDEF* is any hexagon and all the circles are of same radius r with different vertices of the hexagon as their centres as shown in the figure.

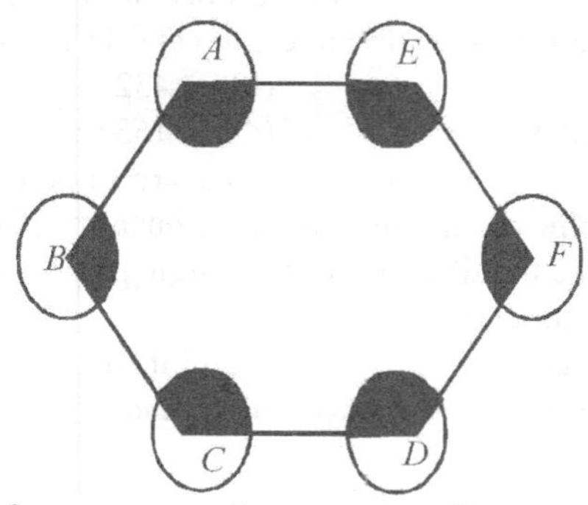

(a) πr^2 (b) $2\pi r^2$ (c) $5\pi r^2/4$ (d) $3\pi r^2/2$

21. A cube is inscribed in a hemisphere of radius *R*, such that four of its vertices lie on the base of the hemisphere and the other four touch the hemispherical surface of the half-sphere. What is the volume of the cube?

(a) $0.25\,R^3$ (b) $0.67\sqrt{\frac{2}{3}}R^3$

(c) $0.5\sqrt{\frac{2}{3}}R^3$ (d) $0.67\,R^3$

22. In the figure given below, *ABCO* represents a quadrant of a circle of radius 10.5 cm with centre *O*. Calculate the area of shaded portion, if *OD* = *DC*.

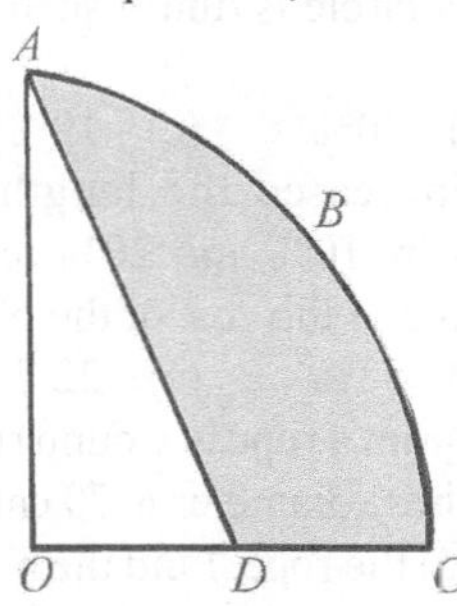

(a) 59 cm^2 (b) 69 cm^2
(c) 79 cm^2 (d) 49 cm^2

23. In the adjoining figure is a park in which shaded area is to be covered by grass. If the rate of covering with grass is ₹ 0.70 per sq. m.

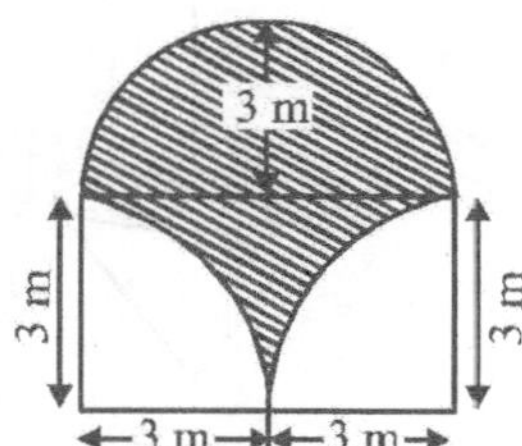

Find the expenditure of covering its field with grass (π = 22/7)
(a) ₹ 12.60 (b) ₹ 6.30
(c) ₹ 9.30 (d) ₹ 10.30

24. *ABCD* is a quadrilateral. The diagonals of *ABCD* intersect at the point P. The area of the triangles *APD* and *BPC* are 27 and 12, respectively. If the areas of the triangles *APB* and *CPD* are equal, then the area of triangle *APB* is
(a) 12 (b) 18
(c) 15 (d) 16

25. Two circles of radius 1 cm touch at point *P*. *A* third circle is drawn through the points *A*, *B* and *C* such that *PA* is the diameter of the first circle and *BC* perpendicular to *AP* is the diameter of the third circle. The radius of the third circle in cm.

(a) $\frac{9}{5}$ (b) $\frac{7}{4}$

(c) $\frac{\sqrt{10}}{2}$ (d) 2

26. A toy is in the shape of a right circular cylinder with a hemisphere on one end and a cone on the other. The height and radius of the cylindrical part are 13 cm and 5 cm respectively. The radii of the hemispherical and conical parts are the same as that of the cylindrical part. Calculate the surface area of the toy if the height of conical part is 12 cm.

(a) 1440 cm^2 (b) 385 cm^2
(c) 1580 cm^2 (d) 770 cm^2

27. A square hole of cross-sectional area 4 cm^2 is drilled across a cube with its length parallel to a side of the cube. If an edge of the cube measures 5 cm, what is the total surface area of the body so formed?

(a) 158 cm^2 (b) 190 cm^2
(c) 166 cm^2 (d) 182 cm^2

28. In the equilateral triangle ABC, $AD = DE = BE$, D and E lies on the AB. If each side of the triangle (i.e., AB, BC and AC) be 6 cm, then the area of the shaded region is:

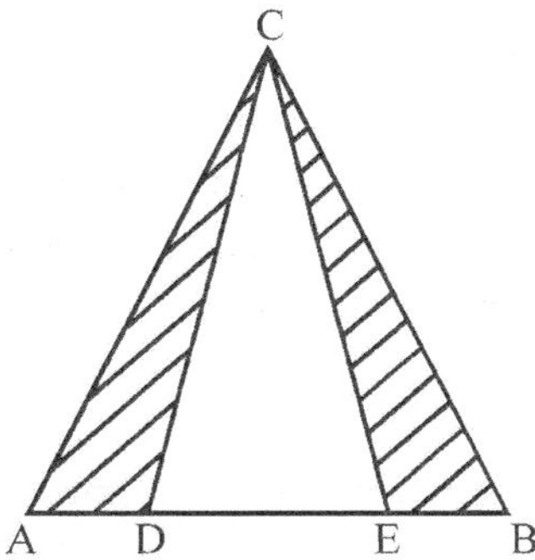

(a) 9 cm^2 (b) $6\sqrt{3}$ cm^2
(c) $5\sqrt{3}$ cm^2 (d) None of these

29. $ABCD$ is a rectangle of dimensions 6 cm × 8 cm. DE and BF are the perpendiculars drawn on the diagonal of the rectangle. What is the ratio of the shaded to that of unshaded region?

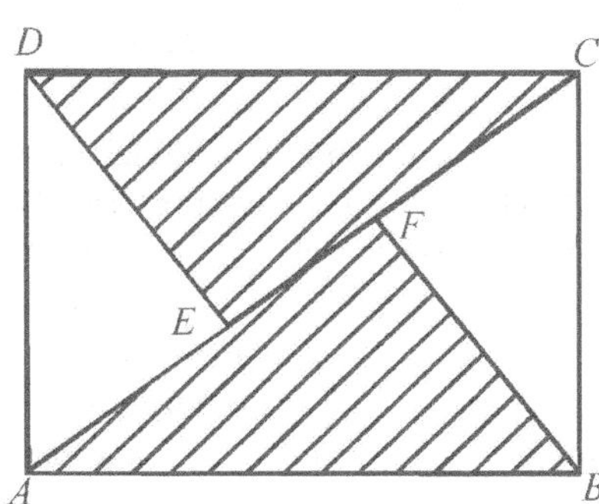

(a) 7 : 3 (b) 16 : 9
(c) $4:3\sqrt{2}$ (d) Data insufficient

30. Two circles touch internally and their centres are O and O' as shown. The sum of their areas is 180π sq. cm. and the distance between their centres is 6 cm.

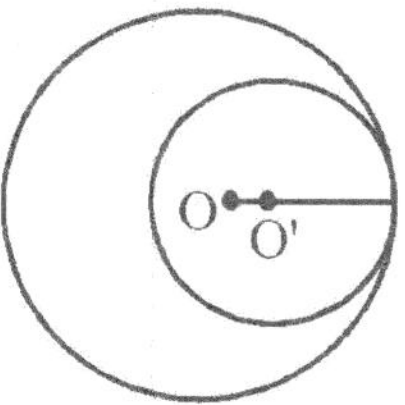

What is the diameter of the larger circle ?

(a) 16 cm (b) 12 cm
(c) 18 cm (d) 24 cm

31. PQRS is the diameter of a circle of radius 6 cm. The lengths PQ, QR and RS are equal. Semi-circles are drawn with PQ and QS as diameters as shown in the figure alongside. Find the ratio of the area of the shaded region to that of the unshaded region.

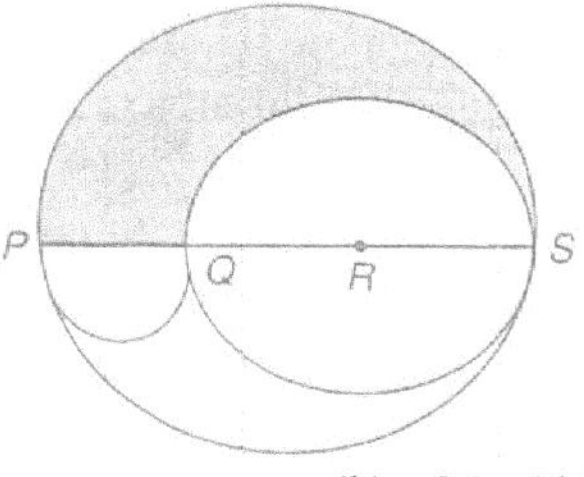

(a) 1: 2 (b) 25 : 121
(c) 5 : 18 (d) 5 : 13

32. The area of a square is 1444 square meters. The breadth of a rectangle is 1/4th the side of the square and the length of the rectangle is thrice the breadth. What is the difference between the area of the square and the area of the rectangle? [*IBPS-PO-2012*]

(a) 1152.38 sq.mtr. (b) 1169.33 sq.mtr
(c) 1181.21 sq.mtr. (d) 1173.25 sq.mtr
(e) None of these

33. The length of the circum-radius of a triangle having sides of lengths 12 cm, 16 cm and 20 cm is [*SSC CGL-2012*]

(a) 15 cm (b) 10 cm
(c) 18 cm (d) 16 cm

34. ABC is a triangle. The medians CD and BE intersect each other at O. Then $\Delta\, ODE : \Delta\, ABC$ is [*SSC CGL-2012*]

(a) 1 : 3 (b) 1 : 4
(c) 1 : 6 (d) 1 : 12

35. If P, R, T are the area of a parallelogram, a rhombus and a triangle standing on the same base and between the same parallels, which of the following is true? [*SSC CGL-2012*]

(a) $R < P < T$ (b) $P > R > T$
(c) $R = P = T$ (d) $R = P = 2T$

36. AB is a diameter of the circumcircle of $\Delta\, APB$; N is the foot of the perpendicular drawn from the point P on AB. If $AP = 8$ cm and $BP = 6$ cm, then the length of BN is [*SSC CGL-2012*]

(a) 3.6 cm (b) 3 cm
(c) 3.4 cm (d) 3.5 cm

37. Two circles with same radius r intersect each other and one passes through the centre of the other. Then the length of the common chord is [SSC CGL-2012]
(a) r (b) $\sqrt{3}r$
(c) $\frac{\sqrt{3}}{2}r$ (d) $\sqrt{5}r$

38. The bisector of $\angle A$ of ΔABC cuts BC at D and the circumcircle of the triangle at E. Then [SSC CGL-2012]
(a) $AB : AC = BD : DC$
(b) $AD : AC = AE : AB$
(c) $AB : AD = AC : AE$
(d) $AB : AD = AE : AC$

39. A metal pipe of negligible thickness has radius 21 cm and length 90 cm. The outer curved surface area of the pipe in square cm is [SSC CGL-2012]
(a) 11880 (b) 11680
(c) 11480 (d) 10080

40. If D is the mid-point of the side BC of ΔABC and the area of ΔABD is 16 cm^2, then the area of ΔABC is [SSC CGL-2012]
(a) 16 cm^2 (b) 24 cm^2
(c) 32 cm^2 (d) 48 cm^2

41. If the inradius of a triangle with perimeter 32 cm is 6 cm, then the area of the triangle in sq. cm is [SSC CGL-2013]
(a) 48 (b) 100
(c) 64 (d) 96

42. If the diagonal of a square is doubled, then its area will be [SSC CGL-2013]
(a) three times (b) four times
(c) same (d) None of these

43. A square is inscribed in a circle of radius 8 cm. The area of the square is [SSC CGL-2013]
(a) 16 square cm (b) 64 square cm
(c) 128 square cm (d) 148 square cm

44. The biggest possible circle is inscribed in a rectangle of length 16 cm and breadth 6 cm. Then its area is [SSC CGL-2013]
(a) 3π cm^2 (b) 4π cm^2
(c) 5π cm^2 (d) 9π cm^2

45. The base of a right pyramid is an equilateral triangle of side 4 cm each. Each slant edge is 5 cm long. The volume of the pyramid is [SSC CGL-2014]
(a) $\frac{4\sqrt{8}}{3}$ cm^3 (b) $\frac{4\sqrt{60}}{3}$ cm^3
(c) $\frac{4\sqrt{59}}{3}$ cm^3 (d) $\frac{4\sqrt{61}}{3}$ cm^3

46. There are two cones. The curved surface area of one is twice that of the other. The slant height of the latter is twice that of the former. The ratio of their radii is [SSC CGL-2014]
(a) 4 : 1 (b) 4 : 3
(c) 3 : 4 (d) 1 : 4

47. In a quadrilateral ABCD, the bisectors of $\angle$A and $\angle$B meet at O. If $\angle C = 70°$ and $\angle D = 130°$, then measure of $\angle$AOB is [SSC CGL-2014]
(a) 40° (b) 60°
(c) 80° (d) 100°

48. In ΔABC, E and D are points on sides AB and AC respectively such that $\angle ABC = \angle ADE$. If AE = 3 cm, AD = 2 cm and EB = 2 cm, then length of DC is [SSC CGL-2014]
(a) 4 cm (b) 4.5 cm
(c) 5.0 cm (d) 5.5 cm

49. In a circle with centre O, AB is a chord, and AP is a tangent to the circle. If $\angle AOB = 140°$, then the measure of $\angle$PAB is [SSC CGL-2014]
(a) 35° (b) 55°
(c) 70° (d) 75°

50. In ΔABC, $\angle A < \angle B$. The altitude to the base divides vertex angle C into two parts C_1 and C_2, with C_2 adjacent to BC. Then
[SSC CGL-2014]
(a) $C_1 + C_2 = A + B$ (b) $C_1 - C_2 = A - B$
(c) $C_1 - C_2 = B - A$ (d) $C_1 + C_2 = B - A$

51. If O is the in-centre of ΔABC; if $\angle BOC = 120°$, then the measure of $\angle$BAC is [SSC CGL-2014]
(a) 30° (b) 60°
(c) 150° (d) 75°

52. Two parallel chords of a circle of diameter 20 cm are 12 cm and 16 cm long. If the chords are in the same side of the centre, then the distance between them is[SSC CGL-2014]
(a) 28 cm (b) 2 cm
(c) 4 cm (d) 8 cm

53. The interior angle of a regular polygon is 140°. The number of sides of that polygon is [SSC CGL-2014]
(a) 9 (b) 8
(c) 7 (d) 6

54. If two circles of radii 9 cm and 4 cm touch externally, then the length of a common tangent is
[SSC CGL-2014]
(a) 5 cm (b) 7 cm
(c) 8 cm (d) 12 cm

55. A wire is bent into the form of a circle, whose area is 154 cm^2. If the same wire is bent into the form of an equilateral triangle, the approximate area of the equilateral triangle is [SSC CGL-2014]
(a) 93.14 cm^2 (b) 90.14 cm^2
(c) 83.14 cm^2 (d) 39.14 cm^2

56. The length of a rectangle is increased by 15% and breadth decreased by 15%. Then the area of the new rectangle is [SSC CGL-2014]
(a) unchanged (b) increased by 2.25%
(c) decreased by 2.25% (d) increased by 15%

Hints & Solutions

Level-I

1. (a) Area of rhombus = side × height
$= 13 \times 20 = 260 \text{ cm}^2$

2. (a) In a circle, circumference = $2\pi r$

Hence, $44 = 2\pi r \quad \therefore r = \frac{44}{2\pi}$

Now, area of circle $= \pi r^2 = \pi \times \frac{44}{2\pi} \times \frac{44}{2\pi} = 154 \text{ m}^2$

3. (a) Let the length and breadth of a rectangle are $9x$ m and $5x$ m respectively.
In a rectangle, area = length × breadth
$\therefore \quad 720 = 9x \times 5x$
or $x^2 = 16 \Rightarrow x = 4$
Thus, length $= 9 \times 4 = 36$ m
and breadth $= 5 \times 4 = 20$ m
Therefore, perimeter of rectangle $= 2(36 + 20) = 112$m

4. (d) Required no. of squares $= \frac{5^2}{1^2} = 25$

5. (c) Let the area of two squares be $9x$ and x respectively.
So, sides of both squares will be
$\sqrt{9x}$ and $\sqrt{x}$ respectively. [since, side = $\sqrt{\text{area}}$]
Now, perimeters of both squares will be
$4 \times \sqrt{9x}$ and $4\sqrt{x}$ respectively.
[since , perimeter = 4 × side]

Thus, ratio of their perimeters $= \frac{4\sqrt{9x}}{4\sqrt{x}} = 3 : 1$

6. (d) Perimeter of the circle = $2\pi r = 2(18 + 26)$

$\Rightarrow \quad 2 \times \frac{22}{7} \times r = 88 \Rightarrow r = 14$

$\therefore$ Area of the circle

$= \pi r^2 = \frac{22}{7} \times 14 \times 14 = 616 \text{ cm}^2$.

7. (a) In a rectangle,

$$\frac{(\text{perimeter})^2}{4} = (\text{diagonal})^2 + 2 \times \text{area}$$

$$\Rightarrow \frac{(14)^2}{4} = 5^2 + 2 \times \text{area}$$

$49 = 25 + 2 \times$ area

$\therefore \text{Area} = \frac{49 - 25}{2} = \frac{24}{2} = 12 \text{ cm}^2$

8. (c) Circumference of circle = Area of circle

or $\pi d = \pi \left(\frac{d}{2}\right)^2$ [where d = diameter]

$\therefore d = 4$

9. (a) In a parallelogram.
Area = Diagonal × length of perpendicular on it.
$= 30 \times 20 = 600 \text{ m}^2$

10. (c) In a triangle,

Area $= \frac{1}{2} \times$ length of perpendicular × base

or $615 = \frac{1}{2} \times$ length of perpendicular × 123

$\therefore$ Length of perpendicular $= \frac{615 \times 2}{123} = 10$ m.

11. (a) Circumference of circular bed = 30 cm

Area of circular bed $= \frac{(30)^2}{4\pi}$

Space for each plant = 4 cm^2
$\therefore$ Required number of plants

$= \frac{(30)^2}{4\pi} \div 4 = 17.89 = 18$ (Approx)

12. (d) Side of square carpet $= \sqrt{\text{Area}} = \sqrt{169} = 13$ m
After cutting of one side,
Measure of one side = 13 – 2 = 11 m
and other side = 13 m (remain same)
$\therefore$ Area of rectangular room $= 13 \times 11 = 143 \text{ m}^2$

13. (a) If area of a circle decreased by x % then the radius of a circle decreases by

$(100 - 10\sqrt{100 - x})\% = (100 - 10\sqrt{100 - 36})\%$

$= (100 - 10\sqrt{64})\%$

$= 100 - 80 = 20\%$

14. (b) Let ABC be the isosceles triangle and AD be the altitude.
Let $AB = AC = x$. Then, $BC = (32 - 2x)$.

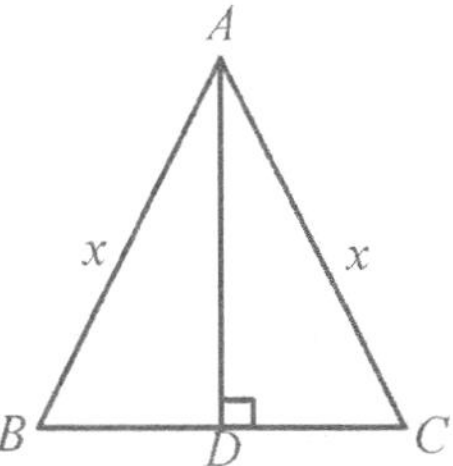

Since, in an isosceles triangle, the altitude bisects the base. So, $BD = DC = (16 - x)$.

In ΔADC, $AC^2 = AD^2 + DC^2$

$\Rightarrow x^2 = (8)^2 + (16 - x)^2$

$\Rightarrow 32x = 320 \Rightarrow x = 10.$

$\therefore BC = (32 - 2x) = (32 - 20)$ cm = 12 cm.

Hence, required area $= \left(\frac{1}{2} \times BC \times AD\right)$

$= \left(\frac{1}{2} \times 12 \times 10\right) \text{cm}^2 = 60 \text{ cm}^2.$

15. (c) Area of field = 576 km^2. Then,

each side of field $= \sqrt{576} = 24$ km

Distance covered by the horse

= Perimeter of square field

= 24 × 4 = 96 km

$\therefore$ Time taken by horse $= \frac{\text{distance}}{\text{speed}} = \frac{96}{12} = 8$ h

16. (b)

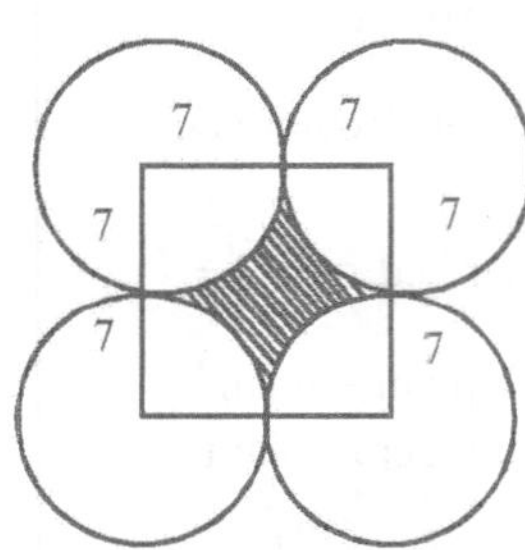

The shaded area gives the required region.

Area of the shaded region

= Area of the square – area of four quadrants of the circles

$= (14)^2 - 4 \times \frac{1}{4} \pi (7)^2$

$= 196 - \frac{22}{7} \times 49 = 196 - 154 = 42 \text{ cm}^2$

17. (b) Perimeter = Distance covered in 8 min.

$= \left(\frac{12000}{60} \times 8\right) \text{m} = 1600 \text{ m}.$

Let length = $3x$ metres and breadth = $2x$ metres.

Then, $2(3x + 2x) = 1600$ or $x = 160$.

$\therefore$ Length = 480 m and Breadth = 320 m.

$\therefore$ Area = (480 × 320) m^2 = 153600 m^2.

18. (c) Length of wire $= 2\pi \times R = \left(2 \times \frac{22}{7} \times 56\right)$ cm = 352 cm.

Side of the square $= \frac{352}{4}$ cm = 88 cm.

Area of the square = (88 × 88) cm^2 = 7744 cm^2.

19. (a) Let the length of the room be ℓ m

Then its, breadth = $\ell/2$

Therefore, $\ell \times \frac{\ell}{2} = \frac{5000}{25}$

or $\ell^2 = 400$

or $\ell = 20$ m

Also, $2\ell h + 2 \times \frac{\ell}{2} \times h = \frac{64800}{240}$

$\Rightarrow 3\ell h = 270$

or $h = \frac{270}{3 \times 20} = \frac{270}{60} = 4.5 \text{m}$

20. (a) Let the edge of the third cube be x cm.

Then, $x^3 + 6^3 + 8^3 = 12^3$

$\Rightarrow x^3 + 216 + 512 = 1728$

$\Rightarrow x^3 = 1000 \Rightarrow x = 10.$

Thus the edge of third cube = 10 cm.

21. (b) Area of the inner curved surface of the well dug

$= [2\pi \times 3.5 \times 22.5] = 2 \times \frac{22}{7} \times 3.5 \times 22.5$

= 44 × 0.5 × 22.5 = 495 sq. m.

$\therefore$ Total cost = 495 × 3 = ₹ 1485.

22. (a) In a cube,

Area = 6 (side)^2

or 150 = 6 (side)^2

$\therefore$ side $= \sqrt{25} = 5$ m

Length of diagonal $= \sqrt{3} \times \text{side} = 5\sqrt{3}$ m

23. (c) Required length = length of the diagonal

$= \sqrt{12^2 + 9^2 + 8^2} = \sqrt{144 + 81 + 64} = \sqrt{289} = 17$ m

24. (c) In a sphere, volume $= \frac{4}{3}\pi r^3$

and surface area $= 4\pi r^2$

According to question, $\frac{4}{3}\pi r^3 \div 4\pi r^2 = 27$

or $r = 27 \times 3 = 81$ cms

25. (a) Let depth of rain be h metre. Then,

volume of water

= area of rectangular field × depth of rain

or 3000 = 500 × 300 × h

$\therefore h = \frac{3000}{500 \times 300} \text{m} = \frac{3000 \times 100}{500 \times 300} \text{cms} = 2$ cms

26. (c) Volume of cylinder = (π × 6 × 6 ×28) cm^3

= (36 × 28) π cm^3.

Volume of each bullet $= \left(\frac{4}{3} \pi \times \frac{3}{4} \times \frac{3}{4} \times \frac{3}{4}\right) \text{cm}^3$

$= \frac{9\pi}{16} \text{cm}^3.$

Number of bullets = $\frac{\text{Volume of cylinder}}{\text{Volume of each bullet}}$

$= \left[(36\times 28)\pi\times\frac{16}{9\pi}\right] = 1792.$

27. **(c)** Let h be the required height then, $\frac{22}{7}\times(60)^2\times h$

$= 30\times 60\times\frac{22}{7}\times(1)^2\times(600)$

$\Rightarrow\ 60\,h = 30\times 600$

$\Rightarrow\ h = 300$ cm = 3 m

28. **(a)** Let radius of the 3rd spherical ball be R,

$\therefore\ \frac{4}{3}\pi\left(\frac{3}{2}\right)^3 = \frac{4}{3}\pi\left(\frac{3}{4}\right)^3 + \frac{4}{3}\pi(1)^3 + \frac{4}{3}\pi R^3$

$\Rightarrow\ R^3 = \left[\left(\frac{3}{2}\right)^3 - \left(\frac{3}{4}\right)^3\right] - 1^3$

$= \frac{27}{8} - \frac{27}{64} - 1 = \frac{125}{64} = \left(\frac{5}{4}\right)^3 \Rightarrow\ R = \frac{5}{4} = 1.25$

$\therefore$ Diameter of the third spherical ball = $1.25\times 2 = 2.5$ cm.

29. **(c)** Let 'A' be the side of bigger cube and 'a' be the side of smaller cube

Surface area of bigger cube = $6\,A^2$

or $384 = 6A^2$

$\therefore\ A = 8$ cm.

Surface area of smaller cube = $6\,a^2$

$96 = 6a^2$

$\therefore\ a = 4$ mm = 0.4 cm

So, Number of small cube = $\frac{\text{Volume of bigger cube}}{\text{Volume of smaller cube}}$

$= \frac{(8)^3}{(0.4)^3} = \frac{512}{0.064} = 8{,}000$

30. **(d)** Volume of the tank = 246.4 litres = 246400 cm^3.

Let the radius of the base be r cm. Then,

$\left(\frac{22}{7}\times r^2\times 400\right) = 246400$

$\Rightarrow r^2 = \left(\frac{246400\times 7}{22\times 400}\right) = 196 \Rightarrow r = 14.$

$\therefore$ Diameter of the base = $2r$ = 28 cm = .28 m

31. **(a)** Total surface area of the remaining solid = Curved surface area of the cylinder + Area of the base + Curved surface area of the cone

$= 2\pi rh + \pi r^2 + \pi r\ell$

$= 2\pi\times 8\times 15 + \pi\times(8)^2 + \pi\times 8\times 17$

$= 240\pi + 64\pi + 136\pi$

$= 440\,\pi\ cm^2$

32. **(d)** $4\pi(r+2)^2 - 4\pi r^2 = 352$

$\Rightarrow (r+2)^2 - r^2 = \left(352\times\frac{7}{22}\times\frac{1}{4}\right) = 28.$

$\Rightarrow (r+2+r)(r+2-r) = 28$

$\Rightarrow 2r+2 = \frac{28}{2} \Rightarrow 2r+2 = 14 \Rightarrow r = 6$ cm

33. **(b)** Volume of material in the sphere

$= \left[\frac{4}{3}\pi\times\left\{(4)^3-(2)^3\right\}\right]cm^3 = \left(\frac{4}{3}\pi\times 56\right)cm^3.$

Let the height of the cone be h cm.

Then, $\frac{1}{3}\pi\times 4\times 4\times h = \left(\frac{4}{3}\pi\times 56\right)$

$\Rightarrow h = \left(\frac{4\times 56}{4\times 4}\right) = 14$ cm.

34. **(b)** Given, playground is rectangular.

Length = 36 m, Breadth = 21 m

Now, perimeter of playground = 2(21 + 36) = 114

Now, poles are fixed along the boundary at a distance 3m.

$\therefore$ Required no. of poles = $\frac{114}{3} = 38$.

35. **(a)** Let width of the field = b m

$\therefore$ length = $2\,b$ m

Now, area of rectangular field = $2b\times b = 2b^2$

Area of square shaped pond = $8\times 8 = 64$

According to the question,

$64 = \frac{1}{8}(2b^2) \Rightarrow b^2 = 64\times 4 \Rightarrow b = 16$ m

$\therefore$ length of the field = $16\times 2 = 32$ m

36. **(a)**

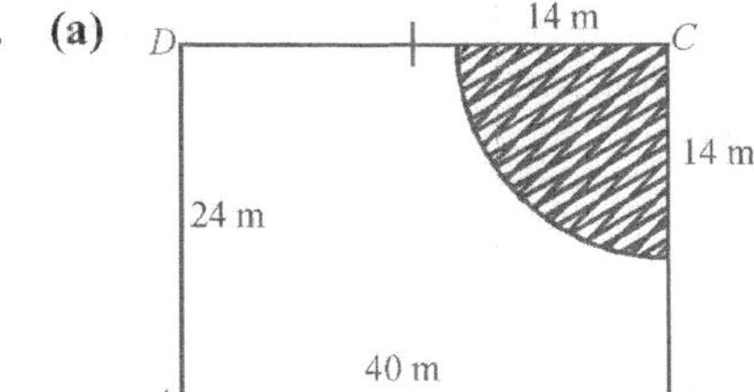

Area of the shaded portion

$= \frac{1}{4}\times\pi(14)^2 = 154\ m^2$

37. **(b)** Let ℓ be the length and b be the breadth of cold storage.

$L = 2B$, $H = 3$ metres

Area of four walls = $2[L\times H + B\times H] = 108$

$\Rightarrow\ 6BH = 108\ \Rightarrow\ B = 6$

$\therefore\ L = 12, B = 6, H = 3$

Volume = $12\times 6\times 3 = 216\ m^3$

38. (c) Surface area of the cube = (6×8^2) sq. ft. = 384 sq. ft.

Quantity of paint required = $\left(\frac{384}{16}\right)$kg = 24 kg.

$\therefore$ Cost of painting = ₹ (36.50 × 24) = ₹ 876.

39. (c) Volume of block = $(6 \times 9 \times 12)$ cm^3 = 648 cm^3.
Side of largest cube = H.C.F. of 6 cm, 9 cm, 12 cm = 3 cm.
Volume of the cube = $(3 \times 3 \times 3) = 27$ cm^3.

$\therefore$ Number of cubes = $\left(\frac{648}{27}\right) = 24.$

40. (d) Circumference of the base of ice-cream cup
= Diameter of the sheet = 28 cm

$2\pi r = 28$

$r = \frac{14}{\pi}$ cm = 4.45 cm

Slant height of cone = radius of the sheet = 14 cm
$\therefore$ $14^2 = (4.45)^2 + h^2$
or $h^2 = 196 - 19.80 = 176.20$
$\therefore$ $h = 13.27$ cm

41. (d) Required no. of squares $= \frac{5^2}{1^2} = 25$

42. (d) Volume of the cone is given by = $1/3 \times \pi r^2 h$
Here, $r = 4.2$ cm, $h = 10.2 - r = 6$ cm
Therefore the volume of the cone = $1/3\, \pi \times (4.2)^2 \times 6$ cm
= 110.88 cm^3

Volume of the hemisphere = $\frac{1}{2} \times \frac{4}{3}\pi r^3 = 155.23$ cm^3

Total volume = 110.88 + 155.232 = 266.112

43. (a) Volume of mud dugout = $10 \times 4.5 \times 3 = 135$ m^3
Let the remaining ground rise by = h m
Then $\{(20 \times 9) - (10 \times 4.5)\}\, h = 135$
$135\, h = 135 \Rightarrow h = 1$ m

44. (c) Consider for an equilateral triangle. Hence ΔABC consists of 4 such triangles with end points on mid pts AB, BC and CA

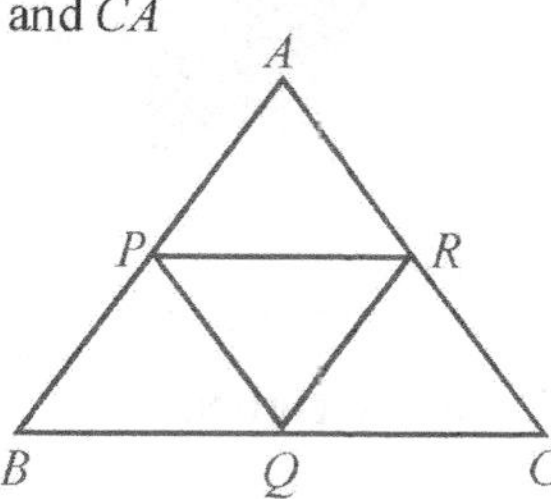

$\Rightarrow \frac{1}{4}\, ar\,(\Delta ABC) = ar\,(\Delta PQR)$

$\Rightarrow ar\,(\Delta PQR) = 5$ sq. units

45. (c) $\frac{\text{Area of uncut portion}}{\text{Area of cut portion}} = \frac{(\pi \times 20 \times 20) - (100\pi)}{(4 \times \pi \times 5 \times 5)}$

$= \frac{300\pi}{100\pi} = \frac{3}{1}$

46. (b) In the figure $\angle ACB$ is 90°
(angle subtended by diameter = 90°)
$AC = 5, AB = 13$
Using pythagoras theorem,

$AB^2 = AC^2 + CB^2 \Rightarrow CB = \sqrt{13^2 - 5^2} = 12$

Area of $\Delta ABC = \frac{1}{2} \times 5 \times 12 = 30$

47. (b)

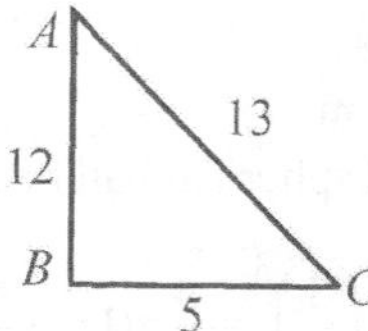

ABC forms a right angled triangle

$\therefore$ Area $= \frac{1}{2} \times 12 \times 5 = 30$

Area of rectangle $= 30 = \ell \times 10$ or $\ell = 3$ units
$\therefore$ Perimeter = 2 (10 + 3) = 26

($\because AE = FD$)

48. (d) (side)^2

$= \left(\frac{1}{2} \times \text{onediagonal}\right)^2 + \left(\frac{1}{2} \times \text{other diagonal}\right)^2$

$13^2 = \left(\frac{1}{2} \times \text{onediagonal}\right)^2 + \left(\frac{1}{2} \times 24\right)^2$

$169 - 144 = \left(\frac{1}{2} \times \text{diagonal}\right)^2$

$25 = \left(\frac{1}{2} \times \text{diagonal}\right)^2$

$5 = \frac{1}{2} \times$ diagonal $\therefore$ diagonal = 10

$\therefore$ Area $= \frac{1}{2} \times 10 \times 24 = 120$ sq. cm.

49. (b) $AC^2 = AB^2 + BC^2 \Rightarrow AC = 10$

We have $r = (A/s)$; $A = \frac{1}{2} \times (6 \times 8) = 24$

$s = (6 + 8 + 10)/2 = 12$
$r = A/s = 24/12 = 2.$

50. (b) Circumference of base $= 2\pi r = 6$ $\therefore$ $r = \frac{3}{\pi}$

51. (c) Volume of spherical shell

$= \frac{4\pi}{3}(R^3 - r^3) = \frac{4\pi}{3}\,(12^2 - 10^3)$

$= \frac{4}{3} \times \pi \times (12 - 10)(12^2 + 12 \times 10 + 10^2)$

$= \frac{4}{3} \times \pi \times 2 \times 364$ cm^3

Weight = volume × density

$= \frac{4}{3} \times \pi \times 364 \times 4.8 = 14.64$ kg

52. **(d)** Let *ABCD* be a square with side = 6 cm. Then the radius of the circle touches the square = 3 cm.
Area of circle = $\pi(r)^2 = 9\pi$ cm^2

53. **(c)** $\because$ Circumference $= \frac{p}{360} = 2\pi R$

54. **(a)** Area of shaded region = Area of equilateral $\Delta ABC - 3$ (Area of sector *AQO*)

$= \frac{\sqrt{3}}{4} \times (2)^2 = 3 \times \frac{60}{360} \times \frac{22}{7} \times (1)^2$

$= \sqrt{3} - \frac{11}{7} = 1.73 - 1.57 = 0.16$ sq. units.

55. **(c)** 2 semicircles = 1 circle with equal radius

So $2\pi r = 132 \Rightarrow 2r = \frac{132}{3.14} = 42$ m diameter

Area of track = Area within external border – Area within internal border.

$\Rightarrow \pi(23^2 - 21^2) + 90 \times 46 - 90 \times 4^2$

$\Rightarrow 88\pi + 360 \Rightarrow 636.3$ m^2

56. **(e)** Circumference = 792

$2\pi r = 792$

$r = \frac{792}{2\pi} = \frac{792 \times 7}{22 \times 2} = 126$ m

57. **(e)** Let the width of rectangle = b

$39 \times b = 1209$

$b = \frac{1209}{39} = 31$ metres.

Perimeter = 2 (39 + 31) = 140 metres.

58. **(d)** Area of field = 3584 m^2

Let the length and breadth be $7x$ and $2x$

Then $7x \times 2x = 3584$ m^2

$14x^2 = 3584$ m^2

$x^2 = 256$

$x = 16$ m

Length = $7x$ = 112 m, Breadth = $2x = 16 \times 2 = 32$ m

Perimeter = $2(l + b) = 2(112 + 32) = 288$ m

59. **(d)** Base = 2 + 2 × altitude

Let, altitude be A

Area of $\Delta = \frac{1}{2} \times$ Base × Altitude

$12 = \frac{1}{2} \times (2 + 2A) \times A$

$12 = A \times (1 + A)$

$12 = A + A^2$

$A^2 + A - 12 = 0$

$(A - 3)(A + 4) = 0$

$A = 3, A = -4$

Altitude = 3 cm

60. **(c)** Volume of sphere = Surface area of sphere

$\frac{4}{3}\pi r^3 = 4\pi r^2$ [where, r → radius]

$\Rightarrow r = 3$

61. **(b)**

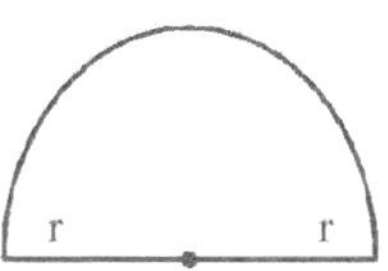

Length of railing to surround
= Length of Arc + Length of diameter

Area of semicircular field = 308

$308 = \frac{1}{2}\pi r^2$

$308 = \frac{1}{2} \times \frac{22}{7} \times r^2$

$\frac{2 \times 308 \times 7}{22} = r^2$

r = 14 m

Length of railing = $\pi r + 2r$

$= \frac{22}{7} \times 14 + 2 \times 14 = 44 + 28 = 72$ m

62. **(b)** According to condition given

Volume of right circular cone = Slant surface area

$\frac{1}{3}\pi r^2 h = \pi r l$ [where, r → radius; h → height; l → slant height]

$\frac{1}{3} rh = l$

$\frac{1}{3} rh = \sqrt{h^2 + r^2}\ [\because l^2 = h^2 + r^2]$

Squaring on both sides

$\frac{1}{9} r^2 h^2 = h^2 + r^2$

Dividing equation by r^2h^2 on both sides

$\frac{1}{9} = \frac{h^2}{r^2h^2} + \frac{r^2}{r^2h^2}$

$\frac{1}{r^2} + \frac{1}{h^2} = \frac{1}{9}$ units

63. (a) Volume of right circular cylinder = Curved surface area of cylinder

$\pi r^2 h = 2\pi rh$ [where, r → radius; h → height]

$\Rightarrow r = 2$ units

64. (c) Radius of cylinder = r units and height = r units

∴ Required ratio $= 2\pi r^2 - 2\pi r^2 = 4\pi r^2$

$: 2\pi r^2 + \pi r^2 = 3\pi r^2 = 4 : 3$

65. (c) Let side of triangle = x

$$\therefore \frac{\sqrt{3}}{4}x^2 = a \quad ...(1)$$

and $\frac{\sqrt{3}}{2}x = b$

$$x = \frac{2b}{\sqrt{3}} \quad ...(2)$$

Putting x in equation (1)

$$\frac{\sqrt{3}}{4}\left(\frac{2b}{\sqrt{3}}\right)^2 = a$$

$$\frac{b^2}{a} = \sqrt{3}$$

66. (b) Volume of sphere $= \frac{4}{3}\pi r^3 = \frac{4}{3}\pi \times 9 \times 9 \times 9$

$= 972\ \pi$ cubic.cm.

If the length of wire be h cm., then

$\pi \times (0.2)^2 \times h = 972\ \pi$

$$\Rightarrow h = \frac{972}{0.2 \times 0.2} = 24300 \text{ cm} = 243 \text{ metre}$$

67. (a) Volume of water flowing from the pipe in 1 minute

$= \pi \times 0.25 \times 0.25 \times 1000$ cu.cm.

Volume of conical vessel

$= \frac{1}{3}\pi \times 15 \times 15 \times 24$ cu.cm.

$$\therefore \text{Required time} = \frac{\pi \times 15 \times 15 \times 24}{3\pi \times 0.25 \times 0.25 \times 1000}$$

= 28 minutes 48 seconds

68. (c) $\frac{V_1}{V_2} = \frac{r_1^2 h_1}{r_2^2 h_2}$

$$\Rightarrow \frac{4}{1} = \frac{25}{16} \times \frac{h_1}{h_2}$$

$$\Rightarrow \frac{h_1}{h_2} = \frac{16 \times 4}{25} = \frac{64}{25}$$

69. (d) AB = BC = CA = $2a$ cm.

∠BAC = ∠ACB = ∠ABC = 60°

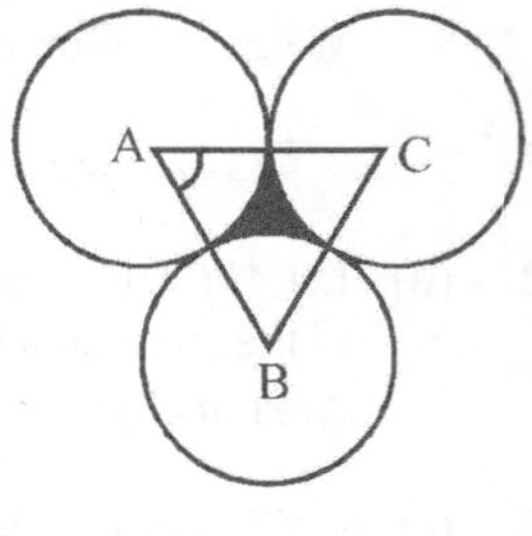

Area of ΔABC $= \frac{\sqrt{3}}{4} \times (\text{side})^2$

$= \frac{\sqrt{3}}{4} \times 4a^2 = \sqrt{3}a^2$ sq.cm.

Area of three sectors

$= 3 \times \frac{60}{360} \times \pi \times a^2$

$= \frac{\pi a^2}{2}$ sq.cm.

Area of the shaded region

$$= \sqrt{3}a^2 - \frac{\pi}{2}a^2 = \left(\frac{2\sqrt{3} - \pi}{2}\right)a^2 \text{ sq.cm.}$$

70. (d) Total surface are of cone $= \pi r(l + r)$

$$S = \frac{22}{7} \times 3 \times \left(\sqrt{3^2 + 4^2} + 3\right)$$

$$= \frac{22}{7} \times 3 \times 8 = \frac{528}{7}$$

S = 75.4 sq. cm

71. (d) Maximum number of boxes $= \frac{800 \times 700 \times 600 \text{ cm}^3}{8 \times 7 \times 6 \text{ cm}^3}$

= 1000000

72. (c) $\pi r_1^2 h_1 = \pi r_2^2 h_2$

$$\frac{r_1}{r_2} = \sqrt{\frac{h_2}{h_1}} = \sqrt{\frac{2}{1}}$$

$r_1 : r_2 = \sqrt{2} : 1$

73. (c) $2\pi r = 22$ cm

$$r = \frac{22 \times 7}{2 \times 22} = \frac{7}{2} \text{ cm}$$

Height, h = 12 cm

Volume of cylinder $= \frac{22}{7} \times \frac{7}{2} \times \frac{7}{2} \times 12 = 462 \text{ cm}^3$

74. (a) Radius of cylinder = Radius of sphere = R

Height of cylinder = 2R

Volume of cylinder $= \pi R^2 \times (2R) = 2\pi R^3$

75. (b) If the height of the godown be h meter, then

$2(15 \times 12) = 2 \times h(15 + 12)$

$\Rightarrow 27h = 15 \times 12$

$$\Rightarrow h = \frac{15 \times 12}{27} = \frac{20}{3} \text{ meter}$$

∴ Volume of the godown

$$= \frac{15 \times 12 \times 20}{3} = 1200 \text{ cu.meter}$$

76. (d)

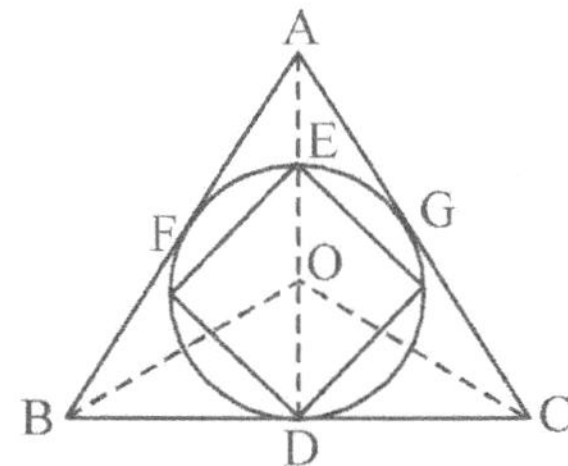

In the given figure ABC is an equilateral Δ of a side with a circle inscribed in it and a square inscribed in the circle. AD, BO and CO are the angle bisectors of $\angle A$, $\angle B$ and $\angle C$ and O is the centre of the circle.
We know that the angle bisector from the vertex of an equilateral triangle is the perpendicular bisector of the opposite side.
AD is the perpendicular bisector of BC.

$$\Rightarrow BD = \frac{a}{2} \text{ and } \angle DBO = \frac{1}{2}\angle B = \frac{1}{2} \times 60° = 30°$$

Now in ΔBOD

$$\tan 30° = \frac{OD}{BD} = \frac{\text{Radius of circle}}{\frac{a}{2}}$$

$$\Rightarrow \text{Radius of circle} = \frac{1}{\sqrt{3}} \times \frac{a}{2} = \frac{a}{2\sqrt{3}}$$

Now in right ΔEDG
$EG^2 + GD^2 = ED^2$ (Pythagoras theorem)

$$2(EG)^2 = (2OD)^2 = \left(\frac{a}{\sqrt{3}}\right)^2 = \frac{a^2}{3}$$

$$\text{Side of the square} = \sqrt{\frac{a^2}{6}} = \frac{a}{\sqrt{6}}$$

Now ar (ΔABC) : ar $(DEFG)$

$$= \frac{\frac{\sqrt{3}}{4}a^2}{\frac{a}{\sqrt{6}} \times \frac{a}{\sqrt{6}}} = \frac{\frac{\sqrt{3}}{4}}{\frac{1}{6}} = 3\sqrt{3} : 2$$

77. (a) Let length = l, breadth = b, height = h.
$l + b + h = 24$ (given) ... (i)
Diagonal of parallellopiped = 15 cm

$\sqrt{l^2 + b^2 + h^2} = 15$ or $l^2 + b^2 + h^2 = 225$
Squaring eqn. (i) on both sides
$l^2 + b^2 + h^2 + 2\,lb + 2bh + 2hl = 576$
$2(lb + bh + hl) = 576 - 225 = 351$
[$\because$ Surface area of parallellopiped = $2(lb + bh + hl)$]

78. (c) Diagonal of a cube = $6\sqrt{3}$

$\sqrt{3}$ × side = $6\sqrt{3}$
$\therefore$ Side of a cube = 6

Surface area of cube = 6 × (side)^2 = 6×6^2
Volume of cube = (side)^3 = $(6)^3$

$$\text{Required ratio} = \frac{6 \times 6^2}{6^3} = \frac{1}{1} \text{ or } 1 : 1$$

79. (b) Length of arc in 18 seconds $= \left(\frac{18}{3600}\right) \times$ circumference

$$= \frac{18}{3600} \times 2 \times \frac{22}{7} \times 35 = 1.1 \text{ cm}$$

80. (a) In a right angled Δ, the length of circumradius is half the length of hypotenuse.
$\therefore$ $H^2 = 6^2 + 8^2$
$H^2 = 36 + 64 \Rightarrow 100$
H = 10 cm

Circumradius = 5 cm

81. (b) Circumradius of a triangle

$$= \frac{abc}{\sqrt{(a+b+c)(a+b-c)(b+c-a)(a+c-b)}}$$

$$= \frac{3 \times 4 \times 5}{\sqrt{(3+4+5)(3+4-5)(4+5-3)(3+5-4)}}$$

$$= \frac{60}{\sqrt{12 \times 2 \times 6 \times 4}} = 2.5 \text{ cm}$$

82. (d) Let the side of square = 'x'
Area of square = x^2

$$\text{New length of rectangle} = \frac{130}{100}x$$

$$\text{New Breadth of rectangle} = \frac{120}{100}x$$

$$\text{Hence, Area of so formed rectangle} = \frac{130}{100} \times \frac{120}{100} \times x^2$$

$$= \frac{156}{100}x^2$$

Therefore, area of rectangle exceeds the area of square by 56%

83. (b) Volume of cubical box = 3.375 m^3

Length of edge of the box = $\sqrt[3]{3.375} = 1.5$ m

84. (c) Angle made by clock in 30 minutes = 180°

$$\therefore \text{Area of sector covered by minute hand} = \frac{\theta}{360°} \times \pi r^2$$

$$= \frac{180°}{360°} \times \frac{22}{7} \times 7 \times 7 = 77 \text{ sq.cm}$$

85. **(b)** Perimeter of a semicircular area = 18 cm

$\Rightarrow \frac{2\pi r}{2} + 2r = 18$

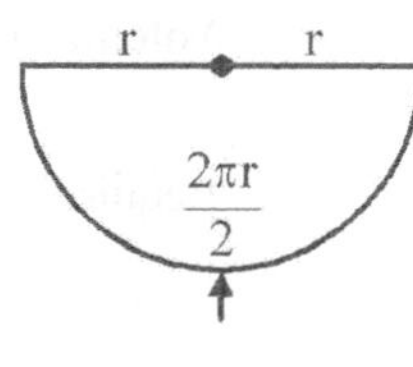

$\Rightarrow r(\pi + 2) = 18$

$r = \frac{18}{\frac{22}{7}+2} = \frac{18 \times 7}{22+14} = 3\frac{1}{2}$ cm

86. **(c)** Circumference = 33 cm

$2\pi r = 33$

$\therefore \quad r = \frac{33 \times 7}{2 \times 22} = \frac{21}{4}$

Volume $= \frac{1}{3}\pi r^2 h$

$= \frac{1}{3} \times \frac{22}{7} \times \frac{21}{4} \times \frac{21}{4} \times 16 = 462$

87. **(b)** Volume of rectangular parallelopiped = 1296

Ratio of edges = 1 : 2 : 3

$\therefore$ x, 2x and 3x are length, breadth and height of parallelopiped respectively.

$x \times 2x \times 3x = 1296$

$\Rightarrow 6x^3 = 1296 \Rightarrow x^3 = 216$

$\Rightarrow x = \sqrt[3]{216} = 6$

Length = 6, Breadth = 12, Height = 18

Required surface area = 2 (lb + bh + hl)

= 2 (6 × 12 + 12 × 18 + 18 × 6) = 792 sq.cm

88. **(a)** Now, T.S.A of pyramid

= ar (ΔABD) + ar (ΔADC) + ar (ΔABC) + ar (ΔBDC)

$\therefore$ T.S.A of pyramid $= \frac{1}{2} \times BD \times AE + \frac{1}{2} \times DC \times AF + \frac{1}{2} \times BC \times AG + \frac{\sqrt{3}}{4} \times (side)^2$

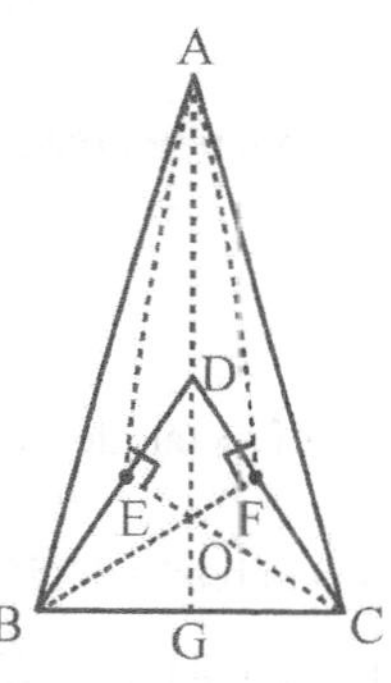

($\therefore AE = AF = AG$ = height of isosceles Δ (h))

$\Rightarrow 270\sqrt{3} = \frac{1}{2} \times h[BD + DC + BC] + \frac{\sqrt{3}}{4}(\text{side})^2$

$\Rightarrow 270\sqrt{3} = \frac{1}{2} \times h[10\sqrt{3} + 10\sqrt{3} + 10\sqrt{3}] + \frac{\sqrt{3}}{4}(10\sqrt{3})^2$

$\Rightarrow 270\sqrt{3} = 15\sqrt{3}h + 75\sqrt{3}$

$\Rightarrow 195\sqrt{3} = 15\sqrt{3}h$

$\Rightarrow h = 13$ cm ...(1)

Now to find height of pyramid (H), we use

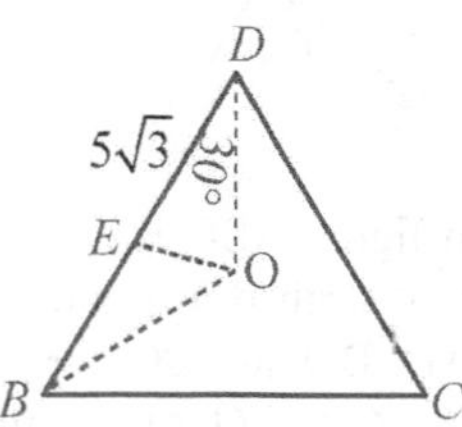

In ΔODE, $\tan 30° = \frac{OE}{ED} = \frac{OE}{5\sqrt{3}}$

$\frac{1}{\sqrt{3}} = \frac{OE}{5\sqrt{3}} \Rightarrow OE = 5$ cm ...(2)

From (1) & (2), we use pythagorals theorem, in Δ AEO

$(AE)^2 = (EO)^2 + (AO)^2$ or $h^2 = (OE)^2 + H^2$

$\Rightarrow (13)^2 - (5)^2 = H^2 \Rightarrow 144 \Rightarrow H = 12$ cm

89. **(c)** Volume of cylinder = 3 × volume of cone

$\pi r_1^2 h = 3 \times \frac{1}{3}\pi r_2^2 h$ (heights are equal)

$r_1 = r_2$

$d_1 = d_2$

90. **(c)**

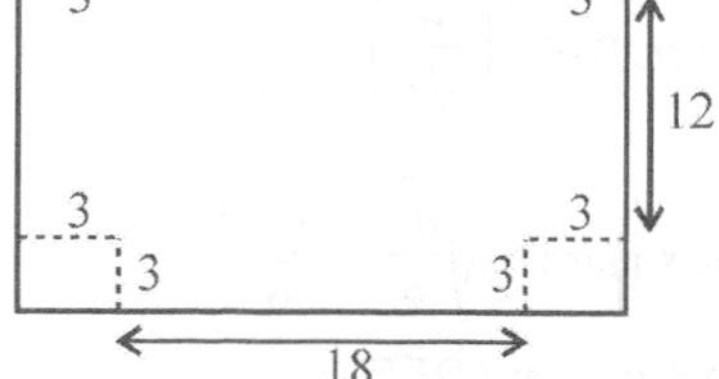

$\ell = 18$ cm, $b = 12$ cm, $h = 3$ cm

$S = 2(\ell h + bh) + \ell b$ {Box is open from upper side}

= 2 (54 + 36) + 216

= 396 cm²

91. **(d)** $(16)^2 + (12)^2 = 400 = (20)^2$

$A = \frac{1}{2} \times 16 \times 12 = 96$ cm²

92. **(a)** Volume of cylinder = volume of sphere (Given)

$\pi r^2 h = \frac{4}{3}\pi r^3$

$h = \frac{4}{3}r$

$h = \frac{4}{3} \times 6\text{cm} = 8$ cm

93. (a) Volume of air in room = 204 m^3
Area of floor × height of room = 204 m^3
Area of floor ×6 = 204 m^3

∴ Area of floor = $\frac{204}{6}$ = 34 m^2

94. (d) Total surface area of cube = 96 cm^2
$6a^2 = 96$ cm^2
$a^2 = 16$ cm^2 $\Rightarrow$ $a = 4$ cm
Now, volume of cube = $a^3 \Rightarrow (4)^3 = 64$ cm^3

95. (d) $\frac{360}{250} = \left(\frac{8}{x}\right)^2$

$\left(\frac{6}{5}\right)^2 = \left(\frac{8}{x}\right)^2$

$x = \frac{20}{3} = 6\frac{2}{3}$ cm

96. (d) $A = \ell b$
$A' = (2\ell)(2b) = 4\ell b = 4A$

% Change = $\frac{4A - A}{A} \times 100 = 300\%$

97. (d) Area of base = $\frac{1}{2} \times r \times a + \frac{1}{2} \times r \times b + \frac{1}{2} \times r \times c$

$= \frac{1}{2} r(a+b+c)$

$= r \times s = 4 \times 14 = 56$ cm^2

[where r = inradius, s = semi-perimeter]
volume of prism = area of base × height
$366 = 56 \times h$
$h = 6.5$ cm [approx]

98. (a) Given,
(Circumference – radius) of circle = 37 cm
$(2\pi r - r) = 37 \Rightarrow r(2\pi - 1) = 37$

$r\left(2 \times \frac{22}{7} - \frac{1}{1}\right) = 37 \Rightarrow r\left(\frac{44-7}{7}\right) = 37$

$r\left(\frac{37}{7}\right) = 37 \Rightarrow r = 7$ cm

Now, Area of circle = $\pi r^2 = \frac{22}{7} \times (7)^2$

$= 22 \times 7 = 154$ sq. cm

99. (c) Sum of interior angles of polygon = $(n - 2) \times 180°$
$(n - 2) \times 180° = 1440$

$n - 2 = \frac{1440}{180} = 8$

$n = 10$
Hence, the number of sides is 10.

100. (a) Consecutive integer = 3, 4 and 5
Smallest side 3 units.

101. (a)

102. (a) $\Delta ABC \sim \Delta PQR$ (given)

$\frac{AB}{PQ} = \frac{BC}{QR} = \frac{AC}{PR}$

(Corresponding sides are proportional)

$\Rightarrow \frac{AB}{PQ} = \frac{BC}{QR} = \frac{AC}{PR} = \frac{AB+BC+AC}{PQ+QR+PR}$

$\Rightarrow \frac{AB+BC+AC}{PQ+QR+PR} = \frac{AB}{PQ} \Rightarrow \frac{\text{Perimeter of ABC}}{\text{Perimeter of PQR}} = \frac{AB}{PQ}$

$\Rightarrow \frac{36}{24} = \frac{AB}{10} \Rightarrow AB = \frac{36 \times 10}{24} \Rightarrow 15$ cm

103. (c)

∵ Radius are equal
Then, CE = DF
CD + DE = DE + EF
CD = EF
EF = 4.5 cm

104. (c) Area of hexagon = $6 \times \frac{\sqrt{3}}{4} a^2 = \frac{3\sqrt{3}}{2} a^2$

or $\frac{9}{2\sqrt{3}} a^2$ sq. units

105. (c) Let length, breadth and height of parallelopiped be l, b and h respectively.
$l + b + h = 24$ cm

$\sqrt{l^2 + b^2 + h^2} = 15$ cm $\Rightarrow l^2 + b^2 + h^2 = 225$ cm^2
$(l + b + h)^2 - 2(lb + hb + lh) = 225$
$(24)^2 - 225 = 2(lb + bh + hl)$
$351 = 2(lb + bh + hl)$
Total surface area is 351 cm^2.

106. (d) Let radius of base of cone be r and height of cylinder be h.
Vol. of cone = Vol. of cylinder

$\frac{1}{3} \pi r^2 \times 24 = \pi \left(\frac{r}{3}\right)^2 \times h$

h = 72 cm

107. (b) Let radius of internal and external circular Plot be r and R respectively.

$2\pi R - 2\pi r = 33$ m

Width of path, $(R - r) = \frac{33 \times 7}{2 \times 22} = \frac{21}{4} = 5.25$m

108. (b) Vol. of cone $= \frac{1}{3}\pi \times (2.5 \text{ cm})^2 \times 11$ cm

Vol. of one sphere $= \frac{4}{3}\pi(0.25\text{cm})^3$

Vol. of all spheres = Vol. of water flows out

$n \times \frac{4}{3}\pi(0.25\text{cm})^3 = \frac{2}{5} \times \frac{\pi}{3} \times (2.5\text{cm})^2 \times 11\text{cm}$

$2n \times \frac{25}{100} \times \frac{25}{100} \times \frac{25}{100} = \frac{1}{5} \times \frac{25}{10} \times \frac{25}{10} \times 11$

n = 440

109. (b) Total cost of plot

= ₹ 630 × 1800

∴ Booking amount

$= \frac{630 \times 1800 \times 45}{100}$

= ₹ 510300

110. (b)

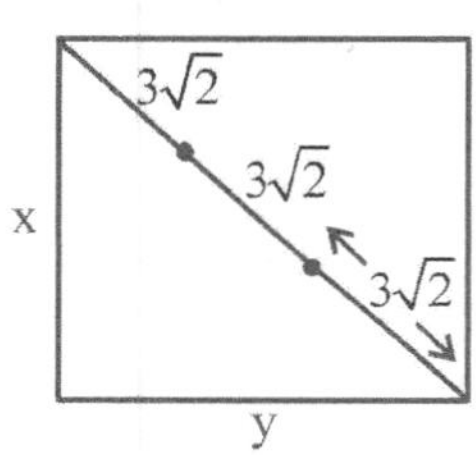

$x^2 + y^2 = (9\sqrt{2})^2$

$2x^2 = 81 \times 9$

x = 9

111. (c) x + x + x + 6 + x + 6 = 84

4x + 12 = 84

x = 18 m

D = 30 m   x = 18

x + 6

18 + 6 = 24

$D^2 = (x + 6)^2 + x^2$

$D^2 = 24^2 + 18^2$

$D^2 = 576 + 324 = 900$

D = 30 m

Base of triangle = 30 m

Height of triangle = x + 6 = 24 m

Area of triangle $= \frac{1}{2} \times 30 \times 24 = 360$ m^2

Level-II

1. (c) Volume of rain that is to be collected

in a pool $= 2 \times 1 \times 10^{10} \times \frac{1}{2}$

$= 10^{10}$ cm $= 10^4$ meter

Volume of pool $= L \times B \times h$

$10^4 = 100 \times 10 \times h$

$h = \frac{10^4}{100 \times 10} = 10$ m.

2. (b) (Volume of solid cylinder) × 0.8 = 8 × Volume of each solid sphere.

$\therefore (\pi \times r_1^2 \times h) \times 0.8 = 8 \times (4/3) \times \pi \times r_2^3$

$\Rightarrow (4^2 \times 22.5 \times 0.8) = 8 \times (4/3) \times r_2^3 \Rightarrow r_2 = 3$ cm.

3. (c) If the radius is diminished by r%, then

Area is diminished by $\left(2r - \frac{r^2}{100}\right)\%$

$= 2 \times 10 - \frac{10^2}{100} = 19\%$

4. (d) Increase in Area $= 10 + 20 + \frac{10 \times 20}{100} = 30 + \frac{200}{100} = 32\%$

Hence, there will be 32% change in the cost of the plot

5. (c) Circumference of the circular face of the cylinder = $2\pi r$

$\Rightarrow \quad 2 \times \frac{22}{7} \times \frac{35}{100} = 2.2$m

Number of revolutions required to lift the bucket by

11 m $= \frac{11}{2.2} = 5$

6. (b)

A B C D, O, a, a

$ABCD$ is square $a^2 = 4 \Rightarrow a = 2$

$ac = BD = 2\sqrt{2}$

perimeters of four triangles

$= AB + BC + CD + DA + 2(AC + BD)$

$= 8 + 2(2\sqrt{2} + 2\sqrt{2}) = 8(1 + \sqrt{2})$

7. (a) $\pi r(r + l) : 3\pi r^2 : 2\pi r(r + h)$

$= \pi \times 1\ (1 + \sqrt{2}) : 3 \times \pi \times 1 : 2 \times \pi \times 1\ (1 + 1)$

$= (\sqrt{2} + 1) : 3 : 4$

8. (c) Let r be the radius of each circle.

Then by given condition,

$\frac{\pi R^2}{2\pi R} = \frac{2\pi R}{\pi R^2} \Rightarrow R^2 = 4 \Rightarrow R = 2$

∴ The length of the side of the square = 8

Now the area covered by 4 coins = $4 \times \pi (2)^2 = 16\pi$
and area of the square = 64

$\therefore$ The area which is not covered by the coins

$= 64 - 16\pi = 16(4 - \pi)$

9. **(b)** $AD = 6.5$

$\therefore AB = 13$ (diameter)

Now $\angle ACB = 90°$ (since the diameter of a circle subtends 90° at the circumference)

So by pythagorus theorem, $CB = 2$ cm.

$\therefore$ area of $\Delta ACB = \frac{1}{2} \times 5 \times 2 = 30$ sq. cm

10. **(b)** Volume of the given ice cuboid = $8 \times 11 \times 2 = 176$

Let the length of the required rod is ℓ.

$\therefore \pi\ell\frac{8^2}{4} = 176 \qquad \therefore \ell = 3.5$ inches

11. **(c)** Surface area of walls = $2(lh + bh)$

$= 2[(12 \times 3) + (4 \times 3)] = 2(36 + 12) = 96\ m^2$

Area of doors = $2 \times 2.5 \times 1.5 = 7.5\ m^2$

Area of window = $2 \times 0.60 = 1.2\ m^2$ and

area of ceiling = $2 \times 4 = m^2$

Area to be coloured = area of walls + area of ceiling – area of doors – area of window

$= 96 + 48 - 7.5 - 1.2 = 135.3\ m^2$

Required cost = $135.3 \times 15 = 2029.5$.

12. **(a)** Let h be the length of water column discharged in 1 hour or 1 minute.

Volume discharged by the 4 pipe = Volumes discharged by the single pipe

$4 \times \pi \times (1.5)^2 \times h = \pi \times (r)^2 \times h$

$\therefore r^2 = 9 \quad \therefore r = 3$

Diameter = 6 inches.

13. **(b)** As per the given conditions,

$$11a^3 = 7 \times \frac{4}{3} \times \pi \times r^3 \quad \therefore \frac{a}{r} = \left(\frac{8}{3}\right)^{1/3}$$

14. **(d)** Let the edge of the cube measure x in.

Then the diameter of the sphere is x in.

Now volume of wood removed

$= \text{volume of cube} - \text{volume of sphere} = \left(x^3 - \frac{\pi x^3}{6}\right) c.$

in By hypothesis, this volume = 35280 c. in.

$\therefore x^3 - \frac{\pi x^3}{6} = 35280 \quad \Rightarrow x^3\left(1 - \frac{22}{7 \times 6}\right) = 35280$

$\Rightarrow x^3(20/42) = 35280 \quad \Rightarrow x = 42.$

15. **(c)** Area of isosceles triangle = $\frac{b}{4}\left(\sqrt{4a^2 - b^2}\right)$

where b is the base and a is any of the equal sides.

Area of the required triangle = $\frac{10}{4}\left(\sqrt{4(8)^2 - (10)^2}\right)$

$= \frac{10}{4}\sqrt{156} = 5\sqrt{39}\ cm^2$

16. **(b)** Area of the shaded portion = Area of quadrant ABC + Area of quadrant ACD – Area of square $ABCD$.

$= \frac{\pi}{2} \times 4^2 - 4^2 = \left(\frac{\pi}{2} - 1\right)4^2 = (\pi - 2)\,8 = 9.12$ sq. cm.

17. **(d)** Volume of pyramid = $\frac{1}{3} \times$ base area $\times$ height

$= \frac{1}{3} \times 6 \times 6 \times 4 = 48$ cc

$\Rightarrow$ Height of slant face $(X) = \sqrt{4^2 + 3^2} = 5$cm.

Area of each slant face = $\frac{1}{2} \times 5 \times 6 = 15$ sq. cm

Area of base = 6×6 sq. cm = 36 sq. cm.

$\Rightarrow$ Total surface area = 4 (15) + 36 sq. cm. = 96 sq. cm.

18. **(c)** $r = \frac{\text{Area of triangle}}{s} = \frac{\Delta}{s}$

$$s = \frac{a+b+c}{2} = \frac{18+24+30}{2} = 36$$

$$\Delta = \sqrt{s(s-a)(s-b)(s-c)}$$

$$\Rightarrow \Delta = \sqrt{36(36-18)(36-24)(36-30)}$$

$$\Delta = \sqrt{36 \times 18 \times 12 \times 6} = 216$$

So, radius of incircle = $\frac{216}{36} = 6$cm.

19. **(b)**

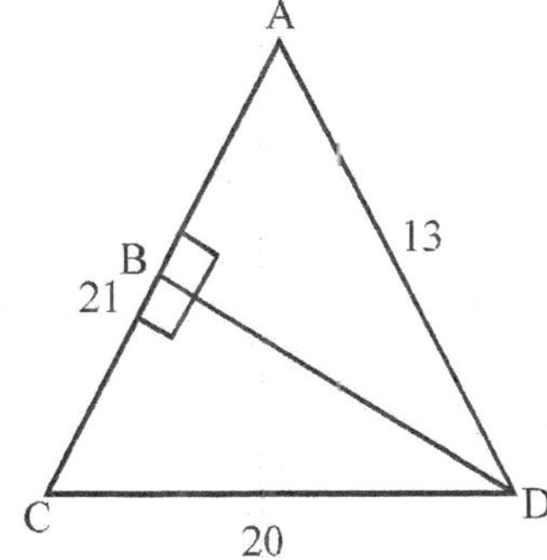

Let the original triangle be = ACD

Longest side = $AC = 21$ cm

In the right angled ΔAED, by Pythagorean triplets, we get $AB = 5$ cm and $BD = 12$ cm

Then, $BC = 21 - 5 = 16$

By Pythagoras theorem,

$BD^2 = CD^2 - BC^2 \Rightarrow BD = 12$ cm

Area of the larger $\Delta BDC = \frac{1}{2} \times 16 \times 12 = 96\ cm^2$

20. (c) Let the internal radius of the cylinder = r
Then, the volume of sphere = Volume of hollow cylinder

$\Rightarrow \dfrac{4\pi.6^3}{3} = \pi h(5^2 - r^2)$

$\Rightarrow \dfrac{864\pi}{3} = 32\pi(25 - r^2)$

$\Rightarrow r^2 = 16 = r = 4$ cm

So thickness of the cylinder = 5 – 4 = 1 cm

21. (b) Sum of interior angles of a hexagon = 720°
6 sectors with same radius r = 2 full circles of same radius.
So area of shaded region $\Rightarrow 2\pi r^2$

22. (b) Let $ABCDEFGH$ be the cube of side a and O be the centre of the hemisphere.

$AC = \sqrt{2}a$

$OD = OC = R$

Let P be the mid-point of AC

$OP = a$

Now in ΔAOC

$$R^2 = a^2 + \frac{a^2}{2} \quad \therefore \quad a = \sqrt{\frac{2}{3}}R$$

$\therefore$ Volume = $a^3 = 0.67\sqrt{\frac{2}{3}}R^3$

23. (a) Since $ABCDOA$ is a quadrant of circle of radius 10.5 cm $OA = OC = r = 10.5$ cm and

$OD = DC = \dfrac{10.5}{2} = 5.25$ cm

Area of shaded portion = (Area of the quadrant) – (Area of ΔAOD)

$$\text{Area} = \frac{\theta}{360}\pi r^2 - \frac{1}{2} \times \text{base} \times \text{height}$$

$$= \frac{90}{360} \times \frac{22}{7} \times (10.5)^2 - \frac{1}{2} \times 5.25 \times 10.25$$

= 86.625 – 27.5625 = 59.06 cm^2.

24. (a) From the fig. the shaded area
= (Area of the rectangle – 2 × quarter of circle) + area of rectangle

$$= \left[\left(3 \times 6 - 2 \times \frac{\pi}{4} \times 3^2\right) + \frac{1}{2}\pi \times 3^2\right] \text{ sq. m}$$

$$= \left[18 - \frac{9\pi}{2} + \frac{9\pi}{2}\right] = 18 \text{ sq. m}$$

$\therefore$ Cost of covering with grass = ₹ $\dfrac{18 \times 70}{100}$

= ₹ $\dfrac{630 \times 2}{100}$ = ₹ 12.60

25. (b) We know ratio of area of triangles divided by diagonals are same.

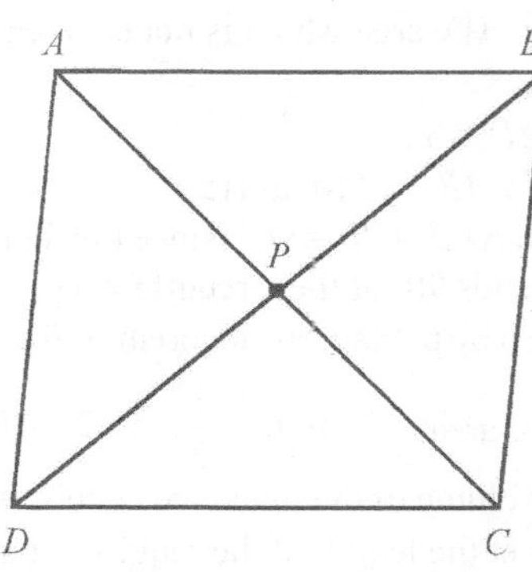

$$\frac{\text{Area of } \Delta APD}{\text{Area of } \Delta APB} = \frac{\text{Area of } \Delta DPC}{\text{Area of } \Delta CPB}$$

$\Rightarrow \dfrac{27}{x} = \dfrac{x}{12}$

$x^2 = 27 \times 12 = 3 \times 3 \times 3 \times 3 \times 2 \times 2$

$x = 18$

26. (d) Radius of cylinder, hemisphere and cone = 5 cm
Height of cylinder = 13 cm
Height of cone = 12 cm

$\ell = \sqrt{h^2 + r^2} = \sqrt{12^2 + 5^2} = 13$

Surface area of toy = $2\pi rh + \dfrac{4\pi r^2}{2} + \pi rL$

$\Rightarrow (2 \times 3.14 \times 5 \times 13) + (2 \times 3.14 \times 25) + (3.14 \times 5 \times 13)$
= 770 cm^2

27. (d) Total surface area of the cube = 6 (Side)^2 = 150
New surface area added
= [Surfaces × side of square cross section × depth]
= 4 × (2 × 5)
Total old surface area to be subtracted = 4 + 4 = 8
Hence net surface area = 150 + 40 – 8 = 182 cm^2

28. (b)

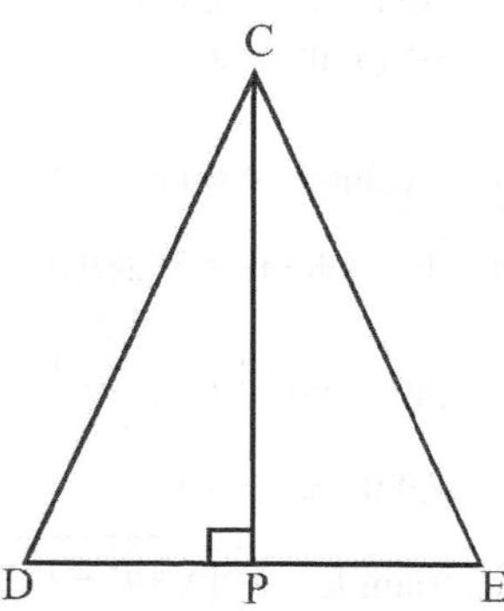

Area of equilateral triangle

$ABC = \dfrac{\sqrt{3}}{4} \times (6)^2 = 9\sqrt{3}$ cm^2

Area of $\Delta ADE = \frac{1}{2} \times DE \times CP$

$= \frac{1}{2} \times 2 \times \frac{\sqrt{3}}{2} \times 6$

$= 3\sqrt{3}$ cm^2

$\therefore$ Area of shaded region $= 9\sqrt{3} - 3\sqrt{3} = 6\sqrt{3}$ cm^2

29. (b) $\frac{\text{Area of } \Delta DAE}{\text{Area of } \Delta DEC} = \frac{\frac{1}{2} \times DE \times AE}{\frac{1}{2} \times DE \times CE}$

$= \frac{AE}{CE} = \frac{(AD)^2}{(DC)^2} = \left(\frac{6}{8}\right)^2 = \frac{9}{16}$

Similarly, in ΔABC,

$\frac{\text{Area of } \Delta BCF}{\text{Area of } \Delta BFA} = \frac{9}{16}$

$\therefore$ The area of shaded to unshaded region $= \frac{16}{9}$

30. (d) $\pi r_1^2 + \pi r_2^2 = 180\pi$ ...(1)

$\Rightarrow r_1^2 + r_2^2 = 180$ and distance between centers i.e.

$= r_1 + r_2 = 6 \Rightarrow r_2 = r_1 - 6$

From the eq. (1), $r_1^2 + (r_1 - 6)^2 = 180$

$\Rightarrow r_1^2 + (r_1^2 - 12r_1 + 36) = 180$

$\Rightarrow 2r_1^2 - 12r_1 + 36 = 180$

$\Rightarrow 2r_1^2 - 12r_1 - 144 = 0$

$\Rightarrow (r_1 - 12)(r_1 + 6) = 0$.

Hence, $r_1 = 12$ cm and $d_1 = 24$ cm.

31. (d) $PQ = QR = RS = \frac{12}{3} = 4$ cm

Area of unshaded region $\Rightarrow \frac{\pi 6^2}{2} + \frac{\pi 4^2}{2}$

$\Rightarrow 18\pi + 8\pi \Rightarrow 26\pi$

Area of shaded region $\Rightarrow \frac{\pi 6^2}{2} - \frac{\pi 4^2}{2}$

$\Rightarrow 18\pi - 8\pi = 10\pi$

Ratio $= \frac{10\pi}{26\pi} \Rightarrow \frac{5}{13} \Rightarrow 5:13$

32. (d) Area of square = 1444 sq. meters

Side of square = $\sqrt{1444} = 38$ m

Breadth of Rectangle $= \frac{1}{4} \times$ side of square

$= \frac{1}{4} \times 38 = 9.5$ m

Length of Rectangle = 3 × breadth
= 3 × 9.5 ⇒ 28.5 m
Area of Rectangle = 28.5 × 9.5 = 270.75 sq. m
Difference in area = 1444 – 270.75
= 1173.25 sq. mtr

33. (b) Circum radius (R) $= \frac{abc}{4 \times \text{Area of triangle}}$

[where a, b and c are sides of triangle]

Area of triangle $= \sqrt{s(s-a)(s-b)(s-c)}$

$\left[\therefore s = \frac{a+b+c}{2} = 24\right]$

Area of triangle $= \sqrt{24 \times 12 \times 8 \times 4} = 8 \times 3 \times 4$ cm^2

$R = \frac{12 \times 16 \times 20}{4 \times 8 \times 3 \times 4} = 10$ cm

34. (d) Area of $\Delta ODE = \frac{1}{2} OK \times DE$

$= \frac{1}{2}\left(\frac{1}{2} BC \times OK\right)$

$= \frac{1}{4}[BC \times (AO - AK)]$

$= \frac{1}{4}\left[BC \times \left(\frac{2}{3}AF - \frac{1}{2}AF\right)\right]$

$= \frac{1}{4} \times \frac{1}{3}\left[\frac{1}{2} AF \times BC\right] = \frac{1}{12}$ area of $\Delta ABC = 1 : 12$

A, D, K, E, O, B, F, C

35. (d) Parallelogram Area = l × b
Rhombus Area = l × b

Triangle Area $= \frac{l \times b}{2}$

Therefore R = P = 2T.

36. (a) Since AB is a diameter. Then ∠APB = 90° (angle in the semicircle)
ΔBPN ~ ΔAPB
So, BN = BP^2 / AB

$BN = \frac{6 \times 6}{10} = 3.6$ cm

37. (b)

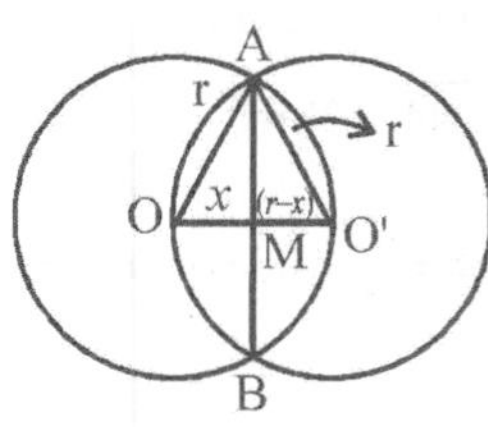

In ΔAOM

$r^2 = AM^2 + x^2$

$AM^2 = r^2 - x^2$ ...(1)

In $\Delta AMO'$

$r^2 = (r - x)^2 + AM^2$

$AM^2 = r^2 - (r - x)^2$ ...(2)

From eqs (1) & (2)

$r^2 - x^2 = r^2 - (r - x)^2$

$\Rightarrow 2rx = r^2$

$\Rightarrow x = \frac{r}{2}$

From eq. (1)

$$AM^2 = r^2 - \left(\frac{r}{2}\right)^2 = \frac{3}{4}r^2$$

$$AM = \frac{\sqrt{3}}{2}r$$

Length of chord AB = 2AM = $2 \times \frac{\sqrt{3}}{2}r = \sqrt{3}r$

38. (a)

39. (a) Curved surface area of cylinder = $2\pi rh$

$$= 2 \times \frac{22}{7} \times 21 \times 90 = 11880 \text{ sq.cm}$$

40. (c) Area of ΔABD = 16 cm^2

Area of ΔABC = 2 × Area of ΔABD [$\because$ In triangle, the midpoint of the opposite side, divides it into two congruent triangles. So their areas are equal and each is half the area of the original triangle]

$\Rightarrow$ 32 cm^2

41. (d) Area of triangle = Inradius × Semi-perimeter

= 6 × 16 = 96 sq. cm

42. (b) Diagonal of a square (d) = $\sqrt{2}$ × side of square (a).

$$d = \sqrt{2}a \Rightarrow a = \frac{d}{\sqrt{2}}$$

Area of square $\Rightarrow a^2 = \frac{d^2}{2}$

Now, diagonal gests doubled

$$a = \frac{(2d)}{\sqrt{2}}$$

Area of square $= \left(\frac{2d}{\sqrt{2}}\right)^2 = 4\left(\frac{d^2}{2}\right)$

$\frac{d^2}{2}$ is area of square

Therefore, area will be four times.

43. (c)

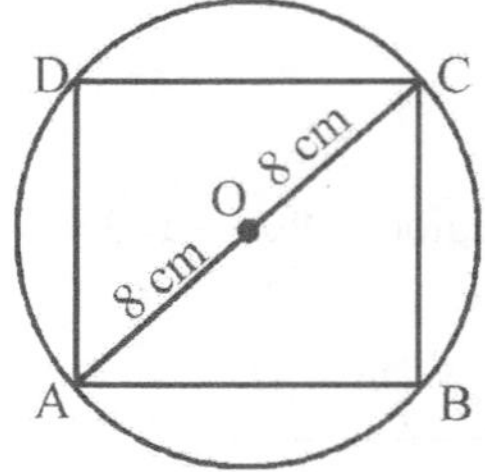

Diagonal of square = Diameter of circle

$\sqrt{2}$ × side of square = 16 cm

Squaring on both sides

$(\sqrt{2} \times \text{sides of square})^2 = 16^2$

$\Rightarrow$ (side of square)^2 = $\frac{16 \times 16}{2}$

$\Rightarrow$ Area of square = 128 sq. cm

44. (d)

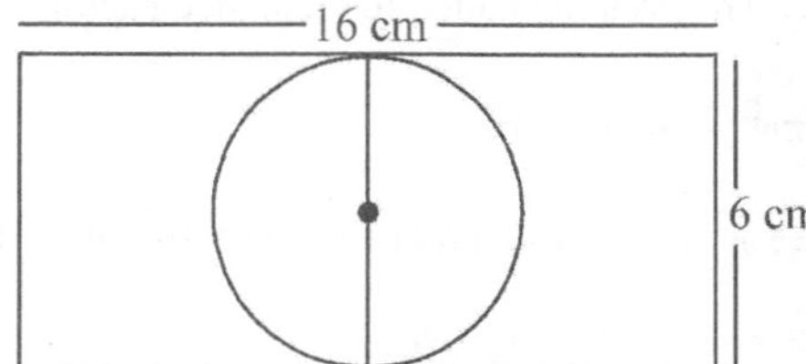

Radius of circle, r = 3 cm

Area of circle, $\pi r^2 = \pi(3)^2 = 9\pi$ cm^2

45. (c) Let ABC be the triangular base of Pyramid.

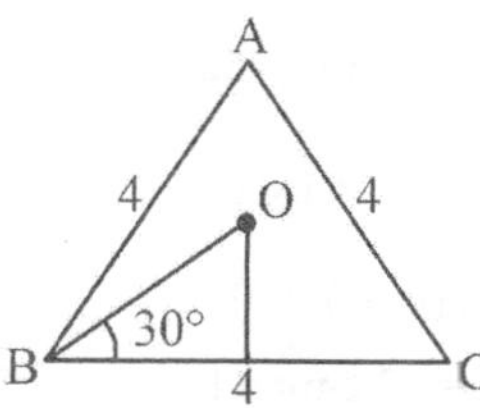

$$\cos 30° = \frac{2}{OB}$$

$$OB = \frac{2}{\cos 30°} = \frac{4}{\sqrt{3}}$$

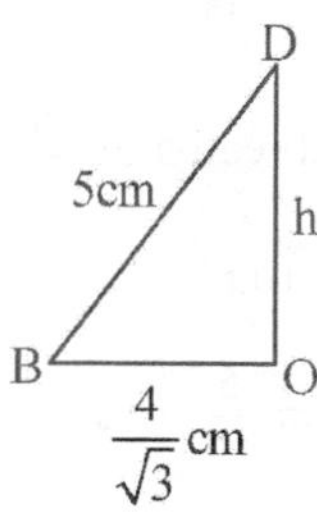

$$\text{(height of pyramid)h} = \sqrt{5^2 - \left(\frac{4}{\sqrt{3}}\right)^2} = \sqrt{\frac{59}{3}}$$

$$\text{Volume of Pyramid} = \frac{1}{3} \times \text{area of base} \times \text{height}$$

$$= \frac{1}{3} \times \frac{\sqrt{3}}{4} \times 4^2 \times \sqrt{\frac{59}{3}}$$

$$= \frac{4\sqrt{59}}{3} \text{cm}^3$$

46. (a) $C_1 = 2C_2$

$\pi r_1 l_1 = 2\pi r_2 l_2$

also, $l_2 = 2l_1$

$\pi r_1 l_1 = 2 \times 2_6 \pi r_2 l_1$

$\frac{r_1}{r_2} = \frac{4}{1}$

47. (d)

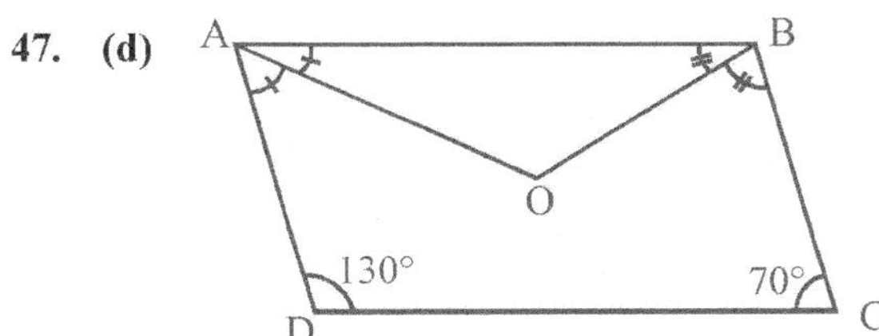

$A + B + C + D = 360$

$A + B = 360 - (130 + 70) = 160°$

$\frac{A}{2} + \frac{B}{2} = 80°$ ...(1)

In Δ AOB,

$\frac{A}{2} + \frac{B}{2} + \angle AOB = 180°$

$\angle AOB = 180° - 80° = 100°$

48. (d)

49. (c)

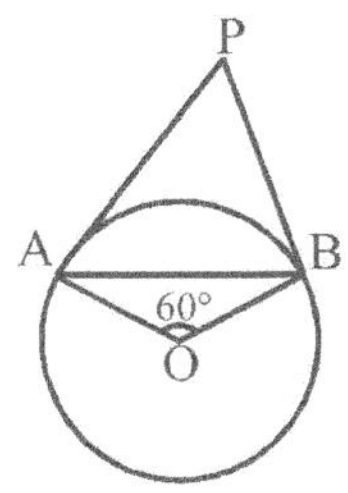

In Δ AOB, $\angle A + \angle B + \angle O = 180°$

$\angle A + \angle B = 180 - 140° = 40°$

$\angle A = \angle B = 20°$ {AO = BO}

$\angle PAO = 90°$

$\angle PAB + \angle BAO = 90°$

$\angle PAB = 90° - 20° = 70°$

50. (c)

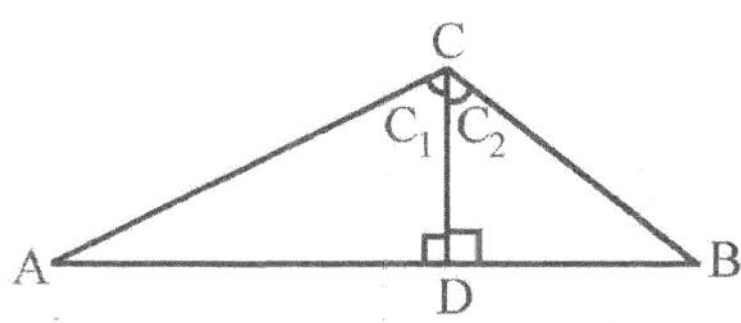

In Δ ADC,

$A + D + C_1 = 180°$; $A + C_1 = 180° - 90° = 90°$

In Δ BDC,

$B + D + C_2 = 180°$; $B + C_2 = 180° - 90° = 90°$

$A + C_1 = B + C_2$

$C_1 - C_2 = B - A$

51. (b)

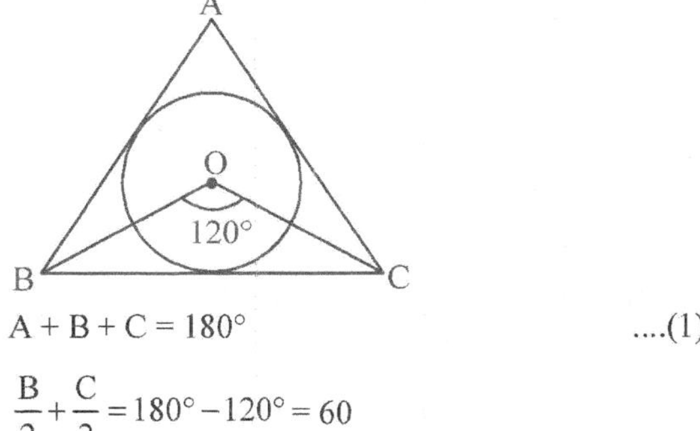

$A + B + C = 180°$ ....(1)

$\frac{B}{2} + \frac{C}{2} = 180° - 120° = 60$

$B + C = 120°$ ...(2)

Using (1) and (2)

$A + 120° = 180°$

$A = 60°$

52. (b)

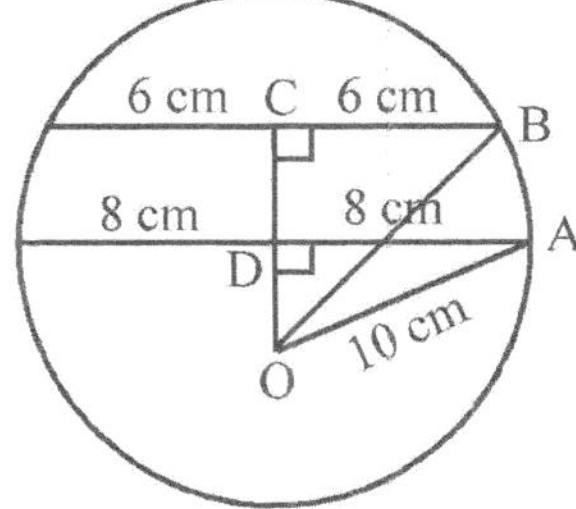

In ΔADO,

$OD = \sqrt{(AO)^2 - AD^2}$

$= \sqrt{100\text{cm}^2 - 64\text{cm}^2} = 6$ cm

In ΔBCO,

$OC = \sqrt{OB^2 - CB^2}$

$= \sqrt{100\text{cm}^2 - 36\text{cm}^2} = 8$ cm

distance between chords = OC – OD = 2 cm

53. (a) Let n be the number of sides.

$(n - 2) \times 180° = 140° \times n$

$180n - 360 = 140\,n$

$40n = 360$

$n = \frac{360}{40} = 9$

54. (d)

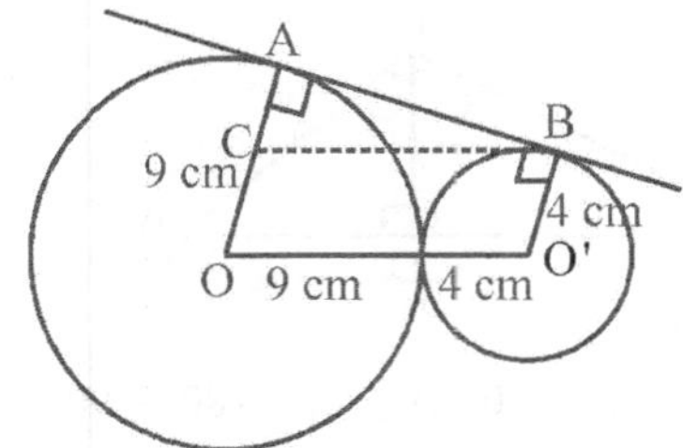

In figure, AC = AO – CO

= 9 cm – 4 cm = 5 cm {CO = BO'}

Also, CB = OO′ = 13 cm

In Δ ABC

$AB = \sqrt{CB^2 - AC^2}$

$= \sqrt{(13cm)^2 - (5cm)^2}$

= 12 cm

55. (a) Let r be the radius of circle.

$\pi r^2 = 154\ cm^2$

$r^2 = \frac{154}{22} \times 7 = 49$

r = 7 cm

length of wire = circumference of circle

$= 2 \times \frac{22}{7} \times 7 = 44$ cm

Now, Perimeter of equilateral triangle = 44 cm

side $= \frac{44}{3}$ cm

Area of equilateral triangle $= \frac{\sqrt{3}}{4} \times \left(\frac{44}{3}\right)^2$

$= \frac{484\sqrt{3}}{9} = 93.14\ cm^2$

Area of equilateral triangle is nearly equal to $93.14\ cm^2$

Hence, option (a) is correct.

56. (c) $A = l \times b$

$A' = \left(l + \frac{15}{100}l\right)\left(b - \frac{15}{100}b\right) = 1.15\,l \times 0.85\,b$

$A' = 0.9775\ A$

$\%\ change = \frac{A - 0.9775A}{A} \times 100 = 2.25\%$

Area decreased as $A' < A$

CHAPTER

20

COORDINATE GEOMETRY

RECTANGULAR COORDINATE AXES

Let XOX' be a horizontal straight line and YOY' be a vertical straight line drawn through a point O in the plane of the paper. Then

the line XOX' is called x-axis

the line YOY' is called y-axis

plane of paper is called xy-plane or cartesian plane.

x-axis and y-axis together are called co-ordinate axes or axis of reference.

The point O is called the origin.

Cartesian Coordinates

Position of any point in a cartesian plane can be described by their cartesian coordinates. The ordered pair of perpendicular distances first from y-axis and second from x-axis of a point P is called cartesian coordinates of P.

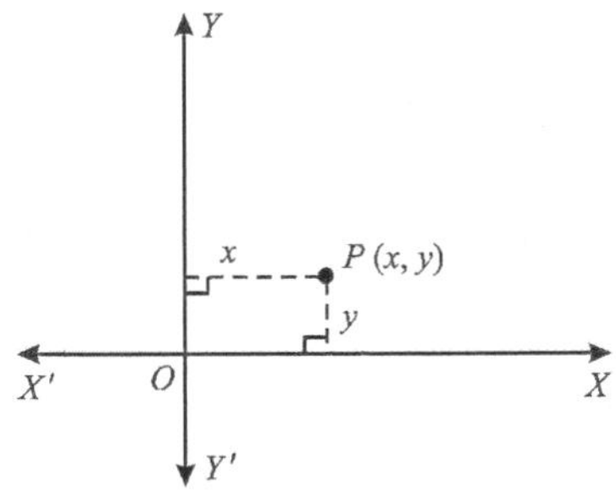

If the cartesian coordinates of point P are (x, y), then x is called abscissa or x-coordinate of P and y is called the ordinate or y-coordinate of point P.

SIGN CONVENTIONS IN THE *xy*-PLANE

(i) All the distances are measured from origin (o).

(ii) All the distances measured along or parallel to x-axis but right side of origin are taken as $+ve$.

(iii) All the distances measured along or parallel to x-axis but left side of origin are taken as $-ve$.

(iv) All the distances measured along or parallel to y-axis but above the origin are taken as $+ve$.

(v) All the distances measure along or parallel to y-axis but below the origin are taken as $-ve$.

According to the Above Sign Conventions

(i) Coordinate of origin is $(0, 0)$

(ii) Coordinate of any point on the x-axis but right side of origin is of the form $(x, 0)$, where $x > 0$.

(iii) Coordinate of any point on the x-axis but left side of origin is of the form $(-x, 0)$, where $x > 0$.

(iv) Coordinate of any point on the y-axis but above the origin is of the form $(0, y)$, where $y > 0$.

(v) Coordinate of any point on the y-axis but below the origin is of the form $(0, -y)$, where $y > 0$.

QUADRANTS OF xy-PLANE AND SIGN OF x AND y-COORDINATE OF A POINT IN DIFFERENT QUADRANTS

x and y-axis divide the xy-plane in four parts. Each part is called a quadrant.

The four quadrants are written as I-quadrant (XOY), II-quadrant (YOX'), III-quadrant ($X'OY'$) and IV-quadrant ($Y'OX$). Each of these quadrants shows the specific quadrant of the xy-plane as shown below:

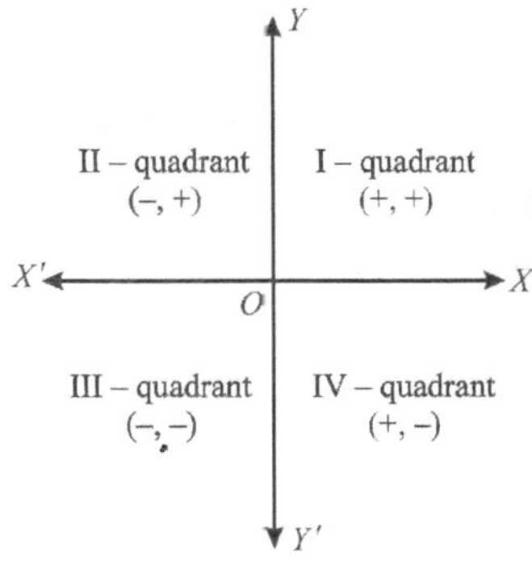

(i) Any of the four quadrants does not includes any part of x or y-axis.

(ii) In the first quadrant both x and y-coordinates of any point are $+ve$.

(iii) In second quadrant x-coordinate of any point is $-ve$ but y-coordinate of any point is $+ve$.

(iv) In third quadrant, both x and y-coordinates of any point are $-ve$.

(v) In fourth quadrant, x-coordinate of any point is $+ve$ but y-coordinate of any point is $-ve$ as shown in the above diagram.

PLOTTING A POINT WHOSE COORDINATES ARE KNOWN

The point can be plotted by measuring its proper distances from both the axes. Thus, any point P whose coordinates are (h, k) can be plotted as follows:

(i) Measure OM equal to h (i.e. x-coordinate of point P) along the x-axis.

(ii) Now perpendicular to OM equal to k.

Mark point P above M such that PM is parallel to y-axis and $PM = k$ (i.e. y-coordinate of point P)

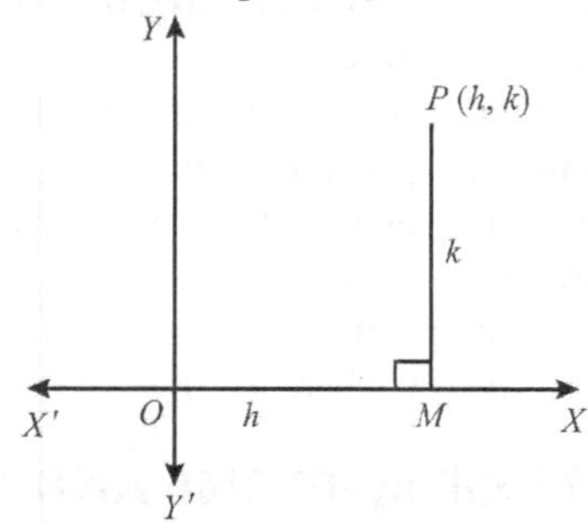

In this chapter, now we shall study to find the distance between two given points, section formula, mid-point formula, slope of a line, angles between two straight lines and equation of a line in different forms etc.

DISTANCE FORMULA

The distance between two points $P(x_1, y_1)$ and $Q(x_2, y_2)$ is given by

$$PQ = \left|\sqrt{(x_1 - x_2)^2 + (y_1 - y_2)^2}\right| \text{ or } \left|\sqrt{(x_2 - x_1)^2 + (y_2 - y_1)^2}\right|$$

Distance of point $P(x, y)$ from the origin $= \left|\sqrt{x^2 + y^2}\right|$

Illustration 1: If distance between the point $(x, 2)$ and $(3, 4)$ is 2, then find the value of x.

Solution:

$$2 = \left|\sqrt{(x-3)^2 + (2-4)^2}\right| \Rightarrow 2 = \left|\sqrt{(x-3)^2 + 4}\right|$$

Squaring both sides

$$4 = (x-3)^2 + 4 \Rightarrow x - 3 = 0 \Rightarrow x = 3$$

Illustration 2: Find the distance between each of the following points :

$A(-6, -1)$ and $B(-6, 11)$

Solution: Here the points are $A(-6, -1)$ and $B(-6, 11)$

By using distance formula, we have

$$AB = \sqrt{\{-6-(-6)\}^2 + \{11-(-1)\}^2} = \sqrt{0^2 + 12^2} = 12$$

Hence, $AB = 12$ units.

SECTION FORMULA

Co-ordinates of a point which divides the line segment joining two points $P(x_1, y_1)$ and $Q(x^2, y^2)$ in the ratio $m_1 : m_2$ are :

(i) $\left(\dfrac{m_1x_2 + m_2x_1}{m_1 + m_2}, \dfrac{m_1y_2 + m_2y_1}{m_1 + m_2}\right)$, for internal division.

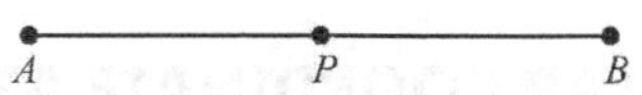

P divides AB internally in the ratio $m : n$

If $m_1 = m_2$, then the point P will be the mid point of PQ whose co-ordinates $= \left(\dfrac{x_1 + x_2}{2}, \dfrac{y_1 + y_2}{2}\right)$

A P B

P is the mid-point of AB

(ii) $\left(\dfrac{m_1x_2 - m_2x_1}{m_1 - m_2}, \dfrac{m_1y_2 - m_2y_1}{m_1 - m_2}\right)$, for external division

n
A B P
m

P divides AB externally in the ratio $m : n$

(iii) When we need to find the ratio in which a point on a line segment divides it, we suppose the required ratio as $k : 1$ or $m/n : 1$.

Note:

(i) Co-ordinates of any point on the line segment joining two points $P(x_1, y_1)$ and $Q(x_2, y_2)$ are

$$\left(\frac{x_1 + \lambda x_2}{1 + \lambda}, \frac{y_1 + \lambda y_2}{1 + \lambda}\right), (\lambda \neq -1)$$

(ii) **Division by axes:** Line segment joining the points (x_1, y_1) and (x^2, y^2) is divided by

(a) x-axis in the ratio $-y_1 / y_2$

(b) y-axis in the ratio $-x / x_2$

If ratio is positive division internally and if ratio is negative division is externally.

(iii) **Division by a line:** Line $ax + by + c = 0$ divides the line joining the points (x_1, y_1) and (x_2, y_2) in the ratio $\left(-\dfrac{ax_1 + by_1 + c}{ax_2 + by_2 + c}\right)$.

Illustration 3: Find the ratio in which the line $3x + 4y = 7$ divides the line segment joining the points $(1, 2)$ and $(-2, 1)$.

Solution: Ratio $= -\dfrac{3(1) + 4(2) - 7}{3(-2) + 4(1) - 7} = -\dfrac{4}{-9} = \dfrac{4}{9} = 4:9$.

Illustration 4: Find the points of trisection of line joining the points A (2, 1) and B (5, 3).

Solution: (2, 1) ←1→←2→ (5, 3)

A P_1 P_2 B

←2→←1→

$$P_1(x, y) = \left(\frac{1\times5+2\times2}{1+2}, \frac{1\times3+2\times1}{1+2}\right) = \left(3, \frac{5}{3}\right)$$

$$P_2(x, y) = \left(\frac{2\times5+1\times2}{2+1}, \frac{2\times3+1\times1}{2+1}\right) = \left(4, \frac{7}{3}\right).$$

Illustration 5: Prove that points A (1, 1), B (– 2, 7) and C (3, – 3) are collinear.

Solution: $AB = \left|\sqrt{(1+2)^2+(1-7)^2}\right| = \left|\sqrt{9+36}\right| = 3\sqrt{5}$

$BC = \left|\sqrt{(-2-3)^2+(7+3)^2}\right| = \left|\sqrt{25+100}\right| = 5\sqrt{5}$

$CA = \left|\sqrt{(3-1)^2+(-3-1)^2}\right| = \left|\sqrt{4+16}\right| = 2\sqrt{5}$

Clearly, $BC = AB + AC$. Hence A, B, C are collinear.

Illustration 6: Find the ratio in which the join of (– 4, 3) and (5, –2) is divided by (i) x-axis (ii) y-axis.

Solution:

(i) x-axis divides the join of (x_1, y_1) and (x_2, y_2) in the ratio of $-y_1 : y_2 = -3 : -2 = 3 : 2$.

(ii) y-axis divides, in the ratio of $-x_1 : x_2 \Rightarrow 4 : 5$.

COORDINATES OF SOME PARTICULAR POINTS

Let A (x_1, y_1), B (x_2, y_2) and C (x_3, y_3) are vertices of any triangle ABC, then

Centroid

Centroid is the point of intersection of the medians of a triangle. Centroid divides each median in the ratio of 2 : 1.

A median is a line segment joining the mid point of a side to its opposite vertex of a triangle.

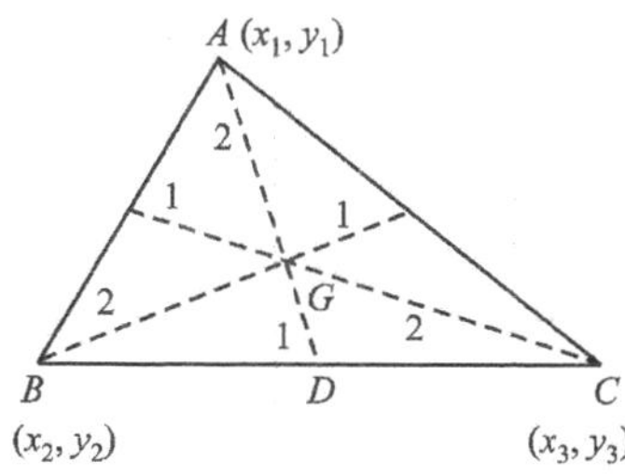

Co-ordinates of centroid, $G = \left(\frac{x_1+x_2+x_3}{3}, \frac{y_1+y_2+y_3}{3}\right)$

Incentre

Incentre is the point of intersection of internal bisectors of the angles of a triangle. Also incentre is the centre of the circle touching all the sides of a triangle.

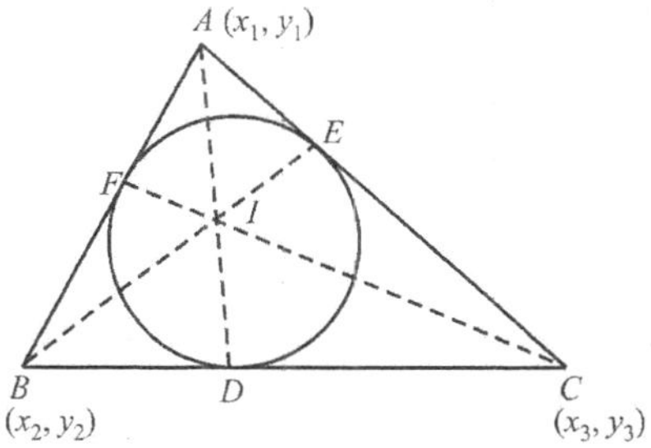

Co-ordinates of incentre,

$$I = \left(\frac{ax_1+bx_2+cx_3}{a+b+c}, \frac{ay_1+by_2+cy_3}{a+b+c}\right),$$

where a, b, c are length of the sides opposite to vertices A, B, C respectively of triangle ABC.

(i) Angle bisector divides the opposite sides in the ratio of the sides included in the angle. For example $\frac{BD}{DC} = \frac{AB}{AC} = \frac{c}{b}$.

(ii) Incentre divides the angle bisectors AD, BE and CF in the ratio $(b + c) : a$, $(c + a) : b$ and $(a + b) : c$ respectively.

Practice Exercise

Level - I

1. If distance between the point $(x, 2)$ and $(3, 4)$ is 2, then the value of $x =$
(a) 0 (b) 2
(c) 3 (d) 4

2. Find the mid-point of the line-segment joining two points $(3, 4)$ and $(5, 12)$.
(a) $(-4, 8)$ (b) $(0, 8)$
(c) $(4, 8)$ (d) $(4, 0)$

3. The mid-point of the line segment joining the points $(-2, 4)$ and $(6, 10)$ is
(a) $(2, 5)$ (b) $(2, 7)$
(c) $(3, 7)$ (d) $(3, 8)$

4. The points $A(-4, -1)$, $B(-2, -4)$, $C(4, 0)$ and $D(2, 3)$ are the vertices of a
(a) Parallelogram (b) Rectangle
(c) Rhombus (d) Square

5. The line $x + y = 4$ divides the line joining the points $(-1, 1)$ and $(5, 7)$ in the ratio
(a) 2 : 1 (b) 1 : 2
(c) 1 : 2 externally (d) None of these

6. If A (3, 5), B (–3, – 4), C (7, 10) are the vertices of a parallelogram taken in the order, then the co-ordinates of the fourth vertex are
(a) $(10, 19)$ (b) $(15, 10)$
(c) $(19, 10)$ (d) $(15, 19)$

7. The centroid of a triangle, whose vertices are $(2, 1)$, $(5, 2)$ and $(3, 4)$ is
(a) $\left(\frac{8}{3}, \frac{7}{3}\right)$ (b) $\left(\frac{10}{3}, \frac{7}{3}\right)$
(c) $\left(-\frac{10}{3}, \frac{7}{3}\right)$ (d) $\left(\frac{10}{3}, -\frac{7}{3}\right)$

8. The incentre of the triangle with vertices $(1, \sqrt{3})$, $(0, 0)$ and $(2, 0)$ is
(a) $\left(1, \frac{\sqrt{3}}{2}\right)$ (b) $\left(\frac{2}{3}, \frac{1}{\sqrt{3}}\right)$
(c) $\left(\frac{2}{3}, \frac{\sqrt{3}}{2}\right)$ (d) $\left(1, \frac{1}{\sqrt{3}}\right)$

9. The centroid of the triangle whose vertices are $(3, 10)$, $(7, 7)$, $(-2, 1)$ is
(a) $(8/3, 6)$ (b) $(6, 8/3)$
(c) $(-4, -7/3)$ (d) None of these

10. The coordinates of the centroid G of a triangle with vertices at $(3, 7)$, $(5, 5)$ and $(-3, 2)$ is
(a) $(10/3, 14/3)$ (b) $(10/3, 10/3)$
(c) $(5/3\ 14/3)$ (d) $(11/3, 10/3)$

11. The coordinates of a point which divides the join of $(5, -5)$ and $(2, -3)$ in the ratio 4 : 3, externally, are:
(a) $(3, 4)$ (b) $(-7, 3)$
(c) $(-7, 9)$ (d) $(8, 3)$

12. Distance between $P(x, y)$ and $Q(3, -6)$ is 10 units and x is positive integer, then $x =$
(a) 3 (b) 9
(c) 7 (d) 11

13. The vertices of a parallelogram in order are $A(1, 2)$, $B(4, y)$, $C(x, 6)$, $D(3, 6)$, then $(x, y) =$
(a) $(6, 3)$ (b) $(3, 6)$
(c) $(5, 6)$ (d) $(1, 4)$

14. The point which divides the line segment joining the points $(7, -6)$ and $(3, 4)$ in ratio 1 : 2 internally lies in the
(a) I quadrant (b) II quadrant
(c) III quadrant (d) IV quadrant

15. How many squares are possible if two of the vertices of a quadrilateral are $(1, 0)$ and $(2, 0)$?
(a) 1 (b) 2
(c) 3 (d) 4

16. In what ratio is the line segment made by the points $(7, 3)$ and $(-4, 5)$ divided by the y-axis?
(a) 2 : 3 (b) 4 : 7
(c) 3 : 5 (d) 7 : 4

17. If the coordinates of the mid-point of the line segment joining the points $(2, 1)$ and $(1, -3)$ is (x, y), then the relation between x and y can be best described by
(a) $3x + 2y = 5$ (b) $6x + y = 8$
(c) $5x - 2y = 4$ (d) $2x - 5y = 4$

18. Points $(4, -1)$, $(6, 0)$, $(7, 2)$ and $(5, 1)$ are joined to be a vertex of a quadrilateral. What will be the structure?
(a) Rhombus (b) Parallelogram
(c) Square (d) Rectangle

19. Find the third vertex of the triangle whose two vertices are $(-3, 1)$ and $(0, -2)$ and the centroid is the origin.
(a) $(2, 3)$ (b) $\left(\frac{-4}{3}, \frac{14}{3}\right)$
(c) $(3, 1)$ (d) $(6, 4)$

20. If the origin gets shifted to $(2, 2)$, then what will be the new coordinates of the point $(4, -2)$?
(a) $(-2, 4)$ (b) $(2, 4)$
(c) $(4, 2)$ (d) $(2, -4)$

21. If the point $R(1, -2)$ divides externally the line segment joining $P(2, 5)$ and Q in the ratio 3 : 4, what will be the coordinates of Q ?
(a) $(-3, 6)$ (b) $(2, -4)$
(c) $(3, 6)$ (d) $(1, 2)$

22. C is the mid-point of PQ, if P is $(4, x)$, C is $(y, -1)$ and Q is $(-2, 4)$, then x and y respectively are
(a) -6 and 1 (b) -6 and 2
(c) 6 and -1 (d) 6 and -2

23. A quadrilateral has the vertices at the points $(-4, 2)$, $(2, 6)$, $(8, 5)$ and $(9, -7)$. Show that the mid-points of the sides of this quadrilateral are the vertices of a parallelogram.
(a) Rectangle (b) Square
(c) Parallelogram (d) Rhombus

24. Find the ratio in which the point $(2, y)$ divides the join of $(-4, 3)$ and $(6, 3)$ and hence find the value of y
(a) $2:3, y=3$ (b) $3:2, y=4$
(c) $3:2, y=3$ (d) $3:2, y=2$

25. If $P\left(\frac{a}{3}, 4\right)$ is the mid-point of the line segment joining the points $Q(-6, 5)$ and $R(-2, 3)$, then the value of a is
(a) -4 (b) -12
(c) 12 (d) -6

26. The ratio in which the line $2x + y - 4 = 0$ divides the line segment joining the points $A(2, -2)$ and $B(3, 7)$ is
(a) 3:7 (b) 4:7
(c) 2:9 (d) 4:9

27. Which of the following points is the nearest to the origin?
(a) $(0, -6)$ (b) $(-8, 0)$
(c) $(-3, -4)$ (d) $(7, 0)$

28. If the points $(1, 1)$, $(-1, -1)$ and $(-\sqrt{3}, k)$ are vertices of a equilateral triangle then the value of k will be :
(a) 1 (b) -1
(c) $\sqrt{3}$ (d) $-\sqrt{3}$

29. The points $(3,0)$, $(-3, 0)$, $(0, -3\sqrt{3})$ are the vertices of
(a) equilateral triangle (b) isosceles triangle
(c) right triangle (d) scalene triangle

30. Ratio in which the line $3x + 4y = 7$ divides the line segment joining the points $(1, 2)$ and $(-2, 1)$ is
(a) 3:5 (b) 4:6
(c) 4:9 (d) None of these

31. If the area of a triangle with vertices $(-3, 0)$, $(3, 0)$ and $(0, k)$ is 9 sq unit, then what is the value of k?
(a) 3 (b) 6
(c) 9 (d) 12

32. The line $y = 0$ divides the line joining the points $(3, -5)$ and $(-4, 7)$ in the ratio
(a) 3:4 (b) 4:5
(c) 5:7 (d) 7:9

33. The line passing through the points $(-2, 8)$ and $(5, 7)$ *[SSC-Sub. Ins.-2012]*
(a) does not cut any axes (b) cuts x-axis only
(c) cuts y-axis only (d) cuts both the axes

Level - II

1. The fourth vertex of a rectangle whose other vertices are $(4, 1)$ $(7, 4)$ and $(13, -2)$ is
(a) $(10, -5)$ (b) $(10, 5)$
(c) $(-10, 5)$ (d) $(-10, -5)$

2. The coordinates of vertices A and B of an equilateral triangle ABC are $(-4, 0)$ and $(4, 0)$ respectively. Which of the following could be coordinates of C
(a) $(0, 2\sqrt{3})$ (b) $(0, 4)$
(c) $(0, 4\sqrt{3})$ (d) $(0, 3)$

3. The three vertices of a parallelogram are $A(3, -4)$, $B(-2, 1)$ and $C(-6, 5)$. Which of the following cannot be the fourth one
(a) $(-1, 0)$ (b) $(7, -8)$
(c) $(1, -5)$ (d) All of these

4. The mid-points of sides of a triangle are $(2, 1)$, $(-1, -3)$ and $(4, 5)$. Then the coordinates of its vertices are:
(a) $(7,9), (-3, -7), (1, 1)$ (b) $(-3, -7), (1, 1), (2, 3)$
(c) $(1, 1), (2, 3), (-5, 8)$ (d) None of these

5. The point whose abscissa is equal to its ordinate and which is equidistant from the points $(1, 0)$ and $(0, 3)$ is
(a) $(1, 1)$ (b) 2,2)
(c) $(3, 3)$ (d) $(4, 4)$

6. If the point dividing internally the line segment joining the points (a, b) and $(5, 7)$ in the ratio 2 : 1 be $(4, 6)$, then
(a) $a=1, b=2$ (b) $a=2, b=-4$
(c) $a=2, b=4$ (d) $a=-2, b=4$

7. The distance of point of intersection of $2X - 3Y + 13 = 0$ and $3X + 7Y - 15 = 0$ from $(4, -5)$, will be
(a) 10 units (b) 12 units
(c) 11 units (d) None of these

8. $A(-2, 4)$ and $B(-5, -3)$ are two points. The coordinates of a point P on Y axis such that $PA = PB$, are
(a) $(3, 4)$ (b) $(0, 9)$
(c) $(9, 0)$ (d) $(0, -1)$

9. The centroid of a triangle formed by $(7, p)$, $(q, -6)$, $(9, 10)$ is $(6, 3)$. Then $p + q$
(a) 6 (b) 5
(c) 7 (d) 8

10. If the three vertices of a rectangle taken in order are the points (2, –2), (8, 4) and (5, 7). The coordinates of the fourth vertex is

(a) (1, 1) (b) (1, –1)
(c) (–1, 1) (d) None of these

11. If $P(1, 2)$, $Q(4, 6)$, $R(5, 7)$ and $S(a, b)$ are the vertices of a parallelogram $PQRS$, then

(a) $a = 2, b = 4$ (b) $a = 3, b = 4$
(c) $a = 2, b = 3$ (d) $a = 3, b = 5$

12. Find the coordinates of the points that trisect the line segment joining (1, –2) and (–3, 4)

(a) $\left(\frac{-1}{3}, 0\right)$ (b) $\left(\frac{-5}{3}, 2\right)$
(c) Both (a) and (b) (d) None of these

13. If the mid-point of the line joining (3, 4) and $(p, 7)$ is (x, y) and $2x + 2y + 1 = 0$, then what will be the value of p?

(a) 15 (b) $\frac{-17}{2}$
(c) –15 (d) $\frac{17}{2}$

14. Two vertices of a triangle are (5, –1) and (–2, 3). If the orthocentre of the triangle is the origin, what will be the coordinates of the third point?

(a) (4, 7) (b) (–4, 7)
(c) (–4, –7) (d) (4, –7)

15. *A* point *P* is equidistant from *A* (3, 1) and *B* (5, 3) and its abscissa is twice its ordinate, then its co-ordinates are.

(a) (2, 1) (b) (1, 2)
(c) (4, 2) (d) (2, 4)

16. If (–1, –1) and (3, –1) are two opposite corners of a square, the other two corners are

(a) (2, 0), (–2, 2) (b) (2, –2), (0, 2)
(c) (3, 0), (4, –2) (d) None of these

17. What is the perimeter of the triangle with vertices $A(-4, 2)$, $B(0, -1)$ and $C(3, 3)$?

(a) $7 + 3\sqrt{2}$ (b) $10 + 5\sqrt{2}$
(c) $11 + 6\sqrt{2}$ (d) $5 + \sqrt{2}$

18. The area (in sq. unit) of the triangle formed by the three graphs of the equations $x = 4$, $y = 3$, and $3x + 4y = 12$, is *[SSC CGL-2012]*

(a) 12 (b) 10
(c) 6 (d) 8

19. The radius of the circumcircle of the triangle made by x-axis, y-axis and $4x + 3y = 12$ is *[SSC CGL-2012]*

(a) 2 unit (b) 2.5 unit
(c) 3 unit (d) 4 unit

20. The total area (in sq. unit) of the triangles formed by the graph of 4x + 5y = 40, x-axis, y-axis and x = 5 and y = 4 is *[SSC CGL-2014]*

(a) 10 (b) 20
(c) 30 (d) 40

Hints & Solutions

Level-I

1. (c) $2=\sqrt{(x-3)^2+(2-4)^2} \Rightarrow 2=\sqrt{(x-3)^2+4}$
Squaring both sides
$4=(x-3)^2+4 \Rightarrow x-3=0 \Rightarrow x=3$

2. (c) Let $A(3,4)$ and $B(5,12)$ be the given points.
Let $C(x,y)$ be the mid-point of AB. Using mid-point formula, we have, $x=\frac{3+5}{2}=4$ and $y=\frac{4+12}{2}=8$
$\therefore$ $C(4, 8)$ are the co-ordinates of the mid-point of the line segment joining two points (3, 4) and (5, 12).

3. (b) 4. (b)

5. (b) Ratio $=-\left(\frac{-1+1-4}{5+7-4}\right)=\frac{1}{2}$

6. (d) Mid point of A (3, 5) and C (7, 10) $= M\left(5,\frac{15}{2}\right)$
$\therefore$ Mid points of $BD = M\left(5,\frac{15}{2}\right)$
$B(-5,-4)$ and $D(x,y)$
$\therefore \frac{-5+x}{2}=5, \; x=10+5=15$
$\frac{-4+y}{2}=\frac{15}{2}, \; y=15+4=19$
Co-ordinates of fourth vertex $D=(15,19)$

7. (b) $x=\frac{2+5+3}{3}=\frac{10}{3}$ and $y=\frac{1+2+4}{3}=\frac{7}{3}$

8. (d) Clearly, the triangle is equilateral.

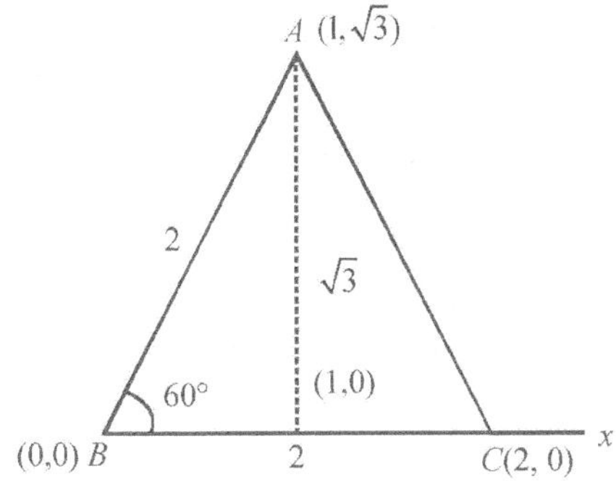

So, the incentre is the same as the centroid.
$\therefore$ Incentre $=\left(\frac{1+0+2}{3},\frac{\sqrt{3}+0+0}{3}\right)=\left(1,\frac{1}{\sqrt{3}}\right)$

9. (a) Centroid $=\left(\frac{3+7-2}{3},\frac{10+7+1}{3}\right)=\left(\frac{8}{3},6\right)$

10. (c) Let G be (X, Y), then $X=\{3+5+(-3)\}/3=5/3$
$Y=(7+5+2)/3=14/3 \Rightarrow G$ is $(5/3, 14/3)$

11. (b) Let the ratio be 4 : 3 or 4/3 : 1.

Now $X=\frac{\frac{4}{3}\times 2-5}{\frac{4}{3}-1}=\frac{\frac{8}{3}-5}{\frac{1}{3}}=\frac{-\frac{7}{3}}{\frac{1}{3}}=-7$

$Y=\frac{\frac{4}{3}x-3+5}{\frac{4}{3}-1}=\frac{\frac{1}{1}}{\frac{1}{3}}=3$. Hence $(-7,3)$

12. (b)

13. (a) Mid-point of AC is $\left(\frac{1+x}{2},\frac{2+6}{2}\right)$ *i.e.,* $\left(\frac{1+x}{2},4\right)$;
Mid-point of BD is $\left(\frac{4-3}{2},\frac{y+5}{2}\right)$
Since for a || gm, diagonals bisect each other
$\therefore \frac{1+x}{2}=\frac{7}{2}$ and $\frac{y+5}{2}=4 \Rightarrow x=6,\; y=3$

14. (d) 15. (c) 16. (d) 17. (b) 18. (a)
19. (c) 20. (d) 21. (c) 22. (a) 23. (c)

24. (c) Let the required ratio be $k:1$
Then, $2=\frac{6k-4\times 1}{k+1} \Rightarrow k=\frac{3}{2}$
$\therefore$ The required ratio is $\frac{3}{2}::1 \Rightarrow 3:2$
Also, $y=\frac{3\times 3+2\times 3}{3+2}=3$

25. (d) 26. (d) 27. (c)

28. (c) The equilateral Δ has its sides equal.
Hence the distance between the vertices should be equal.
$a=\sqrt{2^2+2^2}=\sqrt{(\sqrt{3}+1)^2+k(k-1)^2} \Rightarrow k=\sqrt{3}$

29. (a) Find the three lengths separately
$AB=6,\; BC=\sqrt{3^2+(3\sqrt{3})^2}=6,$
$AC=\sqrt{3^2+(3\sqrt{3})^2}=6$
Hence, the point are the vertices of equilateral triangle.

30. (c) $-\frac{3(1)+4(2)-7}{3(-2)+4(1)-7}=-\frac{4}{-9}=\frac{4}{9}$

31. (a) Let the vertices of the ΔABC be
$A(-3,0)$, $B(3,0)$ and $C(0,k)$.
Given, area is 9
$\Rightarrow 9=\frac{1}{2}\{-3(-k)+1(3k)\}$
$\Rightarrow 18=3k+3k$
$\Rightarrow k=\frac{18}{6}=3$

32. (c) Let $P(x,y)$ be the point of division that divides the line joining $(3,-5)$ and $(-4,7)$ in the ratio of $k:1$
Now, $y=\frac{7k-5}{k+1}$ .... (i)
Since, P lies on $y=0$ or x-axis then, from eq. (i)
$0=\frac{7k-5}{k+1} \Rightarrow 7k=5 \Rightarrow k=\frac{5}{7}$

33. (c)

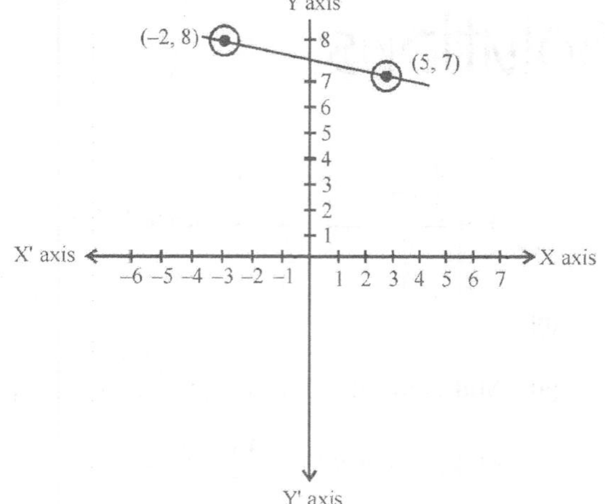

As indicated in the graph, the line passing through the points cuts Y-axis only.

Level-II

1. (a) 2. (c) 3. (d)

4. (a) $\frac{X_1+X_2}{2}=2$, $\frac{X_2+X_3}{2}=-1$, $\frac{X_3+X_1}{2}=4$

$\Rightarrow X_1=7, X_2=-3, X_3=1$

Similarly, y_1, y_2, y_3 can be found

5. (b) Let the point be (X, X), so according to the condition

$(X-1)^2+(X-0)^2=(X-0)^2+(X-3)^2$

$\Rightarrow 2X+1=-6X+9 \Rightarrow X=2$

Hence the point is (2, 2)

6. (c) $\frac{2\times 5+1(a)}{2+1}=4 \Rightarrow a=2$

and $\frac{2\times 7+1(b)}{2+1}=6 \Rightarrow b=4$

7. (b) The point of intersection will be obtained by simultaneously solving the two equations and then by the distance formula, distance can be found.

8. (d) Take points P one by one and see which one (0, – 1) satisfies.

9. (c) By the given condition $\frac{7+q+9}{3}=6$

and $\frac{p-6+10}{3}=3$

$\Rightarrow q=2$ and $p=5$ $\therefore p+q=5+2=7$

10. (c) Let fourth vertex be (x, y), then $\frac{x+8}{2}=\frac{2+5}{2}$

and $\frac{y+4}{2}=\frac{-2+7}{2} \Rightarrow x=-1, y=1$

11. (c) Diagonals cut each other at middle points.

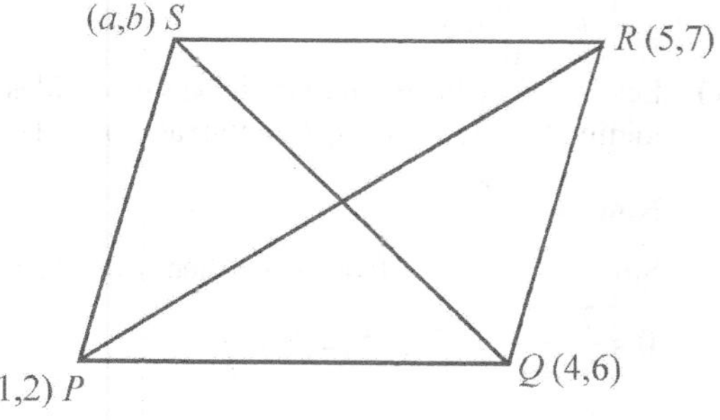

Hence, $\frac{a+4}{2}=\frac{1+5}{2} \Rightarrow a=2$

$\frac{b+6}{2}=\frac{2+7}{2} \Rightarrow b=3$

12. (c) 13. (c) 14. (c)

15. (c) Let the point be $P(2X, X)$. The choices we are left with are (1, 2) and (2, 4).

$AP=\sqrt{(3-2X)^2+(1-X)^2}$,

$BP=\sqrt{(5-2X)^2+(3-X)^2}$

$AP=BP$. (only (4, 2) satisfies)

16. (d) We have the mid-point of diagonal = (1, – 1) which should be the mid point of the other two points as well and which is not satisfied by any given alternative.

17. (b) By using distance formula,

We have,

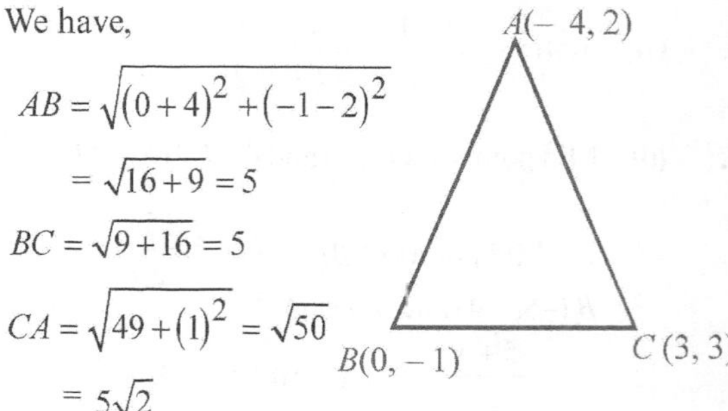

$AB=\sqrt{(0+4)^2+(-1-2)^2}$

$=\sqrt{16+9}=5$

$BC=\sqrt{9+16}=5$

$CA=\sqrt{49+(1)^2}=\sqrt{50}$

$=5\sqrt{2}$

Hence, required perimeter $=AB+BC+CA$

$=10+5\sqrt{2}$

18. (c) $x=4$ ...(1)

$y=3$ ...(2)

$3x+4y=12$ ...(3)

Putting $x=0$ in 3rd equation we get $y=3$

Putting $y=0$ in 3rd equation we get $x=4$

The triangle will be formed by joining the points (3, 0) and (0, 4).

So, base = 3 and altitude = 4

Area $=\frac{1}{2}\times b\times h \Rightarrow \frac{1}{2}\times 3\times 4=6$

19. (b) Putting $x=0$ in $4x+3y=12$ we get $y=4$

Putting $y=0$ in $4x+3y=12$ we get $x=3$

The triangle so formed is right angle triangle with points (0, 0) (4, 0) (0, 3)

So diameter is the hypotenus of triangle $=\sqrt{16+9}$ = 5 unit

radius = 2.5 unit

20. (a)

Area of ΔABC

$=\frac{1}{2}\times BC\times AC$

$=\frac{1}{2}\times(10-5)\times 4$

$=\frac{1}{2}\times 5\times 4$

Area = 10 sq unit.

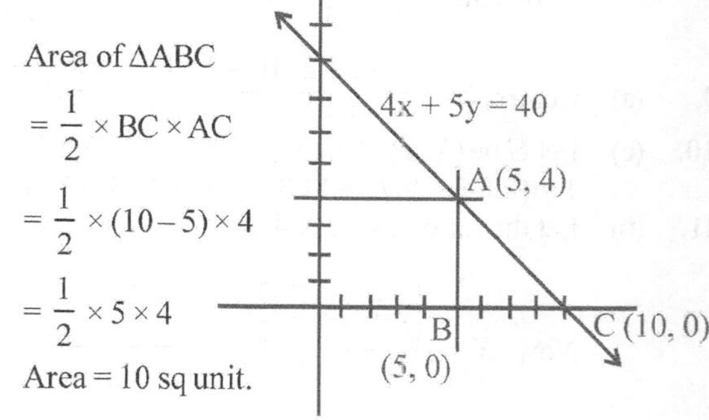

CHAPTER 21

PERMUTATIONS AND COMBINATIONS

FUNDAMENTAL PRINCIPLE OF COUNTING

Multiplication Principle

If an operation can be performed in 'm' different ways; followed by a second operation performed in 'n' different ways, then the two operations in succession can be performed in $m \times n$ ways. This can be extended to any finite number of operations.

Illustration 1: A person wants to go from station P to station R via station Q. There are 4 routes from P to Q and 5 routes from Q to R. In how many ways can he travel from P to R ?

Solution: He can go from P to Q in 4 ways and Q to R in 5 ways.

So number of ways of travel from P to R is $4 \times 5 = 20$.

Illustration 2: A college offers 6 courses in the morning and 4 in the evening. Find the possible number of choices with the student if he wants to study one course in the morning and one in the evening.

Solution: The college has 6 courses in the morning out of which the student can select one course in 6 ways.

In the evening the college has 4 courses out of which the student can select one in 4 ways.

Hence the required number of ways = $6 \times 4 = 24$.

Illustration 3: In how many ways can 5 prizes be distributed among 4 boys when every boy can take one or more prizes ?

Solution: First prize may be given to any one of the 4 boys, hence first prize can be distributed in 4 ways.

Similarly every one of second, third, fourth and fifth prizes can also be given in 4 ways.

$\therefore$ The number of ways of their distribution

$= 4 \times 4 \times 4 \times 4 \times 4 = 4^5 = 1024$

Addition Principle

If an operation can be performed in 'm' different ways and another operation, which is independent of the first operation, can be performed in 'n' different ways. Then either of the two operations can be performed in $(m + n)$ ways. This can be extended to any finite number of independent operations.

Illustration 4: A college offers 6 courses in the morning and 4 in the evening. Find the number of ways a student can select exactly one course, either in the morning or in the evening.

Solution: The college has 6 courses in the morning out of which the student can select one course in 6 ways.

In the evening the college has 4 courses out of which the student can select one in 4 ways.

Hence the required number of ways = $6 + 4 = 10$.

Illustration 5: A person wants to leave station Q. There are 4 routes from station Q to P and 5 routes from Q to R. In how many ways can he travel from the station Q ?

Solution: He can go from Q to P in 4 ways and Q to R in 5 ways. To go from Q to P and Q to R are independent to each other. Hence the person can leave station Q in $4 + 5 = 9$ ways.

FACTORIALS

If n is a natural number then the product of all natural numbers upto n is called factorial n and it is denoted by n ! or $\underline{|n}$

Thus, $n! = n(n-1)(n-2)\ .....\ 3.2.1$

Note that $0! = 1 = 1!$

$$n! = n(n-1)!$$
$$= n(n-1)(n-2)!$$
$$= n(n-1)(n-2)(n-3)!, \text{ etc.}$$

For example $6! = 6 \times 5 \times 4 \times 3 \times 2 \times 1$

But $4! = 4 \times 3 \times 2 \times 1$

$\therefore$ $6! = 6 \times 5 \times 4!$ or $6 \times 5 \times 4 \times 3!$

Remember that

$0! = 1, 1! = 1, 2! = 2, 3! = 6, 4! = 24, 5! = 120, 6! = 720$, etc.

MEANING OF PERMUTATION AND COMBINATION

Each of the different arrangements which can be made by taking some or all of a number of things is called a permutation. Note that in an arrangement, the order in which the things arranged is considerable i.e., arrangement AB and BA of two letters A and B are different because in AB, A is at the first place and B is at the second place from left whereas in BA, B is at the first place and A is at the second place.

The all different arrangements of three letters A, B and C are ABC, ACB, BCA, BAC, CAB and CBA.

Here each of the different arrangements ABC, ACB, BCA, BAC, CAB and CBA is a permutation and number of different arrangement i.e. 6 is the number of permutations.

ABC, *ACB*, *BCA*, *BAC*, *CAB* and *CBA* are different arrangements of three letters *A*, *B* and *C*, because in each arrangement, order in which the letters arranged, is considered. But if the order in which the things are arranged is not considered; then *ABC*, *ACB*, *BCA*, *BAC*, *CAB* and *CBA* are not different but the same. Similarly *AB* and *BA* are not different but the same.

Each of the different selections or groups which can be made by some or all of a number of given things without reference to the order of things in any selection or group is called a combination.

As in selection order in which things are selected is not considered; hence, selections of two letters *AB* and *BA* out of three letters *A*, *B* and *C* are the same. Similarly selections of *BC* and *CB* are the same.

Also selections of *CA* and *AC* are the same.

Hence selection of two letters out of the three letters *A*, *B* and *C* can be made as *AB*, *BC* and *CA* only.

As in arrangements, order in which things are arranged is considered. Hence all arrangements of two letters out of the three letters *A*, *B* and *C* are *AB*, *BA*, *BC*, *CB*, *CA* and *AB*.

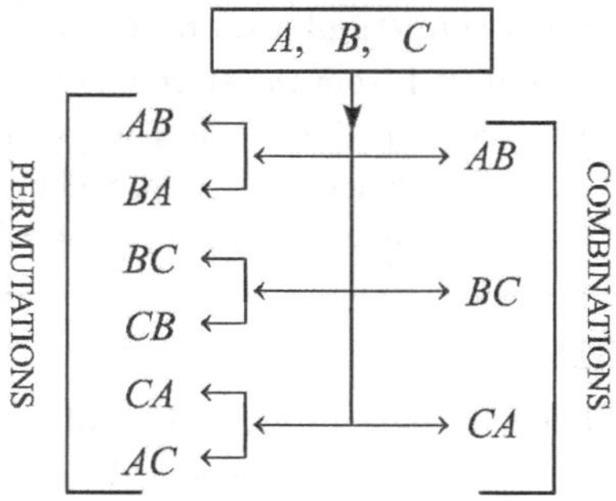

Number of permutations (or arrangements) of two letters out of three letters *A*, *B* and $C = 6$.

Number of combinations (or groups) of two letters out of three letters *A*, *B* and $C = 3$.

Permutations of three different letters *A*, *B* and *C* taken two at a time is also understood as selections of any two different letters *AB*, *BC* or *CA* out of *A*, *B* and *C*, then the selected two letters arranged in two ways as

AB, BA ; BC, CB or CA, AC

Hence using multiplication principle, number of permutations of three different letters *A*, *B* and *C* taken two at a time

= (Number of ways to select any two different letters out of the three given letters) × (Number of arrangements of two selected letters)

$= 3 \times 2 = 6$

Thus permutations means selection of some or all of the given things at a time and then arrangements of selected things. In most of the problems, it is mentioned that the problem is of permutation or combination but in some problems it is not mentioned. In the case where it is not mentioned that problem given is of permutation or combination, you can easily identify the given problem is of permutation or combination using the following classifications of problems:

Problems of Permutations

(i) Problems based on arrangements
(ii) Problems based on standing in a line
(iii) Problems based on seated in a row
(iv) Problems based on digits
(v) Problems based on arrangement letters of a word
(vi) Problems based on rank of a word (in a dictionary)

Problems of Combinations

(i) Problems based on selections or choose
(ii) Problems based on groups or committee
(iii) Problems based on geometry

If in any problem, it is neither mentioned that the problem is of permutation or combination nor does the problem fall in the categories mentioned above for the problems of permutations or problems of combinations, then do you think whether arrangement (i.e. order) is meaningful or not? If arrangement (i.e., order) is considerable in the given problem, then the problem is of permutation otherwise it is of combination. This will be more clear through the following illustrations:

Suppose you have to select three batsmen out of four batsmen B_1, B_2, B_3 and B_4, you can select three batsmen $B_1 B_2 B_3$, $B_2 B_3 B_4$, $B_3 B_4 B_1$ or $B_4 B_1 B_2$.

Here order of selections of three batsmen in any group of three batsmen is not considerable because it does not make any difference in the match.

Hence in the selection process; $B_2 B_3 B_4$, $B_2 B_4 B_3$, $B_3 B_2 B_4$, $B_3 B_4 B_2$, $B_4 B_2 B_3$ and $B_4 B_3 B_2$ all are the same.

But for batting, the order of batting is important.

Therefore for batting; $B_2 B_3 B_4$, $B_2 B_4 B_3$, $B_3 B_2 B_4$, $B_3 B_4 B_2$, $B_4 B_2 B_3$ and $B_4 B_3 B_2$, are different because $B_2 B_3 B_4$ means batsman B_2 batting first then batsman B_3 and then batsman B_4 whereas $B_2 B_4 B_3$ means batsman B_2 batting first then batsman B_4 and then batsman B_3.

COUNTING FORMULA FOR LINEAR PERMUTATIONS

Without Repetition

1. Number of permutations of *n* different things, taking *r* at a time is denoted by ^nP_r or $P(n, r)$, which is given by

$$^nP_r = \frac{n!}{(n-r)!} \ (0 \le r \le n)$$

$$= n(n-1)(n-2) \ldots\ldots (n-r+1),$$

where *n* is a natural number and *r* is a whole number.

2. Number of arrangements of *n* different objects taken all at a time is $^nP_n = n!$

Note:

$$^nP_1 = n, \quad ^nP_r = n.\,^{n-1}P_{r-1}, \quad ^nP_r = (n-r+1).\,^nP_{r-1},$$

$$^nP_n = {^nP_{n-1}}$$

Illustration 6: Find the number of ways in which four persons can sit on six chairs.

Solution: $^6P_4 = 6.5.4.3 = 360$

With Repetition

1. Number of permutations of n things taken all at a time, if out of n things p are alike of one kind, q are alike of second kind, r are alike of a third kind and the rest $n-(p+q+r)$ are all different is
$$\frac{n!}{p!q!r!}$$
2. Number of permutations of n different things taken r at a time when each thing may be repeated any number of times is n^r.

Illustration 7: Find the number of words that can be formed out of the letters of the word COMMITTEE taken all at a time.

Solution: There are 9 letters in the given word in which two T's, two M's and two E's are identical. Hence the required number of words $= \frac{9!}{2!2!2!} = \frac{9!}{(2!)^3} = \frac{9!}{8} = 45360$

NUMBER OF LINEAR PERMUTATIONS UNDER CERTAIN CONDITIONS

1. Number of permutations of n different things taken all together when r particular things are to be placed at some r given places $= {}^{n-r}P_{n-r} = (n-r)!$
2. Number of permutations of n different things taken r at a time when m particular things are to be placed at m given places $= {}^{n-m}P_{r-m}$.
3. Number of permutations of n different things, taken r at a time, when a particular thing is to be always included in each arrangement, is $r.\,{}^{n-1}P_{r-1}$.
4. Number of permutation of n different things, taken r at a time, when m particular thing is never taken in each arrangement is ${}^{n-m}P_r$.
5. Number of permutations of n different things, taken all at a time, when m specified things always come together is
$$m! \times (n-m+1)!$$
6. Number of permutations of n different things, taken all at a time, when m specific things never come together is
$$n! - m! \times (n-m+1)!$$

Illustration 8: How many different words can be formed with the letters of the word 'JAIPUR' which start with '*A*' and end with '*I*'?

Solution: After putting A and I at their respective places (only in one way) we shall arrange the remaining 4 different letters at 4 places in 4! ways. Hence the required number $= 1 \times 4! = 24$.

Illustration 9: How many different 3 letter words can be formed with the letters of word 'JAIPUR' when *A* and *I* are always to be excluded?

Solution: After leaving A and I, we are remained with 4 different letters which are to be used for forming 3 letters words. Hence the required number $= {}^4P_3 = 4 \times 3 \times 2 = 24$.

CIRCULAR PERMUTATIONS

1. Arrangement Around a Circular Table

In circular arrangements, there is no concept of starting point (i.e. starting point is not defined). Hence number of circular permutations of n different things taken all at a time is $(n-1)!$ if clockwise and anti-clockwise order are taken as different.

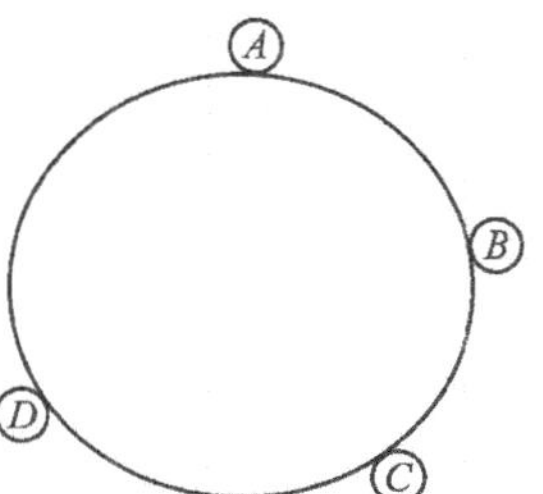

In the case of four persons A, B, C and D sitting around a circular table, then the two arrangements $ABCD$ (in clockwise direction) and $ADCB$ (the same order but in anti-clockwise direction) are different.

Hence the number of arrangements (or ways) in which four different persons can sit around a circular table $= (4-1)! = 3! = 6$.

2. Arrangement of Beads or Flowers (All Different) Around a Circular Necklace or Garland

The number of circular permutations of n different things taken all at a time is $\frac{(n-1)!}{2}$, if clockwise and anti-clockwise order are taken as the same.

If we consider the circular arrangement, if necklace made of four precious stones A, B, C and D; the two arrangements $ABCD$ (in clockwise direction) and $ADCB$ (the same but in anti-clockwise direction) are the same because when we take one arrangement $ABCD$ (in clockwise direction) and then turn the necklace around (front to back), then we get the arrangement $ADCB$ (the same but in anti-clockwise direction). Hence the two arrangements will be considered as one arrangement because the order of the stones is not changing with the change in the side of observation. So in this case, there is no difference between the clockwise and anti-clockwise arrangements.

Therefore number of arrangements of four different stones in the necklace $= \frac{(n-1)!}{2}$.

3. Number of Circular Permutations of *n* Different Things Taken *r* at a Time

Case I: If clockwise and anti-clockwise orders are taken as different, then the required number of circular permutations
$$= \frac{{}^nP_r}{r}.$$

Case II: If clockwise and anti-clockwise orders are taken as same, then the required number of circular permutations
$$= \frac{{}^nP_r}{2r}.$$

4. Restricted Circular Permutations

When there is a restriction in a circular permutation then first of all we shall perform the restricted part of the operation and then perform the remaining part treating it similar to a linear permutation.

Illustration 10: In how many ways can 5 boys and 5 girls be seated at a round table so that no two girls may be together ?

Solution: Leaving one seat vacant between two boys, 5 boys may be seated in 4! ways. Then at remaining 5 seats, 5 girls can sit in 5! ways. Hence the required number = 4! × 5!

Illustration 11: In how many ways can 4 beads out of 6 different beads be strung into a ring ?

Solution: In this case a clockwise and corresponding anticlockwise order will give the same circular permutation. So the required

$$\text{number} = \frac{{}^6P_4}{4.2} = \frac{6.5.4.3}{4.2} = 45.$$

Illustration 12: Find the number of ways in which 10 persons can sit round a circular table so that none of them has the same neighbours in any two arrangements.

Solution: 10 persons can sit round a circular table in 9! ways. But here clockwise and anti-clockwise orders will give the same neighbours. Hence the required number of ways = $\frac{1}{2}9!$.

COUNTING FORMULA FOR COMBINATION

1. Selection of Objects Without Repetition

The number of combinations or selections of n different things taken r at a time is denoted by ${}^nC_r$ or $C\ (n, r)$ or $C\binom{n}{r}$

$$\text{where } {}^nC_r = \frac{n!}{r!(n-r)!};\ (0 \le r \le n)$$

$$= \frac{n(n-1)(n-2)...(n-r+1)}{r(r-1)(r-2)....2.1};$$

where n is a natural number and r is a whole number.

Some Important Results

(i) ${}^nC_n = 1, {}^nC_0 = 1$ (ii) ${}^nC_r = \frac{{}^nP_r}{r!}$

(iii) ${}^nC_r = {}^nC_{n-r}$ (iv) ${}^nC_x = {}^nC_y \Rightarrow x + y = n$

(v) ${}^nC_r + {}^nC_{r-1} = {}^{n+1}C_r$ (vi) ${}^nC_r = \frac{n}{r}.{}^{n-1}C_{r-1}$

(vii) ${}^nC_1 = {}^nC_{n-1} = n$

Illustration 13: If ${}^{20}C_r = {}^{20}C_{r-10}$, then find the value of ${}^{18}C_r$

Solution: ${}^{20}C_r = {}^{20}C_{r-10} \Rightarrow r + (r-10) = 20 \Rightarrow r = 15$

$$\therefore\ {}^{18}C_r = {}^{18}C_{15} = {}^{18}C_3 = \frac{18.17.16}{1.2.3} = 816$$

Illustration 14: How many different 4-letter words can be formed with the letters of the word 'JAIPUR' when A and I are always to be included ?

Solution: Since A and I are always to be included, so first we select 2 letters from the remaining 4, which can be done in ${}^4C_2 = 6$ ways. Now these 4 letters can be arranged in 4! = 24 ways, so the required number = 6 × 24 = 144.

Illustration 15: How many combinations of 4 letters can be made of the letters of the word 'JAIPUR' ?

Solution: Here 4 things are to be selected out of 6 different things.

$$\text{So the number of combinations} = {}^6C_4 = \frac{6.5.4.3}{4.3.2.1} = 15$$

2. Selection of Objects With Repetition

The total number of selections of r things from n different things when each thing may be repeated any number of times is ${}^{n+r-1}C_r$

3. Restricted Selection

(i) Number of combinations of n different things taken r at a time when k particular things always occur is ${}^{n-k}C_{r-k}$.

(ii) Number of combinations of n different things taken r at a time when k particular things never occur is ${}^{n-k}C_r$.

4. Selection From Distinct Objects

Number of ways of selecting at least one thing from n different things is

${}^nC_1 + {}^nC_2 + {}^nC_3 + ...+ {}^nC_n = 2^n - 1.$

This can also be stated as the total number of combination of n different things is $2^n - 1$.

Illustration 16: Ramesh has 6 friends. In how many ways can he invite one or more of them at a dinner ?

Solution: He can invite one, two, three, four, five or six friends at the dinner. So total number of ways of his invitation

$$= {}^6C_1 + {}^6C_2 + {}^6C_4 + {}^6C_5 + {}^6C_6 = 2^6 - 1 = 63$$

5. Selection From Identical Objects

(i) The number of combination of n identical things taking r ($r \le n$) at a time is 1.

(ii) The number of ways of selecting any number r ($0 \le r \le n$) of things out of n identical things is $n + 1$.

(iii) The number of ways to select one or more things out of $(p + q + r)$ things; where p are alike of first kind, q are alike of second kind and r are alike of third kind $= (p + 1)(q + 1)(r + 1) - 1$.

Illustration 17: There are n different books and p copies of each in a library. Find the number of ways in which one or more than one books can be selected.

Solution: Required number of ways

$$= (p + 1)(p + 1)......n \text{ terms} - 1 = (p + 1)^n - 1$$

Illustration 18: A bag contains 3 one ₹ coins, 4 five ₹ coins and 5 ten ₹ coins. How many selection of coins can be formed by taking atleast one coin from the bag?

Solution: There are 3 things of first kind, 4 things of second kind and 5 things of third kind, so the total number of selections $= (3 + 1)(4 + 1)(5 + 1) - 1 = 119$

DIVISION AND DISTRIBUTION OF OBJECTS

1. The number of ways in which $(m + n)$ different things can be divided into two groups which contain m and n things respectively is

$${}^{m+n}C_m\,{}^nC_n = \frac{(m+n)!}{m!n!},\ m \ne n$$

Particular case:

When $m = n$, then total number of ways is

$\dfrac{(2m)!}{(m!)^2}$, when order of groups is considered and

$\dfrac{(2m)!}{2!(m!)^2}$, when order of groups is not considered.

2. The number of ways in which $(m + n + p)$ different things can be divided into three groups which contain m, n and p things respectively is

$$^{m+n+p}C_m \cdot {}^{n+p}C_p \cdot {}^pC_p = \frac{(m+n+p)!}{m!n!p!}, \; m \neq n \neq p$$

Particular case:

When $m = n = p$, then total number of ways is

$\dfrac{(3m)!}{(m!)^3}$, when order of groups is considered and

$\dfrac{(3m)!}{3!(m!)^3}$, when order of groups is not considered.

3. (i) Total number of ways to divide n identical things among r person is $^{n+r-1}C_{r-1}$

(ii) Also total number of ways to divide n identical things among r persons so that each gets atleast one is $^{n-1}C_{r-1}$.

Illustration 19: In how many ways 20 identical mangoes may be divided among 4 persons if each person is to be given at least one mango?

Solution: If each person is to be given at least one mango, then number of ways will be $^{20-1}C_{4-1} = {}^{19}C_3 = 969$.

Illustration 20: In how many ways can a pack of 52 cards be divided in 4 sets, three of them having 17 cards each and fourth just one card?

Solution: Since the cards are to be divided into 4 sets, 3 of them having 17 cards each and 4th just one card, so number of ways $= \dfrac{52!}{1!51!} \cdot \dfrac{51!}{(17!)^3\, 3!} = \dfrac{52!}{(17!)^3\, 3!}$.

IMPORTANT RESULTS ABOUT POINTS

1. If there are n points in a plane of which m ($< n$) are collinear, then

(i) Total number of different straight lines obtained by joining these n points is $^nC_2 - {}^mC_2 + 1$.

(ii) Total number of different triangles formed by joining these n points is $^nC_3 - {}^mC_3$

2. Number of diagonals of a polygon of n sides is $^nC_2 - n$ *i.e.*, $\dfrac{n(n-3)}{2}$.

3. If m parallel lines in a plane are intersected by a family of other n parallel lines, then total number of parallelograms so formed is $^mC_2 \times {}^nC_2$ *i.e.*, $\dfrac{mn(m-1)(n-1)}{4}$.

4. Given n points on the circumference of a circle, then

(i) Number of straight lines obtained by joining these n points $= {}^nC_2$

(ii) Number of triangles obtained by joining these n points $= {}^nC_3$

(iii) Number of quadrilaterals obtained by joining these n points $= {}^nC_4$

Illustration 21: There are 10 points in a plane and 4 of them are collinear. Find the number of straight lines joining any two of them.

Solution: Total number of lines $= {}^{10}C_2 - {}^4C_2 + 1 = 40$.

Illustration 22: If 5 parallel straight lines are intersected by 4 parallel straight lines, then find the number of parallelograms thus formed.

Solution:

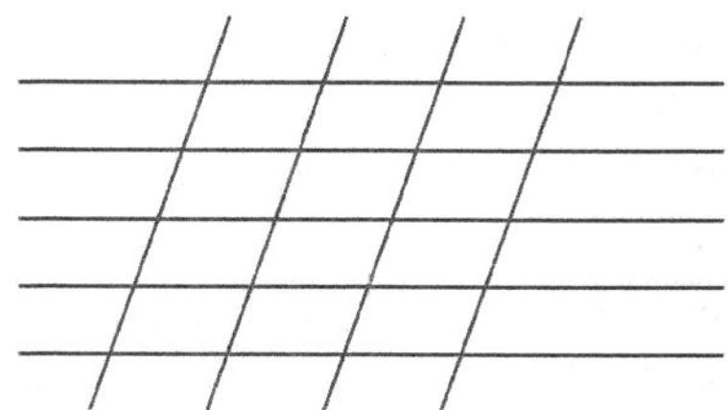

Number of parallelograms $= {}^5C_2 \times {}^4C_2 = 60$.

FINDING THE RANK OF A WORD

We can find the rank of a word out of all the words with or without meaning formed by arranging all the letters of a given word in all possible ways when these words are listed as in a dictionary. You can easily understand the method to find the above mentioned rank by the following illustrations.

Illustration 23: If the letters of the word RACHIT are arranged in all possible ways and these words (with or without meaning) are written as in a dictionary, then find the rank of this word RACHIT.

Solution: The order of the alphabet of RACHIT is A, C, H, I, R, T.

The number of words beginning with A (i.e. the number of words in which A comes at first place) is $^5P_5 = 5!$.

Similarly, number of words beginning with C is $5!$, beginning with H is $5!$ and beginning with I is also $5!$.

So before R, four letters A, C, H, I can occur in $4 \times (5!) = 480$ ways.

Now the word RACHIT happens to be the first word beginning with R. Therefore the rank of this word RACHIT $= 480 + 1 = 481$.

Illustration 24: The letters of the word MODESTY are written in all possible orders and these words (with or without meaning) are listed as in a dictionary then find the rank of the word MODESTY.

Solution:

The order of the alphabet of MODESTY is D, E, M, O, S, T, Y.

Number of words beginning with D is $^6P_6 = 6!$

Number of words beginning with E is $^6P_6 = 6!$

Number of words beginning with MD is $^5P_5 = 5!$

Number of words beginning with ME is $^5P_5 = 5!$

Now the first word start with MO is MODESTY.

Hence rank of the word MODESTY

$= 6! + 6! + 5! + 5! + 5! + 1$

$= 720 + 720 + 120 + 120 + 1$

$= 1681$.

Practice Exercise

Level - I

1. The sum of all the four digit even numbers which can be formed by using the digits 0, 1, 2, 3, 4 and 5 if repetition of digits is allowed is
(a) 1765980 (b) 1756980
(c) 1769580 (d) 1759680

2. How many words beginning with vowels can be formed with the letters of the word EQUATION?
(a) 25200 (b) 15200
(c) 25300 (d) 35200

3. The number of words that can be formed out of the letters of the word COMMITTEE is
(a) $\frac{9!}{(2!)^3}$ (b) $\frac{9!}{(2!)^2}$
(c) $\frac{9!}{2!}$ (d) 9!

4. If $^{10}P_r = 720$, then r is equal to
(a) 4 (b) 2
(c) 3 (d) 1

5. Number of ways in which 12 different balls can be divided into groups of 5, 4 and 3 balls are
(a) $\frac{12!}{5!4!}$ (b) $\frac{12!}{5!4!3!}$
(c) $\frac{12!}{5!4!3!3!}$ (c) None of these

6. How many different letter arrangements can be made from the letter of the word EXTRA in such a way that the vowels are always together?
(a) 48 (b) 60
(c) 40 (d) 30

7. In how many ways can a committee of 5 made out 6 men and 4 women containing atleast one woman?
(a) 246 (b) 222
(c) 186 (c) None of these

8. How many integers greater than 5000 can be formed with the digit 7, 6, 5, 4 and 3, using each digit at most once?
(a) 72 (b) 144
(c) 84 (d) 192

9. Every body in a room shakes hands with every else. If total number of hand-shaken is 66, then number of persons in the room is
(a) 11 (b) 12
(c) 13 (d) 14

10. The number of words from the letters of the words BHARAT in which B and H will never come together, is
(a) 360 (b) 240
(c) 120 (d) None of these

11. A bag contains 3 black, 4 white and 2 red balls, all the balls being different. The number of at most 6 balls containing balls of all the colours is
(a) 42(4!) (b) $2^6 \times 4!$
(c) $(2^6 - 1)(4!)$ (d) None of these

12. How many different ways are possible to arrange the letters of the word "MACHINE" so that the vowels may occupy only the odd positions?
(a) 800 (b) 125
(c) 348 (d) 576

13. If $^nP_r = {}^nP_{r+1}$ and $^nC_r = {}^nC_{r-1}$, then the values of n and r are
(a) 4, 3 (b) 3, 2
(c) 4, 2 (d) None of these

14. If $^nP_r = 720\ {}^nC_r$, then r is equal to
(a) 3 (b) 7
(c) 6 (d) 4

15. In how many ways a hockey team of eleven can be elected from 16 players?
(a) 4368 (b) 4267
(c) 5368 (d) 4166

16. In how many ways can twelve girls be arranged in a row if two particular girls must occupy the end places?
(a) $\frac{10!}{2!}$ (b) 12!
(c) $10! \times 2!$ (d) $\frac{12!}{2!}$

17. To fill a number of vacancies, an employer must hire 3 programmers from among 6 applicants, and 2 managers from among 4 applicants. What is the total number of ways in which she can make her selection?
(a) 1,490 (b) 132
(c) 120 (d) 60

18. A father has 2 apples and 3 pears. Each weekday (Monday through Friday) he gives one of the fruits to his daughter. In how many ways can this be done?
(a) 120 (b) 10
(c) 24 (d) 12

19. If a secretary and a joint secretary are to be selected from a committee of 11 members, then in how many ways can they be selected?

(a) 110 (b) 55
(c) 22 (d) 11

20. On a railway route there are 20 stations. What is the number of different tickets required in order that it may be possible to travel from every station to every other station?

(a) 40 (b) 380
(c) 400 (d) 420

21. If $P(32, 6) = kC(32, 6)$, then what is the value of k?

(a) 6 (b) 32
(c) 120 (d) 720

22. How many straight lines can be formed from 8 non-collinear points on the X-Y plane?

(a) 28 (b) 56
(c) 18 (d) 19860

23. A man has 3 shirts, 4 trousers and 6 ties. What are the number of ways in which he can dress himself with a combination of all the three?

(a) 13 (b) 72
(c) 13!/3!·4!·6! (d) 3!·4!·6!

24. If $({}^{28}C_{2r} : {}^{24}C_{2r-4}) = 225 : 11$. Find the value of r.

(a) 10 (b) 11
(c) 7 (d) 9

25. How many numbers can be formed with the digits 1, 6, 7, 8, 6, 1 so that the odd digits always occupy the odd places.

(a) 15 (b) 12
(c) 18 (d) 20

26. There are 20 people among whom two are sisters. Find the number of ways in which we can arrange them around a circle so that there is exactly one person between the two sisters.

(a) 18! (b) 2!19!
(c) 19! (d) None of these

27. In a company, each employee gives a gift to every other employee. If the number of gifts is 61, then the number of employees in the company is :

(a) 11 (b) 13
(c) 12 (d) 8

28. There are three rooms in a hotel: one single, one double and one for four persons. How many ways are there to house seven persons in these rooms?

(a) 7!/1!2!4! (b) 7!
(c) 7!/3 (d) 7!/3!

29. The digits, from 0 to 9 are written on 10 slips of paper (one digit on each slip) and placed in a box. If three of the slips are drawn and arranged, then the number of possible different arrangements is

(a) 1000 (b) 720
(c) 810 (d) None of these

30. The number of ways in which 7 different books can be given to 5 students if each can receive none, one or more books is

(a) 5^7 (b) 7^5
(c) ${}^{11}C_5$ (d) 12!

31. In how many ways can 13 different alphabets ($a, b, c, \ldots m$) be arranged so that the alphabets f and g never come together?

(a) 13 ! – 12 ! (b) 13 ! – 12! / 2!
(c) 13 ! – 2 × 12 ! (d) None of these

32. Number of ways in which the letters of word GARDEN can be arranged with vowels in alphabetical order, is

(a) 360 (b) 240
(c) 120 (d) 480

33. If 5 parallel straight lines are intersected by 4 parallel straight lines, then the number of parallelograms thus formed is

(a) 20 (b) 60
(c) 101 (d) 126

34. The number of ways in which a couple can sit around a table with 6 guests if the couple take consecutive seat is

(a) 1440 (b) 720
(c) 5040 (d) None of these

35. How many different words beginning with O and ending with E can be formed with the letters of the word ORDINATE, so that the words are beginning with O and ending with E?

(a) 8! (b) 6!
(c) 7! (d) 7!/2!

36. How many 6 digit number can be formed from the digits 1, 2, 3, 4, 5, 6 which are divisible by 4 and digits are not repeated?

(a) 192 (b) 122
(c) 140 (d) 242

37. In how many ways can the eight directors, the vice-chairman and the chairman of a firm be seated at a round-table, if the chairman has to sit between the vice-chairman and the director?

(a) 9! × 2 (b) 2 × 8!
(c) 2 × 7! (d) None of these

38. In how many different ways can the letters of the word 'CREAM' be arranged ? *[SBI Clerk-June-2012]*

(a) 720 (b) 240
(c) 360 (d) 504
(e) None of these

39. Which of the following words can be written in 120 different ways? *[IBPS Clerk-2012]*

(a) STABLE (b) STILL
(c) WATER (d) NOD
(e) DARE

Level - II

1. 5 men and 6 women have to be seated in a straight row so that no two women are together. Find the number of ways this can be done.
(a) 48400 (b) 39600
(c) 9900 (d) 86400

2. The total number of ways in which 8 men and 6 women can be arranged in a line so that no 2 women are together is
(a) 48 (b) ${}^{8}P_{8}.{}^{9}P_{6}$
(c) 8! (84) (d) ${}^{8}C_{8}.{}^{9}C_{8}$

3. The number of different ways in which 8 persons can stand in a row so that between two particular person A and B there are always two person, is
(a) 60 (5!) (b) 15(4!) × (5!)
(c) 4! × 5! (d) None of these

4. From 6 boys and 7 girls a committee of 5 is to be formed so as to include atleast one girl. The number of ways this can be done is
(a) ${}^{13}C_4$ (b) ${}^{6}C_4 . {}^{7}C_1$
(c) $7 . {}^{6}C_4$ (d) ${}^{13}C_5 - {}^{6}C_1$

5. How many different nine digit numbers can be formed from the number 223355888 by rearranging its digits so that the odd digits occupy even positions?
(a) 16 (b) 36
(c) 60 (d) 180

6. If two dices are tossed simultaneously, the number of elements in the resulting sample space is
(a) 6 (b) 8
(c) 36 (d) 24

7. In how many ways can 7 persons stand in the form of a ring?
(a) $P(7, 2)$ (b) 7 !
(c) 6 ! (d) $\frac{7!}{2}$

8. In a football championship 153 matches were played. Every team played one match with each other team. How many teams participated in the championship?
(a) 21 (b) 18
(c) 17 (d) 15

9. If $P(77, 31) = x$ and $C(77, 31) = y$, then which one of the following is correct?
(a) $x = y$ (b) $2x = y$
(c) $77x = 31y$ (d) $x > y$

10. In how many ways can 12 papers be arranged if the best and the worst paper never come together?
(a) 12!/2! (b) 12! – 11!
(c) (12! – 11!)/2 (d) 12! – 2.11!

11. If a team of four persons is to be selected from 8 males and 8 females, then in how many ways can the selections be made to include at least one male.
(a) 1550 (b) 1675
(c) 1725 (d) 1750

12. Letters of the word DIRECTOR are arranged in such a way that all the vowels come together. Find out the total number of ways for making such arrangement.
(a) 4320 (b) 2720
(c) 2160 (d) 1120

13. 4 boys and 2 girls are to be seated in a row in such a way that the two girls are always together. In how many different ways can they be seated?
(a) 1200 (b) 7200
(c) 148 (d) 240

14. In how many ways can 7 Englishmen and 7 Americans sit down at a round table, no 2 Americans being in consecutive positions?
(a) 3628800 (b) 2628800
(c) 3628000 (d) 3328800

15. How many numbers greater than one million can be formed with 2, 3, 0, 3, 4, 2, 3? (repetitions not allowed)
(a) 720 (b) 360
(c) 120 (d) 240

16. 5 Indian and 5 American couples meet at a party & shake hands . If no wife shakes hands with her husband and no Indian wife shakes hands with a male, then the number of hand shakes that takes place in the party is
(a) 95 (b) 110
(c) 135 (d) 150

17. The total number of ways in which letters of the word ACCOST can be arranged so that the two C's never come together will be
(a) 120 (b) 360
(c) 240 (d) 6 ! – 2 !

18. In how many ways can a term of 11 cricketers be chosen from 6 bowlers. 4 wicket keepers and 11 batsmen to give a majority of bastemen if at least 4 bowlers are to be included and there is one wicket keeper?
(a) 27730 (b) 27720
(c) 17720 (d) 26720

19. Three dice are rolled. The number of possible outcomes in which at least one die shows 5 is
(a) 215 (b) 36
(c) 125 (d) 91

20. The sides AB, BC, CA of a traingle ABC have 3, 4 and 5 interior points respectively on them. The total number of triangles that can be constructed by using these points as vertices is

(a) 220 (b) 204
(c) 205 (d) 195

21. If all permutations of the letters of the word AGAIN are arranged as in dictionary, then fiftieth word is

(a) NAAGI (b) NAGAI
(c) NAAIG (d) NAIAG

22. All the words that can be formed using alphabets A, H, L, U and R are written as in a dictionary (no alphabet is repeated). Rank of the word RAHUL is

(a) 71 (b) 72
(c) 73 (d) 74

23. How many new words are possible from the letters of the word PERMUTATION?

(a) $11!/2!$ (b) $(11!/2!) - 1$
(c) $11! - 1$ (d) None of these

24. There are five boys and three girls who are sitting together to discuss a management problem at a round table. In how many ways can they sit around the table so that no two girls are together?

(a) 1220 (b) 1400
(c) 1420 (d) 1440

25. How many 6-digit numbers have all three digits either all odd or all even?

(a) 31,250 (b) 28,125
(c) 15,625 (d) None of these

26. Out of 10 consonants and four vowels, the number of words that can be formed using six consonants and three vowels is

(a) ${}^{10}P_6 \times {}^6P_3$ (b) ${}^{10}C_6 \times {}^6C_3$
(c) ${}^{10}C_6 \times {}^4C_3 \times 9!$ (d) ${}^{10}P_6 \times {}^4P_3$

27. The number of 5 digit numbers that can be made using the digits 1 and 2 and in which at least one digit is different, is

(a) 30 (b) 31
(c) 32 (d) None of these

28. A class photograph has to be taken. The front row consists of 6 girls who are sitting. 20 boys are standing behind. The two corner positions are reserved for the 2 tallest boys. In how many ways can the students be arranged?

(a) $18! \times 1440$ (b) $6! \times 1440$
(c) $18! \times 2! \times 1440$ (d) None of these

29. A,B,C and D are four towns any three of which are non-colinear. Then the number of ways to construct three roads each joining a pair of towns so that the roads do not form a triangle is

(a) 7 (b) 8
(c) 9 (d) More than 9

30. There are 10 points on a line and 11 points on another line, which are parallel to each other. How many triangles can be drawn taking the vertices on any of the line?

(a) 1,050 (b) 2,550
(c) 150 (d) 1,045

31. How many motor vehicle registration number plates can be formed with the digits 1, 2, 3, 4, 5 (No digits being repeated) if it is given that registration number can have 1 to 5 digits?

(a) 100 (b) 120
(c) 325 (d) 205

32. How many different 9-digit numbers can be formed from the number 223355888 by rearranging its digits so that the odd digits occupy even positions?

(a) 120 (b) $9!(2!)^3.3!$
(c) $(4!)(2!)^3.(3!)$ (d) None of these

33. There are 5 different Jeffrey Archer books, 3 different Sidney Sheldon books and 6 different John Grisham books. The number of ways in which at least one book can be given away is

(a) $2^{10} - 1$ (b) $2^{11} - 1$
(c) $2^{12} - 1$ (d) $2^{14} - 1$

34. How many natural numbers not more than 4300 can be formed with the digits 0, 1, 2, 3, 4 (if repetitions are allowed)?

(a) 574 (b) 570
(c) 575 (d) 569

35. The sides of a triangle have 4, 5 and 6 interior points marked on them respectively. The total number of triangles that can be formed using any of these points

(a) 371 (b) 415
(c) 286 (d) 421

36. Total number of ways in which six '+' and four '–' sings can be arranged in a line such that no two '–' sings occur together, is

(a) 35 (b) 18
(c) 15 (d) 42

37. In how many ways can 5 prizes be distributed among 4 boys when every boy can take one or more prizes?

(a) 1024 (b) 625
(c) 120 (d) 600

38. Three dice are rolled. The number of possible outcomes in which at least one die shows 5 is

(a) 215 (b) 36
(c) 125 (d) 91

39. Find the number of numbers between 300 and 3000 that can be formed with the digits 0, 1, 2, 3, 4 and 5, no digit being repeated. *[SBI PO-2011]*

(a) 120 (b) 160
(c) 240 (d) 60
(e) None of these

Hints & Solutions

Level-I

1. (c) | | | | |

Last place can be filled by 0, 2, 4

So total sum = $5 \times 6 \times 6\,(0 + 2 + 4) + 5 \times 6 \times 3 \times 10\,(0 + 1 + 2 + 3 + 4 + 5) + 5 \times 6 \times 3 \times 100\,(0 + 1 + 2 + 3 + 4 + 5) + 6 \times 6 \times 3 \times 1000\,(0 + 1 + 2 + 3 + 4 + 5)$

$= 180 \times 6 + 900 \times 15 + 9000 \times 15 + 10800 \times 15$

$= 1080 + 13500 + 135000 + 1620000 = 1769580$

2. (a) There are 8 letters in the word EQUATION.

A/E/I/O/U							
5 ways	${}^7P_7 = 7! = 5040$						

$\therefore$ Reqd. no. $= 5 \times 5040 = 25200$

3. (a) There are 9 letters in the given word in which two T's, two M's and two E's are identical. Hence the required number of words $= \dfrac{9!}{2!2!2!} = \dfrac{9!}{(2!)^3}$

4. (c) Given, ${}^{10}P_r = 720$

$\therefore \dfrac{10!}{(10-r)!} = 720$

$\therefore 10 \times 9 \times 8 \times \ldots$ to r factors $= 720 = 10 \times 9 \times 8$

$\therefore r = 3$

5. (b) $\dfrac{12!}{5!4!3!}$

6. (a) Considering the two vowels E and A as one letter, the total no. of letters in the word 'EXTRA' is 4 which can be arranged in ${}^4P_4$, i.e. 4! ways and the two vowels can be arranged among themselves in 2! ways.

$\therefore$ reqd. no. $= 4! \times 2! = 4 \times 3 \times 2 \times 1 \times 2 \times 1 = 48$

7. (a) A committee of 5 out of 6 + 4= 10 can be made in ${}^{10}C_5 = 252$ ways.

If no woman is to be included, then number of ways $= {}^5C_5 = 6$

$\therefore$ the required number $= 252 - 6 = 246$

8. (d) 4 digit number | 3 | 4 | 3 | 2 | = 72,

5 digit number = 120

Total = 192

9. (b) If number of persons be n, then total number of hand-shaken $= {}^nC_2 = 66$

$\Rightarrow n\,(n-1) = 132 \Rightarrow (n + 11)\,(n = 12) = 0$

$\therefore n = 12 \qquad (\because n \neq -11)$

10. (b) There are 6 letters in the word BHARAT, 2 of them are identical.

Hence total number of words with these letter = 360

Also the number of words in which B and H come together = 120

$\therefore$ The required number of words $= 360 - 120 = 240$

11. (a) The required number of selections

$= {}^3C_1 \times {}^4C_1 \times {}^2C_1\,({}^6C_3 + {}^6C_2 + {}^6C_0) = 42 \times 4!$

12. (d) MACHINE has 4 consonants and 3 vowels.

The vowels can be placed in position no. 1, 3, 5, 7 ·

$\Rightarrow$ Total number of ways possible $= {}^4P_3 = 24.$

For each of these 24 ways the 4 consonants can occupy the other 4 places in ${}^4P_4$ ways

$\Rightarrow$ Total $= 24 \times 24 = 576$

13. (b) We have, ${}^nP_r = {}^nP_{r+1}$

$\Rightarrow \dfrac{n!}{(n-r)!} = \dfrac{n!}{(n-r-1)!} \Rightarrow \dfrac{1}{(n-r)} = 1$

or $n - r = 1$ ...(1)

Also, ${}^nC_r = {}^nC_{r-1} \Rightarrow r + r - 1 = n$

$\Rightarrow 2r - n = 1$ ...(2)

Solving (1) and (2), we get $r = 2$ and $n = 3$

14. (c) ${}^nP_r = 720\,{}^nC_r$

or $\dfrac{n!}{(n-r)!} = \dfrac{720(n!)}{(n-r)!r!}$

$\Rightarrow r! = 720 = 1 \times 2 \times 3 \times 4 \times 5 \times 6!$

or $r = 6$

15. (a) Total number of ways $= {}^{16}C_{11} = \dfrac{16!}{11! \times 5!} = 4368.$

$= \dfrac{16 \times 15 \times 14 \times 13 \times 12}{5 \times 4 \times 3 \times 2 \times 1} = 4368.$

16. (c) Two particular girls can be arranged in 2! ways and remaining 10 girls can be arranged in 10! ways.

Required no. of ways $= 2! \times 10!$

17. (c) Required no. of the ways $= {}^6C_3 \times {}^4C_2 = 20 \times 6 = 120$

18. (b) Required number of ways $= \dfrac{5!}{2!3!} = 10$.

19. (b) Selection of 2 members out of 11 has ${}^{11}C_2$ number of ways

${}^{11}C_2 = 55$

20. (b) From each railway station, there are 19 different tickets to be issued. There are 20 railway station

So, total number of tickets $= 20 \times 19 = 380.$

21. (d) Since ${}^{32}P_6 = k\,{}^{32}C_6$

$\Rightarrow \dfrac{32!}{(32-6)!} = k.\dfrac{32!}{6!(32-6)!}$

$\Rightarrow k = 6! = 720$

22. **(a)** For a straight line we just need to select 2 points out of the 8 points available. $^8C_2$ would be the number of ways of doing this.

23. **(b)** $^3C_1 \times {}^4C_1 \times {}^6C_1 = 72$

24. **(c)** At $r = 7$, the value becomes
$(28!/14! \times 14!)/(24!/10! \times 14!) \rightarrow 225 : 11$

25. **(c)** The digits are 1, 6, 7, 8, 7, 6, 1. In this seven-digit no. there are four odd places and three even places OEOEOEO. The four odd digits 1, 7, 7, 1 can be arranged in four odd places in [4!/2!×2] = 6 ways [as 1 and 7 are both occurring twice].
The even digits 6, 8, 6 can be arranged in three even places in 3!/2! = 3 ways.
Total no. of ways = 6 × 3 = 18

26. **(d)** First arrange the two sisters around a circle in such a way that there will be one seat vacant between them. [This can be done in 2! ways since the arrangement of the sisters is not circular.]
Then, the other 18 people can be arranged on 18 seats in 18! ways.

27. **(c)** Let the total number of employees in the company be n.

Total number of gifts = $^nC_2 = \dfrac{n(n-1)}{2} = 61$

$\Rightarrow n^2 - n - 132 = 0$ or $(n+11)(n-12) = 0$
or $n = 12$ [– 11 is rejected]

28. **(a)** Choose 1 person for the single room & from the remaining choose 2 people for the double room & from the remaining choose 4 people for the 4 persons room $\rightarrow {}^7C_1 \times {}^6C_2 \times {}^4C_4$.

29. **(b)** $^{10}P_3 = 720$

30. **(a)** Ist book can be given to any of the five students. Similarly other six books also have 5 choices. Hence the total number of ways is 5^7.

31. **(c)** Total possible arrangements = $^{13}P_{13} = 13!$
Total number in which f and g are together
$= 2 \times {}^{12}P_{12} = 2 \times 12!$

32. **(a)** Order of vowels of fixed

∴ required number of ways are $\dfrac{6!}{2!}$

33. **(b)** Number of parallelograms = $^5C_2 \times {}^4C_2 = 60$.

34. **(a)** A couple and 6 guests can be arranged in (7 – 1) ways. But in two people forming the couple can be arranged among themselves in 2! ways.
∴ the required number of ways = 6! × 2! = 1440

35. **(b)** 6! ways, O fixed 1st and E fixed in last.

36. **(a)** For the number to be divisible by 4, the last two digits must be any of 12, 24, 16, 64, 32, 36, 56 and 52. The last two digit places can be filled in 8 ways. Remaining 3 places in $^4P_3$ ways. Hence no. of 5 digit nos. which are divisible by 4 are 24 × 8 = 192.

37. **(b)** Let the vice-chairman and the chairman from 1 unit along with the eight directors, we now have to arrange 9 different units in a circle.
This can be done in 8! ways.
At the same time, the vice-chairman & the chairman can be arranged in two different ways. Therefore, the total number of ways = 2 × 8!.

38. **(e)** C R E A M
1 2 3 4 5
Required number of ways = 5!
= 5 × 4 × 3 × 2 × 1 = 120

39. **(c)** (a) The word STABLE has six distinct letters.
∴ Number of arrangements = 6 !
= 6 × 5 × 4 × 3 × 2 × 1 = 720
(b) The word STILL has five letters in which letter 'L' comes twice.
∴ Number of arrangements
$= \dfrac{5!}{2} = 60$
(c) The word WATER has five distinct letters.
∴ Number of arrangements = 5 ! = 5 × 4 × 3 × 2× 1 = 120
(d) The word 'NOD' has 3 distinct letters.
∴ Number of arrangements = 3 ! = 6
(e) Number of arrangements = 4 ! = 24

Level-II

1. **(d)** Total seats = 5 + 6 = 11.
Arrangement will be : $W\,M\,W\,M\,W\,M\,W\,M\,W\,M\,W$
⇒ Total possible arrangements will be :
$^6P_6 \times {}^5P_5 = 86400$.

2. **(b)** 8 men can sit in a row in $^8P_8$ ways. Then for the 6 women, there are 9 seats to sit
∴ the women can sit in $^9P_6$ ways
∴ total number of ways = $^8P_8 \,.\, {}^9P_6$

3. **(a)** The number of 4 persons including A, B = $^6C_2$
Considering these four as a group, number of arrangements with the other four = 5!
But in each group the number of arrangements
= 2! × 2!
∴ The required number of ways = $^6C_2 \times 5! \times 2! \times 2!$

4. **(d)** From total 13 members 5 can be select as $^{13}C_5$
For at least one girl in the committee, number of ways are $^{13}C_5 - {}^6C_1$

5. **(c)** X-X-X-X-X. The four digits 3, 3, 5,5 can be arranged

at (–) places in $\dfrac{4!}{2!2!} = 6$ ways.

The five digits 2, 2, 8, 8. 8 can be arranged at (X) places

in $\dfrac{5!}{2!3!}$ ways = 10 ways

Total no. of arrangements = $6 \times 10 = 60$ ways

6. **(c)** Number of elements in the sample space
$= 6 \times 6 = 36$
(6, 1), (6, 2), (6, 3), (6, 4), (6, 5), (6, 6)

7. **(c)** Number of ways in which 7 persons can stand in the form of a ring $= (7-1)! = 6!$

8. **(b)** Let total no. of team participated in a championship be n.
Since, every team played one match with each other team.

$$\therefore\ {}^nC_2 = 153 \Rightarrow \frac{n!}{2!(n-2)!} = 153$$

$$\Rightarrow \frac{n(n-1)(n-2)!}{2!(n-2)!} = 153 \Rightarrow \frac{n(n-1)}{2} = 153$$

$\Rightarrow n(n-1) = 306$
$\Rightarrow n^2 - n - 306 = 0$
$\Rightarrow n^2 - 18n + 17n - 306 = 0$
$\Rightarrow n(n-18) + 17(n-18) = 0$
$\Rightarrow n = 18, -17$
n cannot be negative
$\therefore n \neq -17$
$\Rightarrow n = 18$

9. **(d)** As we know
$P(n, r) = r!\ C(n, r)$
$\therefore$ From the question, we have
$x = r!\,(y)$
Here $r = 31$
$\therefore\ x = (31)!.\ y.$

10. **(d)** All arrangements – Arrangements with best and worst paper together $= 12! - 2! \times 11!$.

11. **(d)** $1\text{ m} + 3f = {}^8C_1 \times {}^8C_3 = 8 \times 56 = 448$
$2\text{ m} + 2f = {}^8C_2 \times {}^8C_2 = 28 \times 28 = 784$
$3\text{ m} + 1f = {}^8C_3 \times {}^8C_1 = 56 \times 8 = 448$
$4\text{ m} + 8f = {}^8C_4 \times {}^8C_0 = 70 \times 1 = 70$
Total = 1750

12. **(c)** Taking all vowels (IEO) as a single letter (since they come together) there are six letters among which there are two R.

Hence no. of arrangements $= \frac{6!}{2!} \times 3! = 2160$

There vowels can be arranged in 3! ways among themselves, hence multiplied with 3!.

13. **(d)** Assume the 2 given students to be together (i.e. one).
Now these are five students.
Possible ways of arranging them are $= 5! = 120$
Now they (two girls) can arrange themselves in 2! ways.
Hence total ways $= 120 \times 2 = 240$

14. **(a)** Putting 1 Englishman in a fixed position, the remaining 6 can be arranged in 6! 720 ways, For each such arrangement, there are 7 positions for the 7 Americans and they can be arranged in 7! ways.
Total number of arrangements $= 7! \times 6! = 3628800$

15. **(b)** Required number is greater than 1 million (7 digits).
From given digits, total numbers which can be formed $= 7!$
Number starting from zero $= 6!$
$\Rightarrow$ Required number $= 7! - 6!$
$\because$ Repetition not allowed, so required answer

$$= \frac{7! - 6!}{2!3!} = 360$$

16. **(c)** Total number of hand shakes $= {}^{20}C_2$ of those no Indian female shakes hand with male
$\Rightarrow 5 \times 10 = 50$ hand shakes
No American wife shakes hand with her husband
$= 5 \times 1 = 5$ hand shakes
$\Rightarrow$ total number of hand shakes occurred
$= {}^{20}C_2 - (50 + 5) = 190 - 55 = 135$

17. **(c)** Total number of ways to permute 6 alphabets 2 of which are common $= 6!/2! = 360$.
(1) Treat the two C's as one
$\Rightarrow$ Number of possible ways $= {}^5P_5 = 120$
(b) Number of ways = Total arrangements – Number of arrangements in which they always come together
$= 360 - 120 = 240$.

18. **(b)** 1 wicket keeper from 4 can be selected in

$${}^4C_1 = \frac{4!}{3!.1!} = 4 \text{ ways}$$

If 4 bowlers are chosen then remaining 6 batsmen - can be chosen in ${}^{11}C_6$.

$${}^6C_4.\,{}^{11}C_6 = \frac{6!}{4!.2!} \times \frac{11!}{3!.1!} = \frac{5 \times 6}{2} \times \frac{11 \times 10 \times 9 \times 8 \times 7}{5 \times 4 \times 3 \times 2}$$

$$= 15 \times 14 \times 33 = 6930$$

If we choose 5 bowlers then we have to choose 5 batsmen
$\therefore$ there is no majority.
$\therefore$ Total number of ways $= 4 \times 6930 = 27720$.

19. **(d)** Required number of possible outcomes
= Total number of possible outcomes –
Number of possible outcomes in which 5 does not appear on any dice. (hence 5 possibilities in each throw)
$= 6^3 - 5^3 = 216 - 125 = 91$

20. **(c)** We have in all 12 points. Since, 3 points are used to form a traingle, therefore the total number of traingles including the triangles formed by collinear points on AB, BC and CA is ${}^{12}C_3 = 220$. But this includes the following :

The number of traingles formed by 3 points on $AB = {}^3C_3 = 1$

The number of triangles formed by 4 points on $BC = {}^4C_3 = 4$.

The number of triangles formed by 5 points on $CA = {}^5C_3 = 10$.

Hence, required number of traingles $= 220 - (10 + 4 + 1) = 205$.

21. **(c)** Starting with the letter A, and arranging the other four letters, there are $4! = 24$ words. These are the first 24 words. Then starting with G, and arranging A, A, I, and N in different ways, there are $\frac{4!}{2!1!1!} = \frac{24}{2} = 12$ words.

Hence, total 36 words.

Next, the 37th word starts with I. There are 12 words starting with I. This accounts up to the 48th word. The 49th word is NAAGI. The 50th word is NAAIG.

22. **(d)** No. of words starting with A are $4! = 24$

No. of words starting with H are $4! = 24$

No. of words starting with L are $4! = 24$

These account for 72 words

Next word is RAHLU and the 74th word RAHUL.

23. **(b)** Number of 11 letter words formed from the letter P, E, R, M, U, T, A, I, O, N = 11!/2!.

Number of new words formed = total words – 1 $= 11!/2! - 1$.

24. **(d)** We have no girls together, let us first arrange the 5 boys and after that we can arrange the girls in the space between the boys.

Number of ways of arranging the boys around a circle $= [5 - 1]! = 24$.

Number of ways of arranging the girls would be by placing them in the 5 spaces that are formed between the boys. This can be done in ${}^5P_3$ ways = 60 ways.

Total arrangements $= 24 \times 60 = 1440$.

25. **(b)** When all digits are odd

$5 \times 5 \times 5 \times 5 \times 5 \times 5 = 5^6$

When all digits are even

$4 \times 5 \times 5 \times 5 \times 5 \times 5 \times 5 = 4 \times 5^5$

$5^6 + 4 \times 5^5 = 28125$

26. **(c)** Six consonants and three vowels can be selected from 10 consonants and 4 vowels in ${}^{10}C_6 \times {}^4C_3$ ways. Now, these 9 letters can be arranged in 9! ways. So, required number of words $= {}^{10}C_6 \times {}^4C_3 \times 9!$.

27. **(a)** Total number of numbers without restriction $= 2^5$

Two numbers have all the digits equal. So, the required numbers $= 2^5 - 2 = 30$.

28. **(a)** Two tallest boys can be arranged in 2! ways. Rest 18 can be arranged in 18! ways.

Girls can be arranged in 6! ways.

Total number of ways of arrangement $= 2! \times 18! \times 6!$ $= 18! \times 2 \times 720 = 18! \times 1440$

29. **(d)** To construct 2 roads, three towns can be selected out of 4 in $4 \times 3 \times 2 = 24$ ways.

Now if the third road goes from the third town to the first town, a triangle is formed, and if it goes to the fourth town, a triangle is not formed. So, there are 24 ways to form a triangle and 24 ways of avoiding a triangle.

30. **(d)** For a triangle, two points on one line and one on the other has to be chosen.

No. of ways $= {}^{10}C_2 \times {}^{11}C_1 + {}^{11}C_2 \times {}^{10}C_1 = 1{,}045$.

31. **(c)** Single digit numbers = 5

Two digit numbers $= 5 \times 4 = 20$

Three digit numbers $= 5 \times 4 \times 3 = 60$

Four digit numbers $= 5 \times 4 \times 3 \times 2 = 120$

Five digit numbers $= 5 \times 4 \times 3 \times 2 \times 1 = 120$

Total $= 5 + 20 + 60 + 120 + 120 = 325$

32. **(d)** The odd digits have to occupy even positions. This can be done in $\frac{4!}{2!2!} = 6$ ways

The other digits have to occupy the other positions.

This can be done in $\frac{5!}{3!2!} = 10$ ways

Hence total number of rearrangements possible $= 6 \times 10 = 60$.

33. **(d)** For each book we have two options, give or not give. Thus, we have a total of 2^{14} ways in which the 14 books can be decided upon. Out of this, there would be 1 way in which no book would be given. Thus, the number of ways is $2^{14} - 1$.

34. **(c)** The condition is that we have to count the number of natural numbers not more than 4300.

The total possible numbers with the given digits $= 5 \times 5 \times 5 \times 5 = 625 - 1 = 624$.

Subtract from this the number of natural number greater than 4300 which can be formed from the given digits $= 1 \times 2 \times 5 \times 5 - 1 = 49$.

Hence, the required number of numbers $= 624 - 49$.

35. **(d)** You can form triangles by taking 1 point from each side, or by taking 2 points from any 1 side and the third point from either of the other two sides.

This can be done in: $4 \times 5 \times 6 = {}^4C_2 \times {}^{11}C_1 + {}^5C_2 \times {}^{10}C_1 + {}^6C_2 \times {}^9C_1 = 120 + 66 + 100 + 135 = 421$

36. **(a)** First we write six '+' sings at alternate places i.e., by leaving one place vacant between two successive '+' sings. Now there are 5 places vacant between these sings and these are two places vacant at the ends. If we write 4 '–' sings these 7 places then no two '–' will come together. Hence total number of ways ${}^7C_4 = 35$

37. **(a)** First prize may be given to any one of the 4 boys, hence first prize can be distributed in 4 ways.

Similarly every one of second, third, fourth and fifth prizes can also be given in 4 ways.

∴ the number of ways of their distribution

$= 4 \times 4 \times 4 \times 4 \times 4 = 4^5 = 1024.$

38. **(d)** Required number of possible outcomes

= Total number of possible outcomes – Number of possible outcomes in which 5 does not appear on any dice

$= 6^3 - 5^3 = 91.$

39. **(e)** 3 5 4

— — —

(3 or 4 or 5)

$3 \times 5 \times 4 = 60$

2 5 4 3

— — — —

(1 or 2)

$2 \times 5 \times 4 \times 3 = 120$

Total = 120 + 60 = 180

CHAPTER 22

PROBABILITY

CONCEPT OF PROBABILITY

If you go to buy 10 kg of sugar at ₹ 40 per kg, you can easily find the exact price of your purchase is ₹ 400. On the other hand, the shopkeeper may have a good estimate of the number of kg of sugar that will be sold during the day, but it is impossible to predict the exact amount, because the number of kg of sugar that the consumers will purchase during a day is random.

There are various phenomenon in nature, leading to an outcome, which cannot be predicted in advance. For example, we cannot exactly predict that (i) a head will occur on tossing a coin, (ii) a student will clear the CAT, (iii) India will win the cricket match against Pakistan, etc. But we can measure the amount of certainty of occurrence of an outcome of a phenomenon. This amount of certainty of occurrence of an outcome of a phenomenon is called probability. For example, on tossing a coin certainty of occurrence of each of a head and a tail are the same. Hence amount certainty of occurrence of each of a head and a tail is 50% i.e., $\frac{50}{100}=\frac{1}{2}$. Therefore $\frac{1}{2}$ is the amount of certainty of occurrence of a head (or a tail) on tossing a coin and hence $\frac{1}{2}$ is the probability of occurrence of a head (or a tail) on tossing a coin. On throwing a dice (a dice is a cuboid having one of the numbers 1, 2, 3, 4, 5 and 6 on each of its six faces) certainty of occurrence of each of the numbers 1, 2, 3, 4, 5 and 6 on its top face are the same.

Therefore certainty of occurrence of each of the numbers 1, 2, 3, 4, 5 and 6 is $\frac{1}{6}$.

Therefore $\frac{1}{6}$ is the amount of certainty of occurrence of each of the numbers 1, 2, 3, 4, 5 or 6 on the top face of the dice on throwing the dice and hence $\frac{1}{6}$ is the probability of occurrence of each of the numbers 1, 2, 3, 4, 5, or 6 on the top face of the dice on tossing a dice is $\frac{1}{6}$.

BASIC TERMS

1. **An Experiment:** An action or operation resulting in two or more outcomes is called an experiment. For examples
   (i) Tossing of a coin is an experiment because there are two possible outcomes head and tail.
   (ii) Drawing a card from a pack of 52 cards is an experiment because there are 52 possible outcomes.
2. **Sample Space:** The set of all possible outcomes of an experiment is called the sample space, denoted by S. An element of S is called a sample point. For examples
   (i) In the experiment of tossing a coin, the sample space has two points corresponding to head (H) and Tail (T) i.e., $S\{H, T\}$.
   (ii) When we throw a dice then any one of the numbers 1, 2, 3, 4, 5 and 6 will come up. So the sample space, $S = \{1, 2, 3, 4, 5, 6\}$
3. **An Event:** Any subset of a sample space is an event. For example,
   If we throw a dice then $S = \{1, 2, 3, 4, 5, 6\}$
   Then $A = \{1, 3, 5\}$, $B\ \{2, 4, 6\}$, the null set ϕ and S itself are some events of S, because they all are subsets of set S.
4. **Impossible Event:** The null set ϕ is called the impossible event or null event. For example,
   Getting 7 when a dice is thrown is an impossible or a null event.
5. **Sure Event:** The entire sample space is called sure or certain event. For example,
   Here the event:
   Getting an odd or even number on throwing a dice is a sure event, because the event = S.
6. **Complement of an Event:** The complement of an event A is denoted by $\overline{A}$, A' or A^c, is the set of all sample points of the sample space other than the sample points in A. For example, in the experiment of tossing a fair dice,
   $S = \{1, 2, 3, 4, 5, 6\}$ If $A = \{1, 3, 5, 6\}$, then $A^c = \{2, 4\}$
   Note that $A \cup A^c = S$, $A \cap A^c = \phi$.

7. **Simple (or Elementary) Event:** An event is called a simple event if it is a singleton subset of the sample space S. The singleton subset means the subset having only one element. For example,
   (i) When a coin is tossed, sample space $S = \{H, T\}$
   Let $A = \{H\}$ = the event of occurrence of head and $B = \{T\}$ = the event of occurrence of tail.
   Here A and B are simple events.
   (ii) When a dice is thrown then sample space, $S = \{1, 2, 3, 4, 5, 6\}$
   Let $A = \{5\}$ = the event of occurrence of 5
   $B = \{2\}$ = the event of occurrence of 2
   Here A and B are simple events.
8. **Compound Event:** It is the joint occurrence of two or more simple events. For example,
   The event of at least one head appears when two fair coins are tossed is a compound event,
   $A = \{HT, TH, HH\}$
9. **Equally Likely Events:** A number of simple events are said to be equally likely if there is no reason for one event to occur in preference to any other event. For example,
   In drawing a card from a well shuffled pack of 52 cards, there are 52 outcomes and hence 52 simple events which are equally likely because there is no reason for one event to occur in preference to any other event. For example,

MATHEMATICAL DEFINITION OF PROBABILITY

If an event A consists of m sample points of a sample space S having n elements $(0 \le m \le n)$, then the probability of occurrence of event A, denoted by $P(A)$ is defined to be $\frac{m}{n}$ i.e., $P(A) = \frac{m}{n}$

$$\because \quad 0 \le m \le n \Rightarrow 0 \le \frac{m}{n} \le 1 \Rightarrow 0 \le P(A) \le 1$$

If the event A has m elements, then A' has $(n - m)$ elements.

$$\therefore \; P(A') = \frac{n-m}{n} = 1 - \frac{m}{n} = 1 - P(A)$$

Let $S = \{a_1, a_2, ..., a_n\}$ be the sample space

$P(S) = \frac{n}{n} = 1$, corresponding to the certain event.

$P(\phi) = \frac{0}{n} = 0$, corresponding to the null event ϕ (or impossible event)

If $A_i = \{a_i\}$, $i = 1, 2, ...$ or n; then A_i is the event corresponding to a single sample point a_i, then $P(A_i) = \frac{1}{n}$.

Illustration 1: Two dice are thrown at a time. Find the probability of the followings:

(i) the numbers shown are equal

(ii) the difference of numbers shown is 1

Solution: The sample space in a throw of two dice

$S = \{(1, 1), (1, 2), ..., (1, 6), (2, 1), (2, 2), ..., (2, 6), (3, 1), ..., (3, 6), (4, 1), ..., (4, 6), (5, 1), ..., (5, 6), (6, 1), ..., (6, 6)\}$

$\therefore$ total no. of outcomes, $n(S) = 36$

(i) Here E_1 = the event of showing equal number on both dice

$= \{(1, 1) (2, 2) (3, 3) (4, 4) (5, 5) (6, 6\}$

$$\therefore n(E_1) = 6, \Rightarrow P(E_1) = \frac{n(E_1)}{n(S)} = \frac{6}{36} = \frac{1}{6}$$

(ii) Here E_2 = the event of showing numbers whose difference is 1.

$= \{(1, 2) (2, 1) (2, 3) (3, 2) (3, 4) (4, 3) (4, 5) (5, 4) (5, 6) (6, 5)\}$

$$\therefore \; n(E_2) = 10, \Rightarrow P(E_2) = \frac{n(E_2)}{n(S)} = \frac{10}{36} = \frac{15}{18}.$$

Illustration 2: If three cards are drawn from a pack of 52 cards, what is the chance that all will be queen?

Solution: If the sample space be S, then $n(S)$ = the total number of ways of drawing 3 cards out of 52 cards $= {}^{52}C_3$

Now, if A = the event of drawing three queens, then

$n(A) = {}^4C_3$

$$\therefore \quad P(E) = \frac{n(A)}{n(S)} = \frac{{}^4C_3}{{}^{52}C_3} = \frac{4}{\frac{52 \times 51 \times 50}{3 \times 2}} = \frac{1}{5525}$$

Note that in a pack of playing cards,

Total number of cards: 52(26 red, 26 black)

Four suits: Heart, Diamond, Spade, Club-13 cards of each suit

Court number of cards: 12(4 kings, 4 queens, 4 jacks)

Face number of cards: 16(4 aces, 4 kings, 4 queens, 4 jacks)

Illustration 3: Words are formed with the letters of the word PEACE. Find the probability that 2 E's come together.

Solution: Total number of words which can be formed with the letters of the word P E A C E $= \frac{5!}{2!} = 60$

Number of words in which 2 E's come together $= 4! = 24$

$$\therefore \quad \text{Required prob.} = \frac{24}{60} = \frac{2}{5}.$$

Illustration 4: A and B play a game where each is asked to select a number from 1 to 25. If the two numbers match, both of them win a prize. The probability that they will not win a prize in a single trial is

(a) $\frac{1}{25}$ **(b)** $\frac{24}{25}$

(c) $\frac{2}{25}$ **(d) None of these**

Solution: (b) Total number of possibilities $= 25 \times 25$

Favourable cases for their winning = 25

$$\therefore P(\text{they win a prize}) = \frac{25}{25 \times 25} = \frac{1}{25}$$

$$\therefore P(\text{they will not win a prize}) = = 1 - \frac{1}{25} = \frac{24}{25}$$

ADDITION THEOREM

If A and B are any events in S, then

$$P(A \cup B) = P(A) + P(B) - P(A \cap B)$$

i.e., $$P(A \text{ or } B) = P(A) + P(B) - P(A \text{ and } B)$$

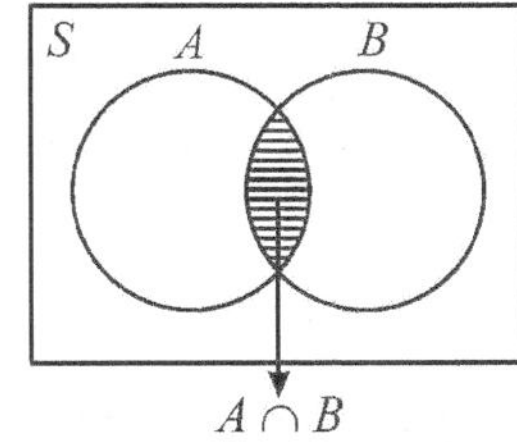

For three events A, B and C in S, we have

$$P(A \cup B \cup C) = P(A) + P(B) + P(C) - P(A \cap B) - P(B \cap C) - P(C \cap A) + P(A \cap B \cap C).$$

Special Addition Rule

If A, B, and C are mutually exclusive, then $P(A \cap B)$, $P(B \cap C)$, $P(C \cap A)$, $P(A \cap B \cap C) = 0$, hence $P(A \cup B) = P(A) + P(B)$ and $P(A \cup B \cup C) = P(A) + P(B) + P(C)$

Illustration 5: A bag contains 6 white, 5 black and 4 red balls. Find the probability of getting either a white or a black ball in a single draw.

Solution: Let A = Event that we get a black ball

Two events A and B are mutually exclusive.

$$P(A) = \frac{^6C_1}{^{15}C_1} = \frac{6}{15},\ P(B) = \frac{^5C_1}{^{15}C_1} = \frac{5}{15}$$

So, $$P(A \cup B) = P(A) + P(B) = \frac{6}{15} + \frac{5}{15} = \frac{11}{15}.$$

Illustration 6: One digit is selected from first 20 positive integers. What is the probability that it is divisible by 3 or 4.

Solution:

Let A = Event that the selected number is divisible by 3

B = Event that the selected number is divisible by 4

Here, the events A and B are not mutually exclusive because 12 is divisible by both 3 and 4.

$$P(A) = \frac{6}{20},\ P(B) = \frac{5}{20},\ P(A \cap B) = \frac{1}{20}$$

$$\therefore\ P(A \cup B) = P(A) + P(B) - P(A \cap B)$$

$$= \frac{6}{20} + \frac{5}{20} - \frac{1}{20} = \frac{10}{20} = \frac{1}{2}.$$

Illustration 7: The probability that at least one of the events A and B occurs is 0.7 and they occur simultaneously with probability 0.2. Then $P(\bar{A}) + P(\bar{B}) =$

(a) 1.8 **(b) 0.6**
(c) 1.1 **(d) 0.4**

Solution: (c) We have $P(A \cup B) = 0.7$ and $P(A \cap B) = 0.2$

Now, $P(A \cup B) = P(A) + P(B) - P(A \cap B)$

$$\Rightarrow P(A) + P(B) = 0.9 \Rightarrow 1 - P(\bar{A}) + 1 - P(\bar{B}) = 0.9$$

$$\Rightarrow P(\bar{A}) + P(\bar{B}) = 1.1$$

INDEPENDENT EVENTS

Two or more events are said to be independent if occurrence or non-occurrence of any of them does not influence the probability of occurrence or non-occurrence of other events.

For example, when two cards are drawn from a pack of 52 playing cards with replacement (i.e., the first card drawn is put back in the pack and then the second card is drawn), then the event of occurrence of a king in the first draw and the event of occurrence of a king in the second draw are independent events because the occurrence or non-occurrence of a king in first draw does not influence the probability of occurrence or non-occurrence of the king in second draw. You can also see that the probability of drawing a king in the second draw is $\frac{4}{52}$ whether a king is drawn in the first draw or not. But if the two cards are drawn without replacement, then the two events are not independent, because in this case probability of drawing a king in the second draw depends on weather a king is drawn in first draw or not. If a king is drawn in first draw, then probability of drawing a king in second draw will be $\frac{3}{51}$ but if a king is not drawn in first draw, then the probability of drawing a king in second draw will be $\frac{4}{51}$.

Illustration 8: A fair coin is tossed repeatedly. If the tail appears on first four tosses, then the probability of the head appearing on the fifth toss equals

(a) $\frac{1}{2}$ **(b)** $\frac{1}{32}$

(c) $\frac{31}{32}$ **(d)** $\frac{1}{5}$

Solution: (a) The event that the fifth toss results a head is independent of the event that the first four tosses results tails.

$\therefore$ Probability of the required event = 1/2.

CONDITIONAL PROBABILITY

Let A and B be two events associated with a random experiment. Then, the probability of occurrence of A under the condition that B has already occurred and $P(B) \neq 0$, is called the conditional probability of occurrence of A when B has already occurred and it is denoted by $P(A/B)$.

Thus, $P(A/B)$ = Probability of occurrence of A, if B has already occurred and $P(B) \neq 0$

$$= \frac{P(A \cap B)}{P(B)} = \frac{\frac{n(A \cap B)}{n(S)}}{\frac{n(B)}{n(S)}} = \frac{n(A \cap B)}{n(B)}$$

Similarly, $P(B/A)$ = Probability of occurrence of B, if A has already occurred and $P(B) \neq 0$

$$= \frac{P(A \cap B)}{P(A)} = \frac{n(A \cap B)}{n(A)}$$

1. Multiplication Theorem on Probability

If A and B are two events associated with a random experiment, then

$$P(A \cap B) = P(A).\,P(B/A), \text{ if } P(A) \neq 0$$

or $$P(A \cap B) = P(B).\,P(A/B), \text{ if } P(B) \neq 0.$$

2. Multiplication Theorem for Independent Events

If A and B are independent events associated with a random experiment, then $P(A/B) = P(A)$ and $P(B/A) = P(B)$

$\therefore \quad P(A \cap B) = P(A)\cdot P(B/A) = P(A).\,P(B)$

i.e., the probability of simultaneous occurrence of two independent events is equal to the product of probability of their individual occurrence.

Extension of multiplication theorem for independent events

If $A_1, A_2, ..., A_n$ are independent events associated with a random experiment, then

$$P(A_1 \cap A_2 \cap A_3 \cap ... \cap A_n) = P(A_1)\,P(A_2)\,...\,P(A_n).$$

3. Probability of Occurrence of at Least One of the *n* Independent Events

If $p_1, p_2, p_3, ..., p_n$ be the probabilities of occurrence of n independent events $A_1, A_2, A_3, ..., A_n$ respectively, then

(i) Probability of happening none of them

$$= P(\bar{A}_1 \cap \bar{A}_2 \cap \bar{A}_3 ..., \cap A_n)$$
$$= P(\bar{A}_1)\,P(\bar{A}_2)\cdot P(\bar{A}_3)\,...\,P(\bar{A}_n)$$
$$= (1-p_1)(1-p_2)(1-p_3)\,...\,(1-p_n)$$

(ii) Probability of happening at least one of them

$$= P(A_1 \cup A_2 \cup A_3 ... \cup A_n)$$
$$= 1 - P\overline{(A_1 \cup A_2 \cup ... \cup A_n)}$$
$$= 1 - P(\bar{A}_1 \cap \bar{A}_2 \cap \bar{A}_3 ... \cap \bar{A}_n)$$
$$= 1 - P(\bar{A}_1)\,P((\bar{A}_2)\,P(\bar{A}_3)\,...\,P(\bar{A}_n)$$
$$= 1 - (1-p_1)(1-p_2)(1-p_3)\,...\,(1-p_n)$$

Illustration 9: A man and his wife appear for an interview for two posts. The probability of the husband's selection is $\frac{1}{7}$ and that of the wife's selection is $\frac{1}{5}$. The probability that only one of them will be selected is

(a) $\frac{6}{7}$ **(b)** $\frac{4}{35}$

(c) $\frac{6}{35}$ **(d)** $\frac{2}{7}$

Solution: (d) Probability that only husband is selected

$$= P(H)\,P(\bar{W}) = \frac{1}{7}\left(1-\frac{1}{5}\right) = \frac{1}{7}\times\frac{4}{5} = \frac{4}{35}$$

Probability that only wife is selected

$$= P(\bar{H})\,P(W) = \left(1-\frac{1}{7}\right)\left(\frac{1}{5}\right) = \frac{6}{7}\times\frac{1}{5} = \frac{6}{35}$$

$\therefore$ Probability that only one of them is selected

$$= \frac{4}{35} + \frac{6}{35} = \frac{10}{35} = \frac{2}{7}$$

Illustration 10: A bag contains 4 red and 4 blue balls. Four balls are drawn one by one from the bag, then find the probability that the drawn balls are in alternate colour.

Solution:

E_1 : Event that first drawn ball is red, second is blue and so on.

E_2 : Event that first drawn ball is blue, second is red and so on.

$$\therefore \quad P(E_1) = \frac{4}{8}\times\frac{4}{7}\times\frac{3}{6}\times\frac{3}{5} \text{ and } P(E_2) = \frac{4}{8}\times\frac{4}{7}\times\frac{3}{6}\times\frac{3}{5}$$

$$P(E) = P(E_1) + P(E_2) = 2\times\frac{4}{8}\cdot\frac{4}{7}\cdot\frac{3}{6}\cdot\frac{3}{5} = \frac{6}{35}$$

Illustration 11: A bag contains 5 red and 4 green balls. Four balls are drawn at random then find the probability that two balls are of red colour and two balls are of green.

Solution:

$n(S)$ = The total number of ways of drawing 4 balls out of total 9 balls = ${}^9C_4$.

If A_1 = The event of drawing 2 red balls out of 5 red balls then $n(A_1) = {}^5C_2$.

A_2 = The event of drawing 2 green balls out of 4 greens balls then $n(A_2) = {}^4C_2$.

Let A = The event of drawing 2 balls are of red colour and 2 balls are of green colour.

$\therefore \quad n(A) = n(A_1)\,.\,n(A_2) = {}^5C_2 \times {}^4C_2$

$$\therefore \quad P(A) = \frac{n(A)}{n(S)} = \frac{{}^5C_2 \times {}^4C_2}{{}^9C_4} = \frac{\frac{5\times4\times4\times3}{2\times2}}{\frac{9\times8\times7\times6}{4\times3\times2}} = \frac{10}{21}$$

Illustration 12: Let A, B, C be 3 independent events such that $P(A) = \frac{1}{3}$, $P(B) = \frac{1}{2}$, $P(C) = \frac{1}{4}$. Then find the probability of exactly 2 events occurring out of 3 events.

Solution: P (exactly two of A, B, C occur)

$$= P(A \cap B) + P(B \cap C) + P(C \cap A) - 3P(A \cap B \cap C)$$
$$= P(A).\,P(B) + P(B).\,P(C) + P(C).\,P(A) - 3P(A).\,P(B).\,P(C)$$
$$= \frac{1}{3}\cdot\frac{1}{2} + \frac{1}{2}\cdot\frac{1}{4} + \frac{1}{4}\cdot\frac{1}{3} - 3\frac{1}{3}\cdot\frac{1}{2}\cdot\frac{1}{4} = \frac{1}{4}.$$

Illustration 13: A bag contains 3 red, 6 white and 7 blue balls. Two balls are drawn one by one. What is the probability that first ball is white and second ball is blue when first drawn ball is not replaced in the bag?

Solution:

Let A = Event of drawing a white ball in first draw

and B = Event of drawing a blue ball in second draw

Here A and B are dependent events.

$$P(A) = \frac{6}{16},\ P\left(\frac{B}{A}\right) = \frac{7}{15}$$

$$P(A \cap B) = P(A)\,.\,P\left(\frac{B}{A}\right) = \frac{6}{16}\times\frac{7}{15} = \frac{7}{40}.$$

Illustration 14: Three coins are tossed together. What is the probability that first shows head, second shows tail and third shows head?

Solution: Let A = The event first coin shows head
B = The event that second coin shows tail
C = The event that third coin shows head

These three events are mutually independent.

So, $P(A \cap B \cap C) = P(A) \,.\, P(B) \,.\, P(C) = \frac{1}{2} \cdot \frac{1}{2} \cdot \frac{1}{2} = \frac{1}{8}$.

Illustration 15: A problem of mathematics is given to three students *A*, *B*, and *C*; whose chances of solving it are 1/2, 1/3, 1/4 respectively. Then find the probability that the problem will be solved.

Solution: Obviously the events of solving the problem by A, B and C are independent.

The problem will be solved if at least one of the three students will solve the problem.

Therefore required probability

$$= 1 - \left[\left(1 - \frac{1}{2}\right)\left(1 - \frac{1}{3}\right)\left(1 - \frac{1}{4}\right)\right] = 1 - \frac{1}{2} \cdot \frac{2}{3} \cdot \frac{3}{4} = \frac{3}{4}$$

Illustration 16: Two dice are thrown simultaneously. Find the probability that the sum of the number appeared on two dice is 8, if it is known that the second dice always exhibits 4.

Solution: Let A be the event of occurrence of 4 always on the second

dice = $\{(1, 4), (2, 4), (3, 4), (4, 4), (5, 4), (6, 4)\}$, $\therefore\ n(A) = 6$

and B be the event of occurrences of such numbers on both dice whose sum is 8 = $\{(2, 6), (3, 5), (4, 4), (5, 3), (6, 2)\}$

Thus, $A \cap B = \{(4, 4)\}$

$\therefore\quad n(A \cap B) = 1$

$$\therefore\quad P\left(\frac{B}{A}\right) = \frac{n(A \cap B)}{n(A)} = \frac{1}{6}.$$

Illustration 17: A coin is tossed thrice. If E be the event of showing at least two heads and F be the event of showing head in the first throw, then find $P\left(\frac{E}{F}\right)$.

Solution:

$S = \{HHH, HHT, HTH, THH, HTT, THT, TTH, TTT\}$
$E = \{HHH, HHT, HTH, THH\}$
$F = \{HHH, HHT, HTH, HTT\}$
$E \cap F = \{HHH, HHT, HTH\}$
$n(E \cap F) = 3,\ n(F) = 4$

$$\therefore \text{ Reqd prob.} = P\left(\frac{E}{F}\right) = \frac{n(E \cap F)}{n(F)} = \frac{3}{4}.$$

Practice Exercise

Level - I

1. Two dice are thrown simultaneously. The probability of obtaining a total score of seven is

(a) $\frac{1}{6}$ (b) $\frac{1}{3}$

(c) $\frac{2}{7}$ (d) $\frac{5}{6}$

2. Four balls are drawn at random from a bag containing 5 white, 4 green and 3 black balls. The probability that exactly two of them are white is

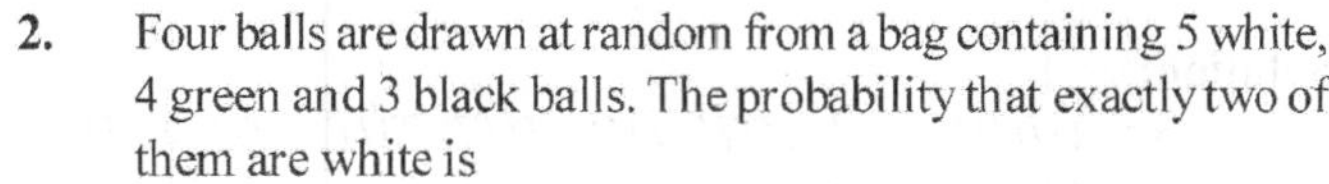

(a) $\frac{14}{33}$ (b) $\frac{7}{16}$

(c) $\frac{18}{33}$ (d) $\frac{9}{16}$

3. Two dice are tossed. The probability that the total score is a prime number is :

(a) $\frac{1}{6}$ (b) $\frac{5}{12}$

(c) $\frac{1}{2}$ (d) $\frac{7}{9}$

4. Anil can kill a bird once in 3 shots. On the assumption that he fires 3 shots, find the probability that the bird is killed.

(a) $\frac{1}{3}$ (b) $\left(\frac{1}{3}\right)^3$

(c) $\frac{19}{27}$ (d) $\frac{8}{9}$

5. If A and B are two independent events with $P(A) = 0.6$, $P(B) = 0.3$, then $P(A' \cap B')$ is equal to :

(a) 0.18 (b) 0.28

(c) 0.82 (d) 0.72

6. The probabilities that A and B will die within a year are p and q respectively, then the probability that only one of them will be alive at the end of the year is -

(a) $p+q$ (b) $p+q-pq$

(c) $p+q+pq$ (d) $p+q-2pq$

7. A pair of dice is thrown thrice. The probability of throwing doublets at least once is

(a) $\frac{1}{36}$ (b) $\frac{25}{216}$

(c) $\frac{125}{216}$ (d) None of these

8. The probability of getting number 5 in throwing a dice is

(a) 1 (b) $\frac{1}{3}$

(c) $\frac{1}{6}$ (d) $\frac{5}{6}$

9. The probability of getting head and tail alternately in three throws of a coin (or a throw of three coins), is

(a) $\frac{1}{8}$ (b) $\frac{1}{4}$

(c) $\frac{1}{3}$ (d) $\frac{3}{8}$

10. A die is thrown once. What is the probability of occurrence of an odd number on the upper face?

(a) $\frac{2}{3}$ (b) $\frac{1}{2}$

(c) $\frac{1}{4}$ (d) $\frac{1}{8}$

11. A die is thrown once. Find the probability that 3 or greater than 3 turns up.

(a) $\frac{1}{2}$ (b) $\frac{1}{3}$

(c) $\frac{1}{4}$ (d) $\frac{2}{3}$

12. Find the probability of getting a multiple of 2 in the throw of a die.

(a) 1/2 (b) 1/4

(c) 1/3 (d) 1/6

13. India and Pakistan play a 5 match test series of hockey, the probability that India wins at least three matches is

(a) $\frac{1}{2}$ (b) $\frac{3}{5}$

(c) $\frac{4}{5}$ (d) None of these

14. The probability that a man can hit a target is 3/4. He tries 5 times. The probability that he will hit the target at least three times is

(a) $\frac{291}{364}$ (b) $\frac{371}{461}$

(c) $\frac{471}{502}$ (d) $\frac{459}{512}$

15. From eighty cards numbered 1 to 80, two cards are selected randomly. The probability that both the cards have the numbers divisible by 4 is given by

(a) $\frac{21}{316}$ (b) $\frac{19}{316}$

(c) $\frac{1}{4}$ (d) None of these

16. The probability of getting sum more than 7 when a pair of dice are thrown is

(a) $\frac{7}{36}$ (b) $\frac{5}{12}$

(c) $\frac{7}{12}$ (d) None of these

17. Two dice are thrown simultaneously then the probability of obtaining a total score of 5 is

(a) $\frac{1}{18}$ (b) $\frac{1}{12}$

(c) $\frac{1}{9}$ (d) None of these

18. The probability that the two digit number formed by digits 1, 2, 3, 4, 5 is divisible by 4 is

(a) $\frac{1}{30}$ (b) $\frac{1}{20}$

(c) $\frac{1}{5}$ (d) None of these

19. Probability of throwing 16 in one throw with three dice is

(a) $\frac{1}{36}$ (b) $\frac{1}{18}$

(c) $\frac{1}{72}$ (d) $\frac{1}{9}$

20. Of a total of 600 bolts, 20% are too large and 10% are too small. The remainder are considered to be suitable. If a bolt is selected at random, the probability that it will be suitable is

(a) $\left(\frac{1}{5}\right)$ (b) $\left(\frac{7}{10}\right)$

(c) $\left(\frac{1}{10}\right)$ (d) $\left(\frac{3}{10}\right)$

21. The probability that in the toss of two dice we obtain the sum 7 or 11 is

(a) $\frac{1}{6}$ (b) $\frac{1}{18}$

(c) $\frac{2}{9}$ (d) $\frac{23}{108}$

22. A card is drawn at random from a pack of 100 cards numbered 1 to 100. The probability of drawing a number which is a square, is

(a) $\frac{1}{10}$ (b) $\frac{1}{100}$

(c) $\frac{9}{10}$ (d) $\frac{90}{100}$

23. The alphabets of word ALLAHABAD are arranged at random. The probability that in the words so formed, all identical alphabets are found together, is

(a) 1/63 (b) 16/17

(c) 5!/9! (d) None of these

24. The probability that Krishna will be alive 10 years hence, is $\frac{7}{15}$ and that Hari will be alive is $\frac{7}{10}$. What is the probability that both Krishna and Hari will be dead 10 years hence ?

(a) $\frac{21}{150}$ (b) $\frac{24}{150}$

(c) $\frac{49}{150}$ (d) $\frac{56}{150}$

25. The probability that in the random arrangement of the letters of the word 'UNIVERSITY', the two I's does not come together is

(a) $\frac{4}{5}$ (b) 1/5

(c) 1/10 (d) 9/10

26. Among 15 players, 8 are batsmen and 7 are bowlers. Find the probability that a team is chosen of 6 batsmen and 5 bowlers:

(a) $\frac{{}^{8}C_6 \times {}^{7}C_5}{{}^{15}C_{11}}$ (b) $\frac{28}{15}$

(c) $\frac{15}{28}$ (d) None of these

27. A four digit number is formed by the digits 1, 2, 3, 4 with no repetition. The probability that the number is odd is

(a) zero (b) $\frac{1}{3}$

(c) $\frac{1}{4}$ (d) None of these

28. X speaks truth in 60% and Y in 50% of the cases. The probability that they contradict each other narrating the same incident is

(a) $\frac{1}{4}$ (b) $\frac{1}{3}$

(c) $\frac{1}{2}$ (d) $\frac{2}{3}$

29. An integer is chosen at random from the numbers 1, 2, ..., 25. The probability that the chosen number is divisible by 3 or 4, is

(a) $\frac{2}{25}$ (b) $\frac{11}{25}$

(c) $\frac{12}{25}$ (d) $\frac{14}{25}$

30. The probability that a leap year will have 53 Friday or 53 Saturday, is

(a) $\frac{2}{7}$ (b) $\frac{3}{7}$

(c) $\frac{4}{7}$ (d) $\frac{1}{7}$

31. An experiment yields 3 mutually exclusive and exhaustive events A, B, C. If $P(A) = 2P(B) = 3P(C)$, then $P(A)$ is equal to

(a) $\frac{1}{11}$ (b) $\frac{2}{11}$

(c) $\frac{3}{11}$ (d) $\frac{6}{11}$

32. If $P(A \cup B) = 0.8$ and $P(A \cap B) = 0.3$, then $P(A') + P(B')$ equals to

(a) 0.3 (b) 0.5

(c) 0.7 (d) 0.9

33. Five coins whose faces are marked 2, 3 are thrown. What is the probability of obtainining a total of 12 ?

(a) $\frac{1}{16}$ (b) $\frac{3}{16}$

(c) $\frac{5}{16}$ (d) $\frac{7}{16}$

34. An aircraft has three engines A, B and C. The aircraft crashes if all the three engines fail. The probabilities of failure are 0.03, 0.02 and 0.05 for engines A, B and C respectively. What is the probability that the aircraft will not crash?

(a) 0.00003 (b) 0.90

(c) 0.99997 (d) 0.90307

35. The probability that a student passes in mathematics is 4/9 and that he passes in physics is 2/5. Assuming that passing in mathematics and physics are independent of each other, what is the probability that he passes in mathematics but fails in physics?

(a) $\frac{4}{15}$ (b) $\frac{8}{45}$

(c) $\frac{26}{45}$ (d) $\frac{19}{45}$

36. From a pack of 52 cards, two cards are drawn, the first being replaced before the second is drawn. What is the probability that the first is a diamond and the second is a king?

(a) $\frac{1}{4}$ (b) $\frac{4}{13}$

(c) $\frac{1}{52}$ (d) $\frac{4}{15}$

37. In a lottery, 16 tickets are sold and 4 prizes are awarded. If a person buys 4 tickets, what is the probability of his winning a prize?

(a) $\frac{4}{16^4}$ (b) $\frac{175}{256}$

(c) $\frac{1}{4}$ (d) $\frac{81}{256}$

38. Each of A and B tosses two coins. What is the probability that they get equal number of heads?

(a) $\frac{3}{16}$ (b) $\frac{5}{16}$

(c) $\frac{4}{16}$ (d) $\frac{6}{16}$

39. The chance of winning the race of the horse A is 1/5 and that of horse B is 1/6. What is the probability that the race will be won by A or B ?

(a) 1/30 (b) 1/3

(c) 11/30 (d) 1/15

40. What is the probability of two persons being born on the same day (ignoring date)?

(a) 1/49 (b) 1/365

(c) 1/7 (d) 2/7

41. The probabilities of two events A and B are given as $P(A) = 0.8$ and $P(B) = 0.7$. What is the minimum value of $P(A \cap B)$?

(a) 0 (b) 0.1

(c) 0.5 (d) 1

42. In tossing three coins at a time, what is the probability of getting at most one head?

(a) $\frac{3}{8}$ (b) $\frac{7}{8}$

(c) $\frac{1}{2}$ (d) $\frac{1}{8}$

43. Two balls are selected from a box containing 2 blue and 7 red balls. What is the probability that at least one ball is blue?

(a) $\frac{2}{9}$ (b) $\frac{7}{9}$

(c) $\frac{5}{12}$ (d) $\frac{7}{12}$

44. The probability of guessing a correct answer is $\frac{x}{12}$. If the probability of not guessing the correct answer is $\frac{2}{3}$, then what is x equal to?

(a) 2 (b) 3

(c) 4 (d) 6

45. A man and his wife appear for an interview for two posts. The probability of the husband's selection is $\frac{1}{7}$ and that of the wife's selection is $\frac{1}{5}$. The probability that only one of them will be selected is

(a) $\frac{6}{7}$ (b) $\frac{4}{35}$

(c) $\frac{6}{35}$ (d) $\frac{2}{7}$

46. The probability that a person will hit a target in shooting practice is 0.3. If he shoots 10 times, the probability that he hits the target is

(a) 1 (b) $1 - (0.7)^{10}$

(c) $(0.7)^{10}$ (d) $(0.3)^{10}$

47. Suppose six coins are tossed simultaneously. Then the probability of getting at least one tail is

(a) $\frac{71}{72}$ (b) $\frac{53}{54}$

(c) $\frac{63}{64}$ (d) $\frac{1}{12}$

48. In a single throw with four dice, the probability of throwing seven is

(a) $\frac{4}{6^4}$ (b) $\frac{8}{6^4}$

(c) $\frac{16}{6^4}$ (d) $\frac{20}{6^4}$

49. Six dice are thrown. The probability that different number will turn up is

(a) $\frac{129}{1296}$ (b) $\frac{1}{54}$

(c) $\frac{5}{324}$ (d) $\frac{5}{54}$

50. If two dice are tossed, find the probability of throwing a total of ten or more.

(a) $\frac{1}{6}$ (b) $\frac{1}{3}$

(c) $\frac{1}{4}$ (d) $\frac{2}{3}$

51. From a pack of 52 cards two are drawn with replacement. The probability, that the first is a diamond and the second is a king, is

(a) 1/26 (b) 17/2704

(c) 1/52 (d) None of these

52. Two cards are selected at random from a deck of 52 playing cards. The probability that both the cards are greater than 2 but less than 9 is

(a) $\frac{46}{221}$ (b) $\frac{63}{221}$

(c) $\frac{81}{221}$ (d) $\frac{93}{221}$

53. If A and B are two independent events such that $P(A)=\frac{1}{2}$ and $P(B)=\frac{1}{5}$, then which is not true?

(a) $P(A\cup B)=\frac{3}{5}$ (b) $P(A/B)=\frac{1}{4}$

(c) $P(A/A\cup B)=\frac{5}{6}$ (d) $P(A\cap B/\bar{A}\cup\bar{B})=0$

54. The probability that a man will live 10 more years is $\frac{1}{4}$ and the probability that his wife will live 10 more years is $\frac{1}{3}$. Then the probability that neither will be alive in 10 years is

(a) $\frac{5}{12}$ (b) $\frac{7}{12}$

(c) $\frac{1}{2}$ (d) $\frac{11}{12}$

55. A and B play a game where each is asked to select a number from 1 to 25. If the two numbers match, both of them win a prize. The probbility that they will not win a prize in a single trial is

(a) $\frac{1}{25}$ (b) $\frac{24}{25}$

(c) $\frac{2}{25}$ (d) None of these

56. The probability of happening an event A in one trial is 0.4. The probability that the event A happens at least once in three independent trials is –

(a) 0.936 (b) 0.216

(c) 0.904 (d) 0.784

57. Find the probability of drawing a jack or an ace from a pack of playing cards.

(a) $\frac{1}{8}$ (b) $\frac{1}{6}$

(c) $\frac{1}{3}$ (d) $\frac{2}{13}$

58. When two dice are thrown, the probability that the difference of the numbers on the dice is 2 or 3 is

(a) $\frac{7}{18}$ (b) $\frac{3}{11}$

(c) $\frac{5}{18}$ (d) $\frac{1}{2}$

59. In shuffling a pack of cards three are accidentally dropped. The probability that the missing cards are of distinct colours is

(a) $\frac{169}{425}$ (b) $\frac{165}{429}$

(c) $\frac{162}{459}$ (d) $\frac{164}{529}$

60. Four persons are selected at random out of 3 men, 2 women and 4 children. The probability that there exactly 2 children in the selection is

(a) $\frac{11}{21}$ (b) $\frac{9}{21}$

(c) $\frac{10}{21}$ (d) None of these

61. It is given that the events A and B are such that

$P(A)=\frac{1}{4}, P(A\mid B)=\frac{1}{2}$ and $P(B\mid A)=\frac{2}{3}$. Then $P(B)$ is

(a) $\frac{1}{6}$ (b) $\frac{1}{3}$

(c) $\frac{2}{3}$ (d) $\frac{1}{2}$

62. A coin is tossed and a dice is rolled. The probability that the coin shows the head and the dice shows 6 is

(a) $\frac{1}{2}$ (b) $\frac{1}{6}$

(c) $\frac{1}{12}$ (d) $\frac{1}{24}$

Level - II

1. If $P(A) = 0.8$, $P(B) = 0.9$, $P(AB) = p$, which one of the following is correct?
(a) $0.72 \leq p \leq 0.8$ (b) $0.7 \leq p \leq 0.8$
(c) $0.72 < p < 0.8$ (d) $0.7 < p < 0.8$

2. A, B, C are three mutually exclusive event associated with a random experiment. Find $P(A)$ if it is given that $P(B) = 3/2\ P(A)$ and $P(C) = 1/2\ P(B)$.
(a) $\frac{4}{13}$ (b) $\frac{2}{3}$
(c) $\frac{12}{13}$ (d) $\frac{1}{13}$

3. The probability that A can solve a problem is $\frac{2}{3}$ and B can solve it is $\frac{3}{4}$. If both attempt the problem, what is the probability that the problem gets solved?
(a) $\frac{11}{12}$ (b) $\frac{7}{12}$
(c) $\frac{5}{12}$ (d) $\frac{9}{12}$

4. A dice is thrown 6 times. If 'getting an odd number' is a 'success', the probability of 5 successes is
(a) $\frac{1}{10}$ (b) $\frac{3}{32}$
(c) $\frac{5}{6}$ (d) $\frac{25}{26}$

5. A bag contains 5 white and 3 black balls, and 4 are successively drawn out and not replaced. What's the chance of getting different colours alternatively?
(a) $\frac{1}{6}$ (b) $\frac{1}{5}$
(c) $\frac{1}{4}$ (d) $\frac{1}{7}$

6. A bag contains 5 white and 7 black balls and a man draws 4 balls at random. The odds against these being all black is
(a) 7 : 92 (b) 92 : 7
(c) 92 : 99 (d) 99 : 92

7. The letters of the word SOCIETY are placed at random in a row. The probability that the three vowels come together is
(a) $\frac{1}{6}$ (b) $\frac{1}{7}$
(c) $\frac{2}{7}$ (d) $\frac{5}{6}$

8. Course materials are sent to students by a distance education institution. The probability that they will send a wrong programme's study material is $\frac{1}{5}$. There is a probability of $\frac{3}{4}$ that the package is damaged in transit, and there is a probability of $\frac{1}{3}$ that there is a short shipment. What is the probability that the complete material for the course arrives without any damage in transit ?
(a) $\frac{4}{5}$ (b) $\frac{8}{60}$
(c) $\frac{8}{15}$ (d) $\frac{4}{20}$

9. A coin is tossed 5 times. What is the probability that head appears an odd number of times?
(a) $\frac{2}{5}$ (b) $\frac{1}{5}$
(c) $\frac{1}{2}$ (d) $\frac{4}{25}$

10. Two dice are tossed. The probability that the total score is a prime number is
(a) $\frac{1}{6}$ (b) $\frac{5}{12}$
(c) $\frac{1}{2}$ (d) $\frac{7}{9}$

11. The probability that the sum of the square of the two numbers, which show up when two fair dice are thrown, is even is
(a) $\frac{3}{7}$ (b) $\frac{4}{7}$
(c) $\frac{5}{7}$ (d) None of these

12. There are 5 pairs of shoes in a cupboard from which 4 shoes are picked at random. The probability that there is at least one pair is
(a) $\frac{8}{21}$ (b) $\frac{11}{21}$
(c) $\frac{13}{21}$ (d) $\frac{12}{31}$

13. The fair dice are thrown. The probability that the number appear are not all distinct is
(a) $\frac{5}{9}$ (b) $\frac{4}{9}$
(c) $\frac{1}{6}$ (d) $\frac{5}{6}$

14. Two dice are thrown simultaneously. What is the probability of obtaining a multiple of 2 on one of them and a multiple of 3 on the other

(a) $\frac{5}{36}$ (b) $\frac{11}{36}$

(c) $\frac{1}{6}$ (d) $\frac{1}{3}$

15. Two dice are thrown at a time, find the probability that the sums of the numbers on the upper faces of the dice are equal to 7.

(a) $\frac{1}{8}$ (b) $\frac{1}{4}$

(c) $\frac{1}{3}$ (d) $\frac{1}{6}$

16. One card is drawn from a well-shuffled pack of 52 cards. What is the probability, that it is not the ace of hearts ?

(a) $\frac{51}{52}$ (b) $\frac{1}{52}$

(c) $\frac{1}{12}$ (d) $\frac{1}{2}$

17. A dice is thrown twice. The probability of getting 4, 5 or 6 in the first throw and 1, 2, 3 or 4 in the second throw is

(a) 1/3 (b) 2/3

(c) 1/2 (d) 1/4

18. Ram and Shyam appear for an interview for two vacancies in an organisation for the same post. The probabilities of their selection are 1/6 and 2/5 respectively. What is the probability that none of them will be selected?

(a) 5/6 (b) 1/5

(c) 1/2 (d) 3/5

19. A class consists of 80 students, 25 of them are girls and 55 are boys. If 10 of them are rich and the remaining poor and also 20 of them are intelligent then the probability of selecting an intelligent rich girl is

(a) $\frac{5}{128}$ (b) $\frac{25}{128}$

(c) $\frac{5}{512}$ (d) None of these

20. If the probability of A to fail in an examination is 0.2 and that for B is 0.3, then probability that either A or B is fail, is :

(a) 0.5 (b) 0.44

(c) 0.8 (d) 0.25

21. The probability of choosing at random a number that is divisible by 6 or 8 from among 1 to 90 is equal to

(a) $\frac{1}{6}$ (b) $\frac{1}{30}$

(c) $\frac{11}{80}$ (d) $\frac{23}{90}$

22. In single cast with two dice the odds against drawing 7 is

(a) 5 (b) $\frac{1}{5}$

(c) 6 (d) $\frac{1}{6}$

23. From a group of 7 men and 4 women a committee of 6 persons is formed. What is the probability that the committee will consist of exactly 2 women?

(a) $\frac{5}{11}$ (b) $\frac{3}{11}$

(c) $\frac{4}{11}$ (d) $\frac{2}{11}$

24. Two numbers a and b are chosen at random from the set of first 30 natural numbers. The probability that $a^2 - b^2$ is divisible by 3 is:

(a) $\frac{37}{87}$ (b) $\frac{47}{87}$

(c) $\frac{17}{29}$ (d) None of these

25. An article manufactured by a company consists of two parts X and Y. In the process of manuifacture of the part X, 9 out of 100 parts may be defective. Similarly, 5 out of 100 are likely to be defective int he manufacture of the part. Y. Calculate the probability that the assembled product will not be defective.

(a) 0.6485 (b) 0.6565

(c) 0.8645 (d) None of these

26. If $P(A) = 3/7$, $P(B) = 1/2$ and $P(A' \cap B') = 1/14$, then are A and B are mutually exclusive events?

(a) No (b) Yes

(c) Either yes or no (d) Cannot be determined

27. One bag contains 4 white balls and 2 black balls. Another bag contains 3 white balls and 5 black balls. If one ball is drawn from each bag, determine the probability that one ball is white and another is black.

(a) 6/24 (b) 5/24

(c) 7/24 (d) 13/24

28. The probability that A can solve a problem is $\frac{2}{3}$ and B can solve it is $\frac{3}{4}$. If both attempt the problem, what is the probability that the problem gets solved?

(a) $\frac{11}{12}$ (b) $\frac{7}{12}$

(c) $\frac{5}{12}$ (d) $\frac{9}{12}$

29. Atul can hit a target 3 times in 6 shots, Bhola can hit the target 2 times in 6 shots and Chandra can hit the 4 times in 4 shots. What is the probability that at least 2 shots (out of 1 shot taken by each one of them) hit the target ?

(a) $\frac{1}{2}$ (b) $\frac{2}{3}$

(c) $\frac{1}{3}$ (d) $\frac{5}{6}$

30. Suppose six coins are tossed simultaneously. Then the probability of getting at least one tail is :

(a) $\frac{71}{72}$ (b) $\frac{53}{54}$

(c) $\frac{63}{64}$ (d) $\frac{1}{12}$

31. Seven digits from the numbers 1, 2, 3, 4, 5, 6, 7, 8, 9 are written in a random order. The probability that this seven digit number is divisible by 9 is

(a) $\frac{2}{9}$ (b) $\frac{7}{36}$

(c) $\frac{1}{9}$ (d) $\frac{7}{12}$

32. A committee of 5 Students is to be chosen from 6 boys and 4 girls. Find the probability that the committee contains exactly 2 girls.

(a) 10/21 (b) 11/21
(c) 12/21 (d) 13/21

33. 4 gentlemen and 4 ladies take seats at random round a table. The probability that they are sitting alternately is

(a) 4/35 (b) 1/70
(c) 2/35 (d) 1/35

34. Two cards are drawn one by one from a pack of cards. The probability of getting first card an ace and second a coloured one is (before drawing second card, first card is not placed again in the pack) :

(a) 1/26 (b) 5/52
(c) 5/221 (d) 4/13

35. Seven people seat themselves indiscriminately at round table. The probability that two distinguished persons will be next to each other is

(a) $\frac{1}{3}$ (b) $\frac{1}{2}$

(c) $\frac{1}{4}$ (d) $\frac{2}{3}$

36. Let $0 < P(A) < 1$, $0 < P(B) < 1$ and $P(A \cup B) = P(A) + P(B) - P(A)P(B)$, then :

(a) $P(B/A) = P(B) - P(A)$

(b) $P(A' \cup B') = P(A') + P(B')$

(c) $P(A \cap B) = P(A')P(B')$

(d) None of these

37. Eleven books, consisting of five Engineering books, four Mathematics books and two Physics books, are arranged in a shelf at random. What is the probability that the books of each kind are all together?

(a) $\frac{5}{1155}$ (b) $\frac{2}{1155}$

(c) $\frac{3}{1155}$ (d) $\frac{1}{1155}$

38. 12 persons are seated around a round table. What is the probability that two particular persons sit together?

(a) $\frac{2}{11}$ (b) $\frac{1}{6}$

(c) $\frac{3}{11}$ (d) $\frac{3}{15}$

39. Two small squares on a chess board are choosen at random. Find the probability that they have a common side:

(a) $\frac{1}{12}$ (b) $\frac{1}{18}$

(c) $\frac{2}{15}$ (d) $\frac{3}{14}$

40. A bag contains 7 blue balls and 5 yellow balls. If two balls are selected at random, what is the probability that **none** is yellow? *[SBI PO-2013]*

(a) $\frac{5}{35}$ (b) $\frac{5}{22}$

(c) $\frac{7}{22}$ (d) $\frac{7}{33}$

(e) $\frac{7}{66}$

41. A die is thrown twice. What is the probability of getting a sum 7 from both the throws? *[SBI PO-2013]*

(a) $\frac{5}{18}$ (b) $\frac{1}{18}$

(c) $\frac{1}{9}$ (d) $\frac{1}{6}$

(e) $\frac{5}{36}$

DIRECTIONS (Qs. 42-46) : ***Study the given information carefully to answer the questions that follow.***

An urn contains 4 green, 5 blue, 2 red and 3 yellow marbles.

42. If four marbles are drawn at random, what is the probability that two are blue and two are red? *[IBPS-PO-2011]*

(a) $\frac{10}{1001}$ (b) $\frac{9}{14}$

(c) $\frac{17}{364}$ (d) $\frac{2}{7}$

(e) None of these

43. If eight marbles are drawn at random, what is the probability that there are equal number of marbles of each colour ? [*IBPS-PO-2011*]

(a) $\frac{4}{7}$ (b) $\frac{361}{728}$

(c) $\frac{60}{1001}$ (d) $\frac{1}{1}$

(e) None of these

44. If two marbles are drawn at random, what is the probability that both are red or at least one is red ? [*IBPS-PO-2011*]

(a) $\frac{26}{91}$ (b) $\frac{1}{7}$

(c) $\frac{199}{364}$ (d) $\frac{133}{191}$

(e) None of these

45. If three marbles are drawn at random, what is the probability that at least one is yellow ? [*IBPS-PO-2011*]

(a) $\frac{1}{3}$ (b) $\frac{199}{364}$

(c) $\frac{165}{364}$ (d) $\frac{3}{11}$

(e) None of these

46. If three marbles are drawn at random, what is the probability that none is green ? [*IBPS-PO-2011*]

(a) $\frac{2}{7}$ (b) $\frac{253}{728}$

(c) $\frac{10}{21}$ (d) $\frac{14}{91}$

(e) None of these

Hints & Solutions

Level-I

1. **(a)** When two are thrown then there are 6 × 6 exhaustive cases ∴ $n = 36$. Let A denote the event "total score of 7" when 2 dice are thrown then A = [(1, 6), (2, 5), (3, 4), (4, 3), (5, 2), (6, 1)].
Thus there are 6 favourable cases.

$\therefore m = 6$ By definition $P(A) = \dfrac{m}{n}$

$\therefore P(A) = \dfrac{6}{36} = \dfrac{1}{6}$.

2. **(a)** ∵ No of ways of drawing 2 white balls from 5 white balls = ${}^5C_2$.
Also, No of ways of drawing 2 other from remaining 7 balls = ${}^7C_2$
Total number of balls = 12

Hence, required probability $= \dfrac{{}^5C_2 \times {}^7C_2}{{}^{12}C_4} = \dfrac{14}{33}$

3. **(b)** Total no. of outcomes when two dice are thrown = $n(S) = 36$ and the possible cases for the event that the sum of numbers on two dice is a prime number, are (1, 1), (1, 2), (1, 4), (1, 6), (2, 1), (2, 3), (2, 5), (3, 2), (3, 4), (4, 1), (4, 3), (5, 1), (5, 6), (6, 1), (6, 5)
Number of outcomes favouring the event = $n(A) = 15$

Required probability $= \dfrac{n(A)}{n(S)} = \dfrac{15}{36} = \dfrac{5}{12}$

4. **(c)** $P(A) = \dfrac{1}{3}$, $P(\bar{A}) = \dfrac{2}{3}$

P (bird killed) = 1 – P (none of 3 shots hit)

$= 1 - \dfrac{2}{3} \times \dfrac{2}{3} \times \dfrac{2}{3} = \dfrac{19}{27}$.

5. **(b)** Since, A and B are independent events
∴ A' and B' are also independent events

$\Rightarrow P(A' \cap B') = P(A').P(B')$

$= (0.4)(0.7) = 0.28$

$[\because P(A') = 1 - P(A),\ P(B') = 1 - P(B)]$

6. **(d)**

7. **(d)** Doublets occur when the numbers thrown are (1, 1), (2, 2), ..., (6, 6). Therefore the probability of a doublet occurring in single throw $= \dfrac{6}{36} = \dfrac{1}{6}$.

The probability of a doublet not occurring at all in three throws $= \left(\dfrac{5}{6}\right)^3 = \dfrac{125}{216}$.

Required probability $= 1 - \dfrac{125}{216} = \dfrac{91}{216}$.

8. **(c)** Required probability = 1/6.

9. **(b)** Total probable ways = 8
Favourable number of ways = HTH, THT

Hence required probability $= \dfrac{2}{8} = \dfrac{1}{4}$

10. **(b)** Any of the six numbers 1,2,3,4,5,6 may appear on the upper face. ∴ $n = 6$
Number of odd numbers = 3, since the odd numbers are 1, 3, 5
∴ $m = 3$.
∴ The required probability

$= \dfrac{\text{number of favourable cases}}{\text{number of all cases}} = \dfrac{m}{n} = \dfrac{3}{6} = \dfrac{1}{2}$

11. **(d)** n = Number of all cases = 6
m = Number of favourable cases = 4 (since the numbers that appear are 3, 4, 5, 6)

∴ The required probability = $p = \dfrac{m}{n} = \dfrac{4}{6} = \dfrac{2}{3}$

12. **(a)** $S = (1, 2, 3, 4, 5, 6)$ ∴ $n(S) = 6$
Let A be the event that the die shows a multiple of 2.
$A = \{2, 4, 6\}$ ∴ $n(A) = 3$

$P(A) = \dfrac{n(A)}{n(S)} = \dfrac{3}{6} = \dfrac{1}{2}$

13. **(a)** India win atleast three matches

$= {}^5C_3\left(\dfrac{1}{2}\right)^5 + {}^5C_4\left(\dfrac{1}{2}\right)^5 + {}^5C_5\left(\dfrac{1}{2}\right)^5 = \left(\dfrac{1}{2}\right)^5(16)$

$= \left(\dfrac{1}{2}\right)$

14. **(d)** Required probability

$= {}^5C_3\left(\dfrac{3}{4}\right)^3\left(\dfrac{1}{4}\right)^2 + {}^5C_4\left(\dfrac{3}{4}\right)^4 \cdot \left(\dfrac{1}{4}\right) + {}^5C_5\left(\dfrac{3}{4}\right)^5 = \dfrac{459}{512}$

15. **(b)** Total no. of numbers divisible by 4 between 1 to 80
$80 = 4 + (n - 1)4$
$80 = 4n$
$\Rightarrow n = 20$

∴ Required probability $= \dfrac{{}^{20}C_2}{{}^{80}C_2} = \dfrac{19}{316}$

16. **(b)** Here $n(S) = 6^2 = 36$

Let E be the event "getting sum more than 7" *i.e.* sum of pair of dice = 8, 9, 10, 11, 12

$$i.e., \ E = \begin{Bmatrix} (2,6), & (3,5), & (4,4), & (5,3), & (6,2), \\ (3,6), & (4,5), & (5,4), & (6,3), & \\ (4,6), & (5,5), & (6,4), & & \\ (5,6), & (6,5), & (6,6) & & \end{Bmatrix}$$

$\therefore \ n(E) = 15$

$\therefore$ Required prob $= \frac{n(E)}{n(S)} = \frac{15}{36} = \frac{5}{12}$

17. **(c)** Number of sample points on throwing two dice
$= 6 \times 6 = 36$

The possible outcomes are (1, 4), (2, 3), (3, 2), (4, 1)

The probability of obtaining a total score of 5 is

$= \frac{4}{6 \times 6} = \frac{1}{9}$.

18. **(c)** Given digits are 1, 2, 3, 4, 5

Total no. of 2 digits numbers formed $= (5)^2 = 25$

Favourable cases are 12, 24, 32, 44, 52

No. of favourable cases = 5

$\therefore$ Required Probability $= \frac{5}{25} = \frac{1}{5}$

19. **(a)** Total no. of cases $= 6^3 = 216$

16 can appear on three dice in following ways

(6, 6, 4), (6, 5, 5), (6, 4, 6), (4, 6, 6), (5, 5, 6), (5, 6, 5).

$\therefore$ No. of favourable cases = 6

Hence, the required probability $= \frac{6}{6^3} = \frac{1}{36}$

20. **(b)** Total number of bolts = 600

Number of too large bolts = 20% of 600

$= \frac{20 \times 600}{100} = 120$

Number of too small bolts = 10% of 600 = 60

Number of suitable bolts = 600 – 120 – 60 = 420

Thus required probability $= \frac{420}{600} = \frac{7}{10}$

21. **(c)** Total possible outcomes = 36

E = Event of getting sum 7

$= \{(1,6), (2,5), (3,4), (4,3), (5,2), (6,1)\}$

F = Event of getting sum 11

$= \{(6,5), (5,6)\}$

Total no. of favourable cases = 6 + 2 = 8

Now required probability

$= \frac{\text{Total favourable cases}}{\text{Total outcomes}} = \frac{8}{36} = \frac{2}{9}$

22. **(a)** $n(S) = 100$

E = square numbers from 1 to 100.

= 1, 4, 9, 16, 25, 36, 49, 64, 81, 100

$n(E) = 10$

$\therefore$ Required probability $= \frac{n(E)}{n(S)} = \frac{10}{100} = \frac{1}{10}$

23. **(a)** (AAAA), (LL), HBD

$$P = \frac{\frac{5!}{9!}}{4!\,2!} = \frac{5! \times 4! \times 2!}{9!} = \frac{24 \times 2}{9 \times 8 \times 7 \times 6} = \frac{1}{63}$$

24. **(b)** The probability that Krishna will be alive 10 years hence, is $\frac{7}{15}$

So, probability that Krishna will be dead 10 years hence, the

$= 1 - \frac{7}{15} = \frac{8}{15}$

Also, probability that Hari will be alive 10 years hence is $\frac{7}{10}$

So, the probability that Hari will be dead 10 years hence, $= 1 - \frac{7}{10} = \frac{3}{10}$

So, the probability that both Krishna and Hari will be dead 10 years hence

$= \frac{8}{15} \times \frac{3}{10} = \frac{24}{150}$

25. **(a)** Total no. of arrangements of the letters of the word UNIVERSITY is $\frac{10!}{2!}$.

No. of arrangements when both I's are together = 9!

So. the no. of ways in which 2 I's do not together

$= \frac{10!}{2!} - 9!$

$\therefore$ Required probability

$$= \frac{\frac{10!}{2!} - 9!}{\frac{10!}{2!}} = \frac{10! - 9!\,2!}{10!}$$

26. **(a)** Total no. of players = 15

Total no. of batsmen = 8

Total no. of bowlers = 7

Total no. of players in the team = 11

$\therefore$ No. of ways to choose a team $= {}^{15}C_{11}$

$\therefore$ No. of way to choose 6 batsmen and 5 bowler

$= {}^{8}C_6 \times {}^{7}C_5$

$\therefore$ Required Probability $= \frac{{}^{8}C_6 \times {}^{7}C_5}{{}^{15}C_{11}}$

27. **(d)** Total number of numbers = 4! = 24
For odd nos. 1 or 3 has to be at unit's place
If 1 is at unit place, then total number of numbers
$= 3! = 6$
And if 3 is at units place, then total number of numbers
$= 3! = 6$
$\therefore$ Total number of odd number = 6 + 6 = 12

$\therefore$ Required probability $= \frac{12}{24} = \frac{1}{2}$

28. **(c)** Required Probability $= P(X).P(\bar{Y}) + P(\bar{X}).P(Y)$

$= \frac{60}{100} \times \frac{50}{100} + \frac{40}{100} \times \frac{50}{100} = \frac{1}{2}$

29. **(c)** $P(3 \cup 4) = P(c) + P(d) - P(3 \cap 4)$

$= \frac{8}{25} + \frac{6}{25} - \frac{2}{25} = \frac{12}{25}$

30. **(b)** In a leap year there are 366 days in which 52 weeks and two days. The combination of 2 days may be : Sun-Mon, Mon-Tue, Tue-Wed, Wed-Thu, Thu-Fri, Fri-Sat, Sat-Sun.

$P(53 \text{ Fri}) = \frac{2}{7}$; $P(53 \text{ Sat}) = \frac{2}{7}$

and $P(53 \text{ Fri and } 53 \text{ Sat}) = \frac{1}{7}$

$\therefore$ $P(53 \text{ Fri or Sat}) = P(53 \text{ Fri}) + P(53 \text{ Sat}) - P(53 \text{ Fri and Sat})$

$= \frac{2}{7} + \frac{2}{7} - \frac{1}{7} = \frac{3}{7}$

31. **(d)** Clearly, $P(A \cup B \cup C) = 1$
$\Rightarrow P(A) + P(B) + P(C) = 1$

$\Rightarrow P(A) + \frac{1}{2}P(A) + \frac{1}{3}P(A) = 1$

$\Rightarrow \frac{11}{6}P(A) = 1$

$\Rightarrow P(A) = \frac{6}{11}$

32. **(d)** Now, $P(A' \cap B') = P(A' \cup B')$
$= 1 - P(A \cup B) = 1 - 0.8 = 0.2$
and $P(A' \cup B') = 1 - P(A \cap B) = 1 - 0.3 = 0.7$
But $P(A' \cup B') = P(A') + P(B') - P(A' \cap B')$
$\Rightarrow 0.7 = P(A') + P(B') - 0.2$
$\Rightarrow P(A') + P(B') = 0.9.$

33. **(c)** Let E be the event of total of 12.
E = (2, 2, 2, 3, 3), (2, 2, 3, 3, 2), (2, 3, 3, 2, 2), (3, 3, 2, 2, 2), (3, 2, 3, 2, 2), (3, 2, 2, 3, 2), (3, 2, 2, 2, 3), (2, 3, 2, 3, 2), (2, 3, 2, 2, 3), (2, 2, 3, 2, 3)
$n(E) = 10$
Sample sapce contain total possibility $= 2^5 = 32$
Hence, $n(s) = 32$

So, $P(E) = \frac{n(E)}{n(S)} = \frac{10}{32} = \frac{5}{16}$

34. **(c)** Since, probabilities of failure for engines A, B and C $P(A)$, $P(B)$ and $P(C)$ are 0.03, 0.02 and 0.05 respectively.
The aircraft will crash only when all the three engine fail. So, probability that it crashes = $P(A).P(B).P(C)$
$= 0.03 \times 0.02 \times 0.05 = 0.00003$
Hence, the probability that the aircraft will not crash,
$= 1 - 0.00003 = 0.99997$

35. **(a)** Probability of passing in mathematics $= \frac{4}{9}$

Probability of passing in physics $= \frac{2}{5}$

Probability of failure in physics $= 1 - \frac{2}{5} = \frac{3}{5}$
Given that both the events are independent.

Required probability $= \frac{4}{9} \times \frac{3}{5} = \frac{4}{15}$

36. **(c)** Probability of getting a diamond, $P(D) = \frac{13}{52} = \frac{1}{4}$

and probability to king, $P(K) = \frac{4}{52} = \frac{1}{13}$

So, required probability $= P(D).P(K)$

$= \frac{1}{4} \times \frac{1}{13} = \frac{1}{52}$

37. **(c)** 16 tickets are sold and 4 prizes are awarded. A person buys 4 tickets, then required probability $= \frac{4}{16} = \frac{1}{4}$

38. **(b)** If both get one head then it is $\frac{1}{4} \times \frac{1}{4}$

and if both get two heads then it is $\frac{1}{2} \times \frac{1}{2}$

$\Rightarrow$ Prob (getting same number of heads) $= \frac{1}{4} \times \frac{1}{4} + \frac{1}{2} \times \frac{1}{2}$

$= \frac{1}{16} + \frac{1}{4} = \frac{5}{16}$

39. **(c)** Let $P(A)$ be the probability that the race will be won by A and $P(B)$ be the probability that the race will be won by B.

$\therefore P(A) = \frac{1}{5}$ and $P(B) = \frac{1}{6}$

$\therefore$ Probability that the race will be won by

A or $B = P(A) + P(B) = \frac{1}{5} + \frac{1}{6} = \frac{11}{30}$

40. **(b)** Required probability $= \frac{365}{365} \times \frac{1}{365} = \frac{1}{365}$.

41. **(c)** As we know $P(A \cup B) \le 1$

$\therefore P(A) + P(B) - P(A \cap B) \le 1$

$\Rightarrow 0.8 + 0.7 - P(A \cap B) \le 1$

$\Rightarrow P(A \cap B) \ge 1.5 - 1$

$\Rightarrow P(A \cap B) \ge 0.5$

Hence, the minimum value of $P(A \cap B)$ is 0.5.

42. **(c)** Possible samples are as follows

$\{HHH, HTH, HHT, THH, TTH, THT, HTT, TTT\}$

Let A be the event of getting one head.

Let B be the event of getting no head.

Favourable outcome for

$A = \{TTH, THT, HTT\}$

Favourable outcome for

$B = \{TTT\}$

Total no. of outcomes = 8

$\therefore P(A) = \frac{3}{8}, P(B) = \frac{1}{8}$

$\therefore$ Required probability = Probability of getting one head + Probability of getting no head

$= P(A) + P(B) = \frac{3}{8} + \frac{1}{8} = \frac{4}{8} = \frac{1}{2}$

43. **(a)** No. of blue balls =2

No. of red balls = 7

Total no. of balls = 9

Required probability

= P (one ball is blue) + P (both ball is blue)

$= \frac{2}{9} \times \frac{7}{8} + \frac{2}{9} \times \frac{1}{8} = \frac{14}{72} + \frac{2}{72} = \frac{16}{72} = \frac{2}{9}$

44. **(c)** Given probability of guessing a correct answer $= \frac{x}{12}$

and probability of not guessing the correct answer $= \frac{2}{3}$

As we know

P (occurence of an event) + P (non-occurence of an event) = 1

$\therefore \frac{x}{12} + \frac{2}{3} = 1 \Rightarrow \frac{x+8}{12} = 1 \Rightarrow x = 12 - 8 = 4$

45. **(d)** Probability that only husband is selected

$= P(H)P(\bar{W}) = \frac{1}{7}\left(1 - \frac{1}{5}\right) = \frac{1}{7} \times \frac{4}{5} = \frac{4}{35}$

Probability that only wife is selected

$= P(\bar{H})P(W) = \left(1 - \frac{1}{7}\right)\left(\frac{1}{5}\right) = \frac{6}{7} \times \frac{1}{5} = \frac{6}{35}$

$\therefore$ Probability that only one of them is selected

$= \frac{4}{35} + \frac{6}{35} = \frac{10}{35} = \frac{2}{7}$

46. **(b)** The probability that the person hits the target = 0.3

$\therefore$ The probability that he does not hit the target in a trial = $1 - 0.3 = 0.7$

$\therefore$ The probability that he does not hit the target in any of the ten trials = $(0.7)^{10}$

$\therefore$ Probability that he hits the target

= Probability that at least one of the trials succeceds

$= 1 - (0.7)^{10}$.

47. **(c)** If six coins are tossed, then the total no. of outcomes $= (2)^6 = 64$

Now, probability of getting no tail $= \frac{1}{64}$

Probability of getting at least one tail

$= 1 - \frac{1}{64} = \frac{63}{64}$

48. **(d)** Total of seven can be obtained in the following ways

1, 1, 1, 4 in $\frac{4!}{3!} = 4$ ways

[there are four objects, three repeated]

Similarly,

1, 1,2, 3 in $\frac{4!}{2!} = 12$ ways

1, 2,2, 2 in $\frac{4!}{3!} = 4$ ways

Hence, required probability $= \frac{4+12+4}{6^4} = \frac{20}{6^4}$

[$\because$ Exhaustive no. of cases $= 6 \times 6 \times 6 \times 6 = 6^4$]

49. **(c)** The number of ways of getting the different number 1, 2, ..., 6 in six dice = 6 !.

Total number of ways $= 6^6$

Hence, required probability $= \frac{6!}{6^6}$

$= \frac{1 \times 2 \times 3 \times 4 \times 5 \times 6}{6^6} = \frac{5}{324}$

50. **(a)** Here the number of favourable cases, consists of throwing 10,11 or 12 with the two dice. The number of ways in which a sum of 10 can be thrown are (4,6), (5,5), (6,4) i.e. 3 ways. The number of ways in which a total of 11 can be thrown are (5, 6), (6,5) i.e. 2 ways. The number of ways in which a total of 12 can be thrown in (6, 6) i.e. 1 way.

m = number of favourable cases = 3 + 2 + 1 = 6

n = Total number of cases = $6 \times 6 = 36$

$\therefore$ Probability $= p = \frac{m}{n} = \frac{6}{36} = \frac{1}{6}$

51. **(c)** Required probability

$= P\text{ (Diamond)} . P\text{ (King)} = \frac{13}{52}.\frac{4}{52} = \frac{1}{52}$

52. **(a)** The cards are of four colours and the number of cards of given description is 24.

The probability $= \frac{24}{52}.\frac{23}{51} = \frac{46}{221}$.

53. **(b)** Since A and B are independent

$\therefore\ P(A \cap B) = P(A).P(B)$

and $P(A/B) = P(A)$

Thus, $P(A/B) = \frac{1}{2}$

Hence, option (b) is not true.

54. **(c)** The probability that a man will not live 10 more years = 3/4 and the probability that his wife will not live 10 more years = 2/3. Then the probability that neither will be alive in 10 years = 3/4 × 2/3 = 1/2

55. **(b)** Total number of possibilities = 25 × 25

Favourable cases for their winning = 25

$\therefore P\text{ (they win a prize)} = \frac{25}{25 \times 25} = \frac{1}{25}$

$\therefore P\text{ (they will not win a prize)} = 1 - \frac{1}{25}$

56. **(d)** Here $P(A) = 0.4$ and $P(\bar{A}) = 0.6$

Probability that A does not happen at all = $(0.6)^3$

Thus required probability = $1 - (0.6)^3 = 0.784$

57. **(d)** As there are four jacks and four aces, the number of favourable cases = 8

$\therefore$ The required probability $p = \frac{8}{52} = \frac{2}{13}$

58. **(a)** The favourable cases are (1, 3), (2, 4), (3,5), (4,6) and (1, 4), (2, 5), (3, 6) and their reversed cases like (3, 1)......

Total number of favourable cases = 2 × 7

$\therefore p = \frac{14}{36} = \frac{7}{18}$

59. **(a)** The first card can be one of the 4 colours, the second can be one of the three and the third can be one of the two. The required probability is therefore

$4 \times \frac{13}{52} \times 3 \times \frac{13}{51} \times 2 \times \frac{13}{50} = \frac{169}{425}$.

60. **(c)** Total number of ways in which 4 persons can be selected out of 3 + 2 + 4 = 9 persons = ${}^9C_4 = 126$

Number of ways in which a selection of 4 contains exactly 2 children = ${}^4C_2 \times {}^5C_2 = 60$

$\therefore$ reqd. prob. $= \frac{60}{126} = \frac{10}{21}$

61. **(b)** $P(A) = 1/4$, $P(A/B) = \frac{1}{2}$, $P(B/A) = 2/3$

By conditional probability,

$P(A \cap B) = P(A)\,P(B/A) = P(B)P(A/B)$

$\Rightarrow \frac{1}{4} \times \frac{2}{3} = P(B) \times \frac{1}{2} \Rightarrow P(B) = \frac{1}{3}$

62. **(c)** Probability of getting a head on tossing a coin $(P_1) = \frac{1}{2}$.

Probability of getting a six on rolling a dice $(P_2) = \frac{1}{6}$.

These two events are independent.

So the probability that the coin shows the head and the dice shows 6 is given by

$P = P_1 \times P_2 = \frac{1}{2} \times \frac{1}{6} = \frac{1}{12}$.

Level-II

1. **(b)** We know,

$P(A \cup B) = P(A) + P(B) - P(A \cap B)$

$\Rightarrow\ 0.8 + 0.9 - p \le 1$

$\Rightarrow\ 1.7 - p \le 1$

$\Rightarrow\ 0.7 \le p$

Now, $P(A) < P(B)$

$\therefore\ P(A \cap B) \le P(A)$

$\Rightarrow\ p \le 0.8$

Hence, $0.7 \le p \le 0.8$

2. **(a)** $P(A) + P(B) + P(C) = 1 \rightarrow 2\,P(B)/3 + P(B) + P(B)/2 = 1 \rightarrow 13P(B)/6 = 1 \rightarrow P(B) = 6/13$. Hence, $P(A) = 4/13$

3. **(a)** The probability that A cannot solve the problem

$= 1 - \frac{2}{3} = \frac{1}{3}$

The probability that B cannot solve the problem

$= 1 - \frac{3}{4} = \frac{1}{4}$

The probability that both A and B cannot solve the problem $= \frac{1}{3} \times \frac{1}{4} = \frac{1}{12}$

$\therefore$ The probability that at least one of A and B can solve the problem $= 1 - \frac{1}{12} = \frac{11}{12}$

$\therefore$ The probability that the problem is solved $= \frac{11}{12}$

4. **(b)** Let A be the event of getting an odd number.

Here, $n(S) = 6$ and

$n(A) = 3$

Probability of getting an odd number $= \frac{3}{6} = \frac{1}{2}$

Hence, probability of not getting an odd number

$=1-\frac{1}{2}=\frac{1}{2}$

Required probability of 5 successes

$= {}^6C_5 \times \left(\frac{1}{2}\right)^5 \times \frac{1}{2} = \frac{3}{32}$

5. **(d)** Total number of balls = 8. Let the first drawn ball is white, so required probability $= \frac{5}{8}\times\frac{3}{7}\times\frac{4}{6}\times\frac{2}{5}=\frac{1}{14}$.

But here we had started with a white ball. When we start with a black ball, the required probability

$= \frac{3}{8}\times\frac{5}{7}\times\frac{2}{6}\times\frac{4}{5}=\frac{1}{14}$.

Since these two cases are mutually exclusive.

Total probability $= \frac{1}{14}+\frac{1}{14}=\frac{2}{14}=\frac{1}{7}$.

6. **(b)** There are 7 + 5 = 12 balls in the bag and the number of ways in which 4 balls can be drawn is ${}^{12}C_4$ and the number of ways of drawing 4 black balls (out of seven) is ${}^7C_4$.

Hence, P (4 black balls)

$= \frac{{}^7C_4}{{}^{12}C_4} = \frac{7.6.5.4}{1.2.3.4}\times\frac{1.2.3.4}{12.11.10.9}=\frac{7}{99}$

Thus the odds against the event 'all black balls' are

$(1-\frac{7}{99}):\frac{7}{99}$: i.e., $\frac{92}{99}:\frac{7}{99}$ or $92:7$.

7. **(b)** The word 'SOCIETY' contains seven distinct letters and they can be arranged at random in a row in ${}^7P_7$ ways, *i.e.*, in 7! = 5040 ways.

Let us now consider those arrangements in which all the three vowels come together. So in this case we have to arrange four letters. S,C,T,Y and a pack of three vowels in a row which can be done in ${}^5P_5$ i.e. 5! = 120 ways.

Also, the three vowels in their pack can be arranged in ${}^3P_3$ i.e. 3! = 6 ways.

Hence, the number of arrangements in which the three vowels come together is 120 × 6 = 720

∴ The probability that the vowels come together

$= \frac{720}{5040}=\frac{1}{7}$

8. **(b)** Probability (sending a correct programme)

$=1-\frac{1}{5}=\frac{4}{5}$

Probability (the packet is not damaged) $=1-\frac{3}{4}=\frac{1}{4}$

Probability (there is no short shipment) $=1-\frac{1}{3}=\frac{2}{3}$

Required probability $=\frac{4}{5}\times\frac{1}{4}\times\frac{2}{3}=\frac{2}{15}=\frac{8}{60}$

9. **(c)** Probability of occurence of head in a toss of a coin is 1/2.

Required probability = Prob [Head appears once] + Prob. [Head appears thrice] + Prob. [Head appears five times]

$= {}^5C_1\left(\frac{1}{2}\right)^5 + {}^5C_3\left(\frac{1}{2}\right)^5 + {}^5C_5\left(\frac{1}{2}\right)^5$

$= \left(\frac{1}{2}\right)^5 [5+10+1] = \frac{16}{32}=\frac{1}{2}$

10. **(b)** Total no. of outcomes when two dice are thrown = $n(S)$ = 36 and the possible cases for the event that the sum of numbers on two dice is a prime number, are (1, 1), (1, 2), (1, 4), (1, 6), (2, 1), (2, 3), (2, 5), (3, 2), (3, 4), (4, 1), (4, 3), (5, 1), (5, 6), (6, 1), (6, 5)

Number of outcomes favouring the event = $n(A)$ = 15

Required probability $=\frac{n(A)}{n(S)}=\frac{15}{36}=\frac{5}{12}$

11. **(d)** Out of 36 possible outcomes the ones which are favourable for the event are

(i) When the numbers are both even and

(ii) When the numbers are both odd. There are six doublets and the pairs. (1, 3), (1, 5), (2, 4), (2, 6) etc. Which make a total of 6 × 3 = 18. The required probability is 1/2.

12. **(c)** There are 5 pairs of shoes and 4 shoes can be picked in 10 × 9 × 8 × 7 ways. Number of ways in which 4 shoes can be picked such that no two are alike = 10 × 8 × 6 × 4.

The required probability $= 1-\frac{10\times8\times6\times4}{10\times9\times8\times7}=\frac{13}{21}$.

13. **(b)** Out of the 6^3 possible outcomes 6.5.5 outcomes will have all distinct numbers.

The probability $= 1-\frac{6.5.4}{6^3}=\frac{4}{9}$.

14. **(b)** Favourable cases for one are there *i.e.*, 2, 4 and 6 and for other are two i.e., 3, 6.

Hence required probability $=\left[\left(\frac{3\times2}{36}\right)2-\frac{1}{36}\right]=\frac{11}{36}$

[As same way happen when dice changes numbers among themselves]

15. **(d)** If a die is thrown, there are 6 equally likely and mutually exclusive cases. Since two dice are thrown, the total number of ways = 6 × 6 = 36. If a sum of 7 is to be obtained from the numbers appearing on the two upper faces, the numbers in the two dice can be (1, 6), (2, 5), (3, 4), (4, 3), (5, 2), (6,1), which are six in number.

∴ Number of favourable cases = m = 6

Total number of cases = 36

∴ The required probability $= p = \frac{m}{n}=\frac{6}{36}=\frac{1}{6}$

16. (a) The ace of hearts can be drawn in only 1 way
($\because$ in a pack of cards there is only one ace of heart)

$P(A)$ = Probability of drawing the ace of hearts $= \frac{1}{52}$.

Hence the probability of not drawing an ace of hearts

$= P(\bar{A}) = 1 - P(A) = 1 - \frac{1}{52} = \frac{51}{52}$

17. (a) Let $P(A)$ and $P(B)$ be the probability of the events of getting 4, 5 or 6 in the first throw and 1, 2, 3 or 4 in the second throw respectively, then

$P(A \text{ and } B) = P(A).P(B) = \frac{1}{2} \times \frac{2}{3} = \frac{1}{3}$

18. (c) Required probability $= \left(1 - \frac{1}{6}\right) \times \left(1 - \frac{2}{5}\right) = \frac{5}{6} \times \frac{3}{5} = \frac{1}{2}$

19. (c) Total 80, Girls = 25, Boys = 55
10 R, 70 P, 20 I

$\frac{1}{4} \times \frac{1}{8} \times \frac{25}{80} = \frac{5}{512}$

20. (b) Given $P(A_f) = 0.2$ and $P(B_f) = 0.3$

Since, A and B are independent events

$\therefore \quad P(A \cap B) = P(A).P(B)$

$= (0.2) \times (0.3) = 0.06$

$\therefore$ Required prob $= P(A \cup B)$

$= P(A) + P(B) - P(A \cap B)$

$= 0.2 + 0.3 - 0.06 = 0.44$

21. (d) Nos. divisible by 6 are 6, 12, 18, ..., 90.
Nos. divisible by 8 are 8, 16, 24, ..., 88.
Now, total no. divisible by 6 = 15
and total no. divisible by 8 = 11
Now, the no. divisible by both 6 and 8 are 24, 48, 72.
So, total no. divisible by both 6 and 8 = 3
$\therefore$ Probability (number divisible by 6 or 8)

$= \frac{15 + 11 - 3}{90} = \frac{23}{90}$

22. (a) Let $E = \{(1, 6), (2, 5), (3, 4), (4, 3), (5, 2), (6, 1)\}$

$\therefore P(E) = \frac{6}{6 \times 6} = \frac{1}{6}$

So, odds against drawing 7

$= \frac{P(\bar{E})}{P(E)} = \frac{1 - \frac{1}{6}}{\frac{1}{6}} = \frac{5}{1}$

23. (a) ${}^6C_2 \times [(7/11) \times (6/10) \times (5/9) \times (4/8) \times (4/7) \times (3/6)]$
$= 5/11$.

24. (b) Out of 30 numbers 2 numbers can be chosen in ${}^{30}C_2$ ways. So, exhaustive number of cases $= {}^{30}C_2 = 435$
Since $a^2 - b^2$ is divisible by 3 if either a and b are divisible by numbers, of cases $= {}^{10}C_2 + {}^{20}C_2 = 235$

Hence, required probability $= \frac{235}{435} = \frac{47}{87}$

25. (c) Required probability
= P (X not defective and Y not defective)

$= P(\bar{X})P(\bar{Y})$

$= \{1 - P(X)\}\{1 - P(Y)\}$

$= \frac{91}{100} \times \frac{95}{100} = \frac{8645}{10000} = 0.8645$

26. (b) A(3/7) B(1/2) X

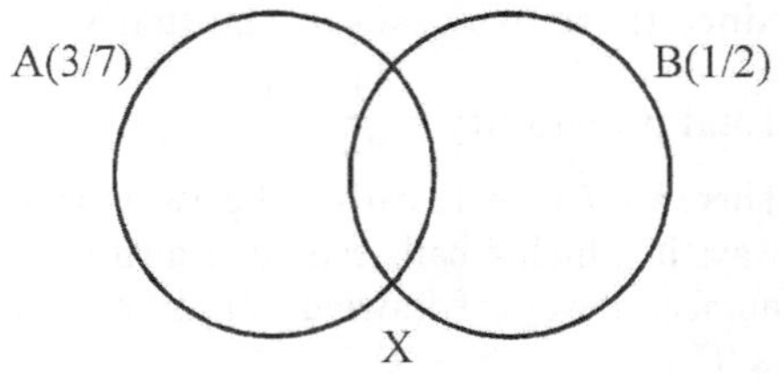

Also $P(A \cup B) = 1 - P(A' \cap B')$

$(3/7) + (1/2) - x = 13/14 \rightarrow x = 0$

Thus, there is no interference between A and B as $P(A \cup B) = x = 0$. Hence, A and B are mutually exclusive.

27. (d) Probability that first ball is white and second black
$= (4/6) \times (5/8) = 5/12$
Probability that first ball is black and second white
$= (2/6) \times (3/8) = 1/8$
These are mutually exclusive events hence the required probability

$P = \frac{5}{12} + \frac{1}{8} = \frac{13}{24}$.

28. (a) The probability that A cannot solve the problem

$= 1 - \frac{2}{3} = \frac{1}{3}$

The probability that B cannot solve the problem

$= 1 - \frac{3}{4} = \frac{1}{4}$

The probability that both A and B cannot solve the problem $= \frac{1}{3} \times \frac{1}{4} = \frac{1}{12}$

$\therefore$ The probability that at least one of A and B can solve the problem $= 1 - \frac{1}{12} = \frac{11}{12}$

$\therefore$ The probability that the problem is solved $= \frac{11}{12}$

29. (b) Chandra hits the target 4 times in 4 shots. Hence, he hits the target definitely.
The required probability, therefore, is given by.
P (both Atul and Bhola hit) + P (Atul hits, Bhola does not hit) + P (Atul does not hit, Bhola hits)

$$= \frac{3}{6}\times\frac{2}{6}+\frac{3}{6}\times\frac{4}{6}+\frac{3}{6}\times\frac{2}{6} = \frac{1}{6}+\frac{1}{3}+\frac{1}{6}=\frac{4}{6}=\frac{2}{3}$$

30. (c) If six coins are tossed, then the total no. of outcomes $= (2)^6 = 64$

Now, probability of getting no tail $= \frac{1}{64}$

Probability of getting at least one tail

$$= 1-\frac{1}{64}=\frac{63}{64}$$

31. (c) A number is divisible by 9, if the sum of its digits is divisible by 9. Here $1 + 2 + 3 + 4 + 5 + 6 + 7 + 8 + 9 = 45$ is divisible by 9.
$\therefore$ the two numbers to be removed should be such that their sum is 9.
$\therefore$ they can be any one of the following pairs (1, 8), (2, 7), (3, 6), (4, 5).
Hence the number of favourable cases = 4
Total number of cases of removing two numbers $= {}^9C_2$

$$\therefore \text{ Required probability} = \frac{4}{{}^9C_2} = \frac{4}{36} = \frac{1}{9}.$$

32. (a) 5 Students can be selected from 10 in ${}^{10}C_5$ ways.

$$\therefore n(S) = {}^{10}C_5 = \frac{10!}{5!.5!} = \frac{10\times9\times8\times7\times6}{5\times4\times3\times2} = 252$$

Let A be the event that the committee includes exactly 2 girls and 3 boys. The two girls. can be selected in ${}^4C_2$ ways and the 3 boys can be selected in ${}^6C_3$ ways.
$\therefore\ n(A) = {}^4C_2 \times {}^6C_3 = 6 \times 20 = 120$

$$\therefore P(A) = \frac{n(A)}{n(S)} = \frac{120}{252} = \frac{10}{21}$$

33. (d) $n(S) = 7!$, $n(E) = (3!) \times (4!)$

$$\therefore P(E) = \frac{(3!)\times(4!)}{7!} = \frac{6}{7\times6\times5} = \frac{1}{15} = \frac{1}{35}$$

34. (c) Let A be the event of getting first card an ace and B be the event of getting second a coloured one. Since, both the events associated with a random experiment.
(i.e. condition of probability)
Therefore, the probability of getting first card an ace

$$P(A) = \frac{4}{52} = \frac{1}{13}$$

and probability of drawing a coloured one in second draw

$$P(B/A) = \frac{15}{51} = \frac{5}{17}$$

(since one card has already been drawn)
Hence, by conditional probability,

$$P(B/A) = \frac{P(A\cap B)}{P(A)}$$

$$\Rightarrow \frac{5}{17} = \frac{P(A\cap B)}{\frac{1}{13}}$$

$$\Rightarrow P(A\cap B) = \frac{5}{17}\times\frac{1}{13} = \frac{5}{221}$$

35. (a) Seven people can seat themselves at a round table in 6! ways. The number of ways in which two distinguished persons will be next to each other = 2 (5) !, Hence, the required probability

$$= \frac{2(5)!}{6!} = \frac{1}{3}$$

36. (d) Given $P(A) + P(B) - P(A)\,P(B) = P(A \cup B)$
Comparing with
$$P(A) + P(B) - P(A \cap B) = P(A \cup B)$$
we get $P(A \cap B) = P(A).P(B)$
$\therefore$ A and B independent events.

37. (d) $\frac{(5!\times 4!\times 2!\times 3!)}{11!} = \frac{24\times2\times6}{11\times10\times9\times8\times7\times6} = 1/1155.$

38. (a)

$$P = \frac{\text{Total no. of ways in which two people sit together}}{\text{Total no. of ways}}$$

$= (10! \times 2!)/11!$

39. (b) The common side could be horizontal or vertical. Accordingly, the number of ways the event can occur is.
$n(E) = 8 \times 7 + 8 \times 7 = 112$
$n(S) = {}^{64}C_2$

$$\text{Required probability} = \frac{2\times8\times7\times2}{64\times63} = \frac{1}{18}$$

40. (c) Total balls = 12
Blue balls = 7
None of two balls are yellow i.e., both balls are blue.

$$\therefore \quad \text{P (both blue balls)} = \frac{7}{12}\times\frac{6}{11} = \frac{7}{22}$$

41. (d) Total possible outcomes when A die is thrown twice = 36
Outcome for getting a sum 7 from both throwns = 6{(2, 6), (2, 5), (3, 4), (4, 3), (5, 2), (6, 1)}

$$\therefore \quad P(E) = \frac{6}{36} = \frac{1}{6}$$

42. (a) According to question,

$$n(S) = {}^{14}C_4 = \frac{14!}{(14-4)!4!} = \frac{14!}{10!4!} \left[\because {}^nC_r = \frac{n!}{(n-r)!r!}\right]$$

$$= \frac{14\times13\times12\times11}{4\times3\times2\times1} = 1001$$

and $n(E) = {}^5C_2 \times {}^2C_2 = \frac{5!}{(5-2)!2!} \times \frac{2!}{(2-2)!2!}$

$$= \frac{5\times4}{2\times1} \times \frac{2\times1}{1\times2\times1} = 10$$

$\therefore$ Required probability $= \frac{n(E)}{n(S)} = \frac{10}{1001}$

43. (c) According to question,

$$n(S) = {}^{14}C_8 = \frac{14!}{(14-8)!8!} \times \frac{14!}{6!8!}$$

$$= \frac{14\times13\times12\times11\times10\times9}{6\times5\times4\times3\times2\times1} = 3003$$

and $n(E) = {}^4C_2 \times {}^5C_2 \times {}^2C_2 \times {}^3C_2$

$$= \frac{4!}{(4-2)!2!} \times \frac{5!}{(5-2)!2!} \times \frac{2!}{(2-2)!2!} \times \frac{3!}{(3-2)!2!}$$

$$= \frac{4!}{2!2!} \times \frac{5!}{3!2!} \times \frac{2!}{0!2!} \times \frac{3!}{1!2!}$$

$$= \frac{4\times3}{2\times1} \times \frac{5\times4}{2\times1} \times \frac{1}{1} \times \frac{3}{1} = 180$$

$\therefore$ Required probability $= \frac{n(E)}{n(S)} = \frac{180}{3003} = \frac{60}{1001}$

44. (e) According to question,

$$n(S) = {}^{14}C_2 = \frac{14!}{(14-2)!2!} = \frac{14\times13}{2\times1} = 91$$

$\therefore$ Probability of at least one red ball

$$= 1 - \frac{{}^{12}C_2}{{}^{14}C_2} = 1 - \frac{66}{91} = \frac{91-66}{91} = \frac{25}{91}$$

45. (b) According to question,

$$n(S) = {}^{14}C_3 = \frac{14!}{(14-3)!3!} = \frac{14\times13\times12}{3\times2\times1} = 364$$

$\therefore$ Required probability

$$= 1 - \frac{{}^{11}C_3}{{}^{14}C_3} = 1 - \frac{165}{364} = \frac{364-165}{364} = \frac{199}{364}$$

46. (e) According to question,

$$n(S) = {}^{14}C_3 = \frac{14!}{(14-3)!3!} = \frac{14\times13\times12}{3\times2\times1} = 364$$

and $n(E) = {}^{10}C_3 = \frac{10!}{(10-3)!3!} = \frac{10\times9\times8}{3\times2\times1} = 120$

$\therefore$ Required probability $= \frac{n(E)}{n(S)} = \frac{120}{364} = \frac{30}{91}$

CHAPTER

23 TRIGONOMETRY AND ITS APPLICATION

ANGLE

Consider a ray $\overrightarrow{OA}$. If this ray rotates about its end point O and takes the position OB, then the angle $\angle AOB$ has been generated. An angle is considered as the figure obtained by rotating a given ray about its end-point. The initial position OA is called the initial side and the final position OB is called terminal side of the angle. The end point O about which the ray rotates is called the vertex of the angle. The measure of an angle is the amount of rotation performed to get the terminal side from initial side.

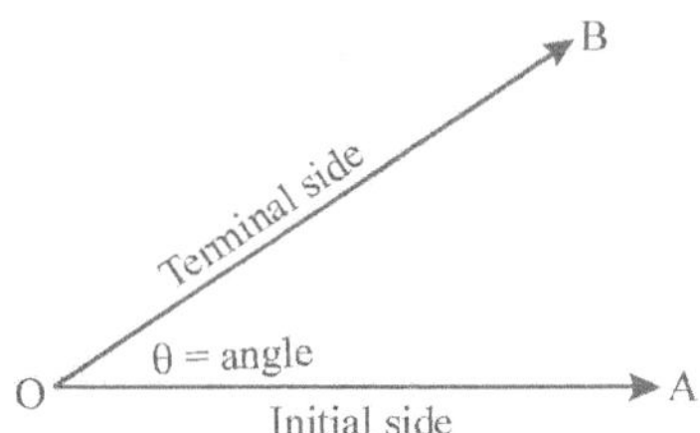

There are several units for measuring angles.

But in this chapter we use degree measure of angle.

Sense of sign of an angle :

The sense of sign of an angle is said to be positive or negative according as the initial side rotates in anticlockwise or clockwise direction respectively to get the terminal side.

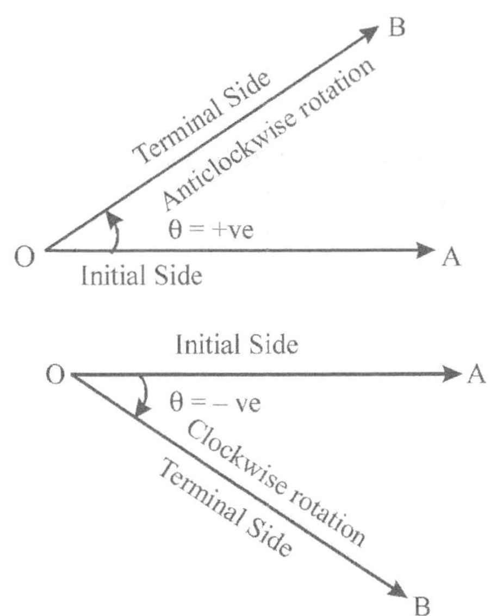

TRIGONOMETRIC RATIOS

In the figure XOX' and YOY' are horizontal and vertical axes respectively.

Horizontal axis $X'OX$ is called X-axis and vertical axis YOY' is called Y-axis.

Let A be a point on OX. Also suppose that the ray OA start rotating in the XY-plane in anti-clockwise direction from the initial position OA about the point O till it reaches its final position OC after some interval of time (See Fig.). Thus, an angle COA is formed with x-axis. Let $\angle COA = \theta$. (θ is a Greek letter, and we read it as "theta"). Draw $CB \perp OX$. Now clearly $\Delta\ CBO$ is a right angled triangled.

In right ΔCBO, OC is the hyponenus. For angle $\theta = \angle\ COA$, BC and OB are called side opposite to angle θ and adjucent side of angle θ respectively.

Let $CB = p$, $OB = b$ and $OC = h$. We define the different ratios between hypotenus, side opposite to angle θ and adjucent side of angle θ as trigonometric ratios for angle θ.

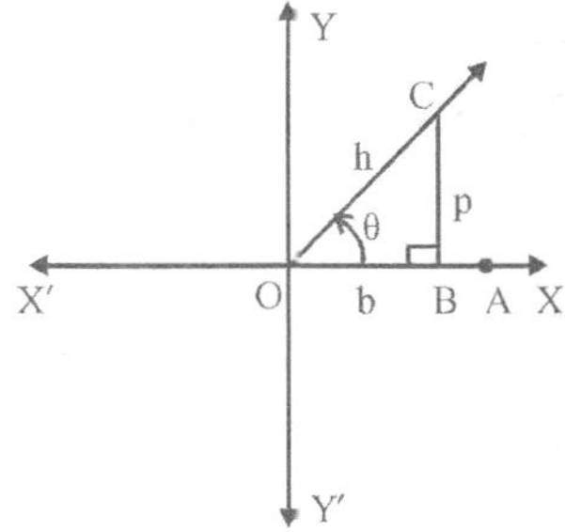

These trigonometrical ratios are :

$$\text{Sine of } \theta = \frac{\text{Side opposite to angle } \theta}{\text{Hypotenuse}} = \frac{CB}{OC} = \frac{p}{h}$$

$$\text{Cosine of } \theta = \frac{\text{Adjacent side to angle } \theta}{\text{Hypotenuse}} = \frac{OB}{OC} = \frac{b}{h}$$

$$\text{Tangent of } \theta = \frac{\text{Side opposite to angle } \theta}{\text{Adjacent side to angle } \theta} = \frac{CB}{OB} = \frac{p}{b}$$

Cotangent of $\theta = \dfrac{\text{Adjacent side to angle } \theta}{\text{Side opposite to angle } \theta} = \dfrac{OB}{CB} = \dfrac{b}{p}$

Secant of $\theta = \dfrac{\text{Hypotenuse}}{\text{Adjacent side to angle } \theta} = \dfrac{OC}{OB} = \dfrac{h}{b}$

Cosecant of $\theta = \dfrac{\text{Hypotenuse}}{\text{Side opposite to angle } \theta} = \dfrac{OC}{CB} = \dfrac{h}{p}$

Sine of θ is abbreviated as sin θ, Cosine of θ is abbreviated as cos θ, Tangent of θ is abbreviated as tan θ, Cotangent of θ is abbreviated as cot θ, Secant of θ is abbreviated as sec θ and Cosecant of θ is abbreviated as cosec θ

Now, throughout the study of trigonometry we shall use only abbreviated form of these trigonometric ratios.

Thus,

$$\sin\theta = \frac{p}{h},\ \cos\theta = \frac{b}{h},\ \tan\theta = \frac{p}{b},\ \cot\theta = \frac{b}{p},$$

$$\sec\theta = \frac{h}{b},\ \text{cosec}\,\theta = \frac{h}{p}$$

Note that sin θ is an abbreviation for "sine of angle θ"and not the product of sin and θ.

VALUE OF TRIGONOMETRIC RATIOS FOR SOME SPECIFIC ANGLES

The values of trigonometric ratios for angles 0°, 30°, 45°, 60° and 90° are quite often used in solving problems in our day-to-day life. Thus the following table is very useful.

IMPORTANT TABLE

(θ)→ Trigonometrical ratio ↓	0°	30°	45°	60°	90°
sin θ	0	$\frac{1}{2}$	$\frac{1}{\sqrt{2}}$	$\frac{\sqrt{3}}{2}$	1
cos θ	1	$\frac{\sqrt{3}}{2}$	$\frac{1}{\sqrt{2}}$	$\frac{1}{2}$	0
tan θ	0	$\frac{1}{\sqrt{3}}$	1	$\sqrt{3}$	not defined
cot θ	Not defined	$\sqrt{3}$	1	$\frac{1}{\sqrt{3}}$	0
cosec θ	Not defined	2	$\sqrt{2}$	$\frac{2}{\sqrt{3}}$	1
sec θ	1	$\frac{2}{\sqrt{3}}$	$\sqrt{2}$	2	not defined

BASIC FORMULAE OR TRIGONOMETRIC IDENTITY

(i) $\sin\theta \cdot \text{cosec}\,\theta = 1$ or $\sin\theta = \dfrac{1}{\text{cosec}\,\theta}$ or $\text{cosec}\,\theta = \dfrac{1}{\sin\theta}$

(ii) $\cos\theta \cdot \sec\theta = 1$ or $\cos\theta = \dfrac{1}{\sec\theta}$ or $\sec\theta = \dfrac{1}{\cos\theta}$

(iii) $\tan\theta \cdot \cot\theta = 1$ or $\cot\theta = \dfrac{1}{\tan\theta}$ or $\tan\theta = \dfrac{1}{\cot\theta}$

(iv) $\sin^2\theta + \cos^2\theta = 1$

or $\cos^2\theta = 1 - \sin^2\theta$ or $\sin^2\theta = 1 - \cos^2\theta$

(v) $\sec^2\theta - \tan^2\theta = 1$

or $\sec^2\theta = 1 + \tan^2\theta$ or $\tan^2\theta = \sec^2\theta - 1$

(vi) $\text{cosec}^2\theta - \cot^2\theta = 1$

or $\text{cosec}^2\theta = 1 + \cot^2\theta$ or $\cot^2\theta = \text{cosec}^2\theta - 1$

(vii) $\tan\theta = \dfrac{\sin\theta}{\cos\theta}$

(viii) $\cot\theta = \dfrac{\cos\theta}{\sin\theta}$

TRIGONOMETRIC RATIOS FOR COMPLEMENTARY ANGLES

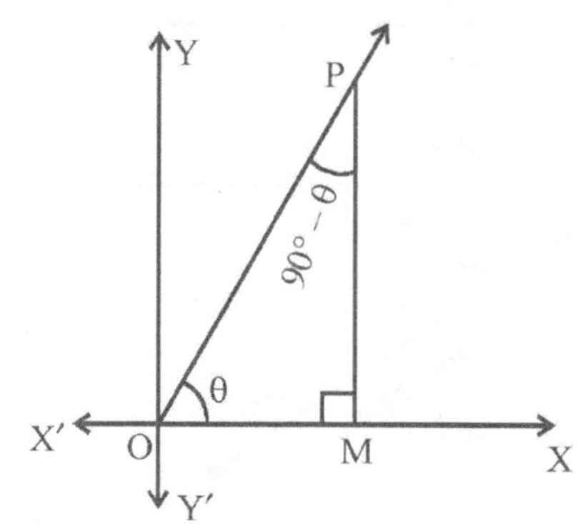

$\sin(90^\circ - \theta) = \frac{OM}{OP} = \cos\theta$, $\cos(90^\circ - \theta) = \frac{PM}{OP} = \sin\theta$,

$\tan(90^\circ - \theta) = \frac{OM}{PM} = \cot\theta$, $\cot(90^\circ - \theta) = \frac{PM}{OM} = \tan\theta$,

$\text{cosec}(90^\circ - \theta) = \frac{OP}{OM} = \sec\theta$ and $\sec(90^\circ - \theta) = \frac{OP}{PM} = \text{cosec}\,\theta$

Illustration 1: If $\tan A = 1$ and $\sin B = \frac{1}{\sqrt{2}}$, find the value of $\cos(A + B)$ where A and B are both acute angles.

Solution : $\tan A = 1 \Rightarrow A = 45^\circ$ and $\sin B = \frac{1}{\sqrt{2}} \Rightarrow B = 45^\circ$

$\therefore \cos(A + B) = \cos(45^\circ + 45^\circ) = \cos 90^\circ = 0$

Illustration 2: If $\sin\theta - \cos\theta = 0$ and $0 < \theta < 90^\circ$, find θ.

Solution :

$$\sin\theta - \cos\theta = 0 \Rightarrow \sin\theta = \cos\theta \Rightarrow \frac{\sin\theta}{\cos\theta} = 1 \Rightarrow \tan\theta = 1$$

But $0 < \theta < 90^\circ \Rightarrow \theta = 45^\circ$

Illustration 3: If A, B and C are interrior angles of a triangle ABC, then show that

$$\sin\left(\frac{B+C}{2}\right) = \cos\frac{A}{2}$$

Solution : In ΔABC,

$$A + B + C = 180^\circ \Rightarrow B + C = 180^\circ - A \Rightarrow \frac{B+C}{2} = 90^\circ - \frac{A}{2}$$

$$\Rightarrow \sin\left(\frac{B+C}{2}\right) = \sin\left(90 - \frac{A}{2}\right) \Rightarrow \sin\left(\frac{B+C}{2}\right) = \cos\frac{A}{2}.$$

Illustration 4: Simplyfy : $\frac{\sin^3\theta + \cos^3\theta}{\sin\theta + \cos\theta} + \sin\theta\cos\theta$

Solution :

$$\frac{\sin^3\theta + \cos^3\theta}{\sin\theta + \cos\theta} + \sin\theta\cos\theta$$

$$= \frac{(\sin\theta + \cos\theta)(\sin^2\theta + \cos^2\theta - \sin\theta\cos\theta)}{(\sin\theta + \cos\theta)} + \sin\theta\cos\theta$$

$$= \sin^2\theta + \cos^2\theta - \sin\theta\cos\theta + \sin\theta\cos\theta$$

$$= \sin^2\theta + \cos^2\theta = 1$$

Illustration 5: Evaluate $\frac{\cos 43^\circ}{\cos 47^\circ} + \frac{\sec 32^\circ}{\text{cosec}\, 58^\circ}$

Solution :

We know that $\cos(90^\circ - \theta) = \sin\theta$

$\sin 47^\circ = \sin(90^\circ - \theta) = \cos 43^\circ$

Also, $\text{cosec}\, 58^\circ = \text{cosec}(90^\circ - 32^\circ) = \cos 32^\circ$

$$\therefore \frac{\cos 43^\circ}{\cos 47^\circ} + \frac{\sec 32^\circ}{\text{cosec} 58^\circ} = \frac{\cos 43^\circ}{\cos 43^\circ} + \frac{\sec 32^\circ}{\sec 32^\circ} = 1 + 1 = 2$$

Illustration 6: Evaluate $\frac{\sec^2 54^\circ - \cot^2 36^\circ}{\text{cosec}^2 57^\circ - \tan^2 33^\circ}$ $+ 2\sin^2 38^\circ \sec^2 52^\circ - \sin^2 45^\circ + \frac{2}{\sqrt{3}} \tan 17^\circ \tan 60^\circ \tan 73^\circ$.

Solution : The given expression is

$$\frac{\sec^2 54^\circ - \cot^2 36^\circ}{\text{cosec}^2 57^\circ - \tan^2 33^\circ} + 2\sin^2 38^\circ \sec^2 52^\circ - \sin^2 45^\circ + \frac{2}{\sqrt{3}} \tan 17^\circ \tan 60^\circ \tan 73^\circ$$

$$= \frac{\sec^2(90^\circ - 36^\circ) - \cot^2 36^\circ}{\text{cosec}^2(90^\circ - 33^\circ) - \tan^2 33^\circ} + 2\sin^2 38^\circ \sec^2 (90^\circ - 38^\circ) - \sin^2 45^\circ + \frac{2}{\sqrt{3}} \tan(90^\circ - 73^\circ) \tan 73^\circ \tan 60^\circ$$

$$= \frac{1}{1} + 2\sin^2 38^\circ \times \frac{1}{\sin^2 38^\circ} - \frac{1}{2} + \frac{2}{\sqrt{3}} \times \frac{1}{\tan 73^\circ} \times \tan 73^\circ \times \sqrt{3}$$

$$[\because \text{cosec}^2\theta - \cot^2\theta = 1,\ \sec^2\theta - \tan^2\theta = 1]$$

$$= 1 + 2 - \frac{1}{2} + 2 = 5 - \frac{1}{2} = \frac{9}{2}$$

ANGLE OF ELEVATION AND ANGLE OF DEPRESSION

Let an observer at the point O is observing an object at the point P. The line OP is called the LINE OF SIGHT of the point P. Let OA be the horizontal line passing through O. O, A and P be in the same vertical plane.

If object P be above the horizontal line OA, then the acute angle AOP, between the line of sight and the horizontal line is known as ANGLE OF ELEVATION of object P. If the object P is below the horizontal line OA then the angle AOP, between the line of sight and the horizontal line is known as ANGLE OF DEPRESSION of object P.

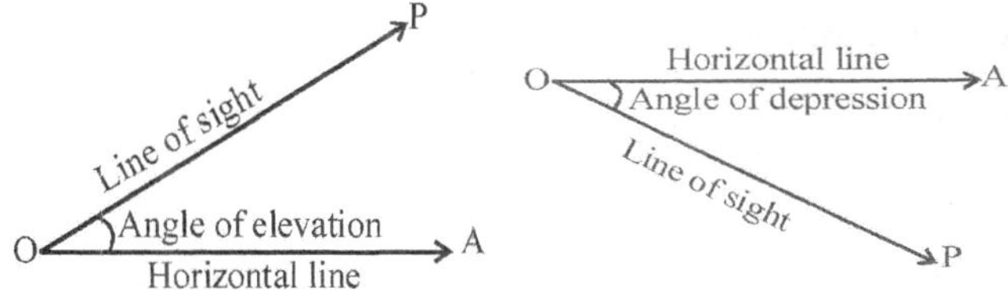

TO FIND THE HEIGHT AND THE DISTANCE OF AN INACCESSIBLE TOWER STANDING ON A HORIZONTAL PLANE

Let AB be a tower and B be its foot. On the horizontal line through B, take two points P and Q. Measure the length PQ.

Let PQ = a.

Let the angles of elevation of the top A of the tower as seen from P and Q be respectively α and β ($\beta > \alpha$), then

$\angle APB = \alpha$, $\angle AQB = \beta$. Let $AB = x$, $BQ = y$.

From right angled Δ ABP, $\tan\alpha = \dfrac{AB}{PB} = \dfrac{x}{a+y}$

$\therefore a + y = x\cot\alpha.$ ....(i)

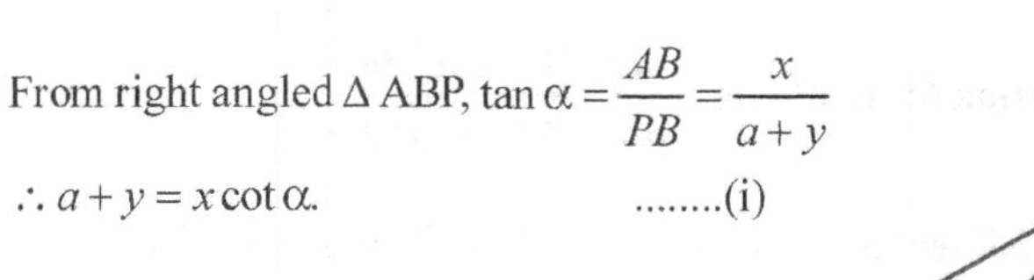

From right angled ΔABQ,

$\tan\beta = \dfrac{AB}{BQ} = \dfrac{x}{y}$

$\therefore\ y = x\cot\beta$ ....(ii)

From equations (i) and (ii),

$\therefore a = x\cot\alpha - x\cot\beta.$

$\Rightarrow x = \dfrac{a}{\cot\alpha - \cot\beta}$

Also $y = x\cot\alpha - a \Rightarrow y = \dfrac{a\cot\alpha}{\cot\alpha - \cot\beta} - a$

$\Rightarrow y = \dfrac{a\cot\alpha - a(\cot\alpha - \text{cct}\beta)}{\cot\alpha - \cot\beta} \Rightarrow y = \dfrac{a\cot\beta}{\cot\alpha - \cot\beta}$

In the above case, P and Q are on the same side of the tower. If the two points are on the opposite sides of the tower then from the adjoining figure, we get

$\tan\alpha = \dfrac{x}{PB}$ or $PB = x\cot\alpha$

and $\tan\beta = \dfrac{x}{BQ}$ or $BQ = x\cot\beta.$

$\therefore a = PB + BQ = x(\cot\alpha + \cot\beta)$

$\therefore x = \dfrac{a}{\cot\alpha + \cot\beta}$

and $y = BQ = x\cot\beta$

Note that : Here, all the lines AP, AQ, AB are in the same plane.

Illustration 7: The angle of elevation of a ladder leaning against a wall is 58°, and the foot of the ladder is 9.6 m from the wall. Find the length of the ladder.

Solution : Let AB be the ladder leaning against a wall OB such that $\angle OAB = 58°$ and OA = 9.6 m

In Δ AOB, we have, $\cos 58° = \dfrac{OA}{AB}$

$\Rightarrow AB = \dfrac{OA}{\cos 58°}$

$\Rightarrow AB = \dfrac{9.6}{0.5299} = 18.11$ m

Illustration 8: A person, standing on the bank of a river, observes that the angle subtended by a tree on the opposite bank is 60° ; when he retreates 20m from the bank, he finds the angle to be 30°. Find the height of the tree and the breadth of the river.

Solution : Let AB be the width of the river and BC be the tree which makes an angle of 60° at a point A on the opposite bank. Let D be the position of the person after retreating 20 m from the bank.

Let AB = x metres and BC = h metres.

From right angled triangles ABC and DBC,

we have $\tan 60° = \dfrac{BC}{AB}$

and $\tan 30° \Rightarrow \sqrt{3} = \dfrac{h}{x}$

and $\dfrac{1}{\sqrt{3}} = \dfrac{h}{x+20} \Rightarrow h = x\sqrt{3}$

and $h = \dfrac{x+20}{\sqrt{3}} \Rightarrow x\sqrt{3} = \dfrac{x+20}{\sqrt{3}}$

$\Rightarrow 3x = x + 20 \Rightarrow x = 10$ m

Putting $x = 10$ in $h = \sqrt{3}\,x$, we get

$h = 10\sqrt{3} = 17.32$ m

Hence, height of the tree = 17.32 m and the breadth of the river = 10 m.

Illustration 9: The angles of elevation of the top of a tower at the top and the foot of a pole of height 10 m are 30° and 60° respectively. Find the height of the tower.

Solution : Let AB and CD be the pole and tower respectively.

Let CD = h

Then $\angle DAC = 60°$ and $\angle DBE = 30°$

Now $\dfrac{CD}{CA} = \tan 60° = \sqrt{3}\ \therefore\ CD = \sqrt{3}\ CA$

$\Rightarrow \dfrac{h}{\sqrt{3}} = CA$

Again $\dfrac{DE}{BE} = \tan 30° = \dfrac{1}{\sqrt{3}}$

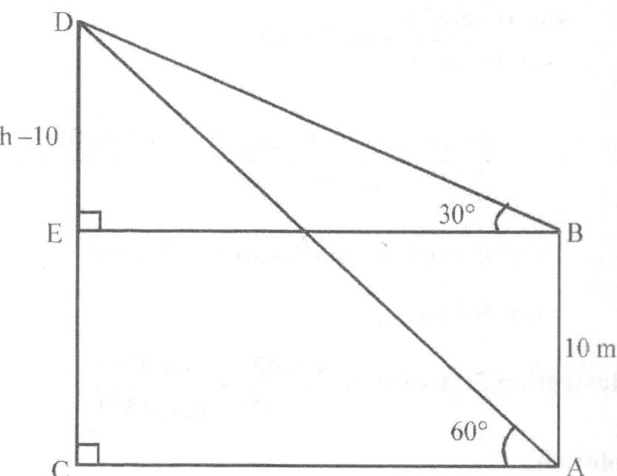

$\therefore (h-10) = \dfrac{BE}{\sqrt{3}} = \dfrac{CA}{\sqrt{3}} = \dfrac{h/\sqrt{3}}{\sqrt{3}} = \dfrac{h}{3}$ [$\because$ BE = CA]

$\Rightarrow 3h - 30 = h \Rightarrow 2h = 30 \Rightarrow h = 15$

Hence, height of the tower = 15 m

Illustration 10: A man is standing on the deck of a ship, which is 8m above water level. He observes the angle of elevation of the top of a hill as 60° and the angle of depression of the base of the hill as 30°. Calculate the distance of the hill from the ship and the height of the hill.

Solution : Let x be the distance of hill from man and $h + 8$ be height of hill which is required.

In rt. ΔACB,

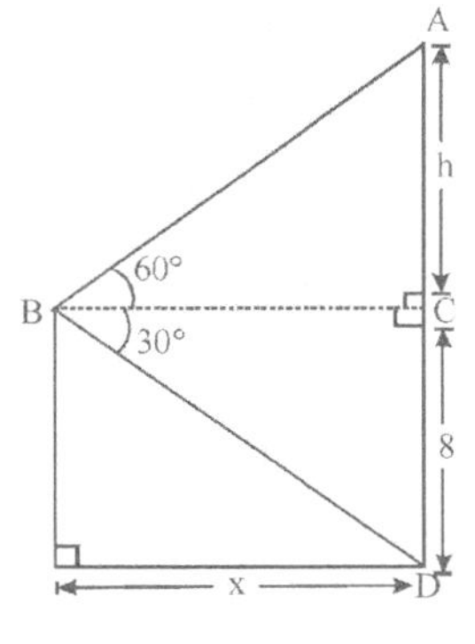

$$\tan 60° = \frac{AC}{BC} = \frac{h}{x}$$

$$\Rightarrow \sqrt{3} = \frac{h}{x}$$

In rt. ΔBCD,

$$\tan 30° = \frac{CD}{BC} = \frac{8}{x}$$

$$\Rightarrow \frac{1}{\sqrt{3}} = \frac{8}{x} \Rightarrow x = 8\sqrt{3}$$

$\therefore$ Height of hill $= h + 8 = \sqrt{3}x + 8 = (\sqrt{3})(8\sqrt{3}) + 8 = 32$ m

Distance of ship from hill $= x = 8\sqrt{3}$ m

Illustration 11: A vertical to stands on a horizontal plane and is surmounted by a vertical flag staff of height 6 meters. At point on the plane, the angle of elevation of the bottom and the top of the flag staff are respectively 30° and 60°. Find the height of tower.

Solution :

Let AB be the tower of height h meter and BC be the height of flag staff surmounted on the tower.

Let the point of the plane be D at a distance m meter from the foot of the tower.

In ΔABD,

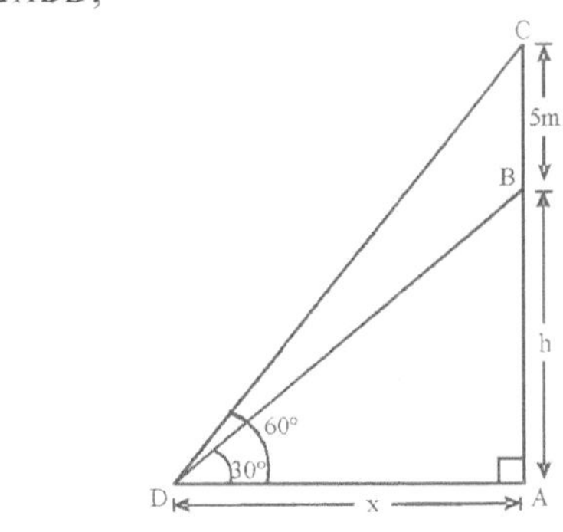

$$\tan 30° = \frac{AB}{BD} \Rightarrow \frac{1}{\sqrt{3}} = \frac{h}{x} \Rightarrow x = \sqrt{3}h \quad \text{..........(i)}$$

In Δ ADC,

$$\tan 60° = \frac{AC}{AD} \Rightarrow \sqrt{3} = \frac{5+h}{x} \Rightarrow x = \frac{5+h}{\sqrt{3}} \quad \text{..........(ii)}$$

From (i) and (ii), $\sqrt{3}h = \dfrac{5+h}{\sqrt{3}} \Rightarrow 3h = 5 + h \Rightarrow 2h = 5$

$$\Rightarrow h = \frac{5}{2} = 2.5 \text{ m}$$

So, the height of tower = 2.5 m

Illustration 12: The angles of depressions of the top and bottom of 8m tall building fron the top of a multistoried building are 30° and 45° respectively. Find the height of multistoried building and the distance between the two buildings.

Solution : Let AB be the multistoried building of height h and let the distance between two buildings be x meters.

$\angle XAC = \angle ACB = 45°$ (Alternate angles)

$\angle XAD = \angle ADE = 30°$ (Alternate angles)

In ΔADE, $\tan 30° = \dfrac{AE}{ED} \Rightarrow \dfrac{1}{\sqrt{3}} = \dfrac{h-8}{x}$

$[\because CB = DE = x]$

$$\Rightarrow x = \sqrt{3}(h-8) \quad \text{..........(i)}$$

In ΔACB,

$$\tan 45° = \frac{h}{x} \Rightarrow 1 = \frac{h}{x} \Rightarrow = \text{h} \quad \text{..........(ii)}$$

From (i) and (ii),

$$\sqrt{3}(h-8) = h \Rightarrow \sqrt{3}h - 8\sqrt{3} = h$$

$$\Rightarrow \sqrt{3}h - h = 8\sqrt{3} \Rightarrow h(\sqrt{3}-1) = 8\sqrt{3}$$

$$\Rightarrow h = \frac{8\sqrt{3}}{\sqrt{3}-1} \times \frac{(\sqrt{3}+1)}{\sqrt{3}+1} \Rightarrow h = \frac{8\sqrt{3}(\sqrt{3}+1)}{2}$$

$$\Rightarrow h = 4\sqrt{3}(\sqrt{3}+1) \Rightarrow h = 4(3+\sqrt{3}) \text{ metres}$$

From (ii), $x = h$

So, $x = 4(3+\sqrt{3})$ metres

Hence, height of multistoried building $= 4(3+\sqrt{3})$ metres

distance between two building $= 4(3+\sqrt{3})$ metres.

Illustration 13: The angle of elevation of an aeroplane from a point on the ground is 45°. After a flight of 15 sec, the elevation changes to 30°. If the aeroplane is flying at a height of 3000 metres, find the speed of the aeroplane.

Solution :

Let the point on the ground is E which is y metres from point B and let after 15 sec. flight it covers x metres distance

In ΔAEB, $\tan 45° = \dfrac{AB}{EB}$

$$\Rightarrow 1 = \frac{3000}{y} \Rightarrow y = 3000\text{m} \quad \text{............(i)}$$

In ΔCED, $\tan 30° = \frac{CD}{ED} \Rightarrow \frac{1}{\sqrt{3}} = \frac{3000}{x+y}$

$(\because AB = CD)$

$\Rightarrow x + y = 3000\sqrt{3}$ .... (ii)

From eqs. (i) and (ii)

$x + 3000 = 3000\sqrt{3} \Rightarrow x = 3000\sqrt{3} - 3000$

$\Rightarrow x = 3000(\sqrt{3}-1) \Rightarrow x = 3000 \times (1.732 - 1)$

$\Rightarrow x = 3000 \times 0.732 \Rightarrow x = 2196\text{m}$

Speed of aeroplane $= \frac{\text{Distance covered}}{\text{Time taken}}$

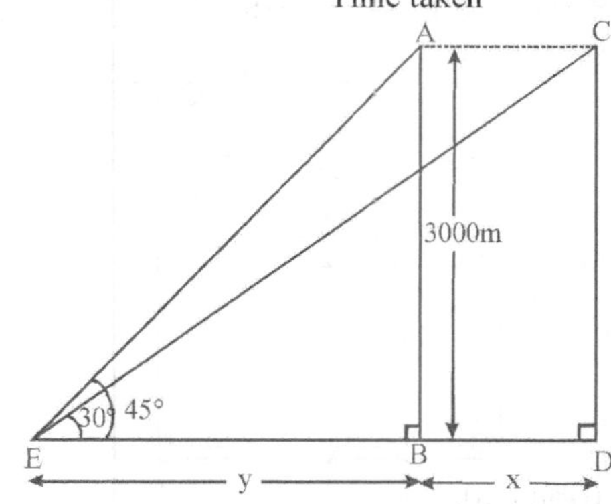

$= \frac{2196}{15} \text{m/sec} = 146.4 \text{ m/sec}$

$= \frac{2196}{15} \times \frac{18}{5} \text{km/hr} = 527.04 \text{ km/hr}$

Hence, the speed of aeroplane is 527.04 km/hr.

Illustration 14: A boy is standing on the ground and flying a kite with 100m of string at an elevation of 30°. Another boy is standing on the roof of a 10m high building and is flying his at an elevation of 45°. Both the boys are on opposite sides of both the kites. Find the length of the string that the second boy must have so that the two kites meet.

Solution :

Let the length of second string be x m.

In ΔABC,

$$\sin 30° = \frac{AC}{AB} \quad \text{or} \quad \frac{1}{2} = \frac{AC}{100} \Rightarrow AC = 50\text{m}$$

In ΔAEF,

$$\sin 45° = \frac{AF}{AE} \Rightarrow \frac{1}{\sqrt{2}} = \frac{AF - FC}{x}$$

$$\Rightarrow \frac{1}{\sqrt{2}} = \frac{50 - 10}{x} \quad [\because AC = 50\text{m}, FC = ED = 10\text{m}]$$

$$\Rightarrow \frac{1}{\sqrt{2}} = \frac{40}{x} \Rightarrow x = 40\sqrt{2}\text{m}$$

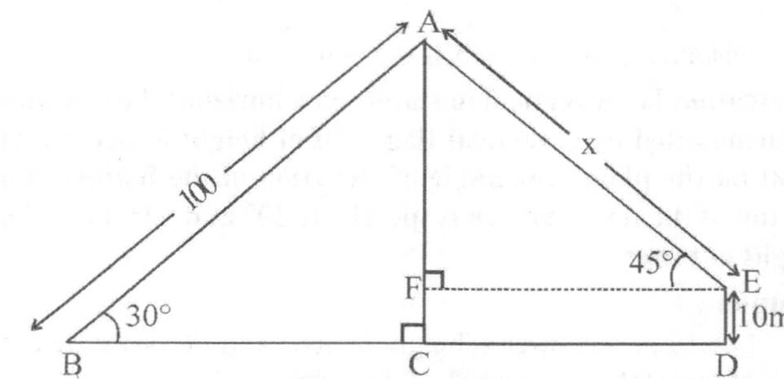

So the length of string that the second boy must have so that the two kites meet $= 40\sqrt{2}\text{m}$

Practice Exercise

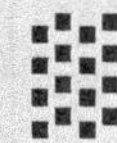

Level - I

1. $\cos 1^\circ . \cos 2^\circ . \cos 3^\circ$ .... $\cos 179^\circ$ is equal to
   (a) -1 (b) 0
   (c) 1 (d) $1/\sqrt{2}$

2. The value of $(\sin^2 7\frac{1}{2}^\circ + \cos^2 7\frac{1}{2}^\circ) - (\sin^2 30^\circ + \cos^2 30^\circ)$ $+ (\sin^2 7^\circ + \sin^2 83^\circ)$ is equal to
   (a) 3 (b) $3\frac{1}{2}$
   (c) 2 (d) 1

3. If $\tan 15^\circ = 2-\sqrt{3}$, then the value of $\cot^2 75^\circ$ is
   (a) $7+\sqrt{3}$ (b) $7-2\sqrt{3}$
   (c) $7-4\sqrt{3}$ (d) $7+4\sqrt{3}$

4. If $b \tan\theta = a$, the value of $\frac{a\sin\theta - b\cos\theta}{a\sin\theta + b\cos\theta}$
   (a) $\frac{a-b}{a^2+b^2}$ (b) $\frac{a+b}{a^2+b^2}$
   (c) $\frac{a^2+b^2}{a^2-b^2}$ (d) $\frac{a^2-b^2}{a^2+b^2}$

5. $\tan 9^\circ \times \tan 27^\circ \times \tan 63^\circ \times \tan 81^\circ =$
   (a) 4 (b) 3
   (c) 2 (d) 1

6. If $\sin\theta = \frac{11}{15}$, then $\cos\theta =$
   (a) $\frac{2\sqrt{26}}{15}$ (b) $\frac{15\sqrt{26}}{2}$
   (c) $\frac{15}{11}$ (d) $\frac{11}{15}$

7. If $\tan\theta = \sqrt{3}$, then $\sec\theta =$
   (a) $\frac{2}{\sqrt{3}}$ (b) 2
   (c) $\sqrt{\frac{3}{2}}$ (d) $\frac{1}{3}$

8. $(1 + \tan^2\theta)/\cos^2\theta =$
   (a) $\frac{1}{\sin^2\theta - \cos^2\theta}$ (b) $\sec^2\theta$
   (c) 1 (d) $\frac{1}{2}$

9. If $4\tan\theta = 3$, then $\left(\frac{4\sin\theta - \cos\theta}{4\sin\theta - \cos\theta}\right)$ is equal to
   (a) $\frac{2}{3}$ (b) $\frac{1}{3}$
   (c) $\frac{1}{2}$ (d) $\frac{3}{4}$

10. $\frac{2\tan 30^\circ}{1+\tan^2 30^\circ} =$
    (a) $\sin 60^\circ$ (b) $\cos 60^\circ$
    (c) $\tan 60^\circ$ (d) $\sin 30^\circ$

11. $\frac{1-\tan^2 45^\circ}{1+\tan^2 45^\circ} =$
    (a) $\tan 90^\circ$ (b) 1
    (c) $\sin 45^\circ$ (d) 0

12. $\sin 2A = 2\sin A$ is true when A =
    (a) 0° (b) 30°
    (c) 45° (d) 60°

13. $\frac{2\tan 30^\circ}{1-\tan^2 30^\circ} =$
    (a) $\cos 60^\circ$ (b) $\sin 60^\circ$
    (c) $\tan 60^\circ$ (d) None of these

14. Given that $\sin\alpha = \frac{1}{2}$ and $\cos\beta = \frac{1}{2}$, then the value of $(\alpha + \beta)$ is
    (a) 0° (b) 30°
    (c) 60° (d) 90°

15. If $\cos 9\alpha = \sin\alpha$ and $9\alpha < 90^\circ$, then the value of $\tan 5\alpha$ is
    (a) $\frac{1}{\sqrt{3}}$ (b) $\sqrt{3}$
    (c) 1 (d) 0

16. $9\sec^2 A - 9\tan^2 A =$
    (a) 1 (b) 9
    (c) 8 (d) 0

17. (sec A + tan A) (1 – sin A) =
(a) sec A (b) sin A
(c) cosec A (d) cos A

18. $\dfrac{1+\tan^2 A}{1+\cot^2 A} =$
(a) $\sec^2 A$ (b) –1
(c) $\cot^2 A$ (d) None of these

19. Which of the following relationship is true ?
(a) sin A/cosec A = cot A
(b) sin A/cos A = tan A
(c) cos A/sin A = sec A
(d) cosec A/sin A = cos A

20. (sin A/tan A) + cos A =
(a) 2 sec A (b) sec A
(c) 2 cosec A (d) 2 cos A

21. If cos (40° + x) = sin 30°, then x is equal to
(a) 20° (b) 30°
(c) 60° (d) 0°

22. If $\sin(A-B)=\dfrac{1}{2}$ and $\cos(A+B)=\dfrac{1}{2}$, then
(a) A = 60°, B = 45° (b) A = 30°, B = 15°
(c) A = 45°, B = 15° (d) A = 60°, B = 30°

23. tan 1° tan 2° tan 3° .... tan 89° is equal to
(a) 1 (b) 0
(c) cannot be determined (d) None of these

24. If $\tan\theta = \dfrac{x\sin\phi}{1-x\cos\phi}$, then $\cot\theta+\cot\phi =$
(a) $\dfrac{\sin\phi}{\sin\theta}$ (b) $\dfrac{1}{x\sin\phi}$
(c) $\dfrac{\sin\theta}{1-\cos\theta}$ (d) $\dfrac{\sin\theta}{1-\cos\phi}$

25. $\tan^2\theta \sin^2\theta$ is equal to
(a) $\tan^2\theta - \sin^2\theta$ (b) $\tan^2\theta + \sin^2\theta$
(c) $\dfrac{\tan^2\theta}{\sin^2\theta}$ (d) $\sin^2\theta \cot^2\theta$

26. If $x = r\sin\alpha\cos\beta$, $y = r\sin\alpha\sin\beta$ and $z = r\cos\alpha$, then
(a) $x^2+y^2+z^2=r^2$ (b) $x^2+y^2-z^2=r^2$
(c) $x^2-y^2+z^2=r^2$ (d) $x^2+y^2-z^2=r^2$

27. If $\dfrac{x}{a}\cos\theta+\dfrac{y}{b}\sin\theta=1$, $\dfrac{x}{a}\sin\theta-\dfrac{y}{b}\cos\theta=1$, then
(a) $x^2+y^2=a^2+b^2$ (b) $\dfrac{x^2}{a^2}+\dfrac{y^2}{b^2}=2$
(c) $a^2x^2+b^2y^2=1$ (d) None of these

28. $\dfrac{\cos 70°}{\sin 20°}+\dfrac{\cos 59°}{\sin 31°}-8\sin^2 30°$ is equal to
(a) 1 (b) –1
(c) 0 (d) 2

29. If $\sin\theta=\dfrac{1}{2}$ and θ is acute, then $(3\cos\theta-4\cos^3\theta)$ is equal to
(a) 0 (b) $\dfrac{1}{2}$
(c) $\dfrac{1}{6}$ (d) –1

30. If tan θ + cot θ = 2, then $\tan^2\theta+\cot^2\theta=$
(a) 4 (b) 3
(c) 2 (d) None of these

31. If $\cos(81°+\theta)=\sin\left(\dfrac{k}{3}-\theta\right)$, then $k=$
(a) 9° (b) 30°
(c) 27° (d) 45°

32. If tan x = sin 45° cos 45° + sin 30°, then x is equal to
(a) 30° (b) 45°
(c) 60° (d) 15°

33. $\cos^2 5° + \cos^2 10° + \cos^2 15° \ldots.. + \cos^2 85° + \cos^2 90°$ is equal to
(a) 10 (b) $\dfrac{19}{2}$
(c) $\dfrac{9}{2}$ (d) $\dfrac{17}{2}$

34. $\dfrac{\sin(90°-\theta)\sin\theta}{\tan\theta}+\sin^2\theta$ is equal to
(a) 2 (b) 1
(c) 0 (d) None of these

35. In the adjoining figure, the length of BC is

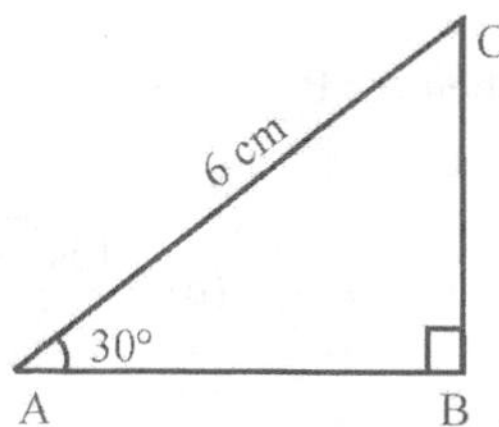

(a) $2\sqrt{3}$ cm (b) $3\sqrt{3}$ cm
(c) $4\sqrt{3}$ cm (d) 3 cm

36. If the angle of depression of an object from a 75 m high tower is 30°, then the distance of the object from the tower is
(a) $25\sqrt{3}$ m (b) $50\sqrt{3}$ m
(c) $75\sqrt{3}$ m (d) 150 m

37. The angle of elevation of the top of a tower at point on the ground is 30°. If on walking 20 metres toward the tower, the angle of elevation become 60°, then the height of the tower is

(a) 10 metre (b) $\frac{10}{\sqrt{3}}$ metre

(c) $10\sqrt{3}$ metre (d) None of these

38. An aeroplane flying horizontally 1 km. above the ground is observed at an elevation of 60° and after 10 seconds the elevation is observed to be 30°. The uniform speed of the aeroplane in km/h is

(a) 240 (b) $240\sqrt{3}$

(c) $60\sqrt{3}$ (d) None of these

39. A 25 m ladder is placed against a vertical wall of a building. The foot of the ladder is 7 m from the base of the building. If the top of the ladder slips 4m, then the foot of the ladder will slide

(a) 5 m (b) 8 m

(c) 9 m (d) 15 m

40. If the length of the shadow of a tower is $\sqrt{3}$ times that of its height, then the angle of elevation of the sun is

(a) 15° (b) 30°

(c) 45° (d) 60°

41. The angles of elevation of the top of a tower from two points at distances *m* and *n* metres are complementary. If the two points and the base of the tower are on the same straight line, then the height of the tower is

(a) $\sqrt{mn}$ (b) mn

(c) $\frac{m}{n}$ (d) None of these

42. The distance between the tops of two trees 20 m and 28 m high is 17 m. The horizontal distance between the two trees is

(a) 9 m (b) 11 m

(c) 15 m (d) 31 m

43. A pole 6 m high casts a shadow $2\sqrt{3}$ m long on the ground, then the sun's elevation is

(a) 60° (b) 45°

(c) 30° (d) 90°

44. The length of a string between a kite and a point on the ground is 85 m. If the string makes an angle θ with level ground such that $\tan\theta = \frac{15}{8}$, how high is the kite ?

(a) 75 m (b) 78.05 m

(c) 226 m (d) None of these

45. A person walking 20 m towards a chimney in a horizontal line through its base observes that its angle of elevation changes from 30° to 45°. The height of chimney is

(a) $\frac{20}{\sqrt{3}+1}$m (b) $\frac{20}{\sqrt{3}-1}$m

(c) $20(\sqrt{3}-1)$m (d) None of these

46. The top of two poles of height 20 m and 14 m are connected by a wire. If the wire makes an angle of 30° with the horizontal, then the length of the wire is

(a) 12 m (b) 10 m

(c) 8 m (d) 6 m

47. Two men standing on opposite sides of a flagstaff measure the angles of the top of the flagstaff is 30° and 60°. If the height of the flagstaff is 20 m, distance between the men is

(a) 46.19 m (b) 40 m

(c) 50 m (d) 30 m

48. If in ΔABC, ∠A = 90°, BC = a, AC = b and AB = c, then the value of tan B + tan C is [*SSC-Sub. Ins.-2012*]

(a) $\frac{b^2}{ac}$ (b) $\frac{a^2}{bc}$

(c) $\frac{c^2}{ab}$ (d) $\frac{a^2+c^2}{b}$

49. A ladder is resting against a wall at height of 10m. If the ladder is inclined with the ground at an angle of 30°, then the distance of the foot of the ladder from the wall is [*SSC-Sub. Ins.-2012*]

(a) $10\sqrt{3}$ m (b) $20\sqrt{3}$ m

(c) $\frac{10}{\sqrt{3}}$m (d) $\frac{20}{\sqrt{3}}$m

50. tan 7° tan 23° tan 60° tan 67° tan 83° is equal to [*SSC-Sub. Ins.-2012*]

(a) $\frac{1}{\sqrt{3}}$ (b) 1

(c) 0 (d) $\sqrt{3}$

51. The value of
(sec θ – cos θ) (cosec θ – sin θ) (tan θ + cot θ) is [*SSC-Sub. Ins.-2012*]

(a) 2 (b) 0

(c) 1 (d) $\frac{3}{2}$

52. If $\tan(\theta_1+\theta_2) = \sqrt{3}$ and $\sec(\theta_1-\theta_2) = \frac{2}{\sqrt{3}}$, then the value of $\sin 2\theta_1 + \tan 3\theta_2$ is equal to [*SSC-Sub. Ins.-2012*]

(Assume that $0 < \theta_1 - \theta_2 < \theta_1 + \theta_2 < 90°$)

(a) 1 (b) 2

(c) 0 (d) 3

53. The value of tan 1° tan 2° tan 3° ... tan 89° is: [SSC-Sub. Ins.-2013]

(a) 1 (b) 2
(c) undefined (d) 0

54. Minimum value of $4\tan^2\theta + 9\cot^2\theta$ is : [SSC-Sub. Ins.-2013]

(a) 12 (b) 1
(c) 6 (d) 13

55. If $\sin\theta - \cos\theta = \frac{1}{2}$ the value of $\sin\theta + \cos\theta$ is : [SSC-Sub. Ins.-2013]

(a) -2 (b) ± 2
(c) $\frac{\sqrt{7}}{2}$ (d) 2

56. If $\operatorname{cosec}\theta - \cot\theta = \frac{7}{2}$, the value of $\operatorname{cosec}\theta$ is : [SSC-Sub. Ins.-2013]

(a) $\frac{47}{28}$ (b) $\frac{51}{28}$
(b) $\frac{53}{28}$ (d) $\frac{49}{28}$

57. From the top of a hill 200 m high, the angle of depression of the top and the bottom of a tower are observed to be 30° and 60°. The height of the tower is (in m) : [SSC-Sub. Ins.-2013]

(a) $\frac{400\sqrt{3}}{3}$ (b) $166\frac{2}{3}$
(c) $133\frac{1}{3}$ (d) $200\sqrt{3}$

58. If $0 \le \theta \le \frac{\pi}{2}$ and $\sec^2\theta + \tan^2\theta = 7$, then θ is [SSC-Sub. Ins.-2014]

(a) $\frac{5\pi}{12}$ radian (b) $\frac{\pi}{3}$ radian
(c) $\frac{\pi}{5}$ radian (d) $\frac{\pi}{6}$ radian

59. The simplest value of $\sin^2 x + 2\tan^2 x - 2\sec^2 x + \cos^2 x$ is [SSC-Sub. Ins.-2014]

(a) 1 (b) 0
(c) –1 (d) 2

60. If $x = a\sin\theta$ and $y = b\tan\theta$ then $\frac{a^2}{x^2} - \frac{b^2}{y^2}$ is [SSC-Sub. Ins.-2014]

(a) 1 (b) 2
(c) 3 (d) 4

61. A kite is flying at a height of 50 metre. If the length of string is 100 metre then the inclination of string to the horizontal ground in degree measure is [SSC-Sub. Ins.-2014]

(a) 90 (b) 60
(c) 45 (d) 30

62. From the top of a light-house at a height 20 metres above sea-level, the angle of depression of a ship is 30°. The distance of the ship from the foot of the light-house is [SSC-Sub. Ins.-2014]

(a) 20 m (b) $20\sqrt{3}$ m
(c) 30 m (d) $30\sqrt{3}$ m

63. If $\frac{2\sin\theta - \cos\theta}{\cos\theta + \sin\theta} = 1$, then value of cot θ is: [SSC 10+2-2012]

(a) $\frac{1}{2}$ (b) $\frac{1}{3}$
(c) 3 (d) 2

64. If $\tan\left(\frac{\pi}{2} - \frac{\theta}{2}\right) = \sqrt{3}$, value of cos θ is: [SSC 10+2-2012]

(a) 0 (b) $\frac{1}{\sqrt{2}}$
(c) $\frac{1}{2}$ (d) 1

65. $2\operatorname{cosec}^2 23° \cot^2 67° - \sin^2 23° - \sin^2 67° - \cot^2 67°$ is equal to [SSC 10+2-2013]

(a) 0 (b) 1
(c) $\sec^2 23°$ (d) $\tan^2 23°$

66. If $5\tan\theta = 4$, then $\frac{5\sin\theta - 3\cos\theta}{5\sin\theta + 2\cos\theta}$ is [SSC 10+2-2013]

(a) $\frac{1}{3}$ (b) $\frac{2}{3}$
(c) $\frac{1}{4}$ (d) $\frac{1}{6}$

67. If $x\sin\theta + y\cos\theta = \sqrt{x^2+y^2}$ and $\frac{\cos^2\theta}{a^2} + \frac{\sin^2\theta}{b^2} = \frac{1}{x^2+y^2}$, then the correct relation is [SSC 10+2-2013]

(a) $\frac{x^2}{a^2} - \frac{y^2}{b^2} = 1$ (b) $\frac{x^2}{b^2} - \frac{y^2}{a^2} = 1$
(c) $\frac{x^2}{a^2} + \frac{y^2}{b^2} = 1$ (d) $\frac{x^2}{b^2} + \frac{y^2}{a^2} = 1$

68. If $2(\cos^2\theta - \sin^2\theta) = 1$ (θ is a positive acute angle), then $\cot\theta$ is equal to [SSC 10+2-2013]

(a) $\sqrt{3}$ (b) $-\sqrt{3}$

(c) $\frac{1}{\sqrt{3}}$ (d) 1

69. The equation $\cos^2\theta = \frac{(x+y)^2}{4xy}$ is only possible when [SSC 10+2-2013]

(a) $x < y$ (b) $x = -y$

(c) $x > y$ (d) $x = y$

70. If $\tan^2\theta - \sin^2\theta = x$, then the value of $\tan^2\theta\sin^2\theta$ is [SSC 10+2-2013]

(a) x (b) 2

(c) $\frac{1}{x}$ (d) 1

71. The value of

$\frac{\sin 25^\circ \cos 65^\circ + \cos 25^\circ \sin 65^\circ}{\tan^2 70^\circ - \operatorname{cosec}^2 20^\circ}$ is [SSC 10+2-2014]

(a) – 1 (b) 0

(c) 1 (d) 2

72. If θ is a positive acute angle and $4\cos^2\theta - 4\cos\theta + 1 = 0$, then the value of $\tan(\theta - 15^\circ)$ is equal to [SSC 10+2-2014]

(a) 0 (b) 1

(c) $\sqrt{3}$ (d) $\frac{1}{\sqrt{3}}$

73. If $(r\cos\theta - \sqrt{3})^2 + (r\sin\theta - 1)^2 = 0$, then the value of $\frac{r\tan\theta + \sec\theta}{r\sec\theta + \tan\theta}$ is equal to [SSC 10+2-2014]

(a) $\frac{4}{5}$ (b) $\frac{5}{4}$

(c) $\sqrt{\frac{3}{4}}$ (d) $\sqrt{\frac{5}{4}}$

74. A vertical pole and a vertical tower are standing on the same level ground. Height of the pole is 10 metres. Form the top of the pole is the angle of elevation of the top of the tower and angle of depression of the foot of the tower are 60° and 30° respectively. The height of the tower is [SSC 10+2-2014]

(a) 20 m (b) 30 m

(c) 40 m (d) 50 m

75. The length of the shadow of a vertical tower on level ground increases by 10 metres when the altitude of the sun changes from 45° to 30°. Then the height of the tower is [SSC 10+2-2014]

(a) $5(\sqrt{3}+1)$ metres (b) $5(\sqrt{3}-1)$ metres

(c) $5\sqrt{3}$ metres (d) $\frac{5}{\sqrt{3}}$ metres

76. The value of $\frac{\tan^2\theta}{1+\tan^2\theta} + \frac{\cot^2\theta}{1+\cot^2\theta}$ is equal to [SSC 10+2-2014]

(a) 0 (b) 1

(c) 2 (d) 3

Level - II

1. $\sin^2\theta + \operatorname{cosec}^2\theta$ is always

(a) greater than 1

(b) less than 1

(c) greater than or equal to 2

(d) equal to 2

2. If $\sin\theta + \cos\theta = a$ and $\frac{\sin\theta + \cos\theta}{\sin\theta\cos\theta} = b$, then

(a) $b = \frac{2a}{a^2-1}$ (b) $a = \frac{2b}{b^2-1}$

(c) $ab = b^2 - 1$ (d) $a + b = 1$

3. If $x = p\sec\theta$ and $y = q\tan\theta$ then

(a) $x^2 - y^2 = p^2q^2$ (b) $x^2q^2 - y^2p^2 = pq$

(c) $x^2q^2 - y^2p^2 = \frac{1}{p^2q^2}$ (d) $x^2q^2 - y^2p^2 = p^2q^2$

4. If $\tan\theta + \sin\theta = m$ and $\tan\theta - \sin\theta = n$, then the value of $m^2 - n^2$ is equal to

(a) $4mn$ (b) $2\sqrt{mn}$

(c) $4\sqrt{mn}$ (d) $2\sqrt{m/n}$

5. The value of expression

$\frac{\sin^2 22^\circ + \sin^2 68^\circ}{\sin^2 22^\circ + \cos^2 68^\circ} + \sin^2 63^\circ + \cos 63^\circ \sin 27^\circ$

(a) 3 (b) 2

(c) 1 (d) 0

6. The value of the expression

$[\operatorname{cosec}(75^\circ + \theta) - \sec(15^\circ - \theta) - \tan(55^\circ + \theta) + \cot(35^\circ - \theta)]$

(a) –1 (b) 0

(c) 1 (d) $\frac{3}{2}$

7. $\sin(45° + \theta) - \cos(45° - \theta)$ is equal to
(a) $2\cos\theta$ (b) 0
(c) $2\sin\theta$ (d) 1

8. $(1 + \tan\theta + \sec\theta)(1 + \cot\theta - \operatorname{cosec}\theta) =$
(a) 0 (b) 1
(c) 2 (d) None of these

9. If $\sin A + \sin^2 A = 1$, then the value of expression $(\cos^2 A + \cos^4 A)$ is
(a) 1 (b) $\frac{1}{2}$
(c) 2 (d) 3

10. If $7\sin^2\theta + 3\cos^2\theta = 4$, then $\sec\theta + \operatorname{cosec}\theta$ is equal to
(a) $\frac{2}{\sqrt{3}} - 2$ (b) $\frac{2}{\sqrt{3}} + 2$
(c) $\frac{2}{\sqrt{3}}$ (d) None of these

11. If $a\cos\theta + b\sin\theta = 4$ and $\sin\theta - b\cos\theta = 3$, then $a^2 + b^2 =$
(a) 7 (b) 12
(c) 25 (d) None of these

12. If $p\sin\theta + q\sin\theta = a$ and $p\cos\theta - q\sin\theta = b$, then $\frac{p+a}{q+b} + \frac{q-b}{p-a}$ is equal to
(a) 1 (b) 2
(c) 0 (d) None of these

13. If $\operatorname{cosec}\theta - \cot\theta = \frac{1}{2}, 0° < \theta < 90°$, then $\cos\theta$ is equal
(a) $\frac{5}{3}$ (b) $\frac{3}{5}$
(c) $-\frac{3}{5}$ (d) $-\frac{5}{3}$

14. If $\cos\theta + \sin\theta = \sqrt{2}\cos\theta$, then $\cos\theta - \sin\theta$ is equal to
(a) $\sqrt{2}\tan\theta$ (b) $\sqrt{2}\sin\theta$
(c) $\frac{\sqrt{2}}{\cos\theta + \sin\theta}$ (d) None of these

15. If $\sec\theta + \tan\theta = x$, then $\sin\theta$ is equal to
(a) $\frac{x^2+1}{2x}$ (b) $\frac{x^2+1}{x^2-1}$
(c) $\frac{x^2-1}{x^2+1}$ (d) $\frac{2x}{x^2-1}$

16. The value of $(1 + \cot\theta - \operatorname{cosec}\theta)(1 + \tan\theta + \sec\theta)$ is
(a) 1 (b) 2
(c) 4 (d) 0

17. $2(\sin^6\theta + \cos^6\theta) - 3(\sin^4\theta + \cos^4\theta)$ is equal to
(a) 0 (b) 3
(c) 1 (d) −1

18. The top of a broken tree has its top touching the ground (shown in the adjoining figure) at a distance of 10m from the bottom. If the angle made by the broken part with ground is 30°, then the length of the broken part is

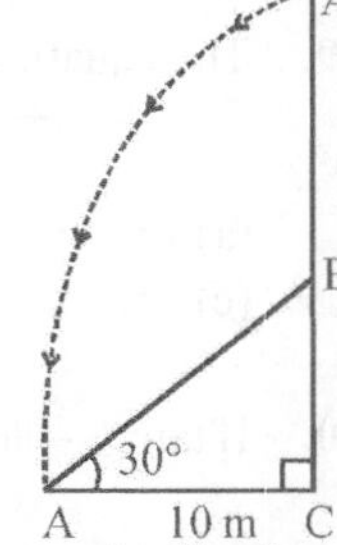

(a) $10\sqrt{3}$ cm
(b) $\frac{20}{\sqrt{3}}$ m
(c) 20 cm
(d) $20\sqrt{3}$ m

19. An aeroplane at a height of 600 m passes vertically above another aeroplane at an instant when their angles of elevation at the same observing point are 60° and 45° respectively. How many metres higher is the one from the other ?
(a) 286.53 m (b) 274.53 m
(c) 253.58 m (d) 263.83 m

20. If a kite is flying at a height of $40\sqrt{3}$ metres from the level ground, attached to a string inclined at 60° to the horizontal, then the length of the string is
(a) 80 m (b) $60\sqrt{3}$ m
(c) $80\sqrt{3}$ m (d) 120 m

21. Two persons are '*a*' metres apart and the height of one is double that of the other. It from the middle point of the line joining their feet, an observer finds the angular elevations of their tops to be complementary, then the height of the shortest persons in metre is
(a) $\frac{a}{4}$ (b) $\frac{a}{\sqrt{2}}$
(c) $a\sqrt{2}$ (d) $\frac{a}{2\sqrt{2}}$

22. The angle of elevation of the top of a rock from the top and foot of 100 m high tower are respectively 30° and 45°. The height of the rock is
(a) $50(3-\sqrt{3})$m (b) $50(3+\sqrt{3})$m
(c) $50\sqrt{3}$m (d) 150 m

23. If $2y\cos\theta = x\sin\theta$ and $2x\sec\theta - y\operatorname{cosec}\theta = 3$, then the relation between x and y is [*SSC CGL-2012*]
(a) $2x^2 + y^2 = 2$ (b) $x^2 + 4y^2 = 4$
(c) $x^2 + 4y^2 = 1$ (d) $4x^2 + y^2 = 4$

24. If $\sec\theta + \tan\theta = \sqrt{3}$, then the positive value of $\sin\theta$ is [SSC CGL-2012]

(a) 0 (b) $\frac{1}{2}$

(c) $\frac{\sqrt{3}}{2}$ (d) 1

25. The radian measure of $63°14'51''$ is [SSC CGL-2012]

(a) $\left(\frac{2811\pi}{8000}\right)^c$ (b) $\left(\frac{3811\pi}{8000}\right)^c$

(c) $\left(\frac{4811\pi}{8000}\right)^c$ (d) $\left(\frac{5811\pi}{8000}\right)^c$

26. In a triangle ABC, $AB = AC$, BA is produced to D in such a manner that $AC = AD$. The circular measure of $\angle BCD$ is [SSC CGL-2012]

(a) $\frac{\pi}{6}$ (b) $\frac{\pi}{3}$

(c) $\frac{2\pi}{3}$ (d) $\frac{\pi}{2}$

27. If $\frac{\cos^4\alpha}{\cos^2\beta} + \frac{\sin^4\alpha}{\sin^2\beta} = 1$, then the value of $\frac{\cos^4\beta}{\cos^2\alpha} + \frac{\sin^4\beta}{\sin^2\alpha}$ is [SSC CGL-2012]

(a) 4 (b) 0

(c) $\frac{1}{8}$ (d) 1

28. $\frac{\sin\theta - \cos\theta + 1}{\sin\theta + \cos\theta - 1}\left(\text{where } \theta \neq \frac{\pi}{2}\right)$ is equal to [SSC CGL-2012]

(a) $\frac{1+\sin\theta}{\cos\theta}$ (b) $\frac{1-\sin\theta}{\cos\theta}$

(c) $\frac{1-\cos\theta}{\sin\theta}$ (d) $\frac{1+\cos\theta}{\sin\theta}$

29. If $\sin^2\alpha = \cos^3\alpha$, then the value of $(\cot^6\alpha - \cot^2\alpha)$ is [SSC CGL-2012]

(a) 1 (b) 0

(c) –1 (d) 2

30. The simplified value of [SSC CGL-2012]
$(1 + \tan\theta + \sec\theta)(1 + \cot\theta - \text{cosec}\,\theta)$ is

(a) –2 (b) 2

(c) 1 (d) –1

31. The angles of elevation of the top of a tower standing on a horizontal plane from two points on a line passing through the foot of the tower at a distance 9 ft and 16 ft respectively are complementary angles. Then the height of the tower is [SSC CGL-2012]

(a) 9 ft (b) 12 ft

(c) 16 ft (d) 144 ft

32. The value of $\frac{\sin 53°}{\cos 37°} \div \frac{\cot 65°}{\tan 25°}$ is [SSC CGL-2013]

(a) 2 (b) 1

(c) 3 (d) 0

33. The value of $\frac{\cos 60° + \sin 60°}{\cos 60° - \sin 60°}$ is [SSC CGL-2013]

(a) – 1 (b) $\sqrt{3} + 2$

(c) $-(2 + \sqrt{3})$ (d) $\sqrt{3} - 2$

34. The value of [SSC CGL-2013]

$\frac{\cot 5°.\cot 10°.\cot 15°.\cot 60°.\cot 75°.\cot 80°.\cot 85°}{(\cos^2 20° + \cos^2 70°) + 2}$ is

(a) $\frac{9}{\sqrt{3}}$ (b) $\frac{1}{9}$

(c) $\frac{1}{\sqrt{3}}$ (d) $\frac{\sqrt{3}}{9}$

35. In a triangle, the angles are in the ratio 2 : 5 : 3. What is the value of the least angle in the radian ? [SSC CGL-2013]

(a) $\frac{\pi}{20}$ (b) $\frac{\pi}{10}$

(c) $\frac{2\pi}{5}$ (d) $\frac{\pi}{5}$

36. If $x = a\cos\theta - b\sin\theta$, $y = b\cos\theta + a\sin\theta$, then find the value of $x^2 + y^2$. [SSC CGL-2013]

(a) a^2 (b) b^2

(c) $\frac{a^2}{b^2}$ (d) $a^2 + b^2$

37. If $\tan\alpha + \cot\alpha = 2$, then the value of $\tan^7\alpha + \cot^7\alpha$ is [SSC CGL-2013]

(a) 2 (b) 16

(c) 64 (d) 128

38. From 125 metre high towers, the angle of depression of a car is 45°. Then how far the car is from the tower ? [SSC CGL-2013]

(a) 125 metre (b) 60 metre

(c) 75 metre (d) 95 metre

39. The value of $\sin^2 1° + \sin^2 2° + \sin^2 3° + ... + \sin^2 89°$ is [SSC CGL-2014]

(a) 22 (b) 44

(c) $22\frac{1}{2}$ (d) $44\frac{1}{2}$

40. The value of $\frac{\cos^3\theta+\sin^3\theta}{\cos\theta+\sin\theta}+\frac{\cos^3\theta-\sin^3\theta}{\cos\theta-\sin\theta}$ is equal to [*SSC CGL-2014*]

(a) -1 (b) 1
(c) 2 (d) 0

41. If $\sin 17° = \frac{x}{y}$ then $\sec 17° - \sin 73°$ is equal to [*SSC CGL-2014*]

(a) $\frac{y}{\sqrt{y^2-x^2}}$ (b) $\frac{y^2}{\left(x\sqrt{y^2-x^2}\right)}$

(c) $\frac{x}{\left(y\sqrt{y^2-x^2}\right)}$ (d) $\frac{x^2}{\left(y\sqrt{y^2-x^2}\right)}$

42. If θ is a positive acute angle and $\operatorname{cosec}\theta + \cot\theta = \sqrt{3}$, then the value of $\operatorname{cosec}\theta$ is [*SSC CGL-2014*]

(a) $\frac{1}{\sqrt{3}}$ (b) $\sqrt{3}$

(c) $\frac{2}{\sqrt{3}}$ (d) 1

43. If $\cos\alpha + \sec\alpha = \sqrt{3}$, then the value of $\cos^3\alpha + \sec^3\alpha$ is [*SSC CGL-2014*]

(a) 2 (b) 1
(c) 0 (d) 4

44. If $\sin\theta + \cos\theta = \sqrt{2}\cos\theta$, then the value of $\cot\theta$ is [*SSC CGL-2014*]

(a) $\sqrt{2}+1$ (b) $\sqrt{2}-1$
(c) $\sqrt{3}-1$ (d) $\sqrt{3}+1$

45. The shadow of a tower standing on a level plane is found to be 30 m longer when the Sun's altitude changes from 60° to 45°. The height of the tower is [*SSC CGL-2014*]

(a) $15(3+\sqrt{3})$ m

(b) $15(\sqrt{3}+1)$ m

(c) $15(\sqrt{3}-1)$ m

(d) $15(3-\sqrt{3})$ m

Hints & Solutions

Level-I

1. **(b)**

2. **(d)** sin 83° = cos7°

$\therefore$ the given expression is 1 – 1 + 1 = 1

3. **(c)** $\cot^2 75° = \left(2-\sqrt{3}\right)^3 = 7-4\sqrt{3}$

4. **(d)** $\tan\theta = \dfrac{a}{b}$

$$\frac{a\sin\theta - b\cos\theta}{a\sin\theta + b\cos\theta} = \frac{a\tan\theta - b}{a\tan\theta + b} = \frac{a^2-b^2}{a^2+b^2}$$

5. **(d)**

6. **(a)** $\cos^2\theta = 1-\sin^2\theta = 1-\left(\dfrac{11}{15}\right)^2 = \dfrac{225-121}{225} = \dfrac{104}{225}$

$$\Rightarrow \cos\theta = \frac{\sqrt{4\times 26}}{15} = \frac{2\sqrt{26}}{15}$$

7. **(b)** $\sec^2\theta = 1 + \tan^2\theta = 1 + 3 = 4 \Rightarrow \sec\theta = 2$

8. **(c)** $(1 + \tan^2\theta)/\cos^2\theta = \sec^2\theta = 1$

9. **(c)** $\dfrac{4\sin\theta-\cos\theta}{4\sin\theta+\cos\theta} = \dfrac{4\dfrac{\sin\theta}{\cos\theta}-1}{4\dfrac{\sin\theta}{\cos\theta}+1} = \dfrac{4\tan\theta-1}{4\tan\theta+1}$

$$= \frac{3-1}{3+1} = \frac{2}{4} = \frac{1}{2}$$

10. **(a)** $\dfrac{2\tan 30°}{1+\tan^2 30°} = \dfrac{2\left(\dfrac{1}{\sqrt{3}}\right)}{1+\left(\dfrac{1}{\sqrt{3}}\right)^2} = \dfrac{\dfrac{2}{\sqrt{3}}}{1+\dfrac{1}{3}} = \dfrac{\dfrac{2}{\sqrt{3}}}{\dfrac{4}{3}} = \dfrac{(2)(3)}{(4)(\sqrt{3})}$

$$= \frac{\sqrt{3}}{2} = \sin 60°$$

11. **(d)** $\dfrac{1-\tan^2 45}{1+\tan^2 45} = \dfrac{1-1}{1+1} = \dfrac{0}{2} = 0$

12. **(a)** sin 2A = 2 sin A

This equation in satisfied for A = 0° and for no other value of A given in options i.e., 30°, 45°, 60°.

Because for A = 0°

sin 2A = sin 2(0°) = sin 0° = 0

and 2 sin A = 2 sin (0°) = 2 (0) = 0

$\therefore$ sin 2A = 2 sin A for the option A = 0°

13. **(c)** $\dfrac{2\tan 30°}{1-\tan^2 30°} = \dfrac{2\left(\dfrac{1}{\sqrt{3}}\right)}{1-\left(\dfrac{1}{\sqrt{3}}\right)^2} = \dfrac{\left(\dfrac{2}{\sqrt{3}}\right)}{1-\dfrac{1}{3}} = \dfrac{\dfrac{2}{\sqrt{3}}}{\dfrac{2}{3}}$

$$= \left(\frac{2}{\sqrt{3}}\right)\left(\frac{3}{2}\right) = \sqrt{3} = \tan 60°$$

14. **(c)** $\cos 9\alpha = \sin\alpha = \cos(90° - \alpha) \Rightarrow 9\alpha = 90° - \alpha$

$\Rightarrow 10\alpha = 90°$

$\Rightarrow 5\alpha = 45° \Rightarrow \tan 5\alpha = \tan 45° = 1$

15. **(b)** $9\sec^2 A - 9\tan^2 A = 9(\sec^2 A - \tan^2 A) = 9(1) = 9$

16. **(c)** $(1 + \tan\theta + \sec\theta)(1 + \cot\theta - \text{cosec}\ \theta)$

$$= \left(1+\frac{\sin\theta}{\cos\theta}+\frac{1}{\cos\theta}\right) = \left(1+\frac{\cos\theta}{\sin\theta}-\frac{1}{\sin\theta}\right)$$

$$= \left(\frac{\cos\theta+\sin\theta+1}{\cos\theta}\right) = \left(\frac{\sin\theta+\sin\theta-1}{\sin\theta}\right)$$

[Use $(a + b)(a - b) = a^2 - b^2$]

$$= \frac{(\cos\theta+\sin\theta)^2-(1)^2}{\cos\theta.\sin\theta}$$

$$= \frac{\cos^2\theta+\sin^2\theta+2\cos\theta\sin\theta-1}{\cos\theta.\sin\theta}$$

$$= \frac{1+2\cos\theta\sin\theta-1}{\cos\theta.\sin\theta} \qquad [\therefore \cos^2\theta + \sin^2\theta = 1]$$

$$= \frac{2\cos\theta\sin\theta}{\cos\theta\sin\theta} = 2$$

17. **(d)** (sec A + tan A) (1 – sin A)

$$= \left(\frac{1}{\cos A}+\frac{\sin A}{\cos A}\right)(1-\sin A)$$

$$= \frac{(1-\sin A)}{\cos A}(1-\sin A) = \frac{1-\sin^2 A}{\cos A} = \frac{\cos^2 A}{\cos A}$$

$[\because \cos^2 A + \sin^2 A = 1 \Rightarrow \cos^2 A = 1 - \sin^2 A]$

= cos A

18. **(d)** $\dfrac{1+\tan^2 A}{1+\cot^2 A} = \dfrac{1+\tan^2 A}{1+\dfrac{1}{\tan^2 A}} = \dfrac{1+\tan^2 A}{\left(\dfrac{\tan^2 A+1}{\tan^2 A}\right)} = \tan^2 A$

19. **(b)** sin A/cos A = tan A

20. **(d)** $\tan A = \dfrac{\sin A}{\cos A} \quad \therefore \dfrac{\sin A}{\tan A} = \cos A$

$$\therefore \frac{\sin A}{\tan A} + \cos A = \cos A + \cos A = 2\cos A$$

21. **(a)** $\cos(40° + x) = \sin 30°$
$\Rightarrow \cos(40° + x) = \cos 60°$
$\Rightarrow 40° + x = 60°$
$\Rightarrow x = 20°$

22. **(c)** $\sin(A - B) = \frac{1}{2}$ and $\cos(A + B) = \frac{1}{2}$
$\Rightarrow \sin(A - B) = \sin 30°$ and $\cos(A + B) = \cos 60°$
$\Rightarrow (A - B) = 30°$ ..... (1)
and $(A + B) = 60°$ ..... (2)
Solving (1) and (2), we get
$a = 45°$ and $B = 15°$

23. **(a)** $\tan 1° \tan 2° \tan 3° \ldots\ldots \tan 89°$
$\Rightarrow (\tan 1°.\tan 89°)(\tan 2° \tan 88°) \ldots\ldots$
..... $(\tan 44° \tan 46°) \tan 45°$
$\Rightarrow (\tan 1°.\cot 1°)(\tan 2° \tan 2°) \ldots\ldots$
..... $(\tan 44° \cot 44°) \times 1$
$\Rightarrow 1 \times 1 \times 1 \times \ldots\ldots \times 1 = 1$

24. **(b)** We have, $\tan\theta = \frac{x\sin\phi}{1 - x\cos\phi}$

$\Rightarrow \frac{1 - x\cos\phi}{x\sin\phi} = \frac{1}{\tan\theta} \Rightarrow \frac{1}{x\sin\phi} - \cot\phi = \cot\theta$

$\Rightarrow \frac{1}{x\sin\phi} = \cot\theta + \cot\phi$

25. **(a)** $\tan^2\theta \sin^2\theta \Rightarrow \frac{\sin^2\theta\sin^2\theta}{\cos^2\theta}$

$\Rightarrow \sin^2\theta\left(\frac{1 - \cos^2\theta}{\cos^2\theta}\right) \Rightarrow \frac{\sin^2\theta}{\cos^2\theta} - \sin^2\theta$

$\Rightarrow \tan^2\theta - \sin^2\theta$

26. **(a)** $x = r\sin\alpha\cos\beta, y = r\sin\alpha\sin\beta, z = r\cos\alpha$
$\therefore x^2 + y^2 + z^2 = r^2\sin^2\alpha\cos^2\beta + r^2\sin^2\alpha\sin^2\beta + r^2\cos^2\alpha$
$= r^2\sin^2\alpha(\cos^2\beta + \sin^2\beta) + r^2\cos^2\alpha$
$= r^2\sin^2\alpha + r^2\cos^2\alpha = r^2$

27. **(b)** $\frac{x}{a}\cos\theta + \frac{y}{b}\sin\theta = 1, \frac{x}{a}\sin\theta - \frac{y}{b}\cos\theta = 1$
Squaring and adding, we get

$$\left(\frac{x}{a}\cos\theta + \frac{y}{b}\sin\theta\right)^2 + \left(\frac{x}{a}\sin\theta - \frac{y}{b}\cos\theta\right)^2 = 1 + 1$$

$\Rightarrow \frac{x^2}{a^2} + \frac{y^2}{b^2} = 2$

28. **(c)** We have, $\frac{\cos 70°}{\sin 20°} + \frac{\cos 59°}{\sin 31°} - 8\sin^2 30°$

$\Rightarrow \frac{\sin 20°}{\sin 20°} + \frac{\sin 31°}{\sin 31°} - 8\left(\frac{1}{2}\right)^2$

[$\because \cos\theta = \sin(90° - \theta)$, $\sin 30° = 1/2$]

$\Rightarrow 1 + 1 - 8 \times \frac{1}{4} \Rightarrow 2 - 2 = 0$

29. **(a)** We have, $\sin\theta = \frac{1}{2} \Rightarrow \theta = 30°$
Now, $3\cos\theta - 4\cos^3\theta$
$\Rightarrow 3\cos 30° - 4\cos^3 30°$

$\Rightarrow 3 \times \frac{\sqrt{3}}{2} - 4 \times \left(\frac{\sqrt{3}}{2}\right)^2 \Rightarrow \frac{3\sqrt{3}}{2} - \frac{4 \times 3\sqrt{3}}{8}$

$\Rightarrow \frac{3\sqrt{3}}{2} - \frac{3\sqrt{3}}{2} = 0$

30. **(c)** We have, $\tan\theta + \cot\theta = 2$
$\Rightarrow (\tan\theta + \cot\theta)^2 = (2)^2$
$\Rightarrow \tan^2\theta + \cot^2\theta + 2\tan\theta\cot\theta = 4$
$\Rightarrow \tan^2\theta + \cot^2\theta + 2 = 4$
$\Rightarrow \tan^2\theta + \cot^2\theta = 2$

31. **(c)** We have, $\cos(81° + \theta) = \sin\left(\frac{k}{3} - \theta\right)$

$\Rightarrow 81° + \theta + \frac{k}{3} - \theta = 90°$

[$\because$ $\cos\alpha = \sin\beta$, then $\alpha + \beta = 90°$]

$\Rightarrow \frac{k}{3} = 90° - 81° \Rightarrow k = 27°$

32. **(b)** We have, $\tan x = \sin 45° \cos 45° + \sin 30°$

$\Rightarrow \tan x = \frac{1}{\sqrt{2}} \times \frac{1}{\sqrt{2}} + \frac{1}{2}$

$\Rightarrow \tan x = \frac{1}{2} + \frac{1}{2} = 1 \Rightarrow x = 45°$

33. **(d)** We have,
$\cos^2 5° + \cos^2 10° + \ldots\ldots + \cos^2 85° + \cos^2 90°$
$= (\cos^2 5° + \cos^2 85°) + (\cos^2 10° + \cos^2 80°)$
..... $\cos^2 45° + \cos^2 90°$
$= (\cos^2 5° + \cos^2 5°) + (\cos^2 10° + \sin^2 10°)$
..... $\left(\frac{1}{\sqrt{2}}\right)^2 + 0$

$= (1 + 1 + \ldots\ldots 8 \text{ times}) + \frac{1}{2} = 8 + \frac{1}{2} = \frac{17}{2}$

34. **(b)** We have, $\frac{\sin(90° - \theta)\sin\theta}{\tan\theta} + \sin^2\theta$

$= \frac{\cos\theta\sin\theta}{\frac{\sin\theta}{\cos\theta}} + \sin^2\theta = \cos^2\theta + \sin^2\theta = 1$

35. **(d)** **Hint:** $\sin 30° = \frac{BC}{AC} \Rightarrow \frac{1}{2} = \frac{BC}{6 \text{ cm}} \Rightarrow BC = 3$ cm.

36. (c) **Hint:** $\tan 30° = \frac{AB}{OB}$

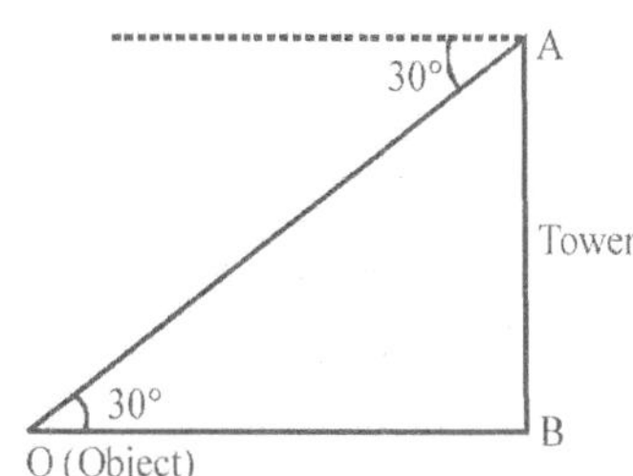

$\Rightarrow \frac{1}{\sqrt{3}} = \frac{75 \text{ m}}{OB}$

$\Rightarrow OB = 75\sqrt{3}$ m

37. (c)

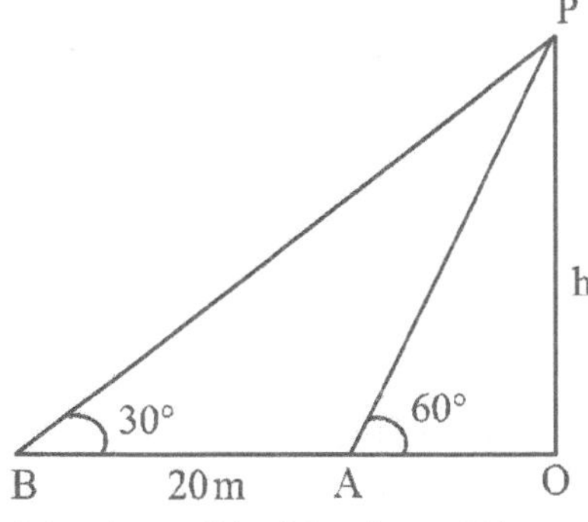

OA = $h \cot 60°$, OB = $h \cot 30°$

OB – OA = 20 = h (cot 30° – cot 60°)

$$\Rightarrow h = \frac{20}{\left(\sqrt{3} - \frac{1}{\sqrt{3}}\right)} = \frac{20\sqrt{3}}{2} = 10\sqrt{3}$$

38. (b) d = H cot 30° – H cot 60°

Time taken = 10 second

$$\text{speed} = \frac{\cot 30° - \cot 60°}{10} \times 60 \times 60 = 240\sqrt{3}$$

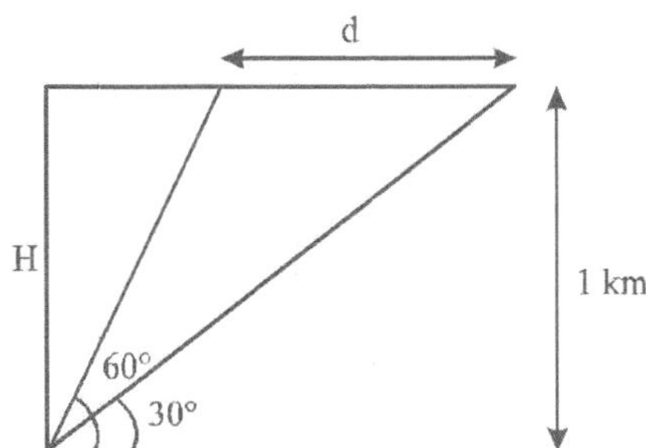

39. (b)

40. (b) **Hint :** Let height of tower (AB) be h metres, then length of its shadow (BC) = $\sqrt{3}\, h$ metres.

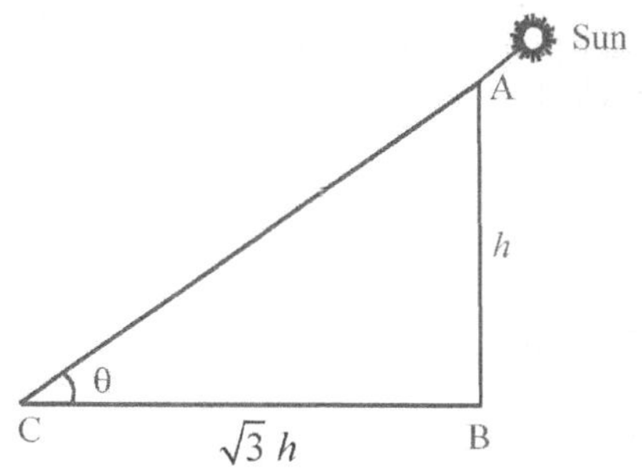

Let angle of elevation be θ,

then $\tan\theta = \frac{h}{\sqrt{3}h} = \frac{1}{\sqrt{3}}$

$\Rightarrow \theta = 30°$

41. (a)

42. (c) $AE^2 = AB^2 - BE^2$

$= 17^2 - 8^2$

$= 289 - 64 = 225$

$\Rightarrow$ AE = 15 m

$\Rightarrow$ CD = 15 m

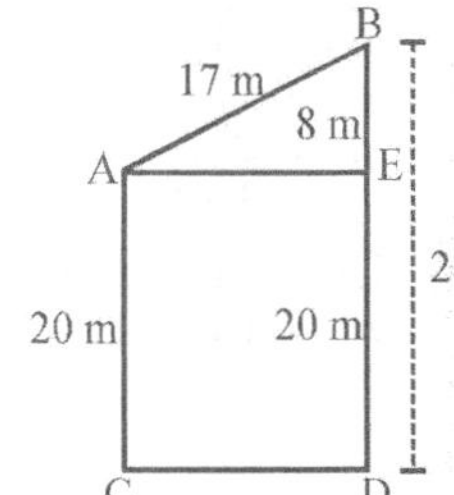

43. (a) $\frac{6}{2\sqrt{3}} = \tan\theta$

$\Rightarrow \tan\theta = \sqrt{3}$

$\Rightarrow \theta = 60°$

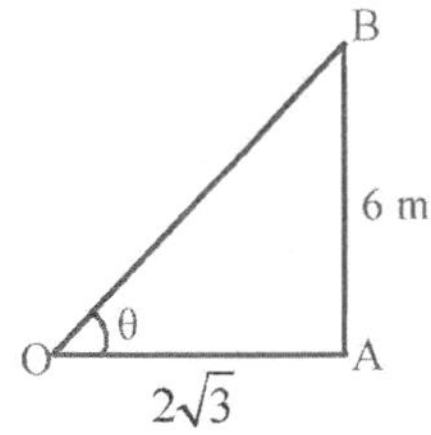

44. (a) Length of the string of the kite AB = 85 m

$\tan\theta = \frac{15}{8}$

$\Rightarrow \cot\theta = \frac{8}{15}$

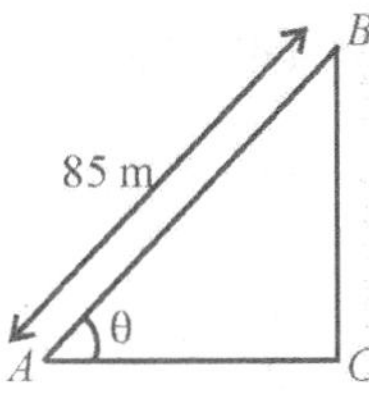

$\Rightarrow \operatorname{cosec}^2\theta - 1 = \frac{64}{225}$

$\Rightarrow \operatorname{cosec}^2\theta = 1 + \frac{64}{225} = \frac{289}{15}$

$\Rightarrow \operatorname{cosec}\theta = \sqrt{\frac{289}{15}} = \frac{17}{15} \Rightarrow \sin\theta = \frac{15}{17}$

In ΔABC, $\sin\theta = \frac{BC}{AB}$

$\Rightarrow \frac{15}{17} = \frac{BC}{85} \Rightarrow BC = 75\text{cm}$

$\therefore$ Height of kite = 75 cm

45. **(b)** Suppose height of the chimney is h metre. Let A and B be the point of observation and distance be $= x$ m.

In ΔACD,

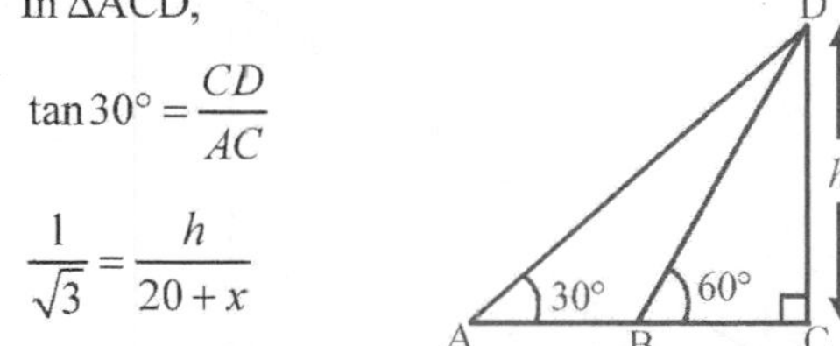

$$\tan 30° = \frac{CD}{AC}$$

$$\Rightarrow \frac{1}{\sqrt{3}} = \frac{h}{20+x}$$

$$\Rightarrow 20 + x = h\sqrt{3}$$

$$\Rightarrow x = h\sqrt{3} - 20 \quad \text{..... (1)}$$

Now, in ΔABC, $\tan 45° = \frac{CD}{BC}$

$$\Rightarrow 1 = \frac{h}{x} \Rightarrow x = h \quad \text{..... (2)}$$

From (1) and (2), we get

$$h = h\sqrt{3} - 20 \Rightarrow h\sqrt{3} - h = 20$$

$$\therefore \quad h = \frac{20}{\sqrt{3}-1}$$

46. **(a)** Here, CD = 20 m [Height of big pole]
AB = 14 m [Height of small pole]

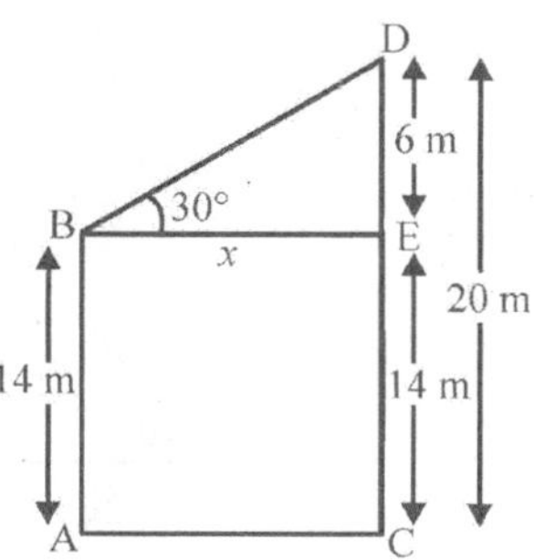

DE = CD – CE ⇒ DE = CD – AB [∵ AB = CE]

$\Rightarrow$ DE = 20 – 14 = 6 m

In ΔBDE, $\sin 30° = \frac{DE}{BD}$

$$\Rightarrow \frac{1}{2} = \frac{6}{BD} \Rightarrow BD = 12 \text{ m}$$

∴ Length of wire = 12 m

47. **(a)** C and D be the position of the men and AB is the height of flagstaff.

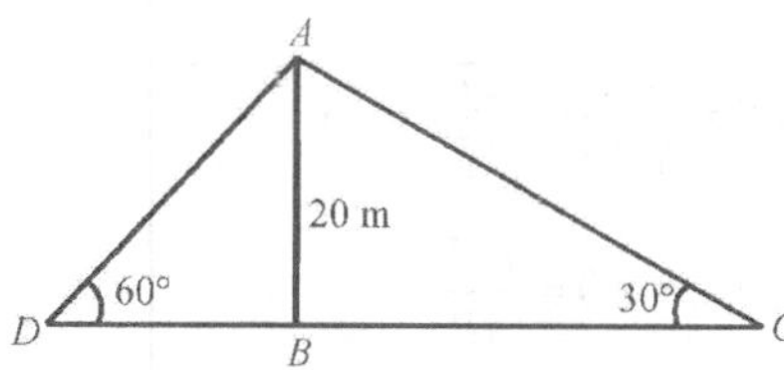

In ΔABC, $\tan 30° = \frac{AB}{BC}$

$$\Rightarrow \frac{1}{\sqrt{3}} = \frac{20}{BC} \Rightarrow BC = 20\sqrt{3}$$

In ΔABD, $\tan 60° = \frac{AB}{BD}$

$$\Rightarrow \sqrt{3} = \frac{20}{BD} \Rightarrow BD = \frac{20}{\sqrt{3}}$$

Distance between the men, CD = BC + BD

$$= 20\sqrt{3} + \frac{20}{\sqrt{3}} + \frac{60+20}{\sqrt{3}} = \frac{80}{\sqrt{3}}$$

$$= \frac{80\sqrt{3}}{3} = \frac{80 \times 1.73}{3} = 46.19 \text{ m}$$

48. **(b)**

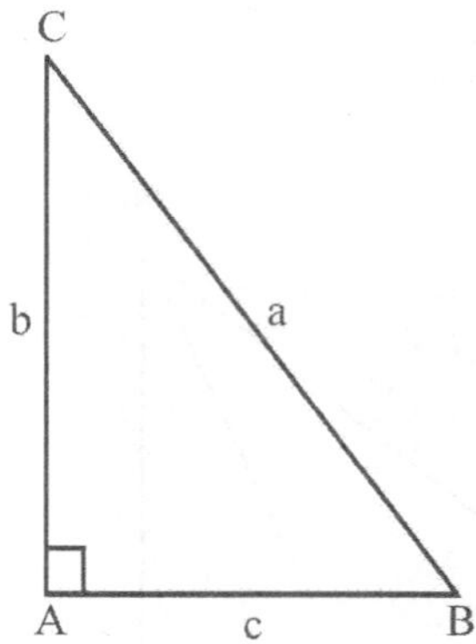

In right angled ΔABC,

$$\tan B = \frac{P}{B} = \frac{b}{c}$$

$$\tan C = \frac{P}{B} = \frac{c}{b}$$

$$\tan B + \tan C = \frac{b}{c} + \frac{c}{b}$$

$$= \frac{b^2 + c^2}{bc} = \frac{a^2}{bc} \qquad [\because a^2 = b^2 + c^2]$$

49. **(a)**

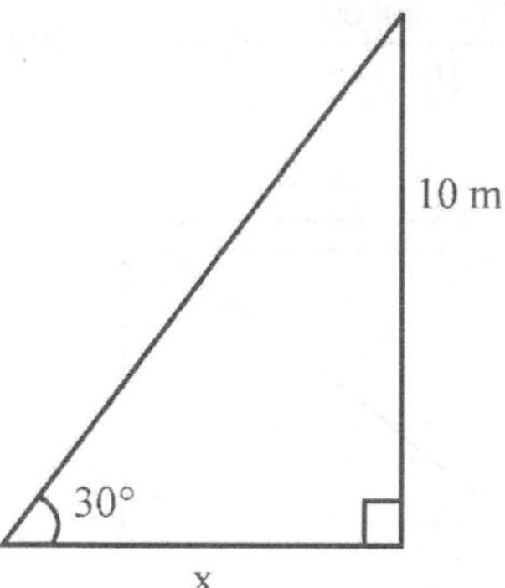

Let ‘x’ be the distance of foot of ladder

$$\tan 30° = \frac{P}{B} = \frac{10}{x}$$

$$\Rightarrow \frac{1}{\sqrt{3}} = \frac{10}{x} \Rightarrow x = 10\sqrt{3} \text{ m}$$

50. **(d)** tan 7° tan23° tan 60° tan 67° tan 83°
= tan (90° – 83°) tan (90° – 67°) tan 60° tan 67° tan 83°
= cot 83° cot 67° tan 60° tan 67° tan 83°
[∵ tan (90° – θ) = cot θ]

$$=\frac{1}{\tan 83^\circ}\times\frac{1}{\tan 67^\circ}\times\tan 60^\circ\times\tan 67^\circ\times\tan 83^\circ$$

$= \tan 60^\circ = \sqrt{3}$

51. **(c)** (sec θ – cos θ) (cosec θ – sin θ) (tan θ + cot θ)

$$\Rightarrow\left(\frac{1}{\cos\theta}-\cos\theta\right)\left(\frac{1}{\sin\theta}-\sin\theta\right)\left(\frac{\sin\theta}{\cos\theta}+\frac{\cos\theta}{\sin\theta}\right)$$

$$\Rightarrow\frac{1-\cos^2\theta}{\cos\theta}\times\frac{1-\sin^2\theta}{\sin\theta}\times\frac{\sin^2\theta+\cos^2\theta}{\sin\theta\cos\theta}$$

$$\Rightarrow\frac{\sin^2\theta}{\cos\theta}\times\frac{\cos^2\theta}{\sin\theta}\times\frac{1}{\sin\theta\cos\theta}\ [\because\sin^2\theta+\cos^2\theta=1]$$

$\Rightarrow 1$

52. **(b)** $\tan(\theta_1+\theta_2)=\sqrt{3}$

or $\tan(\theta_1+\theta_2)=\tan 60^\circ$

$\theta_1+\theta_2=60^\circ$ ... (1)

$\sec(\theta_1-\theta_2)=\frac{2}{\sqrt{3}}$

or $\sec(\theta_1-\theta_2)=\sec 30^\circ$ ... (2)

$\theta_1-\theta_2=30^\circ$

Adding equations (1) & (2)

$\theta_1+\theta_2+\theta_1-\theta_2=90^\circ$

$\theta_1=45^\circ$ & $\theta_2=15^\circ$

Now, sin 2 × 45° + tan 3 × 15°

= sin 90° + tan 45° = 1 + 1 = 2

53. **(a)** tan 1° tan 2° tan 3° ... tan 89°
= tan 1° tan 2°... tan45° ... tan (90 – 2) tan (90 – 1)
= tan 1° tan 2° ... 1 ... cot 2° cot 1°
= (tan 1° cot 1°) (tan 2° cot 2°) ... 1
= 1

54. **(a)** $4\tan^2\theta+9\cot^2\theta$

$\Rightarrow (2\tan\theta)^2+(3\cot\theta)^2$

$(2\tan\theta)^2+(3\cot\theta)^2-12+12$

$=(2\tan\theta-3\cot\theta)^2+12$

∴ Minimum value = 12 because $(2\tan\theta-3\cot\theta)^2\ge 0$

55. **(c)** $\sin\theta-\cos\theta=\frac{1}{2}$

sin θ + cos θ = x.

On squaring and adding.

$2(\sin^2\theta+\cos^2\theta)=\frac{1}{4}+x^2$

$\Rightarrow x^2=2-\frac{1}{4}=\frac{7}{4}$

$\Rightarrow x=\frac{\sqrt{7}}{2}$

56. **(c)** $\operatorname{cosec}\theta-\cot\theta=\frac{7}{2}$ ... (1)

$\operatorname{cosec}^2\theta-\cot^2\theta=1$

$\Rightarrow (\operatorname{cosec}\theta+\cot\theta)(\operatorname{cosec}\theta-\cot\theta)=1$

$\Rightarrow \operatorname{cosec}\theta+\cot\theta=\frac{1}{\operatorname{cosec}-\cot\theta}=\frac{2}{7}$ ...(2)

On adding both equations.

$2\operatorname{cosec}\theta=\frac{7}{2}+\frac{2}{7}$

$=\frac{49+4}{14}=\frac{53}{14}$

$\Rightarrow \operatorname{cosec}\theta=\frac{53}{28}$

57. **(c)**

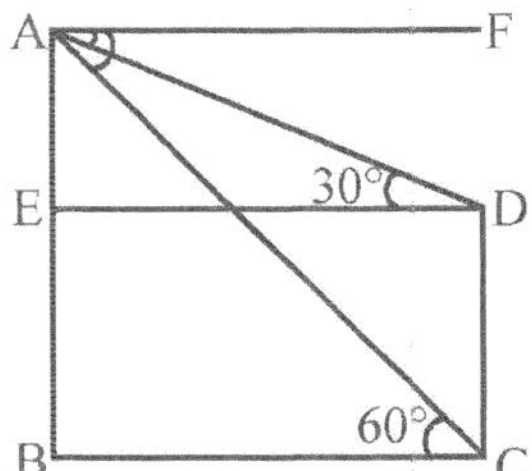

AB = Height of Hill = 200 metre

∠ADE = 30°

∠ACB = 60°

DE = BC = x metre

From ΔABC, $\tan 60^\circ=\frac{AB}{BC}$

$\Rightarrow \sqrt{3}=\frac{200}{x}\Rightarrow x=\frac{200}{\sqrt{3}}$ metre

From ΔAED,

$\tan 30^\circ=\frac{AE}{DE}$

$\Rightarrow \frac{1}{\sqrt{3}}=\frac{AE}{\frac{200}{\sqrt{3}}}\Rightarrow AE=\frac{200}{3}$ meter

$\therefore CD=200-\frac{200}{3}=\frac{400}{3}=133\frac{1}{3}$ metre

58. **(b)** $\sec^2\theta+\tan^2\theta=7$

$1+\tan^2\theta+\tan^2\theta=7$

$\tan^2\theta=\frac{6}{2}=3$

for $0\le\theta\le\frac{\pi}{2}$ $\tan\theta=\sqrt{3}$

$\theta=60^\circ=\frac{\pi}{3}$

59. (c) $\sin^2x + \cos^2x - 2(\sec^2x - \tan^2x)$
$1 - 2(1) = -1$

60. (a) $\dfrac{a^2}{a^2\sin^2\theta} - \dfrac{b2}{b^2\tan^2\theta}$
$\text{cosec}^2\theta - \cot^2\theta = 1$

61. (d) (right triangle: hypotenuse 100 m, opposite side 50 m, angle θ)

$\sin\theta = \dfrac{50\text{m}}{100\text{ m}} = \dfrac{1}{2}$

$\theta = 30°$

62. (b) (right triangle: height 20 m, angle of elevation 30°, base x)

$\tan 30° = \dfrac{20}{x}$

$\dfrac{1}{\sqrt{3}} = \dfrac{20}{x}$

$x = 20\sqrt{3}$

63. (a) $\dfrac{2\sin\theta - \cos\theta}{\cos\theta + \sin\theta} = 1$
Dividing numerator and denominator by cos θ.

$\dfrac{\dfrac{2\sin\theta}{\cos\theta} - \dfrac{\cos\theta}{\cos\theta}}{\dfrac{\cos\theta}{\cos\theta} + \dfrac{\sin\theta}{\cos\theta}} = 1$

$\Rightarrow \dfrac{2\tan\theta - 1}{1 + \tan\theta} = 1$

$\Rightarrow 2\tan\theta - 1 = 1 + \tan\theta \Rightarrow \tan\theta = 2$

Hence, $\cot\theta = \dfrac{1}{\tan\theta} = \dfrac{1}{2}$

64. (c) $\tan\left(\dfrac{\pi}{2} - \dfrac{\theta}{2}\right) = \sqrt{3}$

$\tan\left(\dfrac{\pi}{2} - \dfrac{\theta}{2}\right) = \tan\dfrac{\pi}{3}$

$\dfrac{\pi}{2} - \dfrac{\theta}{2} = \dfrac{\pi}{3} \Rightarrow \dfrac{\theta}{2} = \dfrac{\pi}{2} - \dfrac{\pi}{3} \Rightarrow \theta = \dfrac{\pi}{3}$

$\therefore \cos\theta = \cos\dfrac{\pi}{3} = \dfrac{1}{2}$

65. (c) $\dfrac{2}{\sin^2 23°} \cdot \dfrac{\sin^2 23°}{\cos^2 23°} - (\sin^2 23 + \cos^2 23°) - \tan^2 23°$

$= 2\sec^2 23° - 1 - \tan^2 23°$

$= (\sec^2 23° - 1) + (\sec^2 23° - \tan^2 23°)$

$= \tan^2 23° + 1 = \sec^2 23°$

66. (d) $\dfrac{\dfrac{5\sin\theta - 3\cos\theta}{\cos\theta}}{\dfrac{5\sin\theta + 2\cos\theta}{\cos\theta}} = \dfrac{5\tan\theta - 3}{5\tan\theta + 2} = \dfrac{5\times\dfrac{4}{5} - 3}{5\times\dfrac{4}{5} + 2}$

$= \dfrac{1}{6}$

67. (d) $x\sin\theta + y\cos\theta = \sqrt{x^2 + y^2}$

Put $x = \sin\theta$
$y = \cos\theta$ in the above equation, we have

$\sin^2\theta + \cos^2\theta = \sqrt{\sin^2\theta + \cos^2\theta}$

$\Rightarrow 1 = 1$

$\Rightarrow x = \sin\theta$ & $y = \cos\theta$ is the solution of above equation. Now, on using $x = \sin\theta$ & $y = \cos\theta$ in

$\dfrac{\cos^2\theta}{a^2} + \dfrac{\sin^2\theta}{b^2} = \dfrac{1}{x^2 + y^2}$

$\Rightarrow \dfrac{y^2}{a^2} + \dfrac{x^2}{b^2} = \dfrac{1}{\sin^2\theta + \cos^2\theta}$

$\Rightarrow \dfrac{x^2}{b^2} + \dfrac{y^2}{a^2} = 1$

68. (a) $2\cos 2\theta = 1$

$\cos 2\theta = \dfrac{1}{2}$

$\theta = 30°$

$\cot 30° = \sqrt{3}$

69. (d) $\cos^2\theta = \dfrac{(x+y)^2}{4xy}$

$1 - \cos^2\theta = 1 - \dfrac{(x+y)^2}{4xy} = -\dfrac{(x-y)^2}{4xy}$

$\sin^2\theta$ cannot be – ve
Both $\sin^2\theta$ and $\cos^2\theta$ will be +ve when $x = y$

70. (a) Given $\tan^2\theta - \sin^2\theta = x$

$\dfrac{\sin^2\theta}{\cos^2\theta} - \sin^2\theta = x \Rightarrow \sin^2\theta\left(\dfrac{1}{\cos^2\theta} - 1\right) = x$

$\dfrac{\sin^2\theta(1 - \cos^2\theta)}{\cos^2\theta} \Rightarrow \dfrac{\sin^2\theta}{\cos^2\theta}(1 - \cos^2\theta) = x$

or $\boxed{\tan^2\theta\sin^2\theta = x}$

71. **(a)** $\dfrac{\sin 25° \cos 65° + \cos 25° \sin 65°}{\tan^2 70° - \sec^2 70°}$

$= \dfrac{\sin(25° + 65°)}{-1} = \dfrac{\sin 90°}{(-1)} = -1$

72. **(b)** $4\cos^2\theta - 4\cos\theta + 1 = 0$

$(2\cos\theta - 1)^2 = 0$

or, $2\cos\theta = 1$

$\Rightarrow \cos\theta = \dfrac{1}{2}$

$\theta = 60°$

Hence, the value of $\tan(\theta - 15°) = \tan(60° - 15°)$
$= \tan 45° = 1$

73. **(a)** $\left(r\cos\theta - \sqrt{3}\right)^2 + (r\sin\theta - 1)^2 = 0$

If and only if, $r\cos\theta - \sqrt{3} = 0$, and $r\sin\theta - 1 = 0$

$r\cos\theta = \sqrt{3}$ ...(1)

$r\sin\theta = 1$ ...(2)

dividing (1) by (2)

$\tan\theta = \dfrac{1}{\sqrt{3}}$

$\sec\theta = \sqrt{1 + \tan^2\theta} = \sqrt{1 + \dfrac{1}{3}} = \dfrac{2}{\sqrt{3}}$

Also, $r^2\cos^2\theta + r^2\sin^2\theta = \left(\sqrt{3}\right)^2 + 1^2$

$r^2 = 4 \Rightarrow r = 2$

$$\frac{r\tan\theta + \sec\theta}{r\sec\theta + \tan\theta} = \frac{2 \times \frac{1}{\sqrt{3}} + \frac{2}{\sqrt{3}}}{2 \times \frac{2}{\sqrt{3}} + \frac{1}{\sqrt{3}}} = \frac{4/\sqrt{3}}{5/\sqrt{3}} = \frac{4}{5}$$

74. **(c)**

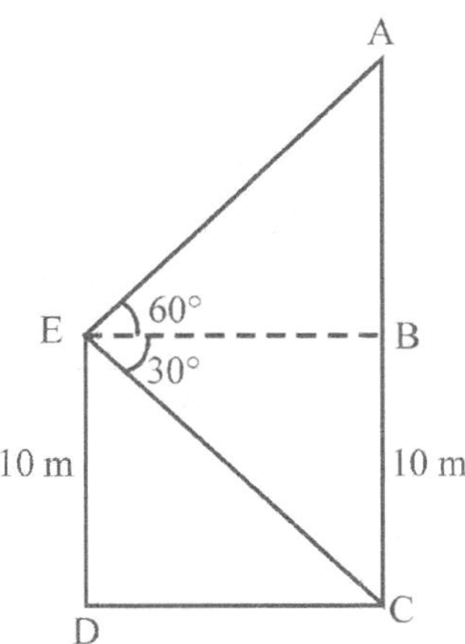

ED = BC = 10 m

In ΔABE, $\tan 60° = \dfrac{AB}{EB}$

$\sqrt{3} = \dfrac{AB}{EB} \Rightarrow AB = \sqrt{3}EB$ ...(1)

In ΔEBC, $\tan 30° = \dfrac{BC}{EB}$

$\dfrac{1}{\sqrt{3}} = \dfrac{10}{EB} \Rightarrow EB = 10\sqrt{3}\text{m}$

Putting value of EB in (1)

$AB = \sqrt{3}\left(10\sqrt{3}\right) = 30\text{m}$

AC = AB + BC = 40 m

75. **(a)**

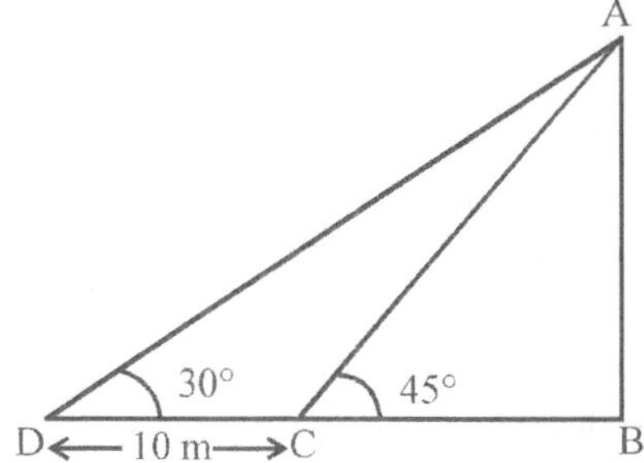

In ΔABC, $\tan 45° = \dfrac{AB}{BC}$

$1 = \dfrac{AB}{BC} \Rightarrow AB = BC$ ...(1)

In ΔABD, $\tan 30° = \dfrac{AB}{BD}$

$\dfrac{1}{\sqrt{3}} = \dfrac{AB}{BD}$

$BD = AB\sqrt{3}$

$DC + BC = AB\sqrt{3}$

$DC = AB\sqrt{3} - AB$ [from (1)]

$10 = AB\left(\sqrt{3} - 1\right)$

$AB = \dfrac{10}{\sqrt{3} - 1} \times \dfrac{\sqrt{3} + 1}{\sqrt{3} + 1} = 5\left(\sqrt{3} + 1\right)\text{m}$

76. **(b)** $\dfrac{\tan^2\theta}{1 + \tan^2\theta} + \dfrac{\cot^2\theta}{1 + \cot^2\theta}$

$= \dfrac{\tan^2\theta}{\sec^2\theta} + \dfrac{\cot^2\theta}{\text{cosec}^2\theta}$

$= \sin^2\theta + \cos^2\theta = 1$

Level-II

1. (c)
2. (a)
3. (d) We know $\sec^2\theta - \tan^2\theta = 1$ and $\sec\theta = \frac{x}{p}$, $\tan\theta = \frac{y}{q}$

$\therefore\ x^2q^2 - p^2y^2 = p^2q^2$

4. (c)
5. (b) $\sin 68° = \sin(90° - 22°) = \cos 22°$ and $\cos 68° = \sin 22°$, $\cos 63° = \sin 27°$

$\therefore$ given expression

$$= \frac{\sin^2 22° + \sin^2 22°}{\cos^2 22° + \sin^2 22°} + \sin^2 63° + \cos 63° = \frac{1}{1} + 1 = 2$$

6. (b) $\sec(15° - \theta) = \sec(90° - 75° - \theta)$
$= \sec(90° - (75° + \theta)) = \text{cosec}(75° + \theta)$
$\cot(35° - \theta) = \cot(90° - 55° - \theta)$
$= \cot(90° - (55° + \theta)) = \tan(55° + \theta)$

$\therefore$ given expression $= \text{cosec}(75° + \theta) - \text{cosec}(75° + \theta) - \tan(55° + \theta) + \tan(55° + \theta) = 0$

7. (b) $\cos(45° - \theta) = \cos(90° - 45° - \theta)$
$= \cos(90° - (45° + \theta)) = \sin(45° + \theta)$
$\therefore$ given expression $= \sin(45° + \theta) - \sin(45° + \theta) = 0$

8. (d) $\sin\alpha = \frac{1}{2} = \sin 30° \Rightarrow \alpha = 30°, \cos\beta = \frac{1}{2} = \cos 60°$

$\Rightarrow \beta = 60°$

$\therefore\ \alpha + \beta = 30° + 60° = 90°$

9. (a) $\sin A + \sin^2 A = 1 \Rightarrow \sin A = 1 - \sin^2 A = \cos^2 A$
$\Rightarrow \sin^2 A = \cos^4 A$
$= 1 - \cos^2 A = \cos^4 A \Rightarrow \cos^2 A + \cos^4 A = 1$

10. (b) $7\sin^2\theta + 3\cos^2\theta = 4$
$\Rightarrow 7\sin^2\theta + 3(1 - \sin^2\theta) = 4$
$\Rightarrow 7\sin^2\theta + 3 - 3\sin^2\theta = 4$

$\Rightarrow 4\sin^2\theta = 1 \Rightarrow \sin^2\theta = \frac{1}{4}$

$\Rightarrow \sin\theta = \frac{1}{2}\quad \therefore\ \theta = 30°$

11. (c) $a\cos\theta + b\sin\theta = 4$ and $a\sin\theta - b\cos\theta = 3$
$\Rightarrow (a\cos\theta + b\sin\theta)^2 = 4^2$ and $(a\sin\theta - b\cos\theta)^2 = 3^2$
Adding, we get
$a^2\cos^2\theta + b^2\sin^2\theta + 2ab\sin\theta\cos\theta + a^2\sin^2\theta + b^2\cos^2\theta - 2ab\sin\theta\cos\theta = 16 + 9$
$\Rightarrow a^2 + b^2 = 25$

12. (c) $p\sin\theta + q\cos\theta = a$ and $p\cos\theta - q\sin\theta = b$
Squaring and adding, we get
$(p\sin\theta + q\cos\theta)^2 + (p\cos\theta - q\sin\theta)^2 = a^2 + b^2$
$\Rightarrow p^2 + q^2 = a^2 + b^2$
$\Rightarrow (p^2 - a^2) + (a^2 - b^2) = 0$
$\Rightarrow (p + a)(p - a) + (q + b)(a - b) = 0$

Dividing both sides by $(p - a)(q + b)$, we get

$$\frac{p+a}{q+b} + \frac{q-b}{p-a} = 0$$

13. (b) We have $\text{cosec}\,\theta - \cot\theta = \frac{1}{2}$ ..... (1)
$\Rightarrow \text{cosec}\,\theta + \cot\theta = 2$ ..... (2)
Adding (1) and (2), we get

$$2\,\text{cosec}\,\theta = \frac{1}{2} + 2 = \frac{5}{2} \Rightarrow \text{cosec}\,\theta = \frac{5}{4}$$

$$\Rightarrow \sin\theta = \frac{4}{5} \Rightarrow \sin^2\theta = \frac{16}{25}$$

$$\Rightarrow 1 - \cos^2 = \frac{16}{25} \Rightarrow \cos^2\theta = 1 - \frac{16}{25} = \frac{9}{25}$$

$$\Rightarrow \cos\theta = \frac{3}{5}$$

14. (b) We have, $\cos\theta + \sin\theta = \sqrt{2}\cos\theta$
$(\cos\theta + \sin\theta)^2 + (\cos\theta - \sin\theta)^2 = 2(\cos^2\theta + \sin^2\theta)$
$\Rightarrow 2\cos^2\theta + (\cos\theta - \sin\theta)^2 = 2\cos^2\theta + 2\sin^2\theta$
$\Rightarrow (\cos\theta - \sin\theta)^2 = 2\sin^2\theta$
$\Rightarrow \cos\theta - \sin\theta) = \sqrt{2}\sin\theta$

15. (c) We have, $\sec\theta + \tan\theta = x$ ..... (1)
$\Rightarrow (\sec\theta + \tan\theta)(\sec\theta - \tan\theta) = x(\sec\theta - \tan\theta)$

$\Rightarrow \sec\theta - \tan\theta = \frac{1}{x}$ ..... (2)

Adding (1) and (2), we get

$$2\sec\theta = x + \frac{1}{x}$$

Substracting (2) from (1), we get

$$2\tan\theta = x - \frac{1}{x}$$

$$\Rightarrow \frac{2\tan\theta}{2\sec\theta} = \frac{x - \frac{1}{x}}{x + \frac{1}{x}} \Rightarrow \sin\theta = \frac{x^2 - 1}{x^2 + 1}$$

16. (b) We have,
$(1 + \cot\theta - \text{cosec}\,\theta)(1 + \tan\theta + \sec\theta)$

$$\Rightarrow \left(1 + \frac{\cos\theta}{\sin\theta} - \frac{1}{\sin\theta}\right)\left(1 + \frac{\sin\theta}{\cos\theta} - \frac{1}{\cos\theta}\right)$$

$$\Rightarrow \left(\frac{\sin\theta + \cos\theta - 1}{\sin\theta}\right)\left(\frac{\cos\theta + \sin\theta - 1}{\cos\theta}\right)$$

$$\Rightarrow \frac{(\sin\theta + \cos\theta)^2 - 1^2}{\sin\theta\cos\theta}$$

$$\Rightarrow \frac{\sin^2\theta + \cos^2\theta + 2\sin\theta\cos\theta - 1}{\sin\theta\cos\theta}$$

$$\Rightarrow \frac{2\sin\theta\cos\theta}{\sin\theta\cos\theta} = 2$$

17. **(d)** We have,

$2(\sin^6\theta + \cos^6\theta) - 3(\sin^4\theta + \cos^4\theta)$

$\Rightarrow 2[(\sin^2\theta + \cos^2\theta)^3 - 3\sin^2\theta\cos^2\theta(\sin^2\theta + \cos^2\theta)] - 3[(\sin^2\theta + \cos^2\theta)^2 - 2\sin^2\theta\cos^2\theta]$

$\Rightarrow 2(1 - 3\sin^2\theta\cos^2\theta) - 3(1 - 2\sin^2\theta\cos^2\theta)$

$\Rightarrow 2 - 6\sin^2\theta\cos^2\theta - 3 + 6\sin^2\theta\cos^2\theta = -1$

18. **(b)** **Hint :** $\cos 30° = \frac{AC}{AB} \Rightarrow \frac{\sqrt{3}}{2} = \frac{10\text{ m}}{AB}$

$\Rightarrow AB = \frac{20}{\sqrt{3}}$ m.

19. **(c)** Let the aeroplanes are at point A and D respectively. Aeroplane A is flying 600m above the ground.

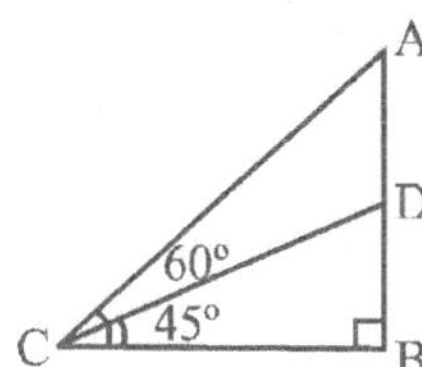

So, AB = 600.

$\angle ACB = 60°$, $\angle DCB = 45°$

From ΔABC, $\frac{AB}{BC} = \tan 60° \Rightarrow BC = \frac{600}{\sqrt{3}} = 200\sqrt{3}$.

From ΔDCB, $\frac{DB}{BC} = \tan 45° \Rightarrow DB = 200\sqrt{3}$.

So, the distance $AD = AB - BD = 600 - 200\sqrt{3}$

$= 200(3 - \sqrt{3}) = 200(3 - 1.7321) = 253.58$m.

20. **(a)** [**Hints:** $\sin 60° = \frac{MP}{OP}$

$\Rightarrow \frac{\sqrt{3}}{2} = \frac{40\sqrt{3}}{OP} \Rightarrow OP = 80$ m.]

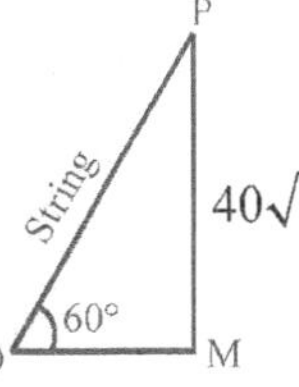

21. **(d)** In ΔCDE,

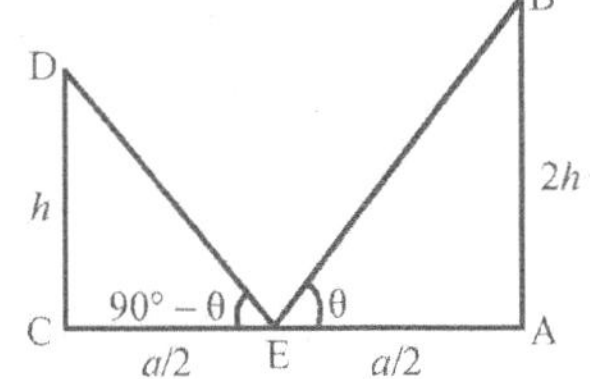

$\tan(90° - \theta) = \frac{h}{a/2}$

$\Rightarrow \cot\theta = \frac{2h}{a}$ ..... (1)

In ΔABE, $\tan\theta = \frac{AB}{EA} = \frac{2h}{a/2}$

$\Rightarrow \tan\theta = \frac{4h}{a}$ ..... (2)

Multiply (1) by (2), we get

$\cot\theta \times \tan\theta = \frac{2h}{a} \times \frac{4h}{a}$

$\Rightarrow 1 = \frac{8h^2}{a^2} \Rightarrow h = \frac{a}{2\sqrt{2}}$

22. **(b)** Let AB be the height of the rock and CD be the height of tower.

CD = BE = 100 m,

AB = H metre

∴ AE = AB – BE = H – 100, CE = BD = x

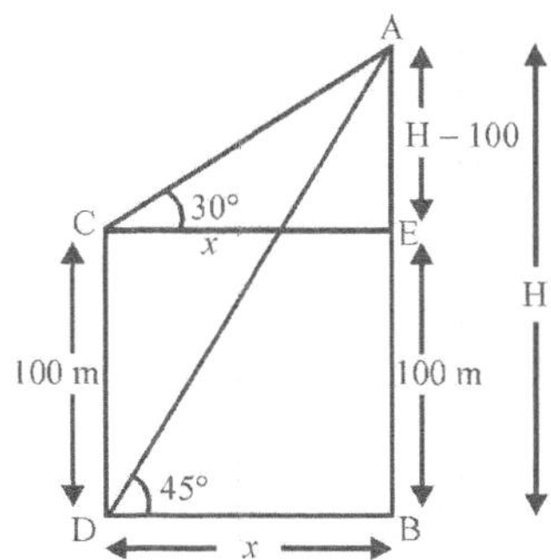

In ΔACE, $\tan 30° = \frac{AE}{CE}$

$\Rightarrow \frac{1}{\sqrt{3}} = \frac{H - 100}{x} \Rightarrow x = \sqrt{3}(H - 100)$ ..... (1)

In ΔABD, $\tan 45° = \frac{AB}{BD}$

$\Rightarrow 1 = \frac{H}{x} \Rightarrow x = H$ ..... (2)

From (1) and (2), we have

$H = \sqrt{3}(H - 100) \Rightarrow H = \sqrt{3}H - 100\sqrt{3}$

$\Rightarrow \sqrt{3}H - H = 100\sqrt{3} \Rightarrow H = \frac{100\sqrt{3}}{\sqrt{3} - 1}$

$\Rightarrow H = \frac{100\sqrt{3}(\sqrt{3} + 1)}{(\sqrt{3} - 1)(\sqrt{3} + 1)} = \frac{100\sqrt{3}(\sqrt{3} + 1)}{2}$

$\Rightarrow H = 50\sqrt{3}(\sqrt{3} + 1) \Rightarrow H = 50(3 + \sqrt{3})$ m

23. **(b)** $2y\cos\theta = x\sin\theta$

$\Rightarrow \sin\theta = \frac{2y}{x}\cos\theta$

And $2x\sec\theta - y\,\text{cosec}\,\theta = 3$

$\Rightarrow 2x\sec\theta - \frac{y}{\sin\theta} = 3$

$\Rightarrow \frac{2x}{\cos\theta} - \frac{yx}{2y\cos\theta} = 3$

$\Rightarrow 3\cos\theta = \frac{3}{2}x \Rightarrow \cos\theta = \frac{x}{2}$

Now $\sin^2\theta + \cos^2\theta = 1$

$\Rightarrow y^2 + \frac{x^2}{4} = 1$

$\Rightarrow 4y^2 + x^2 = 4$

24. (b) $\sec^2\theta - \tan^2\theta = 1$

$(\sec\theta + \tan\theta)(\sec\theta - \tan\theta) = 1$

$$\sqrt{3}(\sec\theta - \tan\theta) = 1 \Rightarrow \sec\theta - \tan\theta = \frac{1}{\sqrt{3}} \quad ...(1)$$

$$\sec\theta + \tan\theta = \sqrt{3} \quad \text{(Given)} \quad ...(2)$$

Adding eqs. (1) and (2)

$$2\sec\theta = \sqrt{3} + \frac{1}{\sqrt{3}} \Rightarrow 2\sec\theta = \frac{4}{\sqrt{3}} \Rightarrow \sec\theta = \frac{2}{\sqrt{3}}$$

$$\therefore \cos\theta = \frac{\sqrt{3}}{2} \quad \left[\because \sec\theta = \frac{1}{\cos\theta}\right]$$

Therefore, $\sin\theta = \sqrt{1-\cos^2\theta}$

$$\Rightarrow \sqrt{1-\frac{3}{4}} = \frac{1}{2}$$

25. (a) $63°\ 14'\left(\frac{51}{60}\right)'$ [1 minute = 60 seconds]

$$\Rightarrow 63°\left[14+\frac{17}{20}\right]' \Rightarrow 63°\left[\frac{297}{20}\right]' \Rightarrow 63° + \frac{297}{20\times 60}$$

[1 degree = 60 minutes]

$$\Rightarrow \left(\frac{75897}{1200}\right)° \Rightarrow \frac{75897}{1200} \times \frac{\pi}{180} \text{radian} \Rightarrow \left(\frac{2811}{8000}\pi\right)^c$$

26. (d) AB = AC

$\therefore \angle ABC = \angle ACB$ ...(1)

[opposite angle of equal sides are equal]

AC = AD

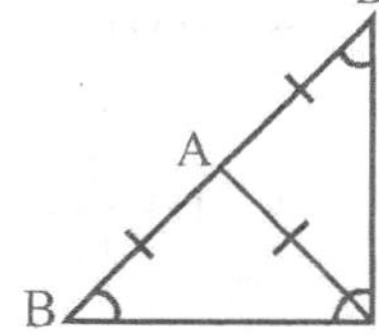

$\therefore \angle ACD = \angle ADC$ ...(2)

In a triangle,

$\angle ABC + \angle ADC + \angle DCB = 180°$

$\angle ABC + \angle ADC + \angle ACB + \angle ACD = 180°$

$2\angle ACB + 2\angle ACD = 180°$ [From eqs. (1) & (2)]

$\therefore \angle BCD = 90°$ or $\pi/2$

27. (d) $\frac{\cos^4\alpha}{\cos^2\beta} + \frac{\sin^4\alpha}{\sin^2\beta} = 1$

$\Rightarrow \cos^4\alpha\sin^2\beta + \sin^4\alpha\cos^2\beta = \cos^2\beta\sin^2\beta$

$\Rightarrow \cos^4\alpha(1-\cos^2\beta) + \cos^2\beta(1-\cos^2\alpha)^2 = \cos^2\beta(1-\cos^2\beta)$

$\Rightarrow \cos^4\alpha - \cos^4\alpha\cos^2\beta + \cos^2\beta - 2\cos^2\alpha\cos^2\beta + \cos^4\alpha\cos^2\beta = \cos^2\beta - \cos^4\beta$

$\Rightarrow \cos^4\alpha - 2\cos^2\alpha\cos^2\beta + \cos^4\beta = 0$

$\Rightarrow (\cos^2\alpha - \cos^2\beta)^2 = 0$

$\Rightarrow \cos^2\alpha = \cos^2\beta$

$\Rightarrow \sin^2\alpha = \sin^2\beta$

Then, $\frac{\cos^4\beta}{\cos^2\alpha} + \frac{\sin^4\beta}{\sin^2\alpha}$

$$\Rightarrow \frac{\cos^2\beta\cos^2\alpha}{\cos^2\alpha} + \frac{\sin^2\beta\sin^2\alpha}{\sin^2\alpha}$$

$\Rightarrow \cos^2\beta + \sin^2\beta = 1$

28. (a) $\frac{\sin\theta - \cos\theta + 1}{\sin\theta + \cos\theta - 1}$

Dividing Numerator and Denominator by $\cos\theta$

$$\Rightarrow \frac{\frac{\sin\theta}{\cos\theta} - \frac{\cos\theta}{\cos\theta} + \frac{1}{\cos\theta}}{\frac{\sin\theta}{\cos\theta} + \frac{\cos\theta}{\cos\theta} - \frac{1}{\cos\theta}} \Rightarrow \frac{\tan\theta - 1 + \sec\theta}{\tan\theta + 1 - \sec\theta}$$

$$\Rightarrow \frac{(\tan\theta + \sec\theta) - (\sec^2\theta - \tan^2\theta)}{\tan\theta - \sec\theta + 1}$$

$$\Rightarrow \frac{(\tan\theta + \sec\theta)[1 - \sec\theta + \tan\theta]}{\tan\theta - \sec\theta + 1} \Rightarrow \tan\theta + \sec\theta$$

$$\Rightarrow \frac{\sin\theta}{\cos\theta} + \frac{1}{\cos\theta} \Rightarrow \frac{1+\sin\theta}{\cos\theta}$$

29. (a) If $\sin^2\alpha = \cos^3\alpha$

$\tan^2\alpha = \cos\alpha$ ...(1)

Now consider, $\cot^6\alpha - \cot^2\alpha$

$$= \frac{1}{\tan^6\alpha} - \frac{1}{\tan^2\alpha} \text{ Since } \cot\alpha = \frac{1}{\tan\alpha}$$

Substituting for $\tan^2\alpha$ with $\cos\alpha$ from (1) above equation will be

$$= \frac{1}{\cos^3\alpha} - \frac{1}{\cos\alpha} = \frac{1-\cos^2\alpha}{\cos^3\alpha} = \frac{\sin^2\alpha}{\cos^3\alpha} = \frac{\tan^2\alpha}{\cos\alpha} = 1$$

30. (b) $(1 + \tan\theta + \sec\theta)(1 + \cot\theta - \operatorname{cosec}\theta)$

$$\Rightarrow \left(1 + \frac{\sin\theta}{\cos\theta} + \frac{1}{\cos\theta}\right)\left(1 + \frac{\cos\theta}{\sin\theta} - \frac{1}{\sin\theta}\right)$$

$$\Rightarrow \left(\frac{\sin\theta + \cos\theta + 1}{\cos\theta}\right)\left(\frac{\sin\theta + \cos\theta - 1}{\sin\theta}\right)$$

$$= \frac{(\sin\theta + \cos\theta)^2 - 1}{\sin\theta\cos\theta} = \frac{\sin^2\theta + \cos^2\theta + 2\sin\theta\cos\theta - 1}{\sin\theta\cos\theta}$$

$$= \frac{2\sin\theta\cos\theta}{\sin\theta\cos\theta} = 2$$

31. (b) In ΔABC

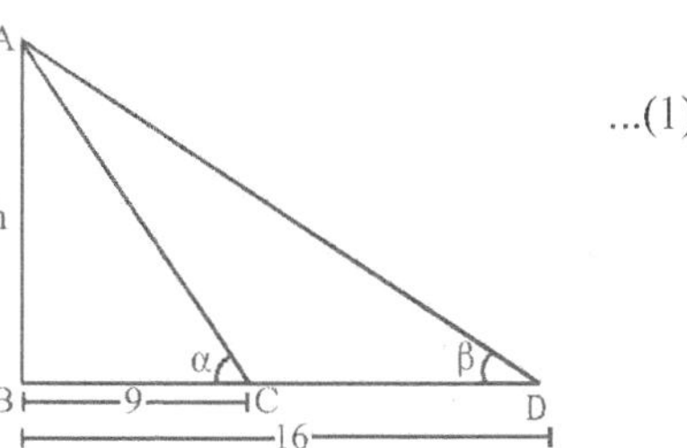

$\tan\alpha = \frac{h}{9}$ ...(1)

In ΔABD

$\tan\beta = \frac{h}{16}$

$\alpha + \beta = 90°$ (given)

$\beta = 90 - \alpha$

since $\tan\beta = \frac{h}{16}$

$\tan(90-\alpha) = \frac{h}{16} \Rightarrow \cot\alpha = \frac{h}{16}$ or $\tan\alpha = \frac{16}{h}$ ...(2)

From eqs. (1) and (2)

$\frac{h}{9} = \frac{16}{h} \Rightarrow h^2 = 16\times 9 \Rightarrow h = 12$ feet.

32. (b) $\frac{\sin 53°}{\cos 37°} \div \frac{\cot 65°}{\tan 25°}$

$\frac{\sin 53°}{\cos 37°} \times \frac{\tan 25°}{\cot 65°} \Rightarrow \frac{\sin 53°}{\cos(90°-53°)} \times \frac{\tan 25°}{\cot(90°-25°)}$

$\Rightarrow \frac{\sin 53°}{\sin 53°} \times \frac{\tan 25°}{\tan 25°} = 1$

[$\because \cos(90° - \theta) = \sin\theta$ and $\cot(90° - \theta) = \tan\theta$]

33 (c) $\frac{\cos 60° + \sin 60°}{\cos 60° - \sin 60°} = \frac{\frac{1}{2} + \frac{\sqrt{3}}{2}}{\frac{1}{2} - \frac{\sqrt{3}}{2}} = \frac{1+\sqrt{3}}{1-\sqrt{3}} \times \frac{1+\sqrt{3}}{1+\sqrt{3}}$

$\Rightarrow \frac{(1+\sqrt{3})^2}{1^2 - (\sqrt{3})^2} = \frac{1+3+2\sqrt{3}}{1-3} = \frac{4+2\sqrt{3}}{-2}$

$\Rightarrow \frac{-2(2+\sqrt{3})}{2} = -(2+\sqrt{3})$

34. (d) $\frac{\cot 5°.\cot 10°.\cot 15°.\cot 60°.\cot 75°.\cot 80°.\cot 85°}{(\cos^2 20 + \cos^2 70°) + 2}$

$\Rightarrow \frac{\cot(90°-85°).\cot(90°-80°).\cot(90°-75°).\cot 60° .\cot 75°.\cot 80° \cot 85°}{(\cos^2(90°-70°) + \cos^2 70°) + 2}$

$\Rightarrow \frac{\cot 60°}{(1+2)} = \frac{\frac{1}{\sqrt{3}}}{3} = \frac{1}{3\sqrt{3}} \times \frac{\sqrt{3}}{\sqrt{3}} = \frac{\sqrt{3}}{9}$

35. (d) Let angles are $2x$, $5x$ and $3x$.

$2x + 5x + 3x = 180°$

(sum of interior angle of triangles is 180°)

$10x = 18°$

$x = 18°$

$\therefore$ Least angle in degree $= 2x = 2 \times 18 = 36°$

In radian $= \frac{\pi}{180°} \times 36° = \frac{\pi}{5}$

36. (d) $x = a\cos\theta - b\sin\theta$

$y = b\cos\theta + a\sin\theta$

$x^2 + y^2 = (a\cos\theta - b\sin\theta)^2 + (b\cos\theta + a\sin\theta)^2$

$\Rightarrow a^2\cos^2\theta + b^2\sin^2\theta - 2ab\cos\theta\sin\theta + b^2\cos^2\theta + a^2\sin^2\theta + 2ab\cos\theta\sin\theta$

$\Rightarrow (a^2 + b^2)\cos^2\theta + (a^2 + b^2)\sin^2\theta$

$\Rightarrow (a^2 + b^2)(\cos^2\theta + \sin^2\theta)$

$\Rightarrow (a^2 + b^2)(1) \Rightarrow a^2 + b^2$

37. (a) $\tan\alpha + \cot\alpha = 2$

$\tan\alpha + \frac{1}{\tan\alpha} = 2 \Rightarrow \tan^2\alpha + 1 = 2\tan\alpha$

$\Rightarrow \tan^2\alpha - 2\tan\alpha + 1 = 0$

$\Rightarrow \tan^2\alpha - \tan\alpha - \tan\alpha + 1 = 0$

$\Rightarrow \tan\alpha(\tan\alpha - 1) - 1(\tan\alpha - 1) = 0$

$(\tan\alpha - 1)(\tan\alpha - 1) = 0$

$\therefore \tan\alpha = 1$

Now, $\tan^7\alpha + \cot^7\alpha \Rightarrow (\tan\alpha)^7 + \frac{1}{(\tan\alpha)^7} = 1 + 1 = 2$

38. (a)

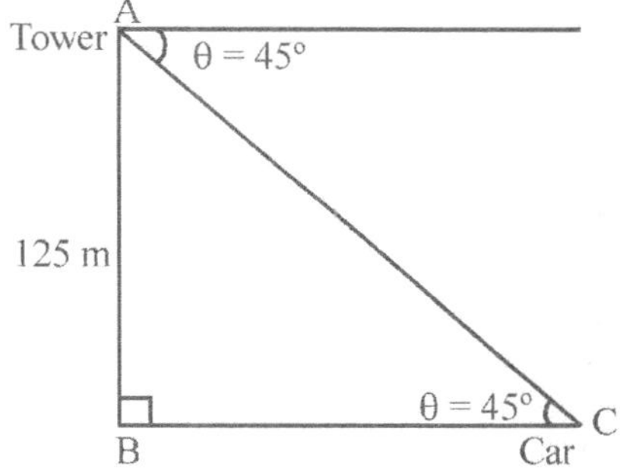

In ΔABC

$\tan\theta = \frac{AB}{BC} \Rightarrow \tan 45° = \frac{125}{BC} \Rightarrow 1 = \frac{125}{BC}$

BC = 125 m

Hence, car is 125 m from the tower.

39. (d) $(\sin^2 1° + \sin^2 89°) + (\sin^2 2° + \sin^2 88°) + ... - (\sin^2 44° + \sin^2 48°) + \sin^2 45°$

$= (\sin^2 1° + \cos^2 1°) + (\sin^2 2° + \cos^2 2°) + ... + (\sin^2 44° + \cos^2 44°) + \sin^2 45°$

$= 1 + 1 + ... + 1$ (44 times) $+ \frac{1}{2}$

$= 44\frac{1}{2}$

40. (c) $\frac{(\cos\theta + \sin\theta)(\cos^2\theta + \sin^2\theta - \sin\theta\cos\theta)}{(\cos\theta + \sin\theta)} + \frac{(\cos\theta - \sin\theta)(\cos^2\theta + \sin^2\theta + \sin\theta\cos\theta)}{(\cos\theta - \sin\theta)}$

$= 2\cos^2\theta + 2\sin^2\theta - \sin\theta\cos\theta + \sin\theta\cos\theta$

$= 2$

41. **(d)** $\sin 17° = \frac{x}{y}$

$\cos 17° = \sqrt{1 - \frac{x^2}{y^2}} = \frac{\sqrt{y^2 - x^2}}{y}$

$\sec 17° - \sin 73°$
$= \sec 17° - \cos 17°$

$= \frac{y}{\sqrt{y^2 - x^2}} - \frac{\sqrt{y^2 - x^2}}{y}$

$= \frac{y^2 - y^2 + x^2}{y\sqrt{y^2 - x^2}} = \frac{x^2}{y\sqrt{y^2 - x^2}}$

42. **(c)** $\operatorname{cosec}\theta + \cot\theta = \sqrt{3}$

$\frac{1}{\sin\theta} + \frac{\cos\theta}{\sin\theta} = \sqrt{3}$

$\frac{1 + \cos\theta}{\sin\theta} = \sqrt{3}$

$\frac{2\cos^2\frac{\theta}{2}}{2\sin\frac{\theta}{2}\cos\frac{\theta}{2}} = \sqrt{3}$

$\cot\frac{\theta}{2} = \sqrt{3}$

$\tan\frac{\theta}{2} = \frac{1}{\sqrt{3}}; \frac{\theta}{2} = 30°; \theta = 60°$

$\operatorname{cosec}\theta = \operatorname{cosec} 60° = \frac{2}{\sqrt{3}}$

43. **(c)** $\cos\alpha + \sec\alpha = \sqrt{3}$
taking cube both sides

$\cos^3\alpha + \sec^3\alpha + 3\cos\alpha\sec\alpha(\cos\alpha + \sec\alpha) = 3\sqrt{3}$

$\cos^3\alpha + \sec^3\alpha + 3\sqrt{3} = 3\sqrt{3}$

$\cos^3\alpha + \sec^3\alpha = 0$

44. **(a)** $\sin\theta + \cos\theta = \sqrt{2}\cos\theta$

$\sin\theta = (\sqrt{2} - 1)\cos\theta$

$\cot\theta = \frac{1}{\sqrt{2} - 1}$

$\cot\theta = \frac{1}{\sqrt{2} - 1} \times \frac{\sqrt{2} + 1}{\sqrt{2} + 1} = \sqrt{2} + 1$

45. **(a)**

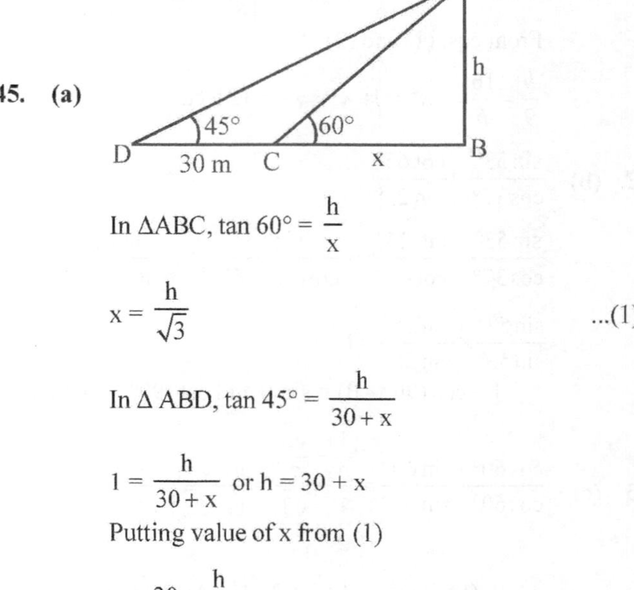

In ΔABC, $\tan 60° = \frac{h}{x}$

$x = \frac{h}{\sqrt{3}}$ ...(1)

In ΔABD, $\tan 45° = \frac{h}{30 + x}$

$1 = \frac{h}{30 + x}$ or $h = 30 + x$

Putting value of x from (1)

$h = 30 + \frac{h}{\sqrt{3}}$

or $h\frac{(\sqrt{3} - 1)}{\sqrt{3}} = 30 \Rightarrow h = 15(3 + \sqrt{3})$ m

CHAPTER 24

DATA INTERPRETATION

CLASSIFICATION OF DATA INTERPRETATION

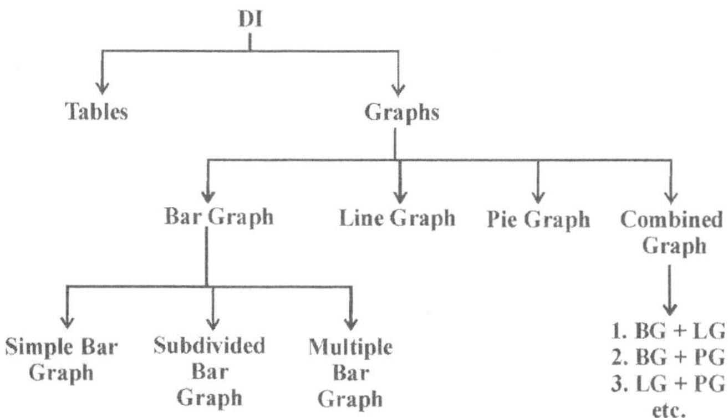

TABLES

A table is one of the easiest way for summarising data.

A statistical table is the logical listing of related quantitative data in vertical columns and horizontal rows of numbers with sufficient explanatory and qualifying words, phrases and statements in the form of titles, heading and notes to make clear the meaning of data.

Remember

$$\text{Average} = \frac{\text{Sum of all items}}{\text{Total number of items}}$$

$$\%\ \text{increase} = \frac{\text{Final value} - \text{Initial value}}{\text{Initial value}} \times 100$$

$$\%\ \text{decrease} = \frac{\text{Initial value} - \text{Final value}}{\text{Initial value}} \times 100$$

$$\%\ \text{change} = \frac{\text{Change in value}}{\text{Initial value}} \times 100$$

GRAPHS

Graphs are a convenient way to represent information. The graphs should be labelled properly to show maximum information.

1. **Bar Graph** : Bar diagram consists of a number of equidistant rectangles. One for each category of the data in which the magnitudes are represented by the length or height of rectangle, whereas width of rectangles are immaterial. Thus, a bar is just one dimensional as only the length of the bar is to be considered and not the width. All the bars drawn in a diagram are generally of uniform width which depends on the number of bars to be constructed and the availability of the space.

   Types of Bar Graphs are :

   (i) **Simple Bar Graph :** It is used to represent only one dependent variable.

   (ii) **Sub-divided Bar Graphs :** These are used to represent the break down of a total into its component bars. A bar is divided into different segments, each segment respresents a given component. Different shades, colours, designs etc. are used to distinguish the various components. An index is given to represent the various components. To compare, the order of various components in the different bars is same.

   (iii) **Multiple Bar Graph (MBG) :** When a combination of inter-related variables are to be represented graphically, multiple bar diagrams are used. These are extended form of simple bar diagrams. In M.B.G. many aspects of the data are presented simultaneously with separated bars or various shades of colours. An index is given to explain the shades or colours used.

2. **Line Graph(LG) :** LG are used to show how a quantity changes, very often the quantity is measured as time changes. If the line goes up, the quantity is increasing and the line goes down, the quantity is decreasing. If the line is horizontal, the quantity is not changing.

3. **Pie Graph(PG) :** is a pictorial representation of numerical data by non-intersecting adjacent sectors of a circle. Sector's area of each sector is proportional to the magnitude of the data represented by the sector.

$$1\%\ \text{of total value} = \frac{360}{100} = 3.6°$$

The % of components parts can be converted to degrees by multiplying 3.6°.

Degree of any component part

$$= \frac{\text{component value}}{\text{total value}} \times 360 .$$

Directions (Qs. 1-5): Study the following graph carefully and answer the questions given below it. The number of students who joined and left the school in the beginning of year for six years, from 1993 to 1998. Initial strength of the school in 1992 = 1500

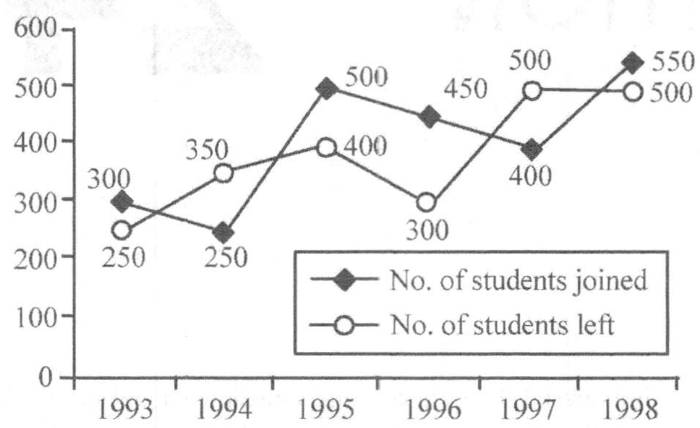

1. What was the increase/decrease in strength of the school from 1994 to 1995?
   (a) Increase by 100 (b) Decrease by 100
   (c) Increase by 200 (d) Decrease by 200
   (e) None of these
2. For which of the following years, the percentage rise/fall in number of students left from the previous year is the **highest**?
   (a) 1994 (b) 1995
   (c) 1996 (d) 1997
   (e) 1998
3. How many students were there in the school during the year, 1996?
   (a) 1495 (b) 1600
   (c) 1550 (d) 1700
   (e) None of these
4. During which of the following pairs of years, the strengths of the school is equal?
   (a) 1994 and 1995 (b) 1995 and 1997
   (c) 1996 and 1998 (d) 1995 and 1998
   (e) 1993 and 1995
5. The number of students in 1996 is **approximately** what per cent of the number of students in 1994?
   (a) 85 (b) 117
   (c) 95 (d) 103
   (e) 108

Directions (Qs. 6-10): Study the following graph carefully and answer the questions given below:

The following graph shows the percentage growth of Branded and Assembled PCs

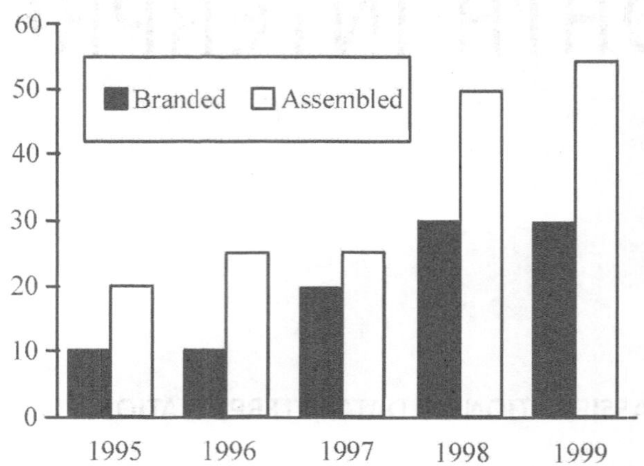

6. What is the average percentage growth of sales of Assembled PCs for the given years?
   (a) 30 (b) 20
   (c) 40 (d) 35
   (e) None of these
7. If the Branded PCs sold in1996 were 100000, how many Branded PCs were sold in 1999?
   (a) 202800 (b) 156000
   (c) 234000 (d) Cannot be determined.
   (e) None of these
8. What is the difference between total Branded and total Assembled PCs sold for the given years?
   (a) 75000 (b) 750000
   (c) 175000 (d) Cannot be determined
   (e) None of these
9. In which year is the difference in the growth between Branded and Assembled PCs lowest'?
   (a) 1995 (b) 1998
   (c) 1999 (d) 1996
   (e) None of these
10. For Assembled PCs sale, which year is the per cent growth the highest compared to previous year'?
    (a) 1999 (b) 1996
    (c) 1998 (d) Cannot be determined
    (e) None of these

Directions (Qs.11-15): Study the chart and give the answer of following questions.

Selling of the car in UK according to the colours

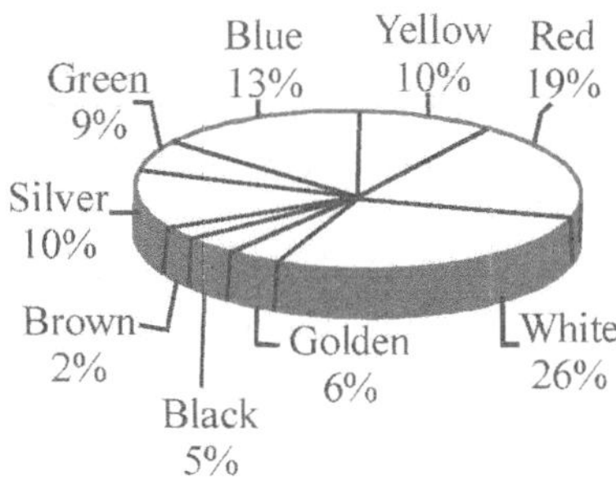

11. 50% of all the cars consisted of which colours of car?
(a) Black, Golden, Blue, Red
(b) Blue, Black, Red, Silver
(c) White, Golden, Blue, Black
(d) White, Blue, Green, Black
(e) None of these

12. Cars of which colour are 20% less popular than white coloured cars?
(a) Black (b) Golden
(c) Red (d) Blue
(e) None of these

13. Cars of which colour are 13% less popular than white cars?
(a) Blue (b) Green
(c) Silver (d) Yellow
(e) None of these

14. Cars of which colour when increased by two per cent and then, combined with that of red cars will make 30 per cent of the total
(a) Gloden (b) Blue
(c) Black (c) Yellow
(e) None of these

15. If in a certain period the total production of all cars was 95400 then, how many more blue cars were sold than green?
(a) 2580 (b) 3618
(c) 2850 (d) 3816
(e) None of these

Solutions

1. **(a)**

2. **(d)** From the graph's inclination, it is clear that the percentage rise/fall is maximum in the year 1997 w.r.t previous year.

3. **(d)** No. of students in 1996 = 1550 + (450 – 300) = 1700

4. **(e)** Strengths of the school in different years

1993	1994	1995	1996	1997	1998
1550	1450	1550	1700	1600	1650

5. **(b)** Reqd. % $= \frac{1700}{1450} \times 100 \approx 117\%$

6. **(d)** Average percentage growth of Assemble PCs

$$= \frac{20+25+25+50+55}{5} = \frac{175}{5} = 35\%$$

7. **(e)** Growth of branded PCs from 1996 to 1999 = 20%

Branded PC's sold in 1999 $= 100000 \times \frac{120}{100} = 1,20,000$

8. **(d)**

9. **(e)** Difference between Assembled and Branded PCs

1995	1996	1997	1998	1999
10%	15%	5%	20%	25%

10. **(c)** Per cent growth of Assembled PCs is

1996	1997	1998	1999
5%	No change	25%	5%

11. **(c)**

12. **(b)**

13. **(a)**

14. **(e)**

15. **(d)**

Practice Exercise

Level - I

DIRECTIONS (Qs.1-5): Study the following table to answer the given questions:

Percentage of marks obtained by seven students in six subjects

Subject (Max, Marks ↓ Students	Eng	His	Com	Math	Science	Econ
	(100)	(100)	(100)	(100)	(100)	(100)
Meera	100	80	50	90	90	60
Subodh	80	70	80	100	80	40
Kunal	90	70	60	90	70	70
Soni	60	60	65	80	80	80
Richu	50	90	62	80	85	95
Irene	40	60	64	70	65	85
Vgay	80	80	35	65	50	75

1. What is the total marks obtained by Meera in all the subject?
(a) 448 (b) 580
(c) 470 (d) 74.67
(e) None of these
2. What is the average marks obtained by these seven students in History? (rounded off to two digits)
(a) 72.86 (b) 27.32
(c) 24.86 (d) 29.14
(e) None of these
3. How many students have got 60% or more marks in all the subjects?
(a) One (b) Two
(c) Three (d) Four
(e) None of these
4. What is the overall percentage of Kunal ?
(a) 64 (b) 65
(c) 75 (d) 64.24
(e) None of these
5. In which subject is the overall percentage the best ?
(a) Maths (b) Economics
(c) History (d) Science
(e) None of these

DIRECTIONS (Qs.6-10): Study the following graph to answer the given questions.

Percent profit earned by two companies over the given years.

$$\%\text{profit} = \frac{\text{Income} - \text{Expenditure}}{\text{Expenditure}} \times 100$$

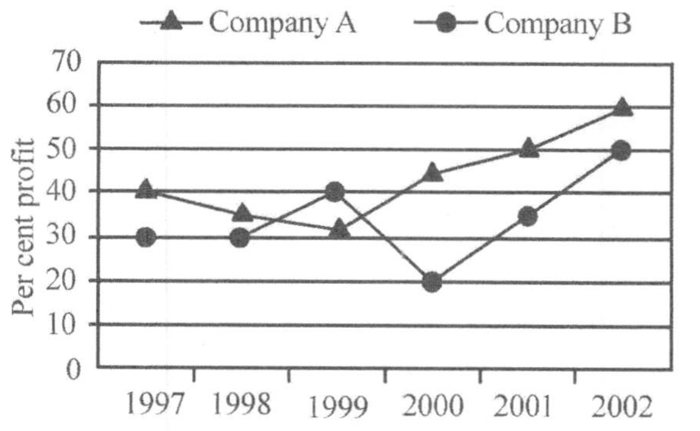

6. If the expenditure of Company B in 2000 was ₹ 200 crores, what was its income?
(a) ₹ 240 crores (b) ₹ 220 crores
(c) ₹ 160 crores (d) Cannot be determined
(e) None ot these
7. If the income of Company A in 2002 was ₹ 600 crores, what was its expenditure?
(a) ₹ 360 crores (b) ₹ 480 crores
(c) ₹ 375 crores (d) Cannot be determined
(e) None ot these
8. If the income of Company B in 1998 was ₹ 200 crores, what was its profit in 1999?
(a) ₹ 21.5 crores (b) ₹ 153 crores
(c) ₹ 46.15 crores (d) Cannot be determined
(e) None of these
9. If the incomes of the two companies in 1998 were equal, what was the ratio of their expenditure?
(a) 1 : 2 (b) 26 : 27
(c) 100 : 67 (d) Cannot be determined
(e) None of these
10. What is the per cent increase in per cent profit for CompanyB from years 2000 to 2001?
(a) 75 (b) 175
(c) 42.86 (d) Cannot be determined
(e) None of these

DIRECTIONS (Qs. 11-15): Read the following graph and answer the questions given below:

Crude oil price (in ₹) per metric tonne

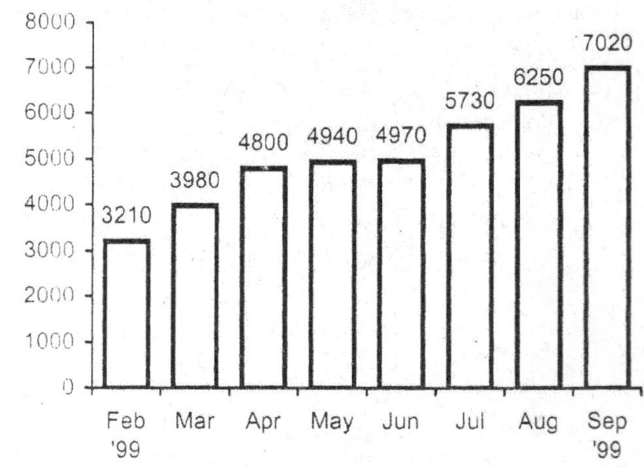

11. How many months experienced more than 10 per cent increase in crude oil price over the earlier month?
(a) 2 (b) 3
(c) 4 (d) 5
(e) None of these

12. Which, month/s experienced more than 10 per cent but less than 20 per cent increase in the price of crude oil over the earlier month?
(a) June and September (b) July and September
(c) April and July (d) March and July
(e) None of these

13. Which month/s had less than one per cent increase in crude oil price over the earlier month?
(a) June only (b) April only
(c) August and April only (d) May only
(e) None of these

14. If in April the crude oil price had been lesser than the given by ₹ 223 per metric tonne then, how much would have been the percentage increase in price over the earlier month?
(a) 10 (b) 12
(c) 14 (d) 16
(e) None of these

15. What is the **approximate** percentage increase in the price of crude oil from February to September?
(a) 100 (b) 80
(c) 130 (d) 120
(e) None of these

DIRECTIONS (Qs. 16-20): Read the following table and answer the questions.

Internet owners in our country (Approximate)

Year	Government	Private
1995-96	3900	–
1996-97	29400	–
1997-98	90000	–
1998-99	230000	12000
1999-2000	520000	120000
2000-2001	1060000	450000
2001-2002	1550000	950000

16. In which period the percentage increase in the total internet owners is least to that over the earlier period?
(a) 1997-98 (b) 1998-99
(c) 1999-2000 (d) 2001-02
(e) None of these

17. What is the total number of fresh internet owners in the period 2001-02?
(a) 54900 (b) 549000
(c) 990000 (d) 99000
(e) None of these

18. What is the proportion of Government internet owners to the Private internet owners in the period 1999-2000?
(a) 13 : 4 (b) 13 : 3
(c) 3 : 13 (d) 4 : 13
(e) None of these

19. What is the **approximate** percentage increase in the Private internet owners in the period 2001-02 over that in the period 1998-99?
(a) 5000 (b) 6000
(c) 7000 (d) 4000
(e) 8000

20. What is the **approximate** percentage of Private internet owners in the total internet owners in 1998-99?
(a) 20 (b) 5
(c) 10 (d) 15
(e) None of these

DIRECTIONS (Qs.21-24): Study the following graph to answer the given questions:

Production of a company (in lakh units) over the given years

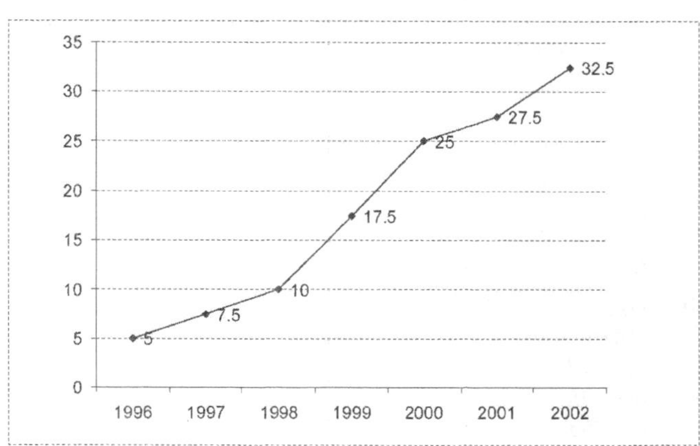

Years

21. The production in 2002 is what per cent of production in 1996?
(a) 650% (b) 550%
(c) 325% (d) 320%
(e) None of these

22. What is the **approximate** average production (in lakhs) for the given years?
(a) 18 (b) 19
(c) 20 (d) 18.5
(e) 17

23. Which of the following is the highest difference in production between two adjacent years?
(a) 5 lakhs (b) 10 lakhs
(c) 9 lakhs (d) 7.5 lakhs
(e) None of these

24. Which year had the highest per cent increase in production over the previous year?
(a) 2000 (b) 1999
(c) 2002 (d) 2007
(e) None of these

DIRECTIONS (Qs. 25 - 29): Study the following table carefully to answer these questions.

Percentage of marks obtained by six students in

Student	Subjects					
	A (out of 60)	B (out of 40)	C (out of 80)	D (out of 50)	E (out of 120)	F (out of 75)
P	80	65	58	68	75	87
Q	55	70	67	74	88	78
R	74	54	72	84	62	76
S	68	76	82	56	72	64
T	75	68	64	72	80	72
U	82	78	75	67	68	82

25. What is the total marks obtained by Student R in Subjects B, D and E?
(a) 200 (b) 138
(c) 168 (d) 156
(e) None of these

26. What is the average marks obtained by all the students in Subject 'C'? (rounded off to the nearest integer)
(a) 56 (b) 58
(c) 54 (d) 70
(e) None of these

27. What is the average percentage of marks obtained by all the students in Subject 'A'?
(a) $73\frac{2}{3}$ (b) $43\frac{2}{5}$
(c) $72\frac{1}{3}$ (d) $48\frac{3}{5}$
(e) None of these

28. Approximately what is the overall percentage of marks obtained by Q in all the subjects?
(a) 77 (b) 72
(c) 78 (d) 70
(e) 74

29. What is the total marks obtained by all the students together in Subject E?
(a) 522 (b) 488
(c) 445 (d) 534
(e) None of these

DIRECTIONS (Qs. 30-34): These questions are based on the graph given below:

Percent profit earned by six companies during 2000 and 2001

Profit = Income – Expenditure

$$\% \text{Profit} = \frac{\text{Income} - \text{Expenditure}}{\text{Expenditure}} \times 100$$

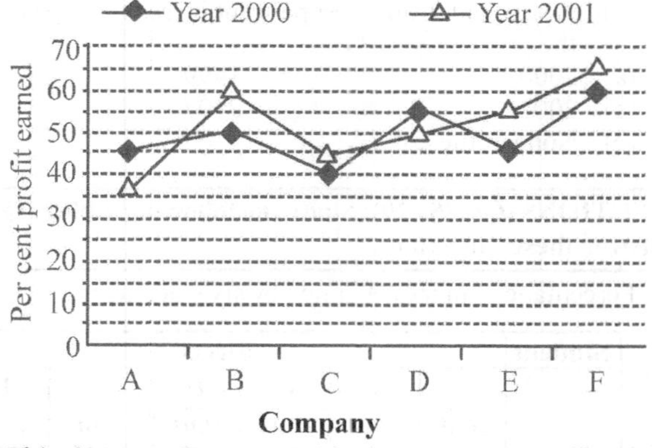

30. If the income of company C in the year 2000 was ₹ 35 lakhs, what was its expenditure in that year?
(a) ₹ 24 lakhs (b) ₹ 21 lakhs
(c) ₹ 25 lakhs (d) Can't be determined
(e) None of these

31. If, in the year 2001, total expenditure of companies B and C was ₹ 48 lakhs, then what was their total income in the same year?
(a) ₹ 32 lakhs (b) ₹ 28.6 lakhs
(c) ₹ 34.2 lakhs (d) Can't be determined
(e) None of these

32. If, in the year 2000, expenditure of Company C was ₹ 32 lakhs, what was the income of the company in the same year?
(a) ₹ 44.2 lakhs (b) ₹ 48.4 lakhs
(c) ₹ 46.4 lakhs (d) ₹ 38 lakhs
(e) None of these

33. If the expenditures of Company E in the years 2000 and 2001 were the same, what was the ratio of the incomes of the company in the same years respectively?
(a) 19 : 21 (b) 11 : 12
(c) 29 : 31 (d) 9 : 11
(e) None of these

34. The income of Company D in the year 2000 was ₹ 31 lakhs. What was the earned profit?
(a) ₹ 11 lakhs (b) ₹ 20 lakhs
(c) ₹ 17 lakhs (d) ₹ 12 lakhs
(e) None of these

DIRECTIONS (Qs. 35-39): Study the following table carefully and answer the questions given below:

Number of students studying in different faculties in seven institutions

	Faculty				
Institution	Arts	Commerce	Science	Engineering	Management
A	125	187	216	98	74
B	96	152	198	157	147
C	144	235	110	164	127
D	165	138	245	66	36
E	215	196	287	86	66
F	184	212	195	112	97
G	255	206	182	138	89

35. What is the percentage of students studying science in the institue G with respect to the total number of students studying in the institute G?
(a) 17.20 (b) 12.70
(c) $21\frac{1}{3}$ (d) $21\frac{2}{3}$
(e) None of these

36. Out of the total students of the institute 'D', **approximately** what percentage of students study Management?
(a) 9 (b) 8 (c) 12 (d) 10 (e) 6

37. The total number of students studying Arts in institutes A, B and C together is **approximately** what per cent of the total number of students studying commerce in institutes D, E, F and G together?
(a) 50 (b) 45 (c) 42 (d) 55 (e) 53

38. What is the percentage of students studying Engineering in institute C with respect to the total students of all institutions studying Engineering? (rounded to the nearest integer)
(a) 19 (b) 20 (c) 18 (d) 21 (e) None of these

39. In which institution, the percentage of students studying Commerce with respect to the total students of the institution is maximum?
(a) F (b) E
(c) C (d) A
(e) None of these

DIRECTIONS (Qs. 40-44) : Study the following graph to answer the given questions.

Production of two companies A & B over the years (Production in lakh units)

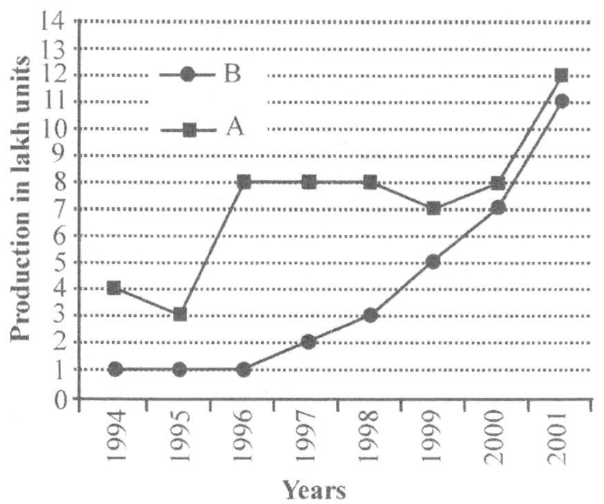

40. For Company A, what is the per cent decrease in production from 1994 to 1995?
(a) 75 (b) 50
(c) 25 (d) 10
(e) None of these

41. In 2001, the production of Company B is **approximately** what per cent of that in 2000?
(a) 60 (b) 157
(c) 192 (d) 50
(e) 92

42. For Company A, in which year is the percentage increase/decrease in the production from the previous year the highest?
(a) 2001 (b) 1995
(c) 1999 (d) 1996
(e) None of these

43. What is the difference in the total production of the two companies for the given years?
(a) 2700000 (b) 3100000
(c) 270000 (d) 310000
(e) None of these

44. Which of the following is the closest average production (in lakh units) of Company B for the given years?
(a) 4.1 (b) 3.5
(c) 4.3 (d) 3.75
(e) 3.9

Directions (Qs. 45-49): Study the following table to answer the given questions.

Centrewise and Postwise number of candidates

Post Specialist Centre	Officer	Clerk	Field Officer	Supervisor	Specialist officer
Bangalore	2000	5000	50	2050	750
Delhi	15000	17000	160	11000	750
Mumbai	17000	19500	70	7000	900
Hyderabad	3500	20000	300	9000	1150
Kolkata	14900	17650	70	1300	1200
Lucknow	11360	15300	30	1500	650
Chennai	9000	11000	95	1650	500

45. In Kolkata, number of Specialist Officers is **approximately** what per cent of Officers?
(a) 8.7 (b) 9
(c) 6.5 (d) 8
(e) 6.9

46. What is the difference between total number of Officers and Clerks?
(a) 29680 (b) 34180
(c) 32690 (d) 28680
(e) None of these

47. In Chennai, the number of Clerks is **approximately** how much per cent more than that of Officers?
(a) 18 (b) 22
(c) 20 (d) 2
(e) 13

48. Which centre has 300% more number of Clerks as compared to those in Bangalore?
(a) Lucknow (b) Mumbai
(c) Hyderabad (d) Chennai
(e) None of these

49. Which centre has the highest number of candidates?
(a) Delhi (b) Kolkata
(c) Hyderabad (d) Mumbai
(e) None of these

DIRECTIONS (Qs. 50-54): Study the following graph carefully and answer the questions given below it :

Per cent profit earned by two companies *A* and *B* over the years 1991 to 1 997

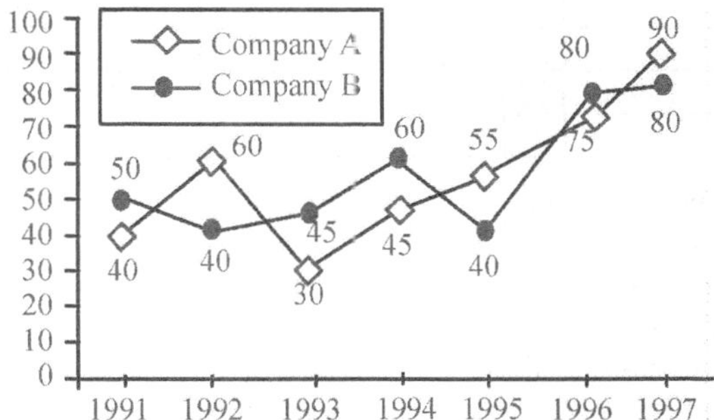

50. Investment of company '*B*' in 1997 is more by 40% than that in the previous year. Income in 1997 was what per cent of the investment in 1996?
(a) 280% (b) 252%
(c) 242% (d) 52%
(e) None of these

51. Average investment of company '*A*' over the years was ₹ 26 lakhs. What was its average income over the years?
(a) ₹ 40.56 lakhs (b) ₹ 41.60 lakhs
(c) ₹ 50.26 lakhs (d) Data inadequate
(e) None of these

52. Income of company '*A*' in 1995 was ₹ 21.7 lakhs. What was the investment ?
(a) ₹ 14.5 lakhs (b) ₹ 15.4 lakhs
(c) ₹ 15.8 lakhs (d) ₹ 14.6 lakhs
(e) None of these

53. Income of company '*A*' in 1995 is equal to the investment of the company '*B*' in 1996. What is the ratio of the investment of company '*A*' in 1995 to the investment of company '*B*' in 1996?

(a) 31 : 36 (b) 31 : 20
(c) 20 : 31 (d) Data inadequate
(e) None of these

54. Investment of company '*B*' in 1993 was ₹ 1540000. What was its income in that year?

(a) ₹ 23.33 lakhs (b) ₹ 22.33 lakhs
(c) ₹ 22. 23 lakhs (d) ₹ 23.23 lakhs
(e) None of these

DIRECTIONS (Qs. 55-59): Study the following table carefully and answer the questions given below it.

A factory was opened in 1994 with certain initial strengths in different units as shown in the table. At the beginning of the subsequent years some of the workers left and some new workers were deployed. No worker left or joined in between. Details are given in the table given below. Study it carefully and answer the questions that follow.

	UNIT									
Year	A		B		C		D		E	
1994 (Initial Strength)	156		132		98		76		125	
	L	J	L	J	L	J	L	J	L	J
1995	12	15	23	32	12	36	6	26	11	13
1996	17	18	16	14	8	19	17	28	11	15
1997	9	20	12	12	17	14	9	16	19	16
1998	32	40	14	17	23	35	12	23	23	14
1999	22	35	11	15	18	25	14	24	32	38
2000	26	32	17	21	13	18	11	19	21	36

Note : L = Left, J = Joined

55. What was the strength of Unit '*B*' in 1998?

(a) 142 (b) 125
(c) 159 (d) 207
(e) None of these

56. In 1999 the strength of workers was maximum in which unit'?

(a) E (b) D
(c) C (d) B
(e) A

57. The strength of workers in unit *C* in 1996 is **approximately** what per cent of the strength in unit *E* in 1997?

(a) 97 (b) 110
(c) 104 (d) 98
(e) 112

58. What was the total strength of workers in all the five units in 1996'?

(a) 647 (b) 570
(c) 690 (d) 697
(e) None of these

59. What was the **approximate** increase/decrease in the strength of the workers in unit *D* in 1998 with respect to its initial strength'?

(a) 47.37% increase (b) 64.47% decrease
(c) 64.47% increase (d) 47.37% decrease
(e) 59.38% increase

DIRECTIONS (Qs. 60-63): Study the following table carefully and answer the questions given below: Per cent marks obtained by 6 students in different subject

	Subject					
Student	Physics (out of 150)	Chemistry (out of 75)	Maths (out of 200)	History (out of 100)	Geography (out of 50)	English (out of 75)
A	77	63	89	55	64	72
B	69	72	71	78	69	66
C	82	78	69	65	75	57
D	73	81	76	67	58	63
E	58	69	54	74	66	75
F	66	57	61	62	71	59

60. What is the total marks obtained by *B* in all the subjects?

(a) 542 (b) 560.5
(c) 425 (d) 459.5
(e) None of these

61. What is the average marks obtained by 6 students in Chemistry out of 75 marks'?

(a) 52.5 (b) 70
(c) 55.5 (d) 62.5
(e) None of these

62. What is the difference in the total marks obtained by *C* in Physics and Chemistry and that obtained by *E* in the same subjects?

(a) 38.75 (b) 33
(c) 42.75 (d) 43
(e) None of these

63. What is the per cent marks obtained by *A* in both Maths and History? Find up to two decimal places.

(a) 72 (b) 77.67
(c) 48 (d) 73.33
(e) None of these

DIRECTIONS (Qs. 64-68): Study the following table carefully to answer the questions given below it.

Percentage of students passed over appeared from six States over the years in an admission test.

State/Year	A	B	C	D	E	F
1996	32	35	37	41	39	29
1997	45	26	29	37	43	37
1998	28	38	22	27	36	42
1999	36	42	38	38	29	45
2000	40	34	26	26	35	30
2001	24	29	33	33	41	36
2002	35	43	40	38	39	28

64. If the number of students appeared from each State in the year 2002 was 5000, **approximately** what was the average number of students qualified?

(a) 1810 (b) 1550
(c) 1380 (d) 1780
(e) 1860

65. If in the year 1998, 18500 students appeared from State '*C*' and 17200 students appeared from State '*E*', what was the total number of students qualified from these two States together?
(a) 10262 (b) 10444
(c) 10536 (d) 10833
(e) None of these

66. If the number of students appeared from State *A* in 1997 was more than that in 1996 by 20%, what was the ratio of numbers qualified in the State *A* in 1996 and 1997 respectively?
(a) 4 : 9 (b) 16 : 27
(c) 5 : 6 (d) 32 : 45
(e) None of these

67. If the numbers of students qualified from State 'D' in 1999 and 2000 were in the ratio of 2: 3 respectively, what was the respective ratio of students appeared in these years?
(a) 13 : 27 (b) 13 : 19
(c) 26 : 57 (d) 19 : 27
(e) None of these

68. If the average number of students qualified from State E for the given years was 532, what was the average number of students appeared?
(a) 1420 (b) 1350
(c) 1422 (d) Cannot be determined
(e) None of these

DIRECTIONS (Qs.69-73): Study the following table carefully to answer these questions.

Production (in lakh tonnes) of product by six companies over the given years.

Year/ Company	1997	1998	1999	2000	2001	2002
A	487	565	648	734	848	765
B	522	378	725	673	729	695
C	746	483	679	499	685	720
D	398	526	498	580	617	732
E	415	680	840	689	780	637
F	632	775	580	720	670	746

69. Production of Company *B* in 1999 was what per cent of the total production of all the companies together for that year (rounded off to the nearest integer)?
(a) 17 (b) 20
(c) 22 (d) 18
(e) None of these

70. During which year was the percentage increase/decrease in production from the previous year the lowest for Company *A*?
(a) 2002 (b) 1998
(c) 2000 (d) 1999
(e) None of these

71. What was the difference between the total productions of companies *E* & *F* (in lakh tonnes) in the given years?
(a) 78 (b) 86
(c) 76 (d) 72
(e) None of these

72. Approximately what was the average production of all the six companies (in lakh tonnes) in the year 1998?
(a) 590 (b) 550
(c) 570 (d) 450
(e) 620

73. What was the percent fall in production of Company '*C*' in 2000 over that in 1999 (rounded off to two digits after decimal)?
(a) 25. 61 (b) 26. 51
(c) 36. 07 (d) 37. 16
(e) None of these

DIRECTIONS (Qs. 74-78): Study the following table carefully and answer the questions given below:

Number of bales of wool processed by 5 woollen mills

	Name of the Company				
Month	Polar	Shephered	Kiwi	Warmwear	Comfy
Jan	900	850	350	1000	850
Feb	800	700	1050	1100	850
March	1050	800	1000	1100	950
April	800	850	850	1100	850
May	950	900	1050	1150	850
Total	4500	4100	4900	5450	4350

74. In the case of which mill is the processing of wool in March the highest percentage of the total processing by that mill during the five month period?
(a) Polar (b) Shephered
(c) Kiwi (d) Warmwear
(e) Comfy

75. The wool-processing by Warmwear in April is what percent of its wool-processing in Janauary?
(a) 91 (b) 110
(c) 115 (d) 10
(e) 11

76. Which of the five mills has the highest ratio of wool processing done in April to that done in February?
(a) Polar (b) Shephered
(c) Kiwi (d) Warmwear
(e) Comfy

77. In the case of which mill is the wool-processing in February and March together the lowest among the five mills processing during the same period?
(a) Comfy (b) Warmwear
(c) Kiwi (d) Shephered
(e) Polar

78. The total of wool-processing done by Kiwi during the given period is approximately what per cent of that done by Shephered?
(a) 80 (b) 87
(c) 8 (d) 108
(e) 120

DIRECTIONS (Qs. 79-83): Study the following graph carefully and answer the questions given below:

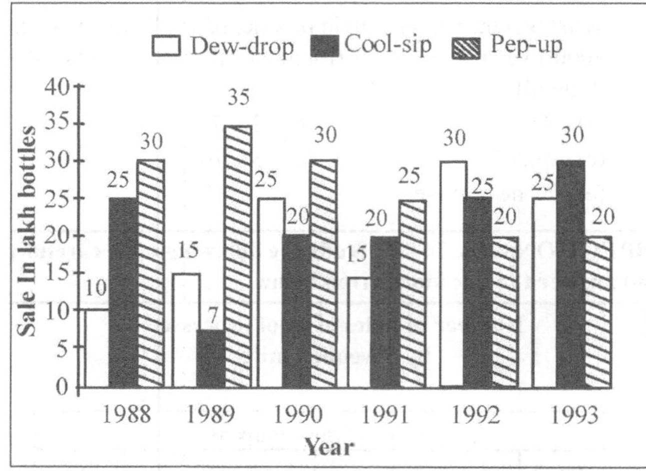

79. In which year was the sale of 'Pep-up' the maximum?
(a) 1990 (b) 1991
(c) 1992 (d) 1993
(e) None of these

80. In the case of which soft drink was the average annual sale maximum in the given period?
(a) Pep-up only (b) Cool-sip only
(c) Dew-drop only (d) Cool-sip and Dew-drop
(e) Pep-up and Dew-drop

81. In the case of Cool-sip drink, what was the **approximate** per cent increase in sale in 1992 over its sale in 1991?
(a) Less than 20 (b) 20-25
(c) 25 (d) 31-35
(e) 36-40

82. In the year 1990, what was the difference between the number of 'Pep-up' and 'Cool-sip' bottles sold?
(a) 50,00,000 (b) 5,00,000
(c) 50,000 (d) 5,000
(e) 10,00,000

83. What was the **approximate** per cent drop in sale of Pep-up in 1990 over its sale in 1989?
(a) 5 (b) 12
(c) 14 (d) 20
(e) None of these

DIRECTIONS (Qs.84-88): Read the following information carefully and answer the questions based on it: In 6 educational years, number of students taking admission and leaving from the 5 different schools which are founded in 1990 are given below

School	A		B		C		D		E	
	Ad	L	Ad	L	Ad	L	Ad	L	Ad	L
1990	1025	–	950	–	1100	–	1500	–	1450	–
1991	230	120	350	150	320	130	340	150	250	125
1992	190	110	225	115	300	150	300	160	280	130
1993	245	100	185	110	260	125	295	120	310	120
1994	280	150	200	90	240	140	320	125	340	110
1995	250	130	240	120	310	180	360	140	325	115

In the above table shown Ad = Admitted, L = Left

84. What is the average number of students studying in all the five schools in 1992?
(a) 1494 (b) 1294
(c) 1590 (d) 1640
(e) None of these

85. What was the number of students studying in school B in 1994?
(a) 2030 (b) 1060
(c) 1445 (d) 1150
(e) None of these

86. Number of students leaving school *C* from the year 1990 to 1995 is **approximately** what percentage of number of students taking admission in the same school and in the same year?
(a) 50% (b) 25%
(c) 48% (d) 36%
(e) 29%

87. What is the difference in the number of students taking admission between the years 1991 and 1995 in school *D* and B?
(a) 514 (b) 1065
(c) 965 (d) 415
(e) None of these

88. In which of the following schools, percentage increase in the number of students from the year 1990 to 1995 is maximum?
(a) A (b) B
(c) C (d) D
(e) E

DIRECTIONS (Qs. 89-93): Study the following table and answer the following questions carefully. Following table shows the percentage population of six states below poverty line and the proportion of male and female

State	Percentage population below proverty line	Proportion of male and female	
		Below poverty line M : F	Above poverty line M : F
A	12	3:02	4:03
B	15	5:07	3:04
C	25	4:05	2:03
D	26	1:02	5:06
E	10	6:05	3:02
F	32	2:03	4:05

89. The total population of state *A* is 3000, then what is the **approximate** no. of females above poverty line in state *A*?
(a) 1150 (b) 2112
(c) 1800 (d) 1950
(e) 2025

90. If the total population of *C* and *D* together is 18000, then what is the total no. of females below poverty line in the above stated states?
(a) 5000 (b) 5500
(c) 4800 (d) Data inadequate
(e) None of these

91. If the population of males below poverty line in state A is 3000 and that in state E is 6000, then what is the ratio of the total population of state A and E?
(a) 3 : 4 (b) 4 : 5
(c) 1 : 2 (d) 2 : 3
(e) None of these

92. If the population of males below poverty line in state B is 500 then what is the total population of that state?
(a) 14400 (b) 6000
(c) 8000 (d) 7600
(e) None of these

93. If in state E population of females above poverty line is 19800 then what is the population of males below poverty line in that state?
(a) 5500 (b) 3000
(c) 2970 (d) Data inadequate
(e) None of these

DIRECTIONS (Qs. 94-98): Study the following graph to answer these questions.

Per cent profit earned by two Companies A & B over the years

Profit = Income – Expenditure

$$\%\text{Profit} = \frac{\text{Profit}}{\text{Expenditure}} \times 100$$

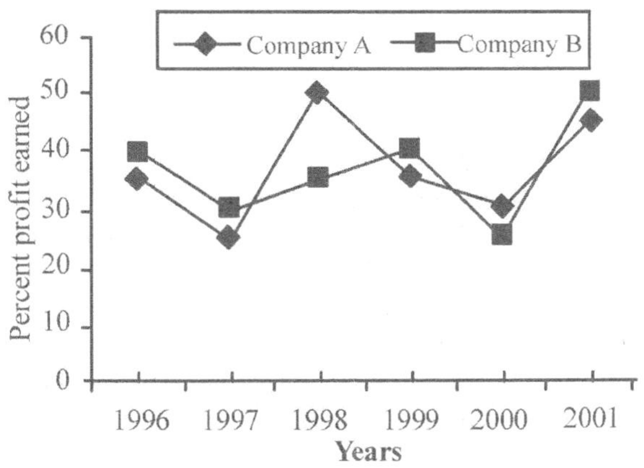

94. If the income of Company 'A' in 1998 was ₹ 1,42,500 what was its expenditure in that year?
(a) ₹ 1,05,000 (b) ₹ 95,500
(c) ₹ 99,500 (d) ₹ 1,05,555
(e) None of these

95. Expenditure of Company 'B' in 1999 was 90% of its expenditure in 1998. Income of Company 'B' in 1999 was what per cent of its income in 1998?
(a) 130.5 (b) $96\frac{2}{3}$
(c) 121.5 (d) $99\frac{1}{3}$
(e) None of these

96. If the expenditure of Company 'A' in 1997 was ₹ 70 lakhs and income of Company A in 1997 was equal to its expenditure in 1998, what was the total income (in ₹ lakh) of the Company A in 1997 & 1998 together?
(a) 175 (b) 131.25
(c) 218.75 (d) Cannot be determined
(e) None of these

97. Expenditure of Company 'B' in years 1996 and 1997 were in the ratio of 5 : 7 respectively. What was the respective ratio of their incomes?
(a) 10 : 13 (b) 8 : 13
(c) 13 : 14 (d) 11 : 14
(e) None of these

98. Total expenditure of Companies A & B together in 2001 was ₹ 13.5 lakhs. What was the total income of the two companies (in ₹ lakh) in that year?
(a) 19.575 (b) 20.25
(c) 19.75 (d) Cannot be determined
(e) None of these

DIRECTIONS (Qs. 99-103): Study the following table carefully to answer these questions.

Percentage of marks obtained by six students in six different subjects

Stu/Sub	Psy (150)	Socio (120)	Econ (80)	Philos (75)	Statis (125)	Geolo (60)
A	72	65	85	65	88	72
B	68	58	74	70	78	54
C	63	73	69	57	68	65
D	56	65	77	61	75	67
E	78	55	82	76	59	74
F	84	70	64	78	82	80

Note: Figures written in bracket under each subject indicate the maximum marks allotted for that subject.

99. What is the difference between the total marks obtained by A in Psychology and Statistics together and the total marks obtained by F in these two subjects together?
(a) 12.5 (b) 6
(c) 10.5 (d) 11.6
(e) None of these

100. What is the average percentage of marks obtained by six students in Sociology?
(a) $77\frac{1}{3}$ (b) $64\frac{1}{5}$
(c) $77\frac{1}{5}$ (d) $64\frac{1}{3}$
(e) None of these

101. What is the average mark obtained by the six students in Philosophy out of 75?
(a) 58.75 (b) 50.875
(c) 67.83 (d) 65.73
(e) None of these

102. What is the total mark obtained by '*A*' in Psychology, Economics and Geology together?
(a) 219.2 (b) 229
(c) 209.8 (d) 229.6
(e) None of these

103. **Approximately** what is the overall percentage of marks obtained by '*C*' in all the subjects together?
(a) 60 (b) 56
(c) 72 (d) 76
(e) 66

DIRECTIONS (Qs. 104-108): Study the following graphs carefully to answer these questions.

Production (in lakh tons) of six units of a company in 2001 and 2002

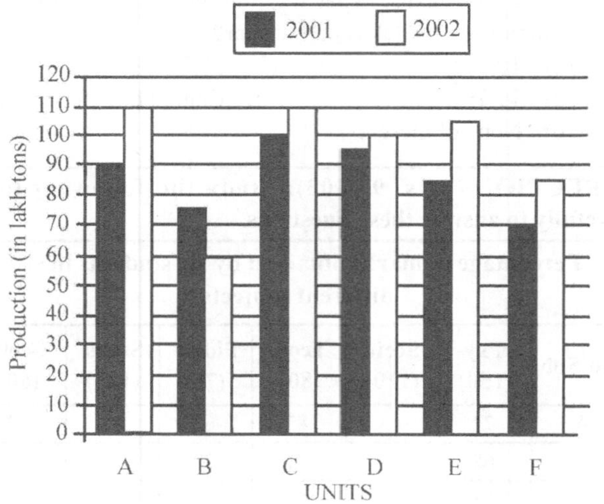

104. What is the average production of all the units (in lakh tons) for the year 2002?
(a) 89 (b) 92
(c) 87 (d) 95
(e) None of these

105. Average production of three units *A, B & C* in 2001 is what per cent of the average production of units *D , E & F* in 2002? (rounded off to two digits after decimal)
(a) 109.43 (b) 90.37
(c) 91.38 (d) 106.43
(e) None of these

106. What is the ratio of total production for two years to-gether for unit *B* to that for *C*?
(a) 17 : 13 (b) 13 : 17
(c) 11 : 13 (d) 19 : 13
(e) None of these

107. Total production for two years together by unit *F* is what per cent of the total production of the two years together by unit *D*? (rounded off to two digits after decimal)
(a) 79.49 (b) 78.49
(c) 78.47 (d) 79.29
(e) None of these

108. What is the total production of units *C, D & E* together for both the years ? (in lakh tons)
(a) 495 (b) 595
(c) 545 (d) 515
(e) None of these

DIRECTIONS (Qs. 109-1113): Study the following graphs carefully and answer the questions that follow. Percentage profit earned by two companies over the given years

$$\%\text{profit} = \frac{\text{Income} - \text{Expenditure}}{\text{Expenditure}} \times 100$$

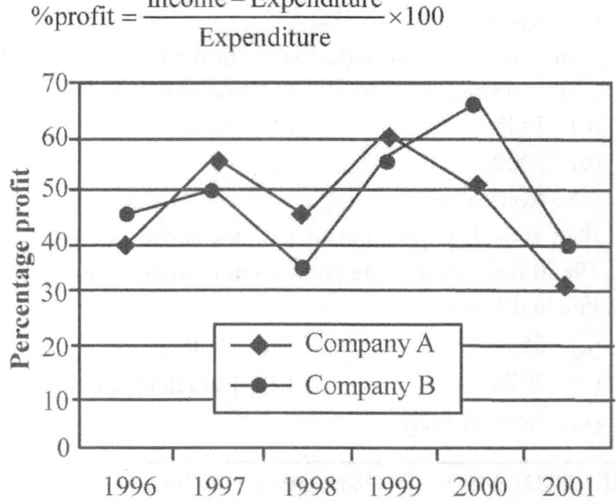

109. If the income of Company A in 1998 was equal to its expenditure in 2000, what was the ratio between Company's expenditures in the years 1998 and 2000 respectively?
(a) 29 : 20 (b) 20 : 29
(c) 19 : 20 (d) Cannot be determined
(e) None of these

110. If the income of Company B in 1999 was ₹ 18.6 lakhs and ratio of incomes of Companies A & B in 1999 was 2:3, what was the expenditure of Company A in 1999 (in ₹ lakhs)?
(a) 12 (b) 12.4
(c) 7.75 (d) 9.75
(e) None of these

111. If the total expenditure of the two Companies in 2001 was ₹ 18 lakhs and expenditures of Companies A & B in that year were in the ratio of 4 : 5 respectively, then what was the income of Company B in that year (in ₹ lakh)?
(a) 8 (b) 10
(c) 10.4 (d) Cannot be determined
(e) None of these

112. If the income of Company A in 1999 was equal to the expenditure of Company B in 2000, then what was the ratio of expenditure of Company A in 1999 to the income of Company B in 2000?
(a) 25 : 66 (b) 66 : 25
(c) 10 : 13 (d) 13 : 10
(e) None of these

113. If the total income of Company A in all the years together was equal to the total expenditure of Company B in all the years together, which was ₹ 265 lakhs, what was the total percentage profit earned by Company A for all the years together?
(a) 45 (b) 37
(c) 52 (d) Cannot be determined
(e) None of these

DIRECTIONS (Qs. 114-118): Study the following table carefully and answer the questions that follow:

The percentage marks obtained by seven students in six different subjects

Subject →	A	B	C	D	E	F
Student ↓	(Out of 75)	(Out of 150)	(Out of 100)	(Out of 50)	(Out of 150)	(Out of 75)
P	85	68	76	92	89	82
Q	78	72	84	80	64	70
R	66	75	79	88	72	66
S	74	62	91	74	70	74
T	90	75	67	68	69	78
V	86	80	69	78	82	80
W	82	68	81	85	76	72

114. What total percentage marks 'R' did secure in all the six subjects together?
(a) 75.73 (b) 74.33
(c) 73.75 (d) 74.75
(e) None of these

115. What is the difference between the marks obtained by 'P' in the subjects 'B', 'D' and 'E' together and by 'T' in the same subjects?
(a) 32.5 (b) 31.5
(c) 37 (d) 34
(e) None of these

116. What is the average of marks obtained by all the students in subject 'B'? (up to two decimal places)
(a) 107.14 (b) 71.4
(c) 114.07 (d) 73.14
(e) None of these

117. What is the average percentage of marks obtained by all the students in the subjects 'C' and 'D' together?
(a) 78 (b) 80.71
(c) 79.43 (d) 77.53
(e) None of these

118. What is the total marks obtained by all the students in subject 'F'?
(a) 422 (b) 398.5
(c) 522 (d) 391.5
(e) None of these

DIRECTIONS (Qs. 119-123): Study the following table carefully and answer the questions that follow:

Investment (in ₹ crores) by six units of XYZ Company from 1996 to 2001

Year → Unit ↓	'96	'97	'98	'99	'00	'01	Total
A	85	132	125	116	142	138	738
B	105	140	145	148	142	144	824
C	114	137	138	136	150	152	827
D	98	125	132	145	158	152	810
E	82	128	141	152	149	165	817
F	108	150	145	156	154	162	875
Total	592	812	826	853	895	913	4891

119. In which of the following years the investment of unit 'C' was **minimum** per cent of the investment of all the companies taken together in the same year?
(a) 1997 (b) 1998
(c) 1999 (d) 2001
(e) None of these

120. In the year 1997 the investment of which of the following units is the maximum per cent of the investment during the given years?
(a) A (b) F
(c) C (d) B
(e) None of these

121. What is the increase per cent in the investment of unit 'D' from 1996 to 1999?
(a) 26.75 (b) 21.55
(c) 21.60 (d) 27.55
(e) None of these

122. How much more/less is the investment by units A, B and C in the year 1998 than the investment by the same three units in the year 1999?
(a) ₹ 10 crores less (b) ₹ 8 crores more
(c) ₹ 8 crores less (d) ₹ 10 crores more
(e) None of these

123. What is the ratio between the total investment of unit A, B and C in the year 1998 and the total investment of units D, E and F in the year 1999?
(a) 36 : 51 (b) 51 : 36
(c) 26 : 43 (d) 43 : 26
(e) None of these

DIRECTIONS (Qs. 124-128) : Study the following table carefully and answer the questions given below.

Percentage of malnourished children in Chile over the years

Year	Tested Number (in thousands)	Percentage of the malnourished		
		Low	Moderate	High
1984	998	12.5	2.9	0.7
1985	1015	12.1	2.7	0.7
1986	1048	12.1	3.0	0.8
1987	1071	11.9	2.5	0.5
1989	1048	10.8	1.8	0.3
1990	1023	10.4	1.6	0.2
1991	1048	10.0	1.4	0.1
1992	1063	3.70	1.1	0.1
1993	1161	7.80	0.9	0.1

124. What is the difference between the total numbers of the malnourished children in the years 1991 and 1986?
(a) 0 (b) 46112
(c) 22008 (d) 41920
(e) None of these

125. In which year was the percentage of the malnourished children the highest?
(a) 1986 (b) 1984
(c) 1985 (d) 1987
(e) None of these

126. Which is true of the following?
(a) Over the years, there was uniform fall in the percentage of high malnourished cases in comparison to the previous year.
(b) Over the years, there was uniform fall in the percentage of moderate malnourished cases in comparison to the previous year.
(c) Over the years, there was uniform fall in the percentage of low malnourished cases in comparison to the previous year.
(d) Over the years, there was no rise in the percentage of high malnourished cases in comparison to the previous year.
(e) Over the years, there was no rise in the percentage of low malnourished cases in comparison to the previous year.

127. The malnutrition level of how many children was high in the year 1987?
(a) 600 (b) 12745
(c) 535 (d) 5355
(e) None of these

128. How many children were malnourished in 1993?
(a) 10,02,168 (b) 1,02,168
(c) 10,216 (d) 1,00,02,168
(e) None of these

DIRECTION (Qs. 129-133) : Study the following graph carefully to answer the question given below it.

Production of paper (in lakh tonnes) by 3 different companies A, B & C over the years

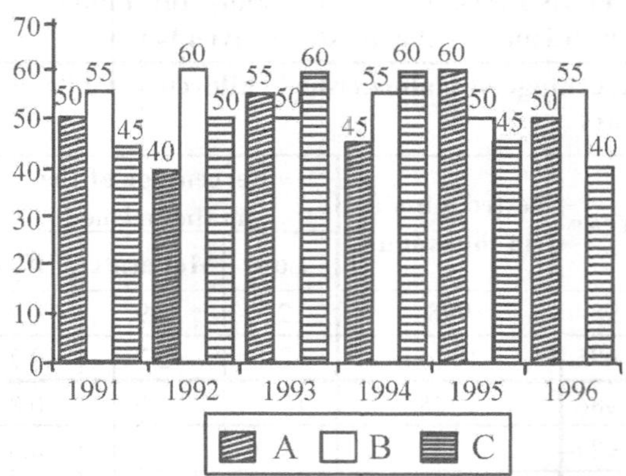

129. What is the difference between the production of company C in 1991and the production of Company A in 1996?
(a) 50,000 tonnes (b) 5,00,00,000 tonnes
(c) 50,00,000 tonnes (d) 5,00,000 tonnes
(e) None of these

130. What is the percentage increase in production of CompanyA from 1992 to 1993?
(a) 37.5 (b) 38.25
(c) 35 (d) 36
(e) None of these

131. For which of the following years the percentage of rise/fall in production from the previous year the **maximum** for Company B?
(a) 1992 (b) 1993
(c) 1994 (d) 1995
(e) 1996

132. The total production of Company C in 1993 and 1994 is what percentage of the total production of Company A in 1991 and 1992?
(a) 95 (b) 90
(c) 110 (d) 115
(e) None of these

133. What is the difference between the average production per year of the company with highest average production and that of the company with lowest average production in lakh tonnes?
(a) 3.17 (b) 4.33
(c) 4.17 (d) 3.33
(e) None of these

DIRECTIONS (Qs. 134-138) : Study the following table carefully and answer the questions given below it.

Fare in rupees for three different types of vehicles

Vehicle	Fare for distance upto					
	2 km	4 km	7 km	10 km	15 km	20 km
Type A	₹ 5.00	₹ 9.00	₹ 13.50	₹ 17.25	₹ 22.25	₹ 26.00
Type B	₹ 7.50	₹ 14.50	₹ 24.25	₹ 33.25	₹ 45.75	₹ 55.75
Type C	₹ 10.00	₹ 19.00	₹ 31.00	₹ 41.50	₹ 56.50	₹ 69.00

Note : Fare per km for intermittent distance is the same.

134. Shiv Kumar has to travel a distance of 15 kms in all. He decides to travel equal distance by each of the three types of vehicles. How much money is to be spent as fare?
(a) ₹ 51.75 (b) ₹ 47.50
(c) ₹ 47.25 (d) ₹ 51.25
(e) None of these

135. Ajit Singh wants to travel a distance of 15 kms. He starts his journey by Type A vehicle. After travelling 6 kms, he changes the vehicle to Type B for the remaining distance. How much money will he be spending in all?
(a) ₹ 42.25 (b) ₹ 36.75
(c) ₹ 40.25 (d) ₹ 42.75
(e) None of these

136. Mr X wants to travel a distance of 8 kms by Type A vehicle. How much more money will be required to be spent if he decides to travel by Type B vehicle instead of Type A?
(a) ₹ 16 (b) ₹ 12.50
(c) ₹ 14 (d) ₹ 13.50
(e) None of these

137. Rita hired a Type B vehicle for travelling a distance of 18 kms. After travelling 5 kms, she changed the vehicle to Type A. Again after travelling 8 kms by Type A vehicle, she changed the vehicle to Type C and completed her journey. How much money did she spend in all?
(a) ₹ 50 (b) ₹ 45.50
(c) ₹ 55 (d) ₹ 50.50
(e) None of these

138. Fare for 14th km by Type C vehicle is equal to the fare for which of the following?
(a) Type B – 1 lth km (b) Type B – 9th km
(c) Type A – 4th km (d) Type C – 8th km
(e) None of these

DIRECTIONS (Qs. 139-143) : Answer these questions on the basis of the information given in the following table.

Production (in lakh tonnes) of six companies over the given years

	1995	1996	1997	1998	1999	2000
A	465	396	524	630	408	650
B	372	482	536	480	512	580
C	694	528	492	575	550	495
D	576	602	387	426	632	518
E	498	551	412	518	647	610
F	507	635	605	600	485	525

139. What is the difference between total productions of Companies A and C for all the given years together?
(a) 2, 61,00,000 tonnes (b) 2,61,900 tonnes
(c) 3,31,00,00 tonnes (d) 3,39,000 tonnes
(e) None of these

140. Approximately, what is the percentage rise/fall in total production of all the Companies together from 1996 to 1997?
(a) 4.5% rise (b) 6% rise
(c) 3.5% fall (d) 7% fall
(e) 7.5% fall

141. During which year is the percentage rise/fall from the previous year in production of company 'F' the highest?
(a) 1999 (b) 2000
(c) 1997 (d) 1996
(e) None of these

142. Production of companies A and B together in 1997 is **approximately** what percentage of the production of companies E and F together in 1998?
(a) 90 (b) 95
(c) 97 (d) 86
(e) 92

143. What is the difference between average production for the given years of companies B and E (in lakh tonnes rounded off to two digits after decimal)?
(a) 56.50 (b) 45.50
(c) 45.67 (d) 55.78
(e) None of these

DIRECTIONS (Qs. 144-148) : Study the following graphs carefully and answer the questions that follow:

Income and Expenditure of Company 'X' during the period 1996 to 2001

Profit / Loss = Income – Expenditure

$$\% \text{ Profit / Loss} = \frac{\text{Income} - \text{Expenditure}}{\text{Expenditure}} \times 100$$

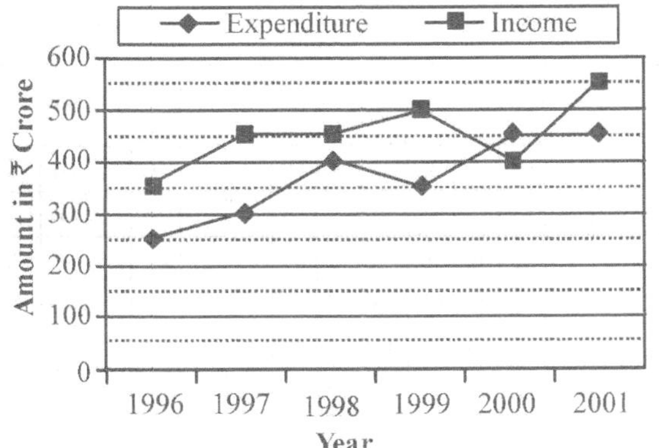

144. What is the average profit earned (in crore ₹) in the given years?
(a) $83\frac{1}{3}$ (b) 600
(c) $113\frac{2}{3}$ (d) 200
(e) None of these

145. What **approximately** is the per cent profit earned during the year 1999?
(a) 48 (b) 43
(c) 52 (d) 49
(e) None of these

146. Which of the following years has the maximum per cent increase/decrease in income from the previous year?
(a) 2000 (b) 1999
(c) 1997 (d) 2001
(e) 1997 & 1999

147. What is the percentage increase in expenditure from 1997 to 1998?
(a) 25 (b) $33\frac{1}{3}$
(c) $33\frac{2}{3}$ (d) 30
(e) None of these

148. What is the average income (in crore ₹) for the given years?
(a) $336\frac{2}{3}$ (b) 280
(c) 450 (d) $366\frac{2}{3}$
(e) None of these

DIRECTIONS (Qs. 149-153): Study the following table carefully to answer these questions.

Distribution of marks obtained by 160 students in each of the three subjects–Hindi, English and Maths– out

Sub/Marks	0-19	20-39	40-59	60-79	80-100
Hindi	12	31	79	30	8
English	21	30	65	42	2
Maths	31	22	34	45	28
Average of three subjects	24	28	68	35	5

149. If the criteria for passing is minimum 40% marks only in Maths, how many students will pass?
(a) 53 (b) 107
(c) 34 (d) 129
(e) None of these

150. If for passing, the student has to obtain minimum 60% marks on average of three subjects, how many students will pass?
(a) 40 (b) 108
(c) 68 (d) 73
(e) None of these

151. If for passing, a student has to obtain 40% marks in any one of the three subjects, what is the minimum number of students who will definitely pass?
(a) 107 (b) 109
(c) 117 (d) 108
(e) None of these

152. How many students will pass in English if minimum passing marks is 40%?
(a) 117 (b) 111
(c) 119 (d) 108
(e) None of these

153. How many students have obtained 20 or more marks in at least one of the three subjects?
(a) 148 (b) 139
(c) 129 (d) Data inadequate
(e) None of these

DIRECTIONS (Qs. 154 – 158) : Study the following tables carefully and answer the questions given below :

Number of cars (In thousands) of different Models and Colours sold in two Metro cities in a year

	Metro M					Metro H				
Type	Colour					Colour				
	Black	Red	Blue	White	Silver	Black	Red	Blue	White	Silver
A	40	25	55	75	15	45	32	40	60	20
B	20	35	60	80	20	30	37	39	81	35
C	35	30	50	90	35	40	42	41	6	37
D	45	40	45	85	40	35	39	37	90	42
E	50	35	35	60	30	50	44	43	77	22
F	55	42	40	65	52	47	34	45	87	17

154. The difference between the white-coloured cars sold in the two metros of which of the following models is the **minimum?**
(a) A (b) C
(c) D (d) F
(e) None of these

155. The total number of blue-coloured cars of Model E and D sold in Metro H is exactly equal to the number of white-coloured cars of which model in Metro M?
(a) B (b) F
(c) C (d) A
(e) None of these

156. What is the difference between the number of blue-colour cars of model 'C' sold in Metro M and number of red colour cars of model 'F' sold in Metro H?
(a) 8,000 (b) 10,000
(c) 12,000 (d) 15,000
(e) None of these

157. The total number of silver-coloured cars sold in Metro H is **approximately** what percentage of that in Metro M?
(a) 130 (b) 140
(c) 90 (d) 100
(e) 110

158. In Metro M the number of cars sold was **maximum** for which of the colour-model combinations?
(a) White - C (b) Blue - B
(c) Silver - B (d) White - D
(e) None of these

DIRECTIONS (Qs. 159 - 163) : Study the following table carefully and answer the questions given below it. Number of candidates from different locations appeared and passed in a competitive examination over the years

Year	Rural		Semi-urban		State capitals		Metropolises	
	App.	Passed	App.	Passed	App.	Passed	App.	Passed
1990	1652	208	7894	2513	5054	1468	9538	3214
1991	1839	317	8562	2933	7164	3248	10158	4018
1992	2153	932	8139	2468	8258	3159	9695	3038
1993	5032	1798	9432	3528	8529	3628	11247	5158
1994	4915	1658	9784	4015	9015	4311	12518	6328
1995	5628	2392	9969	4263	1725	4526	13624	6419

159. For the candidates from which of the following locations was there continuous increase both in appeared and passed?
(a) Semi-urban (b) State capitals
(c) State capital & rural (d) Metropolises
(e) None of these

160. In which of the following years was the percentage passed to appeared candidates from Semi-urban area the least?
(a) 1991 (b) 1993
(c) 1990 (d) 1992
(e) None of these

161. What **approximate** value was the percentage drop in the number of Semi-urban candidates appeared from 1991 to 1992?
(a) 5 (b) 10
(c) 15 (d) 8
(e) 12

162. In 1993 percentage of candidates passed to appeared was **approximately** 35 from which location?
(a) Rural
(b) Rural and metropolises
(c) Semi-urban and metropolises
(d) Rural and semi-urban
(e) None of these

163. The total number of candidates passed from rural in 1993 and semi-urban in 1990 was exactly equal to the total number of candidates passed from State capital in which of the following years?
(a) 1990 (b) 1993
(c) 1994 (d) 1992
(e) None of these

DIRECTIONS (Qs. 164–168) : Study the following table carefully and answer the questions given below:

Marks (out of 50) obtained by five students P, Q, R, S and T in five subjects in five periodical examination of each subject

	Students														
Sub	P					Q					R				
	Periodicals														
	I	II	III	IV	V	I	II	III	IV	V	I	II	III	IV	V
Math	40	30	45	20	35	30	20	35	45	40	30	35	40	45	40
Sc.	30	40	25	30	20	25	45	30	37	28	48	46	31	40	80
His	35	25	15	30	40	33	27	40	34	26	35	45	40	30	35
Geo	45	47	32	39	37	42	43	30	40	25	25	35	48	37	25
Eng	24	28	36	39	43	30	28	37	34	31	26	28	31	30	40

Sub	Students									
	S					T				
						Periodicals				
	I	II	III	IV	V	I	II	III	IV	V
Math	25	35	40	45	30	29	31	39	41	40
Sc.	31	34	38	27	30	44	36	40	30	40
His	34	40	36	42	48	37	43	35	45	40
Geo	39	37	44	40	30	38	39	33	40	40
Eng	31	34	35	45	40	30	30	35	45	40

164. What was the average marks of the five subjects of student Q in the 1st periodical?
(a) 32 (b) 34
(c) 40 (d) 30
(e) None of these

165. What was the total of marks of student T in Science in all the periodicals together?
(a) 160 (b) 180
(c) 190 (d) 140
(e) None of these

166. The average percentage of marks obtained by student P in Maths in the five periodicals was exactly equal to the average percentage of marks obtained by student R in the five periodicals in which of the following subjects?
(a) English
(b) Geography
(c) Science and Geography
(d) Maths
(e) None of these

167. In which of the following subjects was the average percentage of marks obtained by student S the highest?
(a) Maths (b) Science
(c) History (d) Geography
(e) English

168. In which of the periodicals the student P obtained, highest percentage of marks in Geography?
(a) I (b) II
(c) III (d) IV
(e) V

Directions (Qs. 169-173) : Study the following graph carefully and then answer the questions based on it. The percentage of five different types of cars produced by the company during two years is given below.

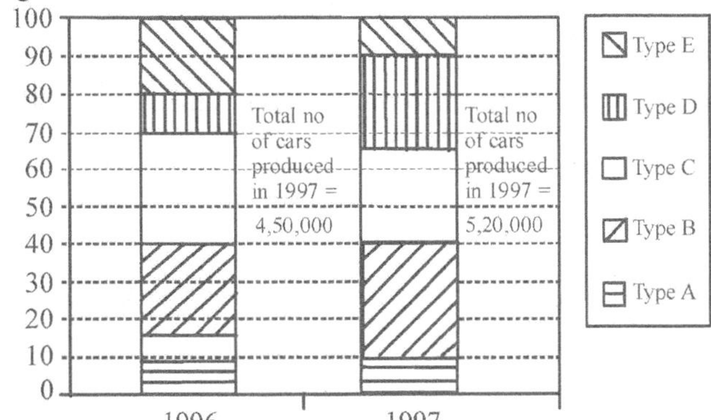

169. What was the difference in the production of C type cars between 1996 and 1997?
(a) 5,000 (b) 7,500
(c) 10,000 (d) 2,500
(e) None of these

170. If 85% of E type cars produced during 1996 and 1997 are being sold by the company, then how many E type cars are left unsold by the company?
(a) 1,42,800 (b) 21,825
(c) 29,100 (d) 25,200
(e) None of these

171. If the number of A type cars manufactured in 1997 was the same as that of 1996, what would have been its **approximate** percentage share in the total production of 1997?
(a) 11 (b) 13
(c) 15 (d) 9
(e) None of these

172. In the case of which of the following types of cars was the percentage increase from 1996 to 1997 the maximum?
(a) A (b) E
(c) D (d) B
(e) C

173. If the percentage production of B type cars in 1997 was the same as that of 1996, what would have been the number of cars produced in 1997?
(a) 1,12,500 (b) 1,20,000
(c) 1,30,000 (d) Data inadequate
(e) None of these

Directions (Qs. 174-178) : Study the following graph carefully and answer the questions given below it:

Imports of 3 companies over the years
₹ in crores

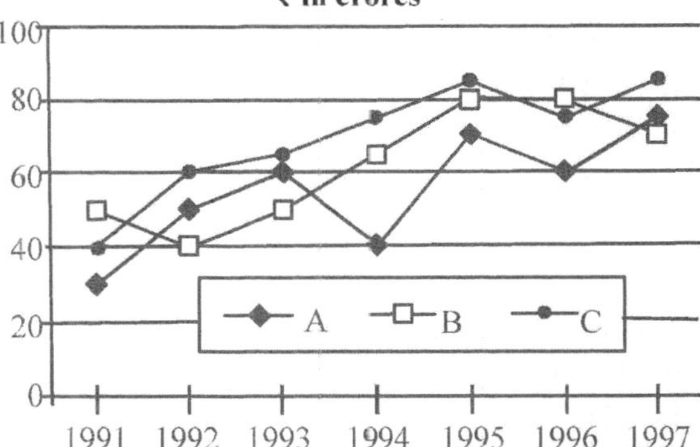

174. In which of the following years, the imports made by Company A was exactly equal to average imports made by it over the given years?
(a) 1992 (b) 1993
(c) 1994 (d) 1995
(e) None of these

175. In which of the following years was the difference between the imports made by Company B and C the maximum?
(a) 1995 (b) 1994
(c) 1991 (d) 1992
(e) None of these

176. In which of the following years was the imports made by Company A exactly half of the total imports made by Company B and C together in that year?
(a) 1992 only (b) 1993 only
(c) 1992 and 1993 (d) 1995 only
(e) None of these

177. What was the percentage increase in imports by Company B from 1992 to 1993?
(a) 10 (b) 25
(c) 40 (d) 20
(e) None of these

178. In which of the following years was the total imports made by all the three companies together the maximum?
(a) 1996 only (b) 1997 only
(c) 1995 only (d) 1995 and 1997 only
(e) None of these

Directions (Qs. 179-183): Study the following table carefully and answer the questions given below it:

Statewise and Disciplinewise Number of Candidates Appeared (App.) and Qualified (Qual.) at a competitive Examination)

State	A.P.		U.P.		Kerala		Orissa		M.P.		W.B.		Total	
Discipline	App.	Qual.	App.	Qual.	App.	Qual.	App.	Qual.	App.	Qual.	App.	Qual.	App.	Qual.
Arts	5420	1840	4980	1690	2450	845	3450	1200	7500	2000	4800	1500	28600	9075
Commerce	8795	2985	6565	2545	3500	2040	4800	2200	8400	2400	7600	2700	39660	14870
Science	6925	2760	8750	3540	4250	2500	4500	1950	6850	3000	8500	3200	39775	16950
Engineering	1080	490	2500	1050	1200	450	1850	850	2500	750	3400	1400	12530	4990
Agriculture	2040	850	1085	455	700	200	450	150	1500	475	1200	500	5775	2130
Total	23060	8425	23880	9280	12100	6035	15050	6350	26750	8625	25500	9300	126340	48015

179. For which of the following disciplines the proportion of qualifying candidates to the appeared candidates from U.P. State is the lowest?
(a) Arts
(b) Commerce
(c) Science
(d) Engineering
(e) Agriculture

180. For which of the pair of States, the qualifying percentage from Agriculture discipline is exactly the same?
(a) A.P. & U.P.
(b) A.P. & West Bengal
(c) U.P & West Bengal
(d) Kerala & Orissa
(e) None of these

181. For which of the following states the percentage of candidates qualified to appeared is the minimum for commerce discipline?
(a) AP (b) UP
(c) Kerala (d) Orissa
(e) MP

182. **Approximately** what is the ratio between total qualifying percentage of UP and that of MP?
(a) 15 : 16 (b) 13 : 14
(c) 14 : 13 (d) 19 : 16
(e) 17 : 16

183. The qualifying percentage for which of the following states is the lowest for Science discipline?
(a) AP (b) UP
(c) Kerala (d) West Bengal
(e) None of these

DIRECTIONS (Qs. 184 -188) : Study the following graph carefully to answer these questions.

The production of fertilizer in lakh tons by different companies for three years 1996, 1997 & 1998

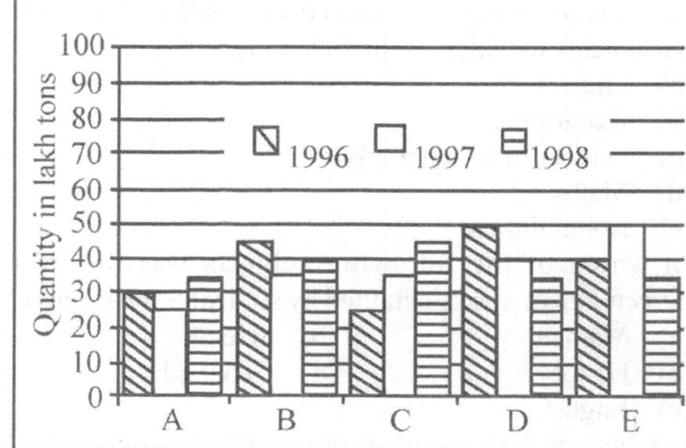

184. The total production by five companies in 1998 is what per cent of the total production by companies B & D in 1996?
(a) 100% (b) 150%
(c) 95% (d) 200%
(e) None of these

185. What is the ratio between average production by Company B in three years to the average production by company C in three years?
(a) 6 : 7 (b) 8 : 7
(c) 7 : 8 (d) 7 : 6
(e) None of these

186. For which of the following companies the rise or fall in production of fertiliser from 1996 to 1997 was the maximum?
(a) A (b) B
(c) C (d) D
(e) E

187. What is the per cent drop in production by Company D from 1996 to 1998?
(a) 30 (b) 43
(c) 50 (d) 35
(e) None of these

188. The average production for three years was maximum for which of the following companies?
(a) B only (b) D only
(c) E only (d) B & D both
(e) D & E both

DIRECTIONS (Qs. 189-193): Study the following table to answer the given questions.

Number of students of different classes of a school playing different games.							
Class → Games ↓	XII	XI	X	IX	VIII	VII	VI
Chess	11	12	5	4	2	2	1
Cricket	38	40	12	17	25	18	20
Basket ball	11	9	7	6	0	0	0
Table Tennis	9	9	21	19	11	9	0
Football	40	27	18	19	12	16	14
Carrom	16	15	8	19	12	16	14
Tennis	8	9	11	5	6	0	0
Badminton	47	39	33	21	19	0	0

189. Approximately what per cent of Class VIII students play Cricket out of the total students playing Cricket?
(a) 13 (b) 4
(c) 25 (d) 15
(e) 17

190. What is the ratio of the students playing Football in Class XI to those in Class X?
(a) 1 : 2 (b) 2 : 5
(c) 2 : 3 (d) 3 : 2
(e) None of these

191. Which game is the most popular?
(a) Badminton (b) Football
(c) Carrom (d) Table Tennis
(e) Cricket

192. Approximately what per cent of Class X students play the Table Tennis out of the total Class X students playing the different given games?
(a) 20 (b) 21
(c) 27 (d) 26
(e) 18

193. Which game has ascending number of students from class IX to XII?
(a) Only Basketball (b) Only Badminton
(c) Chess and Badminton (d) No game
(e) None of these

DIRECTIONS (Qs. 194-198): Study the following graph to answer the given questions:

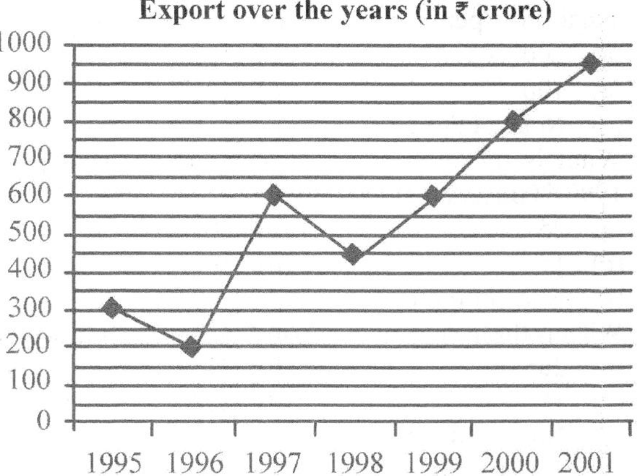

194. Which year has the highest per cent increase/decrease in exports as compared to the preceding year?
(a) 1997 (b) 1998
(c) 2001 (d) Cannot be determined
(e) None of these

195. What is the difference in exports in 1997 and 1998?
(a) ₹ 150 crores (b) ₹ 1500 crores
(c) ₹ 15 crores (d) ₹ 100 crores
(e) None of these

196. What is the per cent increase in exports from the lowest to the highest for the given years?
(a) 750 (b) 475
(c) 950 (d) 375
(e) None of these

197. Exports in 1997 is **approximately** what per cent of that of 1998?
(a) 145 (b) 135
(c) 150 (d) 300
(e) 120

198. What is the total exports (in ₹ crores) for the given years?
(a) 4100 (b) 3700
(c) 3900 (d) 3950
(e) None of these

Directions (Qs. 199-203): Study the following information to answer the given questions.

Income and Expenditure of a company for the given years
Profit = Income – Expenditure

$$\textbf{Per cent profit} = \frac{\textbf{Profit}}{\textbf{Expenditure}} \times 100$$

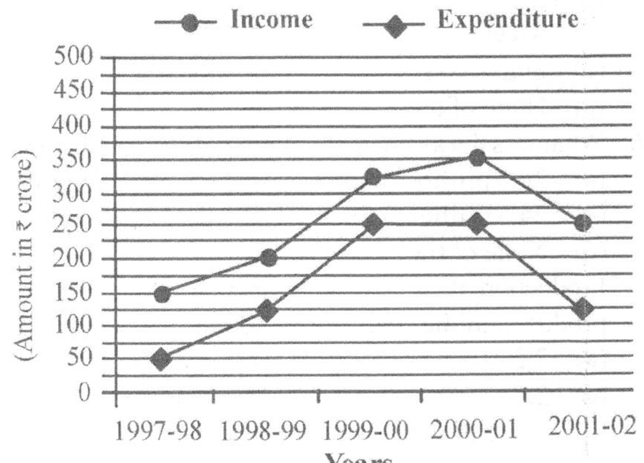

199. What is the per cent profit earned in 2001-2002?
(a) 200 (b) 125
(c) 100 (d) 12.5
(e) None of these

200. What is the difference between the per cent profit in 1999-2000 and that in 2000-2001 ?
(a) 25 (b) 20
(c) 15 (d) 10
(e) None of these

201. What is the average income (in ₹ crore) for the given years?
(a) 265 (b) 2550
(c) 160 (d) 250
(e) None of these

202. Which year is the per cent profit earned the highest?
(a) 2000-2001 (b) 1997-1998
(c) 1999-2000 (d) 2001-2002
(e) None of these

203. What is the average profit earned (in ₹ crore) for the given years?
(a) 95 (b) 160
(c) 70 (d) 19
(e) None of these

DIRECTIONS (Qs. 204-208) : Study the following table carefully and answer the questions given below :

Production of main crops in India (in million tonnes)						
Crops	91 - 92	92 - 93	93 - 94	94 - 95	95 - 96	96 - 97
Pulses	20.5	22.4	24.6	23.5	27.8	28.2
Oilseeds	32.4	34.6	40.8	42.4	46.8	52.4
Rice	80.5	86.4	88.2	92.6	94.2	90.8
Sugercane	140.8	150.2	152.2	160.3	156.4	172.5
Wheat	130.2	138.4	146.8	141.6	152.2	158.4
Coarse grain	45.6	52.8	60.4	62.4	58.2	62.8
Sum	**450**	**484.8**	**513.2**	**522.8**	**535.6**	**565.1**

204. Production of sugarcane in1993 - 94 was **approximately** what percentage of the production of rice in 1992 – 93?
(a) 50 (b) 75
(c) 150 (d) 125
(e) 175

205. Production of what type of crop was going to increase in each year in the given years?
(a) Rice (b) Pulse
(c) Sugarcane (d) Oilseeds
(e) None of these

206. What was the average production of pulse in the given years?
(a) 26.8 million tonnes (b) 20.5 million tonnes
(c) 24.5 million tonnes (d) 22.5 million tonnes
(e) None of these

207. Production of oilseeds was what percentage of the total crops produced in the year 1991 – 92?
(a) 7.2 (b) 8.4
(c) 2.7 (d) 6.4
(e) None of these

208. In which of the following years the total production of oilseeds in the years 1994 – 95, 1995 – 96 and 1996 – 97 was equal to the production of wheat?
(a) 1993 – 94 (b) 1994 – 95
(c) 1996 – 97 (d) 1992 – 93
(e) None of these

DIRECTIONS (Qs. 209-213): Study the following table to answer the given questions.

Average production of six machines for the given years in thousands						
Year	Machine I	Machine II	Machine III	Machine IV	Machine V	Machine VI
1999	620	400	1020	2050	680	980
1998	680	400	1040	2070	670	1000
1997	640	403	1043	2130	680	1020
1996	700	399	1060	1908	690	1060
1995	706	397	1080	1603	685	1200

209. For which machine has there been continuous increase in production from its previous years?
(a) No machine (b) III
(c) IV (d) II
(e) None of these

210. For which year and the machine has the production been highest for the given data?
(a) 1999, IV (b) 1998, IV
(c) 1997, III (d) 1996, IV
(e) None of these

211. Which of the following can be concluded?
(a) As the machine becomes older, the production goes down.
(b) The production goes down in the initial two or three years then it starts improving.
(c) All the fluctuations from one year to the other are in the range of 100.
(d) Each even-numbered machine produces more than the odd-numbered.
(e) None of these

212. Which machine has shown the least fluctuation in production?
(a) I (b) II
(c) V (d) VI
(e) None of these

213. How many machines have production lower than 700 for all the given years?
(a) Nil (b) One
(c) Two (d) Three
(e) None of these

DIRECTIONS (Qs. 214-221) : Read the following table carefully and answer the questions given below.

Highest marks and average marks obtained by students in subjects over the years

The maximum marks in each subject is 100.

	Subjects									
	English		Hindi		Maths		Science		History	
	High	Avg	High	Avg	High	Avg	High	Avg	High	Avg
1992	85	62	75	52	98	65	88	72	72	46
1993	80	70	80	53	94	60	89	70	65	55
1994	82	65	77	54	85	62	95	64	66	58
1995	71	56	84	64	92	68	97	68	68	49
1996	75	52	82	66	91	64	92	75	70	58
1997	82	66	81	57	89	66	98	72	74	62

214. What was the grand average marks of the five subjects in 1996?
(a) 63 (b) 64
(c) 65 (d) 68
(e) None of these

215. The difference in the average marks in History between 1994 and 1995 was exactly equal to the difference in the highest marks in Hindi between which of the following pairs of years?
(a) 1992 and 1995 (b) 1993 and 1995
(c) 1992 and 1996 (d) 1993 and 1997
(e) None of these

216. What was the **approximate** percentage increase in average marks in History from 1992 to 1993?
(a) 20 (b) 25
(c) 24 (d) 16
(e) 18

217. The average highest marks in English in 1992, 1993 and 1996 was exactly equal to the highest marks in Hindi in which of the following years?
(a) 1996 (b) 1997
(c) 1994 (d) 1996
(e) 1993

218. The difference between the highest marks and the average marks in Hindi was maximum in which of the following years?
(a) 1994 (b) 1997
(c) 1995 (d) 1996
(e) 1993

219. The highest marks in Hindi in 1993 was what per cent of the average marks in Mathematics in 1996?
(a) 135 (b) 130
(c) 125 (d) 140
(e) None of these

220. If there were 50 students in 1993, what was the total marks obtained by them in Mathematics?
(a) 2400 (b) 3000
(c) 2500 (d) 3200
(e) None of these

221. The difference between the highest marks in science was maximum between which of the following pairs of years among the given years?
(a) 1992 and 1993 (b) 1992 and 1996
(c) 1996 and 1997 (d) 1992 and 1995
(e) None of these

DIRECTIONS (Qs. 222-228): Study the following graph carefully and answer the questions given below it:

Packaging Materials Used (In tonnes)

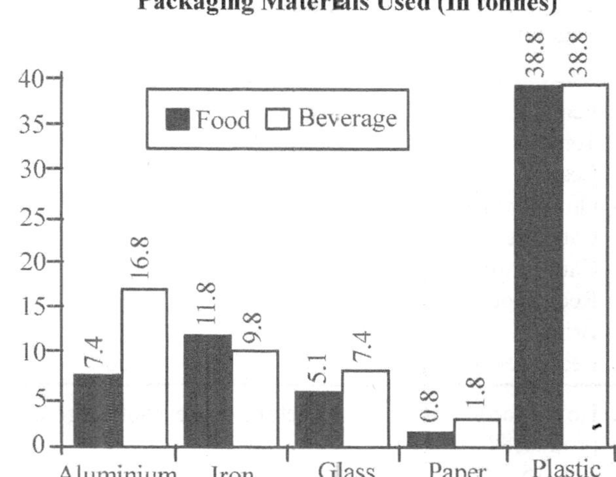

222. What per cent of the total glass packaging material was used for packaging food items?
(a) 40.8 (b) 41.8
(c) 40.7 (d) 41.0
(e) None of these

223. **Approximately** how much per cent more plastic was used than iron for packaging food items?
(a) 32 (b) 320
(c) 33 (d) 325
(e) 225

224. In the case of which one of the following packaging materials used for packing food items and beverages respectively the ratio is 4 : 9?
(a) Glass (b) Paper
(c) Aluminium (d) Iron
(e) None of these

225. What is the ratio between the glass and aluminium packaging materials used for packing beverages?
(a) 17 : 56 (b) 56 : 17
(c) 84 : 37 (d) 37 : 84
(e) None of these

226. **Approximately** what per cent of all the packaging materials used for packing food items was contributed by plastic?
(a) 60 (b) 65
(c) 70 (d) 55
(e) 50

227. **Approximately** what per cent of all the packaging materials used for packing food items and beverages was contributed by plastic and aluminium together?
(a) 60 (b) 70
(c) 80 (d) 65
(e) 75

228. What per cent of all the packaging materials used for packing beverages was contributed by paper? (Find the answer up to two decimal places).
(a) 2.42 (b) 3.41
(c) 2.41 (d) 3.42
(e) None of these

DIRECTIONS (Qs. 229-233): Study the following table carefully and answer the questions given below it:

Area and Production of Different Vegetables

Vegetables	Area (in Hectare)	Production (in Tonnes)
Pea	7200	72792
Tomato	2600	79092
Beans	2100	20895
Onion & Garlic	1500	29490
Cabbage	1700	42670
Cauliflower	700	13790
Root Vegetables	800	18560
Brinjal	300	4500
Leafy Vegetables	2900	28600

229. How many tonnes per hectare were root vegetables produced?
(a) 15.0 (b) 23.2
(c) 19.7 (d) 22.7
(e) None of these

230. In case of how many vegetables the production was more than 20 tonnes per hectare?
(a) 5 (b) 4
(c) 3 (d) 2
(e) None of these

231. What is the ratio between the areas engaged in pea production and onion and garlic production respectively?
(a) 24 : 5 (b) 5 : 24
(c) 23 : 5 (d) 5 : 23
(e) None of these

232. Among the given vegetables in case of how many vegetables the area devoted to production of that vegetables was more than 10 per cent of total areas taken together?
(a) 5 (b) 2
(c) 3 (d) 4
(e) None of these

233. How many more tonnes per hectare cabbage were produced in comparison to cauliflower?
(a) 5.3 (b) 4.5
(c) 3.4 (d) 5.5
(e) None of these

DIRECTIONS (Qs.234-238): Study the pie chart and table carefully to answer the questions that below.

Number of emplyees working in various departments of an organization and the ratio of men to women in the same

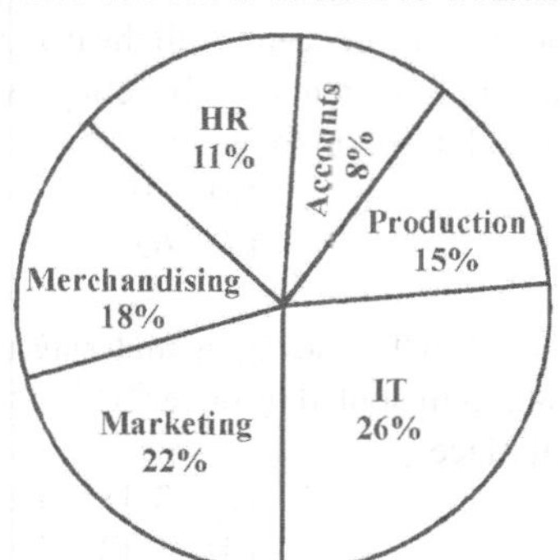

Total number of Employees = 4600
Ratio of men to women

Departments	Men	Women
HR	1	1
Accounts	3	1
Production	3	2
IT	1	3
Marketing	1	1
Merchandising	5	1

234. What is the number of women in the accounts department ?
(a) 86 (b) 102
(c) 80 (d) 92
(e) None of these

235. What is the total number of employees working in the IT department and HR department together ?
(a) 1628 (b) 1742
(c) 1766 (d) 1646
(e) None of these

236. What is the ratio of the total number of men to the total number of women working in all the departments together ?
(a) 63 : 41 (b) 41 : 27
(c) 53 : 47 (d) 27 : 19
(e) None of these

237. The number of women in the merchandising department forms what per cent of the total number of employees in the organization ?
(a) 3% (b) 6%
(c) 1% (d) 12%
(e) None of these

238. What is the ratio of the number of men in the production department to the number of men in the marketing department ?
(a) 7 : 3 (b) 9 : 11
(c) 13 : 7 (d) 11 : 9
(e) None of these

Directions (Qs.239-243): Study the following pie-chart carefully to answer these questions.

Total Expenditure
₹ 60 Lacs

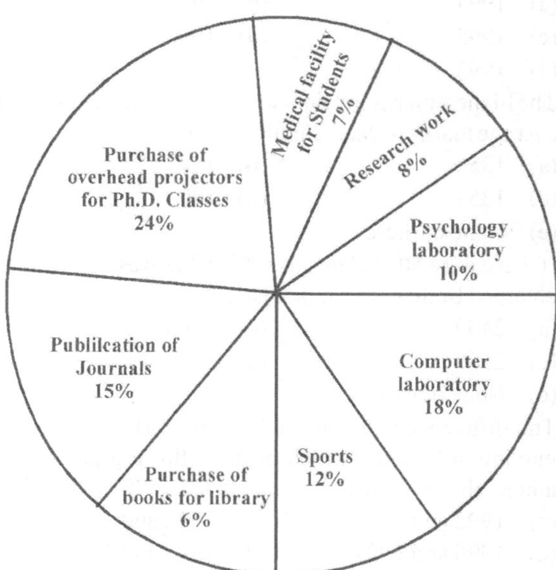

Expenditure of funds by University for various purposes

239. What is the difference between the expenditure mady by university for Publication of Journals and Psychology Laboratory?
(a) ₹ 4 lacs (b) ₹ 3 lacs
(c) ₹ 4.2 lacs (d) ₹ 3.8 lacs
(e) None of these

240. What is the respective ratio between the expenditure made by university on research work and purchase of books for library ?
(a) 4 : 5 (b) 5 : 4
(c) 8 : 3 (d) 8 : 5
(e) None of these

241. What is the total sum of expnditure on Research work, purchase of overhead projectors for Ph.D. Classes and purchase of books for Library together?
(a) ₹ 22.6 lacs (b) ₹ 22.8 lacs
(c) ₹ 23.4 lacs (d) ₹ 20.8 lacs
(e) None of these

242. Which of the following is **definitely true** ?
(a) Ratio between expenditure of university for the purchase of library books and expenditure on computer laboratory is 3 : 1 respectively
(b) Expenditure on medical facilities for students is ₹ 4.6 lacs
(c) Difference between the expenditure on research work and medical facilities for students is ₹ 60,000.
(d) All are true
(e) None of these

243. If the expenditure on purchase of overhead projectors for Ph.D. students is decreased by 7%, what will be the expenditure on the same after the decrease ?
(a) ₹ 1,33,920 (b) ₹ 13,39,200
(c) ₹ 1,02,000 (d) ₹ 1,08,000
(e) None of these

DIRECTIONS (Qs.244-248): The circle graph given here shows the spending of a country on various sports during a particular Year. Study the graph carefully and answer the questions given below it.

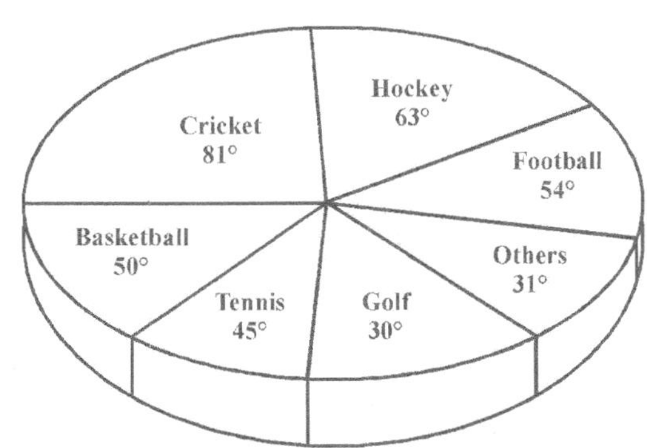

244. What per cent of total expenditure is spent on tennis ?
(a) $12\frac{1}{2}\%$ (b) $22\frac{1}{2}\%$
(c) 25% (d) 45%
(e) 40%

245. How much per cent more is spent on hockey than that on golf?
(a) 27% (b) 35%
(c) 37.5% (d) 75%
(e) 70%

246. If the total amount spent on sports during the year be ₹ 18000000, then the amount spent on basketball exceeds on Tennis by
(a) ₹ 250000 (b) ₹ 360000
(c) ₹ 375000 (d) ₹ 410000
(e) ₹ 30000

247. How much per cent less is spent on football than that on cricket?
(a) $22\frac{2}{9}\%$ (b) 27%
(c) $33\frac{1}{3}\%$ (d) $37\frac{1}{7}\%$
(e) 29%

248. If the total amount spent on sports during the year was ₹ 2 crore, the amount spent on cricket and hockey together was
(a) ₹ 800000 (b) ₹ 8000000
(c) ₹ 12000000 (d) ₹ 16000000
(e) None of these

DIRECTIONS (Qs. 249-251) : Study the information given below and answer the questions that follow :

An article was bought for ₹ 5600. Its price was marked up by 12%. Thereafter it was sold at a discount of 5% on the market price. *[SBI Clerk-June-2012]*

249. What was the market price of the article ?
(a) ₹ 6207/- (b) ₹ 6242/-
(c) ₹ 6292/- (d) ₹ 6192/-
(e) ₹ 6272/--

250. What was the percent profit on the transaction ?
(a) 6.8% (b) 6.3%
(c) 6.4% (d) 6.6%
(e) 6.2%

251. What was the amount of discount given ?
(a) ₹ 319.6 (b) ₹ 303.6
(c) ₹ 306.3 (d) ₹ 313.6
(e) ₹ 316.9

DIRECTIONS (Qs. 252-256) : Study the following graph carefully and answer the questions that follow :

The graph given below represents the number of users of two broadband services A and B across 5 cities P, Q, R, S and T.

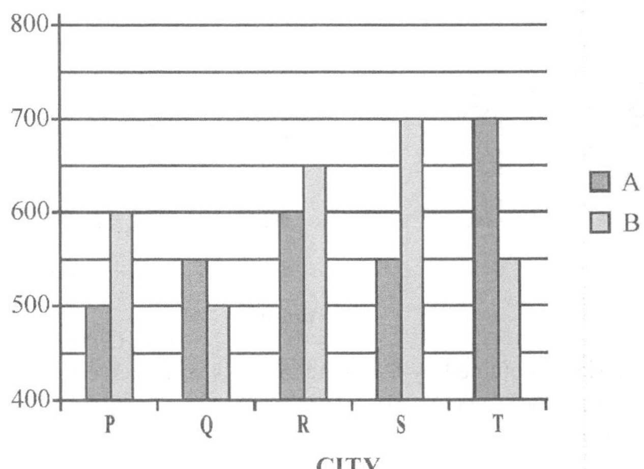

[SBI Clerk-June-2012]

252. What is the total number of users of brand B across all five cities together ?
(a) 2700 (b) 3000
(c) 3100 (d) 2900
(e) 3200

253. The number of users of brand A in city T is what percent of the number of users of brand B in City Q ?
(a) 150 (b) 110
(c) 140 (d) 160
(e) 120

254. What is the average number of users of brand A across all five cities together ?
(a) 560 (b) 570
(c) 580 (d) 590
(e) 550

255. What is the difference between the total number of users of Brand A and B together in city R and the total number of users of brand A and B together in city P ?
(a) 170 (b) 140
(c) 130 (d) 150
(e) 160

256. What is the respective ratio of the number of users of brand A in city P to the number of users of brand B in city S ?
(a) 5 : 7 (b) 4 : 7
(c) 2 : 5 (d) 3 : 4
(e) 5 : 6

DIRECTIONS (Qs. 257-261): *Study the following bar graph carefully to answer the questions.*

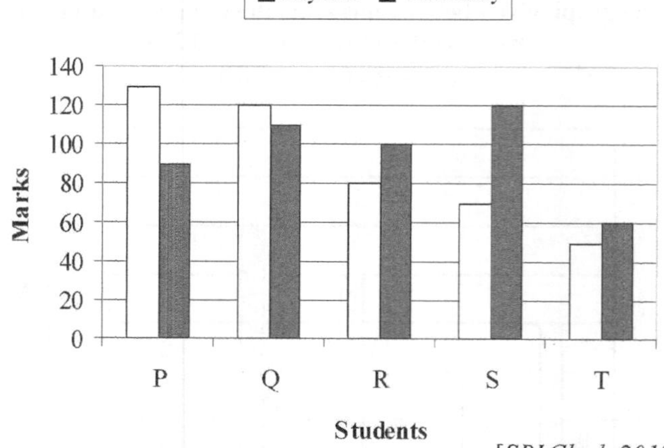

[*SBI Clerk-2012*]

257. Marks obtained by S in Chemistry is what percent of the total marks obtained by all the students in Chemistry ?
(a) 25 (b) 28.5
(c) 35 (d) 31.5
(e) 22

258. If the marks obtained by T in Physics were increased by 14% of the original marks, what would be his new **approximate** percentage in Physics if the maximum marks in Physics were 140 ?
(a) 57 (b) 32
(c) 38 (d) 48
(e) 41

259. Fill in the blank space in order to make the sentence correct as per the given information. Total marks obtained by T in both the subjects together is more than the marks obtained by ________
(a) Q in Chemistry (b) R in Physics
(c) S in Chemistry (d) P in Physics
(e) R in both the subjects together

260. What is the respective ratio between the total marks obtained by P in Physics and Chemistry together to the total marks obtained by T in Physics and Chemistry together ?
(a) 3 : 2 (b) 4 : 3
(c) 5 : 3 (d) 2 : 1
(e) None of these

261. What is the respective ratio between the total marks obtained by Q and S together in Chemistry to the total marks obtained by P and R together in Physics ?
(a) 23 : 25 (b) 23 : 21
(c) 17 : 19 (d) 17 : 23
(e) None of these

DIRECTIONS (Qs. 262-266) : Read the following chart and answer the questions that follows :

The following pie-chart shows the preference of musical instruments of 60,000 people surveyed over whole India.

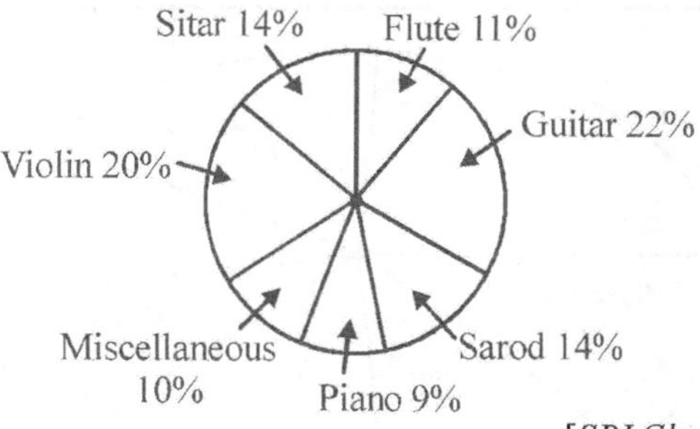

[*SBI Clerk-2014*]

262. If 2100 people be less from the number of people who prefer Flute, the percentage of people who prefer Flute would have been:
(a) 9.5% (b) 6.5%
(c) 7.5% (d) 8.5%
(e) None of these

263. The total number of people who prefer either Sarod or Guitar, is greater than the total number of people who prefer either Violin or Sitar by :
(a) 1200 (b) 1600
(c) 1100 (d) 1400
(e) None of these

264. The number of people who prefer the musical instrument Sarod is :

(a) 7400 (b) 8400
(c) 6400 (d) 8600
(e) None of these

265. If $16\frac{2}{3}\%$ of the people who prefer Piano, would go with the people who prefers Flute, the percentage of people who prefer Flute would have been :

(a) 13.5% (b) 14.5%
(c) 15.5% (d) 12.5%
(e) None of these

266. The number of people who prefer Guitar is greater than the total number of people who prefer either Flute or Piano by :

(a) 1200 (b) 1100
(c) 1300 (d) 1400
(e) None of these

DIRECTIONS (267-271) : The pie-chart given below shows the distribution of workforce by occupational category for country X in 1981 and 1995. Study the chart and answer the questions no. 146 to 150.

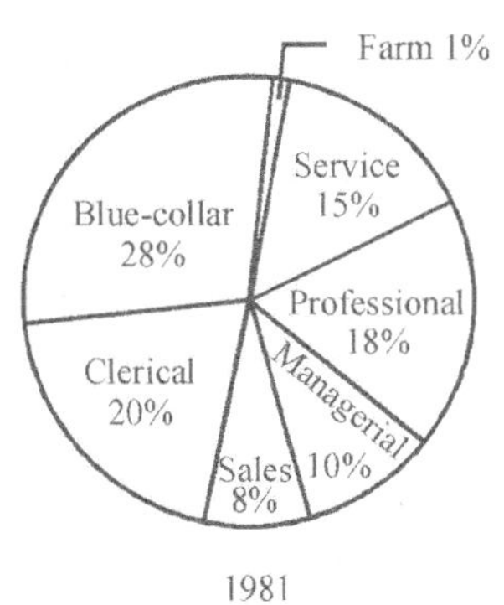

1981
Total workforce
150 million

Farm 2%
Service 16%
Blue-collar 20%
Professional 24%
Clerical 18%
Managerial 13%
Sales 7%

1995
Total workforce
175 million

[*SSC-Sub. Ins.-2012*]

267. In 1981, the number of Sevice workers in the workforce, in millions, was

(a) 15.0 (b) 20.5
(c) 22.5 (d) 28.0

268. In 1981, the number of categories which comprised of more than 25 million workers each, is

(a) two (b) three
(c) four (d) five

269. The ratio of the number of workers in the Professional category in 1981 to the number of such workers in 1995 is

(a) 4 : 9 (b) 5 : 14
(c) 9 : 14 (d) 14 : 9

270. The increase in the number of Clerical workers in the workforce of country X from 1981 to 1995 (in millions) is

(a) 0.75 (b) 1.5
(c) 0.5 (d) 1.25

271. The percentage decrease in the number of Blue-Collar workers in the workforce of country X from 1981 to 1995 is

(a) $42\frac{1}{2}$ (b) 35
(c) 20 (d) $16\frac{2}{3}$

DIRECTIONS (172-175) : The following table shows the productions of food-grains (in million tons) in a state for the period 1999 - 2000 to 2003 - 2004. Read the table and answer the questions.

Production (in million tons)

Year	Wheat	Rice	Barley	Other cereals
1999-2000	680	270	250	450
2000-2001	800	420	440	300
2001-2002	680	350	320	460
2002-2003	720	400	380	500
2003-2004	820	560	410	690

[*SSC-Sub. Ins.-2013*]

272. In 2002 - 2003, the percentage increase in the production of barley as compared to the previous year was:

(a) 14.20 (b) 17.85
(c) 18.75 (d) 7.90

273. During the period 1999 - 2000 to 2003 - 2004, x per cent of the total production is production of wheat. The value of x is about:

(a) 12.6 (b) 37.4
(c) 37.8 (d) 20.2

274. In the year 2003 - 2004, the increase in production was maximum over the previous year for:

(a) Rice (b) Barley
(c) Other cereals (d) Wheat

275. The difference of average production of rice and the average production of barley over the years is :

(a) 50 (b) 60
(c) 80 (d) 40

DIRECTIONS (276-280): Production of three different flavours soft drinks X, Y and Z for a period of six years has been expressed in the following graph. Study the graph and answer the questions.

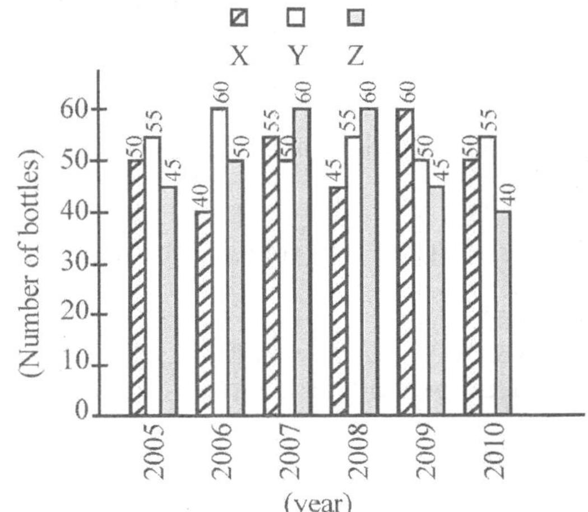

[*SSC-Sub. Ins.-2013*]

276. The approximate decline in the production of flavour Z in 2010 as compared to the production in 2008 is:
(a) 33% (b) 22.5%
(c) 42% (d) 25%

277. The average annual production was maximum in the given period for the flavour:
(a) Y only (b) Z only
(c) X and Z (d) X only

278. What percent of the total production of flavour X in 2005 and 2006 combined is the total production of flavour Z in 2007 and 2008 combined?
(a) 102.25 (b) 115.57
(c) 133.33 (d) 96.67

279. The percentage of rise/fall in production from the previous year is maximum for the flavour Y in this year:
(a) 2007 (b) 2008
(c) 2009 (d) 2006

280. The difference (in lakh bottles) between the average production of flavour X in 2005, 2006, 2007 and the average production of flavour Y in 2008, 2009 and 2010 is :
(a) 2.4 (b) 0.5
(c) 1.5 (d) 5

DIRECTIONS (Qs. 281-284): The bar chart representing the number of first year B.Com. students of St. Xavier's College using different companies' smart phones.

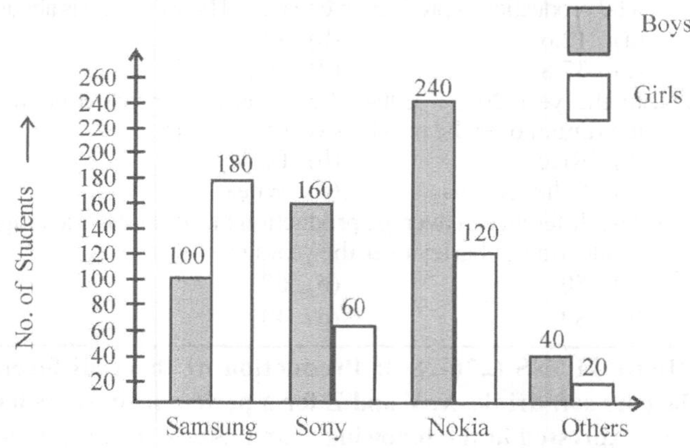

The bar chart representing the no. of students using different smart phones.

[*SSC-Sub. Ins.-2014*]

281. The ratio of the number of boys to the number of girls using the smart phones of Samsung and Sony together is
(a) 12 : 13 (b) 13 : 12
(c) 14 : 11 (d) 11 : 14

282. What percentage of boys are using the smart phones of Samsung ?
(a) 16.52% (b) 17.52%
(c) 18.52% (d) 15.52%

283. What percentage of girls are using the smart phones of Nokia ?
(a) 33.58% (b) 32.58%
(c) 30.58% (d) 31.58%

284. The difference between the total number of students using smart phones of Samsung combined together and the total number of students using smart phone of Sony taken together is
(a) 20 (b) 60
(c) 80 (d) 40

DIRECTIONS (Qs. 285-289) : The following pie-chart shows the monthly expenditure of a family on food, house rent, clothing, education, fuel and miscellaneous.

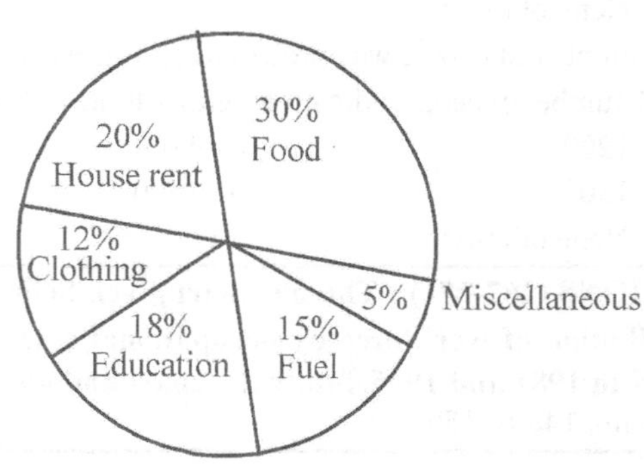

[*SSC-Sub. Ins.-2014*]

285. If the expenditure for food is ₹ 9000, then the expenditure for education is
(a) ₹ 5000 (b) ₹ 5200
(c) ₹ 5400 (d) ₹ 6000

286. The central angle of the sector for the expenditure on fuel (in degrees) is
(a) 50.4 (b) 54
(c) 57.6 (d) 72

287. If the expenditure on fuel is ₹ 3000, then the total expenditure excluding expenditure on house rent and education is
(a) ₹ 11600 (b) ₹ 12000
(c) ₹ 12400 (d) ₹ 12500

288. If the percentage of expenditure on food is x% of the total percentage of expenditure on clothing, education and fuel, then x equals
(a) 66 (b) $66\frac{1}{3}$
(c) $66\frac{2}{3}$ (d) 67

289. Total percentage of expenditure on house rent, clothing and fuel is greater than the percentage of expenditure on food by
(a) 16 (b) 17
(c) 18 (d) 20

290. The following graph represents the maximum and minimum temperature recorded every day in a certain week. The day on which the difference between the maximum and minimum temperature was maximum is [*SSC-MT-2013*]

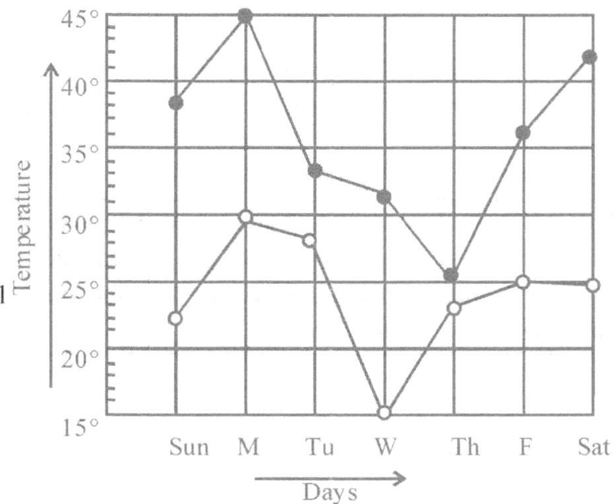

(a) Monday (b) Wednesday
(c) Saturday (d) Sunday

291. Different choices made by a group of 200 students are given below in percentage. The number of students who have taken neither Science nor Commerce is [*SSC-MT-2013*]

Percentage of Students in different streams	
Name of Streams	**Intake Ratio**
Science	29%
Arts	29%
Commerce	31%
Home Science	6%
Others	5%

(a) 40 (b) 80
(c) 120 (d) 60

DIRECTIONS (Qs. 292- 296) : Read the following chart and answer the questions that follows :

The following pie-chart shows the preference of musical instruments of 60,000 people surveyed over whole India.

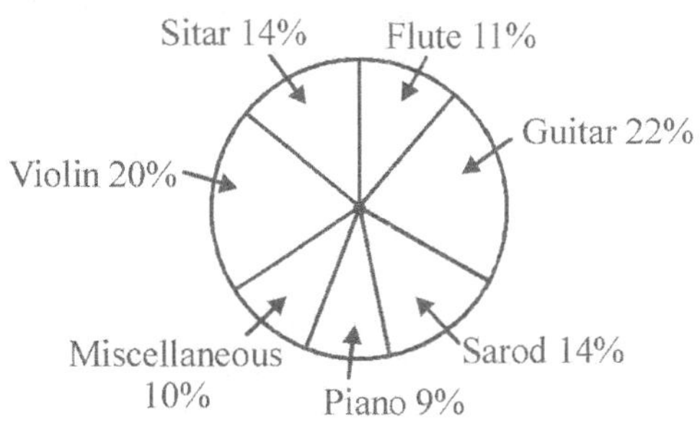

[*SSC 10+2-2012*]

292. If 2100 people be less from the number of people who prefer Flute, the percentage of people who prefer Flute would have been:
(a) 9.5% (b) 6.5%
(c) 7.5% (d) 8.5%

293. The total number of people who prefer either Sarod or Guitar, is greater than the total number of people who prefer either Violin or Sitar by :
(a) 1200 (b) 1600
(c) 1100 (d) 1400

294. The number of people who prefer the musical instrument Sarod is :
(a) 7400 (b) 8400
(c) 6400 (d) 8600

295. If $16\frac{2}{3}\%$ of the people who prefer Piano, would go with the people who prefers Flute, the percentage of people who prefer Flute would have been :
(a) 13.5% (b) 14.5%
(c) 15.5% (d) 12.5%

296. The number of people who prefer Guitar is greater than the total number of people who prefer either Flute or Piano by :
(a) 1200 (b) 1100
(c) 1300 (d) 1400

DIRECTIONS (Qs 297 - 301) : ***Study the graph and answer the questions that follows :***

Circle graph given below shows the expenditure incurred in bringing out a book by a publisher.

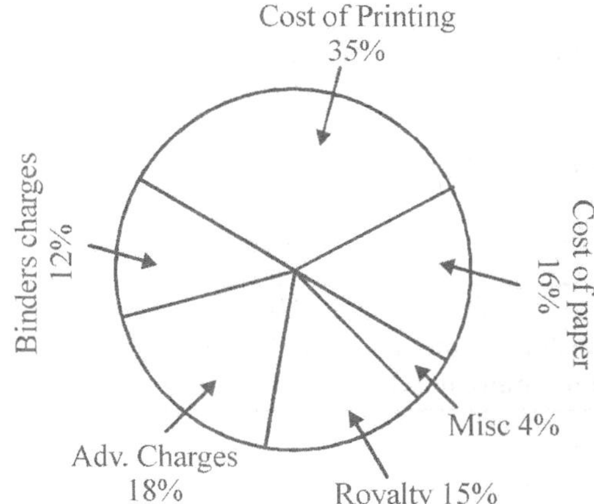

[*SSC 10+2-2012*]

297. The central angle of the sector for the cost of the paper is :
(a) 22.5° (b) 16°
(c) 54.8° (d) 57.6°

298. Royalty on the book is less than the Advertisement charges by:
(a) 3% (b) 25%
(c) 20% (d) $16\frac{2}{3}\%$

299. If 5500 copies are published, Miscellaneous expenditures amounts to ₹ 1848 and publisher's profit is 25%, then marked price of each copy is :
(a) ₹ 12.50 (b) ₹ 10.50
(c) ₹ 10 (d) ₹ 8.40

300. If the cost of printing is ₹ 17,500, the Royalty is:
(a) ₹ 8750 (b) ₹ 6300
(c) ₹ 7500 (d) ₹ 3150

301. If the Miscellaneous charges is ₹ 6,000, the Advertisement charges are:
(a) ₹ 27,000 (b) ₹ 90,000
(c) ₹ 12,000 (d) ₹ 1,333.33

DIRECTIONS (Qs. 302-303): The bar graph shows the marks obtained by a student in an examination out of 100 marks in each subject. Study the diagram answer the questions that follow.

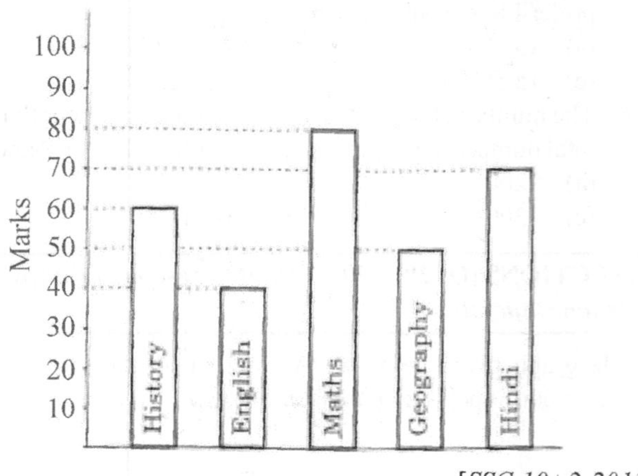

[SSC 10+2-2013]

302. The average marks of Hindi and English is
(a) 60 (b) 65
(c) 50 (d) 55

303. The ratio of the marks of Maths and History is
(a) 4 : 3 (b) 6 : 5
(c) 8 : 5 (d) 3 : 4

DIRECTIONS (Qs. 304-305): Various expenditures incured by a publishing company for publishing a book in 2014 are given below. Study the chart answer the questions that follow.

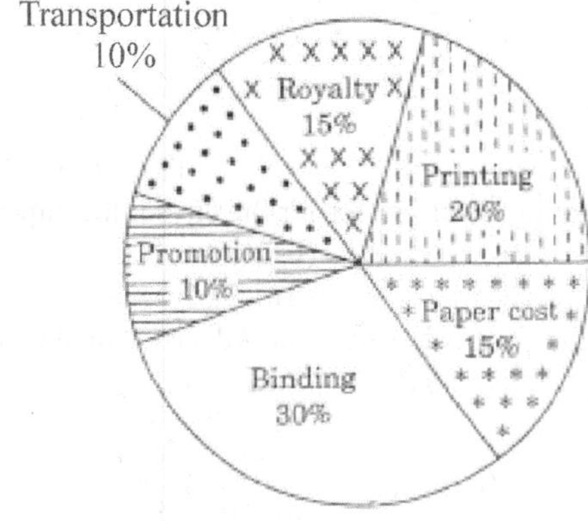

[SSC 10+2-2013]

304. Royalty of a book is less than the printing cost by
(a) 20% (b) 25%
(c) 5% (d) $33\frac{1}{3}\%$

305. Price of a book is 20% above cost price. If the marked price is ₹180, then the cost of paper for a single copy, in ₹, is
(a) 42 (b) 44.25
(c) 36 (d) 22.50

DIRECTIONS (Qs. 306-309): Sales of Books (in thousand numbers) from Six Branches - B1, B2, B3, B4, B5 and B6 of a Publishing Company in 2000 and 2001. Study the graph and answer the question that follow:

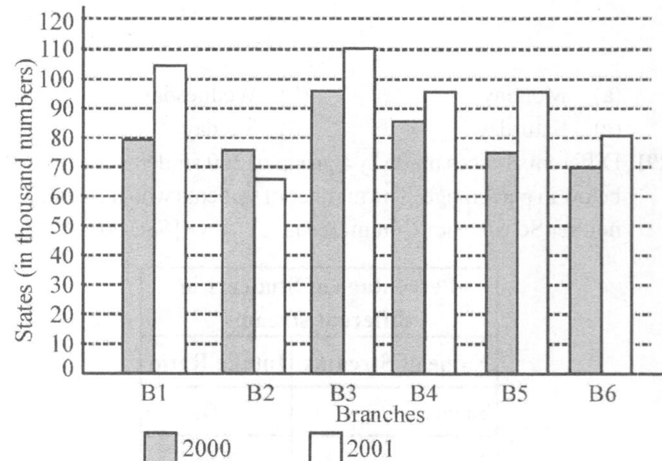

[SSC 10+2-2014]

306. Total sale of branches B1, B3 and B5 together for both the years (in thousand numbers) is
(a) 250 (b) 310
(c) 435 (d) 560

307. Find the ratio of the total sales of branch B2 for both years to the total sales of branch B4 for both years.
(a) 2 : 3 (b) 3 : 5
(c) 4 : 5 (d) 7 : 9

308. Percentage of the average sale of branches B1, B2 and B3 in 2001 and the average sale of branches B1, B3 and B6 in 2000
(a) 87.5 (b) 75
(c) 77.5 (d) 82.5

309. Find the percentage increase in the sales of books of branch B3 in the year 2001 than the branch B2.
(a) 69.2 (b) 50.8
(c) 40.9 (d) 65.7

DIRECTIONS (Qs. 310-314): The diagram shows the age distribution of the patients admitted to a hospital in a particular day. Study the diagram and answer

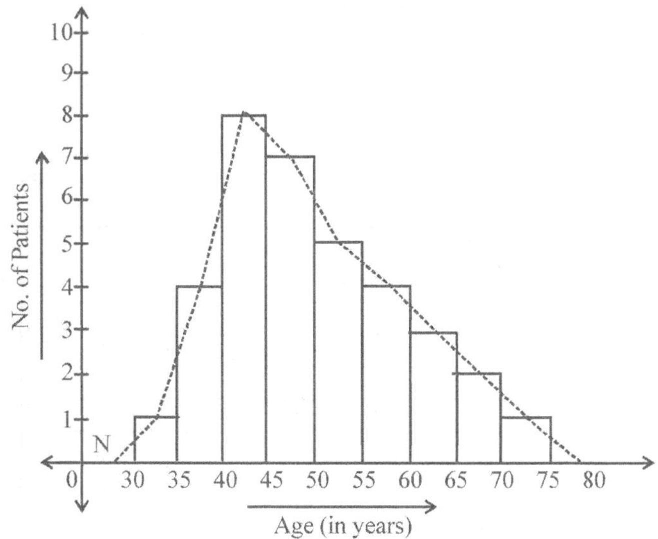

[SSC 10+2-2014]

310. Number of patients of age between 55 years to 60 years, who got admitted to the hospital on that day is

(a) 6 (b) 4
(c) 24 (d) 8

311. Total number of patients of age more than 55 years, who got admitted to the hospital is

(a) 4 (b) 7
(c) 9 (d) 10

312. Number of patients of age more than 40 years and less than 55 years, who got admitted to the hospital on that day is

(a) 20 (b) 30
(c) 15 (d) 12

313. Percentage of patients of age less than 45 years, who got admitted to the hospital on that day is approximately equal to

(a) 14% (b) 20%
(c) 37% (d) 62%

314. About 11% of the patients who got admitted to the hospital on that particular day were of age

(a) either between 35 years and 40 years or between 55 years and 60 years
(b) between 60 years and 65 years
(c) between 35 years and 40 years
(d) between 35 years and 40 years and between 55 years and 60 years

DIRECTIONS (Qs. 315-319): Study the table carefully and answer the given questions.

Number of Pages Printed by 6 Printers in 5 Different Weeks

Week \ Printer	A	B	C	D	E	F
1st	664	618	628	552	638	419
2nd	569	441	519	438	621	537
3rd	440	614	503	527	541	742
4th	256	563	347	651	412	321
5th	717	429	598	582	519	693

[IBPS Clerk-2013]

315. What is the respective ratio between the number of pages printed by Printer B in 2nd week and the number of pages printed by Printer F in 5th week?

(a) 4 : 9 (b) 11 : 13
(c) 9 : 13 (d) 7 : 11
(e) 9 : 11

316. What is the average number of pages printed by all the given printers in 4th week?

(a) 375 (b) 425
(c) 415 (d) 430
(e) 390

317. Which of the following printer printed maximum number of pages in all the given weeks together?

(a) Printer A (b) Printer E
(c) Printer D (d) Printer C
(e) Printer F

318. Number of pages printed by Printer A in 3rd week is what per cent of the total number of pages printed by Printed D in all the given weeks?

(a) 22 (b) 18
(c) 12 (d) 14
(e) 16

319. What is the difference between the total number of pages printed by Printer E in 1st, 2nd and 4th week together and total number of pages printed by Printer C in all the given weeks together?

(a) 952 (b) 878
(c) 924 (d) 934
(e) 918

Level - II

DIRECTIONS (Qs.1-6): Study the following graph to answer the given questions.

Percent profit earned by two companies over the given years

$$\% \text{ Profit} = \frac{\text{Income} - \text{Expenditure}}{\text{Expenditure}} \times 100$$

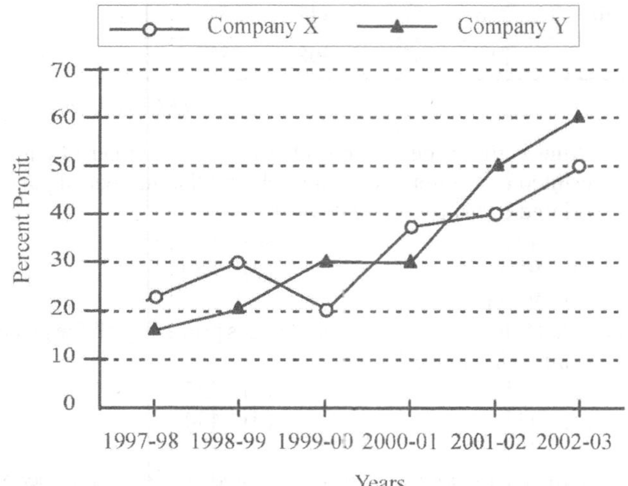

1. If the income of Company X in 1998-99 was equal to the expenditure of Company Y in 2001-2002, what was the ratio of their respective profits?
(a) 13 : 15 (b) 15 : 26
(c) 13 : 26 (d) Cannot be determined
(e) None of these

2. For Company *X*, its income in 2001-2002 was equal to its expenditure in 2002-2003. What was the ratio of its respective incomes in these two years?
(a) 4 : 5 (b) 3 : 4
(c) 2 : 3 (d) Cannot be determined
(e) None of these

3. For Company Y, in which year is the per cent of increase in per cent profit over that of previous year, the highest?
(a) 2002-03 (b) 1999-2000
(c) 2001-02 (d) Cannot be determined
(e) None of these

4. In 1997-98, the expenditure of Company *X* was ₹ 40 crores. What was its income in that year?
(a) ₹ 50 crores (b) ₹ 48 crores
(c) ₹ 46 crores (d) Cannot be determined
(e) None of these

5. What was the difference in the expenditures of the two companies in 1999-2000?
(a) 10 (b) 100
(c) 1000 (d) Cannot be determined
(e) None of these

6. In 2002-03 the income of Company *Y* was ₹ 128 crores. What was its expenditure in that year?
(a) ₹ 76.8 crores (b) ₹ 64 crores
(c) ₹ 48 crores (d) Cannot be determined
(e) None of these

DIRECTIONS (Qs.7-12): Study the following table to answer the given questions:

Production (in crore units) of six companies over the year

Company	Years						Total
	1997	1998	1999	2000	2001	2002	
TP	103	150	105	107	110	132	707
ZIR	75	80	83	86	90	91	505
AVC	300	300	300	360	370	340	1970
CTU	275	280	281	280	285	287	1688
PEN	25	30	35	40	42	45	217
SIO	85	87	89	91	92	96	540
Total	**863**	**927**	**893**	**964**	**989**	**991**	**5627**

7. The production of Company AVC in 2000 is **approximately** what per cent of its average production over the given years?
(a) 300 (b) 110 (c) 136 (d) 18.25 (e) 95

8. For SIO, which year was the per cent increase or decrease in production from the previous year, the highest?
(a) 2001 (b) 1998
(c) 2002 (d) 2000
(e) None of these

9. Which company has less average production in the last three years compared to that of first three years?
(a) No company (b) CTU
(c) ZIR (d) SIO
(e) None of these

10. The total production of the six companies in the first two given years is what per cent of that of last two given years? (round off up to two decimal places)
(a) 87.08 (b) 104.55
(c) 90.40 (d) 10.62
(e) None of these

11. For ZIR, which of the following is the difference between production in 2002 and that in 2001?
(a) 10,00,00,000 (b) 1,00,00,000
(c) 10,00,000 (d) 40,00,000
(e) None of these

12. For how many companies did the production increase every year from that of the previous year?
(a) One (b) Two
(c) Three (d) Four
(e) None of these

DIRECTIONS (Qs. 13-15): These questions are based on the following information. Study the information carefully and answer the questions.

The students of a school have an option to study only Hindi, only Sanskrit or a composite subject Hindi and Sanskrit. Out of, the

175 students in the school, boys and girls are in the ratio of 3 : 4 respectively. 40% of boys have opted for only Hindi. 44% of the students have opted for only Sanskrit. Out of the total number of girls 32% have opted for the composite subject. The number of boys who opted for only Sanskrit and that for composite subject are in the ratio of 2 : 1 respectively.

13. What is the ratio between the number of boys who have opted for only Hindi and the number of girls who have opted for the composite subject respectively?
(a) 15 : 16 (b) 10 : 7
(c) 10 : 9 (d) 11 : 12
(e) None of these

14. How many boys have opted for the composite subject?
(a) 30 (b) 15
(c) 21 (d) 32
(e) None of these

15. How many girls have opted for only Sanskrit?
(a) 72 (b) 47
(c) 51 (d) 77
(e) None of these

DIRECTIONS (Qs. 16-20): Study the following graph carefully to answer these questions.

Annual dividend offered by two companies over the years

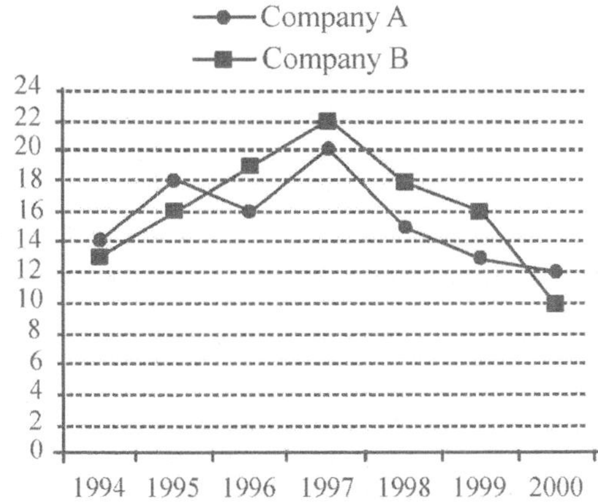

16. Shri Giridhar invested total amount of ₹ 25000 in 1994 for one year in the two companies together and got a total dividend of ₹ 3340. What was the amount invested in Company *A*?
(a) ₹ 12000 (b) ₹ 9000
(c) ₹ 16000 (d) Cannot be determined
(e) None of these

17. Anuja invested ₹ 35000 in Company *B* in 1996. After one year she transferred the entire amount with dividend to Company *A* in 1997 for one year. What amount will be received back by Anuja including dividend?
(a) ₹ 49980 (b) ₹ 49000
(c) ₹ 48300 (d) ₹ 49563.50
(e) None of these

18. An amount of ₹ 18000 was invested in Company *A* in 1997. After one year the same amount was re-invested for one more year. What was the total dividend received at the end of two years?
(a) ₹ 5805 (b) ₹ 7063.20
(c) ₹ 6480 (d) ₹ 6840
(e) None of these

19. Bhushan invested different amounts in Companies *A* and *B* in 2000 in the ratio of 5 : 8. What will be the ratio between the amounts of dividends received from Companies *A* and *B* respectively?
(a) 2 : 3 (b) 5 : 6
(c) 3 : 4 (d) Cannot be determined
(e) None of these

20. In the year 1999, Suraj invested ₹ 56000 in Company *B*. How much more or less dividend would he have received had the amount been invested in Company A?
(a) ₹ 1640 more (b) ₹ 1640 less
(c) ₹ 1860 less (d) ₹ 1680 more
(e) None of these

DIRECTIONS (Qs. 21-25): These questions are based on the following information. Study the information carefully and answer the questions.

Total population of a village is 35000. Out of these 70% are literate. 44% of the total population are females. Out of the total illiterate population, males and females are in the ratio of 28 : 47 respectively.

21. What is the ratio of illiterate to literate females?
(a) 63 : 47 (b) 47 : 63
(c) 16 : 47 (d) 47 : 16
(e) None of these

22. Out of the total literate population what is the ratio of males to females?
(a) 17 : 8 (b) 8 : 17
(c) 9 : 16 (d) 16 : 9
(e) None of these

23. What is the total number of male population?
(a) 15400 (b) 18600
(c) 17800 (d) 19400
(e) None of these

24. If 5% of the male literate population are graduates, how many male graduates are there in the village?
(a) 784 (b) 196
(c) 980 (d) 120
(e) None of these

25. What is the total number of illiterate females?
(a) 6850 (b) 6480
(c) 6580 (d) 8820
(e) None of these

DIRECTIONS (Qs. 26-30): Study the following table carefully to answer these questions.

Number of students studying different disciplines at graduate level from State 'A' over the years

Discipline /year	Arts	Commerce	Science	Agriculture	Medicine	Engineering
1997	2400	3200	4200	840	2350	3180
1998	2250	3500	4820	760	2120	3340
1999	3050	2850	4550	1120	2640	3650
2000	2800	3640	4680	930	1890	3490
2001	2980	3080	5220	780	2260	3280
2002	2770	3800	3950	810	2450	3500

26. Total number of students studying Medicine for all the years together is **approximately** what per cent of those studying Engineering for all the years together?
(a) 60 (b) 67
(c) 72 (d) 75
(e) 73
27. What is the average number of students studying Arts? (Rounded off to an integer)
(a) 2905 (b) 2480
(c) 2308 (d) 2708
(e) None of these
28. For which of the following years, percentage increase/ decrease in the number of students studying Commerce with respect to the previous year is the maximum?
(a) 1998 (b) 1999
(c) 2000 (d) 2001
(e) 2002
29. The number of students studying Agriculture in the year 1999 is what per cent of the total number of students studying rest of the disciplines together during that year? (Rounded off to two digits after decimal)
(a) 6.69 (b) 6.27
(c) 6.82 (d) 6.39
(e) None of these
30. The number of students studying Commerce in 2001 is **approximately** what per cent of the total number of students studying Commerce for all the given years together?
(a) 19 (b) 11
(c) 12 (d) 18
(e) 15

DIRECTIONS (Qs. 31-35): Study the following information to answer the questions given below:

(i) The ratio of the populations of males, females and children 10 years old and above is 11 : 10 : 9 in State 'A'. Out of which 40% males or 8800 are literate, 20% children (10 year old and above) are illiterate while 30% females are literate.

(ii) The number of children below 10 years of age is 10% of the number of females. 5% of the total population of the State are below poverty line and 80% of them are illiterate.

31. What is the number of illiterate persons below the poverty line?
(a) 2480 (b) 3100
(c) 620 (d) Cannot be determined
(e) None of these
32. What is the total population of the State?
(a) 60,000 (b) 62,000
(c) 42,000 (d) 40,000
(e) None of these
33. What is the number of literate children of age 10 years and above?
(a) 14400 (b) 14800
(c) 16200 (d) 12600
(e) None of these
34. Total number of women is what percentage of the total population of the State? (rounded off to two places of decimal)
(a) 28.86 (b) 30.25
(c) 32.86 (d) 32.26
(e) None of these
35. How many women are illiterate?
(a) 20000 (b) 6000
(c) 14400 (d) 16800
(e) None of these

DIRECTIONS (Qs. 36-40) : Study the following graph carefully and answer the questions given below it:

(in thousand tonnes)

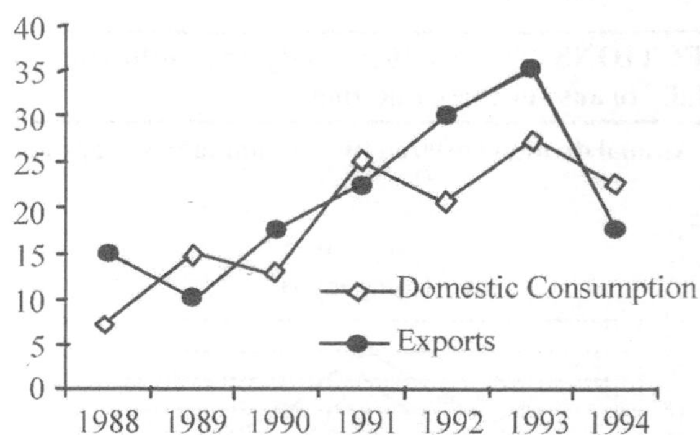

36. In which of the following years was the total of the exports and domestic consumption highest among the given years?
(a) 1991 (b) 1993
(c) 1994 (d) 1992
(e) None of these
37. The difference between the domestic consumption between 1990 and 1991 was exactly equal to the difference in exports between which of the following pairs of years?
(a) 1991 & 1992 (b) 1989 &1991
(c) 1988 &1990 (d) 1991 & 1993
(e) 1992 & 1993
38. If there was 40 per cent increase in the domestic consumption from 1994 to 1995, what would have been the difference between the domestic consumption and exports in 1995 (in tonnes)?
(a) 5,000 (b) 7,000
(c) 8,500 (d) Data inadequate
(e) None of these
39. In which of the following years was the percentage fluctuation (increase/decrease) in exports the maximum from the previous years'?
(a) 1989 (b) 1991
(c) 1994 (d) 1992
(e) 1990

40. What was the difference between the average domestic consumption and the average exports during the given years?
(a) 2,000 tonnes (b) 3,000 tonnes
(c) 2,500 tonnes (d) 4,000 tonnes
(e) None of these

DIRECTIONS (Qs. 41-45) : Read the following table carefully and answer the questions given below it:

Average marks obtained by 20 boys and 20 girls in five subjects from five different schools

Subject	Max Marks	P B	P G	Q B	Q G	R B	R G	S B	S G	T B	T G
Eng	200	85	90	80	75	100	110	65	60	105	110
Hist	100	40	55	45	50	50	55	40	45	65	60
Geo	100	50	40	40	45	60	55	50	55	60	65
Math	200	120	110	95	85	135	130	75	80	130	135
Scien	200	105	125	110	120	125	115	85	90	140	135

In above table, B = Boys and G = Girls

41. What was the total marks obtained by boys in History from school Q?
(a) 900 (b) 1000
(c) 800 (d) 1300
(e) None of these

42. In which of the following subjects did the girls have highest average percentage of marks from all the schools?
(a) Science (b) Geography
(c) English (d) History
(e) Mathematics

43. The pooled average marks of both boys and girls in all the subjects was minimum from which of the following schools?
(a) Q (b) P
(c) T (d) S
(e) R

44. In the case of which of the following schools was total marks obtained by girls in mathematics 100% more than the total marks obtained by boys in History?
(a) R (b) S
(c) P (d) Q
(e) T

45. What was the difference between the total marks obtained in Mathematics by boys from school R and the girls from school S?
(a) Nil (b) 1100
(c) 100 (d) 1200
(e) None of these

DIRECTIONS (Qs. 46-50) : Study the graph carefully and answer the questions given below it.

Per cent profit earned by the two companies A & B over the year

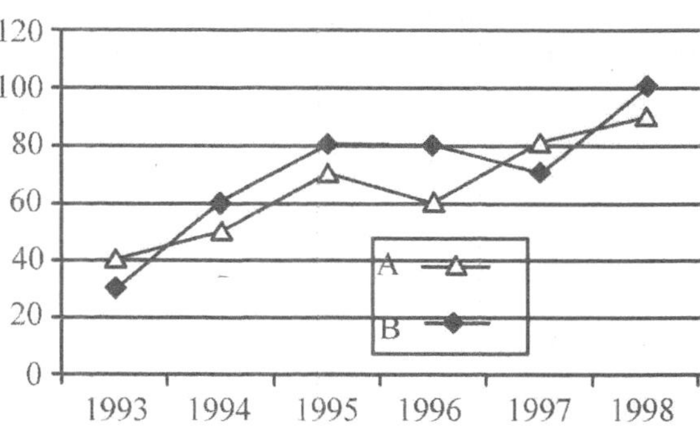

46. If income for Company A in the year 1994 was 35 lakhs what was the expenditure for Company B in the same year?
(a) 123.5 lakhs (b) 128 lakhs
(c) 132 lakhs (d) Data inadequate
(e) None of these

47. The income of Company A in 1996 and the income of Company B in 1997 are equal. What will be the ratio of expenditure of Company A in 1996 to the expenditure of Company B in 1997?
(a) 26 : 7 (b) 37 : 6
(c) 15 : 170 (d) 116 : 17
(e) None of these

48. During which of the following years the ratio of percent profit earned by Company *A* to that of Company *B* was the maximum?
(a) 1993 & 1996 both (b) 1995 & 1997 both
(c) 1993 only (d) 1998 only
(e) None of these

49. If the expenditure of Company B increased by 20% from 1995 to 1996, the income in 1996 will be how many times the income in 1995?
(a) 2.16 times (b) 1.5 times
(c) 1.8 times (d) equal
(e) None of these

50. If the income of Company A in 1996 was ₹ 36 lakhs, what was the expenditure of Company A in 1996?
(a) 22.5 lakhs (b) 28.8 lakhs
(c) 20 lakhs (d) 21.6 lakhs
(e) None of these

DIRECTIONS (Qs.51-55): Study the following pie-chart carefully to answer these questions.

Percentagewise Distribution of Players Who Play Five Different Sports
Total Players are 4200 out of which Female Players are equal to 2000
Total Players = 4200
Percentage of Players who play different sports

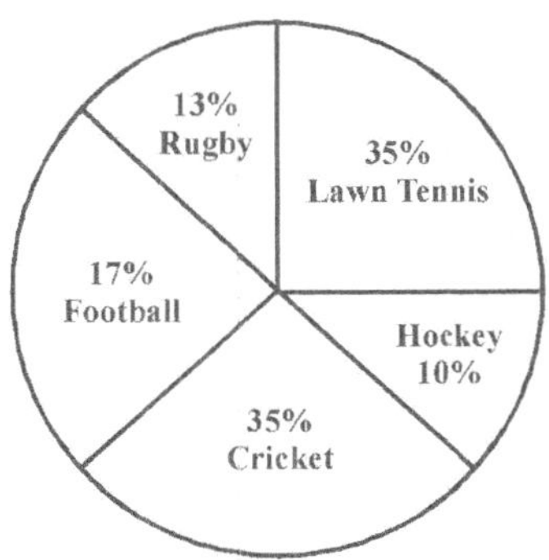

Female Players = 2000
Percentage of Female Players who play different sports

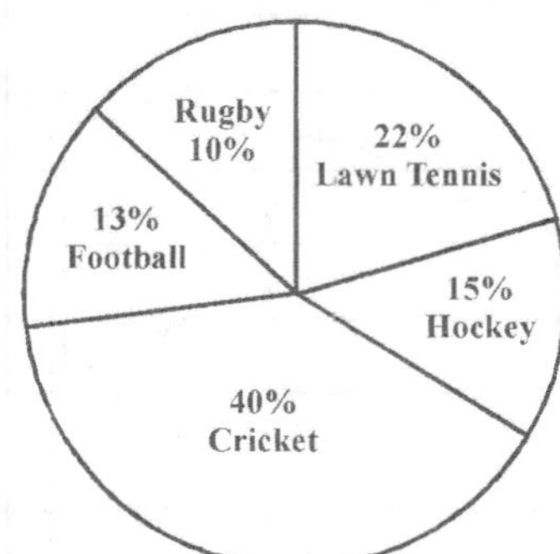

51. What is the average number of players (both male and female) who play football and rugby together?
(a) 620 (b) 357
(c) 230 (d) 630
(e) None of these

52. What is the difference between the number of the female players who play lawn tennis and the number of male players who play rugby?
(a) 94 (b) 84
(c) 220 (d) 240
(e) None of these

53. What is the respective ratio of the number of female players who play cricket and number of male players who play hockey?
(a) 20 : 7 (b) 4 : 21
(c) 20 : 3 (d) 3 : 20
(e) None of these

54. What is the total number of male players who play football, cricket and lawn tennis together ?
(a) 1,724 (b) 1,734
(c) 1,824 (d) 1,964
(e) None of these

55. Number of male players who play rugby is **approximately** what percentage of the total number of players who play lawn tennis ?
(a) 33 (b) 39
(c) 26 (d) 21
(e) 43

DIRECTIONS (Qs.56-60): Study the following pie-chart carefully to answer the questions that follow :

Degree Wise Break-up of Expenditure of a Family in a Month Total Amount Spent in a Month of ₹ 45,800

56. What is the amount spent by the family on commuting ?
(a) ₹ 10,076 (b) ₹ 10,534
(c) ₹ 6,870 (d) ₹ 8,702
(e) None of these

57. What is the respective ratio of amount spent by family on Medicine to the amount spent on Groceries ?
(a) 1 : 2 (b) 13 : 21
(c) 3 : 5 (d) 11 : 23
(e) None of these

58. What is the total amount spent by the family on Entertainment and Shoping together ?
(a) ₹ 9,618 (b) ₹ 13,282
(c) ₹ 13,740 (d) ₹ 11,908
(e) None of these

59. Total amount spent by the family on Groceries. Entertainment and investments together forms **approximately** what per cent of amount spent on Commuting?
(a) 209 (b) 78
(c) 154 (d) 42
(e) 218

60. Amount spent by the family on Medicine forms what per cent of amount spent on shopping ? (rounded off to two digits after decimal)
(a) 43.57 (b) 49.52
(c) 57.89 (d) 61.89
(e) None of these

DIRECTIONS (Qs.61-65): The following pie chart shows the source of funds to be collected by the National Highways Authority of India (NHAI) for its Phase II projects. Study the pie chart and answers the questions that follow.

Sources of Funds to be Arranged by NHAI for Phase II Projects (in ₹ crore)

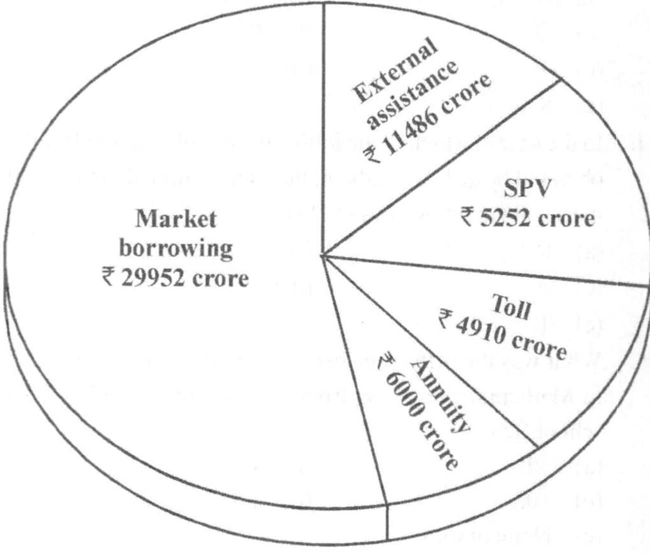

Total Fund = ₹ 57600 crore

61. Near about 20% of the funds are to be arranged through
(a) SPV (b) External assistance
(c) Annually (d) Market borrowing
(e) None of above

62. If NHAI could receive a total of ₹ 9695 crore as external assistance, by what per cent (approaximately) should it increase the market borrowing to arrange for the shortage of funds ?
(a) 4.5% (b) 7.5%
(c) 6% (d) 8%
(e) 7%

63. If the toll is to be collected through an outsourced agency by allowing a maximum 10% commission, how much amount should be permitted to be collected by the outsourced agency so that the project is supported with ₹ 4900 crore?
(a) ₹ 6213 crore (b) ₹ 5827 crore
(c) ₹ 5401 crore (d) ₹ 5316 crore
(e) None of these

64. The central angle corresponding to market borrowing is
(a) 52° (b) 137.8°
(c) 187.2° (d) 192.4°
(e) 100.2°

65. The approximate ratio of the funds to be arranged through toll and that through market borrowing is
(a) 2 : 9 (b) 1 : 6
(c) 3 : 1 (d) 2 : 3
(e) 3 : 2

DIRECTIONS (Qs. 66-71) : Study the following table carefully and answer the questions given below it :

Area and Population of different states

States	Area (in sq kilometres)	Population (in lakhs)
A	6230	1122
B	2540	838
C	8135	649
D	7436	572
E	4893	711
F	3718	286
G	4297	860

[SBI PO-2011]

66. Among the given states, in case of how many states the area of that state was more than 15 per cent of the total areas taken together?
(a) One (b) Three
(c) two (d) Can't say
(e) None of these

67. For which two states the density of population is approximately equal?
(a) No state (b) A and G
(c) D and F (d) C and F
(e) None of these

68. **Approximately** how much more is the density of population of state B in comparison to that of state A?
(a) 15000 (b) 18000
(c) 13000 (d) 14000
(e) 17000

69. In case of how many states the density of population was more than 12 thousand per square kilometre?
(a) Two (b) Five
(c) Three (d) Four
(e) None of these

70. What is the approximate ratio of the areas of state B to the areas of state A and G together.
(a) 1 : 3.8 (b) 1 : 3.5
(c) 1 : 5.2 (d) 1 : 4.5
(e) 1 : 4.1

71. The surface area of a spherical part of a bowl with a flat circular detachable cover, excluding the cover, is 616 sq cm. The area of the cover is 38.5 sq cm. What is the volume of the bowl?
(a) 1339 cm^3 (b) 1430 cm^3
(c) 1570 cm^3 (d) Cannot be determined
(e) None of these

DIRECTION (Qs. 72-75) : Study the following table carefully and answer accordingly :

The distribution of marks (out of 150) obtained by 180 students in each of the five subjects.

Marks → Sub ↓	0-29	30-59	60-89	90-119	120-150
Maths	22	47	74	25	12
Science	39	38	67	22	14
Hindi	19	59	47	36	19
English	24	41	58	34	23
Geography	42	32	52	41	13
Average of five subjects	27	45	60	31	17

[SBI PO-2011]

72. If for passing, the student has to obtain minimum 60% marks in the average of five subjects, how many students will pass?
(a) 108 (b) 58
(c) 48 (d) 72
(e) None of these

73. How many students will pass in Geography if minimum passing marks is 40%?
(a) 74 (b) 106
(c) 96 (d) Can't say
(e) None of these

74. How many students have obtained 60 or more marks in at least one of the five subjects?
(a) 111 (b) 103
(c) 108 (d) 106
(e) Data inadequate

75. If the criteria for distinction is minimum 75% marks in Maths, how many students will get distinction?
(a) 37 (b) 27
(c) 12 (d) Can't say
(e) None of these

76. The no. of students who obtained more than or equal to 40% marks in Science is what per cent less than that of those who scored less than 60% in Hindi?
(a) 17.60% (b) 15.40%
(c) 19.80% (d) 24.30%
(e) None of these

DIRECTIONS (Qs. 77-81) : *Study the following data carefully and answer accordingly.*

Following chart shows the number of students in different universities

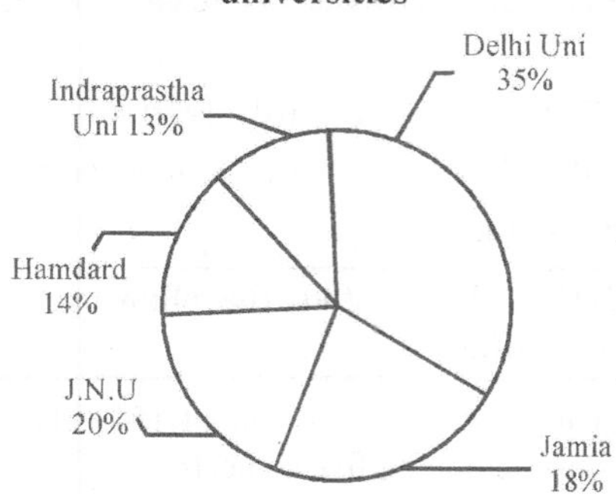

Total no. of students = 120,000

Percentage of listeners of different FM channels in National Capital Region

Universities ↓	FM Channels				
	Radio Mirchi	Radio City	Red FM	FM Gold	Rainbow
Indraprastha	76%	72%	46%	54%	48%
Hamdard	63%	64%	59%	47%	53%
JNU	52%	65%	64%	51%	54%
DU	82%	44%	32%	35%	45%
Jamia	75%	32%	36%	52%	64%

[SBI PO-2011]

77. How many students of JNU listen to Radio city?
(a) 15200 (b) 15600
(c) 14400 (d) 14600
(e) None of these

78. The no. of Indraprastha students listening to Rainbow is what per cent of the no. of Jamia students listening FM Gold?
(a) 65 (b) 56
(c) 68 (d) 58
(e) None of these

79. From which of the following universities, the no. of students liking Red FM is minimum?
(a) Indraprastha (b) Jamia
(c) JNU (d) DU
(e) Hamdard

80. How many students of Indraprastha and Jamia together listen to Red FM?
(a) 12562 (b) 12872
(c) 14952 (d) 14272
(e) None of these

81. Which of the following channels is the most popular among the students of Hamdard and JNU?
(a) Radio Mirchi (b) Radio city
(c) Red FM (d) FM Gold
(e) Rainbow

DIRECTIONS : (Qs. 82-840) : Study the following table carefully and answer accordingly :

The different five-star hotel projects completed by different companies.

Project	No. of rooms	Cost (₹ cr)	Year of completion	Company
A	600	275	1998	P
B	320	210	1999	Q
C	250	250	1999	R
D	400	430	1998	S
E	520	310	2000	T
F	450	400	1998	U
G	500	250	2000	V

[SBI PO-2011]

82. Which project had the minimum cost per room?
(a) A (b) B
(c) G (d) E
(e) D

83. What is the investment in projects to be completed in 1998 as percentage of investment in projects to be completed in all three years?
(a) 41% (b) 52%
(c) 47% (d) 56%
(e) None of these

84. What is the **approximate** average number of rooms that would be built per crore of rupees over the three-year period?
(a) 1.25 (b) 1.70
(c) 0.90 (d) 1.40
(e) 1.55

DIRECTIONS (Qs. 85-89) : Study the following table to answer the given questions :

Each company produces two types of steel. In table I the total production (in lakh tonnes) of both types of steel together of six companies over the years is given. In table II the ratio of production of two types A and B (A : B) is given over the years.

Yrs. → Com. ↓	1997	1998	1999	2000	2001	2002	Total
BS	424	390	258	756	319	427	2574
TIS	339.5	663	812	598	663	782	3857.5
SAI	532	576	364	936	595	665	3668
MPI	620	850	876	1045	1274	1296	5961
ES	612	806	627	406	874	760	4085
LTS	840	836	776	748	384.72	816	4400.72
Total	3367.5	4121	3713	4489	4109.72	4746	24546.22

Table-I

Yrs → Com. ↓	1997	1998	1999	2000	2001	2002
BS	3 : 5	8 : 7	1 : 2	4 : 5	6 : 5	5 : 2
TIS	1 : 2.5	9 : 8	13 : 15	7 : 6	6 : 7	14 : 9
SAI	13 : 15	7 : 9	1 : 3	11 : 7	10 : 7	62 : 71
MPI	41 : 59	17 : 8	100 : 119	53 : 42	24 : 25	7 : 9
ES	2 : 7	12 : 19	6 : 5	3 : 11	10 : 13	19 : 21
LTS	13 : 11	21 : 23	5 : 3	4 : 7	1 : 1.29	7 : 5

Table-II

[SBI PO-2011]

85. The production of steel A by company MPI in 1999 is approximately what per cent of production in 2001?
(a) 51 % (b) 53%
(c) 55% (d) 60%
(e) 64%

86. In 2001, in how many companies production of A type steel is more than that of B type steel?
(a) No company (b) Two
(c) Three (d) Cannot be determined
(e) None of these

87. For how many companies did the production of steel A increase every year and the production of steel B decrease every year together from that of the previous year?
(a) No company (b) One
(c) Two (d) Three
(e) None of these

88. Production of TIS increases by 10% in 2003 and production of SAI decreases by 10% in 2003 in comparison to 2002. If the ratio of production remains the same as in 2002, find the ratio of production of A type to B type steel for both the companies together.
(a) 7026 : 6561 (b) 6561 : 8026
(c) 5226 : 4281 (d) 6066 : 6541
(e) None of these

89. If the profit ratio per unit tonne of steel A and B is 3 : 4 for company BS, what is the actual profit ratio of the company BS for the year 1998 for steel A and B?
(a) 6 : 7 (b) 3 : 4
(c) 8 : 7 (d) Can't be determined
(e) None of these

DIRECTIONS (Qs. 90 & 91) : Contribution of different sources of water to fulfil the requirement in Delhi and consumption of water for different uses by two major sources.

Total consumption of water = 720 million litres

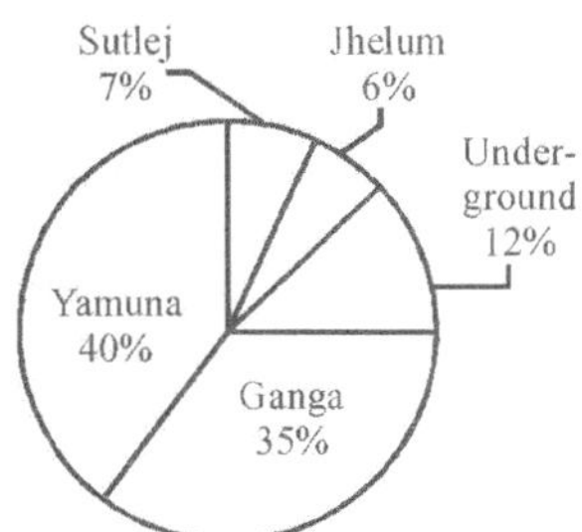

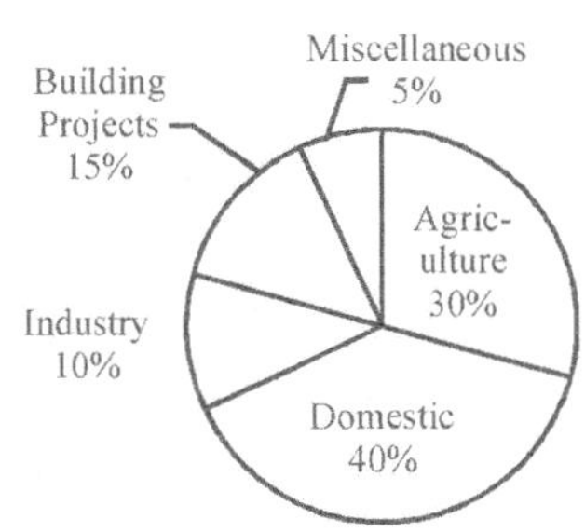

Yamuna

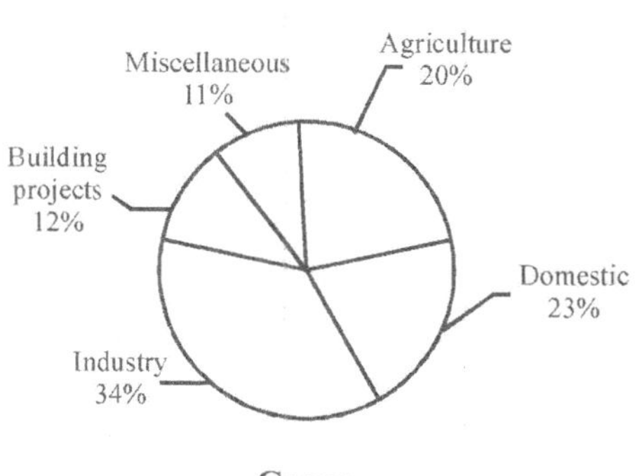

Ganga

[SBI PO-2011]

90. How many litres of Yamuna water is consumed for building projects in Delhi?
(a) 39.9m litre (b) 43.2 m litre
(c) 47.3 m litre (d) 51.9 m litre
(e) None of these

91. What is the ratio of supply of Jehlum water and underground water together to consumption of Yamuna water for domestic purposes and Ganga water for Agriculture purposes together?
(a) 17 : 23 (b) 5 : 8
(c) 17 : 22 (d) 18 : 23
(e) None of these

DIRECTIONS (Qs. 92-96): Study the following information carefully and answer the given questions.

Cost of three different fruits (in rupees per kg. in five different cities)

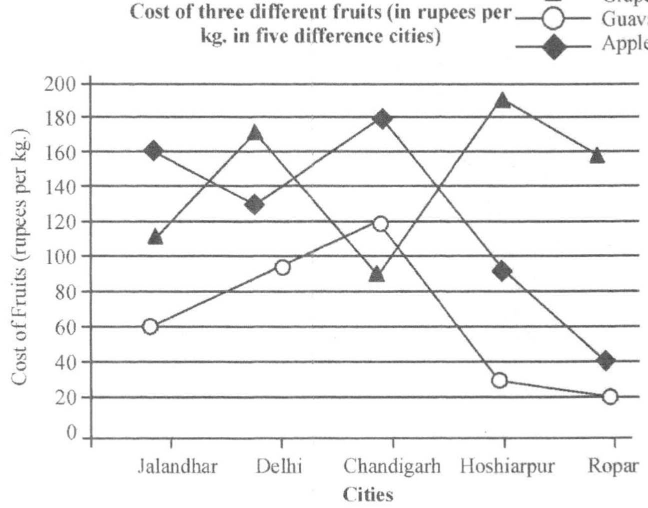

[SBI PO-2013]

92. In which city is the difference between the cost of one kg of apple and cost of one kg of guava second lowest?
(a) Jalandhar (b) Delhi
(c) Chandigarh (d) Hoshiarpur
(e) Ropar

93. Cost of one kg of guava in Jalandhar is **approximately** what percent of the cost of two kgs of grapes in Chandigarh?

(a) 66 (b) 24
(c) 28 (d) 34
(e) 58

94. What total amount will Ram pay to the shopkeeper for purchasing 3 kgs of apples and 2 kgs of guavas in Delhi?

(a) ₹ 530/- (b) ₹ 450/-
(c) ₹ 570/- (b) ₹ 620/-
(e) ₹ 490/-

95. Ravinder had to purchase 45 kgs of grapes from Hoshiarpur. Shopkeeper gave him discount of 4% per kg. What amount did he pay to the shopkeeper after the discount?

(a) ₹ 8,208/- (b) ₹ 8,104/-
(c) ₹ 8,340/- (b) ₹ 8,550/-
(e) ₹ 8,410/-

96. What is the respective ratio between the cost of one kg of apples from Ropar and the cost of one kg of grapes from Chandigarh?

(a) 3 : 2 (b) 2 : 3
(c) $2^2 : 3^2$ (d) $4^2 : 9^2$
(e) $9^2 : 4^2$

DIRECTIONS (Qs. 97-101) : Study the following information carefully and answer the given questions.

Number of students (in thousands) in two different universities in six different years
Number of students

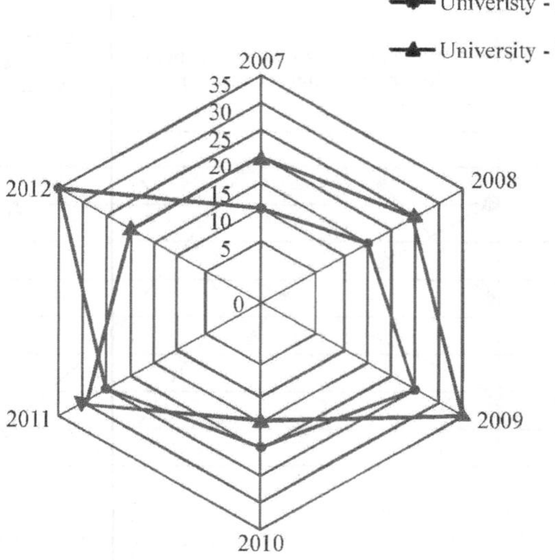

[*SBI PO-2013*]

97. What was the difference between the number of students in university - 1 in the year 2010 and the number of students in university - 2 in the year 2012?

(a) Zero (b) 5,000
(c) 15,000 (d) 10,000
(e) 1,000

98. What is the sum of the number of students in university - 1 in the year 2007 and the number of students in university - 2 in the year 2011 together?

(a) 50000 (b) 55000
(c) 45000 (d) 57000
(e) 40000

99. If 25% of the students in university - 2 in the year 2010 were females, what was the number of male students in the university -2 in the same year?

(a) 11250 (b) 12350
(c) 12500 (d) 11500
(e) 11750

100. What was the present increase in the number of students in University - I in the year 2011 as compared to the previous year?

(a) 135 (b) 15
(c) 115 (d) 25
(e) 35

101. In which year was the difference between the number of students in university - I and the number of students in university - 2 highest?

(a) 2008 (b) 2009
(c) 2010 (d) 2011
(e) 2012

DIRECTIONS (Qs. 102-106) : Study the following information carefully and answer the given questions.

Number of players participating in three different games in five different schools.

Number of players participating in three different games in five different schools

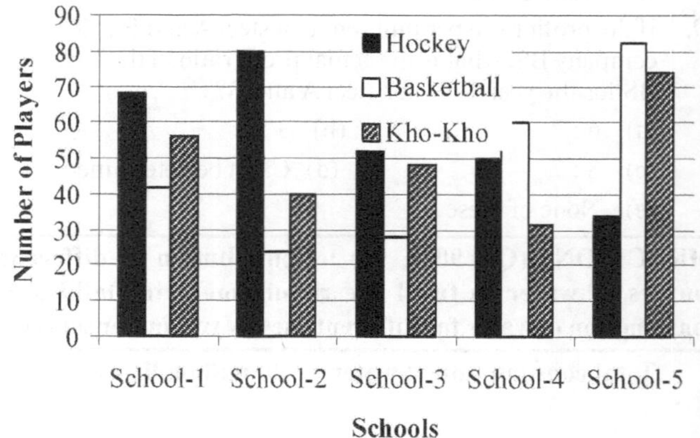

[*SBI PO-2013*]

102. What is the total number of players participating in hockey from all the five schools together?

(a) 324 (b) 288 (c) 342 (d) 284 (e) 248

103. What is the respective ratio between number of players participating in basketball from school-I and the number of players participating in Kho-Kho from school-3?

(a) 5 : 7 (b) 7 : 9 (c) 7 : 8 (d) 9 : 7 (e) 5 : 8

104. In which school is the number of players participating in hockey and basketball together second lowest?
(a) School - 1 (b) School - 2 (c) School - 3
(d) School - 4 (e) School - 5

105. Number of players participating in Kho-Kho from school-4 is what percent of number of players participating in hockey from school-2?
(a) 42 (b) 48 (c) 36 (d) 40 (e) 60

106. 25% of the number of the players particpating in hockey from School-5 are females. What is the number of the hockey players who are males in school-5?
(a) 15 (b) 18 (c) 30 (d) 21 (e) 27

DIRECTIONS (Qs. 107-111): Study the following bar-graph carefully and answer the following questions.

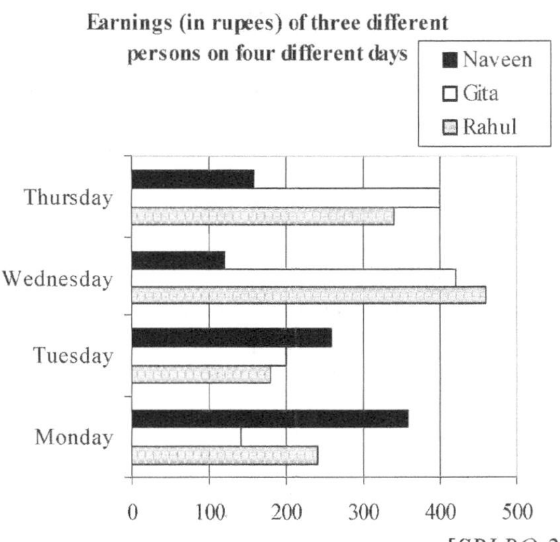

[SBI PO-2013]

107. What is Gita's average earnings over all the days together?
(a) ₹ 285/- (b) ₹ 290/- (c) ₹ 320/-
(d) ₹ 310/- (e) ₹ 315/-

108. What is the total amount earned by Rahul and Naveen together on Tuesday and Thursday together?
(a) ₹ 1,040/- (b) ₹ 1,020/- (c) ₹ 980/-
(d) ₹ 940/- (e) ₹ 860/-

109. Gita donated her earnings of Wednesday to Naveen. What was Naveen's total earnings on Wednesday after Gita's donation?
(a) ₹ 520/- (b) ₹ 550/- (c) ₹ 540/-
(d) ₹ 560/- (e) ₹ 620/-

110. What is the difference between Rahul's earnings on Monday and Gita's earnings on Tuesday?
(a) ₹ 40/- (b) ₹ 20/- (c) ₹ 50/-
(d) ₹ 30/- (e) ₹ 10/-

111. What is the respective ratio between Naveen's earnings on Monday, Wednesday and Thursday?
(a) 7 : 3 : 5 (b) 8 : 6 : 5 (c) 8 : 7 : 4
(d) 9 : 5 : 4 (e) 9 : 3 : 4

DIRECTIONS (Qs. 112-116) : Study the following pie-chart and answer the following questions.

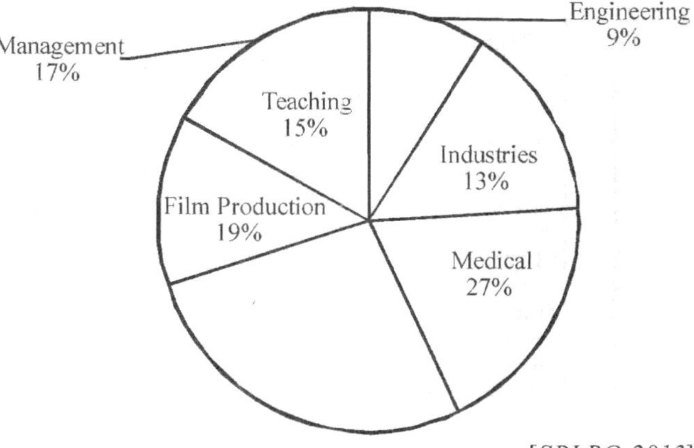

[SBI PO-2013]

112. What is the difference between the total number of employees in teaching and medical profession together and the number of employees in management profession?
(a) 6770 (b) 7700
(c) 6700 (d) 7770
(e) 7670

113. In management profession three-fourth of the number of employees are female. What is the number of male employees in management profession?
(a) 1239 (b) 1143
(c) 1156 (d) 1289
(e) 1139

114. 25% of employees from film production profession went on a strike. What is the number of employees from film production who have not participated in the strike?
(a) 3271 (b) 3819
(c) 3948 (d) 1273
(e) 1246

115. What is the total number of employees in engineering profession and industries together?
(a) 5698 (b) 5884
(c) 5687 (d) 5896
(e) 5487

116. In teaching profession if three-fifth of the teachers are not permanent, what is the number of permanent teachers in the teaching profession?
(a) 1608 (b) 1640
(c) 1764 (d) 1704
(e) 1686

DIRECTIONS (Qs. 117-121): Study the table carefully to answer the questions that follow:

Monthly Bill (in rupees) of landline phone, electricity, laundry and mobile phone paid by three different people in five different months.

Month	Monthly Bills											
	Landline Phone			Electricity			Laundry			Mobile Phone		
	Ravi	Dev	Manu	Ravi	Dev	Manu	Ravi	Dev	Manu	Ravi	Dev	Manu
March	234	190	113	145	245	315	93	323	65	144	234	345
April	124	234	321	270	220	135	151	134	35	164	221	325
May	156	432	211	86	150	98	232	442	132	143	532	332
June	87	123	124	124	150	116	213	324	184	245	134	125
July	221	104	156	235	103	131	413	532	143	324	432	543

[SBI PO-2013]

117. What is the total amount of bill paid by Dev in the month of June for all the four commodities?
(a) ₹ 608/- (b) ₹ 763/-
(c) ₹ 731/- (d) ₹ 683/-
(e) ₹ 674/-

118. What is the average electricity bill paid by Manu over all the five months together?
(a) ₹ 183/- (b) ₹ 149/-
(c) ₹ 159/- (d) ₹ 178/-
(e) ₹ 164/-

119. What is the difference between the mobile phone bill paid by Ravi in the month of May and the laundary bill paid by Dev in the month of March?
(a) ₹ 180/- (b) ₹ 176/-
(c) ₹ 190/- (d) ₹ 167/-
(e) ₹ 196/-

120. In which months respectively did Manu pay the **second highest** mobile phone bill and the **lowest** electricity bill?
(a) April and June
(b) April and May
(c) March and June
(d) March and May
(e) July and May

121. What is the respective ratio between the electricity bill paid by Manu in the month of April and the mobile phone bill paid by Ravi in the month of June?
(a) 27 : 49
(b) 27 : 65
(c) 34 : 49
(d) 135 : 184
(e) 13 : 24

DIRECTIONS (Qs. 122-126): Study the table carefully to answer the questions that follow:

Station Name	Arrival time	Departure time	Halt time (in minutes)	Distance travelled from origin (in km)	No. of Passengers boarding the train at each station
Dadar	Starting point	12.05 am	-	0 km	437
Vasai Road	12.53 am	12.56 am	3 minutes	42 km	378
Surat	4.15 am	4.20 am	5 minutes	257 km	458
Vadodara	6.05 am	6.10 am	5 minutes	386 km	239
Anand Jn.	6.43 am	6.45 am	2 minutes	422 km	290
Nadiad Jn.	7.01 am	7.03 am	2 minutes	440km	132
Ahmedabad	8.00 am	8.20 am	20 minutes	486 km	306
Bhuj	5.40 pm	Ending point	–	977 km	None

[SBI PO-2013]

122. What is the distance travelled by the train from Surat to Nadiad Jn.?
(a) 176 km (b) 188 km
(c) 183 km (d) 193 km
(e) 159 km

123. How much time does the train take to reach Ahmedabad after departing from Anand Jn. (including the halt time)?
(a) 1 hr. 59 min (b) 1 hr. 17 min.
(c) 1 hr. 47 min. (d) 1 hr. 45 min.
(e) 1 hr. 15 min.

124. What is the respective ratio between the number of passengers boarding from Vasai Road and from Ahmedabad in the train?
(a) 21 : 17 (b) 13 : 9
(c) 21 : 19 (d) 15 : 13
(e) 13 : 15

125. If halt time (stopping time) of the train at Vadodara is decreased by 2 minutes and increased by 23 minutes at Ahmedabad. At what time will the train reach Bhuj?

(a) 6.10 am (b) 6.01 pm
(c) 6.05 am (d) 6.50 pm
(e) 6.07 pm

126. Distance between which two stations is second lowest?

(a) Nadiad Jn. to Ahmedabad
(b) Anand Jn. to Nadiad Jn.
(c) Dadar to Vasai Road
(d) Anand Jn. to Vadodara
(e) Vasai Road to Surat

DIRECTIONS (Qs. 127-131): Study the table carefully to answer the questions that follow.

Month	Temperature									
	Bhuj		Sydney		Ontario		Kabul		Beijing	
	Max.	Min.	Max.	Min.	Max.	Min.	Max.	Min.	Max.	Min.
1st September	24	14	12	2	5	1	34	23	12	9
1st October	35	21	5	– 1	15	6	37	30	9	3
1st November	19	8	11	3	4	0	45	36	15	1
1st December	9	2	– 5	– 9	– 11	– 7	31	23	2	– 3
1st January	– 4	– 7	11	– 13	– 14	– 19	20	11	5	– 13

[SBI PO-2013]

127. What is the difference between the maximum temperature of Ontario on 1st November and the minimum temperature of Bhuj on 1st January?

(a) 3°C (b) 18°C
(c) 15°C (d) 9°C
(e) 11°C

128. In which month respectively the maximum temperature of Kabul is second highest and minimum temperature of Sydney is highest?

(a) 1st October and 1st January
(b) 1st October and 1st November
(c) 1st December and 1st January
(d) 1st September and 1st January
(e) 1st December and 1st September

129. In which month (on 1st day) is the difference between maximum temperature and minimum temperature of Bhuj second highest?

(a) 1st September (b) 1st October
(c) 1st November (d) 1st December
(e) 1st January

130. What is the average maximum temperature of Beijing over all the months together?

(a) 8.4°C (b) 9.6°C
(c) 7.6°C (d) 9.2°C
(e) 8.6°C

131. What is the respective ratio between the minimum temperature of Beijing on 1st September and the maximum temperature of Ontario on 1st October?

(a) 3 : 4 (b) 3 : 5
(c) 4 : 5 (d) 1 : 5
(e) 1 : 4

DIRECTIONS (Qs. 132-136): Study the pie-chart and table carefully to answer the questions that follow:

Pie-chart showing percentagewise distribution of cars in four different states
Total cars = 700
Distribution of Cars

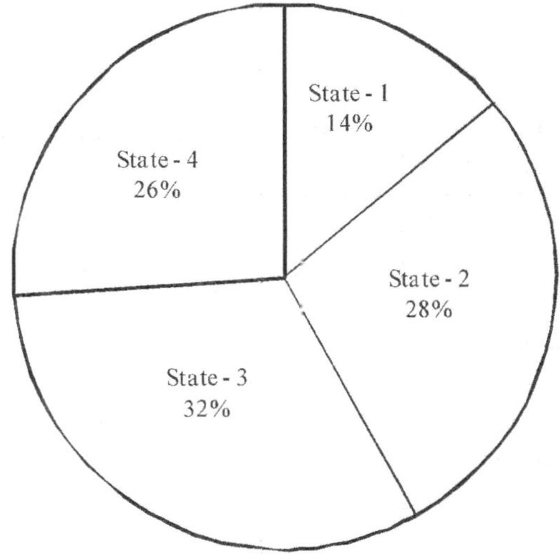

Table showing ratio between diesel and petrol engine cars which are distributed among four different states

State	Diesel Engine Cars	Petrol Engine Cars
State - 1	3	4
State - 2	5	9
State - 3	5	3
State - 4	1	1

[*SBI PO-2013*]

132. What is the difference between the number of diesel engine cars in state-2 and the number of petrol engine cars in state-4?
(a) 159 (b) 21
(c) 28 (d) 34
(e) 161

133. Number of petrol engine cars in state-3 is what percent more than the number of diesel engine cars in state-1?
(a) 100 (b) 200
(c) 300 (d) 125
(e) 225

134. If 25% of diesel engine cars in state-3 are AC and remaining cars are non-AC, what is the number of diesel engine cars in state-3 which are non-AC?
(a) 75 (b) 45
(c) 95 (d) 105
(e) 35

135. What is the difference between the total number of cars in state-3 and the number of petrol engine cars in state-2?
(a) 96 (b) 106
(c) 112 (d) 102
(e) 98

136. What is the average number of petrol engine cars in all the states together?
(a) 86.75 (b) 89.25
(c) 89.75 (d) 86.25
(e) 88.75

DIRECTIONS (Qs. 137-139): Study the information carefully to answer these questions.

In a team there are 240 members (males and females). Two-third of them are males. Fifteen percent of males are graduates. Remaining males are non-graduates. Three-fourth of the females are graduates. Remaining females are non-graduates.

[*SBI PO-2013*]

137. What is the difference between the number of females who are non-graduates and the number of males who are graduates?
(a) 2 (b) 24 (c) 4 (d) 116 (e) 36

138. What is the sum of the number of females who are graduates and the number of males who are non-graduates?
(a) 184 (b) 96 (c) 156 (d) 84 (e) 196

139. What is the ratio between the total number of males and the number of females who are non-graduates? [*SBI PO-2013*]
(a) 6 : 1 (b) 8 : 1 (c) 8 : 3 (d) 5 : 2 (e) 7 : 2

DIRECTIONS (Qs. 140-144) : ***Study the table carefully to answer the questions that follow :***

Number of people visiting six different Super-markets and the percentage of men, women and children visiting those Super-markets.

Names of the Super-markets	Total number of people	Percentage of		
		Men	Women	Children
A	34560	35	55	10
B	65900	37	43	20
C	45640	35	45	20
D	55500	41	26	33
E	42350	6	70	24
F	59650	24	62	14

[*IBPS-PO-2011*]

140. What is the respective ratio of number of women visiting Super-markets A to those visiting Super-market C?
(a) 35 : 37 (b) 245 : 316
(c) 352 : 377 (d) 1041 : 1156
(e) None of these

141. Number of men visiting Super-market D forms approximately what percent of the total number of people visiting all the Super-markets together?
(a) 11 (b) 5.5
(c) 13 (d) 9
(e) 7.5

142. Number of children visiting Super-market C forms what percent of number of children visiting super market F? (rounded off to two digits after decimal)
(a) 91.49 (b) 49.85
(c) 121.71 (d) 109.30
(e) None of these

143. What is the total number of children visiting Super-markets B and D together?
(a) 18515 (b) 28479
(c) 31495 (d) 22308
(e) None of these

144. What is the average number of women visiting all the Super-markets together?
(a) 24823.5 (b) 22388.5
(c) 26432.5 (d) 20988.5
(e) None of these

DIRECTIONS (Qs. 145-149) : ***Study the following pie-chart and answer the questions given below:***

Preferences of students among six beverages in terms of degree of angle in the pie-chart

Total no. of students = 6800

[IBPS-PO-2011]

145. The number of students who prefer Beverage C are approximately what percent of the number of students who prefer Beverage D ?

(a) 7 (b) 12
(c) 18 (d) 22
(e) 29

146. How many students prefer Beverage B and Beverage E together ?

(a) 2312 (b) 2313
(c) 2315 (d) 2318
(e) None of these

147. What is the difference between the total number of students who prefer Beverage A and C together and the total number of students who prefer Beverage D and F together?

(a) 959 (b) 955
(c) 952 (d) 954
(e) None of these

148. The number of students who prefer Beverage E and F together are what percent of the total number of students?

(a) 18 (b) 14
(c) 26 (d) 24
(e) None of these

149. What is the respective ratio between the number of students who prefer Beverage F and the number of students who prefer Beverage A ?

(a) 3 : 11 (b) 3 : 13
(c) 6 : 11 (d) 5 : 11
(e) None of these

DIRECTIONS (Qs. 150-154) : Study the table carefully to answer the questions that follow :

Percentage of Marks Obtained by Different Students in Different Subjects of MBA

Students	SUBJECTS (Maximum Marks)					
	Strategic Management (150)	Brand Management (100)	Compensation Management (150)	Consumer Behaviour (125)	Service Marketing (75)	Training & Development (50)
Anushka	66	75	88	56	56	90
Archit	82	76	84	96	92	88
Arpan	76	66	78	88	72	70
Garvita	90	88	96	76	84	86
Gunit	64	70	68	72	68	74
Pranita	48	56	50	64	64	58

[IBPS-PO-2011]

150. How many marks did Anushka get in all the subjects together?

(a) 369 (b) 463
(c) 558 (d) 496
(e) None of these

151. Who has scored the highest total marks in all the subjects together ?

(a) Archit (b) Gunit
(c) Pranita (d) Garvita
(e) Arpan

152. Marks obtained by Garvita in Brand Management are what percent of marks obtained by Archit in the same subject ? (rounded off to two digits after decimal)

(a) 86.36 (b) 101.71
(c) 115.79 (d) 133.33
(e) None of these

153. How many students have scored the highest marks in more than one subject ?

(a) Three (b) Two
(c) One (d) None
(e) None of these

154. What are the average marks obtained by all students together in Compensation Management ?

(a) 116 (b) 120
(c) 123 (d) 131
(e) None of these

DIRECTIONS (Qs. 155-159) : Study the following graph and answer the questions given below :

No. of students (in thousands) who opted for three different specializations during the given five years in a university

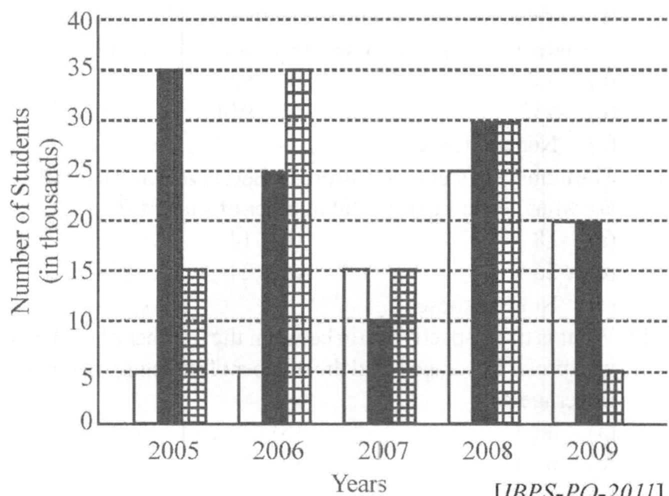

[IBPS-PO-2011]

155. The total number of students who opted for Mathematics in the years 2005 and 2008 together are approximately what percent of the total number of students who opted for all three subjects in the same years ?

(a) 38 (b) 28 (c) 42
(d) 32 (e) 48

156. Out of the total number of students who opted for the given three subjects, in the year 2009, 38% were girls. How many boys opted for Mathematics in the same year ?

(a) 1322
(b) 1332
(c) 1312
(d) Cannot be determined
(e) None of these

157. What is the respective ratio between the number of students who opted for English in the years 2006 and 2008 together and the number of students who opted for Hindi in the year 2005 and 2009 together ?

(a) 11 : 5 (b) 12 : 7 (c) 11 : 7
(d) 12 : 5 (e) None of these

158. If the total number of students in the university in the year 2007 was 455030, then , the total number of students who opted for the given three subjects were approximately what percent of the total students ?

(a) 19 (b) 9 (c) 12
(d) 5 (e) 23

159. What is the total number of students who opted for Hindi and who opted for Mathematics in the years 2006, 2007 and 2009 together ?

(a) 97000 (b) 93000 (c) 85000
(d) 96000 (e) None of these

DIRECTIONS (Qs. 160-164) : Study the following graph and table carefully and answer the questions given below :

Time Taken to Travel (in Hours) by Six Vehicles on Two Different Days

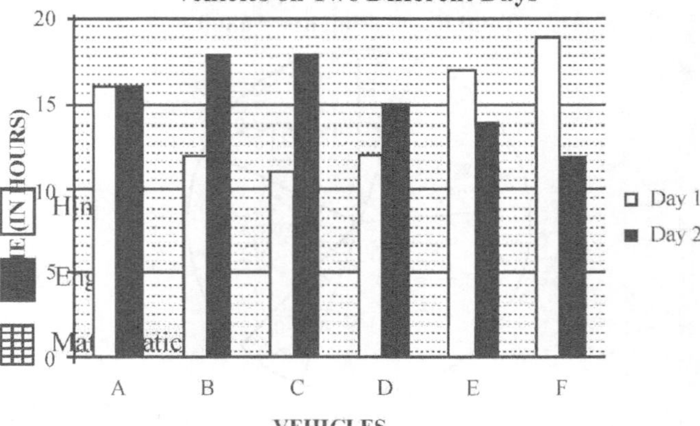

Distance Covered (in Kilometers) by Six Vehicles on Each Day

Vehicle	Day 1	Day 2
A	832	864
B	516	774
C	693	810
D	552	765
E	935	546
F	703	636

[IBPS-PO-2012]

160. Which of the following vehicles travelled at the same speed on both the days ?

(a) Vehicle A (b) Vehicle C
(c) Vehicle F (d) Vehicle B
(e) None of these

161. What was the difference between the speed of vehicle A on day 1 and the speed of vehicle C on the same day ?

(a) 7 km/hr. (b) 12 km/hr.
(c) 11 km/hr. (d) 8 km/hr.
(e) None of these

162. What was the speed of vehicle C on day 2 in terms of meters per second ?

(a) 15.3 (b) 12.8
(c) 11.5 (d) 13.8
(e) None of these

163. The distance travelled by vehicle F on day 2 was approximately what percent of the distance travelled by it on day 1 ?

(a) 80 (b) 65
(c) 85 (d) 95
(e) 90

164. What is the respective ratio between the speeds of vehicle D and vehicle E on day 2 ?

(a) 15:13 (b) 17:13
(c) 13:11 (d) 17:14
(e) None of these

DIRECTIONS (Qs. 165-169) : Study the following pie-chart and table carefully and answer the questions given below:

Percentagewise Distribution of the Number of Mobile Phones Sold by a Shopkeeper During Six Months
Total number of mobile phones sold = 45,000

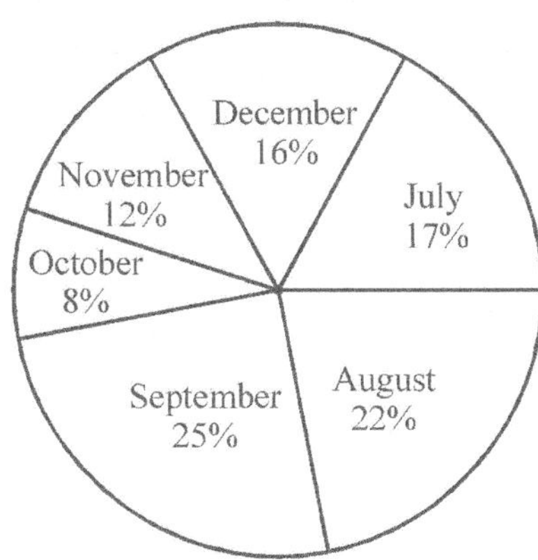

The respective ratio between the number of mobile phones sold of company A and company B during six months

Month	Ratio
July	8:7
August	4:5
September	3:2
October	7:5
November	7:8
December	7:9

[IBPS-PO-2012]

165. What is the respective ratio between the number of mobile phones sold of company B during July and those sold during December of the same company ?
(a) 119:145 (b) 116:135
(c) 119 :135 (d) 119:130
(e) None of these

166. If 35% of the mobile phones sold by company A during November were sold at a discount, how many mobile phones of company A during that month were sold without a discount?
(a) 882 (b) 1635
(c) 1638 (d) 885
(e) None of these

167. If the shopkeeper earned a profit of ₹433/- on each mobile phone sold of company B during October, what was his total profit earned on the mobile phones of that company during the same month ?
(a) ₹ 6,49,900/- (b) ₹ 6,45,900/-
(c) ₹ 6,49,400/- (d) ₹ 6,49,500/-
(e) None of these

168. The number of mobile phones sold of company A during July is approximately what percent of the number of mobile phones sold of company A during December ?
(a) 110 (b) 140
(c) 150 (d) 105
(e) 130

169. What is the total number of mobile phones sold of company B during August and September together ?
(a) 10,000 (b) 15,000
(c) 10,500 (d) 9,500
(e) None of these

DIRECTIONS (Qs. 170-174) : Study the following information and answer the questions that follow :

The premises of a bank are to be renovated. The renovation is in terms of flooring. Certain areas are to be floored either with marble or wood. All rooms/halls and pantry are rectangular. The area to be renovated comprises of a hall for customer transaction measuring 23 m by 29 m, branch manager's room measuring 13 m by 17 m, a pantry measuring 14 m by 13 m, a record keeping cum server room measuring 21m by 13 m and locker area measuring 29 m by 21 m. The total area of the bank is 2000 square meters. The cost of wooden flooring is ₹ 170/- per square meter and the cost of marble flooring is ₹ 190/- per square meter. The locker area, record keeping cum server room and pantry are to be floored with marble. The branch manager's room and the hall for customer transaction are to be floored with wood. No other area is to be renovated in terms of flooring. *[IBPS-PO-2012]*

170. What is the respective ratio of the total cost of wooden flooring to the total cost of marble flooring ?
(a) 1879 : 2527 (b) 1887 : 2386
(c) 1887 : 2527 (d) 1829 : 2527
(e) 1887 : 2351

171. If the four walls and ceiling of the branch managers room (The height of the room is 12 meters) are to be painted at the cost of ₹ 190/- per square meter, how much will be the total cost of renovation of the branch manager's room including the cost of flooring ?
(a) ₹ 1,36,800/- (b) ₹ 2,16,660/-
(c) ₹ 1,78,790/- (d) ₹ 2,11,940/-
(e) None of these

172. If the remaining area of the bank is to be carpeted at the rate of ₹ 110/- per square meter, how much will be the increment in the total cost of renovation of bank premises ?
(a) ₹ 5,820/- (b) ₹ 4,848/-
(c) ₹ 3,689/- (d) ₹ 6,890/-
(e) None of these

173. What is the percentage area of the bank that is not to be renovated ?
(a) 2.2 (b) 2.4
(c) 4.2 (d) 4.4
(e) None of these

174. What is the total cost of renovation of the hall for customer transaction and the locker area ?
(a) ₹ 2,29,100/- (b) ₹ 2,30,206/-
(c) ₹ 2,16,920/- (d) ₹ 2,42,440/-
(e) None of these

DIRECTIONS (Qs. 175-179) : Study the following table carefully and answer the question given below.

Number of People taking Fresh Loans from Different Banks over the Year and the Percentage of Defaulters Amongst them each Year

Year	Bank				
	P	Q	R	S	T
2004	27361	26345	25467	28246	30164
2005	32081	27456	32461	29435	35128
2006	25361	28637	32652	29565	32443
2007	23654	29045	32561	28314	36152
2008	36125	30467	25495	23764	35463
2009	35465	31963	27649	24356	33214
2010	34135	31974	28283	26553	31264

Approximate Percentage of Defaulters Among them

Year	Bank				
	P	Q	R	S	T
2004	12	9	15	13	19
2005	24	8	17	20	23
2006	22	13	16	21	25
2007	18	11	18	22	19
2008	12	10	13	23	18
2009	11	20	11	22	21
2010	9	21	1.2	21	23

[IBPS-PO-2013]

175. Approximately how many people taking a loan from Bank S in the year 2006 were defaulters?

(a) 6490 (b) 6210
(c) 5020 (d) 6550
(e) 5580

176. Approximately what was the difference between the number of defaulters of Bank Q in the year 2004 and 2005?

(a) 175 (b) 125
(c) 190 (d) 205
(e) 140

177. In which of the following years was the number of defaulters of Bank R, the maximum among the given years?

(a) 2005 (b) 2006
(c) 2007 (d) 2010
(e) None of these

178. In which of the following years was the difference in number of people taking loan from Bank P from the previous year the highest?

(a) 2008 (b) 2006
(c) 2007 (d) 2005
(e) None of these

179. Approximately what was the total number of defaulters of Bank T in the years 2007 and 2008 together?

(a) 14110 (b) 13620
(c) 13250 (d) 14670
(e) 15330

DIRECTIONS (Qs. 180-184) : These questions are based on the following data. The distribution of appeared and qualified aspirants in competitive examination from different States.

Total appeard aspirants = 45000 **Total qualified aspirants = 9000**

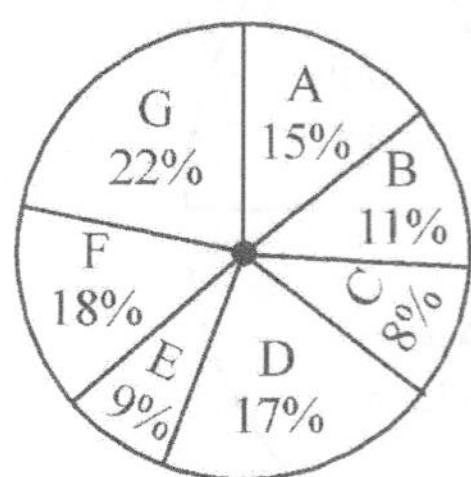

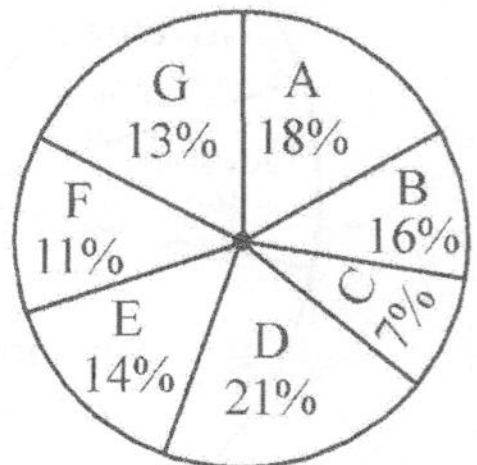

[IBPS-PO-2013]

180. What is the ratio of the number of appeared aspirants from States C and E together to that of the appeared aspirants from States A and F together?

(a) 17:33 (b) 11:13
(c) 13:27 (d) 17:27
(e) None of these

181. In which state the percentage of qualified aspirants to appeared aspirants is the least?

(a) C (b) F
(c) D (d) E
(e) G

182. What is the difference in the number of qualified aspirants in states D and G?

(a) 690 (b) 670
(c) 780 (d) 720
(e) None of these

183. What is the percentage of qualified aspirants with respect to appeared aspirants from states B and C taken together? (Rounded off to two decimal places.)

(a) 23.11 (b) 24.21
(c) 21.24 (d) 23
(e) None of these

184. What is the ratio between number of candidates qualified from States B and D together and the number of candidates appeared from States 'C' respectively?

(a) 8:37 (b) 11:12
(c) 37 : 40 (d) 7 : 37
(e) None of the above

DIRECTIONS (Qs. 185-189) : Study the following table to answer these questions.

Plan of Public Sector Under Various Plans Sector-wise Expenditure out of that total expenditure (in million)

Plan	I	II	III	IV	V	VI
Public sector plan expenditure	19600	46720	85770	2E+05	4E+05	97500
Social service	4180	7440	12960	24620	63720	14035
Education	1530	2730	5890	7860	13360	25240
Health	980	2140	2260	3370	7610	18210
Family Planning	-	20	250	2780	4920	10100
Housing and urban services	330	800	1280	2470	11500	24880
Water supply and sanitation	-	-	1060	4740	10920	39220
Social welfare and related area	1340	1750	2220	3400	15410	22700

[IBPS-PO-2013]

185. In various plans in the ratio of expenditure on public sector, which of the following graphs explain best the expenditure on water supply and sanitation?

(a) 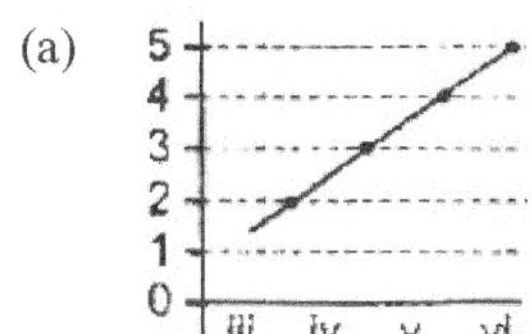

(b) 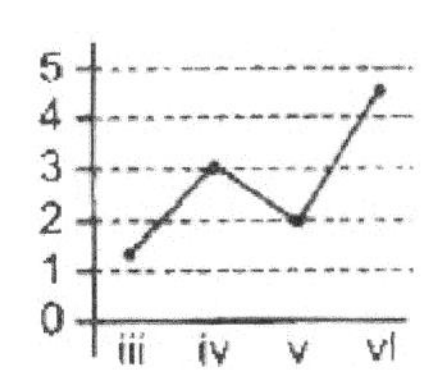

(c) 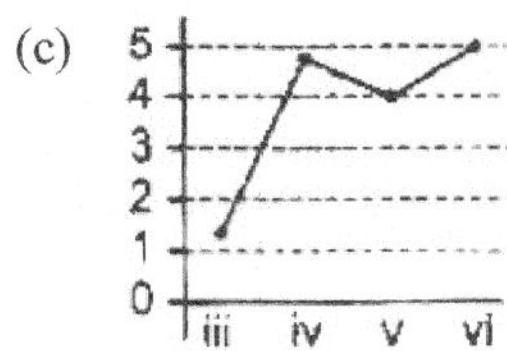

(d) 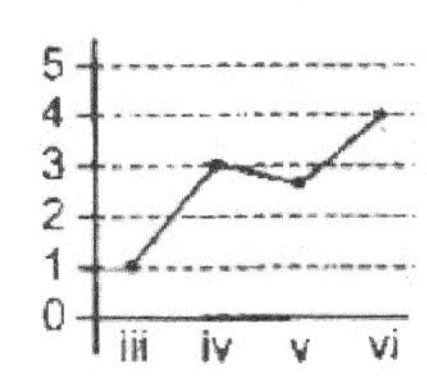

(e) None of these

186. The ratio of public sector expenditure to the expenditure on social services was highest in which plan?

(a) I (b) VI
(c) V (d) II
(e) None of these

187. In the successive plans in the ratio of public sector expenditure there was a continuous decrease in which sector?

(a) In no sector (b) Health
(c) Education (d) Social services
(e) Social welfare and related areas

188. For plan VI out of public sector expenditure, what per cent of expenditure is on Housing and Urban services?

(a) 0.35 (b) 25
(c) 25.5 (d) 2.5
(e) 20.5

189. For all the given plans, what was the difference in expenditure on education and health?

(a) ₹ 220400000 (b) ₹ 224000000
(c) ₹ 22040000000 (d) ₹ 220400000000
(e) None of these

DIRECTIONS (Qs. 190-194) : Study the following information carefully to answer the questions that follow.

There are two trains A and B. Both trains have four different types of coaches viz. General coaches, sleeper coaches, first class coaches and AC coaches. In train A, there are total 700 passengers. Train B has 30% more passengers than train A. 20% of the passengers of train A are in general coaches. One-fourth of the total number of passengers of train A are in AC coaches. 23% of the passengers of train A are in sleeper class coaches. Remaining passengers of train A are in first class coaches. Total number of passengers in AC coaches in both the trains together is 480. 30% of the number of passengers of train B is in sleeper class coaches, 10% of the total passengers of train B are in first class coaches. Remaining passengers of train B are in general class coaches.

[IBPS-PO-2013]

190. What is the ratio of the number of passengers in first class coaches of train A to the number of passengers in sleeper class coaches of train B?

(a) 13 : 7 (b) 7 : 13
(c) 32 : 39 (d) Data inadequate
(e) None of these

191. What is the total number of passengers in the general coaches of train A and the AC coaches of train B together?

(a) 449 (b) 459
(c) 435 (d) 445
(e) None of these

192. What is the difference between the number of passengers in the AC coaches of train A and total number of passengers in sleeper class coaches and first class coaches together of train B?

(a) 199 (b) 178
(c) 187 (d) 179
(e) None of these

193. Total number of passengers in general class coaches in both the trains together is approximately. What percentage of total number of passengers in train B?

(a) 35 (b) 42
(c) 45 (d) 38
(e) 31

194. If cost of per ticket of first class coach ticket is ₹ 450, what total amount will be generated from first class coaches of train A?

(a) ₹ 100080 (b) ₹ 108000
(c) ₹ 100800 (d) ₹ 10800
(e) None of these

DIRECTIONS (Qs. 195-199) : *Study the following pie chart carefully to answer the questions.*

Degree Wise Break-up of Employees Working in Various Departments of an Organization and the ratio of Men to Women

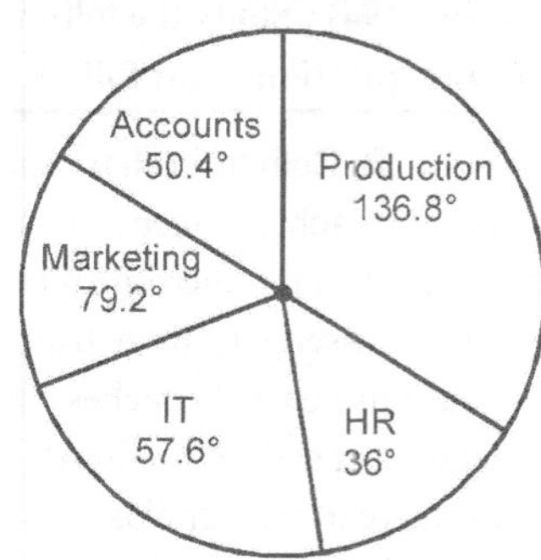

Total number of employees = 3250

Respective Ratio of Men to Women in each Department

Department	Men	Women
Production	4	1
HR	12	13
IT	7	3
Marketing	3	2
Accounts	6	7

[IBPS-PO-2013]

195. What is the number of men working in the Marketing department?
(a) 462 (b) 454
(c) 418 (d) 424
(e) None of these

196. What is the respective ratio of the number of women working in the HR department to the number of men working in the IT department?
(a) 11:12 (b) 17:29
(c) 13:28 (d) 12:35
(e) None of these

197. The number of men working in the production department of the organisation forms what per cent of the total number of employees working in that department?
(a) 88% (b) 90%
(c) 75% (d) 65%
(e) None of these

198. The number of women working in the IT department of the organization forms what per cent of the total number of employees in the organization from all departments together?
(a) 3.2% (b) 4.8%
(c) 6.3% (d) 5.6%
(e) None of these

199. What is the total number of men working in the organization?
(a) 2198 (b) 2147
(c) 2073 (d) 2236
(e) None of these

DIRECTIONS (Qs. 200-204) : The following pie-chart represents the profits earned by a certain company in seven consecutive years. Study the pie-chart carefully and answer the question.

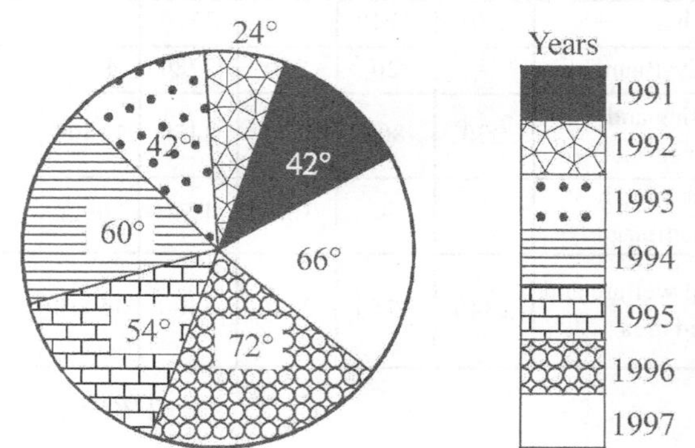

[SSC CGL-2012]

200. If the expenditure in the year 1993 was 30% more than the expenditure in the year 1991, then the income in the year 1993 exceeds the income in the year 1991 by 30% of
(a) the income in the year 1991
(b) the expenditure in the year 1993
(c) the income in the year 1993
(d) the expenditure in the year 1991

201. If x% of the total of profits earned in all the given years is same as the profit earned in the year 1994, then x is
(a) $16\frac{2}{3}$ (b) $33\frac{1}{3}$
(c) $12\frac{1}{2}$ (d) $11\frac{2}{3}$

202. The ratios of expenditures and incomes in the years 1992, 1994 and 1996 are given to be 6 : 5 : 8 and 2 : 3 : 4 respectively. The ratio of the income in the year 1996 to the total expenditure in the years 1992 and 1994 is
(a) 40 : 11 (b) 10 : 7
(c) 20 : 11 (d) 20 : 13

203. The year in which the profit is nearest to the average of the profits earned in all the given years is
(a) 1991 (b) 1995
(c) 1993 (d) 1994

204. If the income in the year 1997 was 5 times the expenditure made in the same year, then the ratio of the profit earned in the year 1991 to the expenditure in the year 1997 was
(a) 11 : 28 (b) 44 : 7
(c) 28 : 11 (d) 7 : 44

DIRECTIONS (Qs. 205-207): The following graph shows the expenditure incurred in bringing a book, by a magazine producer. Study the graph and answer question.

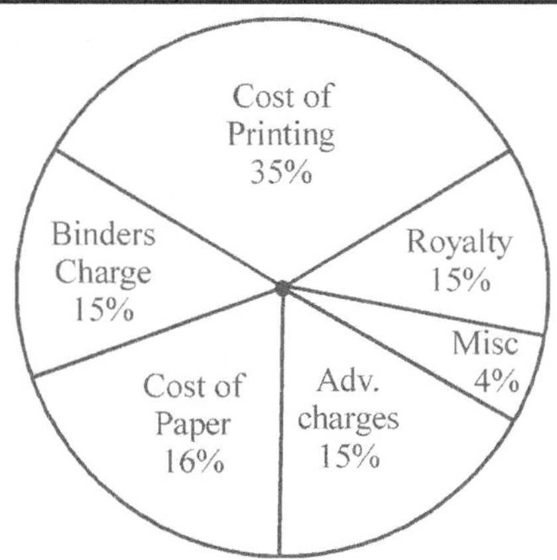

[*SSC CGL-2013*]

205. What should be the central angle of the sector for the cost of the paper ?

(a) 57.6° (b) 54.4°
(c) 56.7° (d) 54.8°

206. If the miscellaneous charges are ₹ 6,000, the cost of paper is

(a) ₹ 12,000 (b) ₹ 18,000
(c) ₹ 15,000 (d) ₹ 24,000

207. If 5500 copies are published, miscellaneous expenditures amount to ₹ 1,848, find the cost price of 1 copy.

(a) ₹ 10.40 (b) ₹ 9.40
(c) ₹ 12.40 (d) ₹ 8.40

DIRECTIONS (Qs. 208-209): The pass percentage for an examination in a school is shown in the adjoining bar diagram, for males and females separately for four years. Study the diagram and answer the question.

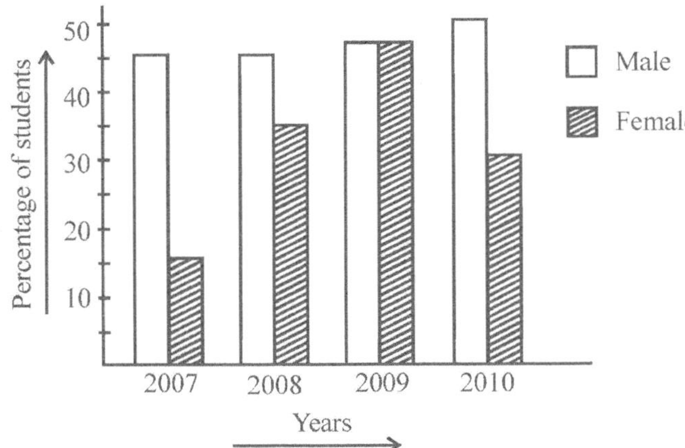

[*SSC CGL-2013*]

208. The maximum percentage of students passed in the year is

(a) 2007 (b) 2008
(c) 2009 (d) 2010

209. The year in which the difference of pass percentage between male and female is maximum, is

(a) 2010 (b) 2009
(c) 2008 (d) 2007

DIRECTIONS (Qs. 210-211): The adjacent histogram shows the average pocket money received by 60 students for a span of one month. Study the diagram and answer the question.

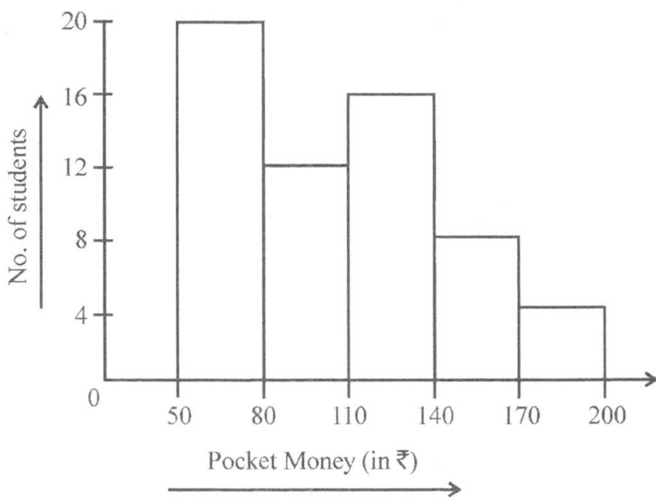

[*SSC CGL-2013*]

210. Maximum number of students received pocket money between

(a) 50 – 80 (b) 140 –170
(c) 80 – 110 (d) 110 – 140

211. The number of students who received pocket money upto ₹ 140 is

(a) 20 (b) 32
(c) 48 (d) 56

DIRECTIONS (Qs. 212-214:)Study the following table and answer

School	No. of students scoring marks less than 50%	Percentage of students scoring marks more than 50%	No. of students appeared
A	240	55	600
B	220	40	400
C	300	20	375
D	280	10	350
E	210	25	300

[*SSC CGL-2014*]

212. The ratio of the total number of students scoring marks less than 50% to that of scoring marks exactly 50% is

(a) 50 : 3 (b) 25 : 2
(c) 25 : 4 (d) 35 : 2

213. Which school has the highest number of students scoring exactly 50% marks?
(a) D (b) E
(c) B (d) A

214. The total number of students scoring 50% or more marks is
(a) 1250 (b) 875
(c) 775 (d) 675

DIRECTIONS (Qs. 215-218): Study the following graph which shows income and expenditure of a company over the years 2005-2009 and answer the questions.

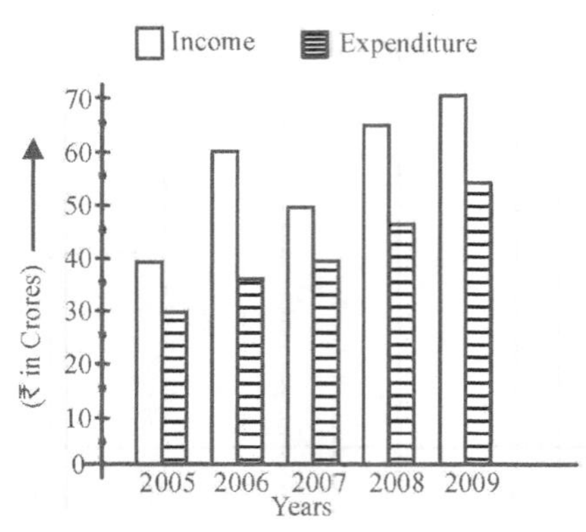

[SSC CGL-2014]

215. The difference in profit (₹ in crores) of the company during 2006 and 2007 is
(a) 10 (b) 15
(c) 20 (d) 25

216. In how many years was the income of the company less than the average income of the given years?
(a) 4 (b) 3
(c) 2 (d) 1

217. The percentage increase in expenditure fo the company from 2007 to 2008 is
(a) 20 (b) 25
(c) 30 (d) 35

218. Profit of the company was maximum in the year
(a) 2009 (b) 2008
(c) 2006 (d) 2005

Hints & Solutions

Level-I

1. **(c)** Total marks obtained by Meera
$= 100 + 80 + 50 + 90 + 90 + 60 = 470$

2. **(a)** Average marks obtained by seven students in History
$$= \frac{80+70+70+60+90+60+80}{7} = 72.86$$

3. **(b)** Only Kunal and Soni got 60% or more marks in all the subjects.

4. **(c)** Average percentage of Kunal
$$= \frac{90+70+60+90+70+70}{6} = 75\%$$

5. **(a)**

6. **(a)** Income of Company B in 2000
$$= 200 \times \frac{120}{100} = ₹\,240 \text{ crores}$$

7. **(c)** Expenditure of Company A in 2002
$$= 600 \times \frac{100}{160} = ₹\,375 \text{ crores}$$

8. **(d)** We can find out the amount of profit in 1998, we do not know the income and expenditure of A and B. therefore, option d is the correct choice.

9. **(b)** Ratio of their expenditure
$$= \frac{100}{135} \times \frac{130}{100} = 26:27$$

10. **(a)** Reqd % increase $= \frac{35-20}{20} \times 100 = 75\%$

11. **(c)** Percentage increase in crude oil price w.r.t. previous month:

March	April	May	June	July	August	Sept
23.98	20.60	2.92	0.61	15.29	9.08	12.32

12. **(b)**

13. **(a)**

14. **(e)** New crude oil price in April
$= (4800 - 223 =)\ ₹\,4577$ per metric tonne
$$\therefore \ \% \text{ increase} = \frac{4577-3980}{3980} \times 100 = 15\%$$

15. **(d)** Reqd % increase
$$= \frac{7020-3210}{3210} \times 100 \approx 120\%$$

16. **(d)** Percentage increase in the total Internet owners

1996-97	1997-98	1998-99	1999-2000	2000-01	2001-02
653.85	206.12	168.89	164.46	135.94	65.56

17. **(c)** Reqd number
$= (1550000 + 950000) - (1060000 + 450000)$
$= 990000$

18. **(b)** Reqd ratio $= 520000 : 120000 = 13 : 3$

19. **(e)** Reqd percentage increase
$$= \frac{950000-12000}{12000} \times 100 \approx 8000\%$$

20. **(b)** Reqd percentage $= \frac{12000}{230000+12000} \times 100 \approx 5\%$

21. **(a)** Production in 1996 = 5 lakh units
Production in 2002 = 32.5 lakh units
$\therefore$ The required percentage
$$= \frac{32.5}{5} \times 100 = 650$$

22. **(a)** Average production
$$= \frac{(5+7.5+10+17.5+25+27.5+32.5)\text{lakhs}}{7}$$
$$= \frac{125}{7} = 17.8 \approx 18 \text{ lakh units}$$

23. **(d)** It is obvious from the graph.

24. **(b)** Per cent increase in 1999 $= \frac{17.5-10}{10} \times 100 = 75$

Per cent increase in 2000 $= \frac{25-17.5}{17.5} \times 100$
$$= \frac{7.5 \times 100}{17.5} = 42.86$$

25. **(b)** Marks obtained by R in B, D and E
$= 54\%$ of $40 + 84\%$ of $50 + 62\%$ of 120
$= 21.60 + 42 + 74.40 = 138.00$

26. **(a)** Average % marks by all the students in C
$$= \frac{58+67+72+82+64+75}{6} = \frac{418}{6} = 69.66\%$$
Average marks = 69.66% of 80 ≈ 56

27. **(c)** Average percentage of marks obtained by all the students in 'A'
$$= \frac{80+55+74+68+75+82}{6} = \frac{434}{6} = 72\frac{1}{3}\%$$

28. **(e)** Total marks obtained by Q in all subjects
$= 55\%$ of $60 + 70\%$ of $40 + 67\%$ of $80 + 74\%$ of $50 + 88\%$ of $120 + 78\%$ of $75 = 33 + 28 + 53.60 + 37 + 105.6 + 58.50 = 315.7$

percentage of marks $= \frac{316}{425} \times 100 \approx 74\%$

29. (d) Total marks obtained by student E in all the subjects $= (75 + 88 + 62 + 72 + 80 + 68)\%$ of $120 = 534$

30. (c) Expenditure of Company C in 2000

$$= 35 \times \frac{100}{140} = ₹\,25 \text{ lakhs}$$

31. (d) Here, the percentage profits of Companies B and C in 2001 were not the same. Therefore, can't be determined is the correct choice.

32. (e) Income of Company C in 2000

$$= 32 \times \frac{140}{100} = ₹\,44.80 \text{ lakhs}$$

33. (c) Reqd ratio = 145:155 = 29 : 31

34. (a) Expenditure of Company D in 2000

$$= 31 \times \frac{100}{155} = ₹\,20 \text{ lakhs}$$

Profit = Income – Expenditure

= 31 – 20 = ₹ 11 lakhs

35. (d) Total no. of students studying in institute G = 225 + 206 + 182 + 138 + 89 = 840

$$\text{Reqd \%} = \frac{182}{840} \times 100 = 21\frac{2}{3}\%$$

36. (e) Total students in institute D = 165 + 138 + 245 + 66 + 36 = 650

$$\text{Reqd \%} = \frac{36}{650} \times 100 \approx 6\%$$

37. (a) $\text{Reqd \%} = \frac{125+96+144}{138+196+212+206} \times 100 = 48.50\%$

38. (b) $\text{Reqd \%} = \frac{164}{821} \times 100 \approx 20\%$

39. (c) Percentage of institute $A = \frac{187}{700} \times 100 = 26.71\%$

$$\% \text{ of institute } B = \frac{152}{750} \times 100 = 20.26\%$$

% of institute C = 30.12 %

% of institute D = 21.23 %

% of institute E = 23.05 %

% of institute F = 26.5 % and G = 23.67 %

Hence, maximum % is in institute C.

40. (c) $\text{Reqd \% decrease} = \frac{4-3}{4} \times 100 = 25\%$

41. (b) $\text{Reqd \%} = \frac{11}{7} \times 100 \approx 157\%$

42. (d) From the graph's slope, it is obvious that the maximum % increase is in the year 1996, i.e., 166.67%.

43. (a) Reqd difference = 58 – 31 = 2700000

44. (e) Average production for Company B $= \frac{31}{8} = 3.9$

45. (d) $\text{Reqd. \%} = \frac{1200}{14900} \times 100 \approx 8\%$

46. (c) Total no. of Officers = 2000 + 15000 + 17000 + 3500 + 14900 + 11360 + 9000 = 72760

Total no. of Clerks = 5000 + 17000 + 19500 + 20000 + 17650 + 15300 + 11000 = 105450

Reqd difference = 105450 – 72760 = 32690

47. (b) Reqd more %

$$= \frac{11000-9000}{9000} \times 100 \approx 22\%$$

48. (c) 300% more means four times the number of Clerks in Bangalore, which is in Hyderabad.

49. (d) No. of candidates in different centres: Bangalore = 3550; Mumbai = 44470; Delhi = 43910; Hyderabad = 33950, Kolkata = 35120; Lucknow = 28840; Chennai = 22245

50. (b) Let the investment of company B in 1996 be ₹ x lakhs.

$\therefore$ Investment of company B in 1997 = ₹ $\frac{7}{5}x$

Income of company B in 1997 $= \frac{9}{5} \times \frac{7}{5}x = \frac{63}{25}x$

$$\therefore \quad \text{Reqd. \%} = \frac{63}{25} \times 100 = 252\%$$

51. (d) Investment for each year is not given.

52. (e) Investment of company A in 1995 $= 21.7 \times \frac{100}{155}$

= ₹ 14 lakhs

53. (c) Let $^{i}95_{(A)} = {}^{e}96_{(B)} = ₹\,x$ lakhs

$$\therefore \quad \text{Regd. ratio} = \frac{x \times \frac{100}{155}}{x} = 20:31$$

54. (b) Income of company B in 1993

$$= 1540000 \times \frac{145}{100} = ₹\,22.33 \text{ lakhs}$$

55. (a) Strength of B in 1998 = 132 + 9 – 2 + 0 + 3 = 142

56. (e) Strength of workers in 1999

A	B	C	D	E
192	146	149	135	125

57. (c) Strength of C in 1996 = 98 + 24 + 11 = 133

Strength of E in 1997 = 125 + 2 + 4 – 3 = 128

$$\therefore \quad \text{Reqd. \%} = \frac{133}{128} \times 100 \approx 104\%$$

58. (e) Total strength of workers in all the five units in 1996 = 160 + 139 + 133 + 107 + 131 = 670.

59. (c) Increase in the strength of workers in D in 1998
$= 20 + 11 + 7 + 11 = 49$
$\therefore$ % increase $= 49/76 \times 100 \approx 64.47\%$

60. (e) Marks obtained by B = 69% of 150 + 72% of 75 + 71% of 200 + 78% of 100 + 69% of 50 + 66% of 75
$= 103.50 + 54 + 142 + 78 + 34.5 + 49.50 = 461.5$

61. (a) Average marks $= \frac{420}{6} \times \frac{75}{100} = 52.5$

62. (c) Difference $= 181.50 - 138.75 = 42.75$

63. (b) % marks obtains by A $= \frac{233 \times 100}{300} = 77.67\%$

64. (e) Since the number of students remain the same for all the states, first of all find the average percentage of passed students,

i.e., $\frac{35+43+40+38+39+28}{6} = 37.17\%$

Reqd no. = 37.17% of 5000 ≈ 1860

65. (a) Reqd no. = 22% of 18500 + 36% of 17200
$= 4070 + 6192 = 10262$

66. (b) Reqd ratio $= \frac{32\% \text{ of } 100}{45\% \text{ of } 120} = 16:27$

67. (c) Reqd ratio $= \frac{2}{3} \times \frac{26}{38} = 26:57$

68. (d) Total no. of students qualified from state E over the years $= 532 \times 7 = 3724$
But that does not lead us anywhere because we can't get the break-up.

69. (d) Total production by all the companies together
$= 648 + 725 + 679 + 498 + 840 + 580 = 3970$

Reqd % $= \frac{725}{3970} \times 100 \simeq 18\%$

70. (a) % increase/decrease for company A 1998 = 16.01%, 1999 = 14.69%, 2000 = 13.27%, 2001 = 15.53%, 2002 = 9.78%

71. (e) Total production of E
$= 415 + 680 + 840 + 689 + 780 + 637 = 4041$
Total production of F
$= 632 + 775 + 580 + 720 + 670 + 746 = 4123$
Reqd difference $= 4123 - 4041 = 82$ lakh tonnes

72. (c) Avg $= \frac{565+378+483+526+680+775}{6}$

$\simeq 570$ lakh tonnes

73. (b)

74. (a) Percentage processing of wool in the month of March by different companies.
Polar = 23.33%. Shephered = 19.51%,
Kiwi = 20.41%, Warmwear = 20.18%
Comfy = 21.84%

75. (b) Reqd % $\frac{1100}{1000} \times 100 = 110\%$

76. (b) If we see the table, we find that only Shepherd shows less value in February in comparison to the month of April So, it gives the maximum ratio.

77. (d) Shephered shows the lowest processing in the month of February and March.

78. (e) Reqd % $= \frac{4900}{4100} \times 100 \approx 120\%$

79. (e) Sale of Pep-up was the maximum in the year 1989.

80. (a) Avg annual sale of Dew-drop

$= \frac{10+15+25+15+30+25}{6} = 20$ lakhs

Avg. annual sale of Cool-sip

$= \frac{25+7+20+20+25+30}{6} = 21.16$ lakhs

Avg. annual sale of Pep-up

$\frac{30+35+30+25+20+20}{6} = 26.66$ lakhs

81. (c) Reqd % $= \frac{25-20}{20} \times 100 = 25\%$

82. (e) Reqd no. $= 30 - 20 = 1000000$

83. (c) Reqd % drop $= \frac{35-30}{35} \times 100 \approx 14\%$

84. (a) Total no. of students studying in all schools in 1992
$= (1025 + 230 + 190 + 950 + 350 + 225 + 1100 + 320 + 300 + 1500 + 340 + 300 + 1450 + 250 + 280) - (120 + 110 + 150 + 115 + 130 + 150 + 150 + 160 + 125 + 130)$
$= 8810 - 1340 = 7470$

$\therefore$ Average $= \frac{7470}{5} = 1494$

85. (c) Number of students studying in school B in 1994
$= 950 + (350 - 150) + (225 - 115) + (185 - 110) + (200 - 90)$
$= 950 + 200 + 110 + 75 + 110 = 1445$

86. (e) Number of students leaving school 'C'
from 1990 to 1995 $= 130 + 150 + 125 + 140 + 180 = 725$
Number of students admitted during the period
$= 1100 + 320 + 300 + 260 + 240 + 310 = 2530$

$\therefore$ Required percentage $= \frac{725}{2530} \times 100 \approx 29\%$

87. (d) Required difference $= (340 + 300 + 295 + 320 + 360) - (350 + 225 + 185 + 200 + 240) = 1615 - 1200 = 415$

88. **(b)** Increase in no. of students in school A
$= (230 - 120) + (190 - 110) + (245 - 100) + (280 - 150) + (250 - 130), = 585$
$\therefore$ % increase from 1990 (1025) to 1995

$$= \frac{585}{1025} \times 100 = 57.07\%$$

Similarly, we can calculate for other schools.
Percentage increases in all schools are given in the following

A	B	C	D	E
57.07%	64.73%	64.09%	61.33%	62.41%

89. **(a)** No. of females above poverty line in state A

$$= 3000 \times (100 - 12)\% \times \frac{3}{7} \approx 1150$$

90. **(d)** Since, we cannot find the population of states C and D separately, we can't find the required value.

91. **(e)** Population of state A below poverty line

$$= 3000 \times \frac{5}{3} = 5000$$

$\therefore$ Total population of state $A = \frac{5000}{12} \times 100$

and the population of state E below poverty line

$$= 6000 \times \frac{11}{6} = 11000$$

Total population of state $E = \frac{11000}{10} \times 100$

Required ratio $= \frac{5}{12} \times \frac{10}{11} = \frac{25}{66}$

92. **(c)** Total population of state $B = 500 \left(\frac{12}{5}\right)\left(\frac{100}{15}\right) = 8000$

93. **(b)** Population of state $E = 19800 \left(\frac{5}{2}\right)\left(\frac{100}{100 - 10}\right) = 55000$

$\therefore$ Population of males below poverty line

$$= 55000 \left(\frac{10}{100}\right)\left(\frac{6}{11}\right) = 3000$$

94-98: As the formula is given in the question, we should follow it to find the solution.

94. **(e)** The percentage profit of company 'A' in 1998 = 50%
Income, = 142500 (given)

$$\text{Expenditure} = 142500 \left(\frac{100}{100 + 50}\right) = 95000$$

95. **(d)** $E_{B99} = \frac{90}{100} E_{B98}$ (given)

$$I_{B99} = \frac{90}{100} \times \frac{140}{100} \times \frac{100}{135} (I_{B98})$$

$$I_{B99} = \frac{280}{300} I_{B98}$$

$$= \frac{280}{3}\% \text{ of } I_{B98} = 93\frac{1}{3}\% \text{ of } I_{B98}$$

96. **(c)** E_{A97} = ₹ 70 lakh

$$\Rightarrow I_{A97} = 70\left(\frac{125}{100}\right) = ₹\ 87.5 \text{ lakh}$$

$I_{A97} = E_{A98}$ = ₹ 87.5 lakh

$$\therefore I_{A98} = 87.5\left(\frac{150}{100}\right) = ₹\ 131.25 \text{ lakh}$$

$\therefore$ reqd value $= I_{A97} + I_{A98}$
$= 87.5 + 131.25 =$ ₹ 218.75 lakh

97. **(a)** We have to find $I_{B96} : I_{B97}$
Given $E_{B96} = 5x$ and $F_{B97} = 7x$

$$I_{B96} = 5x\left(\frac{140}{100}\right) \text{ and } I_{B97} = 7x\left(\frac{130}{100}\right)$$

$$\therefore I_{B96} : I_{B97} = \frac{5x}{7x}\left(\frac{140}{130}\right) = 10 : 13$$

98. **(d)**

99. **(c)** Total marks obtained by A in Psychology and Statistics together

$$\left(\frac{72 \times 150}{100} + \frac{88 \times 125}{100}\right) = 108 + 110 = 218$$

Total marks obtained by F in these two subjects

$$= \left(\frac{84 \times 150}{100} + \frac{82 \times 125}{100}\right) = 126 + 102.5$$

$= 228.5$

$\therefore$ Difference $= (228.5 - 218 =)\ 10.5$

100. **(d)** Average % of marks obtained by six students in Sociology

$$= \frac{(65 + 58 + 73 + 65 + 55 + 70)}{6} = \frac{386}{6}$$

$$= 64\frac{1}{3}$$

101. **(b)** Average marks obtained by 6 students in Philosophy out of 75

$$= \left\{\frac{(65 + 70 + 57 + 61 + 76 + 78 =)\ 407}{6} \times \frac{75}{100}\right\}$$

$= 50.875$

102. **(a)** Required answer

$$= \left(\frac{72 \times 150}{100} + \frac{85 \times 80}{100} + \frac{72 \times 60}{100}\right)$$

$= 108 + 68 + 43.2 = 219.2$

103. (e) Percentage of marks obtained by C in all the subjects together

$$\frac{(63\%\text{ of }150)+(73\%\text{ of }120)+...+(65\%\text{ of }60)}{150+120+....+60}\times 100$$

$$\approx \frac{404}{610}\times 100 \approx 66\%$$

104. (d) $\frac{(110+60+110+100+105+85=)570}{6}$

$= 95$ lakh tons

105. (c) Average production of units A, B and C in 2001 [use white bars]

$$\frac{(90+75+100=)\,265}{3}$$

$\therefore$ Average production of units D, E & F in 2002

$$=\frac{(100+105+85=)290}{3}$$

Required answer $=\frac{265\times 3}{3\times 290}\times 100 = 91.38$

106. (e) Total production by unit B in 2001 and 2002 together $= (75 + 60 =)$ 135 lakh tons

Total production by unit C in 2001 and 2002 together $= (100 + 110 =)$ 210 lakh tons

$\therefore$ Required ratio $= (135 : 210 =)\ 9 : 14$

107. (a) Total production by unit F in year 2001and 2002 together

$= (70 + 85 =)$ 155 lakh tons

Total production by unit D in year 2001 and 2002 together

$= (95 + 100 =)$ 195 lakh tons

$\therefore$ Required percentage $\left(\frac{155}{195}\times 100\right) = 79.487 \approx 79.49$

108. (b) Required total production $= (100 + 110 + 95 + 100 + 85 + 105 =)$ 595 lakh tons

109. (b) $E_{98} : E_{2000} = I_{98}\left(\frac{100}{145}\right) : E_{2000}$

$= 100 : 145\ (\because\ I_{98} = E_{2000})$

$= 20 : 29$

110. (c) According to the given information,

$$\frac{\text{Income of company A in 1999}}{\text{Income of company B in 1999}} = \frac{2}{3}$$

$\Rightarrow$ Income of company A in 1999

$$=\frac{2}{3}\times 18.6$$

$I_{A99} = 12.4$ lakhs

$\Rightarrow\ E_{A99} = 12.4\left(\frac{100}{160}\right) = 7.75$ lakhs

111. (e) Suppose expenditures of A and B in the year 2001 are $4x$ and $5x$ respectively. Then

$4x + 5x = 18$ lakhs

$\therefore\ x = 2$ lakhs; $4x = 8$ lakhs;

$5x = 10$ lakhs

$$I_B = 10\left(\frac{140}{100}\right) = 14 \text{ lakhs}$$

112. (a) $I_{A99} = E_{B2000}$(given)

Now,

$E_{A99} : I_{B2000}$

$$= I_{A99}\left(\frac{100}{160}\right) : E_{B2000}\left(\frac{165}{100}\right)$$

$= 100 \times 100 : 160 \times 165 = 25 : 66$

113. (d) We can't find the expenditure of company A in the given years separately. So we can't find the profit of the company.

114. (c) Marks obtained by R in different subjects

A	B	C	D	E	F
49. 50	112.5	79	44	108	49. 50

Total marks obtained by R out of 600 marks

$= 49.50 + 112.50 + 79 + 44 + 108 + 49.50 = 442.5$

$\therefore$ Required % marks

$$=\frac{442.5\times 100}{600} = 73.75\%$$

115. (b) Marks of P and T in the subjects 'B, 'D' and.'E'

Sub → Students ↓	B	D	E	Total
P	102	46	133.5	281.5
T	112.5	34	103.5	250

Hence required difference

$= 281.5 - 250 = 31.5$

116. (a) Total marks obtained by all the students in subject B

$$=\frac{150\times(68+72+75+62+75+80+68)}{100}$$

$\therefore$ Required average $=\frac{750}{7} = 107.14$

117. (c) $\frac{\text{Total in C} + \text{Total in D}}{1400}\times 100$

$$=\frac{547+565}{14} = 79.43\%$$

118. (d) Required total marks

$$=\frac{75\times(82+70+66+74+78+80+72)}{100}$$

$$=\frac{75\times 522}{100} = 391.5$$

119. (c) Investment per cent of unit C as a fraction of the total investment of all the units in

1996	1997	1998	1999	2000	2001
19.26%	16.87%	16.71%	15.94%	16.76%	16.65%

120. (a) Investment percent in 1997 as a fraction of the total investment in all the given years together of each unit is as follows:

	A	B	C	D	E	F
Investment	132	140	137	125	128	150
Out of	738	824	827	810	817	875
In per cent	17.89%	16.99%	16.57%	15.43%	15.67%	17.14%

121. (e) Required % increase

$$= \frac{(145-98)}{98} \times 100 = 47.96\%$$

122. (b) Investment by units A, B and C in 1998
= 125 + 145 + 138 = 408 crores
Investment by units A, B and C in1999
= 116 + 148 + 136
= 400 crores
Thus, required difference
= 408 – 400 = 8 crores (more)

123. (e) Total investment of units A, B and C in the year 1998
= 125 + 145 + 138 = 408 crores
Investment by the units D, E and F in the years 1999
= 145 + 152 + 156 = 453 crores

Hence required ratio $= \frac{408}{453} = 136 : 151$

124. (b) Total number of the malnourished children in year
(i) 1991 = (10.0 + 1.4 + 0.1)% of 1048000
= 11.5% of 1048000
(ii) 1986 = (12.1 + 3.0 + 0.8)% of 1048000
= 15.9% of 1048000
Hence, the required difference
= (15.9 – 11.5)% of 1048000
= 46112

125. (b)

Year	Percentage of the malnourished
1984	16.1
1985	15.5
1986	15.9
1987	12.9
1989	12.9
1990	12.2
1991	11.5
1992	9.9
1993	8.8

126. (e) Reject (a) and (d) because we see that the percentage of high malnourished cases increases to 0.8 from 0.7 in the year 1986.
Reject (b) because we see that the percentage of moderate malnourished cases increased from 2.7 to 3.0 in the year 1986.
Reject (c) because no such fall is witnesed during the year 1985 to 1986.
Hence, select (e) by elimination.

127. (d) Number of required children
= 0.5% 1071000

$$= \frac{5 \times 1071000}{1000} = 5355$$

128. (b) The required malnourished children in 1993
= (7.8 + 0.9 + 0.1)% of 1161000
= 8.8% of 1161000
= 102168

129. (d) Difference of production of C in 1991and A in 1996 = 5,00,000 tonnes.

130. (a) Percentage increase of A from 1992 to 1993

$$\frac{55-40}{40} \times 100 = 37.5\%$$

131. (b) Percentage rise/fall in production for B

1992	1993	1994	1995	1996
9%	–16.6%	10%	– 9%	10%

Here, the maximum difference is from 1992 to 1993, which is 10. And the second nearest to it is fall or rise of 5. So, undoubtedly the answer is 1993.

132. (e) Percentage production $= \frac{120}{90} \times 100 = 133.3\%$

133. (c) Average production of A = 50
Average production of B = 54.17
Average production of C = 50
Difference of production = 54.17 – 50 = 4.17

134. (d) Distance to be travelled by each type of vehicle

$$= \frac{15}{3} = 5 \text{ km}$$

Since, to travel 5 km by vehicle A, he will pay ₹ 9 for 4 km and for the next 1 km he will have to pay

$$₹ = \frac{13.5-9.00}{(7-4)} \times 1.$$

Similarly, for other cases.

$$\text{Fare by A} = ₹\ 9 + \frac{13.50-9}{7-4} = 9 + 1.50 = ₹\ 10.50$$

Fare by B = $14.50 + \frac{24.25 - 14.50}{7 - 4}$

$= 14.50 + 3.25 = 17.75$

Fare by C = $19 + \frac{31 - 19}{3} = 19 + 4 = 23$

Total fare = 10.50 + 17.75 + 23 = ₹ 51.25

135. (a) Fare byA = $9 + \frac{4.50}{3} \times 2 =$ ₹ 12

Fare by B = $24.25 + \frac{33.25 - 24.25}{3} \times 2 =$ ₹ 30.25

Total fare = 30.25 + 12 = ₹ 42.25

136. (b) Fare for 8 km by A = $13.50 + \frac{17.25 - 13.50}{10 - 7}$

$= 13.50 + \frac{3.75}{3} =$ ₹ 14.75

Fare by B = $24.25 + \frac{33.25 - 24.25}{3} =$ ₹ 27.25

Difference = 27.25 – 14.75 = ₹ 12.50

137. (e) Fare by B for 5 km = 14.50 + 3.25 = ₹ 17.75

Fare by A for 8 km = $13.50 + \frac{17.25 - 13.50}{3}$

= ₹ 14.75

Fare by C for 5 km = $19 + \frac{31 - 19}{3} =$ ₹ 23

Total fare = 17.75 + 14.75 + 23 = 55.50

138. (b) Fare for l4th km by C = $\frac{56.50 - 41.50}{15 - 10} =$ ₹ 3

Fare for 9th km by B = $\frac{33.25 - 24.25}{10 - 7} =$ ₹ 3

139. (a) Total production of
A= 465 + 396 + 524 + 630 + 408 + 650
= 3073 lakh tonnes
C = 694 + 528 + 492 + 575 + 550 + 495
= 3334 lakh tonnes
Hence, required difference = 3334 – 3073
= 261 lakh tonnes

140. (e) Total production of all companies in
1996 = 396 + 482 + 528 + 602 + 551+ 635
= 3194 lakh tonnes
1997 = 524 + 536 + 492 + 387 + 412 + 605
= 2956 lakh tonnes
Hence, required % decrease

$= \frac{3194 - 2956}{3194} \times 100 = 7.451\%$

= 7.5%

141. (d) Percentage rise/fall from the previous year in production of company F are as follows :

1996	1997	1998	1999	2000
24.24%	– 4.72%	– 0.82%	– 19.16%	8.24%

You can give the answer without doing any detailed work. A cursory look will help you detect that the required year is either 1996 or 1999. Again, a step further you get that the rise in production in the year 1996 is more than 20% while the production in 1999 is less than 20%.

142. (b) Production of companies A and B together in 1997 = 524 + 536 = 1050 lakh tonnes
Production of companies E and F together in 1998 = 518 + 600 =1118 lakh tonnes

Hence, required % = $\frac{1060}{1118} \times 100 = 94.81\% \approx = 95\%$

143. (c) Average production of B in the given years (in lakh tonnes)

$= \frac{372 + 482 + 536 + 480 + 512 + 580}{6}$

$= \frac{2962}{6} = 493.66$

Similarly, average production of E in the given years

$= \frac{498 + 551 + 412 + 518 + 647 + 610}{6}$

$= \frac{3236}{6} = 539.33$

Hence, required difference = 539.33 – 493.66 = 45.67 lakh tons

144. (a) We have given profit/loss = Income – Expenditure

Therefore, profit in each of the given years is as follows:

Year	96	97	98	99	00	01
Income	350	450	450	500	400	550
Exp.	250	300	400	350	450	450
Profit	100	150	50	150	– 50	100
in crore ₹						

∴ Average profit

$= \frac{100 + 150 + 50 + 150 - 50 + 100}{6}$

= ₹ 83.33 crore.

145. (b) Profit earned during the year 1999 = ₹ 150 cr
Expenditure during the year 1999 = ₹ 350 cr
Hence, % profit earned in the year 1999

$= \frac{150 \times 100}{350} = 42.85\% \approx 43\%$

146. (d) Per cent increase/decrease in income from the previous year:

1997	1998	1999	2000	2001
28.57%	0%	11.11%	–20%	37.5%

Note : – ve sign indicates fall in income.
you can solve this question merely with the help of the graph.

147. (b) Required % increase

$$= \frac{(400-300)}{300} \times 100 = 33\frac{1}{3}\%$$

148. (c) Average income

$$= \frac{350+450+450+500+400+550}{6} = \frac{2700}{6}$$

= ₹450 crore

149. (b) No.of students who got 0-19 marks in maths = 31
No. of students who got 20-39 marks in Maths = 22; therefore,
no. of students who got less than 40% marks in Maths = 31 + 22 = 53
Hence, no. of students who passed in Maths = 160 – 53 = 107.

150. (a)

Marks	60-79	80-100
Average of three subjects	35	5

Hence, required no. of students = 35 + 5 = 40

151. (a)

	Marks			
Subject ↓	40-59	60-79	80-100	40-100
Hindi	79	30	08	117
English	65	42	02	109
Maths	34	45	28	107

107 is the lowest among 117,109 and 107. Hence, required no. of students = 107.

152. (e) 65 + 42 + 02 = 109

153. (d)

Subject ↓	Marks	
	0-19	No. of those students who obtained 20 or more marks (20-100)
Hindi	12	148
English	21	139
Maths	31	129

Mere this information is not sufficient to obtain the exact number of students who got 20 or more marks in at least one paper.

154. (e) The difference between the white-coloured cars sold is the minimum in B type model.

155. (a) Blue (E + D) = 37 + 43 = 80 = White (B)

156. (e) Reqd. difference = (50 – 34) × 1000 = 16,000

157. (c) Reqd. percentage $= \frac{173}{192} \times 100 \approx 90\%$

158. (a) Colour-model combinations of car in Metro M

Silver-F	White-C	Blue-B	Red-F	Black-F
52	90	60	42	55

159. (e)

160. (d) Our intelligent observation says that the required year can't be 1993, 1994, 1995. Why? Because see the following conclusions:

$$\% \text{ passed to appear} = \frac{\text{Passed}}{\text{Appeared}} \times 100$$

% of passed to appear is least when $\frac{\text{Passed}}{\text{Appeared}}$ is the least

or, $\frac{\text{Passed}}{\text{Appeared}}$ is the most. Now, we do the further calculations mentally. See the following conclusions:

For 1990: $\frac{7894}{2513} \Rightarrow$ Quotient = 3 & Remainder ≈ 300

For 1991: $\frac{8562}{2933} \Rightarrow Q = 3 \;\&\; R \approx 400$

For 1992 : $\frac{8139}{2468} \Rightarrow Q = 3 \;\&\; R \approx 800$

Similarly, for 1993, 1994, 1995, Q is 2.
So, 1992 gives the highest value.

Note: When R is close for close or three years you should go for further calculations and find the exact possible values. But larger difference in R for almost equal divisors gives the option to stop our further calculations, as happened in this case.

161. (a) $\frac{8562-8139}{8562}\times 100=\frac{423}{8562}\times 100\approx\frac{42}{84}\times 10=5$

162. (a) We don't need to calculate the values for each year. Follow as:

For rural area: 35% of 5032 $\approx 35\times 50\approx 1750\approx 1798$

For Semi-urban area : 35% of 9500 $\approx 35\times 95\approx 3300$

Which can't be approximated to 3500.

For State capitals: $35\times 85\approx 3000$

For Metropolises: $35\times 110\approx 3850$

163. (c) 1798 + 2513 = 4311

164. (a) Average marks of Q in 1st periodical

$=\frac{30+25+33+42+30}{5}=\frac{160}{5}=32$

165. (c) Total marks of T in Science

= 44 + 36 + 40 + 30 + 40 = 190

166. (b) Average percentage of marks obtained by P in Marks

$=\frac{80+60+90+40+70}{5}=68\%$

= percentage of marks obtained by student R in Geography.

167. (c) Our observation finds two options which are close to each other. These are History & Geography. When we find the actual value, we find that our answer is History.

Note: You can decide the answer with totalling only. You don't need to calculate the percentage value.

168. (b)

169. (a) Production of C type cars in 1996

= (70 – 40)% of 4,50,000 = 30% of 4,50,000 = 1,35,000

Production of C type cars in 1997

= (65 – 40)% of 5,20,000

= 25% of 5,20,000 = 1,30,000

∴ Required difference = 5,000

170. (e) Production of E type cars in 1996

= (100 – 80)% of 4,50,000

= 20% of 4,50,000 = 90,000

And in 1997 = 10% of 5,20,000 = 52,000

∴ Total production = 90,000 + 52,000 = 1,42,000

∴ Required no. of cars = 15% of 1,42,000 = 21,300

171. (b) Production of A type cars in 1997 = production of A type cars in 1996 (given) = (100 – 85 =) 15% of 4,50,000 = 67,500

∴ Reqd percentage = $\frac{67,500}{5,20,000}\times 100\approx 13$

172. (c) Clearly, by visual inspection D is the desired option.

173. (c) Percentage production of B type cars in 1997 = that in 1996 (given)

= (40 – 15 =) 25% of 5,20,000 = 1,30,000

174. (e) Average imports made by company A

$\frac{30+50+60+40+70+60+75}{7}=\frac{385}{7}=55$

In none of the given years the imports is exactly equal to 55 (crores). Hence, the answer is (e).

175. (d) By visual inspection it is clear that 1992 is the desired year (as the distance between two points is the maximum in 1992.)

176. (a) By mental observation $\left(\text{as } 50=\frac{40+60}{2}\right)$, 1992 only is the desired year. You don't need any calculation. See the year where the point of A lies exactly in the middle of points of B and C.

177. (b) Reqd percentage increase $=\frac{50-40}{40}\times 100=25\%$

178. (c) The total imports (in crores) made by all the three companies together: From the heights of the points we observe that the total heights of three points is the maximum either in 1995 or 1997. If you observe carefully, our clear answer is 1995, but to be sure we find actual values for the two years.

In 1995 = 70 + 80 + 85 = 235.

In 1997 = 75 + 70 + 85 = 230.

Clearly, 1995 is the desired year.

179. (a) UP (Qua/App)

Arts	Commerce	Science	Engg.	Agr.
0.34	0.39	0.4	0.42	0.42

Alternative Approach: $\frac{\text{Qual.}}{\text{App.}}$ should be the least.

$\Rightarrow \frac{\text{App.}}{\text{Qual.}}$ should be the maximum.

Now, for Arts, if we divide (4980 ≈) 5000 by (1690 ≈) 1700 we find the value of quotient near about 3. But in other cases the quotient is just more than 2. So, our answer is Arts.

180. (b)

181. (e) Percentage of students qualified in commerce

A.P.	U.P.	Kerala	Orissa	M.P.
33.9	38.7	58.2	45.8	28.5

182. (d) Qualifying percentage of UP = $\frac{9280}{23880}\times 100=38.86$

Qualifying percentage of MP = $\frac{8625}{26750}\times 100=32.24$

Ratio = 38 : 32 = 19 : 16

183. (d) Qualifying percentage for Science

A.P.	U.P.	W.B.	Kerala	Orissa	M.P.
39.9	40.5	37.7	58.8	43.3	43.8

184. (d) Required percentage

$=\frac{35+40+45+35+35}{45+50}\times 100=\frac{190}{95}\times 100=200$

185. (b) Average production by $B = \frac{45+35+40}{3} = 40$

Average production by $C = \frac{25+35+45}{3} = 35$

Ratio = (40 : 35) 8 : 7

186. (c) Maximum difference is 5 lakh tonnes for three companies C, D & E. So, our answer should be the company for which the production is least in 1996. Because to calculate the % increase or decrease our denominator is the production in 1996.

187. (a) Percentage drop $= \frac{50-35}{50} \times 100 = 30\%$

188. (e) You should not calculate the values to get answer. You can decide by mere visual observation.

189. (d) Total no. of students who play cricket
= 38 + 40 + 12 + 17 + 25 + 18 + 20 = 170

Reqd % $= \frac{25}{170} \times 100 \approx 15\%$

190. (d) Reqd ratio = 27 : 18 = 3 : 2

191. (e)

192. (e) Total Class X students who play different games = 115

Reqd % $= \frac{21}{115} \times 100 \approx 18\%$

193. (e) Basketball and Badminton are the two games which satisfy the conditions.

194. (a) Slope of the line between 1996 and 1997 is the highest. Therefore, in 1997 there is maximum per cent increase in exports as compared to the preceding year.

195. (a) Reqd difference = 600 – 450
= ₹ 150 cr

196. (d) Reqd % increase $= \frac{950-200}{200} \times 100 = 375\%$

197. (b) Reqd % $= \frac{600}{450} \times 100 = \approx 135\%$

198. (c) Total exports = 300 + 200 + 600 + 450 + 600 + 800 + 950 = ₹ 3900 cr

199. (c) Percent profit earned in 2001– 02

$= \frac{250-125}{125} \times 100 = 100\%$

200. (d) Percent profit in 1999-2000

$= \frac{325-250}{250} \times 100 = 30\%$

percent profit in 2000-2001

$\frac{350-250}{250} \times 100 = 40\%$

∴ reqd difference = 40 – 30 = 10%

201. (e) Avg. income

$= \frac{150+200+325+350+250}{5} =$ ₹ 255 cr

202. (b) Year which is having the amount of expenditure minimum and the gap between income and expenditure the maximum. And this condition is fulfilled in 1997- 98.

203. (a) Avg profit $= \frac{100+75+75+100+125}{5} =$ ₹ 95 cr.

204. (e) Required percent $= \frac{152.2}{86.4} \times 100 \approx 175\%$

205. (d)

206. (c) Average production of pulse

$\frac{20.5+22.4+24.6+23.5+27.8+28.2}{6} = \frac{147.0}{6}$

= 24.5 million tonnes

207. (a) Required percentage $= \frac{32.4}{450} \times 100 = 7.2\%$

208. (b) Total production of oilseeds in the given years
= 42.4 + 46.8 + 52.4 = 141.6.
Which is equal to the production of wheat in 1994 - 95.

209. (a)

210. (e) The answer is 1997, Machine IV

211. (e)

212. (b)

213. (c)

214. (a) Average $= \frac{52+66+64+75+58}{5} = \frac{315}{5} = 63.$

215. (a) The difference is 9.

216. (a) Percentage increase $= \frac{55-46}{46} \times 100 \approx 20\%$

217. (e) Average highest marks $= \frac{85+80+75}{3} = \frac{240}{3} = 80.$

218. (e)

219. (c) Required percentage $= \frac{80}{64} \times 100 = 125\%$

220. (b) Marks obtained by students = 50 × 60 = 3000

221. (e) The maximum difference is in the years 1992 & 1997. Since the least value is in 1992 and the highest value is in 1997.

222. (a) Reqd. % $= \frac{5.1}{12.5} \times 100 = 40.8\%$

223. (e) Percentage increase $= \frac{38.8-11.8}{11.8} \times 100 \approx 225\%$

224. (b)

225. (d) Reqd. ratio $= \frac{7.4}{16.8} = 37:84$

226. (a) Reqd. % $= \frac{38.8}{63.9} \times 100 \approx 60\%$

227. (e) Reqd. % $= \frac{101.80}{138.50} \times 100 \approx 75\%$

228. (c) Reqd. % $= \frac{1.8}{74.6} \times 100 = 2.41\%$

229. (b) Root vegetables produce per hectare

$= \frac{18560}{800} = 23.20$

230. (c) Tomato, cabbage, root vegetables

231. (a) Reqd. ratio = 72 : 15 = 24 : 5

232. (d) Total area = 19800 hectares

10% of the total area = 1980 hectares ...(i)

Comparing equation (i) and table, we see that in four types of vegetables, area used for production is more than 10% of the total area.

233. (e) Reqd. number $= \frac{42670}{1700} - \frac{13790}{700}$

$= 25.1 - 19.7 = 5.4$

234. (d) Total number of employees in accounts department

= 8% of 4600 = 368

$\therefore$ Number of women $= \frac{368}{(3+1)} \times 1 = 92$

235. (e) Total number of employers in IT and HR departments

= (26 + 11)% of 4600

$= \frac{37}{100} \times 4600 = 1702$

236. (c) Total number of men in all the departments

$$= \left[\left(\frac{11}{2}\times 1\right)+\left(\frac{8}{4}\times 3\right)+\left(\frac{15}{5}\times 3\right)+\left(\frac{26}{4}\times 1\right)+\left(\frac{22}{2}\times 1\right)+\left(\frac{18}{6}\times 1\right)\right]\% \text{ of } 4600$$

= (5.5 + 6 + 9 + 6.5 + 11 + 15)% of 4600

= 53% of 4600

Total number of women in all the departments

$$= \left[\left(\frac{11}{2}\times 1\right)+\left(\frac{8}{4}\times 1\right)+\left(\frac{15}{5}\times 2\right)+\left(\frac{26}{4}\times 3\right)+\left(\frac{22}{2}\times 1\right)+\left(\frac{18}{6}\times 1\right)\right]\% \text{ of } 4600$$

= (5.5 + 2 + 6 + 19.5 + 11 + 3)% of 4600

= 47% of 4600

Hence, required ratio = 53 : 47

237. (a) Number of women is merchandising department

$= \frac{1}{6}$ of 18% of 4600

Total number employees in the organization = 4600

Then, required percentage

$$= \frac{\frac{1}{6} \text{ of } 18\% \text{ of } 4600}{4600} \times 100\%$$

= 3%

238. (b) Number of men in the production department

$= \frac{3}{5}$ of 15% of 4600

Number of men in marketing department

$= \frac{1}{2}$ of 22% of 4600

Then, required ratio $= \frac{3}{5}$ of 15% of 4600 : $\frac{1}{2}$ of 22% of 4600

$= \frac{3}{5} \times 15 : \frac{1}{2} \times 22 = 9:11$

Note : You are suggested not to calculate these numbers of men or women seperately. Just do as we have done here because it makes your calculation easier.

239. (b) Required difference

= (15 – 10)% of ₹ 60 lacs

$= ₹\left(\frac{5}{100} \times 60\right)$ lacs = ₹ 3 lacs

240. (e) Required ratio = 8 : 6 = 4 : 3

241. (b) Required sum of expenditure

= (8 + 24 + 6)% ₹ 60 lacs $= \left(\frac{38 \times 60}{100}\right)$ lacs

= ₹ 22.8 lacs

242. (c) (a) Ratio = 1 : 3 = Not true

(b) Expenditure on medical facilities $= \frac{7 \times 60}{100}$

= ₹ 4.2 lacs = Not true

(c) Difference between the expenditure on research work and expenditure on research work and medical facilities

$= ₹\left(\frac{8-7}{100} \times 60\right)$ lacs

= 60000 = Definitely true

243. (b) Expenditure on purchase of overhead projectors

$= ₹\left(\frac{24 \times 60}{100}\right)$ lacs

= ₹ 14.40 lacs

After 7% decrease

Now expenditure

$= ₹\left(\frac{93 \times 14.40}{100}\right)$ lacs

$= ₹\ 13,39,200$

244. (a) Percentage of money spent on tennis

$= \left(\frac{45}{360} \times 100\right)\% = 12\frac{1}{2}\%$

245. (d) Degree value of expenditure on hockey = 63°
Degree value of expenditure on golf = 36°

$\therefore$ Required percentage $= \frac{63-36}{36} \times 100 = 75\%$

246. (a) Ammount spent on basketball exceeds that on tennis by

$= ₹\left(\frac{(50-45)}{360} \times 18000000\right) = ₹\ 250000$

247. (c) Degree value of the expenditure
on football = 54°
on cricket = 81°

$\therefore$ Required percentage $= \frac{81-54}{81} \times 100$

$= \frac{27}{81} \times 100 = 33\frac{1}{3}\%$

248. (b) Degree value of expenditure on cricket and hockey
$= (81 + 63)° = 144°$

$\therefore$ Required amount spent on them $= ₹\frac{144}{360} \times 2$ crore

$= 0.8$ crore $= ₹\ 8000000$

249-251.

249. (e) Cost price of article = ₹ 5600

Marked price $= 5600 + 5600 \times \frac{12}{100} = ₹\ 6272$

250. (c) SP $= 6272 - 6272 \times \frac{5}{100} = 5958.4$

Profit% $= \frac{5958.4-5600}{5600} \times 100$

$= \frac{358.4}{56} = 6.4\%$

251. (d) Amount of discount $= 6272 \times \frac{5}{100}$

$= 313.6$

252. (b) Total number of users of brand B across all Five cities
$= 600 + 500 + 650 + 700 + 550 = 3000$

253. (c) $700 = x\%$ of 500

$700 = \frac{x \times 500}{100} \Rightarrow x = \frac{700}{5} = 140$

254. (c) Required average $= \frac{500+550+600+550+700}{5}$

$= 580$

255. (d) Required difference = 1250 – 1100 = 150

256. (a) Required Ratio $= \frac{500}{700} = 5:7$

257. (a) Marks of all student in Chemistry
$= 90 + 110 + 100 + 120 + 60 = 480$

$\frac{\text{Marks obtained by S in Chemistry}}{\text{Marks of all student in Chemistry}} \times 100$

$= \left(\frac{120}{480} \times 100\right)\% = 25\%$

258. (e) Marks obtained by T in Physics = 50
If 14% marks in Physics are increased, new marks
$= 50 + 7 = 57$

New approximate percentage in Physics $= \frac{57}{140} \times 100$

$= 40.71$ or ≈ 41.

259. (b) Total marks obtained by T in both subjects = 50 + 60 = 110
R in Physics = 80

260. (d) Total marks obtained by P in Physics and Chemistry
$= 130 + 90 = 220$
Total marks obtained by T in Physics and Chemistry
$= 50 + 60 = 110$

Ratio $= \frac{220}{110} = 2 : 1$

261. (b) Total marks obtained by Q and S together in Chemistry
$= 110 + 120 = 230$
Total marks obtained by P and R together in Physics
$= 130 + 80 = 210$

Ratio $= \frac{230}{210} = 23 : 21$

262. (c) No. of people who prefer flute = 11% of 60,000

$= \frac{11}{100} \times 60000 = 6600$

2100 people be less from the people who prefer flute.
Therefore, 6600 – 2100 = 4500

Required percentage $= \frac{4500}{60000} \times 100 = 7.5\%$

263. (a) Total number of people who prefer either Sarod or Guitar = 14% of 60000 + 22% of 60000
$\Rightarrow$ 8400 + 13200 = 21600
Total number of people of who prefer violin or Sitar
= 20% of 60000 + 14% of 60000
$\Rightarrow$ 12000 + 8400 = 20400
Required difference = 21600 – 20400 = 1200

264. (b) Required number 14% of 60000

$= \frac{14}{100} \times 60000 = 8400$

265. (d) No. of people who prefer piano = 9% of 60000 = 5400

According to question, $16\frac{2}{3}\%$ no. of the people who prefer piano would go with flute.

Therefore, $\frac{50}{3}\%$ of 5400 = 900

Hence, the required percentage

$$= \frac{900 + 11\% \text{ of } 60000}{60000} \times 100$$

$$= \frac{900 + 6600}{60000} \times 100 = 12.5\%$$

266. (a) No. of people who prefer guitar = 22% of 60000 = 13200

No. of people who prefer Flute or Piano = (11 + 9)% of 60000 = 12000

Required difference = 13200 – 12000 = 1200.

267. (c) In 1981, no. of service workers

= 15% of 150 = 22.5 million

268. (b) In 1981, no. of categories more than 25 million workers i.e. more than 16% of 150.

More than 16% is → Professional, clerical, Blue collar i.e. 3.

269. (c) Ratio of workers to professional in 1981 to 1995

Professional in 1981 → 18%

⇒ 18% of 150 = 27

Professional in 1995 → 24%

⇒ 24% of 175 = 42

$$\text{Ratio} = \frac{27}{42} = \frac{9}{14} \quad \Rightarrow \quad 9 : 14$$

270. (b) Clerical % in country X in 1981 = 20% of 150 = 30

Clerical % in country X in 1985 = 18% of 175 = 31.5

So, increase = 1.5 million

272. (c) % of Blue collar workers in 1981 = 28% of 150 = 42

% of Blue collar workers in 1995

= 20% of 175 = 35

$$\% \text{ decrease} = \frac{42 - 35}{35} \times 100 = 20\%$$

272. (c) Percent increase $= \frac{380 - 320}{320} \times 100 = 18.75$

273. (b) Total production:

Wheat ⇒ 3700 million tonnes

Rice ⇒ 2000 million tonnes

Barley ⇒ 1800 million tonnes

Other cereals ⇒ 2400 million tonnes

Total Production = (3700 + 2000 + 1800 + 2400)

= 9900 million tonns

$$x = \frac{3700}{9900} \times 100 = 37.4\%$$

274. (a) Percentage increase:

$$\text{Rice} = \frac{160}{400} \times 100 = 40$$

$$\text{Barley} = \frac{30}{380} \times 100 = 7.8$$

$$\text{Cereals} = \frac{190}{500} \times 100 = 38$$

$$\text{Wheat} = \frac{100}{720} \times 100 = 13.8$$

275. (d) Required difference

$$= \frac{2000}{5} - \frac{1800}{5} = 400 - 360 = 40 \text{ million tonnes}$$

276. (a) Percentage decrease

$$= \frac{60 - 40}{60} \times 100 = \frac{100}{3} = 33\frac{1}{3}\%$$

277. (a) Average annual production:

Flavour X $\Rightarrow \frac{1}{6} \times 300 = 50$ lakh bottles

Flavour Y $\Rightarrow \frac{1}{6} \times 325 = 54\frac{1}{6}$ lakh bottles

Flavour Z $\Rightarrow \frac{1}{6} \times 300 = 50$ lakh bottles

278. (c) Total production of flavour X in 2005 and 2006 = 90

Total production of flower Z in 2007 and 2008 = 120.

Required percentage

$$= \frac{120}{90} \times 100 = 133.3$$

279 (a) Percentage increase/decrease:

Year 2007 $\Rightarrow \frac{60 - 50}{60} \times 100 \approx 16\%$ decrease

Year 2008 $\Rightarrow \frac{55 - 50}{50} \times 100 = 10\%$ increase

Year 2009 $\Rightarrow \frac{55 - 50}{55} \times 100 \approx 9\%$ decrease

280. (d) Required difference

$$= \frac{1}{3}[(55 + 50 + 55) - (50 + 40 + 55)]$$

$$= \frac{1}{3}(160 - 145) = \frac{15}{3} = 5 \text{ lakh bottles}$$

281. (b) (100 + 160) : (180 + 60)

= 260 : 240 = 13 : 12

282. (c) Required % = $\frac{100}{(100+160+240+40)} \times 100$

$= \frac{100}{540} \times 100 \approx 18.52\%$

283. (d) Required % = $\frac{120}{(180+60+120+20)} \times 100 = 31.58\%$

284. (b) Total students using Samsung = 100 + 180 = 280
Total students using Sony = 160 + 60 = 220
Difference = 280 – 220 = 60

285. (c) Expenditure for education = $\frac{9000}{30} \times 18 = ₹5,400$

286. (b) Central angle of the sector for the expenditure on fuel

$= \frac{360}{100} \times 15 = 54°$

287. (c) Expenditure excluding rent and education

$= \frac{3000}{15} \times (100 - 20 - 18)$

= 200 × 62
= ₹ 12400

288. (c) 30 = x% of (12 + 18 +15)

$30 = \frac{x}{100} \times 45$

$x = \frac{200}{3} = 66\frac{2}{3}$

289. (b) Required difference= (20 + 12 + 15) – 30
= 47 – 30 = 17

290. (c) Saturday

291. (b) Students having both Science and commerce

$= (29+31)\%$ of $200 = \frac{60}{100} \times 200 = 120$

Students who have taken neither science nor commerce
= Total students – Students having both scicence and commerce
= 200 –120 = 80

292. (c) No. of people who prefer flute = 11% of 60,000

$= \frac{11}{100} \times 60000 = 6600$

2100 people be less from the people who prefer flute. Therefore, 6600 – 2100 = 4500

Required percentage = $\frac{4500}{60000} \times 100 = 7.5\%$

293. (a) Total number of people who prefer either Sarod or Guitar = 14% of 60000 + 22% of 60000
⇒ 8400 + 13200 = 21600
Total number of people of who prefer violin or Sitar = 20% of 60000 + 14% of 60000
⇒ 12000 + 8400 = 20400
Required difference = 21600 – 20400 = 1200

294. (b) Required number 14% of 60000

$= \frac{14}{100} \times 60000 = 8400$

295. (d) No. of people who prefer piano = 9% of 60000 = 5400

According to question, $16\frac{2}{3}\%$ no. of the people who prefer piano would go with flute.

Therefore, $\frac{50}{3}\%$ of 5400 = 900

Hence, the required percentage

$= \frac{900 + 11\% \text{ of } 60000}{60000} \times 100$

$= \frac{900 + 6600}{60000} \times 100 = 12.5\%$

296. (a) No. of people who prefer guitar = 22% of 60000 = 13200
No. of people who prefer Flute or Piano = (11 + 9)% of 60000 = 12000
Required difference = 13200 – 12000 = 1200.

297. (d) $\frac{\theta}{360°} \times 100 = 16$

$\theta = \frac{16}{100} \times 360 = \frac{576}{10} = 57.6°$

298. (d) Required difference (in % value)

$= \frac{18-15}{18} \times 100 = \frac{3}{18} \times 100 = 16\frac{2}{3}\%$

299. (b) Given,
Miscellaneous expenditure = ₹ 1848
⇒ 4% of the total expenditure cost for publishing 5500 copies = ₹ 1848
⇒ Total expenditure cost of 5500 copies (i.e. 100%)

$= \frac{₹1848 \times 100}{4} = ₹\ 46200$

⇒ Expenditure cost per copy = $\frac{₹46200}{5500} = ₹\ 8.40$

So, marked price of each copy = ₹ 8.40 + 25% of 8.40
= 8.40 + 2.10 = ₹ 10.50

300. (c) Cost of printing i.e., 35% = 17500

So, Royality i.e., 15% = $\frac{17500}{35} \times 15 = 7500$

301. (a) Miscellaneous charges i.e. 4% of total expenditure = ₹ 6000
So, Advertisement charges i.e., 18%

$= \frac{6000}{4} \times 18 = ₹27000$

302. (d) Marks in Hindi = 70

Marks in English = 40

$\therefore$ Average marks = $\frac{\text{Marks in Hindi+ Marks in English}}{2}$

Average marks = $\frac{70+40}{2} = 55$

303. (a) Marks in Maths = 80

Marks in History = 60

$\therefore$ Ratio of marks of Maths and History = $\frac{80}{60} = 4 : 3$

304. (c) Royalty of book = 15%

Printing of book = 20%

$\therefore$ Royalty of book is less than printing cost by 5%

305. (d) $C.P + \frac{20}{100} \times C.P = 180$

$\frac{6}{5} C.P = 180$

C.P = 150

Paper cost = 15% of C.P

$\frac{15}{100} \times 150 = 22.50$

306. (d) Total sales of branches B1, B3 and B5.

= (80 + 105 + 95 + 110 + 75 + 95) = 560 thousand

307. (d) Required ratio = $\frac{75+65}{85+95} = \frac{140}{180} = \frac{7}{9}$

308. (a) Average sale of B1, B2 and B3 in 2001.

$= \frac{105+65+110}{3} = \frac{280}{3}$

Average sale of B1, B3 and B6 in 2000.

$= \frac{80+95+70}{3} = \frac{245}{3}$

Required % = $\frac{\frac{245}{3}}{\frac{280}{3}} \times 100 = 87.5\%$

309. (a) Sales of books B3 in 2001 = 110 thousand

Sales of books B2 in 2001 = 65 thousand

% increase = $\frac{110-65}{65} \times 100 = 69.2$

310. (b)

311. (d) More than 55 years = 4 + 3 + 2 + 1 = 10

312. (a) Number of patients of age more than 40 years and less than 55 years = 8 + 7 + 5 = 20

312. (c) Total patients = 35

Required % = $\frac{1+4+8}{35} \times 100 \approx 37\%$

314. (a) 11% of 35 = $\frac{11}{100} \times 35 \approx 3.8 \approx 4$

315. (d) Ratio

$= \frac{\text{number of pages printed by printer B in 2nd week}}{\text{number of pages printed by printer F in } 5^{th} \text{ week}}$

$= \frac{441}{693} = \frac{7}{11}$; 7:11

316. (b) Average number of pages printed by all the printer

$= \frac{256+563+347+651+412+321}{6} = 425$

317. (c)

Printer / Week	A	B	C	D	E	F
1st	664	618	628	552	638	419
2nd	569	441	519	438	621	537
3rd	440	614	503	527	541	742
4th	256	563	347	651	412	321
5th	717	429	598	582	519	693
Total up to 5th week	2646	2665	2595	2750	2731	2712

Printer D printed maximum pages.

318. (e) Required percentage (%)

$= \frac{\text{Pages printed by A in 3rd week}}{\text{Total page printed by D from 1st to 5th weeks}} \times 100$

$= \frac{440}{2750} \times 100 = 16\%$

319. (c) Required difference = Total no. of pages printed by printer C in all given weeks – Total no. of pages by E in 1st, 2nd, 4th week

$= 2595 - (638 + 621 + 412) = 924$

Level-II

1. **(e)** Suppose in the year 1998-99 expenditure of Company X = ₹ a

Then, profit earned by Company X in this year = ₹ (30% of 'a')

Hence, income of Company X

= ₹ (130% of a)

Again expenditure of Company Y in 2001-02

$= ₹ \frac{a \times 130}{100}$

Hence, profit earned by company Y in

2001– 02 = ₹ $\frac{a \times 130}{100} \times \frac{50}{100}$

Thus, required ratio $= \dfrac{\frac{30}{100} \times a}{\frac{a \times 130}{100} \times \frac{50}{100}}$

$= \dfrac{30}{100} \times \dfrac{100 \times 100}{130 \times 50} = \dfrac{30}{65} = 6:13$

2. **(c)** $I_{x\ 2001-02} = E_{x\ 2002-03} = \dfrac{I_{x2002-03}}{1.5}$

$I_{x\ 2001-02} : I_{x\ 2002-03} = \dfrac{2}{3} = 2:3$

3. **(c)** Per cent of increase in per cent profit over that of the previous year for the given years is as follows:
Year

1998-99 : $\dfrac{(20-15)}{15} \times 100 = 33.33\%$

1999-00 : $\dfrac{(30-20)}{20} \times 100 = 50\%$

2000-01 : = 0%

2001-02 : $\dfrac{(50-30)}{30} \times 100 = 66\dfrac{2}{3}\%$

2002-03 : $\dfrac{(60-50)}{50} \times 100 = 20\%$

You do not need to do any rough work. See the graph and search for **steep** rise in the line joining the two Δ'S

4. **(b)** Required income
= 120% of ₹ 40 crore
= ₹ 48 crores

5. **(d)** The given graph depicts only the per cent profit earned by the two companies over the given years. Hence, these informations are insufficient to answer the question.

6. **(e)** In 2002-03 profit earned by company *Y* was 60%. Therefore, 160% of expenditure = ₹ 128 crores

Thus, required expenditure $= \dfrac{128}{160} \times 100$

= ₹ 80 crores

7. **(b)** Production of company AVC in 2000 =360 crore units
Average production of AVC over the given years

$= \dfrac{300+300+300+360+370+340}{6} = \dfrac{1970}{6}$

Hence, required per cent $= \dfrac{360 \times 6}{1970} \times 100$

$= 109.64\% \approx 110\%$

8. **(c)** Approximate per cent increase or decrease in production from the previous year for SIO are as follows:

$1998 = \dfrac{2}{85} \times 100 = 2.35\%$

$1999 = \dfrac{2 \times 100}{87} = 2.29\%$

$2000 = \dfrac{2 \times 100}{89} = 2.24\%$

$2001 = \dfrac{1 \times 100}{91} = 1.09\%$

$2002 = \dfrac{4 \times 100}{92} = 4.35\%$

You can solve it with simple rough work. See the difference of produced units between two consecutive years. The difference is maximum for 2001 to 2002, and production during all these years is almost same. Hence, in the year 2002 SIO registered maximum increase in production over the previous year.

9. **(e)** Sum of the productions of the companies in first three years and the last three years in ₹ crore is as follows:

Company	First three years	Last three years
TP	358	349
ZIR	238	267
AVC	900	1070
CTU	836	852
PEN	90	127
SIO	261	279

10. **(c)** Total production of the six companies in first two given years = 863 + 927 = 1790
Again, total production of the six companies in last two given years = 989 + 991 =1980
Therefore, required per cent

$= \dfrac{1790 \times 100}{1980} = 90.40\%$

11. **(b)** The required difference
= (91 – 90) crore units
$= 1 \times 10000000$
= 10000000 units

12. **(c)** Those companies are:
ZIR PEN and SIO

13-15: No. of boys $= \dfrac{3}{7} \times 175 = 75$

No. of girls = 175 – 75 = 100
No. of boys who opt only Hindi
= 40% of 75 = 30
Remaining boys = 75 – 30 = 45
No. of boys who opt only Sanskrit

$= \dfrac{2}{3} \times 45 = 30$

No. of boys who opt composite subjects
= 45 – 30 = 15

Total no. of students who opt only Sanskrit
= 44% of 175 = 77
No. of girls who opt only Sanskrit
= 77 – 30 = 47
No. of girls who opt composite subjects = 32
No. of girls who opt Hindi only
= 100 – (32 + 47) = 21

13. (a) Reqd ratio = 30 : 32 = 15 : 16

14. (b)

15. (b)

16-20: Dividend: At the end of the financial year, a company declares profit. This profit is called dividend. This dividend is expressed in terms of percentage of the nominal value. The share holder receives the dividend on the nominal value of his shares.

Nominal value: The original price of a share is called the nominal value of the share.

16. (b) Suppose Giridhar invested ₹ x in company A.

$$\therefore \quad \frac{x\times 14}{100}+\frac{(25000-x)\times 13}{100}=3340$$

$$\text{or,} \quad \frac{14x}{100}+3250-\frac{13x}{100}=3340$$

$$\text{or,} \quad \frac{x}{100}=90$$

or, x = ₹ 9000.

17. (a) Amount of dividend received by Anuja in 1996 from company B

$$=\frac{35000\times 19}{100}=₹\,6650$$

Total amount invested by Anuja in 1997 in Company A
= 35000 + 6650 = ₹ 41650

$$\text{Reqd amount } = 41650\times\frac{120}{100}=₹\,49980$$

18. (e) $\text{Total dividend } = 18000\times\left(\frac{20}{100}+\frac{15}{100}\right)=₹6300$

19. (c) $\text{Reqd ratio } = \frac{5\times 12}{8\times 10}=3:4$

20. (e) From the graph it is obvious that Suraj will get less dividend in 1999 from company A than from B.
Reqd less amount = 3% of 56000
= ₹ 1680.

21-25: Total population = 35000
Total literate population

$$=\frac{35000\times 70}{100}=24500$$

$$\text{Total females} = \frac{35000\times 44}{100}=15400$$

Total males = (35000 – 15400) = 19600

Total illiterate population = 35000 – 24500
= 10500

$$\text{Total male illiterate} = \frac{28\times 10500}{(28+47)}=3920$$

Total female illiterate = 10500 – 3920 = 6580
Total male literate = (19600 – 3920) = 15680
Total female literate = (15400 – 6580) = 8820

21. (b) (6580 : 8820) = 47 : 63
[140 is common in the ratio.]

22. (d) (15680 : 8820) = 16 : 9
[2 × 7 × 7 × 10 is common in the ratio]

23. (e) 19600

24. (a) $\frac{15680\times 5}{100}=784$

25. (c) Given in the information given above. (= 6580)

26. (b) Total no. of Medicine students = 13710
Total no. of Engineering students = 20440
Required percentage

$$=\left(\frac{13710}{20440}\times 100=\right)\approx\frac{13700}{20400}\times 100=67\%$$

27. (d) Total no. of Arts students over the year = 16250
Total no. of years = 6
∴ average no. of students studying Arts

$$=\left(\frac{16250}{6}\right)\approx 2708$$

28. (c) In this type of questions we do not need to calculate the values for all the years.
By simple comparison we can find out the solution; e.g., For the first three years (1997, 1998, 1999), year 1999 has maximum percentage decrease from the previous year. Now, consider one more year, i.e., year 2000. The difference between the no. of Commerce students for 1998 & 1999 is less than the difference between that for the years 1999 & 2000. Hence, till now year 2000 has maximum percentage increase. Similarly, we can proceed year by year.
[**Note:** For the same difference, or nearly same differences between two pairs of year, the percentage increment/decrement will be more for lesser base value.]

29. (a) Required per cent

$$\frac{1120}{(3050+2850+4550+2640+3650=)16740}\times 100$$

= 6.69%

30. (e) Required per cent

$$\frac{3080\times 100}{(3200+3500+2850+3640+3080+3800=)20070}$$

15.34 ≈ 15

(31-35):

40% of males = 8800

∴ **no. of total males**

$= \frac{8800}{40} \times 100 = 22,000$

Ratio of males, females and children 10 years old and above

=11 : 10 : 9

Hence, **no. of total females**

$= \frac{22,000}{11} \times 10 = 20,000$

No. of total children (10 yrs old and above)

$= \frac{22,000}{11} \times 9 = 18,000$

No. of literate males = 8800

No. of illiterate males

= 22,000 – 8800 =13,200

No. of literate females

$= \frac{20,000 \times 30}{100} = 6,000$

No. of illiterate females

= 20,000 – 6,000 = 14,000

The number of children below 10 years of age = 10% of the number of females

$= \frac{20,000 \times 10}{100} = 2000$

No. of total children

= 18000 + 2000 = 20,000

No. of illiterate children 10 years old and above

$= \frac{18000 \times 20}{100} = 3600$

No. of literate children 10 years old and above

= 18000 – 3600 = 14400

No. of persons below poverty line

= 5% of (22,000 + 20,000 + 20,000)

$= \frac{5 \times 62000}{100} = 3100$

Illiterate persons among these 3100 persons

= 80% of 3100

$= \frac{80 \times 3100}{100} = 2480$

31. **(a)**

32. **(b)**

33. **(a)**

34. **(d)** Required % $= \frac{20,000}{62,000} \times 100$

= 32.26

35. **(e)** 14000

36. **(d)** Evident from the graph itself.

37. **(d)** Domestic consumption in 1990 = 12500

Domestic consumption in 1991 = 25000

Difference= 25000 – 12500

= 12500

= Difference in experts between 1991 & 1993.

38. **(d)** Growth or decline of exports is not given.

39. **(e)** Percentage change in exports is maximum in 1990 (+ 75%).

40. **(c)** Required difference $= \frac{147.5 - 130}{7}$

$= \frac{17.5}{7} \times 1000 = 2,500$ tonnes

41. **(a)** Average marks obtained by 20 boys in History from school $Q = 45$

∴ Total marks = 20 × 45 = 900

42. **(a)** From visual inspection it is clear that Science is the desired subject.

Note : Our visual observation says that it is either Math or Science in which maximum marks has been obtained. So, compare the total of Maths and Science only.

43. **(d)** Total marks obtained by boys and girls in all the subjects:

For school P = (85 + 40 + 50 + 120 + 105) + (90 + 55 + 40 + 110 + 125) = 820

Similarly, for Q = 745, for R = 935, for S = 645 and for T = 1005.

645 is the minimum, so S is the desired school.

Note: From careful observation we find that our answer is school S. The other school nearest to it is either P or Q. But if you compare the marks, P and Q also take lead of at least 100 marks. So, only visual observation gives the result.

44. **(b)** As the no. of boys and girls in the different schools are the same, so for the desired purpose we have to select a certain school in which the average marks of girls in Mathematics be exactly double the average marks of boys in History. By visual inspection (as 80 = 2 × 40), we get that S is the desired school.

45. **(b)** In Mathematics total marks obtained by boys from school R = 135 × 20

By girls from school S = 80 × 20

∴ Reqd difference = (135 – 80 =) 55 × 20 = 1100.

46. **(d)** Incomes-Expenditures of Company A and B cannot be corelated.

47. **(e)** Expenditure of Company A in 1996

$= E_{96(A)} = I_{96(A)}\left[\frac{100}{100+60}\right] = \frac{5}{8} I_{96(A)}$

Expenditure of Company B in 1997

$= E_{97(B)} = I_{97(B)}\left[\frac{100}{100+70}\right] = \frac{10}{17} I_{97(B)}$

Now, $\frac{E_{96(A)}}{E_{97(B)}} = \frac{5}{8} \div \frac{10}{17}$ (Since, $I_{96(A)} = I_{97(B)}$)

$= \frac{5}{8} \times \frac{17}{10} = \frac{17}{16} = 17:16$

48. (c) Ratio $A : B$ is greater than 1 in only 1993 and 1997. It is 1.33 in 1993 and 1.1 in 1997.

49. (e) Suppose $E_{95(B)} = x$
Then $E_{96(B)} = 1.2x$ (Since, $x + 20\%$ of $x = 1.2x$)

Now, $I_{95(B)} = E_{95(B)} \left[\frac{100+80}{100}\right] = 1.8x$

$I_{95(B)} = E_{95(B)} \left[\frac{100+80}{100}\right] = 1.2x(1.8)$

$\therefore \frac{I_{96(B)}}{I_{95(B)}} = \frac{1.2 \times 1.8x}{1.8x} = 1.2$ times

Alternative method : % profits are the same for two years. So if expenditure increases by 20% the income should also increase by 20%. Hence the required ratio

$= \frac{100+20}{100} = 1.2$

50. (a) $E_{96(A)} = I_{96(A)} \left[\frac{100}{100+60}\right]$

$= \frac{36 \text{ lakh} \times 100}{160} = ₹ 22.5$ lakh

51. (d) Average number of players who play Football and Rugby

$= \frac{1}{2}[(14+13)\% \text{ of } 4200]$

$= \frac{1}{2} \times 4200 \times \frac{30}{100} = 630$

52. (a) Number of players who play Rugby

$= 4200 \times \frac{13}{100} = 546$

Number of female players who play Rugby

$= 2000 \times \frac{10}{100} = 200$

$\therefore$ Number of male players who play Rugby
$= 546 - 200 = 346$ play
Number of female players who play Lawn Tennis.

$= 2000 \times \frac{22}{100} = 440$

$\therefore$ Required difference $= 440 - 346 = 94$

53. (c) Number of female cricketers

$= 2000 \times \frac{40}{100} = 800$

Number of male Hockey players

$= \frac{4200 \times 10}{100} - \frac{2000 \times 15}{100}$

$= 420 - 300 = 120$

$\therefore$ Required ratio $= 800 : 120 = 20 : 3$

54. (b) Number of male players who play Football, Cricket and Lawn Tennis
$= (17 + 35 + 25)\%$ of $4200 - (13 + 40 + 22)\%$ of 2000

$= 4200 \times \frac{77}{100} - 2000 \times \frac{75}{100}$

$= 3234 - 1500 = 1734$

55. (a) Number of male players who play Rugby

$= 4200 \times \frac{13}{100} - 200 = 346$

Number of players who play Lawn Tennis

$= 4200 \times \frac{25}{100} = 1050$

$\therefore$ Required percentage

$= \frac{346}{1050} \times 100 \approx 33$

56. (a) The corresponding angle for commuting $= 79.2°$
$\therefore$ $360° = 45800$

$\therefore$ $79.2° = \frac{45800}{360} \times 79.2$

$= ₹ 10,076$

57. (d) Required ratio $= 39.6 : 82.8$
$= 396 : 828 = 11 : 23$

58. (b) Corresponding angle for entertainment and shopping together
$= 36 + 68.4 = 104.4°$
$\because$ $360° = 45800$

$\therefore$ $104.4° \equiv \frac{45800}{360} \times 104.4$

$= ₹ 13,282$

59. (e) Corresponding total angle for groceries, entertainment and investments together
$= 82.8 + 36 + 54 = 172.8$
Corresponding angle for commuting $= 79.2$
$\therefore$ Required percentage

$= \frac{172.8}{79.2} \times 100 \approx 218$

60. (c) Required percentage

$= \frac{39.6}{68.4} \times 100 = 57.89$

61. (b) 20% of the total funds
$= ₹$ (20% of 57600) crore
$= ₹ 11520$ crore $\simeq ₹ 11486$ crore
which is the amount of funds to be arranged through external assistance.

62. **(c)** Shortage of funds arranged through external assistance
= ₹ (11486 – 9695) crore
= ₹ 1791 crore
∴ Increase required in market borrowing
= ₹ 1791 crore
∴ Percentage increase required

$$= \left(\frac{1791}{29952} \times 100\right) = 598\% \simeq 6\%$$

63. **(c)** Amount permitted
= (Funds required from Toll for projects of Phase II) + (10% of these funds)
= ₹ 5401 crore

64. (c) Central angle corresponding to market borrowing

$$= \left(\frac{29952}{57600} \times 360°\right) = 187.2°$$

65. (b) Required ratio

$$= \frac{\text{Fund arranged through toll}}{\text{Fund arranged through market borrowing}}$$

$$= \frac{4910}{29952} = \frac{1}{6.1} \simeq \frac{1}{6} = 1:6$$

66-71 :

Following table can be made easily :
Total area = 37249 sq kilometres

State	Density of population (in thousands)	Approx % area
A	$\frac{112200}{6230} = 18$ (approx)	16.7
B	$\frac{83800}{2540} = 33$ (approx)	6.8
C	$\frac{64900}{8135} = 8$ (approx)	21.8
D	$\frac{572}{7436} = \frac{1}{13} \times 100 = 7.69$ (approx)	20
E	$\frac{71100}{4893} = 14.5$ (approx)	13
F	$\frac{286}{3718} = \frac{1}{13} \times 100 = 7.69$ (approx)	10
G	$\frac{86000}{4297} = 20$ (approx)	11.5

[Note : Density of population is the population per square kilometre.]

66. **(b)** From thae above (column 3) it can be answered. States A, C and D have more than 15 per cent of total areas.

67. **(c)** From column 2 of the above table, it can be observed.

68. **(a)** Required answer = (33 - 18 =) 15 thousand.

69. **(d)** For state A, B, E and G, ie for 4 states.

70. **(e)** Required answer = 2540 : (6230 + 4297 =) 10527 ≈ 1 : 4.1

71. **(d)**

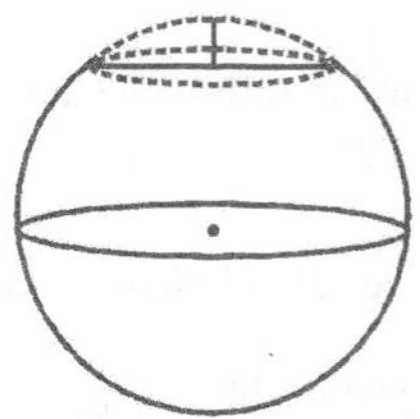

The radius of the spherical part of the bowl can't be determined. Hence volume cannot be calculated.

72. **(c)** 60% of 150 = 90. It means those who obtained either 90 or more than 90 marks in the average of five subjects will be declared as passed. Therefore,
the required number = 31 + 17 = 48

73. **(b)** Reqd no. = 52 + 41 + 13 = 106

74. **(e)** **75.** **(d)**

76. **(a)** No. of students who obtained more than or equal to 40% marks in Science
= 67 + 22 + 14 = 103

No. of students who obtained less than 60% marks in Hindi = 19 + 59 + 47 = 125

$$\text{Reqd } \% = \frac{125-103}{125} \times 100 = 17.60\%$$

77. **(b)** No. of students (of JNU) listening to Radio City

$$= 120000 \times \frac{20}{100} \times \frac{65}{100} = 15600$$

78. **(e)** The no. of Indraprastha students listening to FM Rainbow = 120000 × 13% × 48%
The no. of Jamia students listening to FM Gold = 120000 × 18% × 52%
∴ reqd percentage

$$= \frac{120000 \times 13\% \times 48\%}{120000 \times 18\% \times 52\%} \times 100 = 66.66\%$$

79. **(a)** Indraprastha University

80. **(c)** The no. of Indraprastha students listening to

$$\text{Red FM} = 120000 \times \frac{13}{100} \times \frac{46}{100} = 12 \times 598$$

The no. of Jamia students listening to Red FM

$$= 120000 \times \frac{18}{100} \times \frac{36}{100} = 12 \times 648$$

∴ Total students = (12 × 598) + (12 × 648)
= 12 × 1246 = 14952

81. **(b)** Radio City

82. **(a)** Only for project A, the value of cost per room is less than half crore, whereas in other projects it is either more than or equal to half crore.

83. **(b)** Total investments for all the projects
$= 275 + 210 + 250 + 430 + 310 + 400 + 250 = ₹ 2125$ cr
Investment on those projects which are completed in 1998 = 275 + 430 + 400 = ₹1105 cr
$\therefore$ reqd % $= \frac{1105}{2125} \times 100 = 52\%$

84. **(d)** Total no. of rooms over the three-year period
$= 600 + 320 + 250 + 400 + 520 + 450 + 500 = 3040$
reqd average no. of rooms per crore of rupees
$= \frac{3040}{2125} \approx 1.4$

85. **(e)** $\frac{876\left(\frac{100}{219}\right)}{1274\left(\frac{24}{49}\right)} \times 100 = \frac{400}{624} \times 100 \approx 64\%$

86. **(b)** Using table II, comparing the ratio we find that only BS and SAI have such thing.

87. **(a)** From the table it can be observed.

88. **(e)** In 2003, TIS total production
$= 782\left(\frac{110}{100}\right) = 860.2$
SAI total production $= 665\left(\frac{90}{100}\right) = 598.5$
Total of A-type steel $= \left(\frac{860.2}{23} \times 14\right) + \left(\frac{598.5}{133} \times 62\right)$
$= (523.6) + (279) = 802.6$
Total of B-type steel
$= \left(\frac{860.2}{23} \times 9\right) + \left(\frac{598.5}{133} \times 71\right)$
$= (336.6) + (319.5) = 656.1$
Required ratio = 8026 : 6561

89. **(a)** Required ratio $= \frac{3}{4} \times \frac{8}{7} = \frac{6}{7}$ or 6 : 7

90. **(b)** Required answer
$= 720 \times \frac{40}{100} \times \frac{15}{100} = 43.2$ million litres

91. **(d)** Total supply from Jhelum and underground water
$= 720 \times \frac{(12+6)}{100} = 129.6$ million litres
Consumption of Yamuna water for domestic purposes
$= 720 \times \frac{40}{100} \times \frac{40}{100} = 115.2$
Consumption of Ganga water for agriculture purposes
$= 720 \times \frac{35}{100} \times \frac{20}{100} = 50.4$
$\therefore$ Required ratio
$= \frac{129.6}{(115.2+50.4)} = \frac{18}{23} = 18:23$

92. **(b)** Difference between cost of 1 kg apple and cost of 1 kg guava in 5 cities.
J 160 – 60 = 100
D 130 – 90 = 40
C 180 – 120 = 60
H 90 – 30 = 60
R 40 – 20 = 20
$\therefore$ Cost is second lowest in Delhi.

93. **(d)** Cost of 1 kg guava in Jalandhar = ₹ 60
Cost of 2 kg grapes in chandigarh = ₹ 90 × 2 = ₹ 180
$\% = \frac{60}{180} \times 100 = 33.3 \approx 34\%$

94. **(c)** Cost of 3 kgs apples for Ram = 3 × 130 = ₹ 390
Cost of 2 kgs guavas for Ram = 2 × 90 = ₹ 180
Total cost that Ram pay = 390 + 180 = ₹ 570

95. **(a)** Total cost of 45 kgs grapes from Hoshiarpur
= 45 × 190 = ₹ 8550
After discount 4% Ravinder paid
$= 8550 - \frac{8550 \times 4}{100} = ₹ 8208$

96. **(c)** Cost of 1 kg apples from Ropar :
Cost of 1kg grapes from chandigarh
40 : 90
4 : 9 or $2^2 : 3^2$

97. **(a)** Number of students in university I in 2010 = 20,000
Number of students in university II in 2012 = 20,000
Difference = 20,000 – 20,000 = 0

98. **(e)** Number of students in university I in 2007 = 10,000
Number of students in university II in 2011 = 30,000
Total students = 10,000 + 30,000 = 40,000

99. **(a)** Total no. of students in Uni-2 in 2010 = 15,000
No. of females students = 25% of 15000
$= \frac{25}{100} \times 15,000 = 3750$
No. of male students = 15,000 – 3750 = 11250

100. (d) Number of students in university I in 2011 = 25,000

Number of students in university I in 2010 = 20,000

Increase = 25000 – 20000 = 5000

$$\text{Increase \%} = \frac{5000}{20000} \times 100 = 25\%$$

101. (e) Difference between Number of students in university I & Number of students in university II for the year

2007 – 10,000

2008 – 10,000

2009 – 10,000

2010 – 5,000

2011 – 5,000

2012 – 15,000

$\therefore$ Difference was highest in year 2012.

102. (b) No. of hockey players in all schools

= 68 + 80 + 54 + 50 + 36 = 288

103. (c) No. of basket-ball players in school-1 = 42

No. of Kho-Kho players in school-3 = 48

Ratio = 42 : 48 = 7 : 8

104. (b)

105. (d) No. of Kho-Kho players in school-4 = 32

No. of hockey players in school-2 = 80

$$\% = \frac{32}{80} \times 100 = 40\%$$

106. (e) No. of hockey players in school-5 = 36

No. of females = 25%, No. of males = 75%

$$\text{No. of males} = \frac{36 \times 75}{100} = 27$$

107. (b) Gita's average earning over all the day's

$$= \frac{400 + 420 + 200 + 140}{4} = \frac{1160}{4} = ₹\ 290$$

108. (d) Rahul earned on Tuesday and Thursday

= 180 + 340 = ₹ 520

Naveen earned on Tuesday and Thursday

= 260 + 160 = ₹ 420

Total = 520 + 420 = ₹ 940

109. (c) Gita donated Naveen = ₹ 420

Naveen's total earning on Wednesday

= 420 + 120 = ₹ 540

110. (a) Rahul's earning on Monday = ₹ 240

Gita's earning on Tuesday = ₹ 200

DiH = 240 – 200 = ₹ 40

111. (e) Naveen's earning on Monday, Wednesday and Thrusday

= 360 : 120 : 160

= 9 : 3 : 4

112. (c) No. of employees in teaching and medical = 42%

No. of employees in management = 17%

Difference = 42 – 17 = 25%

$$= \frac{26800 \times 25}{100} = 6700$$

113. (e) In management $\frac{3}{4}$ employees are female

$\therefore$ $\frac{1}{4}$ employees in management are male

$\therefore$ No. of male employees $= \frac{26800 \times 17}{100} \times \frac{1}{4} = 1139$

114. (b) In film production employees on strike = 25%

No. of employees not in the strike = 75%

$\therefore$ No. of employees not in strike

$$= \frac{26800 \times 19}{100} \times \frac{75}{100} = 3819$$

115. (d) Total No. of employees in engineering industries

= 9 + 13 = 22%

$$\therefore \frac{26800 \times 22}{100} = 5896$$

116. (a) $\frac{3}{5}$th of the teachers are not permanent

$\therefore$ $\frac{2}{5}$th of the teachers are permanent

$\therefore$ No. of permanent teachers $= \frac{26800 \times 15}{100} \times \frac{2}{5} = 1608$

117. (c) Total amount paid by Dev in June for all commodities

= 123 + 150 + 324 + 134 = ₹ 731

118. (c) Average electricity bill paid by Manu in all five months

= 315 + 135 + 98 + 116 + 131

$$= \frac{795}{5} = ₹\ 159$$

119. (a) Mobile phone bill paid by Ravi in May = ₹ 143

Loundry bill paid by Dev in March = ₹ 323

Difference = 323 – 143 = ₹ 180

120. (d)

121. (a) Electricity bill paid by Manu in April = 135

Mobile bill paid by Ravi in June = 245

Ratio = 135 : 245 = 27 : 49

122. (c) Distance travelled by train from Surat to Nadiad Jn.

= 440 – 257 = 183 kms

123. (a) Total time taken = 1hr. 35min + 2min + 2min + 20min = 1hr 59 min

124. (a) Ratio between No. of passengers boarding from Vasai Road and from Ahmedabad

= 378 : 306 = 21 : 17

125. (b) Total time increase = 23 – 2 = 21 min.

∴ Train will reach Bhuj at = 5:40PM + 21 min = 6:01 PM

126. (d) Distance between Anand Jn. to Vadodara is second lowest.

127. (e) Difference between the max. temperature of Ontario on 1st Nov. and the min. temperature of Bhuj on 1st Jan. = 4 – (7) = 11°C

128. (b)

129. (c) Difference between max. and min. temperature of Bhuj

Sept. = 10°C

Oct. = 14°C

Nov. = 11°C

Dec. = 7°C

Jan. = 3°C

130. (e) Average of max. temperature of Beijing over all the

months = $\frac{43}{5}$ = 8.6°C

131. (b) Ratio between minimum temperature of Beijing on 1st sept. and maximum temperature of Ontario on 1st Oct. = 9 : 15 = 3 : 5

For (132 –136)

Total Cars = 700

State	Total Cars	Diesel Cars	Petrol Cars
1	98	42	56
2	196	70	126
3	224	140	84
4	182	91	91

132. (b) Difference between No. of diesel cars in S-2 and No. of pertrol cars in S-4

= 91 – 70 = 21

133. (b) Petrol cars in S-3 = 84

Diesel cars in S-1 = 42

$\% = \frac{84}{42}\times 100 = 200$

134. (d) 25% diesel cars in S-3 are AC

∴ 75% diesel cars in S-3 are non-AC

∴ No. of non-AC cars = $\frac{140\times 75}{100} = 105$

135. (e) Difference between total cars in S-3 and petrol cars in S-2

= 224 – 126 = 98

136. (b) Average of petrol cars in all states together

$= \frac{56+126+84+91}{4} = \frac{357}{4} = 89.25$

For (137–139)

Total members = 240

	Males	Females
Total	160	80
Graduates	24	60
Non-graduates	136	20

137. (c) Difference between No. of non-graduates females and no. of graduates males = 24 – 20 = 4.

138. (e) Sum of (graduates females and non-graduates males) = 60 + 136 = 196

139. (b) Ratio between total No. of males and no. of non-grad females

= 160 : 20 = 8 : 1

140. (e) Required ratio = $34560\times\frac{55}{100} : 45640\times\frac{45}{100}$

= 19008 : 20538 = 1056 : 1141

141. (e) Required percentage

$$= \frac{55500\times\frac{41}{100}}{34560+65900+45640+55500+42350+59650}\times 100$$

$$= \frac{22755}{303600}\times 100 = 7.495 \simeq 7.5\%(\text{Approx})$$

142. (d) Required percentage

$$= \frac{45640\times\frac{20}{100}}{59650\times\frac{14}{100}}\times 100 = \frac{9128}{8351}\times 100 = 109.30\%$$

143. (c) Required Number

$$= 65900\times\frac{20}{100} + 55500\times\frac{33}{100} = 13180 + 18315 = 31495$$

144. (a) Required Average Number

$$= \frac{34560\times\frac{55}{100}+65900\times\frac{43}{100}+45640\times\frac{45}{100}+55500\times\frac{26}{100}+42350\times\frac{70}{100}+59650\times\frac{62}{100}}{6}$$

$$= \frac{19008+28337+20538+14430+29645+36983}{6}$$

$$= \frac{148941}{6} = 24823.5$$

145. (b) Required percentage

$$=\frac{14.4^\circ}{122.4^\circ}\times 100 = 11.765 \simeq 12\% (\text{Approx})$$

146. (a) Required number $=\left(\frac{57.6^\circ + 64.8^\circ}{360^\circ}\right)\times 6800$

$$=\frac{122.4^\circ}{360^\circ}\times 6800 = 2312$$

147. (c) Required difference

$$=\frac{(122.4^\circ + 21.6^\circ)-(79.2^\circ + 14.4^\circ)}{360^\circ}\times 6800$$

$$=\frac{144.0^\circ - 93.6^\circ}{360^\circ}\times 6800 = 952$$

148. (d) Required percentage

$$=\frac{64.8^\circ + 21.6^\circ}{360^\circ}\times 100 = \frac{86.4^\circ}{360^\circ}\times 100 = 24\%$$

149. (a) Required Ratio = 21.6° : 79.2° = 3 :11

150. (b) Marks scored by Anushka

$$= 150\times\frac{66}{100} + 100\times\frac{75}{100} + 150\times\frac{88}{100}$$

$$+ 125\times\frac{56}{100} + 75\times\frac{56}{100} + 50\times\frac{90}{100}$$

= 99 + 75 + 132 + 70 + 42 + 45 = 463

151. (d) Marks scored by Archit

$$= 150\times\frac{82}{100} + 100\times\frac{76}{100} + 150\times\frac{84}{100} + 125$$

$$\times\frac{96}{100} + 75\times\frac{92}{100} + 50\times\frac{88}{100}$$

= 123 + 76 + 126 + 120 + 69 + 44 = 558

Marks scored by Arpan

$$= 150\times\frac{76}{100} + 100\times\frac{66}{100} + 150\times\frac{78}{100} + 125$$

$$\times\frac{88}{100} + 75\times\frac{72}{100} + 50\times\frac{70}{100}$$

= 114 + 66 + 117 + 110 + 54 + 35 = 496

Marks scored by Garvita

$$= 150\times\frac{90}{100} + 100\times\frac{88}{100} + 150\times\frac{96}{100} + 125$$

$$\times\frac{76}{100} + 75\times\frac{84}{100} + 50\times\frac{86}{100}$$

= 135 + 88 + 144 + 95 + 63 + 43 = 568

Marks scored by Gunit.

$$= 150\times\frac{64}{100} + 100\times\frac{70}{100} + 150\times\frac{68}{100} + 125$$

$$\times\frac{72}{100} + 75\times\frac{68}{100} + 50\times\frac{74}{100}$$

= 96 + 70 + 102 + 90 + 51 + 37 = 446

Marks scored by Pranita

$$= 150\times\frac{48}{100} + 100\times\frac{56}{100} + 150\times\frac{50}{100} + 125$$

$$\times\frac{64}{100} + 75\times\frac{64}{100} + 50\times\frac{58}{100}$$

= 72 + 56 + 75 + 80 + 48 + 29 = 360

Hence, highest total marks scored by Garvita.

152. (c) Required percentage $=\frac{88}{76}\times 100 = 115.79\%$

153. (b) Two students i.e. Garvita strategic management brand management & compensation management and Archit consumer behaviours service marketing.

154. (a) Required Average Marks

$$=\frac{(88+84+78+96+68+50)\times 150}{100\times 6} = \frac{464\times 150}{600} = 116$$

155. (d) Required percentage

$$=\frac{15000+30000}{5000+35000+15000+25000+30000+30000}\times 100$$

$$=\frac{45000}{140000}\times 100 = 32.14 \simeq 32\% (\text{Approx})$$

156. (e) In 2009

Total number of girls = (20 + 20 + 5)

$$\times\frac{38}{100}\times 1000 = \frac{45\times 38\times 1000}{100} = 17100$$

Total number of boys = 45000 – 17100 = 27900

Total number of boys who opted for Mathematics

$$= 27900\times\frac{5}{45} = 3100$$

157. (a) Required Ratio = (25 + 30) : (5 + 20) = 55 : 25 = 11 : 5

158. (b) Required Percentage = $\frac{(15+10+15)\times1000}{455030}\times100$

$= \frac{40\times1000}{455030}\times100 = 8.79 \simeq 9\%(\text{Approx})$

159. (e) Required total number of students

= (5 + 35 + 15 + 15 + 20 + 5) × 1000

= 95 × 1000 = 95000

160-164.

	Day 1			Day 2		
Vehicle	Time in hr	Distance in km	Speed in km / hr	Time in hr	Distance in km	Speed in km/hr
A	16	832	52	16	864	54
B	12	516	43	18	774	43
C	11	693	63	18	810	45
D	12	552	46	15	765	51
E	16	935	58.4	14	546	39
F	19	703	37	12	636	53

160. (d) Vehicle B.

161. (c) Speed of vehicle A on day 1 = 52 km/hr

Speed of vehicle C on day 1 = 63 km/hr

Difference = 63 – 52 = 11 km / hr

162. (e) Speed of vehicle can day 2 = 45 km/hr

$\Rightarrow\left(45\times\frac{5}{18}\right)\text{m/sec}=12.5\,\text{m/sec}$

163. (e) Percentage

$= \frac{\text{Distance travelled by vehicle F on day 2}}{\text{Distance travelled by vehicle F on day 1}}\times100$

$= \frac{636}{703}\times100\approx\frac{630}{700}\times100\approx90\%$

164. (b) Speed of vehicle D on day 2 = 51

Speed of vehicle E on day 2 = 39

Required ratio = $\frac{51}{39}=\frac{17}{13}$ or 17:13

165-169.

	Total number of Mobiles Sold	Total Number of Mobiles Sold of Company A	Total Number of Mobiles Sold of Company B
July	7650	4080	3570
August	9900	4400	5500
September	11250	6750	4500
October	3600	2100	1500
November	5400	2520	2880
December	7200	3150	4050

165. (c) Number of mobiles sold of company B in July = 3570

Number of mobiles sold of company B in December = 4050

Required ratio = 3570 : 4050 = 119 : 135

166. (c) Total mobiles sold by company A during November = 2520

Total mobiles sold by this company at discount

= 35% of 2520 = 882

Total mobiles sold by company A without discount

= 2520 – 882 = 1638

167. (d) Mobile phones sold of company B during October = 1500

Total profit earned on the mobile phones

= ₹(433 × 1500) = ₹ 6,49,500

168. (e) Number of mobile phones sold of company A during July = 4080

Number of mobile phones sold by company A during December = 3150

Required percentage $= \frac{4080}{3150}\times100 = 129.5 \approx 130\%$

169. (a) Mobile phones sold of company B during August = 5500

Mobile phones sold of company B during September = 4500

Total number of mobile phones = 5500 + 4500 = 10,000

170. (c) Area of customer transaction room

= 23m × 29m = 667 sq.m

Area of branch manager room

= 13m × 17 m = 221 sq. m

Area of Pantry room = 14m × 13m = 182 sq. m

Area of Server room = 21m × 13m = 273 sq. m

Area of locker room = 29 m × 21 m = 609 sq. m

Total cost of wooden flooring = ₹ [(170 × (667 + 221)]

= ₹ (888 × 170)

Total cost of marble flooring

= ₹ [(190 × (182 + 273 + 609)] = ₹ (190 × 1064)

Required Ratio = 888 × 170 : 1064 × 190 = 1887 : 2527

171. (c) Area of 4 walls and ceiling of branch managers room
$= 2\,(lh + bh) + lb = 2\,[17 \times 12 + 13 \times 12] + 13 \times 17$
$= 941$ sq. m
Total cost of renovation = ₹190 × 941 = ₹178790

172. (e) Total area of bank is 2000 sq. m
Total area of bank to be renovated = 1952 sq. m
Remaining Area = 2000 – 1952 = 48 sq. m
Total cost Remaining Area to be carpeted at the rate of ₹110/sq. meter = ₹(48 × 110) = ₹5280

173. (b) Percentage area of bank not to be renovated

$$\Rightarrow \frac{\text{Area bank not be renovated}}{\text{Total area of bank}} \Rightarrow \frac{48}{2000} \times 100 = 2.4\%$$

174. (a) Total cost of hall of customer transaction
= ₹(170 × 667) = ₹ 113,390
Total cost of Locker area = ₹(190 × 609) = ₹115710
Total cost of customer transaction hall + locker area
= ₹(113390 + 115710) = ₹229100

175. (b) Required number of people = 21% of 29565
= 6208.65 = 6210

176. (a) Required difference = 9% of 26345 – 8% of 27456
= 2371.05 – 2196.48 = 174.57 = 175

177. (c) Number of defaulters of Bank R in the year
2004 = 15% of 25467 = 3820.05
2005 = 17% of 32461 = 5518.37
2006 = 16% of 32652 = 5224.32
2007 = 18% of 32561 = 5860.98
2008 = 13% of 25495 = 3314.35
2009 = 11% of 27649 = 3041.39
2010 = 12% of 28283 = 3393.96
Hence, maximum number of defaulters of Bank R is in the year 2007.

178. (a) Difference of number of people taking loan from Bank P from the previous year in the year
2005 = 32081 – 27361 = 4720
2006 = 32081 – 25361 = 6720
2007 = 25361 – 23654 = 1707
2008 = 36125 – 23654 = 12471
2009 = 36125 – 35465 = 660
2010 = 35465 – 34135 = 1330
Hence, the year is 2008.

179. (c) Required number of defaulters
= 19% of 36152 + 18% of 35463
= 6868.88 + 6383.34 = 13252.22 ≈ 13250

180. (a) Required ratio = [45000 × (8 + 9)%] : [45000 × (15 + 18)%] = 17 : 33.

181. (e) Qualified aspirants from C $= \frac{9000 \times 7\%}{45000 \times 8\%} \times 100 = 17.5\%$

From D $= \frac{9000 \times 21\%}{45000 \times 17\%} \times 100 = 24.70\%$

From E $= \frac{9000 \times 14\%}{45000 \times 9\%} \times 100 = 31.11\%$

From F $= \frac{9000 \times 11\%}{45000 \times 18\%} \times 100 = 12.22\%$

From G $= \frac{9000 \times 13\%}{45000 \times 22\%} \times 100 = 11.81\%$

Per cent is least in G.

182. (d) Required difference = 9000 × (21% – 13%)

$= 9000 \times \frac{8}{100} = 720$

183. (b) Required per cent $= \frac{9000 \times (16+7) \times 100}{45000 \times (11+8)} = 24.21\%$

184. (c) Number of candidates qualified from State (B + D)

$= 9000 \times \frac{(16+21)}{100} = 90 \times 37 = 3330$

Number of candidates appeared from state C

$= 45000 \times \frac{8}{100} = 3600$

Required ratio $= \frac{3330}{3600} = 37 : 40$

185. (a) From table we can say that expenditure on water supply and sanitation are increasing in every plan. So, the graph represent in option (a) is best explain of it.

186. (b) By watching table, we see that the ratio of public sector expenditure to the expenditure on social service was highest in the VIth plan.

187. (a) From table, we can say that in all the sector, there is no continuous decrease.

188. (d) Required percentage $= \frac{24880}{975000} \times 100\%$

= 2.551% = 2.5%

189. (c) Total expenditure on education in all the plans
= 1530 + 2730 + 5890 + 7860 + 13360 + 25240
= 56610 million

Total expenditure on health in all the plans

= 980 + 2140 + 2260 + 3370 + 7610 + 18210

= 34570 million

∴ Required difference = (56610 – 34570)

= 22040 million = ₹ 22040 × 1000000

= ₹ 22040000000

Sol. (Q. Nos. 190-194)

Number of Passengers in train A = 700

Number of Passenger in train B = 130% of 700 = 910

Coaches	General	Sleeper	First Class	AC	Total
Train A	140	161	224	175	700
Train B	241	273	91	305	910
Total	381	434	315	480	1610

190. (c) Number of passengers in first class coaches of train A = 224

Number of passengers in sleeper class coaches of train B = 273

∴ Required ratio $= \frac{224}{273} = \frac{7 \times 32}{7 \times 39} = 32:39$

191. (d) Passengers in the general coaches of train A = 140

Passengers in the AC coaches of train B = 305

∴ Total = 140 + 305 = 445

192. (e) The number of passengers in AC coaches of train A = 175

Total number of passengers in sleeper class coaches and first class coaches together of train B

= (273 + 91) = 364

∴ Their difference = 364 – 175 = 189

193. (b) Passengers in general class coach of train A and train B = (140 + 241) = 381

Total number of passengers in train B = 910

∴ Required percentage $= \frac{381}{910} \times 100\% = 41.8\% \approx 42\%$

194. (c) The cost of per ticket of first class coach = ₹ 450

Number of Passenger in first class coaches of Train A = 224.

Total amount = 450 × 224 = ₹ 100800

195. (e) Number of men working in the marketing department

$$3250 \times \frac{79.2^\circ}{360^\circ} \times \frac{3}{5} = 429$$

196. (c) ∴ Required ratio $= \dfrac{3250 \times \frac{36^\circ}{360^\circ} \times \frac{13}{25}}{3250 \times \frac{57.6^\circ}{360^\circ} \times \frac{7}{10}} = 13:28$

197. (e) Number of men working in the production department

$$= 3250 \times \frac{136.8^\circ}{360^\circ} \times \frac{4}{5} = 988$$

Total number of employees in production department

$$3250 \times \frac{136.8^\circ}{360^\circ} = 1235$$

Required percentage $= \frac{988}{1235} \times 100\% = 80\%$

198. (b) Number of women working in IT department

$$= 3250 \times \frac{57.6^\circ}{360^\circ} \times \frac{3}{10} = 156$$

Total number of employees = 3250

Required percentage $\frac{156}{3250} \times 100\% = 4.8\%$

199. (b) Number of men working in accounts department

$$= 3250 \times \frac{50.4^\circ}{360^\circ} \times \frac{6}{13} = 210$$

Number of men working in marketing department

$$= 3250 \times \frac{79.2^\circ}{360^\circ} \times \frac{3}{5} = 429$$

Number of men working in IT department

$$= 3250 \times \frac{57.6^\circ}{360^\circ} \times \frac{7}{10} = 364$$

Number of men working in HR department

$$= 3250 \times \frac{36^\circ}{360^\circ} \times \frac{12}{25} = 156$$

Number of men working in production department

$$= 3250 \times \frac{136.8^\circ}{360^\circ} \times \frac{4}{5} = 988$$

Hence, total number of men working in the organization

= 210 + 429 + 364 + 156 + 988 = 2147

200-204. Profit percentage in given years.

$$1991 \Rightarrow \frac{42}{360} \times 100 = 11.67\%$$

$$1992 \Rightarrow \frac{24}{360} \times 100 = 6.67\%$$

$$1993 \Rightarrow \frac{42}{360} \times 100 = 11.67\%$$

$$1994 \Rightarrow \frac{60}{360} \times 100 = 16.67\%$$

$1995 \Rightarrow \frac{54}{360} \times 100 = 15\%$

$1996 = \frac{72}{360} \times 100 = 20\%$

$1997 = \frac{66}{360} \times 100 = 18.33\%$

200. (d)

201. (a) x % of 100 = 16.67%

$\Rightarrow x = 16\frac{2}{3}\%$

202. (c)

203. (b) Average of year 1995

$= \frac{11.67 + 6.67 + 16.67 + 15 + 20 + 18.33}{7}$

= 14.28% ≈ 15%

204. (c)

205. (a) Central angle of the sector for cost of the paper

$= \frac{\text{Cost of paper }\%}{100} \times 360° = \frac{16}{100} \times 360° = 57.6°$

206. (d) If the cost of paper = ₹ x, then from the given pie-chart

$\frac{\text{Cost of paper}}{\text{Miscellaneous charges}} = \frac{16\%}{4\%}$

$\Rightarrow \frac{x}{6000} = \frac{16}{4} \Rightarrow x = \frac{16 \times 6000}{4} = ₹\ 24000$

207. (d) Let C be the cost price of 1 copy
Cost price of 5500 copies = 5500 C
4% of (5500 C) = 1848

$C = \frac{1848 \times 100}{5500 \times 4} = ₹ 8.40$

208. (c) By observing the graph, we can say that yr. 2009 has maximum percentage of students passed in the year.

209. (d) Year 2007.

210. (a)

211. (c) 20 + 12 + 16 = 48

212. (b) Number of students scoring less than 50%

= (240 + 220 + 300 + 280 + 210) = 1250

Number of student scoring exact 50%

= (30 + 20 + 0 + 35 + 15) = 100

Ratio = 1250 : 100 = 25 : 2

213. (a)

214. (c) Number of students scoring 50% or more marks

= (600 – 240) + (400 – 220) + (375 – 300) + (350 – 280) + (300 – 210)

= 360 + 180 + 75 + 70 + 90 = 775

215. (b) Profit during 2006 = (60 – 35) = 25

Profit during 2007 = (50 – 40) = 10

Difference = 25 – 10 = 15

216. (c) Average Income $= \frac{40 + 60 + 50 + 65 + 70}{5} = \frac{285}{5} = 57$

Income during 2005 and 2007 is less than average

217. (b) Required % $= \frac{50 - 40}{40} \times 100 = 25\%$

218. (c) It is clear from the graph.

CHAPTER 25

DATA SUFFICIENCY

Data sufficiency problems consists of a question and two statements. These statements contain data or information. Questions are based on any topic of the sections Numbers, Arithmetic, Algebra, Geometry, Counting Principles, Trigonometry and Data Interpretation, etc., which we have already studied.

But questions are generally based on the topics of the sections – Numbers, Arithmetic, Algebra and Geometry.

In data sufficiency problems, we have to decide whether the data given in the statements labelled as (1) and (2) are sufficient to solve the given problem. There will be five possible answers –

Option (a), if statement (1) itself is sufficient to answer the question, but statement (2) itself is not;

Option (b), if statement (2) itself is sufficient to answer the equestion, but statement (1) itself is not;

Option (c), if statement (1) and (2) taken together are sufficient to answer the question, even though neither statement by itself is sufficient.

Option (d), if either statement by itself is sufficient to answer the question.

Sometimes, there may be fifth option

Option (e), Both statements together are not sufficient.

Remember

To answer data sufficiency question, we do not actually have to calculate the values. We only have to determine whether the data given in the statements are sufficient for finding the answer.

There are various variations in Data Sufficiency problems :

- Options may be four or five.
- Option may be jumbled.
- One or two options may be new type, while remaining are same.
- There may be three or four statements follow a question.

To attempt a question of data sufficiency, we should follow a systematic approach. A good way is to use a flow chart.

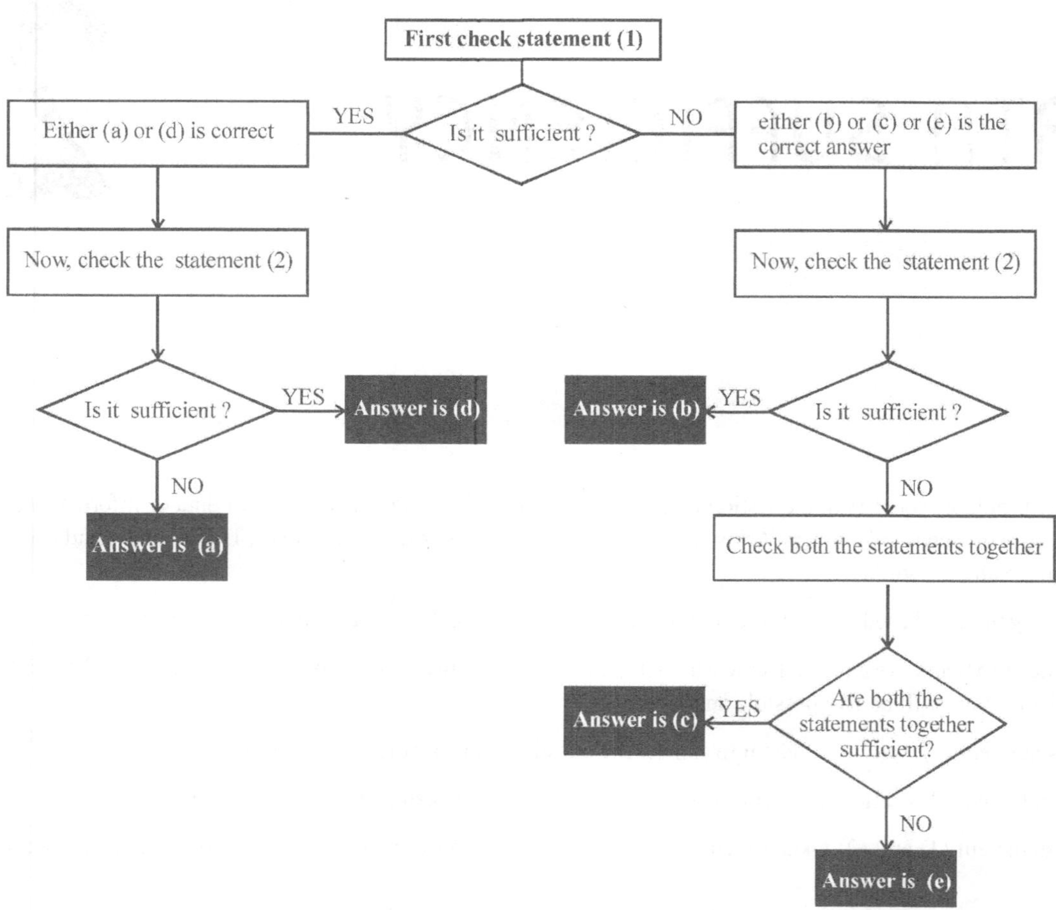

MISCELLANEOUS EXAMPLES

NUMBERS :

Directions : **The following questions accompanied by two statements numbered (I) and (II). You have to decide whether the data provided in the statements are sufficient to answer the questions. Read both the statements and give answer**

(a) If the data in statement I alone are sufficient to answer the questions, while the data in statement II alone are not sufficient to answer questions

(b) If the data in statement II alone are sufficient to answer the question, while the data in statement I alone are not sufficient to answer questions

(c) If the data even in both statements I and II together are not sufficient to answer the question

(d) If the data in both statements I and II together are necessary to answer the questions

Illustration 1:

A, B and C are integers. Is B an even number?

I. (A + B) is an odd number.

II. (C + B) is an odd number.

Solution :

(c) (I) $\Rightarrow$ A + B is odd $\Rightarrow$ either A is odd and B is even or A is even and B is odd.

(II) $\Rightarrow$ C + B is odd $\Rightarrow$ either C is odd and B is even or C is even and B is odd.

Hence, question can't be answered even with the help of two statements. Therefore, both statements together are not sufficient to answer the question.

ARITHMETIC :

Directions : **Each of the questions below consists of a question and two statements numbered (I) and (II) given below it. You have to decide whether the data provided in the statements are sufficient to answer the question. Read both the statements and give answer.**

(a) Statement (I) alone is sufficient, but statement (II) alone is not sufficient.

(b) Statement (II) alone is sufficient, but statement (I) alone is not sufficient.

(c) Both statements together are sufficient.

(d) Both statements together are insufficient.

Illustration 2:

A train crosses another train running in the opposite direction in x seconds. What is the speed of the train?

I. Both the trains have the same length and are running at the same speed.

II. One train crosses a pole in 5 seconds.

Solution :

(d) When the trains are running in opposite direction, time taken to cross each other $= \frac{L_1 + L_2}{S_1 + S_2}$,

where,

L_1 = Length of I train.

S_1 = Speed of I train.

L_2 = Length of II train.

S_2 = Speed of II train.

(I) $\Rightarrow L_1 = L_2 = L$ (let)

$S_1 = S_2 = S$ (let)

$\Rightarrow 2S = \frac{2L}{x} \Rightarrow S = \frac{L}{x}$

(II) $\Rightarrow S = \frac{L}{5}$

Since distance covered (length of train) by the train is not given so, we can't find the value of S even with the help of both statements.

Illustration 3:

Directions : **The question below consists a question and two statements numbered (I) and (II). You have to decide whether the data provided in the statements are sufficient to answer the question. Read both the statements and give answer.**

What was the total compound interest on a sum after three years?

I. The interest after one year was ₹ 100 and the sum was ₹ 1000.

II. The difference between simple and compound interest on a sum of ₹ 1000 at the end of two years was ₹ 10.

(a) Statement (I) ALONE is sufficient, but statement (II) alone is not sufficient.

(b) Statement (II) ALONE is sufficient, but statement (I) alone is not sufficient.

(c) EACH statement ALONE is sufficient.

(d) BOTH statements TOGETHER are sufficient

Solution :

(c) (I) $\Rightarrow$ Interest and sum are given and

$$I = P\left(1 + \frac{r}{100}\right) - P$$

$\Rightarrow$ r can be calculated.

$\Rightarrow$ C.I. after three years can be found.

Hence, statement (I) alone is sufficient to answer the question.

(II) $\Rightarrow$ Since, difference between S.I.

and C.I. $= P \times \left(\frac{r}{100}\right)^n$

$\Rightarrow$ r can be calculated.

Hence, C.I. after three years can be found.

Therefore, statement (II) alone is sufficient to answer the question.

Illustration 4:

Directions : **The following question accompanied by three statements I, II and III. We have to determine which statement(s) is/are sufficient to answer the question.**

What is R's share of profit in a joint venture?

I. Q started a business investing ₹ 80,000/

II. R joined him after 3 months.

III. P joined after 4 months with a capital of ₹ 1, 20,000 and got ₹ 6,000 as his share of profit.

(a) Only (I) & (III) are sufficient.

(b) Only (II) & (III) are sufficient.

(c) All (I), (II) and (III) together are necessary.

(d) Even with all (I), (II) and (III), the answer cannot be arrived at.

Solution :

(d) Since investment amount of R is not given. Therefore, question cannot be answered even with the help of all I, II and III.

Illustration 5:

Directions : **The following question accompanied by three statements I, II and III. We have to determine which statements is/are sufficient to answer the question.**

In how many days can a work be completed by A and B together?

I. A alone can complete the work in 8 days.

II. If A alone works for 5 days and B alone works for 6 days, the work gets completed.

III. B alone can complete the work in 16 days.

(a) Any two of three are sufficient.

(b) II and either I or III are sufficient.

(c) I and II only are sufficient.

(d) II and III only are sufficient.

Solution :

(a) Since any two of three statements give individual work speed of A and B. So, by combining any two of the statements we can easily find the rquired number of days.

Illustration 6:

The following question accompanied by three statements I, II and III. We have to determine which statements is/are sufficient to answer the question.

How many articles were sold?

I. Total profit earned was ₹ 1, 596.

II. Cost price per article was ₹ 632.

III. Selling price per article was ₹ 765.

(a) Only II and III are sufficient.

(b) Only I and II are sufficient.

(c) All I, II and III together are necessary.

(d) Any two of three are sufficient.

Solution :

(c) Let number of articles = N

(II) ⇒ Total C.P. = 632 N

(III) ⇒ Total S.P. = 765 N

(I) ⇒ N(765 – 632) = 1,596

Hence, number of articles can be determined by using all the three statements (I, II and III) together.

Illustration 7:

Directions : **The following question accompanied by two statements numbered (I) and (II). You have to decide whether the data provided in the statements are sufficient to answer the question.**

What is the number x if

I. The LCM of x and 18 is 36.

II. The HCF of x and 18 is 2.

(a) The question can be answered the help of both the statements but not with the help of either statement itself.

(b) The question can not be answered even with the help of both the statements.

(c) The question can be answered with the help of statement II alone

(d) The question can be answered with the help of statement I alone

Solution :

(a) Since product of two numbers = their HCF × LCM

Hence, both the statements together are necessary to answer the question.

Illustration 8:

Directions : **The following question accompanied by two statements numbered (I) and (II). You have to decide whether the data provided in the statements are sufficient to answer the question.**

What is the present age of the Mother?

I. Father's age is eight years more than the Mother's age. Father got married at the age of 28 years.

II. Present age of the Father is 30 years. Four years back the ratio of Mother's age to Father's age was 12 : 13.

(a) The data in statement I alone are sufficient to answer the question, while the data in statement II alone are not sufficient to answer question

(b) The data in statement II alone are sufficient to answer the question, while the data in statement I alone are not sufficient to answer question

(c) The data in statement I alone or in statement II alone are sufficient to answer the question

(d) The data even in both statements I and II together are not sufficient to answer the question

Solution :

(b) Let present age of Mother = x years.

Let present age of Father = y years.

(I) ⇒ y = x + 8

Hence, from (I) we can't be determined the present age of mother.

(II) ⇒ y = 30 years and $\frac{x-4}{30-4}=\frac{12}{13} \Rightarrow \frac{x-4}{26}=\frac{12}{13}$

Hence, present age of mother (x) can be determined by using statement (II) alone.

Illustration 9:

Directions : **The following question accompanied by two statements numbered (I) and (II). You have to decide whether the data provided in the statements are sufficient to answer the question.**

What is the first term of an arithmetic progression of positive integers?

I. Sum of the squares of the first and second term is 116

II. The fifth term is divisible by 7.

(a) The question can be answered with the help of both the statements but not with the help of either statement itself.

(b) The question can not be answered even with the help of both the statements.

(c) The question can be answered with the help of statement II alone

(d) The question can be answered with the help of statement I alone

Solution :

(d) (II) is useless

(I) shows that the two integers on squaring add up to $116 < 11^2$ which means the integers are less than 10. We further find that the numbers are 10 and 4.

Illustration 10:

Directions : **The following question accompanied by two statements numbered (I) and (II). You have to decide whether the data provided in the statements are sufficient to answer the question.**

Navni is three times as efficient as Avantika. In how many days can Avantika alone do the whole work?

I. Both of them together can complete the work in 15 days.
II. Navni alone can do the whole work in 20 days.
(a) The data in statement I alone are sufficient to answer the question, while the data in statement II alone are not sufficient to answer question
(b) The data in statement II alone are sufficient to answer the question, while the data in statement I alone are not sufficient to answer question
(c) The data in statement I alone or in statement II alone are sufficient to answer the question
(d) The data even in both statements I and II together are not sufficient to answer the question

Solution :

(c) Let Navni takes to finish the work = x days
Then Avantika takes to finish the work = 3x days.

(I) $\Rightarrow \frac{1}{x}+\frac{1}{3x}=\frac{1}{15}$

Obviously, x can be determined.
Hence, statement (I) alone is sufficient to answer the question.
(II) $\Rightarrow$ x = 20 days and since, Avantika takes time = 3x days
Hence, statement (II) alone is sufficient to answer the question.

Illustration 11:

Directions : **The following question accompanied by two statements numbered (I) and (II) given below it. You have to decide whether the data provided in the statements are sufficient to answer the question.**

There are 450 boxes to load on a truck. A and B working independently but at the same time they take 30 minutes to load the truck. How long should it take B working by himself to load the truck?

I. A loads twice as many boxes as B.
II. A would take 45 minutes by himself.
(a) Statement I alone is sufficient to answer the question but statement II alone is not sufficient to answer question
(b) Statement II alone is sufficient to answer the question but statement I alone is not sufficient to answer question
(c) You can get the answer from I and II together although neither statement by itself sufficient
(d) Statement I alone or statement II, alone is sufficient

Solution :

(d) (I) $\Rightarrow$ A loaded 300 boxes in 30 minutes and B loaded 150 boxes in 30 minutes. So B should take 90 minutes to load the 450 boxes by himself.
(II) $\Rightarrow$ A loads 10 boxes per minute, hence A loads 300 boxes in 30 minutes. Therefore, B loaded 150 boxes in 30 minutes and hence, B will take 90 minutes to load all the 450 boxes.
$\therefore$ Each statement alone is sufficient to answer the question.

Directions : **The question consists of a question and two statements numbered (I) and (II) given below it. You have to decide whether the data provided in the statements are sufficient to answer the questions. Read both the statements and give answer**

(a) If the data in statement I alone are sufficient to answer the questions, while the data in statement II alone are not sufficient to answer questions
(b) If the data in statement II alone are sufficient to answer the question, while the data in statement I alone are not sufficient to answer questions
(c) If the data in statement I alone or in statement II alone are sufficient to answer the questions
(d) If the data in both statements I and II together are necessary to answer the questions

Illustration 12:

What is the rate of interest p.c.p.a?

I. Difference between compound interest and simple interest on an amount of Rs 10,000 for two years is Rs 225.
II. The amount doubles itself on simple interest in $6\frac{2}{3}$ years.

Solution :

(c) Let rate of interest = r % per year

(I) $\Rightarrow$ Difference between S.I. and C.I. $= P \times \left(\frac{r}{100}\right)^2$

Hence, r can be determined by using statement (I) alone.

(II) $\Rightarrow$ S.I. = Amount = P

and S.I. $= \frac{p \times r \times t}{100}$

$\therefore$ r can be determined by using statement (II) alone.

Illustration -13-15:

Directions : **The following questions accompanied by two statements numbered (I) and (II). You have to decide whether the data provided in the statements are sufficient to answer the questions. Read both the statements and give answer**

(a) If the data in statement I alone are sufficient to answer the questions, while the data in statement II alone are not sufficient to answer questions
(b) If the data in statement II alone are sufficient to answer the question, while the data in statement I alone are not sufficient to answer questions
(c) If the data even in both statements I and II together are not sufficient to answer the question
(d) If the data in both statements I and II together are necessary to answer the questions

Illustration 13 :

Find the number of males and females in a party.

I. The ratio of males and females is 3 : 2

II. The number of males is 50% more than the number of females.

Solution :

(c) Let number of males = M and number of females = F

$(I) \Rightarrow \frac{M}{F} = \frac{3}{2}$

$(II) \Rightarrow M = F\left(1 + \frac{50}{100}\right) = \frac{3}{2}F \Rightarrow \frac{M}{F} = \frac{3}{2}$

Since, both statements give same equations.

So, we can't solve the question even with the help of both statements together.

Illustration 14 :

The towns A, B and C lie on a straight line. C is between A and B. The distance from A to B is 100 miles. How far is it from A to C?

I. The distance from A to B is 25% more than the distance from C to B.

II. The distance from A to C is $\frac{1}{4}$ the distance from C to B.

Solution :

(d) AB = 100 miles and AC + CB = 100 miles.

$(I) \Rightarrow AB = CB\left(1 + \frac{25}{100}\right)$

$\Rightarrow 100 = \frac{5}{4}CB \Rightarrow CB = 80 \text{ miles}$

$(II) \Rightarrow AC = \frac{1}{4}(CB)$

Hence, question can be answered using both statements together.

Illustration 15 :

Plane X flies at r miles per hour from A to B. Plane Y flies at a miles per hour from B to A. Both planes take off at the same time. Which plane flies at a faster rate? Town C is between A and B.

I. C is closer to A than it is to B.

II. Plane X flies over C before plane Y.

Solution :

(c) Since we have no exact idea about distance between A to C and B to C.

Therefore, question can't be answered even with the help of both statements together.

Illustration 16 :

Directions : **The question is accompanied by three statements (I), (II) and (III). You have to determine which statement(s) is/are sufficient/necessary to answer the questions.**

In how many days can 10 women finish a work?

I. 10 men can complete the work in 6 days.

II. 10 men and 10 women together can complete the work in $3\frac{3}{7}$ days.

III. If 10 men work for 3 days and thereafter 10 women replace them, the remaining work is completed in 4 days.

(a) only I and II (b) Any two of the three

(c) only I and III (d) Only II and III

(e) None of these

Solution :

(b) (I) ⇒ One day's work of 10 men = $\frac{1}{6}$th part of whole work.

(II) ⇒ One day's work of 10 men and 10 women = $\frac{7}{24}$th part of the whole work.

(III) ⇒ Let whole work done by 10 men in x days ⇒ work done by 10 men in 3 days = $\frac{3}{x}$th part of the whole work

Remaining work = $1 - \frac{3}{x} \Rightarrow \frac{x-3}{x}$

Hence, any two statements are sufficient to answer the question.

Illustration 17 :

Directions : **The question is accompanied by three statements (I), (II) and (III). You have to determine which statements is/are sufficient/necessary to answer the questions. Read the statement and give answer.**

What is the staff strength of Company 'X'?

I. Male and female employees are in the ratio of 2 : 3 respectively.

II. Of the officer employees 80 % are males.

III. Total number of officers is 132.

(a) I and III only (b) II and either III or I only

(c) All I, II and III (d) Any two of the three

(e) Question cannot be answered even with the information in all the three statements

Solution :

(e) Here, we have no idea of employees in different categories. Therefore, question can't be solved even using all the statements together.

Illustration 18 :

Directions : **The question is accompanied by three statements (I), (II) and (III). You have to determine which statements is/are sufficient/necessary to answer the questions.**

What is the monthly salary of Pravin?

I. Pravin earns ₹ 1, 200 more than Aman.

II. The ratio between Aman's and Vimal's monthly salary is 5 : 3.

III. Vimal earns ₹ 1, 000 less than Aman.

(a) Any two of I, II and III are required
(b) Only I and II are required
(c) Only II and III are required
(d) All I, II and III together are required
(e) None of these

Solution :

(d) Let monthly salary of Pravin = ₹ P and monthly salary of Aman = ₹ A.

(I) $\Rightarrow P = A + 1200$

(II) $\Rightarrow \frac{A}{V} = \frac{5}{3}$, where V is monthly salary of Vimal.

(III) $\Rightarrow A = V + 1000$

Hence, all statements are necessary to answer the question.

Illustration 19 :

Directions : **The question is accompanied by three statements (I), (II) and (III). You have to determine which statements is/are sufficient/necessary to answer the questions.**

What was the amount of profit earned?

I. 10% discount was offered on the labelled price.

II. Had there been no discount, profit would have been 30%.

III. Selling price was more than the cost price by 20%.

(a) I and either II or III
(b) Any two of the three
(c) All I, II and III
(d) Either I or II and III
(e) Question cannot be answered even with the information in all three statements

Solution :

(e) Since we have no idea about labelled price or S.P. Therefore, question can't be answered even with the help of all statements together.

Illustration 20 :

Directions : **The question is accompanied by three statements (I), (II) and (III). You have to determine which statement(s) is/are sufficient/necessary to answer the questions. Read the statement and give answer.**

(a) I only (b) II only
(c) III only (d) I or III only
(e) I or II or III

Three friends X, Y and Z started a partnership business investing money in the ratio of 5 : 4 : 2 respectively for a period of 3 years. What is the amount received by X as the share in the total profit?

I. Total amount invested in the business is ₹ 22,000.

II. Profit was distributed after a period of 2 years.

III. The average amount of profit earned per year is ₹ 2,750.

(a) I and either II or III
(b) Any two of the three
(c) All I, II and III
(d) Either I or II and III
(e) Question cannot be answered even with the information in all three statements

Solution :

(c) When investment ratio is given, the profit of X can be found out if total profit is known. Therefore, statement (III) alone is sufficient.

ALGEBRA :

Directions : **Each of the questions below consists of a question and two statements numbered (I) and (II) given below it. You have to decide whether the data provided in the statements are sufficient to answer the question. Read both the statements and give answer.**

(a) Statement (I) alone is sufficient, but statement (II) alone is not sufficient.
(b) Statement (II) alone is sufficient, but statement (I) alone is not sufficient.
(c) Both statements together are sufficient.
(d) Both statements together are insufficient.

Illustration 21 :

50% of the people in a Steeltown have blue eyes and blond hair. What per cent of the people in Steeltown have blue eyes but do not have blond hair?

I. 70% of the people in Steeltown have blond hair.

II. 60% of the people in Steeltown have blue eyes.

Solution :

(b)

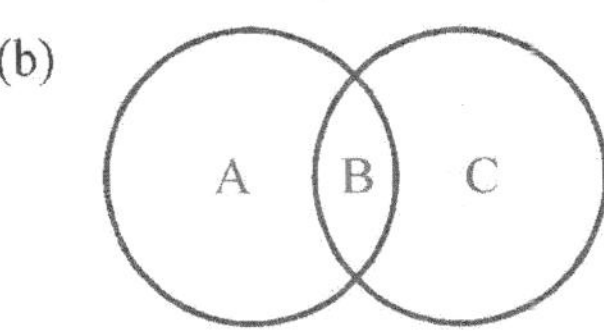

A = People have only blue eyes.

C = People have only blond hair.

B = People have blue eyes and blond hair.

and B = 50%

(I) $\Rightarrow B + C = 70\%$

(II) $\Rightarrow A + B = 60\%$

We have to find 'A'

∴ statement (II) alone is sufficient.

Illustration 22 :

Directions : **The following question accompanied by three statements I, II and III. We have to determine which statements is/are sufficient to answer the question.**

What is the sum of two numbers?

I. The bigger of these two numbers is 6 more than the smaller number.

II. 40% of the smaller number is equal to 30% of the bigger number.

III. The ratio between half of the bigger number and one-third of the smaller number is 2 : 1.

(a) Only II and III are sufficient.
(b) Only I and II are sufficient.
(c) I and either II or III is sufficient.
(d) All, II and III together are sufficient.

Solution :

(c) From the given statements we can make the following equations.

(I) $\Rightarrow y = x + 6$

(II) $\Rightarrow 0.4x = 0.3y \Rightarrow \frac{x}{y} = \frac{3}{4}$

(III) $\Rightarrow \frac{y/2}{x/3} = \frac{2}{1} \Rightarrow \frac{y}{x} = \frac{4}{3} \Rightarrow \frac{x}{y} = \frac{3}{4}$

Obviously, question can be solved by using (I) and either (II) or (III) because equations (II) and (III) are same.

Illustration -23-24:

Directions : **Each of the questions below consists of a question and two statements numbered (I) and (II) given below it. You have to decide whether the data provided in the statements are sufficient to answer the question. Read both the statements and give answer**

(a) If the data in statement I alone are sufficient to answer the question, while the data in statement II alone are not sufficient to answer questions
(b) If the data in statement II alone are sufficient to answer the question, while the data in statement I alone are not sufficient to answer questions
(c) If the data in statement I alone or in statement II alone are sufficient to answer the question
(d) If the data in both statements I and II together are necessary to answer the question

Illustration 23 :

The sum of cubes of three consecutive even numbers is what per cent of the sum of the squares of the same nos.?

I. The average of those nos. is 21.

II. The product of those nos. is 2688.

Solution :

(c) Let numbers are (x –2), x and (x + 2).

(I) $\Rightarrow (x-2) + x + (x+2) = 3 \times 21$

Hence, we can find value of x.

Then, question can be solved using statement (I) alone.

(II) $\Rightarrow x(x-2)(x+2) = 2688$

Obviously, question can be solved using statement (II) alone.

Illustration 24 :

What is the speed of a boat?

I. The boat covers a distance of 48 km in 6 hours while running upstream.

II. It covers the same distance in 4 hours while running downstream.

Solution :

(d) Let the speed of boat = V_b km/h

and the speed of stream = V_s km/h

(I) $\Rightarrow V_b - V_s = 48/6$

$\Rightarrow V_b - V_s = 8$

(II) $\Rightarrow V_b + V_s = 48/4$

$\Rightarrow V_b + V_s = 12$

On solving (I) and (II), we can be determined the value of V_b.

Hence, both the statements together are necessary to answer the question.

GEOMETRY :

Directions : **Each of the questions below consists of a question and two statements numbered (I) and (II) given below it. You have to decide whether the data provided in the statements are sufficient to answer the question. Read both the statements and give answer.**

(a) Statement (I) alone is sufficient, but statement (II) alone is not sufficient.
(b) Statement (II) alone is sufficient, but statement (I) alone is not sufficient.
(c) Both statements together are sufficient.
(d) Both statements together are insufficient.

Illustration 25 :

What is the area of 1 metre wide pathway constructed around all the sides of a rectangular field?

I. Area of the field and the path together is 187 sq. cm.

II. Breadth and the length of the field are in the ratio of 3 : 5 respectively.

Solution :

(c)

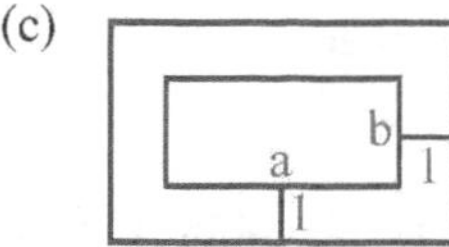

(I) $\Rightarrow (a+2)(b+2) = 187$

(II) $\Rightarrow \frac{b}{a} = \frac{3}{5}$

With the help of both statements, we can calculate 'b'.

Illustration 26 :

Directions : **The following question accompanied by two statements numbered (I) and (II). You have to decide whether the data provided in the statements are sufficient to answer the question.**

Is the figure ABCD a rectangle?

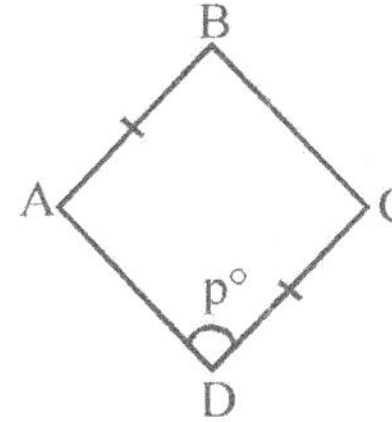

I. p = 90° II. AB = CD

(a) Statement (I) alone is sufficient to answer the question, but statement (II) alone is not sufficient.
(b) Statement (II) alone is sufficient, but but statement (I) alone is not sufficient.
(c) Both statements (I) & (II) together are sufficient, but neither statement alone is sufficient.
(d) Even both statements (I) and (II) together are not sufficient to answer the question.

Solution :

(d) (I) $\Rightarrow$ p = 90°
(II) $\Rightarrow$ AB = CD
Since, we have no idea about other angles and length of sides AD and BC. Therefore, nothing can be said about ABCD. Hence both the statements together are not sufficient to answer the question.

Illustration 27 :

Directions : **The following question accompanied by two statements numbered (I) and (II). You have to decide whether the data provided in the statements are sufficient to answer the question.**

What is the width of a stone slab of length 4 metres and thickness $\frac{1}{16}$th of the length?

I. Volume of the slab is 2 cu metres.
II. Surface area of the slab is 19 sq metres.

(a) Only I is sufficient
(b) Neither I nor II is sufficient
(c) Both I and II are necessary
(d) Either I or II is sufficient but the answers may be different

Solution :

(d) (I) $\Rightarrow l \times b \times h = 2$
Hence, width of slab can be determined by using (I) alone.
(II) $\Rightarrow$ Surface Area = 19 m^2
$\Rightarrow 2(bh + hl + lb) = 19.$
Obviously, question can be answered by using (II) alone.
Hence, each statement alone is sufficient.

Illustration 28 :

Directions : **The following question accompanied by two statements numbered (I) and (II). You have to decide whether the data provided in the statements are sufficient to answer the question.**

What is the area of a circular region X?

I. Area of the square inscribed in the circular region X is 98 sq cm.
II. The perimeter of the circular region X is equal to the perimeter of square Z having area 121 cm^2.

(a) The question can be answered by using one of the statements alone, but cannot be answered using the other statement alone
(b) The question can be answered by using either statement alone
(c) The question can be answered by using both statements together, but cannot be answered using either statement alone
(d) The question cannot be answered even by using both statements together

Solution :

(b) We know that Area of the square inscribed in the circle = $3r^2$, where r is the radius of circle.
(I) $\Rightarrow 3r^2 = 98$
$\Rightarrow$ r can be determined.
$\Rightarrow$ Area of circle can be determined.
(II) $\Rightarrow$ Area of circle = $\frac{4 \times 121}{\pi}$ [since, the area of a square is x cm^2, then area of circle formed by the same perimeter is $\frac{4x}{\pi}$]
Hence, question can be answered by using each statement alone.

Directions : **The following questions accompanied by two statements numbered (I) and (II). You have to decide whether the data provided in the statements are sufficient to answer the questions. Read both the statements and give answer**

(a) If the data in statement I alone are sufficient to answer the questions, while the data in statement II alone are not sufficient to answer questions
(b) If the data in statement II alone are sufficient to answer the question, while the data in statement I alone are not sufficient to answer questions
(c) If the data even in both statements I and II together are not sufficient to answer the question
(d) If the data in both statements I and II together are necessary to answer the questions

Illustration 29 :

What will be the cost of painting the four walls and the roof of a room with length, width and height as 7 metres, 5 metres and 2.5 metres respectively. The room has one door and two windows.

I. Cost of painting per square metre is Rs 72.

II. Area of window is half of the area of the door.

Solution :

(c) Since the dimension of the window and door is not given, therefore, question can't be solved by using both statements together.

Illustration 30 :

***Directions :* The question is accompanied by three statements (I), (II) and (III). You have to determine which statement(s) is/are sufficient/necessary to answer the questions. Read the statement and give answer.**

What is the capacity of a cylindrical tank?

I. The radius of the base is half of its height.

II. The area of the base is 616 sq. metres.

III. The height of the cylinder is 28 metres.

(a) only I and II

(b) only II and III

(c) only I and III

(d) All I, II and III

(e) Any two of the three

Solution :

(e) Capacity of a cylindrical tank = $\pi r^2 h$.

Hence, to find the answer, we need either radius of tank or area of the base and height of the cylinder.

Therefore, any two of three are sufficient to answer it.

Practice Exercise

Level - I

DIRECTIONS (Qs. 1 to 61) : ***Each of the questions below consists of a question and two statements numbered (I) and (II) given below it. You have to decide whether the data provided in the statements are sufficient to answer the question. Read both the statements and give answer***

(a) Statement (I) ALONE is sufficient, but statement (II) alone is not sufficient.

(b) Statement (II) ALONE is sufficient, but statement (I) alone is not sufficient

(c) BOTH statements TOGETHER are sufficient, but NEITHER statement alone is sufficient.

(d) EACH statement ALONE is sufficient

1. How many employees of bank 'X' opted for VRS?
(I) 18% of the 950 officer cadre employees and 6% of the 1100 of all other cadre employees opted for VRS.
(II) 28% of the employees in the age group of 51 to 56 and 17% of the employees in all other age groups opted for VRS.

2. The area of a playground is 1600 square metres. What is its perimeter?
(I) It is a perfect square playground
(II) It costs Rs 3200 to put a fence around the playground at the rate of Rs 20 per metre.

3. Who among P, Q, R, S, T and U teaches Biology? Each person teaches one subject amongst English, Hindi, Biology, History, Geography and Sanskrit.
(I) P and U do not teach Hindi, Biology, History or Geography. Q and S do not teach Biology or Geography. T does not teaches Geography.
(II) P and U teach English and Sanskrit but not necessarily respectively. Similarly, Q and S teach either Hindi or History.

4. What is the value of $m - n \div 37$?
(I) m is the largest possible six-digit number and n is the smallest possible six-digit number.
(II) The diffierence between m and n is known.

5. What is the value of x ?
(I) If 10 is added to x, the result is 34.
(II) If x is multiplied by y, the result is 38.

6. By selling a product for ₹ 100 how much profit was earned?
(I) 20% profit would have been earned if it had been sold for ₹ 90.
(II) The profit was one–third of the purchase price.

7. What was the total compound interest on a sum after three years?
(I) The interest after one year was ₹ 100 and the sum was ₹ 1000.
(II) The difference between simple and compound interest on a sum of Rs. 1000 at the end of two years was ₹ 10.

8. How many hours a day must 4 pumps work to empty a conical tank in 1 day?
(I) 3 pumps working 8 hours a day can empty another tank in 2 days.
(II) The other tank has twice the floor area and one and a half times the depth of the original tank.

9. By selling a product at 20% profit, how much profit was earned?
(I) The difference between cost and selling price is ₹ 40.
(II) The selling price is 120 per cent of the cost price.

10. How much is the average salary of the 30 assembly workers? The foreman is paid a salary of ₹ 12,000.
(I) The total salary paid to the 30 assembly workers and the foreman is ₹ 312,000
(II) The foreman's salary is 120% of the average salary of the 30 assembly workers.

11. A bag contains coins of one-rupee, 50 -paise and 25–paise denominations. The total amount in the bag is Rs.500 . To find the total number of 50–paise coins, which of the following information is sufficient?
(I) The number of the coin is in the ratio 3 : 4 : 5.
(II) The number of one rupee–coins is one– fourth the total number of coins in the bag.

12. Two trains of length 80 m and 100 m are moving in opposite directions on parallel tracks. If they cross each other in 36 second, how much time will they take to pass each other if they move in the same direction?
(I) First one train passes a pole in one minute.
(II) The other train also passes a 120 m long railway bridge in one minute.

13. Kundu and Puttu can do a work in 12 days. In how many days will Puttu complete the work working alone?
(I) Kundu can complete the work in 36 days.
(II) Kundu is half as efficient as Puttu.

14. The speed of a 110 metres long running train 'X' is 45 per cent more than the speed of another 160 meters long train 'Z' running in opposite direction. What is the speed of the train 'Z'?
(I) The two train crossed each other in 6.5 seconds.
(II) The difference between the speed of the two trains was 28 km/hr.

15. A man holding 7 cards in his hand. Four are 'nines' and three are 'fives'. How many cards does he lay on the table?
(I) He lays a card on the table if the number on the card is divisible by 3.
(II) He lays a card on the table if and only if the number n is divisible by 3.

16. What is the percentage of families in the city have telephones ?
(I) 50 % of the families of the city have televisions.
(II) 30% of the television owners of the city have telephones.

17. How many chocolates can Sheena buy if she has to spend 20% of her budget on vegetable and 30% on groceries ?
(I) Sheena has Rs 50 with her.
(II) Each chocolate costs 50 paise.

18. How many litres of a chemical can be stored in a cylindrical tank if the radius of the tank is 5 metres?
(I) The height of the tank is 5 m.
(II) The temperature is 70°F.

19. In ΔABC, find r if AB = 5 and q = 40°

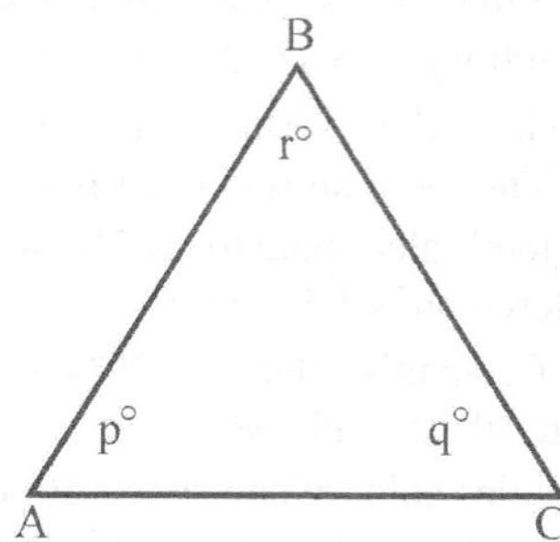

(I) BC = 5
(II) r > p

20. A certain book shop sold 92 copies of Bill Gate's book '*Business at the speed of thought*' during the first day. What precentage of his stock of this book did he sell on that day ?
(I) The total number of books with this title on stock is 230.
(II) There are a total of 1000 books in his store.

21. What selling price should be marked on the article ?
(I) Discount of 5% is to be given and profit percentage should be double the discount. Purchase cost is in the range of ₹ 300 – ₹ 400.
(II) 10% discount is to be allowed and 15% profit is to be obtained on the purchase cost of Rs 200 of the article.

22. At present, how many villagers are voters in village 'X'?
(I) There were 860 voters in village 'X' in the list prepared for the last election.
(II) The present list of village 'X' has 15% more voters than the list for the last election.

23. "You must submit your application within 10 days from the date of release of this advertisement." What is the exact date before which the application must be submitted?
(I) The advertisement was released on 18th February.
(II) It was a leap year.

24. What is Sunil's position in a row of forty students?
(I) There are sixteen students towards the left of Sunil.
(II) There are twenty-three student towards the right of Sunil ?

25. A person bought 2 kg of rice from a shop. But when he entered the next shop he found that the price was less. He calculated that if he bought 6 kg from that shop his average price would be Rs. 20. Then what is the price in the first shop?
(I) The price in the second shop was ₹ 18 per kg.
(II) The difference in the prices was ₹ 8 per kg.

26. Is the number $\frac{M}{3}$ an odd integer? (You may assume that $\frac{M}{3}$ is an integer)
(I) M = 3K, where K is an integer.
(II) M = 6J + 3, where J is an interger.

27. A worker is hired for 6 days. He is paid Rs. 5 more for each day of work than he was paid for the preceding day. How much was he paid for the first day of the work ?
(I) His total wages for 6 days were Rs. 900.
(II) He was paid less than Rs 100 on the first day.

28. How much cardboard will it take to make an open cubical box with no top ?
(I) The area of the bottom of the box is 4 square metres.
(II) The volume of the box is 8 cubic metres.

29. How much minimum marks will be require to pass an examination?
(I) Student A secured 32% marks in that examination and he failed by 1 mark. Student B secured 36% marks in the same examination and his marks was 1 more than the minimum pass marks.
(II) Student A secured 30% of full marks in the examination and he failed by 2 marks. If he had secured 5 more marks his percentage of marks would have been 40%.

30. What number is opposite the face bearing number 5 of the same dice when it is thrown by different persons?
Note that the dice bears the numbers 1 to 6 on different faces of the dice.
(I) When Renu throws the dice on a table, she observes that the visible surfaces bear the numbers 3, 6 and 5, while, when her sister throws the dice, she observes that the visible surfaces are 1, 4 and 5.
(II) When Tulika throws the same dice, she finds herself unable to see the three faces bearing the numbers 1, 3 and 5 while when her brother Shivendra throws the dice he finds himself able to see the three faces bearing the numbers 6, 4 and 5.

31. Find the length of AB?
If ∠YBC = ∠CAX = ∠YOX = 90°

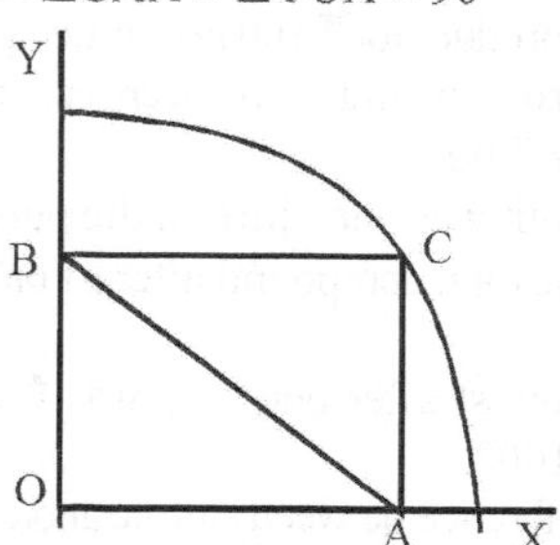

(I) Radius of the Arc is given
(II) OA = 5

32. In a cricket match, the'man of the match' award is given to the player scoring the highest number of runs. In case of a tie, the player (out of those locked in the tie) who has taken the higher number of catches is chosen. Even thereafter if there is a tie, the player (out of those locked in the tie) who has dropped fewer catches is selected. Aakash, Biplab, and Chirag who were contenders for the award dropped at least one catch each. Biplab dropped 2 catches more than Aakash did, scored 50, and took 2 catches. Chirag got two chances to catch and dropped both. Who was the 'man of the match'?
(I) Chirag made 15 runs less than both Aakash and Biplab
(II) The catches dropped by Biplab are 1 more than the catches take by Aakash

33. In figure given below, PAQ is tangent to the circle, with centre O, at a point A.
What would be value of y?

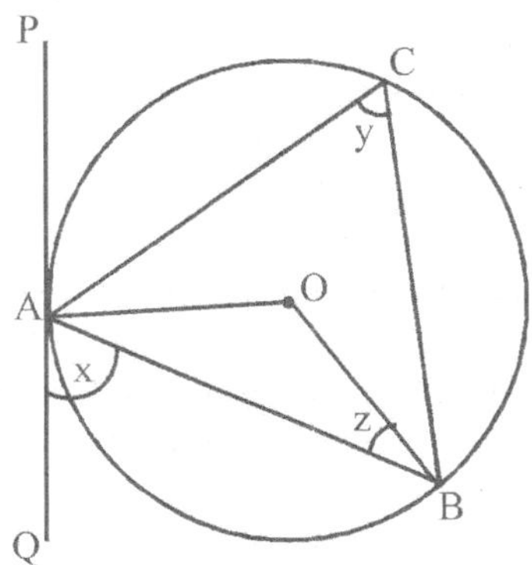

(I) $\angle x = 58°$ (II) $\angle z = 32°$

34. Is $\frac{R}{A^2+B^2+C^2} > \frac{R}{D^2+E^2+F^2}$? It is given that A, B, C, D, E, and R are natural numbers.
(I) A + B + C = D + E + F and among A, B and C, A is the lowest and C is the highest whereas among D, E & F, F is the highest and D is the lowest but F – D < C – A.
(II) $\frac{R}{A^3+B^3+C^3} < \frac{R}{D^3+E^3+F^3}$

35. Is $n^x - n$ divisible by x? It is given that x and n are natural numbers.
(I) Value of n is known.
(II) Value of x is known.

36. A circle C_2 with radius r_2 lies completely within another circle C_1 with radius of r_1. What is the value of r_1?
(I) The difference in the circumference of $C_1$and C_2 is 4 cm.
(II) The difference in the areas of $C_1$and C_2 is $26\frac{8}{11}$ sq. cm.

37. During a five day period, Monday through Friday, the average high temperature was 86 degrees Fahrenheit. What was the high temperature on Friday?
(I) The average high temperature for Monday through Thursday was 87 degrees Fahrenheit
(II) The high temperature on Friday reduced the average high temperature for the week by 1 degree Fahrenheit.

38. How many hours will it take for all the students and teachers together to put up a tent?
(I) There are 4 teachers and 7 students.
(II) All the teachers working together can put up the tent in 5 hours while all the students working together can do so in 3 hours.

39. A sequence of numbers a_1, a_2 .... is given by the rule $a_n^2 = a_{n+1}$. Does 3 appear in the sequence?
(I) $a_1 = 2$
(II) $a_3 = 16$

40. XY is a tangent to the circle with centre O at the point Q. What is the value of $\angle$DPQ?
(I) $\angle DQX = 40°$
(II) $\angle DOQ = 70°$

41. Distance between the two stations X and Y is 470 km. Two trains P and Q starts from X and proceed towards Y at speeds of 40 km/hr and 60 km per hour respectively. An another train R starts from Y and proceeds towards X at a speed of 30 km/hour. At what time, the distance between the train P and Q will be $\frac{2}{3}$rd of the distance between the train P and R.
(I) Train P and Q started at a interval of three hours whereas train R started simultaneous with one of the train among P and Q at 8 am.
(II) At 5 pm both the train P and Q were at the same distance from X.

42. Average weight of four persons is 67 kg. These persons are A, B, C and D. When E is included in the group, average weight decreases by 2 kg. What is the weight of A? A's weight is 4 kg more than B's weight.
(I) When B is displaced by another person F, whose weight is 4 kg more than E, average weight becomes 64 kg.
(II) When D is displaced by another person G, whose weight is 4 kg more than E, average weight becomes 65 kg.

43. H.C.F. of two numbers is 23. Find the smaller of the two numbers.
(I) The product of the two numbers is 6348.
(II) The L.C.M. of the two numbers is 276.

44. What is the value of x ?
(I) 2x + 4 =14
(II) x + y = 7

45. How far is Ranchi from Rohtak ?
(I) Rohtak is 899 km from Ranchi.
(II) There is no air route linking Ranchi and Rohtak directly.

46. What are the dimensions of a certain rectangle ?
(I) The perimeter of the rectangle is 14.
(II) The diagonal of the rectangle is 5.

47. What is the rate of simple interest per annum ?
(I) The sum triples in 20 years at simple interest.
(II) The difference between the sum and the simple interest earned after 10 years is Rs 1000.

48. A train crosses a signal post in X seconds. What is the length of the train?
(I) The train crosses a platform of 100 metres in Y seconds
(II) The train is running at the speed of 80 km/hr.

49. What is the area of a circle?
(I) The circumference of the circle is 308 metres.
(II) The radius of the circle is 28 metres.

50. 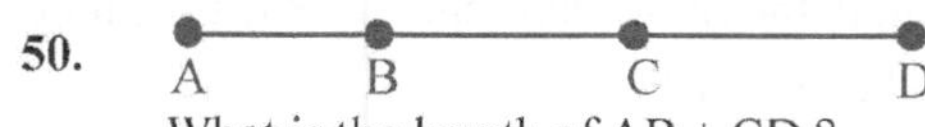

What is the length of AB + CD ?
(I) AD = 30
(II) BC = 10

51. Mahesh's flat is on which floor of the five-floor apartment?
(I) His flat is exactly above Ganesh's flat whose flat is exactly above Nitin's first -floor flat.
(II) Jeevan's flat, which is adjacent to Mahesh's flat, is exactly below Ahmed's flat, who is on fourth floor.

52. Who scored highest among A, B, C, D and E?
(I) B scored more than D, but not as much as C.
(II) E scored more than C, but not more than A.

53. On which date in April was Varun born ?
(I) Varun's mother remembers that Varun was born before nineteenth but after fifteenth.
(II) Varun's sister remembers that Varun was born before seventeenth but after twelfth.

54. Is Bhavana five years old ?
(I) Kalpana is seven years old and is younger than Bhavana.
(II) Bhavana is two years older than Ramesh, who is six years old.

55. Is $\sqrt{x} = 11$?
(I) Square of 11 is not equal to x.
(II) Square of $\sqrt{x}$ is not equal to the square of 11.

56. Is x greater than y?
(I) x is a multiple of y.
(II) $\frac{x}{6} = \frac{y}{3}$

57. What is the average monthly income per family member?
(I) Each male earns Rs 1,250 a month and each female earns Rs 1,050 a month.
(II) Ratio of males to females in the family is 2 : 1

58. What percentage rate of simple interest per annum did Ashok pay to Sudhir ?
(I) Ashok borrowed Rs 8000 from Sudhir for four years.
(II) Ashok returned Rs 8800 to Sudhir at the end of two years and settled the loan.

59. Is $(x^2 - y^2)$ an odd number ?
(I) x and y are integers.
(II) x + y is an odd number.

60. Are x, y and z in A. P. ?
(I) x is greater than y but less than z.
(II) $x = \frac{y+z}{2}$

61. What is the value of 20 percent of x ?
(I) 1/4 of 20 percent of x is 5
(II) 4x = S, 5y = S and y = 80

DIRECTIONS (Qs. 62-122) : ***Each of the questions below consists of a question and two statements numbered (1) and (2) given below it. You have to decide whether the data provided in the statements are sufficient to answer the question. Read both the statements and give answer.***

(a) Statement (1) ALONE is sufficient, but statement (2) alone is not sufficient.
(b) Statement (2) ALONE is sufficient, but statement (1) alone is not sufficient
(c) BOTH statements TOGETHER are sufficient, but NEITHER statement alone is sufficient.
(d) Statements (1) and (2) TOGETHER are NOT sufficient.

62. How many candidates were interviewed everyday by the panel 'A' out of the three panels A, B and C?
(1) The three panels on an average can interview 15 candidates every day.
(2) Out of a total of 45 candidates interviewed everyday by the three panels, the no. of candidates interviewed by panel 'A' is more by 2 than the candidates interviewed by panel 'C' and is less by 1 than the candidates interviewed by panel 'B'.

63. What does "ja" represent in a code language?
Statements :
(1) In that code language "pik ja sok pot" means "you can stay here".
(2) In that code language "ja sok ton te" means you may come here".

64. Among Japanese Yen, Australian Dollar, Hong Kong Dollar and Singapore Dollor, which currency has the lowest value in terms of Indian rupees ?
(1) One rupee is equal to 2.56 Japanese Yen, 0.338 Australian Dollar, and 0.465 Hong Kong Dollar.
(2) Value of Australian Dollar is equal to Singapore Dollar.

65. The book of which subject is at the sixth position from the top in a pile of ten books, including 3 books of History, 3 of Hindi, 2 of Maths and 2 of English ?
(1) Starting from above there is an English book between a History and Math book, a History book between a Math and an English book, a Math book between two Hindi books and two Hindi books between a Maths and a History book.
(2) Counting from bottom the book which is at the fourth position is neither Math nor English book.

66. What is the value of a two-digit number ?
(1) The sum of the two digits is 4.
(2) The difference between the two digits is 2.

67. What was the population of State 'A' in 1999?
(1) Population of the State increases every year by 20% and its population in 1997 was 1,20,000.
(2) Population of State A in 1997 was twice that of State B in the same year.

68. X, Y and Z are integers. Is X an odd number?
(1) An odd number is obtained when X is divided by 5.
(2) (X + Y) is an odd number.

69. What was the cost price of the suitcase purchased by Samir?
(1) Samir got 20 per cent concession on the labelled price.
(2) Samir sold the suitcase for Rs.2000 with 25 per cent profit on the labelled price.

70. What is the speed of the boat in still water?
(1) It takes 2 hours to cover the distance between A and B downstream.
(2) It takes 4 hours to cover the distance between A and B upstream.

71. The area of a square is equal to that of a circle. What is the circumference of the circle?
(1) The diagonal of the square is x inches.
(2) The side of the square is y inches.

72. What is the cost of the laying carpet in a rectangular hall?
(1) Cost of the carpet is ₹ 450 per square metre.
(2) Perimeter of the hall is 50 metres.

73. A starts a business with ₹ 60,000. After 6 months B joins him. What is the profit of B at the end of the year?
(1) The share of the profit is in the ratio 4 : 3
(2) B's capital is ₹ 90,000.

74. How many vinyl squares with sides 5 inches long will be needed to cover the rectangular floor of a room?
(1) The floor is 5 ft long.
(2) The floor is 3 ft wide.

75. If 8A + 9C + 2D + 3I = 106, then what is the value of I?
(1) $6A + 4C - 16D - 4E = 48$
(2) $24C - 13I = 28$

76. Shiva owns 100 shares of stock A and 150 shares of stock B. What is the total value of his stocks?
(1) The value of each share of stock A is twice the value of each share of stock B.
(2) The total value of 4 shares of stock A and 6 shares of stocks B is Rs. 750.

77. Average age of 50 employees working in a departmental store is 30 years. In the next year, ten workers will retire. What will be the average age in the next year?
(1) Retirement age is sixty years.
(2) There are 100 workers in the department.

78. Is the area of ΔABD equal to the area of ΔDBC?

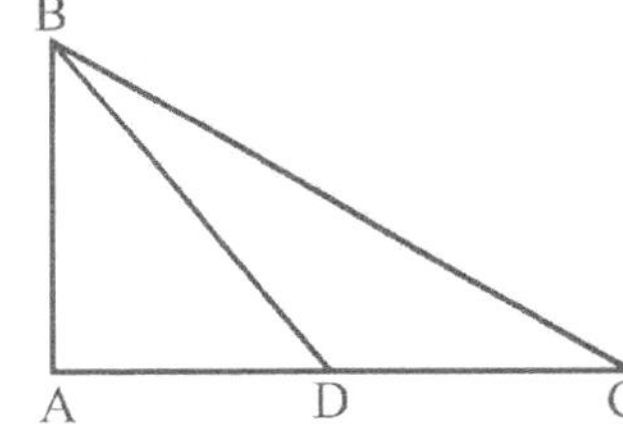

(1) AD = DC
(2) $\angle BAD = 90°$

79. What is the perimeter of rectangle ABCD?

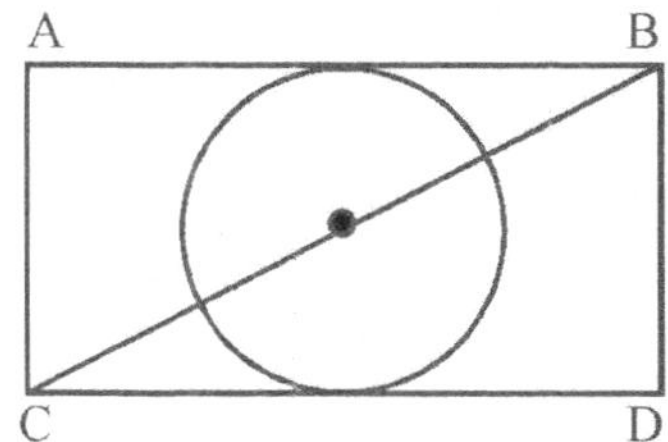

(1) Area of the circle is 78.5 sq cm.
(2) AB = 10 cm.

80. How long will it take for a jeep to travel a distance of 250 km?
(1) The relative speed of the jeep with respect to the car moving in the same direction at 40 kmph is 50 kmph.
(2) The car started at 3.00 am. in the morning.

81. What is value of $p^3 - q^3$.
(1) $p^6 - q^6 = 0$
(2) $q = 0$

82. Is p greater than 1? (You may assume that q is not equal to zero)
(1) $\left(\frac{p}{q}\right)$ is greater than 1
(2) $\left(\frac{1}{q}\right)$ is less than 1.

83. If $a^6 - b^6 = 0$, then what is the value of $a^3 - b^3$?
(1) a is positive
(2) b is greater than 1

84. Is y larger than 1 ?
(1) y is larger than 0 (2) $y^2 - 4 > 0$

85. The pentagon PQRST is inscribed in the circle with centre O. How many degrees is $\angle PQR$?

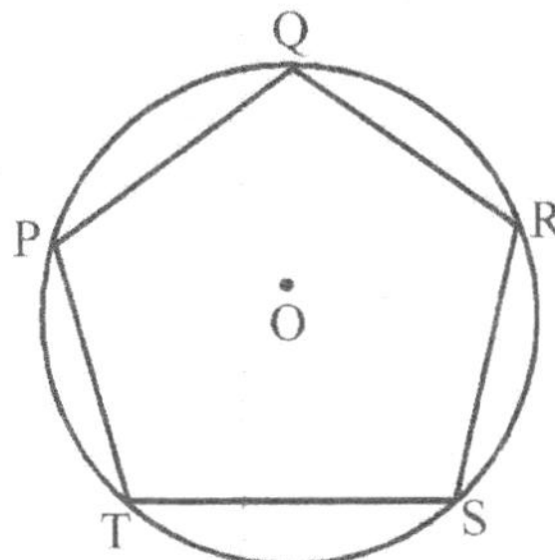

(1) The pentagon PQRST is a regular pentagon.
(2) The radius of the circle is 5 inches.

86. If t is a multiple of prime number S, is t a multiple of S^2?
(1) $S < 4$ (2) $t = 18$

87. Which word in the code language means flower?
(1) 'dem fu la pane' means 'rose flower is beautiful' and 'la quiz' means 'beautiful tree'.
(2) 'dem fu chin' means 'red rose flower' and 'pa chin' means 'red tea'.

88. A, B, C, D and F are seated around a circular table facing at the centre. Who is on the immediate right of B ?
(1) D is between A and F.
(2) C is between B and F.

89. In a one-day cricket match of 50 overs, a team is chasing a target of 300 runs. At one stage the run rate of 4.5 per over was achieved. What is the required run rate after 20th over?
(1) 80 runs were scored in first twenty over.
(2) Five wickets were lost within 20 overs.

90. If both the conveyer belts A and B are used, then they can fill a hopper with iron ore in one hour. How long will it take for the conveyer belt A to fill the hopper without conveyer belt B?
(1) Conveyer belt A moves twice as much iron ore as conveyer belt B.
(2) Conveyer belt B would take more than 3 hours to fill the hopper without belt A.

91. A car, originally, was sold for Rs 2,00,000. After a month, the car was discounted x%, and a month later, the car's price was discounted y%. Is the car's price after the discounts less than Rs 1,75,000?
(1) $y = 10$
(2) $x = 15$

92. If x, y and z are digits, is $x + y + z$ a multiple of 9 ? A digit is one of the integers 0, 1, 2, 3, 4, 5, 6, 7, 8, 9.
(1) The three-digit number xyz is a multiple of 9.
(2) $(x \times y) + z$ is a multiple of 9.

93. What was the combined average attendance per game at the Jawaharlal Nehru Stadium for the months of June and July ?
(1) The total attendance for the month of June was 23100 and the total attendance for the month of July was 25,200
(2) There were 20 games played in June and 22 games played in July.

94. 48 children of a class were asked to sit in rows and columns. How many children are seated in each row?
(1) The number of columns is more than the number of rows.
(2) The number of rows is 3/4 of the number of columns.

95. There are four envelopes E_1, E_2, E_3, E_4 in which one was supposed to put letters L_1, L_2, L_3, L_4 meant for persons C_1, C_2, C_3, C_4, respectively but by mistake the letters got jumbled up and went in wrong envelopes. Now, if C_2 is allowed to open an envelope at random, then how will he identify the envelope containing the letter for him?
(1) L_2 has been put in E_1
(2) The letter belonging to C_3 has gone in the correct envelope

96. A circle has radius r and origin as its centre. Two tangents are drawn from an external point D, d distance away from the origin. what are the angles made by the tangents with the positive X-axis
(1) The co-ordinates of the point D are given
(2) The X-axis bisects one of the tangents

97. Is $\angle BPD > \frac{1}{2}(\angle ABP - \angle APB)$ in the figure shown below? (AP is tangent)

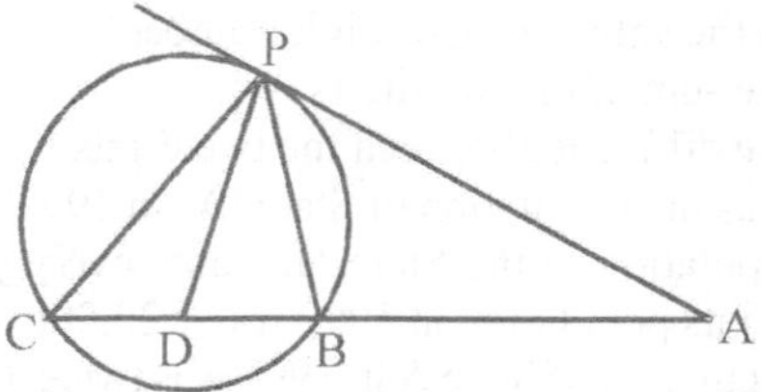

(1) Length of AC is known.
(2) PD is the bisector of $\angle BPC$.

98. In the figure below, is PQ parallel to TU ?

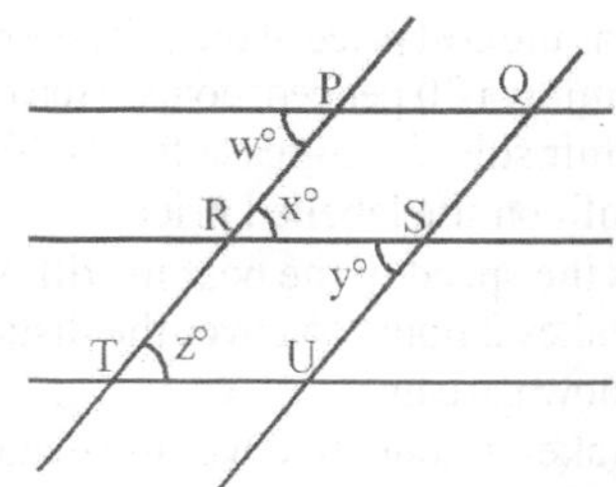

(1) $w = z$
(2) $x = y$

99. In the figure, is PRSU a square?

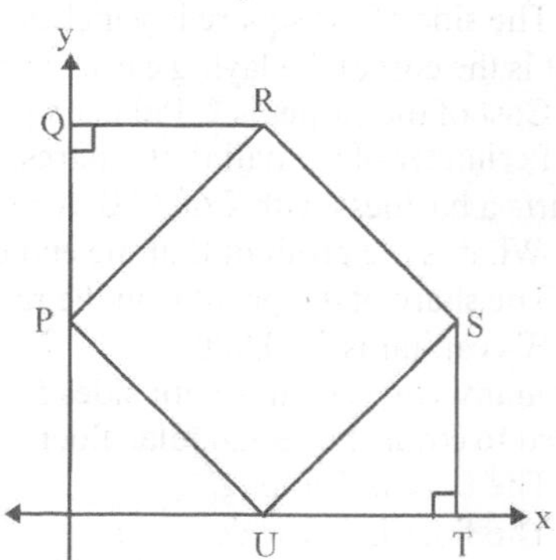

(1) PQ = QR and UT = TS.
(2) PU = RS.

100. Is x an integer?
(1) $\frac{19}{x+1}$ is an integer.
(2) $\frac{x+1}{5}$ is an integer.

101. At a restaurant, Harish left a tip for his waiter equal to 20 percent of his entire dinner check, including tax. What was the amount of the dinner check?
(1) The sum of the dinner check and the tip was Rs. 16.80.
(2) Harish's tip consisted of two bills and four coins.

102. What was the gross income of XYZ Ltd. for the year 2000?
(1) In 1999, the gross income of XYZ Ltd. was Rs. 4300 crore.
(2) The gross income of XYZ Ltd. was 5% greater in 2000 over 1999.

103. If p is a positive number, then what is the value of p ?
(1) $|p - 2| = 1$
(2) $p^2 = 4p - 3$

104. What would be the minimum present age of Raju ? It is given that t_1 years ago the ratio of the age of Raju to that of Renu was R : S and t_2 years hence the ratio will become T : U.

(1) The values of $(t_1 + t_2)$ and T – U are 10 and 1 respectively, and T : U = 2T : 14 (where t_1 and t_2 are natural numbers).

(2) R × U = 42 and T × S = 40.

105. A bill falls due in a year. The creditor agrees to immediate payment of half the amount and defers the rest of the payment for 2 years. What is the amount of the bill?

(1) Creditor gains Rs 40.

(2) Rate of interest is 12.5%.

106. What is Sudha's present age ?

(1) Sudha's present age is five times her son's present age.

(2) Five years ago her age was twenty-five times her son's age that time.

107. What is the height of a right-angled triangle ?

(1) The area of the right-angled triangle is equal to the area of a rectangle whose breadth is 12 cm.

(2) The length of the rectangle is 18 cm.

108. How is 'A' related to 'B' ?

(1) A is married to Y's sister.

(2) B is the name of Y's sister.

109. Towards which direction is C from Q ?

(1) C and Q are opposite to each other. L is equidistant from C and Q.

(2) L is neither towards north-east nor towards south-west of C and Q respectively.

110. When will the sun rise tomorrow ?

(1) The sun will set tomorrow at 6.15 pm.

(2) Tomorrow's day will be of 11 hr 50 min.

111. A certain garden consists only of apple trees, mango trees and orange trees. Which type of tree is the most numerous?

(1) There are 4/5 as many orange trees as there are mango trees.

(2) There are 2/3 as many mango trees as there are apple trees.

112. How many marks did Prakash obtain in Mathematics ?

(1) Prakash secured on an average 55 per cent marks in Mathematics, Physics and Chemistry together.

(2) Prakash secured 10 per cent more than the average in Mathematics.

113. How much was the loss?

(1) The cost is ₹ 300.

(2) The loss is 25 per cent of the selling price.

114. What is the cost of flooring a room ?

(1) The length and breadth of the room is 9 m and 6 m respectively.

(2) The cost of the tiles is ₹ 6 per cm^2.

115. In a certain code language what does 'come' mean ?

(1) 'pit na ja ' means 'come and go' in the code language.

(2) 'na dik sa' means 'you may go ' in the code language.

116. On which day of the week was Navin on leave?

(1) Navin's mother was hospitalised on Thursday.

(2) Navin's brother was on leave on Friday.

117. Five persons are to be seated on a circular table. Who will be seated between Ram and Gita?

(1) Ram will sit on the right of Vinay and on the left of Kamal.

(2) There will be two persons seated between Vinay and Kamal.

118. Which city has the lowest population ?

(1) Bareilly has 20 lakh population, which is less than the population of Bhagalpur.

(2) Patna has population equal to that of Bhagalpur and more than that of Allahabad.

119. In a row of boys facing north who is on the immediate right of Nishikant?

(1) Nishikant is third to the left of Shashikant and third to the right of Ravikant.

(2) Dinanath and Premnath are also in the row but Dinanath is the nearest to Shashikant.

120. What is the difference between the two digits in a two-digit number?

(1) The sum of the two digits is 8.

(2) 1/5 of that number is 15 less than 1/2 of 44.

121. What percentage of X's salary is Y's salary ?

(1) X's salary is 20% of Z's salary.

(2) Y's salary is 45% of Z's salary.

122. How many boys are there in the class ?

(1) The number of boys is 120% of the number of girls in the class.

(2) The number of girls is $\frac{5}{11}$th of the total number of students.

Level - II

DIRECTIONS (Qs. 1 to 34) : ***Each of the questions below consists of a question and two statements numbered (I) and (II) given below it. You have to decide whether the data provided in the statements are sufficient to answer the question. Read both the statements and give answer***

(a) Statement (I) ALONE is sufficient, but statement (II) alone is not sufficient.

(b) Statement (II) ALONE is sufficient, but statement (I) alone is not sufficient

(c) BOTH statements TOGETHER are sufficient, but NEITHER statement alone is sufficient.

(d) EACH statement ALONE is sufficient

1. Out of the four teams A, B, C and D which team is not likely to win as per the opinion poll?
   (I) As per the opinion poll, chances of team C's winning are more than that of team A but not as much as that of team B, whose chances of winning are more than that of team A.
   (II) As per the opinion poll team C's chances of winning are less than that of team B but not less than that of team D, whose chances of winning are more than that of team A.
2. A, B and C are positive integers. Is their product an even number?
   (I) A is an even number.
   (II) The product of A and B is an even number and that of A and C is also an even number.
3. In the diagram shown, is ΔABC isosceles ?

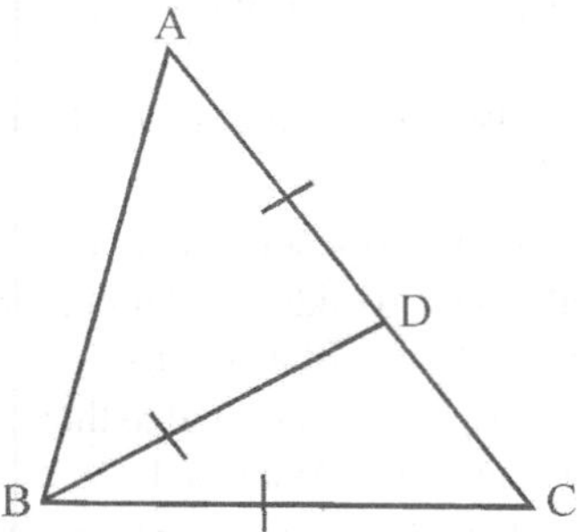

   (I) BC = BD = DA (II) BD bisects $\angle ABC$
4. What was Mr. Srinivasulu's combined income for the year 1994-1999? In 1994, he made ₹ 10,00, 000.
   (I) His average yearly income for the years 1994-1999 was ₹ 12,00,000
   (II) In 1999, his income was ₹ 20,00,0000
5. A fly crawls around the outside of a circle once. A second fly crawls around the outside of a square once. Which fly travels farther?
   (I) The diagonal of the square is equal to the diameter of the circle.
   (II) The fly crawling around the circle takes more time to complete his journey than the fly crawling around the square.
6. In the given figure P, Q and R are centres of three equal circles. What is the area of the shaded portion in the figure shown below ?

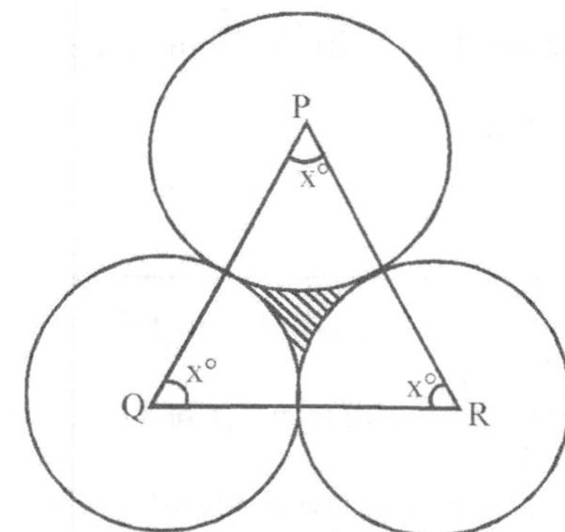

   (I) QR is known.
   (II) The triangle is an equilateral triangle.
7. 3 person were given certain calculations to perform. The calculations were 1 + 1, 1 +1+2, and 1 +2. Their respective answers were 3,3 and 2. How many of them are mathematicians
   (I) Mathematicians can never add two numbers correctly, but they add three numbers correctly
   (II) Whenever the mathematicians add two numbers there is a mistake of +1 or –1
8. An alloy of silver and gold weighs 30 gm in air and 27 gm in water.
   What is the weight of gold in the alloy?
   (I) Both silver and gold lose one-tenth of their weight in water.
   (II) Silver loses 10% of its weight while gold loses 20% of its weight in water respectively.
9. What is the volume of a given cylinder? It is given that the cylinder is within the cube touching all the vertical faces and a cone is inside the cylinder. It is also given that heights are same with the same base.
   (I) Volume of the cone is known.
   (II) Total surface area of the cube is known.
10. Is 1 + (P – 1)! divisible by P?
   (I) P is an even number.
   (II) P is a prime number.
11. N is an integer between 1 and 93. What is the value of N?
   (I) N is both the square of an integer and the cube of an integer.
   (II) The square root of N is divisible by 8.
12. If $x \neq 0, -1$, then is $\frac{1}{x}$ greater than $\frac{1}{x+1}$?
   (I) $x < 1$
   (II) $x > 1$
13. Alka and Madhu put 40 matches on the table and make themselves aware of the game for which the matches are to be used. As per the rules, the winner is the one who takes the last match. Who will won the game ?
   (I) Each players in turn takes 1, 3 or 5 matches.
   (II) Madhu chooses to go first and takes 3 matches.
14. What is the speed of train 'X' in km/h?
   (I) Length of train 'X' is twice that of train 'Y' and speed of Y is 100 m/s.
   (II) Train 'X' passes train 'Y' in 10 seconds when they are running in same direction. Length of train 'X' is 100 metres.
15. Which word in the code language means 'they' ?
   (I) 'ras pak leun' means 'I have pens' and 'pak ras tap' means 'they have pens'.
   (II) 'kip ras tun' means 'boys have cow,' and 'sin tap ras' means 'they have egg.'
16. Is $g - h > 0$?
   (I) $g > h$
   (II) $g^2 > h^2$
17. Which direction is Shashidhar facing ?
   (I) In the early morning Shashidhar was standing in front of a puppet and the shadow of the puppet was falling to the right of Shashidhar.
   (II) In the early morning Shashidhar was standing on the ground. His shadow was falling behind him when he turned to his left.

18. Who is paternal uncle of P ?

(I) P is brother of L, who is daughter of Q, who is sister of N, who is brother of S.

(II) M is brother of K, who is husband of L, who is mother of G, who is sister of P.

DIRECTIONS (Qs. 19-34) : ***Each of the questions below consists of a question and two statements numbered (1) and (2) given below it. You have to decide whether the data provided in the statements are sufficient to answer the question. Read both the statements and give answer.***

(a) Statement (1) ALONE is sufficient, but statement (2) alone is not sufficient.

(b) Statement (2) ALONE is sufficient, but statement (1) alone is not sufficient

(c) BOTH statements TOGETHER are sufficient, but NEITHER statement alone is sufficient.

(d) Statements (1) and (2) TOGETHER are NOT sufficient.

19. A, B, C and D made their project presentation, one on each day, on four consecutive days but not necessarily in that order. On which day did 'C' make his presentation?

(1) The first presentation was made on 23rd, Tuesday and was followed by 'D's presentation.

(2) 'A' did not make his presentation on 25th and one of them made his presentation, between A's and B's.

20. Among the mobile services companies H, A and I which gives maximum talk-time on its coupon of ₹ X ?

(1) Number of subscribers of the company A is more than the number of subscribers of the company I, but not as much as the company H.

(2) Talk time provided by the company I is equal to X minutes, i.e. equal in figure of the price of the coupon. But the same is not true for company A which provides more talk time than company H.

21. What will be the cost of the second necklace?

(1) The cost of the first necklace is $\frac{1}{5}$ more than the second and the cost of the third necklace is $\frac{2}{5}$ more than the second. The total cost of all the three necklaces is Rs 120000.

(2) The cost of the first necklace is $\frac{2}{5}$ more than the second. The cost of the third necklace is the least and total cost of all the three necklaces is Rs. 1,20,000.

22. How many houses on the street are painted blue ?

(1) Houses on the east side with number 122 through 182 are painted blue ?

(2) Houses on the west and east sides have consecutive odd and even number respectively.

23. Is the integer divisible by 3?

(1) the last digit in n is 3.

(2) n + 5 is divisible by 6.

24. Is Mr. 'Y' entitled to get promotion in the month of September 2002?

(1) As per his office rules, the only condition for promotion is completion of 12 years of service in a particular grade on 31st December of every year.

(2) Mr. 'Y' has been working in this office for the last 12 years.

25. ABCD is a square. What is the value of circumference of the inner circle?

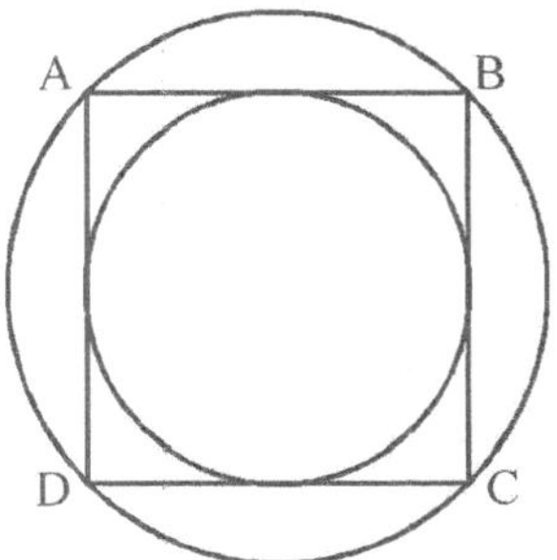

(1) The radius of the outer circle is 10 cm.

(2) The difference between the radii of the two circles is 2.929 cm.

26. What is the distance between D_1 and D_2 ? It is given that a man who can row x km/hr in still water, takes z hrs to row from D_1 to D_2 and back to D_1, in a stream which flows at y km/hr.

(1) Value of $(x^2 - y^2)$ is known.

(2) $(x \div z)$ is known.

27. What is the price of a cake if a man's family consumes 14 cakes in a week ?

(1) If the wages of the man were raised by 12% and the price of cake was raised by 15%, he would gain 60 paise a week, but if his wages were increased by 12% and the price of cake was raised by 16%, he would neither gain nor lose.

(2) If the wages of the man were raised by 4% and the price of cake was also raised by 4%, he would save 80 paise more a week, but if his wages were increased by 8% and the price of cake was also raised by 8%, he would save 160 paise more a week.

28. Is point P located on the circumference of the circle with centre O and radius of length l?

(1) Points P, O and R are the vertices of an equilateral triangle.

(2) Point R is located on the circumference of the circle with centre O and radius of length l.

29. Was China's GDP in 2001 50% higher than that of India? It is given that FEI for a country in a year is the ratio (expressed as a percentage) of its foreign equity inflows to its GDP.

(1) FEI of India in the year 2001 was 0.72 while FEI of China in the year 2001 was 4.80.

(2) China's foreign equity inflows in 2001 were 10 times that to India.

30. The average temperature of 7 days was 44°C. The average temperature of the first three days was 43.8°C. The average temperature of the sixth and the seventh days was 0.1°C more than the average temperature of the first three days. What was the temperature of the fourth day?
   (1) The temperature of fourth day was 0.2°C more than that of fifth day.
   (2) The temperature of fifth day was 0.2°C more than that of sixth day.

31. Is ababab divisible by 222 where a and b are two digits?
   (1) $a + b = 5$.
   (2) b is an even number

32. Is $(X^2 + Y^2 + Z^2)$ even? [X, Y, Z are integers]
   (1) X, Y, Z are consecutive.
   (2) $X^2 + Z^2$ is even.

33. Who among the three friends A, B and C reached the school first?
   (1) A reached the school at 7.15 am five minutes before the bell rang.
   (2) B reached before C, who reached the school before the bell rang.

34. Find the value of algebraic expression $x^3y - \left(\frac{x^3}{y}\right)$.
   (1) $x = 2$
   (2) $y = 1$

Directions (Qs. 35 - 39) : ***Each of the questions below consists of a question and two statements numbered I and II given below it. You have to decide whether the data provided in the statements are sufficient to answer the questions. Read both the statements and:***

Give answer (a) if the data in statement I alone are sufficient to answer the question, while the data in statement II alone are not sufficient to answer the question.

Give answer (b) if the data in statement II alone are sufficient to answer the question, while the data in statement I alone are not sufficient to answer the question.

Give answer (c) if the data either in statement I alone or in statement II alone in sufficient to answer the question.

Give answer (d) if the data even in both the statements I and II together are not sufficient to answer the question.

Give answer (e) if the data in both the statements I and II together are necessary to answer the question.

[SBI PO-2011]

35. What is the height of a triangle?
   I. It is a right-angled triangle.
   II. The area of the triangle is 5 times its base.

36. How much time will Dinesh take to walk a km distance?
   I. The ratio of the speeds at which Dinesh and Ranjay walk is 5 : 6.
   II. The average walking speed of Dinesh and Ranjay is 11 km/hr.

37. Is a two-digit integer 'x' divisible by 12?
   I. When 'x' is divided by 5, the remainder is 2.
   II. When 'x' is divided by 3, the remainder is 1.

38. Is the average of a, b and c equal to b?
   I. $b - a = c - b$
   II. a, b and c are positive integers.

39. What is the monthly income of Rahim?
   I. Total monthly income of Rahim and Suresh is ₹27000, which is 150% of their total monthly expenditure.
   II. The ratio of their monthly expenditures is 5 : 4

Directions (Qs.40-44) : ***The following questions are accompanied by three statements A, B and C. You have to determine which statement(s) is/are necessary/sufficient to answer the question.***

[SBI PO-2011]

40. P, Q and R together invested an amount of ₹42000 in the ratio of 4 : 3 : 7 for different periods of time. What was the amount of profit earned by them individually at the end of one year?
   A. They invested for periods in the ratios of 1 : 2 : 1.
   B. R's profit is ₹ 5500 less than Q's investment.
   C. Total amount of profit at the end of one year is ₹8800.
   (a) Only A and B together
   (b) Only A and C together
   (c) A and either B or C
   (d) All statements are required
   (e) Question can't be answered even after using all the informations

41. What was the sum of the ages of the mother and the son ten years earlier?
   A. Ratio of the mother's present age to the son's age after 10 years is 5 : 3.
   B. The difference between thrice the present age of the mother and five times the present age of the son is 50 years.
   C. The ratio of the age of the mother after 10 years to the age of the son after 16 years is 5 : 3.
   (a) Any two of them
   (b) Any one of them
   (c) B and either A or C
   (d) All statements are required
   (e) Question can't be answered even after using all the informations

42. What will be the cost of fencing a rectangular plot?
   A. Cost of fencing a circular plot whose area is $616m^2$ is ₹ 968.
   B. Perimeter of the rectangular plot is 200 m.
   C. Perimeter of the square whose length is equal to the breadth of the rectangular plot is 20 m.
   (a) Only C (b) A and C together
   (c) A and B together (d) A and either B or C
   (e) Question can't be answered even after using all the informations

43. What is the cost price of an article?

A. After allowing a discount of 10% on marked price the shopkeeper charges ₹ 810.

B. If the shopkeeper does not give the discount the shopkeeper gets a profit of 50%.

C. If the shopkeeper gives only 5% discount on marked price, he will have 42.50% profit.

(a) Any two of them (b) Only A and B together
(c) Only B and C together (d) A and either B or C
(e) Only A

44. What is the rate of interest at which Binod has invested money?

A. The compound interest at this rate of ₹ 2500 in 2 yrs is equal to the simple interest in 3 yrs of ₹ $1716\frac{2}{3}$ at the same rate.

B. The total simple interest on an investment of ₹ 12000 for 3 yrs and ₹10000 for 5 yrs at this rate is ₹5160.

C. In 3 yrs ₹1500 at the same rate becomes ₹1725 by simple interest.

(a) Any of them (b) Either B or C
(c) Only C (d) Only A and either B or C
(e) Any two of them

DIRECTIONS (Q.45) : ***In each of the following questions, a question is followed by information given in three Statements I, II and III. You have to study the question along with the statements and decide the information given in which of the statement(s) is necessary to answer the question.***

[IBPS-PO-2013]

45. What is the rate of interest Percent per annum?

I. An amount doubles itself in 5 yrs on simple interest;

II. Difference between the compound interest and the simple interest earned on a certain amount in two years is ₹ 400.

III. Simple interest earned per annum is ₹ 2000.

(a) Only I (b) II and III
(c) Any two of three (d) I or II and III
(e) Only I or II and III

DIRECTIONS (Q. 46) : ***In each of the following questions, a question is followed by information given in three Statements I, II and III. You have to study the question along with the statements and decide the information given in which of the statement(s) is necessary to answer the question.***

[IBPS-PO-2013]

46. In how many days 10 women can finish the work?

I. 10 men finish the work in 6 days.

II. 10 women and 10 men finish the work in $3\frac{3}{7}$ days.

III. If 10 men work 3 days and after that 10 women are deployed to work for men, the rest work is finished in 4 days.

(a) I and II (b) Any two of three
(c) I and III (d) II and III
(e) None of these

DIRECTIONS (Qs. 47 and 48) : ***In each of the following questions, a question is followed by information given in three Statements I, II and III. You have to study the question along with the statements and decide the information given in which of the statement(s) is necessary to answer the question.***

[IBPS-PO-2013]

47. What is the present age of Sabir?

I. The present age of Sabir is half of his father's age.

II. After five years the ratio of ages of Sabir and his father is 6 : 11.

III. Sabir is younger to his brother by five years.

(a) I and II (b) I and III
(c) II and III (d) All of these
(e) Cannot be determined

48. What is two digit number?

I. The difference between the number and the number formed by interchanging the digit is 27.

II. The difference between two digits is 3.

III. The digit at unit's place is less than that at ten place by 3.

(a) I and II (b) I and either II or III
(c) I and III (d) All of these
(e) None of these

DIRECTIONS (Q. 49) : ***In each of the following questions, a question is followed by information given in three Statements I, II and III. You have to study the question along with the statements and decide the information given in which of the statement(s) is necessary to answer the question.***

[IBPS-PO-2013]

49. What is the cost of flooring the rectangular hall?

I. Length and the breadth of the hall are in the ratio of 3 : 2

II. Length of the hall is 48 m and cost of flooring is ₹ 850 per sq m.

III. Perimeter of the hall is 160 m and cost of flooring is ₹ 850 per sq m.

(a) I and II (b) I and III
(c) Only III (d) I and either II or III
(e) Any two of the three

Hints & Solutions

Level-I

1. (a) Statement (I) gives required information but statement (II) gives information in terms of percentage. Hence (II) is not sufficient.

2. (d) **From (I) :** We can find the side, area and perimeter of square.
From (II) : Since Perimeter × rate of fencing per metre = Total cost (in rupees)
Hence, each statement alone is sufficient.

3. (a) From statement, I, P, U, Q and S do not teach Biology and Geography. From remaining persons R and T, T does not teaches Geography hence T must teaches Biology.
From II, we can't find who among R and T teaches Biology.

4. (a) (I) $\Rightarrow$ m = 999999, n = 100000
$\therefore$ We can find the value of $m - 7 \div 37$
(II) $\Rightarrow$ m – n = known, but neither the value of 'm' is known nor the value of 'n' is known. So, we cannot find the values of $m - n \div 37$.

5. (a) (I) $\Rightarrow x + 10 = 34$
(II) $\Rightarrow xy = 38$
Hence, only statement (1) is sufficient to answer the question.

6. (d) (I) $\Rightarrow$ C.P. can be calculated since

$$\text{C.P.} = \frac{\text{S.P.} \times 100}{(100 + \%\text{profit})}$$

Also, since Profit = S.P. – C.P.
$\therefore$ Profit can be calculated using statement (I) alone.

(II) $\Rightarrow$ $\text{C.P.} = \frac{1}{3} \times \text{Profit}$ and Profit = S.P. – C.P.

Therefore, each statement alone is sufficient to answer the question.

7. (d) $\because$ $\text{Amount} = P\left(1 + \frac{R}{100}\right)^n$

(I) $\Rightarrow$ For P = 1000 and n = 1, A = Rs 1100
$\therefore$ R = 10%
Hence, C.I. after 3 years can be calculated.
$\therefore$ Statement (I) alone is sufficient to answer the question.
Now, since difference between S.I. and C.I. for 2

years $= P \times \left(\frac{R}{100}\right)^2$

$\therefore$ From (II), R can be calculated.
Hence, C.I. after 3 years can be determined.
Therefore, each statement alone is sufficient to answer the question.

8. (c) (I) $\Rightarrow$ Other tank empty in 2(8 + 8 + 8) hours = 48 hours

(II) $\Rightarrow$ Volume of original tank $= \frac{1}{3}$ (volume of other tank)
$\Rightarrow$ Original tank empty in 16 hours
$\Rightarrow$ Required number of hours can be calculated using both the statements together.

9. (a) (I) $\Rightarrow$ Profit = S.P. – C.P. = 40. Hence statement (I) alone is sufficient.
(II) $\Rightarrow$ It is giving same data of question.

10. (d) (I) $\Rightarrow$ Total salary of 30 assembly workers
T_s = Total salary – salary of foreman

and required average $= \frac{T_s}{30}$

(II) $\Rightarrow$ Foreman's salary = 12,000 = 120% of average salary of assembly workers

$\Rightarrow$ Average salary $= \frac{12{,}000 \times 100}{120}$

Hence, each statement alone is sufficient to answer the question.

11. (a) (I) $\Rightarrow$ $3x + 4x\,(0.50) + 5x\,(0.25) = 500$
$\Rightarrow 6.25x = 500 \Rightarrow x = 80$
$\therefore$ The total number of 50-paise coins = 4x = 320
But we can't be solved the question using statement (II).
Hence statement (I) alone is sufficient to answer the question.

12. (d) Given : $V_1 + V_2 = \frac{80 + 100}{36} = 5\,\text{m/s}$

(I) $\Rightarrow$ $V_1 = \frac{80}{60} = \frac{4}{3}\,\text{m/s}$

$\therefore$ $V_2 = 5 - \frac{4}{3} = \frac{11}{3}\,\text{m/s}$

$\therefore$ Required time $= \frac{80 + 100}{\frac{11}{3} - \frac{4}{3}}$ m/s

(II) $\Rightarrow$ $V_2 = \frac{120 + 100}{60} = \frac{11}{3}\,\text{m/s}$

Now, $V_1 = 5 - \frac{11}{3} = \frac{4}{3}\,\text{m/s}$

Hence, required time can be calculated using either statement alone.

13. **(d)** (I) ⇒ Puttu can do the job in $\frac{36\times12}{36-18}$ days

(II) ⇒ Ratio of efficiency for Kundu : Puttu = 1 : 2

∴ ratio of days for Kundu : Puttu = 2 : 1

∴ Puttu will do the job in $12\left(\frac{2+1}{2}\right)$ days and Kundu will do it in $12\left(\frac{2+1}{1}\right)$ days.

Hence, answer can be determined using either statement alone.

14. **(d)** Suppose speed of train Z = x m/s

∴ Speed of train X = x + 45% of

$x = x+\frac{9}{20}x = \frac{29x}{20}$ m/s

Now, (I) ⇒ $\frac{110+160}{6.5} = x+\frac{29x}{20}$

Hence, x can be calculated by statement (I) alone.

(II) ⇒ $\frac{29x}{20} - x = 28\times\frac{5}{18}$

∴ x can also be calculated using statement (II) alone.

15. **(d)** (I) ⇒ He lays 4 cards on the table.

(II) ⇒ He lays 4 cards on the table.

16. **(c)** Let total families in city = x

(I) ⇒ Families having television $= \frac{x}{2}$

(II) ⇒ Families having telephones = 30% of $\frac{x}{2} = \frac{3x}{20}$

∴ Required % $= \frac{3x/20}{x}\times100$

Hence, both statements together are necessary to answer the question.

17. **(c)** (I) ⇒ For chocolates, Sheena has

= 50 – 20% of 50 – 30% of 50 = ₹ 25

(II) ⇒ Number of chocolates

$= \frac{\text{Total remaining rupees}}{\text{cost per chocolate}}$

∴ Both statements together are necessary to answer the question.

18. **(a)** ∵ Storing capacity of tank = volume of tank = πr^2h

Here r is given

and (I) ⇒ height(h)

∴ Statement (I) alone is sufficient to answer the question.

(II) ⇒ There is no need of value of temperature to answer the question.

19. **(a)** (I) ⇒ BC = 5 = AB

⇒ p = q = 40°

⇒ r = 180° – (40 + 40)

Hence, r can be determined by using (I) alone.

(II) ⇒ it is not sufficient to answer the question.

20. **(a)** (I) ⇒ required% = $\frac{92}{230}\times100$

∴ Statement (I) alone is sufficient to answer the question.

(II) ⇒ We have no need of total number of books.

21. **(b)** (I) ⇒ The fixed value of CP is not given, so, SP of the article cannot be determined

(II) ⇒ Let x be SP of an article

$$x\times\frac{9}{100} = \frac{200\times115}{100} = 255.55$$

Hence, we can find the answer using statement II alone.

22. **(c)** (I) ⇒ No. of voters in the last election

(II) ⇒ No. of voters at present = 115% of no. of voters in the last election.

∴ Both statements together are necessary to answer the question.

23. **(a)** (I) ⇒ 10 days from the date of release of advertisement (18th Februrary) means application must be submitted before 27 February.

Hence, statement (I) alone is sufficient to answer the question.

But we can't be determined the exact date using (II).

24. **(d)** (I) ⇒ Sunil is at 17th position from the left end and 40 – 16 = 24th position from the right end.

(II) ⇒ Sunil is at 24th position from the right end and 40 – 23 = 17th position from the left end.

∴ Each statement alone is sufficient to answer the question.

25. **(d)** Let the price in first shop = Rs x per kg

(I) ⇒ $\frac{18\times6+x\times2}{8} = 20$ ⇒ x can be determined.

(II) ⇒ $\frac{6x+2(x+8)}{8} = 20$ ⇒ x can be determined.

∴ Each statement alone is sufficient to answer the question.

26. **(b)** (I) ⇒ $\frac{M}{3} = K$

∴ $\frac{M}{3}$ may be odd or even. Therefore it is not sufficient to answer the question.

(II) ⇒ $\frac{M}{3}$ = 2J + 1, which is always odd because J is an integer.

∴ Statement (II) alone is sufficient to answer the question.

27. **(a)** Suppose he was paid Rs x for the first day.

(I) ⇒ $x+(x+5)+(x+10)+(x+15)+(x+20)+(x+25) = 900$

Hence x can be calculated. Therefore statement (I) alone is sufficient to answer the question.

(II) ⇒ x < 100.

∴ Statement (II) is not sufficient to answer the question because exact figure of wages is not given.

28. (d) (I) $\Rightarrow$ Area of the open cubical box with no top $= 4 \times 5 = 20$ sq m
(II) $\Rightarrow$ Edge of box = 2m. Therefore required area = $5 \times (II)^2 = 20$ sq. m.
Hence we can get answer with statement (I) alone and statement (II) alone.

29. (d) Let the total marks = x
(I) $\Rightarrow$ Minimum pass marks = 32% of x + 1 = 36% of x – 1.
Hence minimum marks can be calculated.
(II) $\Rightarrow$ Minimum pass marks = 30% of x + 2 and 30% of x + 5 = 40% of x
Hence minimum marks can be calculated.
Therefore each statement alone is sufficient for answer the question.

30. (d) (I) $\Rightarrow$ The faces bearing the numbers 1, 3, 4 and 6 are adjacent surfaces of the surface bearing the number 5. Hence 2 is opposite to 5.
(II) $\Rightarrow$ Tulika finds herself able to see the surface bearing the number 2, 4 and 6. While Shivendra can see 6, 4 and 5. From this we can conclude 2 is opposite to 5.

31. (b) Since $\angle YBC = \angle CAX = \angle YOX = 90°$
Hence $AB^2 = OA^2 + OB^2$
therefore we need both the length or any other relation since OACB is a rectangle
Hence $OC = AB$
Therefore radius = AB
Hence, only statement (I) can give us the answer.

32. (d) From statement I we can find the runs made by Aakash, Biplab and Chirag.
From second statement, as catches dropped by Biplab are one more then catches taken by Akash. So Akash has taken atleast 2 catches.
For the man of match, we well require both the statements (I) and (II).

33. (d) Angles in alternate segments are equal i.e.
$\angle ACB = \angle BAQ$
(I) $\Rightarrow \angle BAQ = 58° \Rightarrow \angle ACB = y = 58°$
$\therefore$ Statement (I) alone is sufficient to answer the question.
Now, in ΔAOB, we have OA = OB
[$\because$ Each equal to radius]
$\Rightarrow \angle OBA = \angle OAB$
(II) $\Rightarrow \angle OAB = 32°$
$\Rightarrow \angle AOB = 180° - 32° - 32° = 116°$
$\Rightarrow \angle ACB = \frac{1}{2}\angle AOB = 58°$
Hence, statement (II) alone is sufficient to answer the question.

34. (d) (I) $\Rightarrow A + B + C = D + E + F$
From I, we find that $A^2 + B^2 + C^2$ would be greater than $D^2 + E^2 + F^2$
Hence, $\frac{R}{A^2 + B^2 + C^2} < \frac{R}{D^2 + E^2 + F^2}$
From II, we can also find the answer.
Hence, we can give the answer using each statement alone.

35. (b) Remember that $n^x - n$ is divisible by x if x is a prime number and the divisibility does not depend on the value of n.
$\therefore$ Only statement (II) alone is sufficient to answer the question.

36. (c) (I) $\Rightarrow 2\pi(r_1 - r_2) = 4$
(II) $\Rightarrow \pi(r_1^2 - r_2^2) = 26\frac{8}{11}$
Hence, r_1 can be determined using both statements together.

37. (a) High temperature on Friday = (average high temperature Monday through Friday × 5) – (average high temperature Monday through Thursday × 4)
$\therefore$ Statement (I) alone is sufficient to solve the question.

38. (c) Total time = No. of teachers × time taken by all the teachers working together + No. of students × time taken by all the students working together
$\therefore$ Both statements together are necessary to answer the question.

39. (d) Here $a_1, a_2, a_3, a_4, ...$ are $a_0^2, a_1^2, a_2^2, a_3^2.........$ respectively.
(I) $\Rightarrow a_1 = 2 \Rightarrow a_2 = 4, a_3 = 16$ and so on.
$\therefore$ Statement (I) alone is sufficient to answer the question.
(II) $\Rightarrow a_3 = 16 \Rightarrow a_2 = 4, a_1 = 2, a_0 = \sqrt{2}$
$\therefore$ Statement (II) alone is sufficient to answer the question.

40. (d)

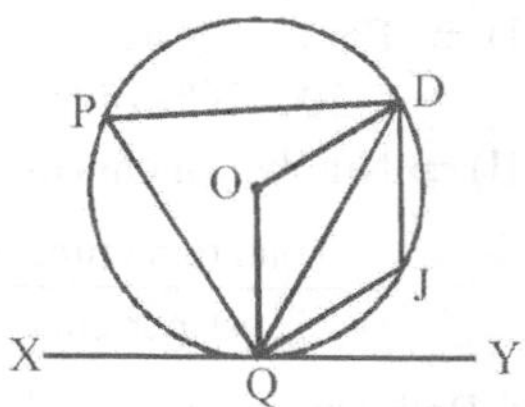

(I) $\Rightarrow \angle DQX = 40° = \angle DJQ$
(Angles in alternate segments of chord DQ)
$\Rightarrow \angle DPQ = 180° - \angle DJQ$
(Opposite angles of cyclic quadrilateral)
Hence statement (I) alone is sufficient to answer the question.
(II) $\Rightarrow \angle DOQ = 70° = 2\angle DPQ$
$\therefore$ Statement (II) alone is sufficient to answer the question.

41. (c) From statement I, we can find the distance between trains P and Q and from statement I and II, we find the distance covered by train R. Hence, both statements will require to give the answer.

42. **(c)** Given, $A + B + C + D = 67 \times 4 = 268$
$A + B + C + D + E = 65 \times 5 = 325$
and $A - B = 4$
$\therefore E = 325 - 268 = 57$
(I) $\Rightarrow F = 57 + 4 = 61$
and $A + C + D = 64 \times 4 - 61 = 195$
(II) $\Rightarrow G = 57 + 4 = 61$
and $A + B + C = 65 \times 4 - 61 = 199$
Solving the above equations, we can get the value of A. Therefore both statements together can solve the question.

43. **(d)** Let the number be 23x and 23y
$\because$ Product of two numbers = Product of their HCF & LCM
$\therefore$ (I) & (II) imply the same thing i.e. $xy = 12$
Now possible value of x & y for which xy is 12 are (1, 12), (2, 6) & (3, 4). Hence we can find the smaller of two numbers by using either statement alone.

44. **(a)** We can find the value of x using the statement (I) alone. While statement (II) alone are not sufficient to answer the question.

45. **(a)** The distance from Ranchi to Rohtak is the same as the distance from Rohtak to Ranchi. So statement (I) alone is sufficient to answer the question completely. Statement (II) has no relevance.

46. **(c)** (I) $\Rightarrow 2x + 2y = 14$ or $x + y = 17$
(II) $\Rightarrow x^2 + y^2 = 5$
We can find the dimensions of certain rectangle using both equation. Therefore, we need both the statements to find the value of x and y.

47. **(a)** $\because I = \frac{P \times R \times T}{100}$ and Amount = P + I
(I) $\Rightarrow$ Amount $= 3P = P + I \Rightarrow I = 2P$
$\therefore R = \frac{I \times 100}{P \times T} = \frac{2P \times 100}{P \times 20} = 10\%$
(II) $\Rightarrow P - I = 1000$; here neither the sum nor interest is given.
$\therefore$ Hence, only statement I alone is sufficient.

48. **(d)** Let the length of the train be 'L' m.
Speed of the train $= \frac{L}{X}$
(I) $\Rightarrow \frac{L}{X} = \frac{L + 100}{Y} \Rightarrow L = \frac{100X}{Y - X}$
(II) $\Rightarrow L = 80 \times \frac{5}{18} X = \frac{200X}{9}$
Therefore, each statement alone is sufficient to answer the question.

49. **(d)** To find area of circle, we must have known radius (r).
(I) $\Rightarrow$ Circumference $= 308 = 2\pi r \Rightarrow r = \frac{308}{2\pi}$
(II) $\Rightarrow$ r is given.
Hence, either statement (I) alone or statement (II) alone is sufficient to answer the question.

50. **(c)** Here, $AD = AB + BC + CD$
$AB + CD = AD - BC$
Hence, both the statements together are necessary to answer the question.

51. **(d)** (I) $\Rightarrow$ 3rd floor - Mahesh, 2nd floor - Ganesh, 1st floor - Nitin
i.e. Mahesh's flat is on the 3rd floor.
(II) $\Rightarrow$ 4th floor – Ahmed
↓
3rd floor – Jeevan – Mahesh
i.e. Mahesh's flat is on the 3rd floor.
Hence, each statement alone is sufficient to answer the question.

52. **(c)** (I) $\Rightarrow C > B > D$
(II) $\Rightarrow A > E > C$
Combining both, we get $A > E > C > B > D$
Hence both statements together are necessary.

53. **(c)** (I) $\Rightarrow$ Varun's birthday is on 16th, 17th or 18th April.
(II) $\Rightarrow$ Varun' s birthday may be 13th, 14th, 15th or 16th April.
From (I) and (II) : Varun's birthday is on 16th April.

54. **(d)** (I) $\Rightarrow$ Kalpana = 7 years < Bhavana
$\Rightarrow$ Bhavana $\neq$ 5 years
(II) $\Rightarrow$ Bhavana = Ramesh + 2 = 6 + 2 = 8 years
Hence, each statement alone is sufficient to answer the question.

55. **(d)** (I) $\Rightarrow x \neq 11^2 \Rightarrow \sqrt{x} \neq 11$
(II) $\Rightarrow (\sqrt{x})^2 \neq 11^2 \Rightarrow \sqrt{x} \neq 11$
Hence, each statement alone is sufficient to answer the question.

56. **(b)** Since the multiple may be a whole number or a fraction.
Therefore, (I) is not sufficient to answer the question.
(II) $\Rightarrow \frac{x}{6} = \frac{y}{3}$
which means $x > y$
Hence, (II) is sufficient to answer the question.

57. **(c)** (I) $\Rightarrow M = 1{,}250$ and $F = 1{,}050$
(II) $\Rightarrow M : F = 2 : 1$
Combining both the statements, we get
Average $= \frac{2K \times 1250 + K \times 1050}{3K}$
$\Rightarrow$ Both statements together are sufficient to answer the question.

58. **(c)** (I) gives P
(II) gives A and T
Now, since $I = A - P$ and $R = \frac{I \times 100}{P \times T}$
Hence, both the statements together are sufficient to answer the question.

59. **(b)** From I, we cannot say that $(x^2 - y^2)$ is odd or even.
Statement (II) $\Rightarrow x + y$ is odd $\Rightarrow$ x is even and y is odd or vice-versa.
$\Rightarrow x - y$ is odd $\Rightarrow (x^2 - y^2)$ is odd

60. (b) (I) $\Rightarrow y < x < z$.
∴ Statement (I) is not sufficient to answer the question.
(II) $\Rightarrow x - y = z - x \Rightarrow$ y, x, z are in A.P.
∴ Only statement (II) is sufficient to answer the question.

61. (d) (I) $\Rightarrow \frac{1}{4}$ of 20% of x = 5 $\Rightarrow$ 20% of x = 20
(II) $\Rightarrow$ S = 400 $\Rightarrow$ x = 100 $\Rightarrow$ 20% of x = 20
∴ Each statement alone is sufficient.

62. (b) (1) There is no data about individual panels. So, it is not sufficient.
(2) $\Rightarrow A = C + 2 = B - 1$ and $A + B + C = 45$. Solving these, we can get the value of A.

63. (d) (1) & (2) $\Rightarrow$ 'ja sok' means 'you here'.
Hence, we cannot be determined the word that means 'ja' even using both the statements together.

64. (c) **From (1) :** Australian Dollar > Hong Kong Dollar > Rupees > Japanese Yen
From (2) : Value of Australian Dollar = Value of Singapore Dollar.
From (1) and (2) : Japanese currency has the lowest value.

65. (a) From statement (1), we can find the proper order of books given in question.
Hence, statement (1) alone is sufficient to answer the question.
(2) $\Rightarrow$ The sixth position from top is occupied by either a History book or a Hindi book. Hence statement II alone is not sufficient.

66. (d) (1) $\Rightarrow x + y = 4$
(2) $\Rightarrow x - y = 2$
Solving above equations, we get x and y. But the numbers may be either 10x + y or 10 y + x. Hence, we can't get the solution.
∴ Both statements together are not sufficient to answer the question.

67. (a) (1) $\Rightarrow$ Population in 1999 = 120% of population in 1998 & Population in 1998 = 120% of 1,20,000.
(2) $\Rightarrow$ We have no data of population of state B
∴ Statement (2) is not sufficient.
Hence, only statement (1) is sufficient to answer the question.

68. (a) (1) $\Rightarrow$ It is sufficient to answer the question because any odd number is divided by any odd number, it gives an odd number.
(2) $\Rightarrow$ Either X or Y is odd $\Rightarrow$ X may be odd or not.
∴ Statement (2) is not sufficient.

69. (c) (2) $\Rightarrow$ Labelled price $= \frac{2000}{125} \times 100$ = Rs.1600

(1) $\Rightarrow$ C.P. of suitcase $= 1600\left(1 - \frac{20}{100}\right)$ = Rs 1280

Hence, both the statements together are necessary to answer the question.

70. (d) Since, $\frac{\text{Speed of the boat}}{\text{Speed of the stream}}$

$= \frac{\text{Upstream time taken + Downstream time taken}}{\text{Upstream time taken – Downstream time taken}}$

Clearly, by using both statements speed of the boat can't be determined.
∴ Both the statements together are not sufficient to answer the question.

71. (d) Given : $\pi r^2 = a^2$
(where r = radius of circle; a = side of square)

$\Rightarrow r = \frac{a}{\sqrt{\pi}}$

Hence, to find the circumference of circle, one should know 'a' the side of square.

(1) $\Rightarrow$ Side of square $= \frac{\text{diagonal}}{\sqrt{2}}$

∴ Statement (1) alone is sufficient.
Also, as stated above, statement (2) alone is sufficient.

72. (d) ∵ Total cost = Area of floor × Cost of carpet per square metre.
∴ Both the statements together are not sufficient to answer the question.

73. (d) By using both statements, we can't be determined the profit of B.
∴ Both statements together are not insufficient to answer the question.

74. (c) ∵ No. of squares $= \frac{\text{Area of floor}}{\text{Area of each square}}$

(1) & (2) $\Rightarrow$ Area of floor
Hence, both statements together are sufficient to answer the question.

75. (d) Both statements together are insufficient because the no. of unknown variables is more than the no. of equations.

76. (c) Suppose the value of each share of Stocks A = ₹ x and the value of each share of stock B = ₹ y. Therefore,
(1) $\Rightarrow x = 2y$
(2) $\Rightarrow 4x + 6y = 750$
Hence, using both statements together we can calculate the total value of his stock.

77. (a) Given : Sum of ages of employees = 1500
(1) $\Rightarrow$ In the next year, sum of ages = 1500 – 10 × 59 and remaining employees = 50 – 10 = 40
Hence, average age in next year can be calculated.
(2) $\Rightarrow$ It is insufficient to answer the question.

78. (c) (2) $\Rightarrow$ Height of ΔABD = Height of ΔBDC = AB
Now, area of

$\Delta ABD = \frac{1}{2} \times \text{height} \times \text{base} = \frac{1}{2} \times AB \times AD$

and area of

$\Delta DBC = \frac{1}{2} \times \text{height} \times \text{base} = \frac{1}{2} \times AB \times DC$

(1) $\Rightarrow$ AD = DC
$\therefore$ area of ΔABD = area of ΔDBC
Hence, we can give the answer using both the statements.

79. (c) (1) $\Rightarrow \pi r^2 = 78.5$
Hence, r can be determined and then breadth of rectangle = 2r
(2) $\Rightarrow$ Length of rectangle
Hence, using statements (1) & (2), perimeter of rectangle can be determined.

80. (a) (1) $\Rightarrow$ Relative speed = Speed of jeep – Speed of car
and time taken by jeep $= \dfrac{\text{Distance}}{\text{Speed of jeep}}$
Hence, statement (1) alone is sufficient to answer the question.
But statement (2) is insufficient to answer the question.

81. (d) (1) $\Rightarrow p^6 - q^6 = 0$
$\Rightarrow (p^3 - q^3)(p^3 + q^3) = 0$
(2) $\Rightarrow$ p = 0
Since the value of $(p^2 + q^3)$ is not given. So, we cannot be determined the answer even using both the statements.

82. (d) (1) $\Rightarrow \dfrac{p}{q} > 1 \Rightarrow p > q$
(2) $\Rightarrow \dfrac{1}{q} < 1 \Rightarrow q > 1$ or q may be negative.
$\therefore$ Using both statements together, nothing can be said about the value of p.

83. (c) $a^6 - b^6 = 0$
$\Rightarrow (a^3 + b^3)(a^3 - b^3) = 0$
(1) and (2) $\Rightarrow a^3 + b^3 > 0$
Hence $a^3 - b^3 = 0$
$\therefore$ Both statements together are necessary to answer the question.

84. (d) (2) $\Rightarrow y^2 > 4 \Rightarrow -2 < y < 2$
(1) $\Rightarrow y > 0 \Rightarrow 0 < y < 2$
Hence, using both the statements we cannot be determined whether y is larger than 1 or smaller than 1.

85. (a) (1) $\Rightarrow$ Pentagon is regular $\Rightarrow$ all angles of pentagon are equal
$\therefore$ Statement (1) alone is sufficient to answer the question.
(2) $\Rightarrow$ It is irrelevant to answer the question.

86. (c) Given, t = nS
From (1) : The value of S will be either 3 or 2
From (2) : t = 18
Hence, we can find the answer using both the statements.

87. (d) (1) $\Rightarrow$ la = beautiful
(2) $\Rightarrow$ chin = red
(1) & (2) $\Rightarrow$ dem fu = rose flower.
$\therefore$ We can't be determined the word that means 'flower'.

88. (d) From 1 and 2 :

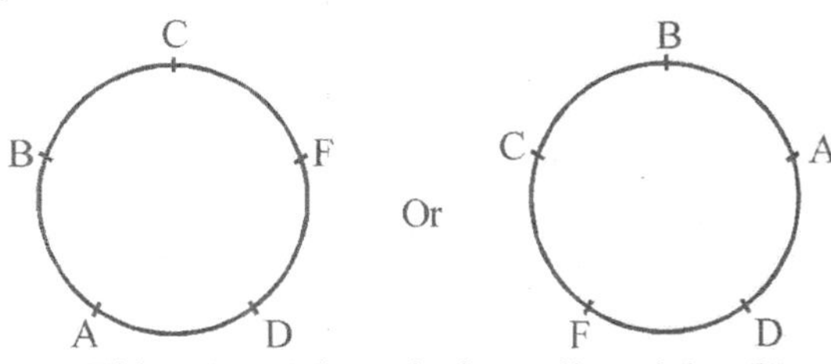

$\therefore$ Either A or C is on the immediate right of B.
$\therefore$ Both statements together are not sufficient to answer the question.

89. (a) From (1) : $\dfrac{300-80}{50-20} = \dfrac{220}{30} = 7\dfrac{1}{3}$ runs per over
(2) is irrelevant to answer the question.
Hence, only statement (1) is sufficient to give answer

90. (a) Suppose volume of the hopper = 3v
From (1) : Conveyer belt A fills 2v of the hopper in one hour.
Conveyer Belt B fills v of the hopper in one hour.
$\therefore$ Conveyer Belt A alone will take $1\dfrac{1}{2}$ hours to fill the hopper.
Hence, statement (1) alone is sufficient to answer the question.

91. (b) (2) $\Rightarrow$ The price of the car after the 1st month = Rs 170000.
Hence only statement (2) is sufficient to answer the question.

92. (a) (1) $\Rightarrow$ xyz is multiple of 9 $\Rightarrow$ x + y + z is multiple of 9.
$\therefore$ Statement (1) alone is sufficient to answer the question.
But (2) is not sufficient to answer the question.

93. (c) From (1) and (2) :
Required average $= \dfrac{23,100 + 25,200}{20 + 22}$
Therefore both statements together are necessary for answer the question.

94. (b) Here, rows × columns = 48 and no. of columns = no. of children in each row
Since, from (2) no. of columns can be calculated. Therefore it is sufficient to answer the question.

95. (a) From (1) we know that L_2 is in E_1 so C_2 will open E_1 from (2) we have no idea where L_2 is . Hence using (1) only we can find the letter of C_2

96. (a) From the question itself, the shape of the figure is known and the only thing remaining is the orientation of the fig. The first statement fixes the position of the point D, which defines the orientation. The second statement gives us four sets of values, which can't answer the query.

97. (b) Since $\angle APB$ and $\angle BCP$ are angles in alternate segments of chord PB.
$\therefore \angle APB = \angle BCP$
(2) $\Rightarrow \angle CPB = 2\angle BPD$

In ΔPCB, side CB has been produced to A, forming exterior angle $\angle ABP$

$\therefore \angle ABP = \angle BCP + \angle CPB$

$\Rightarrow \angle ABP = \angle APB + 2\angle BPD$

$\Rightarrow 2\angle BPD = \angle ABP - \angle APB$

$\Rightarrow \angle BPD = \frac{1}{2}(\angle ABP - \angle APB)$

Hence, we can be determined the answer.

98. **(a)** (1) $\Rightarrow$ w = z, then PQ must be parallel to TU
(2) $\Rightarrow$ PT is parallel to QU. But, it does not determine whether PQ is parallel to TU.
Hence, only statement (1) can give answer the question.

99. **(d)** By using both statements we cannot be determined PR = US.
So, both statements together are not sufficient.

100. **(b)** (1) $\Rightarrow \frac{19}{x+1}$ = an integer, then 19 = (x + 1) × an integer and x + 1 = $\frac{19}{\text{an integer}}$

$\Rightarrow$ x may or may not be an integer.
$\therefore$ Statement (1) alone is not sufficient to answer the question.

(2) $\Rightarrow \frac{x+1}{5}$ = an integer

$\therefore$ x + 1 = a integer × 5 = an integer
$\therefore$ x = an integer – 1 = an integer
Hence, statment (2) alone is sufficient to answer the question.

101. **(a)** (1) $\Rightarrow$ x + 20% of x = Rs. 16.80.
Hence statement (1) alone is sufficient to solve the question.

102. **(c)** Gross income for year 2000 = Gross income for year 1999 + % increase
$\therefore$ Both statements together are necessary to answer the question.

103. **(d)** (1) $\Rightarrow$ p – 2 = 1 or – p + 2 = 1 $\Rightarrow$ p = 3, 1.
From (2) $\Rightarrow P^2 - 4p + 3 = 0 \Rightarrow (p-3)(p-1) = 0$
$\Rightarrow$ p = 3, 1
Hence, using both statements together, we can't be determined the unique value of p.

104. **(c)** (1) $\Rightarrow t_1 + t_2 = 10$; U = 7 and T = 8
$\Rightarrow$ R = 6, S = 5
From (1) and (2) : If $t_1 = 1$ and $t_2 = 9$ then the minimum present ages of Raju and Renu are 31 and 26 years respectively.

105. **(c)** Let the sum be Rs x. Then
From (1) and (2) [half of the sum + present worth of half the sum 2 years hence] – [present worth of the sum due in one year] = Rs 40
Hence, we can find the answer using both the statements.

106. **(c)** (1) $\Rightarrow$ M = 5S
(2) $\Rightarrow$ M – 5 = 25 (S – 5)
Solving these equations, we get the value of M. Therefore, both statements together are required to answer the question.

107. **(d)** (1) $\Rightarrow$ Area of right-angled triangle = 12 × L
(2) $\Rightarrow$ L = 18 cm.
$\therefore$ Using (1) and (2) we can only calculate the area of right-angled triangle. Since we don't know the base of triangle, the height cannot be calculated.

108. **(d)** Here we have not been given the number of sisters of Y in both statements. Hence, we cannot find the answer using both the statements.

109. **(d)** Information given in both the statements does not lead to specific direction. Hence, even both (1) and (2) together are not sufficient.

110. **(c)** Sunrise Time – Sunset time = length of day
$\therefore$ both statements together are necessary to answer the question.

111. **(c)** (1) $\Rightarrow$ orange trees < mango trees
(2) $\Rightarrow$ mango trees < apple trees
Using (1) & (2), we conclude that the most numerous trees are apple trees.
$\therefore$ both statements together are necessary to answer the question.

112. **(d)** (1) $\Rightarrow$ M + P + C = 165%
(2) $\Rightarrow$ Pr $\rightarrow$ M + 10% (average)
$\therefore$ Both statements together are not sufficient to give the marks obtained by Prakash in Match.

113. **(c)** Loss = C.P. – S.P.
(1) $\Rightarrow$ C.P. is given

(2) $\Rightarrow$ Loss = 25% of S.P. $= \frac{1}{4}$ S.P.

Hence, 'Loss' can be calculated using both the statements together.

114. **(c)** $\because$ cost = Area of room × Rate per square area and Area = length × breadth
$\therefore$ Both statements together are necessary to answer the question.

115. **(d)** Using (1) & (2), we get 'na' means 'go'
$\therefore$ either 'pit' or 'ja' means 'come' which cannot be determined.

116. **(d)** Both statements are insufficient to answer the question.

117. **(d)** Since both statements together do not give the correct position of Gita. So, we cannot be determined the answer even using both the statements.

118. **(d)** (1) $\Rightarrow$ Bareilly < Bhagalpur
(2) $\Rightarrow$ Allahabad < Patna = Bhagalpur
$\therefore$ No conclusion can be drawn.

119. **(c)** By using (1) and (2), we get the following order of their positons.

R P N D S

Hence we can find the answer using both the statements.

120. **(b)** Let the two-digit no. be xy, i.e. 10x + y.
(1) $\Rightarrow x + y = 8$

$(2) \Rightarrow \frac{1}{5}(10x+y) = \frac{44}{2} - 15 = 22 - 15 = 7$

$\Rightarrow 10x + y = 35$

$\Rightarrow$ number is 35

$\Rightarrow$ required difference can be determined.

Hence, only statement (2) is sufficient to answer the question.

121. **(c)** (1) $\Rightarrow$ X's salary = 20% of Z's salary

(2) $\Rightarrow$ Y's salary = 45% of Z's salary

$\Rightarrow \frac{\text{X's salary}}{\text{Y's salary}} = \frac{20}{45} = \frac{4}{9}$

Hence, both the statements together are sufficient to answer the question.

122. **(d)** Let the total number of students = N

(2) $\Rightarrow$ Number of Girls $= \frac{5}{11}N$

(1) $\Rightarrow$ Number of Boys = 120% of $\frac{5}{11}N$

Since we do not have the value of N, so, both statements are not sufficient to answer the question.

Level-II

1. **(b)** (I) $\Rightarrow A < C < B$, we have no idea about D.

(II) $\Rightarrow A < D < C < B$

2. **(d)** Since the product of any numbers is an even number if and only if at least one of them is an even number. Now,

(I) $\Rightarrow$ A is an even number $\Rightarrow$ Product of A, B, C is even.

(II) $\Rightarrow A \times B$ is an even number and $A \times C$ is also an even number.

$\Rightarrow$ at least one of them (A, B, C) is even

$\Rightarrow$ their product is an even number.

$\therefore$ Each statement alone is sufficient to answer the question.

3. **(c)** (I) $\Rightarrow \angle ABD = \angle BAD = \theta$ (let)

(II) $\Rightarrow \angle ABD = \angle CBD = \theta$

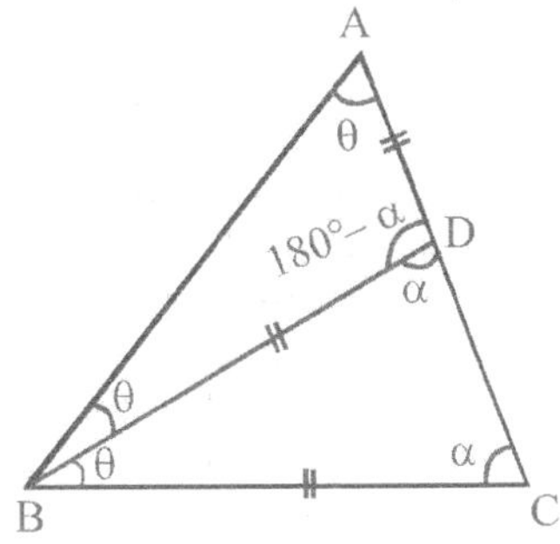

Since BC = BD $\Rightarrow \angle BDC = \angle BCD = \alpha$ (let)

$\therefore$ In ΔABD, $\theta + \theta + (180 - \alpha) = 180°$

$\Rightarrow \alpha = 2\theta$

$\Rightarrow \angle ACB = \angle ABC$

$\Rightarrow \angle AB = \angle AC$

$\Rightarrow \Delta ABC$ is isosceles.

Hence, both statements together are necessary to answer the question.

4. **(a)** (I) $\Rightarrow$ combined income for the years 1994 to 1999 = average yearly income $\times$ total years.

$\therefore$ Statement (I) alone is sufficient to answer the question.

(II) $\Rightarrow$ His salary for 1994-98 is not given, so, it is insufficient to answer the question.

5. **(a)** Let the diameter of circle = d and side of square = a

Then First fly covers the distance = πd

and Second fly covers the distance = 4a

(I) $\Rightarrow a\sqrt{2} = d$

Hence, conclusion can be drawn.

(II) $\Rightarrow$ We have no idea about their speeds. So statement (II) alone is not sufficient.

6. **(a)** Since, all three circles are equal.

$\therefore$ PQ = QR = PR

$\Rightarrow \Delta PQR$ is an equilateral triangle.

And area of shaded portion

= Area of ΔPQR – 3 (area of each sector),

(I) $\Rightarrow$ QR is known.

$\Rightarrow$ side of triangle and radius is known.

Hence, statement (I) alone is sufficient to answer the question.

Now, since all three circles are equal, therefore, statement (II) is restatement of the information given in question.

7. **(d)** From the first statement it gives that mathematician can never add 2 number correctly, but it is quite possible that apart from mathematician, others can also do the same mistake. The same logic is applied for the second statement as mathematician is given. If it is only mathematician then we can answer with the help of both the statements.

8. **(b)** Given G + S = 30

(I) $\Rightarrow$ 90% of G + 90% of S = 27.

i.e. 90G + 80S = 2700 $\Rightarrow$ G + S = 30

Hence, we get the same equation as we have already given in question.

(II) $\Rightarrow$ 90% of G + 80% of S = 27

i.e. 90G + 80G = 2700

Hence, statment (II) alone is sufficient to answer the question.

9. **(d)** Since cone is inside the cylinder which is within the cube

$\Rightarrow$ Diameter of base of cone = height of cone

= diameter of base of cylinder

= height of cylinder = side of cube

(I) $\Rightarrow$ Volume of cone $\Rightarrow$ radius and height of cone

$\Rightarrow$ radius and height of cylinder

Hence, volume of cylinder can be determined.

(II) $\Rightarrow$ Total surface area of cube $\Rightarrow$ side of cube

$\Rightarrow$ radius and height of cylinder

Hence, volume of cylinder can be determined.

$\therefore$ Each statement alone is sufficient to answer the question.

10. (d) (I) ⇒ Because factorial of any number greater than 1 will yield a result of even number. Adding 1 to this number will change it to an odd number. Thus, an odd number will obviously be not divisible by an even number.
(II) ⇒ We get 'yes' as our answer. Because according to Wilson's theorem if P be a prime number, $1 + (P - 1)!$ is divisible by P.

11. (d) (I) ⇒ N = 64 (∵ only 64 is both the square of an integer and the cube of an integer)
∴ Statement (I) alone is sufficient to answer the question.
(II) ⇒ N = 64.
∴ Statement (II) alone is sufficient to answer the question.

12. (b) (I) ⇒ $x < 1$ ⇒ x may be positive or negative.
∴ We can't be determined whether $\frac{1}{x}$ is greater than $\frac{1}{x+1}$.
(II) ⇒ $x > 1$ ⇒ x is positive ⇒ $x < x + 1$
$$\Rightarrow \frac{x}{x(x+1)} = \frac{x+1}{x(x+1)}$$
[dividing both sides by x (x + 1)]
$$\Rightarrow \frac{1}{x+1} < \frac{1}{x}$$
∴ Statement (II) alone is sufficient to answer the question.

13. (c) Every player takes an odd number of matches per play. After the first player goes, there will always be an odd number of matches left. After the second players goes, there will always be an even number of matches left. Therefore, the second player, i.e. Alka, is the winner.

14. (c) (I) ⇒ $L_x = 2L_y$ and $S_y = 100$ m/s
(II) ⇒ $L_x = 100$ m ⇒ $L_y = 50$ m
Now, their relative speed = $S_x - S_y$, where S_x is the speed of train X.
Also, (II) ⇒ 'X' passes 'Y' in 10 sec.
$$\Rightarrow S_x - S_y = \frac{L_x + L_y}{10}$$
Hence, we can find S_x in km/h, using both the statements.

15. (a) **From (I) :** ras pak leun = I have pens ....(i)
pak ras tap = they have pens ....(ii)
(i) and (ii) ⇒ ras pak = have pens ....(iii)
∴ (ii) and (iii) ⇒ tap = they.
From (II) : kip ras tun = boys have cow ....(iv)
sin tap ras = they have egg ....(v)
(iv) and (v) ⇒ ras = have
∴ We cannot be determined the word that means 'they'.
∴ Only statement (I) is sufficient to answer the question, while (II) is not.

16. (a) (I) ⇒ $g - h > 0$
∴ (I) is sufficient to answer the question.
(II) ⇒ g^2 is greater than h^2, but g may not be greater than h. (For example, g might be –3 and h might be 2). Hence, (II) alone is not sufficient to answer the question.

17. (d) (I) ⇒ The sun is to the left of Shashidhar and since it is morning, the left of Shashidhar is East. Hence, he is facing South.
(II) ⇒ Since it is morning, he is facing South.
∴ Each statement alone is sufficient to answer the question.

18. (b) From (I) :
Q(F) – N(M) – S
↓ daughter
P(M) – L(F)
N is maternal uncle of P.
From (II) :
(M)M – K(M) – L(F)
↓ daughter
G(F) – P
M is the paternal uncle of P
Hence we can give the answer using either statement alone.

19. (c) **From (1) :** We get the 1st, 2nd, 3rd and 4th presentation date.
(1) and (2) ⇒ 'A' did not make his presentation on 26th also, because 'D' made presentation on 24th. Hence, 'A' made presentation on 23rd, 'B' on 25th and 'C' on 26th.

20. (d) **From (1) :** Number of subscribers of company gives no indication of company's talk-times on its coupon of different values. Hence, statement (1) alone is not sufficient.
From (2) : A provides more talk-time than H. But still we have no information as to who among A and I provides more talktime. Thus statement (2) also is not sufficient.
Both the informations even together are not sufficient to answer the question.

21. (a) (1) ⇒ Ratio of the costs of first, second and third necklace is 6 : 5 : 7 and total cost is given. Hence the price of second necklace can be calculated.
(2) ⇒ $N_1 = N_2 + \frac{2}{5}$, N_3 is least
and $N_1 + N_2 + N_3 = 120000$
Since we don't know N_3. Therefore statement (2) is not sufficient to answer the question.

22. (c) (2) ⇒ Houses on east side have only even number i.e. 122, 124, 126,......., 180, 182.......
(1) ⇒ Houses with number 122, 124, 126,......., 182 are painted blue.
⇒ 31 houses are painted blue.
∴ Both statements together are necessary to answer the question.

23. (d) (1) ⇒ Last digit of n is 3
⇒ n may be 3, 13, 23, 33, 43, 53, 63, ...
(2) ⇒ n + 5 is divisible by 6
⇒ n may be 1, 7, 13, 19, 25, 31, 37, 43, ...
Hence, we cannot be determined that integer is divisible by 3 even using both the statements.

24. (d) From (1) and (2) it doesn't clear when 12 years were completed. If Y completed 12 years sometime in 2002, he will be entitled to promotion only after Dec. 31st, 2002.
∴ Both statements together are not sufficient to answer the question.

25. (a)

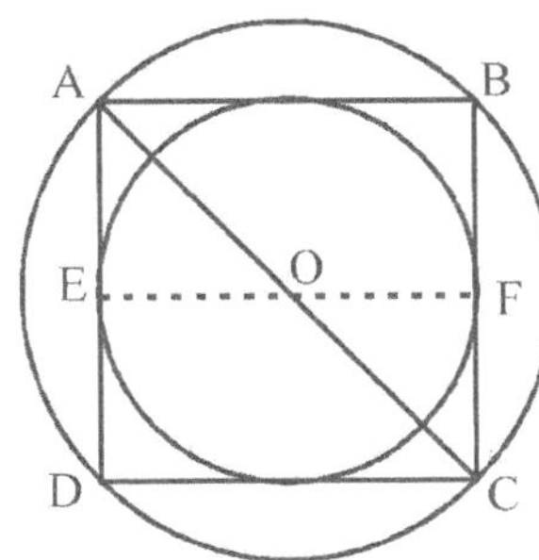

(1) ⇒ OC = 10 cm
∴ Diagonal of square = 20 cm

⇒ Side of square = $\frac{20}{\sqrt{2}} = 10\sqrt{2}$ cm

∴ radius of the inner circle OE = $\frac{10\sqrt{2}}{2} = 5\sqrt{2}$ cm.

Hence circumference of inner circle can be determined. Therefore, statement (1) alone is sufficient to answer the question.
(2) ⇒ It is not sufficient to answer the question.

26. (c) Here, Man's speed upstream = (x – y) km/hr.
Man's speed down stream = (x + y) km/hr
Let the required distance be D km then

$$\frac{D}{(x-y)} + \frac{D}{(x+y)} = z \Rightarrow \frac{D[x+y+x-y]}{(x-y)(x+y)} = z$$

$$\Rightarrow \frac{2Dx}{x^2-y^2} = z \Rightarrow D = \frac{z(x^2-y^2)}{2x} = \frac{x^2-y^2}{2.\frac{x}{z}}$$

∴ To find D, $x^2 - y^2$ and $\frac{x}{z}$ should be known.

Hence, both the statements together are necessary to answer the question.

27. (a) (1) ⇒ 112% of wages of man – 14 × (115% of price of cake) = 0.60.
and 112% of wages of man = 14 × (116% of price of cake)
⇒ Price of a cake can be calculated.
(2) ⇒ 104% of wages of man – 14 × (104% of price of cake) = x + 80
and 108% of wages of man – 14 × (108% of price of cake) = x + 1.6 (where x = initial saving)
∴ We can't be determined the price of a cake.
∴ Statement (1) alone is sufficient to answer the question.

28. (c) (1) ⇒ P might or might not be on the circle.

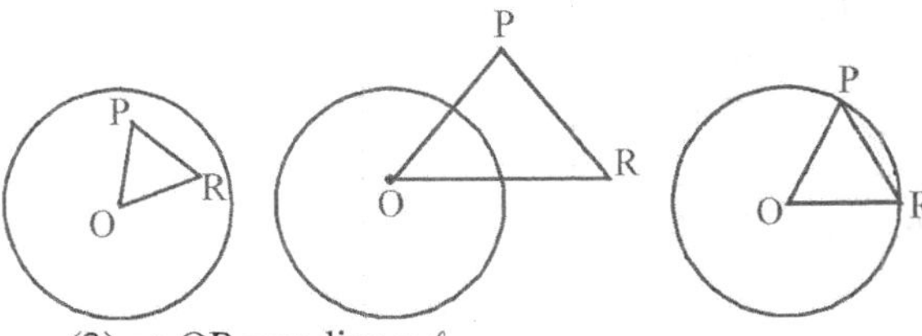

(2) ⇒ OR = radius = ℓ
Now (1) & (2)
⇒ If P, O and R are the vertices of an equilateral triangle and OR = ℓ, then obviously OP = ℓ.
⇒ P is on the circle.
Hence, both statements together are necessary to answer the question.

29. (c) Given : $\text{FEI} = \frac{\text{foreign equity inflows}}{\text{GDP of the country}} \times 100$

$\Rightarrow \text{GDP of a country} = \frac{\text{foreign equity inflows}}{\text{FEI}} \times 100$

Hence, we need FEI and foreign equity infows of a country to get GDP of that country. Here we have been asked about the comparison of GDPs of China and India. Therefore, ratios of foreign equity inflows in the given two countries are sufficient to lead us towards answer.
∴ Statements (1) and (2) together are necessary to answer the question.

30. (a) Given : Total temperature of 7 days = 44×7 = 308°C.
Total temperature of the first three days = 3 × 43.8°C = 131.4°C.
Total tempeature of sixth and seventh days = 2 × (43.8 + 0.1) = 2 × 43.9 = 87.8°C.
Hence, to answer the question, we need a relation between the temperature of fourth and fifth day.
∴ Statement (1) alone is sufficient to answer the question.

31. (b) ∵ ababab = a × 10^5 + b × 10^4 + a × 10^3 + b × 10^2 + a × 10 + b
= a (101010) + b (10101)
= 10101 (10a + b)
Since 10101 is divisible by 111, ababab is always divisible by 111.
⇒ ababab is divisible by 222 only when b is even.
Therefore statement (2) alone is sufficient to answer the question.

32. (d) (1) ⇒ X, Y, Z are consecutive ⇒ Either X, Z are even and Y is odd or X, Z are odd and Y is even.
(2) ⇒ Unique determination can't be possible.
∴ Both statements together are not sufficient to answer the question.

33. (d) (1) ⇒ Bell range at 7.15 + 5 = 7.20 am
(2) ⇒ Both B and C reached the school before 7.20am (using 1). But exact time of reaching can't be determined. Hence, comparison is not possible.

34. (b) (1) ⇒ The value of x is given but we haven't the value of y. Therefore statement (1) is not sufficient to answer the question.
(2) ⇒ y = 1.
∴ Given expression = $x^3 - x^3 = 0$
∴ Statement (2) alone is sufficient to answer the question.

35. (e) From statements I and II, since it is a right-angled triangle, area

$\Rightarrow \frac{1}{2} \times b \times h = 5b \qquad \therefore h = 10$

36. (e) Combining I & II, we get speed of Dinesh = 10 km/hr

$\therefore$ to walk 1 km he takes $\frac{60}{10} = 6$ minutes.

37. (b) From statement II, we find the negative answer since the number is not divisible by 3 then it can't be divisible by 12.

38. (a)

39. (d) Using both the statements, we can find separate expenditures of Rahim and Suresh but can't find their separate incomes.

40. (c) The given question gives the amounts of investment of P, Q and R, which is P = ₹12000, Q = ₹9000 and S = ₹21000. Statement (A) combined with the question's information will give us the ratio of their profits, i.e. 4 : 6 : 7. Now, combine this either with B or C, and profit earned by them can be determined. Therefore, A and either B or C is sufficient.

41. (e) Let the present ages of mother and son be M and S yrs respectively.

A. $\Rightarrow \frac{M}{S+10} = \frac{5}{3}$ or, $3M - 5S = 50$

B. $\Rightarrow 3M - 5S = 50$

C. $\Rightarrow \frac{M+10}{S+16} = \frac{5}{3}$ or, $3M - 5S = 50$

Here, all the three equations are the same. Therefore option (e) is the correct choice.

42. (c) Statement (A) gives the cost of fencing one metre of the plot. Combining this with statement (B), total cost of fencing can be determined.

43. (d) **A.** $\Rightarrow$ Market price of the article

$= \frac{100}{90} \times 810 =$ ₹900

B. $\Rightarrow$ CP of the article (with the help of A)

$= \frac{900 \times 100}{150} =$ ₹600

Now, combining (A) with (C),

SP = 900 × 0.95 = ₹ 855

CP = $855 \times \frac{100}{142.50} =$ ₹600

Hence, A and either B or C are sufficient.

44. (a) Let the rate of interest be r%.

A. $\Rightarrow 2500\left[\left(1+\frac{r}{100}\right)^2 - 1\right] = \frac{5150 \times r \times 3}{3 \times 100}$

B. $\Rightarrow \frac{12000 \times r \times 3}{100} + \frac{10000 \times r \times 5}{100} = 5160$

C. $\Rightarrow r = \frac{1725 - 1500}{3 \times 150} \times 100 = 5\%$

Hence, any one of them is sufficient.

45. (e) From I, If P = 100

A = 200 and SI = 200 – 100 = 100

Rate = $\frac{SI \times 100}{P \times T} = \frac{100 \times 100}{100 \times 5} = 20\%$

From II and III, Rate = $\frac{400 \times 100}{2000 \times 1} = 20\%$

Hence, either I alone or II + III will be sufficient.

46. (b) From I and II, 10 women can finish the work in 1 day

$= \frac{7}{24} - \frac{1}{6} = \frac{7-4}{24} = \frac{1}{8}$

$\therefore$ 10 women can finish the work in 8 days.

From II and III,

Let 10 men can finish the work in x days and 10 women can finish the same work in y days.

Hence, $\frac{1}{x} + \frac{1}{y} = \frac{7}{24}$ ...(1)

and from III – II, $\frac{3}{x} + \frac{4}{y} = 1$ ...(2)

from (1) & (2)

y = 8 days

Again from I and III $\frac{3}{6} + \frac{4}{y} = 1 \Rightarrow y = 8$ days

47. (a) From I, Let present age of Sabir be x yr and age of his father be 2x yrs.

From I and II, $\frac{x+5}{2x+5} = \frac{6}{11} \Rightarrow 12x + 30 = 11x + 55.$

x = 25 yrs.

From I and II, age of Sabir = 25 yrs.

Hence, only from I & II, age of Sabir and his father can be obtained.

48. (e) Let two digit number be 10x + y.

From I, either $x - y = \frac{27}{9} = 3$ or $y - x = \frac{27}{9} = 3$

From II, x – y = 3 or y – x = 3

From III, x – y = 3

Hence, Even by (I) + (II) + (III) we cannot obtain the number.

49. (e) From I and II.

Length = 3x = 48 m

$\therefore$ x = 16

Breadth = 2x = 32 m

Hence, Area of floor = 48 × 32

Cost of flooring = 48 × 32 × 850 = ₹ 1305600

From I and III, 2(l + b) = 160

$\Rightarrow 2(3x + 2x) = 160 \Rightarrow 10x = 160$

$\Rightarrow$ x = 16

$\therefore$ Length = 3 × 16 = 48 m

Breadth = 2 × 16 = 32 m

Cost of flooring = (48 × 32) × 850 = ₹ 1305600

Similarly, from II and III, we can find l = 48 m and b = 32 m

and total cost of flooring = ₹ 1305600

PRACTICE SET-1

DIRECTIONS (Qs. 1-15): *What will come in place of question mark (?) in the following questions ?*

1. 24% of 40% of $\frac{3}{4}$th of 12000 = ?
(a) 868 (b) 864
(c) 866 (d) 862
(e) None of these

2. (82 + 78) × (0.5 + 0.8) = ?
(a) 209 (b) 205
(c) 206 (d) 208
(e) None of these

3. $(15)^2 + (12)^2 - (18)^2 = ?$
(a) 42 (b) 43
(c) 48 (d) 49
(e) None of these

4. 8888 ÷ 8 + 2332 ÷ 2 = ?
(a) 2727 (b) 2772
(c) 2722 (d) 2272
(e) None of these

5. 46813 + 32298 – 13234 – 31112 = ?
(a) 34764 (b) 34767
(c) 34766 (d) 34765
(e) None of these

6. (5994 ÷ 18) × (468 ÷ 18) = ?
(a) 8658 (b) 8656
(c) 8655 (d) 8654
(e) None of these

7. (0.8 × 8800 ÷ 16) × 4 = ?
(a) 1750 (b) 1756
(c) 1766 (d) 1760
(e) None of these

8. $85 \times 12 \div 5 - (2)^3 = (?)^2$
(a) 196 (b) 24
(c) $2\sqrt{14}$ (d) ± 14
(e) $-\sqrt{14}$

9. $(15)^2 - (5)^3 + \sqrt{625} + 44 = (?)^2$
(a) $\sqrt{17}$ (b) 17
(c) $\sqrt{13}$ (d) 13
(e) $\sqrt{269}$

10. $\frac{5}{12} + \frac{5}{36} + \frac{5}{24} = ?$
(a) $\frac{55}{72}$ (b) $\frac{11}{72}$
(c) $\frac{7}{36}$ (d) $\frac{11}{36}$
(e) None of these

11. $?\%$ of $800 + (12)^2 = 504$
(a) 45 (b) 40
(c) 60 (d) 55
(e) None of these

12. $(2 \times 8 \div 4)^4 \div (4 \times 4) \div 5 = ?$
(a) 6.5 (b) 4.2
(c) 3.2 (d) 4.5
(e) None of these

13. 3435 ÷ 3 + 51 = ? × 13
(a) 92 (b) 87
(c) 96 (d) 89
(e) None of these

14. $\frac{7}{12}$ of 48% of 750 = ?
(a) 60 (b) 360
(c) 120 (d) 240
(e) None of these

15. $\frac{8}{21}$ of $189 = (?)^2 \div 2$
(a) 6 (b) 36
(c) $\sqrt{6}$ (d) –36
(e) 12

DIRECTIONS (Qs. 16-20) : *What should come in place of the question mark (?) in the following number series?*

16. 2 12 60 240 720 ?
(a) 1445 (b) 1440
(c) 1420 (d) 1444
(e) 1448

17. 80 48 64 56 60 ?
(a) 57 (b) 68
(c) 54 (d) 62
(e) 58

18. 32 38 50 68 92 ?
(a) 118 (b) 128
(c) 124 (d) 122
(e) 116

19. 55000 11000 2200 440 88 ?
(a) 16.4 (b) 18.3
(c) 17.6 (d) 14.5
(e) None of these

20. 5 30 120 600 ?
(a) 3600 (b) 1500
(c) 1800 (d) 2000
(e) None of these

21. The average age of father and son at present is 27 years. Also, the difference between their ages is 30 years. What is the respective ratio between the present age of father and the present age of son ?
(a) 8 : 3 (b) 7 : 2
(c) 2 : 7 (d) 3 : 8
(e) None of these

22. The ratio of the present ages of Tarun and Varun is 3 : 7. After 4 years Varun's age will be 39 years. What was Tarun's age 4 years ago ?
(a) 12 years (b) 13 years
(c) 19 years (d) 18 years
(e) None of these

23. A truck covers a distance of 420 km in a certain interval of time at a speed of 70 km/hr. What is the average speed of a bike that travels a distance of 36 km less than the truck in the same time ?
(a) 62 km/hr (b) 64 km/hr
(c) 66 km/hr (d) 68 km/hr
(e) None of these

24. Kiran sold an item for ₹ 8,160 and incurred a loss of 15%. At what price should she have sold the item to have a profit of 25%?
(a) ₹ 12,200 (b) ₹ 12,300
(c) ₹ 13,000 (d) Cannot be determined
(e) None of these

25. Sunita scored 56 marks in Hindi, 42 marks in Science, 63 marks in Maths, 94 marks in Social Science and 61 marks in English. The maximum marks of each subject is 110. How much approximate percentage of marks did she get?
(a) 68 (b) 51
(c) 57 (d) 62
(e) 48

26. The average of five numbers is 50.8. The average of the first and the second number is 29. The average of the fourth and the fifth number is 66.5. What is the third number ?
(a) 65 (b) 73
(c) 61 (d) 77
(e) None of these

27. The length of a rectangle is 15 cm which is 6 cm less than the diameter of a circle. What is the area of the circle?
(a) 346.5 sq cm (b) 173.25 sq cm
(c) 156 sq cm (d) 132 sq cm
(e) None of these

28. 12 women can complete a piece of work in 5 days, whereas 3 women and 9 children together can complete the same piece of work in 10 days. In how many days can 36 children complete the same piece of work?
(a) 10 days (b) 2 days
(c) 5 days (d) 8 days
(e) None of these

29. What is the area of the circle whose circumference is 264 cm?
(a) 5544 sq cm (b) 5454 sq cm
(c) 5540 sq cm (d) 5548 sq cm
(e) None of these

30. The average speed of a tractor is two-fifths the average speed of a car. The car covers 450 km in 6 hours. How much distance will the tractor cover in 8 hours?
(a) 210 km (b) 240 km
(c) 420 km (d) 480 km
(e) None of these

31. The simple interest accrued in 3 years on a principal of ₹ 25,000 is three-twentieths the principal. What is the rate of simple interest pcpa?
(a) 5 (b) 4
(c) 6 (d) 3
(e) None of these

32. A train crossed a platform in 43 seconds. The length of the train is 170 metres. What is the speed of the train?
(a) 133 km./hr (b) 243 km./hr
(c) 265 km./hr (d) Cannot be determined
(e) None of these

33. The angles in a triangle are in a ratio of 19 : 10 : 7. What is the sum of twice the smallest angle and the largest angle?
(a) 165° (b) 185°
(c) 155° (d) 175°
(e) None of these

34. Out of certain sum, $\frac{1}{3}$rd is invested at 6%, $\frac{1}{6}$th at 12% and the rest at 16%. If the simple interest for 4 years from all these investments amounts to ₹ 2400, find the original sum.
(a) ₹ 4000 (b) ₹ 5000
(c) ₹ 6000 (d) ₹ 4500
(e) None of these

35. The length of a rectangular field is thrice its width. Inside the field there is a square pond $3\sqrt{3}$ m long. If the area of the pond is $\frac{1}{9}$ of the area of field, what is the length of the field ?
(a) 27 m (b) 25 m
(c) 20 m (d) 22 m
(e) None of these

DIRECTIONS (Qs. 36–40): *Study the table carefully to answer the questions that follow :*

Number of Employees (in Thousands) working is six different companies in six years.

Years	Company					
	K	L	M	N	P	R
2009	3.4	2.4	2.8	3.9	4.2	5.6
2010	1.2	3.4	3.0	4.1	1.9	4.5
2011	5.4	4.9	3.7	3.4	2.7	5.7
2012	6.3	4.7	5.5	5.3	5.6	2.9
2013	11.0	5.9	5.7	6.7	4.8	6.7
2014	7.2	7.4	5.9	9.6	6.4	4.8

36. What was the respective ratio between the employees working in company R is year 2010 and employees working in company L in year 2009 ?
(a) 8 : 15 (b) 15 : 8
(c) 5 : 8 (d) 13 : 9
(e) None of these

37. . What was the total number of employees working in company L and R together in year 2010 ?
(a) 7800 (b) 7840
(c) 7000 (d) 7900
(e) None of these

38. In which company were the total number of employees working in year 2012 the second highest?
(a) R (b) N
(c) L (d) K
(e) P

39. Number of employees working in company K in year 2014 was what percentage of all the employees working in company N in all the years together ?
    (a) 25 (b) 30
    (c) 22 (d) 31
    (e) 35
40. What was the average number of employees working in year 2012 in all the companies together ?
    (a) 5,05,000 (b) 50,050
    (c) 50,500 (d) 5,500
    (e) None of these

DIRECTIONS (Qs. 41-45): *Study the following pie-chart carefully to answer these questions.*

Total number of passengers = 8500
Percentage of passengers

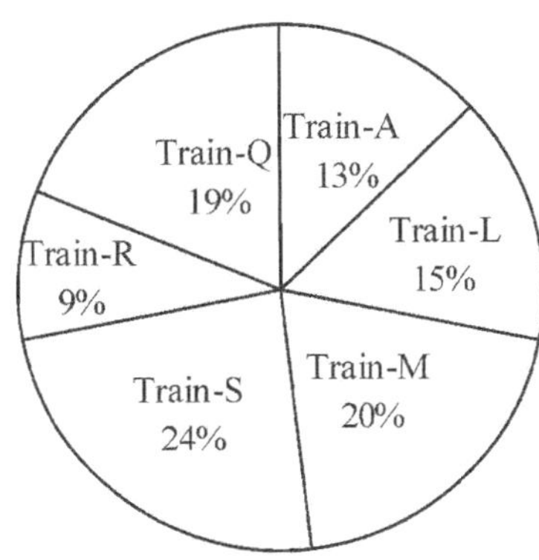

41. What was the **approximate** average number of passenger in Train-S, Train-M and Train-L together?
    (a) 1521 (b) 1641 (c) 1651 (d) 1671 (e) 1691
42. If in Train-R 34 per cent of the passengers are females and 26 per cent are children, what is the number of males in that train?
    (a) 306 (b) 316
    (c) 308 (d) 318
    (e) None of these
43. The number of passengers in Train-Q is approximately what percentage of the total number of passengers to Train-A and Train-R?
    (a) 90 (b) 70 (c) 75 (d) 80 (e) 86
44. Which train has the second highest number of passengers?
    (a) A (b) Q (c) S (d) M (e) L
45. How many more per cent (approximately) number of passengers are there in Train-M as compared to the number of passengers in Train-L?
    (a) 29 (b) 49 (c) 43 (d) 33 (e) 39

DIRECTIONS (Qs. 46-50): *Each of the questions below consists of a question and three statements denoted A, B and C are given below it. You have to study the questions and all the three statements and decide whether the question can be answered with any one or two of the statements or all the statements are required to answer the question.*

46. What is R's share of profit in a joint venture?
    A. Q started business investing ₹ 80,000/-
    B. R joined him after 3 months.
    C. P joined after 4 months with a capital of ₹ 1,20,000 and got ₹ 6,000 as his share of profit.
    (a) Only A and C are required
    (b) Only B and C are required
    (c) All A, B and C together are required
    (d) Even with all A, B and C the answer cannot be arrived
    (e) None of these
47. What is the area of a right angled triangle?
    A. The perimeter of the triangle is 30 cm.
    B. The ratio between the base and the height of the triangle is 5 : 12.
    C. The area of the triangle is equal to the area of a rectangle of length 10 cms.
    (a) Only B and C together are required
    (b) Only A and B together are required
    (c) Only either A or B and C together are required
    (d) Only A and C together are required
    (e) None of these
48. What will be sum of two numbers?
    A. Among the two numbers, the bigger number is greater than the smaller number by 6.
    B. 40% of the smaller number is equal to 30% of the bigger number.
    C. The ratio between half of the bigger number and $\frac{1}{3}$rd of the smaller number is 2 : 1.
    (a) Only B and C together are necessary
    (b) Only A and B together are necessary
    (c) Out of A, B and C any two together are necessary
    (d) All three A, B and C together are necessary
    (e) None of these
49. How much profit did Mahesh earn on the cost price of an article by selling it?
    A. He got 15% discount on the marked price at the time of purchase.
    B. He sold it for ₹ 3060.
    C. He earned 2% profit on the marked price.
    (a) Only A and B both together are necessary.
    (b) Only B and C both together are necessary.
    (c) Only A or C and B together are necessary.
    (d) Even A, B and C all together are not sufficient to answer the question.
    (e) All three A, B and C together are necessary.
50. How many marks did Arun secure in English?
    A. The average marks obtained by Arun in four subjects including English is 60.
    B. The total marks obtained by him in English and Mathematics together is 170.
    C. The total marks obtained by him in Mathematics and Science together is 180.
    (a) All three A, B and C together are necessary.
    (b) Only A and B together are necessary
    (c) Only B and C together are necessary.
    (d) Only A and C together are necessary.
    (e) None of these

SOLUTIONS

1. (b) $? = \frac{24}{100} \times \frac{40}{100} \times \frac{3}{4} \times 12000 = 864$

2. (d) $(82 + 78) \times (0.5 + 0.8)$

$= 160 \times 1.3 = 208$

3. (e) $? = (15)^2 + (12)^2 - (18)^2$

$= 225 + 144 - 324 = 45$

4. (e) $\frac{8888}{8} + \frac{2332}{2} = 1111 + 1166 = 2277$

5. (d) $79111 - 44346 = 34765$

6. (a) $\left(\frac{5994}{18}\right) \times \left(\frac{468}{18}\right) = 333 \times 26 = 8658$

7. (d) $(0.8 \times 550) \times 4 = ?$

$440 \times 4 = ?$

$? = 1760$

8. (d) $85 \times \frac{12}{5} - 8 = (?)^2$

$17 \times 12 - 8 = ?^2$

$196 = ?^2$

$? = \pm 14$

9. (d) $225 - 125 + 25 + 44 = (?)^2$

$\therefore \quad ? = \sqrt{169} = 13$

10. (a) $? = \frac{5}{12} + \frac{5}{36} + \frac{5}{24} = \frac{30 + 10 + 15}{72} = \frac{55}{72}$

11. (a) $\frac{?}{100} \times 800 + 144 = 504$

$? \times 8 = 360$

$? = 45$

12. (c) $\left(\frac{2 \times 8}{4}\right)^4 \div (4 \times 4) \div 5 = ?$

$\frac{256}{16 \times 5} = ?$

$? = 3.2$

13. (a) $\frac{3435}{3} + 51 = ? \times 13$

$? = \frac{1196}{13} = 92$

14. (e) $\frac{7}{12} \times \frac{48}{100} \times 750 = ?$

$\frac{28}{100} \times 750 = 210$

15. (e) $\frac{8}{21} \times 189 = \frac{(?)^2}{2}$

$(?)^2 = 72 \times 2 = 144$

$? = 12$

16. (b) 2 12 60 240 720 1440

×6 ×5 ×4 ×3 ×2

17. (e) 80 48 64 56 60 58

-32, $\frac{+32}{2}$, $\frac{-32}{4}$, $\frac{+32}{8}$, $\frac{-32}{16}$

18. (d) 32 38 50 68 92 122

$+(6\times1)$ $+(6\times2)$ $+(6\times3)$ $+(6\times4)$ $+(6\times5)$

19. (c) 55000 11000 2200 440 88 17.6

÷5 ÷5 ÷5 ÷5 ÷5

20. (a) $5 \times 2 = 10,\ 10 \times 3 = 30,\ 30 \times 4 = 120,\ 120 \times 5 = 600,$

$600 \times 6 = 3600$

21. (b) Let the present ages of father and son be x and y years respectively.

Now, according to the question,

$\frac{x + y}{2} = 27$

$x + y = 54$ .... (i)

and $x - y = 30$ .... (ii)

On solving equation (i) and (ii), we have

$x = 42$ and $y = 12$

$\therefore$ Required ratio $= x : y = 42 : 12 = 7 : 2$

22. (e) Varun's present age = 39 – 4 = 35 years

So, Tarun's present age = 15 years

Tarun's age 4 years ago = 11 years

23. (b) Time taken by truck to cover 420 km at 70 km/hr

$= \frac{420}{70} = 6$ hours.

Speed of bike which covers (420 – 36) km in 6 hours

$= \frac{384}{6} = 64$ km/hr

24. (e) Let the cost price of the item be ₹ x.

Kiran sold it for ₹ 8160 at a loss of 15%.

So, $0.85x = 8160$

$x = \frac{8160}{0.85}$

To gain a profit of 25%, it should be sold at $1.25x$,

that is, $1.25 \times \frac{8160}{0.85} =$ ₹12000

25. (c) Total of maximum marks in 5 subjects = 110 × 5 = 550
Marks scored by Sunita = 56 + 42 + 63 + 94 + 61 = 316

Percentage of marks scored $= \frac{316}{550} \times 100 = 57.45\%$

26. (e) Sum of all 5 numbers = 50.8 × 5 = 254

Sum of 1st and 2nd number = 29 × 2 = 58

Sum of 4th and 5th number = 66.5 × 2 = 133

∴ 3rd number = 254 – 58 – 133 = 63

27. (a) Diameter of the circle = 15 + 6 = 21 cm

Area of the circle $= \frac{22}{7} \times \frac{21}{2} \times \frac{21}{2} = 346.5 \text{ cm}^2$

28. (c) 12 women complete the work in 5 days.

So, 1 women in 1 day works $= \frac{1}{5 \times 12} = \frac{1}{60}$

3 women in 10 days works $= \frac{3 \times 10}{60} = \frac{1}{2}$

So, 9 children complete $\frac{1}{2}$ work in 10 days.

1 child in 1 day $= \frac{1}{2 \times 10 \times 9}$

36 children in 1 day $= \frac{36}{2 \times 10 \times 9} = \frac{1}{5}$

So, 36 children will complete the work in 5 days.

29. (a) Circumference = $2\pi r = 264$

$r = \frac{264}{2\pi} = \frac{132}{\pi}$

Area of circle $= \pi r^2 = \pi \times \frac{132}{\pi} \times \frac{132}{\pi}$

$= \frac{132 \times 132}{22} \times 7 = 5544$ sq cm

30. (b) Average speed of the car $= \frac{450}{6} = 75$ km/h

Average speed of the tractor $= 75 \times \frac{2}{5} = 30$ km/h

Distance covered by the tractor = 30 × 8 = 240 km

31. (a) $I = 25000 \times \frac{3}{20} = 3750 \qquad r = \frac{3750 \times 100}{25000 \times 3} = 5\%$

32. (d) Length of the platform is not given

33. (a) The smallest angle $= \frac{7 \times 180}{(19+10+7)} = 35°$

Largest angle $= \frac{19 \times 180}{(19+10+7)} = 95°$

∴ Required sum = 2 × 35° + 95° = 165°

34. (b) Total S.I. $= \frac{\frac{P}{3} \times 6 \times 4 + \frac{P}{6} \times 12 \times 4 + \frac{P}{2} \times 16 \times 4}{100} = \frac{48P}{100}$

$P = \frac{2400 \times 100}{48} =$ ₹5000

35. (a) Let l be the length and b be the width.

Area of field = $l \times b = 3b^2$

Area of pond $= (3\sqrt{3})^2 = 27$

Area of field = 9 × 27

$3b^2 = 9 \times 27$

$b^2 = 81 \Rightarrow b = 9$ m

$l = 27$ m

36. (b) Required ratio $=\frac{4.5}{2.4}=15:8$

37. (d) Total number of employees

$= (3.4 + 4.5) \times 1000$

$= 7.9 \times 1000 = 7900$

38. (e) $\because$ K = 6.3, L = 4.7, M = 5.5, N = 5.3, $\boxed{P = 5.6}$, R = 2.9

$\therefore$ Required company = P

39. (c) $\therefore$ Required per cent

$$=\left(\frac{7.2}{3.9+4.1+3.4+5.3+6.7+9.6}\right)\times 100$$

$$=\frac{7.2\times 100}{33}\% = 21.82\% \approx 22\% \text{ (Approx.)}$$

40. (e) $\therefore$ Required average

$$=\frac{6.3+4.7+5.5+5.3+5.6+2.9}{6}\times 1000$$

$$=\frac{30.3\times 1000}{6}=\frac{30300}{6}=5050$$

41. (d) Required average number of passengers

$$=\frac{1}{3}[(24+20+15)\% \text{ of } 8500]$$

$$=\frac{1}{3}\times\frac{8500\times 59}{100}\approx 1671$$

42. (a) Number of passengers in Train R $=\frac{8500\times 9}{100}=765$

$\therefore$ Number of males $=(100-34-26)\%$ of 765

$$=\frac{765\times 40}{100}=306$$

43. (e) Required per cent $=\frac{19}{(13+9)}\times 100\approx 86$

44. (d) It is clear from the pie - chart

45. (d) Required per cent $=\frac{20-15}{15}\times 100\approx 33$

46. (d) The question cannot be answered because R's share in investment is not given.

47. (b) Hypotenuse $=\sqrt{5^2+12^2}=\sqrt{25+144}=\sqrt{169}=13$

Base : Height : Hypotenuse = 5 : 12 : 13

Base + Height + Hypotenuse = 30 cm

$\therefore$ Base $=\frac{5}{5+12+13}\times 30=5$ cm

Height $=\frac{12}{5+12+13}\times 30=12$ cm

Area $=\frac{1}{2}\times\text{base}\times\text{height}=\frac{1}{2}\times 5\times 12=30\text{ cm}^2$

So only statements A and B are required.

48. (b) A. $x-y=6$

B. $0.4y=0.3x$

$\frac{x}{y}=\frac{4}{3}$; C. $\frac{x}{2}.\frac{y}{3}=2:1$

$\frac{x}{y}\times\frac{3}{2}=\frac{2}{1}\Rightarrow\frac{x}{y}=\frac{4}{3}$

B and C give the same expression / information and hence are equivalent.

$x=\frac{4}{3}y\Rightarrow x-y=6$

$\frac{4}{3}y-y=6\Rightarrow\frac{y}{3}=6$

$y=18$ and $x=\frac{4}{3}\times 18=24$

So statements A and B together are necessary.

49. (e) Let the marked price be ₹x

A. cost price $=(1-0.15)x=$ ₹$0.85x$

B. S.P. = ₹3060

C. Profit = 2% of $x=0.02x$

Profit earned on the cost price $=\frac{0.02x}{0.85x}\times 100\approx 2.35\%$

$0.02x=3060-0.85x$

$0.87x=3060$ or $x=\frac{3060}{0.87}$

Actual profit $=0.02x=0.02\times\frac{3060}{0.87}=$ ₹70.34

So all 3 statements are required.

50. (e) A. Total marks in 4 subjects including English $=4\times 60=240$

B. Total marks in English and Maths = 170

C. Total marks in Maths and Science = 180

The question can't be answered because nothing has been said about the marks in the fourth subject. Also, there are four unknowns but only three equations can be formed with given data.

PRACTICE SET-2

DIRECTIONS (Qs. 1-5): *Each of the questions below consists of a question and three statements denoted A, B and C are given below it. You have to study the questions and all the three statements and decide whether the question can be answered with any one or two of the statements or all the statements are required to answer the question.*

1. What is R's share of profit in a joint venture?
   A. Q started business investing ₹ 80,000/
   B. R joined after 3 months
   C. P joined after 4 months with a capital of ₹ 1,20,000 and got ₹ 6,000 as his share of profit.
   (a) Only A and C are required
   (b) Only B and C are required
   (c) All A, B and C together are required
   (d) Even with all A, B and C the answer cannot be arrived at
   (e) None of these
2. What is the area of a right angled triangle?
   A. The perimeter of the triangle is 30 cm.
   B. The ratio between the base and the height of the triangle is 5 : 12.
   C. The area of the triangle is equal to the area of a rectangle of length 10 cms.
   (a) Only B and C together are required
   (b) Only A and B together are required
   (c) Only either A or B and C together are required
   (d) Only A and C together are required
   (e) None of these
3. What will be sum of two numbers?
   A. Among the two numbers, the bigger number is greater than the smaller number by 6.
   B. 40% of the smaller number is equal to 30% of the bigger number.
   C. The ratio between half of the bigger number and $\frac{1}{3}$rd of the smaller number is 2 : 1
   (a) Only B and C together are necessary
   (b) Only A and B together are necessary
   (c) Out of A, B and C any two together are necessary
   (d) All three A, B and C together are necessary
   (e) None of these
4. How much profit did Mahesh earn on the cost price of an article by selling it?
   A. He got 15% discount on the market price at the time of purchase.
   B. He sold it for ₹ 3060
   C. He earned 2% profit on the marked price
   (a) Only A and B both together are necessary
   (b) Only B and C both together are necessary
   (c) Only A or C and B together are necessary
   (d) Even A, B and C all together the question
   (e) All three A, B and C together are necessary
5. How much marks did Arun secure in English?
   A. The average marks obtained by Arun in four subjects including English is 60.
   B. The total marks obtained by him in English and Mathematics together is 170.
   C. The total marks obtained by him in Mathematics and Science together is 180.
   (a) All three A, B and C together are necessary
   (b) Only A and B together are necessary
   (c) Only B and C together are necessary
   (d) Only A and C together are necessary
   (e) None of these
6. In an exam, the average was found to be x marks. After deducting computational error, the average marks of 94 candidates got reduced from 84 to 64. The average thus came down by 18.8 marks. The numbers of candidates who took the exam were:
   (a) 100 (b) 90
   (c) 110 (d) 105
7. A bag contains 25 paise, 50 paise and 1 rupee coins. There are 220 coins in all and the total amount in the bag is ₹ 160. If there are thrice as many 1 rupee coins as there are 25paise coins, then what is the number of 50paise coins?
   (a) 60 (b) 40
   (c) 120 (d) 80
8. The area of a square is 196 sq cm whose side is half the radius of a circle. The circumference of the circle is equal to breadth of a rectangle, if perimeter of the rectangle is 712 cm. What is the length of the rectangle?
   (a) 196 cm (b) 186 cm
   (c) 180 cm (d) 190 cm
   (e) None of these
9. The respective ratio between the speeds of a car, a jeep and a tractor is 3 : 5 : 2. The speed of the jeep is 250% the speed of the tractor which covers 360 km in 12 h. What is the average speed of car and jeep together ?
   (a) 60 km/h (b) 75 km/h
   (c) 40 km/h (d) Cannot be determined
   (e) None of these

10. A man gets a simple interest of ₹1000 on a certain principal at the rate of 5% per annum in 4 yrs. What compound interest will the man get on twice the principal in two years at the same rate ?
(a) ₹1050 (b) ₹1005
(c) ₹11025 (d) ₹10125
(e) None of these

DIRECTIONS (Qs. 11-15) *Study the following table carefully and answer the questions given below*

Units of Garments Manufactured By Various Factories over the Years (units in Lakhs)

Year \ Factory	P	Q	R	S	T
2009	62.25	60.18	58.50	51.35	59.48
2010	63.36	62.16	60.00	59.50	59.75
2011	63.59	64.25	63.15	62.30	63.15
2012	64.26	64.70	64.20	63.75	64.18
2013	68.76	65.85	67.15	63.35	67.25
2014	70.00	68.32	68.20	70.15	70.35

11. What is the diference between the total number of units of garments manufactured by the various factories in the year 2011 and the total number of units of garments manufactured by various factories in the year 2013?
(a) ₹15.94 lakh (b) ₹18.22 lakh
(c) ₹17.92 lakh (d) ₹15.92 lakh
(e) None of these

12. What are the total units of garments manufactured by the Factory Q over the years?
(a) 38456000 (b) 34564000
(c) 36845000 (d) 38546000
(e) None of these

13. What is the approximate per cent increase in the units of garments manufactured by Factory T in the year 2011 from the previous year?
(a) 3 (b) 10
(c) 12 (d) 8
(e) 6

14. The units of garments manufactured by Factory S in the year 2010 are approximately, what per cent of the units of garments manufactured by the same factory in the year 2014?
(a) 88 (b) 85
(c) 79 (d) 75
(e) 71

15. What is the respective ratio of the units of garments manufactured by the Factory P in the year 2009 to the units of garments manufactured by Factory R in the same year?
(a) 29 : 26 (b) 28 : 27
(c) 27 : 26 (d) 29 : 27
(e) None of these

DIRECTIONS (Qs. 16-20) *What should come in place of question mark (?) in the following number series?*

16. 4, 8, 24 ,60, ? 224
(a) 178 (b) 309
(c) 96 (d) 141
(e) None of these

17. 8000, 1600,320,64,12.8 ?
(a) 2.56 (b) 3.5
(c) 3.2 (d) 2.98
(e) None of these

18. 1, 9, 125, 49 , 729, 121 ?
(a) 2197 (b) 729
(c) 125 (d) 1
(e) None of these

19. 6, 18, 90, 630, ? 62370
(a) 4900 (b) 4800
(c) 5400 (d) 5600
(e) None of these

20. 1250,500, 200 ?, 32, 12.8, 5.12
(a) 86 (b) 82
(c) 85 (d) 88
(e) None of these

21. A car covers its journey at the speed of 80 km/hour in 10 hours. If the same distance is to be covered in 4 hours, by how much the speed of car will have to increase?
(a) 8 km/hr (b) 10 km/hr
(c) 12 km/hr (d) 16 km/hr
(e) None of these

22. A grocer purchased 2 kg. of rice at the rate of ₹ 15 per kg. and 3 kg. of rice of the rate of ₹ 13 per kg. At what price per kg. should he sell the mixture to earn $33\frac{1}{3}\%$ profit on the cost price ?
(a) ₹28.00 (b) ₹20.00
(c) ₹18.40 (d) ₹17.40
(e) None of these

23. 12 men take 36 days to do a work while 12 women complete $\frac{3}{4}$th of the same work in 36 days. In how many days 10 men and 8 women together will complete the same work?
(a) 6 (b) 27
(c) 12 (d) Data inadequate
(e) None of these

24. The difference between a two digit number and the number obtained by interchanging the positions of its digits is 36. What is the difference between the two digits of that number ?
(a) 4 (b) 9
(c) 3 (d) Cannot be determined
(e) None of these

25. Deepa bought a calculator with 30% discount on the listed price. Had she not got the discount, she would have paid ₹ 82.50 extra. At what price did she buy the calculators?
(a) ₹192.50 (b) ₹275
(c) ₹117.85 (d) Cannot be determined
(e) None of these

DIRECTIONS (Qs. 26-30): *Study the following pie graph carefully and answer the questions given below.*

A survey conducted on 5800 villagers staying in various villages and having various favourite fruits

Favourite fruits **People staying in various villages**

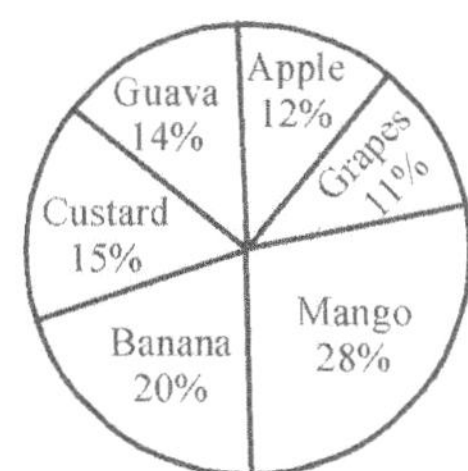

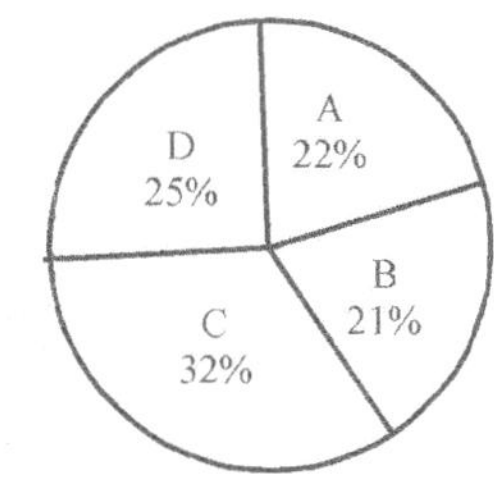

26. Mango is the favourite fruit of 50% of the people from village C. People having their favourite fruit as mango from village C form approximately, what percent of the people having their favourite fruit as mango from all the villages together ?
(a) 48 (b) 53 (c) 61 (d) 57 (e) 45

27. 20% of the people from village D have banana as their favourite fruit and 12% of the people from the same village have guava as their favourite fruit. How many people from that village like other fruits?
(a) 764 (b) 896
(c) 874 (d) 968
(e) None of these

28. How many people in all have custard as their favourite fruit ?
(a) 850 (b) 864
(c) 870 (d) 812
(e) None of these

29. 50% of the people from village B have banana as their favourite fruit. How many people from other villages have the same favourite fruit ?
(a) 1160 (b) 551
(c) 1020 (d) 609
(e) None of these

30. What is the total number of people having their favourite fruit as apple and grapes together ?
(a) 1334 (b) 1286
(c) 1300 (d) 1420
(e) None of these

DIRECTIONS (Qs. 31-36) : *The piechart given below shows the distribution of work force by occupational category for country X in 1981 and 1995. Study the chart and answer the questions.*

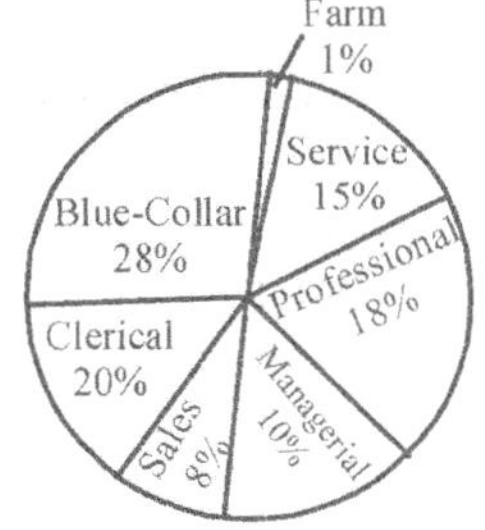

Total workforce 150 million

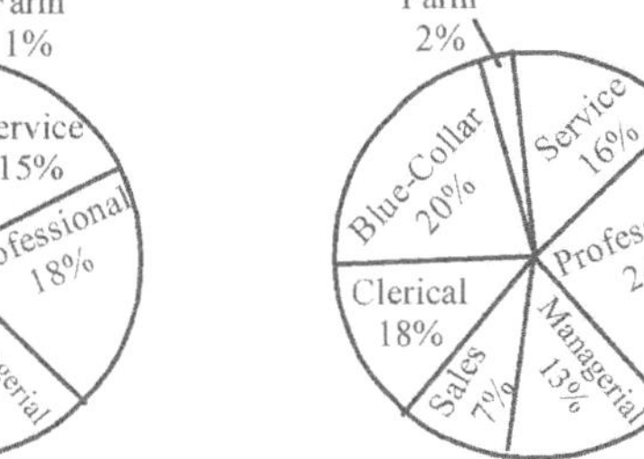

Total workforce 175 million

31. The increase in the number of Clerical workers in the workforce of country X from 1981 to 1995 (in millions) is
(a) 0.5 (b) 1.25
(c) 0.75 (d) 1.5

32. The percentage decrease in the number of Blue – Collar workers in the workforce of country X from 1981 to 1995 is
(a) 20 (b) $16\frac{2}{3}$
(c) $42\frac{1}{2}$ (d) 35

33. In 1981, the number of Service workers in the workforce in millions, was
(a) 22.5 (b) 28.0
(c) 15.0 (d) 20.5

34. In 1981, the number of categories which comprised of more than 25 million workers each, is
(a) four (b) five
(c) two (d) three

35. The ratio of the number of workers in the Professional category in 1981 to the number of such workers in 1995 is
(a) 3 : 4 (b) 14 : 9
(c) 4 : 9 (d) 5 : 14

36. The price of a jewel, passing through three hands, rises on the whole by 65%. If the first and the second sellers earned 20% and 25% profit respectively, find the percentage profit earned by the third seller.
(a) 10% (b) 20%
(c) 30% (d) 40%

37. The salaries of A, B, C are in the ratio 2 : 3 : 5. If the increments of 15%, 10% and 20% are allowed respectively in their salaries, then what will be the new ratio of their salaries?
(a) 3 : 3 : 10 (b) 10 : 11 : 20
(c) 23 : 33 : 60 (d) Cannot be determined

38. If the ages of P and R are added to twice the age of Q, the total becomes 59 years. If the ages of Q and R are added to thrice the age of P, the total becomes 68 years. And if the age of P is added to thrice the age of Q and thrice the age of R, the total becomes 108 years. What is the age of P?
(a) 15 years (b) 19 years
(c) 17 years (d) 12 years

39. 12 men can complete a piece of work in 4 days, while 15 women can complete the same work in 4 days. 6 men start working on the job and after working for 2 days, all of them stopped working. How many women should be put on the job to complete the remaining work, if it is to be completed in 3 days?
(a) 15 (b) 18
(c) 22 (d) Data inadequate

40. Rajesh gave ₹ 1200 on loan. Some amount he gave at 4% per annum on simple interest and remaining at 5% per annum on simple interest. After two years, he got ₹ 110 as interest. Then the amounts given at 4% and 5% per annum on simple interest are, respectively
(a) ₹500, ₹700 (b) ₹400, ₹800
(c) ₹900, ₹300 (d) ₹1100, ₹1100

DIRECTIONS (Qs. 41-45) : *Study the following data carefully and answer the questions given below.*

Out of a total number of commuters commuting daily in a city, 17171 commuters commute only by trains. 7359 commuters commute only by bikes and 22077 commuters commute only by buses. 14718 commuters commute only by their private cars and 4906 commuters commute only by autos. 7359 commuters commute only by taxis. 26983 commuters commute by buses as well as trains. 9812 commuters commute only by autos as well as trains. 12265 commuters commute only by buses as well as autos.

41. The total number of commuters commuting by trains forms what per cent of the total number of commuters commuting daily?
(a) 22% (b) 44%
(c) 14% (d) 36%
(e) None fo these

42. The total number of commuters commuting by autos forms what per cent of the total number of commuters commuting daily?
(a) 10% (b) 18%
(c) 22% (d) 4%
(e) None of these

43. The total number of commuters commuting by bikes and taxis together forms what per cent of the total number of commuters commuting daily?
(a) 12% (b) 6%
(c) 8% (d) 16%
(e) None of these

44. What is the total number of commuters in the city commuting daily?
(a) 122650 (b) 126250
(c) 162250 (d) 152260
(e) None of these

45. The number of commuters commuting only by buses forms what per cent of the total number of commuters commuting daily ?
(a) 40% (b) 22%
(c) 32% (d) 18%
(e) None of these

DIRECTIONS (Qs. 46-50) : *Study the following graph and table to answer the given questions.*

Percentage of Different Types of Employees in an Organization

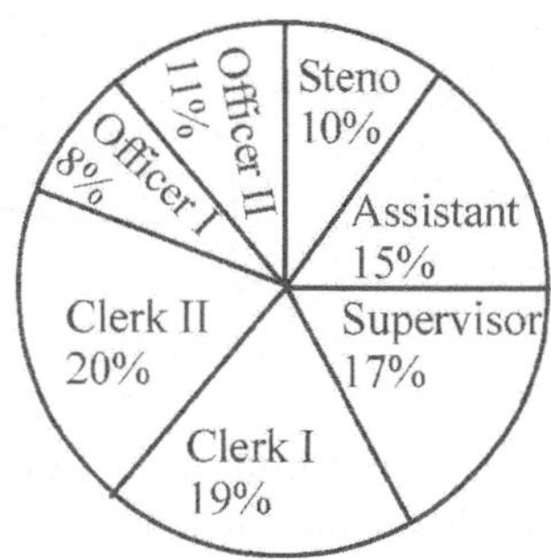

Total number of employees = 7000

Employees	Out of these per cent of	
	Direct Recruit	Promotees
Steno	30	70
Assistant	40	60
Supervisor	50	50
Clerk I	90	10
Clerk II	30	70
Officer I	90	10
Officer II	70	30

46. What is the difference between direct recruits and promotee assistant?
(a) 210 (b) 280
(c) 180 (d) 110
(e) None of these

47. The promotee clerk I is approximately what per cent of that of direct recruit clerk I?
(a) 10% (b) 9%
(c) 11% (d) 10.50%
(e) None of these

48. How many employees are supervisors?
(a) 1050 (b) 1019
(c) 1109 (d) 1290
(e) None of these

49. How many total direct recruits among all types of employees are there ?
(a) 4000 (b) 3885
(c) 3000 (d) 3115
(e) None of these

50. Which types of employees have maximum number of direct recruits?
(a) Clerk I and officer I (b) Officer I
(c) Clerk I (d) Clerk II
(e) None of these

SOLUTIONS

1. (d) The question cannot be answered because R's share in investment is not given.

2. (b) Hypotenuse = $\sqrt{5^2+12^2}$

$=\sqrt{25+144}=\sqrt{169}=13$

Base : Height : Hypotensuse

= 5 : 12 : 13

Base + Height + Hypotenuse = 30 cm

$\therefore$ Base $=\frac{12}{5+12+13}\times30=5$ cm

Height $=\frac{12}{5+12+13}\times30=12$ cm

Area $=\frac{1}{2}\times$ base $\times$ height

$=\frac{1}{2}\times5\times12=30\text{ cm}^2$

3. (b) A. $x-y=6$

B. $0.4y=0.3x$;

$\frac{x}{y}=\frac{4}{3}$

C. $\frac{x}{y}:\frac{y}{3}=2:1$;

$\frac{x}{y}\times\frac{3}{2}=\frac{2}{1}$; $\frac{x}{y}=\frac{4}{3}$

Statement B and C give the same expression/information and hence are equivalent.

$x=\frac{4}{3}y$

$x-y=6$

$\frac{4}{3}y-y=6$

$\frac{y}{3}=6$; $y=18$

$x=\frac{4}{3}\times18=24$

4. (e) Let the marked price be ₹ x

A. cost price $=(1-0.15)x$
$=$ ₹ $0.85x$

B. S.P. = ₹ 3060

C. Profit = 2% of $x=0.02x$

Profit earned on the cost price

$\frac{0.02x}{0.85x}\times100\approx2.35\%$

$0.02x=3060-0.85x$

$0.87x=3060$

$x=\frac{3060}{0.87}$

Actual profit $=0.02x$

$=0.02\times\frac{3060}{0.87}=$ ₹ 70.34.

5. (e) A. Total marks in 4 subjects including English $=4\times60=240$

B. Total marks in English and Maths = 170

C. Total marks in Maths and Science = 180

The question can't be answered because nothing has been said about the marks in the fourth subject. Also, there are four unknowns but only three equations can be formed with given data.

6. (a) $\frac{(84-64)\times94}{18.8}=100$

7. (a) The no. of coins of 1 rupee = 3x and 25p = x.

Conventionally, we can solve this using equations as follows.

$A+B+C=220$ ...(1)

$A=3C$ ...(2)

$A+0.5B+0.25C=160$ ...(3)

We have a situation with 3 equations and 3 unknowns and we can solve for

A (no. of 1 rupee coins),

B (no. of 50 paise coins)

and C (no. of 25 paise coins)

However, a much smarter approach would be to go through the options. If we check option (a) – no. of 50 paise coins = 60 we would get the number of 1 rupee coins as 120 and the number of 25 paise coins as 40.

$120\times1+60\times0.5+40\times0.25=160$

This fits the conditions perfectly and is hence the correct answer.

8. (c) Area of square $=(a)^2=196$

$a=\sqrt{196}=14$ cm

Radius of a circle $=14\times2=28$ cm

$\therefore$ Circumference $=\frac{22}{7}\times2\times28=176$ cm

Now, according to question,

$b=176$ cm

Also, $2(l+b)=712$

$2(l + 176) = 712$

$l + 176 = 356$

$\therefore \quad l = 356 - 176$

$\therefore \quad l = 180$ cm

9. (a) Speed of tractor = $\frac{360}{12} = 30$ km/h

Speed of jeep = $30 \times 250\% = 75$ km/h

Speed of car = $\frac{3}{5} \times 75 = 45$ km/h

Average speed of car and jeep together = $\frac{75+45}{2}$ = 60 km/h

10. (e) Principal = $\frac{1000 \times 100}{5 \times 4}$ = ₹ 5000

Compound interest = $10000\left[\left(1+\frac{5}{100}\right)^2 - 1\right]$

$= 10000 \times \frac{41}{400}$ = ₹ 1025

11. (d) Total number of units of garments manufactured by the various factories in:

Year 2011 $\rightarrow (63.59 + 64.25 + 63.15 + 62.30 + 63.15)$ lakh $= 316.44$ lakh

Year 2013 $\rightarrow (68.76 + 65.85 + 67.15 + 63.35 + 67.25)$ lakh $= 332.36$ lakh

$\therefore$ Required difference $= (332.36 - 316.44)$ lakh $= 15.92$ lakh

12. (d) Total units of garments manufactured by the factory Q

$= (60.18 + 62.16 + 64.25 + 64.70 + 65.85 + 68.32)$ lakh

$= 385.46$ lakh

$= 38546000$

13. (e) Percentage increase

$\frac{63.15 - 59.75}{59.75} \times 100 = \frac{3.4 \times 100}{59.75} \approx 6$

14. (b) Required percentage = $\frac{59.5}{70.15} \times 100 \approx 85\%$

15. (a) Required ratio = 62.25 : 58.50 = 249 : 235

16. (e)

4, 8, 24, 60, ? [124], 224

$+2^2$, $+4^2$, $+6^2$, $+8^2$, $+10^2$

Hence, the answer will be 124.

17. (a)

8000, 1600, 320, 64, 12.8, ? [2.56]

÷5, ÷5, ÷5, ÷5, ÷5

Hence, the answer will be 2.56.

18. (a) $1^3 \rightarrow 3^2 \rightarrow 5^3 \rightarrow 7^2 \rightarrow 9^3 \rightarrow 11^2 \rightarrow 13^3$

So 1, 9, 125, 49, 729, 121, [2197]

19. (e) The given pattern is × 3, × 5, × 7, × 9, × 11

So, the missing term is 630 + 9 = 5670

20. (e) The pattern is × 0.4

So, the missing term is 200 × 0.4 = 80

21. (e) Initial speed = 80 km/hr

Total distance = 80 × 10 = 800 km.

New speed = $\frac{800}{4}$ = 200 km/hr

Increase in speed

= 200 – 80 = 120 km/hr.

22. (c) Mixture : 2 kg of rice at ₹ 15/kg + 3 kg of rice at ₹ 13/kg

Total weight = 2 + 3 = 5 kg

Total cost price

= (2 × 15) + (3 × 13)

= 30 + 39 = ₹ 69

Cost price per kg of the mixture $\frac{69}{5}$ = ₹ 13.80

Selling price to get $33\frac{1}{3}\%$ profit

$= 100 + 33\frac{1}{3} \times$ ₹ 13.80

$= \frac{400}{3 \times 100} \times$ ₹ 13.80

$= \frac{4}{3} \times 13.80 =$ ₹ 18.40

23. (b) In 36 days 12 men can do 1 complete work.

In 36 days 12 women can do $\frac{3}{4}$th of the work

Since time and the no. of persons is the same in both cases,

1 women's daily work = $\frac{3}{4}$th of 1 man's daily work

8 women's daily work

$= \frac{3}{4} \times 8 = 6$ men's daily work

(10 men + 8 womens daily work)

= (10 men + 6 men)

= 16 men's daily work.

12 men can do the work is 36 days

$\therefore$ 16 men can do the work is $36 \times \frac{12}{16} = 27$ days.

24. (a) Let the unit's digit be x and ten's digits be y.

Also let $y > x$

∴ Number = $10y + x$ and number obtained by interchanging the digits = $10x + y$

∴ $10y + x - 10x - y = 36$

$\Rightarrow 9y - 9x = 36$

$\Rightarrow 9(y - x) = 36$

$\Rightarrow y - x = 4$

25. (a) Let the listed price be ₹ x.

∴ Discount = 30% of x

$= \frac{30x}{100} = ₹\frac{3x}{10}$

According to the question,

$\frac{3x}{10} = 82.5$

$\Rightarrow x = \frac{82.5 \times 10}{3} = ₹\ 275$

∴ Required cost price of calculator = 70% of 275

$= ₹\ \frac{70 \times 275}{100} = ₹\ 192.50$

26. (d) Number of people liking mango from all the villages

$= \frac{28}{100} \times 5800 = 1624$

Number of people liking mango from villages C

$= \frac{50}{100} \times 1856 = 928$

Hence, required percentage $= \frac{928}{1624} \times 100 = 57$

27. (e) Percentage of people liking banana and guava from village D = 32

Hence, number of people who like other fruits

$= 68\% \text{ of } 1450 = \frac{68}{100} \times 1450 = 986$

28. (c) Number of people who like custard

$= 15\% \text{ of } 5800 = \frac{15}{100} \times 5800 = 870$

29. (b) Number of people who like banana from all villages

$= \frac{201}{100} \times 5800 = 1160$

Number of people liking bananas from village B

$= \frac{50}{100} \times 1218 = 609$

Hence, required number of people = 1160 – 609 = 551

30. (a) Number of people who like apple and grapes together

= (12 + 11)% of 5800

$= \frac{23}{100} \times 5800 = 1334$

31. (d) Number of clerical workers in 1981

$= \frac{150 \times 20}{100} = 30$ million

Number of clerical workers in 1995

$= \frac{175 \times 18}{100} = 31.5$ million

Difference = 31.5 – 30 = 1.5 million

32. (b) Number of blue-collar workers in 1981

$= \frac{150 \times 28}{100} = 42$ million

Number of blue-collar workers in 1991

$= \frac{175 \times 20}{100} = 35$ million

Percentage decrease

$= \frac{42 - 35}{42} \times 100$

$= \frac{50}{3} = 16\frac{2}{3}$

33. (a) Services workers in 1981

$= \frac{150 \times 15}{100}$

= 22.5 million

34. (d) $\frac{150 \times x}{100} = 25$

$\Rightarrow x = \frac{25 \times 100}{150} = 16\frac{2}{3}\%$

Required answer = blue-collar, professional and clerical workers.

35. (a) Required ratio

18 : 24

3 : 4

36. (a) Let the original price of the jewel be ₹ P and let the profit earned by the third seller be x%.

Then, (100 + x)% of 125% of 120% of P = 165% of P

$\Rightarrow \left[\frac{(100 + x)}{100} \times \frac{125}{100} \times \frac{120}{100} \times P\right] = \left(\frac{165}{100} \times P\right)$

$\Rightarrow (100 + x) = \left(\frac{165 \times 100 \times 100}{125 \times 120}\right) = 110$

$\Rightarrow x = 10\%$

OR

$100 \xrightarrow[+65]{65\% \uparrow} 165$

$100 \xrightarrow[+20]{20\% \uparrow} 120 \xrightarrow[30]{25\% \uparrow} 150$

So, third seller profit $= \frac{165 - 150}{150} \times 100 = 10\%$

37. (c) Let A = 2k, B = 3k and C = 5k.

A's new salary $= \frac{115}{100}$ of $2k = \left(\frac{115}{100} \times 2k\right) = \frac{23}{10}k$

B's new salary $= \frac{110}{100}$ of $3k = \left(\frac{110}{100} \times 3k\right) = \frac{33}{10}k$

C's new salary $= \frac{120}{100}$ of $5k = \left(\frac{120}{100} \times 5k\right) = 6k$

$\therefore$ New ratio $= \frac{23k}{10} : \frac{33k}{10} : 6k = 23 : 33 : 60.$

38. (d) $P + R + 2Q = 59$;
$Q + R + 3P = 68$
and $P + 3(Q + R) = 108$
Solving the above two equations, we get P = 12 years.

39. (a) 1 man's 1 day's work $= \frac{1}{48}$;

1 woman's 1 day's work $= \frac{1}{60}$.

6 men's 2 day's work $= \left(\frac{6}{48} \times 2\right) = \frac{1}{4}$.

Remaining work $= \left(1 - \frac{1}{4}\right) = \frac{3}{4}$.

Now, $\frac{1}{60}$ work is done in 1 day by 1 woman.

So, $\frac{3}{4}$ work will be done in 3 days by

$\left(60 \times \frac{3}{4} \times \frac{1}{3}\right) = 15$ women.

40. (a) Let the amount given 4% per annum be ₹x.
then, amount given at 5% per annum = ₹(1200 – x)

Now, $\frac{x \times 4 \times 2}{100} + \frac{(1200 - x) \times 5 \times 2}{100} = 110$

$\Rightarrow$ x = ₹500
And, the amount given at 5% per annum
= ₹(1200 – x) = ₹(1200 – 500) = ₹700

41. (b) Total number of commuters
= 17171 + 7359 + 22077 + 14718 + 4906 + 7359 + 26983 + 9812 + 12265 = 122650
Number of commuters commuting by train
= 17171 + 26983 + 9812 = 53966

Hence, required percentage $= \frac{53966}{122650} \times 100\% = 44\%$

42. (c) Total number of commuters commuting by autos
= (4906 + 9812 + 12265) = 26983

$\therefore$ Required percentage $= \frac{26983}{122650} \times 100\% = 22\%$

43. (a) Total number of commuters commuting by bikes and taxies = 7359 + 7359 = 14718

Hence, required percentage $= \frac{14718}{122650} \times 100\% = 12\%$

44. (a) Total number of commuters = 122650

45. (d) Number of commuters commuting by buses only = 22077

$\therefore$ Required percentage $= \frac{22077}{122650} \times 100\% = 18\%$

46. (a) Total number of assistants = 15% of 7000

$= \frac{15}{100} \times 7000 = 1050$

Number of direct recruits = 40% of 1050

$= \frac{40}{100} \times 1050 = 420$

Now, number of promotee assistants = 1050 – 420 = 630
Hence, required difference = 630 – 420 = 210

47. (c) Number of clerk I = 19% of 7000 $= \frac{19}{100} \times 7000 = 1330$

Number of direct recruits = 90% of 1330

$= \frac{90}{100} \times 1330 = 1197$

Now, number of promotees = 1330 – 1197 = 133

Hence, required percentage = $= \frac{133}{1197} \times 100\% = 11\%$

48. (e) Number of supervisors = 17% of 7000 $= \frac{17}{100} \times 7000$
= 1190

49. (b)

Number of direct recruits	Type of employees
Steno	$\frac{10}{100} \times \frac{30}{100} \times 7000 = 210$
Assistant	$\frac{15}{100} \times \frac{40}{100} \times 7000 = 420$
Supervisor	$\frac{17}{100} \times \frac{50}{100} \times 7000 = 595$
Clerk I	$\frac{19}{100} \times \frac{90}{100} \times 7000 = 1197$
Clerk II	$\frac{20}{100} \times \frac{30}{100} \times 7000 = 420$
Officer I	$\frac{8}{100} \times \frac{90}{100} \times 7000 = 504$
OfficerII	$\frac{11}{100} \times \frac{70}{100} \times 7000 = 539$

Total direct recruits
= 210 + 420 + 595 + 1197 + 420 + 504 + 539 = 3885

50. (c) From the table, it is clear that clerk I has highest number of direct recruits.

■■■

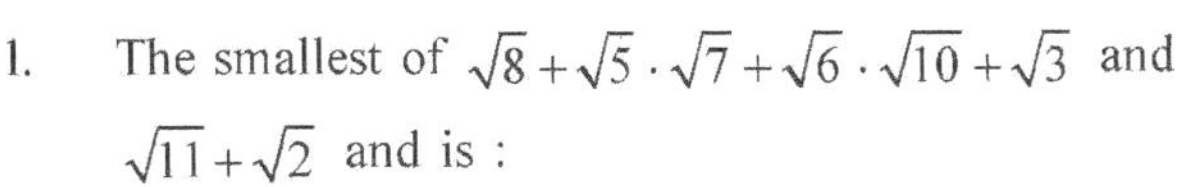

PRACTICE SET-3

1. The smallest of $\sqrt{8}+\sqrt{5}\cdot\sqrt{7}+\sqrt{6}\cdot\sqrt{10}+\sqrt{3}$ and $\sqrt{11}+\sqrt{2}$ and is :
   (a) $\sqrt{8}+\sqrt{5}$ (b) $\sqrt{7}+\sqrt{6}$
   (c) $\sqrt{10}+\sqrt{3}$ (d) $\sqrt{11}+\sqrt{2}$
2. Find the highest power of 5 in 100!.
   (a) 19 (b) 22
   (c) 25 (d) None of these
3. A contract is to be completed in 46 days and 117 men were set to work, each working 8 hours a day. After 33 days, 4/7 of the work is completed. How many additional men may be employed so that the work may be completed in time, each man now working 9 hours a day ?
   (a) 80 (b) 81
   (c) 82 (d) 83
4. A water tank has three taps A, B and C. A fills four buckets in 24 minutes, B fills 8 buckets in 1 hour and C fills 2 buckets in 20 minutes. If all the taps are opened together a full tank is emptied in 2 hours. If a bucket can hold 5 litres of water, what is the capacity of the tank?
   (a) 120 litres (b) 240 litres
   (c) 180 litres (d) 60 litres
5. A can do 50% more work as B can do in the same time. B alone can do a piece of work in 20 hours. A, with the help of B, can finish the same work in how many hours ?
   (a) 12 (b) 8
   (c) $13\frac{1}{3}$ (d) $5\frac{1}{2}$
6. Base of a right pyramid is a square. length of diagonal of the base is $24\sqrt{2}$ is the volume of the pyramid is 1728 cu.m. its height is
   (a) 7 m (b) 8 m
   (c) 9 m (d) 10 m
7. Which one set of letters when sequentially placed at the gaps in the given letter series shall complete it ?
   __ ab__ b __ aba __ __ abab
   (a) a ab aa (b) bb aa b
   (c) ab aa b (d) a aa ba

DIRECTIONS (Qs. 8) : *In this question, a series is given, with one term missing. Choose the correct alternative from the given ones that will complete the series.*

8. 121, 144, 289, 324, 529, 576, $\frac{?}{}$
   (a) 961 (b) 841
   (c) 900 (d) 729
9. A trader marked his goods at 20% above the cost price. He sold half the stock at the marked price, one quarter at a discount of 20% on the marked price and the rest at a discount of 40% on the marked price. His total gain is
   (a) 2% (b) 4.5%
   (c) 13.5% (d) 15%
10. A trader wants 10% profit on the selling price of a product whereas his expenses amount to 15% on sales. What should be his rate of mark up on an article costing ₹ 9?
    (a) 20% (b) $66\frac{2}{3}\%$
    (c) 30% (d) $\frac{100}{3}\%$
11. A manufacturer sells goods to an agent at a profit of 20%. The agent's wholesale price to a shopkeeper is at a profit of 10% and the shopkeeper retails his goods at a profit of 12%. Find the retailer's price of an article which had cost the manufacturer ₹ 25.
    (a) ₹ 37 (b) ₹ 40
    (c) ₹ 44 (d) ₹ 46
12. A cistern cotains 50 litres of water, 5 liters of water is taken out of it and replaced by wine. The process is repeated again. Find the proportion of wine and water in the resulting mixture.
    (a) 1 : 4 (b) 41 : 50
    (c) 19 : 81 (d) 81 : 19
13. A jar full of whisky contains 40% alcohol. A part of this whisky is replaced by another containing 19% alcohol and now the percentage of alcohol was found to be 26%. The quantity of whisky replaced is:
    (a) $\frac{1}{3}$ (b) $\frac{2}{3}$
    (c) $\frac{2}{5}$ (d) $\frac{3}{5}$
14. The average age of boys in class is 16.66, while the average age of girls is 18.75. Thus the average age of all the 40 students of the class is 17.5. If the difference between the no. of boys and girls is 8, then the no. of girls in the class is:
    (a) 12 (b) 16
    (c) 18 (d) Data insufficient
15. In a school, the average age of students is 6 years, and the average age of 12 teachers is 40 years. If the average age of the combined group of all the teachers and student is 7 years, then the number of student is :
    (a) 396 (b) 400
    (c) 408 (d) 416

16. An article is listed at ₹ 65. A customer bought this article for ₹ 56.16 and got two successive discounts of which the first one is 10%. The other rate of discount of this scheme that was allowed by the shopkeeper was
(a) 3% (b) 4%
(c) 6% (d) 2%

17. A's income is $6\frac{1}{4}$% more than B's. How much % is B's less than A's ?
(a) 5.89% (b) 4.78%
(c) 2.39% (d) None of these

18. A train leaves station X at 5 a.m. and reaches station Y at 9 a.m. Another train leaves station Y at 7 a.m. and reaches station X at 10: 30 a.m. At what time do the two trains cross each other ?
(a) 7:36 am (b) 7:56 am
(c) 8:36 am (d) 8:56 am

19. An athlete runs to and fro between points A and B at a speed of 10 km/h. A second athlete simultaneously runs from point B to A and back at a speed of 15 km/h. If they cross each other 12 min after the start, after how much time will they cross each other?
(a) 18 min (b) 24 min
(c) 36 min (d) 48 min

20. If $x^{\frac{1}{3}} + y^{\frac{1}{3}} = z^{\frac{1}{3}}$, then $(x + y - z)^3 + 27\,xyz$ equals :
(a) –1 (b) 1
(c) 0 (d) 27

21. Karnal took ₹ 6800 as a loan which along with interest is to be repaid in two equal annual instalments. If the rate of Interest is $12\frac{1}{2}$% compounded annually, then the value of each instalment is
(a) ₹ 8100 (b) ₹ 4150
(c) ₹ 4050 (d) ₹ 4000

22. If $a+\frac{1}{b} = 1$ and $b+\frac{1}{c} = 1$, then $c+\frac{1}{a}$ is equal to
(a) 0 (b) $\frac{1}{2}$
(c) 1 (d) 2

23. The area of the triangle formed by the line $5x - 3y + 15 = 0$ with coordinate axes is
(a) 15 sq. units (b) 5 sq. units
(c) 8 sq. units (d) $\frac{15}{2}$ sq. units

24. The value of k for which the lines $x + 2y - 9 = 0$ and $kx + 4y + 5 = 0$ are parallel, is
(a) $k = 2$ (b) $k = 1$
(c) $k = -1$ (d) $k = -2$

25. If sec θ + tan θ = = p, then cos θ is
(a) $\frac{p^2+1}{p^2-1}$ (b) $\frac{p^2-1}{(p^2+1)^2}$
(c) $\frac{2p}{p^2+1}$ (d) $\frac{4p^2}{(p^2+1)^2}$

26. If tan A + sin A = p and tan A – sin A = q, them
(a) $p^2 + q^2 = 4\sqrt{pq}$ (b) $p^2 - q^2 = 4\sqrt{pq}$
(c) $p^2 - q^2 = \sqrt{pq}$ (b) $p^2 - q^2 = 2\sqrt{pq}$

27. The area of circle whose radius is 6 cm is trisected by two concentric circles. The radius of the smallest circle is
(a) $2\sqrt{3}$ cm (b) $2\sqrt{6}$ cm
(c) 2 cm (d) 3 cm

28. PQ and RS are two parallel chords of a circle with centre C such that PQ = 8 cm RS = 16 cm. If the chords are on the same side of the centre and the distance between them is 4 cm, then the radius of the circle is
(a) $3\sqrt{2}$ cm (b) $3\sqrt{5}$
(c) $4\sqrt{5}$ cm (d) $5\sqrt{5}$

29. If $ABCD$ is a parallelogram and E, F are the centroids of ΔABC and $\Delta\ BCD$ respectively, then EF is equal to
(a) AE (b) BE
(c) CE (d) DE

30. If the angle of elevation of the Sun changes from 30° to 45°, the length of the shadow of a pillar decreases by 20 metres. The height of the pillar is
(a) $20\ (\sqrt{3}-1)$ m (b) $20\ (\sqrt{3}+1)$ m
(c) $10\ (\sqrt{3}-1)$ (d) $10\ (\sqrt{3}+1)$ m

31. If cosec 39° = x, the value of $\frac{1}{\text{cosec}^2 51°}\sin^2 39° + \tan^2 51° - \frac{1}{\sin^2 51° \sec^2 39°}$ is
(a) $\sqrt{x^2-1}$ (b) $\sqrt{1-x^2}$
(c) $x^2 - 1$ (d) $1 - x^2$

32. If 3 cot θ = 4, then $\left(\frac{5\sin\theta+3\cos\theta}{5\sin\theta-3\cos\theta}\right)$ is equal to
(a) $\frac{1}{9}$ (b) $\frac{1}{3}$
(c) 3 (d) 9

33. If tan x + cot x =3, then $\sec^2$ + $\text{cosec}^2\, x$ is equal to
(a) 3 (b) 9
(c) 12 (d) 15

34. The greatest value of sin θ + cos θ is equal to
(a) 0 (b) 1
(c) 2 (d) $\sqrt{2}$

35. In the following figure, O is the centre of the circle and $\angle ABO = 30°$, find $\angle ACB$.

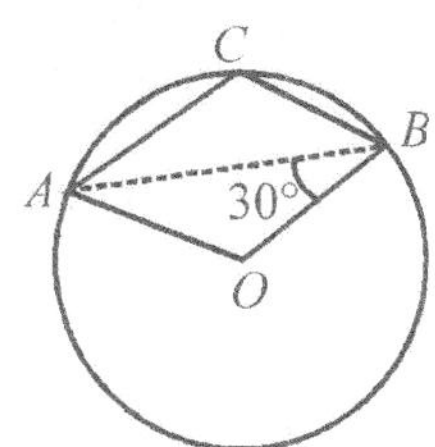

(a) 60° (b) 120°
(c) 75° (c) 90°

36. A right triangle with sides 3 cm, 4 cm and 5 cm is rotated about the side of 3 cm to form a cone. The volume of the cone so formed is
(a) 16π cm^3 (b) 12π cm^3
(c) 15π cm^3 (d) 20π cm^3

37. If $x = 3+2\sqrt{2}$ then the value of $\sqrt{x} - \frac{1}{\sqrt{x}}$ is
(a) 1 (b) $2\sqrt{2}$
(c) 2 (d) $3\sqrt{3}$

38. If $a = 0.25$ $b = -0.05$, $c = 0.5$, then the value of
$\frac{a^2 - b^2 - c^2 - 2bc}{a^2 + b^2 - 2ab - c^2}$ is
(a) $\frac{7}{8}$ (b) $\frac{14}{17}$
(c) 1 (d) $\frac{25}{16}$

39. If $x^2 - 3x + 1 = 0$, then the value of $x + \frac{1}{x}$ is
(a) 0 (b) 1
(c) 2 (d) 3

40. A chord of length 16 cm is drawn in a circle of radius 10 cm. The distance of the chord from the centre of the circle is
(a) 6 cm (b) 9 cm
(c) 12 cm (d) 8 cm

41. Each interior angle of a regular polygon is 18° more than eight times an exterior angle. The number of sides of the polygon is
(a) 10 (b) 15
(c) 20 (d) 25

42. The ratio of the outer and the inner perimeter of a circular path is 23 : 22. If the path is 5 m wide, the diameter of the inner circle is $\left(\text{use } \pi = \frac{22}{7}\right)$
(a) 110 m (b) 55 m
(c) 220 m (d) 230 m

43. A cistern has two pipes. One can fill it with water in 8 hours and other can empty it in 5 hours. In how many hours will the cistern be emptied if both the pipes are opened together when $\frac{3}{4}$ of the cistern is already full of water?
(a) $13\frac{1}{3}$ hours (b) 10 hours
(c) 6 hours (d) $3\frac{1}{3}$ hours

DIRECTIONS (Qs. 44-46) : *Study the following table carefully and answer the questions given below.*

Number of Students of Different Classes of a School Playing Difference Games

	XII	XI	X	IX	VIII	VII	VI
Chess	11	12	5	4	2	2	1
Cricket	38	40	12	17	25	18	20
Basket Ball	11	9	7	6	0	0	0
Table Tennis	9	9	21	19	11	9	0
Football	40	27	18	19	12	16	14
Carrom	16	15	8	19	12	16	14
Tennis	8	9	11	5	6	0	0
Badminton	47	39	33	21	19	0	0

44. Approximately what per cent of VIII class students play Cricket out of the total student playing Cricket?
(a) 13 (b) 4
(c) 25 (d) 15

45. What is the respective ratio of XI and X class students playing Football?
(a) 1 : 2 (b) 2 : 5
(c) 2 : 3 (d) 3 : 2

46. Which game is the most popular?
(a) Badminton (b) Cricket
(c) Carrom (d) Table Tennis

DIRECTIONS (Qs. 47-50) : *The following graph shows the demand and production of cotton by 5 companies A, B, C, D and E. Study the graph and answer the questions.*

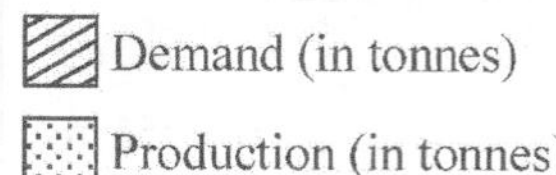

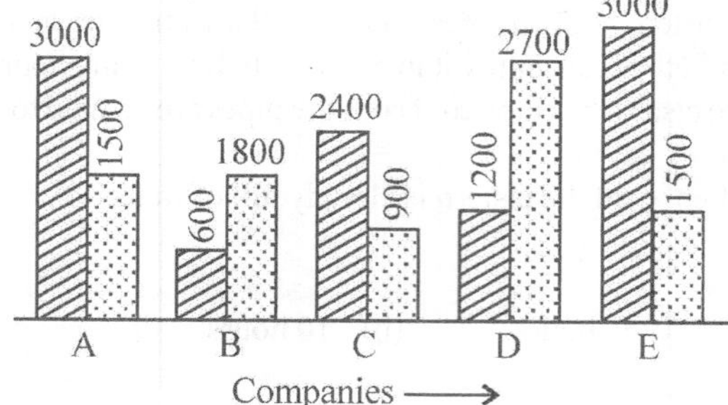

47. What is the ratio of companies having more demand than production to those having more production than demand?
    (a) 2 : 3 (b) 4 : 1
    (c) 3 : 2 (d) 1 : 4
48. What is the difference (in tonnes) between average demand and average production of the five companies taken together?
    (a) 320 (b) 420
    (c) 2100 (d) 1050
49. The production of company D is how many times that of the production of the company A?
    (a) 1.8 (b) 1.5
    (c) 0.5 (d) 0.4
50. The demand for company B is what percent of the demand for comapany C?
    (a) 1.5 (b) 2.5
    (c) 25 (d) 30

SOLUTIONS

1. (d) Here, $(\sqrt{8}+\sqrt{5})^2 = (\sqrt{8})^2 + (\sqrt{5})^2 + 2\times\sqrt{8}\times\sqrt{5}$

$= 8 + 5 + 2\times\sqrt{8\times5} = 13+2\sqrt{40}$

Similarly,

$(\sqrt{7}+\sqrt{6})^2 = 7+6+2\times\sqrt{7\times6} = 13+2\sqrt{42}$,

$(\sqrt{10}+\sqrt{3})^2 = 10+3+2\times\sqrt{10\times3}$

$= 13 + 2\sqrt{30}$, $(\sqrt{11}+\sqrt{2})^2$

$= 11 + 2 + 2\sqrt{11\times2}$

$= 13 + 2\sqrt{22}$

Clearly, $13 + 2\sqrt{22}$ is the smallest among these.

$\therefore$ $\sqrt{11}+\sqrt{2}$ is the smallest.

2. (d) Divide 100 successively by 5 and keep on writing the quotient and then find the summation of all the quotient this summation will give us the highest power of 5 in 100!.

	100
5	20
5	4

Sum of all the quotient is 20 + 4 = 24, hence highest power of 5 in 100! *Is* 24.

Alternately

Required highest power of 5 is

$\left[\frac{100}{5^1}\right]+\left[\frac{100}{5^2}\right] = 20 + 4 = 24$ hence highest power of 5 in 100! *Is* 24.

3. (b) Let x additional men employed.

117 men were supposed to finish the whole work in $46\times8=368$ hours.

But 117 men completed $\frac{4}{7}$ of the work in $33\times8=264$ hours

$\therefore$ 117 men could complete the work in 462 hours.

Now $(117+x)$ men are supposed to do $\frac{3}{7}$ of the work, working 9 hours a day, in $13\times9=117$ hours, so as to finish the work in time.

i.e., $(117+x)$ men are supposed to complete the whole work in $117\times\frac{7}{3}=273$ hours.

$\therefore$ $(117+x)\times273=117\times462$

$\Rightarrow$ $(117+x)\times7=3\times462$

$\Rightarrow$ $x+117=3\times66=198 \Rightarrow x=81$

$\therefore$ Required number of additional men to finish the work in time = 81.

4. (b) Tap A fills 4 buckets ($4\times5=20$ litres) in 24 min.

In 1 hour tap A fills $\frac{20}{24}\times60 = 50$ litres

In 1 hour tap B fills $= 8\times5 = 40$ litres

In 1 hour tap C fills $\frac{2\times5}{20}\times60 = 30$ litres

If they open together they would fill $50+40+30=120$ litres in one hour but full tank is emptied in 2 hours

So, tank capacity would be $120\times2=240$ litres.

5. (b) B alone can do a work in 20 hours.

$\therefore$ A alone can do $\frac{3}{2}$ of the work in 20 hours.

i.e., A alone can do the same work in $\frac{40}{3}$ hours

$\therefore$ (A + B)'s one hour's work $= \frac{3}{40}+\frac{1}{20}=\frac{5}{40}=\frac{1}{8}$

$\Rightarrow$ A and B together can finish the whole work in 8 hours.

6. (c) Area of base $= \frac{1}{2}\times$ (diagonal)^2

$= \frac{1}{2}\times24\sqrt{2}\times24\sqrt{2} = 576$ sq. metre.

$\therefore$ Volume of pyramid $= \frac{1}{3}\times$ height $\times$ area of base

$\Rightarrow$ $1728 = \frac{1}{3}\times h\times576$

$\Rightarrow h = \frac{1728\times3}{576} = 9$ metre

7. (d) $\boxed{a}$ a b $\boxed{a}$ b / $\boxed{a}$ a b a

$\boxed{b}$/$\boxed{a}$ a b a b

8. (b) $11^2, 12^2, 17^2, 18^2, 23^2, 24^2, 29^2, 30^2$

9. (a) Let C.P. of whole stock = ₹ 100.

Then, marked price of whole stock = ₹ 120.

M.P. of $\frac{1}{2}$ stock = ₹ 60, M.P. of $\frac{1}{4}$ stock = ₹ 30.

$\therefore$ Total S.P. = ₹ [60 + (80% of 30) + (60% of 30)]
= ₹ (60 + 24 + 18) = ₹ 102

Hence, gain% = (102 – 100)% = 2%.

10. (d) Let the SP of the article be ₹ x

Expenses = 15% of x = ₹ 0.15x

Profit = 10% of x = ₹ 0.10x

CP = ₹ 9 (given)

Therefore, $9 + 0.15x + 0.1x = x \Rightarrow x = 12$

$\therefore$ % increase for marked price $= \frac{12-9}{9} \times 100$

$= \frac{100}{3}\%$

11. (a) Retailer's price = 112 % of 110% of (120 % of 25)

$= \frac{112}{100} \times \frac{110}{100} \times \frac{120}{100} \times 25 =$ ₹ 36.96 ≈ ₹ 37

12. (c) Amount of water left = 50 × 9/10 × 9/10 = 40.5 litres. Hence, wine = 9.5 litres. Ratio of wine and water = 19 : 81.

13. (b) By the rule of alligation, we have :

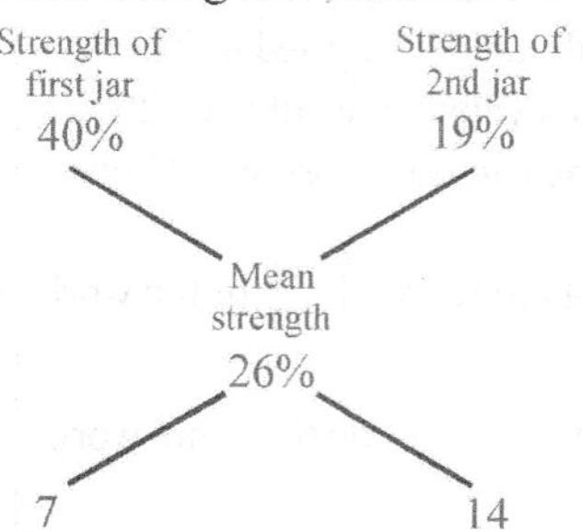

So, ratio of 1st and 2nd quantities = 7 : 14 = 1 : 2.

$\therefore$ Required quantity replaced $= \frac{2}{3}$.

14. (d)

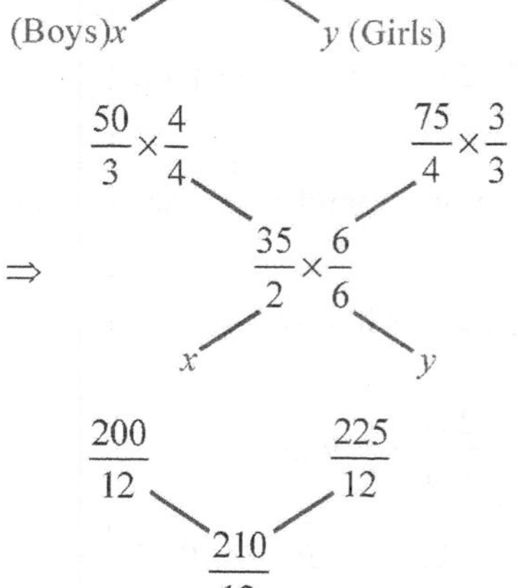

$\Rightarrow$ Thus the number of girls = 16 and number of boys = 24

15. (a) Let the number of students be n. Then,

$7 = \frac{n \times 6 + 12 \times 40}{n + 12}$

$\Rightarrow$ 7n + 84 = 6n + 480

$\Rightarrow$ n = 480 – 84 = 396

16. (b) Price of the article after first discount

65 – 6.5 = ₹ 58.5

Therefore, the second discount

$= \frac{58.5 - 56.16}{58.5} \times 100 = 4\%$

17. (a) B's income is $\left(\frac{6\frac{1}{4}}{100 + 6\frac{1}{4}} \times 100\right)\%$ less than A's

i.e., $\left(\frac{6.25}{106.25} \times 100\right)\% = 5.89\%$ less than A's income

18. (b) Let the distance between X and Y be x km. Then, the speed of A is $\frac{x}{4}$ km/h and that of B is $\frac{2x}{7}$ km/h.

$\frac{2x}{7}$ km/h • X ←— x km —→ Y • $\frac{x}{4}$ km/h

Relative speeds of the trains

$= \left(\frac{x}{4} + \frac{2x}{7}\right) = \frac{15x}{28}$ km / h

The distance between the trains at 7 a.m. $= x - \frac{x}{2} = \frac{x}{2}$ km

Hence, time taken to cross each other

$= \frac{\frac{x}{2}}{\frac{15x}{28}} = \frac{x}{2} \times \frac{28}{15x} = \frac{14}{15} \times 60 = 56$ min

Thus, both of them meet at 7 : 56 a.m.

19. (c) 10 km/h A ——→ B 15 km/h

Both the athlete are crossing each other after 12 minutes which means the distance between them is 5 km. It will be easy to go through the ratio of the speed which is 2 : 3. The answer is 36 minutes.

20. (c) $x^{\frac{1}{3}} + y^{\frac{1}{3}} = z^{\frac{1}{3}}$ ...... (i)

Taking cube both sides,

$\left(x^{\frac{1}{3}} + y^{\frac{1}{3}}\right)^3 = z$

$\Rightarrow (x + y + 3\, x^{\frac{1}{3}}.y^{\frac{1}{3}}\left(x^{\frac{1}{3}} + y^{\frac{1}{3}}\right) = z$

[$\because (a + b)^3 = a^3 + b^3 + 3ab\,(a + b)$]

$\Rightarrow x + y - z = -3.\, x^{\frac{1}{3}}.y^{\frac{1}{3}}.z^{\frac{1}{3}}$ ....(ii)

[From equation (i)]

$\therefore (x + y - z)^3 + 27\,xyz$

$= \left(-3x^{\frac{1}{3}}.y^{\frac{1}{3}}.z^{\frac{1}{3}}\right)^3 + 27\,xyz$

[From equation (ii)]

$= -27xyz + 27xyz = 0$

21. (c) Let the annual instalment be ₹ x

By formula $A = P\left(1 + \frac{R}{T}\right)^T$

$A = P_1\left(1 + \frac{25}{200}\right)$

$\Rightarrow x = P_1 \times \frac{9}{8}$

$\Rightarrow P_1 = \frac{8}{9}x$

Similarly, $= P_2 = \frac{64}{81}x$

$P_1 + P_2 = 6800$

$\Rightarrow \frac{8}{9}x + \frac{64}{81}x = 6800$

$\Rightarrow \frac{72x + 64x}{81} = 6800$

$\Rightarrow \frac{136x}{81} = 6800$

$\Rightarrow x = \frac{6800 \times 81}{136} =$ ₹ 4050

22. (c) $a + \frac{1}{b} = 1 \Rightarrow a = 1 - \frac{1}{b} = \frac{b-1}{b}$

$\Rightarrow \frac{1}{a} = \frac{b}{b-1}$ and

$b + \frac{1}{c} = 1 \Rightarrow \frac{1}{c} = 1 - b \Rightarrow c = \frac{1}{1-b}$

$\therefore c + \frac{1}{a} = \frac{1}{1-b} + \frac{b}{b-1}$

$= \frac{1}{1-b} - \frac{b}{1-b} = \frac{1-b}{1-b} = 1$

23. (d) $\frac{5x}{-15} + \frac{3y}{15} = 1 \Rightarrow \frac{x}{-3} + \frac{y}{5} = 1$

$\therefore$ area of $\Delta = \frac{1}{2} \times 3 \times 5 = \frac{15}{2}$

24. (a) Since, the given lines are parallel, we have

$\frac{1}{k} = \frac{2}{4} \Rightarrow k = 2$

25. (c) $\sec\theta + \tan\theta = p$ or $\frac{1}{\cos\theta} + \frac{\sin\theta}{\cos\theta} = p$

$\Rightarrow 1 + \sin\theta = p\cos\theta$

On squaring both sides, we get

$1 + \sin^2\theta + 2\sin\theta = p^2\cos^2\theta$

$1 + 1 - \cos^2\theta + 2\sin\theta$

$= p^2\cos^2\theta$

$2 + 2\sin\theta - \cos^2\theta = p^2\cos^2\theta$

$2(1 + \sin\theta) - \cos^2\theta = p^2\cos^2\theta$

$2 \times p\cos\theta - \cos^2\theta = p^2\cos^2\theta$

$2p\cos\theta = p^2\cos^2\theta + \cos^2\theta$

$\therefore 2p\cos\theta = \cos^2\theta\,(1 + p^2)$

$\Rightarrow \cos\theta = \frac{2p}{1+p^2}$

26. (b) $\tan A + \sin A = p$

$\therefore \tan^2 A + \sin^2 A + 2\tan A\sin A = p^2$ ...(i)

Again, $\tan A - \sin A = q$

$\tan^2 A + \sin^2 A - 2\tan A\sin A = q^2$ ...(ii)

On subtracting Eq. (ii) from Eq. (i), we get

$p^2 - q^2 = 4\tan A\sin A$

$\therefore \tan A\sin A = \frac{p^2 - q^2}{4}$

Also, $(\tan A - \sin A)(\tan A + \sin A) = pq$

$\Rightarrow \tan^2 A - \sin^2 A = pq$

$\Rightarrow \frac{\sin^2 A}{\cos^2 A} - \sin^2 A = pq$

$\Rightarrow \frac{\sin^2 A(1 - \cos^2 A)}{\cos^2 A} = pq$

$\Rightarrow \frac{\sin^2 A\sin^2 A}{\cos^2 A} = pq$

$\Rightarrow \tan^2 A\sin^2 A = pq$

$\therefore \tan A\sin A = \sqrt{pq}$

Also, $\tan A\sin A = \frac{p^2 - q^2}{4}$

$\therefore \sqrt{pq} = \frac{p^2 - q^2}{4}$

$\therefore p^2 - q^2 = 4\sqrt{pq}$

27. (a) Area of original circle
$= \pi \times (6)^2 = 36\pi \text{ cm}^2$
After trisection, the area of the smallest circle
$= \frac{1}{3} \times 36\pi$
$= 12\pi = \pi \times (2\sqrt{3})^2$
$\therefore$ Required radius $= 2\sqrt{3}$ cm

28. (c) In ΔSBO, $r^2 = x^2 + 8^2$ ...(i)
In ΔQAO, $r^2 = (4 + x)^2 + 4^2$ ...(ii)
From Eqs. (i) and (ii), $x^2 + 64 = 16 + x^2 + 8x + 16$

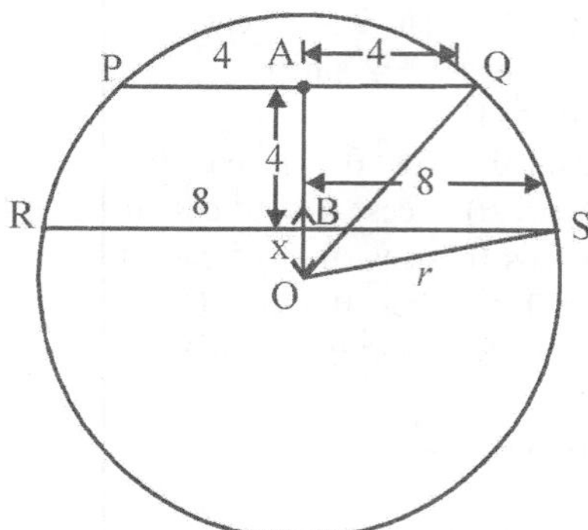

$\Rightarrow$ $8x = 32 \Rightarrow x = 4$
$\Rightarrow$ From Eq. (i), $r = \sqrt{16+64} = \sqrt{80} = 4\sqrt{5}$ cm

29. (a) E and F are the centroids of ΔABD and ΔBCD, so
$\frac{AE}{EO} = \frac{CF}{OF} = \frac{2}{1}$
$AE = CF = 2x$ (say)

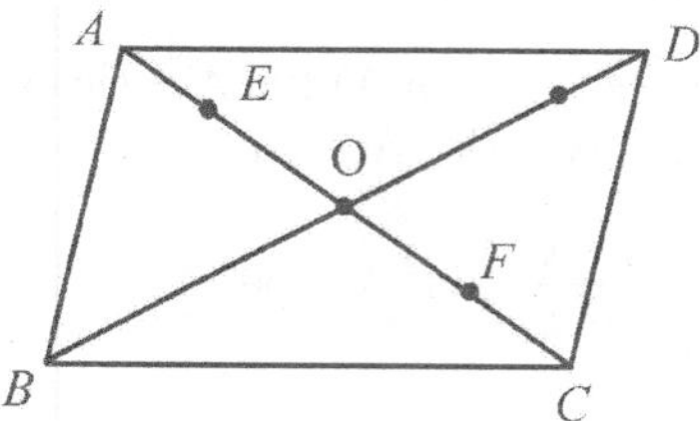

Then, $EO = x = OF$
$\Rightarrow$ $EF = 2x$
$\therefore$ $AE = \frac{1}{2} EF = CF$

30. (d)

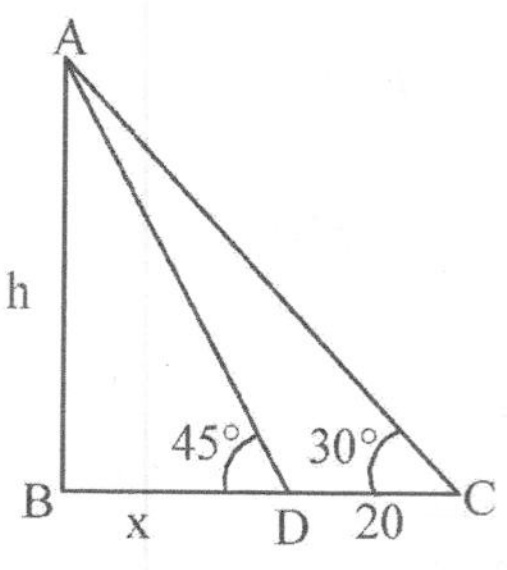

Let AB be a pillar of height h metre.
If BC = length of shadow = $x + 20$,
DC = 20m
From ΔABD,
$\tan 45° = \frac{h}{x} \Rightarrow h = x$ ..... (i)
From ΔABC,
$\tan 30° = \frac{AB}{BC} \Rightarrow \frac{1}{\sqrt{3}} = \frac{h}{x+20}$
$\Rightarrow \frac{1}{\sqrt{3}} = \frac{h}{h+20} \Rightarrow \sqrt{3}h = h + 20$
$\Rightarrow \left(\sqrt{3}-1\right)h = 20 \Rightarrow h = \frac{20}{\sqrt{3}-1}$
$= \frac{20}{\sqrt{3}-1} \times \frac{\sqrt{3}+1}{\sqrt{3}+1}$
$= \frac{20\left(\sqrt{3}+1\right)}{2} = 10\left(\sqrt{3}+1\right)$ metre

31. (c) $\frac{1}{\text{cosec}^2 51°} + \sin^2 39° + \tan^2 51°$
$- \frac{1}{\sin^2 51°.\sec^2 39°}$
$= \sin^2 51°. + \sin^2 39° + \tan^2 (90° - 39°)$
$- \frac{1}{\sin^2(90°-39°).\sec^2 39°}$
$= \cos^2 39° + \sin^2 39° + \cot^2 39°$
$- \frac{1}{\cos^2 39°.\sec^2 39°}$
[$\because \sin (90° - \theta) = \cos \theta$),
$\tan (90° - \theta) = \cot \theta$]
$= 1 + \cot^2 39° - 1$
$= \text{cosec}^2 39° - 1 = x^2 - 1$

32. (d) $3 \cot \theta = 4$
$\therefore \cot \theta = \frac{4}{3}$
Now, $\frac{5\sin\theta + 3\cos\theta}{5\sin\theta - 3\cos\theta}$
$= \frac{\frac{5\sin\theta+3\cos\theta}{\sin\theta}}{\frac{5\sin\theta-3\cos\theta}{\sin\theta}}$
$= \frac{5+3\cot\theta}{5-3\cot\theta} = \frac{5+3\times\frac{4}{3}}{5-3\times\frac{4}{3}} = 9$

33. (b) $\tan x + \cot x = 3$

Squaring on both the sides, we get

$(\tan x + \cot x)^2 = 9$

$\Rightarrow \tan^2 x + \cot^2 x + 2 = 9$

$\Rightarrow (\sec^2 x - 1) + (\operatorname{cosec}^2 x - 1) = 7$

$\therefore \sec^2 x + \operatorname{cosec}^2 x = 9$

34. (d) The greatest value of $(\sin\theta + \cos\theta)$ is $\sqrt{2}$ at $\theta = \frac{\pi}{4}$.

35. (b)

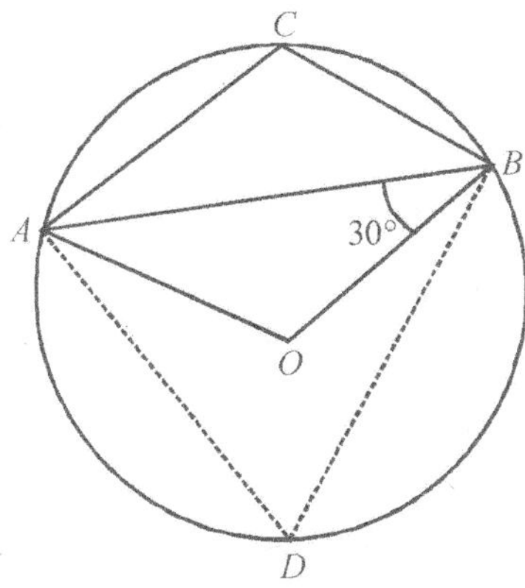

OB = OA = radius of the circle

$\angle AOB = 180 - (30 + 30)$ {Sum of angles of triangle = 180°}

$\Rightarrow 120°$

Then $\angle ADB = \frac{120}{2} = 60°$, because the angle subtended by a chord at the centre is twice of what it can subtend at the circumference. Again, $ACBD$ is a cyclic quadrilateral;

So $\angle ACB = 180° - 60° = 120°$ (because opposite angles of cyclic quadrilateral are supplementary).

36 (a) $\because h = 3$ cm, and $r = 4$ cm

$\therefore$ Volume of cone $= \frac{1}{3}\pi r^2 h = \frac{1}{3}\pi \times 16 \times 3 = 16\pi \text{ cm}^3$

37. (c) $\left(\sqrt{x} - \frac{1}{\sqrt{x}}\right)^2 = x + \frac{1}{x} - 2$

$= (3 + 2\sqrt{2}) + \frac{1}{3 + 2\sqrt{2}} - 2$

$= (3 + 2\sqrt{2}) + \frac{(3 - 2\sqrt{2})}{(3 + 2\sqrt{2})(3 - 2\sqrt{2})} - 2$

$= (3 + 2\sqrt{2}) + \frac{(3 - 2\sqrt{2})}{9 - 8} - 2$

$= 3 + 2\sqrt{2} + 3 - 2\sqrt{2} - 2 = 4$

$\therefore \sqrt{x} - \frac{1}{\sqrt{x}} = \sqrt{4} = 2$

38. (a) $\frac{a^2 - b^2 - c^2 - 2bc}{a^2 + b^2 - 2ab - c^2}$

$= \frac{a^2 - (b^2 + c^2 + 2bc)}{(a^2 + b^2 - 2ab) - c^2} = \frac{a^2 - (b + c)^2}{(a - b)^2 - c^2}$

$= \frac{(a + b + c)(a - b - c)}{(a - b + c)(a - b - c)}$

$= \frac{a + b + c}{a - b + c} = \frac{0.25 - 0.05 + 0.5}{0.25 + 0.05 + 0.5}$

$= \frac{0.7}{0.8} = \frac{7}{8}$

39. (d) $x^2 - 3x + 1 = 0$

$\therefore x = \frac{3 \pm \sqrt{9 - 4}}{2} = \frac{3 \pm \sqrt{5}}{2}$

$\left[\text{If } ax^2 + bx + c = 0, \text{ then } x = \frac{-b \pm \sqrt{b^2 - 4ac}}{2a}\right]$

Let $x = \frac{3 + \sqrt{5}}{2}$

$\therefore \frac{1}{x} = \frac{2}{3 + \sqrt{5}} = \frac{2(3 - \sqrt{5})}{(3 + \sqrt{5})(3 - \sqrt{5})}$

$= \frac{2(3 - \sqrt{5})}{9 - 5} = \frac{3 - \sqrt{5}}{2}$

$\therefore x + \frac{1}{x} = \frac{3 + \sqrt{5}}{2} + \frac{3 - \sqrt{5}}{2}$

$= \frac{3 + \sqrt{5} + 3 - \sqrt{5}}{2} = \frac{6}{2} = 3$

When, $x = \frac{3 - \sqrt{5}}{2}, \frac{1}{x} = \frac{3 + \sqrt{5}}{2}$

$\therefore x + \frac{1}{x} = \frac{3 - \sqrt{5}}{2} + \frac{3 + \sqrt{5}}{2} = 3$

40. (a)

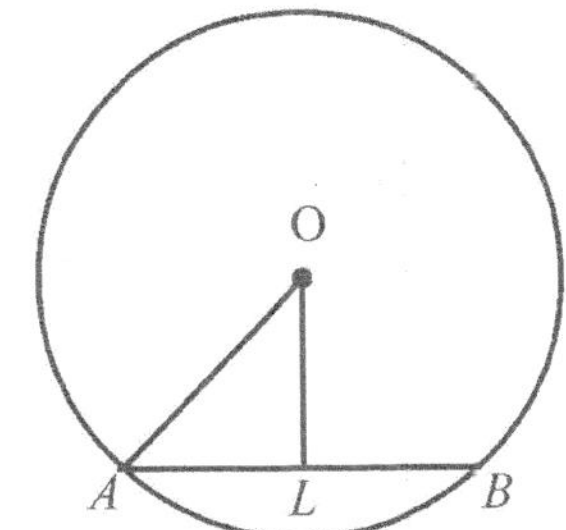

Let O be the centre of the circle and AB be the chord.

Draw $OL \perp AB$

Then, $AL = \frac{1}{2} \times AB = \left(\frac{1}{2} \times 16\right)$ cm = 8 cm and OA = 10 cm

$OL^2 = OA^2 - AL^2 = (10)^2 - (8)^2$

$= (100 - 64) = 36 = OL = \sqrt{36} = 6$ cm

41. (c) If the number of sides of the regular polygon be n, then

$\therefore \left(\frac{2n-4}{n}\right) \times 90° = 8 \times \frac{4 \times 90°}{n} + 18$

$\Rightarrow \left(\frac{2n-4}{n}\right) \times 5 = \frac{160}{n} + 1$

$\Rightarrow 10n - 20 = 160 + n$

$\Rightarrow 10n - n = 180$

$\Rightarrow 9n = 180$

$\Rightarrow n = 20$

42. (c) $\therefore \frac{2\pi r_2}{2\pi r_1} = \frac{23}{22}$

$\Rightarrow \frac{r_2}{r_1} = \frac{23}{22}$

Let $r_1 = 22 \times x$

and $r_2 = 23 \times x$

$\therefore r_1 - r_2 = 5$

Hence, the diameter will be 110 × 2 = 220

$23x - 22x = 5$

$x = 5$

$\therefore$ Inner radius = 22 × 5 = 110 m

43. (b) Part of cistern emptied in 1 hour

$= \frac{1}{5} - \frac{1}{8} = \frac{8-5}{40} = \frac{3}{40}$

$\because \frac{3}{40}$ part is emptied in 1 hour.

$\because \frac{3}{4}$ part is emptied n

$\frac{40}{3} \times \frac{3}{5} = 10$ hours.

44. (d) Total number of students playing cricket

= 38 + 40 + 12 + 17 + 25 + 18 + 20 = 170

Required percentage = $\frac{25}{170} \times 100 = 14.706 \approx 15\%$

45. (d) Required ratio = 27 : 18 = 3 : 2

46. (b) Looking at the table, we find Cricket is the most popular game.

47. (c) Required ratio = 3 : 2

48. (b) Required difference

$= \left(\frac{3000+600+2400+1200+3300}{5}\right)$

$- \left(\frac{1500+1800+900+2700+1500}{5}\right)$

= 2100 – 1680 = 420 tonnes

49. (a) Required value = $\frac{2700}{1500} = 1.8$

50. (c) Required percentage = $\frac{600}{2400} \times 100 = 25\%$

■■■

PRACTICE SET-4

1. What is the value of $\sin^3 60° \cot 30° - 2\sec^2 45° + 3\cos 60° \tan 45° - \tan^2 60°$?

(a) $\frac{35}{8}$ (b) $-\frac{35}{8}$

(c) $-\frac{11}{8}$ (d) $\frac{11}{8}$

2. If $\tan\theta = -\frac{1}{\sqrt{7}}$, then $\frac{\text{cosec}^2\theta - \sec^2\theta}{\text{cosec}^2\theta + \sec^2\theta} = ?$

(a) $-\frac{3}{4}$ (b) $-\frac{2}{3}$

(c) $\frac{2}{3}$ (d) $\frac{3}{4}$

3. If $7\sin^2 x + 3\cos^2 x = 4$, $0 < x < 90°$, then the value of $\tan x$ is

(a) $\sqrt{3}$ (b) 1

(c) $\frac{\sqrt{3}}{2}$ (d) $\frac{1}{\sqrt{3}}$

4. The angle of elevation of a tower at a point is 45°. After going 40 m towards the foot of the tower, the angle of elevation of the tower becomes 60°, Find the height of the tower.

(a) $\frac{40\sqrt{3}}{\sqrt{3}+1}$m (b) $40\sqrt{3}$ m

(c) $\frac{40\sqrt{3}}{\sqrt{3}-1}$m (d) None of these

5. If a regular hexagon is inscribed in a cricle of radius r, find the perimeter of the hexagon.

(a) $5r$ (b) $6r$

(c) $8r$ (d) $9r$

6. The external bisector of ∠B and ∠C of ΔABC (where AB and AC extended to E and F respectively) meet at point P. If ∠BAC = 100°, then the measure of ∠BPC is

(a) 50° (b) 80°

(c) 40° (d) 100°

7. The sum of the H.C.F . and L.C.M. of two number is 680 and the L.C.M. is 84 times the H.C.F. If one of the numbers is 56, the other is :

(a) 84 (b) 12

(c) 8 (d) 96

8. A number when divided by 119 leaves remainder 19. If the same number is divided by 17, the remainder will be

(a) 12 (b) 10

(c) 7 (d) 2

9. If $x = 2^{\frac{1}{3}} + 2^{-\frac{1}{3}}$, then the value of $2x^3 - 6x$ will be

(a) 5 (b) 6

(c) 8 (d) 10

10. If 10 men or 20 boys can make 260 mats in 20 days, then how many mats will be made by 8 men and 4 boys in 20 days?

(a) 260 (b) 240

(c) 280 (d) 520

11. A cuboidal water tank contains 216 litres of water. Its depth is $\frac{1}{3}$ of its length and breadth is $\frac{1}{2}$ of $\frac{1}{3}$ of the difference between length and depth. The length of the tank is :

(a) 72 dm (b) 18 dm

(c) 6 dm (d) 2 cm

12. Triangle PQR circumscribes a circle with centre O and radius r cm such that ∠PQR = 90°. If PQ = 3 cm, QR = 4 cm, then the value of r is :

(a) 2 (b) 1.5

(c) 2.5 (d) 1

13. An article was sold at 16% gain. Had it been sold for ₹ 200 more, the gain would have been 20%. Then the cost price of the article is :

(a) ₹ 5000 (b) ₹ 4800

(c) ₹ 4500 (d) ₹ 5200

14. The marked price of a T.V. is ₹ 16,000. After two successive discounts it is sold for ₹ 11,400. If the first discount is 5%, then the rate of second discount isf

(a) 15% (b) 20%

(c) 30% (d) 25%

15. If the numerator of a fraction is increased by 25% and the denominater is doubled, the fraction thus obtained is 5/9, What is the original fraction?

(a) 2/3 (b) 4/9

(c) 8/9 (d) None of these

16. If A : B = 3 : 4, B : C = 5 : 7 and C : D = 8 : 9, then, the ratio A : D is :

(a) 3 : 7 (b) 7 : 3

(c) 21 : 10 (d) 10 : 21

17. A, B and C can complete a work in 10, 12 and 15 days respectively. They started the work together. But A left the work before 5 days of its completion. B also left the work 2 days after A left. In how many days was the work completed ?

(a) 4 (b) 5

(c) 7 (d) 8

18. The areas of a square and a rectangle are equal. The length of the rectangle is greater than the length of a side of the square by 5 cm and the breadth is less than the length of the side of the square by 3 cm. The perimeter of the rectangle is :
(a) 17 cm (b) 26 cm
(c) 30 cm (d) 34 cm

19. A began business with ₹ 45000 and was joined after wards by ₹ 54000. After how many months did B join if the profits at the end of the year were divide in ratio 2 : 1 ?
(a) 4 (b) 5
(c) 6 (d) 7

20. A man gets ₹ 13 more by selling an article at a profit of $12\frac{1}{2}\%$ than selling it at a loss of $12\frac{1}{2}\%$. The cost price of the article is :
(a) ₹25.50 (b) ₹38
(c) ₹52 (d) ₹65

21. If 50% of $(x - y)$ = 30% of $(x + y)$, then what per cent of x is y ?
(a) 25% (b) $33\frac{1}{3}\%$
(c) 40% (d) 400%

22. The population of a town is 8000. If the males increased by 6 % and the females by 10 %, the population will be 8600. Find the number of females in the town.
(a) 1,000 (b) 2,000
(c) 3,000 (d) 5,000

23. A sum of ₹ 1500 is lent out in two parts in such a way that the simple interest on one part at 10% per annum for 5 years is equal to that on another part at 12.5% per annum for 4 years. The sum lent out at 12.5% is :
(a) ₹500 (b) ₹1000
(c) ₹750 (d) ₹1250

24. The average weight of 15 students in a class increses by 1.5 kg when one of the students weighting 40 kg is replaced by a new student. What is the weight (in kg) of the new student ?
(a) 64.5 (b) 56
(c) 60 (d) 62.5

25. The average of 30 results is 20 and the average of other 20 results is 30. what is the average of all the results ?
(a) 24 (b) 48
(c) 25 (d) 50

26. A car can cover a certain distance in $4\frac{1}{2}$ hours. If the speed is increased by 5 km/hour, it would take $\frac{1}{2}$ hour less to cover the same distance. Find the slower speed of the car.
(a) 50 km / hour (b) 40 km / hour
(c) 45 km / hour (d) 60 km / hour

27. A solid wooden toy is in the shape of a right circular cone mounted on a hemisphere. If the radius of the hemisphere is 4.2 cm and the total height of the toy is 10.2 cm, find the volume of the wooden toy (nearly)
(a) 104 cm^3 (b) 162 cm^3
(c) 427 cm^3 (d) 266 cm^3

28. The greatest among
$\sqrt{7}-\sqrt{5},\ \sqrt{5}-\sqrt{3},\ \sqrt{9}-\sqrt{7}.\sqrt{11}-\sqrt{9}$ is
(a) $\sqrt{7}-\sqrt{5}$ (b) $\sqrt{5}-\sqrt{3}$
(c) $\sqrt{9}-\sqrt{7}$ (d) $\sqrt{11}-\sqrt{9}$

29. If $x = 3 + \sqrt{8}$, then $x^2 + \frac{1}{x^2}$ is equal to
(a) 38 (b) 36
(c) 34 (d) 30

30. From four corners of a square sheet of side 4 cm, four pieces, each in the shape of arc of a circle with radius 2 cm, arc cut out. The area of the remaining portion is :
(a) $(8 - \pi)$ sq.cm. (b) $(16 - 4\pi)$ sq.cm
(c) $(16 - 8\pi)$ sq.cm. (d) $(4 - 2\pi)$ sq.cm

31 The area of a circle inscribed in a square of area 2 m^2 is
(a) $\frac{\pi}{4}m^2$ (b) $\frac{\pi}{2}m^2$
(c) πm^2 (d) $2\pi m^2$

32. If $x = a - b$, $y = b - c$, $z = c - a$. then the numerical value of the algebraic expression $x^3 + y^3 + z^3 - 3xyz$ will be
(a) $a + b + c$ (b) 0
(c) $4(a + b + c)$ (d) $3abc$

33. If $x : y = 2 : 1$, then $(x^2 - y^2) : (x^2 + y^2)$ is
(a) 3 : 5 (b) 5 : 3
(c) 4 : 5 (d) 5 : 6

34. The sides of a triangle are in the ratio 3 : 4 : 6. The triangle is :
(a) acute angled
(b) right angled
(c) obtuse angled
(d) either acute angled or right angled

35. If ABC is a quarter circle and a circle is inscribed in it and if AB = 1 cm, find the radius of smaller circle.

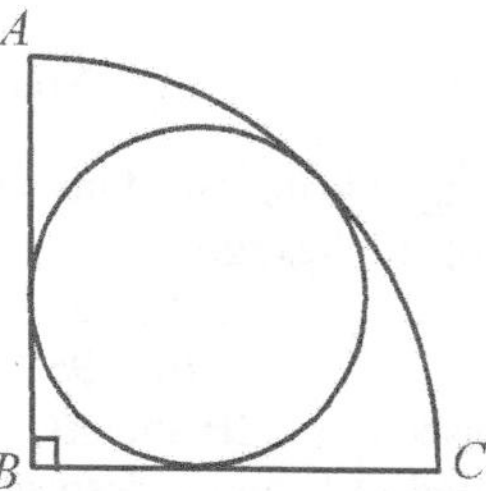

(a) $(\sqrt{2}-1)$ cm (b) $\left(\sqrt{2}-\frac{1}{2}\right)$ cm
(c) $\frac{(\sqrt{2}+1)}{2}$ (d) $(1-2\sqrt{2})$ cm

36. The value of $\left(\frac{2+\sqrt{3}}{2-\sqrt{3}}+\frac{2-\sqrt{3}}{2+\sqrt{3}}+\frac{\sqrt{3}-1}{\sqrt{3}+1}\right)$ is
(a) $2-\sqrt{3}$ (b) $2+\sqrt{3}$
(c) $16-\sqrt{3}$ (d) $40-\sqrt{3}$

37. The simplified value of $\sqrt{5+\sqrt{11+\sqrt{19+\sqrt{29+\sqrt{49}}}}}$ is

(a) 3 (b) 2
(c) 4 (d) 6

38. A tree breaks due to storm and the broken part bends so that the top of the tree touches the ground making an angle 30° with it. The distance between the feet of the tree to the point where to touches the ground in 8 m. Find the height of the tree.

(a) 8 m (b) $8\sqrt{3}$ m
(c) 24 m (d) 12 m

39. The least fraction to be subtracted from the expression

$$\frac{3\frac{1}{4}-\frac{4}{5} of \frac{5}{6}}{4\frac{1}{3}\div\frac{1}{5}-\left(\frac{3}{10}+21\frac{1}{5}\right)}$$ to make it an integer.

(a) $\frac{1}{2}$ (b) $\frac{5}{6}$
(c) $\frac{1}{4}$ (d) $\frac{3}{10}$

40. Two solid cylinders of radii 4 cm and 5 cm and lengths 6 cm and 4 cm respectively are recast into cylindrical disc of thickness 1 cm. The radius of the disc is

(a) 7 cm (b) 14 cm
(c) 21 cm (d) 28 cm

41. ABC is an equilateral triangle of side 2 cm. With A, B, C as centres and radius 1 cm, three arcs are drawn. The area of the region within the triangle bounded by the three arcs is

(a) $\left(3\sqrt{3}-\frac{\pi}{2}\right)$ cm^2 (b) $\left(\sqrt{3}-\frac{3\pi}{2}\right)$ cm^2
(c) $\left(\sqrt{3}-\frac{\pi}{2}\right)$ cm^2 (d) $\left(\frac{\pi}{2}-\sqrt{3}\right)$ cm^2

DIRECTIONS (Qs. 42-45) : *Study the following graph carefully and answer the questions given below.*

Total Number of Boys and Girls in Various Colleges Number in Thousands

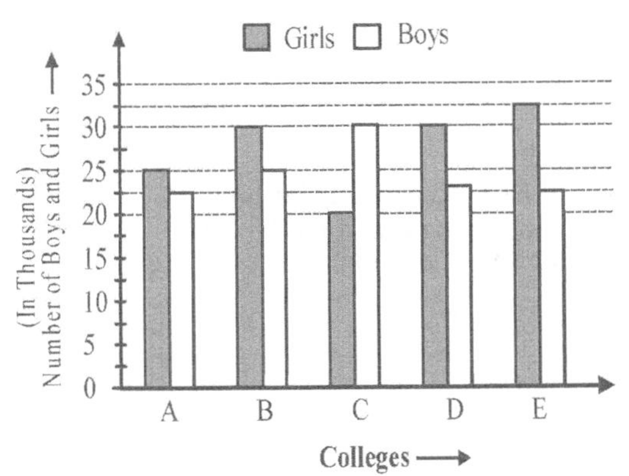

42. What is the average number of girls from all the colleges together ?

(a) 25000 (b) 27500
(c) 27000 (d) 25500

43. The total number of girls from colleges D and E together are approximately what per cent of the total number of girls from colleges A, B and C together ?

(a) 83% (b) 75%
(c) 70% (d) 88%

44. What is the difference between the total number of girls and the total number of boys from all the colleges together?

(a) 13500 (b) 14000
(c) 15500 (d) None of these

45. What is the respective ratio of the number of boys from college D to the number of boys from college B ?

(a) 10 : 9 (b) 8 : 9
(c) 9 : 10 (d) 9 : 8

DIRECTIONS (Qs. 46-49) : *Study the following graph and answer the questions. Number on the top of a bar is the number of TVs.*

Demand and Production of Colour T.Vs of Five Companies for January 2006

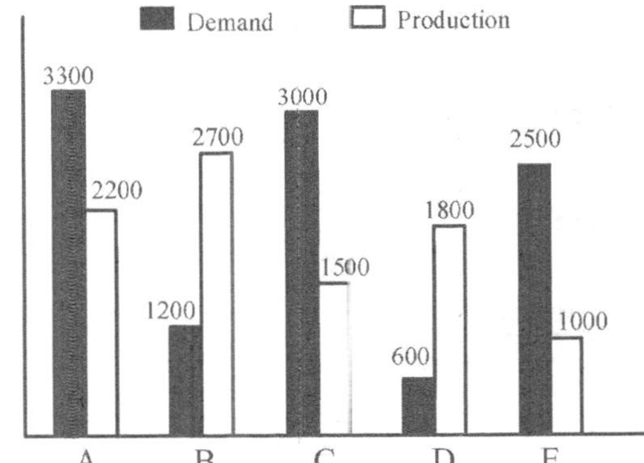

46. What is the ratio of the companies having more demand than production to the companies having more production than demand ?

(a) 2 : 3 (b) 4 : 1
(c) 2 : 2 (d) 3 : 2

47. What is the difference between average demand and average production of the five companies taken together?

(a) 1400 (b) 400
(c) 280 (d) 138

48. Demand of company D is approximately what per cent of demand of company E ?

(a) 12% (b) 20%
(c) 24% (d) 30%

49. What is the ratio of average demand to average production of comapanies B and D ?

(a) 1 : 5 (b) 2 : 5
(c) 3 : 5 (d) 4 : 5

50. What is the ratio of demand and production of comapany E to demand and production of company C ?

(a) 1 : 1 (b) 1 : 2
(c) 2 : 1 (d) 7 : 9

SOLUTIONS

1. (b) $\sin^3 60° \cot 30° - 2\sec^2 45° + 3\cos 60° \tan 45° - \tan^2 60°$

$$= \left(\frac{\sqrt{3}}{2}\right)^3 .\sqrt{3} - 2.(\sqrt{2})^2 + 3.\frac{1}{2}.1 - (\sqrt{3})^2$$

$$= \frac{3\sqrt{3}}{8} \times \sqrt{3} - 2 \times 2 + \frac{3}{2} - 3 = \frac{9}{8} - 4 + \frac{3}{2} - 3$$

$$= \frac{9 - 32 + 12 - 24}{8} = \frac{21 - 56}{8} = -\frac{35}{8}$$

2. (d) $\tan\theta = \frac{BC}{AB} = \frac{1}{\sqrt{7}}$

$AC^2 = AB^2 + BC^2 = 7k^2 + k^2 = 8k^2$

$\therefore AC = \sqrt{8k^2} = 2\sqrt{2}k$

$$\sec\theta = \frac{AC}{AB} = \frac{2\sqrt{2}k}{\sqrt{7}k} = \frac{2\sqrt{2}}{\sqrt{7}},$$

$$\text{cosec}\,\theta = \frac{AB}{BC} = \frac{2\sqrt{2}k}{\sqrt{k}} = 2\sqrt{2}$$

$$\therefore \frac{\text{cosec}^2\theta - \sec^2\theta}{\text{cosec}^2\theta + \sec^2\theta}$$

$$= \frac{(2\sqrt{2})^2 - \left[\frac{2\sqrt{2}}{\sqrt{7}}\right]^2}{(2\sqrt{2})^2 + \left[\frac{2\sqrt{2}}{\sqrt{7}}\right]^2} = \left[\frac{8 - \frac{8}{7}}{8 + \frac{8}{7}}\right] = \frac{48}{64} = \frac{3}{4}$$

3. (d) $7\sin^2 x + 3\cos^2 x = 4$

or $7\sin^2 x + 3(1 - \sin^2 x) = 4$

or $4\sin^2 x + 3 = 4$

or $\sin^2 x = \frac{1}{4} \Rightarrow \sin x = \frac{1}{2} = \sin 30°$

or $x = 30°$

$\therefore \tan x = \tan 30° = \frac{1}{\sqrt{3}}$

4. (c)

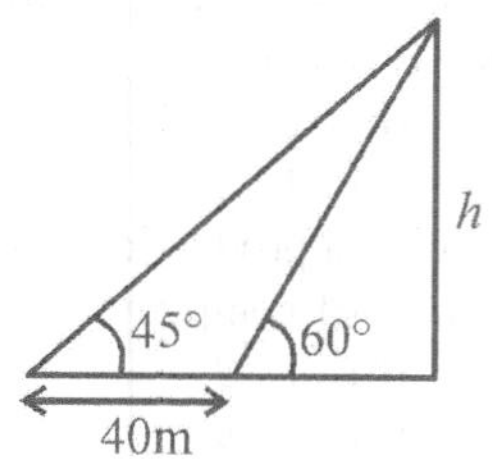

Here, $h = ?$, $x = 40$

$\theta_1 = 60°$, $\theta_2 = 45°$

$$\therefore h = \frac{40}{\cot 45° - \cot 60°} = \frac{40}{1 - \frac{1}{\sqrt{3}}} = \frac{40\sqrt{3}}{\sqrt{3} - 1}$$

$$\therefore h = \frac{40\sqrt{3}}{\sqrt{3} - 1}\text{ m}$$

5. (b) Here, $OA = OB = AB = r$

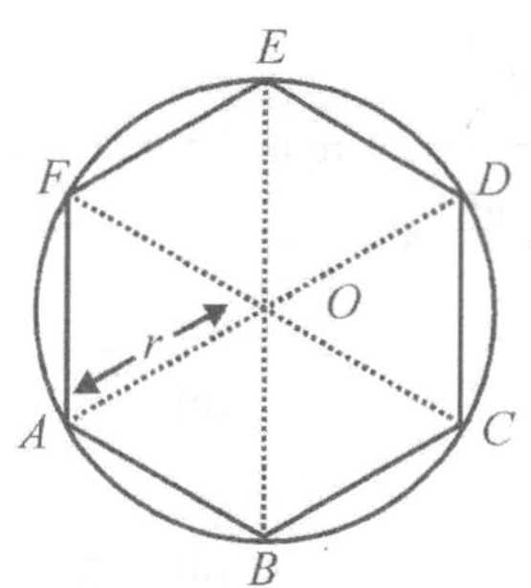

$\therefore$ Perimeter of hexagon $= 6 \times AB = 6r$

6. (c)

In ΔABC,

$\angle A = x$, $\angle B = y$; $\angle C = z$

In ΔPBC,

$\angle PBC + \angle PCB + \angle BPC = 180°$

$\Rightarrow \frac{1}{2}\angle EBC + \frac{1}{2}\angle FCB + \angle BPC = 180°$

$\angle EBC + \angle FCB + 2\angle BPC = 360°$

$\Rightarrow (180° - y) + (180° - z) + 2\angle BPC = 360°$

$\Rightarrow 360° - (y + z) + 2\angle BPC = 360°$

$\Rightarrow 2\angle BPC = y + z$

$\Rightarrow 2\angle BPC = 180° - x = 180° - \angle BAC$

$\therefore \angle BPC = 90° - \frac{1}{2}\angle BAC$

$= 90° - 50 = 40°$

7. (d) Let HCF be h and LCM be l.

Then, $l = 84h$ and $l + h = 680$

$\Rightarrow 84h + h = 680$

$\Rightarrow h = \frac{680}{85} = 8$

$\therefore l = 680 - 8 = 672$

$\therefore$ Other number $= \frac{672 \times 8}{56} = 96$

8. (d) When we divide the number by 119, remainder = 19
$\because$ 119 is exactly divisible by 17
$\therefore$ the required remainder = remainder obtained by dividing 19 by 17 = 2

9. (a) $x = 2^{1/3} + 2^{-1/3}$
On cubing both sides, we get
$x^3 = 2 + 2^{-1} + 3(2^{1/3} + 2^{-1/3})$
$\Rightarrow \ x^3 = \frac{4+1}{2} + 3x$
$\Rightarrow \ 2x^3 = 5 + 6x$
$\Rightarrow \ 2x^3 - 6x = 5$

10. (a) 10 men = 20 boys
$\therefore$ 1 man = 2 boys
$\therefore$ 8 men + 4 boys
= (16 + 4) boys = 20 boys
Hence, 8 men and 4 boys will make 260 mats in 20 days.

11. (b) Let the length of the tank be x cm.
$\therefore$ Depth = $\frac{x}{3}$
Breadth = $\frac{1}{2} \times \frac{1}{3} \times \left(x - \frac{x}{3}\right) = \frac{x}{9}$
Now, $x \times \frac{x}{3} \times \frac{x}{9} = 216 \times 1000$
$\Rightarrow \ x^3 = 27 \times 216 \times 1000$
$\Rightarrow \ x = (27 \times 216 \times 1000)^{1/3}$
$\Rightarrow \ x = 3 \times 6 \times 10$
= 180 cm = 18 dm

12. (d)

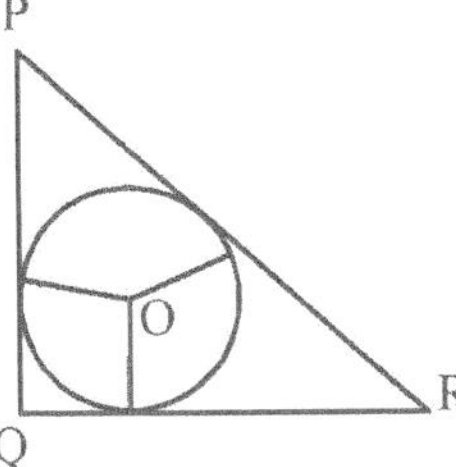

$PR^2 = PQ^2 + QR^2$
$3^2 + 4^2 = 25$
$\therefore$ PR = $\sqrt{25}$ = 5 cm
$r = \frac{\text{Area of triangle}}{\text{Semi- perimeter of triangle}}$
$\frac{\frac{1}{2} \times 3 \times 4}{\frac{3+4+5}{2}} = \frac{6}{6} = 1$ cm

13. (a) If the C.P. of article be ₹ x then
$x \times \frac{116}{100} + 200 = \frac{x \times 120}{100}$
$\Rightarrow \ x \times \frac{4}{100} = 200$
$\Rightarrow \ x = \frac{200 \times 100}{4} = ₹ \ 5000$

14. (d) After a discount of 5%
SP = $\frac{95 \times 16000}{100}$ = ₹ 15200
Let the second discount be x%.
x% of 15200 = (15200 – 11400)
$\Rightarrow \ \frac{x \times 15200}{100} = 3800$
$\Rightarrow \ x = \frac{3800 \times 100}{15200} = 25$
$\therefore$ Second discount = 25%

15. (c) Let the original fraction be $\frac{x}{y}$, Then,
$\frac{\frac{125x}{100}}{2y} = \frac{5}{9} \Rightarrow \frac{125x}{100 \times 2y} = \frac{5}{9} \Rightarrow \frac{x}{y} = \frac{5 \times 100 \times 2}{125 \times 9} = \frac{8}{9}$

16. (d) $\frac{A}{B} \times \frac{B}{C} \times \frac{C}{D} = \frac{3}{4} \times \frac{5}{7} \times \frac{8}{9}$
$\Rightarrow \ \frac{A}{D} = \frac{10}{21}$

17. (c) Let the work be completed in x days
According to the question,
$\frac{x-5}{10} + \frac{x-3}{12} + \frac{x}{15} = 1$
$\Rightarrow \ \frac{6x - 30 + 5x - 15 + 4x}{60} = 1$
$\Rightarrow \ 15x - 45 = 60$
$\Rightarrow \ 15x - 105 \Rightarrow \quad x = \frac{105}{15} = 7$
Hence, the work will be completed in 7 days.

18. (d) Let the side of square be x cm.
$\therefore$ Length of rectangle
= $(x + 5)$ cm
Breadth of rectangle
$(x - 3)$ cm
As given,
$(x + 5)(x - 3) = x^2$
$\Rightarrow \ x^2 + 5x - 3x - 15 = x^2$
$\Rightarrow \ 2x = 15$
$\therefore$ Perimeter of rectangle
= $2(x + 5 + x - 3)$
= $2(2x + 2)$
= $2(15 + 2)$ = 34 cm

19. (d) Let B remained in business for x months.
Ratio of equavalent capitals
= 45000 × 12 : 54000 × x
= 10 : x
$\therefore \ \frac{10}{x} = \frac{2}{1}$
$\Rightarrow \ 2x = 10 \Rightarrow x = 5$
Clearly, B joined after (12 – 5) = 7 months.

20. (c) Let the C.P of article be ₹ x.
According to the question,

$$\left(100+\frac{25}{2}\right)\% \text{ of } x - \left(100-\frac{25}{2}\right)\% \text{ of } x = 13$$

$$\Rightarrow \frac{x}{100}\left(100+\frac{25}{2}-100+\frac{25}{2}\right) = 13$$

$$\Rightarrow \frac{x}{100}\times 25 = 13$$

$\Rightarrow x = 13 \times 4 = 52$

21. (a) $\frac{30}{100}(x+y) = \frac{50}{100}(x-y)$

$\Rightarrow 30x + 30y = 50x - 50y$

$\Rightarrow 80y = 20x$

$\Rightarrow y = \frac{20}{80}x = \frac{1}{4}x$

$\Rightarrow$ y is 25% of x.

22. (c) Let the population of females be x.

Then 110 % of x + 106 % of (8000 – x) = 8600

or, $\frac{110x}{100} + \frac{106(8000-x)}{100} = 8600$

or, $x(110-106) = 8600\times 100 - 8000\times 106$

$\therefore x = \frac{8600\times100-8000\times106}{110-106} = \frac{12,000}{4} = 3,000$

By Method of Alligation

Average % of increase $= \frac{600}{8000}\times 100 = \frac{15}{2} = 7.5\,\%$

Now,

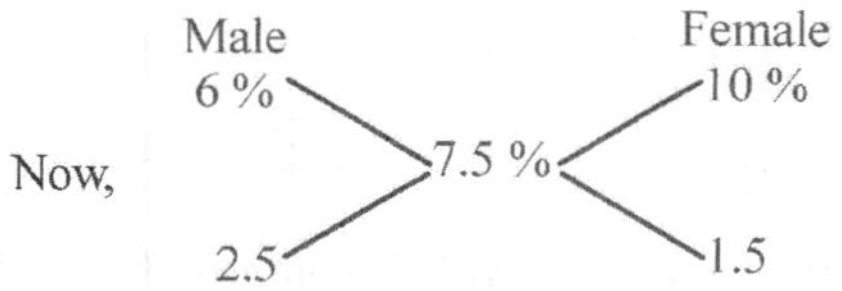

$\therefore$ Male : Female = 2.5 : 1.5 = 5 : 3

$\therefore$ The population of females $= \frac{8000}{5+3}\times 3 = 3000$

23. (c) Let the sum lent out at 12.5% be ₹ x

Sum lent out at 10%

$= 1500 - x$

Now, $\frac{(1500-x)\times10\times5}{100}$

$= \frac{x\times12.5\times4}{100}$

$\Rightarrow 50\,(1500-x) = 50x$

$\Rightarrow 2x = 1500$

$\Rightarrow x = \frac{1500}{2} =$ ₹750

24. (d) Weight of the new student

$= (40 + 15 \times 1.5)$ kg

$= (40 + 22.5)$ kg

$= 62.5$ kg

25. (a) Required average

$$= \frac{20\times30+20\times30}{30+20} = \frac{600+600}{50} = \frac{1200}{50} = 24$$

26. (b) Let the initial speed of the car be x kmph and the distane be y km.

$\therefore\quad y = \frac{9}{2}x$ .... (i)

and, $y = 4\,(x+5)$ ....(ii)

$\therefore\quad \frac{9}{2}x = 4\,(x+5)$

$\Rightarrow 9x = 8x + 40$

$\Rightarrow x = 40$ kmph

27. (d)

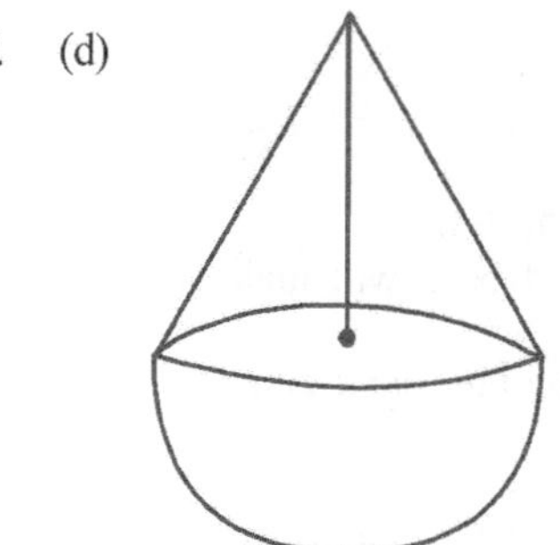

Radius of cone = 4.2 cm

Height of cone = 10.2 – 4.2 = 6 cm

Volume of the toy

Volume of cone + Volume of hemisphere

$$= \frac{1}{3}\pi(4.2)^2\times6+\frac{2}{3}\pi(4.2)^3 = \frac{1}{3}\pi(4.2)^2\,(6+2\times4.2)$$

$$= \frac{1}{3}\times\frac{22}{7}\times 4.2\times4.2\times14.4 = 266 \text{ cu.cm.}$$

28. (b) $\frac{1}{\sqrt{7}-\sqrt{5}}$

$$= \frac{\sqrt{7}+\sqrt{5}}{(\sqrt{7}-\sqrt{5})(\sqrt{7}+\sqrt{5})} = \frac{\sqrt{7}+\sqrt{5}}{7-5} = \frac{\sqrt{7}+\sqrt{5}}{2},$$

$$\frac{1}{\sqrt{5}-\sqrt{3}} = \frac{\sqrt{5}+\sqrt{3}}{(\sqrt{5}-\sqrt{3})(\sqrt{5}+\sqrt{3})}$$

$$= \frac{\sqrt{5}+\sqrt{3}}{5-3} = \frac{\sqrt{5}+\sqrt{3}}{2}$$

Similarly,

$$\frac{1}{\sqrt{9}-\sqrt{7}} = \frac{\sqrt{9}+\sqrt{7}}{2}$$

$$\frac{1}{\sqrt{11}-\sqrt{9}} = \frac{\sqrt{11}+\sqrt{9}}{2}$$

Clearly, $\frac{\sqrt{5}+\sqrt{3}}{2}$ is the smallest.

$\therefore\ \frac{1}{\sqrt{5}-\sqrt{3}}$ is the smallest.

$\therefore\ \sqrt{5}-\sqrt{3}$ is the greatest.

29. (c) $x = 3+\sqrt{8}$

$\therefore \frac{1}{x} = \frac{1}{3+\sqrt{8}} = \frac{3-\sqrt{8}}{(3+\sqrt{8})(3-\sqrt{8})}$

$= \frac{3-\sqrt{8}}{9-8} = 3-\sqrt{8}$

Now, $x^2 + \frac{1}{x^2} = \left(x+\frac{1}{x}\right)^2 - 2$

$= (3+\sqrt{8}+3-\sqrt{8})^2 - 2 = 36 - 2 = 34$

30. (b)

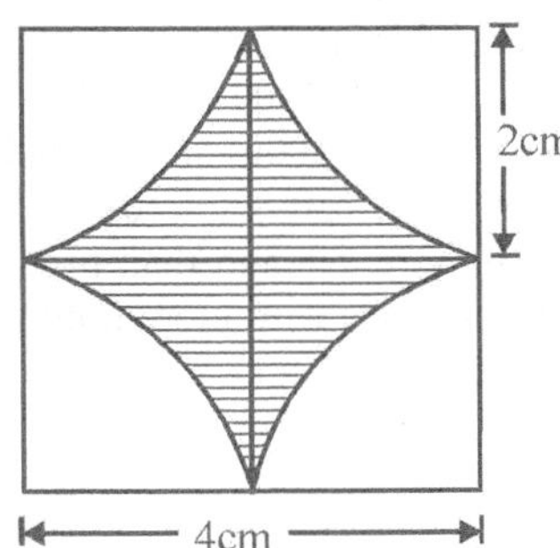

Area of sectors $= \pi r^2$
$= 4\pi$ sq. cm
Area of square $= 4 \times 4 = 16$ sq.cm.
Area of the remaining portion
$(16 - 4\pi)$ sq.cm

31. (b) Side of square $= \sqrt{2}$ metre
Radius of in-circle

$\frac{\sqrt{2}}{2} = \frac{1}{\sqrt{2}}$ metre

Area of the circle $= \pi r^2$

$= \pi \times \frac{1}{2} = \frac{\pi}{2}$ sq. metre.

32. (b) $x + y + z = a - b + b - c + c - a = 0$
$\therefore x^3 + y^3 + z^3 - 3xyz = 0$

33. (a) Here, $\frac{x}{y} = \frac{2}{1} \Rightarrow \frac{x^2}{y^2} = \frac{4}{1}$

$\therefore \frac{x^2 - y^2}{x^2 + y^2} = \frac{\frac{x^2}{y^2} - 1}{\frac{x^2}{y^2} + 1} = \frac{4-1}{4+1} = \frac{3}{5}$

34. (c) Let the sides of the triangle be $3x$, $4x$ and $6x$ units.
Clearly, $(3x)^2 + (4x)^2 < (6x)^2$
$\therefore$ The triangle will be obtuse angled.

35. (a) Let the radius of the smaller circle be x cm, then
$OP = OQ = PB = x$ cm
and $OB = (1 - x)$ cm [$\therefore BR = 1$ cm]
In ΔPOB, $(1 - x)^2 = x^2 + x^2$
$\Rightarrow x^2 + 2x - 1 = 0$

$\Rightarrow x = \frac{-2 \pm \sqrt{4-4(-1)}}{2} = \frac{-2 \pm 2\sqrt{2}}{2} = -1 \pm \sqrt{2}$

$\therefore x = (\sqrt{2}-1)$ cm or $-(\sqrt{2}+1)$ cm

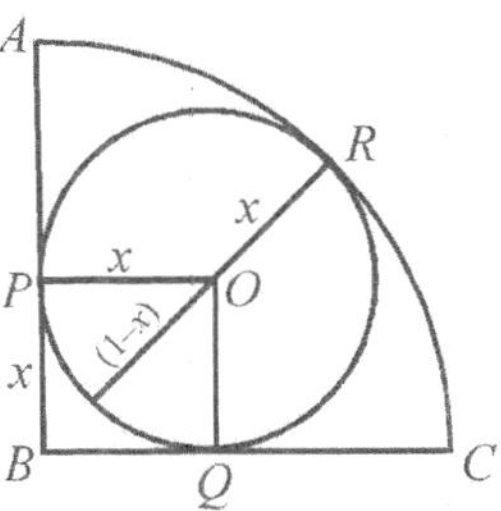

36. (c) $\frac{2+\sqrt{3}}{2-\sqrt{3}} + \frac{2-\sqrt{3}}{2+\sqrt{3}} + \frac{\sqrt{3}-1}{\sqrt{3}+1}$

$= \frac{(2+\sqrt{3})}{(2-\sqrt{3})} \times \frac{(2+\sqrt{3})}{(2+\sqrt{3})} + \frac{(2-\sqrt{3})}{(2+\sqrt{3})} + \frac{(2-\sqrt{3})}{(2-\sqrt{3})}$

$\times \frac{(\sqrt{3}-1)(\sqrt{3}-1)}{(\sqrt{3}+1)(\sqrt{3}-1)}$

$= \frac{(2+\sqrt{3})^2}{4-3} + \frac{(2-\sqrt{3})^2}{4-3} + \frac{(\sqrt{3}-1)^2}{3-1}$

$= 4 + 3 + 4\sqrt{3} + 4 + 3 - 4$

$\sqrt{3} + \frac{3+1-2\sqrt{3}}{2}$

$= 14 + 2 - \sqrt{3} = 16 - \sqrt{3}$

37. (a) Expression

$= \sqrt{5+\sqrt{11+\sqrt{19+\sqrt{29+7}}}} = \sqrt{5+\sqrt{11+\sqrt{19+6}}}$

$= \sqrt{5+\sqrt{11+5}} = \sqrt{5+4} = \sqrt{9} = 3$

38. (b) In ΔABC, $\tan 30° = \frac{AB}{BC}$

$\frac{1}{\sqrt{3}} = \frac{AB}{8}$ $AB = \frac{8}{\sqrt{3}}$ m

Again, $\cos 30° = \frac{BC}{AC}$

$\frac{\sqrt{3}}{2} = \frac{8}{AC}$

$AC = \frac{16}{\sqrt{3}}$

$DB = AB + AD$
$= AB + AC$ $(\because AD = AC)$

$= \frac{8}{\sqrt{3}} + \frac{16}{\sqrt{3}} = \frac{24}{\sqrt{3}} = 8\sqrt{3}$ m

39. (a) Experssion

$= \frac{\frac{13}{4} - \frac{4}{5} \times \frac{5}{6}}{\frac{13}{3} \times 5 - \left(\frac{3}{10} + \frac{106}{5}\right)} = \frac{\frac{13}{4} - \frac{2}{3}}{\frac{65}{3} - \frac{3}{10} - \frac{106}{5}}$

$$= \frac{\frac{39-8}{12}}{\frac{650-9-636}{30}} = \frac{31}{12}\times\frac{30}{5} = \frac{31}{2} = 15\frac{1}{2}$$

∴ Required answer

$$= 15\frac{1}{2} - 15 = \frac{1}{2}$$

40. (b) Sum of the volumes of two cylinders

$$= \pi r_1^2 h_1 + \pi r_2^2 h_2$$

$$= \frac{22}{7}(4\times4\times6+5\times5\times4)$$

$$= \frac{22}{7}(96+100) = \frac{22}{7}\times196 = 616\text{ cm}^3$$

Let the radius of the disc be r cm.

∴ $\pi r^2 \times 1 = 616$

$\Rightarrow \frac{22}{7}\times r^2 = 616$

$\Rightarrow r^2 = \frac{616\times7}{22} = 196$

$\Rightarrow r = \sqrt{196} = 14$ cm

41. (c)

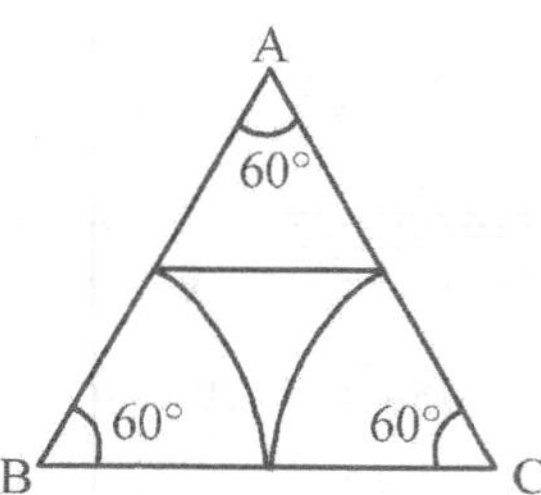

Each angle of the triangle = 60°

Required area of the three sectors $= 3\times\frac{60}{360}\times\pi(1)^2$

$$= \frac{\pi}{2}\text{ cm}^2$$

Area of region bounded by 3 arcs $= \left(\sqrt{3} - \frac{\pi}{2}\right)\text{cm}^2$

42. (b) Total number of girls
= (25 + 30 + 20 + 30 + 32.5) thousands
= 137.5 × 1000 = 137500

Average number of girls $= \frac{137500}{5} = 27500$

43. (a) Total number of girls in colleges D and E
= (30 + 32.5) thousands
= (62.5 × 1000) = 62500
Total number of girls in colleges A,B and C
= (25 + 30 + 20) thousands
(75 × 1000) = 75000

Required percentage $= \left(\frac{62500}{75000}\times100\right) = 83.3\%$

≈ 83%

44. (d) Total number of boys
= (22.5 + 25 + 30 + 22.5 + 22.5) thousands
= (122.5 × 1000) = 122500
Total number of girls = 137500.
Required difference = (137500 – 122500) = 15000.

45. (c) Required ratio = (22.5 × 1000) : (25 × 1000)
= 225 : 250 = 9 : 10

46. (d) Companies with more demand than production are A, C and E, Companies with more production than demand are B and D.
∴ Required ratio = 3 : 2

47. (c) Average demand

$$= \frac{3300+1200+3000+600+2500}{5}$$

$$= \frac{10600}{5} = 2120$$

Average production

$$= \frac{2200+2700+1500+1800+1000}{5}$$

$$= \frac{9200}{5} = 1840$$

∴ Required difference
= 2120 – 1840 = 280

48. (c) Required percentage $= \frac{600}{2500}\times100 = 24$

49. (b) Average demand of companies B and D

$$= \frac{1200+600}{2} = \frac{1800}{2} = 900$$

Average production of companies B and D

$$= \frac{2700+1800}{2} = 2250$$

∴ Required ratio = 900 : 2250 = 2 : 5

50. (d) Required ratio, $\frac{2500+1000}{3000+1500} = \frac{7}{9}$

PRACTICE SET-5

DIRECTIONS (Qs. 1-5) : *Study the following information carefully to answer the questions given below it.*

In a school of 2500 students, all the students have enrolled for different games viz. hockey, table-tennis, badminton, football, cricket, chess and carrom. The respective ratio of girls to boys in the school is 3 : 2. 20% of the boys play only cricket. 25% of the girls play table tennis, badminton and carrom only. 26.8% of the boys play only football. The number of girls playing only cricket is 175% of the boys playing the same. The respective ratio of girls and boys playing only chess is 12 : 11. 25.7% of the boys play hockey, table-tennis and carrom only. One-fourth of the girls play only badminton. The remaining girls play football and hockey only. The remaining boys play only chess.

1. How many students play more than one game ?
   (a) 850 (b) 862
   (c) 732 (d) 671
   (e) None of these
2. The Total number of students playing hockey is what per cent of the total number of students in the school ?
   (a) 25.7% (b) 10.8%
   (c) 14.28% (d) 21.14%
   (e) None of these
3. What is the respective ratio of total number of boys playing chess to the total number of girls playing badminton ?
   (a) 11 : 30 (b) 13 : 32
   (c) 9 : 29 (d) 13 : 29
   (e) None of these
4. What is the total number of students playing football, cricket and table tennis ?
   (a) 1300 (b) 1550
   (c) 1450 (d) 1650
   (e) None of these
5. How many students play carrom ?
   (a) 475 (b) 600
   (c) 538 (d) 482
   (e) None of these

DIRECTIONS (Qs. 6-10) : *Study the following graph carefully and answer the questions given below it :*

Account of income and expenditure (in crores) of seven companies in the year 2014

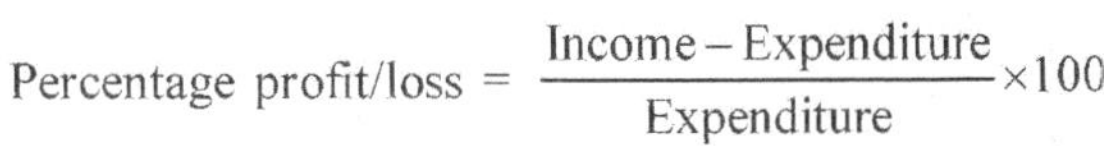

$$\text{Percentage profit/loss} = \frac{\text{Income} - \text{Expenditure}}{\text{Expenditure}} \times 100$$

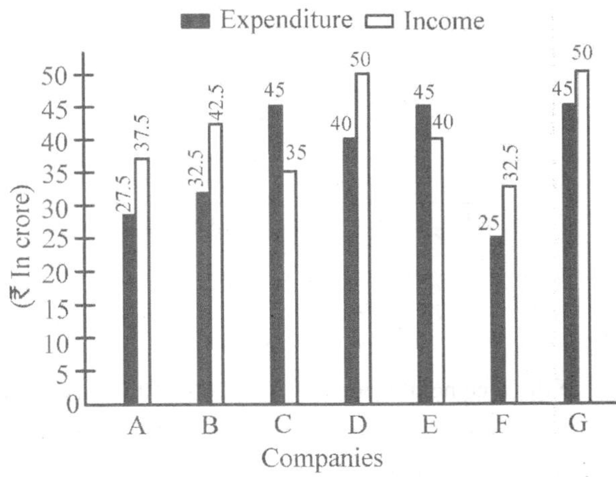

6. Approximately what is the percentage of profit/loss that Companies C and D had together ?
   (a) 10% profit (b) 12% profit
   (c) 10% loss (d) 12% loss
   (e) There was no loss or profit
7. If the income of Company A had increased by 10% in year 2014 from year 2013 and profit earned in 2013 was 20% what was its expenditure in 2013 ? (The value upto two decimal places in crores)
   (a) 36.36 (b) 32.32
   (c) 30.30 (d) Can't be determined
   (e) None of these
8. If the expenditure of Company G in 2014 was 20% more than its expenditure in the year 2013 and the Company has earned a profit of 10% in 2013. What was the company income in 2013 in crores ₹ ?
   (a) 37.5 (b) 41.25
   (c) 34.09 (d) Cannot be determined
   (e) None of these
9. Which of the following Companies had the highest percentage of profit/loss in 2014 ?
   (a) B (b) C
   (c) F (d) A
   (e) None of these
10. What is the approximate percentage of profit earned by all the Companies together in 2014 ?
    (a) 11 (b) 11.5
    (c) 10.5 (d) 12
    (e) 12.5

DIRECTIONS (Qs. 11-16) : *Study the following graph to answer the given questions.*

Per cent profit earned by two companies over the given years

$$\%\text{profit} = \frac{\text{Income} - \text{Expenditure}}{\text{Expenditure}} \times 100$$

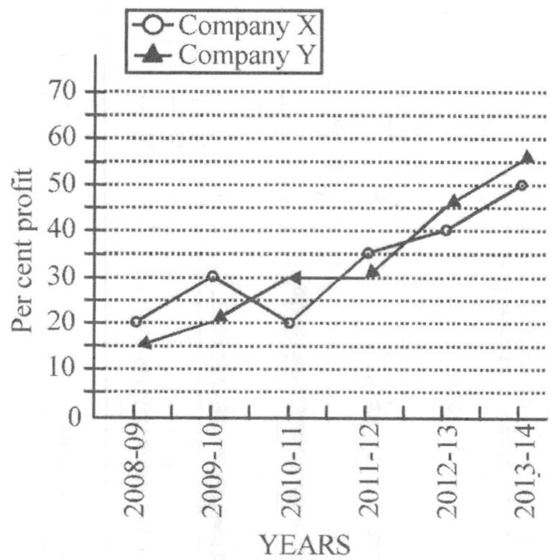

11. If the income of Company X in 2009-10 was equal to the expenditure of Company Y in 2012-2013, What was the ratio of their respective profits ?
    (a) 13 : 15 (b) 15 : 26
    (c) 13 : 26 (d) Cannot be determined
    (e) None of these
12. For Company X, its income in 2012-2013 was equal to its expenditure in 2013-2014. What was the ratio of its respective incomes in these two years ?
    (a) 4 : 5 (b) 3 : 4
    (c) 2 : 3 (d) Cannot be determined
    (e) None of these
13. For Company Y, which year is the percent of increase in per cent profit over that of previous year the highest ?
    (a) 2013-2014 (b) 2010-2011
    (c) 2012-2013 (d) Cannot be determined
    (e) None of these
14. In 2008-09, the expenditure of Company its income in that year ?
    (a) ₹ 50 crores (b) ₹ 48 crores
    (c) ₹ 46 crores (d) Cannot be determined
    (e) None of these
15. What was the difference in expenditure of the two companies in 2010-2011 ?
    (a) 10 (b) 100
    (c) 1000 (d) Cannot be determined
    (e) None of these
16. In 2013-2014, the income of Company 4 was ₹ 128 crores. What was its expenditure in that year ?
    (a) ₹ 76.8 crores (b) ₹ 64 crores
    (c) ₹ 48 crores (d) Cannot be determined
    (e) None of these

DIRECTIONS (Qs. 17-20) : *The following questions are based on the stacked bar graph given below :*

Sales of various precious stones in India for the period of 2009-10 to 2013-2014

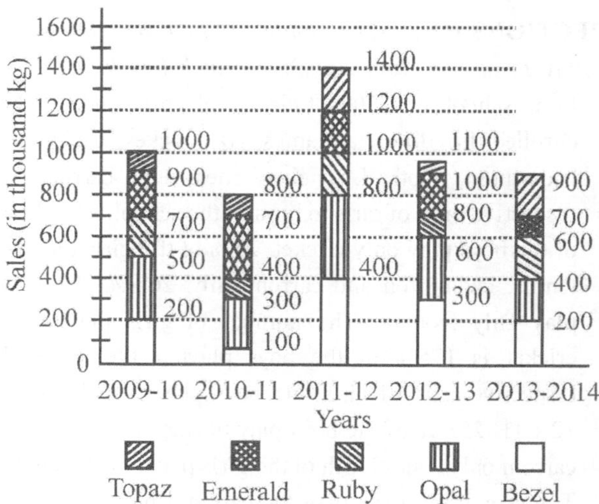

17. What is the total sales of ruby as a per cent of the total sales of precious stones for the given period ?
    (a) 17.3%
    (b) 19.23%
    (c) 23.1%
    (d) Cannot be determined
    (e) None of these
18. By what per cent is the average annual sales of opal for the given period more than the sales of Emerald in 2012-13 ?
    (a) 120% (b) 50%
    (c) 25% (d) 40%
    (e) None of these
19. For how many years is the sales of Topaz as a percentage of the total sales of precious stones less than the of Bezel ?
    (a) One (b) Two
    (c) Three (d) Four
    (e) None of these
20. If the sales of Bezel increased from 2008-09 to 2009-10 by 25% and increased from 2013-2014 to 2014-15 by 50%, then what is the difference between the sales of Bezel in 2008-09 and that in 2014-15 (in kg)?
    (a) 50000 (b) 100000
    (c) 140000 (d) 160000
    (e) None of these

DIRECTIONS (Qs. 21-25) : *Study the following graph carefully to answer these questions.*

Details about the distribution of employees and expenditure of an organization (distributed proportionately across the departments)

Annual Expenditure on different items

Total Expenditure = ₹12 crores

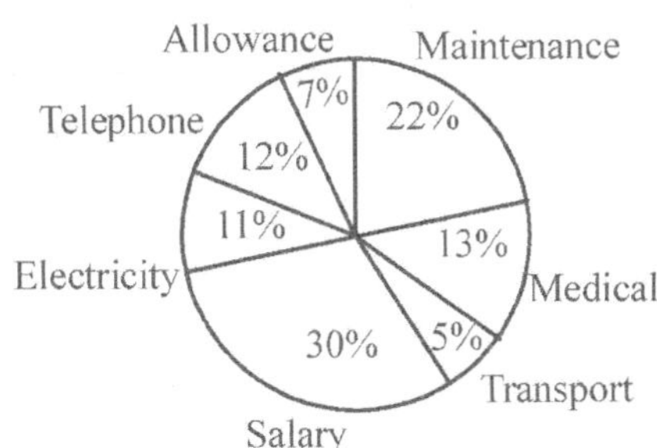

Departmentwise distribution of employees

Total number of employees = 1200

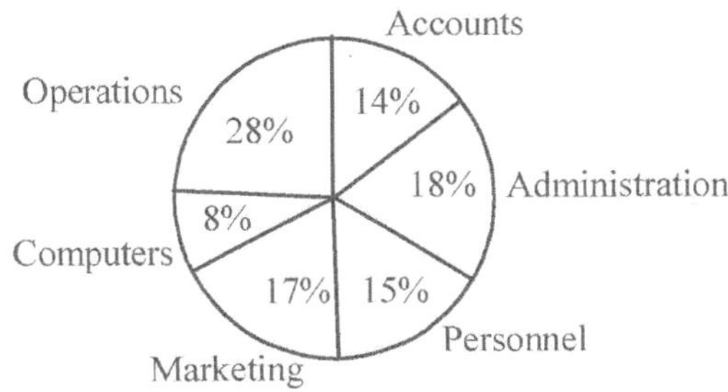

21. What was the total expenditure on Accounts Department?
(a) ₹ 16.8 lakhs (b) ₹ 1,680 lakhs
(c) ₹ 18.6 millions (d) ₹ 16.8 millions
(e) None of these

22. What was per employee expenditure on Medical ?
(a) ₹ 12,000 (b) ₹ 13,000
(c) ₹ 12,500 (d) ₹ 13,500
(e) None of these

23. What was the total expenditure on salary of employees in Marketing Department ?
(a) ₹ 6.12 lakhs (b) ₹ 61.2 lakhs
(c) ₹ 6.12 millions (d) ₹ 176 lakhs
(e) None of these

24. What was amount spent on electricity ?
(a) ₹ 13.2 millions (b) ₹ 13.2 lakhs
(c) ₹ 26 millions (d) ₹ 12.6 lakhs
(e) None of these

25. What was the expenditure on telephone for employees in Computer Department ?
(a) ₹ 11.52 lakhs (b) ₹ 11.52 millions
(c) ₹ 10.72 lakhs (d) ₹ 10.72 millions
(e) None of these

DIRECTIONS (Qs. 26-30) : *Study the following tables carefully and answer the questions given below :*

Number of Cars (in thousands) of Different Models and Colours sold in two Metro Cities in a year

	Metro M Type Colour					Metro H Colour				
	Black	Red	Blue	White	Silver	Black	Red	Blue	White	Silver
A	40	25	55	75	15	45	32	40	60	20
B	20	35	60	80	20	30	37	39	81	35
C	35	30	50	90	35	40	42	41	86	37
D	45	40	45	85	40	35	39	37	90	42
E	50	35	35	60	30	50	44	43	77	22
F	55	42	40	75	52	47	34	45	87	17

26. The total number of silver coloured cars sold in Metro H is approximately what percentage to that of Metro M?
(a) 130 (b) 140
(c) 90 (d) 100
(e) 110

27. What is the difference between the numbers of blue colour cars model C sold in Metro M and number of red colour cars model F sold in Metro H ?
(a) 8,000 (b) 10,000
(c) 12,000 (d) 15,000
(e) None of these

28. In Metro M the number of cars sold was maximum for which of the colour-model combination?
(a) White-C (b) Blue-B
(c) Silver-B (d) White-D
(e) None of these

29. The total number of blue coloured cars of Model E and D sold in Metro H is exactly equal to the number of white coloured cars of which model in Metro M ?
(a) B (b) F
(c) C (d) A
(e) None of these

30. The difference between the white coloured cars sold between the two metros of which of the following models in the minimum ?
(a) A (b) C
(c) D (d) F
(e) None of these

DIRECTIONS (Qs. 31-35) : *Study the following table carefully and answer the questions given below.*

Quantity of Rice Produced by Various States Over the Years (Quantity in Tonnes)

State	Year					
	2009	**2010**	**2011**	**2012**	**2013**	**2014**
A	1500	1480	1620	1700	1540	1650
B	1250	1190	1400	1450	1320	1380
C	1160	1190	1310	1300	1340	1360
D	1522	1500	1480	1590	1630	1580
E	1440	1350	1430	1280	1380	1400
F	1600	1620	15[illegible]0	1610	1580	1590

31. In which state, has the production of rice increased continuously over the years ?
(a) A (b) B
(c) C (d) D
(e) None of these

32. In which year, was the production of rice the highest in all the states together ?
(a) 2014 (b) 2009
(c) 2012 (d) 2013
(e) None of these

33. Which state produced the lowest quantity of rice over the years ?
(a) E (b) D
(c) C (d) A
(e) None of these

34. What is the respective ratio of the average quantity of rice produced by State D to the average quantity of rice prduced by State F over the years ?
(a) 66 : 791 (b) 310 : 317
(c) 138 : 155 (d) 276 : 317
(e) None of these

35. Rice produced by State C in the year 2012 in approximately what per cent of the rice produced by state A in the same year ?
(a) 82 (b) 72
(c) 88 (d) 76
(e) 69

DIRECTIONS (Qs. 36-41) : *Study the following table to answer the given questions :*

Producton (in crore units) of six companies over the years

Company	Years						Total
	2009	2010	2011	2012	2013	2014	
TP	103	150	105	107	110	132	707
ZIR	75	80	83	86	90	91	505
AVC	300	300	300	360	370	340	1970
CTU	275	280	281	280	285	287	1688
PEN	25	30	35	40	42	45	217
SIO	85	87	89	91	92	96	540
Total	863	927	893	964	989	991	5627

36. The Production of Company AVC in 2012 is approximately what per cent of its average production over the given year?
(a) 300 (b) 110
(c) 135 (d) 18.25
(e) 95

37. For SIO, which year was the per cent increase or decrease in production from the previous year the highest ?
(a) 2013 (b) 2010
(c) 2014 (d) 2012
(e) None of these

38. Which Company has less average production in the three years compared to that of first three years ?
(a) No company (b) CTU
(c) ZIR (d) SID
(e) No of these

39. The total prodution of the six companies in the first two given years is what per cent of that of last two given years ? (round off upto two decimal places)
(a) 87.08 (b) 104.55
(c) 90.40 (d) 10.62
(e) None of these

40. For ZIR, Which of the following is the difference in production in 2014 and 2013 ?
(a) 10,00,00,000 (b) 1,00,00,000
(c) 10,00,000 (d) 40,00,000
(e) None of these

41. For how many comapanies did the production increase every year from that of the previous year?
(a) One (b) Two
(c) Three (d) Four
(e) None of these

DIRECTIONS (Qs. 42-45) : *Study the graph carefully to answer the questions that follow.*

$$\text{Per cent profit} = \frac{\text{Income} - \text{Expenditure}}{\text{Expenditure}} \times 100$$

Percent profit made by two companies over the years

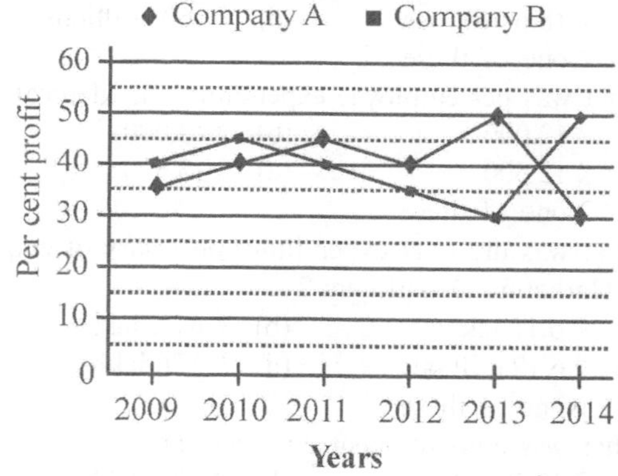

42. If the amounts of profit earned by company A in the year 2012 was ₹ 1.5 lakh, what was its expenditure in that year?
(a) ₹ 1.96 lakh (b) ₹ 2.64 lakh
(c) ₹ 1.27 lakh (d) ₹ 3.75 lakh
(e) None of these

43. What is the respective ratio of the amount of profit earned by companies A and B in the year 2014?
(a) 2 : 3 (b) 4 : 7
(c) 11 : 15 (d) Cannot be determined
(e) None of these

44. If in the year 2009 the expenditure incurred by companies A and B was the same, what was respective ratio of the income of companies A and B in that year ?
(a) 27 : 28 (b) 14 : 23
(c) 13 : 19 (d) Cannot be determined
(e) None of these

45. What is the average per cent profit earned by company B over all the years together ?
(a) $19\frac{1}{3}$ (b) $24\frac{1}{6}$
(c) $24\frac{1}{3}$ (d) $37\frac{1}{6}$
(e) None of these

DIRECTIONS (Qs. 46-50) : *Study the following graph carefully to answer the question given below it.*

Production of paper by 3 different companies A, B, & C over the years

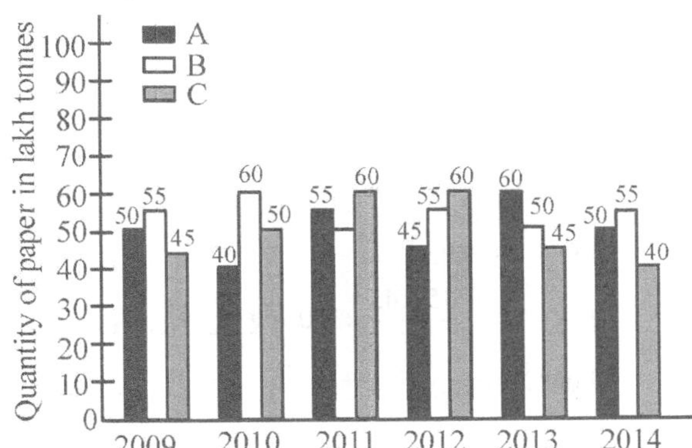

46. What is the difference between the production of company C in 2009 and the prodution of company A in 2014 ?
(a) 50,000 tonnes (b) 5,00,00000 tonnes
(c) 50,00,000 tonnes (d) 5,00,000 tonnes
(e) None of these

47. What is the percentage increase in production of company 'A' from 2010 to 2011 ?
(a) 37.5 (b) 38.25
(c) 35 (d) 36
(e) None of these

48. For which of the following years the percentage of rise/ fall in production from the previous year is the maximum for company B ?
(a) 2010 (b) 2011
(c) 2012 (d) 2013
(e) 2014

49. The total production of comapany C in 2011 and 2012 is what percentage of the total production of company A in 2009 and 2010 ?
(a) 95 (b) 90
(c) 110 (d) 115
(e) 133.33

50. What is the difference between the average production per year of the company with the highest average production and the company with the lowest average production in lakh tonnes ?
(a) 3.17 (b) 4.33
(c) 4.17 (d) 3.33
(e) None of these

SOLUTIONS

Sol. (1-5)

Number of girls $= \frac{3}{5} \times 2500 = 1500$

Number of boys $= \frac{2}{5} \times 2500 = 1000$

Number of boys playing cricket only $= \frac{20}{100} \times 1000 = 200$

Number of girls playing table-tennis, badminton and carrom only $= \frac{25}{100} \times 1500 = 375$

Number of boys playing football only

$= \frac{26.8}{100} \times 1000 = 268$

Number of girls playing cricket only $= \frac{175}{100} \times 200 = 350$

Number of boys playing hockey, table-tennis and carrom only $= \frac{25.7}{100} \times 1000 = 257$

Number of girls playing badminton only $= \frac{1}{4} \times 1500 = 375$

Number of girls playing chess only $= \frac{12}{11} \times 275 = 300$

Number of girls playing football and hockey only
$= 1500 - (375 + 350 + 375 + 300)$
$= 1500 - 1400 = 100$

Boys play chess only $= 1000 - (200 + 268 + 257)$
$= 1000 - 725 = 275$

1. (c) Number of students playing more than one game
$= 375 + 257 + 100 = 732$

2. (c) Number of students playing hockey
$= 257 + 100 = 357$

Hence, required percentage $= \left(\frac{357}{2500} \times 100\right) = 14.28\%$

3. (a) Required ratio $= 275 : 375 \times 2 = 11 : 30$

4. (b) Required number of students
$= 200 + 375 + 268 + 350 + 257 + 100 = 1550$

5. (e) Number of students playing carrom
$= 375 + 257 = 632$

6. (e) Total expenditure of Companies C and D
$= 45 + 40 =$ ₹ 85 crores
Total incomes of Companies
C and D $= (35 + 50)$ crores
$=$ ₹ 85 crores
Clearly, there is no profit or loss.

7. (e) Income of Company A is 2013

$=$ ₹ $\frac{100}{110} \times 37.5$ crores $=$ ₹ 34.09 crores

Let expenditure in 2013 be ₹ x crores

$\therefore 20 = \frac{34.09 - x}{x} \times 100$

$\Rightarrow 0.2x = 34.09 - x$

$\Rightarrow x = \frac{34.09}{1.2} = 28.40$ crores

8. (b) Expenditure of Company G in 2013 $= \frac{100}{120} \times 45$
$=$ ₹ 37.5 crores
Let its income in 2013 be I crores

$\therefore 10 = \frac{I - 37.5}{37.5} \times 100$

$\Rightarrow 3.75 = I - 37.5$

$\Rightarrow I = 37.5 + 3.75 =$ ₹ 41.25 crores

9. (d) For Company B
Profit %

$= \frac{42.5 - 32.5}{32.5} \times 100 = \frac{1000}{32.5} =$ ₹ 30.77 crores

For Company C

Loss % $= \frac{35 - 45}{45} \times 100 =$ ₹ 22.2 crores

For Company F

Profit % $= \frac{32.5 - 25}{25} \times 100 =$ ₹ 30 crores

For Company A

Profit % $= \frac{37.5 - 27.5}{27.5} \times 100 =$ ₹36.36 crore

10. (c) Total expenditure
$= 27.5 + 32.5 + 45 + 40 + 45 + 25 + 45$
$=$ ₹ 260 crores
Total income
$= 37.5 + 42.5 + 35 + 50 + 40 + 32.5 + 50$
$=$ ₹ 287.5 crores
$\therefore$ Required Profit %

$= \frac{287.5 - 260}{260} \times 100 = 10.57$

11. (d) Cannot be determined because expenditure of company X income of company Y data not given.

12. (c) For company X, % profit in 2012 - 2013 = 40
Let the expenditure be ₹100 crores.
$\therefore$ Income = ₹ 140 crores
$\therefore$ Expenditure in 2013-2014
= Income of company X. In 2012-2013 = ₹ 140 crores.
% Profit = 50

$\therefore$ Income $= 140 \times \frac{150}{100} =$ ₹ 210 crores.

$\therefore$ The required ratio $= 140 : 210 = 2 : 3$

13. (c) % Increase during

$2009\text{-}10 = \frac{5\times100}{15} = 33\frac{1}{2}\%$

$2010\text{-}11 = \frac{10\times100}{20} = 50\%$

$2012\text{-}2013 = \frac{20\times100}{30} = \boxed{66\frac{2}{3}\%}$

$2013\text{-}2014 = \frac{10\times100}{50} = 20\%$

14. (b) Let the income be ₹ x crores

$\therefore\ 20 = \frac{x-40}{40}\times100$

Or, $x - 40 = \frac{20\times40}{100} = 8$

Or, $x = 40 + 8 = 48$

∴ The required income = ₹ 48 crores

15. (d)

16. (e) Let the expenditure be ₹ x crores.

$\therefore\quad 60 = \frac{128-x}{x}\times100$

or, $\frac{128-x}{x} = \frac{60}{100} = \frac{3}{5}$

or, $128\times5 - 5x = 3x$

or, $8x = 128\times5$

or, $x = \frac{128\times5}{8} = 80$

∴ The required expenditure = ₹ 80 crores

17. (a) Total sales of all the stones = 1000 + 800 + 1400 + 1100 + 900 = 5200

Total sales of ruby = 200 + 100 + 200 + 200 + 200 = 900

Now, required percentage $= \frac{900}{5200}\times100 = 17.3\%$

18. (d) Total sales of opal = 300 + 200 + 400 + 300 + 200 = 1400

Average annual sales of opal $= \frac{1400}{5} = 280$

and, sales of Emerald in 2012-13 = 200

∴ Required percentage $= \frac{280-200}{200}\times100 = 40\%$

19. (c) It is clear from the chart.

20. (c) Sales to Bezel in 2008-09

$= \frac{200000}{1.25} = 160000$ kg

Sales of Bezel in 2014-2015 = 200000 × 1.5 = 300000 kg

Now, required increase = 300000 – 160000 = 140000 kg

21. (d) Total expenditure on Accounts department

= 14% of ₹ 12 crores = ₹ $\frac{14\times12}{100}$ crores

= ₹ 1.68 crores = ₹ 16.8 millions

22. (b) Total expenditure on medical = 13% of ₹ 12 crores

= ₹ $\frac{13\times12}{100}$ crores = ₹ 1.56 crores = ₹ 15600000

∴ Expenditure per employee

$= \frac{15600000}{1200}$ = ₹ 13000

23. (c) Total expenditure on salary of employees in marketing department

= 17% of 30 of ₹ 12 crores

$= \frac{17}{100}\times\frac{30}{100}\times120000000$ = ₹ 6120000

= ₹ 6.12 millions

24. (e) Amount spent on electricity

= 11% of ₹ 12 crores

= ₹ $\frac{11\times120}{100}$ millions = ₹ 13.2 millions

25. (a) Expenditure on telephone in computer department

= 12% of 8% of ₹ 12 crores

= ₹ $\frac{12}{100}\times\frac{8}{100}\times120$ millions

= ₹ 1.152 millions = ₹ 11.52 lakhs

26. (c) No. of silver coloured cars sold,

In Metro H = 20 + 35 + 37 + 42 + 22 + 17 = 173

In Metro M = 15 + 20 + 35 + 40 + 30 + 52 = 192

$\frac{173}{192}\times100 \approx 90\%$

27. (e) Blue colour cars model C sold in Metro M = 50,000

Red colour cars model F sold in Metro H = 34,000

Difference = 50000 – 34000 = 16000

28. (a) 90,000 is the largest no. of cars sold in Metro M and it is for white colour Model C.

29. (a) Total no. of blue coloured car of Model E and D sold in Metro H = 43 + 37 = 80

In Metro M, 80 white coloured Model B was sold.

30. (e) Difference between the white coloured cars sold between the two Metros :

A : 75 – 60 = 15

B : 81 – 80 = 1

C : 90 – 85 = 5

D : 90 – 86 = 4

E : 77 – 60 = 17

F : 87 – 75 = 12

The minimum difference is 1 and it is for Model B.

31. (e)

32. (a)

Year	Production of rice in the states together (in tonnes)
2009	1500 + 1250 + 1160 + 1522 + 1440 + 1600 = 8472
2010	1480 + 1190 + 1190 + 1500 + 1350 + 1620 = 8330
2011	1620 + 1400 + 1310 + 1480 + 1430 + 1510 = 8750
2012	1700 + 1450 + 1300 + 1590 + 1280 + 1610 = 8930
2013	1540 + 1320 + 1340 + 1630 + 1380 + 1580 = 8790
2014	1650 + 1380 + 1360 + 1580 + 1400 + 1590 = 8960

33. (c)

State	Production rice over the years
A	9490 tonnes
B	7990 tonnes
C	7660 tonnes
D	9300 tonnes
E	8280 tonnes
F	9510 tonnes

Hence, the State C produced lowest quantity of rice.

34. (b) Required ratio $= \frac{9300}{6} : \frac{9510}{6} = 1550 : 1585$

$= 310 : 317$

35. (d) Required ratio $= \frac{1300}{1700} \times 100 \approx 76$

36. (b) Average production of company AVC

$= \frac{1970}{6} = \frac{985}{3}$ crore units

Production of company AVC in 2012 = 360 crore units

∴ The required percentage

$= \frac{360}{\frac{985}{3}} \times 100 = \frac{360 \times 300}{985} = 109.6 \approx 110$

37. (c) It is obvious from the table as the production increased from 92 to 96 crore units.

38. (e) For company TP, average production in the last three years

$= \frac{107 + 110 + 132}{3} = 116.3$ crore units.

Obvisously, its average production in the last three years is less compared to the first three years.

39. (c) The total production of the six companies in the last two years = 989 + 991

= 1980 crore units

The total production of the six companies in the first two years

= 863 + 927 = 1790 crore units

∴ The required percentage

$= \frac{1790}{1980} \times 100 = 90.40$

40. (b) The required difference

= (91 – 90) crore units

= 1 crore units

41. (c) The production increased every year from that of the previous year for the companies ZIR, PEN, SIO.

42. (d) Suppose the expenditure of company A in year 2012 was ₹ x lakh.

$\therefore\ 40 = \frac{1.5}{x} \times 100$

$\Rightarrow 40x = 150 \Rightarrow x = \frac{150}{40} = 3.75$ lakh

43. (d) Data insufficient

44. (a) Suppose in 2009, expenditure by company A and B each was ₹ l lakh.

For company A, $35 = \frac{l_1 - l}{l} \times 100$

$\Rightarrow 135l = 100l_1$ ....(i)

For company B,

$40 = \frac{l_2 - l}{l} \times 100$

$\Rightarrow 140l = 100l_2$

From Eqs. (i) and (ii), we get ...(ii)

$\frac{l_1}{l_2} = \frac{135}{140} = \frac{27}{28} = 27 : 28$

45. (e) Average percentage profit earned by company B over the years

$= \frac{40 + 45 + 40 + 35 + 30 + 45}{6} = \frac{235}{6} = 39\frac{1}{6}\%$

46. (d) Production of company C in 2009 = 45 lakh tonnes.
Production of company A in 2014 = 50 lakh tonnes

∴ Required difference = 50 – 45 = 5 lakh tonnes

47. (a) Required percentage

$= \frac{55 - 40}{40} \times 100 = \frac{75}{2} = 37.5\%$

48. (b)

49. (b) Total production of company C in 2011 and 2012
= 120 lakh tonnes

Total production of company A in 2009 and 2010
= 90 lakh tonnes

∴ Required percentage $= \frac{120}{90} \times 100 = 133\frac{1}{3}\%$

50. (c) Average production of company A

$= \frac{50 + 40 + 55 + 45 + 60 + 50}{6}$

$= \frac{300}{6} = 50$ lakh tonnes

Average production of company B

$= \frac{55 + 60 + 50 + 55 + 50 + 55}{6}$

$= \frac{325}{6} = 54.17$ lakh tonnes

Average production of company C

$= \frac{45 + 50 + 60 + 60 + 45 + 40}{6} = \frac{300}{6}$

= 50 lakh tonnes

∴ Required difference

= 54.17 – 50 = 4.17 lakh tonnes.

■■■